Fungi and their Role in Sustainable Development: Current Perspectives

Praveen Gehlot • Joginder Singh
Editors

Fungi and their Role in Sustainable Development: Current Perspectives

Springer

Editors
Praveen Gehlot
Department of Botany
Jai Narain Vyas University
Jodhpur, Rajasthan, India

Joginder Singh
School of Bioengineering & Biosciences
Lovely Professional University
Phagwara, Punjab, India

ISBN 978-981-13-0392-0   ISBN 978-981-13-0393-7 (eBook)
https://doi.org/10.1007/978-981-13-0393-7

Library of Congress Control Number: 2018949363

This Springer imprint is published by the registered company Springer Nature Singapore Pte Ltd.
The registered company address is: 152 Beach Road, #21-01/04 Gateway East, Singapore 189721, Singapore

Contents

Part I Sustainable Cultivation and Conservation Strategies of Fungi

1 Cultivation, Conservation and Medicinal Significance of Macrofungi 3
S. K. Singh and Rakesh Pathak

2 Fungi as Biocontrol Agent: An Alternate to Chemicals 23
Sangeeta Singh, Shiwani Bhatnagar, Sunil Choudhary, Bindu Nirwan, and Kuldeep Sharma

3 Incredible Role of Fungi in Various Fields for Sustainable Development 35
Tulika Mishra and Gunjan Mukherjee

4 The Biological Promises of Endophytic *Muscodor* Species 51
Mahiti Gupta and Vineet Meshram

5 Slime Moulds: The Tiny Charmers 75
Anubha Pathak and Sharda Vaidya

6 Fungal Biotechnology: Role and Aspects 91
Debjani Mukherjee, Simranjeet Singh, Mohit Kumar, Vijay Kumar, Shivika Datta, and Daljeet Singh Dhanjal

Part II Sustainable Aspects of Fungi in Agriculture

7 Fungal Endophytes: Role in Sustainable Agriculture 107
Pratibha Vyas and Anu Bansal

8 Fungal Endophytes and Their Secondary Metabolites: Role in Sustainable Agriculture 121
Anamika, Samiksha Joshi, Manvika Sahgal, Sharda Sahu, and Anil Prakash

9 **Agricultural Important Microorganisms: From Rhizosphere to Bioformulation as Biological Control Weapons for Sustainable Agriculture** ... 147
Gunjan Mukherjee, Pannalal Dey, and Sunny Dhiman

10 **Recruit the Plant Pathogen for Weed Management: Bioherbicide – A Sustainable Strategy** ... 159
Manish Mathur and Praveen Gehlot

11 **Mycopesticides: Fungal Based Pesticides for Sustainable Agriculture** ... 183
S. Bhuvaneswari Manivel and G. Subashini Rajkumar

12 **Multifactorial Role of Arbuscular Mycorrhizae in Agroecosystem** ... 205
U. N. Bhale, S. A. Bansode, and Simranjeet Singh

13 **Strategic Approaches for Applications of Entomopathogenic Fungi to Counter Insecticide Resistance in Agriculturally Important Insect Pests** ... 221
V. Ambethgar

14 **Arbuscular Mycorrhizal Fungi As Phosphate Fertilizer for Crop Plants and Their Role in Bioremediation of Heavy Metals** ... 255
Arun Arya, Shalini Ojha, and Simranjeet Singh

15 **Fungal Community in Mitigating Impacts of Drought in Plants** ... 267
Richa Raghuwanshi

16 **Role of Fungi As Biocontrol Agents for the Control of Plant Diseases in Sustainable Agriculture** ... 283
D. U. Gawai

17 **Fungal Disease Management in Chickpea: Current Status and Future Prospects** ... 293
Sarvjeet Kukreja, Neha Salaria, Kajal Thakur, and Umesh Goutam

18 **Recent Approaches for Late Blight Disease Management of Potato Caused by *Phytophthora infestans*** ... 311
Umesh Goutam, Kajal Thakur, Neha Salaria, and Sarvjeet Kukreja

Part III Fungi for sustainable Industrial and Environmental Aspects

19 **Recent Advances and Industrial Applications of Microbial Xylanases: A Review** ... 329
Sunny Dhiman and Gunjan Mukherjee

20 **Bio-valorization of Dairy Whey for Bioethanol by Stress-Tolerant Yeast** ... 349
Deepansh Sharma, Mehak Manzoor, Parul Yadav, Jagdip Singh Sohal, Gajender Kumar Aseri, and Neeraj Khare

21 **Nutritional Quality Attributes of Edible Gasteroid Wild Mushroom *Astraeus hygrometricus*** ... 367
Mundamoole Pavithra, Kandikere R. Sridhar, and Ammatanda A. Greeshma

22 **Biotechnology of Fungal Lipases** ... 383
Gautam Kumar Meghwanshi and Abhishek Vashishtha

23 **Biopigments: Fungal Pigments** ... 413
Sharmila Tirumale and Nazir Ahmad Wani

24 **Bioethanol Production from Renewable Biomass by Yeast** ... 427
Ajay Kumar, Ranjan Deb, and Joginder Singh

25 **Fungi Inhabiting in Hypersaline Conditions: An Insight** ... 449
Abhishek Vashishtha and Gautam Kumar Meghwanshi

26 **Carbon Sequestration and the Significance of Soil Fungi in the Process** ... 467
Santosh Kumar Mehar and S. Sundaramoorthy

27 **Fungal Infections and Intervention in Diabetic Complication** ... 483
Amarish Kumar Sharma and Anjana Rana Sharma

Part IV Bioconversion Technologies of Fungi

28 **Arbuscular Mycorrhizal Fungi: Effects on Secondary Metabolite Production in Medicinal Plants** ... 507
Devendra K. Pandey, Prabhjot Kaur, and Abhijit Dey

29 **Endophytic Fungi: Carrier of Potential Antioxidants** ... 539
Jyoti Goutam, Ranjana Singh, Rajyoganandh S. Vijayaraman, and Mukesh Meena

30 **Current Perspectives of Endophytic Fungi in Sustainable Development** ... 553
Rashmi Mishra and V. Venkateswara Sarma

31 **Marine Fungi for Sustainable Development** ... 585
V. Venkateswara Sarma

32 **Fungi-Mediated Biodeterioration of Household Materials, Libraries, Cultural Heritage and Its Control** ... 597
Bhupendra Koul and Hina Upadhyay

33 **Co-cultivation Strategies to Induce De Novo Synthesis of Novel Chemical Scaffolds from Cryptic Secondary Metabolite Gene Clusters** 617
Dharmesh Harwani, Jyotsna Begani, and Jyoti Lakhani

Part V Modern Biotechnological Interventions of Fungi

34 **Computational Approaches to Understand the Genome and Protein Sequences of Fungi** 635
Atul Kumar Upadhyay and Gaurav Sharma

35 **Bioprospecting of Endophytic Fungi for Bioactive Compounds** 651
C. Ganesh Kumar and Poornima Mongolla

36 **Bioactive Compounds from Endophytic Fungi** 699
Vijayalakshmi Selvakumar and A. Panneerselvam

37 **Bioactive Potential of Nonconventional Edible Wild Mushroom *Amanita*** 719
Ammatanda A. Greeshma, Kandikere R. Sridhar, Mundamoole Pavithra, and Kaori Tomita-Yokotani

38 **Fungus *Monascus*-Fermented Red Yeast Rice (RYR): Natural Therapeutic Statin Source or Mycotoxin?** 739
Seema Patel and Nadeem Akhtar

39 **Nanoparticles from Fungi (Myconanoparticles)** 753
G. Subashini and S. Bhuvaneswari

About the Editors

Dr. Praveen Gehlot is Associate Professor, Department of Botany, Jai Narain Vyas University, Jodhpur, India. He did his Ph.D. from JNV University Jodhpur, under the able guidance of Prof. D.K. Purohit. He has about 18 years of teaching and research experience. He has been an awardee of Young Scientist under FAST Track Scheme by Department of Science and Technology, New Delhi, and later worked as a Pool-Officer of Council of Scientific and Industrial Research, New Delhi. Earlier, he taught at Budha Institute of Technology and Science Research College, Jodhpur, and Government Bangor PG College, Pali, Rajasthan, and subsequently served as an Associate Professor and Head, Department of Microbiology, Maharaja Ganga Singh University, Bikaner, Rajasthan. Dr. Gehlot has successfully completed three major research projects sponsored by UGC, DST and CSIR. He has published more than 75 research papers, book chapters and review articles in journals of international and national repute and has edited three books entitled *Pyrenomycetes Fungi, Arbuscular Mycorrhizal Fungi and Microbes: In Action* published by AgroBios, India. He attended several international and national seminars, symposia, conferences and chaired technical sessions. He is an active member of various scientific societies, including Mycological Society of India, Society of Mycology and Plant Pathology, India, Association of Microbiologists of India, and Indian Science Congress Association.

Dr. Joginder Singh is presently working as an Associate Professor at the Department of Biotechnology, Lovely Professional University, Punjab, India. Previously, he has worked as Young Scientist at Microbial Biotechnology and Biofertilizer Laboratory, Department of Botany, Jai Narain Vyas University in Department of Science and Technology, Govt. of India. His research interest is to explore different efficient strategies towards bioremediation and phytoremediation of pollutants from water and soil. Presently, he is working on designing and development of cleanup technology (biofilters) for in-situ bioremediation of textile industrial effluents. He is an active member of various scientific societies and organizations, including

Association of Microbiologists of India, Indian Society of Salinity Research Scientists, Indian Society for Radiation Biology, and European Federation of Biotechnology. Till now, he has published more than 60 research and review articles in peer-reviewed journals and authored or co-authored numerous books and book chapters.

Part I
Sustainable Cultivation and Conservation Strategies of Fungi

Chapter 1
Cultivation, Conservation and Medicinal Significance of Macrofungi

S. K. Singh and Rakesh Pathak

Abstract Mushrooms are saprophytic macrofungi. Due to differences in sexuality patterns, mostly diploid pure mycelia cultures are raised as tissue culture on suitable cereal grains. Cultivation techniques of *Pleurotus* spp., *Calocybe indica* and *Agaricus bisporus* (long, short and indoor composting methods) with spawn and substrate preparation are discussed. Improved methodologies of mushroom culture preservation for a short or long period with advantages and limitations are discussed. Anticancer, anti-heart attack, anti-HIV, hypoglycaemic, hypotensive, hepato- and nephroprotective, antioxidant, immunomodulatory, cardiovascular, respiratory, anti-hepatotoxic, pharmacological properties (antifungal, antibacterial and antiviral), and production of novel bioactive molecules by mushrooms are highlighted.

Keywords Mushrooms · Mycelium conservation · Medicinal properties · Bioactive molecules

1.1 Introduction

Mushrooms are saprophytic macrofungi. Unlike plants they lack chlorophyll and draw nutrition from dead and decaying organic materials. The mycelium obtains nutrition by penetrating substrate and at maturity or under unfavourable conditions forms fruiting bodies of various shapes, sizes and colour. Generally, mushroom fruit body consists of pileus and stipe, and a few mushrooms have veil or annulus, cup or volva and perform vital functions during the life cycle of the fungus. The pileus of the fruit body produces profuse spores of microscopic size upon maturity. These spores are dispersed and fall in soil/substrate. Under favourable conditions of moisture and temperature, these spores germinates to produces mycelia on suitable substrate. This mycelium continues to grow, mate by several means and undergo fructification to complete its life cycle. Due to saprophytic nutritional habit,

S. K. Singh (✉) · R. Pathak
ICAR-Central Arid Zone Research Institute, Jodhpur, Rajasthan, India

P. Gehlot, J. Singh (eds.), *Fungi and their Role in Sustainable Development: Current Perspectives*, https://doi.org/10.1007/978-981-13-0393-7_1

mushrooms usually grow in shady moist places on dead organic matter, decaying plant remains, wood logs, etc. Therefore, mushrooms grow in ample in forests, fields and meadows. However, some mushrooms are not edible and can be highly poisonous, and commonly referred to as 'toadstools' are objects of fear and distrust and have led to certain amount of inhibition towards mushroom eating. Mushroom cultivation being an indoor activity is highly remunerative as it requires low investment, relatively small space than any other agricultural activity and has simple cultivation technology.

Out of 10,000 macrofungi, only 2000 have been reported growing from different parts of the globe. Since India has rich floral diversity than any other country of its size, it is expected to have great fungal wealth and species diversity (Watling and Gregory 1980). Indian mycologists especially working on fleshy fungi have either described or reported several species from different parts of the country without mentioning their edibility. This was probably due to their major interest in survey and taxonomic studies of the fleshy fungi rather than their economic importance. All the reported fungi are not edible (Verma et al. 2003).

Purkayastha and Chandra (1976) published a book entitled *Indian Edible Mushrooms* in which 122 edible species have been recorded, of which 105 species are briefly described. In the 1970s two national symposia on survey and cultivation of macrofungi were organized in Srinagar (Atal et al. 1978; Kaul and Kapoor 1987). Manjula (1983) enlisted 115 genera and 538 species of agaricoid and boletoid fungi from India and Nepal. Purkayastha and Chandra (1985) published a manual of *Indian Edible Mushrooms* and described more than 430 edible fungi along with their characters and distribution in India. Bhawani Devi and Nair updated the list of Indian *Agaricales* in 1987. Dhar and Singh (1989) presented an account of biodiversity in Indian mushrooms during FAO workshop on Mushroom Biology and Mushroom Products held in Sydney.

This chapter deals with cultivation, conservation and medicinal significance of macrofungi to underline their role in sustainable development.

1.2 Cultivation

The mushroom cultivation involves several steps and is summarized as under.

1.2.1 Pure Culture Preparation

Mushroom cultures are required for variety of purposes such as preparation of quality spawn, conservation of genetic resources and maintenance of true to the type mushroom strains. The mushroom pure cultures are usually raised from tissue culture as it is expected to have diploid number of chromosomes. The young mushroom fruit bodies are first surface sterilized using surface disinfectants such as mercuric chloride or sodium hypochlorite. The fruit body is then longitudinally cut into two pieces with the help of sterilized scalpel, and a small piece from the collar region is carefully transferred onto a suitable culture media such as malt extract agar (MEA) and potato dextrose agar (PDA) prepared in Petri plates and incubated in a BOD incubator at 22–28 °C for 5–7 days. The growing mycelium from the margins of the tissue is carefully transferred onto test tubes that have sterilized culture media and further incubated at suitable temperatures in an incubator for about 14–20 days to raise mycelial cultures. Alternatively, pure mushroom cultures can also be obtained from spores. Most of the mushroom fruit bodies bear millions of spores which remained covered with pileus. The spore prints can be obtained by placing mature fruit body onto a sterilized paper or sterilized Petri plate.

Diluted spores are transferred to suitable presterilized culture medium in test tubes or Petri plates and inoculated at favourable temperatures for 10–14 days to obtain pure culture. A number of culture media such as malt extract agar, potato dextrose agar, oatmeal agar, compost extract agar, wheat extract agar, etc. are often used for preparation of mushroom pure cultures.

1.2.2 Spawn Preparation

Millions of minute spores are produced in mushroom fruit bodies which cannot be used directly as seed for mushroom production under controlled conditions. Moreover different mushrooms have different sexuality patterns, and every spore is not essentially fertile. Therefore, mushroom spawn is normally raised as tissue culture on suitable cereal grains. The spawn is defined as mushroom mycelia multiplied on a suitable culture medium. The spawn consists of mushroom mycelia and a suitable nutritive medium, which provides food to mushroom during active growing phase. Culpeper described spawn as mycelium of mushroom. Until the late nineteenth century, spawn used to be gathered from wild. Before the advent of modern spawn production technology using cereal grain spawn, virgin spawn from pastures and meadows; flake spawn by breaking of beds; mill-track spawn using cattle and horse dung, manure and soil; etc. were used to grow mushrooms.

The Pennsylvania State University did pioneering research in standardizing spawn-making process using cereal grains and hold two patents. Subsequently licences under the patent technology were accessible to other mushroom spawn laboratories. The process for the production of grain spawn and the fundamentals have not changed since then; you still need a starter culture, cereal grain. The grains are sterilized and cooled, and the product is grown out. It is no secret that anyone can make spawn, just as anyone can grow mushrooms. The traditional spawn laboratories throughout the world use cereal grains as base material for spawn preparation. The spawn preparation steps are as follows.

1.2.2.1 Substrate Preparation

Various types of cereal grains such as wheat, bajra, rye, jowar and agricultural and wooden waste are used for mushroom spawn production. Good substrate should not contain inhibitory compounds; large surface area of substrate should be available for fungal colonization and should provide essential nutrients required by mushroom mycelium to grow. Edible quality grains without any pesticide treatment or infestation (fungal or insect) should be used to avoid contamination and retarded growth.

For spawn production of oyster, white button and paddy straw mushrooms, cereal grains are the most suitable substrate, whereas for black ear and shiitake mushrooms prefer sawdust-based substrates being wood-rotting fungi. To prepare grain spawn, the grains are washed in clean drinking water thrice to remove admixtures of straw, soil and seed of other grasses. These cleaned and washed cereal grains are then boiled for approximately 20 min. Excess water from the boiled grains is then removed by drying the grains in shed for few hours. These moist grains are mixed with 2% calcium sulphate and 0.5% calcium carbonate. The application of these two chemicals not only provides nutrition to the growing mycelium but also overcomes the problem of stickiness, lump formation and maintains favourable pH of the grains near neutral to support the growth of the mushroom mycelium.

1.2.2.2 Mother Spawn

About 350 g boiled and treated grains are filled in half kg bottles up to 2/3 volume and plugged with nonabsorbent cotton. These bottles are then autoclaved at 22 lb p.s.i. pressure at 126 °C for 1.5–2 h. After cooling a piece of growing mycelium is aseptically inserted onto grains in bottles, and inoculated bottles are incubated at 22–30°C as per the most favourable temperature for the mycelia growth of the mushroom to be cultivated for 2–3 weeks. The commercial spawn is prepared from mother spawn after complete colonization of mycelium of mother spawn on cereal grains. The most suitable temperatures of mother spawn incubation for *Agaricus* and *Pleurotus* spp. are between 22 and 25 °C and for *Calocybe* and *Volvariella* spp. is around 30 °C.

1.2.2.3 Commercial Spawn

The commercial spawn is prepared in the same way as described under mother spawn, but instead of glass bottles, double-sealed polypropylene bags are used for multiplication of mushroom mycelium. Ideally, 500 g of boiled wheat grains treated with calcium carbonate and calcium sulphate are filled in each polypropylene bag and packed by inserting cotton in the polypropylene ring and sterilized in an autoclave at 126 °C for a period of approx. 2 h. The bags are left overnight and then inoculated with about 50–60 grains from prepared mother spawn and are mixed thoroughly and incubated in a BOD incubator at 22–30 °C as per the requirement of the mushroom strain. During incubation, the mushroom mycelium starts spreading in all directions and covers the entire bag during 2–3 weeks. Ready commercial spawn is stored at 4 °C until used.

*1.2.3 Oyster Mushrooms (***Pleurotus** *spp.)*

The *Pleurotus* is generally referred to as 'oyster mushroom' or 'dhingri' in India. Oyster mushrooms are 100% vegetarian and are considered as highly nutritive food material. The *Pleurotus* species are rich source of vitamin C and B complex, protein contents with calcium, sodium, potassium, phosphorus, iron, etc. As compared to other vegetables, *Pleurotus* contains more than ten times niacin content. The pleurotin which is a polycyclic compound isolated from *P. griseus* exhibited antibiotic properties. This is the most popular subtropical variety and stands next only to the button mushroom in world production. It utilizes agro-wastes without composting for producing protein-rich food. This variety is most suitable for production in India because it grows on a variety of cereal straws, agricultural and forest wastes and wide range of temperatures. Its conversion rate is the highest (BE up to 100%). It is the most suitable source of revenue generation in rural areas and a potential source of self-employment due to low cost of production, easy dehydration, choice of species and resilience to a wide range of temperatures.

Across the globe more than 20 species of *Pleurotus*, viz. *P. australis*, *P. cystidiosus*, *P. cornucopiae*, *P. columbinus*, *P. djamor*, *P. eryngii*, *P. flabellatus*, *P. florida*, *P. fossulatus*, *P. ostreatus*, *P. sajor-caju*, *P. sapidus*, *P. opuntiae*, *P. platypus*, *P. tuber-regium*, *P. purpureo-olivaceus*, *P. populinus*, *P. levis*, *P. membranaceus* and *P. yuccae*, are grown on a wide range of substrates, viz. wheat straw, sugarcane bagasse, paddy straw, cotton waste, stalks/leaves of bajra, jowar, jute, maize and peanut shells, sunflower stalks, used tea leaves and other industrial by-products and wastes such as apple pomace, paper mill sludge's, coffee waste, etc.

Generally, cellulose-rich substrates support the growth of *Pleurotus* species as the mycelium is saprophytic and grows on various types of substrates. The substrate needs to be sterilized for the development of *Pleurotus* mycelium as other competitor moulds compete with *Pleurotus* and antagonize the growth by secreting toxic metabolites but for limited availability of food. Therefore, in order to make the substrate selective for the growth of *Pleurotus* mycelium, the most popularly used method is chemical sterilization. About 10–12 kg of substrate is soaked in 100 l of water in plastic or iron drums containing carbendazim (7.5 g) and formaldehyde (125 ml) overnight and covered with polythene to eliminate or reduce the growth of the competitor moulds prior to spawning of *Pleurotus*. Carbendazim percolates gradually and sticks to straw killing microorganisms, whereas formaldehyde being volatile releases poisonous fumes and thereby eliminates living organisms from straw in about 18 h. The excess water is then drained off and the straw spread on the floor and then filled in polythene bags of about 4–5 kg capacity. For larger unit, instead of drums, bigger cemented blocks with more water holding capacity can be engaged for chemical treatment of straw. The quantity of the fungicide for chemical sterilization of substrate is based on per hundred litres of water. Alternatively the wheat straw can be sterilized by soaking the straw in hot water 65–70 °C for 1–2 h.

The rate of spawn mixed is 2–3% vis-à-vis wet weight of the substrate by thorough mixing or by layer spawning method. Spawned substrate is filled in polythene bags and tightly closed. About 25–30 holes should be made on all sides of each bag with 2 side cuts at the bottom to drain excess water. The spawned bags are kept in incubation room at 25 °C for mycelial growth in a closed room. During this period fresh air and diffused light are not required, and higher CO_2 environment is helpful in early spawn run. Depending on the substrate and the *Pleurotus* species used, the mycelial run is often completed in 18–22 days. Once the mycelial run is complete, a thick mycelial mat is formed indicating that the macrofungi are ready to fruit. The polythene coverings of the bags are completely removed, and the blocks are arranged on shelves and watered at least twice daily. During this period diffused light of low intensity is required. This can be achieved from sunlight entering through window or 1–2 bulbs for 3–4 h daily. High relative humidity during cropping period should be maintained (between 70 and 80%). The fruiting initials start within 3–5 days, and the fruiting bodies mature within a week. The mature fruit bodies should be regularly harvested and used fresh as vegetable or dried to be used later.

*1.2.4 White Button Mushroom (***Agaricus bisporus***)*

A. bisporus requires well composted materials for its growth. Composting is a complex microbial process which involves chain of microbial succession during composting. The process begins with piling up of pre-wetted ingredients into heap to start with mesophilic flora to develop and degrade the straw. As the temperature of the pile rises, mesophilic microbes are replaced by thermophilous bacteria and later replaced by actinomycetes and help in the decomposition of the substrates. Among different thermophilic fungi, *Scytalidium thermophilum* is the most important in terms of compost production, selectivity and nutrition of *A. bisporus* (Vijay et al. 1997). Composting is essentially anaerobic process, and the mixture is often turned to expose every portion of the substrate for aeration. A variety of base material is used for composting of *A. bisporus*, i.e. wheat straw, paddy straw, barley, oat, maize stalks, soybean stalks, mustard stalks, etc. A little quantity of nitrogen is also added to initiate fermentation; it is supplemented with nitrogen and carbohydrate sources such as animal manures, urea, calcium ammonium nitrate, ammonium sulphate and molasses; concentrated meals such as wheat and rice bran, soybean meal, castor meal, etc.; and mineral sources, i.e. gypsum superphosphates, etc. A number of formulations are available for compost preparation for long, short and indoor composting.

Long method of composting requires sterilization of the compositing area with 2% formaldehyde solution. Wheat straw and other ingredients are spread onto the composting yard and wetted with water thoroughly till it absorbs sufficient moisture for 1–2 days. Other ingredients, i.e. wheat bran, chicken manure and fertilizers except gypsum, are mixed with water and covered with polythene sheet and kept for 1 day. At zero days both the ingredient and straw are mixed, and pile is made and left for 5 days. The temperature in the heap rises up to 70 °C in 1–2 days which favours multiplication of the thermophilic microorganisms that eventually degrade the substrate. On the sixth day, first turning is done by dismantling the pile, reconstructing it to facilitate aerobic fermentation of the composting mixture. On the 10th day, second turning is done by breaking the pile to release excess production of ammonia; on the 13th day with third turning, the pile is again turned, and gypsum is added. Subsequently 4th, 5th, 6th and 7th turnings are done on 16th, 19th, 22nd and 25th day, respectively. Well-prepared compost is of dark brown colour with pH ranging 7.2–7.8 and moisture 65–67 devoid of any foul smell of ammonia. On the 28th day, the bags are filled with compost, and spawning is done.

Short method of composting consists of two parts. One is outdoor composting (phase I) for a period of 10–20 days followed by pasteurization and conditioning with free air circulation inside an insulated room under controlled set of conditions (phase II) which lasts for 7 days. Phase I is like long method of composting which includes wetting of ingredients followed by breaking of stack after 2 days, and addition of water to dry portions and again an aerobic stack is made. The ingredients are allowed to ferment under uncontrolled conditions to achieve partial decomposition of manure, whereas phase II has two stages: pasteurization and control environment. During pasteurization phase, the compost exhibits greater difference in temperature in peripheral and central zones. The temperature above 55 °C is harmful at which microorganisms are killed. It is usually done at 57 °C for 6 h. The second phase is conditioning, which has two steps, one is pre-pasteurization and another is post-pasteurization controlling. The whole compost masses are brought to temperature between 45 and 52 °C for the growth of thermophilic flora. The excess ammonia is released in the atmosphere, whereas post-pasteurization condition regenerates the lost thermophilic organisms. During this phase sufficient oxygen is supplied to the compost to maintain aerobic conditions. Indoor composting is an improvement over long and short traditional composting methods which involve environment problems. The total duration is greatly reduced to around 10 days.

Spawning of the compost is done immediately after composting. The grain spawn of *A. bisporus* is mixed with the compost under aseptic conditions in beds or polythene bags and compressed hard. Polythene bags are closed to prevent loss of moisture. The spawn is directly mixed with compost thoroughly at 0.5–0.7% and kept under dark room conditions at high CO_2 concentration. During spawn run no fresh air is supplied. The mushroom mycelium covers entire bags in about 22–25 days. These bags are then opened, and a casing layer of 3–4 cm thickness is applied to facilitate fruiting. There are large numbers of casing materials used and applied for the purpose. Due to its water retention capacity of peat, it is one of the most desirable casing materials. Alternatively, composted coir pith, composted farmyard manure, spent mushroom compost, etc. can also be used. An ideal casing material should have a neutral pH with low EC. The casing layer should be kept moist. It takes about a week for complete case run; at this stage air temperatures are lowered to 15–17 °C and RH around 80–85%, which induces pin heads within 7–10 days. The crop is regularly watered, and mushrooms are harvested for 6–8 weeks.

*1.2.5 Milky Mushroom (***Calocybe indica***)*

The only mushroom partially adopted by the farmers is the milky mushroom (*Calocybe indica*) which was first cultivated by Purkayastha and Nayak (1981) and later refined by Doshi et al. (1989, 1993). Its cultivation on wheat/paddy straw supplemented with various organic additives gives about 70% conservation at 25–35 °C and is catching the imagination of farmers/consumers of west Bengal, Karnataka, Tamil Nadu, Maharashtra and Uttarakhand, but it needs sustained extension support. This is a tropical mushroom and has a good scope as it is pure white, grows well in high temperature ranging from 25–37 °C and is suitable for tropical regions. The effect of different substrates and casing materials on the growth and yield of *C. indica* has been studied by Ruhul et al. (2010). Its keeping quality is excellent and doesn't turn brown. The production technology is very simple. This mushroom is suitable for chutney and pickle production.

1.3 Conservation

Not much organized efforts have been made to conserve the available mushroom biodiversity ex situ, which require their extensive collection, isolation, pure cultures, identifications and conservation in well-established gene banks as regional, national and even international facilities. Such a coordinated effort for mushroom genetic resource conservation has become necessary. In view of the threats like over-exploitation of wild mushroom and restriction imposed on germplasm exchange due to patenting, IPR issues, etc., mushroom scientists need to lay more emphasis on isolating pure culture of wild mushrooms and deposit them in well-recognized regional and/or national gene bank for their long-term conservation and future exploitation.

When a new genus or a species of microbe is discovered and described, it is generally deposited in established germplasm bank. This ensures availability of the organism for use in future. In addition, microbial strains of industrial importance are preserved and patented, and their availability became restricted (Jong and Birmingham 1991). Presently there is no authentic on-the-spot examination; method is accessible to certify spawn quality; a method of preserving selected strains tested and proved desirable is of primary importance. The preservation is not a simple and routine process as it seems to be, due to degeneration during the preparation or maintenance. This gradually leads to the loss of desired traits resulting in reduced recovery, retarded growth and low production and productivity (Chang and Miles 1989; Stadelmann 1986).

Spores produced through asexual processes, i.e. heterothallic or secondary homothallic species, have genetic differences (Miles and Chang 2004). Although spores from primary homothallic species are principally genetically similar, they still exhibit genetic variations, for example, *Volvariella* (Chang et al. 1981). To ensure fruiting from single spore of heterothallic species requires mating tests, and therefore, usually, diploid vegetative mycelia of elite strain are conserved (Snell 1984). Once the pure mycelial cultures are prepared, they can be preserved by different methods as per the requirements of a culture bank, a spawn laboratory and/or a research organization. Nevertheless, availability of the required equipments and funds becomes a major limiting factor for mushroom culture preservation for a relatively a long period.

1.3.1 Subculturing

The viability of any mushroom culture becomes limited because of continuous mycelial growth onto a suitable culture medium under favourable conditions and eventually exhaustion of nutrients. Thus, frequent subculturing is essential to maintain any mushroom in its active growing state. Kaur et al. (2011) observed that the mycelial growth of *A. bisporus* declined in linear growth when stored for a long period but remained static with reduced incubation. Fortnightly subculturing of *A. bisporus* (strain U-3) showed a rapid and persistent growth when grown as control for 16 days. The mycelial cultures can be raised on agar slants in culture bottles and/or test tubes and can be stored at ambient temperatures in a laboratory for about a month or so. The duration of storage can be enhanced by placing the mycelial cultures at 4 °C in a refrigerator or cold room (Singh et al. 2001a). *Volvariella volvacea* and *Calocybe indica* are incubated at 30–32 °C for 10 and 15 days, respectively. Both these mushrooms are subcultured after every 2 months, whereas strains of *Agaricus*, *Lentinula* and *Pleurotus* species can survive in a refrigerator at 4 °C for a longer period and are subcultured after every 4–6 months.

Degeneration and variation in subcultures may be identified by closely observing mycelial cultures. The most conspicuous symptoms are thin and weak mycelium, slow sectors of growth, fluffy but still may exhibit usual growth. Mushroom mycelia exhibiting slow growth and/or fluffy mycelium can carry virus or form stroma and reduce yields (Chang and Miles 1989). Culture tubes of *Volvariella* spp. form chlamydospores; those are light to dark brown in colour. Cultures of *Volvariella* spp. with more chlamydospores are a clear indicative of vigorous growth and potential for higher yields. Constant subculturing of stock cultures tends to degenerate or mutate and gradually leads to reduction in mushroom-forming ability and desired traits, or from genetic recombination and selection in continuous field, cultivation of re-established culture is relatively common in the spawn produced from cultures maintained by these methods (Chang and Hayes 2013). Such conventional procedures of conservation of living fungi are time-consuming, costly and risky. Ultimately, repeated subculturing may end in maintaining a culture which is diverse from the original one (Smith and Onions 1994). The disadvantages of frequent subculturing include loss of desirable traits, chances of airborne contamination, insect infestation, regular technical supervision by a specialist, labour-intensive and time-consuming process, etc.

1.3.2 Storage Under Mineral Oil

Mineral oil preservation envisages protection to the actively growing mycelial cultures by preventing dehydration and lowering metabolic activity and growth through reduced oxygen tension. In this storage method, mineral oil is autoclaved for 20 min at 22 p.s.i. consecutively for 2 days and allowed to cool at ambient temperatures. Thereafter mycelia cultures prepared in test tubes are filled in this presterilized mineral oil covering the culture and the media which disconnect the contact of growing mycelium with the external environment. Basically, shorter slants require less oil to cover them. Alternatively, mycelial discs from growing culture in a Petri plate or wheat grain from mother or commercial spawn can be plunged into sterilized mineral oil and may be stored at 4 °C in a refrigerator (Singh et al. 2001a).

In juxtaposition with preservation of the culture at 4 °C in a refrigerator, it is an efficient way of conserving mycelial cultures. Retrieval of mycelia is easy by using grain spawn or disc with the help of sterilized needle and draining off as much oil as possible and streaking the inoculum onto agar in plates or tubes. Tilting the culture plate facilitates drainage; the excess of mineral oil can also be soaked with the help of a sterilized filter paper. Subculture at the time of retrieval shows retarded growth rate, and a second subculture is expected to re-establish the culture.

Contamination by airborne spore and retarded growth on retrieval are two disadvantages of mineral oil storage. The method involves cutting 5-mm-diameter discs from fully grown cultures and storing them at ambient temperature in test tubes in liquid paraffin and plugged with nonabsorbent cotton. The cultures remained viable for 8 years.

1.3.3 Water Storage

Boeswinkel (1976) reported that water storage of 650 plant pathogens belonging to the *Phycomycetes*, *Ascomycetes*, *Fungi Imperfecti* and *Basidiomycetes* could remain viable for 7 years. For water storage, the mycelia cultures are raised in Petri plates onto a suitable culture media, and small bits of 5 mm diameter are transferred aseptically to precooled and sterilized McCartney bottles containing distilled water; the lids are screwed down tightly and are stored at ambient temperatures.

Most of the mushroom cultures except a few like *Volvariella* spp. and *Calocybe indica* can be stored by this method. Demineralized water without any nutrition works better. Revival of culture is by removal of a block and placing the mycelium on a suitable growth medium. Survival of fungal cultures stored in water is reported for 2–5 years period satisfactorily at IMI, Kew, Survey, UK (Smith and Kolkowski 1996). Freire et al. (2016) evaluated the viability, contamination and morphological changes of endophytic fungi maintained under different preservation methods.

1.3.4 Lyophilization

Lyophilization is a freeze-drying process for long-term conservation of spore-bearing fungi. During the process mushroom mycelia and/or spores are frozen to minus temperatures, and water present inside the cells is withdrawn by sublimation in a vacuum by creating reduced pressure. The main advantages of lyophilization are prolonged stability and relatively longer storage (Jong et al. 1984).

Pure cultures are first raised on suitable culture medium and then mixed in equal volume of culture suspension with suspending media containing 100 g skimmed milk or 7.5 g trypticase soy broth with 100 g sucrose and 50 g bovine serum albumin in 1 l distilled water. The basidiospores of mushroom can be dehydrated in glass ampoules by the following freeze-drying protocol. The glass ampoules are oven-sterilized at higher temperatures for 1 h, and then cotton is plugged at the neck of each ampoule and autoclaved for 15 min. at 126 °C at 22 lb. p.s.i. Mushroom mycelia or spore culture suspension is mixed with skimmed milk or other suitable medium and poured at 0.2 ml/autoclaved ampoule, and a representative sample of each culture suspension to be preserved is suitable diluted, and prefreezing viable count is determined. All the ampoules with spore suspension are kept in a freezer (−70 °C) for a few hours, and when the temperature of the freeze chamber drops down to approximately −40 °C, the frozen ampoules are then placed in the freeze-dryer chamber, and vacuum is created. Primary drying is accomplished at −40 °C in 4 h. Subsequently, the temperature is raised in 10 °C increments keeping at each temperature for at least half an hour and at 20 °C for 1 h, and vacuum is released. These ampoules are stored at −20–80 °C in a deep freezer for 24 h and then dried for 2–3 h, and vacuum is released. Cotton plugs are then pushed inside, and slight constrictions are made in the glass ampoules above the cotton plug in a way that the ampoule is not closed. The ampoules are then attached to the freeze-dryer through rubber tubes for secondary drying under vacuum at 20 °C for 2 h and sealed while remain attached to the freeze-dryer itself with the help of a LPG gas-oxygen gas torch. To check the vacuum environment inside the ampoule, a representative ampoule is tested using vacuum tester. If the sealing is perfect, a spark is seen inside the ampoule upon lighting the torch near ampoule. To check the postfreezing count, any ampoule may be selected at random and break open to check the viability before finally storing the ampoules for longer durations. If vacuum is perfect, then viability of most of the organisms does not decrease significantly upon freeze-drying, and lyophilized cultures can be stored in a refrigerator for several years satisfactorily. While retrieving the lyophilized culture, it should be re-hydrated in sterile distilled water for half an hour so as to absorb moisture before inoculating on a suitable culture medium.

Tan et al. (1991) demonstrated that hyphal cooling at the rate of –1 °C/min to temperatures of – 45 °C and then −75 °C produced fully freeze-dried mycelia. Freeze-drying was performed for 2 h at −40 °C, 20 h at −2 °C and 8 h at 20 °C, resulting in residual moisture content of 2%. Hyphae of *Ascomycetes* and *Basidiomycetes* have been reported to survive freeze-drying. A new lyophilization protocol has been developed by lyophilization of mushroom mycelium multiplied on pearl millet grains instead of culture suspension in glass ampoules to improve survival rates reaching almost 100% (Singh et al. 2004a). Croan et al. (1999) reported successful lyophilization of basidiomycetes using trehalose solution for freeze-drying. Lyophilized spores of dictyostelids could remain viable for 30 years (Raper 1984).

1.3.5 Storage at −70 °C

Glycerol and dimethyl sulfoxide (DMSO) are potential cryoprotectants. An aqueous solution of about 5–10% glycerol is sterilized by moist heat sterilization in an autoclaving at 22 lbs for 15–20 min. Otherwise, DMSO is sterilized by filtration using 0.22-micron Teflon filter. The culture suspensions in glycerol or DMSO are prepared, and aliquots are dispersed in small vials and then stored at −70 °C. Many national and international culture repositories are preserving macrofungi cultures this way suitably for many years. Nakasone et al. (2004) suggested that repeating freezing and thawing reduces viability of the cultures.

1.3.6 Conservation in Liquid Nitrogen

Liquid nitrogen storage of microbial cultures is considered to be the best method of culture preservation (Kirsop and Doyle 1991; Singh et al. 2004b). Sub-zero temperatures are known to reduce metabolic activities of the living cells. When the water inside the cells is completely frozen, biochemical reactions and cell metabolism are suspended (Franks 1981). Although some cell metabolism continues below −70 °C, recrystallization of ice and ice crystal growth can occur at temperature above −139 °C (Morris 1981), and this can cause injury to living cells during storage. Change of physical state from liquid to solid results in increases of volume of water by 10% and creates mechanical stress inside the cells (Grout and Morris 1987), whereas at −196 °C no recrystallization or movement of molecule has been reported, and complete dormancy is induced, during which the organism does not undergo any phenotypic or genotypic change, provided sufficient care is taken during freezing of cultures and thawing to recover from preserved culture. This method can be applied to both sporulating and mycelial cultures. The recoveries of temperature-sensitive strains have been enhanced by standardization of the technique for individual strain and that have previously failed (Morris et al. 1988; Singh et al. 2004b).

The temperature of nitrogen gas at its vapour phase is 139–140 °C temperature and that of liquid phase is −196 °C. The mycelia suspension of mushrooms in cryoprotectant like glycerol (10–15%) or DMSO that can sustain ultra-low temperature is prepared and distributed in aliquots of 0.5 ml–1 ml in plastic screw cap cryovials that can withstand ultra-low temperature. Programmed cooling at 1–10 °C per minute is perfect. Alternatively, culture vials are kept in a mechanical freezer maintained at −70 °C for 60 min and plunged into liquid nitrogen. Within 24 h the viability checks before and after freezing of a culture are performed to ensure long-term preservation of the cultures. Rapid thawing of cultures at 37 °C improves viability (Singh et al. 2001b).

San Antonio (1979) found that cryogenic storage did not affect culture viability and mushroom production for 9 years. Elliott and Challen (1979) stored 1012 cultures of *Agaricus bisporus* and related species and reported 95% recovery rate. Morphological and physiological characters remained unchanged after cryogenic storage of eight commercial mushroom cultures of *A. brunnescens* (*A. bisporus*) preserved in liquid nitrogen for 10 years (Jodon et al. 1982). Challen and Elliott (1986) reported successful preservation of species of *Agaricus*, *Coprinus*, *Lentinula*, *Pleurotus*, *Schizophyllum*, *Tremella* and *Polyporus* except *Volvariella* in 10% aqueous glycerol solution. They also reported that 10% aqueous DMSO solution gave constantly reliable retrieval of *V. volvacea*. Chen (1987) reported that 122 strains from 42 *Basidiomycetes* species including important edible mushrooms survived better with slow cooling (at 1 °C/min.) than rapid freezing. Slow freezing and rapid thawing generally gave the highest viability count. Ohmasa et al. (1996) recorded mycelial growth of *Flammulina velutipes* cultures stored in liquid nitrogen for 7 years. Singh et al. (2004b) preserved mushroom mycelium multiplied on wheat grains instead of mycelial disc in liquid nitrogen and tested survival, yield and genetic stability of 11 edible mushroom stock cultures. They modified cryopreservation protocols, gave experimental demonstration of genetic stability of stock cultures and validated the use of liquid nitrogen cryopreservation for long-term preservation of mushroom cultures. Nevertheless, cryopreservation with liquid nitrogen, using DMSO as cryoprotectant, has been reported not the most appropriate one for *A. blazei* preservation (Colauto et al. 2012).

1.3.7 Granular Medium

Xiang (1991) developed an economically substitute granular structure culture medium in place of cereal grain medium as medium for the preservation of mushroom strains. The ingredients of granular structure culture medium are sawdust (72%), wheat flour (20%), soybean powder (5.5%), complex additives (2%) and adhesive (0.5%). These ingredients are sterilized in 500 ml glass jars, and mushroom mycelium is multiplied and subsequently preserved at 2–4 °C. The viability of the cultures can be retained for 5 years. When the spawn is prepared for cultivation by this new method, less inoculum needs to be removed each time from the specific 500 ml jar. Removed inoculum is then transferred and reproduced into spawn. Each time the inoculum is taken, it has to be done at a temperature of 2–4 °C in a sterile environment. Using this method, a 500 ml jar of the preserved strain can provide the original and pure inoculum for a long time. Alternatively, fresh granular medium is replaced into the jar, and mycelium grows upwards from the jar bottom to top when kept at 20 °C to culture the inoculum. When the mycelial growth reaches the surface, the jar is immediately returned to 2–4 °C. This cycle can be continued for many years. This method is simple, practical and keeps rejuvenating mycelium in the preserved mushroom strain.

1.3.8 Storage in Mechanical Freezer

The cessation of biological activity at ultra-low temperatures enhances viability of the preserved cultures. The ultra-low temperature mechanical freezers operating efficiently at −140 °C or −150 °C are now available in the market. Glass transition temperature of water below −130 °C suspends all enzyme activity and thermally driven reactions. Therefore living cells, spores or frozen propagules may be stored indefinitely below −130 °C. The mycelia culture suspension with cryoprotectant is prepared and gradually cooled, i.e. 20 °C, and then at −70 °C for a few hours and finally below −130 °C in freezer. The culture preservation in mechanical freezer is at par with that of liquid nitrogen. Nevertheless, mechanical freezers are run on electricity and therefore are not very successful where electricity supply is not continuous, and on-the-spot repairs are unattainable.

The preference of the preservation method depends upon the purpose, requirement, expertise, equipment, funds and facilities. It is recommended that each mushroom strain may be preserved and maintained by short and/or long method by more than one method. Conservation of edible mushrooms can be best done in mineral oil and liquid nitrogen conditions. Standardization of slow cooling and rapid thawing protocols may maximize survival of edible mushrooms. Once any mycelia culture of a mushroom has been successfully frozen, it eventually results in indefinite conservation of mushroom because of cessation of all biological and biochemical activities (Grout and Morris 1987).

1.4 Medicinal Significance

Mushrooms contain 20–35% protein with high digestibility which is higher than most vegetables due to low calories, low fat and high K:Na ratio; mushrooms are the dietician's choice for those with obesity and hypertension and suffering from hyperacidity and constipation. With low calorie, no sugars and starch and high protein, they are set to be 'delight of the diabetic'. Alkaline ash content coupled with high fibre present in mushrooms is suited to those with hyperacidity and constipation (Rai and Arumuganathan 2005).

Besides the medicinal significance, the pharmacological research confirms antifungal, antibacterial, antioxidant and antiviral properties of mushrooms (Wani et al. 2010). Nutritionally important chemical components, namely, proteins, amino acids, carbohydrates, fat, ash, fibre, antioxidants, vitamins, flavour and taste compounds and contents of wild-grown mushrooms, have been reviewed by Wang et al. (2014).

Ganoderma lucidum, called Rishi mushroom in Japan and Ling Zia in China, is the most priced medicinal mushroom in the Chinese and Japanese system of medicine but, of late, found appreciable acceptance throughout the world including the developed countries. It has been found to grow on trees as a phytopathogenic fungus during monsoon season and found associated with wide range of hosts including *Prosopis* spp., *Acacia* spp., *Azadirachta indica*, etc. causing economic losses. Its fruiting bodies have gained wide popularity due to rare medicinal property throughout the world particularly in Japan, China and the USA. It is reported to possess a plethora of health and medicinal benefits including significant anticancer, anti-heart attack and anti-HIV activities and hypoglycaemic, hypotensive, hepato- and nephroprotective and antioxidant properties (Kim et al. 1996; Liu et al. 1995; Chang 1995, 1996). It is consumed as a tonic and supplement due to its medicinal properties like antitumour, immunomodulatory, anti-platelet, aggregation, cardiovascular, respiratory, anti-hepatotoxic, anti-HIV, hypoglycaemic and anti-nociceptive effects. There is a big herbal market in American and European countries for this mushroom due to its health benefits. It is a tropical mushroom fruit under high temperature (30–38 °C) and high humidity. Its cultivation technology has been developed by the Directorate of Mushroom Research, Solan, on sawdust substrate (Rai 2003). The most important bioactive polysaccharide from *G. lucidum* β-1-3, β-1-6, D-glucan is high molecular weight polysaccharide and has exhibited significant antitumour activity. The medicinal component of *G. lucidum* such as polysaccharides, triterpenes, organic germanium, etc. and their function have been described by Cheng and Buswell (1996).

Lentinula edodes generally referred to as Shiitake is considered as medicinal food mainly in China, Japan and other Asian countries (Yap and Ng 2001). Its cultivation technology has been standardized by Singh et al. (2008). This mushroom has got as an excellent potential to boost the economy of farmers due to its medicinal applications. It possesses anticancerous, anti-hepatitis, cholesterol-lowering, antihypertensive, antiviral and libido-increasing properties. A bioactive metabolite from *L. edodes* called lentinan has been reported to be useful in the treatment of lung, liver, ovarian and stomach cancers (Hazama et al. 1995). It is known to possess strong immunomodulatory properties. The methanolic extract of Shiitake mushroom confers protection against hydrogen peroxide-induced cytotoxicity in peripheral blood mononuclear cells (Kuppusamy et al. 2009).

Phellorinia species is a popular mushroom in Rajasthan, Delhi, Haryana, Punjab and Uttar Pradesh (Doshi and Sharma 1997). This mushroom is collected in tonnes and after one or two good monsoon rains from different zones of Rajasthan even from sand dunes and sold in the market as fresh as well as in dried form. The rural people have identified the richness of calcium through experience and confirmed scientifically later. The species of *Phellorinia* mixed with pure ghee in semisolid state is given to person suffering from bone crack. It is also given to pregnant women with local Ayurvedic preparations. Attempts have been made to domesticate *Phellorinia* spp. under ICAR ad hoc project by Dr. Anila Doshi and co-workers. Bioactive components of *Podaxis pistillaris* exhibited antibacterial activity (Al-Fatimi et al. 2006). Further production technology has not been standardized for *Podaxis pistillaris*.

A lot of *Pleurotus* species have been found associated with dead and living tree bark. The cultivation technology of *P. sajor-caju*, *P. sapidus*, *P. florida* and *P. citrinopileatus* and *P. flabellatus* has been standardized on wheat straw-based substrate (Anonymous 2009). Fractionation extract of edible mushroom *Volvariella volvacea*, *Agaricus bisporus* and *Calocybe indica* yields nicotinic acid carboxylic acid and ergosterol and correlated with the beneficial health effects as food (Mallavadhani et al. 2006). Belova and Denisova (2005) studied the possibilities of use of white-rot wood-destroying fungi for agricultural and industrial waste utilization as a possible source of energy. Valencia et al. (2006) studied the biological quality of protein from three strains of *Pleurotus* spp. Fruit bodies of all the strains possessed high-protein content, approximately 27% with very high-protein digestibility up to 98%. Sarangi et al. (2006) identified three natural proteoglycans from *Pleurotus ostreatus* mycelia as immune modulators and anticancer agent. Matsubara et al. (2006) demonstrated the potential of white-rot fungi in the bioremediation of contaminated land. *P. eryngii* extracts alleviate the decrease in the trabecular bone mineral density and had significant role in bone metabolism when tested on rats (Kim et al. 2006).

Yi et al. (2006) carried out studies on dietary fibre in *P. tuber-regium*. Gambato et al. (2016) evaluated mushroom productivity and antioxidant properties of edible macrofungi *Pleurotus albidus* and *Pycnoporus sanguineus* and found that the substrate ingredients affected mushroom production and chemical composition. Ren et al. (2014) tested antioxidant and antibacterial activities of eight edible mushrooms. They reported that the aqueous extracts of all the mushrooms demonstrated radical scavenging activities and *P. australis* with the highest antioxidant activity.

The novel antibacterial compounds from *Agaricus bisporus* and *Pleurotus sajor-caju* were found the most sensitive against *Escherichia coli* enterobacter spp. and *Pseudomonas* spp. (Tambekar et al. 2006). Puttaraju et al. (2006) studied the antioxidant activity in 23 indigenous edible mushrooms. The two mushrooms *Termitomyces heimii* and *T. mummiformis* displayed the maximum antioxidant activity potential due to gallic acid, gentisic acid, protocatechuic acid and tannic acid and developed an important database for the studies of mushroom for preparation of mushroom-based nutraceutics. Loganathan et al. (2009) performed a relative study on the anticancer, antioxidant and antimicrobial property of *A. bisporus*. Antioxidant activity of three species of wild mushroom *Cantharellus* has also been reported from Northwestern Himalaya (Kumari et al. 2011). Guo et al. (2012) found the macrofungi species *Boletus edulis*, *Boletus regius*, *Suillus bovinus*, *Thelephora ganbajun* and *Volvariella volvacea* exhibited the highest antioxidant capacities and total phenolic contents. They identified and quantified gallic, homogentisic, protocatechuic and *p*-hydroxybenzoic acid as bioactive compounds which contribute significantly to the antioxidant capacities of these macrofungi.

Cordyceps sinensis is native of Himalayan Mountains in India, Tibet and Nepal at an altitude of 3000–5000 m. It is entomopathogenic in nature and lives as a parasite on the larvae of *Hepialus armoricanus* and grows in soil and canalizes caterpillar and forms a fruiting body at the head of the larva. It has got antioxidant antitumourous and immune-stimulating properties (Singh et al. 2007). *Grifola frondosa* popularly known as Maitake has been shown to be effective in regulating blood pressure, constipation and diabetes. Its extract has shown to kill AIDS virus and anti-HIV activity (Kim et al. 1996). PSK, a top selling anticancer drug extracted from *Coriolus versicolor*, accounts for more than 25% of Japan total sales of anticancer drug (Cheng and Buswell 1996).

References

Al-Fatimi MAA, Julich WD, Jansen R, Lindequist U (2006) Bioactive components of the traditionally used mushroom *Podaxis pistillaris*. Evid Based Complement Alternat Med 3(1):87–92

Anonymous (2009) Annual Report-ICAR-Central Arid Zone Research Institute, Jodhpur, Rajasthan, p 156

Atal CK, Bhat BK, Kaul TN (1978) First national symposium on survey and cultivation of edible mushrooms in India, RRL, Srinagar (J & K), September 22–24

Belova NV, Denisova NP (2005) White rot wood destroying fungi and possibility of their use for waste utilization. Biotekhnologiya 4:55–58

Bhawani D, Nair MC (1987) List of Indian Agaricales. In: Proceedings of international conference on science and cultivation technology of edible fungi, RRL, Srinagar (J & K), pp 371–374

Boeswinkel HJ (1976) Storage of fungal cultures in water. Trans Br Mycol Soc 66:183–185

Challen MP, Elliott TJ (1986) Polypropylene straw ampoules for the storage of microorganisms in liquid nitrogen. J Microbiol Methods 5:11

Chang ST (1995) Ganoderma: the leader in production and technology of mushroom nutraceuticals. In: Kim BK, Kim IH, Kim YS (eds) Recent advances in *Ganoderma lucidum* research. The Pharmaceutical Society of Korea, Seoul, pp 43–52

Chang ST (1996) Mushroom research and development equality and mutual benefit. In: Royse DJ (ed) Mushroom biology and mushroom products. Pennsylvania State University, State College, pp 1–10

Chang ST, Hayes WA (2013) The biology and cultivation of edible mushrooms. Academic, New York, p 842

Chang ST, Miles PG (1989) Edible mushrooms and their cultivation. CBS Delhi, New Delhi

Chang ST, Miles PG, Wali CC (1981) A study of monosporous isolates of *Volvariella volvacea*. Mushroom Sci 2:603

Chen YY (1987) The preservation of basidiomycetes culture dry ultra-low temperature freezing. Acta Myco Sin 6(2):110

Cheng ST, Buswell JA (1996) Mushroom nutraceuticals. World J Microbiol Biotechnol 12:473–476

Colauto NB, Eira AF, Linde GA (2012) Cryopreservation of *Agaricus blazei* in liquid nitrogen using DMSO as cryoprotectant. Biosci J 28(6):1034–1037

Croan SC, Burdsall HH Jr, Rentmeester MR (1999) Preservation of tropical wood-inhibiting Basidiomycetes. Mycologia 91:908–916

Dhar BL, Singh SK (1989) Current status of mushroom genetic resources in India. FAO workshop on diversity, conservation and utilization of mushroom genetic resources in Asia Pacific Region, Sydney, Australia

Doshi A, Sharma SS (1997) Wild mushroom of Rajasthan. In: Rai RD, Dhar BL, Verma RN (eds) Advances in mushroom biology and production. Mushroom Society of India, Solan, pp 193–203

Doshi A, Sidana N, Chakravarti BP (1989) Cultivation of summer mushroom *Calocybe indica* P&C in Rajasthan. Mushroom Sci 12:395–400

Doshi A, Sharma SS, Trivedi A (1993) A promising edible mushroom from the tropics *Calocybe indica* P&C. Mushroom Info 86:14–22

Elliott TJ, Challen MP (1979) The storage of mushroom strains in liquid nitrogen. The Glasshouse Crops Research Institute, Annual Report, pp 194

Franks F (1981) Biophysics and biochemistry of low temperatures and freezing. In: Morris GJ, Clarkes A (eds) Effects of low temperature of biological membranes. Academic, London, pp 3–19

Freire AKL, Bentes AS, Sampaio IL, Lima AM et al (2016) Availability and morphological characteristics of endophytic fungi held in different methods of preservation. Sci Res Essays 11(6):76–79

Gambato G, Todescato K, Pavão EM, Scortegagna A et al (2016) Evaluation of productivity and antioxidant profile of solid-state cultivated macrofungi *Pleurotus albidus* and *Pycnoporus sanguineus*. Bioresour Technol 207:46–51

Grout BWW, Morris GJ (eds) (1987) The effect of low temperatures on biological systems. Edward Arnold, London

Guo YJ, Deng GF, Xu XR, Wu S et al (2012) Antioxidant capacities, phenolic compounds and polysaccharide contents of 49 edible macro-fungi. Food Funct 3:1195–1205

Hazama S, Oka M, Yoshino S, Lizuka N et al (1995) Clinical effects and immunological analysis of intra-abdominal and intra pleural injection of lentinan for malignant ascites and pleural effusion of gastric carcinoma. Cancer Chemother 22:1595–1597

Jodon MH, Royse DJ, Jong SC (1982) Productivity of *Agaricus brunnescens* stock culture following 5, 7 and 10 year storage periods in liquid nitrogen. Cryobiology 19:602

Jong SC, Birmingham JM (1991) Patent development in mushroom biotechnology. In: Maher MJ (ed) Mushroom science, science and cultivation of edible fungi. Balkema/Illus Maps, Rotterdam/Brookfield

Jong SC, Levy A, Stevenson RW (1984) In: Kocur M, da Silva E (eds) Proceedings of the fourth international conference on culture collection. World federation for culture collections, London, pp 125–136

Kaul TN, Kapoor BM (1987) Proceeding of international conference on science and cultivation technology of edible fungi, RRL, Srinagar (J & K)

Kaur L, Dhanda S, Sodhi HS, Kapoor S et al (2011) Storage and preservation of temperate mushroom cultures, *Agaricus bisporus* and *Pleurotus florida*. Indian J Microbiol 51(2):234–238

Kim BK, Kim HW, Chi EC (1996) Medicinal efficiency of *Ganoderma lucidum* (XV). Anti HIV activity of *Ganoderma lucidum*. In: Royse DJ (ed) Mushroom biology and mushroom products. Pennsylvania State University, State College, pp 187–194

Kim SW, Kim HG, Lee BE, Hwang HH et al (2006) Effects of mushroom, *Pleurotus eryngii*, extracts on bone metabolism. Clin Nutr 25(1):166–170
Kirsop BE, Doyle A (1991) Maintenance of microorganisms: a manual of laboratory methods, 2nd edn. Academic, London, p 308
Kumari D, Sudhakara RM, Upadhyay RC (2011) Antioxidant activity of three species of wild mushroom genus *Cantharellus* collected from North Western Himalaya. Int J Agric Biol 13:415–418
Kuppusamy UR, Chang YL, Mahmood AA, Indran M et al (2009) *Lentinula edodes* (shiitake) extract protects against hydrogen peroxide induced cytotoxicity in peripheral blood nuclear cells. Indian J Biochem Biophys 46:161–165
Liu F, Doi VEC, Chang ST (1995) Anti-tumor components and culture filtrates from *Trichoderma* spp. World J Microbiol Biotechnol 11:486–490
Loganathan JK, Venkata KV, Shenbhagaraman R, Kaviyarasan V (2009) Comparative study on the antioxidant, anticancer and antimicrobial property of *Agaricus bisporus* (JE Lange) Imbach before and after boiling. Afr J Biotechnol 8(4):654–661
Mallavadhani UV, Sudhakar AVS, Satyanarayana KVS, Mahapatra A et al (2006) Chemical and analytical screening of some edible mushrooms. Food Chem 95(1):58–64
Manjula B (1983) A revised list of agaricoid and boletoid basidiomycetes from India and Nepal. Proc Indian Acad Sci (Plant Sci) 92(2):81–213
Matsubara M, Lynch JM, De Leij FAAM (2006) A simple screening procedure for selecting fungi with potential for use in the bioremediation of contaminated land. Enzyme Microb Technol 39(7):1365–1372
Miles PG, Chang ST (2004) Mushrooms: cultivation, nutritional value, medicinal effect, and environmental impact, 2nd edn. CRC Press, Washington, DC, p 480
Morris GJ (1981) Cryopreservation: an introduction to cryopreservation in culture collections, vol 27. Culture Centre of Algae and Protozoa, Cumbria
Morris GJ, Smith D, Coulson GE (1988) A comparative study of the morphology of hyphae during freezing with the viability upon thawing of 20 species of fungi. J Gen Microbiol 134:2897–2906
Nakasone KK, Peterson SW, Jong SC (2004) Preservation and distribution fungal cultures. In: Mueller GM, Bills GF, Foster MS (eds) Biodiversity of fungi. Elsevier Academic Press, Amsterdam, pp 37–47
Ohmasa M, Tsunoda M, Babasaki K, Hiraide M et al (1996) Fruit body production of test cultures of *Flammulina velutipes* preserved for seven years by freezing at three different temperature. Mycoscience 37:449–454
Purkayastha RP, Chandra A (1976) A new technique for in vitro production of *Calocybe indica* as edible mushroom from India. Mushroom J 40:1–2
Purkayastha RP, Chandra A (1985) Manual of Indian edible mushrooms. Today and Tomorrow's Printers and Publishers, New Delhi, p 266
Purkayastha RP, Nayak D (1981) Development of cultivation method and analysis of proteins of a promising edible mushroom *Calocybe indica* P&C. Mushroom Sci 11(2):697–713
Puttaraju NG, Venkateshaiah SU, Dharmesh SM, Urs SMN et al (2006) Antioxidant activity of indigenous edible mushrooms. J Agric Food Chem 54(26):9764–9772
Rai RD (2003) Successful cultivation of the medicinal mushroom Reishi (*Ganoderma lucidum*) in India. Mushroom Res 12(2):87–91
Rai RD, Arumuganathan T (2005) Post harvest handling of the fresh mushrooms. In: Rai RD, Upadhyay RC, Sharma SR (eds) Frontiers in mushroom biotechnology. National Research Centre for Mushroom, Solan, pp 365–375
Raper KB (1984) Thedictyostelids. Princeton University Press, Princeton, p 453
Ren L, Hemar Y, Perera CO, Lewis G et al (2014) Antibacterial and antioxidant activities of aqueous extracts of eight edible mushrooms. Bioact Carbohydr Diet Fibre 3:41–51
Ruhul A, Abul K, Nuhu A, Lee TS (2010) Effect of different substrates and casing materials on the growth and yield of *Calocybe indica*. Mycobiology 38(2):97–101

San Antonio JP (1979) Stability of spawn stocks of the cultivated mushrooms stored for nine years in liquid nitrogen (–160°C) to (–196°C). Mushroom Sci 1:103
Sarangi I, Ghosh D, Bhutia SK, Mallick SK et al (2006) Anti-tumor and immunomodulating effects of *Pleurotus ostreatus* mycelia derived proteoglycans. Int Immunopharmacol 6(8):1287–1297
Singh SK, Verma RN, Upadhyay RC, Yadav MC (2001a) Studies on improved methods of mushroom germplasm conservation. Indian J Plant Genet Resour 14:55–59
Singh SK, Upadhyay RC, Verma RN (2001b) Effect of cryoprotectants on preservation of mycelial cultures of edible mushrooms. Mushroom Res 10(2):67–72
Singh SK, Upadhyay RC, Yadav MC, Tiwari M (2004a) Development of a novel lyophilization protocol for preservation of mushroom mycelial cultures. Curr Sci 87(5):568–570
Singh SK, Upadhyay RC, Kamal S, Tewari M (2004b) Mushroom cryopreservation and its effect on survival, yield and genetic stability. Cryo Letters 25:23–32
Singh RP, Verma RC, Arora RK, Mishra KK et al (2007) Medicinal mushrooms of Uttaranchal with reference to *Ganoderma auricularia* and *Cordyceps sinensis*. In: Rai RD, Singh SK, Yadav MC, Tewari RP (eds) Mushroom biology and biotechnology. Mushroom Society of India, Solan, pp 321–324
Singh SK, Yadav MC, Rai RD, Tewari RP (2008) Enhancing production of medicinal mushroom shiitake. ICAR News 14(2):18
Smith D, Kolkowski J (1996) Fungi. In: Hunter Cevera JC, Belt A (eds) Maintaining cultures for biotechnology and industry. Academic Press, New York, pp 251–269
Smith D, Onions AHS (1994) The preservation and maintenance of living fungi, IMI Technical Handbooks No. 2, 2nd edn. CAB International, Wallingford, p 122
Snell JJS (1984) General introduction to maintenance methods. In: Krisop BE, JJS S (eds) Maintenance of microorganisms: a manual of laboratory methods. Academic Press, London, p 11
Stadelmann R (1986) Preservation and degeneration of mushroom strains. Mushroom J 158:41
Tambekar DH, Sonar TP, Khodke MV, Khante BS (2006) The novel antibacterials from two edible mushrooms: *Agaricus bisporus* and *Pleurotus sajor caju*. Int J Pharmacol 2(5):582–585
Tan CS, van Ingen CW, Stalpers JA (1991) Freeze-drying of fungal hyphae and stability of the product. Genetics and breeding of Agaricus. In: van Griensven LJD (ed) Proceedings of the first international seminar on Mushroom Science, Mushroom Experimental Station, Horst, Netherlands, 14–17 May 1991. Wageningen, Netherlands
Valencia DTG, Castelan VR, Garin AME, Leal LH (2006) Biological quality of proteins from three strains of *Pleurotus* spp. Food Chem 94(4):494–497
Verma RN, Upadhyay RC, Yadav MC, Singh SK et al (2003) Genetic resources of commercial mushrooms, their conservation, characterization and improvement. In: Upadhyay RC, Singh SK, Rai RD (eds) Current vistas in mushroom biology and production. Mushroom Society of India, National Research Centre for Mushroom, Solan, pp 1–9
Vijay B, Sharma SR, Lakhanpal TN (1997) Mycoflora of *Agaricus bisporus* compost. In: Rai RD, Dhar BL, Verma RN (eds) Advances in mushroom biology and production. Mushroom Society of India, National Research Centre for Mushroom, Solan, pp 139–159
Walting R, Gregory NM (1980) Larger fungi from Kashmir. Nova Hedwigia 32:494–564
Wang XM, Zhang J, Wub LH, Zhao YL et al (2014) A mini-review of chemical composition and nutritional value of edible wild-grown mushroom from China. Food Chem 151:279–285
Wani BA, Bodha RH, Wani AH (2010) Nutritional and medicinal importance of mushrooms. J Med Plant Res 4(24):2598–2604
Xiang Y (1991) A new granular structure medium for spawn manufacture and the preservation of strains. In: Maher MJ (ed) Mushroom science, science and cultivation of edible fungi, vol 1. Balkema/Illus Maps, Rotterdam/Brookfield, pp 123–124
Yap AT, Ng ML (2001) An improved method for the isolation of lentinan from the edible and medicinal shiitake mushroom *Lentinus edodes* (Berk.) Sing. (Agaricomycetideae). Int J Med Mushrooms 3:9–19
Yi J, Chen TQ, Wu JZ (2006) Studies on dietary fiber in mushrooms. J Fungal Res 4(2):49–53

Chapter 2
Fungi as Biocontrol Agent: An Alternate to Chemicals

Sangeeta Singh, Shiwani Bhatnagar, Sunil Choudhary, Bindu Nirwan, and Kuldeep Sharma

Abstract To increase production, we are using fungicides indiscriminately leading to much negative effect on humans, animals and environment. An alternative to these fungicide is application of biological control agents which not only helps directly in management of diseases below economic threshold level but also have many folds beneficial effect on growth and production. Out of these biocontrol agent's fungus plays very important role. These fungi are ubiquitous in nature, and many strains are present within the species making it more specific against insects and diseases. They are self-sustainable since spores are the means by which the infection occurs, which are produced in large numbers and are produced continuously as long as the growth conditions for it remain favourable. Thus, cost of application is also reduced. Moreover, their handing and application are also convenient, and they neither cause any harmful effect to humans and livestock nor cause any other environmental issues. The main advantage is that they readily fit into the integrated management programmes.

Keywords Fungi · Biocontrol

2.1 Introduction

To increase production, predominantly where production methods were intensified to increase agricultural output and where different crop species were introduced in a new agroecological zones, the use of chemical pesticides has increased rapidly (Jatala 1986). Synthetic pesticides used in agriculture can have negative impact on humans and the natural environment as Geiger et al. (2010) reported that the use of insecticides and fungicides had not only consistent undesirable effects on biodiversity but they reduce the opportunities for biological pest control (Geiger et al. 2010). Studies have shown that chemical pesticides remain in the atmosphere and soil and

S. Singh (✉) · S. Bhatnagar · S. Choudhary · B. Nirwan · K. Sharma
Forest Protection Division, Arid Forest Research Institute, Jodhpur, Rajasthan, India

P. Gehlot, J. Singh (eds.), *Fungi and their Role in Sustainable Development: Current Perspectives*, https://doi.org/10.1007/978-981-13-0393-7_2

enter in water channel polluting the environment. According to a report, 15% of chemical fungicide is used against 1% of the biological products among agricultural control measures (Fravel 2005).

The damage caused to the crops due to plant diseases can be reduced by application of biological control agent which is an environment-friendly strategy for management of diseases (Murali et al. 2012). The microorganisms present in the soil are an efficient indicator of soil health and are the most diverse and abundant group of organisms on Earth. Although a large number of microbes are considered as pests of plants and animals, for some group, these, in fact, are beneficial, and without their presence, it will be very difficult to sustain human life. Biological control of plant pathogens using natural resources is a widely accepted practice of management for sustainable agriculture (Azcón-Aguilar and Barea 1997) in which not only beneficial microorganism but their genes and other products like secondary metabolites are being used which reduced the damage caused by plant pathogens and promote positive response in terms of plant growth (Vinale et al. 2008). India has a vast potential for biopesticides (Gupta and Dikshit 2010). Globally also biopesticides are being used for managing insect pests and diseases (Kumar 2012). Although biological control is less spectacular and takes more time than most physical or chemical controls, it is usually also more stable and longer lasting due to its self-sustainability (Baker and Cook 1974).

Fungi are a diverse group of organisms and are ubiquitous in nature. It is present everywhere, in almost every environment on Earth. Most of the fungi are parasitic in nature causing direct or indirect impairment to various eukaryotes, including plants and insects. The fungal biological control agents offer considerable potential for insect, disease and weed control (Butt and Copping 2000). Some fungal endophytes have also reported to be good biocontrol agents (Mejía et al. 2008). The most important feature of biological control agent is that it should be highly host specific. The existence of many strains within species of plant pathogen such as rust fungi necessitates selecting the most efficient one, which can reduce the adverse effect of the target weed (Sheppard 2003).

In this chapter, we will discuss about some fungi which play very important role in ecological balance. The important feature of fungi is that they can be mass multiplied in large quantities on artificial media and their storage is also very easy. Moreover, their method of application is also very convenient using conventional spray which is nontoxic if inhaled and does not have any environmental issues.

2.2 Fungi as Fungicide

Plant pathogenic fungi are the causal agents of the many detrimental infectious diseases in different plants and trees and thereby cause considerable yield losses all over the world. Rhizosphere of plants has association of many microorganism which shows antagonistic potential against soilborne pathogens (Cook and Baker 1983; Elad et al. 1986). Seed treatment with biological control agents can protect the seeds from the seed- and soilborne fungal pathogens like damping-off. The interactions among the plant, the pathogen, the biocontrol agent, the microbial community on and around the plant and the physical environment lead to inhibition of disease.

Some fungi are obligate parasites which require the living host to grow and reproduce, but majority of them are saprophytic which doesn't need any host and can survive in the soil, water or air. Although a large number of fungicides are present in the market which can manage these fungal pathogens at economic threshold level, the insignificant use of these fungicides not only is responsible for changing the genetic makeup of the pathogens, making them more resistant to these chemicals, but also pollutes environment. Fungicide of biological origin is used nowadays to manage these fungal pathogens. Not only seed- and soilborne pathogens but foliar pathogens are also now being managed by this alternative means of chemical fungicide. Biological control of soilborne plant pathogens is a potential substitute to the use of chemical pesticides, which have already been proven to be harmful to the environment (Chet and Inbar 1994). *Trichoderma harzianum* is one of the most studied commercial biocontrol agents which have been reported to control the foliar pathogens *Botrytis cinerea*, *Pseudoperonospora cubensis*, *Sclerotinia sclerotiorum*, *Sphaerotheca fusca*, *Fusarium*, *Pythium* and *Rhizoctonia* (syn. *S. fuliginea*) (Elad 2000).

The fungal plant pathogens are parasitized by *Trichoderma* spp. The hyphal branches of the parasite extend in the direction of the target host pathogen where it coils around the fungal hyphae and attaches itself with the help of appressorium-like bodies and punctures the host mycelium (Chet and Baker 1981; Goldrnan et al. 1994). In another study for management of *Ganoderma* root rot disease, the use of *T. harzianum* has been suggested as a potential antagonist (Nirwan et al. 2016). Degradation of plant pathogenic fungi by *Trichoderma harzianum* was earlier studied by Elad et al. (1882) where they have reported that *Trichoderma harzianum* showed high activity of enzymes like β-1, 3-glucanase, chitinase, protease and lipase when grown with *S. rolfsii* in media. Another study indicates that *Trichoderma* shows chelation and reduction activities, thereby helping in biocontrol of plant pathogens (Altomare et al. 1999). Approximately 40 fungal species have been so far tested and reported as the potential biocontrol agents and natural antagonists of powdery mildews (Kiss 2003).

The association of vesicular-arbuscular mycorrhizal (VAM) with the root of plants has also been reported to reduce the damage caused by soilborne plant pathogens, although here the protective role played is by indirect method by the improvement of plant nutrition, compensation for pathogen damage and competition for photosynthates or colonization/infection sites and not by directly killing the pathogen (Azcón-Aguilar and Barea 1997). *Pythium aphanidermatum* which causes damping-off disease in *Aquilaria agallocha* seedlings was managed by application of *Glomus fasciculatum*, which not only helped directly by reducing the percentage of disease incidence but also indirectly by significantly increasing host plant height, total biomass and dry matter (Tabin et al. 2009). Similarly, *Coniothyrium minitans* can be used for biological control *Sclerotinia sclerotiorum,* a widespread pathogen which substantially reduces the yield of many crops since it is a naturally occurring parasite of the pathogen (De Vrije et al. 2001).

Sometimes hypovirulent strains of the parasitic fungus are also used to control the virulent strains. Chestnut blight disease caused by *Cryphonectria parasitica* was successfully controlled by using its hypovirulent strains (Lee et al. 2006). Similarly, *Helicobasidium mompa* also contains adsRNA fragment which is associated with hypovirulence in it (Ikeda et al. 2003). The typical characteristic of hypovirulent strains is the presence of dsRNA virus in the cytoplasm. Transmission of double-stranded RNA mycoviruses from hypovirulent strains to virulent strains renders the virulent strains hypovirulent and makes it unable to infect or spread further (Sneh 1998). Thus, nonpathogenic strains of various fungi could be used for the development of biocontrol preparations. List of some fungi as potential biocontrol agent which is produced commercially is given below:

2.3 Fungi as Nematicide

Among the microorganisms that parasitize or prey on nematodes or reduce nematode populations by their antagonistic behaviour, fungi hold significant positions, and some of them have shown great potential as biocontrol agents (Siddiqui and Mahmood 1996). The fungi and the nematode live in close association in rhizosphere; nematodes affect the quality and quantity of the exudates which may act as a signal for microorganism that are either antagonistic or had mutualistic relationship. The antagonistic fungi continuously destroy nematodes either by trapping it or by producing metabolites.

A range of specialist and generalist microorganism in the rhizosphere attacks plant-parasitic nematodes (Kerry 2000). Although there is a long list of fungi preying nematodes, the most important are *Paecilomyces*, *Verticillium*, *Hirsutella*, *Nematophthora*, *Arthrobotrys*, *Drechmeria*, *Fusarium* and *Monacrosporium*. *Verticillium chlamydosporium* was also reported to be a promising nematophagous fungus for management of some cyst and root-knot nematodes (De Leij and Kerry 1991). The mode of action is either by trapping in the hyphal network of the fungi or by producing secondary metabolites or nematicides. *Paecilomyces lilacinus* has been reported to significantly reduce the tuber and root galling caused by *M. incognita* in potato (Jatala 1985). The *Trichoderma* strains have been reported to colonize *M. javanica*-separated eggs and second-stage juveniles' in vitro condition and have also been reported to penetrate the egg masses (Sharon et al. 2001).

The toxicity of culture filtrate of different fungi on *Meloidogyne* larvae has been studied earlier (Mani and Sethi 1984; Dahiya and Singh 1985). The toxic culture filtrate of the nematophagous fungus *Paecilomyces lilacinus* not only kills the adult but also destroys the eggs of *Meloidogyne arenaria*. The mechanism involved in the toxin activity was found to be neurotropic one since a reversible effect was observed when the treatment period was less than 48 hours. The secondary metabolites of nematode-trapping fungi having antimicrobial and nematicidal activities were screened against few nematodes, and the results gave three new antimicrobial metabolites from cultures of five *Arthrobotrys* strains. The compounds exhibited nematicidal activities towards *Caenorhabditis elegans* and *Meloidogyne incognita* (Anke et al. 1995). The endophytes, *Phomopsis phaseoli*, and four strains of *Melanconium betulinum* isolated from leaf of a tropical tree and twigs of *Betula pendula* and *B. pubescens*, respectively, produce 3-hydroxypropionic acid which showed selective nematicidal activity against the plant-parasitic nematode *Meloidogyne incognita* with LD_{50} values of 12.5–15 μg/ml (Schwarz et al. 2004). Further, it was suggested that enhanced proteolytic activity of the antagonist may be essential for the biological control of the nematodes (Sharon et al. 2001).

2.4 Fungi as Weedicide

Weed problem is one of the major threats during cultivation of crops. Many of the weed problems are linked with extensive agricultural practices in which biological controls are often only practical method of management. Australia is world's leader in the development and implementation of practical weed biological control (Julien and white 1997). In biological weed control, living organisms used to reduce weed populations include insects, nematodes, bacteria or fungi. Parasitic fungi are used extensively as biocontrol agents. Fungal pathogens are either non-host specific and can infect a wide range of plant species or are host specific and restricted to one or few host species, and thus they can be a very effective biological control agent. *Colletotrichum truncatum* is one of the bioherbicidal fungi which was effectively used to manage *Sesbania exaltata* in filed condition (Boyette et al. 1993). There are certain criteria which are considered when a pathogen is tested as biological control of weed, that is, the taxonomy, life cycle, pathogenicity testing and host specificity should be tested. The mechanism of action may be, by reducing the leaf surface, thus affecting the photosynthetic ability. A rust fungus, pathogen for rush skeleton weed, has been used to eliminate it from thousands of acres of rangeland in the West. Prospects for the control of *Cyperus rotundus*, *Lantana camara*, *Portulaca oleracea*, *Mikania micrantha*, *Cassia* spp., *Parthenium hysterophorus*, *Heliotropium indicum* and *Commelina benghalensis*, using an augmentative (mycoherbicide) approach, appear to be good (Evans 1987).

2.5 Fungi as Biopesticides

Pesticide may be defined as any substance that intends to prevent, destroy or repel any pest. Over the past 50 years, crop protection has relied heavily on synthetic chemical pesticides (Chandler et al. 2011). The use of pesticides is widespread, and there are many examples of poisoning with pesticides. The application of pesticides is also at very greater rate which is ecologically unacceptable. Though the use of pesticides in agriculture has slightly reduced, from 948 million pounds (2000) to 877 million pounds (2007), i.e. about 1% per year, still billion pounds of poisonous chemicals are deliberately introduced into the environment every year (http://www.panna.org/blog/long-last-epa-releases-pesticide-use-statistics/2016). The ill effects of synthetic chemicals have compelled researchers to search for some alternative methods. Compounds of biological origin having various biological activities are the need of the hour, and in the future, there will be social pressure to incorporated biopesticides as a tool to manage pest as these are safe to humans as well as nontarget organisms. This review outlines the current knowledge on the potential use of fungi as biopesticides in global management of pests.

Agriculture and forests serve as an important resource for sustaining global economic, environmental and social system. To attain the global challenge of securing high-quality yields of agricultural and forestry produce, chemical means of plant/forest protection occupy the primary place in integrated pest management and diseases of plants. But it is well known that pesticides cause toxicity to humans and warm-blooded animals, and also a number of factors are threatening the effectiveness and continued use of these agents. Therefore, an environment-friendly alternative is required. One such alternative is biopesticides which are effective, biodegradable and do not leave any harmful effect on environment.

Biochemical substances that manage pests below economic threshold level by non-hazardous/nontoxic mechanisms and are present naturally are called biopesticides. These are derived from living organisms (natural enemies) or their products or byproducts and can be used in combination with other substances including chemical pesticides and thus become part of bio-intensive integrated pest management programme. Although the use of plant extracts as natural biopesticides dates back to seventeenth century (http://www.bpia.org/history-of-biopesticides), the use of fungi as biopesticide was established for the first time by an Italian entomologist Agostino Bassi (1835). He observed that the fungus which was later known as *Beauveria bassiana* (name given in his honour) causes infection in silkworm (Callow et al. 2003). Many fungal pathogens act as natural control agent of insect species, and, also, they are well suited to be developed as biopesticides (Thomas and Read 2007).

Shi and Feng (2004a, b) bioassayed ten isolates of *Beauveria bassiana*, *Metarhizium anisopliae* and *Paecilomyces fumosoroseus* to check their lethal effects on the eggs of the carmine spider mite, *Tetranychus cinnabarinus*. Results confirmed the ovicidal activity of the three fungal species and suggested the feasibility to search for more ovicidal isolates from fungal species that may serve as biocontrol agents against spider mites such as *T. cinnabarinus*. Two isolates of entomopathogenic fungi, *Beauveria bassiana* SG8702 and *Paecilomyces fumosoroseus* Pfr153, were also bioassayed against *T. cinnabarinus* eggs (Shi and Feng 2004a, b). Moreover, it was reported by Wraight et al. (2000) that the microbes *B. bassiana* and *P. fumosoroseus* can manage nymphal whiteflies.

Entomopathogenic fungi belonging to the order *Hypocreales* (class: *Sordariomycetes*) have been used against *Bactericera cockerelli* (Sulc) (Hemiptera: Triozidae), a potato psyllid, (Lacey et al. 2011). The commercial mycoinsecticide 'Boverin' based on *B. bassiana* with reduced doses of trichlorfon has been used to suppress the second-generation outbreaks of *Cydia pomonella* L. (Ferron 1971). Duarte et al. (2009) demonstrated the impact of the fungus *Neozygites floridana* on the tomato red spider mite, *Tetranychus evansi* Baker and Pritchard, in the field as well as under screen houses throughout four crop cycles of tomato and nightshade in Piracicaba, SP, Brazil.

Out of the seven strains of entomopathogenic fungi against *Ceratitis capitata* adults, aqueous suspensions of conidia of five strains were found to be most effective (Castillo et al. 2000). Among these five, the extract from *M. anisopliae* was the most toxic, resulting in mortality of the insect of about 90%. The entomopathogenic fungi are even used in combination with botanical extract to increase their efficacy since there is only one obstacle in employing entomopathogenic fungi for insect pest management, that is, they may kill their host too slowly (Thomas and Read 2007). Islam et al. (2010) conducted the compatibility test of the entomopathogenic fungus *Beauveria bassiana* (Balsamo) Vuillemin with neem against sweet potato whitefly, *Bemisia tabaci* (Gennadius) (Hemiptera: Aleyrodidae), on eggplant (Table 2.1).

2.6 Advantages of Biological Control Agents

An efficient biocontrol agent should be genetically stable, effective even at very low concentrations and easy to mass produce in culture on low-cost media and have a broad host range for being effective to manage a wide range of pathogens together. Moreover, it should be able to quash the fungistatic effect of soil. The most important part is its being self-sustainable which reduces the cost of application. Other advantages of the bioagents are having no residual problem since certain fungicides have either fungicidal or fungistatic effect and thus inhibit germination of entomopathogenic fungi but have little or no effect on their virulence against target insects. The only disadvantage of fungal bioagents is that they are highly specific particularly in case of insect pest management. As compared to bacteria, the spores or sclerotia of fungal biocontrol agents are more stable than the bacterial spores. Moreover, fungi develop into epidemic on host more potentially, and this strategy helps it to manage insect pests and weed population (Jackson 1997). The area under organic cultivation (crops) in India is estimated to be around one lac hectare. Besides, there are lakhs of hectare of forest area being certified as organic (Gupta and Dikshit 2010). Since the last so many years, people are becoming aware of consequences of synthetic pesticide, and thus demand of organic cultivation is growing. These all suggest that there is wide scope for the biopesticide industries. The biopesticides work as efficiently as the conventional pesticides when used as a component in integrated pest management (IPM) practices, especially for crops like fruits, vegetables, nuts and flowers (Kumar 2012).

Table 2.1 List of fungal biocontrol agents and their target weeds, pathogens, nematodes and insect pest

Fungal biocontrol agents	Target	References
Cercospora rodmanii	Water hyacinth (*Eichhornia crassipes*)	Butt and Copping (2000)
Ampelomyces quisqualis	*Erysiphales* (powdery mildews)	
C. gloeosporioides	*Cuscuta chinensis*, *C. australis* in soybeans	Butt and Copping (2000)
Puccinia chondrillina	Rush skeleton weed in *Chondrilla juncea*	Emge et al. (1981)
Trichoderma lignorum and *T. viride*	*Rhizoctonia solani*	Weindling (1932)
	Damping-off of citrus seedlings	
Trichoderma harzianum	*Sclerotinia sclerotiorum*—a soilborne plant pathogen attacking soybean	Inbar et al. (1996)
Pythium oligandrum	*Pythium ultimum*	Butt and Copping (2000)
Gliocladium virens	Damping-off and root pathogens	Butt and Copping (2000)
Phlebiopsis (*Peniophora gigantea*)	*Heterobasidion annosum*	Butt and Copping (2000)
Trichoderma harzianum	*Meloidogyne incognita*	Haseeb et al. (2005)
Metarhizium anisopliae	Locusts, grasshoppers, cockroaches and termites, major pests	Sandhu et al. (1993)
Beauveria bassiana	Insect species, causing white muscardine disease, Colorado potato beetle, the codling moth, and several genera of termites, American bollworm *Helicoverpa armigera*	Jain et al. (2008)
Verticillium lecanii	*Trialeurodes vaporariorum*	Hamlen (1979)
	Control whitefly and several aphid's species, including the green peach aphids (*Myzus persicae*)	
Nomuraea rileyi	Insect species belonging to Lepidoptera including *Spodoptera litura* and some to Coleoptera are very susceptible	Ignoffo (1981)
Paecilomyces lilacinus	Root-knot and cyst nematodes	Seryczynska and Bajan (1975)
Paecilomyces fumosoroseus	Controlling the nymphs of whitefly	Kim et al. (2002)

2.7 Disadvantages

One of the most important disadvantages of biopesticides is that it has a relatively short shelf life that can be for a few weeks and is highly sensitive to the environmental conditions. But, this constrain can be neutralized to several years depending on storage conditions. The other disadvantage is selectivity of biopesticides which may necessitate the use of conventional pesticides to control other pests, which can lead to compatibility issues. Moreover, the farmers want a quick response which can be attained only by synthetic fungicides, since the biocontrol agents are alive, and they require some time for augmentation.

References

Altomare C, Norvell WA, Björkma T, Harman GE (1999) Solubilization of phosphates and micronutrients by the plant-growth-promoting and biocontrol fungus *Trichoderma harzianum*. Appl Environ Microbiol 65(7):2926–2933

Anke H, Stadler M, Mayer A, Sterner O (1995) Secondary metabolites with nematicidal and antimicrobial activity from nematophagous fungi and ascomycetes. Can J Bot 73:932–939

Azcón-Aguilar C, Barea JM (1997) Arbuscular mycorrhizas and biological control of soil-borne plant pathogens–an overview of the mechanisms involved. Mycorrhiza 6(6):457–464

Baker KF, Cook RJ (1974) Biological control of plant pathogens. WH Freeman and Company, San Francisco

Bassi A (1835) English translation by Yarrow PJ. 1958. On the mark disease, calcinacccio or muscardine, a disease that affects silk worms. Ainsworth GC, Yarrow PJ (ed) APS Phytopathol. Classics 10:1–49

Boyette CD, Quimby PC Jr, Bryson CT, Egley GH, Fulgham FE (1993) Biological control of hemp sesbania (*Sesbania exaltata*) under field conditions with *Colletotrichum truncatum* formulated in an invert emulsion. Weed Sci:497–500

Butt TM, Copping LG (2000) Fungal biological control agents. Pestic Outlook 11(5):186–191

Callow M, Drmanac R, Drmanac S, Callow MJ, Drmanac Radoje T (2003) Universal selective genome amplification and universal genotyping system. US Patent Application 10:608–293

Castillo MA, Moya P, Hernández E, Primo-Yufera E (2000) Susceptibility of *Ceratitis capitata* Wiedemann (Diptera: Tephritidae) to entomopathogenic fungi and their extracts. Biol Control 19(3):274–282

Chandler D, Bailey AS, Tatchell GM, Davidson G, Greaves J, Grant WP (2011) The development, regulation and use of biopesticides for integrated pest management. Philos Trans R Soc Lond Ser B Biol Sci 366(1573):1987

Chet H, Baker R (1981) Isolation and biocontrol potential of *Trichoderma hamatum* from soil naturally suppressive to *Rhizoctonia solarti*. Phytopathology 71:286–290

Chet I, Inbar J (1994) Biological control of fungal pathogens. Appl Biochem Biotechnol 48(1):37–43

Cook RJ, Baker KF (1983) The nature and practice of biological control of plant pathogens. American Phytopathological Society, St. Paul

Dahiya JS, Singh DP (1985) Inhibitory effects of *Aspergillus niger* culture filtrate on mortality and hatching of larvae of *Meloidogyne* sps. Plant Soil 86(1):145–146

de Leij FAAM, Kerry BR (1991) The nematophagous fungus Verticillium chlamydosporium as a potential biocontrol agent for Meloidogyne arenaria. Revue de Ne'matologie 14:157–164

De Vrije T, Antoine N, Buitelaar RM, Bruckner S, Dissevelt M, Durand A, Gerlagh M, Jones EE, Lüth P, Oostra J, Ravensberg WJ (2001) The fungal biocontrol agent *Coniothyrium minitans*: production by solid-state fermentation, application and marketing. Appl Microbiol Biotechnol 56(1):58–68

Duarte VS, Silva RA, Wekesa VW, Rizzato FB, Dias CTS, Delalibera I (2009) Impact of natural epizootics of the fungal pathogen *Neozygites floridana* (Zygomycetes: Entomophthorales) on population dynamics of *Tetranychus evansi* (Acari: Tetranychidae) in tomato and nightshade. Biol Control 51(1):81–90

Elad Y (2000) Biological control of foliar pathogens by means of *Trichoderma harzianum* and potential modes of action. Crop Prot 19(8):709–714

Elad Y, Zvieli Y, Chet I (1986) Biological control of *Macrophomina phaseolina* (Tassi) Goid by *Trichoderma harzianum*. Crop Prot 5(4):288–292

Emge RG, Melching JS, Kingsolver CH (1981) Epidemiology of *Puccinia chondnllina*, a rust pathogen for the biological control of rush skeleton weed in the United States. Phytopathology 7:839–843

Evans HC (1987) Fungal pathogens of some subtropical and tropical weeds and the possibilities for biological control. Biocontrol News Inf 8(1):7–30

Ferron P (1971) Influence of relative humidity on development of fungal infection caused by *Beauveria bassiana*. Entomol Exp Appl 14:57–76

Fravel DR (2005) Commercialization and implementation of biocontrol. Annu Rev Phytopathol 43:337–359

Geiger F, Bengtsson J, Berendse F, Weisser WW, Emmerson M, Morales MB, Ceryngier P, Liira J, Tscharntke T, Winqvist C, Eggers S (2010) Persistent negative effects of pesticides on biodiversity and biological control potential on European farmland. Basic Appl Ecol 11(2):97–105

Goldrnan GH, Vasseur V, Contreras R, van Montagu M (1994) Sequence analysis and expression studies of a gene encoding a novel serine + alanine-rich protein in *Trichoderma harzianum*. Gene 144:113–117

Gupta S, Dikshit AK (2010) Biopesticides: an ecofriendly approach for pest control. J Biopest 3(1):186–188

Hamlen RA (1979) Biological control of insects and mites on European greenhouse crops: research and commercial implementation. Proce Florida State Hort Soci 92:367–368

Haseeb A, Ahmad V, Shukla PK (2005) Comparative efficacy of pesticides, bio-control agents and botanicals against Meloidogyne incognita-fusarium oxysporum disease complex on Vigna mungo. Ann Plant Prot Sci 13(2):434–437

Ignoffo CM (1981) The fungus *Nomuraea rileyi* as a microbial insecticide: fungi. In: Burges HD (ed) Microbial control of pests and plant diseases. Academic Press, London, pp 513–538

Ikeda KI, Nakamura H, Matsumoto N (2003) Mycelial incompatibility operative in pairings between single basidiospore isolates of *Helicobasidium mompa*. Mycol Res 107(7):847–853

Inbar J, Menendez A, Chet I (1996) Hyphal interaction between *Trichoderma harzianum* and *Sclerotinia sclerotiorum* and its role in biological control. Soil Biol Biochem 28:757–763

Islam M, Castle SJ, Ren S (2010) Compatibility of the insect pathogenic fungus *Beauveria bassiana* with neem against sweet potato whitefly, *Bemisia tabaci*, on eggplant. Entomol Exp Appl 134(1):28–34

Jackson MA (1997) Optimizing nutritional conditions for the liquid culture production of effective fungal biological control agents. J Ind Microbiol Biotechnol 19(3):180–187

Jain N, Rana IS, Kanojiya A, Sandhu SS (2008) Characterization of *Beauveria bassiana* strains based on protease and lipase activity and their role in pathogenicity. J Basic Appl Mycol I-II:18–22

Jatala P (1985) Biological control of nematodes. An advanced treatise on Meloidogyne. Biol Control 1:303–308

Jatala P (1986) Biological control of plant-parasitic nematodes. Annu Rev Phytopathol 24(1):453–489

Julien MH, White G (1997) Biological control of weeds: theory and practical application. Canberra, Australian Centre for International Agricultural Research

Kerry BR (2000) Rhizosphere interactions and the exploitation of microbial agents for the biological control of plant-parasitic nematodes. Annu Rev Phytopathol 38(1):423–441

Kim JJ, Lee MH, Yoon CS, Kim HS, Yoo JK, Kim KC (2002) Control of cotton aphid and greenhouse whitefly with a fungal pathogen. J Nat Inst Agric Sci Technol:7–14

Kiss L (2003) A review of fungal antagonists of powdery mildews and their potential as biocontrol agents. Pest Manag Sci 59(4):475–483

Kumar S (2012) Biopesticides: a need for food and environmental safety. J Biofertil Biopestic 3(4):1–3

Lacey LA, Liu TX, Buchman JL, Munyaneza JE, Goolsby JA, Horton DR (2011) Entomopathogenic fungi (Hypocreales) for control of potato psyllid, *Bactericera cockerelli* (Šulc)(Hemiptera: Triozidae) in an area endemic for zebra chip disease of potato. Biol Control 56:271–278

Lee SH, Lee S, Choi D, Lee YW, Yun SH (2006) Identification of the down-regulated genes in a mat1-2-deleted strain of Gibberella zeae, using cDNA subtraction and microarray analysis. Fungal Genet Biol 43:295–310

Mani A, Sethi CL (1984) Effect of culture filtrates of *Fusarium oxysporum* f. Sp. ciceri and *Fusarium solani* on hatching and juvenile mobility of *Meloidogyne incognita*. Nematropica 14(2):139–144

Mejía LC, Rojas EI, Maynard Z, Van Bael S, Arnold AE, Hebbar P, Samuels GJ, Robbins N, Herre EA (2008) Endophytic fungi as biocontrol agents of *Theobroma cacao* pathogens. Biol Control 46(1):4–14

Murali M, Amruthesh KN, Sudisha J, Niranjana SR, Shetty HS (2012) Screening for plant growth promoting fungi and their ability for growth promotion and induction of resistance in pearl millet against downy mildew disease. J Phytology 4(5):1

Nirwan B, Choudhary S, Sharma K, Singh S (2016) *In vitro* studies on management of root rot disease caused by *Ganoderma lucidum* in on *Prosopis cineraria*. Curr Life Sci 2(4):118–126

Sandhu SS, Rajak RC, Agarwal GP (1993) Studies on prolonged storage of *Beauveria bassiana* conidia: effects of temperature and relative humidity on conidial viability and virulence against chickpea borer *Helicoverpa armigera*. Biocont Sci Technol 3:47–53

Schwarz M, Köpcke B, Weber RW, Sterner O, Anke H (2004) 3-Hydroxypropionic acid as a nematicidal principle in endophytic fungi. Phytochemistry 65(15):2239–2245

Seryczynska H, Bajan C (1975) Defensive reactions of L3, L4 larvae of the Colorado beetle to the insecticidal fungi *Paecilomyces farinosus* (Dicks) Brown et Smith, *Paecilomyces fumosoroseus* (Wize), *Beauveria bassiana* (Bols/Vuill.) (Fungi Imperfecti: Moniliales). Bulletin de l'Academie Polonaise des Sciences 23(4):267–271

Sharon E, Bar-Eyal M, Chet I, Herrera-Estrella A, Kleifel O, Spiegel Y (2001) Biological control of the root-knot nematode *Meloidogyne javanica* by *Trichoderma harzianum*. Phytopathology 91(7):687–693

Sheppard AW (2003) Prioritising agents based on predicted efficacy: beyond the lottery approach. CRC for Australian weed management technical series, vol 7, pp 11–22

Shi W, Feng M (2004a) Ovicidal activity of two fungal pathogens (Hyphomycetes) against *Tetranychus cinnabarinus* (Acarina: Tetranychidae). Chin Sci Bull 49(3):263–267

Shi WB, Feng MG (2004b) Lethal effect of *Beauveria bassiana, Metarhizium anisopliae*, and *Paecilomyces fumosoroseus* on the eggs of *Tetranychus cinnabarinus* (Acari: Tetranychidae) with a description of a mite egg bioassay system. Biol Control 30(2):165–173

Siddiqui ZA, Mahmood I (1996) Biological control of plant parasitic nematodes by fungi: a review. Bioresour Technol 58(3):229–239

Sneh B (1998) Use of non-pathogenic or hypovirulent fungal strains to protect plants against closely related fungal pathogens. Biotechnol Adv 16(1):1–32

Tabin T, Arunachalam A, Shrivastava K, Arunachalam K (2009) Effect of arbuscular mycorrhizal fungi on damping-off disease in Aquilaria agallocha Roxb. Seedlings. Trop Ecol 50(2):243

Thomas MB, Read AF (2007) Can fungal biopesticides control malaria? Nat Rev Microbiol 5(5):377

Vinale F, Sivasithamparam K, Ghisalberti EL, Marra R, Woo SL, Lorito M (2008) *Trichoderma*–plant–pathogen interactions. Soil Biol Biochem 40(1):1–10

Weindling R (1932) Trichoderma lignorum as a parasite of other soil fungi. Phytopathology 22(8):837–845

Wraight SP, Carruthers RI, Jaronski ST, Bradley CA, Garza CJ, Galani-Wraight S (2000) Evaluation of the entomopathogenic fungi *Beauveria bassiana* and *Paecilomyces fumosoroseus* for microbial control of the silver leaf whitefly, *Bemisia argentifolii*. Biol Control 17:203–217

Chapter 3
Incredible Role of Fungi in Various Fields for Sustainable Development

Tulika Mishra and Gunjan Mukherjee

Abstract Fungi are leaving its remarkable contribution in human race since a long back. History dictates its utilization in one of the oldest civilizations, Chinese civilization. It has been and still playing a remarkable role in many fields. Fungi are playing implausible role in food and feed providing many dietary supplements. Its role in impacting our daily life cannot be denied, as it is now used in food and feed industry. The most common fungus *Agaricus bisporus* is one of the examples of most consumable fungi across the world. Utilization of baker's yeast is another such example. Not only this, its many species are utilized for the production of pharmaceutical proteins, immunosuppressive agents, and antitumor agents and in other medical fields. Fungi have extended its helping hand to sustain agriculture too. Its role as biofertilizer and as biocontrol agent enhances its utilization and importance. Fungi are playing significant role in bioremediation of heavy metals and other pesticides also. In addition fungi like *Phlebia radiata* or *Poria subvermispora* are utilized in paper industry. There are fields where fungus is showing good results at the starting level and needs to be explored for its utilization at higher scale. This chapter is about putting some light on significance of fungi in various fields.

Keywords Fungi · Agriculture · Food and feed · Bioremediation · Paper industry · Medical

3.1 Introduction

Since time immemorial, fungi have been an important part of human race. Its historical usage in various form is well documented, whether its use as poison by Agrippina for poisoning of Roman emperor Claudius (Ramsbottom 1972) or use as a food as shown in the Chapel of Plaincourault (Wasson 1968). The human race is utilizing nature's gift for various products like bread, cheese, wine, etc. Earlier, the mycology used to focus only toward those areas where fungi are causing

T. Mishra · G. Mukherjee (✉)
Department of Biotechnology, Chandigarh University, Mohali, Punjab, India

P. Gehlot, J. Singh (eds.), *Fungi and their Role in Sustainable Development: Current Perspectives*, https://doi.org/10.1007/978-981-13-0393-7_3

damage, but with the passage of time, change has been observed. New era of fungi started with World War I, where utilization of microbial world took a leap. The breakthrough was discovery of penicillin by Alexander Fleming. Continuous efforts in the field have now resulted into many beneficial products including enzymes, antibiotics, vitamins, etc. There are many industrial processes like production of polyhydric alcohols, enzymes, pigments, etc., where contribution of fungi cannot be denied.

Fungi kingdom is providing enormous diversity with approximately 70,000 known species. Most of the known fungi are filamentous, differing from yeast in their morphology, development, and metabolic complexity. Ascomycetes and basidiomycetes have entered in the various fields providing their benefits to human race (Berdy 2005). The involvement of fungal enzymes in various industries, like food, animal feed, baking, and detergent textiles, is usually from the filamentous fungi (Guimaraes et al. 2006). Enzymes like phosphatases are used in biological process including cell cycle, differentiation, etc. Phosphatases are classified as alkaline phosphatases (used in ELISA assay, blotting sequencing, etc.) and acid phosphatases (used in animal feed). Esters and anhydrides of phosphoric acid are hydrolyzed by phosphatase. Amylase is another important enzyme that is used in food, baking, detergents, and drinks. α-Amylases and glucoamylases help in conversion of starch to glucose (Pandey et al. 2000). Pectinases are another important enzyme produced by fungi and used in various wine and fruit juices. Pectinases act on pectin, which is a polymeric substance present in fruit lamella and cell walls. Pectinases break the fruit cell wall, thereby helping in juice extraction (Sathyanarayana and Panda 2003).

Large bulk of lignocellulose is produced on earth as these are the polysaccharides of plants. But its chunk does not accumulate on the earth due to lignocellulolytic enzymes produced by many microorganisms. Among the various microorganisms, fungi play a very important role. Degradation of lignocellulose involves various enzymes including ligninases represented by laccases and peroxidases and hemicellulases represented by xylanases and cellulases (Sajith et al. 2016). Fungi have been also used in various medical fields like production of mammalian gene, human interferon, human hemoglobin gene, etc. Since prebiblical time, fungi are used in food also, as they possess high content of protein, have all essential amino acids and high vitamin content, and are low in fat. Thirteen thousand years ago, inhabitants of Chile used to consume wild mushroom (Rojas and Mansur 1995). Records have shown the use of fermented foods in Sumeria and Babylon (Elander and Lowe 1994). There are many other records that show the usage of fungus kingdom in many fields even before the birth of Christ. With the passage of time and development of molecular techniques, new ways have been developed to use various fungal strains. Many strains have acquired the status of GRAS. Fungi are playing miraculous roles in many fields, and in this chapter light is thrown on various fields where fungi are playing an incredible role.

3.2 Fungi in Food and Feed

There are many species of mushroom that are consumed in a processed form or even directly as a replacement of nonvegetarian diet for the rich source of protein. There are approximately 100 species of mushroom that are edible, but only few are extensively used. The most common of them is *Agaricus bisporus* commonly called as button mushroom or Portobello mushrooms, used as vegetable, as salad, in soups, etc. (Beelman et al. 2003). There are many Asian fungi including straw mushrooms (*Volvariella volvacea*), oyster mushrooms (*Pleurotus ostreatus*), shiitakes (*Lentinula edodes*), and enokitake (*Flammulina* spp.) that are commonly found in market. After *Agaricus bisporus*, *Lentinula edodes* (Shiitake mushroom) is the second most cultivated variety. These are the rich source of essential and nonessential amino acids. Oyster mushroom (*Pleurotus ostreatus*) is also one of the common varieties which is a saprobic fungus that grows on dead trees in nature (Cohen et al. 2002). Truffle (*Tuber melanosporum*) is a European mushroom that is nearly white when young and as it matures it turns darker. Its aroma is very strong and is used as flavoring agent. *Tremella fuciformis*, commonly called as silver ear, is widely used in China and regarded as Chinese delicacy and also used as herb against tuberculosis, common cold, and high blood pressure (Cheung 1996).

Utilization of baker's yeast for numerous products is very old and common method for various products. Baker's yeast is a fungal strain of *Saccharomyces cerevisiae* and is used by mixing with bread dough for the sugar fermentation, resulting in the leavening effect. Other product like Quorn is produced by the fermentation of *Fusarium venenatum* and is used as mycoprotein supplement. Use of *Saccharomyces cerevisiae* in brewing and wine-making industries is very well known. In Japan, *Aspergillus oryzae* is used for the production of "sake" by inoculating rice with them (Yoshizawa 1995). Fungus plays an important role in converting starch to simple sugar. *Penicillium, P. camemberti*, are used for the preparation of cheese. Eastern culture has utilized mycelial fungi (*Aspergillus oryzae*) for the formation of "soya sauce." Enzymes like pectinases and xylanase from the fungal origin are used to clarify juice and feed formulation, respectively.

The most popular ruminant diet fungal DFM (direct-fed microbial) has been produced from various strains of fungi including Saccharomyces *cerevisiae* (Martin and Nibs 1992) and *Aspergillus oryzae*. There are many other recombinant proteins that are used in food and feed industry, but they showed low yields, and interest has been moved toward the food-grade proteins that are in use which are of fungal origin (Archer 2000). Recombinant protein from *Aspergillus oryzae* producing proteinase from *Rhizomucor miehei* is approved by FDA for cheese production (Pariza and Johnson 2001). Fungal secondary metabolites are also used as food colorant. *Monascus purpureus* has been traditionally used in the production of red wine and also used to prepare koji or ang-kak (red rice) (Li et al. 1998). *Phaffia rhodozyma*, source of astaxanthin (carotenoid), is used to impart pink color in salmonids. The carotenoid pigment astaxanthin has important applications in the nutraceutical, cosmetics, food, and feed industries. *Termitomyces* which is a filamentous basidiomycota is a producer of extracellular glucosidases and is capable of hydrolyzing the polysaccharides and thus used for softening and leavening of bread and also in clarification of non-citrus fruit juice (Wei and Yao 2003).

3.3 Role of Fungi in Sustainable Agriculture

3.3.1 Fungal Biofertilizers

Biofertilizers are formed from the microbial inocula or collection of various microorganisms that supports the plant growth and helps in increased crop yield (Fuentes-Ramirez and Caballero-Mellado 2005). These microorganisms help in faster uptake of nutrients from the soil and solubilization of phosphorous and are also able to fix atmospheric nitrogen (Malik et al. 2005). Arbuscular mycorrhizal fungi (AMF) are one such type of biofertilizer and are abundant in agriculture soil (Khan 2006). It has been reported to increase the crop yield to a greater extent by enhanced nutrient uptake, by stimulation of plant growth hormone, and by faster decomposition of organic residues (Wani and Lee 2002). While providing so much benefit to plants, these fungi take sugars from the plant for their sustainability, and this type of symbiotic relationship is known as mycorrhiza. There are seven types of mycorrhiza but ectomycorrhizae (ECM) and arbuscular mycorrhizae (AM) are the dominant ones. These help in faster nutrient and water uptake and also provide protection to plant from other pathogens, but there are a lesser number of plants that form association with AM and ECM (Siddiqui and Pichtel 2008).

There are reports showing the association between ECM fungi and forest trees that includes willow, poplar, eucalyptus, pine, etc. (Rinaldi et al. 2008). Most of the ECM fungi that are allied with forest trees are basidiomycetes i.e. *Pisolithus* sp., *Lactarius* sp., *Rhizopogon* sp., and *Amanita* sp. (Buyck et al. 2008; Rinaldi et al. 2008). These ECM fungi help in tree growth faster than that of nonmycorrhizal roots by enhancing faster nutrient and water uptake, making trees even drought tolerant of *Pisolithus tinctorius* which is one of the most commonly used ECM (Gentili and Jumpponen 2006). Another important mycorrhiza is an arbuscular mycorrhiza (AM). Here the fungus in used is *Endomycorrhizae*. Endomycorrhizal fungi penetrate the root cortical cells and form arbuscular vesicles and hence are known as vesicular arbuscular mycorrhiza (VAM). AM fungi belong to nine genera: *Acaulospora*, *Archaeospora*, *Entrophospora*, *Gerdemannia*, *Geosiphon*, *Gigaspora*, *Glomus*, *Paraglomus*, and *Scutellospora* (Das et al. 2007). AM fungi are found in most of the agricultural and natural ecosystem. AM fungi help in plant growth and nutrition by helping plant for faster uptake of nutrients and water especially when it is available in a lesser quantity and making plant drought tolerant, metal tolerant, and temperature tolerant, improving soil quality, and reducing pathogen attack (Chen 2006).

In addition to this, there are other fungal species that are acting as biofertilizer following the same mechanism, including *Penicillium* species, viz., *Penicillium bilaiae*, *Penicillium radicum*, and *P. italicum* (Wakelin et al. 2004; El-Azouni 2008). Species of *Aspergillus* include *A. flavus*, *A. niger*, and *A. terreus* (Akintokun et al. 2007). *Trichoderma* species are used as both biofertilizer and biofungicide.

3.3.2 Fungi as Biocontrol Agent

The overuse of chemical pesticides and their detrimental effects on environment made the researchers to turn toward the safe eco-friendly technology that can control pest without alarming effects (Calhelha et al. 2006; Haggag and Mohamed 2007). These have led to a new world of usage of biocontrol agents. It's a need of time to control plant diseases and maintain quality of food, feed, and fiber around the world. Biocontrol agents refer to the usage of natural product that can be extracted or fermented from various sources that can control other harmful organism or natural enemies. Fungi are playing a remarkable and appreciable role to work as biocontrol agent. Fungi are generally genetically stable; show better destruction of host; are ubiquitous; have high persistence, good dispersal efficiency, and ease of culture maintenance; are effective at low concentration and effective against wide range of pathogens; have resistance to pesticides; and are nontoxic to humans. These are the factors which make them suitable biocontrol agents (Irtwange 2006). Fungi follow various mechanisms to work as biocontrol agent, and it includes direct antagonism, antibiosis, or competition. Direct antagonism is the death of pathogen by other microorganisms, and this phenomenon is called hyperparasitism, and the fungi that follow this mechanism are termed as mycoparasites (Bakers and cook 1974). One of the fungi that has shown this mechanism is *Ampelomyces quisqualis* that is used against powdery mildews. It infects to form fruiting bodies within powdery mildews hyphae, conidiophores, and cleistothecia. Hyperparasitism reduces the growth and eventually kills the mildew colony (Kiss 2003). Antibiosis is another mechanism that involves a procedure of secretion of antimicrobial compounds by antagonist fungi to kill or suppress pathogenic fungi within its growth area. *Trichoderma* species produces many secondary metabolites, including trichoviridin, gliotoxin, viridian, harzianic acid, etc., that possess antibiotic activity and when combined with various cell wall-degrading enzymes produces inhibitory effects against plant pathogens (Woo and Lorito 2007; Vinale et al. 2008). *Trichoderma harzianum* against *Sclerotinia clerotiorum* is an example of such kind of mechanism. *Sclerotinia clerotiorum* is a soilborne plant pathogen that attacks many economically important crops. Starvation is one of the major causes of death among the various microorganisms. Competition for the limiting nutrients results in the competitive mechanisms. Some of the *Trichoderma* sp. produces highly efficient siderophores that chelate iron making it unavailable for pathogenic fungi, and it is observed that it helped in the control of *Pythium* by *Trichoderma* (Chet and Inbar 1994).

Likewise, there are many other fungi that are working as mycofungicides. Few examples are *Trichoderma harzianum* against *Botrytis cinerea*, *Fusarium*, *Pythium*, and *Rhizoctonia* (Khetan 2001); *Chaetomium globosum* and *C. cupreum* can be used against *Fusarium*, *Phytophthora*, and *Pythium* (Soytong et al. 2001); *Ampelomyces quisqualis* against powdery mildew (Liang et al. 2007); *Coniothyrium minitans* – against *Sclerotinia* (Whipps et al. 2008); and *Gliocladium virens* effective against soilborne pathogens (Viterbo et al. 2007). Besides these, there are many other fungi also that are used as biocontrol agent.

3.4 Fungi in Medical Field

History flashes a light on the microbial biodiversity for producing many important products that are in use with medical perspective. Fungi are one of them that are used for commercial production of many biochemical agents that are helping human race. It has been estimated that in total, approx. there exist 1.5 million fungal species, out of which 70,000 are known. Most of them are filamentous fungi and are known for production of secondary metabolites and various enzymes (Berdy 1995). From the known antibiotics, 22% could be produced from the filamentous fungi (Berdy 1995; Strohl 1997). Many antibiotics include tetracycline, penicillin, and cephalosporin; many immunosuppressive agents like cyclosporine A and hypocholesterolemic agents like lovastatin and pravastatin are reported to be produced from various fungi. The first secondary metabolite mycophenolic acid is not in use, but its ester 2-morpholinoethylester is commercialized as immunosuppressive agent. The rice fermented with *Monascus purpureus* used as Chinese's traditional medicine contains monascorubramine and rubropunctamine (Juzlova et al. 1996). It has been reported that *Ascomycetes* and *Fungi Imperfecti* are producing approx. 6400 compounds, *Basidiomycetes* or mushrooms are producing 200 active compounds, and *Myxomycetes* are touching the range of 60 compounds (Berdy 2005). These metabolites are exhibiting various activities and playing a major role in many medical applications.

3.4.1 Antibiotics

In literal words, antibiotic is a term used to define a substance that is against life. This term came into limelight after the discovery of *Penicillin* by Alexander Fleming and was isolated from blue green mold *Penicillium notatum*. Likewise, there were tetracycline and cephalosporin, and list increased as the days passes by. Most commonly, *Aspergillales* group found to produce antibiotics. Reports are available that show that natural penicillin G and biosynthetic penicillin V have made up their market for $4.4 billion and semisynthetic antibiotic like cephalosporins has made their market even beyond $11 billion (Jose et al. 2003).

3.4.2 Pharmaceutical Proteins

Yeast is the single-celled eukaryotic fungus that is commonly used for the production of recombinant proteins and does not show problems dealing with fold and need for glycosylation as observed in *E. coli.* They exhibit higher yield, stable production strains, cost-effectiveness, high-density growth, suitability for production of isotopically labeled protein, rapid growth in chemically defined media, and product processing similar to mammalian cells, can handle S-S-rich proteins, can assist protein folding, and can glycosylate proteins (Demain and Vaishnav 2009). The two most commonly yeast strains include *S. cerevisiae* and *P. pastoris*. *S. cerevisiae* comes under the GRAS and has been considered as safer for the production of recombinant proteins. There are many genes that have been cloned and expressed in *S. cerevisiae*. It includes human interferon (Hitzeman et al. 1983), human epidermal growth factor (Brake et al. 1984), and human hemoglobin (Strohl 1997). Production of the first safe hepatitis B vaccine was only possible due to the *S. cerevisiae* (Miyanohara et al. 1983). *P. pastoris* is used to produce many mammalian recombinant proteins like insulin precursor at the rate of 1.5 g/L (Wang et al. 2001), 4 g/L of intracellular interleukin 2 (Cregg et al. 1993), and 6 g/L of tumor necrosis factor (Dale et al. 1999). *P. pastoris* showed better results for the production of many recombinant proteins than that of *S. cerevisiae*. *S. cerevisiae* produced serum albumin at the rate of 0.15 g/L, whereas in *P. pastoris*, the titer was 10 g/L (Nevalainen et al. 2005).

Another important fungus includes filamentous fungi such as *A. niger* which are striking hosts for recombinant DNA technology. They have the ability to secrete elevated levels of bioactive proteins with posttranslational processing such as glycosylation. *A. niger* excretes 25 g/L of glucoamylase (Ward et al. 2006). Utilizing the plasmids, foreign gene can be incorporated in the chromosomes of filamentous fungi that provide long-term stability also.

3.4.3 Immunosuppressive Agents

Immunosuppression involves the reduction of activity of immune system or efficiency of immune system. Immunosuppresants are used to control severe manifestations of allergic response, of autoimmune disorders, or in the transplantation. Cyclosporine A is discovered from mold, *Tolypocladium niveum*, and it acts as antifungal peptide (Borel et al. 1976). Its discovery has facilitated the organ transplant. Another important agent includes ester of mycophenolic acid, mycophenolate mofetil (CellCept), produced from several species of *Penicillium* and was helpful in kidney transplants. Antamides are cyclic peptides, produced from *Amanita phalloides* spp.; they play a role in induction of cell necrosis; collutellin A is another cyclic octapeptide produced from *Colletotrichum dematium* and plays a role in reduction of IL-2 production (Thell et al. 2014).

3.4.4 Antitumor Agents

Natural polyketides have taken most of the attention for various biological activities. There are some of the fungal polyketide that are famous for their anticancer potential, and they belong to statin family. Statins are known for the cholesterol synthesis inhibitors and are used to treat hypercholesterolemia and cardiovascular diseases. Several in vitro studies showed that naturally derived compactin, lovastatin, and pravastatin have anticancer potential also. Compactin inhibited acute myeloid leukemia (AML) cells with a full inhibitory concentration (IC100) of 2.6 μM (Sharma et al. 1984); lovastatin and synthetic simvastatin inhibited the growth of AML cells (Newman et al. 1997). Further lovastatin has shown a potential activity against various lung cancer cell lines (Maksimova et al. 2008), ovarian cancer cell lines (Martirosyan et al. 2010), and breast cancer cell line, liver cancer cell line, and cervical cancer HeLa cell line (Mahmoud et al. 2012). Another small polyketide chain terrein, produced from *A. terreus*, showed the potential activity against breast cancer MCF-7 cell line, pancreatic cell line (PANC-1), and liver cancer cell lines (HepG2) (Liao et al. 2012). Asperlin, isolated from *A. nidulans*, reduces cell proliferation and induces G2/M cell cycle arrest in the human cervical carcinoma HeLa cell line (He et al. 2011). Sequoiamonascin A is a polyketides with spiro-ring structure derived from *A. parasiticus* and found to be selectively cytotoxic against leukemia cell lines and melanoma cell lines; it also showed potential activity against breast cancer MCF-7, lung cancer NCI-H460, and central nervous system (CNS) cancer SF-268 cell lines (Stierle et al. 2003). Another compound penisimplicissin derived from *T. pinophilus* showed a cytotoxic activity against leukemia HL-60 and CCRF-CEM cell lines (Buommino et al. 2004). There are many other nitrogen-containing compounds and terpenoids derived from various fungi shown to possess anticancer activity, Taxol being the most remarkable. Taxol also known as paclitaxel is the most common and acceptable drug against cancer. Earlier, it was derived from bark of yew tree *Taxus brevifolia*. Later it was found that Taxol was also produced by the fungus *Taxomyces andreanae* (Strobel et al. 1997).

3.4.5 Hypocholesterolemic Agents

There are natural polyketides derived from the statin family and known for the cholesterol synthesis inhibitors and used to treat hypercholesterolemia and cardiovascular diseases. Lovastatin, pravastatin, etc. act as inhibitors of 3-hydroxy-3-methylglutarylcoenzyme A reductase, the regulatory and rate-limiting enzyme of cholesterol biosynthesis in the liver. Statins are boon to the cholesterol patients. Statins are found to be produced by various fungal strains including *Penicillium brevicompactum*, *Penicillium citrinum*, *Monascus ruber*, and *Aspergillus terreus*.

3.4.6 *Mycotoxins*

Different species of *Claviceps* are producing ergot alkaloids that are used for migraine, cerebral circulatory diseases, bleeding after the child birth, and also prevention of pregnancy (Bentley 1997; Vining and Taber 1979).

3.5 Fungi in Paper Industry

Like any other field, fungi play a remarkable role in paper and pulp industry. In the eucalyptus wood, there is tendency of forming pitch which is a low molecular weight oleophilic material that deposits during paper manufacturing and causes holes in the paper. Fungi are used to remove free and esterified steroid ketones and thus prevent the formation of pitches. There is report showing that 1–2-week treatment with either *Phlebia radiata* or *Poria subvermispora* helped in the removal of sterols and very lesser percentage of wood loss (Martinez-Inogi et al. 2001). During the paper manufacturing, biobleaching is also performed that helps in reduction of environmental pollution; in terms of lesser usage of chemicals for bleaching and thus reducing effluent toxicity, this is done by utilizing enzymes of fungal origin like xylanases. By utilizing biobleaching improvements in paper brightness, kappa index, breaking length, and burst index have been observed (Jimenez et al. 1997). In the conventional treatment of paper, chlorine is used, but by adopting biobleaching through recombinant xylanases, usage of chlorine is also reduced (Sandal et al. 1997). It has been found that fungi can be used to biopulp the wood. The natural microflora of the wood is reduced by debarking, chipping, and steaming, but when the material is inoculated with biopulping fungi and ventilated with filtered and humidified air, it reduces the loss of energy by 30% and increases mill output by 30% and also increases the fineness and strength of paper (Shukla et al. 2004). The combination of fungal mycelium and conventional fibers is used while making the paper from the recycled fibers (Yamanak et al. 1992). Utilization of fungi in paper industry has remarkably enhanced the quality of paper and reduced the energy consumption and deleterious environmental impacts.

3.6 Fungi for Bioremediation

Rapid industrialization, population explosion, and fast development of today's world have gifted environment with enormous amount of problem of chemical and solid waste management. Due to human activities, environment is now burdened with recalcitrant compounds like polyaromatic hydrocarbons (PAHs), polychlorinated biphenyls (PCBs), polychlorinated dibenzo-p-dioxins (PCDDs), polychlorinated dibenzofurans (PCDFs), heavy metals, etc. (Deshmukh et al. 2016). There are many physicochemical methods that are available to treat such compounds, but they are not proving to be as promising as they should be. Bioremediation may provide an answer to such problems, which is a natural phenomenon and is eco-friendly also. Bioremediation is a process of usage of suitable microbes in the polluted system, which by using their metabolism degrades these pollutants. Various species of fungi are involved in this process as they can sustain life in variety of environment. There are reports showing the growth of fungi even in the effluent treatment plants (Badia-Fabregat et al. 2015; Zhang et al. 2013). Various species of white rot fungi *Phanerochaete chrysosporium*, *Trametes versicolor*, *Bjerkandera adusta*, and *Pleurotus* sp. are found to produce laccase and peroxidase enzymes and thereby helpful in transformation of pollutant like pesticides (Rodríguez-Rodríguez et al. 2013). Various fungal groups, viz., *Coriolus versicolor*, *Hirschioporus larincinus*, *Inonotus hispidus*, *Phanerochaete chrysosporium*, and *Phlebia tremellosa*, are found to possess ability for decolorization in the dye effluent (Jebapriya and Gnanadoss 2013). There are marine fungi that are involved in biotransformation of persistent organic pollutants (POPs). Pentachlorophenol is one of the POPs which has been transformed by marine-derived fungus *Trichoderma harzianum* (Vacondio et al. 2015), while marine-derived fungi including *Mucor*, *Aspergillus*, and *Penicillium* showed bioremediation for water-soluble crude oil fractions (Hickey 2013).

Fungi have found to have ability of bioremediation of various pollutant including textile dyes, petroleum hydrocarbons, pulp and paper industry effluents, leather tanning effluents, PAHs, pesticides, and PPCPs. *Aspergillus*, *Curvularia*, *Acrimonium*, and *Pythium* are reported to have metal tolerance ability (Sousa et al. 2014). *Pleurotus ostreatus* (white rot fungi) and *T. versicolor* (basidiomycota) have been reported to have potential to degrade PAHs in solid-state fermentation (Rosales et al. 2013). *Aspergillus*, *Curvularia*, *Drechslera*, *Fusarium*, *Lasiodiplodia*, *Mucor*, *Penicillium*, *Rhizopus*, and *Trichoderma* have been reported to efficiently bioremediate PAHs on account of production of lipase enzyme (Llado et al. 2013; Balaji et al. 2014). *Aspergillus terreus* has shown the complete degradation of toxic compounds, pesticide chlorpyrifos, within 24 h of incubation (Silambarasan and Abraham 2013). *A. flavus* and *A. niger* are found to reduce chromium, thereby reducing its harmful impact (Bennett et al. 2013). There are many other fungi that are playing important role in bioremediation of harmful compounds.

Besides all these applications, fungi are known to produce various pigments. Fungi like *Aspergillus*, *Fusarium*, *Penicillium*, *Monascus*, *Trichoderma*, *and Laetiporus* are reported to produce quinones, anthraquinones, rubropuntamine, rubropuntatin, ankaflavin, monascin, β-carotene, and many other pigments responsible for various colors, viz., red, purple, yellow, brown, orange, and green. In addition to providing natural colors, these pigments possess many therapeutic applications like immune modulators, anticancer, antioxidant, antiproliferative, etc.

3.7 Conclusion

The role of fungi in various fields emphasizes that more study is needed to use this nature's magical bullet. In addition to the above fields, the successful usage of their enzymes in various industries cannot be denied. Laccase enzyme produced by ascomyceteous, deuteromyceteous, and basidiomyceteous fungi is utilized in delignification and pulp bleaching and bioremediation; xylanases from fungi are also successfully utilized in various industries. There are many other enzymes produced from various fungal species that are in use in detergents, starch, drinks, food, textile, animal feed, baking, pulp and paper, leather, and chemical and biomedical products. In this chapter, we have tried to bring together various fields where various fungi and its products are explored for benefits to human race. The chapter elaborated various aspects of medical field, food and feed, paper and pulp industry, and bioremediation where fungi are playing an outstanding role and can further show a future prospect where more of the strains can be used. Metabolites produced by various fungi are showing potential activity against various cell lines, which can further be explored for future clinical trials. Similarly, various strains have potential for bioremediation of heavy metals, POPs, etc., which can be further utilized at large scale.

References

Akintokun AK, Akande GA, Akintokun PO, Popoola TOS, Babalola AO (2007) Solubilization on insoluble phosphate by organic acid-producing fungi isolated from Nigerian soil. Int J Soil Sci 2:301–307

Archer D (2000) Filamentous fungi as microbial cell factories for food use. Curr Opin Biotech 11:478–483

Badia-Fabregat M, Lucas D, Gros M, Rodríguez-Moraz S, Barceló D, Caminal G, Vicent T (2015) Identification of some factors affecting pharmaceutical active compounds (PhACs) removal in real wastewater case study of fungal treatment of reverse osmosis concentrate. J Haz Mater 283:663–671

Bakers KF, Cook RJ (1974) Biological control of plant pathogens. American Phytopathological Society, St. Paul, p 433

Balaji V, Arulazhagan P, Ebenezer P (2014) Enzymatic bioremediation of polyaromatic hydrocarbons by fungal consortia enriched from petroleum contaminated soil and oil seeds. J Environ Biol 35:521–529

Beelman RB, Royse D, Chikthimmah N (2003) Bioactive components in button mushroom *Agaricus bisporus* (j.lge) imbach (*Agaricomycetideae*) of nutritional, medicinal, or biological importance (review). Int J Med Mushroom 5:321–337

Bennett RM, Cordero PRF, Bautista GS, Dedeles GR (2013) Reduction of hexavalent chromium using fungi and bacteria isolated from contaminated soil and water samples. Chem Ecol 29:320–328

Bentley R (1997) Microbial secondary metabolites play important roles in medicine: prospects to discovery of new drugs. Perspect Biol Med 40:364–394

Berdy J (1995) Are actinomycetes exhausted as a source of secondary metabolites? In: Debanov VG (ed) Biotehnologija (ISSN 0234-2758) Proceedings of the 9th international symposium on the biology of Actinomycetes, Moscow 1994, pp 13–34
Berdy J (2005) Bioactive microbial metabolites. J Antibiot 58(1):1–26
Borel JF, Feurer C, Gabler HU, Stahelin H (1976) Biological effects of cyclosporin A: a new anti-lymphocytic agent. Agents Action 6:468–475
Brake AJ, Merryweather JP, Coit DG, Heberlein UA, Masiarz FR, Mullenbach GT, Urdea MS (1984) a-Factor-directed synthesis and secretion of mature foreign proteins in Saccharomyces cerevisiae. Proc Natl Acad Sci U S A 81:4642–4646
Buommino E, Nicoletti R, Gaeta GM, Orlando M, Ciavatta ML, Baroni A, Tufano MA (2004) 3-*O*-methylfunicone, a secondary metabolite produced by *Penicillium pinophilum*, induces growth arrest and apoptosis in HeLa cells. Cell Proliferat 37:413–426
Buyck B, Hofstetter V, Eberhardt U, Verbeken A, Kauff F (2008) Walking the thin line between *Russula* and *Lactarius*: the dilemma of *Russula* subject. *Orchricompactae*. Fungal Divers 28:15–40
Calhelha RC, Andrade JV, Ferreira IC, Estevinho LM (2006) Toxicity effects of fungicide residues on the wine-producing process. Food Microbiol 23:393–398
Chen JH (2006) The combined use of chemical and organic fertilizers and/or biofertilizer for crop growth and soil fertility. International workshop on sustained management of the soil-rhizosphere system for efficient crop production and fertilizer use, 16–20 October 2006 Land Development Department, Bangkok 10900, Thailand, pp 1–11
Chet I, Inbar J (1994) Biological control of fungal pathogens. Appl Biochem Biotechno 48:37–43
Cheung CKP (1996) The hypocholesterolemic effect of two edible mushrooms: *Aurzcularia auricula* (tree-ear) and *Tremella fuciformis* (white jelly-leaf) in hypercholesterolemic rat. Nutr Res 16:1721–1725
Cohen R, Persky L, Hadar YPL (2002) Biotechnological applications and potential of wood-degrading mushrooms of the genus *Pleurotus*. App Microbiol Biotech 58:582–594
Cregg JM, Vedvick TS, Raschke WC (1993) Recent advances in the expression of foreign genes in *Pichia pastoris*. Biotechnology 11:905–910
Dale C, Allen A, Fogarty S (1999) *Pichia pastoris*: a eukaryotic system for the large-scale production of biopharmaceuticals. BioPharm 12:36–42
Das A, Prasad R, Srivastava A, Giang HP, Bhatnagar K, Varma A (2007) Fungal siderophores: structure, functions and regulation. In: Varma A, Chincholkar SB (eds) Soil biology, Microbial siderophores, vol 12. Springer, Berlin/Heidelberg, pp 1–42
Demain AL, Vaishnav P (2009) Production of recombinant proteins by microbes and higher organisms. Biotechnolo Adva 27:297–306
Deshmukh R, Khardenavis AA, Purohit HJ (2016) Diverse metabolic capacities of Fungi for bioremediation. Indian J Microbio. https://doi.org/10.1007/s12088-016-0584-6
Elander RP, Lowe DA (1994) Fungal biotechnology: an overview. In: Handbook of applied mycology. Marcel Dekker, New York, pp 1–34
El-Azouni IM (2008) Effect of phosphate solubilizing fungi on growth and nutrient uptake of soybean (Glycine Max L.) plants. J App Sci Res 4:592–598
Fuentes-Ramirez LE, Caballero-Mellado J (2005) Bacterial biofertilizers. In: Siddiqui A (ed) PGPR: biocontrol and biofertilization. Springer, Dordrecht, pp 143–172
Gentili F, Jumpponen A (2006) Potential and possible uses of bacterial and fungal biofertilizers. In: Rai MK (ed) Handbook of microbial biofertilizers. Food products press, New York, pp 1–28
Guimaraes LHS, Peixoto-Nogueira SC, Michelin M, Rizzatti ACS, Sandrim VC, Zanoelo FF, Aquino ACM, Junior AB, Polizeli MLT (2006) Screening of filamentous fungi for production of enzymes of biotechnological interest. Braz J Microbio 37:474–480
Haggag WM, Mohamed HAA (2007) Biotechnological aspects of microorganisms used in plant biological control. Am Euras J Sust Agric 1:7–12

He L, Nan MH, Oh HC, Kim YH, Jang JH, Erikson RL, Ahn JS, Kim BY (2011) Asperlin induces G2/M arrest through ROS generation and ATM pathway in human cervical carcinoma cells. Biochem Biophys Res Commun 409:489–493

Hickey P (2013) Toxicity of water soluble fractions of crude oil on some bacteria and fungi isolated from marine water. Am J Anim Res 3:24–29

Hitzeman RA, Leung DW, Perry LJ, Kohr WJ, Levine HL, Goeddel DW (1983) Secretion of human interferons by yeast. Science 219:620–625

Irtwange VS (2006) Application of biological control agents in pre- and postharvest operations. Agric Eng Int CIGR J Inv Overv 3:1–12

Jebapriya GR, Gnanadoss JJ (2013) Bioremediation of textile dye using white-rot fungi: a review. Int J Curr Res Rev 5:1–13

Jimenez L, Martinez C, Perez I, Lopez F (1997) Biobleaching procedures for pulp from agricultural residues using Phanerochaete chrysosporium and enzymes. Process Biochem 32:297–304

Jose L, Adrio AE, Arnold L and Demain. 2003. Fungal biotechnology. Int Microbiol 6: 191–199

Juzlova P, Martinkova L, Kren V (1996) Secondary metabolites of the fungus Monascus: a review. J Ind Microbiol 16:163–170

Khan AG (2006) Mycorrhizoremediation-an enhanced form of phytoremediation. J Zhejiang Univ Sci B 7:503–514

Khetan SK (2001) Microbial pest control. Marcel Dekker, Inc, New York/Basel, p 300

Kiss L (2003) A review of fungal antagonists of powdery mildews and their potential as biocontrol agents. Pest Manag Sci 59:475–483

Li C, Zhu Y, Wang Y (1998) *Monascus purpureus* fermented rice (red yeast rice): a natural food product that lowers blood cholesterol in animal models of hypercholesterolemia. Nutr Res 18:1–81

Liang C, Yang J, Kovács GM, Szentiványi O, Li B, Xu XM, Kiss L (2007) Genetic diversity of *Ampelomyces mycoparasites* isolated from different powdery mildew species in China inferred from analyses of rDNA ITS sequences. Fungal Divers 24:225–240

Liao WY, Shen CN, Lin LH, Yang YL, Han HY, Chen JW, Kuo SC, Wu SH, Liaw CC (2012) Asperjinone, a nor-neolignan, and terrein, a suppressor of ABCG2-expressing breast cancer cells, from thermophilic *Aspergillus terreus*. J Nat Prod 75:630–635

Llado S, Covino S, Solanas AM, Vinas M, Petruccioli M, Dannibale A (2013) Comparative assessment of bioremediation approaches to highly recalcitrant PAH degradation in a real industrial polluted soil. J Hazard Mater 248–249:407–414

Mahmoud AM, Al-Abd AM, Lightfoot DA, El-Shemy HA (2012) Anti-cancer characteristics of mevinolin against three different solid tumor cell lines was not solely p53-dependent. J Enzym Inhib Med Ch 27:673–679

Maksimova E, Yie TA, Rom WN (2008) *In vitro* mechanisms of lovastatin on lung cancer cell lines as a potential chemopreventive agent. Lung 186:45–54

Malik KA, Hafeez FY, Mirza MS, Hameed S, Rasul G, Bilal R (2005) Rhizospheric plant – microbe interactions for sustainable agriculture. In: Wang Y-P, Lin M, Tian Z-X, Elmerich C, Newton WE (eds) Biological nitrogen fixation, sustainable agriculture and the environment. Springer, Berlin, pp 257–260

Martin SA, Nibs DJ (1992) Effect of direct-fed microbials on rumen microbial fermentation. J Dairy Sci 75:1736

Martinez-Inogi MJ, Gutierrez A, del Rio JC, Martinez MJ, Martinez AT (2001) Time course of fungal removal of lipophilic extractives from Eucalyptus globulus wood. J Biotechnol 84:119–126

Martirosyan A, Clendening JW, Goard CA, Penn LZ (2010) Lovastatin induces apoptosis of ovarian cancer cells and synergizes with doxorubicin: potential therapeutic relevance. BMC Cancer 10:103

Miyanohara A, Toh-e A, Nozai C, Hamada F, Ohtomo N, Matsubara K (1983) Expression of hepatitis B surface antigen gene in yeast. Proc Natl Acad Sci U S A 80:1–5

Nevalainen KMH, SJT V, Bergquist PL (2005) Heterologous protein expression in filamentous fungi. Trends Biotech 23(9):468–474

Newman A, Clutterbuck RD, Powles RL, Catovsky D, Millar JL (1997) A comparison of the effect of the 3-hydroxy-3-methylglutaryl coenzyme A (HMG-CoA) reductase inhibitors simvastatin, lovastatin and pravastatin on leukaemic and normal bone marrow progenitors. Leuk Lymphoma 24:533–537

Pandey A, Nigam P, Socol CR, Socol VT, Singh D, Mohan R (2000) Advances in microbial amylases – review. Biotechnol Appl Biochem 31:135–152

Pariza MW, Johnson EA (2001) Evaluating the safety of microbial enzyme preparations used in food processing: update for a new century. Regul Toxicol Pharmacol 33:173–186

Ramsbottom J (1972) Mushrooms and toadstools, 6th edn. Collins, London 306 pp

Rinaldi AC, Comandini O, Kuyper TW (2008) Ectomycorrhizal fungal diversity: separating the wheat from the chaff. Fungal Divers 33:1–45

Rodríguez-Rodríguez CE, Castro-Gutie'rrez V, Chin-Pampillo JS, Ruiz-Hidalgo K (2013) On-farm biopurification systems: role of white-rot fungi in depuration of pesticide-containing wastewaters. FEMS Microbiol Lett 345:1–12

Rojas C, Mansur E (1995) Ecuador: informaciones generales sobre productos non madereros en Ecuador. In: Memoria, consulta de expertos sobre productos forestales no madereros para America Latina y el Caribe. FAO Regional Office for Latin America and the Caribbean (Serie Forestal #1), Santiago, pp 208–223

Rosales E, Pazos M, Angeles Sanromán M (2013) Feasibility of solid-state fermentation using spent fungi-substrate in the biodegradation of PAHs. Clean Soil Air Water 41:610–615

Sajith S, Priji P, Sreedevi S, Benjamin S (2016) An overview on fungal cellulases with an industrial perspective. J Nutr Food Sci 6:1–10

Sandal T, Kofod LV, Kauppinen MS, Andersen LN (1997) An enzyme with xylanase activity. Patent WO/1997/027293

Sathyanarayana N, Panda GT (2003) Purification and biochemical properties of microbial pectinases- a review. Process Biochem 38:987–996

Sharma S, Hulcher F, Dodge W (1984) Effects of cholesterol synthesis inhibitor on the proliferation of avian granulocytic and monocytic progenitor cells and primary acute monocytic leukemia cells. J Nutr 2:65–69

Shukla OP, Rai UN, Subramanyam SV (2004) Biopulping and biobleaching: an energy and environmental saving technology for Indian pulp and paper industry. 10(2) (Reproduced from the archives if Environ News ISEB of India)

Siddiqui ZA, Pichtel J (2008) Mycorrhixae: an overview. In: Siddiqui ZA, Akhtar MS, Futai K (eds) Mycorrhizae: sustainable agriculture and forestry. Springer, Dordrecht, pp 1–36

Silambarasan S, Abraham J (2013) Ecofriendly method for bioremediation of chlorpyrifos from agricultural soil by novel fungus *Aspergillus terreus* JAS1. Water Air Soil Pollut 224:1369

Sousa NR, Ramos MA, Marques APGC, Castro PML (2014) A genotype dependent-response to cadmium contamination in soil is displayed by Pinus pinaster in symbiosis with different mycorrhizal fungi. Appl Soil Ecol 76:7–13

Soytong K, Kanokmadhakul S, Kukongviriyapa V, Isobe M (2001) Application of *Chaetomium* species (Ketomium®) as a new broad spectrum biological fungicide for plant disease control: a review article. Fungal Divers 7:1–15

Stierle DB, Stierle AA, Bugni T (2003) Sequoiamonascins A–D: novel anticancer metabolites isolated from a redwood endophyte. J Org Chem 68:4966–4969

Strobel GA, Torczynski R, Bollon A (1997) *Acremonium* sp.– a leucinostatin A producing endophyte of European yew (*Taxus baccata*). Plant Sci 128:97–108

Strohl WR (1997) Industrial antibiotics: today and the future. In: Strohl WR (ed) Biotechnology of antibiotics, 2nd edn. Marcel Dekker, New York, pp 1–47

Thell K, Hellinger R, Schabbauer G, Gruber CW (2014) Immunosuppressive peptides and their therapeutic applications. Drug Discov Today 19(5):645–653

Vacondio B, Birolli WG, Ferreira IM, Seleghim MH, Gonçalves S, Vasconcellos SP, Porto AL (2015) Biodegradation of pentachlorophenol by marine-derived fungus *Trichoderma harzianum* CBMAI 1677 isolated from ascidian Didemnun ligulum. Biocatal Agric Biotechnol 4:266–275

Vinale F, Sivasithamparam K, Ghisalberti EL, Marra R, Woo SL, Lorito M (2008) *Trichoderma* plant pathogen interactions. Soil Biol Biochem 40:1–10
Vining LC, Taber WA (1979) Ergot alkaloids. In: Rose AH (ed) Economic microbiology secondary products of metabolism, vol 3. Academic, London, pp 389–420
Viterbo A, Inbar J, Hadar Y, Chet I (2007) Plant disease biocontrol and induced resistance via fungal mycoparasites. In: Kubicek CP, Druzhinina IS (eds) Environmental and microbial relationships. The Mycota IV, 2nd edn. Springer, Berlin/Heidelberg, pp 127–146
Wakelin SA, Werren PR, Ryder HM (2004) Phosphate solubilization by *Penicillium* spp. closely associated with wheat root. Biol Fertil Soils 40:36–43
Wang Y, Liang ZH, Zhang YS, Yao SY, Xu YG, Tang YH, Zhu SQ, Cui DF, Feng YM (2001) Human insulin from a precursor overexpressed in the methylotrophic yeast Pichia pastoris and a simple procedure for purifying the expression product. Biotechnol Bioeng 73(1):74–79
Wani SP, Lee KK (2002) Biofertilizers for sustaining cereal crops production. In: Kannaiyan S (ed) Biotechnology of biofertilizers. Narosa publishing house, New Delhi, pp 50–64
Ward OP, Qin WM, Hanjoon JD, Singh EJYA (2006) Physiology and biotechnology of *Aspergillus*. Adv Appl Microbiol 58:1–75
Wasson RG (1968) Soma. divine mushroom of immortality. Harcourt Brace Ivanovich Inc., New York 380 pp
Wei TZ, Yao YJ (2003) Literature review of *Termitomyces* species in China. Fungal Sci 22:39–54
Whipps JM, Sreenivasaprasad S, Muthumeenakshi S, Rogers CW, Challen MP (2008) Use of *Coniothyrium minitans* as a biocontrol agent and some molecular aspects of sclerotial mycoparasitism. Eur J Plant Pathol 121:323–330
Woo LS, Lorito M (2007) Exploiting the interactions between fungal antagonists, pathogens and the plant for biocontrol. In: Vurro M, Gressel J (eds) Novel biotechnologies for biocontrol agent enhancement and management. Springer, Dordrecht, pp 107–130
Yamanak S, Kikuchi R, Ajinomoto KK (1992) Manufacture of reinforced fibres with fungi. Japanese patent 04112794
Yoshizawa K (1995) Sake No Kagaku (Science of alcoholic beverages). Asakura, Tokyo, Japan
Zhang Y, Xie J, Liu M, Tian Z, He Z, van Nostrand JD, Ren L, Zhou J, Yang M (2013) Microbial community functional structure in response to antibiotics in pharmaceutical wastewater treatment systems. Water Res 47:6298–6308

Chapter 4
The Biological Promises of Endophytic *Muscodor* Species

Mahiti Gupta and Vineet Meshram

Abstract With increasing human population and limited land for agriculture, it has become a herculean task to provide food security to over 7 billion people around the world. The majority of the food produced is lost in postharvest losses. The present approaches to manage the current situation appear to be ineffective and unsustainable. Further, the development of resistance in microbes against the current arsenal of drugs and the destructive effects of methyl bromide and sulfur dioxide on the ozone layer has worsened the situation. This calls for an urgent need to explore alternate avenues for management of postharvest losses. Besides various chemical approaches, exploration of natural resources for finding out new anti-infective and biocontrol agents appears to be a plausible and sustainable solution for management of postharvest losses. *Muscodor* is a genus of sterile endophytic fungi which has the remarkable property to produce a mixture of volatile organic compounds (VOCs) which are lethal against a number of plant and human pathogenic bacteria and fungi, nematodes, and moths. Further, the VOCs of *Muscodor* spp. have also shown promising application as biocontrol agent and in management of human waste. Recently, the extrolites of *Muscodor* species have also shown promising antimicrobial, anti-obesity, antihyperuricemic, and antioxidant activity. Hence, the current chapter embodies the potential uses of volatiles and other extrolites produced by *Muscodor* species and their possible application in agriculture and pharmaceutical industries.

Keywords Endophytic fungi · *Muscodor* · Biocides · Mycofumigation biocontrol agent

---

M. Gupta
Department of Biotechnology, Lovely Professional University, Phagwara, Punjab, India

V. Meshram (✉)
Department of Plant Pathology and Weed Research, Agriculture Research Organization, The Volcani Centre, Rishon Lezion, Israel
e-mail: Vineet@volcani.agri.gov.il

P. Gehlot, J. Singh (eds.), *Fungi and their Role in Sustainable Development: Current Perspectives*, https://doi.org/10.1007/978-981-13-0393-7_4

4.1 Introduction

There are more than 300,000 plant species that exist on earth, and each plant harbors a suite of microorganisms which can be classified as pathogens, epiphytes, and endophytes. Endophytes are group of highly diverse, polyphyletic microorganisms which colonize the intra- or intercellular parts of the plant for at least a portion of their lives without causing obvious infections (Petrini 1991; Bacon and White 2000; Strobel and Daisy 2003; Kusari et al. 2013). They are ubiquitous in nature and occupy all the niches on the earth. Once the endophyte enters the internal tissue of the host, they assume the latent phase for their entire life cycle or for an extended duration (Aly et al. 2011; Kaul et al. 2012). Their relationship with the host plant ranges from symbiotic, benign commensals, decomposers to latent pathogens (Promputha et al. 2007). Fungal endophytes are more frequently encountered as compared to bacterial counterparts. Endophytic fungi are hyperdiverse, and it is estimated that more than 1.5 million species may exist (Arnold et al. 2000). Under selective pressure of biotic and abiotic stresses, endophytes undergo a constant process of strain development and are engaged in a bi-, tri-, or multipartite interaction with their host to produce value-added natural products which may directly or indirectly be used as therapeutic agents (Kusari et al. 2009; Aly et al. 2011; Qadri et al. 2013). The genetic recombination of the endophytes with the host plant enables them to mimic the biological properties of their host and produce analogous bioactive metabolites like paclitaxel (Stierle et al. 1993), podophyllotoxin (Puri et al. 2006), deoxypodophyllotoxin (Kusari et al. 2009), and camptothecin (Puri et al. 2005). Thus, the endophytic fungi exhibit biological adeptness related to their ecological functions which immensely contribute to their profound metabolic proficiency thereby making them a lucrative resource for exploring natural bioactive compounds (Qadri et al. 2013).

Muscodor are volatile-producing sterile endophytic fungi that have been isolated from various ecological niches (Zhang et al. 2010; Meshram et al. 2016). The VOCs produced by *Muscodor* spp. exhibit a broad-spectrum antibacterial, antifungal, insecticidal, and nematicidal activity (Strobel 2006a, b, 2011). The volatile antimicrobials produced by *Muscodor* have been utilized as a mycofumigation/biofumigation agent to control the postharvest losses caused due to decay of fruits and vegetables and damping off of plant. Apart from this, the VOCs of *Muscodor* species have also been utilized for control of building molds and sewage treatment (Mercier and Manker 2005; Gabler et al. 2006; Mercier and Jimnez 2007; Mercier et al. 2007). Furthermore, *Muscodor* species also produce biologically active soluble compounds (Fig. 4.1) (Kapoor and Saxena 2016; Qadri et al. 2016). In this chapter, the role and impact of secondary metabolites produced by *Muscodor* species has been discussed along with their possible biotechnological applications.

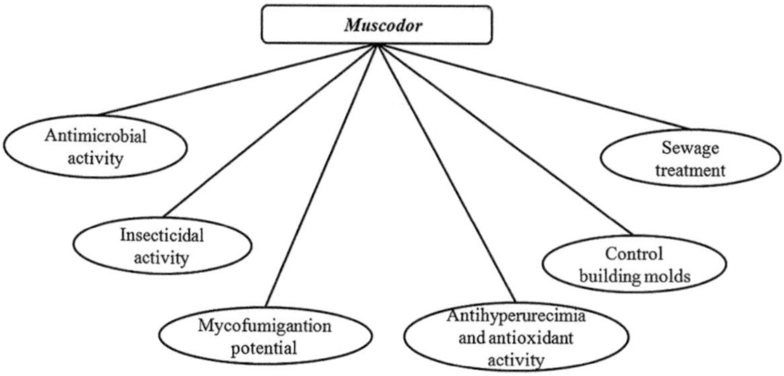

Fig. 4.1 Intervention areas of *Muscodor* species

4.2 Discovery of *Muscodor*

The discovery of *Muscodor* species can be refereed as a scientific accident. The genus *Muscodor* was erected with the isolation of isolate cz620 from the stems of *Cinnamomum zeylanicum* by Prof. Gary Strobel, Department of Plant Sciences and Plant Pathology, Montana State University, USA, during his forest forays in Honduras, Central America, in the late 1990s (Woropong et al. 2001). The Petri dishes containing the limbs of *C. zeylanicum* were kept in tightly sealed plastic boxes to eliminate the invasion of mites and fastidious microorganisms. After few days, fungal colonies were observed germinating from the plant samples kept in the plastic boxes. The hyphae of the germinating fungal colonies were individually transferred to the fresh potato dextrose agar plate and again incubated. After 2 days of incubation, it was observed that, among the transferred fungal colonies, only one fungus was able to grow, whereas the growth of the rest of the fungi was suppressed. It was thought that the limiting oxygen supply in plastic boxes could have killed the emerging fungi, but very soon it became evident that this was due to the production of volatile antibiotics by the alive fungal species. Thus, a hypothesis was given which suggested that the alive fungus has wide range of biological activities which is attributed to the VOCs produced by it. Further, on the basis of morphological features, gas chemistry, and molecular taxonomy of the isolate, a separate genus *Muscodor* belonging to family Xylariaceae was erected. The isolate was designated as *Muscodor albus* cz620 based on its odor and white color (Strobel et al. 2001; Strobel 2006a, b, 2011).

4.3 Geographical Distribution of *Muscodor* Species

Muscodor species hold enormous potential to be used as a mycofumigation agent by the virtue of their VOCs. It has attracted researchers from all over the world to look for newer biotypes of this genus. Since the discovery of *M. albus* cz620, the genus is getting expanded with the discovery of newer biotypes from different hosts and geographical locations covering four continents of the world (Suwannarach et al. 2013; Meshram et al. 2016). Since the discovery of *M. albus* cz620 in late 1990s from cinnamon plant, 18 different species have been discovered in this genus from different host and ecological niches of the world (Table 4.1). They are *M. roseus* A3–5 (Worapong et al. 2002), *M. vitigenus* P-15 (Daisy et al. 2002), *M. crispans* B-23 (Mitchell et al. 2008), *M. yucatanensis* B110 (Gonzalez et al. 2009), *M. fengyangensis* ZJLQ070 (Zhang et al. 2010), *M. cinnamomi* CMU-Cib461 (Suwannarach et al. 2010), *M. sutura* CA22-D (Kudalkar et al. 2012), *M. oryzae* CMU-WR2, *M. suthepensis* CMU-Cib462, *M. musae* CMU-MU3, *M. equiseti* CMU-M2 (Suwannarach et al. 2013), *M. kashayum* #16 AMLWLS (Meshram et al. 2013), *M. darjeelingensis* #1 CCSTITD (Saxena et al. 2014), *M. strobelii* #6610 (Meshram et al. 2014), *M. tigerii* #2 CCSTITD (Saxena et al. 2015), *M. ghoomensis* #6 CCSTITD, *M. indica* #6(b) CCSTITD (Meshram et al. 2016), and *M. heveae* RTM5IV3 (Siri-udom et al. 2016). Apart from these, several strains of *M. albus* have been isolated across the globe including Australia, China, Cuba, Ecuador, India, Indonesia, and the USA (Ezra et al. 2004a; Atmosukarto et al. 2005; Strobel et al. 2007; Banerjee et al. 2010, 2013; Yuan et al. 2011; Banguela-Castillo et al. 2015). Recently, the existence of *M. vitigenus* and *M. equiseti* has been observed in the leaves of rubber tree (Siri-udom et al. 2016). Geographically, *Muscodor* species have been isolated from Asia, Australia, and North and South America. The conserved rain forest of Southeast Asia appears to be the most suitable habitat for colonization of *Muscodor* species since maximum number of novel isolates have been obtained from these habitats. Similarly, maximum number of isolates was obtained from the cinnamon plant (Suwwannarach et al. 2013; Meshram et al. 2014, 2016; Saxena et al. 2014, 2015). Most of *Muscodor* species exhibited strong antibacterial and antifungal activity, whereas *M. vitigenus* displayed nematocidal activity, and *M. yucatanensis* showed phytoinhibitory activity (Kudalkar et al. 2012; Meshram et al. 2014).

4.4 Gas Chemistry of *Muscodor* Species

Muscodor species produces a mixture of low-molecular-weight VOCs which exhibited broad-spectrum antimicrobial, nematocidal, and insecticidal activity. The biological activities of *Muscodor* spp. are attributed to the synergistic effect of these mixtures of VOCs secreted into the head space of the culture (Strobel 2006a, b, 2011; Alpha et al. 2015). Production of VOCs is dependent on availability of

Table 4.1 Biological and culture characteristics of *Muscodor* species

Muscodor sp.	Host plant	Location	Hyphal growth	Pigment	Bioactivity	Reference
M. albus	*Cinnamomum zeylanicum*	Honduras, Central America	Rope-like	None	Antibacterial, antifungal	Worapong et al. (2001)
M. roseus	*Grevillea pteridifolia*	Northern Australia	Rope-like, forming erumpent pie-shaped sectors	Light rose	Antibacterial, antifungal	Worapong et al. (2002)
M. vitigenus	*Paullinia paullinioides*	Bahuaja-Sonene National Park, Peru	Rope-like	None	Insecticidal activity	Daisy et al. (2002)
M. crispans	*Ananas ananassoides*	Bolivian Amazon, South America	Rope-like with cauliflower-like bodies	Pink in light	Antibacterial, antifungal	Mitchell et al. (2008, 2010)
M. yucatanensis	*Bursera sirnaruha*	Yucatan Peninsula, Mexico	Rope-like with coiled hyphae with swollen cells	None	Antifungal phytoinhibitory	Gonzalez et al. (2009); Macías-Rubalcava et al. (2010)
M. fengyangensis	*Actinidia chinensis*	Zhejiang Province, China	Rope-like with coiled hyphae	Yellow	Antibacterial, antifungal	Zhang et al. (2010)
M. cinnamomi	*Cinnamomum bejolghota*	Chiang Mai Province, Thailand	Rope-like with cauliflower-like bodies	Pale orange (in light)	Antibacterial, antifungal	Suwannarach et al. (2010, 2015a, b)
M. sutura	*Prestonia trifidi*	Colombia, South America	Rope-like bands extracellular bodies	Reddish in dark	Antifungal	Kudalkar et al. (2012)
M. equiseti	*Equisetum debile*	Chiang Mai Province, Thailand	Rope-like with coils structure and swollen cell	None	Antibacterial, antifungal	Suwannarach et al. (2013)
M. musae	*Musa acuminata*	Chiang Mai Province, Thailand	Rope-like with coils structure	None	Antibacterial, antifungal	Suwannarach et al. (2013)
M. oryzae	*Oryza rufipogon*	Chiang Mai Province, Thailand	Rope-like with coils structure	Pale orange	Antibacterial, antifungal	Suwannarach et al. (2013)
M. suthepensis	*Cinnamomum bejolghota*	Chiang Mai Province, Thailand	Rope-like with coils structure	Pale pink (in light)	Antibacterial, antifungal	Suwannarach et al. (2013)

(continued)

Table 4.1 (continued)

Muscodor sp.	Host plant	Location	Hyphal growth	Pigment	Bioactivity	Reference
M. kashayum	*Aegle marmelos*	Kerala, India	Fused rope-like hyphal strands	None	Antibacterial, antifungal	Meshram et al. (2013)
M. darjeelingensis	*Cinnamomum camphora*	West Bengal, India	Rope-like with cauliflower-like bodies	None	Antibacterial, antifungal	Saxena et al. (2014)
M. strobelii	*Cinnamomum zeylanicum*	Karnataka, India	Rope-like, slimy; Zinnia- and bud-like bodies	Pale yellow	Antibacterial, antifungal	Meshram et al. (2014)
M. tigerii	*Cinnamomum camphora*	West Bengal, India	Rope-like with structure coils and nondescript structures	Brown on maturation	Antibacterial, antifungal	Saxena et al. (2015)
M. heveae	*Hevea brasiliensis*	Nong Bua Lamphu Province, Thailand	Rope-like with coils structure	Pale yellow	Antibacterial, antifungal	Siri-udom et al. (2016)
M. ghoomensis	*Cinnamomum camphora*	West Bengal, India	Rope-like with structure coils and grape-like structure	Pale yellow	Inhibitory effect	Meshram et al. (2016)
M. indica	*Cinnamomum camphora*	West Bengal, India	Rope-like with coils structure	Pale yellow	Inhibitory effect	Meshram et al. (2016)

carbohydrates, surrounding temperature and water content. With increase or decrease in temperature, the activity of *Muscodor* is lost (Ezra et al. 2004b; Corcuff et al. 2011; Braun et al. 2012). Solid-phase microextraction-gas chromatography mass spectroscopy (SPME-GC/MS) has been exclusively used for qualitative and quantitative analysis of VOCs produced by different *Muscodor* spp. But now, proton transfer reaction mass spectroscopy (PTR-MS) has been used for determining the fungal volatile composition because of its high speed, accuracy, and sensitivity (detection limit in ppb). Apart from this, its small size makes it possible to monitor the VOCs in field itself (Ezra et al. 2004b). The VOCs emitted by *Muscodor* spp. constitutes both aliphatic and aromatic heterocyclic compounds which predominantly belongs to the chemical class of alcohols, amines, acids, esters, ketones, certain hydrocarbons, etc. The aromatic compounds produced by *Muscodor* spp. are majorly azulene, naphthalene, and sesquiterpenes (Daisy et al. 2002; Zhang et al. 2010; Meshram et al. 2014, 2016; Alpha et al. 2015). The VOCs analysis indicated that the major volatile emitted by *M. albus*, *M. crispans*, *M. equiseti*, *M. musae*, *and M. sutura* in their head space is 2-methylpropanoic acid, whereas VOCs composition of *M. yucatanensis*, *M. roseus*, and *M. vitigenus* is unique and different from above mentioned *Muscodor* species (Strobel et al. 2001; Mitchell et al. 2010; Kudalkar et al. 2012; Suwannarach et al. 2013). Similarly, *M. fengyangensis* produces azulene and β-phellandrene as their premier VOCs (Zhang et al. 2010). The Indian *Muscodor* isolates *M. kashayum* and *M. darjeelingensis* predominantly produce β-Bisabolol and 2,6-Bis(1,1-dimethylethyl)-4-(1-oxopropyl)phenol, whereas *M. strobelii*, *M. tigerii*, *M. ghoomensis*, and *M. indica* emitted 4-octadecylmorpholine as the chief contributor of their gas chemistry (Meshram 2013, 2014, 2016; Saxena et al. 2014, 2015) (Table 4.2). *Muscodor tigerii* produces campesterol and stigmasterol which possess anticancer property, whereas it also emits phytol and squalene which have reported antibacterial properties (Saxena et al. 2015). *Muscodor ghoomensis* isolated from *C. camphora* produced a geranyl linalool isomer in its head space which is an analogous volatile oil originally isolated from aerial parts of the camphor tree, thus validating the theory that endophytic fungi mimic the property of the host by producing analogous bioactive substances (Meshram et al. 2016). The VOCs produced by *Muscodor* also act as a chemotaxonomic marker. The partial ITS DNA sequence of *M. sutura* was similar to that of *M. vitigenus*, but on the basis of its volatile gas chemistry, *M. sutura* was introduced as a new species of genus *Muscodor* (Kudalkar et al. 2012).

4.5 VOCs Stress Bioassay

VOCs stress assay is used as a selection tool for isolation of new *Muscodor* species. Further, the same assay is also employed in screening VOCs for their antifungal and antibacterial activity (Ezra et al. 2004a; Meshram et al. 2013). In this assay, compartments are made in the Petri dish by removing the agar strip from the center to inhibit the movement of any diffusible metabolite. One side of the Petri dish is

Table 4.2 Major volatile organic compounds (VOCs) produced by various *Muscodor* species

Muscodor sp.	Major VOCs produced
M. albus[a]	1-butanol, 3-methyl-, acetate; naphthalene; azulene; ethanol
M. roseus[b]	Ethyl 2-butenoate and 1,2,4-trimethylbenzene
M. vitigenus[c]	Naphthalene
M. crispans[d]	Propanoic acid, 2-methyl-, methyl ester; ethanol; ethyl acetate
M. yucatanensis[e]	1R,4S,7S,11R-2,2,4,8-tetramethyltricyclo[5.3.1.0(4,11)]undec-8-ene; caryophyllene
M. fengyangensis[f]	Propanoic acid, 2-methyl-; β-phellandrene; azulene
M. cinnamomi[g]	Ethyl 2-methylpropanoate
M. sutura[h]	Propanoic acid, 2-methyl-, methyl ester; thujopsene; octadecanoic acid
M. equiseti[i]	2-methylpropanoic acid
M. musae[i]	2-methylpropanoic acid
M. oryzae[i]	3-methylbutan-1-ol
M. suthepensis[i]	2-methylpropanoic acid
M. kashayum[j]	β-Bisabolol; 2,6-bis(1,1-dimethylethyl)-4-(1-oxopropyl)phenol;
M. darjeelingensis[k]	2, 6-bis (1, 1-dimethylethyl)-4-(1-oxopropyl)phenol; 1, 6-dioxacyclododecane-7, 12-dione
M. strobelii[l]	4-octadecylmorpholine; tetraoxapropellan; aspidofractinine-3-methanol
M. tigerii[m]	4-octadecylmorpholine; 1-tetradecanamine,N,N-dimethyl
M. heveae[n]	3-methylbutan-1-ol; 3-methylbutyl acetate; azulene
M. ghoomensis[o]	4-octadecylmorpholine; 1-nonadecamine-N.N-dimethyl
M. indica[o]	1, 6-dioxacyclododecane-7,12-dione; 4-octadecylmorpholine; squalene; pogostol

[a]Worapong et al. (2001)
[b]Worapong et al. (2002)
[c]Daisy et al. (2002)
[d]Mitchell et al. (2008)
[e]Gonzalez et al. (2009)
[f]Zhang et al. (2010)
[g]Suwannarach et al. (2010)
[h]Kudalkar et al. (2012)
[i]Suwannarach et al. (2013)
[j]Meshram et al. (2013)
[k]Saxena et al. (2014)
[l]Meshram et al. (2014)
[m]Saxena et al. 2015
[n]Siri-udom et al. (2016)
[o]Meshram et al. (2016)

inoculated with actively growing *Muscodor* species and incubated (4–5 days) for the formation of volatile atmosphere inside the chamber. Further, on the remaining chambers, plant samples were inoculated and incubated for next few days till the endophytic fungi emerges from the host tissue. The VOCs inhibit the growth of fastidious microorganism and allows the growth of volatile-tolerant *Muscodor*-related species (Mitchell et al. 2010; Meshram et al. 2013; Saxena et al. 2015). Several *Muscodor* isolates including *M. crispans*, *M. sutura*, *M. kashayum*, and *M.*

strobelli have been isolated using this technique (Mitchell et al. 2008; Kudalkar et al. 2012; Meshram et al. 2013, 2014).

For screening the biological activity of *Muscodor* species, the growth of the pathogenic fungi and bacteria is monitored in presence and absence of VOCs in the head space of the Petri dish using VOCs stress bioassay as discussed above. From *M. albus* to *M. heveae*, their antifungal and antibacterial activity was determined by same assay (Mitchell et al. 2010; Meshram et al. 2013; Suwannarach et al. 2013). Further, a microcup assay is devised to determine the effect of artificial mixture of VOCs on pathogenic microorganisms. A microcup (cap of microcentrifuge tube) having varying amount of VOCs is placed in the center of the Petri dish, and the test organism is inoculated around the periphery and incubated for 48–72 h. The growth of the test organism was observed with respect to its growth in absence of the VOCs mixture. The inhibitory concentration (IC_{50}) is calculated by dividing the amount of VOCs required to cause 50% death of the microorganism. This assay was used to determine the effect of VOCs produced by *M. albus* on pathogenic bacteria and fungi and *M. crispans on Mycobacterium tuberculosis* (Strobel et al. 2001; Mitchell et al. 2010).

4.6 Antimicrobial Activity of *Muscodor* spp.

Muscodor species exhibit antimicrobial activity against an array of plant and human pathogenic microorganisms by the virtue of the VOCs produced by it (Meshram et al. 2013, 2016). The VOCs produced by *Muscodor* species completely kills or inhibits the growth of plant pathogens like *Alternaria*, *Aspergillus*, *Cercospora*, *Fusarium*, and *Rhizoctonia*, whereas it also kills human pathogens including *Candida*, *Staphylococcus*, *Pseudomonas*, and *E.coli*. *Muscodor* species including *M. albus*, *M. crispans*, *M. equiseti*, *M. fengyangensis*, *M. musae*, *M. kashayum*, and *M. heveae* exhibited both antifungal and antibacterial activity, whereas *M. sutura* and *M. yucatanensis* only showed antifungal activity (Suwannarach et al. 2013; Meshram et al. 2014; Siri-udom et al. 2016). Further, *M. ghoomensis* and *M. indica* do not completely kill the pathogen but only inhibit its growth (Meshram et al. 2016). *M. albus*, *M. crispans*, and *M. sutura* completely inhibited the growth of two of the most important pathogens, i.e., *Pythium* and *Phytophthora* species, whereas *M. kashayum*, *M. suthepensis*, and *M. equiseti* also checked the growth of *Fusarium* spp. that are potential plant pathogen leading to huge crop loss (Table 4.3, Fig. 4.2). Similarly, the VOCs of *M. crispans* completely inhibited the growth of drug-resistant *Mycobacterium tuberculosis* (Mitchell et al. 2010; Krajaejun et al. 2012). The bioactivity of the volatile antimicrobial produced by *Muscodor* species is greatly affected by the substrate. The effectiveness and quality of the VOCs produced by *Muscodor* species is greatly influenced by the composition of the medium used to support the growth of fungus. In case of *M. albus*, sugar-enriched medium more effectively inhibit the growth of *Fusarium*, *Pythium*, and *Rhizoctonia* (Ezra and Strobel 2003). Recently, the mode of action of VOCs produced by *M. albus* was

Table 4.3 Antimicrobial activity of volatile organic compounds (VOCs) produced by various *Muscodor* species

Muscodor sp.	Antimicrobial activity									
	A. flavus	*B. cinerea*	*C. beticola*	*C. gloeosporioides*	*F. oxysporum*	*F. solani*	*R. solani*	*C. albicans*	*E. coli*	*S. aureus*
M. albus	ND	ND	+	ND		+	+++	+++	+++	+++
M. crispans	ND	+++	ND	+++	+++	++	+++	+++	+++	ND
M. fengyangensis	ND	+++	ND	ND	++	++	+++	ND	+++	ND
M. sutura	ND	+++	+++	+++	ND	++	+++	–	–	ND
M. equiseti	++	+++	ND	+++	+++	+++	+++	+++	+++	+++
M. musae	++	++	ND	+++	++	+++	+++	+++	+++	+++
M. oryzae	+++	+++	ND	+++	++	+++	+++	+++	+++	+++
M. suthepensis	+++	+++	ND	+++	+++	+++	+++	+++	+++	+++
M. kashayum	+	++	+++	+++	+++	ND	+++	+++	+++	+++
M. darjeelingensis	+	+	++	+	+	+	+	++	–	++
M. strobelii	ND	+++	++	+	+	ND	+	+++	+++	+++
M. tigerii	+	+	+++	–	+	+	++	++	ND	++
M. heveae	+++	ND	ND	+++	+	+++	+++	ND	ND	+++
M. ghoomensis	+	+	++	+	ND	+	+	+	ND	+
M. indica	+	+	++	+	ND	+	+	+	ND	++

+, poor activity; ++, strong activity (60–80% inhibition); +++, very strong activity (80–100% inhibition); –, no activity; *ND*, refers to "not done"
Data collated from the protologue publications

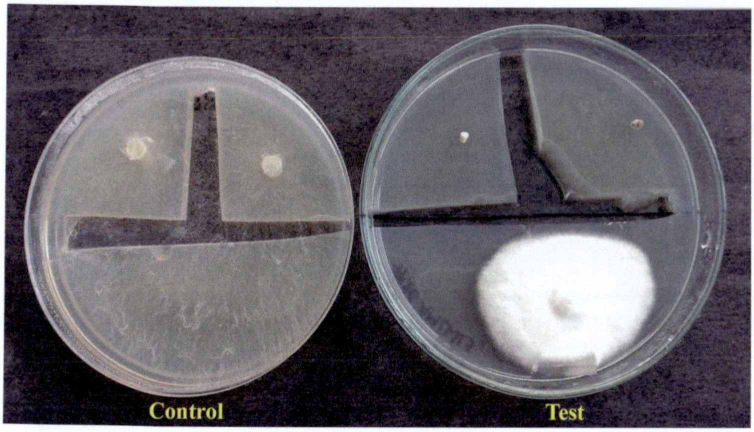

Fig. 4.2 Antimicrobial activity of a *Muscodor* species against *Rhizoctonia solani* in a VOCs stress bioassay. Inoculation with *Muscodor* (right, lower colony) inhibited the growth of *R. solani* (inoculation in the upper quarter) as compared to the control (left). (Pictures were taken 72 h post VOCs exposure)

examined by a series of genetic screen and biochemical assays, and it was found that VOCs induce microbial killing via DNA damage (Alpha et al. 2015).

4.7 Insecticidal Activity of *Muscodor* spp.

The hazardous effect of chemical fumigants like sulfur dioxide and methyl bromide has led to the exploration of alternative biofumigants for quarantine security of fruits with reduced side effects. A possible alternative is the biologically generated fumigant could provide sustainable protection of the fruits and vegetables by inhibiting or checking the attack of insects and pests which degrade the product. Potato tuber moth, *Phthorimaea operculella*, is a widespread pest of tuber family in many regions of the world. The shortcomings encountered by the conventional approaches for controlling this moth have led to the exploitation of *M. albus* as a biofumigant. About 30 g of the rye grain formulation of *M. albus* was found to induce 90% mortality in the adult moth, whereas the same amount of formulation led to the reduction of larval growth by 72% (Lacey and Neven 2006). Further, the VOCs produced by *M. albus* also controlled the growth of codling moth, *Cydia pomonella.* Exposure to VOCs of *M. albus* for 3 days has led to the 81% mortality of the moth. Similarly, mycofumigation with *M. albus* for 3 days led to 86% of larva mortality (Lacey et al. 2009). The mycofumigation potential of *M. albus* has also been exploited for controlling western cherry fruit fly (*Rhagoletis indifferens*). The volatile exposure for a fortnight has led to 86% reduction in the fly emergence (Yee et al. 2009). Thus, mycofumigation with *M. albus* could be a novel and sustainable approach for control of notorious insects.

4.8 Nematicidal and Nematostatic Activity of *Muscodor* spp.

Plant-parasitic nematodes are the vital crop pathogens that cause humongous economic loss of around 100 billion dollar annually. Despite adopting several precautionary strategies like crop rotation, planting resistant varieties, and application of chemical fungicides, not much of a success has been achieved in controlling nematodes. Further, the hazardous effect of chemical fungicides on ozone layer has led to their restricted use. Owing to their low side effects on the environment, there has been a growing interest to use biological methods for control of pests as compared to their synthetic counterparts (Gu et al. 2007). Under greenhouse conditions, VOCs of *M. albus* were found to induce both nematicidal and nematostatic effect over the tested four nematodes including *Meloidogyne chitwoodi*, *Meloidogyne hapla*, *Paratrichodorus allius*, and *Pratylenchus penetrans*. Treatment with *M. albus* induces 82.7% mortality for *P. allius*, 82.1% for *P. penetrans*, and 95% for *M. chitwoodi*, whereas only 21.9% of *M. hapla* died in the test condition. Further, the treatment with *M. albus* caused 100% reduction of all four nematodes in soil, whereas in case of root hairs the reduction varied from 56–100%, respectively (Riga et al. 2008). Further, a biorational mixture of *M. albus* volatiles containing 2-methyl acetate and propanoic acid 2-methyl-, 2-methyl propyl ester and butanol, 2-methyl-1-butanol protected tomatoes from the root-knot nematode *M. incognita* (Grimme et al. 2007).

Apart from *M. albus*, the VOCs produced by *M. vitigenus* have also been found to effectively repel the wheat sawfly *Cephus cinctus* under in vitro condition in a Y tube bioassay test. *Muscodor vitigenus* produces naphthalene as its principal volatile. The artificial mixture of naphthalene (flow rate 12 ng/h) successfully repelled two third (>70%) of the sawfly on successive treatment in a Y tube bioassay, whereas the naphthalene produced by *M. vitigenus* exhibited better insect repellent property with above 81% efficacy (Daisy et al. 2002).

4.9 Control of Building Molds by *Muscodor* spp.

The dampness in the house promotes microbial growth that deteriorates the indoor air quality leading to certain health problems among the people living in that environment which includes asthma, headache, dermatological symptoms, and gastrointestinal disorders. The condition is known as sick building syndrome and is generally caused by the presence of molds like *Aspergillus*, *Penicillium*, and *Stachybotrys* that grow on building materials. The volatiles of *M. albus* have shown promising results in controlling the growth of sick building molds. To determine its curative effect, gypsum walls (both dry and moist, respectively) were fumigated with rye grain formulation of *M. albus* (5 and 20 g) and then infected (10^5–10^6 cfu/cm^2) with *Cladosporium cladosporioides*, *A. niger*, and *S. chartarum*. Upon incubation, a huge reduction was observed in walls fumigated with 5 g *M. albus* formulation, whereas no detectable amount of fungal population was present in fumigated dry walls as compared to non-fumigated walls with 20 g *M. albus*. Further, in case of moist walls, *Cladosporium* was easily eliminated, whereas longer exposure of VOCs caused five log reductions in *Aspergillus* population. The major VOCs responsible for this curative treatment were isobutyric acid, 2-methyl-1-butanol, and isobutanol. The mycofumigation property of *Muscodor* species could be an important tool for controlling building molds in areas that are difficult to reach and where the use of liquid detergents is undesirable (Mercier and Jimnez 2007; Strobel 2011).

4.10 Mycofumigation with *Muscodor* Species for Control of Postharvest Losses

The major cause of economic losses in postharvest storage is the spoilage of food product by bacterial and fungal infections. Harvesting of crops leads to epidermal lesions which act as a gateway for opportunistic pathogens. Several strategies like storage at low treatment and treatment with fungicides has reduced the risk of rotting during shipment. However, the over use of fungicides has led to the development of resistance among them. Further, certain pesticides cause hazardous environmental effects resulting in their discontinuation from commercial use. Synthetic fungicides provide immediate results, but it is not a sustainable solution for the problem; that is why the use of alternative means such as physicochemical methods, biofumigation, and microbial antagonists has been extensively studied (Mercier and Jimenez 2004; Mercier and Smilanick 2005; Ramim et al. 2005; Schotsmans et al. 2008; Strobel 2011). Mycofumigation is a term generally applied to the exploitation of gas-producing fungi for control of pathogens by production of volatile antibiotics (Stinson et al. 2003). Over the last few decades, many volatile-producing fungi including *Trichoderma*, *Phomopsis*, *Nodulisporium*, and *Hypoxylon* spp. have been reported to control microbial pathogenesis, but none of them has a comprehensive spectrum of antimicrobial activity as that of *Muscodor* species (Mitchell et al. 2010; Singh et al. 2011; Gomes et al. 2015; Pena et al. 2016).

Mycofumigation with *Muscodor* is done by using rye grain formulation where the actively growing fungus is colonized in sterile rye seeds and subsequently dried using desiccators. Further, the rye grains are packed in sachet. The rye grain formulations are hydrated just before the use. The hydration causes release of volatiles which are then spilled over the fruit and provide shielding effect. The release of volatiles like isobutyric acid is a good indicator of antifungal activity for biofumigation formulations. The volatiles are active even at −31 °C. Studies have suggested that 1 g of *Muscodor* formulation is effective for protecting 100 g of fruits up to 2 months in cold storage. The major advantage with mycofumigation is that there is no direct contact between fruit and the fungus; secondly it does not require any manual handling. Additionally, the VOCs produced are degraded rapidly thus reducing the chances of toxicological effects (Mercier and Jimenez 2004; Schnabel and Mercier 2006; Mercier et al. 2009; Braun et al. 2012). Following are few examples where mycofumigation with *Muscodor* was performed:

4.10.1 Control of Postharvest Gray Mold of Table Grapes

Botrytis cinerea is both a friend and foe to the grape growers because on one hand it leads to perfect sweetening of vine and secretion of resveratrol that adds to its medicinal value while on the contrary it causes gray mold disease (Gabler et al. 2006). Mycofumigation with rye grain formulations of *M. albus* was found to control the growth of *B. cinerea* on grapes. Mycofumigation with 50 and 100 g of rye grain formulation of *M. albus* completely hindered the growth of *B. cinerea* at 20 °C and 5 °C for 7 days, respectively. Further, it was also observed that when the fumigation was done within 3–24 h post inoculation, there was no longer viability of *B. cinerea* on grapes after 7 days at 20 °C, and the grapes can be stored for 3 more days at same condition without any further infections. Apart from this, a significant reduction in the gray mold incidence was observed when clamshell boxes and polyethylene cluster bags consisting of *M. albus* formulation (20 g) was used. The development of gray mold incidences in *M. albus* treated formulation reduced to 1% and 6.7% as compared to their respective control after 7 days incubation at 15 °C (Gabler et al. 2006). Further, an integrated approach comprising of chemical and biological fumigation potentially checked the growth of *B. cinerea*. The cocktail of ozone and *M. albus* (5000 μl/l) reduced the infection rate of *B. cinerea* to 10% from 91%. Further, the effect of this cocktail was also effective at low temperature (5 °C) where it brought down the infection rate from 31% to just 3.4%. Although the mixture of ozone and *M. albus* was less effective than commercially used synthetic fungicide sulfur dioxide (1.1%), but this could be an alternative to chemical fungicide particularly in organic farming (Gabler et al. 2010).

4.10.2 Control of Fungal Decay of Apple and Peaches

Fungal pathogens like *B. cinerea* and *Penicillium expansum* lead to development of gray and blue mold in pome fruits, whereas *Monilinia fructicola* causes brown rot in peaches. These pathogens through their conidia infest air, water, and packaging machineries. Most of the fungicides that were used to control these infections are withdrawn from market owing to their toxicological effects. Thus, the need of new alternative has opened pathways for utilizing curative activity of *M. albus*. Mycofumigation with *M. albus* shielded apples from gray and blue mold infection. Rye grain formulation of *M. albus* provided complete safety from *B. cinerea* and *P. expansum* infection up to 14 days. Similarly, fumigation with *M. albus* also protected peaches from brown-rot decay. The growth of infectious pathogen *M. fructicola* was also completely hindered by *M. albus* for 10 days post inoculation. Further, a pad delivery system of *M. albus* was used for management of brown rot of peaches in shipping cartons. Fumigation (at 1–2 °C) with pads containing 50 or 200 g of *M. albus* per pack in a carton containing 14 kg of peaches was found to effectively reduce the incidences of brown rot in Coronet and Red Globe peaches (Schnabel and Mercier 2006).

4.10.3 Control of Green Mold and Sour Rot of Lemon

The green mold of citrus is caused by *Penicillium digitatum*, whereas the sour rot of lemon is caused by *Geotrichum citri-aurantii*. These pathogens have become resistant to the current armamentarium of the fungicides including imazalil and thiabendazole due to their overuse. Mycofumigation with *M. albus* controlled green mold and sour rot of citrus. Mycofumigation with *M albus* killed *P. digitatum* for 3 days in an 11 l plastic box. Further, a test was conducted to see the effect of mycofumigation in harvested citrus fruit assisted by presence of ethylene. It was observed that mycofumigation brought down the incidences of green mold from 89% to only 26% when tested in an 11 m^3 room supplemented with 5 ppm ethylene at 20 °C, and it also had no adverse effect on color development. Mycofumigation with *M. albus* for 72 h also reduced the incidences of sour rot from 39% to 2.4% when incubated immediately after fungal lesion (Mercier and Smilanick 2005). Apart from this, *M. suthepensis* CMU-Cib462 exhibited a potential dose dependence control of green mold on tangerine fruit (Suwannarach et al. 2015b). Recently, a new *Muscodor* isolate, *Muscodor* LGMF1254, prevented citrus black spot disease by controlling the growth and pycnidia formation of causal organism *Phyllosticta citricarpa* (Pena et al. 2016). Thus mycofumigation with *Muscodor* spp. could be a handy tool for controlling the decay of citrus in storage rooms and shipping packages.

4.10.4 Control of Smut

Smut fungi like *Tilletia* have been associated with various kinds of bunts including Karnal bunt of wheat and kernel bunt of rice. Fumigation with *M. albus* leads to the non-germination of teliospores of *Tilletia horrida*, *Tilletia indica*, and *Tilletia tritici*. Further, it was found that fumigation with *M. albus* can also be utilized for controlling the seedling infecting smut in furrow or seed treatment (Goates and Mercier 2009). In greenhouse experiments, *M. albus* has also been found to control *Ustilago hordei* in barley without disturbing the growth of the plant (Strobel et al. 2001).

4.10.5 Control of Soilborne Diseases

Fungal pathogens like *Rhizoctonia*, *Pythium*, and *Phytophthora* cause root rot and damping off in plants resulting in poor plant growth. Further, the withdrawal of methyl bromide which was used to manage the obnoxious pathogens has worsened the situation. Mycofumigation with *Muscodor* could be an effective substitution of methyl bromide (Zidack et al. 2002, 2003). Mycofumigation with *Muscodor* has exhibited control from soil pathogens including *Aphanomyces cochlioides*,

Rhizoctonia solani, *Pythium ultimum*, *Phytophthora capsici*, and *Verticillium dahliae* (Stinson et al. 2003; Mercier and Manker 2005; Grimme et al. 2007; Camp et al. 2008; Worapong and Strobel 2009). The rye grain formulation of *M. albus* mixed with soil completely inhibited the growth of *R. solani* thereby providing control to damping off of broccoli seedlings. Further, it was also observed that *M. albus* completely control the formation of root rot of bell pepper caused by *P. capsici.* Mycofumigation with *M. albus* was effective between 4 and 22 °C. During the study, it was also noticed that *M. albus* enhances plant growth by eliminating the harmful microorganisms that often contaminate the soil (Mercier and Manker 2005; Mercier and Jimenez 2009). Further, *Muscodor cinnamomi* completely inhibited the growth of damping-off pathogen *R. solani* (Suwannarach et al. 2012). Recently, *M. cinnamomi* was found to control the growth of *R. solani* in tomato plant thereby increasing its root length along with its shoot and root dry mass (Suwannarach et al. 2015a). Fumigation with *M. albus* and *M. roseus* leads to decrease in the disease severity in sugar beet and eggplant caused by *A. cochlioides*, *P. ultimum*, *R. solani*, and *V. dahliae* (Stinson et al. 2003). Later on, it was found that a small amount of biorational mixture of *M. albus* is required (2 and 0.75 μl/cm^3, respectively) for controlling damping off of sugar beet caused by *Rhizoctonia* and *Pythium* (Grimme et al. 2007). Another isolalate *M. albus* MFC2 controlled the root rot caused by *P. ultimum* in *Brassica oleracea* (Worapong and Strobel 2009). Mycofumigation with *M. albus* also controlled the *Phytophthora* blight of sweet pepper and butternet squash (Camp et al. 2008).

4.10.6 **Muscodor** *in Treatment of Human and Animal Wastes*

The VOCs produced by *M. albus* are effective in controlling the growth of *E. coli* and other bacteria that are found in human wastes (Strobel 2006a). This inherent property of *Muscodor* species has been harnessed by a US-based company, Phillips Environmental Products, Belgrade, MT, to solve the problem of degradation of human wastes products in pit toilets or portable toilets. Initially, a technology was developed to support the growth of *M. albus* in these conditions. Further, a fungal formulation comprising a mixture of *M. albus* and *Fusarium culmorum* is developed and is placed in disposable/wag bags that trap human wastes in biotoilets and decomposes them. Both these fungi are compatible with each other. *M. albus* carries bacterial decontamination, whereas *F. culmorum* degrades waste. Thus, this technology significantly contribute in decomposition of solid or liquid waste in places where public has been removed from sanitation facility. This technology will be helpful to military organizations, mountaineers, and many national parks (Patent no. US 7341862 B2; US 7858362 B2).

4.11 Lead Molecules Produced by *Muscodor* Species

Muscodor has gained popularity because of its volatile antimicrobial which has possible applications as a biocontrol agent. However, not much has been done to investigate the biological properties of the secreted secondary metabolites (extrolites) produced in the culture broth by various *Muscodor* species. The very first attempt in this direction was made by Macias-Rubalcava et al. (2010) where they have evaluated the inhibitory activity of the organic and mycelial extract of *M. yucatanensis* on plant pathogenic fungi. Both the extracts exhibited a dose-dependent inhibitory activity on phytopathogenic fungi *Colletotrichum* sp., *Guignardia mangiferae* and *Phomopsis* sp. The extracts of *M. yucatanensis* also retarded the growth of tomato, amaranth, and barnyard grass. In both the cases, organic extract has better activity as compared to mycelial extracts (Macias-Rubalcava et al. 2010). Apart from this, a new strain of *Muscodor yucatanensis* Ni30 isolated from Central America produced brefeldin A, whereas its mutant EV1 produced ergosterol and xylaguaianol C in their culture broth (Qadri et al. 2016). Recently, *M. cinnamomi* was found to produce indole-3-acetic acid (45.36 ± 2.4 μg/ml) in its culture medium. The production of phytohormone indole-3-acetic acid supported the root and coleoptile elongation and increased seed germination. Further, the isolate also solubilized various toxic metals including calcium, cobalt, cadmium, lead, and zinc and was also found tolerant to certain herbicides and insecticides including methomyl and 2,4-D-dimethylammonium (Suwannarach et al. 2015a).

Indian *Muscodor* isolates have been tested for their antibacterial potential against a spectrum of Gram-positive and Gram-negative bacteria including *Staphylococcus aureus*, *S. epidermidis*, *Pseudomonas aeruginosa*, *P. putida*, *Bacillus subtilis*, and *E.coli*. Among the tested seven *Muscodor* species, the ethyl acetate extract of *M. indica* exhibited maximum antibacterial activity followed by *M. ghoomensis* and *M. camphora*. The antibacterial activity exhibited by these isolates was predominantly against Gram-positive bacteria comprising three isolates of *S. aureus* and one isolate of *S. epidermidis* and only 1 g negative bacteria, i.e., *P. aeruginosa* (Boparai et al. 2015). Further, a mild pancreatic lipase inhibitory activity was observed in the culture broth of *M. darjeelingensis* (Gupta et al. 2015). Recently, xanthine oxidase inhibitory (antigout/antihyperuricemia) activity was reported in the chloroform extract of Indian *Muscodor* isolates. Among seven isolates, three isolates (*M. darjeelingensis*, *M. tigerii*, and *M. kashayum*) exhibited >75% inhibition, whereas three isolates (*M. strobelli*, *M. ghoomensis*, and *M. camphora*) showed >35% inhibition of xanthine oxidase. Additionally, the Indian *Muscodor* isolates also exhibited strong antioxidant properties. Maximum antioxidant activity via free radical scavenging was observed in *M. indica* (71%) followed by *M. strobelli* (69%), *M. darjeelingensis* (68%), and *M. kashayum* (65%) (Kapoor and Saxena 2016). Thus *Muscodor* appears to be a putative source of novel bioactive secondary metabolites which can have possible application in agro and pharmaceutical industries.

4.12 Commercial Products of *Muscodor* or Products in Pipeline

On September 25, 2005, the US Environmental Protection Agency Office of Pesticide Programs has partially permitted AgraQuest, Inc., Davis, CA, USA, to use products of *M. albus* strain QST 20799 as biopesticide. Further, in the same year, United States Patent and Trademark Office has issued a trademark to *Muscodor* (s.no. 78658729) in the category of pharmaceutical products. The product of *M. albus* strain QST 20799 has been registered as 20799® by AgraQuest, Inc., with their end product names as Andante™, Arabesque™, and Glissade™. Arabesque is used for controlling postharvest and tuber disease-producing fungi and bacteria, whereas Andante and Glissade have been used as a soil biofumigant for controlling root rot, damping off, and wilt disease-producing bacteria and fungi. Additionally, Andante also control nematodes (US Environmental Protection Agency Office of Pesticide Programs 2005). Further, a number of patents have been granted to various companies which are developing different products from *Muscodor* species having their use in biofumigation, nematode control, management of human and animal waste, control of *M. tuberculosis*, and oral hygiene (Table 4.4).

Table 4.4 Patents granted for various applications of *Muscodor* species

Patent title	Patent no	Applicants	Assignee	Field of the invention
Novel endophytic fungi and methods of use	CA 2443295 C (Apr 11, 2002–24 Oct, 2002)	Gary Strobel, Julien Mercier, Denise Carol Manker	AgraQuest, Inc., Davis, CA, USA	Isolation of volatile antibiotics producing fungi having biological activity having antimicrobial and insecticidal and nematicidal activity
Methods and compositions relating to insect repellents from a novel endophytic fungus	US 7267975 B2 (Oct 15, 2003–Sep 11, 2007)	Gary Strobel, Bryn Daisy	Montana State University, Bozeman, MT, USA	Discovery of *Muscodor vitigenus*, naphthalene production and its use as an insect repellent
Application *of Muscodor albus* to control harmful microbes in human and animal wastes	US 7341862 B2 (Mar 17, 2004–Mar 11, 2008)	Gary Allan Strobel, David Ezra	Montana State University, Bozeman, MT, USA	Discovery of new *Muscodor albus* and control of human and animal waste by its VOCs

(continued)

Table 4.4 (continued)

Patent title	Patent no	Applicants	Assignee	Field of the invention
Method of using endophytic fungi to decontaminate and decompose human and animal wastes	US 7858362 B2 (May 25, 2005–Dec 28, 2010)	Brian J. Phillips, Gary Allan Strobel, Emilie Dirkse, David Ezra, Uvidelio Castillo	Montana State University, Bozeman, MT USA; Phillips Environmental Products, Inc., Belgrade, USA	Use of *Fusarium culmorum* and *Muscodor albus* in waste treatment (biotoilets)
Antimicrobial compositions and related methods of use	CA 2760150 A1 (Apr 27, 2010–Nov 11, 2010)	Niranjan Gandhi, Victoria Palmer Skebba, Gary Strobel	Jeneil Biosurfactant Company Llc., Waxahachie, TX, USA	Discovery of *M. crispans* and use of its uses as food and flavoring agent, in control of *M. tuberculosis*, oral hygiene product/ mouthwash
Compounds derived from *Muscodor* fungi	US 0120058058 A1 (May 11, 2010–Mar 8, 2012)	Jorge I. Jimenez, Jonathan S. Margolis, John Kenneth Baird, Sarah F. Lego	AgraQuest, Inc., Davis, CA, USA	Novel compounds from *Muscodor*, their composition and uses as an antimicrobial and pest control agent including insects and nematodes
Muscodor albus strain producing volatile organic compounds and methods of use	US 20140086879 A1 (Sep 23, 2013–Mar 27, 2014)	Gary Strobel, Vu Phong Bui, Hai Su, Phyllis Himmel, Pamela Marrone, Lijuan Xing, Sarah Lewis	Marrone Bio Innovations, Inc., Davis, CA, USA	Production of VOCs from *M. albus* and their use as pest control
Compositions of volatile organic compounds and method of use thereof	US 8968798 B2 (Apr 9, 2013–Mar. 3, 2015)	Wayne A. Green, Gary Strobel	Synthetic Genomics, Inc., La Jolla, CA, USA	Discovery of *M. stobelii* and use its use as antimicrobial agent

4.13 Concluding Remarks

The bioactive metabolites produced by *Muscodor* species have potential biotechnological applications with a broader market value beyond those in food and agriculture industries. The available studies on *Muscodor* have only scratched the surface. Nevertheless, several roadblocks remain before their potential can be exploited for commercial purposes. The VOCs are emitted in small quantities, making them difficult to characterize and study. The life cycle of *Muscodor* still remains an untold story.

Acknowledgment Authors are thankful to Dr. Gary Strobel, Prof. Emeritus (Retd.), Department of Plant Sciences and Plant Pathology, Montana State University, Bozeman, USA, for all his kind guidance and support over the years. We dedicate this book chapter to him.

References

Alpha JC, Campos M, Wagner JC, Strobel AS (2015) Mycofumigation by the volatile organic compound-producing fungus *Muscodor albus* induces bacterial cell death through DNA damage. Appl Environ Microbiol 81:1147–1156

Aly AH, Debbab A, Proksch P (2011) Fungal endophytes: unique plant inhabitants with great promises. Appl Microbiol Biotechnol 90(6):1829–1845

Arnold AE, Maynard Z, Gilbert GS, Coley PD, Kursar TA (2000) Are tropical fungal endophytes hyperdiverse? Ecol Lett 3:267–274

Atmosukarto I, Castillo U, Hess WM, Sears J, Strobel G (2005) Isolation and characterization of *Muscodor albus* I-41.3s, a volatile antibiotic producing fungus. Plant Sci 169:854–861

Bacon CW, White JF (2000) Microbial endophytes. Marcel Dekker, New York

Banerjee D, Strobel G, Geary B, Sears J, Erza D, Liarzi O, Coombs J (2010) *Muscodor albus* strain GBA, an endophytic fungus of *Ginkgo biloba* from United States of America, produces volatile antimicrobials. Mycology 1:179–186

Banerjee D, Pandey A, Jana M, Strobel G (2013) *Muscodor albus* MOW12 an endophyte of *Piper nigrum* L. (Piperaceae) collected from North East India produces volatile antimicrobials. Indian J Microbiol 54:27–32

Banguela-Castillo AB, Gonzalez PLR, Marey MP, Tanaka FAO, Moraes MCB, Rodriguez LH, Cabrera RI (2015) Crop Prot 78:284–292

Boparai JK, Saxena S, Meshram V (2015) *In vitro* antimicrobial potential of Indian *Muscodor* species. J Basic Appl Mycol 11:22–25

Braun G, Vailati M, Prange R, Bevis E (2012) *Muscodor albus* volatiles control toxigenic fungi under controlled atmosphere (CA) storage conditions. Int J Mol Sci 13:15848–15858

Camp AR, Dillard HR, Smart CD (2008) Efficacy of *Muscodor albus* for the control of *Phytophthora* blight of sweet pepper and butternut squash. Plant Dis 92:1488–1492

Corcuff R, Mercier J, Tweddell R, Arul J (2011) Effect of water activity on the production of volatile organic compounds by *Muscodor albus* and their effect on three pathogens in stored potato. Fungal Biol 115:220–227

Daisy B, Strobel G, Ezra D, Castillo U, Baird G, Hess WM (2002) *Muscodor vitigenus* anam. sp. nov., an endophyte from *Paullinia paullinioides*. Mycotaxon 84:39–50

Ezra D, Strobel GA (2003) Effect of substrate on the bioactivity of volatile antimicrobials produced by *Muscodor albus*. Plant Sci 165:1229–1238

Ezra D, Jasper J, Rogers T, Knighton B, Grimsrud E, Strobel G (2004a) Proton transfer reaction-mass spectrometry as a technique to measure volatile emissions of *Muscodor albus*. Plant Sci 166:1471–1477

Ezra D, Hess WM, Strobel GA (2004b) New endophytic isolates of *Muscodor albus*, a volatile antibiotic producing fungus. Microbiology 150:4023–4031

Gabler MF, Fassel R, Mercier J, Smilanick JL (2006) Influence of temperature, inoculation interval, and dosage on biofumigation with *Muscodor albus* to control postharvest gray mold on grapes. Plant Dis 90:1019–1025

Gabler MF, Mercier J, Jimenez JI, Smilanick JL (2010) Integration of continuous biofumigation with *Muscodor albus* with pre-cooling fumigation with ozone or sulfur dioxide to control postharvest gray mold of table grapes. Postharvest Biol Technol 55:78–84

Goates BJ, Mercier J (2009) Effect of biofumigation with volatiles from *Muscodor albus* on the viability of *Tilletia* spp. teliospores. Can J Microbiol 55:203–206

Gomes AAM, Queiroz MV, Pereira OL (2015) Mycofumigation for the biological control of post-harvest diseases in fruits and vegetables: a review. Austin J Biotechnol Bioeng 2:2–8
Gonzalez MC, Anaya AL, Glenn AE, Macias-Rubalcava ML, Hernandez-Bautista BE, Hanlin RT (2009) *Muscodor yucatanensis*, a new endophytic ascomycete from Mexican chakah, *Bursera simaruba*. Mycotaxon 110:363–372
Grimme E, Zidack NK, Sikora RA, Strobel GA, Jacobsen BJ (2007) Comparison of *Muscodor albus* volatiles with a biorational mixture for control of seedling diseases of sugar beet and root-knot nematode on tomato. Plant Dis 91:220–225
Gu YQ, Mo MH, Zhou JP, Zou CS, Zhang KQ (2007) Evaluation and identification of potential organic nematicidal volatiles from soil bacteria. Soil Biol Biochem 39:2567–2575
Gupta M, Saxena S, Goyal D (2015) Potential pancreatic lipase inhibitory activity of an endophytic *Penicillium* species. J Enzyme Inhib Med Chem 30(1):15–21
Kapoor N, Saxena S (2016) Xanthine oxidase inhibitory and antioxidant potential of Indian *Muscodor* species. 3 Biotech 6:248
Kaul S, Gupta S, Ahmed M, Dhar MK (2012) Endophytic fungi from medicinal plants: a treasure hunt for bioactive metabolites. Phytochem Rev 11:487–505
Krajaejun T, Lowhnoo T, Yingyong W, Rujirawat T, Fucharoen S, Strobel G (2012) *In vitro* antimicrobial activity of volatile organic compounds from *Muscodor crispans* against the pathogenic oomycete *Pythium insidiosum*. Southeast Asian J Trop Med Public Health 43:1474–1483
Kudalkar P, Strobel G, Riyaz-Ul-Hassan S, Geary B, Sears J (2012) *Muscodor sutura*, a novel endophytic fungus with volatile antibiotic activities. Mycoscience 53:319–325
Kusari S, Lamshoft M, Spiteller M (2009) *Aspergillus fumigatus* Fresenius, an endophytic fungus from *Juniperus communis* L Horstmann as a novel source of the anticancer pro drug deoxy-podophyllotoxin. J Appl Microbiol 107:1019–1030
Kusari S, Pandey SP, Spiteller M (2013) Untapped mutualistic paradigms linking host plant and endophytic fungal production of similar bioactive secondary metabolites. Phytochemistry 91:81–87
Lacey LA, Neven LG (2006) The potential of the fungus, *Muscodor albus*, as a microbial control agent of potato tuber moth (Lepidoptera: Gelechiidae) in stored potatoes. J Invertebr Pathol 91:195–198
Lacey LA, Horton DR, Jones DC, Headrick HL, Neven LG (2009) Efficacy of the biofumigant fungus *Muscodor albus* (Ascomycota: Xylariales) for control of codling moth (Lepidoptera: Tortricidae) in simulated storage conditions. J Econ Entomol 102:43–49
Macias-Rubalcava ML, Hernandez-Bautista BE, Oropeza F, Duarte G, Gonzalez MC, Glenn AE, Hanlin RT, Anaya AL (2010) Allelochemical effects of volatile compounds and organic extracts from *Muscodor yucatanensis,* a tropical endophytic fungus from *Bursera simaruba*. J Chem Ecol 36:1122–1131
Mercier J, Jimenez JI (2004) Control of fungal decay of apples and peaches by the biofumigant fungus *Muscodor albus*. Postharvest Biol Technol 31:1–8
Mercier J, Jimenez JI (2007) Potential of the volatile-producing fungus *Muscodor albus* for control of building molds. Can J Microbiol 53:404–410
Mercier J, Jimenez JI (2009) Demonstration of the biofumigation activity of *Muscodor albus* against *Rhizoctonia solani* in soil and potting mix. BioControl 54:797–805
Mercier J, Manker DC (2005) Biocontrol of soil-borne diseases and plant growth enhancement in greenhouse soilless mix by the volatile-producing fungus *Muscodor albus*. Crop Prot 24:355–362
Mercier J, Smilanick JL (2005) Control of green mold and sour rot of stored lemon by biofumigation with *Muscodor albus*. Biol Control 32:401–407
Mercier J, Jimenez-Santamaria JI, Guerra PT (2007) Development of the volatile producing Fungus *Muscodor albus* Worapong, Strobel and Hess as a novel antimicrobial biofumigant. Rev Mex Fitopatol 25:173–179

Mercier J, Lego SF, Smilanick JL (2009) In-package use of *Muscodor albus* volatile-generating sachets and modified atmosphere liners for decay control in organic table grapes under commercial conditions. Fruits 65:31–38

Meshram V, Kapoor N, Saxena S (2013) *Muscodor kashayum* sp. nov.–a new volatile antimicrobial producing endophytic fungus. Mycology 4:196–204

Meshram V, Saxena S, Kapoor N (2014) *Muscodor strobelii*, a new endophytic species from South India. Mycotaxon 128:93–104

Meshram V, Gupta M, Saxena S (2016) *Muscodor ghoomensis* and *Muscodor indica*: new endophytic species based on morphological features, molecular and volatile organic analysis from Northeast India. Sydowia 67:133–146

Mitchell AM, Strobel GA, Hess WM, Vargas PN, Ezra D (2008) *Muscodor crispans*, a novel endophyte from *Ananas ananassoides* in the Bolivian Amazon. Fungal Divers 31:37–43

Mitchell AM, Strobel GA, Moore E, Robison R, Sears J (2010) Volatile antimicrobials from *Muscodor crispans,* a novel endophytic fungus. Microbiology 156:270–277

Pena LC, Jung LF, Savi DC, Servienski A, Aluizio R, Goulin EH, Galli-Terasawa LV, Maia BH, Annies V, Franco CR, Glienke C, Kava V (2016) A *Muscodor* strain isolated from *Citrus sinensis* and its production of volatile organic compounds inhibiting *Phyllosticta citricarpa* growth. J Plant Dis Prot 15:1–12

Petrini O (1991) *Fungal endophytes of tree leaves*. In: Fokkema NJ, Heuvel VD (eds) Microbial ecology of leaves. Cambridge University Press, Cambridge, pp 185–187

Promputtha I, Lumyong S, Dhanasekaran V, Huge E, Mackenzie C, Hyde K, Jeewon R (2007) Phylogenetic evaluation of whether endophytes become saprotrophs at host senescence. Microb Ecol 53(4):579–590

Puri SC, Verma V, Amna T, Qazi GN, Spiteller M (2005) An endophytic fungus from *Nothapodytes foetida* that produces camptothecin. J Nat Prod 68:1717–1719

Puri SC, Nazir A, Chawla R, Arora R, Riyaz-Ul-Hasan S, Amna T, Ahmed B, Verma V, Singh S, Sagar R, Sharma A, Kumar R, Sharma RK, Qazi GN (2006) The endophytic fungus *Trametes hirsuta* as a novel alternative source of podophyllotoxin and related aryl tetralin lignans. J Biotechnol 122:494–510

Qadri M, Johri S, Shah BA, Khajuria A, Sidiq T, Lattoo SK, Abdin MZ, Hasan SRL (2013) Identification and bioactive potential of endophytic fungi isolated from selected plants of the Western Himalayas. Springer Plus 2(1):8–14

Qadri M, Nalli Y, Jain SK, Chaubey A, Ali A, Strobel GA, Vishwakarma RA, Riyaz-Ul-Hassan S (2016) An insight into the secondary metabolism of *Muscodor yucatanensis*: small-molecule epigenetic modifiers induce expression of secondary metabolism-related genes and production of new metabolites in the endophyte. Microb Ecol. https://doi.org/10.1007/s00248-016-0901-y

Ramin AA, Braun PG, Prange RK, Delong JM (2005) *In vitro* effects of *Muscodor albus* and three volatile components on growth of selected postharvest microorganisms. Hortscience 40:2109–2114

Riga E, Lacey LA, Guerra N (2008) *Muscodor albus*, a potential biocontrol agent against plant-parasitic nematodes of economically important vegetable crops in Washington State, USA. Biol Control 45:380–385

Saxena S, Meshram V, Kapoor N (2014) *Muscodor darjeelingensis*, a new endophytic fungus of *Cinnamomum camphora* collected from northeastern Himalayas. Sydowia 66:55–67

Saxena S, Meshram V, Kapoor N (2015) *Muscodor tigerii* sp. nov. volatile antibiotic producing endophytic fungus from the Northeastern Himalayas. Ann Microbiol 65:47–57

Schnabel G, Mercier J (2006) Use of a *Muscodor albus* pad delivery system for the management of brown rot of peach in shipping cartons. Postharvest Biol Technol 42:121–123

Schotsmans WC, Braun G, Delong JM, Prange RK (2008) Temperature and controlled atmosphere effects on efficacy of *Muscodor albus* as a biofumigant. Biol Control 44:101–110

Singh SK, Strobel GA, Knighton B, Geary B, Sears J, Ezra D (2011) An endophytic *Phomopsis* sp. possessing bioactivity and fuel potential with its volatile organic compounds. Microb Ecol 61:729–739

Siri-udom S, Suwannarach N, Lumyong S (2016) Existence of *Muscodor vitigenus, M. equiseti* and *M. heveae* sp. nov. in leaves of the rubber tree (*Hevea brasiliensis* Müll.Arg.), and their biocontrol potential. Ann Microbiol 66:437. https://doi.org/10.1007/s13213-015-1126-x

Stierle A, Strobel G, Stierle D (1993) Taxol and taxane production by *Taxomyces andreanae*, an endophytic fungus of pacific yew. Science 260:214–216

Stinson AM, Zidack NK, Strobel GA, Jacobsen BJ (2003) Mycofumigation with *Muscodor albus* and *Muscodor roseus* for control of seedling diseases of sugar beet and *Verticillium* wilt of eggplant. Plant Dis 87:1349–1354

Strobel G (2006a) *Muscodor albus* and its biological promise. J Ind Microbiol Biotechnol 33:514–522

Strobel G (2006b) Harnessing endophytes for industrial microbiology. Curr Opin Microbiol 9:240–244

Strobel G (2011) *Muscodor* species- endophytes with biological promise. Phytochem Rev 10:165–172

Strobel G, Daisy B (2003) Bioprospecting for microbial endophytes and their natural products. Microbiol Mol Biol Rev 67:491–502

Strobel GA, Dirkse E, Sears J, Markworth C (2001) Volatile antimicrobials from *Muscodor albus*, a novel endophytic fungus. Microbiology 147:2943–2950

Strobel G, Kluck K, Hess WM, Sears J, Ezra D, Vargas PN (2007) *Muscodor albus* E-6, an endophyte of *Guazuma ulmifolia,* making volatile antibiotics: isolation, characterization and experimental establishment in the host plant. Microbiology 153:2613–2620

Suwannarach N, Bussaban B, Hyde KD, Lumyong S (2010) *Muscodor cinnamomi*, a new endophytic species from *Cinnamomum bejolghota*. Mycotaxon 114:15–23

Suwannarach N, Kumla J, Bussaban B, Lumyong S (2012) Biocontrol of *Rhizoctonia solani* AG-2, the causal agent of damping off by *Muscodor cinnamomi* CMU-Cib 461. World J Microbiol Biotechnol 28:3171–3177

Suwannarach N, Kumla J, Bussaban B, Hyde KD, Matsui K, Lumyong S (2013) Molecular and morphological evidence support four new species in the genus *Muscodor* from northern Thailand. Ann Microbiol 63:1341–1351

Suwannarach N, Kumla J, Matsui K, Lumyong S (2015a) Characterization and efficacy of *Muscodor cinnamomi* in promoting plant growth and controlling *Rhizoctonia* root rot in tomatoes. Biol Control 90:25–33

Suwannarach N, Bussaban B, Nuangmek W, Pithakpol W, Jirawattanakul B, Matsui K, Lumyong S (2015b) Evaluation of *Muscodor suthepensis* CMU-Cib462 as a postharvest biofumigant for tangerine fruit rot caused by *Penicillium digitatum.* J Sci Food Agric 96:339–345

US Environmental Protection Agency Office of Pesticide Programs (2005) Biopesticide Registration Action Document. *Muscodor albus* QST 20799(QST 20799® technical) 1–44

Worapong J, Strobel G (2009) Biocontrol of a root rot of kale by *Muscodor albus* strain MFC2. BioControl 54:301–306

Worapong J, Strobel GA, Ford EJ, Li JY, Baird G, Hess WM (2001) *Muscodor albus* anam. nov., an endophyte from *Cinnamomum zeylanicum*. Mycotaxon 79:67–79

Worapong J, Strobel GA, Daisy B, Castillo U, Baird G, Hess WM (2002) *Muscodor roseus* anam. sp. nov., an endophyte from *Grevillea pteridifolia*. Mycotaxon 81:463–475

Yee WL, Lacey LA, Bishop BJB (2009) Pupal mortality and adult emergence of western cherry fruit fly (Diptera: Tephritidae) exposed to the fungus *Muscodor albus* (Xylariales: Xylariaceae). J Econ Entomol 102:2041–2047

Yuan ZL, Su ZZ, Mao LJ, Peng YQ, Yang GM, Lin FC, Zhang CL (2011) Distinctive endophytic fungal assemblage in stems of wild rice *(Oryza granulata)* in China with special reference to two species of *Muscodor* (Xylariaceae). J Microbiol 49:15–23

Zhang CL, Wang GP, Mao LJ, Komon-Zelazowska M, Yuan ZL, Lin FC, Druzhinina IS, Kubicek CP (2010) *Muscodor fengyangensis* sp. nov. from Southeast China: morphology, physiology and production of volatile compounds. Fungal Biol 114:797–808

Zidack N, Grimme E, Erza D, Stinson A, Strobel G, Jacobsen B (2002) Development of mycofumigation for control of soil-borne plant pathogens. Department of Plant Science and Plant Pathology, Montana State University, Bozeman

Zidack N, Grimme E, Erza D, Stinson A, Strobel G, Jacobsen B (2003) Progress in the development of mycofumigation for control of soil-borne plant diseases. Department of Plant Science and Plant Pathology, Montana State University, Bozeman

Chapter 5
Slime Moulds: The Tiny Charmers

Anubha Pathak and Sharda Vaidya

Abstract Slime moulds are the special organisms exhibiting characters similar to lower animal groups on one side and with fungi on the other side. They show a plasmodial stage in the life cycle with amoeboid movement which is similar to protozoans. They produce sporocarp-bearing spores within similar to fungi. Because of this, they were sometimes considered as "animalcules", or they gained the name "slime moulds". Today they are separated from both these groups and are thought to form a separate group. But even today these organisms are studied by the mycologists. There are basically two types of slime moulds – cellular and plasmodial. In this chapter, we will primarily discuss the plasmodial slime moulds. They typically belong to the group Myxomycetes.

5.1 Introduction

A number of Indian mycologists have studied Myxomycetes from different parts of India; Manoharachay and Nagaraju (2016) have taken a review of their work. Myxomycetes represent a small group of eukaryotic organisms with nearly 60 genera and around 2000 species all over the world. About 500 species are expected to be present in India. The prominent mycologists and their contributions are Agnihothrudu 1956, 1958, 1959a, b, 1961, 1968; Bhide et al. 1987; Indira 1968a, b, 1975; Lakhanpal and Mukherji 1981; Manoharachary et al. 2012; Manoharachary and Rajithasri 2015; Ranade et al. 2012; Mishra 1980; Kadam 2010; and Kadam and Vaidya 2011, 2015.

A. Pathak · S. Vaidya (✉)
Department of Botany, Smt. C.H.M. College, Ulhasnagar, Maharashtra, India

P. Gehlot, J. Singh (eds.), *Fungi and their Role in Sustainable Development: Current Perspectives*, https://doi.org/10.1007/978-981-13-0393-7_5

Hemitrichia cerpula

Physarum mellius

Didumium squamulosum

Diderma effusum

Arcyria denudata

Trichia decipens

In Indian climate, the slime moulds can be found from June to November depending upon the duration of rainfall. The growth can be observed during bright sunlight for 4–5 days after a continuous rainfall of about the same period. The growth is peculiar and can be located on decaying logs, leaves or sometimes even on the living plant parts such as grass leaves, leaf bases of coconut, tree trunks, etc. The collection should be done very carefully by lifting the upper moist decaying matter. If the pieces of decaying leaves or branches are covered with white, yellowish or variously coloured, reticulate plasmodial mass, the slime moulds can be located. Normally the collection is done when the fruiting bodies are mature.

5.2 Morphology

The initial stages in the life cycle of Myxomycetes members show plasmodium when the favourable conditions for growth are present. If the unfavourable conditions prevail, it turns to develop fructifications.

5.2.1 *Plasmodium*

It is the vegetative phase that exhibits amoeboid movement and can cover sizable distances in the field. The growth can be seen as veins. This helps in its propagation to wide areas in the same field. Plasmodium is a naked mass of cytoplasm covered by thin plasma membrane. The nutrition of plasmodium is holozoic and saprophytic, and it can survive on bacteria, filamentous fungi, spores, etc. The cytoplasm shows the streaming movements that can be observed because of the pigments that it contains. The cytoplasmic streaming continues in opposite directions. In 1962 and 1963, Alexopoulos described four types of plasmodia.

5.2.1.1 Phaneroplasmodium

It is granular, with conspicuous thick veins, comparatively stationary, with jellified ectoplasm, spreading in the fanlike manner as a sheet of protoplasm. It is very common in the order Physarales.

5.2.1.2 Aphanoplasmodium

It does not possess granules and is difficult to observe at young stage. Jellified ectoplasm is present only in large veins. The growing fanlike structure is formed in thin strands forming an open network. It is common in the order Stemonitales.

5.2.1.3 Protoplasmodium

It is always microscopic, granular, without the growth of advancing fan like growth. It lacks the features such as channels, veins, reticulum, cytoplasmic streaming, etc., but the granules clearly depict random movement. It is found in orders Echinosteliales and Liceales.

5.2.1.4 Intermediate Type

It grows in the fanlike manner and possesses granular cytoplasm similar to phaneroplasmodium. It shows the presence of jellified ectoplasm only in the largest veins similar to aphanoplasmodium. The random movement of granules is similar to protoplasmodium. Hence it is the type intermediate between the three types. It is commonly found in the order Trichiales (Stephenson and Stempen 2000).

But these studies have to be confirmed by further analysis.

Phaneroplasmodium

Aphanoplasmodium

Intermediate type

5.3 Fructifications

At the end of favourable season, the plasmodium starts shrinking. The horizontally spreading plasmodium aggregates into fructifications that are small, 1–2 mm in height, vertical structures. They bear spores either externally (Genus *Ceratiomyxa*) or internally (all other genera). There are basically four types of fructifications.

5.3.1 Plasmodiocarp

The plasmodium shrinks into few large veins forming a clear colourful network. It develops peridium containing spores within, e.g. *Hemitrichia*. It may also occur as a discrete, erect, spherical or elongated fruiting body.

5.3.2 Sporangium

The protoplasm aggregates in minute, individual sporangia which may be stalked or sessile. Normally many sporangia develop collectively in small patches. It is because the multicellular plasmodial stage merely shows compact aggregation. The plasmodium covers a wider area. The plasmodium in this entire area develops fruiting bodies. Unlike higher eukaryotes, there is no differentiation of tissues for the formation of sporangia. The sporangia may be rounded, elongated, ovoid, fusiform, etc. They may be variously coloured.

5.3.3 *Aethalia*

When the entire protoplasm is converted into a large fructification, it is regarded as aethalia. The fructification is cushion-like and is covered with peridium, e.g. *Fuligo*.

5.3.4 *Pseudoaethalia*

In this type of fructification, the individual sporangia are fused together. The walls between them can be observed or can be rudimentary (Lakhanpal and Mukherji 1981).

5.4 Parts of Fructification

5.4.1 *Stipe*

It is the stalk of sporangium. It may be short or long, straight or bent, thick or thin or concolorous or with different colours than the sporangium, tapering towards the tip or may be with uniform thickness, smooth or with ridges.

5.4.2 *Hypothallus*

It is the membrane-like layer produced by the aggregating protoplasm near the base of the stipe. Presence or absence of hypothallus is an important feature in the classification.

5.4.3 *Peridium*

The noncellular wall of sporangium formed by the secretory materials of the protoplasm that encloses the spores is called peridium. It may be thick, thin or fugacious (*Arcyria*). It may remain till dehiscence (*Physarum*) or may disappear before it (*Stemonitis*). It may be variously coloured and with or without metallic lustre.

5.4.4 *Columella*

The extension of stipe penetrating the sporangium is called columella. It may be present or absent. It may be calcareous (Physarales) or noncalcareous (Stemonitales).

5.4.5 *Capillitium*

These are threadlike structures interspersed with spores in the fructification of some members of Myxomycetes. Their presence/absence, nature determines important criteria in classification. They may be calcareous or noncalcareous, may be with or without nodes, branched or unbranched, smooth or spirally coiled, coloured, hyaline or white or elastic or non-elastic.

5.4.6 *Spores*

They are generally one-celled, globose or subglobose. The size of the spore may vary from less than 5 μ to more than 15 μ. They may be brown, violet or hyaline. The spore wall may be smooth, reticulate or verrucose (Thind 1977).

5.5 Role of Slime Moulds in Sustainable Development

The members of Myxomycetes play a variety of roles in the environment and can be employed in diverse ways for the benefit of human beings.

5.5.1 *Biodegradation*

Since the plasmodia crawl along the substrate, they utilize fungi, bacteria and spores as their food exhibiting phagotrophic mode of nutrition. Because of this, they are known as microbial predators (Keller et al. 2008). But plasmodia also show other modes of nutrition. One of the modes is saprophytic one for which it has to secrete the enzymes extracellularly. The enzymes include cellulase, protease, etc. The enzymes are responsible for degrading the substrate. This can be observed in the following photographs. The leaves of *Ficus religiosa* possessing growth of slime mould, *Physarum cinereum*, were kept in moist chamber for 3 months. The leaf tissue degraded where the growth of plasmodium was more. The enzymes and substances secreted by the plasmodium on the substrate might be responsible for this.

The biodegradative activities are important in nature in recycling of minerals and in removal of debris of the dead and decaying substances.

Growth of slime moulds on the leaf

Portion of leaf deteriorated after three months

5.5.2 *Enzymes*

The slime moulds normally show holozoic nutrition. But the recent studies have proved that the cultures and the plasma membrane possess cellulolytic, lignolytic and agarolytic activity. Moitra and Nishi (1991, 1993) detected β-galactosidase activity from the plasma membrane and the extracellular fraction of culture medium by gel filtration, ion exchange and hydrophobic chromatography. The molecular weight of the enzyme was found to be 65 kD. Optimum pH was found to be 4.5, and optimum temperature was found to be 55 °C. The enzyme activity was also found in extracellular fraction after attaining the maximal growth. The activity of this enzyme was very little during the log phase of growth. The activity increased when the plasmodia were grown in glucose-free media. The enzyme was effective on laminarin, lichenan and other glucosides. It was inhibited by D-glucono-1,5-lactone, N-Bromosuccinimide and Hg^{+2}. The distribution of enzyme activities changed according to culture conditions and the stage of plasmodial growth. In mannitol, the medium differentiated into spherules covered by a hard cell wall. B-glucosidase activity is much decreased. When plasmodia were incubated in the synthetic media, the enzyme activity in the membrane fraction becomes low. It increases if the medium is supplemented by the addition of tryptone or yeast extract. These substances contain glucans, which are used by the plasmodia as carbon source. Purified β-glucoside had highest hydrolytic activity against laminarin and lichenan. It was inhibited by 1,5-glucono-D-lactone. It means chemically that it is exo-β-1,3-glucanase. It is common in bacteria, fungi and higher plants. Plasmodia utilize β-1,3-glucan as carbon source. Hence it can replace glucose. B-1,3-glucan is common in the natural environment around plasmodia. Macabago and de la Cruz (Macabago and dela Cruz 2014) cultured 18 species of Myxomycetes for 4 weeks, 7 of which developed plasmodia, 10 developed amoeboflagellates. These were preserved in 15% glycerol for 3 months at 5 °C. The plasmodia were tested for their activity against starch and protein degradation. They proved to be amylase and protease enzymes. These enzymes are very important for industries.

The tetramic acid (pyrrolidine-2,4-dione) ring system forms the main structural unit of many natural products. This moiety is found as a 3-acyl derivative. Fuligorubin A, a yellow pigment produced from *Fuligo septica*, consists of hydroxybenzyl substituent at C-5 position of tetramic acid ring which has been shown to possess antibiotic and cytotoxic activity. One of its derivatives has also been found from the plasmodium of *Leocarpus fragilis* (Mohalid et al. 2015).

5.5.3 Cellulolytic and Lignolytic Activity

Keller et al. (2008) have demonstrated the extracellular activity of cellulase, amylase, protease and agarase in different cultures of Myxomycetes members. They cultured five members of Myxomycetes on different media. When the plasmodia developed the clear zones on specific substrates, the secretion of respective enzymes was proved. Clearing zones and floating with 1% Lugol's iodine on starch agar indicated the presence of amylase enzyme. Clearing zones on carboxymethyl cellulose agar medium and flooding with 1% congo red were the indicators of cellulose enzyme. The clearing zones on skimmed milk agar showed the protease activity of plasmodia. The soft water agar showed liquefaction because of the presence of agarase enzymes. Legendary mycologist Dr. Indira Kalyansundaram had worked on this aspect. If these enzymes are extracted from the cultures, they can be a good source for industrial utilization.

The members of Myxomycetes, viz. *Physarum viride*, *Physarum polycephalum*, *Physarum echinosporum* and *Physarum cinereum*, were cultured on water agar, and 35 °C and 95% moisture content were maintained. As the growth of plasmodium progressed, the media started becoming semisolid from solid state. Ultimately some portion of media became hyaline by the end of the week as it can be seen in the following photographs.

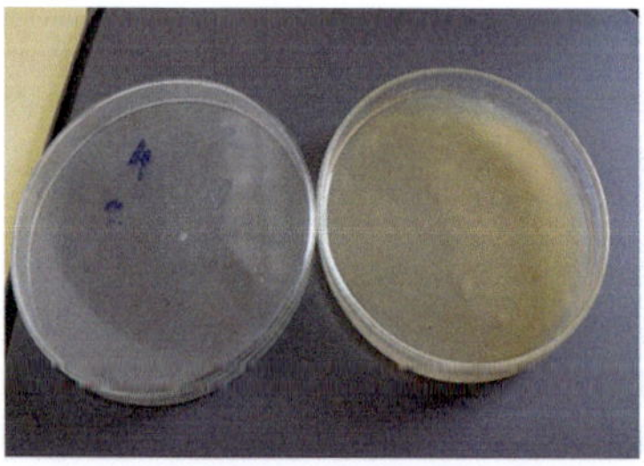

The medium before inoculation of slime mould

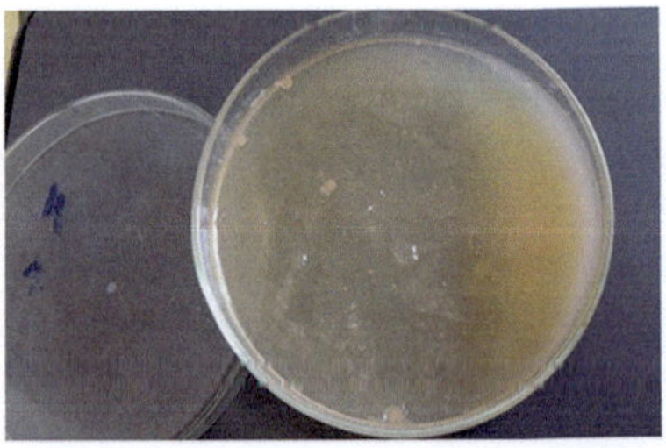

The dissolved patch of the medium after one week

5.5.4 Medicinal Value and Extraction of Pigments

The plasmodia of Myxomycetes are generally white, yellowish creamish or rarely brightly coloured. But the fruiting bodies are generally brightly coloured. These are the sources of pigments that can give a number of natural colours. The cultures and the fruiting bodies are also the sources of a number of secondary metabolites such as quinones, polyenes, triterpenoids, glycosides, lactones, sterols, etc. (Ishibashi 2009; Shintani et al. 2003; Misono et al. 2009; Dembitsky et al. 2010).

More than hundred secondary metabolites have been isolated from the members of Myxomycetes. These include lipids, fatty acid amides and their derivatives, alkaloids, amino acids, peptides, naphthaquinone pigments and terpenoids (Dembitsky et al. 2005). Crude extracts of plasmodium cultured with *Escherichia coli* showed antibiotic activity against it. Crude extracts of *Physarum melleum* also exhibited antibiotic properties against *Bacillus subtilis*. There were two components, viz. melleumin A and B, that did not show this property. The yellow pigment isolated from the crude extracts of plasmodium may be responsible for the antibiosis (Nakatani et al. 2005). The problem with Myxomycetes was the quantity. The microscopic body and the limited occurrence created the limitation in the process. Because of this, sufficient quantity is not available from the natural resources. If the cultures are maintained and manipulated, this problem can be solved. Arcyriaflavin A obtained from *Arcyria denudata* and *Lycogala epidendrum* showed moderate antibiotic antifungal activity (Steglich 1989; Fröde et al. 1994; Hoyosa et al. 2005). It also inhibits signals in Wnt cell signaling pathway.

Schroeder and Mallette (Shroeder and Mallette 1973) cultured *Physarum gyrosum* on glucose yeast agar (pH 6.5) with *E. coli*. The extracts of plasmodia were the sources of many antibiotics. One of them was purified by them which revealed the antibiotic activity against *B. cereus*. Similar results against different organisms were shown by Taylor and Mallette (1976) and Considine and Mallette (1965).

Some of the secondary metabolites isolated from Myxomycetes are under clinical trials. Most of them are indolo-carbazoles, e.g. arcyriarubins, arcyriaflavins, arcyroxepins, arcyriacyanins, lycogallic acid, lycogarubins, etc.

Wild-type plasmodium of *Didymium iridis* is brown in colour. But two colour mutants obtained from different isolates were found. Each of these, in double-recessive condition, results in cream-coloured plasmodium. The mutants were complementary in heterozygous state but non-complementary in diploid plasmodial heterokaryon. The colour was observed because of a red pigment produced that may be related to sporulation (Collins 1969).

Racoczy (1998) observed two prominent changes in the plasmodia of *Physarum polycephalum*. The yellow plasmodia were exposed to white light for a short duration. They became colourless. He called this phenomenon as photobleaching. The plasmodia regained yellow colour when placed in the dark. The same plasmodia if exposed for a longer duration developed orange colouration. The orange pigments were isolated and characterized. They showed resemblance to carotenoids.

5.5.5 *Maintaining the Ecological Balance*

These organisms utilize bacteria and fungi as their foods that may otherwise supersede the number of other organisms in the soil. This helps in controlling the number of these organisms in the soil. It thus helps in establishing the ecological balance in the soil among the various soil organisms. The myxamoebae mainly engulf the spores or the entire organism irrespective of its size (Olive 1975). Because of this, they are described as secondary saprotrophs by Adl and Gupta (Adl and Gupta 2006). The fungi trophic activity may be used as a biocontrol agent (Adl 2003) against plant pathogens in the soil. But there are limitations to this as the Myxomycetes members that do not penetrate the hyphae and are nonspecific in action. Hence they may reduce the number of beneficial fungi in the soil.

Fuligo septica has large aethalium spreading to long distances. It contains millions of spores. These spores and yellow-coloured plasmodium are used to study hyperaccumulation of zinc, heavy metal by Stijve and Andrey (Stijve and Andrey 1999) and Zhulidov et al. (Zhulidov et al. 2002). It is also found to accumulate barium, calcium, iron, manganese and strontium. This property can be used in bioremediation of polluted soils.

The Myxomycetes growing on trees exhibit four categories according to pH range to which they adapt, viz. 3–4.5, 4.6–6.0, 6.1–7.5 and 7.6–10.0. There are few species that adapt to wide pH range (Everhart et al. 2008, 2009).

The species diversity on a tree or vine can become a pollution indicator especially to acid rain, air pollutants produced by forest fire, etc. (Kilgore et al. 2009). But this knowledge is incomplete. There should be enough data collected for trees and tree species and the food for Myxomycetes, i.e. bacteria, fungi, etc. available in those conditions.

5.5.6 *Pathogenicity*

Some of the slime moulds are found to grow on the living parts of the plants such as the bark, stem, leaves, e.g. *Diachea leucopodia*, *Fuligo septica*, *Mucilago crustacea*, *Physarum cinereum*, *Stemonitis*, etc. But their penetration in the host tissues is not observed. It is assumed that the plasmodium moves along the stem or bark and develops the fruiting bodies there. (Kellar and Everhart 2010).

5.5.7 *Myxomycetes and Ageing Research*

Cummins and Rusch (1968) and Ljubimova et al. (2008) performed experiments to determine the life cycle, senescence and nuclear behaviour in plasmodia of *Didymium iridis* and *Physarum cinereum*. They found that ageing or longevity is under the nuclear control and cytological factors are not responsible for it.

5.5.8 Cancer Studies

Cummins and Rusch (1968) and Ljubimova et al. (2008) also showed that *Physarum polycephalum* proved to be the best model organism for cancer studies. The plasmodium of *Physarum polycephalum* behaves as a single celled giant amoeba. But growth, mitosis, differentiation, cytokinesis and karyokinesis are separate in each cell. The synchronous divisions may be due to biological clock in the nucleus or may be caused by some cytological factor. This fact helps in understanding uncontrolled cell division in cancer. It may be that the mitotic synchrony is triggered inside the nucleus and is transferred to the nucleus just before the mitosis. Arcyriacyanin A, from *Arcyria nutans* (Steglich 1989), showed inhibitory activity against human cancer cell lines and against the enzymes, protein kinase C and protein tyrosine kinase (Hibino and Choshi 2002). Since the proteins play important role in regulation of cell growth and differentiation which is encoded by many oncogenes, it is important to study why protein inhibitors are produced. Studies by Murase et al. (2000) have proved that though high dosages are required, arcyriacyanin A or its precursors may be useful in inhibition of cancer cell growth.

5.5.9 Drugs for Cancer

Physarum polycephalum yields a nontoxic, non-immunogenic, biodegradable, nanoconjugate drug delivery system for the drug called Polycefin. It is produced from purified β-L-malic acid. It is then modified for directed delivery of morpholino antisense oligonucleotides (gene silencing therapy), antibodies and antitumour drugs in certain tumour cells. When fluorescent-labelled Polycefin is injected in the tail vein of the mouse, it accumulated in the breast and brain tumour cells (Ljubimova et al. 2008).

5.5.10 Use of Slime Moulds in Biological Robots

Nakagaki et al. (2000) performed an experiment demonstrating primitive brain power or primitive intelligence. They placed the plasmodium of *Physarum polycephalum* in two petri plates, one with oat flakes at two points and the other without it. The plate without oat flakes showed regular spread up of plasmodium in the petri plate. In the plate containing the oat flakes, the plasmodium moved towards flakes with the shortest possible distance.

Admitzky and Jones (2008) and Admitzky (2009) tried to develop the first biological, amorphous robot (non-silicon plasmobot). They used the plasmodium of *Physarum polycephalum* as the motive source. The robot was a six-legged structure having a six-pointed star on the top. The plasmodium was grown on this star. The legs of the robot were connected to circuit and remotely to the computer. The light source was attached to one of the points of the circuit. The plasmodium showed movement either to or away from the light. This movement controlled the robot. But these studies are yet very primitive and need much more contribution.

5.5.11 Use in Space Technology

Members of Myxomycetes are now studied for their uses and cultural behaviour in space as the growth of plasmodia is typical. The plasmodium of *Physarum polycephalum* is used in American-German and Russian Biosatellite Kosmos – 1129 and 1179. It showed that migration of plasmodium and protoplasmic streaming were maintained under microgravity conditions (Tairbekov et al. 1984). A joint experiment in 1986 between German spacelab and NASA (USA) used plasmodium of *Physarum polycephalum* as a model to study contraction behaviour and gravity response. It showed the sensitivity to both, gravity and light. Weightlessness experiments in space confirmed the validity of zero gravity (Block et al. 1986).

5.5.12 Myxomycetes Spores as Aeroallergens

The spores of Myxomycetes are tiny and can be liberated during their handling. Spores of many species are reported to cause rhinitis or asthma (Rockwell et al. 1989). RAST and skin testing of extracts of *Fuligo septica* proved to be aeroallergens. Though there are meagre studies in this field, the species producing large number of spores are expected to be allergic.

5.5.13 Myxomycetes as Human Food Source

In the state of Veracruz, Mexico, the young, immature fruiting bodies and plasmodia of *Fuligo septica* and *Reticularia lycoperdon* are fried-like eggs with onions and pepper and eaten on a tortilla (Lopez et al. 1982; Villareal 1983). The plasmodium of *Fuligo septica* is called “caca de luna”. The meaning of this is the excrement of the moon.

References

Adl MS (2003) The ecology of soil decomposition. CABI Publishing, Wallingford

Adl MS, Gupta VVSR (2006) Protists in soil ecology and forest nutrient cycling. Can J For Res 36:1805–1817

Admitzky A (2009) *Physarum* boats: if Plasmodium sailed, it would never leave port. Appl Bionics Biomech. https://doi.org/10.1080/11762320902863890 Online

Admitzky A, Jones J (2008) Towards *Physarum* robots: computing and manipulating on water surface. J Bionic Eng 5:348–357
Agnihothrudu V (1956) Some slime molds of Southern India IV and V. J Indian Bot Soc 35(27,37):210–221
Agnihothrudu V (1968) Some slime moulds of Southern India. 10. Sydowia 22:179–182
Agnihothrudu V (1958) Notes on Fungi from NorthEast India II. An undescribed myxomycetes from Assam. J Indian Bot Soc 37:499–503
Agnihothrudu V (1959a) A list of Indian myxomycetes. J Madras Univ B:120
Agnihothrudu V (1959b) Notes on Fungi from Northeast India IV. J Indian Bot Soc 38:418–491
Agnihothrudu V (1961) Some slime molds of Southern India 10. Sydowia 22:179–182
Bhide VP, Sathe AV, Pande A, Patwardhan PG, Rao VG (1987) Fungi of Maharashtra. Suppl. I. Maharashtra Association for the Cultivation of Science, Pune
Block I, Briegleb W, Sobick V, Wohlfarth-Bettermann KE (1986) Contraction behaviour and protoplasmic streaming in the slime mold *Physarum polycephalum* (*Physarum* kinetics). In: Proceedings of Norderney Symposium on Scientific Results of the German Spacelab Mission D1. Institut für Zytologie. Universität Bonn, Bonn, pp 408–418
Collins OR (1969) Complementation between two color mutants in a true slime mold, *Didymium iridis*. Genetics 63:93–102
Considine JM, Mallette MF (1965) Production and partial purification of antibiotic materials formed by *Physarum gyrosum*. Appl Microbiol 13(3):464–468
Cummins JE, Rusch HP (1968) Natural synchrony in the slime molds. Endeavour 27:124–129
Dembitsky VM, Razenka T, Spizek J, Hanus LO (2005) Secondary metabolites of slime molds. (Myxomycetes). Phytochemistry 66:747–769
Dembitsky VM, Razenka T, Spizek J, Hanus LO (2010) Secondary metabolites of slime molds. Theory Biosci 129(1):15–23
Everhart SE, Keller HW, Ely JS (2008) Influence of bark pH on the occurrence and distribution of tree canopy Myxomycetes species. Mycologia 100:191–204
Everhart SE, Ely JS, Keller HW (2009) Evaluation of a tree canopy epiphytes and bark characteristics associated with the presence of Corticolous Myxomycetes. Botany 87:509–517
Fröde R, Hinze C, Josten I, Schmidt B, Steffian B, Steglich W (1994) Isolation and synthesis of 3,4-bis(indol-3-yl)pyrrole-2,5-dicarboxylic acid derivatives from the slime mold, *Lycogala epidendrum*. Tetrahedron Lett 35:1689–1690
Hibino S, Choshi T (2002) Simple indole alkaloids and those with a non-rearranged Monoterpenoid unit. Nat Prod Rep 19:148–180
Hoyosa T, Yamamoto Y, Uehara Y, Hoyashi M, Komiyama K, Ishibashi M (2005) New cytotoxic bis indole alkaloids with protein tyrosine kinase inhibitory activity from a Myxomycetes, *Lycogala epidendrum*. Bioorg Med Chem Lett 15:2776–2780
Indira PU (1968a) Some slime molds from Southern India VIII. J Indian Bot Soc 47:155–186
Indira PU (1968b) Some slime molds from Southern India IX-distribution, habitat and variation. J Indian Bot Soc 47:330–340
Indira PU (1975) Some slime molds from southern India XI. Kavaka 3:41–54
Ishibashi M (2009) Study of myxomycetes as a source bioactive natural products. Chem Pharma Bull (Tokyo) 57(8):894–895
Kadam P (2010) Study of biodiversity of myxomycetes from Badlapur and Allied Area. M. Sc. Thesis. University of Mumbai
Kadam P, Vaidya S (2011) Biodiversity of slime molds from Badlapur River and Barvi Dam. In: Proceedings of National Conference on "Forests for the Future of Man" held at Rathnam College, Bhandup, Mumbai on 16–17 Sept 2011, pp 268–275
Kadam P, Vaidya S (2015) Taxonomy of Plasmodial Myxomycetes. Proceedings of National Conference on "New and Emerging Trends in Bioinformatics and Taxonomy" Held at B. N. Bandodkar College, Thane, Mumbai.14–15 Jan 2015
Keller HW, Everhart SE (2010) Importance of Myxomycetes in biological research and teaching. Fungi 3(1):13–27

Keller HW, Kilgore CM, Everhart SE, Carmack GJ, Crabtree CD, Scarborough AR (2008) Myxomycete plasmodia and fruiting bodies: unusual occurrences and user friendly study techniques. Fungi 1(1):24–37
Kilgore CM, Keller HW, Ely JS (2009) Aerial reproductive structures on vascular plants as a microhabitat for Myxomycetes. Mycologia 101:303–317
Lakhanpal TN, Mukherji KG (1981) Taxonomy of the Indian myxomycetes. A. R. Gantner- Verlag. K. G., Vaduz
Ljubimova JYM, Fujita NM, Khazenzon BS, Lee S, Wachmann-Hogiu S, Farkas DL, Black KL, Holler E (2008) Nanoconjugate based on polymalic acid for tumor targeting. Chem Biol Interact 171:195–203
Lopez A, Garcia H, Herrador JL (1982) Nuevos Registros de Hongos Cosmetibles de la Region del Cofre de Porte, Estado de Veracruz. (Abstract). P.30. in Primer Congresso National de Mycologia. Sociedad Mexicana de Mycologia, Xalapa
Macabago SAB, dela Cruz TEE (2014) Preservation and extracellular enzyme production of myxomycetes from Lubang Island, Occidental Mindoro, Philippines. Philipp Sci Lett 7(2):331–336
Manoharachary C, Rajithasri AB (2015) Interesting slime molds from Telangana State, India. J Indian Bot Soc 94(3 and 4):286–288
Manoharachary C, Kunwar IK, Tilak KVBR (2012) Some myxomycetes from Andhra Pradesh, India. J Indian Bot Soc 9(4):427–429
Manoharachay C, Nagaraju D (2016) Myxomycetes: the forgotten fungi-like living organisms from India. Kavaka 47:35–41
Mishra RL (1980) Myxomycetes and soil fungi of Maharashtra. Ph. D. Thesis. Pune University, Pune-411007
Misono Y, Ishikawa Y, Yamamoto Y, Hayashi MDL, Komiyama K, Ishibashi M (2009) Dihydrolindbladiones, three new naphthaquinone pigments from a myxomycetes *Lindbladia tubulina*. Chem Pharm Bull 57(8):894–895
Mohalid N, Hanzah AS, Shaameri Z (2015) Chemical exploration of 4-Hydroxybenzylated-3-Gantian Acid Tetramik. Malas J Anal Sci 19(2):359–362
Moitra M, Nishi A (1993) Purification and partial characterization of β-galactosidase from plasmodial membrane and culture medium of *Physarum polycephalum*. J Gen Microbiol 139:1635–1641
Morita M, Nishi A (1991) Glycoproteins and enzymes associated with the plasmodial membrane and slime layer of *Physarum polycephalum*. J Gen Appl Microbiol 37:93–109
Murase M, Watanabe K, Yoshida T, Topinga S (2000) A new concise synthesis of arcyriacyanin A and its unique inhibitory activity against a panel of human cancer cell lines. Chem Pharm Bull 48:81–84
Nakagaki T, Yamada H, Toh A (2000) Intelligence: maze solving by an amoeboid organism. Nature 407:470
Nakatani S, Kamata K, Sato M, Onuki H, Hirota H, Matsumoto J, Ishibashi M (2005) Melleumin A, a novel peptide lactone isolated from the cultured myxomycetes, *Physarum melleum*. Tetrahedron Lett 46:267–271
Olive LS (1975) The mycetozoans. Academic, New York
Racoczy L (1998) Plasmodial pigmentation of the acellular slime mold, *Physarum polycephalum* in relation to the irradiation period. Pol J Environ Stud 7(6):337–342
Ranade VD, Korade ST, Jagtap AV, Randive KR (2012) Checklist of myxomycetes from India. Mycosphere 3:358–390
Rockwell WJ, Collins RP, Santilli J (1989) Fuligo, a myxomycetes, an allergen. J Allergy Clin Immunol 83:266
Shintani A, Yamazaki H, Yamamoto Y, Ahmed F, Ishibashi M (2003) Cribarione C, a naphthaquinone pigment from myxomycete *Cribaria meylanii*. J Nat Proc 66(7):999–1001
Shroeder HR, Mallette MF (1973) Isolation and purification of antibiotic material from *Physarum gyrosum*. Antimicrob Agents Chemother 4(2):160–166

Steglich W (1989) Slime molds (Myxomycetes) as a source of new biochemically active metabolites. Pure Appl Chem 61:281–288
Stephenson SL, Stempen H (2000) Myxomycets: a handbook of slime molds. Timber Press, Portland
Stijve T, Andrey D (1999) Accumulation of various metals by *Fuligo septica* (L) Wiggers and by some other slime molds (Myxomycetes). Aust Mycoogist 18(2):23–26
Tairbekov MG, Belina SI, Lairand DB, Budnitzky AA, Lendev VV (1984) Plasmodium of myxomycetes as the object of the investigation in gravitation biology. Iswestiya Akademii Nauk USSR Seriya. Biologitscheskaya 2:198–209
Taylor RL, Mallette MF (1976) Growth of *Physarum gyrosum* on agar plates and in liquid culture. Antimicrob Agents Chemother 10(4):613–617
Thind KS (1977) The myxomycetes of India. I. C. A. R, New Dehli
Villareal L (1983) Algunas Especies de Myxomycetes no Registradas del Estado Veracruz. Boletin de la Sociedad Mexicana de Mycologia 18:153–164
Zhulidov DA, Robarts RD, Zhulidov AV, Zhulidova OV, Markelov DA, Rusanov VA, Headley JV (2002) Zinc accumulation by the slime mold *Fuligo septica* (L) Wiggers in the former soviet union and North Korea. J Environ Qual 31:1038–1042

Chapter 6
Fungal Biotechnology: Role and Aspects

Debjani Mukherjee, Simranjeet Singh, Mohit Kumar, Vijay Kumar, Shivika Datta, and Daljeet Singh Dhanjal

Abstract Sustainability has become the prominent necessity in every human event in today's time, which can initiate from the household and leads to this planet earth. The Earth Summit organized by the United Nations has played the significant role in creating awareness about sustainable development to retort against the natural calamities, due to over-exploitation of natural resources and exponential growth of the human population for centuries. For the quest of sustainability, fungi have emerged as a suitable candidate. Fungi play a pivotal role in fundamental and modern processes of biotechnology. Nowadays many processes such as baking, brewing and the synthesis of alcohols, antibiotics, enzymes, organic acid as well as the additional pharmaceutical product are carried out using fungal bioproducts. Due to recent advances in genomics and rDNA technology, yeast and fungi have attained the forefront position because of their present industrial purposes. In general, the term "mycotechnology" is used, which states about the various roles of fungi with the addition to its impact on biotechnology as well as economy.

Fungi play the significant role in sustaining the health and terrestrial ecosystem. During disastrous event which leads to disruption of the earth ecosystem, fungi prepare themselves to prevail in the future. The aid of fungal population in sustaining the environment is showing promising result. About 90% of plant grows in symbiosis with fungi such as vesicular-arbuscular mycorrhizal (VAM) fungi, mycorrhizae, out of which *Glomus* is the most exploited genera (Van der Heijden et al. 1998). Fungi persisting on this earth have widespread complex relationship

---

D. Mukherjee (✉)
Department of Civil and Environmental Engineering, University of Windsor, Windsor, ON, Canada

S. Singh · D. S. Dhanjal
Department of Biotechnology, Lovely Professional University, Phagwara, Punjab, India

M. Kumar
IDS Infotech Ltd, Mohali, India

V. Kumar
Department of Chemistry, Lovely Professional University, Phagwara, Punjab, India

S. Datta
Department of Zoology, Lovely Professional University, Phagwara, Punjab, India

P. Gehlot, J. Singh (eds.), *Fungi and their Role in Sustainable Development: Current Perspectives*, https://doi.org/10.1007/978-981-13-0393-7_6

among the range of microbes, which can be arthropods, bacteria and nematodes. The dwelling zone of these fungi is named as the "rhizosphere".

A fungus belongs to the group of Eukaryotes, which consists of microbes like moulds and yeast along with more familiar mushrooms. All of them are categorized under the kingdom *Fungi*. The fungi are omnipresent in every environment and play critical role in complex biological processes. Thus, it can work as decomposers, which aid in nutrient cycles, exclusively as symbiont as well as saprotrophs, in disintegrating the organic constituents into inorganic constituents, which gets retraced in the anabolic pathways of metabolic activities taking place in organism as well as plants.

Keywords Fungal biotechnology · Fungal enzymes · Bioprocessing · Fermentation · Secondary metabolites

6.1 Introduction and Definitions

Fungi belong to the lower class of eukaryotes according to contemporary biologist and sometimes are also regarded as the fifth kingdom on the basis of the mode of nutrition. Fungi project different types of enzymes into their surrounding and after the action of these enzymes engulf the pre-treated food. With assorted morphology, ecology and physiology, many eminent fungi still have an adverse effect on the well-being of humans as they are involved in different plant diseases (like blights, rusts, smuts and wilts) and biodeterioration (like mildews and rots) as well as pathogenic to animal (by producing mycoses and toxins). Range of fungi starts from the micro-sized moulds and yeasts to macro-sized truffles as well as mushroom. Large number of macro-sized species of fungi are believed to be delicate; thus they are grown and accumulated for human uses such as food or its supplements, whereas there are micro-sized fungi, comprising of genera like *Aspergillus*, *Penicillium* and *Saccharomyces*, which have positive influence in context to human activities as their regulated metabolisms are utilized for synthesizing enzymes as well as metabolites. These abilities have made fungi one of the foundation stones in modern biotechnology.

There are numerous ways by which the term biotechnology can be explained. The Spinks Commission, UK, was the first group to give a formal explanation: "Biotechnology is the application of biological organisms, systems, or processes to manufacturing and service industries". On the other hand, the European Federation has quoted the similar definition but with a broad range of aspect: "the integrated use of biochemistry, microbiology and engineering sciences in order to achieve technological (industrial) application of the capabilities of micro-organisms, cultured tissue cells, and parts thereof". Further, the National Institute of Health and Food and Drug Administration, USA, defined: "Biotechnology is the application of biological systems and organisms to technical and industrial processes. The technologies included in this area include genetic selection, in vitro modification of genetic material, e.g. recombinant DNA, gene splicing, cell fusion, hybridoma technology etc." and other novel techniques for modifying genetic material of living organisms (Bennett et al. 1997).

For better understanding, the meaning of biotechnology can be comprehended and incorporated to the fermentation processes which are employed for producing wine and penicillin. Thus, the term "mycotechnology" needs to be incorporated with various biotechnological practices, both old and new era, that rely on fungal products and processes.

6.2 Premodern Fungal Technology

The term "modernism" separates the twentieth century by violating the tradition set in the nineteenth century. Like in art, abstraction superseded by representation; in architecture, functionalization gets superseded by ornamentation; in literature, new style forms got superseded by conventional narrative. Daily the new applications are discovered for basic science to revolutionize the living standards of the people. Adjectives such as "premodern", "modern" and "postmodern" are employed as descriptive terms for assessing the massive number of procedures as well as products which involves fungal biotechnology.

For millennia, the bread, beer, wine, koji and various fermented food along beverages have been integral part of human regime, but they have lost their relic (Table 6.1). Historical documents have made it clear that the individual knows about microbes like moulds and yeasts on the basis of the function. Pasteur on visualizing the microbes under the microscope during sugar fermentation process, considered the fermentation as "organized ferment", whereas amendment taking place in the solution without the traces of any microbes were regarded as "unorganized ferment". Later on, it was found that unorganized ferments were the metabolites which were synthesized by organized ferment. Kuhne coined the word "enzyme". Finally, "enzyme" was predominantly for unorganized ferments irrespective of the microbe producing which can be bacteria, fungi or yeast (Lutman 1929).

Table 6.1 Premodern examples of mycotechnology (Gray 1970; Chang and Hayes 1978; Hesseltine 1983)

Process/substrate for Asian food fermentations	Microorganisms involved
Ang-kak	*Monascus purpureus*
Miso	*Aspergillus oryzae*
Ontjam	*Neurospora crassa*
Soy sauce	*Aspergillus oryzae, Aspergillus sojae*
Tempeh	*Rhizopus niveus*
Brewing and baking	*Saccharomyces cerevisiae, Saccharomyces carlsbergensis*
Mould-ripened cheeses	*Penicillium roqueforti, Penicillium camemberti*
Mushroom cultivation	*Agaricus bisporus, Auricularia* sp., *Flammulina velutipes, Lentinus edodes, Pleurotus* sp., *Volvariella volvacea*

6.3 Modern Fungal Technology

Alcohols, enzymes, organic acids and different pharmaceutical products synthesized by fungi are the key for the advancement of the modern technology. Some of the chief industrial products produced by fermentation process along with synthesizing species are illustrated in Table 6.2.

Traditional enzyme, diastase (also known as amylase), is one of the most exploited enzymes which were isolated from the barley, and it is now applied for manufacturing beer. In 1894, Jokichi Takamine realized about the industrial importance of enzymes produced by these moulds and fungi. Takamine used *Aspergillus oryzae*, a Japanese koji mould, for synthesizing the diastase, for which he inoculated the spores of the mould onto the wheat bran/steamed rice and allowed it to proliferate so as to form a thin layer over it. In 1894, successive patents were filed, in which he protected the processes involved but advised the new role of diastase (partially purified) as a malting enzyme. Additionally, he also suggested that diastase can aid digestive enzyme for treating dyspepsia (Takamine 1894; Underkofler 1954). During the early period of the twentieth century, analogous procedures were established for various other enzymes. In 1983, there were 30 classes of enzyme which were being commercially exploited, out of which half of them were of fungal origin. Due to exponential exploitation of these commercial enzymes, Godfrey and West compiled the summary of these enzymes in 1996.

Table 6.2 List of examples of modern industrial products by mycotechnology

Industrial product	Fungi
Antibiotics/pharmaceutical product	
Penicillins	*Penicillium chrysogenum*
Cephalosporin	*Cephalosporium acremonium*
Cyclosporin	*Tolypocladium inflatum*
Ergot alkaloids	*Claviceps purpurea*
Griseofulvin	*Penicillium griseofulvin*
Mevalonin	*Aspergillus terreus*
Enzymes	
a-Amylases	*A. niger*, *A. oryzae*
Cellulase	*Humicola insolens*, *Penicillium funiculosum*, *Trichoderma viride*
Glucoamylases	*Aspergillus phoenicis*, *Rhizopus delemar*, *R. niveus*
Glucose oxidase	*A. niger*
Invertase	*A. niger*, *A. oryzae*
Laccase	*Coriolus versicolor*
Pectinase	*A. niger*, *A. oryzae*, *Humicola insolens*
Proteinases	*A. oryzae*, *Aspergillus melleus*, *R. delemar*
Rennin (microbial)	*Mucor miehei*, *M. pusillus*
Organic acids	
Citric acid	*A. niger*
Itaconic acid	*A. terreus*

Another product synthesized by these filamentous fungi which is the centre of attraction in modern biotechnology is the citric acid. Formerly, the citric acid was isolated from the citrus fruit, but at end of the nineteenth century, it was enlightened that these filamentous fungi were responsible for citric acid production. Pfizer, Brooklyn, New York, USA, developed the conventional process which gained worldwide recognition for its application in beverage and food industry. Further, these processes are utilized to produce antifoaming agents, cosmetics, detergents, tablets, textile treatment and preservatives for storing blood. Different approaches of modern fermentation technology were improved focussing on improving the yield of citric acid by amending the growing conduction and by exploiting the submerged process for enhancing the process of product recovery (Crueger and Crueger 1982).

The turning point took place in the industrial microbiology when penicillin was discovered, and further derivatives of penicillin were named as "wonder drug" (Wainwright 1990). Exploration of secondary metabolites with additional antimicrobial activity was elicited after the penicillin discovery. Beside that invigorating research on physiology of fungi, fermentation technology and development of industrial strain was taking place. During 1940–1950, varied number of antibiotics was discovered, and this time is also regarded as the "golden era of antibiotics". Great success was achieved by Selman Waksman and his colleagues, who were working in Rutgers University and Merck Corporation, New Jersey, as they were the only one to screen out the antimicrobial metabolites. Most of the antibiotics found by Waksman group were synthesized by soil isolates, notably actinomycetes. As moulds and actinomycetes form the complex filamentous network, it emerged as a challenge for chemical engineers to remodel the industrial-level fermentation method perfectly for both batch and continuous procedure (Smith and Berry 1975; Demain 1981). During the revolution of rDNA technology, minor alteration in genetic material of microbes was done in order to synthesize the genetically engineered product of microbial origin.

During the golden era of antibiotics, where research was focusing mainly on finding novel drugs, the laboratory in Japan under the supervision of Umezawa was screening for the microbes with antidiarrhoeal, antihypertensive, antimutagenic, antitumour and immune-stimulant potentials (Umezawa 1982). Cyclosporin (immune-suppressant) and mevalonins (antihypertensive) are the pharmaceutical product which are acquired from the filamentous fungi (Von Wartburg and Trabor 1986; Monghan and Tkacz 1990).

6.4 Postmodern Fungal Technology

The advent of rDNA technology has transformed biology. Under the term "postmodern", mycotechnology insinuates to recent improvements procreated by embracing the techniques like gene splicing along with the additional post-rDNA techniques other than the traditional industrial techniques. Few examples of this hybrid mix are depicted in Table 6.3. The regulation of heterologous protein

Table 6.3 Postmodern example of mycotechnology

Development	Examples
Expression of heterologous genes	Fungal, plant and mammalian
Amplification of homologous genes	Antibiotic pathways and enzymes
Manipulation of secondary pathways	New semi-synthetic antibiotics and hybrid antibiotics
Large-scale genomics	*Aspergillus nidulans*, *Neurospora crassa*, *Magnaporthe grisea* and *Phytophthora infestans*
DNA chips	High-density DNA arrays for screening gene expression
"Mining" fungal biodiversity of new pharmaceutical	Sampling environmental DNA genomics-based screening

synthesized by the filamentous fungi has attained substantial apperception. Presently, the chief host fungi are *Aspergillus nidulans*, *A. niger*, *A. oryzae* as well as *Trichoderma reesei* (Davies 1991), whereas *Neurospora crassa* is under evaluation (Rasmussen-Wilson et al. 1997). Formerly, researchers wanted to synthesize the mammalian protein that is also in high amount and were successful in expressing bovine prochymosin (also known as rennin) in different species (Davies 1991). Moreover, the expression of mammalian protein was less than the expected level. The molecular stages of secretion pathway of fungi, the PTM (post-translational modification) of metabolic proteins and the discharging of these proteins into environment via hyphae are only limited to textual level (Wosten et al. 1991). Thus, extensive research at the molecular level is required to get the insight about the fungal gene expression and mode of secretion.

Molecular analysis also acquiesced the secondary metabolic pathways. Certainly, penicillin family was the first family of antibiotic which gained the profit from innovative approaches. On cloning the gene encoding for isopenicillin N synthase, it was uncovered that various gene involved in the pathway were present in cluster; thus it accelerated the isolation process (Skatrud 1991). The strains showing high yield were determined to have multiple copies of the gene which code for the main enzyme involved in the penicillin pathway. In few instances, the researchers were able to engineer the fraction of the metabolic pathway with the help of atypical precursor or the host organism, hence exaggerating chemical diversity of the nature (Skatrud 1992).

Molecular analysis has also benefited the group of secondary metabolites named polyketides. Till date, extensive research is done in actinomycetes targeting either mixing or matching of the polyketide synthases (Kao et al. 1994). Moreover, these strategies have benefited fungi a lot. The variation spawned by employing the diverse initiating units, by amending the oxidation as well as stereochemistry of chemical during elongation and by inducing different post-polyketide amendment which resulted in the synthesis of different theoretic molecules. By exploiting the genetically altered polyketide synthases, one is even able to synthesize the artificial natural product.

The advances in the genetic transformation techniques have enabled to amend the fungal strain and provide them the potency as some of the species lacks sexual as well as parasexual cycles (Esser 1997). Enhancement in the conventional fermentation by fungi is observed along with that genetic modification that has enabled to amend the fungi to perform specific function. Incorporation of both homologous and heterologous gene into fungal host has improved the yield and properties of the enzyme. Now, the enzyme could be synthesized on the varied substrate and at different optimum temperatures (Kinghorn and Lucena 1994).

Even traditional approach of cultivating the mushroom got the boost by the latest mycotechnology. Most of the appealing species are difficult to culture; thus they are needed to amass in their sporadic phase. Model species like *Coprinus cinereus* (Pukkila 1993) and *Schizophyllum commune* (Raper and Horton 1993) were chosen to evaluate the genetic foundation responsible for formation of fruiting body. It has been forecasted that to obtain the commercially available improved strain of mushroom, the same breeding techniques using molecular tools are needed. However, it is believed that isolating the gene responsible for the development of mushroom in in vitro models will provide the perception of fruiting in the exotic species.

The automation in DNA sequencing has made it possible to sequence the genome, and this genomic is transforming the whole biology. The bacterium *Haemophilus influenza* Rd was the microbes whose whole genome was sequenced (Fleischmann et al. 1995); although 7 other genomes have been sequenced beside it, 100 species are under evaluation. The first eukaryotic species, which got whole genome sequenced, is yeast, *Saccharomyces cerevisiae* (Dujon 1996). *S. cerevisiae* comprises 16 chromosomes containing 12,067,266 bp excluding the repeats. Various yeast genes show similarity with mammalian gene; thus their functionality is assessed to study about cancer as well as other human diseases (Botstein et al. 1997). In contrast, the open reading frames recognized via the sequencing result went into vain as they didn't display any phenotypic characteristics.

Genomic project started with the sequencing of the yeasts *Candida albicans* and *Schizosaccharomyces pombe* and some filamentous fungi, such as *Aspergillus nidulans* and *Neurospora crassa* (Bennett 1997). The significant amount of data like genetic as well as biochemical makes it feasible to understand the correlation between sequence and phenotypic traits.

Microbial genomics assures to reform not only the fundamental biology, but it gears up the drug discovery procedure by regulating the clinical experiments. New possibilities arise on finding the correlation among the few gene sets with the drug potential. By developing a DNA microchip which encompasses the fixed oligonucleotide genomic sequence of the yeast onto silica chip with the help of photolithography, probes are employed to assess the intensity of the gene expression on treating it with the suitable reagent. Diagnostic based on genome is more liable in association with spatial chemistry as well as miniature assay. DNA chips accelerate the identification process of drug suitable for the target. Genomic analysis enables us to explore novel bioactive compounds from the environmental samples and unexplored areas, and it also includes the marine and macro-fungi (Blanchard and Hood 1996). Mycotechnology works according to the scientific frontiers in the same framework of biotechnology. New development in scientific frontiers leads to development of new techniques in the field of agriculture, industry and medicine. Molecular biology has revolutionized and procreated the social as well as political controversies. On the other hand, sometimes regulatory issues get masked by the scientific matters. However, economic element is not the new aspect in the industrial microbiology but is taken in consideration during setting up of new biotechnology start-up company. This altogether has generated the huge amount of literature concerning the financing, government regulations, intellectual property as well as safety (Moses and Cape 1991).

6.5 Role of Fungi in Agriculture and Soil Science

6.5.1 The Role of Arbuscular Mycorrhizal Fungi in Agro- and Natural Ecosystem

Symbiosis is the key which determines the growth and development of the plant. Arbuscular mycorrhizal fungi (AMF) are the most common variety of the mycorrhizal fungi. Advantages of these AMF are they aid in translocating the nutrients from the soil, maintaining soil integrity, guarding the plant from stress due to drought and pathogen persisting in the soil as well as generating diversity in the plants. AMF belong to the order of class *Glomales*, which presently encompasses six genera. AMF comprises the special structure in their root which is named as arbuscules, which help them in transferring of the nutrients among the fungus and plant (Dodd 2000).

6.5.2 Plant Nutrient Uptake by AMF

Phosphorus, element which is immobile in soil and important nutrient for growth and proliferation of plant, requires mycorrhiza. On incorporating the AMF in the plant-growing mixture, the plant growth system gets boosted as it was deprived of indigenous inoculum as in sterilized soil. The result can be visualized by assessing the branched root system of Chenopodiaceae and Brassicaceae family. On growing these plants in container, symbiosis will be effective if nutrition balance and environmental conditions like light are properly regulated (Dodd 2000).

6.5.3 Alleviation of Environmental Stress by AMF

AMF aids in elevating the resistance against drought not only in the plants growing in the arid but also in the area where drought prevails for short duration. The reason proposed is that it helps in nutrient absorption. AMF also induce the tolerance against different heavy metals like Cd (cadmium). The AMF metal-resistant strain which was isolated from the heavy metal-contaminated site proves its adaptation against the heavy metals (Dodd 2000).

6.5.4 Sustainable Production Using AMF

Under natural conditions, the young seeds germinate and get plug into the hyphae of AMF, which possess the ability to go under the soil and adhere to different plants. Due to this ability, AMF has gained a lot of attention. Moreover, AMF requires less carbon in comparison to plants for photosynthesis as it has the pre-existing mycelium. However, agriculture permits only those AMF which have the potential to synthesize new ERM for their survival (Dodd 2000).

Soil microbes form the colonies over the root of higher plants and develop the symbiotic relation. Nodulating bacteria play a significant role during nitrogen fixation (such as rhizobia over legumes); symbiotic relation between the root and fungi is called mycorrhizae, which is essential for phosphorus uptake. Mycorrhizae are omnipresent in nature and can be isolated from any kind of soil. Chenopodiaceae, Cruciferae, Cyperaceae, Juncaceae and Proteaceae are the few main families which form the vascular system with the fungi (Bolan 1991). The adaptation of plant with surrounding microbes is stated as rhizosphere, which is the narrow region of soil nearby the roots (Johansson et al. 2004). Elevated microbial activity is observed in rhizosphere as seepage of organic constituents from the root takes place. The concept of rhizosphere has been extended on incorporating fungal constituents responsible for symbiosis, and it is termed as “mycorrhizosphere”. Thus, this progression in nature has directed us towards the traditional concentrated management to low-cost, sustainable crop cultivation (Johansson et al. 2004).

6.5.5 Effect of AM Fungi on Mycorrhizosphere Bacteria

AM fungi directly and indirectly affect the bacteria colonizing on the roots of plants. Direct interaction involves the pH change which is stimulated by fungi and creates the competition for nutrient consumption. However, indirect interaction influences the growth of plant, soil texture as well as exudation from the roots (Mishra et al. 2016; Johansson et al. 2004).

6.5.6 Relevance of Mycorrhizosphere Interactions to Sustainable Agriculture

AMF is considered to amend the phosphorus nutrition, and along with that it will extemporize disease resistance in the target plant. Thus, the mycorrhizal fungi are essential for sustaining the agriculture land where the nutrient inputs are very low. AM mycelia in collaboration with bacteria/fungi aid in mobilizing the nutrients (Johansson et al. 2004). In nodulating legumes, the N-fixing bacteria and AMF work synergistically. The N-fixation gene in *Burkholderia* species that is present in the hyphae of AMF has been determined and advised that it may enhance N supply by fixing the atmospheric N for mycorrhizal plants.

6.5.7 Degradation of Pesticides by White-Rot Fungi and Its Relationship with Ligninolytic Potential

The white-rot fungi comprise those fungi which have potential to degrade the lignin (polyphenolic polymer) present in the lignocellulosic substrate. White-rot got its name from the appearance of the wood which is infected by fungi, and leaching out of substrate during lignin removal causes the white-rot appearance. Basidiomycetes are most exploited white-rot fungi [13].

6.5.8 The Ligninolytic System of White-Rot Fungi

White-rot fungi possess the ability to secrete more than one extracellular enzyme, having the potential to degrade the lignin. Along with these enzymes fusion of different processes will result in mineralization of the lignin, thus also known as the lignin-modifying enzymes (LMEs) (Orth and Tien 1995). These three enzymes are heme-containing glycosylated peroxidases, lignin peroxidase (LiP, E.C. 1.11.1.14) and Mn-dependent peroxidase (MnP, E.C. 1.11.1.13) and a copper-containing phenoloxidase, laccase (Lac, E.C. 1.10.3.2) (Thurston 1994).

The peroxidase enzyme synthesized during the white-rot has the ability to degrade different aromatic xenobiotics, which involve polyaromatic hydrocarbons (PAHs), polychlorinated biphenyls, pentachlorophenol and group of pesticides (Bending et al. 2002). The ligninolytic enzyme has the ability to disintegrate the C-C as well as C-O bonds leading to depolymerization of the lignin, which subsequently leads towards cleaving of aliphatic/aromatic components (Hammell 1997). White-rot has been associated with Poly R-478 degradation to 3–5 rings of PAHs. The ability of ligninolytic enzyme for degrading mono-aromatic xenobiotic is still unclear (Bending et al. 2002). The chief objective of this experimentation was to evaluate the potency of white-rot fungi to disintegrate the mono-aromatic compounds and to establish the characteristic value of ligninolytic enzyme acquired from fungi to disintegrate the pesticides. Nine different types of fungi namely, *Agrocybe semiorbicularis, Auricularia auricula, Coriolus versicolor, Dichomitus squalens, Flammulina velutipes, Hypholoma fasciculare, Phanerochaete velutina, Pleurotus ostreatus* and *Stereum hirsutum* having the white-rot traits were selected for investigation. The assessment was done to find the correlation between the potency of fungi in degrading the pesticides. Even the mechanism responsible for ligninolytic ability is still unknown (Bending et al. 2002).

6.5.9 *Degradation of Atrazine by Soil Fungi for a Sustainable Environment*

Microbial disintegration is the main process which provides the insight about the pesticide behaviour inside the soil (Singh et al. 2017a, b; Singh et al. 2016; Kumar et al. 2015). Various studies were conducted to observe the potential of pure isolates to degrade the ^{14}C-ethylamino- and 14C-ring-labelled atrazine (2-chloro-4-ethylamino-6-isopropylamino-s-triazine). The chief objective of the experimentation was to uncover the potential of fungi isolated from the soil to degrade the atrazine. *Aspergillus fumigatus* is reported to degrade simazine, which on dealkylation leads to the formation of 2-chloro-4-amino-6-ethylamino-s-triazine. Successive catabolization of the intermediate molecule by the processes like dealkylation and dehalogenation leads to the synthesis of ammelide (Kaufman and Blake 1970; Kearney et al. 1965). A bioassay of oat seedling was done to assess the ability of fungi isolated from the soil to degrade the atrazine, simazine and propazine into non-toxic or medium toxic constituents. In another research, the cultures were inoculated into sandy loam of Lakeland procured in the unperforated pot. The culture solution inoculated was having different growth times as follows: 0, 3, 6, 12, 36 or 72 days. The composition of culture solution was 5 mg/l of atrazine, simazine or propazine dissolved in the basal medium. The comparison between culture solution containing s-triazine and sterile s-triazine was done for each culturing time. Whole experiment was executed in four replicas. After 3 weeks of growing period, the seedlings of oats were harvested, and their fresh weigh was assessed and illustrated in percent to that of the sterile control. Basal medium comprising of ^{14} C-ring- and chain-labelled (^{14} C ethyl) atrazine with the concentration of 5 ppm was inoculated with microbes and incubated at 24 °C inside a closed flask. Results were depicted as percent for initial radioactivity of ^{14}C. Cells were collected, filtered and washed with 0.85% NaCl at the end. *A. fumigatus* was readily purified by the simazine, whereas partial detoxification of atrazine and minute detoxification took place in the case of propazine. Different fungal strains also showed the analogous result. As a conclusion for this study, microbes were able to disintegrate the herbicide s-triazine. On associating the result of in vitro study, N-dealkylation method was found as the effective mechanism for s-triazine detoxification by the fungi isolated from the soil (Kaufman and Blake 1970).

6.6 Conclusion

The exploration of the green endophytic growth of fungi persisting in the soil and water in symbiosis with marine creatures, functioning as nutrient translocation unit or the factory of synthesizing the bioactive molecules, is a major perspective. The exploitation is being done in far-fetched niches like remote ocean and hypoxic region. The microbes have become the part of our life, operating as bio-protector, bio-fertilizers, bio-remediators as well as drug producers. Fungi being the conventional fermentation workhorses, for the processes like brewing, baking and synthesizing the antibiotics, enzymes, organic acids and various other pharmaceutical products, are economically important. Thus, fungi have become the frontiers of molecular biotechnology. These micro-sized eukaryotes have become the vital model for understanding the basic science and synthesizing the commercial product. Genomes of yeast act as platform for functional genomics as well as the DNA microchip method. In coming time, the researchers will keep on exploiting these filamentous fungi and yeast to gain knowledge about their physiology, different biochemical pathways, mechanism of developing resistance and production of secondary metabolites in order to build a strong foundation for developmental research.

References

Bending GD, Friloux M, Walker A (2002) Degradation of contrasting pesticides by white rot fungi and its relationship with ligninolytic potential. FEMS Microbiol Lett 212(1):59–63
Bennett JW (1997) White paper: genomics for filamentous fungi. Fungal Genet Biol 21:3–7
Bennett JW, Heller F, Case CL (1997) Bio-tech-nol-o-gy,n.: the many definitions of biotechnology. Soc Indust Microbiol News 47:240–243
Blanchard AP, Hood L (1996) Sequence to array: probing the genome's secrets. Nat Biotech 14:1649
Bolan NS (1991) A critical review on the role of mycorrhizal fungi in the uptake of phosphorus by plants. Plant Soil 134(2):189–207
Botstein D, Cherfitz SA, Cherry JM (1997) Yeast as a model organisms. Science 277:1259–1260
Chang ST, Hayes WA (1978) The biology and cultivation of edible mushrooms. Academic Press, New York
Crueger W, Crueger A (1982) Biotechnology: a textbook of industrial microbiology. Sinauer, Sunderland
Davies RW (1991) Expression of heterologous genes in filamentous fungi. In: Peberdy JR, Caten CE, Ogden JE, Bennett JW (eds) Applied molecular genetics of fungi. Cambridge University Press, Cambridge, pp 103–117
Demain AL (1981) Industrial microbiology. Science 214:987–991
Dodd JC (2000) The role of arbuscular mycorrhizal fungi in agro-and natural ecosystems. Outlook on Agriculture 29(1):55–62
Dujon B (1996) The yeast genome project; what did we learn? Trends Genet 12:263–270
Esser K (1997) Fungal genetics: from fundamental research to biotechnology. Prog Bot 58:3–38
Fleischmann RD, Adams MD, White O, Clayton RA, Kirkness EF et al (1995) Whole-genome random sequencing and assembly of *Haemophilus influenzae* Rd. Science 269:496–512
Godfrey T, West S (eds) (1996) Industrial enzymology, 2nd edn. Stockton Press, New York
Gray WD (1970) The use of fungi as food and in food processing. CRC Crit Rev Food Technol 1:1–225
Hammell KE (1997) Fungal degradation of lignin. In: Cadisch G, Giller KE (eds) Driven by nature, plant litter quality and decomposition. CAB International, Wallingford, pp 33–45
Hesseltine CW (1983) Microbiology of oriental fermented foods. Annu Rev Microbiol 37:575–592
Johansson JF, Paul LR, Finlay RD (2004) Microbial interactions in the mycorrhizosphere and their significance for sustainable agriculture. FEMS Microbiol Ecol 48(1):1–13
Kao CM, Katz L, Khosla C (1994) Engineered biosynthesis of a complete macrolactone in a heterologous host. Science 265:509–512
Kaufman DD, Blake J (1970) Degradation of atrazine by soil fungi. Soil Biol Biochem 2(2):73–80
Kearney PC, Kaufman DD, Sheets TJ (1965) Metabolites of simazine by Aspergillus fumigatus. J Agric Fd Chem 13:369–373

Kinghorn JR, Lucena N (1994) Biotechnology of filamentous fungi. In: Iberghin L, Frontali L, Sensi P (eds) ECB6: proceedings of the 6th European congress on biotechnology. Elsevier, pp 277–286
Kumar V, Singh S, Singh J, Upadhyay N (2015) Potential of plant growth promoting traits by bacteria isolated from heavy metal contaminated soils. Bull Environ Contam Toxicol 94(6):807–814
Lutman BF (1929) Microbiology. McGraw Hill, New York
Mishra V, Gupta A, Kaur P, Singh S, Singh N, Gehlot P, Singh J (2016) Synergistic effects of Arbuscular mycorrhizal fungi and plant growth promoting rhizobacteria in bioremediation of iron contaminated soils. Int J Phytoremediation 18(7):697–703
Monghan RL, Tkacz JS (1990) Bioactive microbial products: focus upon mechanisms of action. Annu Rev Microbiol 44:271–301
Moses V, Cape RE (eds) (1991) Biotechnology: the science and the business. Harwood Academic Publishers, London
Orth AB, Tien M (1995) Biotechnology of lignin degradation. In: Esser K, Lemke PA (eds) The Mycota. II. Genetics and biotechnology. Springer, New York/Berlin/Heidelberg, pp 287–302
Pukkila PJ (1993) Methods of genetic manipulation in Coprinus cinereus. In: Chang ST, Buswell JA, Miles PG (eds) Genetics and breeding of edible mushrooms. Gordon and Breach, Amsterdam, pp 249–264
Raper CA, Horton JS (1993) A strategy for isolating mushroom-inducing genes in edible basidiomycetes. In: Chang ST, Buswell JA, Miles PG (eds) Genetics and breeding of edible mushrooms. Gordon and Breach, Amsterdam, pp 285–295
Rasmussen-Wilson SJ, Palas JS, Wolf VJ, Taft CS, Selitrennikoff CP (1997) Expression of a plant protein by Neurospora crassa. Appl Environ Microbiol 63:3488–3493
Singh S, Singh N, Kumar V, Datta S, Wani AB, Singh D, Singh K, Singh J (2016) Toxicity, monitoring and biodegradation of the fungicide carbendazim. Environ Chem Lett 14(3):317–329
Singh S, Kumar V, Chauhan A, Datta S, Wani AB, Singh N, Singh J (2017a) Toxicity, degradation and analysis of the herbicide atrazine. Environ Chem Lett:1–27
Singh S, Kumar V, Upadhyay N, Singh J, Singla S, Datta S (2017b) Efficient biodegradation of acephate by *Pseudomonas pseudoalcaligenes* PS-5 in the presence and absence of heavy metal ions [Cu (II) and Fe (III)], and humic acid 3. Biotech 7(4):262
Skatrud PI (1991) Molecular biology of the beta-lactam producing fungi. In: Bennett JW, Lasure LS (eds) More gene manipulations in fungi. Academic Press, San Diego, pp 364–395
Skatrud PI (1992) Genetic engineering of beta-lactam antibiotic biosynthetic genes in filamentous fungi. Tibtech 10:17–20
Takamine J (1894) United States Patents 525,820 and 525,823
Thurston CF (1994) The structure and function of fungal laccases. Microbiology 140:19–26
Umezawa SI (1982) Low-molecular weight enzyme inhibitors of molecular origin. Annu Rev Microbiol 36:75–99
Underkofler, L.A. 1954. Fungal amylolytic enzymes. In: Underkofler, L.A., Hickey, R.J. (Eds.), Industrial Fermentations. Chemical Publishing Company, New York, pp. 94–118
Van der Heijden MGA, Klironomos JN, Ursic M, Moutoglis P, Streitwolf-Engel R, Boller T, Wiemken A, Sanders IR (1998) Mycorrhizal fungal diversity determines plant biodiversity, eco- system variability and productivity. Nature 396:69–72
Von Wartburg A, Trabor R (1986) Chemistry of the natural cyclosporin metabolites. Prog Alleg 38:28–45
Wainwright M (1990) Miracle cure-the story of antibiotics. Blackwell, Oxford
Wosten H, Moukha S, Sietsma J, Wessels J (1991) Localization of growth and secretion of proteins in Aspergillus Niger. J Gen Microbiol 137:2017–2023

Part II
Sustainable Aspects of Fungi in Agriculture

Chapter 7
Fungal Endophytes: Role in Sustainable Agriculture

Pratibha Vyas and Anu Bansal

Abstract In view of escalating cost and pollution related with chemical fertilizers and pesticides, interest has increased to find alternative methods of fertilization and control of pests. Fungal endophytes residing symbiotically inside the plant tissues play an important role in the growth promotion and resistance to various biotic and abiotic stresses and diseases in plants. They also produce phytohormones, antimicrobial compounds, and many agrochemical bioactive metabolites. These endophytes hold huge potential to be used as safe and cost-effective alternative to chemical pesticides and fertilizers in view of their wide range of plant growth-promoting activities. The present chapter describes the role of endophytic fungi in the agriculture sector.

Keywords Fungal endophytes · Sustainable agriculture · Phytohormones · Plant growth promotion · Phosphate solubilization

7.1 Introduction

Endophytes can be defined as the microorganism living within the tissues of plant without causing them probable disease symptoms. Microbial endophytes have developed a mutualistic association with their plant hosts, wherein the sufficient nutrients and habitation for survival of the endophytes are provided by their host plant. Compounds produced by endophytes help host plants to resist different biotic and abiotic stresses and help them to survive under extreme conditions.

P. Vyas (✉)
Department of Microbiology, School of Bioengineering and Biosciences, Lovely Professional University, Phagwara, Punjab, India

A. Bansal
Department of Biochemistry, School of Bioengineering and Biosciences, Lovely Professional University, Phagwara, Punjab, India

P. Gehlot, J. Singh (eds.), *Fungi and their Role in Sustainable Development: Current Perspectives*, https://doi.org/10.1007/978-981-13-0393-7_7

Fungal endophytes are an important source of compounds which can be exploited in the field of agriculture (Zhao et al. 2011). They promote the growth of plant and improve soil fertility by virtue of their capability to produce plant growth-promoting compounds. They also have received great attention in terms of their ability to produce a wide range of bioactive compounds with medical potential. These fungi have been known to produce bioactive compounds with antimicrobial, anticancer, cytotoxic, and insecticidal properties. Similar kind of bioactive compounds are produced by many fungi which have also been reported to be produced by their host plants. Endophytic fungi producing bioactive compounds could also be beneficial to the pharmaceutical industry for a commercial exploitation.

In view of escalating antibiotic resistance of pathogenic microorganisms, the requirement for new antimicrobial agents and interest in natural methods of pathogen control have increased. Fungal endophytes are the largest group of microorganisms producing secondary metabolites. They are known to produce extracellular enzymes like cellulases, proteinase, lipases, and esterases. Metabolite products like amines and amides produced by endophytes have found to be toxic to insects but not to mammals. Fungal endophytes are a significant component of plant micro-ecosystems and have been found in large number of plant species examined. They have been reported to solubilize insoluble phosphates and produce plant growth-promoting hormones including auxins, cytokinins, and gibberellins. Furthermore, they provide protection to plants against pathogens by producing antagonistic compounds, inducing host defense mechanisms, or providing competition for nutrients and colonization sites.

7.2 Fungal Endophytes from Medicinal Plants

Endophytic fungi have been found to be associated with almost all investigated plants. They play widespread ecological roles and interact with host plants in different ways (Fig. 7.1).

Raviraja (2005) isolated *Curvularia clavata*, *C. lunata*, *F. oxysporum*, and *C. pallescens* fungi from bark, stem, and leaf tissues of medicinal plants from Western Ghats of India. In a study carried out by Gautam et al. (2013), fungi belonging to *Aspergillus niger*, *A. flavus*, *A. nidulans*, *Colletotrichum*, *Curvularia*, *Cladosporium*, *Penicillium chrysogenum*, *P. citrinum*, *Phoma*, and *Rhizopus* were isolated from leaves, stem, and petiole samples of *Cannabis sativa*. Likewise, various other endophytes have been reported from different medicinal plants (Table 7.1).

Fig. 7.1 Isolated fungal endophytes from medicinal plant samples: (**a**) *Adhatoda vasica*, (**b**) *Zanthoxylum alatum*, (**c**) *Tinospora cordifolia*, (**d**) endophytic fungi growing around the vicinity of plant sample, (**e–f**) pure cultures of fungal endophytes

7.3 Fungal Endophytes

7.3.1 Fungal Endophytes in Sustainable Agriculture

In order to increase plant productivity, usage of more than 100 million tons of chemical fertilizers annually has been estimated (Glick et al. 1999). Due to the adverse effects of using chemical fertilizers on the environment and their high production expense, the focus on the research on microbial inoculants has increased. Endophytic fungi perform various symbiotic associations with plants and enrich plant progression by various direct and indirect mechanisms (Fig. 7.2). They are essential components of sustainable agriculture in view of their ability to enhance plant growth and yield and increase plant fitness by providing biotic and abiotic stress tolerance (Barka et al. 2002; Tanaka et al. 2005; Vega et al. 2008). In addition, they release various secondary metabolites minimizing the effect of pathogens and also induce host plant defenses against phytopathogens (Giménez et al. 2007; Gao et al. 2010).

Table 7.1 Fungal endophytes from medicinal plants

Endophytic fungus	Plant	References
Acremonium, Chaetomium, Cylindrocarpon Paecilomyces, Trichoderma	*Actinidia macrosperma*	Lu et al. (2012)
Alternaria, Colletotrichum, Aspergillus, Fusarium, Gliocladium, and *Cunninghamella*	*Malus sieboldii*	Cai and Wang (2012)
Alternaria, Fusarium, Cladosporium, Colletotrichum, Leptosphaeria, Paraconiothyrium, Pyrenochaeta, Stephanonectria	*Holcoglossum rupestre* and *H. flavescens*	Tan et al. (2012)
Chaetomium, Alternaria, Cercophora, Fusarium, Hypoxylon, Nigrospora, Cladosporium, Thielavia, Schizophyllum, Gibberella	*Cannabis sativa, Cedrus deodara, Pinus roxburghii, Picrorhiza kurroa, Withania somnifera, Abies pindrow*	Qadri et al. (2013)
Alternaria tenuissima, Aspergillus fumigates, A. japonicas, A. niger, A. repens, Curvularia pallescens, Fusarium solani, F. semitectum, Phoma hedericola, and *Drechslera australien*	*Ricinus communis*	Sandhu et al. (2014)
Phomopsis	*Brucea javanica*	Liang et al. (2014)
Pestalotiopsis, Phomopsis, Aspergillus, Xylaria, Nectria, Penicillium, and *Fusarium*	*Myrcia guianensis*	dos Banhos et al. (2014)
Acremonium, Aspergillus niger, Cladosporium, Curvularia lunata, C. brachyspora, Penicillium species	*Urginea indica*	Shiva Kameshwari et al. (2015)
Nigrospora, Fusarium sp.	*Crescentia cujete*	Prabukumar et al. (2015)
Ramichloridium cerophilum	Chinese cabbage	Xie et al. (2016)

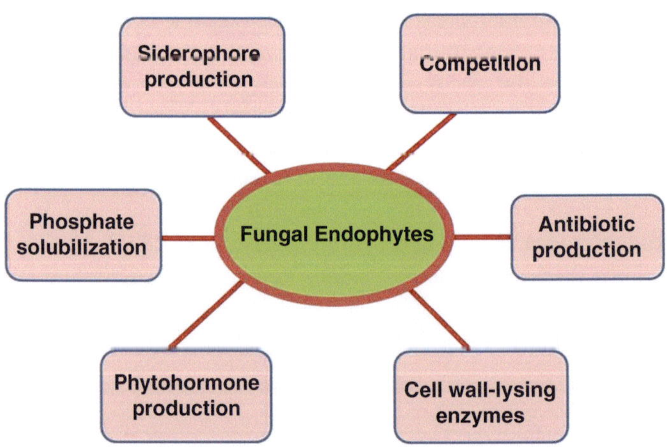

Fig. 7.2 Mechanisms used by fungal endophytes for plant growth promotion

7.3.2 Phytohormone Production

Phytohormones help in the regulation of plant enlargement and plant responses toward the biotic stress. Fungal endophytes are studied to promote plant growth by producing different plant hormones. Among phytohormones, auxins constitute an important group influencing different cellular functions and regulating plant growth. In response to light and gravity, they help in orientating the growth of root and shoot, differentiating vascular tissues, and initiating lateral and adventitious roots (Frankenberger and Arshad 1995; Patten and Glick 1996; Cecchetti et al. 2008). Auxins that have been found involved in host-parasite interactions are mainly indole-3-acetic acid (IAA) (Hamill 1993; Gutierrez et al. 2009). *Fusarium tricinctum* and *Alternaria alternate* isolated from leaves of potato (*Solanum nigrum*) produced IAA and also enhanced the growth of rice plants (Khan et al. 2015a, b).

Adverse effects of abiotic stress can be overcome by exploiting gibberellins produced by endophytic fungi, essential for increasing plant growth and biomass production under stressed environment (Khan et al. 2015a, b). *Galactomyces geotrichum* isolated from *Trapa japonica* has been reported for IAA and gibberellin production (Waqas et al. 2014b). Two strains *Phoma glomerata* and *Penicillium* sp. from cucumber roots have been found to synthesize gibberellic acid and IAA. Inoculating these strains in cucumber plants under drought stress, has shown a significant increase in plant biomass, growth parameters, and assimilation of essential nutrients and reduced sodium toxicity. Stress modulation is also ensured through the alteration in the level of jasmonic acid, downregulation of abscisic acid, and increased salicylic acid content (Waqas et al. 2012). Similar effects were observed by the same research group working on cucumber endophyte *Paecilomyces formosus* (Khan et al. 2012). *Fusarium proliferatum* from *Physalis alkekengi* have shown strong plant growth-promoting activity due to its ability to produce gibberellins (Rim et al. 2005). Bioactive gibberellic acids were also found to be produced by species of *Aspergillus*, *Cladosporium*, and *Talaromyces* from soybean (Hamayun et al. 2009; Khan et al. 2011a, b).

Abscisic acid, jasmonic acid, and salicylic acid act as defense signaling substances and thereby respond to abiotic stress stimuli. Protective secondary metabolites and defense-related protein biosynthesis are induced by jasmonic acid, which responds to abiotic and biotic stresses (Balbi and Devoto 2008). Salicylic acid responds to abiotic and biotic stress by initiating induced systemic resistance against fungi by microorganisms that tend to promote plant growth. Endophytic fungi *Phoma* and *Penicillium* produced gibberellins and indoleacetic acid and significantly promoted the shoot and growth attributes of rice during stresses like salinity and drought (Waqas et al. 2012). *Galactomyces geotrichum* from *Trapa japonica* has been studied to produce large amount of jasmonic acid, thereby inducing systemic resistance in soybean plants (Waqas et al. 2014b).

Penicillium citrinum and *Aspergillus terreus* inoculated in sunflower plants showed alteration in the levels of salicylic acid and jasmonic acid of the plants (Waqas et al. 2015). Three different fungal endophytes isolated from the medicinal plant of *Asclepias sinaica* identified as *Penicillium chrysogenum*, *Alternaria alternata*, and sterile hyphae produced extracellular enzymes including amylase, pectinase, cellulase, gelatinase, xylanase, and tyrosinase. The fungal isolates promoted the root growth as a result of ammonia and IAA production (Fouda et al. 2015).

7.3.3 Biocontrol Agents

Pathogenic microorganisms that affect plant health are a foremost threat to crop production. Instead of using synthetic chemicals, biological control is an effective and eco-friendly alternative against plant pathogens. Both in medicine and agriculture, fungal infections have emerged as a serious problem. On annual basis, filamentous fungi among several fungal species cause major economic losses (Pennisi 2001). Moreover, a substantial loss of postharvest food leads to fungal contamination during storage (Muńoz et al. 2013). Sustainability in agriculture in an eco-friendly way can be achieved by management strategies involving endophytic symbiosis that reduce excessive fungicide use.

Some chemical compounds used in plant defenses produced by endophytic fungi are earlier thought to be produced by their host plant. In perennial ryegrass, wide studies had been done to know the chemical basis of insect resistance in plant-endophyte mutualisms. Indole diterpenes, ergot alkaloids, and peramine have been found as the three major classes of secondary metabolites (Rowan et al. 1986). Terpenes and alkaloids act as a highly toxic substance against phytophagous insects and mammalian herbivores (Clay and Cheplick 1989; Popay et al. 1990). The imperfect stage of the fungus *Pezicula cinnamomea* is *Cryptosporiopsis quercina*, isolated as an endophyte from medicinal plant *Tripterygium wilfordii* that exhibited antifungal activity against *Candida albican* (Strobel et al. 1999). Two compounds with antifungal activity were isolated from *C. quercina* which have been identified as "Cryptocandin" and "Cryptocin." Cryptocin has shown the potential activity against *Pyricularia oryzae* and large numbers of other plant pathogenic fungi.

Several studies on inoculation of plants with commonly occurring fungal endophytes such as *Fusarium verticillioides* (Lee et al. 2009); *Acremonium zeae* (Poling et al. 2008); *Colletotrichum magna* (Redman et al. 1999); *Colletotrichum* sp., *Fusarium nectria* sp., and *Xylaria* sp. (Arnold et al. 2003); and *Colletotrichum gloeosporioides*, *Clonostachys rosea*, and *Botryosphaeria ribis* (Mejía et al. 2008) have described the mitigation of pathogenic diseases. Similarly, fungal endophytes identified as *Trichoderma*, *Nigrospora*, and *Curvularia* from an essential medicinal plant of Assam, *Rauwolfia serpentina*, showed antagonistic activity against *Fusarium oxysporum* and *Phytophthora* spp. (Li et al. 2000; Doley and Jha 2010). *Talaromyces flavus* from *Sonneratia apetala* synthesize a new non-sesquiterpene peroxides which can be used in managing certain plant diseases (Li et al. 2011).

Similarly, *Aspergillus niger* and *A. flavus* isolated as endophytes from *Cannabis sativa* have been shown to inhibit *Colletotrichum gloeosporioides* and *Curvularia lunata*, two common plant pathogens (Gautam et al. 2013).

Plant growth enhancement and decreased effect of stem rot of sunflower have been shown by *Aspergillus terreus* and *Penicillium citrinum* (Waqas et al. 2015). During disease incidence, plant inoculated with the fungal endophyte enhanced the growth by modifying the responses related to host plant defense. *Hypoxylon* sp. which is an endophytic fungus isolated from *Persea indica* produced volatile organic compounds, including 1,8-cineole and 1-methyl-1,4-cyclohexadiene. These volatile organic compounds showed antimicrobial activity against four pathogenic fungi including *Botrytis*, *Cercospora*, *Phytophthora*, and *Sclerotinia*, clearly signifying the role of these compounds in the survivability of the endophytes in their host plants (Tomsheck et al. 2010). The study showed that fungal sourcing of 1,8-cineole and other volatile compounds produced by *Hypoxylon* sp. has greatly expanded potential applications in the field of industry and medicine.

In addition, fungal endophytes have been reported to have insecticidal activities. For the first time, *Phomopsis oblonga* was reported against the beetle *Physocnemum brevilineum* on elm tree (Webber 1981). Endophytic fungus *P. oblonga* produced ergot alkaloids and mycotoxins that control vector *P. brevilineum* of pathogen *Ceratocystis ulmi*, a causative agent of Dutch elm disease, and thereby reduce spread of disease. An insecticidal property against the bowl fly larvae is exhibited potentially by nodulisporic acids (novel indole diterpines) (Demain 2000). *Muscodor vitigenus*, isolated as an endophyte from liana plant (*Paullina paullinioides*), is reported to produce naphthalene that is majorly used as active ingredient in mothballs (Daisy et al. 2002).

7.3.4 Phosphate Solubilization

Phosphorus (P) mineral is considered as an essential nutrient for the growth of plant. Phosphorus affects the structure of a plant at cellular level, encourages growth, and hastens maturity. Plants which lack mineral phosphorus shows stunted growth, wilting of leaves, delayed maturity, and reduced yield (Loria and Sawyer 2005). Most of the soils are insufficient in P availability to plants (Richardson 2001; Fernández et al. 2007). According to the data on available P in the Indian soils, approximately 2% of the area is rich in P (Hasan 1994). Soluble inorganic phosphates are used as chemical fertilizers. After adding to soil, a maximum portion of it gets converted into the forms that plants are not able to consume (Richardson 1994; Fernández et al. 2007). pH and soil type largely determine the fixation and precipitation of P. Fixation of P in acidic soil is done by free oxides and hydroxides of aluminum and iron, while in the alkaline soils, it gets fixed by calcium (Goldstein 1986; Halford 1997). The availability of soil P to the plants is mainly mediated by the endophytic microorganisms. Endophytic fungi like *Aspergillus niger*, *Penicillium sclerotiorum*, *P. chrysogenum*, and *Fusarium oxysporum* isolated from *Camellia sinensis* growing in Assam, India, showed auxin production, phosphate solubilization, potassium solubilization, and zinc solubilization (Nath et al. 2012a, b). Likewise, *Fusarium verticillioides* and *Humicola* sp. have been shown to solubilize phosphates under salt stress (Radhakrishnan et al. 2015). Similarly, *Trichoderma pseudokoningii* isolated from tomato roots from central Himalaya exhibited plant growth-promoting activities of phosphate solubilization and synthesis of auxins, siderophores, HCN, and ammonia (Chadha et al. 2015).

7.3.5 Siderophore Production

Iron having redox activity behaves as a cofactor of many enzymes and is therefore considered as an essential plant micronutrient. Iron occurs in the form of ferric hydroxide which is insoluble in nature. Due to this, it limits the plant growth even in the soil rich with iron. Siderophores are iron-binding molecules with low molecular weight less than 1 kDa produced by microorganisms under low-iron conditions. Siderophores help in iron uptake by binding Fe^{3+} with high affinity (Neilands 1981; Glick et al. 1999). Many mycorrhizal and endophytic fungi produce siderophores. Its production by pathogenic fungi is typically associated with virulence. Foliar endophytic fungi of Scots pine (*Pinus sylvestris* L.) and Labrador tea (*Rhododendron tomentosum* Harmaja) have been shown to produce ferricrocin siderophore (Kajula et al. 2010). Likewise, an endophytic fungus *Acremonium sclerotigenum* inhabiting *Terminalia bellerica Roxb* has been shown to produce siderophore and also inhibit pathogenic microorganisms (Prathyusha et al. 2015). Seven endophytic fungi belonging to *Colletotrichum*, *Lasiodiplodia*, and *Fusarium* showed siderophore zone more than 30 mm on CAS agar (Aramsirirujiwet et al. 2016).

7.3.6 Agriculturally Important Enzyme Production

Degradation of the dead biomasses by microorganisms is a major step in bringing the utilized nutrients back to the ecosystem. Fungi play an important role in carbon and nitrogen cycling. Many studies have shown that endophytes have important role in biodegradation of the litter of its host plants (Osono 2006; Korkama-Rajala et al. 2008; Promputtha et al. 2010). Endophytic fungi produce extracellular hydrolases including cellulase, laccase, pectinase, phosphatase, lipase, xylanase, and proteinase as a resistance mechanism against pathogenic invasion and to obtain nutrition from host. Enzymes like amylase, cellulase, and laccase which are hydrolytic in nature are also of major interest with their various industrial applications. Endophytic fungi differ in their ability to decompose organic components, including lignin, cellulose, and hemicelluloses (He et al. 2012). Endophytic fungi of five ethno-medicinal plants from the forests of Meghalaya exhibited the production of amylase, cellulase, protease, lipase, and xylanase (Bhagobaty and Joshi 2012).

Endophytic fungi like *Fusarium*, *Penicillium*, *Phoma*, *Acremonium*, *Nigrospora*, *Pestalotiopsis*, *Phomopsis*, *Tetraploa*, and *Xylaria* from *Opuntia ficus-indica* have shown their utmost ability for defilement in biotechnological processes which involves production of enzymes like pectinase, cellulase, xylanase, and protease. An endophyte, *Acremonium zeae*, isolated from maize produced the enzyme hemicellulase extracellularly, which may be used in the bioconversion of lignocellulosic biomass into fermentable sugars (Bischoff et al. 2009).

Chitin, a constituent of the insect's exoskeleton, crustacean's shells, and fungal cell wall, is a linear homopolymer of β-1,4 N-acetylglucosamine. Fungal chitinases have vital role in the biome as they help in degradation and cycling of carbon and nitrogen from chitin molecule. Many fungal endophytes isolated from leaves of trees grown in the forests of Western Ghats, Southern India, have shown the production of chitinases (Rajulu et al. 2011). Bioremediation, an important approach in the reduction of wastes, mainly rely on biological processes for the breakdown of different pollutants. Reduced crop growth, yield, and quality have been caused by a phenolic allelochemical, cinnamic acid, widely found in continuous cropping soils. *Phomopsis liquidambari*, an endophytic fungus, effectively catabolizes cinnamic acid to styrene (Xie and Dai 2015).

7.3.7 Plant Growth Promotion

Plant growth and yield are promoted either directly or indirectly by plant growth-promoting microorganisms (Klöepper et al. 1989; Glick 1995). Fungi have been known to show plant growth-promoting activities and plant growth promotion. Overall performance and health of plants improve due to increased availability of limiting nutrients and increase in the ability of infected plants to defend them. Growth of host plant has been enhanced tremendously by endophytic *Piriformospora indica* through its root colonization (Waller et al. 2005; Varma et al. 2012). Fungus *Piriformospora indica* that forms colonies around plant root, isolated from rhizosphere of *Prosopis juliflora* and *Ziziphus nummularia*, has shown to promote growth during its symbiotic relationship with a broad spectrum of plants (Varma et al. 1999, 2012). Various studies had been done to know the molecular mechanisms used by *P. indica* to promote growth and biomass production of various plant species (Lee et al. 2011). Increased tillering and root growth in tall fescue inoculated with endophytic *Neotyphodium coenophialum* have been demonstrated by Schardl and Phillips (1997). Fresh biomass increase up to 50% in *Artemisia annua* L. (Varma et al. 1999; Franken 2012), and better growth of *Brassica oleracea* var. *capitata* resulted after inoculation with *P. indica* (Kumari et al. 2003). Pocasangre et al. (2000) observed similar outcome showing enhanced biomass of tissue-cultured plants treated with endophytic *Fusarium oxysporum* as compared to control plants. Plants of *Lolium multiflorum* infected with endophyte were found to have more vegetative tillers and more root and seed biomass than plants without endophyte infection. A fungal endophyte of perennial ryegrass named *Neotyphodium lolii* produces bioactive alkaloids thereby improving grass fitness. Effects in growth and physiology of grasses are observed hosting *Neotyphodium* species. Impact of *N. lolii* on plant growth and photosynthesis was found to be endophyte concentration independent in the plant studied in Spiering et al. 2006 which clarifies that the endophyte mycelium is not an energy outlet to the plant. After co-cultivating *Bacopa monniera* with *P. indica*, enhancement of biomass and antioxidant activity in *Bacopa monniera* was reported by Prasad et al. (2013). Gibberellic acid and indole-3-acetic acid are synthesized by endophytic fungi *Aspergillus fumigatus*, *Paecilomyces* sp., *Penicillium* sp., *Phoma glomerata*, *Chrysosporium pseudomerdarium*, and *P. formosus*. These fungi play a vital role in both mutant- and wild-type rice by promoting shoot length, chlorophyll contents, and biomass (Waqas et al. 2014a, b). Inoculation of *P. indica* in Chinese cabbage plant resulted in twofold higher fresh weight as compared with uninoculated plants. Auxin level in roots was also found twofold higher in fungus-inoculated plant as compared to uncolonized controls. Likewise, *P. citrinum* and *A. terreus* enhanced the sunflower growth (*Helianthus annuus* L.) and disease resistivity against the stem rot caused by *Sclerotium rolfsii* (Waqas et al. 2015). Endophytic *Absidia* and *Cylindrocladium* were tested to govern their effect on rice plant growth. The results showed that the plants inoculated with both fungal isolates showed increases in plant height and fresh and dry weight significantly (Atugala and Deshappriya 2015).

7.4 Conclusion

Extensive use of chemicals for increasing agriculture productivity has disturbed the delicate ecological balance leading to the buildup of pesticide resistance among pathogens and health risks to mankind. There has been an ever-increasing interest in finding eco-friendly and safe approaches to increase agriculture productivity. Fungal endophytes are important components of sustainable agriculture in view of their ability to produce phytohormones and solubilize phosphates, siderophore production, inhibiting plant pathogens, and promoting plant growth. Keeping in view the various beneficial activities carried out by fungal endophytes, the number of investigations carried out has shown a sharp increase to study plant-fungus associations. In recent years, the use of genetically modified endophytes is focused for research to improve plant yields and defensive properties.

References

Aramsirirujiwet Y, Gumlangmak C, Kitpreechavanich V (2016) Studies on antagonistic effect against plant pathogenic fungi from endophytic fungi isolated from *hottuynia cordata* thunb. and screening for siderophore and Indole-3-acetic acid production. KKU Res J 21(1):55–66

Arnold AE, Mejía LC, Kyllo D, Rojas EI, Maynard Z, Robbins N, Herre EA (2003) Fungal endophytes limit pathogen damage in a tropical tree. Proc Natl Acad Sci USA 100:15649–15654

Atugala DM, Deshappriya N (2015) Effect of endophytic fungi on plant growth and blast disease incidence of two traditional rice varieties. J Natl Sci Found Sri Lanka 43(2):173–187
Balbi V, Devoto A (2008) Jasmonate signaling network in *Arabidopsis thaliana*: crucial regulatory nodes and new physiological scenarios. New Phytol 177:301–318
Barka AE, Gognies S, Nowak J, Audran J-C, Belarbi A (2002) Inhibitory effect of endophyte bacteria on *Botrytis cinerea* and its influence to promote the grapevine growth. Biol Control 24:135–142
Bhagobaty RK, Joshi SR (2012) Enzymatic activity of fungi endophytic on five medicinal plant species of the Pristine sacred forests of Meghalaya, India. Biotechnol Bioprocess Eng 17:33–40
Bischoff KM, Wicklow DT, Jordan DB, de Rezende ST, Liu S, Hughes SR, Rich JO (2009) Extracellular hemicellulolytic enzymes from the maize endophyte *Acremonium zeae*. Curr Microbiol 58:499–503
Cai G, Wang X (2012) Isolation, identification and bioactivity of endophytic fungi from medicinal plant *Malus sieboldii*. ZhongguoZhong Yao ZaZhi 37(5):564–568
Cecchetti V, Altamura MM, Falasca G, Costantino P, Cardarelli M (2008) Auxin regulates *Arabidopsis* anther dehiscence, pollen maturation, and filament elongation. Plant Cell 20:1760–1774
Chadha N, Prasad R, Varma A (2015) Plant promoting activities of fungal endophytes associated with tomato roots from central Himalaya, India and their interaction with *Piriformospora indica*. Int J Pharm Bio Sci 6:333–343
Clay K, Cheplick GP (1989) Effect of ergot alkaloids from fungal endophyt-infected grasses on fall armyworm (Spodoptera frugiperda). J Chem Ecol 15:169–182
Daisy BH, Strobel GA, Castilo U, Ezra D, Scars J, Weaver D et al (2002) Nepthalene- an insect repellent, is produced by Muscodorvitigenus, a novel endophytic fungus. Microbiology 148:3737–3741
Demain AL (2000) Microbial natural products: a past with a future. In: Wringley SK, Hayes MA, Thomas R, Chrystal EJT, Nicholson N (eds) Biodiversity: a new leads for pharmaceutical and agrochemical industries. The Royal Society of Chemistry, Cambridge, pp 3–16
Doley P, Jha DK (2010) Endophytic fungal assemblages from ethnomedicinal plant *Rauwolfia serpentina* (L) Benth. J Plant Pathol Microbiol 40(1):44–48
dos Banhos EF, de Souza AQL, de Andrade JC, Souza ADL, Koolen HHF, Albuquerque PM (2014) Endophytic fungi from *Myrcia guianensis* at the Brazilian Amazon: distribution and bioactivity. Braz J Microbiol 45:153–161
Fernández LA, Zalba P, Gomez MA, Sagardoy MA (2007) Phosphate-solubilization activity of bacterial strains in soil and their effect on soybean growth under greenhouse conditions. Biol Fertil oils 43:805–809
Fouda AH, Hassan SED, Eid AM, Ewais EED (2015) Biotechnological applications of fungal endophytes associated with medicinal plant *Asclepias sinaica* (Bioss.). Ann Agric Sci 60:95–104
Franken P (2012) The plant strengthening root endophyte *Piriformospora indica*: potential application and the biology behind. Appl Microbiol Biotechnol 96:1455–1464
Frankenberger WT, Arshad M (1995) Phytohormones in soils: microbial production and function. Marcel Dekker, Inc, NewYork, p 503
Gao FK, Dai CC, Liu XZ (2010) Mechanisms of fungal endophytes in plant protection against pathogens. Afr J Microbiol Res 4:1346–1351
Gautam AK, Kant M, Thakur Y (2013) Isolation of endophytic fungi from *Cannabis sativa* and study their antifungal potential. Arch Phytopathol Plant Protect 46:627–635
Giménez C, Cabrera R, Reina M, Coloma-González A (2007) Fungal endophytes and their role in plant protection. Curr Org Chem 11:707–720
Glick BR (1995) The enhancement of plant growth by free-living bacteria. Can J Microbiol 41:109–117
Glick BR, Patten CL, Holguin G, Penrose DM (1999) Biochemical and genetic mechanisms used by plant growth promoting bacteria. Imperial College Press, London, p 267
Goldstein AH (1986) Bacterial solubilization of mineral phosphates: historical perspectives and future prospects. Am J Altern Agric 1:51–57
Gutierrez OA, Wubben MJ, Howard M, Roberts B, Hanlon E, Wilkinson JR (2009) The role of phytohormones ethylene and auxin in plant-nematode interactions. Russ J Plant Physiol 56:1–5

Halford ICR (1997) Soil phosphorus: its measurement, and its uptake by plants. Aust J Soil Res 35:227–239

Hamayun M, Khan SA, Ahmad N, Tang DS, Kang SM, Na CI, Sohn EY, Hwang YH, Shin DH, Lee BH et al (2009) *Cladosporium sphaerospermum* as a new plant growth-promoting endophyte from the roots of *Glycine max* (L.) Merr. World J Microbiol Biotechnol 25:627–632

Hamill JD (1993) Alterations in auxin and cytokinin metabolism of higher plants due to expression of specific genes from pathogenic bacteria: a review. Aust J Plant Physiol 20:405–423

Hasan R (1994) Phosphorus fertility status of soils in India. In: Dev G (ed) Phosphorus researches in India. Potash and phosphate of Canada-India Programme, Gurgaon, pp 7–13

He X, Han G, Lin Y, Tian X, Xiang C, Tian Q, Wang F, He Z (2012) Diversity and decomposition potential of endophytes in leaves of a *Cinnamomum camphora* plantation in China. Ecol Res 27:273–284

Kajula M, Tejesvi MV, Kolehmainen S, Mäkinen A, Hokkanen J, Mattila S, Pirttilä AM (2010) The siderophore ferricrocin produced by specific foliar endophytic fungi *in vitro*. Fungal Biol 114:248–254

Khan AL, Hamayun M, Kim YH, Kang SM, Lee IJ (2011a) Ameliorative symbiosis of endophyte (*Penicillium funiculosum* LHL06) under salt stress elevated plant growth of *Glycine max* L. Plant Physiol Biochem 49:852–861

Khan AL, Hamayun M, Kim YH, Kang SM, Lee JH, Lee IJ (2011b) Gibberellins producing endophytic *Aspergillus fumigatus* sp. LH02 influenced endogenous phytohormonal levels, isoflavonoids production and plant growth in salinity stress. Proc Biochem 46:440–447

Khan AL, Hamayun M, Kang SM, Kim YH, Jung HY, Lee JH, Lee IJ (2012) Endophytic fungal association via gibberellins and indole acetic acid can improve plant growth under abiotic stress: An example of *Paecilomyces formosus* LHL10. BMC Microbiol 12:3

Khan AR, Ullah I, Waqas M, Shahzad R, Hong SJ, Park GS, Jung BK, Lee IJ, Shin JH (2015a) Plant growth-promoting potential of endophytic fungi isolated from Solanumnigrum leaves. World J Microbiol Biotechnol 31(9):1461–1466

Khan AL, Hussain J, Al-Harrasi A, Al-Rawahi A, Lee IJ (2015b) Endophytic fungi: resource for gibberellins and crop abiotic stress resistance. Crit Rev Biotechnol 35:62–74

Klöepper JW, Lifshitz R, Zablotowicz RM (1989) Free living bacterial inocula for enhancing crop productivity. Trends Biotechnol 7:39–44

Korkama-Rajala T, Müller MM, Pennanen T (2008) Decomposition and fungi of needle litter from slow- and fast-growing Norway spruce (*Picea abies*) clones. Microb Ecol 56:76–89

Kumari R, Yadav HK, Bhoon YK, Varma A (2003) Colonization of cruciferous plants by *Piriformospora indica*. Curr Sci 85:1648

Lee K, Pan JJ, May G (2009) Endophytic *Fusarium verticillioides* reduces disease severity caused by *Ustilago maydis* on maize. FEMS Microbiol Lett 299:31–37

Lee YC, Johnson JM, Chien C-T, Sun C, Cai D, Lou L, Oelmüller R, Yeh K-W (2011) Growth promotion of Chinese cabbage and *Arabidopsis* by *Piriformospora indica* is not stimulated by mycelium-synthesized auxin. MPMI 24:421–431

Li JY, Strobel GA, Harper JK, Lobkovsky E, Clardy J (2000) Cryptocin, a potent tetramic acid antimycotic from the endophytic fungus *Cryptosporiopsis quercina*. Org Lett 2:767–770

Li H, Huang H, Shao C, Huang H, Jiang J, Zhu X, Liu Y, Liu L, Lu Y, Li M, Lin Y, She Z (2011) Cytotoxic norsesquiterpene peroxides from the endophytic fungus *Talaromyces flavus* isolated from the mangrove plant *Sonneratia apetala*. J Nat Prod 74:1230–1235

Liang ZN, Zhu H, Lai KP, Chen L (2014) Isolation of endophytic fungi from medicinal plant *Brucea javanica* and their microbial inhibition activity. Zhong Yao Cai 37(4):564–568

Loria ER, Sawyer JE (2005) Extractable soil phosphorus and inorganic nitrogen following application of raw and anaerobically digested swine manure. Agron J 97:879–885

Lu Y, Chen C, Chen H, Zhang J, Chen W (2012) Isolation and identification of endophytic fungi from *Actinidia macrosperma* and investigation of their bioactivities. Evid Based Compl Altern Med, Article ID 382742

Mejía LC, Rojas EI, Maynard Z, Bael SV, Arnold AE, Hebbar P, Samuels GJ, Robbins N, Herre EA (2008) Endophytic fungi as biocontrol agents of *Theobroma* cacao pathogens. Biol Control 46:4–14

Muñoz A, Gandía M, Harries E, Carmona L, Read ND, Marcos JF (2013) Understanding the mechanism of action of cellpenetrating antifungal peptides using the rationally designed hexapeptide PAF26 as a model. Fungal Biol Rev 26:146–155

Nath R, Sharma GD, Barooah M (2012a) Efficiency of tricalcium phosphate solubilization by two different endophytic *Penicillium* sp. isolated from tea (*Camellia sinensis* L.). *European*. J Exp Biol 2(4):1354–1358

Nath R, Sharma GD, Barooah M (2012b) Plant growth promoting endophytic fungi isolated from tea (*Camellia sinensis* L.) shrubs of Assam, India. Appl Ecol Environ Res 13(3):877–891

Neilands JB (1981) Iron absorption and transport in microorganisms. Annu Rev Nutr 1:27–46

Osono T (2006) Role of phyllosphere fungi of forest trees in the development of decomposer fungal communities and decomposition processes of leaf litter. Can J Microbiol 52:701–716

Patten CL, Glick BR (1996) Bacterial biosynthesis of indole-3- acetic acid. Can J Microbiol 42:207–220

Pennisi E (2001) The push to pit genomics against fungal pathogens. Science 292:2273–2274

Pocasangre LE, Sikora RA, Vilich V, Schuster RP (2000) Survey of banana endophytic fungi from Central America and screening for biological control of *Radopholus similis*. Acta Hortic 531:283–289

Poling SM, Wicklow DT, Rogers KD, Gloer JB (2008) *Acremonium zeae*, a protective endophyte of maize, produces dihydroresorcylide and 7-hydroxydihydroresorcylides. J Agric Food Chem 56:3006–3009

Popay AJ, Prestidge RA, Rowan DD, Dymock JJ (1990) The role of *Acremonia lolii* mycotoxins in insect resistance of perennial ryegrass (*Lolium perenne*). In: Quisenberry SS, Joost RE (eds) Proceedings of the 1st international symposium on *Acremonium*/grass interactions. Louisiana Agriculture Experiment Station, Baton Rouge

Prabukumar S, Rajkuberan C, Ravindran K, Sivaramakrishnan S (2015) Isolation and characterization of endophytic fungi from medicinal plant *Crescentia cujete* L. and their antibacterial, antioxidant and anticancer properties. Int J Pharm Pharm Sci 7:316–321

Prasad R, Kamal S, Sharma PK, Oelmüller R, Varma A (2013) Root endophyte *Piriformospora indica* DSM 11827 alters plant morphology, enhances biomass and antioxidant activity of medicinal plant *Bacopa monniera*. J Basic Microbiol 53:1016–1024

Prathyusha P, Rajitha Sri AB, Ashokvardhan T, Satya Prasad K (2015) Antimicrobial and Siderophore Activity of the Endophytic Fungus *Acremonium sclerotigenum* Inhabiting *Terminalia bellerica* Roxb. Int J Pharma Sci Rev Res 16:84–87

Promputtha I, Hyde KD, McKenzie EHC, Peberdy JF, Lumyong S (2010) Can leaf degrading enzymes provide evidence that endophytic fungi becoming saprobes? Fungal Divers 41:89–99

Qadri M, Johri S, Shah BA, Khajuria A, Sidiq T, Lattoo SK, Abdin MZ, Riyaz-Ul-Hassan S (2013) Identification and bioactive potential of endophytic fungi isolated from selected plants of the Western Himalayas. Springerplus 2:8

Radhakrishnan R, Khan AL, Kang SM, Lee I-J (2015) A comparative study of phosphate solubilization and the host plant growth promotion ability of *Fusarium verticillioides* RK01 and *Humicola* sp KNU01 under salt stress. Ann Microbiol 65:585–593

Rajulu MBG, Thirunavukkarasu N, Suryanarayanan TS, Ravishankar JP, Gueddari NEE, Moerschbacher BM (2011) Chitinolytic enzymes from endophytic fungi. Fungal Divers 47:43–53

Raviraja NS (2005) Fungal endophytes in five medicinal plant species from Kudremukh Range, Western Ghats of India. J Basic Microbiol 45:230–235

Redman RS, Freeman S, Clifton DR, Morrel J, Brown G, Rodriguez RJ (1999) Biochemical analysis of plant protection afforded by a nonpathogenic endophytic mutant of *Colletotrichum magna*. Plant Physiol 119:795–804

Richardson AE (1994) Soil microorganisms and phosphorus availability. In: Pankhurst CE, Doube BM, Grupta VVSR, Grace PR (eds) Soil biota, management in sustainable farming systems. CSIRO, Melbourne, pp 50–62

Richardson AE (2001) Prospects for using soil microorganisms to improve the acquisition of phosphorus by plants. Aust J Plant Physiol 28:897–906

Rim SO, Lee JH, Choi WY, Hwang SK, Suh SJ, Lee IJ, Rhee IK, Kim JG (2005) *Fusarium proliferatum* KGL0401 as a new gibberellin-producing fungus. J Microbiol Biotechnol 15:809–814
Rowan DD, Hunt MB, Gaynor DL (1986) Peramine, a novel insect feeding deterrent from ryegrass infected with the endophyte *Acremonium loliae*. J Chem Soc Chem Commun 12:935–936
Sandhu SS, Kumar K, Aharwal RP (2014) Isolation and identification of endophytic fungi from *Ricinus communis* Linn. and their antibacterial activity. Int J Res Pharm Chem 4(3):611–618
Schardl CL, Phillips TD (1997) Grass endophytes. Plant Dis 81:5
Shiva Kameshwari MN, Mohana B, Thara Saraswathi K (2015) Isolation and identification of endophytic fungi from *Urginea indica*, a medicinal plant from diverse regions of South India. Int J Latest Res Sci Technol 4:75–80
Spiering MJ, Greer DH, Schmid J (2006) Effects of the fungal endophyte, *Neotyphodium lolii*, on net photosynthesis and growth rates of perennial ryegrass (Loliumperenne) are independent of In Planta endophyte concentration. Ann Bot 98(2):379–387
Strobel GA, Milln RV, Condron M, Teplow DB, Hess WM (1999) Cryptocandin- a potent antimycotoic from endophytic fungus *Cryptosporiopsis quercina*. Microbiology 145:1919–1926
Tan XM, Chen XM, Wang CL, Jin XH, Cui JL, Chen J, Guo SX, Zhao LF (2012) Isolation and identification of endophytic fungi in roots of nine *Holcoglossum* plants (Orchidaceae) collected from Yunnan, Guangxi, and Hainan provinces of China. Curr Microbiol 64(2):140–147
Tanaka A, Tapper BA, Popay A, Parker EJ, Scott B (2005) A symbiosis expressed non-ribosomal peptide synthetase from a mutualistic fungal endophyte of perennial ryegrass confers protection to the symbiotum from insect herbivory. Mol Microbiol 57:1036–1050
Tomsheck AR, Strobel GA, Booth E, Geary B, Spakowicz D, Knighton B, Floerchinger C, Sears J, Liarzi O, Ezra D (2010) *Hypoxylon* sp., an endophyte of *Persea indica*, producing 1,8-cineole and other bioactive volatiles with fuel potential. Microb Ecol 60(4):903–914
Varma A, Verma S, Sudha Sahay N, Buttehorn B, Franken P (1999) *Piriformospora indica*, a cultivable plant-growth-promoting root endophyte. Appl Environ Microbiol 65:2741–2744
Varma A, Bakshi B, Lou B, Hartmann A, Oelmueller R (2012) *Piriformospora indica*: a novel plant growth-promoting mycorrhizal fungus. Agric Res 1:117–131
Vega FE, Posada F, Aime MC, Pava-Ripoll M, Infante F, Rehner SA (2008) Entomopathogenic fungal endophytes. Biol Control 46:72–82
Waller F, Achatz B, Baltruschat H, Fodor J, Becker K, Fischer M, Heier T, Hückelhoven R et al (2005) The endophytic fungus *Piriformospora indica* reprograms barley to salt-stress tolerance, disease resistance, and higher yield. Proc Natl Acad Sci USA 102:13386–13391
Waqas M, Khan AL, Kamran M, Hamayun M, Kang SM, Kim YH, Lee IJ (2012) Endophytic fungi produce gibberellins and indoleacetic acid and promotes host-plant growth during stress. Molecules 17:10754–10773
Waqas M, Khan AL, Lee IJ (2014a) Bioactive chemical constituents produced by endophytes and effects on rice plant growth. J Plant Interact 9:478–487
Waqas M, Khan AL, Kang SM, Kim YH, Lee IJ (2014b) Phytohormone-producing fungal endophytes and hardwood-derived biochar interact to ameliorate heavy metal stress in soybeans. Biol Fertil Soils 50:1155–1167
Waqas M, Khan AL, Hamayun M, Shahzad R, Kang S-M, Kim J-G, Lee I-J (2015) Endophytic fungi promote plant growth and mitigate the adverse effects of stem rot: an example of *Penicillium citrinum* and *Aspergillus terreus*. J Plant Interact 10:280–287
Webber J (1981) A natural control of Dutch elm disease. Nature 292:449–451
Xie X-G, Dai C-C (2015) Biodegradation of a model allelochemical cinnamic acid by a novel endophytic fungus *Phomopsis liquidambari*. Int Biodeterior Biodegrad 104:498–507
Xie L, Usui E, Narisawa K (2016) An endophytic fungus, *Ramichloridium cerophilum*, promotes growth of a non-mycorrhizal plant, Chinese cabbage. Afr J Biotechnol 15:1299–1305
Zhao J, Shan T, Mou Y, Zhou L (2011) Plant-derived bioactive compounds produced by endophytic fungi. Mini-Rev Med Chem 11(2):159–168

Chapter 8
Fungal Endophytes and Their Secondary Metabolites: Role in Sustainable Agriculture

Anamika, Samiksha Joshi, Manvika Sahgal, Sharda Sahu, and Anil Prakash

Abstract In today's constantly changing scenario, there is an increase in the use of novel and useful bioactive compounds for solving myriad of problems mankind faces, viz. appearance of drug-resistant bacteria, emergence of life-threatening viruses, increasing incidences of fungal infections in the world's population and problems in eliminating food scarcity from some areas of the globe to help human populations. Fungal endophytes though not extensively studied yet are potent source of novel natural products useful in industry, agriculture and medicine. Each of the 300,000 plant species existing on earth is host to one or more endophytes. Till date about one tenth of an estimated one million plant species have been studied for fungal endophytes which are considerably diverse. This chapter deals with the range of bioactive metabolites produced by the fungal endophytes studied so far with emphasis on those useful in increasing food production.

Keywords Endophytic fungi · Bioactive metabolites · Volatile compounds · Tripartite interaction · Signalling molecules

8.1 Functional Grouping of Plant-Endophyte Interaction

Plant-endophyte interactions can be mutualistic, symbiotic and parasitic. Various factors such as endophytic infection pattern, its transmission mode, genetic background and environmental conditions (Saikkonen et al. 1998; Rodriguez and Redman 2008) influence plant-endophyte interaction. Endophytes are transmitted vertically (systemic) and horizontally (nonsystemic). Vertically transmitted endophytes are mutualistic, whereas those transmitted horizontally depict antagonism to the host (Schardl et al. 1991; Saikkonen et al. 1998). It is observed that endophytic

Anamika · S. Joshi · M. Sahgal (✉)
Department of Microbiology, G. B. Pant University of Agriculture & Technology, Pantnagar, Uttarakhand, India

S. Sahu · A. Prakash
Department of Microbiology, Barkatullah University, Bhopal, India

P. Gehlot, J. Singh (eds.), *Fungi and their Role in Sustainable Development: Current Perspectives*, https://doi.org/10.1007/978-981-13-0393-7_8

fungi become pathogenic upon leaf ageing or senescence and thereafter become more widespread causing external infections. Genetic makeup of the host determines the relationship of fungi with the host (Redman et al. 2001; Unterseher and Schnittler 2010). Therefore, the type of interaction between plant and fungi is determined by the differential expression of fungal genes in response to plant host or vice versa. Thus, symbiosis may be positive, negative or neutral based on minor genetic variations in the genomes of both partners (Moricca and Ragazzi 2008). Environmental variations (Moricca and Ragazzi 2008), imbalance in nutrient exchange (Kogel et al. 2006) and physiological stress and senescence (Halmschlager et al. 1993) also affect the plant-fungi interaction. Many a time expression of disease in host plant infected by endophyte is due to miscommunication rather than aggressive pathogenicity (Rai and Agarkar 2014). Plants colonised with endophytic fungi are found to be well adapted to both abiotic and biotic stresses and tolerant towards high temperature and salinity (Arnold et al. 2003; Zhang et al. 2006; Akello et al. 2007). Tolerance of plants with endophytic colonisation towards biotic stress is attributed to production of secondary metabolite by fungal partner (Saikkonen et al. 1998; Zhang et al. 2006; Tan and Zou 2001; Rocha et al. 2011). In conclusion, mutualistic interactions between endophyte and plant host are under environmental, physiological and genetic control (Kogel et al. 2006). Natural products of endophytes are antimicrobial and protect host plant against various microbial phytopathogens (Gunatilaka 2006). The production of flavonoids and other phenolic antioxidants was increased in endophyte-colonised plants (Herrera-Carillo et al. 2009; Huang et al. 2007; Torres et al. 2009). ROS produced by fungal hyphae oxidises and denatures the host cell causing leakage of nutrients across plant cells and their subsequent absorption by endophyte (White and Torres 2010). Isobenzofurans (Harper et al. 2003), isobenzofuranones (Strobel et al. 2002) and carbohydrate and mannitol (Richardson et al. 1992) are also produced by endophytes and enhance stress tolerance in host plants (Huang et al. 2007; White and Torres 2010).

Stress tolerance conferred by endophytes is habitat-adapted phenomenon. *Dichanthelium lanuginosum* dominantly colonised by *Curvularia protuberata* provides heat tolerance to the plant and allows it to grow in geothermal soils at Yellowstone National Park, USA. When grown in axenic conditions and exposed to heat stress >38 °C, both the symbiotic partners died (Redman et al. 2002). It is also observed that *Leymus mollis* becomes salt tolerant upon colonisation with *Fusarium culmorum*. Thus, many times symbiotic association of fungi with plant helps it to sustain stress (Rodriguez and Redman 2008).

Endophytic colonisation modifies biochemical pathways of host plant leading to enhanced production of growth hormones, viz. cytokinins, indole-3- acetonitrile and indole-3-acetic acid (Tan and Zou 2001; Zhang et al. 2006). Sometimes nutrient absorption by the host is also increased. The metal sequestration, chelation and degradation abilities of endophytes help plant survive in soils contaminated with metals (Weyens et al. 2009). Thus endophytic colonisation enhances fitness and competitive abilities of the plant host. On the other hand, plant hosts protect endophytes from desiccation, provide nutrient support and spatial structure (Saikkonen et al. 1998) and allow their transfer to the next generation of hosts (Rudgers and

Strauss 2004; Faeth and Fagan 2002). Plants also provide essential nutrients required for growth, self-defence and completing life cycle of the endophytes (Metz et al. 2000; Strobel et al. 2002).

Fungal endophytes play a pivotal role in degradation of a dead host plant (Oses et al. 2008) and thus are important part of nutrient cycling (Zhang et al. 2006; Boberg et al. 2011; Parfitt et al. 2010; Sun et al. 2011; Vega et al. 2010).

8.2 Role of Secondary Metabolites Within the Host

Tan and Zou (2001) reviewed the function of metabolites from endophytic fungi and emphasised that they play significant role in ecological interactions. The plants in which roots are colonised by endophytes synthesise various plant growth-promoting substances and phytohormones and hence grow faster than noninfected ones (Petrini 1991; Tudzynski and Sharon 2002; Tudzynski 1997). They also depict enhanced tolerance to environmental stresses (Schulz et al. 1999). These plants show better ecological adaptation due to production of antimicrobials against phytopathogens (Schulz et al. 1995, 2002) and predators (Azevedo et al. 2000; Liu et al. 2001). About 80% of endophytic fungi produce bioactive compounds with antimicrobial and herbicidal properties (Schulz et al. 2002). The grasses infected with *Neotyphodium/Epichloë* produced alkaloids (Leuchtmann 1992). The maize plants infected with *Fusarium verticillioides* produced fumonisin (White et al. 2000; Schardl and Phillips 1997; Miller 2001). It is well established that secondary metabolites have an ecological significance and play a major role within the host. Demain (1980) presumed that fungus which is an efficient producer of metabolites *in vitro* must also play an important role in ecology. Several workers studied role of secondary metabolites during endophyte-host interaction through dual culture assay in which endophytes were confronted with their host callus, for example, *Lamium purpureum* with *Coniothyrium palmarum*, *Phaseolus vulgaris* with *Fusarium* sp. and *Teucrium scorodonia* with *Phomopsis* sp. (Peters et al. 1998a; Gotz et al. 2000), the callus degenerated following necrosis without coming in contact with fungi indicating that metabolites secreted by fungal endophytes into the medium were toxic to the callus. The similar response is generated when the callus is confronted with pathogen (Peipp and Sonnenbichler 1992). The necrosis was also observed in callus during various interactions upon addition of endophytic extracts into the growth medium (Peters et al. 1998b). In addition to culture extracts, purified secondary metabolites from *Coniothyrium palmarum* (*palmarumycins*; Krohn et al. 1994, 1997) and *Phomopsis* sp. (*phomopsins* and *biarylethers*; Krohn et al. 1996) were equally toxic to host and non-host seedlings (Peters et al. 1998b) indicating that their action is non-specific. Almost 94% of the secondary metabolites extracted from endophytic cultures depict non-specificity (Krohn et al. 1994, 1996, 1997; Schulz et al. 1999, 2002).

The secondary metabolites of non-balansiaceous endophytic fungi are structurally diverse (Schulz et al. 2002). The concentration of secondary metabolites secreted

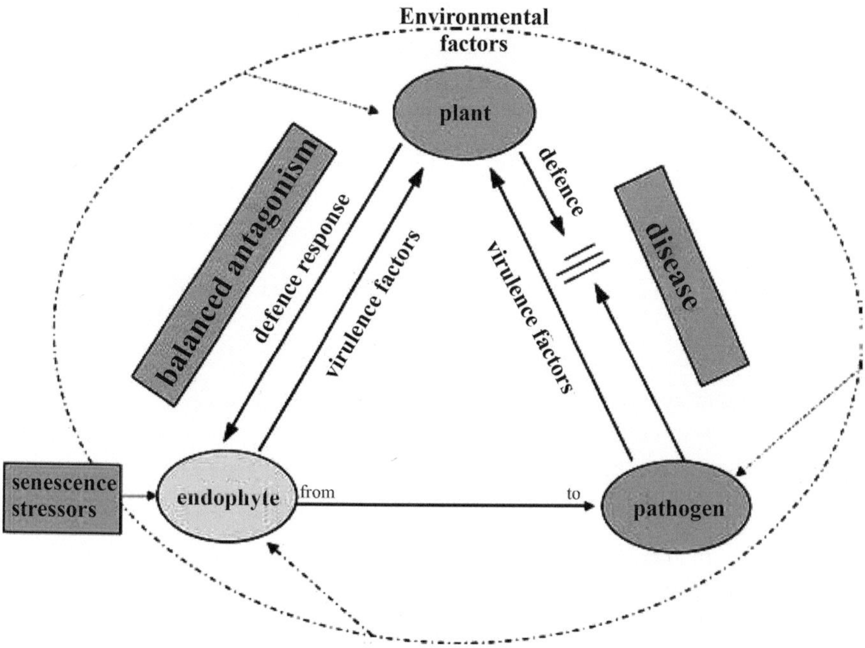

Fig. 8.1 A balance of antagonisms between endophytic virulence and plant defence response results in asymptomatic colonisation (Schulz and Boyle 2005)

in vivo is inadequate for the expression of severe disease but helps in its survival in an ecosystem. This has been validated through histological studies (Boyle et al. 2001; Deckert et al. 2001) which indicate limited colonisation of the above-ground organs. Probably, a fine balance between host-endophyte interactions hinders pathogenesis. There is no disease development as long as there is balance between fungal virulence and host defence response (Fig. 8.1; Schulz and Boyle 2005).

Peter and co-workers (1998a, b) revealed that culture extracts of endophytic fungi depict algicidal and herbicidal action. They tested the effect of endophyte culture extract on the oxygen production in *Chlorella fusca* and observed that extracts with anti-algal property inhibited photosynthesis and not respiration. Similarly, inhibition of photosynthesis was also observed in banana and maize colonised by *Fusarium verticillioides* and *Colletotrichum musae*, respectively (Pinto et al. 2000).

8.3 Root Endophytes and Nutrient Uptake

Root endophytes are abundant, phenotypically plastic and taxonomically diverse like foliar endophytes. Their ecological role is similar to saprotrophic, pathogenic and mycorrhizal fungi present in soil. In the late 1990s, significance of root-associated fungal endophytes in nutrient uptake has been recognised (Jumpponen and Trappe 1998). Endophytic fungi belonging to order *Sebacinales* are mycorrhizal as well as endophyte of several angiosperms (Selosse et al. 2009), but their role in nutrient uptake was not known until Ngwene and co-workers (2016) isolated, characterised and identified *Piriformospora indica* in the order *Sebacinales* from Indian desert. This endophytic fungus showed plant growth-promoting properties and exhibited tolerance to stress as well as ability to solubilise significant amounts of phosphorous (P) from inorganic sources such as Ca_3 $(PO_4)_2$ and rock phosphate (Ngwene et al. 2016). They further demonstrated that P solubilisation in this fungus is not due to enzymatic activities but rather lowering pH of the medium. All genes involved in P solubilisation were repressed in the presence of higher amounts of inorganic phosphorus and expressed when organic phosphorous compound phytate was added to the medium. The solubilisation of phosphorous was also studied at transcriptome level. There was accumulation of RNA from P-solubilising genes, but no intra- or extracellular enzymatic activity was detected. Moreover all genes involved in P solubilisation were repressed in higher amounts of inorganic phosphorus and were expressed when phytate was added to the medium.

8.4 Fungal Endophytes as Source of Bioactive Products

Bacteria and fungi are most common microbes existing as endophytes. However, there is no evidence of other microbial forms, such as mycoplasmas and archaebacteria existing as endophytes. Of these, the most frequently isolated endophytes are the fungi. Fungal endophytes represent a relatively untapped source of microbial diversity that produces novel natural products for exploitation in agriculture, industry and medicine. It is noteworthy that, of approximately 300,000 plant species existing on earth, each is host to one or more endophytes. The endophytic biology of very few plants has been studied in detail. Thus, there are enormous possibilities for the recovery of novel fungal forms, taxa and biotypes. Only about 10% of one million different fungal species have been described as yet (Strobel and Daisy 2003). Presently major challenges mankind faces are environmental degradation, loss of biodiversity and spoilage of land and water. Secondary metabolites may play major role in nature. Endophytes secrete a large number and diversity of bioactive compounds possessing a range of activities. Major categories of fungal natural products are antibiotics, antioxidants, insecticidal, antidiabetic and immunosuppressive agents belonging to various chemical classes.

8.4.1 Amines and Amide Alkaloids

Acremonium, grass endophytes and *Neotyphodium* (earlier known as *Epichloë*) secrete secondary metabolites belonging to amines and amides (Glenn et al. 1996; Schardl and Phillips 1997). Peramine, a pyrrolopyrazine alkaloid, is produced by *Neotyphodium coenophialum*, *N. lolli*, *Epichloë festucae* and *E. typhina* in culture as well as *in planta* when associated with stem, and leaf of tall fescue, ryegrass and other grasses is toxic to insects like Argentine stem weevil without any harmful effects on mammals (Rowan and latch 1994; Dew et al. 1990). An endophytic *Phoma* sp. produces 1-*N*-methylalbonoursin inhibiting ras farnesyl transferase (Ishii et al. 2000). *Rhinocladiella* sp., endophyte of perennial vine *Tripterygium wilfordii*, produces cytochalasin E (Wagenaar et al. 2000). A novel metabolite phomopsichalasin, a cytochalasin with an isoindolone moiety fused to a 13-membered tri cyclic ring, is produced by an endophyte *Phomopsis* sp., recovered from twigs of *Salix gracilistyla* var. *melanostachys* (Horn et al. 1996). This metabolite possesses antimicrobial activity against various bacterial and fungal species, for example, *Bacillus subtilis*, *Salmonella gallinarum*, *Staphylococcus aureus* and *Candida albicans*, and human pathogenic yeast.

8.4.2 Indole Derivatives

The culture of *Neotyphodium* secretes indole alkaloids such as agroclavine, chanoclavine and elymoclavine (Powell and Petroski 1992), toxic to a few insects and mammals (Schardl and Phillips 1997). *Phomopsis* sp., isolated from the bark of living *Cavendishia pubescens* tree (Bills et al. 1992), and *Aspergillus flavus* indole derivatives such as tremorgenic aspalitrems A and C are also secreted by an endophytic fungi. Endophytic fungi like *Aureobasidium pullulans*, *Acremonium coenophialum*, *Colletotrichum* sp. and *Epicoccum purpurascens* present in *Artemisia annua* produce indole-3-acetic acid. *Hypoxylon serpens* strains, isolated from tobacco, produce indole-3-acetonitrile and cytokinins along with IAA (Petrini et al. 1991). Recently a novel indole derivative, 6-isoprenylindole-3-carboxylic acid, was characterised from *Colletotrichum* sp. endophytic with *A. annua*. It is antagonistic against pathogenic bacteria as well as fungi. It shows moderate antibacterial activity against *Bacillus subtilis*, *Pseudomonas* sp., *Sarcina lutea* and *Staphylococcus aureus*. It also depicts antifungal activity against phytopathogenic fungi: *Gaeumannomyces graminis* var. *tritici*, *Phytophthora capsici* and *Rhizoctonia cerealis* (Lu et al. 2000).

8.4.3 Pyrrolizidines

These are alkaloid compounds with pyrrolizidine as a central structure. These are produced by plants in families Asteraceae, Boraginaceae, Convolvulaceae, Fabaceae, Lamiaceae, Orchidaceae and Poaceae as a self-defence against insect pests. Grasses infected with endophytic fungi secrete lolines, the saturated 1-aminopyrrolizidine. It is reported that lolines were neither produced in noninfected grasses nor endophytic cultures *in vitro* (Schardl and Phillips 1997). They were secreted only during interaction of endophyte with plant host. Lolines are broad-spectrum insecticides, acting both as feeding deterrents and metabolic toxins. The allelopathic properties of host grasses are attributed to certain loline analogues secreted by endophytes (Bush et al. 1997). For example, *Festuca arundinacea* infected with *N. coenophialum* and *F. pratensis* infected with *N. uncinatum* characteristically secrete lolines. Lolines are less toxic as compared to ergot and indole diterpene alkaloids.

8.4.4 Sesquiterpenes

Several sesquiterpenes have been identified from the endophytic fungi associated with different plants. For example, 2α-hydroxydimeninol and pestalotiopsins A–C have been identified from *Pestalotiopsis* spp. endophytic with *T. brevifolia* (Pulici et al. 1996a, b). Several sesquiterpenes show toxicity against insect pests. The hydroheptelidic and heptelidic acid isolated from *Phyllosticta* sp., a fungal endophyte of *Abies balsamea*, is toxic to larvae of spruce budworm (*Choristoneura fumiferana*) (Calhoun et al. 1992). Two new benzofuran-carrying normonoterpene derivatives, toxic to spruce budworm larvae, have been characterised from fungus, endophytic on *Gaultheria procumbens* (Findlay et al. 1997). *Epichloë typhina*, endophytic on *Phleum pratense*, produces sesquiterpenes chokols A–G, fungitoxic to *Cladosporium phlei*, a leaf spot disease pathogen (Koshino et al. 1989).

8.4.5 Steroids

Fungus *Colletotrichum* sp., an endophyte of *A. annua*, secretes steroids such as ergosterol, 3β,5α,6β-trihydroxyergosta-7,22-diene, 3β-hydroxyergosta-5-ene, 3-oxoergosta-4,6,8(14),22-tetraene, 3β-hydroxy-5α,8α-epidioxyergosta-6,22-diene, 3β-hydroxy-5α,8α-epidioxyergosta-6,9(11),22-triene, 3-oxoergosta-4-ene, 3β,5α-dihydroxy-6β- acetoxyergosta-722-diene and 3β,5α-dihydroxy-6β-phenylacetoxyergosta-7, 22-diene. The above-mentioned metabolites were antagonistic to crop pathogens, viz. *Gaeumannomyces graminis* var. *tritici*, *Helminthosporium sativum*, *Phytophthora capsici* and *Rhizoctonia cerealis* (Lu et al. 2000).

8.4.6 Terpenoids

Plant hosts colonised by fungi are known to secrete terpenoids mainly sesqui- and diterpenes.

8.4.7 *Diterpenes*

The endophyte of balsam fir, *Abies balsamea*, secretes insecticidal toxins with pimarane diterpene framework (Findlay 1995). Diterpene subglutinol A and B have been characterised from *Fusarium subglutinans*, endophytic on *Tripterygium wilfordii*, a perennial twining vine (Lee et al. 1995). An unidentified fungus isolated from *Daphnopsis americana* growing in Guanacaste, Costa Rica, produced guanacastepene, a novel diterpenoid that is antibacterial against methicillin-resistant *Staphylococcus aureus* and vancomycin-resistant *Enterococcus faecium* (Singh et al. 2000). Taxol, an efficient anticancer diterpene originally characterised from the inner bark of *Taxus brevifolia*, a pacific yew growing in Montana, USA, is (Wani et al. 1971; Georg et al. 1994) produced by fungus *Taxomyces andreanae* endophytic on plant *T. brevifolia*. Fungus *T. andreanae* produced Taxol in vitro (Stierle et al. 1993). Since then, diverse endophytic fungi isolated from *T. brevifolia* (Stierle et al. 1995), *T. wallichiana* (Strobel et al. 1996), *T. yunnanensis* (Qiu et al. 1994), *T. baccata* (Caruso et al. 2000), *T. mairei* (Wang et al. 2000), *Taxodium distichum* (Li et al. 1996), *Torreya grandifolia* (Li et al. 1998) and *Wollemia nobilis* (Strobel et al. 1997) were also reported to produce Taxol or taxane derivatives.

8.4.8 *Isocoumarin Derivatives*

Several isocoumarin derivatives produced by endophyte-colonised plants exhibit biocontrol potential. *Pezicula* sp. produces (*R*)-mellein (Schulz et al. 1995) that is strongly algicidal, fungicidal and herbicidal. *E. typhina* an endophyte from *P. pratense* produces an isocoumarin compound gamahorin (Koshino et al. 1992). The isocoumarin-like metabolites, identified from endophytes of conifers, were toxic to spruce budworm larvae and cells (Findlay et al. 1995a, b).

8.4.9 *Quinones*

Rugulosin, an insecticidal compound secreted by *Hormonema dematioides*, an endophytic fungus of balsam fir is a quinone (Calhoun et al. 1992). Similarly, a quinone 1′,5′-trihydroxy-3′,4′dihydro-1′*H*-[2,4′]binaphthalenyl-1,4,2′-trione toxic to spruce budworm larvae is secreted by an unidentified endophyte of *Larix laricina* needle (Findlay et al. 1997). Further, altersolanol A, a highly hydroxylated quinone, exhibiting antibacterial properties is secreted by *Alternaria* spp. and *Phoma* sp. when in association with plants (Yang et al. 1994). An endophytic *Coniothyrium* sp. produces preussomerin N_1, palmarumycin CP_{4a} and palmarumycin CP_5 that are ras farnesyl protein transferase inhibitors (Krohn et al. 2001).

8.4.10 Flavonoids

Blue grass (*Poa ampla*) infected with an endophyte produces tricin a flavone glycosides, toxic to mosquito larvae (Ju et al. 1998).

8.4.11 Peptides

Cryptocandin, a cyclopeptide with potent antifungal property, is produced by *Cryptosporiopsis* cf. *quercina* endophyte of redwood produces (Strobel et al. 1999). Cyclopeptides echinocandins A, B, D and H are produced by *Aspergillus rugulosus*, *A. nidulans* var. *echinulatus*, *Cryptosporiopsis* sp. and *Pezicula* sp. endophytic on *Pinus sylvestris* and *Fagus sylvatica* (Benz et al. 1974; Traber et al. 1979). Leucinostatin A, an oligopeptide with anticancer, phytotoxic and antifungal properties, originally identified from *Penicillium lilacinum* (Arai et al. 1973) is also detected from culture of *Acremonium* sp., a fungal endophyte of *Taxus baccata* (Strobel et al. 1997).

8.4.12 Phenolic Compounds

Endophytes are known to produce biologically active phenolic compounds. For example, *Phoma* sp. produces 2-hydroxy-6-methylbenzoic acid, a compound with antibacterial activity (Yang et al. 1994). *Pezicula* sp. strain 553, a tree endophyte, produces 2-methoxy-4-hydroxy-6-methoxymethyl benzaldehyde, which exhibited antifungal property against *Cladosporium cucumerinum*, a phytopathogen (Schulz et al. 1995). Tyrosol, *p*-hydroxyphenylacetic acid, *p*-hydroxybenzoic acid and *cis*- and *trans*-*p*-coumaric acids are antifungal phenolic acids isolated from stromata of *E. typhina*, an endophyte of *P. pratense* (Koshino et al. 1988). Colletotric acid, an antimicrobial tridepside, is secreted by *Colletotrichum gloeosporioides*, a fungal endophyte of *Artemisia mongolica* (Zou et al. 2000). *Cytonaema* sp., an endophyte recovered from *Quercus* sp., secretes cytonic acids A and B (Guo et al. 2000).

8.4.13 Phenylpropanoids and Lignans

Phenylpropanoid and lignan compounds were identified from stromata of fungus *Epichloë typhina*, endophytic on *Phleum pratense* (Koshino et al. 1988). Several lignans are crucial for establishing fungus-plant interaction (Chapela et al. 1991). For example, coniferin and syringing, two monolignol glucosides, are critical in establishing interaction between fungal species and *Xylariaceae* species.

8.4.14 Aliphatic Compounds

Various endophytic fungi secrete aliphatic compounds, e.g. stromata of *E. typhina* endophytic on *P. pratense* (Koshino et al. 1989), and an endophyte of the eastern larch secretes esoteric metabolite exhibiting antibacterial activity against *Pseudomonas aeruginosa*, *Staphylococcus aureus* and *Vibrio salmonicida* (Findlay et al. 1997). Aliphatic compounds like phomodiol and homoproline B are produced by *Phomopsis* spp., an endophyte of genus *Salix* and non-willow plants (Horn et al. 1996).

8.4.15 Chlorinated Metabolites

Three chlorinated metabolites have been isolated from cultures of balsam fir needle endophyte *Phyllosticta* sp. strain 76 (Calhoun et al. 1992) and tree endophytes *Pezicula* sp. and *P. livida* strain 1156 (Schulz et al. 1995), respectively.

8.4.16 Miscellaneous

Pentaketide, an antifungal metabolite, is produced by *Fusarium* sp., an endophyte of *Selaginella pallescens* stem (Brady and Clardy 2000). *Aspergillus parasiticus*, an endophyte of redwood, produces sequoiatones A and B possessing antitumor properties (Stierle et al. 1999). *Pestalotiopsis* spp., endophytic fungi associated with *Taxus brevifolia*, produce monoterpenes and C-methylated acetogenins (Pulici et al. 1997). The culture of a *Chaetomella acutisea* (MF5686), an endophytic fungus, secretes chaetomellic acids A and B that specifically inhibit farnesyl protein transferase (FPTase) (Lingham et al. 1993).

8.5 Volatile Organic Compounds

Fungal volatile organic compounds (VOCs) are derived from both primary and secondary metabolic pathways (Korpi et al. 2009). They are low-molecular-weight lipophilic compounds that are readily vaporised at 0.01 KPa and 20 °C (Pagans et al. 2006). VOCs act as signalling molecules in short- and long-distance intercellular or organismal interactions as they can move from the point of secretion to distant places through air, liquids and porous soils and hence are also known as "infochemicals". Approximately 250 volatile organic compounds have been identified from fungi. These are placed in different classes and occur as a mixture of aldehydes, alcohols, cyclohexanes, simple hydrocarbons, heterocycles, indole and their derivatives, ketones, phenols, thioalcohols, thioesters and benzene derivatives (Chiron and Michelot 2005; Ortiz-Castro et al. 2009; Korpi et al. 2009). Generally fungi emit mixture of VOCs. The qualitative and quantitative composition of VOCs is dependent on fungal species and environmental conditions provided for its growth (Effmert et al. 2012). VOCs produced by majority of fungal groups are antibiotics, and a few root strongly affects atmospheric chemistry, terrestrial carbon dynamics and belowground ecology. Endophytic fungi belonging to *Ascomycota* lineages and *Xylariaceae* family are capable of producing VOCs (Lee et al. 2009). A few basidiomycetous members also produce VOCs. MVOCs can serve as alternative to bactericides, fungicides and harmful pesticides and hence can play key role in sustainable agriculture. The volatile metabolites are involved in microbial interactions. Within few ecosystems, bacterial or fungal VOCs also stimulate biotic aggregations (Davis et al. 2013) which in turn prompt coevolution. Production of MVOCs sometimes leads to complex interactions at trophic level. They act as chemical windows providing insights of molecular regulation of microbial activities (Korpi et al. 2009; Liang et al. 2008; Thorn and Greenman 2012) and also as environmental marker for bacteria and fungi.

8.5.1 VOCs of **Muscodor** *Species*

Genus enters into obligate endophytes and exhibits comprehensive spectrum of antimicrobial activity. There is variation in the volatile organic compounds produced by *Muscodor* species depending on the plant species on which it is a symbiont. A number of *Muscodor* species associated with different plant species produce various VOCs like azulene, aromadendrene, β-caryophyllene, 2-methylfuran, naphthalene, tetrahydrofuran, α-phellandrene, β-phellandrene and 2-pentylfuran when growing on plant hosts such as *Actinidia chinensis*, *Ananas ananassoides*, *Ginkgo biloba* and *Myristica fragrans* (Macias-Rubalcava et al. 2010; Yuan et al. 2012). A *Muscodor* species, *Phialocephala fortinii*, produces *β*-caryophyllene, a VOC.

8.5.2 VOCs of Truffles

Truffles produce more than 200 VOCs during different growth stages, i.e. presymbiotic, symbiotic (mycelial as well as mycorrhizal) and reproductive stage (Splivallo et al. 2011). *Tuber magnatum*, *T. melanosporum* and *T. borchii* are three model species used for studying the profile of volatiles from truffles. A volatile compound, 2-octenal, has been identified in culture filtrates of *T. borchii*, *T. indicum* and *T. melanosporum*. Other compounds like 2-methylbutanal, dimethyl sulphide, DMDS and 3-methylbutanal have been found in most truffles studied till now, whereas 2-methyl,4,5-dihydrothiophene has been found only from fruiting bodies of *T. borchii*. Volatile compound bis(methylthio)methane along with methyl(methylthio) methyldisulphide, dimethyl sulphide, methanethiol, benzothiazole and terpenoids including limonene, guaiene, p-cymene, carveol and cumene hydroperoxide contribute to aroma of white truffle.

8.5.3 VOCs Emitted by Other Fungi

Fungi belonging to genera *Epichloë* (Schiestl et al. 2006), *Puccinia* and *Uromyces* (rust fungi) (Kaiser 2006), *Tuber* spp.(truffles) (Splivallo et al. 2011) and *Trichoderma* sp. that are soil saprophyte *and* mushroom sporocarps secrete volatile compounds (Yuan et al. 2012; Fraatz and Zom 2010). In addition fungus *Ceratocystis fimbriata* isolated from building materials, diseased plants and wood (Hung et al. 2013) also secretes volatile organic compounds. An endophytic fungus *Phleum pratense* isolated from *Epichloë typhina* is reported to produce sesquiterpenes, chokols A–G, which are toxic to *Cladosporium phlei*, a leaf spot disease pathogen (Kumar and Kaushik 2012). *Phaeosphaeria nodorum*, a leaf endophytic fungus of plum (*Prunus domestica*), produces several volatiles identified as 2-propyl-1-ol, acetic acid, ethyl acetate, 3-methylbutan-1-ol and 2-propenenitrile. These volatiles were antagonistic against *Monilinia fructicola* and inhibited the growth of fungus by disintegrating hyphal content (Pimenta et al. 2012). The major volatile compounds from *Ampelomyces* sp. and *Cladosporium* sp. are *m*-cresol and methyl benzoate, respectively. These elicit induced systemic resistance in *Arabidopsis* against *Pseudomonas syringae* pv. *tomato* DC3000 (Naznin et al. 2014). *Fomes fomentarius* is reported to emit 1-octen-3-ol, 3-octanone and 3-octanol, all inducing behavioural changes in *Bolitophagus reticulatus*, a vorous beetle during olfactometer bioassays (Holighaus et al. 2014). *Penicillium expansum* secreted styrene and 3-methylanisole, volatile metabolites that significantly reduced attraction of pine weevil's (*Hylobius abietis*) towards cut pieces of Scots pine twigs, whereas 3-methylanisole reduced the attraction of male weevil to pine twigs (Azeem et al. 2013). Several microbial volatile compounds have been reported to show plant growth promotory influence. For example, VOCs of *Trichoderma viride*, isobutyl alcohol, isopentyl alcohol, 3-methylbutanal, α-bergamotene, bicyclogermacrene, farnesene, geranylacetone, β-sesquiphellandrene, valencence, α-ylangene and zingiberene (Hung et al. 2013; Muller et al. 2013), enhanced plant height and plant vigour and prompted lateral rooting and early flowering in *Arabidopsis*. It is reported that VOC, 1-octen-3-ol, enhanced plant resistance to the necrotrophic fungal pathogen *Botrytis cinerea* by inducing defence signalling cascades (Contreras-Cornejo et al. 2014; Kishimoto et al. 2007). Similarly, volatiles of *Alternaria alternata*, *Penicillium aurantiogriseum* and *Penicillium charlesii* promote growth and starch accumulation in several plant species (Ezquer et al. 2010). A non-pathogenic strain of *Fusarium oxysporum*, MSA35, associated with ectosymbiotic bacteria secretes volatiles that are involved in growth promotion of lettuce (*Lactuca sativa*) (Minerdi et al. 2009, 2011). This volatile compound was identified as β-caryophyllene, a sesquiterpene produced by the ectosymbiotic bacterial species (Minerdi et al. 2011).

The endophytic fungus NRRL50072, isolated from *Eucryphia cordifolia* in northern Patagonia, produces volatile compounds like ethyl acetate, heptanes, 3-methylbutan-1-ol, 2 methylhexanoate and 2 pentene with potential for use as myco-diesel (Griffin et al. 2010). *Streptomyces alboflavus* emit volatiles like 1, 4-dimethyladamantane, 3-methylisoborneol and 1, 2, 3-trimethylbenzene which exhibit inhibition against storage fungi *Aspergillus flavus*, *A. niger*, *A. ochraceus*, *F. moniliforme* and *Penicillium citrinum* in vitro.

Saprophytic species *Stropharia rugosoannulata* emits α-muurolene, 3-phenylpropan-1-olandγ-selinene. Yeast strain *Saccharomyces cerevisiae* CR-1 releases 3-methylbutan-1-ol, 2-methyl butan-1-ol, 2-phenyl ethanol and ethyl octanoate as volatile organic compounds that inhibited the production of enzymes involved in morphogenesis in turn inhibiting the mycelial development (Fialho et al. 2011). *Marasmius crinis-equi* produced antifungal volatiles 3-oxo-β-ionol and 2-phenyl-3,4,5,6-tetramethylpyridine (Su et al. 2012).

8.6 MVOCs: Advantage Over Chemical Pesticides

Microbial volatile organic compounds are effective at low concentrations, biodegradable and nonhazardous as compared to synthetic pesticides or fertilisers. The biocontrol potential of MVOCs is just beginning to be demonstrated and exploited (Song and Ryu 2013). There is a need to perform more physiological, molecular and field studies to demonstrate the full potential of MVOCs in sustainable crop protection. MVOCs can be formulated for safe handling, storing and delivery to fields. A few volatile metabolites of microbes may be structurally novel but pose health hazard to humans upon long-term exposure (Korpi et al. 2009). Thus, profiling of endophyte-origin VOCs for toxicity to human health is mandatory for registration and certification standards. For example, 1-octen-3-ol, a volatile compound detected in mould and mushroom odours, is cytotoxic to human embryonic stem cells (Inamdar et al. 2012). Similarly, the biosecurity of volatiles used as fumigants to control fruit- or soil-borne disease should also be monitored as there are instances when endophytes themselves are found to be latent plant pathogens.

8.7 Fungal Endophytes Producing Anti-insect Metabolites

Fungi that infect and kills insect play fascinating role in agriculture. The use of agrochemicals reduced the attack of insect pests and phytopathogenic microbes but at the same time poses a high risk to field workers and consumers. Moreover, the excess use of chemical fungicides and pesticides is harmful as well as economically unviable. Biological control of pests and diseases using entomopathogenic and antagonistic microorganisms is an alternative to reduce or eliminate the use of agrochemicals in agriculture. In the early 1980s, there were first reports of endophytic microorganisms, mainly fungi, playing an important role inside plants and their presence in plant host resulting in the reduction of insect attacks. There are more than 100 fungal species depicting insect pathogenic property. They are diverse and fall in broad taxonomic groups ranging from chytridiomycetes to basidiomycetes (Madelin 1963).

8.7.1 Anti-insect Metabolites

The anthraquinone metabolite, rugulosin, isolated for the first time from endophyte *A. balsamea* and later from *Picea scopiformis* and *Picea glauca* resulted in growth rate reductions of *Choristoneura fumiferana*, *Lambdina fiscellaria* and *Zeiraphera canadensis* (Sumarah et al. 2008). Three sesquiterpene lactones, heptelidic acid (HA) and their derivatives HA chlorohydrins and hydro-HA isolated from *A. balsamea*, were shown to be toxic to *C. fumiferana* larvae in concentrations ranging from 5 to 15 μM. Several toxic isocoumarins and related metabolites were isolated and characterised from *Conoplea elegantula* endophytic on *P. mariana* (Findlay et al. 1995). Recently, these compounds were also isolated from *Mycosphaerella* spp. endophytic on *Picea rubens* (Crous et al. 2007). *L. laricina* was the source of two novel metabolites, out of which one was toxic to *C. fumiferana* and was probably produced through pentaketide dimerisation biogenesis, whereas another was antimicrobial (Findlay et al. 1997). *P. mariana*, an endophyte producing terreic acid, showed moderate toxicity to *C. fumiferana* cells and larvae (Findlay et al. 1996). An endophyte form *P. glauca* DAOM 221611 was reported to produce macrocyclic antibiotic vermiculin showing toxicity to *C. fumiferana* and cytotoxic to HeLa cells (Horakova and Betina 1976).

8.7.2 Endophyte-Herbivore Interactions

Fungal endophytes have been reported to provide protection against herbivorous insects (Azevedo et al. 2000), plant diseases (Arnold et al. 2003; O'Hanlon et al. 2012) or plant parasitic nematodes (Waweru et al. 2014). Clavicipitaceous fungus which forms an intimate association with grasses produces specific alkaloid compounds that are toxic to insects. Most species of entomopathogenic fungi belong to two divisions – *Ascomycota* and *Zygomycota* (Roy et al. 2006). Biological control of insect pests of crop plants was reported for the first time when higher mortality and reduction in larval population of the European corn borer (*Ostrinia nubilalis* Hbn.) were observed in corn plants endophytically colonised by *Beauveria bassiana* (Balsamo-Crivelli) Vuillemin (BB). Similarly, in corn and sorghum plants inoculated with BB isolates, there was a reduction in tunnelling symptoms induced by lepidopteran larvae (*Ostrinia nubilalis* or *Sesamia calamistis*) (Bing and Lewis 1991; Reddy et al. 2009; Cherry et al. 1999, 2004). The banana plants treated with

BB isolate were highly vulnerable to root borer attack. Mycosis was observed in both egg and adult (Akello et al. 2008). The gall wasp *Iraella luteipes* larva abundance in opium poppy treated with endophytic BB isolate was reduced up to 73% (Quesada-Moraga et al. 2006). Further it was reported that larvae of *Liriomyza huidobrensis*, a leaf mining fly, could not develop properly and pupation was less on common bean plants with endophytic colonisation of BB isolates. Moreover, in inoculated plants, the emergence of adults from pupae was reduced (Akutse et al. 2013).

There was significant reduction in reproduction rate in two aphid species (*Aphis fabae* and *Acyrthosiphon pisum*) colonising the leaves of faba plants growing from seeds pre-soaked in a spore suspension of *Beauveria bassiana* and *Metarhizium anisopliae* (de Faria and Wraight 2007: Castrillo et al. 2011). The possible mode of action is degradation of insect cuticle and host body component as there is existence of various enzymes like chitinases, lipases and proteases within the genome of *Metarhizium* and *Beauveria* (St Leger et al. 1996; Pedrini et al. 2013; Schrank and Vainstein 2010). The infection strategies are well documented and understood (Small and Bidochka 2005; Ortiz-Urquiza and Keyhani 2013). The secretion and overexpression of the most important virulence factor, protease Pr1A (Schrank and Vainstein 2010), from the entomopathogenic fungi *Metarhizium anisopliae* are dependent on the host; it colonises dependent regulation. However, destruxins were detected in cowpea plants endophytically colonised by *Metarhizium robertsii* (Golo et al. 2014). *T. trachyspermus*, an endophytic fungus isolated from *W. somnifera*, is known to produce hydrolytic enzyme like chitinase and shows insecticidal and fungicidal properties with other plant growth attributes (unpublished data). Thus focus of future studies should be mainly on demonstrating that anti-insect effect is due to the fungus or its metabolites and not by fungus-mediated changes in host plant metabolism.

8.7.3 Mechanisms of Insect Control Displayed by Endophytic Fungi

The toxin production by endophytes is well correlated with their ability to repel insects. Thus, several toxins produced by endophytic fungi confer protection to host against different herbivores. The production of toxins leaves the plant unpalatable to various pests like aphids, beetles and grasshoppers (Carroll 1988; Clay 1988a, b). For the first time, Bacon et al. (1977) established a correlation between toxicity of *F. arundinacea* against herbivorous mammals and its colonisation with fungi *Epichloë typhina*. Bacon and Hills (1996) in their review described important toxins of *L. perenne*, which include ergot alkaloids, ergopeptine, clavine and neurotoxins like lolitrems. The neurotoxins produced by several endophytes are precursors of various other toxins, for example, neurotoxins secreted by *A. lolii* are precursors of toxins like paxilline. Prestidge and Gallagher (1988) reported the reduction in insect (of *Listronotus bonariensis*) attacks on *Lolium perenne* plants infected with fungus *A. lolii* due to production of highly toxic, lolitrem B by fungus. As soon as insect feeds on *L. perenne* plants infected with *A. lolli*, lolitrem B enters into insect gut. The growth of insect was retarded finally leading to its death. The grasses infected with *Acremonium* spp. and *E. typhina* produce alkaloids, mainly peramine and ergovaline. *Festuca longifolia* infected with *E. typhina* produced peramine, ergovaline and lolitrem B. Insect species, *Rhopalosiphum padi* and *Schizaphis graminum*, do not attack grasses containing the alkaloid loline in their tissues but attack those containing ergovaline. Hence, loline was sensitive whereas ergoline insensitive against both insect species. The methanolic extracts of *F. arundinacea*, infected with *A. coenophialum*, contain lolines of fungal origin, which alter weight and feeding behaviour of insect pests. The weight and behaviour of *S. frugiperda* and *O. nubilalis* were reduced and altered, respectively, when fed with diets amended with extracts containing loline derivatives (Riedell et al. 1991). There was reduction in attacks of *Popillia japonica*, Japanese beetle, on *Lolium* and *Festuca* colonised by *Acremonium* spp. due to the production of alkaloids by plants (Patterson et al. 1992). The production and level of toxins in endophytic fungi are influenced by seasonal variations. It has been reported that changes in temperature modify the levels of toxin peramine, produced in *L. perenne* by *A. lolii* (Breen 1992). The antixenosis towards the aphid *S. graminum* is dependent on the extent of colonisation and concentration of peramine. Fungal alkaloids deter the insect attack on plants by altering their behaviour. Two compounds, ergotamine and ergovaline, made the plant resistant against insect attack (Ball et al. 1997). Recently, Miles et al. (1998) observed that *Neotyphodium* sp. produces N-formilonine and an analogue of paxilline in the host *Echinopogon ovatus* which depict insecticidal activity against various insects including *L. bonariensis*. There is considerable specificity between certain endophytic fungi and host plants. Hence there is occurrence of physiological races in endophytic fungi (Leuchtmann 1992).

8.8 Recent Developments

The commercial production of bioactive compounds from endophytes is still in its infancy due to lack of proper isolations and characterisation (Kusari et al. 2014). Modern genomic studies like genome sequencing, comparative genomics, microarray, next-generation sequencing, metagenomics, metatranscriptomics involving metaomics and comparative studies can improve unravelling of grey areas of endophytic interaction between plants and microbes (Kaul et al. 2016). Greenfield et al. (2015) reported a novel method for surface sterilisation of multiple plant tissue samples simultaneously in bulk, based on 24 perforated Falcon™ tubes, as surface sterilisation is very time-consuming and limiting factor for large number of samples. Some endophytic bacteria are capable to degrade toxic compounds, tolerate high heavy metal concentrations and enhance plant tolerance to xenobiotics and thus used in bioremediation (Dourado et al. (2015) utilising mechanisms like biosorption, intracellular sequestration, efflux transporter and extracellular chelation (Haferburg and Kothe 2007). In addition these produce highly resistant biodegradable plastic similar to conventional plastic like polyhydroxyalkanoate (PHA) and polyhydroxybutyric acid (PHB). *M. extorquens* was genetically modified to increase PHA and PHB production from methanolic substrate (Hofer et al. 2011). Endophytes can be used to synthesise nanomaterials on a large scale. These exhibit antimicrobial activity and find application in biomedicine. Verma et al. (2011) synthesised extracellular AgNPs and intracellular gold nanoparticles (AuNPs), especially nanotriangles (GNT) with size range of 20–35 nm using endophytic fungus, *A. clavatus*. Parsa et al. (2016) studied biopriming effect of different endophytic fungi with several bean cultivars and suggested endophytic *A. pullulans* offers significant potential to enhance common bean production as part of integrated pest management programmes. Another useful field of interest for healthy and improved plant growth is the biofilm formation in endophytic organisms. There is distant possibility of interaction among endophytes on account of limited space and physical barriers in the endophytic environment of plant tissues. In vitro production of beneficial endophytes and preparation of their biofilmed inocula are relevant researchable issues to achieve enhanced productivity in any agroecosystem. The traditional methods of plant inoculation with monocultures or consortia of potent microbes do not show similar effect as achieved by using microbial biofilm (Bandara et al. 2006), so a lot of hidden information and query are still under deep depth regarding the use and applications of endophytic microbes.

8.9 Conclusion

The fungal endophytes are a rich source of novel secondary metabolites. The biology of toxicity of some compounds *in planta* is of utility to improve the tolerance to insect pests and provide disease resistance. The use of endophytic insect pathogenic fungi (EIPF) as biological control agents requires extensive research for its exploitation in a myriad of positive ways. Further studies on the taxonomy and ecology of these endophytic fungi need to be processed. It is likely that systematic exploration and characterisation of MVOC for their ecological roles will unravel novel mechanisms regulating diverse biological processes critical to plant health and offer tangible practical solutions for agricultural and environmental issues.

Acknowledgements Authors Sharda Sahu and Anil Prakash gratefully acknowledge the financial support provided by DBT-Builder Programme, Barkatullah University, Bhopal. Manvika Sahgal acknowledges the financial assistance provided by ICAR-GOI and Director Experiment Station, GBPUAT, Pantnagar-India.

References

Akello J, Dubois T, Gold CS, Coyne D, Nakavuma J, Paparu P (2007) *Beauveria bassiana* (Balsamo) Vuillemin as an endophyte in tissue culture banana (*Musa* spp.). J Invertebr Pathol 96(1):34–42

Akello J, Dubois T, Coyne D, Kyamanywa S (2008) Endophytic *Beauveria bassiana* in banana (*Musa* spp.) reduces banana weevil (*Cosmopolites sordidus*) fitness and damage. Crop Prot 27(11):1437–1441

Akutse KS, Maniania NK, KKM F, Van den Berg J, Ekesi S (2013) Endophytic colonization of Vicia faba and Phaseolus vulgaris (Fabaceae) by fungal pathogens and their effects on the life-history parameters of Liriomyza huidobrensis (Diptera: Agromyzidae). Fungal Ecol 6(4):293–301

Arai T, Mikami Y, Fukushima K, Utsumi T, Yazawa K (1973) A new antibiotic, leucinostatin, derived from Penicillium lilacinum. J Antibiot 26(3):157–161

Arnold AE, Mejía LC, Kyllo D, Rojas EI, Maynard Z, Robbins N, Herre EA (2003) Fungal endophytes limit pathogen damage in a tropical tree. Proc Natl Acad Sci 100(26):15649–15654

Azeem M, Rajarao GK, Terenius O, Nordlander G, Nordenhem H, Nagahama K, Borg-Karlson AK (2013) A fungal metabolite masks the host plant odor for the pine weevil (Hylobius abietis). Fungal Ecol 13:103–111

Azevedo JL, Maccheroni W Jr, Pereira JO, de Araújo WL (2000) Endophytic microorganisms: a review on insect control and recent advances on tropical plants. Electron J Biotechnol 3(1):15–16

Bacon CW, Hills NS (1996) Symptomless grass endophytes: products of coevolutionary symbioses and their role in the ecological adaptation of grasses. In: Redlin SC, Carris LM (eds) Endophytic fungi in grasses and woody plants. American Phytopathologycal Society Press, St. Paul, pp 155–178

Bacon CW, Porter JK, Robins JD, Luttrell ES (1977) Epichloë typhina from toxic tall fescue grasses. Appl Environ Microbiol 34:576–581

Ball OJP, Miles CO, Prestidge RA (1997) Ergopeptide alkaloids and Neotyphodium lolli-mediated resistance in perennial ryegrass against adult Heteronychus arator (Coleoptera: Scarabaeidae). J Econ Entomol 90:1382–1391

Bandara WM, Seneviratne G, Kulasooriya SA (2006) Interactions among endophytic bacteria and fungi: effects and potentials. J Biosci 31(5):645–650

Benz F, Knüsel F, Nüesch J, Treichler H, Voser W, Nyfeler R, Keller-Schierlein W (1974) Stoffwechselprodukte von Mikroorganismen 143. Mitteilung. Echinocandin B, ein neuartiges Polypeptid-Antibioticum aus Aspergillus nidulans var. echinulatus: Isolierung und Bausteine. Helv Chim Acta 57(8):2459–2477

Bills GF, Giacobbe RA, Lee SH, Peláez F, Tkacz JS (1992) Tremorgenic mycotoxins, paspalitrem A and C, from a tropical Phomopsis. Mycol Res 96(11):977–983

Bing LA, Lewis LC (1991) Suppression of Ostrinia nubilalis (Hübner) (Lepidoptera: Pyralidae) by endophytic *Beauveria bassiana (Balsamo) Vuillemin*. Environ Entomol 20:1207–1211

Boberg JB, Ihrmark K, Lindahl BD (2011) Decomposing capacity of fungi commonly detected in Pinus sylvestris needle litter. Fungal Ecol 4(1):110–114

Boyle R, McLean S, Foley W, Davies NW, Peacock EJ, Moore B (2001) Metabolites of dietary 1, 8-cineole in the male koala (Phascolarctos cinereus). Comp Biochem Physiol C: Toxicol Pharmacol 129(4):385–395

Brady SF, Clardy J (2000) CR377, a new pentaketide antifungal agent isolated from an endophytic fungus. J Nat Prod 63(10):1447–1448

Breen JP (1992) Temperature and seasonal effects on expression of *Acremonium* endophyte-enhanced resistance to *Schizaphis graminum (Homoptera: Aphididae)*. Environ Entomol 21:68–74

Bush LP, Wilkinson HH, Schardl CL (1997) Bioprotective alkaloids of grass-fungal endophyte symbioses. Plant Physiol 114(1):1

Calhoun LA, Findlay JA, Miller JD, Whitney NJ (1992) Metabolites toxic to spruce budworm from balsam fir needle endophytes. Mycol Res 96(4):281–286
Carroll G (1988) Fungal endophytes in stems and leaves: from latent pathogen to mutualistic symbiont. Ecology 69(1):2–9
Caruso M, Colombo AL, Fedeli L, Pavesi A, Quaroni S, Saracchi M, Ventrella G (2000) Isolation of endophytic fungi and actinomycetes taxane producers. Ann Microbiol 50(1):3–14
Castrillo LA, Griggs MH, Ranger CM, Reding ME, Vandenberg JD (2011) Virulence of commercial strains of *Beauveria bassiana* and *Metarhizium brunneum* (*Ascomycota: Hypocreales*) against adult *Xylosandrus germanus* (*Coleoptera: Curculionidae*) and impact on brood. Biol Control 58(2):121–126
Chapela IH, Petrini O, Hagmann L (1991) Monolignol glucosides as specific recognition messengers in fungus-plant symbioses. Physiol Mol Plant Pathol 39(4):289–298
Cherry AJ, Lomer CJ, Djegui D, Schulthess F (1999) Pathogen incidence and their potential as microbial control agents in IPM of maize stem borers in West Africa. BioControl 44(3):301–327
Cherry AJ, Banito A, Djegui D, Lomer C (2004) Suppression of the stem-borer *Sesamia calamistis (Lepidoptera; Noctuidae)* in maize following seed dressing, topical application and stem injection with African isolates of *Beauveria bassiana*. Int J Pest Manag 50(1):67–73
Chiron N, Michelot D (2005) Odeurs des champignons: chimie et rôle dans les interactions biotiques-une revue. Cryptogam Mycol 26(4):299–364
Clay K (1988a) Fungal endophytes of grasses. A defensive mutualism between plants and fungi. Ecology 69:10–16
Clay K (1988b) Fungal endophytes of grasses. Their potential as biocontrol agents. Mycol Res 92:1–12
Contreras-Cornejo HA, Macías-Rodríguez L, Herrera-Estrella A, López-Bucio J (2014) The 4-phosphopantetheinyl transferase of *Trichoderma virens* plays a role in plant protection against *Botrytis cinerea* through volatile organic compound emission. Plant Soil 379(1–2):261–274
Crous PW, Braun U, Groenewald JZ (2007) Mycosphaerella is polyphyletic. Stud Mycol 58:1–32
Davis TS, Crippen TL, Hofstetter RW, Tomberlin JK (2013) Microbial volatile emissions as insect semiochemicals. J Chem Ecol 39(7):840–859
de Faria MR, Wraight SP (2007) Mycoinsecticides and mycoacaricides: a comprehensive list with worldwide coverage and international classification of formulation types. Biol Control 43(3):237–256
Deckert RJ, Melville LH, Peterson RL (2001) Structural features of a Lophodermium endophyte during the cryptic life-cycle phase in the foliage of *Pinus strobus*. Mycol Res 105(8):991–997
Demain AL (1980) Microbial production of primary metabolites. Naturwissenschaften 67(12):582–587
Dew RK, Boissonneault GA, Gay N, Boling JA, Cross RJ, Cohen DA (1990) The effect of the endophyte (*Acremonium coenophialum*) and associated toxin (s) of tall fescue on serum titer response to immunization and spleen cell flow cytometry analysis and response to mitogens. Vet Immunol Immunopathol 26(3):285–295
Dourado MN, Neves AAC, Santos DS, Araújo WL (2015) Biotechnological and agronomic potential of endophytic pink-pigmented methylotrophic methylobacterium spp. Hindawi Publishing Corporation. BioMed Res Int 2015:909016 19 pages
Effmert U, Kalderás J, Warnke R, Piechulla B (2012) Volatile mediated interactions between bacteria and fungi in the soil. J Chem Ecol 38(6):665–703
Ezquer I, Li J, Ovecka M, Baroja-Fernández E, Muñoz FJ, Montero M, Etxeberria E (2010) Microbial volatile emissions promote accumulation of exceptionally high levels of starch in leaves in mono-and dicotyledonous plants. Plant Cell Physiol 51(10):1674–1693
Faeth SH, Fagan WF (2002) Fungal endophytes: common host plant symbionts but uncommon mutualists. Integr Comp Biol 42(2):360–368
Fialho MB, Ferreira LFR, Monteiro RTR, Pascholati SF (2011) Antimicrobial volatile organic compounds affect morphogenesis-related enzymes in *Guignardia citricarpa*, causal agent of citrus black spot. Biocontrol Sci Tech 21(7):797–807

Findlay JA, Li G, Penner PE, Miller JD (1995a) Novel diterpenoid insect toxins from a conifer endophyte. J Nat Prod 58(2):197–200

Findlay JA, Buthelezi S, Lavoie R, Peña-Rodriguez L, Miller JD (1995b) Bioactive isocoumarins and related metabolites from conifer endophytes. J Patural Prod 58(11):1759–1766

Findlay S, Carreiro M, Krischik V, Jones CG (1996) Effects of damage to living plants on leaf litter quality. Ecol Appl 6(1):269–275

Findlay JA, Buthelezi S, Li GQ, Seveck M, Miller JD (1997) Insect toxins from an endophytic fungus from wintergreen. J Nat Prod 60:1214–1215

Fraatz MA, Zorn H (2010) Fungal flavors. In: Esser K, Hofrichter M (eds) The mycota X: industrial applications. Springer, Heidelberg, pp 249–268

Georg GI, Chen TT, Ojima I, Vyas DM (1994) Taxane anticancer agents. Basic science current status, American Chemical Society. Symposium Series no. 583. American Chemical Society, Washington, DC

Glenn AE, Bacon CW, Price R, Hanlin RT (1996) Molecular phylogeny of *Acremonium* and its taxonomic implications. Mycologia 88:369–383

Golo PS, Gardner DR, Grilley MM, Takemoto JY, Krasnoff SB, Pires MS, Roberts DW (2014) Production of destruxins from *Metarhizium* spp. fungi in artificial medium and in endophytically colonized cowpea plants. PLoS One 9(8):e104946

Götz M, Dammann U, Schulz B, Boyle C (2000) Influence of colonization of the apoplast by endophytic fungi on the nutrient status of the host plant. Abstracts, third international congress on symbiosis, Marburg, p 78

Greenfield M, Pareja R, Ortiz V, María I, Jiménez G, Fernando E, Vega, Parsa S (2015) A novel method to scale up fungal endophyte isolations. Biocontrol Sci Tech 25(10):1208–1212

Griffin MA, Spakowicz DJ, Gianoulis TA, Strobel SA (2010) Volatile organic compound production by organisms in the genus Ascocoryne and a re-evaluation of myco-diesel production by NRRL 50072. Microbiology 156(12):3814–3829

Gunatilaka AL (2006) Natural products from plant-associated microorganisms: distribution, structural diversity, bioactivity, and implications of their occurrence. J Nat Prod 69(3):509–526

Guo LD, Hyde KD, Liew ECY (2000) Identification of endophytic fungi from Livistona chinensis based on morphology and rDNA sequences. New Phytol 147(3):617–630

Haferburg G, Kothe E (2007) Microbes and metals: interactions in the environment. J Basic Microbiol 47(6):453–467

Halmschlager E, Butin H, Donaubauer E (1993) Endophytic fungi in leaves and twigs of Quercus petraea. Eur J For Pathol 23(1):51–63

Harper JK, Arif AM, Ford EJ, Strobel GA, Porco JA, Tomer DP, Grant DM (2003) Pestacin: a 1, 3-dihydro isobenzofuran from Pestalotiopsis microspora possessing antioxidant and antimycotic activities. Tetrahedron 59(14):2471–2476

Herrera-Carillo Z, Torres MS, Singh AP, Vorsa N, Gianfagna T, Meyer W, White Jr JF (2009) Phenolic, flavonoid and antioxidant profiling for cool-season grasses with and without endophyte. In: Proceedings of the 18th annual rutgers turfgrass symposium, vol 12, January 2009

Hofer P, Vermette P, Groleau D (2011) Introducing a new bioengineered bug: *Methylobacterium extorquens* tuned as a microbial bioplastic factory. Bioeng Bug 2(2):71–79

Holighaus G, Weißbecker B, von Fragstein M, Schütz S (2014) Ubiquitous eight-carbon volatiles of fungi are infochemicals for a specialist fungivore. Chemoecology 24(2):57–66

Horakova K, Betina V (1976) Cytotoxic activity of macrocyclic metabolites from fungi. Neoplasma 24(1):21–27

Horn WS, Simmonds MS, Schwartz RE, Blaney WM (1996) Variation in production of Phomodiol and Phomopsolide B by *Phomopsis* spp. Mycologia 88:588–595

Huang Z, Wang B, Eaves DH, Shikany JM, Pace RD (2007) Phenolic compound profile of selected vegetables frequently consumed by African Americans in the Southeast United States. Food Chem 103(4):1395–1402

Hung R, Lee S, Bennett JW (2013) Arabidopsis thaliana as a model system for testing the effect of Trichoderma volatile organic compounds. Fungal Ecol 6(1):19–26

Inamdar AA, Moore JC, Cohen RI, Bennett JW (2012) A model to evaluate the cytotoxicity of the fungal volatile organic compound 1-octen-3-ol in human embryonic stem cells. Mycopathologia 173:13–20

Ishii T, Hayashi K, Hida T, Yamamoto Y, Nozaki Y (2000) TAN-1813, a novel Ras-farnesyltransferase inhibitor produced by *Phoma* sp. J Antibiot 53(8):765–778

Ju Y, Sacalis JN, Still CC (1998) Bioactive flavonoids from endophyte-infected blue grass (Poa ampla). J Agric Food Chem 46:3785–3788

Jumpponen ARI, Trappe JM (1998) Dark septate endophytes: a review of facultative biotrophic root-colonizing fungi. New Phytol 140(2):295–310

Kaiser R (2006) Flowers and fungi use scents to mimic each other. Science 311:806–807

Kaul S, Sharma T, Dhar MK (2016) Omics tools for better understanding the plant–endophyte interactions. Front Plant Sci. https://doi.org/10.3389/fpls.2016.00955

Kishimoto K, Matsui K, Ozawa R, Takabayashi J (2007) Volatile 1-octen-3-ol induces a defensive response in *Arabidopsis thaliana*. J Gen Plant Pathol 73(1):35–37

Kogel KH, Franken P, Hückelhoven R (2006) Endophyte or parasite–what decides? Curr Opin Plant Biol 9(4):358–363

Korpi A, Jarnberg J, Pasanen A-L (2009) Microbial volatile organic compounds. Crit Rev Toxicol 39:139–193

Koshino H, Terada SI, Yoshihara T, Sakamura S, Shimanuki T, Sato T, Tajimi A (1988) Three phenolic acid derivatives from stromata of *Epichloë typhina* on Phleum pratense. Phytochemistry 27(5):1333–1338

Koshino T, Yoshihara S, Sakamura T, Shimanuki T, Sato A, Tajimi (1989) A ring B aromatic sterol from stromata of *Epichloë typhina*. Phytochemistry 28:771–772

Koshino H, Yoshihara T, Okuno M, Sakamura S, Tajimi A, Shimanuki T (1992) Gamahonolides A, B, and gamahorin, novel antifungal compounds from stromata of *Epichloë typhina* on Phleum pratense. Biosci Biotechnol Biochem 56(7):1096–1099

Krohn K, Michel A, Flörke U, Aust HJ, Draeger S, Schulz B (1994) Biologically active metabolites from fungi, 5. Palmarumycins C1–C16 from Coniothyrium sp.: isolation, structure elucidation, and biological activity. Eur J Org Chem 1994(11):1099–1108

Krohn K, Michel A, Romer E, Florke U, Aust HJ, Draeger S, Schulz B (1996) Biologically active secondary metabolites from fungi 6: phomosines AC. Two new biaryl ethers and one new arylbenzyl ether from *Phomopsis* sp. Nat Prod Lett 8:43–48

Krohn K, Bahramsari R, Flörke U, Ludewig K, Kliche-Spory C, Michel A, Antus S (1997) Dihydroisocoumarins from fungi: isolation, structure elucidation, circular dichroism and biological activity. Phytochemistry 45(2):313–320

Krohn K, Florke U, John M, Root N, Steingrover K, Aust HJ, Draeger S, Schulz B, Antus S, Simonyi M, Zsila F (2001) Biologically active metabolites from fungi part 16. New preussomerins J, K and L from an endophytic fungus: structure elucidation, crystal structure analysis and determination of absolute configuration by CD calculations. Tetrahedron 57:4343–4348

Kumar S, Kaushik N (2012) Metabolites of endophytic fungi as novel source of biofungicide: a review. Phytochem Rev 11(4):507–522

Kusari S, Singh S, Jayabaskaran C (2014) Biotechnological potential of plant-associated endophytic fungi: hope versus hype. Trends Biotechnol 32:6

Lee JC, Lobokovsky NB, Strobel GA, Clardy JC (1995) Subglutinol A and B: immunosuppresive compounds from the endophytic fungus *Fusarium subglutinans*. J Org Chem 60:7076–7077

Lee SO, Kim HY, Choi GJ, Lee HB, Jang KS, Choi YH, Kim JC (2009) Mycofumigation with *Oxyporus latemarginatus* EF069 for control of postharvest apple decay and Rhizoctonia root rot on moth orchid. J Appl Microbiol 106(4):1213–1219

Leuchtmann A (1992) Systematics, distribution, and host specificity of grass endophytes. Nat Toxins 1(3):150–162

Li JY, Strobel G, Sidhu R, Hess WM, Ford EJ (1996) Endophytic taxol-producing fungi from bald cypress, *Taxodium distichum*. Microbiology 142(8):2223–2226

Li JY, Sidhu RS, Ford EJ, Long DM, Hess WM, Strobel GA (1998) The induction of taxol production in the endophytic fungus—Periconia sp from Torreya grandifolia. J Ind Microbiol Biotech 20:259–264

Liang B, Lehmann J, Solomon D, Sohi S, Thies JE, Skjemstad JO, Wirick S (2008) Stability of biomass-derived black carbon in soils. Geochim Cosmochim Acta 72(24):6069–6078

Lingham RB, Silverman KC, Bills GF, Cascales C, Sanchez M, Jenkins RG, Mochales S (1993) Chaetomella acutiseta produces chaetomellic acids A and B which are reversible inhibitors of farnesyl-protien transferase. Appl Microbiol Biotechnol 40(2–3):370–374

Liu CH, Zou WX, Lu H, Tan RX (2001) Antifungal activity of Artemisia annua endophyte cultures against phytopathogenic fungi. J Biotechnol 88(3):277–282

Lu H, Zou WX, Meng JC, Hu J, Tan RX (2000) New bioactive metabolites produced by *Colletotrichum* sp., an endophytic fungus in Artemisia annua. Plant Sci 151(1):67–73

Macías-Rubalcava ML, Hernández-Bautista BE, Oropeza F, Duarte G, González MC, Glenn AE, Anaya AL (2010) Allelochemical effects of volatile compounds and organic extracts from *Muscodor yucatanensis*, a tropical endophytic fungus from *Bursera simaruba*. J Chem Ecol 36(10):1122–1131

Madelin MF (1963) Diseases caused by hyphomycetous fungi. In: Steinhaus E (ed) Insect pathology. An advanced treatise, vol 2. Academic, New York, pp 233–272

Metz AM, Haddad A, Worapong J, Long DM, Ford EJ, Hess WM, Strobel GA (2000) Induction of the sexual stage of *Pestalotiopsis microspora*, a taxol-producing fungus. Microbiology 146(8):2079–2089

Miller JD (2001) Factors that affect the occurrence of fumonisin. Environ Health Perspect 109(Suppl 2):321

Minerdi D, Bossi S, Gullino ML, Garibaldi A (2009) Volatile organic compounds: a potential direct long-distance mechanism for antagonistic action of *Fusarium oxysporum* strain MSA 35. Environ Microbiol 11:844–854

Minerdi D, Bossi S, Maffei ME, Gullino ML, Garibaldi A (2011) Fusarium oxysporum and its bacterial consortium promote lettuce growth and expansin A5 gene expression through microbial volatile organic compound (MVOC) emission. FEMS Microbiol Ecol 76(2):342–351

Moricca S, Ragazzi A (2008) Fungal endophytes in Mediterranean oak forests: a lesson from *Discula quercina*. Phytopathology 98(4):380–386

Müller A, Faubert P, Hagen M, zu Castell W, Polle A, Schnitzler JP, Rosenkranz M (2013) Volatile profiles of fungi–chemotyping of species and ecological functions. Fungal Genet Biol 54:25–33

Naznin HA, Kiyohara D, Kimura M, Miyazawa M, Shimizu M, Hyakumachi M (2014) Systemic resistance induced by volatile organic compounds emitted by plant growth-promoting fungi in Arabidopsis thaliana. PLoS One 9(1):e86882

Ngwene B, Boukail S, Söllner L, Franken P, Andrade-Linares DR (2016) Phosphate utilization by the fungal root endophyte *Piriformospora indica*. Plant Soil 405(1–2):231–241

O'Hanlon KA, Knorr K, Jørgensen LN, Nicolaisen M, Boelt B (2012) Exploring the potential of symbiotic fungal endophytes in cereal disease suppression. Biol Control 63(2):69–78

Ortíz-Castro R, Contreras-Cornejo HA, Macías-Rodríguez L, López-Bucio J (2009) The role of microbial signals in plant growth and development. Plant Signal Behav 4(8):701–712

Ortiz-Urquiza A, Keyhani NO (2013) Action on the surface: entomopathogenic fungi versus the insect cuticle. Insects 4(3):357–374

Oses R, Valenzuela S, Freer J, Sanfuentes E, Rodriguez J (2008) Fungal endophytes in xylem of healthy Chilean trees and their possible role in early wood decay. Fungal Divers 33(7):77–86

Pagans E, Font X, Sánchez A (2006) Emission of volatile organic compounds from composting of different solid wastes: abatement by biofiltration. J Hazard Mater 131(1):179–186

Parfitt D, Hunt J, Dockrell D, Rogers HJ, Boddy L (2010) Do all trees carry the seeds of their own destruction? PCR reveals numerous wood decay fungi latently present in sapwood of a wide range of angiosperm trees. Fungal Ecol 3(4):338–346

Parsa S, García-Lemos AM, Castillo K, Ortiz V, López-Lavalle LAB, Braun J, Vega FE (2016) Fungal endophytes in germinated seeds of the common bean, *Phaseolus vulgaris*. Fungal Biol 120(5):783–790

Patterson CG, Potter DA, Fannin FF (1992) Feeding deterrency of alkaloids from endophyte-infected grasses to Japanese beetle grubs. Entomol Exp Appl 61:285

Pedrini N, Ortiz-Urquiza A, Zhang S, Keyhani NO (2013) Targeting of insect epicuticular lipids by the entomopathogenic fungus *Beauveria bassiana*: hydrocarbon oxidation within the context of a host-pathogen interaction. Front Microbiol 4:24

Peipp H, Sonnenbichler J (1992) Secondary fungal metabolites and their biological activities, II. Occurrence of antibiotic compounds in cultures of *Armillaria ostoyae* growing in the presence of an antagonistic fungus or host plant cells. Biol Chem Hoppe Seyler 373(2):675–684

Peters S, Dammeyer B, Schulz B (1998a) Endophyte-host interactions. I. Plant defense reactions to endophytic and pathogenic fungi. Symbiosis 25(1–3):193–211

Peters S, Draeger S, Aust HJ, Schulz B (1998b) Interactions in dual cultures of endophytic fungi with host and nonhost plant calli. Mycologia 90:360–367

Petrini O (1991) Fungal endophytes of tree leaves. In: Microbial ecology of leaves. Springer, New York, pp 179–197

Pimenta RS, Moreira da Silva JF, Buyer JS, Janisiewicz WJ (2012) Endophytic fungi from plums (Prunus domestica) and their antifungal activity against Monilinia fructicola. J Food Prot 75(10):1883–1889

Pinto C, Rodrigues LS, Azevedo JL, Pereira JO, Carneiro Vieira ML, Labate CA (2000) Symptomless infection of banana and maize by endophytic fungi impairs photosynthetic efficiency. New Phytol 147(3):609–615

Powell RG, Petroski RJ (1992) Alkaloid toxins in endophyte-infected grasses. Nat Toxins 1(3):163–170

Prestidge RA, Gallagher RT (1988) Endophyte conifers resistance to ryegrass: argentine stem weevil larval studies. Ecol Entomol 13:429–435

Pulici M, Sugawara F, Koshino H, Uzawa J, Yoshida S, Lobkovsky E, Clardy J (1996a) Pestalotiopsins A and B: new caryophyllenes from an endophytic fungus of Taxus brevifolia. J Org Chem 61(6):2122–2124

Pulici M, Sugawara F, Koshino H, Uzawa J, Yoshida S, Lobkovsky E, Clardy J (1996b) A new isodrimeninol from Pestalotiopsis sp. J Nat Prod 59(1):47–48

Pulici M, Sugawara F, Koshino H, Okada G, Esumi Y, Uzawa J, Yoshida S (1997) Metabolites of *Pestalotiopsis* spp., endophytic fungi of *Taxus brevifolia*. Phytochemistry 46(2):313–319

Qiu DY, Huang MJ, Fang XH, Zhu C, Zhu ZQ (1994) Isolation of an endophytic fungus associated with *Taxus yunnanensis*. Acta Mycol Sin 13:314

Quesada-Moraga E, Landa BB, Muñoz-Ledesma J, Jiménez-Diáz RM, Santiago-Alvarez C (2006) Endophytic colonisation of *opium poppy*, *Papaver somniferum*, by an entomopathogenic *Beauveria bassiana* strain. Mycopathologia 161(5):323–329

Rai M, Agarkar G (2014) Plant–fungal interactions: what triggers the fungi to switch among lifestyles? Crit Rev Microbiol 42(3):428–438

Reddy NP, Khan APA, Devi UK, Sharma HC, Reineke A (2009) Treatment of millet crop plant (Sorghum bicolor) with the entomopathogenic fungus (*Beauveria bassiana*) to combat infestation by the stem borer, Chilo partellus Swinhoe (Lepidoptera: Pyralidae). J Asia Pac Entomol 12(4):221–226

Redman RS, Dunigan DD, Rodriguez RJ (2001) Fungal symbiosis from mutualism to parasitism: who controls the outcome, host or invader? New Phytol 151(3):705–716

Redman RS, Sheehan KB, Stout RG, Rodriguez RJ, Henson JM (2002) Thermotolerance generated by plant/fungal symbiosis. Science 298(5598):1581–1581

Richardson MD, Chapman GW, Hoveland CS, Bacon CW (1992) Sugar alcohols in endophyte-infected tall fescue under drought. Crop Sci 32(4):1060–1061

Riedell WE, Kieckhefer RE, Petroski RJ, Powell RG (1991) Naturally-occurring and synthetic loline alkaloid derivatives: insect feeding behaviour modification and toxicity. J Entomol Sci 26:122–129

Rocha AC, Garcia D, Uetanabaro AP, Carneiro RT, Araújo IS, Mattos CR, Góes-Neto A (2011) Foliar endophytic fungi from Hevea brasiliensis and their antagonism on Microcyclus ulei. Fungal Divers 47(1):75–84

Rodriguez R, Redman R (2008) More than 400 million years of evolution and some plants still can't make it on their own: plant stress tolerance via fungal symbiosis. J Exp Bot 59:1109–1114

Rowan DD, Latch GCM (1994) In: Bacon CW, White JF Jr (eds) Biotechnology of endophytic fungi of grasses. CRC Press, Boca Raton, pp 169–183

Roy HE, Steinkraus DC, Eilenberg J, Hajek AE, Pell JK (2006) Bizarre interactions and endgames: entomopathogenic fungi and their arthropod hosts. Annu Rev Entomol 51:331–357

Rudgers JA, Strauss SY (2004) A selection mosaic in the facultative mutualism between ants and wild cotton. Proc R Soc Lond B Biol Sci 271(1556):2481–2488

Saikkonen K, Faeth SH, Helander M, Sullivan TJ (1998) Fungal endophytes: a continuum of interactions with host plants. Annu Rev Ecol Syst 29(1):319–343

Schardl CL, Phillips TD (1997) Protective grass endophytes: where are they from and where are they going? Plant Dis 81(5):430–438

Schardl CL, Liu JS, White JF, Finkel RA, An Z, Siegel MR (1991) Molecular phylogenetic relationships of nonpathogenic grass mycosymbionts and *clavicipitaceous* plant pathogens. Plant Syst Evol 178(1):27–41

Schiestl FP, Steinebrunner F, Schulz C, Von Reuss S, Francke W, Weymuth C, Leuchtmann A (2006) Evolution of 'pollinator'-attracting signals in fungi. Biol Lett 2(3):401–404

Schrank A, Vainstein MH (2010) Metarhizium anisopliae enzymes and toxins. Toxicon 56(7):1267–1274

Schulz B, Boyle C (2005) The endophytic continuum. Mycol Res 109(6):661–686

Schulz B, Sucker J, Aust HJ, Krohn K, Ludewig K, Jones PG, Döring D (1995) Biologically active secondary metabolites of endophytic Pezicula species. Mycol Res 99(8):1007–1015

Schulz B, Römmert AK, Dammann U, Aust HJ, Strack D (1999) The endophyte-host interaction: a balanced antagonism? Mycol Res 103(10):1275–1283

Schulz B, Boyle C, Draeger S, Römmert AK, Krohn K (2002) Endophytic fungi: a source of novel biologically active secondary metabolites** Paper presented at the British Mycological Society symposium on Fungal Bioactive Compounds, held at the University of Wales Swansea on 22–27 April 2001. Mycol Res 106(9)996–1004

Selosse MA, Dubois MP, Alvarez N (2009) Do Sebacinales commonly associate with plant roots as endophytes? Mycol Res 113(10):1062–1069

Singh MP, Janso JE, Luckman SW, Brady SF, Clardy J, Greenstein M, Maiese WM (2000) Biological activity of guanacastepene, a novel diterpenoid antibiotic produced by an unidentified fungus CR115. J Antibiot 53(3):256–261

Small CLN, Bidochka MJ (2005) Up-regulation of Pr1, a subtilisin-like protease, during conidiation in the insect pathogen Metarhizium anisopliae. Mycol Res 109(3):307–313

Song GC, Ryu CM (2013) Two volatile organic compounds trigger plant self-defense against a bacterial pathogen and a sucking insect in cucumber under open field conditions. Int J Mol Sci 14:9803–9819

Splivallo R, Ottonello S, Mello A, Karlovsky P (2011) Truffle volatiles: from chemical ecology to aroma biosynthesis. New Phytol 189:688–699

St Leger RJ, Joshi L, Bidochka MJ, Roberts DW (1996) Construction of an improved mycoinsecticide overexpressing a toxic protease. Proc Natl Acad Sci 93(13):6349–6354

Stierle A, Strobel G, Stierle D (1993) Taxol and taxane production by *Taxomyces andreanae,* an endophytic fungus of Pacific yew. Science-New York Then Washington 260:214–214

Stierle A, Strobel G, Stierle D, Grothaus P, Bignami G (1995) The search for a taxol-producing microorganism among the endophytic fungi of the Pacific yew, Taxus brevifolia. J Nat Prod 58(9):1315–1324

Stierle AA, Stierle DB, Bugni T (1999) Sequoiatones A and B: novel antitumor metabolites isolated from a redwood endophyte. J Org Chem 64(15):5479–5484
Strobel G, Daisy B (2003) Bioprospecting for microbial endophytes and their natural products. Microbiol Mol Biol Rev 4:491–502
Strobel G, Yang X, Sears J, Kramer R, Sidhu RS, Hess WM (1996) Taxol from Pestalotiopsis microspora, an endophytic fungus of Taxus wallichiana. Microbiology 142(2):435–440
Strobel GA, Torczynski R, Bollon A (1997) Acremonium sp.—a leucinostatin A producing endophyte of European yew (Taxus baccata). Plant Sci 128(1):97–108
Strobel GA, Miller RV, Martinez-Miller C, Condron MM, Teplow DB, Hess WM (1999) Cryptocandin, a potent antimycotic from the endophytic fungus Cryptosporiopsis cf. quercina. Microbiology 145(8):1919–1926
Strobel G, Ford E, Worapong J, Harper JK, Arif AM, Grant DM, Chau RMW (2002) Isopestacin, an isobenzofuranone from Pestalotiopsis microspora, possessing antifungal and antioxidant activities. Phytochemistry 60(2):179–183
Su M, Hung R, Bennett JW (2012) Fungal volatile organic compounds: a review with emphasis on their biotechnological potential. Fungal Biol Rev 26:73–83
Sumarah MW, Puniani E, Blackwell BA, Miller JD (2008) Characterization of polyketide metabolites from foliar endophytes of Picea glauca. J Nat Prod 71(8):1393–1398
Sun X, Guo LD, Hyde KD (2011) Community composition of endophytic fungi in Acer truncatum and their role in decomposition. Fungal Divers 47(1):85–95
Tan RX, Zou WX (2001) Endophytes: a rich source of functional metabolites. Nat Prod Rep 18(4):448–459
Thorn RMS, Greenman J (2012) Microbial volatile compounds in health and disease conditions. J Breath Res 6(2):024001
Torres MS, Singh AP, Shah S, Herrera-Carrillo Z, Gianfagna T, White Jr JF, Vorsa N (2009) LC–MS–MS identification and quantification of phenolics in symbiotic tall fescue. In: Proceedings of the 18th Annual Rutgers Turfgrass Symposium, vol 12, New Brunswick, New Jersey: Rutgers University, Center for Interdisciplinary Studies in Turfgrass Science, January 2009
Traber R, Keller-Juslén C, Loosli HR, Kuhn M, Von Wartburg A (1979) Cyclopeptid-Antibiotika aus Aspergillus-Arten. Struktur der Echinocandine C und D. Helv Chim Acta 62(4):1252–1267
Tudzynski B (1997) Fungal phytohormones in pathogenic and mutualistic associations. In: Plant relationships. Springer, Berlin/Heidelberg, pp 167–184
Tudzynski B, Sharon A (2002) Biosynthesis, biological role and application of fungal phytohormones. In: Industrial applications. Springer, Berlin/Heidelberg, pp 183–211
Unterseher M, Schnittler M (2010) Dilution-to-extinction cultivation of leaf-inhabiting endophytic fungi in beech (Fagus sylvatica L.)–different cultivation techniques influence fungal biodiversity assessment. Mycol Res 113(5):645–654
Vega FE, Simpkins A, Aime MC, Posada F, Peterson SW, Rehner SA, Arnold AE (2010) Fungal endophyte diversity in coffee plants from Colombia, Hawai'i, Mexico and Puerto Rico. Fungal Ecol 3(3):122–138
Verma VC, Singh SK, Solanki R, Prakash S (2011) Biofabrication of anisotropic gold nanotriangles using extract of endophytic *Aspergillus clavatus* as a dual functional reductant and stabilizer. Nanoscale Res Lett 6:16–22
Wagenaar MM, Corwin J, Strobel G, Clardy J (2000) Three new cytochalasins produced by an endophytic fungus in the genus Rhinocladiella. J Nat Prod 63(12):1692–1695
Wang J, Li G, Lu H, Zheng Z, Huang Y, Su W (2000) Taxol from Tubercularia sp. strain TF5, an endophytic fungus of Taxus mairei. FEMS Microbiol Lett 193:249–253
Wani MC, Taylor HL, Wall ME, Coggon P, McPhail AT (1971). Plant antitumor agents. VI. The isolation and structure of taxol, a novel antileukemia and antitumor agent from Taxus brevifolia. Journal of the American Chemical Society, 93(9): 2325–2327
Waweru B, Turoop L, Kahangi E, Coyne D, Dubois T (2014) Non-pathogenic fusarium oxysporum endophytes provide field control of nematodes, improving yield of banana (Musa sp.). Biol Control 74:82–88

Weyens N, van der Lelie D, Taghavi S, Vangronsveld J (2009) Phytoremediation: plant–endophyte partnerships take the challenge. Curr Opin Biotechnol 20(2):248–254

White JF, Torres MS (2010) Is plant endophyte-mediated defensive mutualism the result of oxidative stress protection? Physiol Plant 138(4):440–446

White Jr JF, Reddy PV, Bacon CW (2000) Biotrophic endophytes of grasses: a systematic appraisal. In: Microbial endophytes. Marcel Dekker, New York, pp 49–62

Yang X, Strobel G, Stierle A, Hess WM, Lee J, Clardy J (1994) A fungal endophyte-tree relationship: Phoma sp. in Taxus wallichiana. Plant Sci 102(1):1–9

Yuan J, Raza W, Shen Q, Huang Q (2012) Antifungal activity of Bacillus amyloliquefaciens NJN-6 volatile compounds against Fusarium oxysporum f. sp. cubense. Appl Environ Microbiol 78(16):5942–5944

Zou WX, Meng JC, Lu H, Chen GX, Shi GX, Zhang TY, Tan RX (2000) Metabolites of *Colletotrichum gloeosporioides*, an endophytic fungus in Artemisia mongolica. J Nat Prod 63(11):1529–1530

Chapter 9
Agricultural Important Microorganisms: From Rhizosphere to Bioformulation as Biological Control Weapons for Sustainable Agriculture

Gunjan Mukherjee, Pannalal Dey, and Sunny Dhiman

Abstract Fungi are an integral part of the soil ecology. Rhizosphere holds foremost importance as a primitive site for microbial colonization and operation. A significant number of organic compounds including secretions, sloughed off cells, lysates and exudates are secreted by actively growing roots into the rhizosphere. As a matter of fact, the microbial activity in the rhizosphere is reckoned to be upraised and significantly distinctive in comparison to microbes existing in the bulk soil. Over the years agriculturally important microorganisms (AIM) have witnessed a significant utilization in a broad range of agroecosystems including both immanent and artificial circumstances in diverse applications ranging from nutrient supply, bioremediation, biocontrol and rehabilitation of degraded lands. The successful development of AIM in stressed ecosystem poses many challenges. The adverse effect to the environment due to indiscriminate use of chemical pesticides is of great concern, and hence development of alternate control strategies such as biological control as a substitute for chemicals or as a key component in integrated disease management system is gaining momentum. Biological control agents are usually target specific, and by using these agents in conjunction with fungicides, the level of fungicide applied can be reduced. The role of microorganisms to inhibit phytopathogen and possibility of bioformulation of next-generation products for agriculture market has been discussed.

Keywords *Trichoderma* · Rhizosphere · Agriculturally important microorganisms · Bioformulation · Next-generation products

---

G. Mukherjee (✉)
Department of Biotechnology, Chandigarh University, Mohali, Punjab, India

The Energy and Resources Institute, India Habitat Centre, New Delhi, India

P. Dey
The Energy and Resources Institute, India Habitat Centre, New Delhi, India

S. Dhiman
Department of Biotechnology, University Institute of Biotechnology,
Chandigarh University, Mohali, Punjab, India

P. Gehlot, J. Singh (eds.), *Fungi and their Role in Sustainable Development: Current Perspectives*, https://doi.org/10.1007/978-981-13-0393-7_9

9.1 Introduction

Fungi hold a significant position amongst various microbes present in the soil. Many important plant pathogens and plant growth enhancing microorganisms (ecto- and endomycorrhizae) and biocontrol agents are fungi. Saprobic fungi occupy a significant spot amongst the fungal species in soil. They have been documented to play a significant role in the disintegration of plant structural polymers such as cellulose, hemicellulose, chitin and lignin, thereby bestowing to the sustentation of the global carbon cycle. Furthermore such catabolic operations make fungi amenable to grow upon relatively cheaper substrates. Thus owing to this property coupled with their potential to generate commercially pivotal organic molecules and enzymes, the biotechnological utilization of filamentous fungi has witnessed a significant rise.

Hiltner originally coined the term rhizosphere in 1904. Rhizosphere can be categorized into ecto- and endorhizosphere. The word endorhizosphere describes the multilayered microenvironment consisting of a mucoid layer on the root surface, the epidermal layer of the root tissue including the root hairs and the cortical cells (Bolton et al. 1993). Rhizosphere holds foremost importance as a primitive site for microbial colonization and operation. A significant number of organic compounds including secretions, sloughed off cells, lysates and exudates are secreted by actively growing roots into the rhizosphere (Lynch and Whipps 1990; Bowen and Rovira 1991).

Apparently microbial activity in the rhizosphere is reckoned to be upraised and significantly distinctive in comparison to microbes existing in the bulk soil.

A detailed knowledge of rhizosphere microbiota also paves a way to ennoble disease restraining microflora in the rhizosphere. For the past so many years, scientific aspiration has been centralized on structure of microbial communities in the rhizosphere assessed by cultivation-based studies. Such studies have revealed that microbial divergence in rhizosphere is humongous and there are significant deviances in bacterial community structures between bulk (non-rhizosphere) soil and rhizosphere soil.

9.2 Soil Microorganisms and Their Potential Applications in Agriculture

Sorensen 1997 documented that rhizosphere zone is featured by eminent rates of oxygen consumption as a consequence of root and microbial respiration. The oxygen level in the rhizosphere soil atmosphere may either increase or decrease in comparison to the oxygen level of the surrounding soil.

Over the years agriculturally important microorganisms (AIM) have witnessed a significant utilization in a broad range of agroecosystems including both immanent and artificial circumstances in diverse applications ranging from nutrient supply, bioremediation, biocontrol and rehabilitation of degraded lands. The successful development of AIM in stressed ecosystem poses many challenges. The last two decades of research resulted in many successful approaches to select tolerant strains for nitrogen fixation, P solubilization, plant growth promotion and biocontrol. There has been also good progress in formulation technology suitable for dryland and wetland ecosystems with the emerging application of molecular biology techniques.

A plant exhibiting active transpiration evicts huge quantities of water from the soil. Papendick and Campbell 1975 documented that the water potential in rhizosphere soil can reach up to a level more than 1 MPa owing to the extent of water supply from the soil around to the rhizosphere. The rhizosphere soil is relatively drier than the surrounding bulk soil during day hours. This is in contrast to the fact that rhizosphere soil in few terrestrial ecosystems can boast relatively higher water volume in comparison to the surrounding soil at night owing to the phenomenon of "hydraulic redistribution" (Caldwell and Richards 1989). Hinsinger et al. 2003 documented that plant roots alter the rhizosphere pH by extruding protons via H + -ATPase in epidermal cells.

9.3 Edaphic Environment and Community Ecology of the Rhizosphere

Traditional agricultural measures and plant pathogens enhance the susceptibility of a crop to disease. Moreover, the edaphic environment is transfigured with eminent inorganic nutrient accessibility and low-diversity carbon inputs associated with conventional agricultural systems. This profoundly sways substrate, habitat availability and microbial community dynamics (Hoitink and Boehm 1999). Besides that, such environmental alterations in consortium with short rotations are the root of several soilborne diseases. The use of inoculation in conjunction with salutary biological control organisms possessing the ability to colonize the rhizosphere shows a great degree of potential as a possible measure to lessen the occurrence of plant disease (Cook 1993).

9.4 Rhizosphere Microflora: Direct and Indirect Effects on Plant Growth

Shaik and Nusrath (1987) observed *Trichoderma viride* and *Aspergillus niger* as a part of microflora of wilt-resistant cultivar, while susceptible cultivar showed a predominance of *Fusarium udum* and *Fusarium* spp. during all the stages of plant growth.

Plant growth-promoting rhizobacterial strains belonging to fluorescent pseudomonas were explored from the rhizosphere of rice and sugarcane by Kumar et al. (2002). Amongst 40 strains that were identified as *Pseudomonas fluorescens, 18* exhibited strong antifungal activity against *Rhizoctonia bataticola* and *Fusarium oxysporum* especially through the production of antifungal metabolites. Genotyping of *P. fluorescens* strains was made by PCR-RAPD analysis, since differentiation by biochemical methods was limited.

9.5 The Composition of Rhizosphere Microbial Communities

Plant species may play a vital role in exploring the structure of rhizosphere microbial communities (Stephan et al. 2000), with both positive and negative effects on different microbial groups. The recent advancements in molecular tools and techniques to identify soil microorganisms have paved a way to surpass the small subset of culturable soil microorganisms and begin to elucidate populations and communities of microbes belowground. It is progressively common to characterize complex microbial communities genotypically utilizing the small subunit 16S ribosomal DNA gene, a highly conserved region that is quintessential for low homologous gene transfer and a nice reflection of overall phylogenetic relatedness. Fingerprinting methods such as T-RFLP and DGGE can be utilized for partial analysis of a collection of 16S genes, for detailed analysis by sequencing the entire populations or communities to create libraries.

9.5.1 Pathogenic Communities

Ansari and Dhirendra (1986) have studied rhizosphere and rhizoplane mycoflora of barley infected with *Ustilago hordei* and discussed the certain biochemical changes that occur due to infection. Higher fungal population and number of fungal species were encountered in the infected plants in comparison to their healthy counterparts. Gopinath et al. (1987) have reported the colonization of *Fusarium* sp. in sorghum seeds and their significance, and they came with the conclusion that 30 high-yielding cultivars of sorghum analysed showed severe infection of *Fusarium moniliforme*, *F. oxysporum*, *F. semitectum* and *F. solani*; *F. semitectum* infected the embryonic tissue in 93% seeds, while *F. semitectum* and *F. solani* colonized the embryo in 8 and 5% seeds, respectively. But *F. oxysporum* did not colonize the embryo. Hee (1991) has studied the selection and identification of antagonistic rhizobacteria in relation to controlling soilborne diseases of vegetables. 926 isolates of rhizobacteria were obtained from the soils of 22 plant species using three selective media. Pathogenic bacteria were dual cultured with 10 species of important soilborne plant pathogenic fungi, respectively, and measured their antagonism by their inhibition zone; the population density of rhizobacteria in the same field was different according to the crop species planted, and the isolation frequency of the antagonistic bacteria from the species of plant was also markedly different according to the fields or regions where the soils were collected for the effective isolation of rhizobacteria; M523 and King's B media were more suitable than D+ medium of 926 rhizobacteria isolated from the soils with 22 plant species. Amongst these, 63 isolates were selected which were found to be antagonistic to *Phytophthora capsici* while 54 isolates antagonistic to *Rhizoctonia solani* and 17 isolates antagonistic to *Fusarium oxysporum* f.sp. *lycopersici*, respectively. Of these, one isolate RB 173 was finally selected as the most effective antagonist to the nine species of soilborne plant pathogenic fungi and identified as *Pseudomonas fluorescens*. Tissues and cellular location of cross-reactive antigen shared by *Fusarium oxysporum*, soybean roots and *Bradyrhizobium japonicum* was carried out by Chakraborty et al. (1995). They have pre-inoculated the seeds of soybean cultivars with *B. japonicum* and discussed cross protection mechanism involved in reduction of disease severity. The plant rhizosphere is an important zone where many microorganisms both friends and foe exit (Mathur et al. 2004). The microflora associated with a plant rhizosphere is generally influenced by the soil type, pH and temperature. Stem rot caused by *Rhizoctonia solani*, one of the diseases of chillis, becomes very severe and destructive under the favourable conditions. The fungus also colonized seed, leaves and fruits of chilli. Tiwari and Singh (2004) analysed potential use of rhizosphere microorganisms of chillis for the management of stem rot disease.

9.5.2 Non-pathogenic Communities

The mode of antagonisms of *Trichoderma viride* against *Alternaria triticina* causing leaf blight of wheat and dual culture interaction in vitro revealed that mycelial strand of *T. viride* coiled around the hyphae of the test pathogen forming a ropelike structure and finally inhibited the growth of *A. triticina* in vitro (Parveen and Vijay 2004).

Rhizosphere of healthy pigeon pea plant was heavily colonized by *Aspergillus niger*, *Penicillium* sp., *Trichoderma viride* and *Gliocladium virens*. Resident *Trichoderma* and *Gliocladium* were highly antagonistic to the pathogen (*Fusarium udum*). *T. viride* formed loops and coiling and ruptured the cell wall of the pathogen. Mechanism of parasitism between *F. udum* and *G. virens* resulted in twisting, air bubbling and disintegration of pathogen hyphae, while *T. harzianum* causes shrinkage and coagulation of cytoplasm of pathogen hyphae (Pandey and Upadhyay 2000). Two species of *Aspergilli* and ten other fungi were isolated from rhizosphere mycoflora of onion (*Allium cepa*). *Aspergilli* in general were dominant contributing 38.59% to the total mycoflora by Kallurmath and Rajasab (2000). *A. niger* and *A. flavus* were dominant on onion bulbs with the progress of their maturity. Rhizosphere colonization is one of the primitive steps in the pathogenesis of soilborne microorganisms. This may also prove crucial for the action of microbial inoculants utilized as biofertilizers, biopesticides, phytostimulators and bioremediators.

9.6 Biocontrol Agents (BCAs)

The adverse effect to the environment due to indiscriminate use of chemical pesticides is of great concern, and hence development of alternate control strategies such as biological control as substitute for chemicals or as a key component in integrated disease management system is gaining momentum (Mathivannan et al. 2006; Mukhopadhyay 2009). Biological control has been developed as an academic discipline during the 1970s and is now a mature science supported by both the public and private sectors (Mukhopadhyay 2009) which involves the utilization of salutary organisms, their genes and/or gene products, such as metabolites, that lessens the negative sways of plant pathogens and promote positive responses by the plant (Vinale et al. 2008)). Biological control is the purposeful utilization of introduced or resident living organisms other than disease-resistant host plants to restrain the activities and populations of one or more plant pathogens. *Trichoderma* is the most widely exploited fungal genus as biocontrol agent (BCA) in the field of agriculture for the management of crop diseases caused by a vast range of fungal pathogens.

9.6.1 **Trichoderma** *Species: Most Potent Antagonistic Fungi*

Species of the fungal genus *Trichoderma* are typically soil dwellers, existing as anamorphs belonging to the subdivision *Deuteromycotina* (fungi imperfecti) (Hawksworth, et al. 1983). As a rule of thumb, *Trichoderma* species are fast-growing fungi which commonly occur in a variety of soil types, such as agricultural, prairie, forest, salt marsh and desert soils in all climatic zones (Domsch et al. 1980).

Species of *Trichoderma* were documented to dominate the rhizosphere of established tea bushes in a descriptive study conducted from various tea-growing locations in India, and the population of *Trichoderma* spp. showed less variation (Pandey and Upadhyay 2000). *Trichoderma* species were also isolated and analysed for their diversity by Chakraborty et al. (2010a, b, c) from different rhizosphere soils collected from various locations of North Bengal region.

9.6.2 In Vitro Inhibition of Phytopathogens and Biological Control of Plant Diseases

A huge number of reports have been documented on the potential of *Trichoderma* species to antagonize a broad range of commercially important plant pathogens coupled with their potential to lessen the occurrence of disease caused by these pathogens in a huge variety of crops. The utilization of biological control agents puts forward themselves as an alternate to the use of chemicals for pest and disease control. Biological control has certain advantages over the utilization of chemicals which include relatively higher public acceptance, reduced risk of chemical residue and reduced contamination of the environment. Pandey and Upadhyay (2000) reported that rhizosphere of healthy pigeon pea plant was heavily colonized by resident *Trichoderma* and *Gliocladium* which were highly antagonistic to the pathogen. *T. viride* formed loops and coiling and ruptured the cell wall of the pathogen. Mechanism of parasitism between *Fusarium udum* and *G. virens* resulted in twisting, air bubbling and disintegration of pathogen hyphae, while *T. harzianum* causes severe vacuolation, shrinkage and coagulation of cytoplasm of pathogen hyphae.

In vitro evaluation of antagonists revealed that 38.42% and 32.78% inhibition of *Rhizoctonia solani* was achieved by the application of *Trichoderma longibrachiatum* and *T. harzianum*, respectively (Sharma and Gupta 2003). Parveen and Vijay (2004) reported the mode of antagonisms of *Trichoderma viride* against *Alternaria triticina* causing leaf blight of wheat which was studied in vitro by employing dual culture technique. *T. viride* inhibited the growth of the pathogen; its mycelial strands coiled around the hyphae of the test pathogen forming a ropelike structure and finally disintegrating the test pathogen, *A. triticina*. Singh et al. (2004) evaluated *Trichoderma viride*, *Trichoderma harzianum*, *Gliocladium virens* and *Aspergillus nidulans* as seed, soil and combined seed and soil treatment for the control of tomato wilt caused by *Fusarium oxysporum* f. sp. *lycopersici* in greenhouse. *Trichoderma viride*, *Trichoderma harzianum* and *Gliocladium virens* as seed treatment at 10 g/kg seed were effective in controlling seedling mortality up to 85% and were at par with carbendazim. Sharma et al. (2004) reported that *Aspergillus niger*, *Trichoderma harzianum*, *Trichoderma viride* and *Penicillium aurantiogriseum* and the bacterium (B1) and *Bacillus subtilis* were isolated from the rhizosphere, while *A. nidulans* var *cristatus*, *Drechslera spicifera*, *Gliocladium virens*, *Fusarium solani*, *Fusarium moniliforme*, *Fusarium oxysporum* and the bacteria (B2) were isolated from rhizoplane. Amongst the various rhizospheric microorganisms, *Trichoderma viride* and from rhizoplane microorganisms *Gliocladium virens* and bacterium (B2) proved effective against *Fusarium oxysporum* Schlecht under experimental condition. The spore of *F. oxysporum* Schlecht germinated minimum in association with rhizospheric *Trichoderma viride*. The rhizoplane microorganisms *G. virens* and bacterium (B2) exhibited minimum spore germination of *Fusarium oxysporum*.

9.6.3 *Bioformulations of* Trichoderma

Species of *Trichoderma* exhibiting good biological control activity have also proved to be preferably ideal to studies. These species compete well for food and site, grow well on root surfaces, synthesize a wide range of antibiotics and act as mycoparasites by utilizing an enzyme system having potential of attacking a broad range of plant pathogenic fungi. Those *Trichoderma* isolates that have exhibited good biological control activity most frequently belong to one of the four species aggregates: *T. hamatum*, *T. harzianum*, *T. koningii* and *T viride*. Biological control activity of these isolates has been demonstrated in vitro and glasshouse studies where the environment is controlled and in numerous field trials (Table 9.1).

Table 9.1 Examples of the successful field control of phytopathogens by *Trichoderma* species

Trichoderma	Crop	Pathogen	References
Trichoderma	Citrus trees, kiwi fruit vines, pine trees	*Armillaria* species	Bliss (1951), Cutler and Hill (1994), and Munnecke (1972)
Trichoderma	Apple, strawberry, kiwi fruit	*Botrytis cinerea*	Sutton and Peng (1993) and Tronsmo and Raa (1977)
Trichoderma	Stone fruit and other crops	*Chondrostereum purpureum*	Dubos and Ricard (1974) and Meyer and Plaskowitz (1989)
Trichoderma	Tomato, bean, iris, sugar beet, cotton	*Sclerotium rolfsii*	Elad et al. (1980), Latunde-Dada (1993), and Upadhyay and Mukhopadhyay (1986)
Trichoderma	Tomato	*Fusarium oxysporum*	Marois et al. (1981) and Sivan (1987)
Trichoderma	Apples	*Neonectria galligena*	Corke and Hunter (1979)
Trichoderma	Sugar beet	*Phoma betae*	Grondona et al. (1992)
Trichoderma	Strawberry, cucumber, potato, tomato, cotton	*Rhizoctonia solani*	Beagle-Ristaino and Papavizas (1985), Chet (1987), and Lewis and Papvizas (1980)
Trichoderma	Chrysanthemum	*Sclerotinia sclerotiorum*	Delgado De Kallman and Arbelaez Torres (1990)
Trichoderma	Onion	*Sclerotium cepivorum*	Abd-El-Moity and Shatla (1981)
Trichoderma	Maize, melon	*Macrophomina phaseolina*	Elad and Chet (1986)
T. harzianum T-39, T. atroviride P1	Bean, tomato, pepper, tobacco, lettuce,	*Botrytis cinerea*	De Meyer et al. (1998)
Trichoderma GT3-2	Cucumber	Green mottle mosaic virus	Lo et al. (1998)
T. harzianum T-22	Tomato	*Alternaria solani*	Seaman (2003)
T. asperellum T-203	Cucumber	*Colletotrichum orbiculare*	Koike (2001)
T. harzianum	Apple	*Phytophthora capsici*	Ahmed et al. (2000)
Trichoderma virens	Gladiolus	*Fusarium oxysporum*	Mishra et al. (2005)
Trichoderma harzianum	Chickpea	*Fusarium oxysporum*	Mukhopadhyay (1992)

However, despite extensive research over the last 70 years on the biological control capabilities of *Trichoderma* species, few isolates have been commercialized. Three examples of where preparations of *Trichoderma* have been commercialized include the marketing of *Trichoderma* biocontrol agents by Binab Corporation (Sigtuna, Sweden) and the use of *Trichoderma*-based biofungicide products (Agrimm Technologies Limited, New Zealand) in the New Zealand horticultural industry for the control of a range of plant pathogens and the marketing of a preparation called TRICHODEX (Abbot Laboratories, Australia) for the control of Botrytis bunch rot of grapes.

Some of the currently available commercial bioformulations of *Trichoderma* worldwide are Biofungus (Belgium), Bineb-T (Sweden, UK), Planterbox (USA), Rootpro, TRICHODEX, Trichoderma 2000 (Israel), Supresivit (Denmark), Trichopel and Trichodowels (New Zealand) and talc-based formulations – Bio-Cure-F, Biogourd, Funginil, Echoderma, Trieco, Trishul and Trichodermin-6 (Indian market).

9.7 Future Prospects

The fact that there are relatively few examples of commercialization can be attributed to the lack of consistency observed in the control of phytopathogens by *Trichoderma* species. In an attempt to address this problem, current investigation in the field of biological control is now being centralized on understanding how disease control is achieved and how the factors that affect its efficiency can be optimized. For instance, research is being focused more towards understanding the mode of action of *Trichoderma* biological control agents, with a view to enhancing biological control activity via either mutation or genetic manipulation of genes associated with biological control activity (Hayes et al. 1993). Research conducted at HortResearch, using 14 of the 50 strains of *Trichoderma*, has found that those that produced high quantities of the antimicrobial secondary metabolite 6-penryl-a-pyrone (PAP) and other active compounds exhibited the greatest biological control activity (Robert Hill pen comm.). Also, direct injection of kiwi fruit vines with either PAP (extracted from one of the *Trichoderma* isolates) or synthetic 6-amyl-cr-pyrone increased the survival rate of vines when under natural *Armillaria*-induced disease conditions. However, injection with formulations containing propagules of *Trichoderma* was more effective. Moreover, when pastes containing propagules of *Trichoderma* were applied directly to areas of infection, vines were completely healed, even in situations where as much as four fifths of the vascular cambium had been destroyed. The spread of *Armillaria* within kiwi fruit orchards was also inhibited when formulations, consisting of mixed populations of *Trichoderma* strains, were used to coat old tree stumps within the orchard or added to barrier trenches (physical barriers separating kiwi fruit vines from infectious *Armillaria* sites) or the soil (Cutler and Hill 1994). The same *Trichoderma* strains were also tested in pine tree field trials. The survival and vigour of pine trees were determined after 15 months. Those trees which had had their seedling root systems immersed in a slurry containing mixed populations of *Trichoderma* strains prior to planting had both a higher survival rate and were significantly more vigorous when compared to trees that were not treated (Cutler and Hill 1994). In addition to controlling *Armillaria*-induced disease, these same *Trichoderma* strains have also provided good control of *Chondrostereum purpureum* (silver leaf)-induced disease of pip fruit, stone fruit and *Leucadendron* as well as *Corticium rolfsii* (sclerotium) disease in capsicum.

There have been many benefits to the plants through *Trichoderma* applications which include marked improvement in plant growth, development and yield.

The use of modern techniques is helpful to study and identify molecular activities of fungus exploiting interaction with other microbes and the plant (Hanhong 2011; Mukherjee et al. 2013). To utilize for biotechnological improvement, ideal biological characters can be identified from selected species (Hermosa et al. 2012; Mukherjee et al. 2013). Next-generation products for higher yield and greater safety for agriculture market can be developed based on such information.

References

Abd-El-Moity TH, Shatla MN (1981) Biological control of white rot disease of onion (*Sclerotium cepivorum*) by *Trichoderma harzianum*. Phytopathol Z 100:29–35

Ahmed AS, Sanchez CP, Candela ME (2000) Evaluation of induction of systemic resistance in pepper plants (*Capsicum annuum*) to *Phytophthora capsici* using *Trichoderma harzianum* and its relation with capsidiol accumulation. Eur J Plant Pathol 106:817–824

Ansari AR, Dhirendra P (1986) Rhizosphere and rhizoplane mycoflora of berley infected with *Ustilago haordei*. Indian Phytophatol 39(3):380–384

Beagle-Ristaino JE, Papavizas GC (1985) Biological control of *Rhizoctonia* stem canker and black scurf of potato. Phytopathology 75:560–564

Bliss DE (1951) The destruction of *Armillaria mellea* in citrus soils. Phytopathology 41:665–683

Bolton B, Fredrickson JK, Elliott FE (1993) Microbial ecology of the rhizosphere. In: Metting FB Jr (ed) Soil microbial ecology: applications in agricultural and environmental management. Marcel Dekker, Inc, New York, pp 27–63

Bowen GD, Rovira AD (1991) The rhizosphere, the hidden half. In: Waisel Y, Eshel A, Kafkafi U (eds) Plant roots – the hidden half. Marcel Dekker Inc, New York, pp 641–649

Caldwell MM, Richards JH (1989) Hydraulic lift – water efflux from upper roots improves effectiveness of water-uptake by deep roots. Oecologia 79:1–5

Chakraborty BN, Chakraborty U, Saha A (1995) Defence strategies of tea (*Camellia sinensis*) against fungal pathogens. In: Daniel M, Purkayastha RP (eds) Handbook of phytoalexin metabolism and action. Marcel Dekker Inc, New York, pp 485–502

Chakraborty BN, Chakraborty U, Saha A, Sunar K, Dey PL (2010a) Evaluation of phosphate solubilizers from soils of North Bengal and their diversity analysis. World J Agri Sci 6(2):195–200

Chakraborty BN, Chakraborty U, Saha A, Dey PL, Sunar K (2010b) Molecular characterization of *Trichoderma viride* and *Trichoderma harzianum* isolated from soils of North Bengal based on rDNA markers and analysis of their PCR-RAPD profiles. Global J Biochem Biotechnol 5(1):55–61

Chakraborty BN, Chakraborty U, Dey PL, Sunar K (2010c) Phylogenetic relationship of *Trichoderma* isolates of North Bengal based on sequence analysis of ITS region of rDNA. J Appl Sci Res 6(10):1477–1482

Chet I (1987) *Trichoderma*- application, mode of action and potential as a biocontrol agent of soilborne plant pathogenic fungi. In: Chet I (ed) Innovative approaches to plant disease control. Wiley, New York, pp 137–160

Cook RJ (1993) Making greater use of introduced microorganisms for biological-control of plant-pathogens. Annu Rev Phytopathol 31:53–80

Corke ATK, Hunter T (1979) Biocontrol of *Nectria galligena* infection of pruning wounds on apple shoots. J Hortic Sci 54:47–55

Cutler HG, Hill RA (1994) Natural fungicides and their delivery systems as alternatives to synthetics. In: Wilson CL, Wisniewski ME (eds) Biological control of post harvest disease. Theory and practice. CRC Press Inc, Boca Raton, pp 135–152

De Meyer G, Bigirimana J, Elad Y, Hofte M (1998) Induced systemic resistance in *Trichoderma harzianum* T39 biocontrol of *Botrytis cinerea*. Eur J Plant Pathol 104:79–286
Delgado De Kallman L, Arbelaez Torres G (1990) Control of *Sclerotinia clerotiorum* (Lib) deary in *Chrysanthemum* and snapbean with different isolates of *Trichoderma* and with fungicides. Agronomia Colombiana 7:33–39
Domsch KH, Gams W, Anderson TH (1980) Compendium of soil fungi, vol I. Academic, London
Dubos B, Ricard JL (1974) Curative treatment of peach trees against silver leaf disease(*Stereum purpureum*) with *Trichoderma viride* preparations. Plant Dis Rep 58:147–150
Elad Y, Chet I, Karan J (1980) *Trichoderma harzianum*: A biocontrol agent effective against *Sclerorium rotfsii* and *Rhizoctonia solani*. Phytopathology 70:119–121
Elad Y, Zvieli Y, Chet I (1986) Biological control of *Macrophomina phaseolina* (Tassi) Goid by *Trichoderma harzianum*. Crop Prot 5:288–292
Gopinath A, Shekhar SH and Prakash HS (1987) Colonization of *Fusarium* species in sorghum seeds and their significance. Indian Phytopathology 40:181–185.
Grondona I, Perez de Algaba A, Monte E, Garcia-Acha I (1992) Biological control of sugarbeet diseases caused by *Phoma betae*: greenhouse and field tests. Brlerin Organisation Internationale de Lutte Biologique contre les Animaux et les Plantes Nuisiblesl Section Regionale Ouest Palearctique 15:39–41
Hanhong B (2011) *Trichoderma* species as abiotic and biotic stress quenchers in plants. Res J Biotechnol 6(3):73–79
Hawksworth DL, Sutton BC, Ainsworth GC (1983) Ainsworth and Bisby's dictionary of fungi, 7th edn. Commonwealth Mycological Institute, Kew
Hayes CK, Harman GE, Woo SL, Gullino ML, Lorito M (1993) Methods of electrophoretic karyotyping of filamentous fungi in the genus *Trichoderma*. Anal Biochem 209:176–182
Hee R (1991) Selection and identification on antagonistic rhizobacteria for controlling soil borne diseases of vegetables. Res Rep Rural Dev. ADM (Suweon) 33 (Icrop prot): 10–19
Hermosa R, Viterbo A, Chet I, Monte E (2012) Plant-beneficial effects of *Trichoderma* and of its genes. Microbiology 158(1):17–25
Hiltner L (1904) Uber neuere Erfahrungen und Probleme auf dem Gebiete der Bodenbakteriologie unter besonderer Berücksichtigung der Gründüngung und Brache. Arb DLG 98:59–78
Hinsinger P, Plassard C, Tang C, Jaillard B (2003) Origins of root mediated pH changes in the rhizosphere and their environment constraints – a review. Plant Soil 248:43–59
Hoitink HAJ, Boehm J (1999) Biocontrol within the context of soil microbial communities: A substrate-dependent phenomenon. In: Webster RK (ed) Annual review of phytopathology, vol 37. Annual Reviews Inc, Palo Alto, pp 427–446
Kallurmath RS, Rajasab AH (2000) Rhizosphere Mycoflora of onion. Indian J Mycol Plant Pathol 30(3):339–342
Koike N (2001) Induction of systemic resistance in cucumber against several diseases by plant growth-promoting fungi: lignification and superoxide generation. Eu J Plant Pathol 107:523–533
Kumar NR, Arasu TV, Gunasekaran P (2002) Genotyping of antifungal compounds producing plant growth promoting rhizobacteria, *Pseudomonas fluorescens*. Curr Sci 82(12):1463–1466
Latunde-Dada AO (1993) Biological control of southern blight disease of tomato caused by *Sclerotium rolfsii* with simplified mycelial formulations of *Trichoderma koningii*. Plant Pathol 42:522–529
Lewis JA, Papvizas GC (1980) Integrated control of *Rhizoctonia* fruit rot of cucumber. Phytopathology 70:85–89
Lo CT, Nelson EB, Hayes CK, Harman GE (1998) G. E. Ecological studies of transformed *Trichoderma harzianum* strain 1295-22 in the rhizosphere and on the phylloplane of creeping bentgrass. Phytopathology 88:129–136
Lynch JM, Whipps JM (1990) Substrate flow in the rhizosphere. Plant Soil 129:1–10
Marois JJ, Mitchell DJ, Sonoda RM (1981) Biological control of *Fusarium* crown rot of tomato under field conditions. Phyropathology 71:1257–1260

Mathivannan N, Manibhushanrao K, Murugesan K (2006) Biological control of Plant pathogens. In: Anand N (ed) Recent trends in biochemical research. University of Madras, Chennai, pp 275–323
Mathur K, Gurjar RBS, Jhamaria SL (2004) Rhizosphere Mycoflora of Chilli crop. Infected with *Rhizoctonia solani*. Indian J Mycol Plant Pathol 34(2):312–313
Meyer RJ, Plaskowitz JS (1989) Scanning electron microscopy of conidia and conidial matrix of *Trichoderma*. Mycologia 81:312–317
Mishra PK, Mukhopadhyay AN, Fox RTV (2005) Integrated and biological control of gladiolus corm rot and wilt caused by *Fusarium oxysporum* f.sp. *gladioli*. Ann App Biol 137(3):361–364
Mukherjee PK, Horwitz BA, Herrera-Estrella A, Schmoll M, Kenerley CM (2013) *Trichoderma* research in the genome era. Annu Rev Phytopathol 51:105–129
Mukhopadhyay AN (1992) Integrated control of Chickpea wilt complex by *Trichoderma* and chemical methods in India. Int J Pest Management 38(4):372–375
Mukhopadhyay AN (2009) Challenges, changes and choices before *Trichoderma* based biopesticides. J Mycol Plant Pathol 39(3):542
Munnecke DE (1972) Factors affecting the efficacy of fungicides in soil. Annu Rev Phytopathol 10:375–398
Pandey KK, Upadhyay JP (2000) Microbial population from rhizosphere and non-rhizosphere soil of pigeon pea. J Mycol Plant Pathol 30(1):7–10
Papendick RI, Campbell GS (1975) Water potential in the rhizosphere and plant and methods of measurement and experimental control. In: Bruehl GW (ed) Biology and control of soil-born plant pathogens. American Phytopathological Society, St. Paul, pp 34–49
Parveen S, Vijay RK (2004) Antagonism by *Trichoderma viride* against leaf blight pathogen of wheat. J Mycol Plant Pathol 34(2):220–222
Seaman A (2003) Efficacy of OMRI-approved products for tomato foliar disease control. N Y State Integr Pest Manag Program Publ 129:164–167
Shaik I, Nusrath M (1987) Vertical variation in the rhizosphere and rhizosphere mycoflora of *Cajanus cajan* with special reference to wilt diseases. Ind J Bot 10:126–129
Sharma M, Gupta SK (2003) Ecofriendly methods for the management of root-rot and web blight of French bean. J Mycol Plant Pathol 33(3):345–361
Sharma RN, Champawat RS, Gaur RB (2004) Efficacy of Rhizospheric and Rhizoplane microflora on spore germination in *Fusarium oxysporum*, the luciant of *Jojoba* wilt. J Mycol Plant Pathol 34(2):579–580
Singh F, Hooda I, Sindhan GS (2004) Biological control of tomato wilt caused by *Fusarium oxysporum* f. sp. *lycopersici*. J Mycol Plant Pathol 34(2):568–570
Sivan A (1987) Biological control of *Fusarium* crown rot of tomato by *Trichoderma harzianum* under field conditions. Plant Dis 71:587–592
Sorensen J (1997) The rhizosphere as a habitat for soil microorganisms. In: van Elsas JD, Trevors JT, EMH W (eds) Modern soil microbiology. Marcel-Dekker, Inc, New York, pp 21–45
Stephan A, Meyer AH, Schmid B (2000) Plant diversity affects culturable soil bacteria in experimental grassland communities. J Ecol 88:988–998
Sutton JC, Peng G (1993) Biocontrol of *Botrytis cinerea* in strawberry leaves. Phytopathology 83:615–621
Tiwari RKS, Singh A (2004) Efficacy of fungicides of *Rhizoctonia solani* and *Sclerotium rolfsii* and their effect on *Trichoderma harzianum* and *Rhizobium leguminosarum*. J Mycol Plant Pathol 34(2):212–224
Tronsmo A, Raa J (1977) Antagonistic action of *Trichoderma pseudokoningii* against the apple pathogen *Botrytis cinerea*. Phytopathol Z 89:216–220
Upadhyay JP, Mukhopadhyay AN (1986) Biological control of *Sclerotium rolfsii* by *Trichoderma harzianum* in sugarbeet. Trop Pest Manag 32(21):5–220
Vinale F, Sivasithamparam K, Ghisalberti EL, Marra R, Woo SL, Lorito M (2008) Review *Trichoderma*- plant pathogen interactions. Soil Biol Biochem 40:1–10

Chapter 10
Recruit the Plant Pathogen for Weed Management: Bioherbicide – A Sustainable Strategy

Manish Mathur and Praveen Gehlot

Abstract Innovative and sustainable methods of pathogen controls are the paramount in modern-day integrated disease management (IDM) approaches. Weeds, although representing the diversity of habitat, provide many ecological services, but at the same time, they are problematic and produce intense competition for our crops or other economic plants. Herbicides are largely being utilized for their control with some known advantages and disadvantages as well. Bioherbicides are the biological agent that utilizes for control of weeds. In present chapter, different attributes of bioherbicides like their basic terminology, advantages and disadvantages, approaches for combining them with other management tools, their implementation at field level, prior steps for research and product development, present status, market, attributes of some marketed products, types of their formulations, factors affecting efficacy of bioherbicides, and their mode of actions were discussed. Further weed management by means of utilization of biological agents was gauge through land use vise, viz., croplands, pasture/grazing lands, and water bodies. Frequency distribution analysis of such weed biological agent was carried out for the first time.

Keywords Bioherbicide · Herbicide resistance · *Colletotrichum* · *Phoma* · *Sclerotinia* · *Xanthomonas* · *Pseudomonas*

10.1 Weed and Their Dual Roles

Weeds are the unwanted plants that are unintentionally sown and occupied the larger spaces at various habitats. Within the plant community, they play dual roles. They perform many vital ecological functions like providing habitat for insect and animal, restoring the biodiversity and habitat, checking soil erosion, replenishing organic matter, and feeding and restoring soil life and are also useful in carbon

M. Mathur (✉)
ICAR-Central Arid Zone Research Institute, Jodhpur, India

P. Gehlot
Department of Botany, Jai Narain Vyas University, Jodhpur, India

P. Gehlot, J. Singh (eds.), *Fungi and their Role in Sustainable Development: Current Perspectives*, https://doi.org/10.1007/978-981-13-0393-7_10

Table 10.1 Potential crop yield loss due to weeds in India (Rao et al. 2014)

Crop	Yield loss (%)	Crop	Yield loss (%)	Crop	Yield loss (%)
Chickpea	10–50	Pea	10–50	Cotton	40–60
Pearl millet	16–65	Finger millet	50	Pigeon pea	20–30
Green gram	10–45	Potato	20–30	Groundnut	30–80
Rice	10–100	Horse gram	30	Sorghum	45–69
Jute	30–70	Soybean	10–100	Lentil	30–35
Sugarcane	25–50	Maize	3040	Vegetable	30–40

sequestration. At the same time, they are also very problematic as they compete for light, nutrient, moisture, and space with crops and plantations and allelopathic effect, serve as collateral and alternate host for plant pathogens, and restrict the air circulation around the crop. Thus with reference to sustaining human food demands, weeds are regarded as pest as they produce both bottom-up and top-down competitions for other plant species, consequently reducing the yield and value of our crops and plantation (Anjea et al. 2013). They are the major threats for our livestock health and impose severe competition for our important economic plant species. At global level, around 1800 weed species are affecting crop production (9.7%) that costs $ 32 billion economic losses (Li et al. 2003; Chutia et al. 2007). In India weed is causing 1980 Rs crores economic loss (Rao et al. 2014) with 10–100% yield reduction in different crops (Table 10.1).

Weeds are more proliferating then our crops and such invasive capacity are associated with their traits like r-selection species (producing larger progeny), deep root system, drought and frost hardy and high nutrient utilization efficiency then crops. They spread through various agencies like weed, animal, water, and air. Global spread (spread in different countries) of weed species at agricultural and at pasture-/rangelands is depicted in Figs. 10.1 and 10.2, respectively. *Amaranthus hybridus*, *A. spinosus*, *Avena fatua*, *Chenopodium album*, *Convolvulus arvensis*, *Cynodon dactylon*, *Cyperus esculentus*, *Cyperus rotundus*, *Digitaria sanguinalis*, *Echinochloa crus-galli*, *Eleusine indica*, *Paspalum conjugatum*, *Portulaca oleracea*, *Rottboellia*, and *Sorghum halepense* are the major weed species of agricultural lands, and among them *Cyperus rotundus* is the most problematic species spreading in 92 different countries followed by *Portulaca oleracea* and *Cynodon dactylon.* Similarly *Ageratum* spp., *Agropyron repens*, *Anagallis arvensis*, *Argemone mexicana*, *Bidens pilosa*, *Brachiaria mutica*, *Capsella bursa-pastoris*, *Cenchrus echinatus*, *Chromolaena odorata*, *Cirsium arvense*, *Commelina benghalensis*, *Cyperus difformis*, *Cyperus iria*, *Dactyloctenium aegyptium*, *Digitaria adscendens*, *Eclipta prostrata*, *Euphorbia hirta*, *Galinsoga parviflora*, *Heliotropium indicum*, *Lantana camara*, *Lolium temulentum*, *Mikania cordata*, *Mimosa* spp., *Panicum maximum*, *Panicum repens*, *Paspalum dilatatum*, *Pennisetum clandestinum*, *Pennisetum purpureum*, *Polygonum convolvulus*, *Salvinia auriculata*, *Setaria verticillata*, *Setaria viridis*, *Sida acuta*, *Solanum nigrum*, *Sonchus oleraceus*, *Spergula arvensis*, *Stellaria media*, *Striga lutea*, *Tribulus terrestris*, and *Xanthium strumarium* are dominating at pasturelands (Bailey 2014).

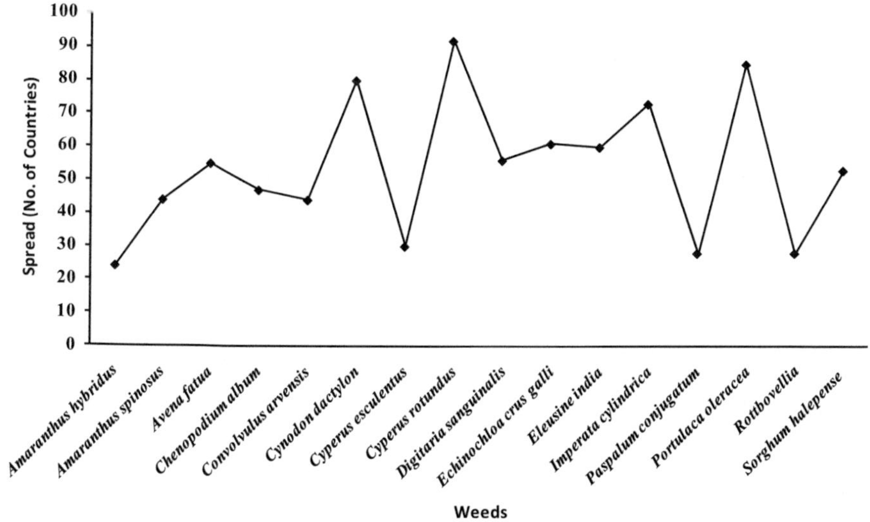

Fig. 10.1 Global spread of weed in agricultural lands. (Prepared with the help of text by Bailey 2014)

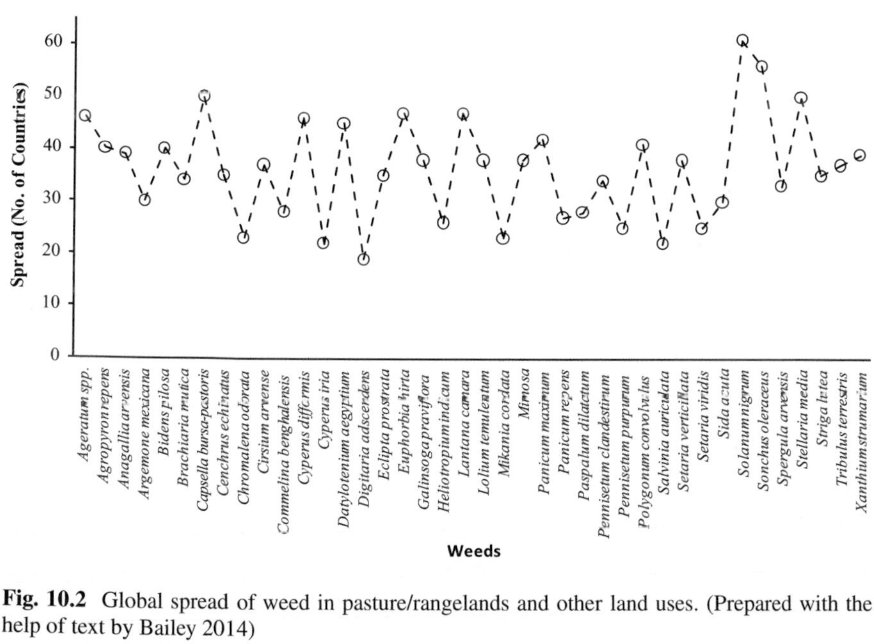

Fig. 10.2 Global spread of weed in pasture/rangelands and other land uses. (Prepared with the help of text by Bailey 2014)

There are varieties of mechanism available for weed control led by intensive use of synthetic herbicides (Gordeau et al. 2016) costing $14 billion annually (Kiely et al. 2004). Despite of their rapid and effective mode of actions, herbicides suffers with several disadvantages pertains to (1) many of them are non-biodegradable and having long resistant period, (2) slight to moderate toxicity. (3) with irrigation or rainfall they leached into underground water and causing problems for flora and fauna and (4) grazing of such treated plants by herbivores facilitate the passed up of such toxic substance in to the food chain and that ultimately harmful for animal and humans (Jurado et al. 2011).

Therefore, novel and sustainable weed control/management strategies are required that keep the weed population below economic threshold level (Bajwa 2014). Integrated weed management (IWM) are very much effective for arresting the weed population below certain threshold level. IWM provides a better cost-effective and environmentally safe procedure for their control (Zimdahl 2004). One step ahead, now the researchers are linking the weed-crop interactions with competition types and associated traits. Biological weed control by means of living entities, predominantly with fungal pathogens, has the greater possibility for sustainable crop production and weed management (Aneja 2009).

10.2 Sustainable Weed Management

Sustainable crop production is an important aspect to ensure the global food demands and environmental safety. With respect to crop protection, sustainable disease management typically involves an integration of various individual component actions, which include:

- Raising of nursery material or seeds that have the capacity to avoid harmful pathogens.
- Use of integration of vertical and horizontal resistance, disease deployment, and multiline varieties.
- Adopting innovating cultural practices like early- and late-sowing varieties, crop rotations, etc. Such practices will inhibit pathogen survival and their development.
- Integration of traditional as well as innovative disease management practices.

Under weed management approach, sustainable strategy deals with minimizing the open niches for weeds in cropping systems. Bajwa (2014) reviewed the sustainable weed management strategies in relation to conservation agriculture. He has listed modified tillage, improved cultural practices, mulching and soil cover, biological weed control (bioherbicides), allelopathy, crop nutrition, and integrated weed management (crop rotation, crop residue management, adjustment of sowing time, preventive measures, herbicide management, seed rate adjustment, planting geometry adjustment, etc.).

Sustainable weed management can be categorized into preseason planning, preventive practices, and control tactics. Such categories provide knowledge of weed type, controlling weed spread, and their sustainable control.

10.3 Bioherbicides

The term bioherbicide refers to herbicide that relies on natural living agents such as fungi, bacteria, viruses, protozoans, and nematodes (Anjea et al. 2013). The Conseil international de la langue francaise in 1971 defined bioherbicides as the material deliberately introduced to reduce weed population that are environmentally safe also. Boyetchko and Peng (2004) and Bailey (2014) suggested that bioherbicides are phytopathogenic microorganism or microbial phytotoxins useful for weed control that are applied and worked similarly as conventional herbicides. When the organism used is a fungus, the product is termed as mycoherbicide. However, the use of pathogens other than fungi as bioherbicides is limited. Major traits of bioherbicides are provided by Bailey (2014); he described them on the basis of their host specificity, crop tolerance, efficacy, environmental fate, temperature and moisture spectrum, mode of action, and toxicology.

10.3.1 Advantages and Disadvantages of Bioherbicides

With comparison to chemical herbicides, bioherbicides provided many advantages, and these include:

- Environmental friendly, and they generally minimize the residue buildup. Thus, this approach recues the negative environmental impacts.
- Provides targeted weed control.
- Safer for beneficial plant or man and nontarget species.
- Avoids other problems which are associated with chemical agents like herbicide resistance breaking (Charudattan 2001).
- Lower production cost as compared to the chemical agents.
- Can be worked at microlevel quantity and decompose rapidly as well.
- Highly fulfill the objectives of integrated disease management programs.

Thus, desirable characteristics of a bioherbicide included the following traits:

- Specific host range
- Easy to use
- Genetic constancy
- Cost-effective mass production with high shelf life
- Rapid and predictable mode of actions
- Besides these advantages, bioherbicides also possessed several disadvantages which are related with environmental, biological, technical, and commercial restrictions (Altman et al. 1990; Bailey et al. 1998; Kempenaar and Scheepens 1999; Wheeler and Center 2001; Auld et al. 2003 and Pacanoski 2015):
- Their modes of action are depending on environmental conditions.
- High degree of host specificity and its resistance types.

- Technical restrictions are related with their development and large-scale productions.
- Market competence, patent protection, confidence, and adjustment are their commercial restrictions.

10.3.2 Approaches for Combining Bioherbicide with Other Management Tools

Under the era of sustainable development, this innovative management tool can also be visualized with its incorporation with other tactics. Scharer and Collins (2012) have partitioned these combinations under the vertical and horizontal integration. Vertical integration is related with the control tactics against a single weed species, while the other one is related with the control of many weed species in one crop. Horizontal approach chiefly comprises the mixture of microbial herbicides with chemical or mechanical methods.

With respect to vertical integration, Muller-Scharer and Collins (2012) further described purpose-specific approaches, ecological integration, and physiological integration. Under the purpose-specific approach, the level of control is selected based on the context-specific requirements, that is, involvement of different management tools applied at different sites. For example, chemical agents may be adopted to avoid new weed infestations that are currently under spreading; however, biological control may be used for long-term control of large and established infestation (Yang et al. 2000). Synergistic effects of acifluorfen and bentazon provided significant control in several weed/pathogen combinations (*Senna obtusifolia-Alternaria cassia*, *Aeschynomene virginica-Colletotrichum gloeosporioides*, *Sesbania exaltata-Colletotrichum truncatum*, and *Desmodium tortuosum-Fusarium lateritium*). Similarly synergistic interaction between *Myrothecium verrucaria* and glyphosate provides better control of *Brunnichia ovata* and *Campsis radicans*.

Under the ecological integration, different tactics are applied at the same time on the same infestation. Such integrations provide environmental effective and holistic weed management (Muller and Reiger 1998; Wymore et al. 1987; DiTommaso et al. 1996; Sheppard 1996). Physiological integration relies on synergistic changes in weed biochemistry, brought by sublethal effects of herbicides and by biological control agents. Herbicides (synergists) are known to facilitate pathogen (or biological agent) entry and its initial establishment within the host.

10.3.3 Implementation of Bioherbicide at Field

At the field level, bioherbicide can be implemented through either by introduction of foreign pathogenic organisms (classical approach) or by augmentative bioherbicide strategies (Fig. 10.3) for biological weed control. With respect to augmentative

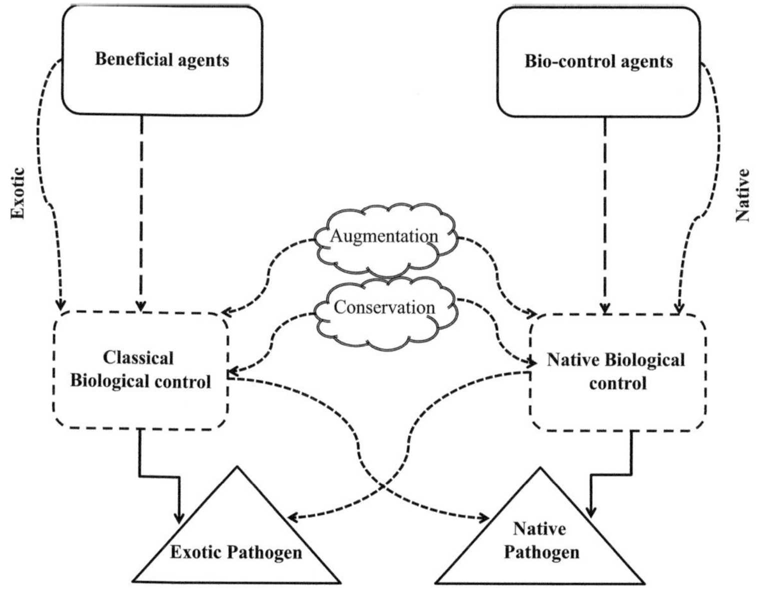

Fig. 10.3 Strategies for implementing the biological control of weed by pathogens

approach, pathogenic or biological agents are already present (native or introduced), and their population can be enhanced by mass rearing (Dagno et al. 2012). Hasan and Ayres (1990) termed these approaches as inoculative and inundative strategy, respectively.

With respect to initial size of the inoculums released, the origin of the control organism used and the ecosystem where the biological control program implemented Muller-Scharer and Collins (2012) provided three basic approaches used for bioherbicide application, and these includes inundation which are related to the application or introduction of bioagent materials (fungal spores or bacterial suspensions) at concentrations that would not normally occur in nature (Harding and Raizada 2015), system management approach (i.e., conservation, i.e., protection or maintenance of existing population of biocontrol agents), and classical approach (i.e., inoculation) under which natural pathogen of species is released that has the ability to reduce the weed species in an entire ecosystem (Shaw et al. 2009). In both classical and inundative approaches, the agent inoculums can be used as liquid sprays or solid granules (Caldwell et al. 2012).

Classical-type approach is generally more frequently utilized for biological weed control compared to conservation approach. This approach is mainly directed toward the control of exotic weeds that have invaded, spread, and established. Weed control is accomplished through the introduction and release of highly host-specific pathogens. Such bioagent is established and reproduced on weed host and gradually suppressed its population. Examples of such weed-pathogen interaction are *Ageratina riparia-Entyloma ageratinae*, *Acacia saligna-Uromycladium tepperianum*,

Chondrilla juncea-Puccinia chondrillina, *Centaurea solstitialis-Puccinia jaceae* var. *solstitialis*, and *Passiflora tripartita-Septoria passiflorae*. High initial cost, high host specificity of BCA, and difficulties in post-release controls of BCA are the major drawbacks of classical method. Further, increase in BCA populations and subsequent effective weed control by them are also linearly associated with environmental conditions (El-Sayed 2005).

With respect to inundative approach, native plant pathogens are isolated from weed and are cultured for producing large quantity of infective material (Kremer 2005). Such agents eradicate the problematic weed before it reaches to economic damage threshold (Tinaudo et al. 2010; Pacanoski 2015). Under this tactic, the weed pathogen is applied alike herbicide. However, faith of bio-inoculums is entirely depending on environmental conditions. Weed pathogen examples of this approach are *Aeschynomene virginica-Colletotrichum gloeosporioides* f. sp. *aeschynomene*, *Morrenia odorata-Phytophthora palmivora*, *Cyperus* spp.-*Puccinia canaliculata*, *Isatis tinctoria-Puccinia thlapeos*, and *Taraxacum-Scleotinia minor*.

10.3.4 Prior Steps for Research and Product of Bioherbicides

Many important aspects should be considered before the research and marketable product of bioherbicides, and these include:

1. Threat imposed by the weed
2. Succession stage of the weed (establishment/climax):
    (a) With consideration to its establishment, is the eradication or containment feasible?
    (b) Survival strategies adopted by weed (Deevy curve and R/K selection).
3. What are the ecosystem services provided by the weed (provisional, cultural, or regulatory)? For example, medicinal values, dune plantation, soil binder, broom making, etc.
4. Availability of other biocontrol methods and legislation support by the government
5. Density and spatial extent and type of weed
6. Availability of possible biocontrol agents

Various prior considerations for the development of a bioherbicide are depicted in Fig. 10.4. Major steps in developing a bioherbicide are:

1. Check that a bioherbicide product is needed and that there is sufficient industry and commercial backing to proceed.
2. Searching for suitable pathogens (if not previously explored).
3. Identification of highly pathogenic (disease-causing) isolates for weed species.
4. Develop an efficient methodology for their mass production with high shelf life.

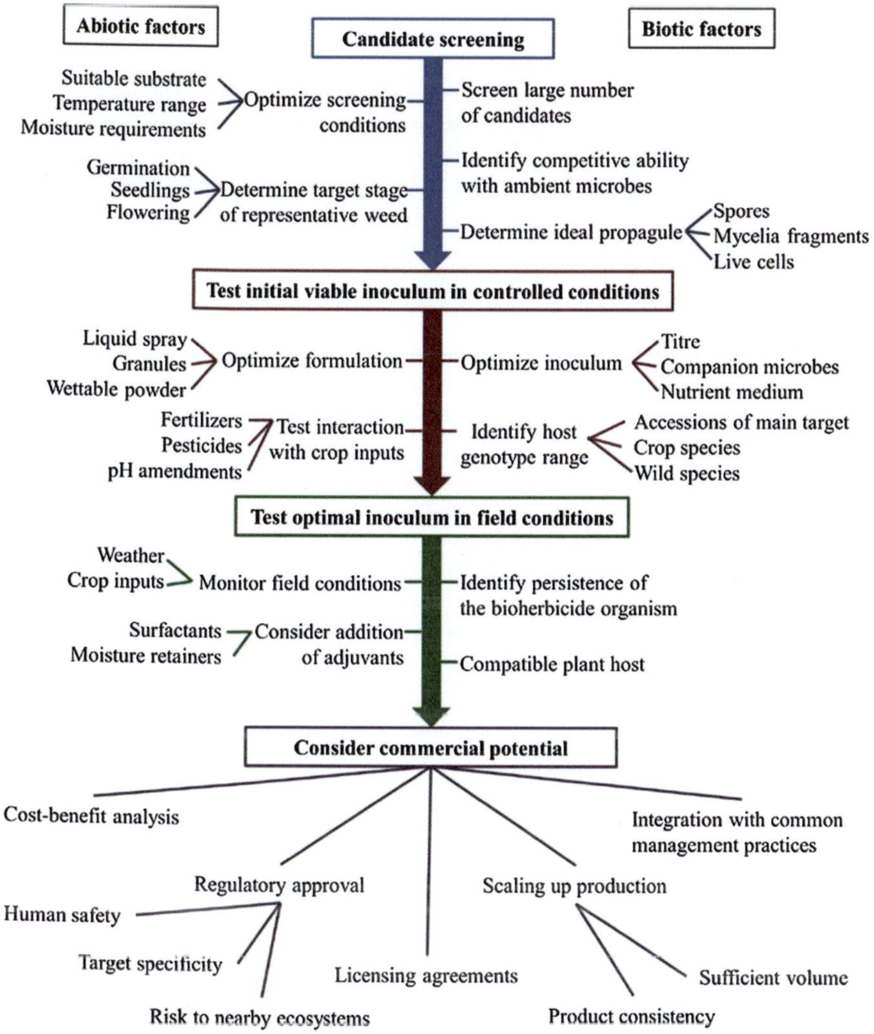

Fig. 10.4 Important attributes associated with the development of a bioherbicide. (Adopted with kind permission of the author Manish N Raizada)

5. Complete information on suitable conditions for infection and disease development.
6. Knowledge of pathogen host specificity and virulence efficacy with other hosts.
7. Development of most suitable formulation and their application technology.
8. Field trials and improve formulation if necessary.
9. Obtain registration for the product, and market and distribute the product.

10.3.5 Present Status

In this chapter we have reviewed many available resources from different researchers/agencies from different geographical locations and observed that 132 different weed species have evaluated for their biological control specifically with bioherbicides (Table 10.2). Land use vise can be categorized under the cropland with 77 weeds, grazing land/pasture/forestlands (50), and water bodies (5). Seventy-seven different cropland weeds were evaluated with one hundred bioagents. Biocontrol of 50 grazing land/pasture/forest weed species were evaluated with 71 bioagents, and similarly 5 water weeds were evaluated with 15 biocontrol agents. *Alternanthera philoxeroides*, *Eichhornia crassipes*, *Hydrilla verticillata*, *Mimosa pigra*, and *Salvinia molesta* are the weed species and were evaluated with *Alternaria eichhorniae*, *Alternaria eichhorniae isolate 5*, *Cercospora echii*, *Cercospora rodmanii* (ABG-5003), *Cercospora piaropi*, *Cyrtobagous salviniae*, *Diabole cubensis*, *Fusarium culmorum*, *Mycoleptodiscus terrestris*, *Neochetina bruchi*, *Neochetina eichhorniae*, *Nimbya alternantherae*, *Paulinia acuminata*, *Phoma hedericola*, and *Sphaerulina mimosae-pigrae*. Among these waterweeds, *Eichhornia crassipes* were evaluated with maximum seven bioherbicide agents. With reference to the weed of croplands, frequency distribution analysis suggested that the number of use of bioherbicides for control of 77 weeds is not following normal distribution (Fig. 10.5). Control of most of the weeds is gauge with using one bioherbicide only. However, control of *Taraxacum officinale* has been visualized with six different bioherbicides like *Phoma herbarum*, *Phoma macrostoma*, *Puccinia* sp., *Ramularia inaequale*, *Sclerotinia minor (Sarritor)*, and *Sphaerotheca fuliginea* followed by *Striga hermonthica* and *Abutilon theophrasti* weeds. Among the bioherbicide agent, *Myrothecium verrucaria* (*Brunnichia ovata*, *Euphorbia maculata*, *Euphorbia supine*, *Portulaca oleracea*, *Pueraria lobata*, and *Trianthema portulacastrum*) and *Arbuscular mycorrhizal* fungi (*Amaranthus retroflexus*, *Chenopodium album*, *Digitaria sanguinalis*, *Echinochloa crus-galli*, Setaria viridis, and *Sinapis arvensis*) were approached for the control of six different weeds. Details of other weed-bioherbicide complexes are presented in Table 10.2. Similar frequency distribution result also revealed for grazing/pasture/forestland weeds. Eight different bioherbicide agents (*Alternaria cirsinoxia*, *Phomopsis cirsii*, *Phoma nebulosa*, *Phyllosticta cirsii*, *Pseudomonas syringae* pv. *tagetis*, *Puccinia punctiformis*, *Sclerotinia sclerotiorum*, and *Stagonospora* sp.) were approached for the control of *Cirsium arvense* weed (Fig. 10.6).

10.3.6 Market of Bioherbicides

Market report (http://www.marketsandmarkets.com/Market-Reports/bioherbicides-market-175213366.html?gclid=EAIaIQobChMI2NHRofaw1gIVlgRoCh1C6A8nEAAYASAAEgJnQ_D_BwE) evaluated the bioherbicide value at USD 698.7 million

Table 10.2 Examined weed species for their control using plant pathogens

S. no	Weed name	Pathogen use for control
1	*Ageratina riparia*	*Entyloma ageratinae*
2	*Abutilon theophrasti*	*Colletotrichum coccodes* **(Velgo)**
3	*Acacia saligna*	*Uromycladium tepperianum*
4	*Acacia* spp.	*Cylindrobasidium laeve* **(Stumpout)**
5	*Acroptilon repens*	*Subanguina picridis*
6	*Aegilops cylindrica*	*Pseudomonas* and *Xanthomonas* spp.
7	*Aeschynomene virginica*	*Colletotrichum gloeosporioides* f. sp. *aeschynomene* **(Collego)**
8	*Ageratina adenophora*	*Phaeoramularia eupatorii-odorati*
9	*Ailanthus altissima*	*Aecidium ailanthi, Coleosporium* sp.; *Fusarium oxysporum* f. sp. *perniciosum*; *Verticillium albo-atrum*; *Phoma destructiva*; *Phoma hedericola*; *Mycelia sterile*; *Phoma nebulosa*; *Phomopsis cirsii*
10	*Alternanthera philoxeroides*	*Nimbya alternantherae*
11	*Amaranthus hybridus*	*Phomopsis amaranthicola*; *Alternaria cassiae*; *Colletotrichum dematium*; and *Fusarium udum*
12	*Amaranthus retroflexus*	*Albugo amaranthi, Arbuscular mycorrhizal fungi*
13	*Amaranthus* spp.	*Phomopsis amaranthicola*; *Alternaria alternata*; *Trematophoma lignicola*
14	*Ambrosia artemisiifolia*	*Albugo tragopogonis*
15	*Ambrosia trifida*	*Protomyces gravidus*
16	*Anoda cristata*	*Alternaria macrospora*
17	*Arceuthobium tsugense*	*Neonectria neomacrospora*
18	*Artemisia vulgaris*	*Puccinia absinthii*; *Erysiphe artemisiae*; *Septoria artemisiae*; and *Mycovellosiella ferruginea*
19	*Atriplex patula*	*Passalora dubia*
20	*Avena fatua*	*Fusarium avenaceum*; *Fusarium culmorum*; *Drechslera avenacea*; and *Puccinia coronata* f. sp. *avenae*
21	*Baccharis halimifolia*	*Puccina evadens*
22	*Broad-leaved trees*	*Chondrostereum purpureum* **(BioChon)**
23	*Bromus diandrus*	*Pseudomonas trivialis strain X33D*
24	*Bromus tectorum*	*Pseudomonas fluroescens strain D7*; *Pyrenophora seminiperda*; *Ustilago bullata*
25	*Brunnichia ovata*	*Neonectria neomacrospora* and *Myrothecium verrucaria*
26	*Carduus tenuiflorus*	*Puccinia carduorum*
27	*Calystegia sepium*	*Stagonospora convolvuli*
28	*Campsis radicans*	*Neonectria neomacrospora, Myrothecium verrucaria*
29	*Cannabis sativa*	*Puccinia carduorum, Fusarium oxysporum* var. *cannabis*
30	*Carduus pycnocephalus*	*Puccinia cardui-pycnocephali*
31	*Carduus thoermeri*	*Puccinia carduorum*
32	*Cassia obtusifolia*	*Alternaria cassiae* **(CASST)**
33	*Cassia occidentalis*	*Alternaria cassiae*

(continued)

Table 10.2 (continued)

S. no	Weed name	Pathogen use for control
34	*Centaurea cyanus*	*Puccinia cyani*, *Puccinia jacea*, and *Puccinia absinthii*
35	*Centaurea diffusa*	*Fusarium oxysporum*
36	*Centaurea maculosa*	*Fusarium oxysporum*
37	*Chenopodium album*	*Ascochyta caulina*, *Monodidymaria chenopodii* and *Ascochya chenopodii*, *Ascochya caulina*, *Arbuscular mycorrhizal fungi*
38	*Chondrilla juncea*	*Puccinia chondrillina*
39	*Chromolaena odorata*	*Apion brunneonigrum*; *Cecidochares connexa*
40	*Chrysanthemoides monilifera* ssp. *monilifera*	*Endophyllum osteospermi*
41	*Cirsium arvense*	*Phyllosticta cirsii*; *Stagonospora sp.*; *Alternaria cirsinoxia*; and *Pseudomonas syringae* pv. *tagetis*, *Sclerotinia sclerotiorum*, *Puccinia punctiformis*, *Pseudomonas syringae* pv. *tagetis*, and *Phomopsis cirsii*
42	*Clematis vistalba*	*Phoma clematidina*
43	*Clidemia hirta*	*Colletotrichum gloeosporioides* f. sp. *clidemiae*
44	*Convolvulus arvensis*	*Stagonospora convolvuli*, *Phomopsis convolvulus*, *Erysiphe convolvuli*
45	*Cirsium arvense*	*Puccinia punctiformis*; *Stenella kansensis*; *Ramularia cynarae*; and *Erysiphe cichoracearum*
46	*Crotalaria spectabilis*	*Phomopsis amaranthicola*; *Alternaria cassiae*; *Colletotrichum dematium*; *Fusarium udum*; and *Ascochyta cirsii*
47	*Cryptostegia grandiflora*	*Maravalia cryptostegiae*
48	*Cucurbita texana*	*Fusarium solani* f. sp. *cucurbitae*
49	*Cuscuta* spp.	*Colletotricum gloeosporioides* f. sp. *cuscutae* **(Lubao)**, *Alternaria destruens*
50	*Cyperus esculentus*	*Puccinia canaliculata* **(Dr. BioSedge)**
51	*Cyperus rotundus*	*Dactylaria higginsii* and *Cercospora caricis*, *Puccinia romagnoliana*
52	*Cytisus scoparius*	*Fusarium tumidum*
53	*Damasonium minus*	*Plectosporium alismatis*
54	*Datura stramonium*	*Colletotrichum truncatum*
55	*Digitaria sanguinalis*	*Curvularia eragrostidis*, *Curvularia intermedia*, *Arbuscular mycorrhizal fungi*
56	*Dodder*	*Alternaria destruens* **(Smolder)**
57	*Echinochloa crus-galli*	*Exserohilum fusiform*, *Eucalyptus tereticornis*, and *Arbuscular mycorrhizal fungi*
58	*Echium plantagineum*	*Cercospora echii*
59	*Eichhornia crassipes*	*Neochetina eichhorniae* and *Neochetina bruch*, *Cercospora echii*, and *Alternaria eichhorniae*, *Alternaria eichhorniae isolate 5*, and *Cercospora piaropi*
60	*Elytrigia repens*	*Ascochyta agropyrina* var. *nana*
61	*Erythroxylum coca*	*Fusarium oxysporum* f. sp. *erythroxyli*

(continued)

Table 10.2 (continued)

S. no	Weed name	Pathogen use for control
62	*Euphorbia escula*	*Uromyces scutellatus*
63	*Euphorbia heterophylla*	*Bipolaris euphorbia*, *Sphaceloma poinsettiae*
64	*Euphorbia maculata*	*Myrothecium verrucaria*
65	*Euphorbia supina*	*Myrothecium verrucaria*
66	*Euphorbia helioscopia*	*Melampsora helioscopae* and *Sphaerotheca euphorbiae*
67	*Fallopia convolvulus*	*Puccinia polygoni*
68	*Galega officinalis*	*Uromyces galegae*
69	*Galeopsis tetrahit*	*Septora galeopsidis*
70	*Galinsoga ciliata*	*Colletotrichum gloeosporioides*
71	*Galium spurium*	*Plectosporium tabacinum*
72	*Galium aparine*	*Puccinia punctata*
73	*Gaultheria shallon*	*Phoma exigua* and *Valdensina heterodoxa*
74	*Hakea sericea*	*Colletotrichum gloeosporioides* **(Hatatak)**
75	*Hedychium gardnerianum*	*Ralstonia solanacearum*
76	*Heliotropium europaeum*	*Uromyces heliotropii*
77	*Hydrilla verticillata*	*Fusarium culmorum* and *Mycoleptodiscus terrestris*
78	*Hypericum androsaemum*	*Melampsora hypericorum*
79	*Imperata cylindrica*	*Colletotrichum caudatum*; *Bipolaris sacchari* and *Drechslera gigantea*
80	*Isatis tinctoria*	*Puccinia thlaspeos*
81	*Lantana camara*	*Mycovellosiella lantanae* var. *lantanae* and *Corynespora cassiicola* f. sp. *lantanae*, *Ophiomyia lantanae*, *Teleonemia scrupulosa*, *Diastema tigris*, and *Salbia haemorrhoidalis*
82	*Malva pusilla*	*Colletotrichum gloeosporioides* **(Biomal)**
83	*Matricaria perforata*	*Colletotrichum truncatum*
84	*Mentha arvensis*	*Uromyces viciae-craccae* and *Puccinia menthae*
85	*Miconia calvescens*	*Ditylenchus drepanocercus*
86	*Microstegium vimineum*	*Bipolaris* species
87	*Mikania micrantha*	*Puccinia spegazzinii*
88	*Mimosa pigra*	*Sphaerulina mimosae-pigrae* and *Diabole cubensis*
89	*Morrenia odorata*	*Phytophthora palmivora* **(DeVine)**
90	*Nassella neesiana*	*Uromyces pencanus*
91	*Orobanche cumana*	*Fusarium oxysporum* f. sp. *orthoceras*
92	*Opuntia* spp.	*Dactylopius ceylonicus*; *Dactylopius confusus*; *Dactylopius opuntiae*; and *Opuntia elatior*
93	*Orobanche crenata*	*Pseudomonas fluorescens Pf7–9*
94	*Orobanche aegyptiaca*	*Fusarium solani*
95	*Orobanche foetida*	*Pseudomonas fluorescens Pf7–10*
96	*Orobanche ramosa*	*Fusarium oxysporum*

(continued)

Table 10.2 (continued)

S. no	Weed name	Pathogen use for control
97	*Orobanche* spp.	*Fusarium oxysporum* f. sp. *erythroxyli*
98	*Papaver somniferum*	*Pleospora papaveracea*
99	*Parthenium hysterophorus*	*Cladosporium* sp. *(MCPL-461)*, *Puccinia abrupta* var. *partheniicola*, *Zygogramma bicolarata*
100	*Passiflora tripartita*	*Septoria passiflorae*
101	*Poa annua*	*Xanthomonas campestris* **(Camperico)**
102	*Polygonum aviculare*	*Erysiphe polygoni*; *Uromyces polygoni-avicularis*; and *Puccinia punctata*
103	*Portulaca oleracea*	*Myrothecium verrucaria*, *Dichotomophthora portulacaceae*
104	*Prunus serotina*	*Chondrostereum purpureum* **(BioChon)**
105	*Pueraria lobata*	*Myrothecium verrucaria*
106	*Ranunculus acris*	*Sclerotinia sclerotiorum*
107	*Raphanus raphanistrum*	*Hyaloperonospora parasitica*; *Pseudomona fluorescens*
108	*Rottboellia cochinchinensis*	*Sporisorium ophiuri* and *Colletotrichum graminicola*
109	*Rubus* spp.	*Phragmidium violaceum*
110	*Sagittaria* spp.	*Rhynchosporium alismatis*
111	*Salsola kali*	*Uromyces salsolae*
112	*Salvinia molesta*	*Paulinia acuminata* and *Cyrtobagous salviniae*
113	*Schinus terebinthifolius*	*Neofusicoccum batangarum*
114	*Senecio vulgaris*	*Puccinia lagenophorae*
115	*Senna obtusifolia*	*Phomopsis amaranthicola*; *Alternaria cassiae*; *Colletotrichum dematium*, *Fusarium udum*, *Colletotrichum gloeosporioides*, *Alternaria cassiae*
116	*Sesbania exaltata*	*Colletotrichum truncatum* **(Coltru)**
117	*Setaria viridis*	Arbuscular mycorrhizal fungi
118	*Sinapis arvensis*	Arbuscular mycorrhizal fungi
119	*Solanum ptychanthum*	*Colletotrichum coccodes*
120	*Solanum viarum*	*Ralstonia solanacearum*
121	*Sonchus arvensis*	*Coleosporium sonchi* and *Erysiphe cichoracearum*
122	*Sorghum halepense*	*Sporisorium cruentum*
123	*Sphenoclea zeylanica*	*Alternaria* sp. and *Colletotrichum gloeosporioides*
124	*Setaria viridis*	*Drechslera gigantea*; *Exserohilum rostratum*; *Exserohilum longirostratum*; and *Pyricularia setariae*
125	*Striga hermonthica*	*Fusarium oxysporum*; *Pseudomonas fluorescens* and *Pseudomonas putida*, *Fusarium nygamai*; *Fusarium oxysporum* and *Fusarium semitectum* var. *majus*
126	*Taraxacum officinale*	*Phoma herbarum*; *Phoma macrostoma* and *Sclerotinia minor*, *Ramularis inaequale*; *Puccinia sp.* and *Sphaerotheca fuliginea*
127	*Trianthema portulacastrum*	*Gibbago trianthema*, *Myrothecium verrucaria*

(continued)

Table 10.2 (continued)

S. no	Weed name	Pathogen use for control
128	*Ulex europaeus*	*Chondrostereum purpureum* and *Fusarium tumidum*
129	*Vicia cracca*	*Erysiphe pisi*
130	*Xanthium occidentale*	*Puccinia xanthii*
131	*Xanthium* spp.	*Alternaria zinniae*
132	*Xanthium strumarium*	*Alternaria helianthi, Curvularia lunata*

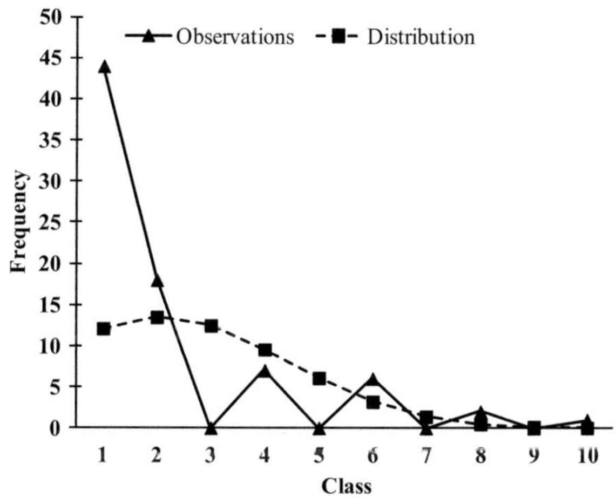

Fig. 10.5 Frequency distribution of screening of the weed with respect to the number of pathogens for their controls at croplands

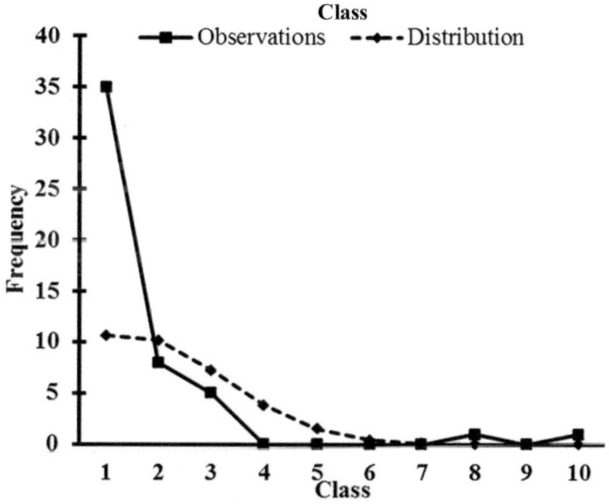

Fig. 10.6 Frequency distribution of screening of the weed with respect to the number of pathogens for their controls at grazing lands

in 2015 projected to grow at a compound annual growth rate (CAGR) of 14.5% from 2016, to reach USD 1573.7 million by 2021. Review of literature suggested that at global level, 17 different mycoherbicide formulations/products have been registered; of these, 8 were registered in the USA, 4 in Canada, 2 in South Africa, and 1 each in the Netherlands, Japan, and China (Aneja et al. 2013). The bioherbicide market can be segmented as follows:

Base type	Segment type
On the basis of source	Microbial, biochemical, and others (plant phytotoxic residues and other botanical extracts)
On the basis of application	Seed, soil, foliar, and postharvest
On the basis of application	Agricultural crop type and nonagricultural crop type
On the basis of formulation	Granular, liquid, and others (pellets and dust/powder form)
On the basis of region	North America, Europe, Asia-Pacific, and Latin America

Ash (2010) pointed out that the interest in bioherbicide research has remained consistent as there were 509 papers published from 1987 to 2009 (23 papers/year); he further indicated low success rate in biopesticide research with consideration of registration and commercialization.

A bioherbicide product/formulation is regarded as verifiable success when it is having regularity in their registration, in their commercialization, and in their utilization Charudattan (2005). That means bioherbicides that were registered but not commercialized for various reasons are not considered successful. He also concluded that that there have been five variable successes: *Colletotrichum gloeosporioides* f. sp. *aeschynomene*, *Phytophthora palmivora*, *Xanthomonas campestris*, *Chondrostereum purpureum*, and *Acremonium* sp. Bailey and Falk (2011) point out that when we compare the rate of success to the total number of project, only 8.1% were verifiable successes, 19.4% were uncertain (i.e., registered but no commercialized), and 72.5% were ineffective. A list of product names and their weed-pathogen complex and formulation type is depicted in Table 10.3.

Bioherbicide market can be segmented based on crop type, by application and by geography. Based on crop type, market can be categorized as cultivated crops, tree plantations, and others. Based on application type, their market can be classified into cereals and pulses, oil seeds, fruits and vegetables, and turf and ornamental grass. Based on geography, the market is segmented into North America, Europe, Asia-Pacific, and Rest of the World. Andermatt Biocontrol AG, BioHerbicides Australia, Bayer CropScience AG, Camson Biotechnologies Ltd., Hindustan Biotech, ISAGRO Agrochemicals Pvt. Ltd., Jiangsu Dongbao Agrochemical Co., Ltd., MycoLogic Inc., Marrone Bio Innovations, and Valent BioSciences Corp. are the major agencies for production of bioherbicides.

Bioherbicide traits like environmentally safe, low research and development costs, integrations with other farming and disease management practices, and growing demand for food safety and quality are associated with their market demands. However, factors such as low consumer adoption, risk of secreted metab-

Table 10.3 Bioherbicide product names, weed-pathogen complexes, and their formulation types

Product name	Weed	Pathogen	Formulation type
ABG-5003	*Eichhornia crassipes*	*Cercospora rodmanii*	Wettable powder
Acremonium diospyri	*Diospyros virginiana*	*Acremonium diospyri*	Liquid-conidial suspension
BioChon	*Prunus serotina*	*Chondrostereum purpureum*	Liquid-mycelial suspension in water
Biomal	*Malva pusilla*	*Colletotrichum gloeosporioides*	Mallet wettable powder
Casst	*Cassia* spp.	*Alternaria cassiae*	Solid
Chontol	*Alnus* species	*Chondrostereum purpureum*	Liquid-spray emulsion and paste
Collego	*Aeschynomene virginica*	*Colletotrichum gloeosporioides* f. sp. *aeschynomene*	Solid-wettable powder
DeVine	*Morrenia odorata*	*Phtophthora palmivora*	Liquid-conidial suspension
Dr. BioSedge	*Cyperus esculentus*	*Puccinia canaliculata*	Liquid-emulsified suspension
Hakatak	*Hakea gummosis*	*Colletotrichum acutatum*	Liquid-conidial suspension
Hakatak	*Hakea sericea*	*Colletotrichum acutatum*	Granular dry conidia
LockDown	*Aeschynomene virginica*	*Colletotrichum gloeosporioides* f. sp. *aeschynomene*	Solid-wettable powder
Lubao	*Cascuta chinensis* and *Cuscuta australis*	*Colletotrichum gloeosporioides* f. sp. *cuscutae*	Liquid-conidial suspension
MycoTech	*Deciduous tree species*	*Chondrostereum purpureum*	Liquid-paste
Sarritor	*Taraxacum officinale*	*Sclerotinia minor*	Solid-granular
Smolder	*Cuscuta* sp.	*Alternaria destruens*	Liquid-conidial suspension
SolviNix	*Solanum viarum*	*Tobacco mild green mosaic virus*	Liquid-foliar spray
Stumpout	*Poa annua*	*Cylindrobasidium laeve*	Liquid-oil suspension
Velgo	*Abutilon theophrasti*	*Colletotrichum coccodes*	Wettable powder
Woad Warrior	*Isatis tinctoria*	*Puccinia thlaspeos*	Powder

olites by the microbes, high costs, and low availability are restraining the growth of the bioherbicide market. Commercial successes of bioherbicides are related with their cost; unique, simple, and effective market strategy; and knowledge of their registration cost (Bailey and Falk 2011).

10.3.7 Attributes of Some Marketed Products

Marketed bioherbicide products are depicted in Table 10.3. Besides the established potential of bioherbicide, surpassingly at global level, only 18 bioherbicide products are currently available in the market. The first bioherbicides were marketed in the 1980s, and since then, a large number of bioherbicides have been evaluated for their efficiency against many weeds, but their commercial products are very less. Details of different bioherbicide products can be found in papers of Bailey and Falk (2011), Aneja et al. (2013), Harding and Raizada (2015), and Cordeau et al. (2016).

Most commercial biological weed formulations are based on fungi (mycoherbicides), and the fungal species like *Colletotrichum gloeosporioides* f. sp. *malvae*, *Phoma herbarum*, *Phoma macrostoma*, *Phoma chenopodicola*, *Sclerotinia minor*, *Sclerotinia sclerotiorum*, *Puccinia thlaspeos*, *Alternaria destruens strain 059*, *Phytophthora palmivora*, *Myrothecium verrucaria*, and *Fusarium* spp. are the promising bioherbicides. Bacteria like *Pseudomonas fluorescens* and *Xanthomonas campestris* (strains, referred to as BRG100) have been recognized to have suppressive activity on the *Setaria viridis* (Caldwell et al. 2012). Similarly strain JT-P482 of *X. campestris* pv. *poae* was registered in Japan under the product name Camperico to control *Poa annua* weed. Among the virus, tobacco mild green mosaic tobamovirus and Araujia mosaic virus are investigated for the control of *Solanum viarum* and *Araujia hortorum*, respectively. A patent on the former biological control agent has been granted in 2015 (EPA 2015).

10.3.8 Types of Bioherbicide Formulation

The main types of formulations currently in use include emulsions, organosilicone surfactants, hydrophilic polymers, and alginate-, starch-, or cellulose-encapsulated granules, all of which have their own advantages and disadvantages (Hallet 2005). These formulations facilitate the predisposal of pathogen on weed species and ultimately disease development (Kremer 2005). A typical bioherbicide formulation is fundamentally the combinations of the active ingredient – the biological propagules with a carrier or solvent – and an adjuvant (Chutia et al. 2007). Bioherbicides can be prepared either by liquid or by solid formulation approaches. Fluid or formulation suspended in an aqueous solution has generally been utilized for the foliage portion of the weed, while solid formulations are utilized for biological agents that acted at soil level or are incorporated into the soil. A specific Pesta formulation consisted wheat gluten mixed with biological agent. This material is used with *Fusarium oxysporum* to control sicklepod (Rosskopf et al. 1999). Liquid and solid are the two basic formulations.

Liquid formulation	Solid formulation
Water suspension of the spores with small amount (like 0.1% v/v) of wetting agent like Tween 20 (polyoxyethylene sorbitan monolaurate) is an example of simple's liquid formulations. Principally a liquid formulation contains the aqueous oil or polymer-based product, oil suspension emulsion, inverted emulsion, etc. (Boyetchko and Peng 2004; Chutia et al. 2007). Water is the basic bioherbicide delivery system that contains the agent propagules (Connick et al. 1990). Adjutant application in the bioherbicide formulation enhances the weed mortality. Polymers, simple emulsions (vegetable oil emulsions 10% oil and 1% of emulsifying agent), invert emulsions (continuous oil phase with water droplet), and water in oil in water emulsion are the innovative liquid formulation approaches (Auld et al. 2003)	Solid formulations are effectively acting as preemergence applicant, attacking the weed seedling. Such formulations contain dried propagules; therefore they may have a longer shelf life than liquid-based formulation and thus are very important for a commercial product. Calcium alginate, wheat flour, kaolin clay, water-absorbent starch, hydrated silica, and sorbitol are utilized by various researchers (Auld et al. 2003). This type of formulation is more recommended for the weed of pasture, rangeland, and other natural ecosystems. However such formulations require suitable moisture conditions for fungal growth and infection
	In some circumstances, unique formulations can also be prepared and applied. Likely, Gohbara and Tsukamoto (1999) provided a floating formulation in which microorganisms are coated with invert emulsion and a further coating of a low bulk gravity powder

Boyetchko et al. (1998) categorized the challenges that have hampered the bioherbicide advancement, and these are pertaining to biological, environmental, technological, and commercial.

10.3.9 Factors Affecting Efficacy of Bioherbicides

Dew duration and dew period temperature, plant growth stage, conidial/spore or plant-pathogen concentration, and adjutants are the major factors that are associated with bioherbicide efficacy (Rosskopf et al. 2005 and Cai and Gu 2016). Appropriate humidity is the pre-request for germination and establishment of bioherbicide agent on weed (Kremer 2005; Rosskopf et al. 2005), while the long dew period is essential for infection on the aerial surfaces of the target host (Auld et al. 2003). Available soil moisture can be linearly linked with bioherbicide efficiency. Abu-Dieyeh and Watson (2009) reported that the use of a jute fabric cover to reduce soil water loss significantly enhanced the granular *Sclerotinia minor* efficacy for controlling *Taraxacum* spp., *Trifolium repens*, *Plantago major*, *Glechoma hederacea*, and *Polygonum aviculare*. The commercial success of the bioherbicide product also depends on their shelf lives, and lower shelf lives' trait of the microorganism imposed a drawback on their commercial development. Similarly earlier Boyette et al. (1993) suggested that inverted oil emulsion application can also be utilized for checking moisture loss, and this tactic improves the 100% efficacy of *Colletotrichum truncatum* for the control of *Sesbania exaltata*.

Charudattan (2001) advocated the prior fine-tune of spray droplet size, droplet retention and distribution, spray application volume, and the equipment used, and such prior visualization can enhance the bioherbicide efficiency. Singh et al. (2002) suggested that adjuvant used in the solution, droplet size and speed, surface properties, and morphology of weed and its biotype affect the efficiency of bioherbicides. Smaller droplet sizes of *Colletotrichum truncatum* ensure greater efficacy in controlling *Matricaria perforate* (Doll et al. 2005). Further, the use of compressed air rather than carbon dioxide can elevate the efficacy (Charudattan 2001). Chadramohan et al. (2002) have concluded that a combination of multiple pathogens can provide effective weed management, and according to them, combination of *Alternaria cassiae*, *Phomopsis amaranthicola*, and *Colletotrichum dematium* greatly controls *Amaranthus* spp., *Senna obtusifolia*, and *Crotalaria spectabilis* weeds. Similarly Chandramohan and Charudattan (2003) also reported that a mixture *Drechslera gigantea- Exserohilum longirostratum-Exserohilum rostratum* greatly suppressed the growth of *Crotalaria* and other six weeds in citrus groves in Florida.

Types of formulations that are using organosilicone surfactants, hydrophilic polymers, and emulsions are also associated with bioherbicide efficacy (Charudattan 2001). Emulsions are related with predisposing factor, while organosilicone surfactants, such as Silwet L-77, facilitate direct entry of bacterial cells and small spores into weed tissues. Tiourebaev et al. (1999) suggested the impact of amino acid extraction on virulence of bioherbicides, and according to them, valine excretion by mutants of *Fusarium oxysporum* controlled *Cannabis sativa* by 70–90% compared to 25% by a wild-type isolate.

10.3.10 Mode of Action

Their mode of actions is similar to plant-pathogen interactions and mechanism and allelopathy (Harding and Raizada 2015 and Cordeu et al. 2016). A successful bioherbicide should be compatible with target weed along with having suitable chemical weapons. The biological agent first utilizes enzyme or other chemicals to degrade the weed cell wall, and this facilitates the entry of biological agents. Secondly the agents could be any phyto-metabolites and peptides that act as toxins that interfere with weed metabolisms (Stergiopolus et al. 2013). Ghorbani et al. (2005) have listed the various factors affecting the efficiency of plant pathogen used in biological weed control, and these are related with (1) biotic environment (virulence, density of biocontrol agents, and weed growth stage), (2) physical environment (temperature, moisture, dew period, dew temperature, wind, and light), and (3) soil environment (soil nutrients, soil reaction, and soil microorganisms).

References

Abu-Dieyeh MH, Watson AK (2009) Increasing the efficacy and extending the effective application period of a granular turf bioherbicide by covering with jute fabric. Weed Technol 23:524–530

Altman J, Neate S, Rovira AD (1990) Herbicide pathogens interaction and Mycoherbicides as alternative strategies for weed control. In: Hoagland RE (ed) Microbes and microbial products as herbicides, ACS Symposium Series 439. American Chemical Society, Washington, DC, pp 240–259

Aneja KR (2009) Biotechnology: an alternative novel strategy in agriculture to control weeds resistant to conventional herbicides. In: Lawrence R, Gulati AK, Abraham G (eds) Antimicrobial resistance from emerging threats to reality. Narosa Publishing House, New Delhi, pp 160–173

Aneja KR, Kumar V, Jiloha P, Kaur M, Sharma C, Surain P, Dhiman R, Aneja A (2013) Potential bioherbicides: Indian perspectives. In: Salar RK (ed) Biotechnology: prospects and applications. Springer, New Delhi, pp 197–215. https://doi.org/10.1007/978-81-322-1683-4_15

Ash GJ (2010) The science, art and business of sucessful bioherbicides. Biol Control 52:230–240

Auld BA, Hetherington SD, Smith HE (2003) Advances in bioherbicide formulation. Weed Biol Manag 3(2):61–67

Bailey KL (2014) The bioherbicide approach to weed control using plant pathogen. In: Abrol P, Dharam PA (eds) Integrated pest management: current concepts and ecological perspective. Elsevier, New York, pp 245–266

Bailey KL, Falk S (2011) Turning research on microbial bioherbicides into commercial product-A Phoma Story. Pest Technol 5:73–79

Bailey BA, Hebbar PK, Strem M, Lumsden RD, Darlington LC, Connick WJ, Daigle DJ, Lumsde RD (1998) Formulation of Fusarium oxysporum f. Sp erythroxyli for Biocontrol of Erythroxylum coca var coca. Weed Sci 46(6):682–689

Bajwa AA (2014) Sustainable weed management in conservation agriculture. Crop Prot 65:105–113

Boyetchko SM, Peng G (2004) Challenges and strategies for development of Mycoherbicides. In: Arora DK (ed) Fungal biotechnology in agricultural, food, and environmental applications. Marcel Dekker, New York, pp 211–221

Boyetchko S, Pederson E, Punja Z, Reddy M (1998) Formulations of biopesticides. In: Hall FR, Menn JJ (eds) Methods in biotechnology, vol. 5: biopesticides: use and delivery. Humana Press, Totowa, pp 487–508

Boyette CD, Quimby PC Jr, Bryson CT, Egley GT, Fulgham FE (1993) Biological control of hemp sesbania (Sesbania exaltata) under field conditions with Colletotrichum truncatum formulated in emulsion. Weed Sci 41:497–500

Cai X, Gu M (2016) Bioherbicides in organic horticulture. Horticulture 2:1–10. https://doi.org/10.3390/horticulturae2020003

Caldwell CJ, Hynes RK, Boyechko SM, Korber DR (2012) Colonization and bioherbicidal activity on green foxtail by Pseudomonas fluorescens BRG100 in a pesta formulation. Can J Microbiol 58:1–9

Chadramohan S, Charudattan R, Sonoda RM, Singh M (2002) Field evaluation of a fungal mixture for the control of seven weedy grasses. Weed Sci 50:204–213

Chandramohan S, Charudattan R (2003) A multiple-pathogen system for bioherbicidal control of several weeds. Biocontrol Sci Tech 13:199–205

Charudattan R (2001) Biological control of weeds by means of plant pathogens: significance for integrated weed management in modern agro-ecology. BioControl 46(2):229–260

Charudattan R (2005) Use of plant pathogens as bioherbicides to manage weeds in horticultural crops. Proc Fla State Hortic Soc 118:208–214

Chutia M, Mahanta JJ, Bhattacharyya N, Bhuyan M, Boruah P, Sharma TC (2007) Microbial herbicides for weed management: prospects, progress and constraints. Plant Pathol J 6:200–218

Connick W, Lewis JA, Quimby PC (1990) Formulation of biocontrol agents for use of plant pathology. In: Baker RR, Dunn PE (eds) New direction in biological control: alternative for suppressing agricultural Pest and diseases. Academic Press Inc, New York, pp 345–272

Cordeu S, Triolet M, Wayman S, Steinberg C, Guillemin JP (2016) Bioherbicides: dead in water? A review of the existing products of integrated weed management. Crop Prot 87:44–49

Dagno K, Lahlai R, Diourter M, Jijakli MH (2012) Present status of the development of mycoherbicides against water hyacinth: success and challenges. A review. Biotechnol Agron Soc Environ 16(3):360–368

DiTommaso A, Watson AK, Hallett SG (1996) Infection by the fungal pathogen Colletotrichum coccodes affects velvetleaf (Abutilon theophrasti)-soybean competition in the field. Weed Sci 44(4):924–933

Doll DA, Sojka PE, Hallett SG (2005) Effect of nozzle type and pressure on the efficacy of spray applications of the bioherbicidal fungus Microsphaeropsis amaranthi. Weed Technol 19:918–923

El-Sayed W (2005) Biological control of weeds with pathogens: current status and future trends. ZPflanzenkrankh Pflanzenschutz 112:209–221

EPA (2015) Biopesticides registration action document: tobacco mild green mosaic Tobamovirus strain U2. PC Code: 056705. United States Environmental Protection Agency. Available at: http://www.regulations.gov/#!documentDetail;D=EPA-HQ-OPP-2013-0759-0017

Ghorbani R, Leifert C, Seel W (2005) Biological control of weeds with antagonistics plant pathogen. Adv Agron 86:191–225

Gohbara M, Tsukamoto H (1999) Inventors Waterborne microbial pesticides. European Patent Application PCT/JP97/04800

Hallet SG (2005) Where are the bioherbicides? Weed Sci 53:404–415

Harding DP, Raizada NM (2015) Controlling weeds with fungi, bacteria and viruses: a review. Front Plant Sci 6:659. https://doi.org/10.3389/fpls.2015.00659

Hasan S, Ayres PG (1990) The control of weeds through fungi: principles and prospects. New Phytol 115:201–222

Jurado AS, Fernandes MAS, Videira RA, Peixoto FP, Vicente JAF (2011) Herbicides: the face and the reverse of the coin an. An in vitro approach to the toxicity of herbicides in non-target organism. In: Andreas K (ed) Herbicides and environment. InTech, Rijeka, pp 33–34

Kempenaar C, Scheepens PC (1999) Dutch case studies showing the success and limitations of biological weed control. In: Pallet K (ed) Brighton conference on weeds. The British Crop Protection Council, Brighton, pp 297–302

Kiely T, Donaldson D, Grube A (2004) Pesticides industry sales and usage, 2000 and 2001 market estimates. US Environmental Protection Agency Office of Pesticide Programs, Washington, DC

Kremer RJ (2005) The role of bioherbicides in Weed Management. Biopesticides Int 1(3–4):127–141

Li Y, Sun Z, Zhuang X, Xu L, Chen S, Li M (2003) Research progress on microbial herbicides. Crop Prot 47:247–252

Müller-Schärer H, Rieger S (1998) Epidemic spread of the rust fungus Puccinia lagenophorae and its impact on the competitive ability of Senecio vulgaris in celeriac during early development. Biocontrol Sci Tech 8(1):59–72

Pacanoski Z (2015) Bioherbicides. Herbicides, physiology of action and safety. Intech Open Science. https://doi.org/10.5772/61528

Rao AN, Wani SP, Ladha JK (2014) Weed management research in India – an analysis of the past and outlook for future. In: Souvenir (1989–2014), DWR Publication No. 18. Directorate of Weed Research, Jabalpur, pp 1–26

Rosskopf EN, Charudattan R, Kadir JB (1999) Use of plant pathogen in weed control. In: Handbook of biological control. Academic Press, New York, pp 891–918

Rosskopf EN, Yandoc CB, Charudattan R, DeValerio JT (2005) Influence of epidemiological factors on the bioherbicidal efficacy of Phomopsis amaranthicola on Amaranthus hybridus. Plant Dis 89:1295–1300

Scharer MH, Collins RA (2012) Integrated weed management. Encyclopedia of Environmental Management https://doi.org/10.1081/E-EEM-120046243
Shaw RH, Bryner S, Tanner R (2009) The life history and host range of the Japanese knotweed psyllid, Aphalara itadori Shinji: potentially the first classical biological weed control agent for the European Union. Biol Control 49:105–113
Sheppard AW (1996) The interaction between natural enemies and inter-specific plant competition in the control of invasive pasture weeds. In: Moran VC, Hoffman JH (eds) IX International symposium on biological control of weeds. University of Cape Town, Stellenbosch, pp 19–26
Singh M, Siyun T, Sharma SD (2002) Adjuvants enhance weed control efficacy of foliar applied Diuron. Weed Technol 16(1):74–78
Stergiopoulos I, Collemare J, Mehrabi R, De Wit P (2013) Phytototoxic secondary metabolites and peptides produce by plant pathogenic Dothideomycete fungi. Ferm Microbiol Rev 37(1):67–93
Tinaudo V, Barberi P, Giovannetti M, Heijden MGA (2010) Mycorrhizal fungi suppress aggressive agricultural weeds. Plant Soil 333:7–20
Tiourebaev KS, Nelson S, Zidak NK, Kaleyva GT, Pilgeram AL, Anderson TW, Sands DC (1999) Amino acid excretion enhances virulence of bioherbicides. In: Spencer NR (ed) Proceedings of the X international symposium on biological control of weeds. Montana State University, Bozeman, pp 295–299
Wheeler GS, Center TD (2001) Impact of the biological control agent Hydrellia pakistanae (Dipetra: Ephydridae) on the submersed aquatic weed Hydrilla verticillata (Hydrocharitaceae). Biol Control 21(2):168–181
Wymore LA, Watson AK, Gotlieb AR (1987) Interaction between Colletotrichum coccodes and thidiazuron for control of velvetleaf (Abutilon theophrasti). Weed Sci 35(3):377–383
Yang YK, Kim SO, Chung HS, Lee YH (2000) Use of Colletotrichum graminicola KA001 to control barnyard grass. Plant Dis 84(1):55–59
Zimdahl RL (2004) Weed- crop competition. Blackwell Publishing Professional, Iowa, p P229

Chapter 11
Mycopesticides: Fungal Based Pesticides for Sustainable Agriculture

S. Bhuvaneswari Manivel and G. Subashini Rajkumar

Abstract In agriculture, the increasing demand for introducing chemical fertilizer has increased resistance to insecticide and gave a great impact to progress the synthesis of new forms of insect pest control. The usage of chemical fertilizer is reduced by applying mycopesticides into the field. The world's largest countries depend on agriculture. India is one of the developing countries that mainly depend on agricultural resources. In an environment, mycobiocontrol is rapidly used to destroy most of the insects. It mainly focuses in reducing disease and crop damage. Their way of action appears little tedious but their resistance can be highly developed to be a biopesticide. It was found that fungi are used as a selective pesticide in earlier days. The recent progress in the field of mycobiocontrol shows a bright future for further developments.

Keywords Biocontrol · Biopesticicdes · Entomopathogenic fungi · *Trichoderma*

11.1 Introduction

In India, a variety of cereals, oil seeds, pulses, vegetables, and horticultural crops are cultivated; hence, it is considered as the largest country that depends on agriculture. The introduction of chemical fertilizers increases upto 30% in food grain production in the country since 1980. Many factors were responsible for this, such as an intensive cultivation of high yielding crop varieties, change in cropping pattern, mono-cropping, and extensive use of chemical fertilizers. The major problem for agriculture productivity is caused by fungi, and in India it was nearly 22 to 25% crop damage every year.

To resolve these problems we need some eco-friendly biocontrol agents. So, these methods diminish the dependence on synthetic agrochemicals. In biocontrol activity, bacteria and fungi are concerned as biological control agents. It is suggested

S. Bhuvaneswari Manivel (✉) · G. Subashini Rajkumar
PG & Research Department of Microbiology, Shrimati Indira Gandhi college, Trichy, India

P. Gehlot, J. Singh (eds.), *Fungi and their Role in Sustainable Development: Current Perspectives*, https://doi.org/10.1007/978-981-13-0393-7_11

as the chemical control of fungal phytopathogens. Since 1981, fungi are enrolled as an effective biological control.

Pesticides are defined as the substance used to avert, demolish, or resist any pest from insects to microorganism. It prevents pest disease, which causes damage to crops and other property. Mycopesticides are one of the organic pesticides. It is a best alternative to chemical pesticides due to their high specificity and their enhanced safety to human, pest, and insects. Some of the most common fungus such as *Lecanicillium lecanii, Beauveria bassinia*, and *Trichoderma* species has been widely used in that field. In this competitive environment, organic farming justified the all concern in the population. Mycobiocontrol is a biological process, which reduces disease and crop damage.

11.2 History of Mycopesticide

Last few decades, chemical insecticides became the major tool that farmers utilized to control the damage caused by pests and diseases. For over hundred years, scientists studied various species of entomopathogenic fungi, and the role of those organisms in the environment has been very specific. In the early 1800s, entomopathogenic fungi used in silkworm industry in France were distressed by *Beauveria bassi* along with *Trichoderma.* Mycobiocontrol substances are highly used to reduce the plant pathogen like *Selerotium*, *Rhizoctonia*, and *Phytophthora*. The three important species of fungi are *Trichoderma viride*, *Trichoderma harzianum*, and *Trichoderma koningii*. Most recently, DNA sequences analysis revealed the variations among the entomopathogenic fungi with others. It plays various beneficial roles to plant growth and also to plant endophytes.

Hirsutella thompsonii was the first mycoinsecticide registered in USA in 1981 in the name Mycar. The insect control of this fungi widely covers a special division of Eumycota such as Mastigomycotina, Zygomycotina, Ascomycotina, and Deuteromycotina.

11.3 Role of Pest and Pesticide in Agriculture

Pest is defined as any organism that interferes in some way with human welfare or activities. A pesticide is any substance or mixer of substances intended for destroying, preventing any pest. It can be classified as ideal pesticide and nonideal pesticide (Fig. 11.1).

"Ideal pesticides" have a narrow spectrum that kill only target organisms. It does not move around in the environment, but nonideal pesticides" have a broad spectrum that kill more than the target. It can persist or it can degrade into other compounds

Fig. 11.1 Various pests in environment

that can be more dangerous. It moves around in the environment. Generally, it was grouped by its target organism as insecticide, herbicide, fungicide, or rodenticide. Worldwide nearly 85% of the pesticides were used in agriculture (Mallikarjuna et al., 2004).

During pre 1940s the first-generation pesticides are inorganics (minerals) such as lead, mercury, and arsenic. It is very persistent and bioaccumulate in environment. But organics (botanicals) are plant-derived compounds. They break down readily. During post 1940s (second-generation pesticides) synthetic botanicals were made by altering natural botanicals, e.g., Dichloro Diphenyl Trichloro Ethane (DDT). Agricultural, veterinary, domestic, and institutional areas are the different places where pesticides are used as various formulations such as liquid, gel, paste, etc. They can be stored in various containers such as metal flasks, bottles, etc., and their concentration ranges from 2% to 80%.

Infection by fungal pathogens in the field affects the health of humans, especially when the fungus produces toxic residues in or on consumable products. There are 200 major varieties of pests of economic importance belonging to insect orders Coleoptera, Diptera, Hemiptera, Pidoptera, and Orthoptera. *Helicovera armigera* is a dreadful pest of cotton, pea, peanut, sorghum, millet, tomato, etc., and it has been estimated to cause a loss of over Rs. 10 billion per year in India due to crop loss. The fungal diseases such as wilt, rots, cankers, and mildews cause considerable losses in yield and reduce the value of agricultural commodities. The problems due to insect pests and fungal pathogens of crop plants cause heavy losses to agricultural production of the country and also disturb the national economy.

11.4 Classification of Chemical Pesticides

Frequently used pesticides to kill specific pests are listed as follows.

- The common insecticides such as organochlorines, organophosphates, and carbamates and insect repellents as diethyltoluamide and citronella
- Weed killers (e.g., paraquat, glyphosate, and propanil) are otherwise called as herbicides
- For killing mold or fungi, fungicides are used and sometimes they are used as wood preservatives

The chemical pesticides can also occur as gas or vapor at room temperature. They are termed as fumigants. Thay are highly toxic to humans and animals.

• For killing mice, rats, and other rodents, rodenticides are used, e.g., warfarines

11.4.1 Dichloro Diphenyl Trichloro Ethane

DDT was discovered in 1939. It is an insecticide also known as 1,1,1-trichloro-2,2-bis-(p-chlorophenyl) ethane. In 1948, Paul Muller received Nobel Prize for medicine and physiology. In 1941, the product entered the swiss market, chiefly in marine animals and fishes. DDT can be easily engrossed by some plants, animals, and humans. It is stored in adipose tissues of humans and animals as it is fat soluble. By eating contaminated fish and shellfish, DDT can be easily exposed to humans and in infants through breast milk (Table 11.1).

Table 11.1 Advantages and disadvantages of DDT

S. No.	Advantages	Disadvantages	Health effects
1	Low volatility	Persistence in the environment	Paresthesia of tongue, lips, and face
2	Chemical stability	Bioconcentration	Irritability, dizziness, vertigo, tremor
3	Lipid solubility	Biomagnification in food chain	Hypertrophy of hepatocytes
4	Slow rate of biotransformation	Profound effects on wild life ("silent spring")	Hepatic tumors
5	Slow rate of degradation	–	Low rate of absorption through the skin

11.5 Chemical Pest Control

In modern agriculture, the fungicides and insecticides played a vital role in routine agricultural practices due to their rapid action and inexpensive nature. In India, insecticides are used by about 80% followed by 10% of fungicides, 7% of herbicides, and 3% of other chemicals. Among the total pesticides used in the country, more than 60% of the pesticides, worth of nearly USD 630 million, is consumed annually by the agriculture sector.

11.5.1 Chemical Fungicides

First generation fungicides are inorganic chemicals like copper and tin. The organic chemicals such as organotins and quinines are second generation fungicides. Third generation fungicides are also organic such as carboxamides and 2-aminopyrimidines that penetrated the plant tissue and controlled well-known infections. Fourth generation fungicides were biodegradable in natural situations and are selective in mode of action to reduce nontarget effect.

In 1948, pesticide usage started in India, and for controlling malarial disease DDT was highly recommended. Hence, DDT was used on more than three hundred different agricultural commodities as a general insecticide. Among the inorganic pesticides, DDT enjoyed a product life of 50 years all over the world before it got banned as a result of its harmful effects.

11.6 Biological Pest Control

11.6.1 Mycopesticides

The term Myco means Mushroom or fungus. Pesticide is referred as the substance that prevents, destroys, and diminishes any pest. The pests are of various types such as insects – structural, human disease vectors, biting, agricultural, microbial – bacterial and fungal pathogens related to human and plant health, and rodents.

11.6.2 Sources of Mycobiocontrol Agents

- Since 1960, mycoinsecticides have been urbanized worldwide.
- They have a wide host range.

- They exhibit high particularity between species.
- Example *Metarhiziuum aniopliae.*
- *Beauveria bassiana* is the commercial mycoinsecticide.
- There is a huge potential for genetic improvement of fungi.

11.6.3 Types of Mycopesticides

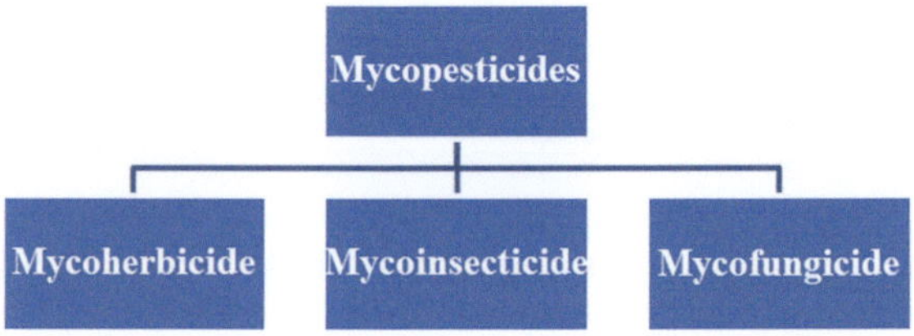

11.6.3.1 Mycoherbicide

- It breaks the condition that need to be evident for infection to occur in plants.
- To suppress weed growth it releases phytopathogens.
- *Collectrichum gloesporioides* and *Phytophthora palmivoraa* are the common species.
- Other types of fungi cause rusts and mildews.

11.6.3.2 Mycoinsecticide

- When a fungus is used as insecticide, it is called mycoinsecticides.
- The common fungal insecticides are *Beauveria bassiana* and *Metarhizium anisopliae*.
- The fungal hyphae were used to inactivate the host by the release of insectotoxins in spray form.

11.6.3.3 Mycofungicide

- In the host large amount of resting spores like *Sclerotia* are parasitized by another fungi.
- It causes the cell wall breakdown by liberating lytic enzymes
- Chemotaxis, recognition, attachment, and penetration are the four main stages of this process.
- The most commonly used fungus is *Trichoderma.*

11.6.4 Development of Mycopesticide

The steps involved in the development of mycopesticide are as follows.

11.6.4.1 Strain Isolation

The soil dilution method is commonly used for the isolation of both mycoparasitic and entomopathogenic fungi (Goettel and Inglis, 1996). The baiting technique called the pre-colonized plate method in which field inoculum was placed on to the host fungus was used for the isolation of epiphytic mycoparasites. Alternatively, mycosed larvae from fields may be collected to isolate highly virulent entomopathogens causing epidemics in particular insect host (Deshpande and Tuor, 2002). Screening and isolation of fungi producing extracellularly chitinase is usually done on media containing chitin (Fig. 11.2).

For the assortment or assessment of chitinolytic microorganisms Ramirez et al. (2004) proposed a simple method in a liquid medium as a carbon source on the basis of colloidal chitin stained with Remazol Brilliant Blue R. The general procedure for collecting and cultivating entomopathogens is mentioned in Fig. 11.3.

11.6.4.2 Habitat Association

Initially, it was followed that the organisms are randomly selected from soil and plant surfaces to prospect zones of inhibition on culture media. Healthy plants and their rhizosphere were examined secondly. Isolation of biocontrol organisms such as mycoparasite species is due to some disease suppressive agricultural fields. Bidochka et al. (2001) reported that the isolation of the entomopathogenic fungi was affected by the locality from where the soil was taken and also by the method of isolation selected. Generally, soil samples collected from the rhizosphere of

Fig. 11.2 Strategy for the development of mycopesticide

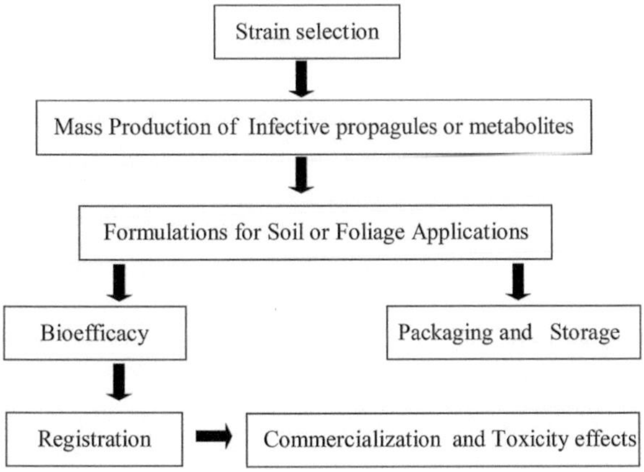

Fig. 11.3 Collecting and Cultivating Entomopathogens

crops where chemical insecticides are not routinely sprayed one can expect more number of isolates that had high infectivity against insect pest. For instance, *Metarhizium* isolates from fields of tomato and other vegetables heavily sprayed with chemicals were less virulent to *H. armigera* than the isolates from custard apple field rarely sprayed with chemicals (Nahar et al., 2003). This can also be correlated with the natural insect flora of the host plant, not necessarily pest that affects the virulence of entomopathogenic fungi. The chitinolytic fungi like *Myrothecium, Trichoderma, Beauveria, Penicillium, and Aspergillus* were isolated from rhizospheral soil samples, chitinous wastes, and marine environments.

11.6.4.3 Criteria for Strain Selection

The capability to synthesize differential levels of cuticle degrading and mucolytic enzymes plays vital criteria in strain selection. Interestingly, spore characteristics such as the size, viability, production, speed of germination, and relative hyphal growths are important for strain selection too (Liu et al., 2003).

The environmental conditions like temperature, pH, and water activity affect the growth and in turn efficacy of biocontrol agents. Thus, the ability of biocontrol agents to withstand fluctuations in the environmental conditions requires consideration in strain selection. Certain other criteria suggested by Antal et al. (2000) that also governed the success of the biocontrol preparation were field performance studies, genetic stability, productivity, and stability of conidia in storage, stability in formulation, field persistence, mammalian safety, low environmental impact, and capacity to recycle in the environment.

11.6.4.4 Strain Improvement

- In the development of biocontrol agents, genetic improvement plays the vital role.
- To understand the contribution of diverse genes in virulence entomopathogenic as well as mycoparasitic fungi, mutagenesis was employed for killing process as well as for the strain improvement.
- The precious tool for strain improvement is protoplast fusion.
- To produce hybrids biocontrol agent against *Beauveria sulfurescens*.

11.6.4.5 Preservation of Fungi

The fungal spores are normally stored at -20°C or -80°C using skimmed milk, water, or mineral oil. Spores remained viable when stored with water or mineral oil and preserved at -80°C. Alternatively, fungal isolates can be stored for long period as dry alginate pellets (biomass plus sterile wheat bran and aqueous sodium alginate) at -20°C. The fungal conidia can be stored in carriers such as mineral oils, wheat bran, talc powder, silica, etc., to retain the viability and virulence of the conidia, which is an important requisite for the performance of the biocontrol preparation. Also, to retain the virulence, regular serial passage of fungus through the insect hosts for entomopathogenesis and for mycoparasitic in dual culture are essential (Chavan et al., 2006).

11.6.5 Uses of Mycopesticides

- Biopesticides have an inferior toxicity, improved safety, as well as a high efficiency.
- The total worldwide pesticide market has been raised to 0.98%.
- In pest management, it shows a large fraction.

11.6.6 Benefits of Mycopesticides

- Narrow bandwidth of target specificity using cloned ecotypes.
- Self-replicating so it is inexpensive to manufacture.
- Low toxicity and no ecotoxicity compared to chemical pesticides.
- Collected from native or local stock to target local pests.

11.6.7 Concerns of Mycopesticides

- Some insects are beneficial at one stage and a pest at another.
- Caterpillars eating agricultural crops.
- Adult butterflies serve as beneficial pollinators.

11.6.8 Challenges of Mycopesticides

- DNA and Patent protection to Identify and File Delivery System or Trade Secret Process.
- Permitting the field trials by EPA & Clemson Pesticide Permitting.
- Industrial development for process development and scaling from parallel models
- License and distribute to manufacture or license active ingredient.

11.6.9 Fungi as a Bio Control Agent

In 1930s, it was first showed that fungal pathogens of plants could be infected or parasitized by other fungi (mycoparasites). Since then number of mycoparasitic strains were isolated from their natural habitats and studied for their biocontrol efficacy against several plant pathogenic fungi in phylosphere or rhizosphere. Biotrophic mycoparasites like *Sporidesmium sclerotivorum* require a persistent contact with or occupation of living cells of the host *Sclerotinia*. Whereas necrotrophic mycoparasites kill the host cells often in advance of contact and penetration. Majority of the necrotrophic mycoparasites were developed as biocontrol agents due to their saprophytic nature, less specialized mode of action, and broad host range. *Trichoderma* species (necrotrophic mycoparasites) were developed into several commercial products (Harman, 2000). In addition, mycoparasites inhibit plant pathogenic fungi by producing toxins and enzymes. Most of the biocontrol preparations were developed using mycoparasitic fungal strains. Thus, it concludes that entamopathogens act as a biocontrol agent for insect test.

However, fungi enter the host only through the cuticle but bacteria and virus infect insects through their digestive tract. Then, the fungal biomass can be synthesized easily and delivered in a variety of formulations that act as direct contact sprays, foliage sprays, or granules. Further advantages are their mammalian safety and minimal impacts on nontarget insects due to relative host specificity of different isolates of the same species, no toxic residues, environmentally safe, and no known resistance in insect community. Another important aspect in the interactions of fungi with insect through their entry into the host only through the cuticle is the possibility of identifying a strain that has capability to show dual pathogenesis possibly mediated through specific hydrolytic enzymes (Shah and Pell 2003) (Fig. 11.4).

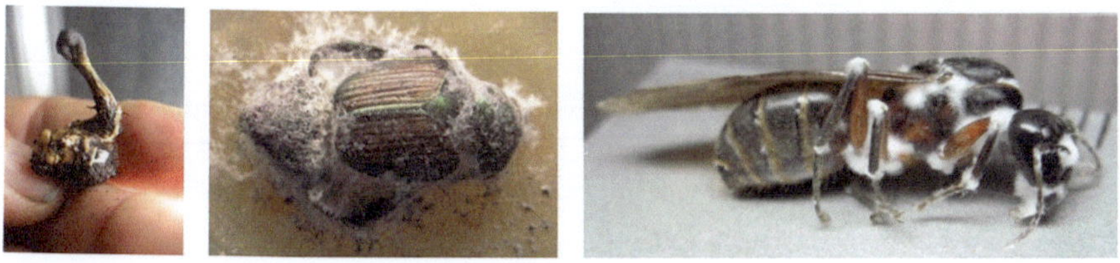

Fig. 11.4 Mycopest associated with different insects in their cuticle

11.6.10 Entomopathogenic Fungi

Most of the entomopathogens belong to the Deuteromycetes and Entomophthorales. Around 700 species of these fungi are pathogenic to insects. But, some of them have limited host range, for example, *Beauveria bassiana* are highly pathogenic to insects. Biopesticide has very tedious action unlike chemical pesticides; hence, host resistance was not well developed.

11.6.11 Biomanagement of Pest by Entomopathogenic Fungi

The word entomon indicates insect. It means microorganism which arise in insects. This organism shows uniqueness in their activity to control pests which cause damage to humans. One of the important species of mycobiocontrol of insect pest such as *Beauveria bassiana* one of the filamentous fungi which is called as fungi imperfection. This is highly host specific. It is generally found in all soils, and it behaves as a pathogen to numerous insects and cause muscardine disease (Sandhu and Vikrant 2004). It also controls white fly, culex mosquito (McNeil Donald Jr 2005). The fungal spores are sprayed on infected parts of the plant in powder form. Thus, it behaves as a specific bioinsecticides.

11.6.11.1 *Verticillium lecanii*

Verticillium lecanii is a most common fungus, mainly found in humid environment. It was highly effective biological control agent against *Trialeurodes vaporariorum*. This fungus destroys the nymphs and adults present in the underside of leaf by their filamentous mycelium. *Verticillium lecanii* have ability to control whitefly and several aphids species. *Vertcillium lecaii* was considered as an important parasite in nematode populations of some plants. *Verticillium chlamydosporium* has a broad host range, and it act as an effective biological control agent.

11.6.11.2 *Metarhizium*

Metarhizium anisopliae is a very effective pathogen for mycobiocontrol of various insect pests. The entire mechanism of this fungi has been trailed on *Eutectona machaeralis* and found to be an important mycobiocontrol agent of plant pest (Sandhu et al. 2000).

11.6.11.3 *Nomuraea*

Nomuraea rileyi is a dimorphic hyphomycete entomopathogenic fungi. It is lethal to various insects. It shows unique host specificity. It is widely used in insect pest management due to their ecofriendliness. Its route of entry and development was well documented in various hosts such as *Heliothis zea*, *Psedoplusia*, and *Bombyx mori*.

11.6.11.4 Paecilomyces

Paecilomyces is a dreadful nematophagous fungus which causes disease in the nematodes. Thus, it can be used as a bionematicide to control nematodes when it is introduced into the soil (Tables 11.2 and 11.3).

11.6.11.5 *Paecilomyces fumosoroseus*

In whiteflies, it is the most important natural invader. It produces a disease called "yellow muscardine." It grows well in humid conditions and it mostly found on the leaf surface. *Paecilomyces fumosoroseus* is one of the best control agents in nymphs of whiteflies. Fungal mycelia shows the feathery appearance in the nymphs. It is a most potential control agent to eliminate mosquito.

Table 11.2 Different types of organisms and their biological action

Infective host	Microbes	Biological activity
Insects as parasites	*Trichogramma chilonis*	They live inside or outside of the host
Insects as predators	*Chrysoperla carnea*	Insects which kill the prey
Microorganisms (bacteria fungi, viruses)	Bacteria – *Bacillus thuringiensis Verticillium, Metarhizium, and Beauveria* viruses, nuclear polyhedrosis virus	Cause disease in pests

Table 11.3 Various bioactive products derived from entomopathogenic fungi

Product	Fungus	Biological action
Mycotal pelicide biologic bio 1020 Brocaril	*Verticillium lecanii Paecilomyces lilacinus Metarhizium anisopliae Beauveria bassiana*	Fungal pesticide effective against nematode mycelium as pesticide
Botanigard	*Beauveria bassiana*	Pesticide produced from mycelium. Effective against whiteflies

11.6.12 Role of Entomopathogenic Fungus

The insect killing fungi are fastidious microorganisms, having the ability to cause disease in insects. It occupies the largest group of insect pathogens among microorganisms. Entomopathogens are the promising mycobiocontrol agents for numerous crop pests. Various species of this fungi belonging to the order *Homoptera, Lepidoptera, Hymenoptera, Diptera,* and *Coleoptera* are susceptible to various fungal infections. These fungi comprise a wide range of 750 species and act as great potential mycobiocontrol agents.

11.6.12.1 Mechanism of Infection Process

The fungi enter the cuticle and finally reach the mouth parts of the insect. It shows specific mode of infection. In intestine, the fungal spores do not germinate and finally it is released in the faeces. The insect death was due to various factors: mechanical damage by the tissue penetration, toxin production, and nutrient depletion.

11.6.12.2 Cuticle Attached with Conidia

The initial adherence of an organism into the host is either by means of wind or water. Then the *Beavearia bassiana* spores have an outer layer made of network of fascicles. The rodlets have hydrophobic forces that involve the attachment of spores and cuticles. The conidia and insect cuticle binding with the help of proteins are present in the fungal conidial surfaces. Various factors such as water availability, nutrients, oxygen, temperature and moisture play a vital role in the attachment of microbe into the host.

11.6.12.3 Formation of an Infection Structure

The host cuticle entry by fungi is by forming an appressorium into the host. The cuticle has outer epicuticle and inner procutile. The fatty acids, lipids, and sterols are present in the epicuticles and it lacks chitin. The procuticle is mainly composed of protein matrix associated with chitin fibrils along with quinones and lipids. Surface topography and biochemical components are the host factors that influence the formation of appressorium.

11.6.12.4 Invasion of the Cuticle

Entamopathogenic fungi uptake nutrients from insect body through the cuticle for their growth and multiplication. Mechanical pressure and enzymatic degradation are used for the host invasion. Chitinases, proteases, lipases, and esterases are the extracellular enzymes that are involved in the degradation of insect cuticle (Bhattacharyya et al. 2004).

11.6.12.4.1 Fungal Infective Process

Spore present on the surface

⇩

Germ tube formation

⇩

Excreating enzymes (Proteases, Chitanases, Lipases) are released

⇩

Penetrate the insect cuticle

⇩

Blastospores formed

⇩

Fungal spores spreaded into the Haemolymph

11.6.12.5 Production of Toxins

Huge amounts of toxic compounds are produced by *Beaveria bassiana* inside the host. *Beauveria bassiana* produced many toxins likewise Beauverolides, Bassianolide, Destruxins, and Beauvericin. The toxins lyse the insect tissues and impart negative effect on the host. Destruxins highly inhibit the insect function.

11.6.12.6 Advantages of Mycobiocontrol

The advantages of fungi as mycobiocontrol are as follows:

- Pest control specificity should be very higher.
- Beneficial insect predators and parasites are unaffected when harmful insect pests gets controled.
- The negative effect on mammals causes reduction in environmental hazard.
- It provides prolonged pest control.
- The presence of mycopesticide in the environment gives long-term benefits on pest suppression.
- High protection on biodiversity.

11.6.12.7 Disadvantages of Mycobiocontrol

- Short shelf life of spores.
- Mycopesticide need more than 2 weeks to kill the insects, but chemical pesticides take only few hours.
- High relative humidity enhances the action of mycopesticide.
- The additional control agents are needed due to high specific activity.
- The production cost is very high.
- Necessitates of cold storage upto 80% humidity.
- Entomopathogenic fungal efficiency varies in host population.
- To retain long-term benefits various methods to be optimized.
- Immunosuppressive people are at risk.

11.6.13 Trichoderma *an Efficient Mycobiopesticides*

11.6.13.1 *Trichoderma* – A Challenged Fungus

These mushrooms belonging to *Trichoderma* genus have been known since 1920 for their ability to function as biocontrol agents (BCAs) against plant pathogens. *Trichoderma* has numerous interactions mainly *Trichoderma harzianum* and *Trichoderma* strain atroviride strain T22 plant P1 crops and fungal pathogens in soil (Woo et al., 2006). The different species of this *Trichoderma* are known not only for the control of plant diseases but also for its ability to increase plant growth, reproductive capacity, growth potential in favorable conditions, competition use of nutrients, strong aggressiveness against phytopathogenic fungi, and effective in supporting plant growth and improving defense mechanisms (Tripathi et al. 2013, Keswani et al., 2014). These properties have made ubiquitous *Trichoderma* genus capable of growing in larger and high population density environments (Chaverri et al., 2011).

Trichoderma spp. It was well known for its ability to produce numerous antibiotics that inhibit the growth of pathogens and are used as effective biocontrol agents. The first theoretical test was the fight against the diseases of *Trichoderma* cultures against *Armillaria belle* in citrus. It also effectively monitors a large number of plant pathogens such as *Pythium, Sclerotium, Fusarium, Phytophthora, Rhizoctonia, and Galumannomyces*. Now a days the mass production was done by three main species such as *Trichoderma viridae, Trichoderma harzianum and Trichoderma koningii*.

11.6.13.2 *Trichoderma* Spp. Is Omnipresent

Trichoderma is a soil fungus that reproduces sexually. It is found in woody and herbaceous plants. In nature, the vegetative forms of fungi persist often as clone heterarióticos. The competitive context is rapidly growing, strong opportunistic invaders, prolific spore producers, and strong antibiotic producers (Gal-Hemed et al., 2011). These properties make many dominant and ecological *Trichoderma* strains to serve ubiquitous in all climatic zones. Currently, isolated *Trichoderma marini* have been characterized by evaluating their potential use as halotolerant biocontrol agents.

11.6.13.3 *Trichoderma* as a Chemical Fertilizer

In agriculture, the production of modern microbial pesticides is more familiar in all agricultural sectors. Today, there are more than 50 different *Trichoderma* farm products produced in different countries and sold to farmers for the best yields in different crops. Currently, *Trichoderma* spp. is considered a relatively new type of biological control agent (BCAs). Recently, *T. harzianum* is used as an active agent in a wide range of commercially available biopesticides (Lorito et al., 2010).

The biological control of various strains of *Trichoderma* used to control a wide range of plant pathogens. *Trichoderma* acts as a model organism in several areas such as the molecular pathology of plants, phytosanitary, and microbial ecology. Due to their different substrates using this capacity, they are the dominant flora in the soil in different environments, such as agriculture, forests, pastures, and desert landscapes in a wide range of climatic zones. Isolated native *Trichoderma spp*. can often be isolated and characterized for biological control activity so they relate to the integrated fight. The *Trichoderma* genus is spread all over the world and can easily soil isolated by a series of dilution technique. They are usually called Deuteromycotine because they lack sexual reproduction mode. In cultivated soils, a faster growth rate and the production of conid conidias showing different cultural characteristics are reported. It acts as a more effective biological control agent.

11.6.14 Biocontrol Mechanism Trichoderma

11.6.14.1 Induced Systemic Resistance

One of the most common mechanisms of plants against plant pathogens is called induced systemic resistance (ISR). When primary pathogenic infection begins in plants that are able to produce an immune response, known as systemic acquired resistance. The pathogenic expression is due to the production of beta-1, 3 glucanase and quitinase and acidic or basal endocytic activity involves the activation of SAR. Trichoderms produce various low molecular weight proteins and compounds, and these compounds are responsible for the production of ethylene and other hypersensitive defense mechanisms in plants.

11.6.14.2 Fungistasis

When fungal growth is inhibited on the floor, a process known as soil fungistase. *Trichoderma* grow well on the ground as it can withstand many toxic compounds.

11.6.14.3 The Production of Enzymes That Degrade the Cell Wall

Trichoderma species produce few β-1, 3-glucanase quitinase; this is the key enzyme for cell wall lysis during interactions against pathogenic fungal plant.

11.6.14.4 The Antagonism of Mushrooms

Trichoderma and other plant fungi exude waste from the host plant. It can easily invade plant roots due to the release of exudate from plants that greatly benefit from its symbiotic association with plants.

11.6.14.5 Mycoplasmic Activity of Other Fungi

Trichoderma can easily colonize other mushrooms by secretion of a digestive enzyme that decomposes chitin into its cell wall. Then you can easily extract the nutrients from them.

11.6.14.6 Secondary Fungal Metabolites

Fungal enzymes have the ability to degrade cellulose and chitin because it is a better production of extracellular proteins. Over 100 different metabolites are produced by different fungal strains with antibiotic activity. Several secondary metabolites produced by *Trichoderma* as hydrocyanic acid, aldehydes, and ketones play an important role in controlling plant pathogens.

11.6.15 Mass Production of Trichoderma

In India, the inoculum of the base vehicle was developed at the University of Tamil Nadu, Coimbatore (Jeyarajan and Nakkeeran 2000). Therefore, based on the inoculum vector is formed by both solid state fermentation and the liquid state.

11.6.15.1 Solid State Fermentation

In this fermentation, various solid substrates such as maize, rye, sorghum, and millet are used for Trichoderma's mass production. These granules are dampened, sterilized, and inoculated with *Trichoderma spp.* and incubated for 10–15 days. After incubation, the organism produces dark green spores that cover the whole grain. Then these grains can be powdered and mixed in a sterilized vehicle to improve their efficiency for long periods. Finally, the processed product can be used to treat seeds and soil. This type of fermentation is used for commercial production levels. The main disadvantage of this technique is that it is very tedious and takes time.

11.6.15.2 Fermentation Liquid

In this type of fermentation, *Trichoderma* has grown on a shaker containing liquid medium. Then talcum powder is mixed with fermented medium in 1:9 ratio. For high production of biomass the following fermentation conditions such as temperature, pH, controlled aeration and foam control were maintained. Then maximum biomass was produced within a short time Organic Control Project (PDBC), Bangalore, has revealed that the maximum amount of biomass can be obtained *T. viride* within 96 hours of fermentation in a fermenter under optimum conditions (Prasad and Rangeshwaran 2000).

11.6.16 Application in the Field of Trichoderma

Appropriate substrate selection is the first step in mass production of an organism. The substrate must be available at a low price. For the mass production of *Trichoderma*, agro-food wastes such as wheat bran, manure, maize flour, sorghum, and vegetable waste are used. For the preparation of the average composition, all waste materials are dried and are grounded to a fine powder and then packed in sterile bags (Misra, 2005). The substrate must contain at least 45% moisture and sterilized twice at 121 °C. *Trichoderma sp.* that has been inoculated in multiplying product waste quickly reaches the maximum count. Then 10 ml of *Trichoderma* spore suspension was inoculated in sterile bags and incubated at 28 ° C. After incubation, the inoculated bags were transferred to trays and sterilized plastic clean and covered with a thin plastic film. Finally, the substrate is mixed with the carrier to maintain the vitality of *Trichoderma.* Wheat straw, tea (wastes), lime (FYM), sarmiento grain sorghum (Jowar), and corn flour are the most commonly used vector.

Various methods are used for bio fungicide fields for the success of treating plant diseases. The mode of administration of bio fungicide on plants plays a very important role in controlling infections. The most common methods of applying trichoderma are bio seed pruning, plumage seedlings, soil application, and leaf sprays.

11.6.16.1 Seed Treatment

In this method, the seed is coated with *Trichoderma* dry powder just before sowing. For commercial purposes, dry pathogenic dust for plants is used from 3 to 10 g per kg of seed, depending on the seed size. This is one of the most common and effective methods. *Harzianum, T. virens*, and *T. viride* were effective in protecting seeds inhibiting *R. solani* and *Pythium* sp.

11.6.16.2 Seed Biopriming

In this method, the seeds are treated with *Trichoderma* and incubated in hot, humidity conditions just before applying to the field. This technique has enormous advantages because the simple coating of the seeds result in uniformly and rapidly growing seedlings. *Trichoderma* cones germinate on the seed surface and form a layer around bioprimadas seeds and tolerate adverse conditions.

11.6.16.3 Root Treatment

Singh and Zaidi (2002) reported that roots of young seedlings can immerse themselves with fungal spores or cellular suspension antagonists in nursery beds or biocontrol dive root suspended prior to transplantation. This procedure is commonly used for horticultural crops, grape rice under the main field.

11.6.16.4 Soil Treatment

The application of ground biological control agents and other artificial means can be done before or when planting to control a wide range of terrestrial pathogenic fungi. This type of introduction is suitable for nurseries.

11.6.16.5 Air Spraying

Trichoderma was actually applied to the aerial parts of the plant to prevent deterioration in the bushes and trees caused *by* plant pathogens. Therefore, they act as an effective biocontrol agent.

11.7 Conclusions

Mycopesticides play a better role in the fight against some remarkable plant pathogens. The main disadvantages of this biocontrol are due to the lack of knowledge and knowledge of the rhizosphere ecology and the use of in vitro antagonism for the selection of biological control agents. But the advantages of this method are very great.

Trichoderma spp. It is one of the mushrooms commonly found in the soil that is saprophytic in nature. It acts as a potential biological control agent because of its ability to reduce disease caused by pathogenic plant fungi.

References

Antal ZS, Manczinger L, Szakacs GY, Tengerdy RP, Ferenczy L (2000) Influence of environmental parameters on biocontrol Trichoderma strains. Mycol Res 104:545–549

Bhattacharyya AC, Samal, Kar S (2004) Entomophagous fungus in pest management. News Letter 5:12

Bidochka MJ, Kamp AM, Lavender TM, Dekoning J, Amritha de Croos JN (2001) Appl Environ Microbiol 67:1335–1342

Chavan S, Kulkarni M, Deshpande MV (2006) Annu Rev Plant Pathol 4(83):35–39

Chavcrri P, Gazis R, Samuels GJ (2011) *Trichoderma amazonicum,* a new endophytic species on *Hevea brasiliensis* and *H. guianensis* from the Amazon basin. Mycologia 103:139–151

Deshpande MV, Tuor U (2002) In: Proceedings of the III Asia Pacific Crop Protection Conference, Suchak's Consultancy Services (ed.), New Delhi, India, pp 56–59

Gal-Hemed I, Atanasova L, Komon-Zelazowska M, Druzhinina IS, Viterbo A, Yarden O (2011) Marine isolates of *Trichoderma* spp. as potential halotolerant agents of biological control for arid-zone agriculture. Appl Environ Microbiol 77:5100–5109

Goettel MS, Inglis GD (1996) Manual of techniques in insect pathology. Academic Press, Sandiego, pp 213–249

Harman GE (2000) Myth and dogmas of biocontrol; changes in perceptions derived from research on *Trichoderma harzianum* T-22. Plant Dis 84:377–393

Jeyarajan, R. and Nakkeeran, S. (2000). Exploitation of microorganisms and viruses as biocontrol agents for crop disease management. In: Biocontrol potential and their exploitation in sustainable agriculture, (Ed. Upadhyay et al.,) Kluwer Academic/Plenum Publishers, Boston, pp 95–116

Keswani C, Mishra S, Sarma B, Singh S, Singh H (2014) Unraveling the efficient applications of secondary metabolites of various *Trichoderma* spp. Appl Microbiol Biotechnol 98:533–544

Liu BL, Kao PM, Tzeng YM, Feng KC (2003) Production of chitinase from Verticillium lecanii F091 using submerged fermentation. Enzym Microb Technol 33:410–415

Lorito M, Woo SL, Harman GE, Monte E (2010) Translational research on *Trichoderma*: from omics to the field. Annu Rev Phytopathol 48:395–417

Mallikarjuna N, Kranthi KR, Jadhav DR, Kranthi S, Chandra S (2004) Influence of foliar chemical compounds on the development of *Spodoptera litura* (fab.) in interspecific derivatives of groundnut. J Appl Entomol 128:321–328

G McNeil Donald Jr . 2005. Fungus fatal to mosquito may aid global war on malaria. *The New York Times*

Misra HP (2005) Fungi uses in pest management. Orissa Review:53–56

Nahar P, Kulye M, Yadav P, Hassani M, Tuor U, Keller S, Deshpande MV (2003) Insect resistance to transgenic Bt crops: lessons from the laboratory and field. J Mycol Pl Pathol 33:372–377

Prasad RD, Rangeshwaran R (2000) Effect of soil application of a granular formulation of *Trichoderma harzianum* on *Rhizoctonia solani* incited seed rot and damping off of chickpea. J Mycol Pl Pathol 30:216–220

Ramirez MG, Rojas Avelizapa LI, Rojas Avelizapa NG, Cruz Camarillo R (2004) Colloidal chitin stained with Remazol brilliant blue RR, a useful substrate to select chitinolytic microorganisms and to evaluate chitinases. J Microbiol Meths 56:213–219

Sandhu SS, Vikrant P (2004) Myco-insecticides: control of insect pests: In Microbial diversity: Opportunities & Challenges 3:47–53

Sandhu SS, Rajak RC, Hasija SK (2000) Potential of entomopathogens for the biological management of medically important pest: progress and prospect. In: Glimpses in Plant Sciences 2000:110–117

Shah PA, Pell JK (2003) Fungi as biological control agents. Appl Microbiol Biotechnol 61(5–6):413–423

Tripathi P, Singh PC, Mishra A, Puneet S, Chauhan DS, Thakur R, Deo B, Tripathi R (2013) *Trichoderma*: a significant fungus for agriculture and environmental cleanup. Clean Techn Environ Policy 15:541–550

Woo SL, Scala F, Ruocco M, Lorito M (2006) The molecular biology of the interactions between *Trichoderma* spp., pathogenic fungi, and plants. Phytopathology 96:181–185

Chapter 12
Multifactorial Role of Arbuscular Mycorrhizae in Agroecosystem

U. N. Bhale, S. A. Bansode, and Simranjeet Singh

Abstract Arbuscular mycorrhizal fungi (AMF) are naturally occurring organisms and associated with most of the plant families (90%). The main mechanism of AMF is the uptake of nutrients and water from the soil when colonized and through hyphae glomalin (biological glue) produced. AMF are tolerant to different environmental conditions. However, AMF also are in microbial activity. AMF are predictable biocontrol agents in disease management and in plant health. In the agricultural point of view, AMF improved nutrition and enhanced plant growth. In the recent years of organic and sustainable products, reduction in chemical fertilizers application and biological control of plant pathogens are a goal of governments, producers and food safety organizations; AMF, in addition to other benefits and microorganism can access this kind of production. Some important soil-borne phytopathogenic diseases are controlled by AMF especially *Glomus* species. Some antagonists' microbes could also obstruct with AMF fungi and positive interaction with other microorganisms for biomass and yield. AMF have multifaceted approaches in the different agroecosystem. Therefore this article presents an overview of current knowledge on mycorrhiza and their potential benefits to agriculture ecosystem.

Keywords Mycorrhizae · Nutrients · Biofertilizer · Bioprotectant · Sustainable agriculture

---

U. N. Bhale (✉) · S. A. Bansode
Department of Botany, Arts, Science and Commerce College,
Naldurg, Tq. Tuljapur, Osmanabad, Maharashtra 413602, India

S. Singh
Department of Biotechnology, Lovely Professional University, Phagwara, Punjab, India

P. Gehlot, J. Singh (eds.), *Fungi and their Role in Sustainable Development: Current Perspectives*, https://doi.org/10.1007/978-981-13-0393-7_12

12.1 Introduction

Mycorrhizae are 50 times more efficient than non-mycorrhizal plants as a function to access moisture, nutrients, and collection system for their host plants. Arbuscular mycorrhizae gear up productivity and plant fitness unswervingly through escalating uptake of insoluble micronutrients and "P" phosphorus. Arbuscular mycorrhizal fungi (AMF) are ecologically significant to the majority of vascular plants. It is quite easier to separate the list of mycorhizae from others (Harley and Smith 1983). McGonigle (1988) found that AMF colonize up to 37% in a field trial of 78 surveys, whereas Lekberg and Koide (2005) observed that the increase colonization is 23% in 290 greenhouse and field studies. AMF also increase seed yield than the controlled groups of flax seeds, and it depends on status of nutrient, management, and type of soil (Thompson 1994).

The other beneficial role of AMF is to control root pathogens and their hormonal production that has higher potential to withstand synergistic interaction and water stress. Management practices, such as tillage rotation, influence mycorrhizal activity via phosphorus fertilization and cropping. During establishment, it ensures effective symbiosis and rapid colonization to plants. Recent advancements in research on physiology of ecology and mycorrhizae lead to the important function of mycorrhizae in the agroecosystem and how to manage practices under the influence to increase efficacy of mycorrhizal symbiosis (Table 12.1).

12.2 Mycorrhizae, Plant, and Soil Ecosystem Interaction

12.2.1 Colonization and Symbiosis

The growths of AMF fungi proceeding to root colonization are known as presymbiosis which mainly consists of three stages: (1) germination of spores, (2) growth of hyphae, and (3) recognition of host and formation of appressorium (Douds and Nagahashi 2000). Usually, spores of AMF are multinucleate resting structures and thick walled (Wright 2005). They also germinate in appropriate conditions of the temperature, phosphorus concentration Co_2, soil matrix and pH (Douds and Nagahashi 2000). The spore germination is not under the direct control of host plant and has been germinated both in soil condition and in vitro under experimental conditions in the dearth of plants. Colonization of AMF is higher in poor nutrient soils and lower in the glut of phosphate enriched fertilizer (Vivekanandan and Fixen 1991; Read et al. 1976). They enable hyphal growths which have chemotactic abilities (Sbrana and Giovannetti 2005). Gianinazzi-Pearson (1996) observed that arbuscules are the exchange for nutrients, water phosphorus, and carbon. It also reduces sensitivity and increases resistance against soil pathogens to toxic substances in their host plants. But today various other practices of agriculture lead to the obliteration of these symbiotic or beneficial associations.

Table 12.1 Some phytopathogenic diseases controlled by AMF

Crop	Disease	Phytopathogens	References
Tomato (*Lycopersicon esculentum*)	Damping-off	*Pythium aphanidermatum*	Jallali and Chand (1987)
Tomato (*Lycopersicon esculentum*)	Verticillium wilt	*Verticillium dahliae*	Karagiannidis et al. (2002)
Tomato (*Lycopersicon esculentum*)	Root-knot nematode	*Meloidogyne incognita*	Momotaz et al. (2015)
Asparagus (*Asparagus officinalis*) and *French* bean (*Phaseolus vulgaris*)	Root rot	*Fusarium oxysporum*	Matsubara et al. (2002)
French bean (*Phaseolus vulgaris*)	Root rot	*Rhizoctonia solani*	Neeraj (2011)
Asparagus (*Asparagus officinalis*)	Root rot	*Helicobasidium mompa*	Kasiamdari et al. (2002)
Mung bean (*Vigna radiata*)	Root and stem rots	*Rhizoctonia solani*	Kjoller and Rosendahl (1996)
Onion (*Allium cepa*)	White rot onion	*Sclerotium cepivorum*	(Torres-Barragán et al. 1996)
Pea (*Pisum sativum*)	Root rot	*Aphanomyces euteiches*	Larsen and Bodkar (2001)
Brinjal (*Solanum melongena*)	*Fusarium wilt*	*Fusarium solani*	Ojha et al. (2012)
Cotton (*Gossypium herbaceum*)	*Verticillium wilt*	*Verticillium* sp.	Kobra et al. (2011)
Sesame (*Sesamum indicum*)	Root and wilt rot	*Fusarium oxysporum* f. sp. sesame, *Macrophomina phaseolina*	Ziedan et al. (2011)
Chili (*Capsicum annuum*)	Root knot	*Meloidogyne incognita*	Chaudhary and Kaul (2013)
Soybean (*Glycine max*)	Red crown rot	*Cylindrocladium parasiticum*	Gao et al. (2012)
Bean (*Vigna mungo* and *Vigna radiata*)	Dry root rot	*Macrophomina phaseolina*	Chandra et al. (2007)
Bean (*Vigna mungo*)	Root knot nematode	*Meloidogyne incognita*	Sankaranarayanan and Sundarababu (2009)
Apple (*Malus domestica*)	Stem brown canker	*Botryosphaeria* sp.	Krishna et al. (2010)
Chrysanthemum (*Chrysanthemum carinatum*)	Phytoplasma	*Candidatus phytoplasmas asteris*	Sampo et al. 2012

12.2.2 *Nutrient Uptake and Exchange*

The hyphae of mycorrhizae are much more efficient than other plant roots. Via diffusion method, phosphorus moves to the roots, and in the uptake mechanism of the phosphorus, the hyphae of AMF decrease the desired distance. The inflow rate of phosphorus is six times higher in mycorrhizae as compared to root hairs. The major advantage of AMF to plants is to uptake nutrients increasingly. This uptake increases storage, surface area of soil contact, movement of nutrients and modification of the root environment into mycorrhizae. (Bolan 1991). AMF also increase uptake of other soil nutrients in plant like Cu, P, and Zn. The increased uptake is only because of the increased surface area of hyphae. In high-phosphorus soil, there is a decrease in colonization of AMF which also mediates other micronutrients such as Zn and Cu in plants (Pacovsky et al. 1986). The effect of AMF on plants micronutrients simply relies on the nutrient level of the soil (Liu et al. 2000; Lambert et al. 1979).

12.2.3 Phosphorus Fertility

Phosphorus (P) is one of the major nutrients, which is necessary for growth of plant. Phosphorus uptake from agricultural soil is mediated by arbuscular mycorrhizal fungi (AMF) directly to roots of plant. The activity of mycorrhizal colonization is inhibited by the excess of phosphorus, and in limited conditions its activity is quite high. Symbiotic association of AMF and low phosphorus results in the increase uptake of phosphorus and enhance crop yield (Grant et al. 2005). The main function of vesicular-arbuscular mycorrhizal (VAM) fungi is to provide phosphorus to roots of plan roots through phosphate transporters present in the hyphal membrane of the plant. The networks of extra radical and filamentous hyphae of AMF assist in the uptake of freely available phosphates. AMF result in the hydrolysis of organic phosphates which are present in soil and supply soluble phosphates to their host plant through hyphae (Bagyaraj et al. 2015).

12.2.4 Glomalin in Soil

Glomalin is a glycoprotein which is released by AM fungus after death of the fungus to quantify glomalin-related soil protein (GRSP). It contains 2–5% Fe, 4–6% O, 0.03–0.1% P, 36–59% C, 33–49% H, and 3–5% N (Wright et al. 1999). It is a good protector of hyphae and helps in the stabilization of soil aggregation. AMF hypha forms glomalin which is a reddish brown in appearance (Rilling et al. 2005). Due to its high adhesiveness and hydrophobicity, it plays a major role as cementing material with the particles of soil and acts as a highly stable form of organic carbon storage which represents an important fraction of the total organic matter present in the soil. AMF should be controlled by the community composition and abundance of AMF present in the soil.

12.3 Organic Farming

In accordance with the ecological and socioeconomic conditions, organic agriculture encourages the use of alternative technologies and various other agronomic practices which are carried out by using organic techniques, tools, and materials. Communities of soil microbes have been considered an important factor for the success in organic farming and to functionalize agroecosystems. In the soil microbial communities, AMF are a major component in the functioning of agroecosystems, soil fertility, soil microbiota, and plant nutrition. Thus, soil microbial communities such as AMF are usually lower in conventional type as compared to organic farming systems. The increasing concentration of AMF in soil results in the nutrient uptake of phosphate ions. In comparing the farming systems, i.e., organic farming systems (OFSs) and conventional farming systems (CFSs), OFSs promote enhanced soil biodiversity (Mäder et al. 2000), alleviate environmental stress (Altieri 2002), and enhance soil biodiversity (Mäder et al. 2000) and soil structure formation (Wright et al. 1999). Organic farming systems (OFSs) offer a unique context to know the exact relationship between biogeochemistry and soil biology (Cavagnaro et al. 2006). Soil microbial communities are considered an important factor for the success in organic farming and functioning of agroecosystems (Gosling et al. 2006). In organic farming the role of AMF is most likely to bring disease attack by AM-colonized roots, protection against pest, and to enhance productivity.

12.3.1 *Host Habit of Mycorrhizae*

Crop plants differ in the dependence and extent on AMF for nutrient uptake. Various plants may be obligate mycotrophs, while others are facultative mycotrophs (Smith and Read 2002). Crop plant root factors such as growth rate response to soil conditions, root hair abundance and length, surface area, and exudations determine the dependency on arbuscular mycorrhizal fungus for uptake of nutrients (Smith and Read 2002). Crops like flax (*Linum usitatissimum*) and corn (*Zea mays*) solely depend on arbuscular mycorrhizal fungus to fulfill their phosphorus requirements (Thingstrup et al. 1999; Plenchette 1983), and beans (Fabaceae), potatoes (*Solanum tuberosum*), and legumes (Leguminosae) also significantly benefitted from mycorrhizae. Oat (*Avena sativa*), barley (*Hordeum* pp.) and wheat (*Triticums* pp.) benefitted from VAM symbiosis but not solely depend on conditions of high soil fertility. In the different families of Brassicaceae and Polygonaceae, they do not form any symbiotic association with arbuscular mycorrhizal fungus (Harley and Smith 1983); it includes mustard (*Brassica juncea*), buckwheat (*Fagopyrum esculentum*), beets (*Beta vulgaris*) and canola (*Brassica napus*).

12.3.2 Mycorrhizal Fungi as Biofertilizer

AMF directly control the growth of its host plant by providing nutrition through plant roots (Smith and Read 2008). In addition, some extracellular secretions from fungal hyphae (Ravnskov et al. 1999) directly affect microflora of the mycorrhizosphere and also play a very significant role in soil aggregation (Rillig and Mummey 2006). VAM fungi have some unique structures known as arbuscules and vesicles, and they increase the length of the root 100-fold of the normal because of which the root reaches up to moist soil and helps plants to absorb available nutrients such as copper, molybdenum, phosphorus and zinc. AM fungal species cover sheath around the root and enhance the tolerance of seedling to drought, high temperature, pathogens, and soil acidity conditions. Treatment of *Aspergillus niger* fermentation along with AMF was added in the treated medium with sugar beet waste which increases the growth of lucerne (alfalfa) (Rodriguez et al. 1999). It was reported that *Azolla*, *Azotobacter*, *Azospirillum*, biofertilizers, mycorrhizal fungi, and phosphate-solubilizing microorganisms have experimentally proven and successfully been used experimentally in guinea grass plant (George et al. 1999).

12.4 Management of Mycorrhizal Biotechnology

12.4.1 Biocontrol of Soil-Borne Phytopathogens

Biocontrol agents can easily grow in lab conditions than AMF, and in combination with AMF, they act as powerful synergistic controls of pathogens. There is abrupt decrease in the harmful effects of soil-borne pathogens when AMF root is colonized into the plants (Gerdemann 1968, 1974), and it was also observed on bacteria nematodes, stramenopiles, and various fungi (Whipps 2004). In association of the clover plants with the *Glomus*, the infection cv. Sonja was controlled by *Pythium ultimum* (Carlsen et al. 2008). AMF compete with soil-borne pathogens and host roots within the mycorrhizosphere. Larsen and Bodker (Larsen and Bødker 2001) studied the use of fatty acid profiles and decrease in energy and biomass reserves of co-occupying pea roots in both *Aphanomyces euteiches* and *G. mosseae*. Cordier et al. (1996) demonstrated that *G. mosseae* and *Phytophthora nicotianae* are not occupied at the same time on tomato root tissues. A decrease in the quantity of mycorrhizal colonization by diverse plant pathogens has been also reported (Krishna and Bagyaraj 1983) representing the probable occurrence of competitive interactions. Because of these competitive interactions, the AM fungus is inoculated ahead of pathogen to favor the efficiency of biocontrol agent. AMF have not been reported for biological control; they are only used to reduce various symptoms of disease such as damping-off, wilt, root rot, and yellowing disease (Whipps 2004). Even though the research concerning biocontrol is widespread, only some studies have plainly considered interactions of arbuscular fungus. Other optimistic effects of AM fungal colonization on bacteria are the direct synergistic effects on mycorrhizal colonization itself of roots, and it is due to antagonistic effects of bacteria and AMF on competing pathogens (Azcon-Aguilar and Barea 1996; Budi et al. 1999). The role of AMF is to control intra-radical disease symptoms and the proliferation of soil-borne pathogens.

12.4.2 Mycorrhizal Fungi as Pest Control

Some studies reveal that AMF reduce the damage to roots caused by pathogenic nematodes. It also suppresses growth of nematode in roots and root galling of banana (*Musa* spp.) with inoculation of *Glomus mosseae* (Jaizme-Vega et al. 1997). AMF species increases the growth of coffee (*Coffea arabica*) when coming in contact with the nematode (Vaast et al. 1998) *Acaulospora mellea* and *Glomus clarum* which in combination reduces the egg production of the nematode *Meloidogyne incognita* in *Pratylenchus coffeae*, and infection was prevented by inoculating *Glomus fasciculatum* (Nagesh et al. 1999). Another study was demonstrated to show the susceptibility of the nematode *Meloidogyne arenaria* to peanut (*Arachis hypogaea*) (Carling et al. 1996) and of tomato to the nematode *Meloidogyne incognita* inoculated with AMF which results in 24% increase in shoot weight compared with non-mycorrhizal plants.

12.4.3 Management of Weeds to Increase Mycorrhizal Activity

Many common weed species are AMF hosts (Table 12.2). This includes wild oats (*Avena fatua*), Canada thistle (*Cirsium arvense*), dandelion (*Taraxacum officinale*), cleavers (*Galium aparine*), green foxtail (*Setaria viridis*), and chickweed (*Stellaria media*). Weeds in the families of Brassicaceae, Polygonaceae, Chenopodiaceae, and Amaranthaceae are non-mycorrhizal. They include lamb's quarters (*Chenopodium album*), canola (*Brassica napus*), kochia (*Kochia scoparia*), redroot pigweed (*Amaranthus retroflexus*), wild buckwheat (*Polygonum convolvulus*), and stinkweed (*Thlaspi arvense*). There are various evidences that AMF enhance effective symbiosis with the crop plant thus maintaining a diverse AMF population and was also found that the advantages of AMF to maize yield from keeping up a different weed cover species which outweighed any yield penalty because of rivalry (Feldman and Boyle 1998). Indigenous AMF host species may give a powerful bridge to AMF in the middle of cropping periods and contrasted a winter wheat cover-up with a dandelion cover and found that dandelion delivered higher AMF colonization, P take-up, and yield in the accompanying maize crop (Kabir and Koide 2000). In addition to the already mentioned role of strigolactones as molecule of signals for AM symbiotic interaction foundation, they additionally act in the rhizosphere as

Table 12.2 Multifaceted positive approaches of AMF in agroecosystem

Ecological aspect	Benefits to host plants
Agro-system stability	Improved plant root design (Bhattacharya et al. 2002), soil quality improvement by reestablishment of soil microflora (Nottingham et al. 2013), soil chemistry (Morovvat et al. 2012), and soil matrix (Rillig 2004)
Enhanced nutrient uptake	Improves the absorption of phosphate (Karandashov and Bucher 2005) along with other macronutrients like N (Othira et al. 2012), K (Porras-Soriano et al. 2009), Ca and Mg (Clark 2002), and S (Schreiner 2007) and micronutrients like Zn (Gao et al. 2007), Cu (Carvalho et al. 2006), Fe (Farzaneh et al. 2011), and Mn (Rivera-Becerril et al. 2013)
Stress resistance	Water scarcity (Wu et al. 2007), flooded condition (Muthukumar et al. 1997), sewage water utilization (Arriagada et al. 2009), radioactive metal stress in soil (Berreck and Haselwandter 2001) and heavy metal pollution (Leyval et al. 1997), and saline environments (Garcia and Mendoza 2007)
Improved carbohydrate level	Photosynthetic activity (Estrada-Luna et al. 2000)
Plant production and improvement in value added products	Increased shoot and root biomass (Aboul-Nasr 1996), productivity (Shinde et al. 2013; Bhale et al.2014; Bansode et al. 2014), synthesis of alkaloids (Karthikeyan et al. 2008), lycopene (Ordookhani et al. 2010), and essential oil (Karagiannidis et al. 2002)
Resistance to pathogens	Crop protection against pathogens (Berta et al. 2005) and reduced nematode effect (Jothi et al. 2005)
Crop management and cost reduction	Reduced use of chemical fertilizers (Nedorost and Pokluda 2012) and fungicides (Bailey and Safir 1978), improvement in yields in tissue culture (Jefwa et al. 2010) and transplantation shocks (Barrows and Roncadori 1977)
Habitat restoration	Biogeochemical cycling (Johnson et al. 2006)
Degraded ecosystems	Degraded area (Barea et al. 2011; Jeffries and Barca 2012)

identification signals for root parasitic plants of the family Orobanchaceae, including the genera *Phelipanche*, *Striga*, *and Orobanche* (López- Ráez et al. 2011; Bouwmeester et al. 2007).

Mycorrhizal fungi play an essential role in sustainable agriculture which is the urgent need of the modern era. Mycorrhizal fungi act as an effective biofertilizer and bioprotectant. These improve the plant vigor and soil quality. They also functionalized crucial role in plant nutrient, i.e., N, P, K, S, Mg, Fe, Zn, and Cu uptake, diversity and productivity of plant, biotic (root pathogen) and abiotic (drought, salinity, heavy metal) stress resistance, microbial diversity and population in mycor rhizosphere, soil aggregation and soil structure, reduction of nutrient leaching, weed suppression, and stimulating soil biological activity.

It is evidenced from the earlier studies and literature that symbiotic association of mycorrhiza plays an important role in different ecosystems and they must be considered as a necessary factor for promoting plant health, productivity, and soil fertility.

12.5 Interaction Between AMF and Other Organisms

Triple interaction of soil fungus and plant increases plant resistance to diseases, biological nitrogen fixation, plant resistance to drought, photosynthesis rates, and lower concentrations of harmful elements such as cadmium and arsenic in plant tissues as well as improves soil physical properties which ultimately result in better growth. Total dry weight along with alkaline and acid phosphatase activities of the roots was enhanced as compared to control in *Vetiveria zizanioides* when inoculated with *G. fasciculatum*, *Glomus aggregatum*, *G. intraradices*, and *G. mosseae* (Ratti et al. 2002). Manjunath et al. (1981) recorded the effects of *Beijerinckia mobilis*, *Aspergillus niger*, and *Glomus fasciculatum* singly and in combination of growth of onion where *G. fasciculatus*- or *B. mobilis*-inoculated plants improved dry weight. AMF along with plant growth-promoting rhizobacteria (PGPR) positively influence shoot potassium content, antioxidant activity, and fruit lycopene in tomato (Ordookhani et al. 2010). Berta et al. (2005) reported that *Pseudomonas fluorescens* and *Glomus mosseae* suppress *Rhizoctonia solani* root rot in tomato. Ozgonen et al. (2010) stated that AMF could effectively be used to enhance yield by reducing stem rot caused by *Sclerotium rolfsii* Sacc. Masunaka et al. (2011) inoculated *Lotus japonicus* a legume plant with plant growth-promoting fungi (*Trichoderma koningi*, *Fusarium equiseti*, and *Penicillium simplicissimum*) which develops symbiotic association rather than parasitism. Hage-Ahmed et al. (2009) found decreased germination of plant pathogenic fungi *Fusarium oxysporum* sp. *lycopersici* (Fol) in rhizosphere soil of *Glomus mosseae* treated with tomato than non-mycorrhizal plant. Vázquez et al. (2000) tested positive effects of *Azospirillum*, *Trichoderma*, and *Pseudomonas* upon mycorrhizal infection (*Glomus mosseae*, *Glomus deserticola*, and indigenous AMF) in maize plants. Garcia-Romera et al. (1998) found that the combination of some strains of *Fusarium* sp. with *G. mosseae* led to growth enhancement in soybean plants, along with shoot dry weight. The AMF have a vital role in the decrease of plant pathogens such as *Phytophthora* sp. and *Rhizoctonia solani* (Cordier et al. 1996; Trotta et al. 1996). Tarafdar and Marschner (1995) reported that by dual inoculation of *Aspergillus fumigatus* and *Glomus mosseae* have shown enhanced shoot, root dry weight, and root length in wheat (*Triticum aestivum* L.).

In combination of inoculation *Frankia* and AMF in *Casuarina*, there is threshold increase in the number of nodules, levels of N and P, total dry weight of shoots and roots, etc. (Vasanthakrishna et al. 1994). Harmonious effect on plant growth was observed and studied when both *Glomus fasciculatum* and *Azotobacter chroococcum* were inoculated in tomato plant (Bagyaraj and Menge 1978). There is steady reduction of disease symptoms which has been described for fungal pathogens, such as *Aphanomyces* spp. *Botrytis fabae*, *Fusarium oxysporum*, *P. splendens*, *P. vignae*, *Chalara (Thielaviopsis) basicola*, *R. bataticola*, *Sclerotium rolfsii*, *Pythium ultimum*, *Gaeumannomyces graminis* var. *tritici*, *P.cactorum*, *Dothiorella gregaria*,

Phytophthora parasitica, and *Ganoderma pseudoferreum*; bacteria such as *Ralstonia solanacearum*, *Rhizoctonia solani*, *Pseudomonas syringae*; and nematodes such as *Radopholus similis*, *Pratylenchus brachyurus*, *M. hapla*, *T. vulgaris*, *M. javanica*, *M. incognita*, *Tylenchulus semipenetrans*, and *Meloidogyne arenaria* (Bagyaraj and Chawla 2012). The study of plants inoculated with arbuscular mycorrhizal fungus and biocontrol agents requires special attention for the reason that these are capable to act as fungal antagonists which interfere with AMF.

12.6 Future Approach

1. Development of various factors of mycorrhizal technology leading to enhance functioning.
2. Arbuscular mycorrhizal fungus has high potential in the restoration of disturbed land and low fertility soil; therefore, there is a need to clear understanding of the effects of environmental changes on the AM fungal species.
3. Research priority relates to influence mycorrhizal effectiveness and is evident that there are conflicts.
4. It will improve the AMF and their involvement in interactions with environmental condition.
5. There will be the need of field-based studies.
6. It will be a strong understanding to the percentage disease incidence controlled by AMF.

12.7 Conclusion

Mycorrhizae are an essential element of successful low soil test phosphorus production systems. Encouragement of mycorrhizal symbiosis has great potential to benefit modern agricultural systems. Interest in food safety and environmental conservation has recently increased, and eco-friendly agricultural practices have been recognized as an important food safety and farming system by minimizing the use of mineral fertilizers and synthetic biocides. It is a cost-effective, high-productivity, and low-input farming system. Besides all these, mycorrhizal fungi also act as biocontrol agents and protect the host plant root from various soil-borne plant pathogens such as bacteria, fungi, and nematodes by using physical, chemical, and biological mechanisms. Mycorrhizal fungi play a very vital role in sustainable agriculture as an effective biofertilizer, and bioprotectant and interact with other microorganisms which is the urgent need of modern era.

References

Aboul-Nasr A (1996) Effects of vesicular-arbuscular mycorrhiza on *Tageteserecta* and *Zinnia elegans*. Mycorrhiza 6:61–64

Altieri MA (2002) Agroecology: the sciences of natural resource management for poor farmers in marginal environments. Agric Ecosyst Environ 93:1–24

Arriagada C, Pacheco P, Pereira G, Machuca A, Alvear M, Ocampo JA (2009) Effect of arbuscular mycorrhizal fungal inoculation on *Eucalyptus globulus* seedlings and some soil enzyme activities under application of sewage sludge amendment. J Soil Sci Plant Nutr 9(2):89–101

Azcon-Aguilar C, Barea JM (1996) Arbuscular mycorrhizas and biological control of soil-borne plant pathogens – an overview of the mechanisms involved. Mycorrhiza 6:457–464

Bailey JE, Safir OR (1978) Effect of benomyl on soybean endomycorrhizae. Phytopathology 68:1810–1812

Bagyaraj DJ, Sharma MP, Maiti D (2015) Phosphorus nutrition of crops through arbuscular mycorrhizal fungi. Curr Sci 108(7):1288–1293

Bagyaraj DJ, Menge JA (1978) Interaction between a VA mycorrhiza and *Azotobacter* and their effects on the rhizosphere microflora and plant growth. New Phytol 80:567–573

Bagyaraj DJ, Chawla G (2012) Status and prospects for enhancing the uptake of antagonistic organisms for nematode management in India. NBAII Publication, Bangalore, pp 74–89

Barrows JB, Roncadori RW (1977) Endomycorrhizal synthesis by *Gigaspora margarita* in poinsettia. Mycologia 69(6):1173–1184

Barea JM, Palenzuela J, Cornejo P, Sánchez-Castro I, Navarro-Femándcz C, Lopéz-García A, Estrada B, Azcón R, Ferrol N, Azcón-Aguilar C (2011) Ecological and functional roles of mycorrhizas in semi-arid ecosystems of Southeast Spain. J Arid Environ 75:1292–1301

Bhattacharya PM, Paul AK, Saha J, Chaudhuri S (2002) Changes in the root development pattern of bamboo and sweet orange plants upon arbuscular mycorrhization. Mycorrhiza News 14(1):15–18

Berreck M, Haselwandter K (2001) Effect of the arbuscular mycorrhizal symbiosis upon uptake of cesium and other cations by plants. Mycorrhiza 10:275–280

Berta G, Sampo S, Gamalero E, Massa N, Lemanceau P (2005) Suppression of *Rhizoctonia* root-rot of tomato by *Glomus mossae* BEG12 and *Pseudomonas fluorescens*A6RI is associated with their effect on the pathogen growth and on the root morphogenesis. Eur J Plant Pathol 111(3):279–288

Bhale UN, Sawant VS, Bansode SA (2014) River sedimentary soils retort to biomass production of Soybean *(Glycine max*) and arbuscular mycorrhizal fungal (AMF) status. Int J Adv Lif Sci 6(5):510–515

Bansode SA, Sawant VS, Bhale UN (2014) Biomass production of sunflower plant and pretense of Arbuscular Mycorrhizal Fungi (AMF) in newly renovated agricultural land. Flora and Fauna 20(2):191–195

Bolan NS (1991) A critical review of the role of mycorrhizae fungi in the uptake of phosphorus byplants. Plant Soil 134:189–207

Bouwmeester HJ, Roux E, López-Ráez JA, Beeard G (2007) Rhizosphere communication of plants, parasitie plants and AM fungi. Trends Plant Sci 12:224–230

Budi SW, van Tuinen D, Martinotti G, Gianinazzi S (1999) Isolation from the Sorghum bicolor mycorrhizosphere of a bacterium compatible with arbuscular mycorrhiza development and antagonistic towards soil borne fungal pathogens. Appl Environ Microbiol 65:5148–5150

Carling DE, Roncadori RW, Hussey RS (1996) Interactions of arbuscular mycorrhizae, Meloidogynearenaria, and phosphorus fertilization on peanut. Mycorrhiza 6:9–13

Carlsen SCK, Understrup A, Fomsgaard IS, Mortensen AG, Ravnskov S (2008) Flavonoidsin roots of white clover: interaction of arbuscular mycorrhizal fungi and a pathogenic fungus. Plant Soil 302:33–43

Cordier C, Gianinazzi S, Gianinazzi-Pearson V (1996) Colonisation patterns of root tissues by *Phytophthora nicotianae* var. *parasitica* related to reduced disease in mycorrhizal tomato. Plant Soil 185:223–232

Carvalho L, Caçador I, Martins-Loução M (2006) Arbuscular mycorrhizal fungi enhance root cadmium and copper accumulation in the roots of the salt marsh plant *Aster tripolium* L. Plant Soil 285(1–2):161–169

Cavagnaro TR, Jackson LE, Six J, Ferris H, Goyal S, Asami D, Scow KM (2006) Arbuscular mycorrhizas, microbial communities, nutrient availability, and soil aggregates in organic tomato production. Plant Soil 282:209–225

Chandra S, Khare V, Kehri HK (2007) Evaluation of Arbuscular Mycorrhizal Fungi against *Macrophomina phaseolina* causing dry root-rot of urd and mung bean. Indian Phytopathology 60(1):42–47

Chaudhary KK, Kaul RK (2013) Efficacy of *Pasteuria penetrans* and various oil seed cakes in management of *Meloidogyne incognita* in Chilli pepper (*Capsicum annuum* L.). J Agr Sci Tech 15:617–626

Clark RB (2002) Differences among mycorrhizal fungi for mineral uptake per root length of switch grass grown in acidic soil. J Plant Nutr 25(8):1753–1772

Douds DD, Nagahashi G (2000) Signalling and recognition events prior to colonisation of Rootsby arbuscular mycorrhizal Fungi. In: Podila G, Douds DD (eds) Current advances in mycorrhizae research. APS Press, Minnesota, pp 11–18

Estrada-Luna AA, Davies FT Jr, Egilla JN (2000) Mycorrhizal fungi enhancement of growth and gas exchange of micropropagated guava plantlets (*Psidium guajava* L.) during ex vitro acclimatization and plant establishment. Mycorrhiza 10:1–8

Farzaneh M, Vierheilig H, Lössl A, Kaul HP (2011) Arbuscular mycorrhiza enhances nutrient uptake in chickpea. Plant Soil Environ 57(10):465–470

Feldman F, Boyle C (1998) Weed-mediated stability of arbuscular mycorrhizal fungi effectiveness in maize monocultures. J Appl Bot 73:1–5

Gao X, Kuyper TW, Zou C, Zhang F, Hoffland E (2007) Mycorrhizal responsiveness of aerobic rice genotypes is negatively correlated with their zinc uptake when non mycorrhizal. Plant Soil 290:283–291

Gao X, Lu X, Wu M, Zhang H, Pan R, Tian J, Li S, Liao H (2012) Co-inoculation with rhizobia and AMF inhibited soybean red crown rot: from field study to plant defense-related gene expression analysis. Public Libr Sci one 7(3):e33977

Garcia VI, Mendoza RE (2007) Arbuscular mycorrhizal fungi and plant symbiosis in a saline-sodic soil. Mycorrhiza 17:167–174

Garcia-Romera I, Garcia-Garrido JM, Martin J, Fracchia S, Mujica MT, Godeas A, Ocampo JA (1998) Interaction between saprophytic *Fusarium* strains and arbuscular mycorrhizas of soybean plants. Symbiosis 24:235–246

Gerdemann J (1968) Vesicular arbuscular mycorrhiza and plant growth. Annu Rev Phytopathol 6:397–418

Gerdemann JW (1974) Vesicula-arbuscular mycorrhiza. Academic Press, NY

George S, Pillai GR, Pushpakomari R (1999) Influence of biofertilisers on productivity of Guinea grass intercropped in coconut gardens. Indian J Agron 43:622–627

Gianinazzi-Pearson V (1996) Plant cell responses to arbuscular mycorrhizae fungi: getting to the roots of symbiosis. The Plant Cell 8:1871–1883

Gosling P, Hodge A, Goodlass G, Bending GD (2006) Arbuscular mycorrhizal fungi and organic farming. Agric Ecosyst Environ 113:17–35

Grant C, Bittman S, Montreal M, Plenchette C, Morel C (2005) Soil and fertilizer phosphorus: effects on plant P supply and mycorrhizal development. Can J Plant Sci 85:3–14

Hage-Ahmed K, Chobot V, Postl W, Voglgruber A, Hadacek F, Steinkellner S (2009) Alteration of plant metabolites and root exudate-mediated interactions by pathogenic and mycorrhizal fungi in tomato. international symposium “Root Research and Applications” (RootRAP), Boku–Vienna, Austria, 2–4 Sept, pp 1–4

Harley JL, Smith SE (1983) Mycorrhizal symbiosis. Academic Press, London
Jaizme-Vega MC, Tenoury P, Pinochet J, Jaumot M (1997) Interactions between the root-knot nematode *Meloidogyne incognita* and *Glomus mosseae*in banana. Plant Soil 196:27–35
Jallali BL, Chand H (1987) In: Mahadevan et al (eds) Proceedings of iSI Asian conference on Mycorrhizae at Madras, pp 209–214
Jefwa J, Vanlauwe B, Coyne D, Van Asten P, Gaidashova S, Rurangwa E, Mwashasha M, Elsen A (2010) Benefits and potential use of Arbuscular Mycorrhizal Fungi (AMF) in banana and plantain (*Musa* spp.) systems in Africa. In: Dubois T et al (eds) Proceedings international conference on Banana & Plantain in Africa, Acta Hort, 879. ISHS, The Hague, pp 479–486
Jeffries P, Barca JM (2012) Arbuscular Mycorrhiza-a key component of sustainable plant-soil ecosystems. In: Hock B (ed) The Mycota. Springer, Berlín/Heidelberg, pp 51–75
Johnson D, Leake JR, Read DJ (2006) Role of arbuscular mycorrhizal fungi in carbon and nutrient cycling in grassland. In: Gadd GM (ed) Fungi in biogeochemical cycles. Cambridge University Press, Cambridge, pp 129–150
Jothi G, Babu RS, Rajendren G (2005) Biomanagement of nematodes by mycorrhiza – a review. Agric Rev 26(4):249–260
Kabir Z, Koide RT (2000) The effect of dandelion or a cover crop on mycorrhiza inoculum potential, soil aggregation and yield of maize. Agric Ecosyst Environ 78:167–174
Karagiannidis N, Bletsos F, Stavropoulos N (2002) Effect of Verticillium wilt (*VerticilliumdahliaeK*leb.) and mycorrhiza (*Glomus mosseae*) on root colonization, growth and nutrient uptake in tomato and egg plant seedlings. Sci Hortic 94:145–156
Karandashov V, Bucher M (2005) Symbiotic phosphate transport in arbuscular mycorrhizas. Trends Plant Sci 10(1):22–29
Karthikeyan B, Jaleel CA, Changxing Z, Joe MM, Srimannarayan J, Deiveekasundaram M (2008) The effect of AM fungi and phosphorous level on the biomass yield and ajmalicine production in *Catharanthus roseus*. Eur Asia J BioSci 2(3):26–33
Kasiamdarim RS, Smithm SE, Smith FA, Scott ES (2002) Influence of the mycorrhizal fungus, *Glomus coronatum*, and soil phosphorus on infection and disease caused by binucleate*Rhizoctonia* and *Rhizoctoniasolani* on mung bean (*Vigna radiata*). Plant Soil 238:235–244
Kjøller R, Rosendahl S (1996) The Presence of the Arbuscular Mycorrhizal Fungus Glomus intraradices Influences Enzymatic Activities of the Root Pathogen Aphanomyces euteiches in Pea Roots. *Mycorrhiza* 6(6):487–491
Kobra N, Jalil K, Youbert G (2011) *Arbuscular mycorrhizal* fungi and biological control of *Verticillium* wilted cotton plants. Arch Phytopathol Plant Protect 44(10):933–942
Krishna H, Das B, Attri BL, Grover M, Ahmed N (2010) Suppression of *Botryosphaeria* canker of apple by arbuscular mycorrhizal fungi. Crop Protect 29:1049–1054
Krishna KR, Bagyaraj DJ (1983) Interaction between *Glomus fasciculatum* and *Sclerotium rolfsii* in peanut. Can J Bot 61:2349–2351
Lambert DH, Baker DE, Cole H Jr (1979) The role of mycorrhizae in the interactions of phosphorus with zinc, copper and other elements. Soil Sci Soc Am J 43:976–980
Larsen J, Bødker L (2001) Interactions between pea root-inhabiting fungi examined using signature fatty acids. New Phytol 149:487–493
Lekberg Y, Koide RT (2005) Is plant performance limited by an abundance of arbuscular mycorrhizal fungi? a meta-analysis of studies published between 1988–2003. New Phytol 168:189–204
Leyval C, Turnau K, Haselwandter K (1997) Effect of heavy metal pollution on mycorrhizal colonization and function: physiological, ecological and applied aspects. Mycorrhiza 7:139–153
Liu A, Hamel C, Hamilton RI, Ma BL, Smith DL (2000) Acquisition of Cu, Zn, Mn, and Fe bymycorrhizal maize (*Zea mays* L) grown in soil at different P and micronutrient levels. Mycorrhiza 9:331–336
López-Ráez JA, Charnikhova T, Femández I, Bouwrneester H, Pozo MJ (2011) Arbuscular mycorrhizal syrnbiosis decreases strigolactone production in tomato. J Plant Physiol 168:294–297

Mader P, Edenhofer S, Boller T, Wiemken A, Niggli U (2000) Arbuscular mycorrhizae in a long-term field trial comparing low-input (organic, biological) and high-input (conventional) farming systems in a crop rotation. Biol Fertil Soils 31:150–156

Manjunath BA, Mohan R, Bagyaraj DJ (1981) Interaction between *Beijerinckiamobilis*, *Aspergillusniger* and *Glomusfasciculatus* and their effects on growth of onion. New Phytol 87:723–727

Matsubara YI, Hasegawa N, Fukui H (2002) Incidence of Fusarium root rot in asparagus seedlings infected with arbuscular mycorrhizal fungus as affected by several soil amendments. J Jpn Soc Hortic Sci 71:370–374

Masunaka A, Hyakumachi M, Takenaka S (2011) Plant growth–promoting fungus, *Trichoderma koningi* suppresses isoflavonoid phytoalexin vestitol production for colonization on/in the roots of *Lotus japonicus*. Microbes Environ 26(2):128–134

McGonigle TP (1988) A numerical analysis of published field trials with vesicular-arbuscular mycorrhizal fungi. Funct Ecol 2:473–478

Momotaz R, Alam MM, Islam MN, Alam KM, Rahman MZ (2015) Management of the root-knot nematode of tomato by inoculation with Arbuscular Mycorrhizal Fungi. Int J Sustain Crop Prod 10(2):48–54

Morovvat A, Ronaghi A, Zarei M, Emadi M, Heidarianpour MB, Gholami L (2012) Effect of arbuscular mycorrhiza fungi application on distribution of phosphorus forms in rhizosphere soils of sunflower (*Helianthus annuus* L.). Int J Agric Sci Res Technol 2(2):77–82

Muthukumar T, Udaiyan K, Karthikeyan A, Manian S (1997) Influence of native endomycorrhiza, soil flooding and nurse plant on mycorrhizal status and growth of purple nutsedge (*Cyperus rotundus* L.). Agric Ecosyst Environ 61:51–58

Nagesh M, Reddy PP, Kumar MVV, Nagaraju BM (1999) Studies on correlation between *Glomus fasciculatum* spore density, root colonization and *Meloidogyne incognita* infection on *Lycopersicon esculentum*. J Plant Dis Protect 106:82–87

Nedorost L, Pokluda R (2012) Effects of arbuscular Mycorrhizal fungi on tomato yield and nutrient uptake under different fertilization levels. Acta Universitatis Agriculturae et Silviculturae Mendelianae Brunensis 60(8):181–186

Neeraj SK (2011) Organic amendments to soil inoculated *arbuscular mycorrhizal* fungi and *Pseudomonas fluorescens* treatments reduce the development of root-rot disease and enhance the yield of *Phaseolus vulgaris* L. Eur J Soil Biol 47:288–295

Nottingham AT, Turner BL, Winter K, Chamberlain PM, Stott A, Tanner EV (2013) Root and arbuscular mycorrhizal mycelial interactions with soil microorganisms in lowland tropical forest. FEMS Microbiol Ecol 85(1):37–50

Ojha S, Chakraborty M, Chatterjee NC (2012) Influence of salicylic acid and *Glomusfasciculatum*on *Fusarial*wilt of tomato and brinjal. Arch Phytopathol Plant Protect 45(13):1599–1609

Ordookhani K, Khavazi K, Moezzi A, Rejali F (2010) Influence of PGPR and AMF on antioxidant activity, lycopene and potassium contents in tomato. Afr J Agric Res 5(10):1108–1116

Othira JO, Omolo JO, Wachira FN, Onek LA (2012) Effectiveness of arbuscular mycorrhizal fungi in protection of maize (*Zea mays* L.) against witchweed (*Striga hermonthica* Del Benth) infestation. J Agric Biotechnol Sustain Dev 4(3):37–44

Ozgonen H, Akgul DS, Erkilic A (2010) The effects of arbuscular mycorrhizal fungi on yield and stem rot caused by *Sclerotiumrolfsii*Sacc. in peanut. Afr J Agric Res 5(2):128–132

Pavkovsky RS, Bethlenlfalvay GJ, Paul EA (1986) Comparisons between P-fertilized and mycorrhizal plants. Crop Sci 26:151–156

Plenchette C (1983) Growth responses of several plant species to mycorrhizae in a soil of moderate P-fertility. Plant Soil 70:199–209

Porras-Soriano A, Soriano-Martín ML, Porras-Piedra A, Azcón R (2009) Arbuscular mycorrhizal fungi increased growth, nutrient uptake and tolerance to salinity in olive trees under nursery conditions. J Plant Physiol 166(13):1350–1359

Ratti N, Gautam SP, Verma HN (2002) Impact of four *Glomus* species on the growth, oil content, P content and phosphatase activity of *Vetiveria zizanioides*. Indian Phytopathology 55(4):434–437

Rillig MC (2004) Arbuscular mycorrhizae, glomalin, and soil aggregation. Can J Soil Sci 84:355–363

Rilling MC, Lutgen ER, Ramsey PW, Klironomos JN, Gannon JE (2005) Microbiota accompanying different arbuscular mycorrhizal fungal isolates influence soil aggregation. Pedobiologis 49:251–259

Rillig MC, Mummey DL (2006) Mycorrhizas and soil structure. New Phytol 171:41–53

Rivera-Becerril F, Juárez-Vázquez LV, Hernández-Cervantes SC, Acevedo-Sandoval OA, Vela-Correa G, Cruz-Chávez E, Moreno-Espíndola IP, Esquivel-Herrera A, De León-González F (2013) Impacts of manganese mining activity on the environment: interactions among soil, plants, and arbuscular mycorrhiza. Arch Environ Contam Toxicol 64(2):219–227

Rodriguez R, Vassilev N, Azcon R (1999) Increases in growth and nutrient uptake of alfalfa grown in soil amended with microbially-treated sugar beet waste. Appl Soil Ecol 11(1):9–15

Ravnskov S, Larsen J, Olsson PA, Jakobsen I (1999) Effects of various organic compounds growth and phosphorus uptake of an arbuscular mycorrhizal fungus. New Phytol 141:517–524

Read DJ, Koucheki HK, Hodgson J (1976) Vesicular-arbuscular mycorrhiza in natural vegetation systems. New Phytol 77:641–653

Sankaranarayanan C, Sundarababu R (2009) Reciprocal influence of Arbuscular Mycorrhizal Fungus and root knot nematode and interaction effects on Blackgram. Nematol Medit 37:197–202

Sampo S, Nadia M, Simone C, Urnuberta D, Dominica B, Christina M, Graziella B (2012) Effect of two Am fungi on phytoplsama infection in the model plant *Chrysantemum carinatum*. Agric Food Sci 21:39–51

Sbrana C, Giovannetti M (2005) Chemotropism in the arbuscular mycorrhizal fungus *Glomus mosseae*. Mycorrhiza 15:539–545

Schreiner RP (2007) Effects of native and nonnative arbuscular mycorrhizal fungi on growth and nutrient uptake of 'Pinot noir' (*Vitis vinifera* L.) in two soils with contrasting levels of phosphorus. Appl Soil Ecol 36:205–215

Shinde SK, Shinde BP, Patale SW (2013) The alleviation of salt stress by the activity of AM fungi in growth and productivity of onion (*Allium cepa* L.) plant. Int J Life Sci Farma Res 3(1):11–15

Smith SE, Read DJ (2002) Mycorrhizal symbiosis. Academic Press, London

Smith SE, Read DJ (2008) Mycorrhizal symbiosis, 3rd edn. Academic Press, Amsterdam

Tarafdar JC, Marschner H (1995) Dual inoculation with *Aspergillus fumigatus* and *Glomus mosseae* enhances biomass production and nutrient uptake in wheat (*Triticum aestivum* L.) supplied with organic phosphorus as Na-phytate. Plant Soil 173:97–102

Thingstrup I, Rubaek G, Sibbensen E, Jakobsen I (1999) Flax (*Linum usitatissimum* L.) depends on arbuscular mycorrhizal fungi for growth and P uptake at intermediate but not high soil P levels in the field. Plant Soil 203:37–46

Thompson JP (1994) Inoculation with vesicular-arbuscular mycorrhizal fungi from cropped soil overcomes long-fallow disorder of linseed (*Linum usitatissium* L.) by improving P and Zn uptake. Soil Biol Biochem 26(9):1133–1143

Torres-Barragán A, Zavalc-Tamcjia E, Gonzalcz-Chavcz C, Fcrrcra-Ccrrato R (1996) The use of arbuscular mycorrhizae to control onion white rot (*Sclerotium cepivorum*) under field conditions. Mycorrhiza 6:253–257

Trotta A, Varese GC, Gnavi E, Fusconi A, Sampo S, Berta G (1996) Interactions between the soil-borne root pathogen *Phytophthora nicotianae* var. *parasitica* and the arbuscular mycorrhizal fungus *Glomus mosseae* in tomato plants. Plant Soil 185:199–209

Vaast P, Caswell-Chen EP, Zasoski RJ (1998) Influeences of a root-lesion nematode, *Pratylenchus coffeae*, and two arbuscular mycorrhizal fungi, *Acaulospora mellea* and *Glomus clarumon* coffee (*Coffea Arabica* L.). Biol Fertil Soils 26:130–135

Vázquez MM, César S, Azcón R, Barea JM (2000) Interactions between arbuscular mycorrhizal fungi and other microbial inoculants (*Azospirillum, Pseudomonas, Trichoderma*) and their effects on microbial population and enzyme activities in the rhizosphere of maize plants. Appl Soil Ecol 15(3):261–272

Vasanthakrishna M, Muthanna MB, Bagyaraj DJ (1994) Succession of vesicular arbuscular mycorrhizal fungi associated with *Casuarina equisetifoliaL.* Ann For 2:123–126

Vivekanandan M, Fixen PE (1991) Cropping systems effects on Mycorrhizal colonization, early growth, and Phosphorus uptake. Soil Sci Soc Am J 55:136–140

Whipps JM (2004) Prospects and limitations for mycorrhizas in biocontrol of root pathogens. Can J Bot 82:1198–1227

Wright SF (2005) Management of Arbuscular Mycorrhizal Fungi. In: Zobel RW, Wright SF (eds) Roots and soil management: interactions between roots and the soil. American Society of Agronomy, Madison, pp 183–197

Wright SF, Starr JL, Paltineanu IC (1999) Changes in aggregate stability and concentration of glomalin during tillage management transition. Soil Sci Soc Am J 63(6):1825–1829

Wu QS, Zou YN, Xia RX, Wang MY (2007) Five Glomus species affect water relations of *Citrus tangerine* during drought stress. Bot Stud 48:147–154

Ziedan E, Sayed EI, Mostafa M, Sahab A (2011) Application of mycorrhizae for controlling root diseases of sesame. J Plant Protect Res 51(4):355–361

Chapter 13
Strategic Approaches for Applications of Entomopathogenic Fungi to Counter Insecticide Resistance in Agriculturally Important Insect Pests

V. Ambethgar

Abstract Insect attack is a serious agricultural problem, both in the field and during storage, leading to substantial yield losses and reduced product quality. Annually, insect pests destroy about 25% of food crops worldwide. Intense use of insecticides in pest management systems has caused several ecological complications. Of them, the development of resistance to insecticides amongst the target insect pests is the major impediment in agricultural pest management programme. To date, more than 645 species of phytophagous insects and mite pests have developed high degree of resistance to several classes of insecticides mediated through a complex of biological and physiochemical mechanisms mediated via a complex of enzymatic induction. Resistance management strategies employing combination of insecticides or alternation of insecticides are suggested to improve the sustainability and to prevent/delay the development of resistance, but with potentialities of cross resistance between the currently approved insecticides, practical options are few. Naturally occurring insect pathogenic fungi are widely prevalent amongst insect species in diverse agroecosystems. Many species of entomopathogenic strains derived from Hyphomycetes and Zygomycetes classes have potential to control a range of pest insects with their unique contact action. Novel strategies employing myco-insecticides derived from fungi are advantageous in dealing with insecticide resistance management (IRM). The Hyphomycete species, viz. *Beauveria bassiana* (Balsamo) Vuillemin, *Metarhizium anisopliae* (Metschnikoff) Sorokin, *Nomuraea rileyi* (Farlow) Samson, *Paecilomyces fumosoroseus* (Wize) Brown and Smith, *Lecanicillium (=Cephalosporium = Verticillium) lecanii* (Zimmermann) Viegas and *Hirsutella thompsonii* Fisher, and Zygomycete species, viz. *Entomophthora virulenta* Hall and Dunn, *Erynia neoaphidis* Remaudiere and Hennebert and *Zoophthora radicans* (Brefeld) Batko, are being employed to counter insect resistance to insecticides against diverse insect pests. Many of these fungi induce quick mortality of target pests by inhibiting enzymatic detoxification mechanisms which in turn predispose

V. Ambethgar (✉)
Horticultural College and Research Institute for Women, Tamil Nadu Agricultural University, Tiruchirappalli, Tamil Nadu, India

P. Gehlot, J. Singh (eds.), *Fungi and their Role in Sustainable Development: Current Perspectives*, https://doi.org/10.1007/978-981-13-0393-7_13

the target pest insects for fungal infection. Instances of joint application of promising isolates of fungal entomopathogens with selective insecticides have been employed against a diverse suite of pests to slash down the selection pressure to insecticides and to overcome concurrent resistance risks in target pests. Integration of promising strains of insect pathogenic fungi with selective insecticides tends to improve the pest control efficiency, besides decrease the amount of insecticides required and minimize the risks associated with hazards involved in environmental and health issues, which further render delayed expression of insecticide resistance in insect pests. This review focuses on the potentials for employing some promising stains of fungal entomopathogens for complementing existing insect pest management measures with insecticide resistance management for agriculturally important insect pests.

Keywords Insecticide resistance management · Entomopathogenic fungi · Agriculture

13.1 Introduction

Insect pest menace is a serious agricultural problem that destabilizes crop productivity on a global basis. The rich biodiversity of agroecosystems provides congenial environments for proliferation of insect herbivores under diverse cropped ecosystems (Ranjekar et al. 2003). A spectrum of insect pests belonging to diverse orders damages various crops both in the field and during storage, leading to reduced yield and product quality. The most practical means of achieving greater production of agricultural crops is to mitigate the losses associated with pest insects. In agriculture, the total production loss due to pest infestations is estimated at 14–51% in different crops (Sharma et al. 2001). Overall, herbivorous insects alone destroy about 25% of food crops annually (David 2008). Arthropod pests not only inflict direct loss to the agricultural produce but also induce indirect loss due to their role as vectors of crop diseases (Ahmad et al. 2002), which pose additional monetary loss incurring towards the cost of pesticides applied for vector control. *Agro Magazine* points out that overall losses of agricultural production are estimated between 10% and 100% when no insecticides are used (Riley and Sparks 2006). In systems based on the use of insecticides and non-chemical control methods, losses due to insect attack come to an estimated 13%, whereas in natural ecosystems, roughly 10% of all produced crops are lost annually.

Over-reliance on broad-spectrum pesticides comes under severe criticism from different parts of the world after the publication of *Silent Spring* in 1962 by Rachel Carson. Since then, an alternative ecofriendly strategy for the management of noxious insect pests has been searched to reduce harmful effects of chemical insecticides on humanity. In agriculture, intensive use of pesticides ends up with rapid development of resistance in pest insects (Umina 2007; Silva et al. 2012), which often threatens the effectiveness and sustainability of pest management programme in agriculture. Repeated applications of the same insecticide over multiple generations of pest induce resistant individuals to reproduce, resulting in flare-up of resistant pest populations that can no longer be controlled with that insecticide even at increased doses (Riley and Sparks 2006). In pest control programme where insecticides are used frequently, resistance management practices must be considered to preserve ecological balance, besides ensuring safety towards nontarget organisms and human health (Deedat 1994; Sharma et al. 2001). While aiming to achieve sustainable crop productivity, it is important to adopt safer pest control strategies in order to mitigate insecticide resistance development in target pests. Furthermore, with strict legislation of industrialized nations on maximum permitted levels of pesticide residues on food safety aspects, it is extremely important to research alternative methods for pest control.

In recent years, crop protection based on biological control of crop pests with microbial pathogens has been recognized as a valuable tool in pest management (Bhattacharya et al. 2003). The appropriate use of environment-friendly microbial pesticides can play a significant role in sustainable crop production by providing a stable pest management programme. Currently, resurgence of interest persists to integrate microbial compounds to counter insecticide-induced resistance in insects. Amongst the microbial entities, employing fungi with selective insecticides would be an ideal option to mitigate insecticide resistance pressures in insect pests (Ambethgar 2009). In nature, entomopathogenic fungi are widely dispersed amongst insect species in diverse agroecosystems (Agarwal 1990). Fungi belonging to the Hyphomycete species, viz. *Beauveria bassiana* (Balsamo) Vuillemin, *Metarhizium anisopliae* (Metsch) Sorokin, *Nomuraea rileyi* (Farlow) Samson, *Paecilomyces fumosoroseus* (Wize) Brown and Smith, *Lecanicillium (=Verticillium) lecanii* (Zimmermann) Viegas and *Hirsutella thompsonii* Fisher, and Zygomycete species, viz. *Entomophthora virulenta* Hall and Dunn, *Erynia neoaphidis* Remaudiere and Hennebert and *Zoophthora radicans* (Brefeld) Batko, have been employed to counter insecticide resistance in many agricultural pests (Butt and Brownbridge 1997; Chandler et al. 2000). Entomopathogenic fungi alone amongst the microbes have unique features to invade insect host by simple contact mechanism via direct penetration of the cuticle (Ambethgar 2001; Olivera and Neves 2004; Jaramillo et al. 2005). However, unlike instant killing action of insecticides, fungi requires extended time between the application and the death of the host, which is the main obstacles with the use of fungi for biological pest control programme (Butt and Brownbridge 1997; Jacobson et al. 2001. Maintaining high level of virulence is another important aspect for proving efficiency of fungi (Batista Filho and Leite 1995). Accordingly, efforts have been made to describe new approaches for mitigating insecticide resistance in insect pests through integrating fungal entomopathogens with commonly used insecticides and/or botanical pesticides. This paper also examines the potential utility of some promising entomopathogenic fungal species and discusses the challenges ahead with insecticide resistance management programme of certain agriculturally important insect pests. Keeping all these issues in view, I appeal that field consultants and pest management policymakers use this resource information to combat the risk of insect resistance to insecticides for change towards sustainability in pest management system.

13.2 Trends in Pesticides Used for Insect Pest Management

Globally, about 50% of all pesticides produced are used for pest management in agriculture (Deedat 1994), and the demand for pesticides continuously persists. According to Sharma et al. (2001), about 95% of the insecticides produced are used for crop protection programme. Deedat (1994) estimated that in the absence of pesticides, about two-third of all crop yield providing food for millions of people would be lost. For example, from the 1960s rice crop became a major consumer of pesticides with the introduction of high-yielding varieties following the "Green Revolution" (Regupathy 1995). In India, instances of injudicious application of pesticides in crop plants especially cotton, pulses, vegetables and fruit crops resulted in development of pesticide resistance amongst the key pests (Regupathy 1995; David 2008). The massive scale use of chemical pesticides in many field situations has triggered the resistance development in target pests, besides disrupting the biocontrol mechanisms. Besides, several cases of insecticide resistance have become a worldwide threat to commercial agriculture (Sundaramurthy and Gahukar 1998). In the past, numerous brands of insecticides in almost all chemical classes have been successfully employed for control of many serious agricultural pests (David 2008). However, indiscriminate use of such toxic pesticides has induced several impediments in the evolution of insecticide resistance in a range of agricultural pest insects, mites, fungi, bacteria, nematodes and weeds in diverse agroecosystems (Luckmann and Metcalf 1982). However, arthropod pests (insects and mites) have acquired remarkable capacity to develop resistance to all types of insecticides including inorganic and microbial pesticides (Deedat 1994).

13.3 Insecticide Resistance Research: Overview

The Insecticide Resistance Action Committee (IRAC) defines "resistance to insecticide is a heritable change in the sensitivity of a pest population that is reflected in the repeated failure of a product to achieve the expected level of control when used according to the label recommendation for that pest species" (Russell 1999). The magnitude of resistance in insect population was relatively insignificant for the farmers in the pest-affected areas (Russell 1999). The first published data of insecticide resistance was reported in the State of Washington in a short notice when orchardists faced problems in controlling San Jose scale, *Quadraspidiotus perniciosus* (Comstock), and codling moth, *Cydia pomonella* (L.) in *Garden and Forest* in 1897. With the intensive use of synthetic organic insecticides, resistant insect species were met with mainly from houseflies, mosquitoes, granary weevils and bedbugs (www.pesticideresistance.com), which created concern for agriculture and public health in the early 1950s. In later periods, more cases of insect resistance to insecticides grew exponentially. By the end of 2006, there were 645 specific cases of agricultural insecticide resistance, affecting 316 compounds across the world (David 2008; www.pesticideresistance.com).

13.4 Insecticide Resistance Mechanisms

The genetic and biochemical mechanisms of insecticide resistance that persists in arthropod population are generally regulated via specific gene or combination of genes present. The popular expression of "Functions create organs and heredity determines the change in offspring" is emphasized as the basis of evolution. From a philosophical point of view, the scientific community generally accepted the evolutionary theory of **Darwin** based on **the** emergentism *The Origin of Species* in 1859. From the perspectives of insect pest management, specific group of insecticides are formulated to kill the target pests, but many of these toxicants are rarely effective as the residual population might possess peculiar behavioural trait that tends to avoid the direct exposures to insecticides to detoxify the insecticide in target pests (Kranthi et al. 1997). If the surviving insect population mates, it passes these resistance traits to their offspring, which will contain fewer susceptible individuals. Eventually, in the subsequent insect generation, the entire population may become resistant, if the offspring were continuously exposed to the same toxicants repeatedly over prolonged periods. Figure 13.1 depicts the generalized mechanism of insecticide resistance build-up in insects. Resistance in insect pests occurs through complex mechanisms, especially via metabolic, physical, physiological, behavioural and biochemical mechanism or combination of two or more mechanisms (Soderhall and Smith 1986.

According to Brogdon and McAllister (1998), resistance is governed by inherited ability of an individual insect to survive to a concentration of insecticide that is

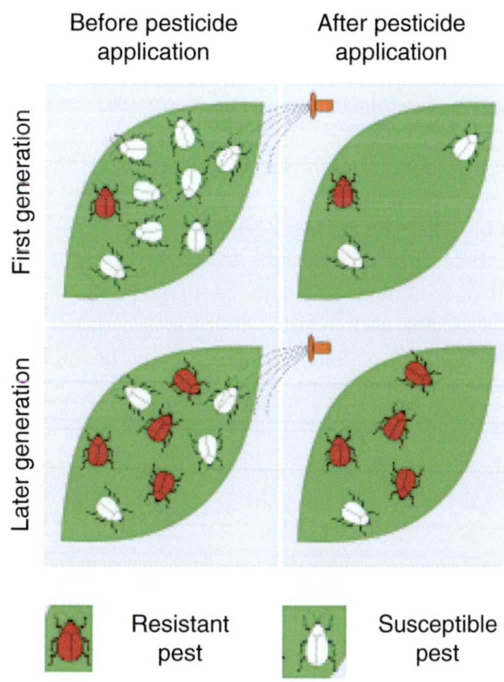

Fig. 13.1 Development of resistance in insects before and after pesticidal application (*Source*: Wikipedia)

lethal to other individuals that lack this gene. Usually, an insect inherits this resistance gene from its predecessors. In insects, resistance development is effected by complex of factors including the rate or dose, frequency of application of insecticide and certain pest characteristics (Regupathy 1995; Duraimurugan and Regupathy 2005). Small arthropod pests, especially the phytophagous acari and hemipterous insects, have identical physiological traits that contribute to rapid resistance development due to their limited dispersal capacity, short lifespans, high reproductive capacity with a lot of offspring, exposure of overlapping generations to a pesticide or sublethal doses of insecticides (Metcalf 1982; Ahmad et al. 1999). Insecticide-resistant individuals are seldom prevalent before exposure to any insecticide. However, with repeated use of insecticides, insects tend to develop resistance because of the frequent exposures (Georghiou and Taylor 1986).

13.5 Types of Insecticide Resistance

In nature, insect species tend to have evolved with complex modes of resistance to insecticides. For example, *esterase resistance* evolved through metabolic mechanism confers strong resistance to organophosphate compounds because of the overproduction of carboxylesterases sequestered to degrade insecticide esters before the toxicants reach the target sites of the nervous system (Metcalf 1982). Similarly, "modified acetylcholinesterase (MACE)", involved in target-site mechanism, confers immunity to the dimethyl carbamate (Ahmad et al. 1995). The knock-down target-site mechanisms *kdr* and super-*kdr* confer strong resistance against pyrethroid insecticides. Some insects adapt to *class resistance* which occurs in pest populations that develop resistance to any specific class of insecticides such as organophosphates (or), carbamates (or) and pyrethroids. The complex nature of *cross resistance* occurs when resistance to one insecticide confers resistance to another insecticide, which is illustrated in Fig. 13.2. Insect population that possess cross resistance is hard to manage with even most powerful insecticides (Olson et al. 2000; Zhang et al. 2000).

Insects that possess *multiple resistance* are endowed with independent mechanisms leading to resistance to different chemical families (e.g., carbamates, organophosphates and pyrethroids). On the other hand, insects that tend to *tolerate* a particular insecticide possess specific physiological or behavioural adaptations for increased survivorship to specific baseline toxicity.

13.6 Occurrence of Resistance in Insects to Insecticides

A diverse suite of arthropod pests developed resistance to different classes of chemical insecticides (Georghiou and Taylor 1986), with as many as 645 case histories of resistance met with insects belonging to Lepidopteran, Hemipteran and

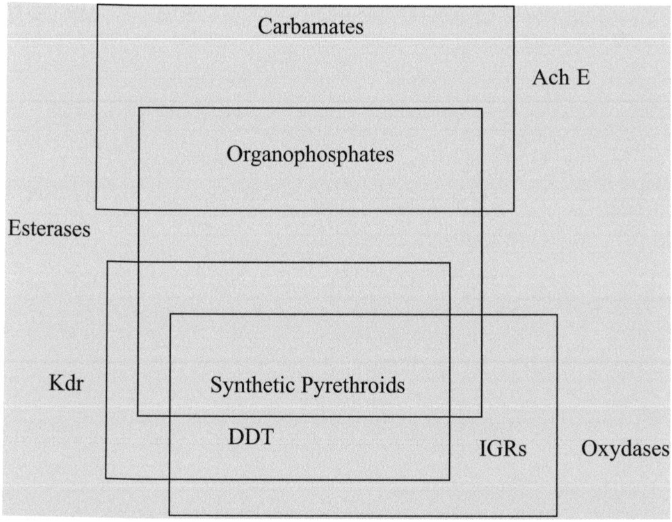

Fig. 13.2 Cross-linking resistance mechanism in different classes of insecticides
IGRs insect growth regulators, *Kdr* knock-down resistance, *AchE* acetylcholine esterase

Thysanopteran orders and mite pests of several being reported (Regupathy and Paramjothi 1977; Sharma et al. 2001). Many species of phytophagous pest insects, especially defoliators (caterpillars, beetles) and sap feeders (aphids, mealybugs, leafhoppers, plant hoppers, whiteflies and thrips), have developed resistance to several groups of insecticides (Regupathy and Paramjothi 1977), including currently available newer insecticides. Sharma et al. (2001) found that maximum instances of resistance encountered in order of severity were with organophosphates (250), synthetic pyrethroids (156), carbamates (154) and others including chlorinated hydrocarbons (85) (Sharma et al. 2001). The gravity of resistance development amongst the organisms also vary, which may occur so rapidly in some species, more slowly in certain groups and not at all in others (Sharma et al. 2001). Globally, there exist a number of case history pertaining to insecticide-induced resistance in insects (e.g., Georghiou and Taylor 1986; Olson et al. 2000). Some major examples of arthropod pests resistant to commonly used insecticides are provided in Table 13.1. Insect pests of every crop production system including field crops (cereals, pulses, oilseeds, cotton, tobacco) and horticultural crops (vegetables, fruits and ornamentals) have been prone to resistance to several insecticides (Ahmad et al. 1995; Kranthi et al. 1997). The two important polyphagous pests, viz. American fruitworm (*Helicoverpa armigera* Hub.) and tobacco caterpillar (*Spodoptera litura* Fab.), have shown extreme level of resistance to several groups of insecticides (Regupathy 1995; Ahmad et al. 2002). Field populations of Colorado potato beetle, *Leptinotarsa decemlineata* (Say), developed cross resistance to several insecticides as reported by Olson et al. (2000). Similarly, in many cruciferous vegetables, the worms of diamondback moth (*Plutella xylostella*) have shown high degree of resistance to any effective insecticide recommended to its management (Sarfraz and Keddie 2005).

Table 13.1 Some important examples of insect/mite pests resistant to commonly used insecticides in agroecosystem

Insect/mite pests		Resistance built to				
Order/common name	Scientific name	OC	OP	Pyr	Car	Mechanism
Coleoptera						
Colorado potato beetle	*Leptinotarsa decemlineata*	+	+	–	+	x, M
Coffee berry borer	*Hypothenemus hampei*	+	+	–	+	x, M
Singhara beetle	*Galerucella birmanica*	+	–	–	–	x
Granary weevil	*Sitophilus granarius*	+	+	–	–	x
Sawtoothed grain beetle	*Sitophilus oryzae*	+	+	–	–	x
Red flour beetle	*Tribolium castaneum*	+	–	–	+	x
Diptera						
Fruit fly	*Drosophila melanogaster*	+	–	–	–	–
Leaf miner	*Liriomyza trifolii*	+	–	–	–	–
Heteroptera						
White-tailed mealy bug	*Ferrisia virgata*	+	–	+	–	x, kdr
Tea mosquito bug	*Helopeltis antonii*	–	–	+	–	kdr
Tea mosquito bug	*Helopeltis theivora*	–	–	+	–	kdr
Tarnished plant bug	*Lygus lineolaris*	+	+	+	–	M, kdr
Homoptera						
Cotton leafhopper	*Amrasca bigutula*	–	–	+	–	kdr
Cotton leafhopper	*Amrasca devastans*	+	–	–	–	–
Mango leafhopper	*Idioscopus* species	–	+	+	–	kdr
Cotton aphid	*Aphis gossypii*	+	–	–	–	–
Cotton whitefly	*Bemisia tabaci*	+	+	–	+	x, M
Glasshouse whitefly	*Trialeurodes vaporariorum*	+	–	–	–	x, M
Mustard aphid	*Lipaphis erysimi*	+	+	–	–	–
Potato aphid	*Myzus persicae*	–	+	+	–	kdr
Tobacco aphid	*Myzus nicotianae*	–	+	–	–	x
Rice brown plant hopper	*Nilaparvata lugens*	+	+	+		x, kdr
Pear psylla	*Psylla pyricola*	+	+	–	–	x, M
Lepidoptera						
Rice leaffolder	*Cnaphalocrocis medinalis*	+	+	–	–	x
Rice leaffolder	*Marasmia patnalis*	+	+	–	–	x
Maize corn borer	*Heliothis zea*	+	+	–	–	–
Tobacco budworm	*Heliothis virescens*	+	+	+	+	x, M/kdr
Cotton bollworm	*Helicoverpa armigera*	+	+	–	+	x, M
Spotted bollworm	*Earias vitella*	–	+	–	–	x
Leaf feeder	*Hellula undalis*	–	+	–	–	–
Brinjal fruit borer	*Leucinodes orbonalis*	+	–	–	–	–
Black-headed caterpillar	*Opisina arenosella*	–	+	–	–	–
Diamondback moth	*Plutella xylostella*	+	+	+	–	x, kdr
Beet armyworm	*Spodoptera exigua*	–	+	+	–	kdr
Fall armyworm	*Spodoptera furigperda*	+	+	–	–	x, M

(continued)

Table 13.1 (continued)

Insect/mite pests		Resistance built to				
Order/common name	Scientific name	OC	OP	Pyr	Car	Mechanism
Tobacco caterpillar	*Spodoptera litura*	+	+	+	–	x, M
Thysanoptera						
Tobacco thrips	*Thrips tabaci*	+	–	+	–	–
Western flower thrips	*Frankliniella occidentalis*	–	–	+	+	x, M
Acarina						
Two-spotted spider mite	*Tetranychus urticae*	+	+	–	–	–

OC organochlorides, *OP* organophosphorus, *Car* carbamates, *Pyr* pyrethroids, + and – indicate development and nondevelopment of resistance, respectively, *X* cross resistance, *M* multiple resistance, *kdr* knock-down resistance

Besides, insecticide resistance problems were reported on several sap-feeding pests including whitefly (*Bemisia tabaci*), peach aphid (*Myzus persicae*), cotton aphid (*Aphis gossypii*) and mustard aphid (*Lipaphis erysimi*) in diverse crops. Pyrethroid resistance in tarnished plant bug, *Lygus lineolaris* (Palisot de Beauvois), was first reported in the delta region of Mississippi in 1993 and later found to spread widely throughout Arkansas, Louisiana and Mississippi; it included multiple resistance to organophosphates, carbamates and cyclodienes (Leland and Behle 2004).

13.7 Existing IRM Strategies

Insecticide resistance management forms a complementary towards integrated pest management (IPM) technique that should be implemented in order to prevent or delay the development of resistance (Regupathy 1995). The development of insecticide resistance management strategies requires a comprehensive knowledge of mechanisms. Generally, insecticide mixture and alternating of insecticides are useful strategies to slash down insecticide resistance. However, in the occurrence of cross resistance even with the new molecule insecticides (Leland and Behle 2004), practical options to alleviate resistance problems are very few. Currently, "high dose or refuge" strategy is practically adaptable practice to solve the risk of resistance (Regupathy 1995; Ahmad et al. 2002; Leland and Behle 2004). The high dose strategy, which is usually targeted against the heterozygous insects, can effectively alter any resistance traits into physiologically susceptible population (Ranjekar et al. 2003). But, for resource-poor farmers, this strategy is uneconomical especially where commodity quality standards through restricted application of registered pesticides should be maintained. In the past, alternative resistance management strategies involving IPM practices combining several tactics including biological methods were recommended to reduce the frequency of insecticide applications (Pedigo 2002) and to eliminate resistant pest populations (Sarfraz and Keddie 2005).

13.8 Need for Developing Alternate Management Strategies

For sustainable insect pest management, the requirement of new molecular insecticides with specific modes of action is becoming increasingly necessary. On the other hand, excessive use of same pesticide or frequent use of such pesticides leads to development of resistance amongst the target pests. Furthermore, insecticide mixtures adopted elsewhere to minimize the insect damage not only leave enormous toxic residues in commodities but also decimate beneficial natural enemies of target pests in a particular cropped ecosystem. This situation has resulted in complete failure of pest control, which even forced the farmers to commit suicide because of extreme debts incurred for crop protection (Sharma et al. 2001; Feng and Xiao 2005). Misuse of pesticides creates pack of complications. Conversely, increased emphasis on the development of novel IPM strategies is suggested to obviate insecticide resistance problems.

13.9 Entomopathogenic Fungi to Counter the Risk of Insecticide Resistance

Entomopathogenic fungi have great potential as mycoinsecticide agents against diverse insect pests in agriculture (Hajek and Leger 1994). The use of entomopathogenic fungi and their products as mycoinsecticides is one of the important components of IPM (Jayaraj et al. 1985; Narayanasamy 1991). Many promising species of entomopathogenic fungi have been successfully integrated with chemical control (Laird 1962; Bajan et al. 1977). Examples of entomopathogenic fungi recommended for management of agriculturally important pests (Agarwal 1990; Narayanasamy et al. 1995; Ambethgar 2009) include pathogens from classes Hyphomycetes and Zygomycetes (Table 13.2).

Many potential fungal candidates in the class Deuteromycotina, viz. *Beauveria bassiana*, *Metarhizium anisopliae*, *Nomuraea rileyi*, *Paecilomyces fumosoroseus*

Table 13.2 Major taxa of entomopathogenic fungi for insect control in agroecosystem

Fungal taxa	Genus	Propagule	Examples of target species
Class: Hyphomycetes	*Beauveria*	Conidia	Beetles, plant hoppers, caterpillars
(Deuteromycotina)	*Metarhizium*	Conidia	Beetles, hoppers, bugs, termites
	Nomuraea	Conidia	Caterpillars of Lepidoptera
	Paecilomyces	Conidia	Whiteflies, leafhoppers
	Lecanicillium	Conidia	Aphids, whiteflies, scales
Class: Zygomycetes	*Conidiobolus*	Conidia	Aphids, plant hoppers, moths
(Zygomycotina)	*Entomophthora*	Conidia	Grasshoppers, caterpillars
	Neozygites	Conidia	Thrips, mites
	Zoophthora	Conidia	Aphids, hoppers, caterpillars

and *Lecanicillium lecanii*, in combination with certain registered insecticides were reported to synergize overall control efficacy and reduced the development of resistance (Balikai and Sattigi 2000).

13.10 Infection Mechanisms of Entomopathogenic Fungi

Insect exoskeleton is a structurally complex barrier for invasion fungus to penetrate and establish infection (Clarkson and Charnley 1996). Here, the thin outer layer called epicuticle is enveloped with impermeable waxy coating. The thick procuticle comprises mainly chitin fibrils embedded in a proteinaceous matrix. The penetration of fungi through the massive barrier presented by the insect cuticle is reported to be due to the synchronized action of mechanical pressure and enzymic degradation, which paves the way for successful establishment infection. The infection process of a fungus in insect is progressed through a series of systematic metabolic events that are crucial for the successful establishment of mycosis.

(a) Adhesion of fungal propagules to the cuticle
(b) Germination of conidial spores on the cuticle
(c) Formation of appressorium and infection peg
(d) Penetration of fungus into the cuticle
(e) Proliferation of blastospores in haemocoel
(f) Production of mycotoxins to cause paralysis of host insect
(g) Death of the host insect

Entomopathogenic fungi infect insects by breaching the host cuticle (Fig. 13.3), while other groups of pathogens such as bacteria and viruses are necessarily to be ingested by host insects to be infective (Beauvais et al. 1989; Agarwal 1990). The insect cuticle is the protective barrier, while entomofungi secrete a complex of cuticle-degrading enzymes to disintegrate respective cuticle elements to allow the fungus penetrate into the cuticle (Cadatal and Gabriel 1970; Clark et al. 1982). The infection process of fungi involves mechanical pressure followed by solubilization of cuticle with cuticle-degrading enzymes (Chandler et al. 2000; Hiromori and Nishigaki 2001), which facilitate the entry of fungus in to haemocoel, where the vegetative hyphae produce a complex of toxic metabolites and peptides. Usually, the fungi *M. anisopliae* and *B. bassiana* release copious cuticle-lysing enzymes such as chymoelastase protease or Pr1, esterases and chitinases to degrade the host cuticle (St. Leger et al. 1996). These toxins further tend to cause paralysis to the host insect (Bajan et al. 1977). At this stage, the host insect shows sluggishness, reduced feeding and insensitivity to external stimuli (Bidochka and Khachatourian 1988; Bidochka and Hajeck 1998).

Under appropriate favourable weather conditions, especially with high humidity, the vegetative mycelia proliferate inside the cadaver in haemocoel by utilizing nutrients (Hajeck and St. Leger 1994), destroy host cells and eventually cause death of the host insect (Clarkson and Charnley 1996; St. Leger et al. 1998). Under unfa-

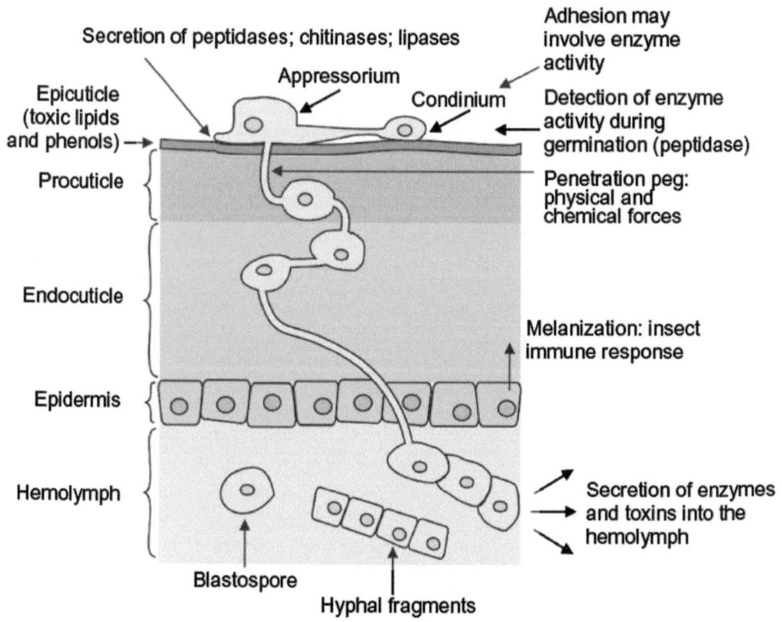

Fig. 13.3 Imaginary sectional view of integument carrying fungal infection in insects (*Source*: Wikipedia)

vourable conditions, especially under parched dry weather, the vegetative mycelia live in the form of dormant filaments inside the cadaver (Silva et al. 1993; Sandhu et al. 2001). On the return of favourable weather conditions, the dormant mycelia mature to form conidiophores which re-penetrate the cuticle from inside to out and produce numerous aerial conidia over the cadavers which spread horizontally to infect next-generation target species (Riba et al. 1983; Sandhu et al. 2001). The entire infection cycle of fungal pathogen-induced mortality increases with the fungal infection age, with the average times from condyle attachment to death of host that takes less than 10 days (Santos et al. 2007).

13.11 Potential Benefits of Fungus-Insecticide Combinations in IRM

Application of fungi with insecticides forms an ideal alternative measure for insecticide resistance management (Quntela and McCoy 1998; Hoy 1999), because many herbivore pests easily develop resistance to insecticides. This strategy produces multiple advantages as it requires decreased dose of insecticides and delayed expression of resistance and gives residue-free produce (Purwar and Sachan 2006). Joint application of a fungal entomopathogen with sublethal doses of selective insecticide is a useful method to mitigate the development of insecticide resistance in pest population (Ramaraje Urs et al. 1967). In addition, prelethal effects of fungal infection, including feeding propensity and fecundity, could substantially reduce or prevent further crop damage. In this context, clarification on the influence of toxicants on the native population of fungal entomopathogens is necessary.

The IPM system strongly encourages sequential application of insecticides and entomopathogenic fungi rather than simultaneous applications. Recent studies have indicated that insecticides, especially at sublethal doses, often synergize the speed of fungal infection (Quntela and McCoy 1998). Many authors have reviewed the effects of fungal entomopathogen and insecticide combinations in two-in-one tank mix method (Steinhaus 1958; Franz 1961; Laird 1962; Hall 1963; Fargues 1975). Majority of the investigations proved additive role of fungi with multiple advantages over sole use of insecticides in agricultural production systems. Initially, the pathogen tends to induce the pest to more susceptible to the pesticide. On the other hand, the insecticidal toxicant weakens the target pest sufficiently to make it more susceptible for infection by the pathogen. For example, the additive effects of fungal-insecticidal combinations with reduced gravity of resistance have been worked out with tremendous successes (Steinhaus 1996; Ericsson et al. 2007).

13.12 Fungal Infection and Immune Reactions in Insects

The mechanism in relation to interactions between breakdown of resistance and fungal infection is complex to understand. According to Bidochka and Khachatourian (1988), immune traits in insects prevent a pathogen to infect the non-permissive host insects. Wago (1995) indicated that the insect immune reaction is activated in two phases, viz. cellular and noncellular immunity. In cellular response, haemocytes react to a foreign body through degranulation, phagocytosis and/or encapsulation (Gotz and Boman 1985; Gunnarsson and Lackie 1985). But, during the fungal infection process, haemocytic encapsulation is identified by some structural components of the fungal cell wall composed of chitin and β-1,3-glucan (Beauvais et al. 1989). In case of noncellular immunity, also called as "adaptive humoral immunity", certain specific proteins are induced in response to a "non-self" elicitor or the activation of certain immune protein (Wheeler et al. 1993; Bidochka and Hajeck 1998). The mechanisms of resistance to insecticide occur in response to enzymic action, especially monooxygenases and the esterases (Soderhall and Smith 1986). Many candidate fungi in conjunction with insecticides discharge powerful enzymes to debilitate the physiological state of host insect which in turn predisposes the host insects to pathogenic infection by inducing stress effects (Kanost et al. 1990). The sublethal concentrations of insecticide readily alter the immune reactions of insect by targeting the humoral defence mechanism (Bidochka and Hajeck 1998), besides enhancing the efficacy and lifespan of insecticides where resistance attained maximum levels (Feng et al. 2004). Therefore, insecticide resistance management largely depending on the speed of fungal infection and the extent of resistance built up prevailed amongst the insect population.

13.13 The Concept of Joint Action

Joint action between bio-inoculants and chemical pesticides has been investigated by many researchers (Hurpin and Robert 1968; Roberts and Campbell 1977; Thurston et al. 1993). The main objective of this concept is to obtain synergistic action out of simultaneous application of insecticide and fungus agent, in which the insecticide performs as stress inducer to the target insects, while the fungus acts as the major control agent (Thurston et al. 1993), which invades the host rapidly because of stress created by the toxicants. Documentation on the joint action of agrochemicals with microbial insecticides is gradually accumulating. Roberts and Campbell (1977) summarized the data on the interactions of microbial-chemical pesticides with varied nature of susceptibility. A majority of these experiments have been performed under in vitro environment using insecticide-amended nutrient media of both solid and liquid state (Hall 1963; Roberts and Campbell 1977). However, employing in vitro mycelial growth as criterion for inhibition factors and evaluating these principles with experiments in open in vivo environment are entirely different aspects for comparison of their effectiveness (Gardner et al. 1979; Clark et al. 1982). However, the additive advantages of compatibility between pathogens and pesticides have been evaluated in numerous situations with a view to slash down resistance problems in certain arthropod pests (Benz 1971; Anderson et al. 1989). Many of these experiments clearly proved that the pesticides compatible with pathogens under controlled in vitro conditions also shown synergistic action under the open field conditions (Fargues 1975; Anderson et al. 1989).

13.14 Interactive Effects of Entomopathogenic Fungi and Insecticides

Currently available new molecule insecticides are compatible with many fungal entomopathogens. For example, viability assays using *M. anisopliae* in the presence of imidacloprid demonstrated normal growth and development of the fungus without causing any impediments on the biological parameters and vitality of the organism (Anderson and Roberts 1983; Dayakar et al. 2000; Damodar et al. 2008). Chemical pesticides including insecticides and phytochemicals have been integrated with entomopathogenic fungi to mitigate the resistance development in insect pests (Prasad 1989; Mietkiewski and Gorski 1995; Ambethgar 2003; Damodar et al. 2008; Sabbour 2013). Many researchers have tested certain chemical substances as *stress inducer* in combination with entomopathogens to improve overall pest control efficacy (Anderson et al. 1989; Quntela and McCoy 1998) and also examined the elements responsible for synergistic action (Anderson et al. 1989; Hassan and Charnley 1989; Boucias et al. 1996; Kaakeh et al. 1997; Quntela and McCoy 1997).

In any agroecosystem, insecticides are an epizootically vital component, because insecticides may synergize or antagonize disease progression in insects (Jacobson et al. 2001). For instance, Ferron (1971) demonstrated that reduced doses of insecticides with spores of *Beauveria tenella* Sacc. (McLeod) increased mortality of *Melolontha melolontha* L. larvae leading to desirable control of the pest. In another study, Quntela and McCoy (1997) observed reduced larval mobility of root weevil (*Diaprepes abbreviatus* L.) treated with *M. anisopliae* and *B. bassiana* was associated with synergistic effects induced due to the action of imidacloprid. Agrochemicals produce differential effects on pathobiological process, viz. survival, growth and metamorphosis of fungal strains (Vanninen and Hokkanen 1988; Anderson et al. 1989), depending on the biopesticide products and prevailing environmental factors (Anderson and Roberts 1983; Todorova et al. 1998; Ambethgar et al. 2009).

Insecticidal activity of phyto-products, especially derivatives of neem (*Azadirachta indica* A. Juss.) as "natural insecticides/green pesticides", has been well documented against many agricultural pests in organic crop cultivation (Gupta et al. 1999; Bhattacharya et al. 2003; David 2008). The usefulness of *neem oil* as adjuvants to enhance the persistence of entomopathogenic fungi has also been demonstrated which significantly increased the effectiveness of *B. bassiana* against crop pests (Akbar et al. 2005; Depieri et al. 2005). The approaches using a combination of phytochemicals and fungal conidia affected multiple benefits as it ensures rapid mortality of crop pests. For example, Gupta et al. (1999) found that active principles of neem increased persistence of fungal conidia apart from inducing pest mortality rates and it reduced resistance development when used as combination sprays (Gupta et al. 2002). However, while integrating this strategy, careful vigil and diligent caution are needed on the aspects of resurgence of minor pests, effect on non-target organisms and control efficiency on target pests, in order to achieve desirable pest management.

13.15 Factors Influencing Joint Action

Insecticides have the tendency to debilitate the physiological strength of insects, which leads to hasten the infectivity of entomopathogens (Mietkiewski et al. 1991; Farenhorst et al. 2010). The differential effects of pathogenicity under the exposure of pesticides shown under in vitro laboratory and in vivo field conditions are of complex nature, which are influenced by interaction between biotic and abiotic factors. Prior research have indicated that pesticides applied in soil strongly impair the dynamics of fungal entomopathogens and other beneficial soil microorganisms, which are epizootiologically important elements to establish infections in target pests (Mietkiewski et al. 1997; Dutt and Balasubramanian 2002). Besides the epizootiological factors, type of formulations, carrier materials, emulsifying agents, dosage and edaphological and physiological condition of host plants are reported to influence the joint action of entomofungi and insecticides (Todorova et al. 1998;

Jacobson et al. 2001). The complexities of biotic and abiotic factors are difficult to discriminate in fields. However, a thorough laboratory and field experiments could predict the effects of pesticides on entomopathogenic fungi (Mohan et al. 2007; Ambethgar et al. 2011). Prior investigations indicated that in IPM systems, selective pesticides which were proven to be innocuous to fungal formulations may be used in conjunction with other components.

13.16 Insecticide Resistance Management with Entomopathogenic Fungi: Case Studies

Knowledge of fungus-insect association helps to integrate fungal and insecticidal formulation against pestiferous insects in agriculture. In fields, fungal epizootics provide additional fitness cost for the insects. For example, fungal infection not only restricts the spread of resistance in insects but also slows down the speed of insecticide resistance development in the long run (Hall and Dunn 1959; Hall 1981). Besides, combination treatment of selective fungi and insecticides also tends to reduce the expression of insecticide resistance and enhance the persistence of fungal propagules in treated fields. Many potential fungal entomopathogens including *B. bassiana* and *M. anisopliae* have been employed to overcome insecticide resistance in certain agriculturally important polyphagous crop pests (Mietkiewski et al. 1997; Gupta et al. 1999). Manipulation of technologies to enhance the virulence of fungal propagules can be used to improve the commercial effectiveness of fungal-based control methods. This review addresses on the effective utilization of entomopathogenic fungi in insecticide resistance management with some selected examples of case studies.

13.16.1 Beauveria bassiana *(Balsamo) Vuillemin*

The fungal pathogen *Beauveria bassiana* (Balsamo) Vuillemin is commonly called as "white muscardine fungus", a potential control agent of many insect species of Exopterygota and Endopterygota orders and mite pests (Islam and Omar 2012; Sanjaya et al. 2015). Its development as biocontrol agent has been of considerable interest with insecticide resistance in crop production systems (Stewart et al. 1997; Zhao et al. 2000). Many researchers have proven the synergistic interactions of several insecticides in sublethal doses with *B. bassiana* preparations in insect control (Kaakeh et al. 1997; Gardner and Kinard 1998; Quntela and McCoy 1998; Lacey et al. 1999; Ramakrishnan et al. 1999; Furlong and Groden 2001; Gupta et al. 2002; Ying et al. 2003). Some successful examples of *B. bassiana* utilized in insecticide resistance management programme of crop pests are presented in Table 13.3.

Table 13.3 Examples of synergistic interaction of *B. bassiana* and sublethal dose of some insecticides

Product	Insecticide	Target pest/ organism	Testing method	References
Boverin WP	Trichlorophon	*Cydia pomonella* L.	In vitro foliar spray	Ferron (1971) and Gardner and Kinard (1998)
Aqueous suspension (AS)	Chlorpyrifos	*Ostrinia nubilalis* Hub.	Spray in corn field	Foschi and Grassi (1985)
	Carbofuran	*Ostrinia nubilalis* Hub.	Spray in corn field	Lewis et al. (1996) and Steinhaus (1996)
	Imidacloprid	*Leptinotarsa decemlineata*	Spray in potato field	Anderson et al. (1989)
	Imidacloprid	*Leptinotarsa decemlineata*	In vitro soil treatment	Furlong and Groden (2001)
	Triazine (IGR)	*Leptinotarsa decemlineata*	In vitro leaf disc test	Furlong and Groden (2001)
	Imidacloprid	*Reticulitermes flavipes*	In vitro leaf disc test	Boucias et al. (1996)
	Imidacloprid	*Blissus leucopterus* Say	In vitro soil treatment	Studdert and Kaya (1990)
	Imidacloprid	*Spodoptera exigua* Hub.	In vitro soil treatment	Krueger et al. (1991)
	Imidacloprid	*Spodoptera litura*	In vitro leaf disc test	Jayanthi and Padmavthamma (2001)
Mycotrol WP	Imidacloprid	*Lygus lineolaris*	In vitro leaf disc test	Steinkraus and Tugwell (1997)
	Imidacloprid	*Diaprepes abbreviatus*	In vitro soil treatment	Quntela and McCoy (1998)
	Imidacloprid	*Diaprepes abbreviatus*	In vitro soil treatment	Maurao et al. (2003)
	Imidacloprid	Mole cricket	In vitro soil treatment	Thompson and Brandenburg (2006)
	Imidacloprid	*Atta sexdens rubropilosa*	In vitro soil treatment	Santos et al. (2007)
	Imidacloprid	*Popillia japonica*	In vitro soil treatment	Prabhu et al. (2007)
	Neonicotinoids	*Popillia japonica*	Spray in greenhouse	Morales-Rodriguez and Peck (2009)

(continued)

Table 13.3 (continued)

Product	Insecticide	Target pest/ organism	Testing method	References
Aqueous suspension (AS)	Imidacloprid	*Bemisia argentifolii*	Spray in cucumber	James and Elzen (2001) and Mohan et al. (2007)
	Imidacloprid	*Trialeurodes vaporariorum*	Foliar spray	Feng et al. (2004)
	Fenvalerate	*Diaprepes abbreviatus*	In vitro soil treatment	Jayanthi and Padmavthamma (2001)
	Triademefon	*Galleria mellonella*	Laboratory screening	Mietkiewski et al. (1997)
	Cypermethrin	*Hypothenemus hampei*	Spray in coffee field	Vyas et al. (1990)
	Thiamethoxam	*Spodoptera litura*	Spray in groundnut field	de Oliveira et al. (2003)
Aqueous suspension (AS)	Dimethoate	*Spodoptera litura*	Laboratory screening	Padmini Palem et al. (2010)
	Indoxacarb	*Plutella xylostella*	Laboratory screening	Tian and Feng (2006)
	Chlorpyrifos	*Cnaphalocrocis medinalis*	Laboratory screening	Ambethgar (2009)
	Dicofol	*Tetranychus* spp.	Laboratory screening	Shi et al. (2005)

13.16.2 Metarhizium anisopliae *(Metchnikoff) Sorokin*

The green muscardine fungus, *Metarhizium anisopliae* (Metchnikoff) Sorokin, is widely distributed in soil and has very broad range of crop pests including elm bark beetle, plant hoppers, coconut leaf beetle, rhinoceros beetle, onion thrips, storage cowpea, white grub, cattle tick and also termite species (Latch and Falloon 1976; Moorhouse et al. 1992). Synergistic effects of *M. anisopliae* in combination with sublethal doses of certain insecticides have been documented earlier (Barbosa and Moreira 1982). Examples with utilization of *M. anisopliae* in insecticide resistance management against some agriculturally important crop pests are presented in Table 13.4.

13.16.3 Nomuraea rileyi *(Farlow) Samson*

The fungus, *Nomuraea rileyi*, is highly pathogenic to larvae of Noctuidae family of the order Lepidoptera, including several agriculturally important pests. Many researchers have demonstrated on the synergistic reaction of *N. rileyi* to many commonly used insecticides through paper disc bioassay technique (Ignoffo et al. 1975), in vitro pathogenicity test (Silva et al. 1993) and field screening methods (Vimala

Table 13.4 Examples of synergistic interaction of *M. anisopliae* products and sublethal dose of insecticides

Product	Insecticide	Target pest/organism	Testing method	References
Aqueous suspension (AS)	Chlorpyrifos	*Ostrinia nubilalis* Hub.	Laboratory screening	Riba et al. (1983) and Aguda et al. (1988)
	Chlorpyrifos	*Otiorhynchus sulcatus*	Laboratory screening	Moorhouse et al. (1992)
	Imidacloprid	*Anomala cuprea*	In vitro soil treatment	Hiromori and Nishigaki (1998)
	Imidacloprid	*Diaprepes abbreviatus*	In vitro soil treatment	Quntela and McCoy (1998)
	Thiamethoxam	*Adoryphorus couloni*	In vitro soil treatment	Hiromori and Nishigaki (1998)
	Acetamiprid	Conidial viability test	Laboratory screening	Neves et al. (2001) and Ye et al. (2005)
	Imidacloprid	*Blattella germanica*	Laboratory screening	Kaakeh et al. (1997)
	Chlorpyrifos	*Blattella germanica*	Laboratory screening	Pachamuthu and Kamble (2000)
	Boric acid	*Blattella germanica*	Laboratory screening	Zurek et al. (2002)
	Quinalphos	*Spodoptera litura*	In vitro foliar spray	Dayakar et al. (2002)
	Imidacloprid	*Cyrtomenus bergi*	Laboratory screening	Jaramillo et al. (2005)
	Deltamethrin	Conidial viability test	Laboratory screening	Bahiense et al. (2006)
	Spinosad	Conidial viability test	Laboratory screening	Rachappa et al. (2007)
	Imidacloprid	*Popillia japonica*	In vitro soil treatment	Morales-Rodriguez and Peck (2009)
	Pyriproxyfen	Conidial viability test	Laboratory screening	Rashid et al. (2010)
	Imidacloprid	*Aedes aegypti*	Laboratory screening	Paula et al. (2011) and Johnson et al. (1992)
	Teflubenzuron	*Schistocerca gregaria*	Laboratory screening	Joshi et al. (1992)and Seyoum (2001)
	IGR Nomolt	*Schistocerca gregaria*	Laboratory screening	Johnson et al. (1992)
	Dimilin	*Manduca sexta*	Laboratory screening	Hassan and Charnley (1989)
	Dimilin	*Manduca sexta*	Laboratory screening	Butt and Brownbridge (1997)
	Flufenoxuron	*Manduca sexta*	Laboratory screening	Hiromori and Nishigaki (1998)

Table 13.5 Examples of synergistic interaction of *N. rileyi* products and sublethal dose of insecticides in resistance management

Product	Insecticide	Target pest/ organism	Testing method	References
Aqueous suspension (AS)	OP insecticides	*Spodoptera litura*	In vitro paper disc	Ignoffo et al. (1975)
	Permethrin	*Trichoplusia ni*	Laboratory screening	Silva et al. (1993)
	Diflubenzuron	*Heliothis zea*	Laboratory screening	Silva et al. (1993)
	NSKE	*Spodoptera litura*	Laboratory screening	Vimala Devi and Prasad (1996)
	Pongamia oil	*Spodoptera litura*	Laboratory screening	Vimala Devi and Prasad (1996)
	Endosulfan	*Plutella xylostella*	Laboratory screening	Gopalakrishnan and Mohan (2002)
	Acephate	*Spodoptera litura*	Laboratory screening	Manjula and Krishna Murthy (2005)

Devi and Prasad 1996). Examples of synergistic reaction of *N. rileyi* with commonly used insecticides against some important crop pests are presented in Table 13.5. In all field efficacy experiments, selective application of insecticides at sublethal doses coinciding with onset of *N. rileyi* natural infection proved with enhanced larval mortality.

13.16.4 **Paecilomyces** ***species***

The soil-dwelling fungal genus *Paecilomyces* is one of the virulent biological control agents of several agriculturally important pests including nematodes (James 2003; Kubilay and Gokee 2004). Some examples of synergistic action rendered through *Paecilomyces farinosus* to various agrochemicals in the presence of target host insects has been shown in Table 13.6.

13.16.5 **Lecanicillium (=Cephalosporium = Verticillium)** ***)*** **lecanii** ***(Zimmermann) Veigas***

The white halo fungus, *Lecanicillium lecanii*, previously known as *Cephalosporium lecanii Verticillium lecanii* (Zimmermann) Veigas is highly infective to a wide range of glasshouse pests, especially to pests possessing piercing and sucking mouth parts (Hall and Dunn 1959; Hall 1981). For example, *L. lecanii* effectively controlled potential crop pests such as coffee green bug, *Coccus viridis* (Santharam et al.

Table 13.6 Examples of synergistic interaction of *Paecilomyces* species and sublethal dose of insecticides in resistance management

Product	Insecticide	Target pest/organism	Testing method	References
Aqueous suspension (AS)	Azadirachtin	*Macrosiphoniella sanborni*	Laboratory screening	Lindquist (1993)
	OP insecticides	*Galleria mellonella*	Paper disc assay	Mietkiewski et al. (1997)
	Azadirachtin	*Aphis gossypii*	Laboratory screening	James (2003)
	Neonicotinoids	Several agriculture pests	Laboratory screening	Kubilay and Gokee (2004)
	Acetamiprid	Poisoned food method	Laboratory screening	Oliveira and Neves (2005)
	Imidacloprid	*Trialeurodes vaporariorum*	Greenhouse screening	Feng et al. (2004)
	Chlorpyrifos	*Plutella xylostella*	Field screening	Damodar et al. (2008)

1977), and banana aphid, *Pentalonia nigronervosa* forma *typica* Coq. (Regupathy and Paramjothi 1977). However, previous investigations on compatibility of chemical insecticides with *L. lecanii* have found varying results. Some examples of synergistic interaction of *L. lecanii* and sublethal dose of insecticides are indicated in Table 13.7. In these studies, the combination product derived out of fungal biocontrol agents and insecticides may have reduced the resistance development in target pests.

13.16.6 Hirsutella thompsonii *Fisher*

The genus *Hirsutella* primarily infective to mite pests also causes natural epizootics on a number of arthropod pests including insects and nematode pests (McCoy et al. 1988) in diverse cropped ecosystems. This acro-pathogenic fungus was originally identified by Fisher, who isolated it from the citrus rust mite, *Phyllocoptruta oleivora*, in Florida. The mite-specific fungal species, *Hirsutella thompsonii*, can cause spectacular natural epizootics amongst mite populations infesting citrus, blueberry, coconut and tomato crops under hot, humid weather conditions. Candidates of *Hirsutella* are considered to be one of the key natural enemies of various mite pests (Chandler et al. 2000). The compatibility of *H. thompsonii* has been tested with several compounds of insecticides including sulphur, dicofol, dichlorvos and imidacloprid with varying degree of sensitivity (McCoy 1981). However, Sreerama Kumar and Singh (2000) opined that insecticides inflicted mild inhibition to the fungus than fungicides. Combined application of neem-based oil formulation containing azadirachtin 0.03% EC at 500 ml and myco-acaricide Bio-Catch WP formulation at 1000 g in 200 l of spray fluid was reported to contain eriophyid mite, *Aceria*

Table 13.7 Examples of synergistic interaction of *L. lecanii* and sublethal dose of insecticides in resistance management

Product	Insecticide	Target pest/ organism	Testing method	References
Aqueous suspension (AS)	Pirimicarb	Aphids	Tank mixing	Hall (1981)
	Dichlorvos	Coffee bug, *Coccus viridis*	Laboratory screening	Easwaramoorthy and Jayaraj (1977)
	Carbaryl	Coffee bug, *C. viridis*	Laboratory screening	Santharam et al. (1977)
	Phosphamidon	Banana aphids, *P. nigronervosa*	Field screening	Regupathy and Paramjothi (1977)
	Fenthion	Coffee bug, *C. viridis*	Field screening	Easwaramoorthy et al. (1978)
	Imidacloprid	Aphids	Laboratory screening	Moino and Alves (1998)
	Nicotinoids	*Bemisia tabaci* nymphs	Laboratory screening	Andrew et al. (2005)
	Imidacloprid	Aphids	Laboratory screening	Cuthberston et al. (2005)
	Thiamethoxam	Coffee bug, *C. viridis*	Laboratory screening	Senthilkumar and Regupathy (2007)
	Imidacloprid	*B. tabaci* nymphs	Laboratory screening	Andrew et al. (2008)
	Teflubenzuron	*B. tabaci* nymphs	Laboratory screening	Cuthberston et al. (2008)
	Buprofezin	*B. tabaci* nymphs	Glasshouse screening	Cuthberston et al. (2010)

(*Eriophyes*) *guerreronis*, population in coconut plantations. Ramarethinam et al. (2000) reported that imidacloprid at 100, 200 and 500 ppm was synergized by inducing germination of *H. thompsonii*. Similarly, Dara and Hountondji (2001) reported that combination of subnormal concentration of imidacloprid and *H. thompsonii* rendered mortality greater than 95% of cassava green mite, *Mononychellus tanajoa* (Bondar).

13.17 Ecological Implications

Biodiversity in agroecosystems delivers significant ecosystem services to an array of native biological control agents specifically the entomopathogenic fungi. However, the virulence of these fungi inducing infectivity largely depends on viability of conidia which determine epizootics of native organisms (Mietkiewski et al. 1991; Vanninen 1995; Bidochka and Hajeck 1998; Oliveira and Neves 2005). Fungus propagules that invade non-host insects germinate poorly with aborted infection (Alizadeh et al. 2007). Propagules produced from current season infection can survive longer either within the cadavers or in soil to infect later generations of pest populations. However, agrochemicals applied to control crop pests and pathogens in the same habitat tend to synergize or antagonize towards the resident organisms (Majchrowiez and Poprawski 1993) depending on the prevailing ecological conditions. In this context, Olmert and Kenneth (1974) stressed the necessity of clarifying the influence of pesticides on specific biocontrol fungi in order to integrate any pesticide insect resistance management. For example, the anamorphic taxa *Beauveria bassiana* and *Metarhizium anisopliae*, Hypocreales (Ascomycota), were reported to tolerate diverse classes of crop protection chemicals and have a tendency to persist within the host cadavers (Latch and Falloon 1976). Many researchers have clarified that fungal infectivity is seldom affected at sublethal concentrations of pesticides (Vanninen and Hokkanen 1988; Anderson et al. 1989; Butt and Brownbridge 1997). Therefore, pesticides that are harmful in the controlled in vitro environment do not necessarily exhibit the same reaction under open field conditions, which fluctuate time to time.

13.18 Environmental Implications

Entomopathogenic fungi are amongst the natural enemies of pests in agroecosystems, especially for future conservation biological control programme. Conservation of native organisms is an important component of biological control strategy in which farming practices and environmental manipulations are adopted to enhance the living conditions for specific natural enemies of pests. However, in order to manipulate the environment for the benefit of populations of the entomopathogens, knowledge of fundamental aspects of the ecology of the fungi considered is necessary. The complexity of environmental factors in relation to natural occurrence of fungi, their population dynamics in agroecosystems and interactions with other organisms and agronomical practices determines the insecticidal tolerance of various biocontrol fungi. Furthermore, current pest control is focused towards integration of biocontrol and chemical strategies with a view to mitigate the insecticide resistance problem vis-a-vis for protection of environment-related issues.

13.19 Strategic Approaches

Biopesticides based on entomopathogenic fungi have attained wider utility as an environment benign approach across the world, because of their extensive use in agriculture, horticulture and forestry for the past three decades. Fungi were also exploited as a component of resistance management in IPM system. Adoption of the following farm manipulative practices may help the farmers for effective implementation of insecticide resistance management programme:

- Preservation of naturally occurring myco-pathogens is needed in order to re-establish them in field environment in order to abate concurrent occurrence of resistance build-up in field populations of insect pests.
- Screening of newer insecticides for safety aspects to beneficial microorganisms is to be considered in favour of potential viability and pathogenicity to target pests.
- Appropriate laboratory studies with confirmative field experiments are essential for manipulating compatibility between introduced fungi and insecticides.
- Research on the genetic improvement of biocontrol fungi for locating insecticide-resistant/insecticide-tolerant strains are needed to counter the insect resistance to insecticides.

In this context, biotechnological innovations like genetic engineering may be helpful for strain improvement of fungal entomopathogens for tolerance to exposure of agrochemicals. In this direction, Daboussi et al. (1989) for the first time introduced genetic transfer systems for building pesticide resistance mechanisms. Subsequently, Sandhu et al. (2001) demonstrated genetic transformation in *B. bassiana* using β-tubulin gene of *Neurospora crassa* encoding resistance to benomyl fungicide, to compare the generations of resistant strains of *B. bassiana* and their infectivity to larvae of *Helicoverpa armigera*. Gene manipulations that render overproduction of a cuticle-degrading protease have also been shown to induce the fungus to provide rapid killing of host insects (St.Leger et al. 1996). Similar such studies with selected strains of fungal entomopathogens and insecticides would result in the development of mutant strains with improved virulence against major arthropod pests in order to augment their compatibility with other pest control options for sustainable insecticide resistance management programme.

13.20 Conclusion

Prior documented evidences with synergism between an array of insecticides and fungal entomopathogens are enormous across the world as evident from the foregoing discussion. The instillation of novel technology for joint action of subnormal doses of insecticides with promising entomopathogens is largely suggested for insecticide resistance management. In this context, compatible insecticide potential to induce physiological fitness of fungi to cause infections in host insect is to be identified. Specific isolates and strains of candidate fungi such as *B. bassiana*, *M. anisopliae*, *P. farinosus*, *P. fumosoroseus* and *V. lecanii* in both in vitro and in vivo environments were reported to cause synergistic action with many selected classes of insecticides. Joint application of fungi with insecticides as tank mix strategy is largely employed to slash down the risk of resistance amongst the target pests of crops, because individual applications of either insecticides or fungal formulation in separate occasions are cumbersome and impractical to the farmer. Low doses of toxicants in combination with fungal biocontrol agents not only serve to reduce or delay the selection pressure, but such combinations also induce high level of mortality amongst genetically resistant pest population and reduce the frequency of insecticide applications.

Still, basic researches are needed to explore the practical implications for exploiting fungal entomopathogens as a component in insecticide resistance management programme. Currently there is a great interest in using combination interventions with distinct modes of action as management strategy, not only to control resistant insect population but to delay the selection of novel resistance, which indicates a potential role for fungi with other categories of insecticides. Such research could enable the development of novel IPM strategies that would sustain the useful lifespan of current insecticide-based interventions and maximize control in the face of emerging insecticide resistance. With the advancement in molecular biology and genetic engineering tools, it is possible to explore tolerance mechanism through strain improvement of fungi, which would further help to abate resistance development in insects. From the foregoing discussion, it is evident that deployment of fungal entomopathogens and their products will continue to render vital role as a component in insecticide resistance management programme.

References

Agarwal GP (1990) Entomogenous fungi in India and management of insect pests. Indian Phytopathol 43(2):131–142

Aguda RM, Rombach MC, Roberts DW (1988) Effect of pesticides on germination and growth of three fungi of rice insects. Int Rice Res Newsl 13(6):39–40

Ahmad M, Arif MI, Ahmad M (1995) Monitoring insecticide resistance in *Helicoverpa armigera* (Lepidoptera:Noctuidae) in Pakistan. J Econ Entomol 88(4):771–776

Ahmad M, Arif MI, Ahmad M (1999) Patterns of resistance to organophosphate insecticides in field populations of *Helicoverpa armigera* in Pakistan. Pestic Sci 55(6):626–632

Ahmad M, Arif MI, Ahmad Z, Denholm I (2002) Cotton white fly (*Bemisia tabaci*) resistance to organophosphate and pyrethroids insecticides in Pakistan. Pest Manag Sci 58(2):203–208

Akbar W, Jeffrey CI, James RN, Thomas MI (2005) Efficacy of *Beauveria bassiana* for red flour beetle when applied with plant essential oils or in mineral oil and organosilicone carriers. J Econ Entomol 98:683–688

Alizadeh A, Amin-Samih M, Khezri M, Saberi-Riseh R (2007) Compatibility of *Beauveria bassiana* (Bals.) Vuill. with several pesticides. Int J Agric Biol 9(1):31–34

Ambethgar V (2001) Exploitation of entomogenous fungi in biological control of crop pests. In: Upadhyay RK, Mukerji KG, Chamola BP (eds) Biocontrol potential and its exploitation in sustainable agriculture, vol. 2, insect pests. Kluwer Academic/Plenum Publishers, New York, pp 39–55

Ambethgar V (2003) Investigations on the development of mycoinsecticide formulations based on *Beauveria bassiana* (Bals.) Vuill. for the management of rice leaf folder, *Cnaphalocrocis medinalis* Guenee. Ph.D thesis submitted to TNAU, Coimbatore, India

Ambethgar V (2009) Potential of entomopathogenic fungi in insecticide resistance management (IRM): a review. J Biopest 2(2):177–193

Ambethgar V, Swamiappan M, Rabindra RJ, Rabindran R (2009) Biological compatibility of *Beauveria bassiana* (Balsamo) Vuillemin isolate with different insecticides and neem formulations commonly used in rice pest management. J Biol Control 23(1):11–15

Ambethgar V, Swamiappan M, Rabindran R (2011) Field assessment of aqueous suspension of *Zoophthora radicans* conidia and low application rates of imidacloprid for control of diamondback moth, *Plutella xylostella* on cauliflower in Tamil Nadu, India. In: Proceedings of the sixth international workshop on management of the diamondback moth and other crucifer insect pests, 21–25 March, 2011, Kasetsart University, Kamphaeng Saen Campus, Nakhon Pathom, p 23

Anderson TE, Roberts DW (1983) Compatibility of *Beauveria bassiana* isolate with insecticide formulations used in Colorado potato beetle (Coleoptera: Chrysomelidae) control. J Econ Entomol 76:1437–1441

Anderson TE, Hajek AE, Roberts DW, Preisler HK, Robertson JL (1989) Colorado potato beetle (Coleoptera: Chrysomelidae): effects of combinations of *Beauveria bassiana* with insecticides. J Econ Entomol 82:83–89

Andrew CGS, Keith FAW, Carola C (2005) Compatibility of the entomopathogenic fungus, *Lecanicillium muscarium* and insecticides for eradication of sweetpotato whitefly, *Bemisia tabaci*. Mycopathologia 160(1):35–41

Andrew CGS, Blackburn LF, Northing P, Luo W, Cannon RJC, Keith FAW (2008) Further compatibility tests of the entomopathogenic fungus, *Lecanicillium muscarium* with conventional insecticide products for control of sweet potato whitefly, *Bemisia tabaci* on poinsettia plants. Insect Sci 15(4):355–360

Bahiense TC, Fernandes EKK, Bittencourt VREP (2006) Compatibility of fungus *Metarhizium anisopliae* and deltamethrin to control a resistant strain of *Boophilus microplus* tick. Vet Parasitol 141:319–324

Bajan C, Kmitowa K, Wojciechowska M (1977) The effect of Endofos 50 and its active substance: Chlorfenvinphos on the growth and pathogenicity of entomopathogenic fungi. Pol Ecol Stud 3:65–77

Balikai RA, Sattigi HN (2000) Management of sunflower head borer by using bioinsecticide alone and in combination with insecticides. Agric Sci Dig 20(2):110–111

Barbosa FR, Moreira WA (1982) In vitro effect of six insecticides on *Metarhizium anisopliae* (Metsch.) Sorokin. An Soc Entomol Brasil 11(2):245–250

Batista Filho A, Leite LG (1995) Enhanced activity of *Beauveria bassiana* (Bals.)Vuill. associated with mineral oil against *Cosmopolites sordidus* (Germar) adults. Ann Soc Entomol Braz 24:405–408

Beauvais A, Latge JP, Vey A, Pervost MC (1989) The role of surface components of the entomopathogenic fungus, *Entomophaga aulicae* in the cellular immune response of *Galleria mellonella* (Lepidoptera). J Gen Microbiol 135:489–498

Benz G (1971) Synergism of microorganisms and chemical insecticides. In: Burges HD, Hussey NW (eds) Microbial control of insects and mites. Academic, New York, pp 327–355

Bhattacharya AK, Mondal P, Ramamurthy VV, Srivastava RP (2003) *Beauveria bassiana*: a potential bioagent for innovative integrated pest management programme. In: Srivastava RP (ed) Biopesticides and bioagents in integrated pest management of agricultural crops. International Book Distributing Co, Lucknow 860pp

Bidochka MJ, Hajeck AE (1998) A nonpermissive entomophthoralean fungal infection increases activation of insect prophenoloxidase. J Invertebr Pathol 72:231–238

Bidochka MJ, Khachatourian GG (1988) N-acetyl-D-glucosamine mediated regulation of extracellular protease in the entomopathogenic fungus, *Beauveria bassiana*. Appl Environ Microbiol 54:2699–2704

Boucias DG, Stokes C, Storey G, Pendland JC (1996) The effects of imidacloprid on the termite *Reticulitermes flavipes* and its interaction with the mycopathogen, *Beauveria bassiana*. Pflanzenshutz-Nachr Bayer 49:105–151

Brogdon WG, McAllister JC (1998) Insecticide resistance and vector control. Emerg Infect Dis 4(4):605–613

Butt TM, Brownbridge M (1997) Fungal pathogens of thrips. In: Lewis T (ed) Thrips as crop pests. CAB International, Wallingford, pp 399–433

Cadatal TD, Gabriel BP (1970) Effect of chemical pesticides on the development of fungi pathogenic to some rice insects. Philipp Entomol 1:379–395

Chandler D, Davidson G, Pell JD, Ball BV, Shaw K, Sunderland KD (2000) Fungal biocontrol of Acari. Biocontrol Sci Tech 10:357–384

Clark RA, Casagrande RA, Wallace DB (1982) Influence of pesticides on *Beauveria bassiana* a pathogen of the Colorado potato beetle. Environ Entomol 11:67–70

Clarkson JM, Charnley AK (1996) New insights to the mechanisms of fungal pathogens in insects. Trends Microbiol 4:197–203

Cuthberston AGS, Blackburn LF, Northing P, Luo W, Cannon RJC, Walters KFA (2008) Further compatibility tests of the entomopathogenic fungus, *Lecanicillium muscarium* with conventional insecticide products for control of sweetpotato whitefly, *Bemisia tabaci* on poinsettia plants. Insect Sci 15(4):355–360

Cuthberston AGS, Blackburn LF, Northing P, Luo W, Cannon RJC, Walters KFA (2010) Chemical compatibility testing of the entomopathogenic fungus, *Lecanicillium muscarium* to control *Bemisia tabaci* in glasshouse environment. Int J Environ Sci Technol 7(2):405–409

Daboussi MJ, Djeballi A, Gerlinger C, Blaiau PL, Bouvier I, Cassan M, Lebrun MH, Parisot D, Brygoo.Y. (1989) Transformation of seven species of filamentous fungi using the nitrate reductase gene of *Aspergillus nidulans*. Curr Genet 15:453–456

Damodar JS, Kalyanasundaram M, Suganthi A, Karuppuchamy P (2008) Management of diamond back moth, *Plutella xylostella* (L.) using entomopathogenic fungi in combination with reduced doses of newer molecules of insecticides. Pest Manag Econ Zool 16(1):77–82

Dara SK, Hountondji FCC (2001) Effect of formulated imidachloprid on two mite pathogens, *Neozygites floridana* (Zygomycotina: Zygomycetes) and *Hirsutella thompsonii* (Deuteromycotina: Hyphomycetes). Insect Sci Appl 21(2):133–138

David BV (2008) Biotechnological approaches in IPM and their impact on environment. J Biopest 1(1):1–5

Dayakar S, Kanaujia KR, Rathore RRS (2002) Compatibility of entomogenous fungi with commonly used insecticides for the management of *Spotoptera litura* (Fabricius). In: Maniyan K (ed) Microbials in insect pest management. Bharathiyar University, Coimbatore, pp 51–52

Dayakar S, Kanaujia KR and Rathore RRS (2002). Compatibility of entomogenous fungi with commonly used insecticides for the management of Spotoptera litura (Fabricius). In: Maniyan K (ed.) Microbials in Insect Pest Management, Bharathiyar University, Coimbatore, India. pp 51- 52.

de Oliveira CN, Neves PMOJ, Kawazoe LS (2003) Compatibility between the entomopathogenic fungus *Beauveria bassiana* and insecticides used in coffee plantations. Sci Agric 60(4):663–667

Deedat YD (1994) Problems associated with the use of pesticides: an overview. Insect Sci Appl 15(3):247–251

Depieri RA, Martinez SS, Menezes AO Jr (2005) Compatibility of the fungus, *Beauveria bassiana* (Bals.) Vuill. (Deuteromycetes) with extracts of neem seeds and leaves and the emulsible oil. Neotrop Entomol 34:601–606

Duraimurugan P, Regupathy A (2005) Joint action of *Beauveria bassiana* (Balsamo) Vuillemin and synthetic pyrethroids against *Helicoverpa armigera* (Hubner) (Noctuidae: Lepidoptera). J Food Agric Environ 3:65–67

Dutt MS, Balasubramanian R (2002) In vitro bioefficacy of *Beauveria bassiana* isolate (FB-12) and its combination with chemical insecticides against cabbage diamond back moth, *Plutella xylostella* on cabbage. In: Deshmukh AM (ed) Biotechnology in agricultural industry and environmental microbilogists society, Karad, 70–72pp

Easwaramoorthy S, Jayaraj S (1977) Effect of certain insecticides and fungicides on the growth of the coffee green bug fungus, *Cephalosporium lecanii* Zimm. Madras Agric J 64(4):243–246

Easwaramoorthy S, Regupathy A, Santharam G, Jayaraj S (1978) The effect of subnormal concentrations of insecticides in combination with the fungal pathogen, *Cephalosporium lecanii* Zimm. in the control of coffee green scale, *Coccus viridis* green. Z Angew Entomol 86:161–166

Ericsson J, Kabaluk JT, Goetell MS, Myers JH (2007) Spinosad interacts synergistically with the insect pathogen, *Metarhizium anisopliae* against two exotic wireworms, *Agriotes lineatus* and *Agriotes obscurus* (Coleoptera: Elateridae). J Econ Entomol 100:31–38

Farenhorst M, Knols BGJ, Thomas MB, Howard AF, Takken W, Rowland M, N'Guessan RK (2010) Synergy in efficacy of fungal entomopathogens and permethrin against West African insecticide-resistant mosquitoes. Public Libr Sci One 5(8)

Fargues J (1975) Elude experimentale dans la nature de l utilisation combinee de *Beauveria bassiana* et d' insecticides a dose reduite contre *Leptinotarsa decemlineata*. Annal Zool Ecol Anim 7:247–264

Feng MG, Xiao P (2005) Time-concentration-mortality modeling of the synergistic interaction of *Beauveria bassiana* and imidacloprid against *Nilaparvata lugens*. Pest Manag Sci 61:363–370

Feng MG, Chen B, Ying SH (2004) Trials of *Beauveria bassiana*, *Paecilomyces fumosoroseus* and imidacloprid for the management of *Trialeurodes vaporariorum* (Homoptera: Aleyrodidae) on greenhouse lettuce. Biocontrol Sci Tech 14(6):531–544

Ferron P (1971) Modification of the development of *Beauveria tenella* mycosis in *Melolontha melolontha* larvae by means of reduced doses of organophosphorus insecticides. Entomol Exp Appl 14:457–466

Foschi S, Grassi S (1985) Results of treatment with *Beauveria bassiana* (Balsamo) Vuill and with *Metarhizium anisopliae* (Metsch.) Sorok on *Ostrinia nubilalis* Hub. Difesa delle Piante 8:301–308

Franz JM (1961) Biological control of pest insects in Europe. Annu Rev Entomol 6:183–200

Furlong MJ, Groden E (2001) Evaluation of synergistic interactions between the Colorado potato beetle (Coleoptera: Chrysomelidae), pathogen *Beauveria bassiana* and the insecticides, imidacloprid and crysomazine. J Econ Entomol 94:344–356

Gardner WA, Kinard DJ (1998) *In vitro* germination and growth response of two entomogenous fungi to imidacloprid. J Entomol Sci 33:322–324

Gardner WA, Sutton RM, Noblet R (1979) Evaluation of the effects of six selected pesticides on the growth of *Nomuraea rileyi* and *Beauveria bassiana* in broth cultures. J Georgia Entomol Soc 16(2):106–113

Georghiou GP, Taylor CE (1986) Factors influencing the evolution of resistance. In: Pesticide resistance strategies and tactics for management. National Academy Press, Washington, DC, pp 157–169

Gopalakrishnan C, Mohan S (2002) Field efficacy of *Nomuraea rileyi* (Farlow) Samson against certain lepidopterous insect pests of cabbage. In: Proceedings of the symposium on Biological Control of Lepidopterous Pest, July 17–18, 2002, Bangalore, pp 247–250

Gotz P, Boman HG (1985) Insect immunity. In: Kerkut GA, Gilbert LI (eds) Comprehensive insect physiology, biochemistry and pharmacology. Pergamon, Oxford, pp 453–485

Gunnarsson SGS, Lackie AM (1985) Hemocytic aggregation in *Scistocerca gregaria* and *Periplaneta americana* as a response to injected substances of microbial origin. J Invertebr Pathol 46:312–319

Gupta P, Paul MS, Sharma SN (1999) Studies on compatibility of white muscardine fungus, *Beauveria bassiana* with some neem products. Indian Phytopathol 52:278–280

Gupta RBL, Sharma S, Yadava CPS (2002) Compatibility of two entomofungi *Metarhizium anisopliae* and *Beauveria bassiana* with certain fungicides, insecticides and organic manures. Indian J Entomol 64:48–52

Hajeck AE, St. Leger R (1994) Interactions between fungal pathogens and insect hosts. Annu Rev Entomol 39:293–322

Hall IM (1963) Microbial control. In: Steinhaus EA (ed) Insect pathology, an advanced treatise, vol 2. Academic, New York, pp 477–511

Hall RA (1981) Laboratory studies on the effects of fungicides, acaricides and insecticides on the entomopathogenic fungus, *Verticillium lecanii*. Entomol Exp Appl 29:39–48

Hall IM, Dunn PH (1959) The effect of certain insecticides and fungicides on fungi pathogenic to the spotted alfalfa aphid. J Econ Entomol 52:28–29

Hassan AEM, Charnley AK (1989) Ultra structural study of the penetration by *Metarhizium anisopliae* through dimilin affected cuticle of *Manduca sexta*. J Invertebr Pathol 54:117–124

Hiromori H, Nishigaki J (1998) Joint action of an entomopathogenic fungi (*Metarhizium anisopliae*) with synthetic insecticides against the scarab beetle *Anomala cuprea* (Coleoptera: Scarabaeidae) larvae. Appl Entomol Zool 33(1):77–84

Hiromori H, Nishigaki J (2001) Factor analysis of synergistic effect between the entomopathogenic fungus, *Metarhizium anisopliae* and synthetic insecticides. Appl Entomol Zool 36(2):231–236

Hoy MA (1999) Myths, models and mitigation of resistance to pesticides. In: Denholm I, Pickett JA, Devonshire AL (eds) Insecticide resistance: from mechanisms to management. CABI Publishing, New York, pp 111–119

Hurpin B, Robert P (1968) Experiments on simultaneous infections of the common cockchafer, *Melolontha melolontha*. J Invertebr Pathol 11:203–213

Ignoffo CM, Hosteller DL, Garcia C, Pinnell RE (1975) Sensitivity of the entomopathogenic fungus, *Nomuraea rileyi* to chemical pesticides used on soybean. Environ Entomol 4:765–768

Islam MT, Omar DB (2012) Combined effect of *Beauveria bassiana* with neem on virulence of insect in case of two application approaches. J Anim Plant Sci 22(1):77–82

Jacobson RJ, Chandler D, Fenlon J, Russell KM (2001) Compatibility of *Beauveria bassiana* (Balsamo) Vuillemin with *Amblyseius cucumaris* Oudemans (Acarina: Phytoseiidae) to control *Frankliniella occidentalis* Pergande (Thysanoptera: Thripidae) on cucumber plants. Biocontrol Sci Tech 11:391–400

James RR (2003) Combining azadirachtin and *Paecilomyces fumosoroseus* (Deuteromycotina: Hyphomycetes) to control *Bemisia argentifolii* (Homoptera: Aleyrodidae). J Econ Entomol 96:25–30

James RR, Elzen GW (2001) Antagonism between *Beauveria bassiana* and imidacloprid when combined for *Bemisia argentifolii* (Homoptera: Aleyrodidae) control. J Econ Entomol 94:357–361

Jaramillo J, Borgemeister C, Ebssa L, Gaigl A, Tobon R, Zimmermann G (2005) Effect of combined application of *Metarhizium anisopliae* (Metsch) Sorokin (Deuteromycotina: Hyphomycetes) strain CIAT 224 and different dosage of imidacloprid on the subterranean burrower bug *Cyrtomenus bergi* Froeschner (Hemiptera: Cydnidae). Biol Control 34:12–20

Jayanthi PDK, Padmavthamma K (2001) Joint action of microbial and chemical insecticides on *Spodoptera litura* (Fab.) (Lepidoptera: Noctuidae). J Trop Agric 39(2):142–144

Jayaraj S, Rabindra RJ, Narayanan K (1985) Development and use of microbial agents for the control of *Heliothis* spp. (Lepidoptera: Noctuidae) in India. In: International workshop on "biological control of heliothis: increasing the effectiveness of natural enemies", 11–15, November, New Delhi

Jia M, Cao G, Li Y, Tu X, Wang G, Nong X, Whitman DW, Zhang Z (2016) Biochemical basis of synergism between pathogenic fungus *Metarhizium anisopliae* and insecticide chlorantraniliprole in *Locusta migratoria* (Meyen). Sci Rep 28424(6):1–15

Johnson DL, Goettel MS, Bradley C, van der Paauw H, Maiga B (1992) Field trials with the entomopathogenic fungus, *Beauveria bassiana* against grasshoppers in Mali, West Africa, July, 1990. In: Lomer CJ, Prior C (eds) Biological control of locust and grasshoppers. CAB International, Wallingford, pp 296–310

Joshi L, Charnley AK, Arnold G, Brain P, Bateman R (1992) Synergism between entomopathogenic fungi, *Metarhizum* spp., and the benzoylphenyl urea insecticide, teflubenzuron against the desert locust, *Schistocerca gregaria*. In: Proceedings, brighten crop protection conference, pest and diseases, British Crop Protection Council, Farham, pp 369–374

Kaakeh W, Reid BL, Bohnert TJ, Bennett GW (1997) Toxicity of imidacloprid in the German cockroach (Dictyoptera: Blattellidae) and the synergism between imidacloprid and *Metarhizium anisopliae* (imperfect fungi: Hyphomycetes). J Econ Entomol 90(2):473–482

Kanost MR, Kawooya JK, Law JH, Ryan RO, Van Heusden MC, Ziegler R (1990) Insect hemolymph protein. Adv Insect Physiol 22:299–396

Kranthi KR, Armes NJ, Rao GV, Raj S, Sundaramoorthy VT (1997) Seasonal dynamics of metabolic mechanisms mediating pyrethroid resistance in *Helicoverpa armigera* in Central India. Pestic Sci 50:91–98

Krueger SR, Nechols JR, Ramoska WA (1991) Infection of chinch bug, *Blissus leucopterus leucopterus* (Hemiptera: Lygaeidae) adults from *Beauveria bassiana* (Deuteromycotina: Hyphomycetes) conidia in soil under controlled temperature and moisture conditions. J Invertebr Pathol 58:19–26

Kubilay M, Gokee A (2004) Effects of selected pesticides used against glasshouse tomato pests on colony growth and conidial germination of *Paecilomyces fumosoroseus*. Biol Control 3:398–404

Lacey LA, Horton DR, Chauvin RL, Stocker JM (1999) Comparative efficacy of *Beauveria bassiana*, *Bacillus thuringiensis* and aldicarbfor control of Colorado potato beetle in an irrigated desert agroecosystem and their effects on biodiversity. Entomol Exp Appl 93:189–200

Laird M (1962) *Coelomomyces* fungi, an important group of mosquito parasites. In: WHO/BEL 1; issued by World Health Organization. Switzerland, Geneva

Latch GCM, Falloon RF (1976) Studies on the use of *Metarhizium anisopliae* to control *Oryctes rhinoceros*. Entomophaga 21:39–48

Leland JE, Behle RW (2004) Formulations of entomopathogenic fungus *Beauveria bassiana* with resistance to UV degradation for control of tarnished plant bug, *Lygus lineolaris*. In: Beltwide Cotton conferences, San Antonio, TX, January 5–9, pp 1800–1808

Lewis LC, Berry EC, Obrycki JJ, Bing L (1996) Aptness of insecticides (*Bacillus thuringiensis* and carbofuran) with endophytic *Beauveria bassiana* in suppressing larval populations of the European corn borer. Agric Ecosyst Environ 57(1):27–34

Lindquist RK (1993) Integrated insect, mite and disease management programmes on greenhouse crops: pesticides and application methods. In: Robb K, Hall J (eds) Proceedings, 9th conference on Insect and Disease Management in Ornamentals, Del Mar, California. Society of American Florists, Alexandria, pp 38–44

Luckmann WH, Metcalf RL (1982) The pest management concept. In: Metcalf RL, Luckman WH (eds) Introduction to insect pest management, 2nd edn. Wiley, New York, pp 1–31

Majchrowiez I, Poprawski TJ (1993) Effects in vitro of nine fungicides on growth of entomopathogenic fungi. Biocontrol Sci Tech 3:321–336

Manjula K, Krishna Murthy KVM (2005) Combined effect of *Nomuraea rileyi* (Farlow) Samson and insecticides against third instar larvae of *Spodoptera litura* Fab. Indian J Plant Prot 33(1):141–142

Maurao SA, Zanancio VE, Jose C (2003) Selectivity of insecticides and fungicides to the entomopathogenic fungus *Beauveria bassiana*. Neotrop Entomol 32:103–106

McCoy CW (1981) Pest control by the fungus *Hirsutella thompsonii*. In: Burges HD (ed) Microbial control of pest and plant diseases 1970–1980. Academic, London, pp 499–512

McCoy CW, Samson RA, Boucias DG (1988) Entomogenous fungi. In: Ignoffo C (ed) CRC microbial insecticides. CRC Press, Orlando, pp 156–236

Metcalf RL (1982) Insecticides in pest management. In: Metcalf RL, Luckman WH (eds) Introduction to pest management, 2nd edn. Wiley, New York, pp 217–277

Mietkiewski R, Gorski R (1995) Growth of selected entomopathogenic fungi species and isolates on media containing insecticides. Acta Mycol 30:27–33

Mietkiewski RT, Zurek M, Tkaczuk C, Balazy S (1991) Occurrence of entomopathogenic fungi in arable soil, forest soil and leaf litter (in polish). Pol Ann Agric Ser E-Plant Prot 21:61–68

Mietkiewski RT, Pell JK, Clark SJ (1997) Influence of pesticide use on the natural occurrence of entomopathogenic fungi in arable soils in the UK: field and laboratory comparisons. Biocontrol Sci Tech 7:565–575

Mohan MC, Narasimha P, Reddy NP, Devi UK, Kongara R, Sharma HC (2007) Growth and insect assays of *Beauveria bassiana* with neem to test their compatibility and synergism. Biocontrol Sci Tech 17:1059–1069

Moino JRA, Alves SB (1998) Efeito de imidacloprid e fipronil sobre *Beauveria bassiana* (Bals.) Vuill. E *Metarhizium anisopliae* (Metsch.) Sorok. e no comportamento de limpeza de *Heterotermes tenuis* (Hagen). An Soc Entomol Braz 27:611–619

Moorhouse ER, Gillespie AT, Sellers EK, Charnley AK (1992) Influence of fungicides and insecticides on the entomogenous fungus, *Metarhizium anisopliae*, a pathogen of the vine weevil, *Otiorhynchus sulcatus*. Biocontrol Sci Tech 2:49–58

Morales-Rodriguez A, Peck DC (2009) Synergies between biological and neonicotinoid insecticides for the curative control of the white grubs *Amphimallon majale* and *Popillia japonica*. Biol Control 51(1):169–180

Narayanasamy P (1991) Development and use of myco-insecticide from indigenous fungal pathogens against the brown planthopper (Nilaparvata lugens Stal.) problem in rice. In: Second Year annual report of Mycoinsectide Scheme. Financed by Ministry of environment and forest, New Delhi, pp 14–25

Narayanasamy P, Ramesh V, Ambethgar V, Gunasekaran (1995) Infection and extra cellular chitinase activity of *Pandora delphacis* associated with the rice brown planthopper, *Nilaparvata lugens*. Cytobios 84:111–125

Neves PMOJ, Hirose E, Tehujo PT, Moino A Jr (2001) Compatibility of entomopathogenic fungi with neonicotinoid insecticides. Neotrop Entomol 30:263–268

Oliveira RC, Neves PMOJ (2005) Compatibility of *Beauveria bassiana* with acaricides. Neotrop Entomol 33:353–358

Olivera RC, Neves PMOJ (2004) Biological control compatibility of *Beauveria bassiana* with acaricides. Neotrop Entomol 33:353–358

Olmert I, Kenneth RG (1974) Sensitivity of entomopathogenic fungi, *Beauveria bassiana, Verticillium lecanii* and *Verticillium* sp. to fungicides and insecticides. Environ Entomol 3:33–39

Olson ER, Dively GP, Nelson JO (2000) Baseline susceptibility to imidacloprid and cross resistance patterns in Colorado potato beetle (Coleoptera:Chrysomelidae) populations. J Econ Entomol 93:447–458

Pachamuthu P, Kamble ST (2000) In vivo study on combined toxicity of *Metarhizium anisopliae* (Deuteromycotina: Hyphomycetes) strain ESC-1 with sublethal doses of chlorpyrifos, propetamphos and cyfluthrin against German cockroach (Dictyoptera: Blattellidae). J Econ Entomol 93(1):60–70

Padmini Palem PC, Padmaja V, Vadlapudi V (2010) Evaluation of *Beauveria bassiana* isolates for tolerance to organophosphorus pesticides by in vitro and SDS-page analysis. Arch Appl Sci Res 2(2):202–210

Paula AR, Carolino AT, Paula CO, Samuels RI (2011) The combination of the entomopathogenic fungus *Metarhizium anisopliae* with the insecticide Imidacloprid increases virulence against the dengue vector *Aedes aegypti* (Diptera: Culicidae). Parasit Vectors 4(8):1–8

Pedigo LP (2002) Entomology and pest management, 4th edn. Prentice Hall, Upper Saddle River 742 pp

Pevling R, Weyrich J (1992) Effects of neem oil, *B. bassiana* and Dieldrin on non target tenebrionid beetle in desert zone of the Republic of Niger. In: Biological control of locusts and gras hoppers: proceedings of the workshop held at International Institute of Tropical Agriculture: Cotonou, Republic of Nenin, April–May 1991, CABI, Wallingford, pp 321–336

Prabhu T, Srikanth J, Santhalakshmi G (2007) Compatibility of selected pesticides with three entomopathogenic fungi of sugarcane pests. J Biol Control 21(1):73–82

Prasad VD (1989) Studies on certain entomopathogens of *Heliothis armigera* Hub. and *Spodoptera litura* Fab. and their interaction with some botanicals. Ph.D. thesis, Tamil Nadu Agricultural University Coimbatore, 187pp
Purwar JP, Sachan GC (2006) Synergistic effect of entomogenous fungi on some insecticides against Bihar hairy caterpillar, *Spilosoma oblique* (Lepidoptera: Arctiidae). Microbiol Res 161:38–42
Quntela ED, McCoy CW (1997) Pathogenicity enhancement of *Metarhizium anisopliae* and *Beauveria bassiana* to first instar of *Diaprepes abreviatus* (Coleoptera: Curculionidae) with sub-lethal doses of imidachloprid. Environ Entomol 26:1173–1182
Quntela ED, McCoy CW (1998) Conidial attachment of *Metarhizium anisopliae* and *Beauveria bassiana* to the larvae cuticle of *Diaprepes abbreviatus* (Coleoptera: Curculionidae) treated with imidacloprid. J Invertebr Pathol 72:220–230
Rachappa V, Lingappa S, Patil RK (2007) Effect of agrochemicals on growth and sporulation of *Metarhizium anisopliae* (Metschnikoff) Sorokin. Karnataka J Agric Sci 20(2):410–413
Ramakrishnan R, Suiter DR, Nakatsu CH, Humber RA, Bennett GW (1999) Imidacloprid-enhanced *Reticulitermes flavipes* (Isoptera: Rhinotermitidae) susceptibility to the entomopathogen, *Metarhizium anisopliae*. J Econ Entomol 92:1125–1132
Ramaraje Urs NV, Govindu HC, Shivashankara Shatry KS (1967) The effect of certain insecticides on the entomogenous fungi, *Beauveria bassiana* and *Metarhizium anisopliae*. J Invertebr Pathol 9:398–403
Ramarethinam S, Marimuthu S, Murugesan NV (2000) Studies on the effect of entomopathogens *Hirsutella thompsonii, Verticillium lecanii* and *Paecilomyces* sp. and a neem derivative in the control of coconut mite, Aceria (Eriophyes) guerreronis. Pestology 24(5):5–7
Ranjekar PK, Patanka A, Gupta V, Bhatnagar R, Bentur J, Anand KP (2003) Genetic engineering of crop plants for insect resistance. Curr Sci 84(3):321–329
Rashid M, Baghdadi A, Sheikhi A, Pourian HR, Gazavi M (2010) Compatibility of *Metarhizium anisopliae* (Ascomycota: Hypocreales) with several insecticides. J Plant Prot Res 50(1):22–26
Regupathy A (1995) Problems and prospects of management of insecticide resistance in Helicoverpa armigera in India. In: Constable GA, Forrester NW (eds) Challenging the future: proceedings of world cotton conference–I, 14–17 February, Brisbane, Australia, CSIRO, Melbourne, pp 556–562
Regupathy A, Paramjothi G (1977) Evaluation of insecticides and white halo fungus. *Cephalosporium lecanii* Zimm. against banana aphid, *Pentalonia nigronervosa* Coq. Pesticides 13(9):35–37
Riba G, Marcandier S, Richard G, Larget T (1983) Susceptibility of maize pyralid (*Ostrinia nubilalis*) (Lepidoptera: Pyralidae) to entomopathogenic Hyphomycetes. Entomophaga 28(1):55–64
Riley DG, Sparks AS (2006) Insecticide resistance management for diamondback moth in Georgia. The University of Georgia, College of Agricultural and Environmental Sciences and the U.S. Department of Agriculture Cooperating, Circular No.899, pp 1–11
Roberts DW, Campbell AS (1977) Stability of entomopathogenic fungi. Entomol Soc Am 10:19–76
Russell EP (1999) The strange career of DDT: experts, federal capacity, and environmentalism in world war II. Technol Cult 40:770–796
Sabbour M (2013) Efficacy of entomopathogenic fungi alone or in combination with inorganic insecticides for protecting a broad bean against certain coleopteran stored products beetles in Egypt. Glob J Biodivers Sci Manag 3:182–187
Sandhu SS, Unkles SE, Rajak RC, Kinghorn JR (2001) Generation of benomyl resistant *Beauveria bassiana* strains and their infectivity against *Helicoverpa armigera*. Biocontrol Sci Tech 11:245–250
Sanjaya Y, Ocampo VR, Caoili BL (2015) Infection process of entomopathogenic fungi *Beauveria bassiana* in the *Tetrancyhus kanzawai* (Kishida) (Tetranychidae: Acarina). Arthropods 4(3):90–97

Santharam G, Easwaramoorthy S, Regupathy A, Jayaraj S (1977) Possibility of increasing the pathogenicity of *Cephalosporium lecanii* on coffee green bug *Coccus viridis* during summer. J Plant Crop 5(2):121–126

Santos AV, de Olivera BL, Samuels RI (2007) Selection of entomopathogenic fungi for use in combination with sub-lethal doses of imidacloprid: perspectives for the control of the leaf-cutting ant, *Atta sexdens rubropilosa* Forel (Hymenoptera: Formicidae). Mycopathologia 163:233–240

Sarfraz M, Keddie BA (2005) Conserving the efficacy of insecticides against *Plutella xylostella* (L.) (Lepidoptera: Plutellidae). J Appl Entomol 129:149–157

Schmutterer H (1990) Properties and potential of natural pesticides from the neem tree, *Azadirachta indica*. Annu Rev Entomol 35:197–271

Senthilkumar CM, Regupathy A (2007) Laboratory studies and field assessment on the compatibility and combined efficacy of neonicotinoids with *Lecanellium (=Verticillium) lecanii* (Zimmermann) Viegas in the control of coffee green scale, *Coccus viridis* (Green) (Hemiptera: Coccidae). National conference on applied zoology and sustainable development, July, 13–14 2007, Indian Institute of Chemical Technology, Hyderabad. Abstract 38 p

Seyoum E (2001) The synergistic effects of *Metarhizium anisopliae* (Metchnikof) with the acyl urea insecticides teflubenzuron and diflubenzuron for *Schistocerca gregaria* (Orthoptera: Acrididae). Ethiop J Sci 24(1):113–125

Sharma HC, Sharma KK, Seetharama N, Ortiz R (2001) Genetic transformation of crop plants: risk and opportunities for the rural poor. Curr Sci 80(12):1495–1508

Shi WB, Jiang Y, Feng MG (2005) Compatibility of ten acaricides with *Beauveria bassiana* and enhancement of fungal infection to *Tetranychus cinnabarinus* (Acari: Tetranychidae) eggs by sublethal application rates of pyridaben. Appl Entomol Zool 40(4):659–666

Silva LDA, Silva RFP, Heineck MA (1993) In vitro evaluation of the effect of different insecticides on the sporulation of the fungus, *Nomuraea rileyi* (Farlow) Samson. An Soc Entomol Brazil 22:99–103

Silva AX, Jander G, Samaniego H, Ramsey JS, Figueroa CC (2012) Insecticide resistance mechanisms in the green peach aphid Myzus persicae (Hemiptera: Aphididae) I: a transcriptomic survey. PLoS One., www.plosone.org 7:1–14

Soderhall K, Smith VJ (1986) The prophenoloxidase activating system: the biochemistry of its activation and role in arthropod cellular immunity, with special reference to crustaceans. In: Brehelin M (ed) Immunity in invertebrates. Springer, New York, pp 208–223

Sreerama Kumar P, Singh SP (2000) *Hirsutella thompsonii*: the best biological control option for the management of the coconut mite in India. Indian Coconut J 9:11–17

St. Leger RJ, Joshi L, Bidochka MJ, Roberts DW (1996) Construction of an improved mycoinsecticide overexpressing a toxic protease. Proc Natl Acad Sci USA 93:6349–6354

St. Leger RJ, Joshi L, Roberts DW (1998) Ambient pH is a major determinant in the expression of cuticle degrading enzymes and hydrophobin by *Metarhizium anisopliae*. Appl Environ Microbiol 64:709–713

Steinhaus DC (1958) Stress as a factor in insect disease. In: Proceedings of 10th international congress of entomology 4:725–730

Steinhaus DC (1996) Control of tarnished plant bug with *Beauveria bassiana* and in interaction with imidacloprid. In: Beltwide cotton conferences, 9–12 January, Nashville, pp 888–889

Steinkraus DC, Tugwell NP (1997) *Beauveria bassiana* (Deuteromycotina: Moniliales) effects on *Lygus lineolaris* (Hemiptera: Miridae). J Entomol Sc 32:79–90

Stewart JG, Kennedy GC, Sturz AV (1997) Incidence of insecticide resistance in populations of Colorado potato beetle, *Leptinotarsa decemlineata* (Say) (Coleoptera:Chrysomelidae), on Prince Edward Island. Can Entomol 129:21–26

Studdert JP, Kaya HK (1990) Water potential, temperature, and clay coating of *Beauveria bassiana* conidia: effect on *Spodoptera exigua* pupal mortality in two soil types. J Invertebr Pathol 56:327–336

Sundaramurthy VT, Gahukar RT (1998) Integrated management of cotton insect pests in India. Outlook Agric 27(4):261–269

Thompson SR, Brandenburg RL (2006) Effect of combining imidacloprid and diatomaceous earth with *Beauveria bassiana* on mole cricket (Orthoptera: Gryllotalpidae) mortality. Biol Microb Control 90(6):1948–1954

Thurston GS, Kaya HK, Burlando TM, Harrison RE (1993) Milky disease bacterium as a stressor to increase susceptibility of scarabaeid larvae to an entomopathogenic nematode. J Invertebr Pathol 61:167–172

Tian L, Feng MG (2006) Evaluation of the time-concentration-mortality responses of *Plutella xylostella* larvae to the interaction of *Beauveria bassiana* with nereistoxin analogue insecticide. Pest Manag Sci 62:69–76

Todorova SI, Coderre D, Duchesne RM, Cote JC (1998) Compatibility of *Beauveria bassiana* with selected fungicides and herbicides. Biol Control 27(2):427–433

Umina PA (2007) Pyrethroid resistance discovered in a major agricultural pest in southern Australia: the redlegged earth mite (Acari: Penthaleidae). Pest Manag Sci 63:1185–1190

Vanninen I (1995) Distribution and occurrence of four entomopathogenic fungi in Finland: effect of geographical locations, habitat type and soil type. Mycol Res 100:93–101

Vanninen I, Hokkanen H (1988) Effect of pesticides on four species of entomopathogenic fungi in vitro. Ann Agric Fenniae 27:159–169

Vimala Devi PS, Prasad YG (1996) Compatibility of oils and antifeedants of plant origin with the entomopathogenic fungus, *Nomuraea rileyi*. J Invertebr Pathol 68:91–93

Vyas RY, Yadav DN, Patel RJ (1990) Compatibility of *Beauveria brongniartii* with some pesticides used in groundnut pest management. Ann Biol 6:21–26

Wago H (1995) Host defense reactions of insects. J Appl Entomol Zool 39:1–14

Wheeler MB, Stuart GS, Hapner KD (1993) Agglutinin mediated opsonization of fungal Blastospores in *Melanoplus differentialis* (Insecta). J Insect Physiol 39:477–483 www.psticideresistance.com

Ye SD, Dun YH, Feng MG (2005) Time and concentration dependent interactions of *Beauveria bassiana* with sublethal rates of imidacloprid against the aphid pests *Macrosiphoniella sanborni* and *Myzus persicae*. Ann Appl Biol 146:459–468

Ying SH, Feng MG, Xu ST (2003) Field efficacy of emulsifiable suspensions of *Beauveria bassiana* conidia for control of Myzus persicae population on cabbage in China. J Appl Ecol 14:545–548

Zhang Y, Zhang W, Han X, Li X (2000) Biochemical and physiological mechanisms of insecticide resistance in *Helicoverpa armigera* (Hübner). Acta Entomol Sin 43(3):247–253

Zhao JZ, Bishop BA, Grafius EJ (2000) Inheritance and synergism of resistance to imidacloprid in the Colorado potato beetle (Coleoptera: Chrysomelidae). J Econ Entomol 93:1508–1514

Zurek L, Watson DW, Schal C (2002) Synergism between *Metarhizium anisopliae* (Deuteromycotina: Hyphomycetes) and boric acid against the German cockroach (Dictyoptera: Blattellidae). Biol Control 23(3):296–302

Chapter 14
Arbuscular Mycorrhizal Fungi As Phosphate Fertilizer for Crop Plants and Their Role in Bioremediation of Heavy Metals

Arun Arya, Shalini Ojha, and Simranjeet Singh

Abstract Endophytic fungi are universally present in plants. These fungi form a symbiotic association and help the plants in different ways. These fungi also help in protection against insect attack and herbivory. Many root-associated fungi form an endophytic symbiotic association which assists in providing phosphorus, other nutrients, and water balance. The members of Glomerales colonize cortex of roots and help in better plant growth. More than one fungus may be associated, and this helps in bioremediation of pollutant heavy metals. Their association with plants is as old as the evolution of angiosperms millions of years back.

Keywords Arbuscular mycorrhizal fungi · Crops · *Glomus* · Heavy metals · Mushroom · Phytoremediation

14.1 Introduction

Arbuscular mycorrhizal fungi (AMF) facilitate in the uptake of the nutrients for the benefit of the plant, working as an extension of the root system of the plant. Approximately, 85% of the angiospermic plants work in association with fungi. On colonizing over the root, the hyphal structure of fungi starts developing and it extends itself into the soil. The large area of the soil is covered by these hyphal structures in comparison with the root hairs and aids in the absorption of nutrients like P, S, and Zn from the soil which are less mobile. Fungal hyphae facilitate nutrient transport towards the root that was oozed by the root cells. Elevated nutrient level prompts the disease resistance in plants and averts water stress. AMF mainly sustain in the soil, which causes and alters the balance of soil microbes. AMF predominantly elevate the growth of soil microbes and constrain the proliferation of plant pathogen, which as a resultant reduces the stress due to disease over the plant/crop.

A. Arya (✉) · S. Ojha · S. Singh
Department of Botany, Faculty of Science, The Maharaja Sayajirao University of Baroda, Vadodara, India

P. Gehlot, J. Singh (eds.), *Fungi and their Role in Sustainable Development: Current Perspectives*, https://doi.org/10.1007/978-981-13-0393-7_14

Hyphae of the AMF stabilize the soil by secreting glomalin, a glue-like material, which aids in wrapping of soil particle into small clumps which keep the soil intact. These soil clumps augment a number of pores within the structure of soil. These augmented pores facilitate the soil to hold large amount of air, which is essential for both microbial activity and roots, which subsequently leads to increase in absorption and water-retaining capacity during heavy rains.

Because of the symbiotic nature, AMF are dependent on the roots of the plants for carbon/sugar source which helps them to proliferate and reproduce. Eventually, both fungi and plants favor each other as they grow in symbiosis. By supporting AMF population in soil, we can improve the soil fertility. Almost every plant species has mycorrhizal association with one or more species of fungi (Molina et al. 1992). All conifer species have proved to be mycorrhizal in their natural habitat (Okland and Eilertsen 1993); this association is of ectomycorrhizal fungi. Of over 500 species of angiosperms that have been studied so far, 70% are obligatory mycorrhizal and the other 12% are facultatively mycorrhizal in that they may not form mycorrhizae (Trappe 1987). Mycorrhizosphere is found to help in disease control. Mycorrhizal association makes the plant more tolerant to soil toxicants and thus maintain a higher level of growth and health than non-mycorrhizal plants (Davis et al. 1993).

14.2 Fungi: Role in Litter Decomposition

Fungi are stated to be the most exuberant organism which decomposes the waste material and eminent part of food web of the soil (Rhodes 2012) and also provides the nutrition to the other microbes residing in the soil. The lands of forest are roofed with the leaves' debris of the previous season, which cannot flourish on their own, as they are tough, undigestible, and have nutrients clenched within themselves. The microbe which has the ability to decompose the debris of leaves is fungus, predominantly, the mycelium. Mycelium, in fungi is a vegetative part, which can be discerned by fine white threadlike growth, is oozing out of the deceased leaves, wood, etc. Undeniably, fungi is the microbe which possesses the ability to degrade the wood. The mycelial part of fungi secretes the extracellular enzyme, which has the ability to degrade the cellulose as well as lignin, which are the integral parts of the plant fibers. During this decomposition process of leaves and wood, the organic-rich material is formed, named humus. These consortiums of microbes start consuming the varied substrates present around and increase the degradation rate when the essential nutrients like N, P, and K and inorganic nutrients are present in abundant amount (Rhodes 2013). *Aspergillus* and molds are reported to be the decomposer of material like cellulose, hemicellulose, pectin, starch, and other polymers of sugar. Few of the *Aspergillus* spp. are also recorded which possess the ability to disintegrate the intractable substrates like chitin, fats, keratin, and oil. Anthropogenic substrates, like cotton, linen, paper, and jute, are promptly decomposed by the molds, and this process is known as biodeterioration. In 1969, Florence (also known

Firenze) got flooded; from there 74% of isolates were reported of *Aspergillus versi color*, which isolated from the ravaged Ghirlandaio fresco of the church Ognissanti (Rhodes 2013).

14.3 Fungi As Symbiont

In different parts of the world, extensive use of fertilizers containing phosphorus (P) in the large amount has led to the accumulation of P in different types of soil. Plants which have the capability to absorb P via roots only cease the growth of AMF which adhere to their roots by blocking the sugar/nutrient flow to the fungi. With the passage of time, this issue has become the serious concern as it has declined population of many valuable fungal species in the soil. The soil in which fertilizer contains the P is not used; their AMF help in the plant growth via symbiosis. Moreover, if we stop using the fertilizer containing P, one can save wealth, subside the environmental effect because of the mining as well as refine the P, and one can even reduce the chances of P discharge to the surface water.

AMF hyphae perform two main functions: first, they aid as an organ which absorbs nutrient, and second, they adhere to newly formed roots, as plowing disrupts the network of fungal hyphae with soil and hinders both the functions. To prove this, researchers are conducted and the study found that planting the seeds under non-plowing area facilitates in rapid adherence of the AMF; additionally, it elevates the absorption of P as to the seedling cultivated in the plowing region of the soil.

Monoculturing of one crop dominates the development of particular fungi, which have the ability to grow in symbiosis with that crop, and leads to reduction in a number of other AMF species. As a fact, continuous monoculturing of one crop with the same AMF species leads to reduction in yield. Different researches have illustrated that crops grow promptly when they have been colonized by different species of AMF. Thus, for attaining the diverse communities of AMF in soil, rotation of different types of crop should be followed. In 1996, Maiti et al. reported about the AMF which adhere on the roots of rice which previously were residing on the roots of weeds, whereas inter-cropping of rice and red gram caused the substantial growth of the AMF (Maiti et al. 1996a, b).

There are few crops which suppress the colonization of AMF on their root. The crops belong to the family Brassica, comprising brocculi, cabbage, cauliflower, spinach, sugar beet, lupine, and rapeseed. If these crops are grown in rotation, they will suppress the growth of AMF in the soil. Thus, for sustaining the AMF number while cultivating these crops, interspersed mixture of complement crop is required to facilitate the growth of AMF.

Researchers have illustrated that soils which have been managed for a long time encompass diverse AMF population, but increased number of these AMF will only be possible if fungus-friendly approaches are employed. An investigation in Rodale Institute demonstrates the one-time crop cover or one no plowing cycle, elevating the amount of AMF for colonizing over the succeeding crop. Furthermore, after 8 years of time shifting from traditional towards the organic management in the Rodale Institute's Farming Systems Trial, it has been realized that fungal spores are in large numbers in organic plot in contrast to plot which was traditionally farmed.

14.4 Contaminated Land and Bioremediation

For effective mycoremediation, there is a need to select the suitable fungal species, which has the ability to degrade specific pollutant by following the simple procedure of screening (Matsubara et al. 2006). A comprehensive review enlightening the role of fungi in degrading the organic contaminants till 2006 is documented (Singh 2006). Another article named Mycelium Running: How Mushrooms Can Help Save the World (Stamets 2005) provides information for cultivating mushroom and how to exploit these mushrooms for remediation procedure, like remediation of accidental spills of chemical toxins and oil at pilot scale.

Bioremediation can be stated as the procedure which disintegrates the contaminants present in the soil as well as water with the help of microbes (Rhodes 2013). There are two approaches of bioremediation, i.e., in situ and ex situ. The main difference between these two approaches is their treatment; on site is known as in situ, whereas treatment done after physical removing of the material from the on site is regarded as the ex situ. Ex situ treatment approaches are cost-effective, as chemical and incineration are required for remediating the contaminants from the soil. On the other, hand if the treatment is done on the target site, the total expenses are less than the ex situ approach. On the contrary, this chemical treatment process simply moves these contaminants from one site to another, actually eradicating it, whereas incineration has challenges of its own as it leads to the formation of dioxin and requires high energy. Bioremediation methods aid in degrading the toxic organic compound such as industrial waste, oil spills, and pesticides at the level of molecular by transforming these toxic compounds to nontoxic compounds. Bioremediation's firm goal is to mineralize the pollutants by transforming them into carbon dioxide, water, nitrogen gas, hydrochloric acid, etc. Decomposition of the heavy metal and radioactive ions is difficult, but they are transformed to low-soluble form, i.e., by rendering their oxidation state from U(IV) to UO_2 (Singh et al. 2014) into less harmful state which could be physically removed with the help of mycoremediation/phytoremediation, which encompasses the cultivation of fungus and plants together.

14.5 Bioremediation Using Fungi (Mycoremediation)

White-rot fungi got its name because it secretes the enzyme, which decomposes lignin and cellulose, and gives white appearance because of the undissolved cellulose, whereas brown-rot fungi only degrades the cellulose and leaves the deposits of lignin which give brownish appearance. These fungi lead to the formation of mixed, cubical crack as well as shrinkage of wood, generally seen in the confer trees (Stamets 2005). Approximately, 30% of the literature-related bioremediation, with help of fungi, is associated with white-rot fungus (Singh 2006). There are certain sets of mechanism involved which make white-rot fungi unique from the other fungi and bacteria employed for bioremediation. For bacteria to degrade the particular pollutant, they need to be adapted for synthesizing the specific enzyme which can complete the task of degradation. This technology has its own limitation, as the bacteria are adapted to disintegrate the contaminants to certain level (Adenipekun and Lawal 2012). Many different organic molecules, including the untraceable as well as enduring constituents like PAHs, are vulnerable to erratic degrees via the numerous strains of white-rot fungi with the ability to degrade it (Singh 2006). *Phanerochaete chrysosporium*, the white-rot fungi, is a superlative bioremediation model which is efficient in disintegrating the insoluble-toxic constituents than any other bacteria or fungi. Nowadays, redox reactions take place in varied conditions, like the type of pollutant, its degree, and site where it's taking place. Different white-rot fungi named *Pleurotus ostreatus*, *Trametes versicolor*, *Bjerkandera adusta*, *Lentinula edodes*, *Irpex lacteus*, *Agaricus bisporus*, *Pleurotus tuber-regium*, and *Pleurotus pulmonarius* have been reported to disintegrate importunate xenobiotic compounds (Singh 2006; Adenipekun and Lawal 2012). The soil contaminated with crude oil is decomposed mixing it with the lignocellulosic substrate such as corn cob and saw dust, which aid in the proliferation of fungal species in the soil (Lang 1995). Other toxic constituents like dioxins, pesticides, phenols, polychlorinated biphenyls, chlorophenols, pulp and paper mill effluents, dyestuffs, and heavy metals have been reported for successful disintegration by the white-rot fungus (Singh 2006). Planning is being done to deploy the fungal species at bioremediation site contaminated with complex amalgam of PAHs such as crude oil, coal tar, as well as creosote. Moreover, heavy metals along with mediators like 2,2′-azinobis-(3-ethylbenzothiazoline-6-sulfonate) and vanillin have been found to induce disintegration of the benzo[*a*]pyrene with the help of *Pleurotus ostreatus*. In experiment, by increasing the Cu concentration to 15 mM, the degradation activity was recorded to be 74.2%, whereas further declination was observed on increasing the Cu concentration. On the other, when 5 mM vanillin was added, the degradation rate increased to 83.6% (Bhattacharya et al. 2014). Thus, it has been predicted that by employing the vanillin (by-product of lignin) will enhance the mycoremediation process with the help of white-rot fungi in the field. In 2011, Isikhuemhen et al. determined the ability of *L. squarrosulus* to degrade the corn stalks after 30 days, whereas the maximum activity of lignocellulolytic enzyme was attained after 6 days of cultivation, confirmed by production of exopolysaccharides. Therefore, *L. squarrosulus* can be employed in industry for the pretreatment as well as delignification of the lingo-cellulose waste. The key points make the white-rot fungi the ideal preference as it secretes variety of compounds that can be an extracellular low-specificity enzyme; hence, it can act on the substrate like lignin and its analog (Adenipekun and Lawal 2012). The list of enzyme which has been documented for lignin degradation encompasses hydrogen peroxidase, laccase, lignin peroxidase, and manganese peroxidase (Kirk and Farrell 1987), though there are no lignolytic fungi which show these three types of enzyme activity.

AMF enhance the uptake of Zn and reduce the Cd uptake toward the shoot when the concentration of the heavy metals in the soil is low. Infection due to mycorrhizal fungi elevates the root/shoot barrier against the heavy metals; sometime it also induces the heavy metal resistance among the plants (Dehn and Schuepp 1990). Maize mycorrhiza augments the Cu uptake, whereas soybean mycorrhiza augments the uptake of the heavy metals like Cu, Fe, Mn, and Zn. It has been determined in the addition of heavy metal in both crops which declined the concentration of the nutrients. Cr treatment induced the maximum inhibition as compared to the others (Wang and Chao 1989).

In 2007, Krishnamurthi et al. conducted the study to evaluate the solid phase distribution, mobility of Cu, and phytoavailability on wheat durum (*Triticum durum*) by growing it on contaminated as well as non-contaminated Italian soil. The location of selected sites was changed because of land use and current vegetation. For studying the distribution of Cu in the solid phase, the vineyard area of cultivation was also selected as it showed the high concentration of Cu about 132–253 mg kg^{-1}. Eight-step selected sequential procedure was employed for the extraction of solid phase Cu fraction from the soil. In results it was found that Cu was promptly associated on the binding sites of organic molecule approximately 62.6–74.8%. Furthermore, the DTPA-TEA as well as NH_4Cl extraction procedure was employed for the assessment of phytoavailability of Cu via durum wheat in the greenhouse using two types of soils. Due to limited translocation in the plant, most of Cu was retained within the root region, in which regression coefficient was found to be 0.960 and p-value was 0.0001 among the Cu content of plant vs Cu accumulated with metal-fulvate complex, where result obtained directed toward metal-fulvate complex is to be responsible for phytoavailability of Cu. Although the polluted soil showed high proportion of Cu, i.e., about 77% which were associated with binding site of organic molecule, the nonpolluted soil showed the proportion of Cu, i.e., 21.3, which illustrated the reason behind the high phytoavailability of Cu. When leeks were inoculated with *Glomus mosseae*, it increased the Cu uptake (Gildon and Tinker 1983).

For the fractionation of the metalloids like Ar and Se, unique sequence of the extraction chemicals is required as different oxidation states will be attained for these metals in the soil. Different fractionation assessment steps are performed under the redox condition, in which the initial oxidation state is amended. Even, one can question the results of extraction procedure, as they are suitable for evaluating the metalloids bioavailability under varied environmental conditions.

14.6 Comparison of Mycoremediation Using Mushroom and AM Fungi

For employing the white-rot fungi for the bioremediation purpose, there is a need to get the proper insight of the subjects like physiology, chemistry, enzymology, genetics, molecular biology, ecology, as well as genetic engineering. Four different approaches have been developed (Lamar and White 2001): benchtop treatment, on-site pilot scale assessment, inoculum preparation, and industrial-scale application. Different types of substrates like alfalfa, bagasse of sugarcane, bark, canola meal, corn cobs, exhausted compost of mushroom, fish oil, grain and sawdust nutrient-fortified mixture, okra, peat, pulp of coffee and sugar beet, rice, sawdust, stems and wood of annual plant, wheat straw, wood chips, cyclodextrins, as well as the surfactants can be utilized for the synthesis of the inoculum either on-site or off-site or in the form of mixture with the polluted soil to enhance the degradation process (Singh 2006). As it is difficult to regulate the ideal nitrogen/carbon ratio over the varied substrate, overcoming these inhibiting effects can affect the efficacy of fungi during bioremediation process. Therefore, the fungi inoculum is coated with agarose, alginate, carrageenan, chitosan, gelatin, etc. in the shape of beads, which give better result in comparison to inoculum grown over the bulk substrate.

The approach is named as "encapsulation," taken from the mushroom-propagating industry, as it facilitates in preserving the inoculum in viable stage and also aids in maximum contaminant degradation. Although this encapsulation approach increases the efficacy and survival rate of the fungal species, solid-state fermentation is the chief source of fungal inoculum, as it increases the success rate in the first phase so that technical amendments can be engineered in the second stage. The third and fourth stage is largely dependent on the procedure of remediation-encompassing process like evaluation, optimization, maintenance, and continuation of the complete process. Even the niche microbes give the tough competition to this mycoremediation process; still there are limitations because of the absence of ideal procedures which can eradicate these hindrances. Only few patents are available which are related to the subject of bioremediation with the help of white-rot fungi (Singh 2006).

Biosorption, one of the rapid as well as reversible processes, is employed form removing the toxic heavy metals from the contaminated solution. Different environmental factors affect the chemical interaction between the surface of bacteria and metal during biosorption. The factors responsible are ionic strength, pH, temperature, and presence of other impurities like organic constituent or heavy metal. The age of bacteria is also determining factors for the adsorption capacity of heavy metal. Various factors like pH, nature of the surface of the absorbent, number of sorption site and charge, as well as nature of Me-L species are present in the solution which can affect the sorption of trace elements over components of soil when biological and inorganic ligands are present. The number of natural organic constituents is higher in rhizospheric soil as compared to the bulk soil of root-soil interface. Large fraction of the plant production is due to the C flow in the rhizosphere. About 20% of the accumulated C via photosynthesis can be removed through the roots. For the rhizosphere region, root adhesive is discharged which comprises of both high and low molecular weight constituents. The utmost high molecular weight constituents are ectoenzymes, mucilage, and polysaccharide; on the other hand, the prime low molecular weight constituent comprises amino acids, carbohydrates, organic acids, peptides, and phenolic compounds. Microbes synthesize a variety of extracellular enzymes which can interact with metal complex present in the solution, which encompasses organic acids, pigments, polysaccharides, and siderophores. Mycorrhizal and free-living fungi, plant roots, and lichens are known for producing citric as well as oxalic acids.

Bioremediation is a process in which microbes are used for treating the contaminated soil. AMF (arbuscular mycorrhizal fungi), also known as root-colonizing symbiosis microbes, are usually enmeshed in phytoremediation process, which involves plant for remediating the soil. Phytoremediation encompasses the different groups of methods which employ plant for containment, devastation, or extraction purpose (EPA 2000). In recent years, these methods have attained considerable appreciation for being conventional, cost-effective, as well as non-biological method. With regard to the contaminant, the different methods of phytoremediation are employed. In each instance, vegetation process declines the infiltration of water as well as erosion. Heavy metals are generally disintegrated via extraction (also known as phytoextraction) from the soil; beside this, they are immobilized as non-toxic constituent (also known as phytostabilization). AMF has been found to reduce the toxicity due to metals in the plant via declining the translocation rate from root to shoot (Levyval et al. 1997; Jeffries et al. 2003). Moreover, this aids the plant to survive in the site which is heavily contaminated with the heavy metals and could be employed as a comparison to immobilization method. Phytoextraction involves those plants which have the ability to accumulate a large amount of heavy metals; further these plants can be cultivated and discarded, or heavy metals could be extracted. Thus, due to this reason, plants acquire the ability to accumulate the metals, generally plants of *Brassicaceae* family, also these plants are determined to be non-mycorrhizal. Furthermore, the other plants synthesize a large amount of biomass and are regarded as mycotrophic. Pollutants of organic nature like polycyclic aromatic hydrocarbons (PAHs) are converted or disintegrated via activity of microbes, which is usually observed around the roots (also referred as rhizodegradation). Still it is unconfirmed that this elevated disintegration within the rhizosphere is because of the active molecules oozed out from the plant, which can be enzyme, surfactant or physiochemical change, or elevated microbial activity. Other suggested mechanism for the disintegration of the organic contaminants involves the plant metabolism. Moreover, the above process is of significant importance for PAHs (Binet et al. 2000a, b) degradation and could not be conferred here. PAH degradation involves the exudates from root, as well as root-accumulated microbial population such as AMF. A case was studied in which they assessed the different methods for improving the soil quality and in which they studied the non-inoculated soil, soil inoculated with the one type of exotic AMF, and soil inoculated with the mixture of indigenous AMF (arbuscular mycorrhizal fungi). Out of the soil, inoculated with the mixture of indigenous, AMF were found to be effective. Due to the presence of AMF nodules, it enhanced the plant growth, water infiltration ability, and soil aeration because of the aggregation of the soil. Phosphorus uptake has also increased on inoculating the native AMF (Jeffries et al. 2003).

14.7 Conclusion

Eventually, on analyzing all aspects of fungi and its scope for the degrading the intractable, persisting, highly toxic contaminant like TNT (2,4,6-trinitrotoluene) (http://www.defmin.fi/files/2461/Steffen_Kari.pdf), nerve gases as well as sarin (Stamets 2005) was observed on contaminated site. On inoculating the oyster mushroom, *Pleurotus ostreatus*, to the land polluted with the diesel oil, it was found that the mycelial growth degraded the 95% of PAHs into nontoxic constituent after the 4 weeks of inoculation. Microbes which naturally residing in the soil in association with fungi degrade most of the organic contaminants to CO_2 and H_2O, i.e., complete mineralization. In a report about cargo ship spill in 2007, in which 58,000 gallons of fuel was spilled on the sideways of the shoreline of San Francisco, human hair-knitted mat was employed as sponge for soaking the spilled oil. And these mats were then exposed to the oyster mushrooms along with straws: as resultant the mushroom disintegrates the oil, and after a few weeks' time, soil was cleansed enough that it could be utilized for roadside landscaping. For degrading the toxic aromatic contaminants of petroleum and pesticides containing chlorine, the wood-degrading fungi are employed (Rhodes 2013). The procedure in which mycelia act as a filter removes the toxic constituent and microbes through the water from soil and is regarded as mycofiltration. The confined definition of "mycoremediation" was stated by Paul Stamets who was asked to form a "Mycological Response Team" which aimed to use fungi to recycle and regain the quality of healthy soil in the area where the following incident of contamination such as chemical leaks, oil spills, and radiation egress has taken place. Proposal has been made to grow the mushroom for mycoremediation purpose and diagnose them if they are safe for human consumption after the mycoremediation process (Kulshreshtha et al. 2014). Indeed, everything depends on the contaminant nature, where heavy metals emerge as a problem (because they get adsorbed and accumulated in the mushroom); on the other hand, organic contaminant disintegrates without imparting the toxicity. In conclusion, the land which is contaminated as well as unfit for the agricultural purpose can be cleaned via AMF. It also elevates yield of crop as well as its nutritious value. These AMF are found to be the potent degrader of PAHs compounds. Thus, AMF can be utilized as bioassay for assessing soil quality and toxicity, because of its sensitivity against the wide range of contaminants. Further, they can also confirm about the adequate quality of soil which has been established after process of bioremediation.

References

Adenipekun CO, Lawal R (2012) Uses of mushrooms in bioremediation: a review. Biotechnol Mol Biol Rev 7(3):62–68

Bhattacharya S, Angayarkanni J, Das A, Palaniswamy M (2014) Mycoremediation of Benzo[*a*] pyrene by *Pleurotusostreatus* in the presence of heavy metals and mediators. Biotechnology 4(2):205–211

Binet P, Portal JM, Leyval C (2000a) Fate of polycyclic aromatic hydrocarbons (PAH) in the rhizosphere and mycorrhizosphere of ryegrass. Plant Soil 227:207–213

Binet P, Portal JM, Leyval C (2000b) Dissipation of 3–6-ring polycyclic aromatic hydrocarbons in the rhizosphere of ryegrass. Soil Biol Biochem 32:2011–2017

Davis FT, Potter JR, Linderman RG (1993) Drought resistance of mycorrhizal pepper plants independent of leaf P concentration-response in gas exchange and water relations. Plant Physiol 87:45–53

Dehn B, Schuepp H (1990) Influence of VA mycorrhizae on the uptake and distribution of heavy metals in plants. Agric Ecosyst Environ 29(1–4):79–83

EPA (2000) Introduction to phytoremediation. Environmental Protection Agency, Cincinatti, p 70

Gildon A, Tinker PB (1983) Interactions of vesicular arbuscular mycorrhizal infection and heavy metals in plants. the effects of heavy metals on the development of vesicular-arbuscular mycorrhizas. New Phytol 95:247–261

Isikhuemhen OS et al (2011) The tropical white rot fungus, *Lentinussquarrosulus*Mont: lignocellolytic enzymes activities and sugar release from cornstalks under solid state fermentation. World J Microbiol Biotechnol 28(5):1961–1966

Jeffries P, Gianinazzi S, Perotto S, Turnau K, Barea JM (2003) The contribution of arbuscular mycorrhizal fungi in sustainable maintenance of plant health and soil fertility. Biol Fertil Soils 37:1–16

Kirk TK, Farrell RL (1987) Enzymatic combination–microbial degradation of lignin. Annul Rev Microbiol 41:465–505

Kulshreshtha S, Mathue N, Bhatnagar P (2014) Mushroom as a product and their role in Mycoremediation. AMB Express 4:29. https://doi.org/10.1186/s13568-014-0029-8 http://www.amb-express.com/content/4/1/29

Lamar RT, White RB (2001) Mycoremediation: commercial status and recent developments. In: Magar VS et al (eds) Proceedings of the sixth international symposium on in situ and on-site bioremediation, 6, San Diego, pp 263–278

Lang E (1995) Interaction of white rot fungi and micro-organisms leading to biodegradation of soil pollutants. In: Van de Brink WJ et al (eds) Proceedings of the 5th international FZK/TNo conference on contaminated soil, 30th October–5 November 1995, Maustrient, 95. Contaminated Soils, Netherlands, pp 1277–1278

Levyval C, Turnau K, Haselwandler K (1997) Effect of heavy metal pollution on mycorrhizal colonization and function: physiological, ecological and plied aspect. Mycorrhiza 7:139–153

Maiti D, Variar M, Singh RK (1996a) Perpetuation of native VAM fungi under monocropped, rainfed upland agro-ecosystem. Mycorrhiza News 8(1):7–9

Maiti D, Variar M, Singh RK, Singh CV, Saha J (1996b) Enhancing VA mycorrhizal association in rainfed upland rice (Oryza sativa L) based cropping system. Mycorrhiza News 8(4):11–14

Matsubara et al (2006) A simple screening procedure for selecting fungi with potential for use in the bioremediation of contaminated land. Enzym Microb Technol 39(7):1365–1372

Molina R, Massicotte H, Trappe JM (1992) Specificity phenomenon in mycorrhizal symbiosis: community-ecological and practical implications. In: Allen ME (ed) Mycorrhizal functioning; an integrative plant fungal process. Chapman and Hall, New York, pp 357–423

Okland RH, Eilertsen O (1993) Vegetation environment relationship of boreal coniferous forests in the Sdhomfjell area, Gjerstad, S. Norway. Sommerfeltia 16:1–254

Rhodes CJ (2012) Feeding and healing the world: through regenerative agriculture and permaculture. Sci Prog 95(4):345–446

Rhodes CJ (2013) Applications of bioremediation and phytoremediation. Sci Prog 96(4):417–427

Singh H (2006) Mycoremediation: fungal bioremediation. Wiley, Hoboken

Singh G et al (2014) Reoxidation of biogenic reduced uranium: a challenge toward bioremediation. Crit Rev Environ Sci Technol 44(4):391–415

Stamets P (2005) Mycelium running*: how mushrooms can help save the world.* Ten Speed Press, Berkley

Trappe JM (1987) Phylogenetic and ecologic aspects of mycotrophy in the angiosperms from an evolutionary standpoint. In: Safir GR (ed) Ecophysiology of VA mycorrhizal plants. CRC Press, Florida

Wang YP, Chao CC (1989) The relationship of heavy metals- mycorrhizal fungi. IV effects of addition of heavy metals on uptake of micronutrients by vesicular arbuscular mycorrhizal crops in sand culture. J Agric For 38(1):105–119

Chapter 15
Fungal Community in Mitigating Impacts of Drought in Plants

Richa Raghuwanshi

Abstract Drought is a serious and common threat to plant survival worldwide and is getting intensified due to global warming and decreasing water levels thereby possessing challenges on food security. The responses of plants to changing environment are complex. All adaption and acclimatization strategies are at physiological cost at the organism level which can affect ecosystem functioning at large. Mycorrhizae and endophytes are representative symbiotic association of the plants and fungi and are capable of modulating the physiological response of plants to water stress and overcome yield barrier. Water and nutrients available to plant are determined by the rhizospheric water potential which can be manifested to some extent by the AMF. The fungal community forming symbiosis with plants may exert their effect through phytohormones production, solubilization of nutrients, and induction of pathogen resistance or increasing abiotic stress tolerance through increased antioxidant levels in plants. While research supports the fungal endophytes and mycorrhizae as an ecofriendly alternative to combat drought stress, a better perceptive of physiological effects of these microbes to stress can develop a stronger and resilient agroecosystem.

Keywords Endophytes · Mycorrhizae · Drought stress

15.1 Introduction

Plant evolution and their geographic distribution have principally been governed by different abiotic environmental stresses that have adversely affected their growth and development. The major climatic factor which impacts plant growth is rainfall. The world's shrinking water supply and increasing food security issues are raising challenges to the scientific world to fill the gap of high food demand with low resources. Limited agricultural land and issues like water scarcity, decreased soil productivity due to enhanced soil degradation and decrease in soil water table and

R. Raghuwanshi (✉)
Department of Botany, MMV, Banaras Hindu University, Varanasi, India

P. Gehlot, J. Singh (eds.), *Fungi and their Role in Sustainable Development: Current Perspectives*, https://doi.org/10.1007/978-981-13-0393-7_15

nutrient depletion need serious thought. Water stress is the condition where the plant water potential and turgor gets decreased enough to inhibit normal plant function. The loss of farmable area because of abiotic stress straightforwardly influences the crop yield. India is amongst the most vulnerable water stress-inclined nations in the world. Of all the twentieth-century normal perils, water stress brought on the greatest negative effect. Drought/water stress is defined as "A period of inadequate or no rainfall over extended time creating soil moisture deficit and hydrological imbalances", and in context to plants, "water stress is the condition where plant water potential and turgor are decreased enough to inhibit normal plant function" (Hsiao et al. 1974).

The symbiotic association of plants with the fungal community was formed 400 million years ago during the Ordovician period when plants became terrestrial in nature (Pirozynski and Malloch 1975; Redecker 2000; Simon et al. 1993). Plants during the process of terrestrialization were able to adapt to a wide range of biotic and abiotic stress by forming a mutual symbiosis with microorganisms (Tanaka et al. 2006). Within plants, a huge diversity is seen in the microbial community which consists of archaea, bacteria, fungi and Protista (Hardoim et al. 2015). The fungal community has been harbouring plants growing in natural community forming a symbiotic association wherein both partners get benefited besides empowering plants to survive, grow and reproduce more efficiently even under stress conditions (Sylvia and Williams 1992). The plant roots and fungal symbionts have coevolved over 450 million (Smith and Read 2008) years and are highly interrelated in their growth-promoting activities. Mycorrhizal fungi are grouped under the phylum Glomeromycota and have been forming symbiotic association with root systems of 80% of terrestrial plants. Studies done by Francis and Read (1995), Johnson et al. (1997) and Graham and Eissenstat (1998) report the parasitic behaviour of AM fungi.

Not only the mycorrhizal symbiosis but also the mutual relationships among plants and endophytes in relation to stress tolerance have increased attention in the recent years (Mei and Flinn 2010). Researchers estimate that approximately 300,000 plant species harbour endophytic fungal species (Strobel and Daisy 2003), and there might be more than 1 million endophytes of fungi (Ganley et al. 2004) present globally. A high percent colonization of higher plants by endophytic fungi signifies it as a crucial component in fungal biodiversity. The lifestyle fungi exhibited in host depend upon the host genotype, its physiological status and environmental factors. Endophytes in both pathogenic and non-pathogenic form can be invariably isolated from healthy plant cells, signifying that the fungal community can live in either non-pathogenic lifestyles or can infect and remain silent in plants (Schulz et al. 1999).

The initial phase of plant-fungal interaction whether it is a pathogen or a symbiont is the same for most of the fungi, and the type of lifestyle exhibited by the fungi is a post-colonization phenomenon and is governed by genetic talk between the fungi and host. A fungal pathogen is observed to switch to a symbiotic lifestyle in genetically divergent host as changing the host from cucurbits to solanaceous species and even in many times in different cultivars of a species (e.g. tomato), revealing the importance of the dialogue between the fungi and host in directing the lifestyles exhibited (Redman et al. 2001).

The plant suppresses the plant immune response towards fungi by creating an oxidative burst which inactivates the defence response of plants, and the fungi thereby successfully establish a mutualistic relation with the plant as reported by Tanaka et al. (2006) and Nanda et al. (2010). Fungal endophytes help plants to adapt to a particular habitat and improve their performance and defence under stress conditions. They secrete various novel secondary metabolites that also include the volatile organic compounds.

The endophytes are protected in plants from desiccation and nutrients deprivation and are also many times vertically transmitted to the offspring (Schulz 2006; Aly et al. 2011) and easily spread in the vegetatively reproducing plants. Multifarious signalling is involved in plant symbioses with either the fungal community (mycorrhizae) or the bacterial community involving the *Agrobacterium* or *Rhizobium*, and it is apparent that adaptive symbiosis ensures a symbiotically conferred fitness benefits to plants. While global warming and changing climatic conditions have shifted the research interest at ecosystem dynamics, the influence of climate on microbial community and its effect on plant system need to be understood better. Although the complexities underlying in the switching of lifestyle by fungal community is less understood, they have been looked as novel agents in promoting stress tolerance in plants.

15.2 Systematic Position

Endophytes as defined by Hallmann et al. (1997) are 'Microorganisms which can be obtained from surface-sterilized plant cells or extracted from the plant and which do not show any pathogenic symptoms or are asymptomatic'. However, it does not take into account the endophytes that do not grow in vitro, and, second, it doesn't distinguish latent pathogens from endophytes which are important constituent of the microbiome community (Hardoim et al. 2015; Mercado-Blanco 2015; Card et al. 2016). To qualify a microorganism as an endophytic, it must be successfully reintroduced into disinfected plantlets resulting in pathogenicity and thereby fulfilling Koch's postulates. Fungal endophytes have been categorized (Rodriguez et al. 2009) in two major groups: one being the clavicipitaceous endophytes and the other being the non-clavicipitaceous endophytes. Clavicipitaceous fungi belong to the division Ascomycetes and Clavicipitaceae family. They usually infect grasses and weeds (Schardl et al. 2004). The clavicipitaceous endophytes were also referred as 'class 1 endophytes' (Rodriguez et al. 2009). The non-clavicipitaceous plant-infecting endophytes embody many other families of Ascomycetes. Endophytic fungi though being biotrophic do not form structures for nutrient absorption which enter host cells; rather they remain intimately attached to the plant cell wall and absorb the leaked nutrients from host tissues and develop intercellularly in plants. Clavicipitalean endophytes produce ROS (Torres et al. 2009) which modify host cell membranes and facilitating the leaking of minerals into the free apoplastic region which is absorbed by fungal hyphae.

Schussler et al. (2001) have placed AMF in phylum Glomeromycota based on small subunit rRNA gene sequences. The phylum Glomeromycota has three classes, viz. Glomeromycetes, Archaeosporomycetes and Paraglomeromycetesand five orders, viz. Glomerales, Diversisporales, Gigasporales, Paraglomeralesand Archaeosporales, with 14 families and 26 genera (Stürmer 2012). The ubiquitous genera of AMF include the *Glomus*, *Gigaspora*, *Entrophospora*, *Scutellospora* and *Acaulospora*.

15.3 Endophytes in Different Ecosystems

Fungal endophytes have been known to protect plants from stress, increase their resilience, and help plants to adapt to new habitats and therefore establishment of ecosystems (Schulz 2006; Strobel and Daisy 2003; Rodriguez et al. 2008; Friesen 2013). Endophytic fungi which are predominantly ascomycetous are ubiquitous and have been isolated from all groups of plants starting from the lower forms like mosses and ferns to the higher ones including the conifers and the angiospermic plants (Arnold 2007) evolved under varied ecosystems like the hot deserts in the arid regions, cold regions of the Arctic tundra, mangroves, forests lying in temperate regions and tropical regions, grasslands, savannahs and crop lands. The diversity of endophytic microorganisms has been found to decrease from the tropics as we move towards the northern boreal forests. The diverse spectrum of endophytes residing within a plant changes with space, time and function. Some fungal endophytes in ecosystems also commence the biological decay of a dead or senescing host plant, recycling the nutrient (Strobel and Daisy 2003; Vega et al. 2010; Aly et al. 2011; Boberg et al. 2011).

Trade-offs and links between plants and mycorrhizal community have been responsible for a wide range of processes in ecosystems. Mycorrhizal fungi have played a fundamental role in establishment of plants and ecosystems. They empowered plants with fitness under adverse biotic and abiotic stress conditions, primarily through improved nutrient cycling and modulating the soil architecture. Mycorrhizal symbiosis has come up as vital technique in improving the ability of plants to counteract drought and confer normal growth and promote sustainability not only of the agricultural fields but also of the degraded land areas and natural arid and semiarid areas, where multiple environmental stresses exist which hinder plant growth (Gianinazzi et al. 2010; Barea et al. 2011).

15.4 Role of Endophytes in Plant Growth Promotion and Drought Tolerance

Plant growth and development are negatively affected by a number of abiotic stress factors like nutrient limitation, drought, salination and altered soil pH values and temperature. While the biotic stresses are easy to overcome by human interventions, the abiotic stresses mainly being governed by a number of environmental factors acting together are difficult to overcome and are responsible for up to 50% yield loss of the foremost important crop plants (Bray et al. 2000). Drought, salinity and extreme temperatures, which are often interconnected, result in oxidative burst and cellular damage (Wang et al. 2003). For example, water stress and salinization, which are apparent osmotic stress, disrupt the ion distribution and metabolism in the cell (Zhu 2001a, b). Elevation in temperature leads to extensive denaturation and aggregation of proteins in the cell, which leads to cell death.

Endophytes exhibiting wide host benefits and potential to make over the productivity of agricultural crops in a sustainable manner are areas of current research. Global agricultural production is subjected to increasing environmental constraints predominantly drought which frequently combines with heat and salinity stress leaving a high magnitude of impact on plants. *P. indica* is known for plant growth promotion in various plants (Franken 2012). An extensive phytohormone signalling is involved in plant growth promotional phenomena, which lead to increased plant growth and biomass. Colonization of plants by *P. indica* shows salt tolerance, and this attribute is well demonstrated in tobacco (*Nicotiana*) and members of Poaceae family like barley (*Hordeum vulgare* L.), wheat (*Triticum*) and rice (*Oryza sativa*) which involve the detoxification of ROS by initiating production of antioxidants and enhancing the photosynthetic efficiency (Johnson et al. 2014). Under low temperature and higher nutrient input, *P. indica* triggers flowering earlier and increases grain yield indicating its potential as an effective endophyte promoting growth. Commercial product of *P. indica* is 'ROOTONIC' (trade name) and is currently under the field trials in India (Shrivastava and Verma 2014). Many endophytes isolated from wild varieties of barley (*H. murinum* sub sp. *murinum* L.) not only show benefits against biotic stresses like pest and disease resistance but also show increased grain yield under nutrient-deficient conditions (Murphy et al. 2015). Such endophytes can have great impact in reducing fertilizer inputs while maintaining yields. Fungal endophytes are a chief component of ecosystems that make their hosts compatible to thrive under adverse environments. Several studies have shown that endophytes may be profitable to plants harnessing in stressed conditions, like the presence of pollutants, heavy metals (Alquisira et al. 2017), organic compounds with chlorinated aromatic molecules (Becerra-Castro et al. 2013), high salinity (Aly et al. 2011), elevated temperature and drought (Sinha and Raghuwanshi 2016), by basically controlling the oxidative stress. Reports are well documented (Rodriguez et al. 2009) on enhanced tolerance of plants harbouring fungal endophytes towards water stress, ionic stress, soil temperature and resistance towards parasitic fungi and herbivores. With growing issues on climate change and its repercussions on agriculture, knowledge on fungal endophyte conferring drought tolerance has become gradually more important. By altering the plants physiological and biochemical responses to stress, which in turn affect the plant morphology and development, the fungal endophytes may induce drought avoidance, tolerance and/or recovery from drought in their hosts. Tolerance of the host *Festuca arundinacea* against a hydrocarbon mixture was enhanced by endophytic fungi *Lewia* (Cruz-Hernández et al. 2013). The endophytic fungi *Lewia* sp. belonging to Ascomycotina and family Pleosporaceae has been isolated from *Limonium tetragonum* and *Phragmites australis*, which grow in high salt conditions (Khalmuratova et al. 2015).

Less availability of water to the agricultural fields and consequent increase in soil salinization have become a growing challenge in the agricultural sector in many parts of the world (Egamberdieva et al. 2008; Egamberdieva and Lugtenberg 2014) and are expected to raise due to climate change effects. Mechanisms adopted by plants to combat water deficit conditions comprise of a complex series of biochemical network including osmotic regulation, formation of antioxidants and altered stomatal activity along with genetic modifications (Griffiths and Parry 2002). Endophytes which can even switch to pathogenic lifestyle like *Colletotrichum* species, too, confer drought tolerance in plants (Redman et al. 2001, 2002a, b). *Colletotrichum magna* and *C. protuberata* are well reported for water stress tolerance in wheat (*Triticum* sp.), tomato (*Solanum lycopersicum*) and watermelon (*Citrullus lanatus*) plants. This may also reflect the importance of fungal community in evolution of plants onto land ca. 400 million years ago when plant water relation was one of the most tricky stress to overcome, and fungal symbionts played a significant role in conferring drought tolerance in both monocots and dicots (Pirozynski and Malloch 1975). Tolerance to salt and drought stress can be enhanced in rice varieties unadapted to these stresses (Redman et al. 2011) by endophytic fungi. Plants inhabited by endophytic fungi had enhanced drought tolerance as they bargained water consumption less by 20–30% while promoting plant growth, yield and biomass. They have also been conferred to induce cold tolerance in plants grown in growth chambers (Redman et al. 2002a, b), although the mechanism needs to be worked out.

Drought stress in nature is often accompanied by heat stress. Plants cope with the increased temperature and heat stress by overexpression of heat-shock proteins and activation of antioxidant mechanism, which help in regulating the cell osmotic potential and integrity of membrane lipids (Iba 2002). Certain endophytes have a high degree of plasticity in moving and communicating between genetically distant plants in such a manner that both organisms are able to survive under varied environmental constrains under which either of them are unable to on their own. This strategy has provided endophytes a chance to evolve and expand their habitat range by processes like spreading of endophyte-inhabited plant parts (rhizomes, seed, seed coats). Symbionts adapted to a particular habitat have a wide host range and confer tolerance in plants and play an important role in plant establishment and resilience. The non-clavicipitaceous endophytes have been reported for providing tolerance to plants under habitat-specific selective pressures such as temperature, soil pH and salinity (Rodriguez et al. 2009). Grass species growing on coastal areas harbour symbiotic fungal endophytes which strengthen them against salinity and heat (Rodriguez et al. 2008). These endophytes have been reported to induce salt tolerance in several noncoastal plants. *Leymus mollis* (dunegrass), plants harbouring endophyte *Fusarium culmorum*, flourished on the coastal beaches of the USA, which are areas of high salinization. Salt-sensitive plants not growing in coastal areas were able to develop salt tolerance when inoculated with the endophyte *Fusarium culmorum* (Rodriguez et al. 2008). Similarly studies have demonstrated that endophytic fungi also aid heat tolerance. A good example of it is the *Dichanthelium lanuginosum* (panic grass), which grows at Yellowstone National Park with soil showing geothermal properties, which forms a mutualistic interaction with the fungal endophyte *Curvularia protuberate*, which bestow thermal tolerance (Redman et al. 2002a, b). A study conducted on 200 plants of *D. lanuginosum* plants reported the presence of symbiotic fungus *Curvularia protuberata* (Redman et al. 2002a, b) in them. None of the plant parts like roots, leaves and seed coats analysed were free of the endophyte. *D. lanuginosum* inhabiting geothermal soils where temperature may rise up to 57 °C grows as small clusters suggesting that the symbioses have played an adaptive role and *C. protuberata* contributes to the heat tolerance and growth of *D. lanuginosum.* Remarkable increase in heat tolerance was observed in plants which had endophytic fungi as they tolerated heat up to 65 °C and yet survived, whereas neither the endophyte nor the plant tolerated soil temperatures higher than 38 °C (M'arquez et al. 2007) alone. The effect however was not confined solely to the endophyte plant symbiosis as it was discovered that the endophyte harbours a virus and thereby a tripartite mutualism was maintained and the virus was an important component in bestowing heat tolerance to host plants, as the virus-free endophytes were heat-sensitive (M'arquez et al. 2007). This showed the significance of interactions among the microbiomes in plant evolution, and it can be looked forward in conferring abiotic stress tolerance in plants under the changing climatic conditions. Mostly these endophytes promote specific tolerance related to the habitat they are adapted to. As mostly the endophytes confer stress tolerance in plants irrespective of their habitat of isolation (Rodriguez et al. 2008), they can be exploited in the restoration programme of degraded lands and in arid and semiarid areas.

15.5 Endophyte Response Towards Oxidative Stress

Plant exposure to unfavourable environmental conditions can augment the reactive oxygen species production, e.g. $^{1}O_2$, $O_2^{\cdot -}$, H_2O_2 and $OH^{\cdot}$. The ROS are known to cause various destructive processes resulting in cellular damage as they interact with a number of cellular molecules and interfere with the normal cellular metabolic activities. Antioxidants are proved beneficial substances that increase plant tolerance and diminish the effects of biotic and abiotic stresses (Sharma and Dubey 2005). The antioxidant defence system in plants consists of enzymatic as well as non-enzymatic components such as low-molecular-mass antioxidants (ascorbate, carotenoids, glutathione) and enzymes like superoxide dismutase (SOD), peroxidise (POX), catalase (CAT) and ascorbate peroxidase (APX) that scavenge the ROS generated during stress (Apel and Hirt 2004). Endophytes enhance stress tolerance in plants by production antioxidant compounds (Malinowski and Belesky 2006; Yuan et al. 2009).

An increase in ROS has been observed after endophytic infection in response to which plants produce antioxidants. Studies on tall fescue, fine fescues and perennial ryegrass showed an increase in flavonoid content and antioxidants of phenolic nature (Herrera-Carillo et al. 2009; Malinowski et al. 2005). A positive correlation was found between the phenolic content and antioxidant activity of 292 fungal endophytes signifying the effect of endophytic antioxidants (Huang et al. 2007) like phenolic acids, tannins, flavonoids, hydroxyanthraquinones and terpenoids. *Pestalotiopsis microspora*, a fungal endophyte well known for digesting polyurethane, produces antioxidants, pestacin and isopestacin, which hunt the free radicals like superoxide, hydroxyl-free radicals, etc. (Strobel and Daisy 2003). Fungal endophytes besides producing the complex secondary metabolites also produce simple carbohydrate compounds like the fungal sugar alcohol mannitol which have good antioxidant activity (Jennings et al. 1998). A higher glucose and fructose level was reported in the apoplastic spaces of tall fescue plants inhabiting *Neotyphodium coenophialum* over the non-endophytic plants (Richardson et al. 1992). Similar observations were reported when the fungus *Alternaria alternata* exhibited an endophytic lifestyle in tobacco plant (Jennings et al. 1998). It has been speculated that endophyte antioxidants of all forms help in overcoming the overall oxidative burst and uplift the tolerance level of the host plants. Endophytes are therefore used to increase the defensive properties in crop plants by enhancing the antioxidants level which makes the plants tolerant to the oxidative burst created in plants by biotic and abiotic factors.

15.6 Arbuscular Mycorrhizal Fungi (AMF) in Plant Growth Promotion

The mycorrhizal fungus is dependent on the host as it fulfils its carbon requirement from the host plant (Schussler et al. 2001) and in return, provides nutritional benefits to the plant, thereby protecting host plants, against biotic and abiotic stress like drought (Augé 2001). AMF colonization promotes the host growth during adverse conditions like water stress by altering nutrient acquisition and hydration through increased water use efficiency (Simpson and Daft 1990). The intraradical mycelium of mycorrhizal fungi, residing in the root cortex of plants, helps in improving water relations, photosynthesis rates and drought responses. Several mechanisms are evolved by AMF to help plants survive in stressful conditions. These mechanisms include enhanced growth, prevention of nutrient deficiency and ion toxicity, osmotic adjustment, enhancing the activities of antioxidants, prevention of oxidative damage and improving photosynthesis and water status.

15.7 Enhanced Growth by Improved Root Architecture

AM symbiosis leads to augmented plant biomass and altered root-shoot: root length-leaf area ratios, enhanced plant growth and nutrient uptake ratios (Al-Karaki et al. 2004). AM formation results in an ecological niche where the roots become more accessible to water resources (Ruiz-Lozano 2003), as the fine fungal hyphae are capable of penetrating the soil pores which are otherwise out of the reach to the thick root hairs. Studies have also reported as altered root architecture in response to symbiosis which aids in improving the uptake of water at low level of soil moisture. Besides this the extraradical fungal hyphae is able to explore water and nutrients beyond the depletion zone.

15.8 Prevention of Nutrient Deficiency

AM symbiosis enhances the nutrients uptake like phosphate, nitrogen, sulphur or even more trace elements like copper and zinc (Tian et al. 2010). Nutrient mobility is declined under drought conditions. Mobility of phosphorous gets decreased in arid soils, and AM symbiosis becomes helpful in upgrading the P acquisition by plants and consequently improving the root growth which increases the water content of host plants (Augé et al. 2007). On the earth crust, most of the phosphorus is orthophosphate form, which is directly absorbed at the plant root epidermal cells and root hairs. Indirect entry into the plant is through the fungal-root interface by the external AM hyphae (Manoharachary 2001). AM colonization has been observed to be connected with the increase in activities of certain enzymes that help in hydrolysis and mobilization of nutrients. Often, P forms complexes with Ca and Mg rendering P unavailable for uptake. Higher acid phosphatase activity in mycorrhizosphere enables increased hydrolysis and mobilization of P resulting in higher uptake and has been positively correlated to soil water content (Chethan Kumar et al. 2008). Thus, as AM colonization improves the nutrient acquisition and the root architecture, which indirectly affects the soil architecture in the mycorrhizosphere, and thus the relation partly alleviates plant drought stress (Wu et al. 2011). Colonization by AMF enhances both the nutritional status and N assimilation rate of drought-stressed plant (Ruiz-Lozano 2003). Increased N assimilation may result from an immediate hyphal uptake of NO_3^- or NH_4^+ (Cardoso and Kuyper 2006) which may promote protein synthesis in AM plants over non-AM plants.

15.9 Soil Aggregates Formation

AMF hyphae play a basic role in the adjustment of soil complex aggregates by increasing the soil nutrient uptake by plants (Miller and Jastrow 1990). The exudation of extracellular polysaccharides like glomalin helps in entangling soil particles with the hyphae network (Treseder and Turner 2007). The polysaccharide glomalin, having binding property, adheres the soil particles, viz. silt, sand and clay together, which form soil aggregates (Miller and Jastrow 1990), thereby viewed as key for ecosystem functioning (Koide and Mosse 2004), because they play a crucial role in soil fertility, stability and biodiversity within plant communities.

15.10 Improved Water Status

The mechanism proposed for drought tolerance in AM plants has been credited to the increase in osmotic potential adjustments (Augé et al. 1986) by AM colonized plants due to increased water uptake, which further improves the stomatal conductivity and transpiration rate (Allen and Boosalis 1983) and is also supported by better mineral nutritional status particularly phosphorus (Hodge et al. 2010). Thus, overall the leaves are in a better hydrated state after AM colonization (Boomsma and Vyn 2008).

Besides this the other mechanisms proposed include the changes in plant vascular architecture which smoothen the plant hydraulic conductance and the non-hydraulic root signals wherein the altered cytokinin and auxin content (Boomsma and Vyn 2008) plays an important role.

15.11 Higher Photosynthetic Efficiency

AM plants perform higher rate of photosynthesis than their non-mycorrhizal counterparts, as AMF effect on stomatal conductance. AM symbiosis increases the units of photosynthesis (Ruiz-Sánchez et al. 2010) so as to increase the rates of photosynthetic storage and export in the plants (Augé 2001). An increase in chlorophyll content of AM plants has been observed compared to their non-mycorrhizal plants (Fan and Liu 2011). The higher photosynthetic rates linked with mycorrhization benefit the plants through an increase in concentration of soluble sugars and photosynthetic byproducts in the leaf symplasm (Porcel and Ruiz-Lozano 2004) which also act as osmoprotectant under water stress.

15.12 Osmotic Adjustment

Proline which is an important organic osmoprotectant is found to be alleviated in AM plants exposed to drought stress. Plant maintains better turgor via sugar accumulation in response to drought. Higher turgor in root helps in augmented root growth and elongation thereby enhancing nutrient water uptake (Studer et al. 2007).

15.13 Antioxidant Enzymes Production

AM symbiosis provokes a more potent oxidants scavenging system in host plants and minimizes the damage at the cellular level under water stress condition. AM symbiotic plants faced to water stress shows lower lipid peroxidation than non-mycorrhizal plants (Porcel and Ruiz-Lozano 2004). Porcel et al. (2003) stated that AMF protects host plants against oxidative damage by increments of enzymatic as well as non-enzymatic antioxidants. Increased enzymatic antioxidants, i.e. catalase (CAT) and peroxidase (POX), alleviating ROS breakage are some of the ways of protecting AM plant organisms under oxidative stress (Zhu et al. 2011).

15.14 Future Prospects

Fungal symbiosis, which has been playing a critical role in plant terrestrialization and evolution, can also play a critical role in meeting current global challenges, which embodies environmental deterioration, ecosystem conservation and efforts towards sustainable agriculture and food security. The modern breeding/domestication/selection programmes aimed at improving agricultural outputs under adverse conditions and low fertilization should also consider features like mycorrhization and the endophytic microflora during which are otherwise not considered during breeding. Unravelling the contribution of fungal community to the nutritional quality and drought resistance becomes a priority in the era of climate change and growing population. However, issues related to the fungal community still remains to be understood before they can be exploited in meeting the challenges of the growing population.

References

Al-Karaki G, Mc Michael B, Zak J (2004) Field response of wheat to arbuscular mycorrhizal fungi and drought stress. Mycorrhiza 14:263–269

Allen MF, Boosalis MG (1983) Effect of two species of VA mycorrhizal fungi on drought tolerance of winter wheat. New Phytol 93:67–76

Alquisira CM, Rojas MG, Márquez HG, Sepúlveda TV (2017) Improved growth and control of oxidative stress in plants of *Festuca arundinacea* exposed to hydrocarbons by the endophytic fungus *Lewia* sp. Plant Soil 411:347–358

Aly AH, Debbab A, Proksch P (2011) Fungal endophytes: unique plant inhabitants with great promises. Appl Microbiol Biot 90:1829–1845

Apel K, Hirt H (2004) Reactive oxygen species: metabolism, oxidative stress, and signal transduction. Ann Rev Plant Biol 55:373–399

Arnold AE (2007) Understanding the diversity of foliar endophytic fungi: progress, challenges, and frontiers. Fungal Biol Rev 21:51–66

Augé RM (2001) Water relations, drought and vesicular-arbuscular mycorrhizal symbiosis. Mycorrhiza 11:3–42

Augé RM, Schekel KA, Wample RL (1986) Osmotic adjustment in leaves of VA mycorrhizal non-mycorrhizal rose plants in response to drought stress. Plant Physiol 82:765–770

Augé RM, Toler HD, Moore JL, Cho K, Saxton AM (2007) Comparing contributions of soil versus root colonization to variations in stomatal behaviour and soil drying in mycorrhizal *Sorghum bicolor* and *Cucurbita pepo*. J Plant Physiol 164:1289–1299

Barea JM, Palenzuela J, Cornejo P, Sánchez-Castro I, Navarro-Fernández C, Lopéz-García A et al (2011) Ecological and functional roles of mycorrhizas in semi-arid ecosystems of Southeast Spain. J Arid Environ 75:1292–1301

Becerra-Castro C, Prieto-Fernández Á, Kidd P, Weyens N, Rodríguez-Garrido B, Touceda-González M, Acea MJ, Vangronsveld J (2013) Improving performance of *Cytisus striatus* on substrates contaminated with hexachlorocyclohexane (HCH) isomers using bacterial inoculants: developing a phytoremediation strategy. Plant Soil 362:247–260

Boberg JB, Ihrmark K, Lindahl BD (2011) Decomposing capacity of fungi commonly detected in *Pinus sylvestris* needle litter. Fungal Ecol 4:110–114

Boomsma CR, Vyn TJ (2008) Maize drought tolerance: potential improvements through arbuscular mycorrhizal symbiosis. Field Crop Res 108:14–31

Bray EA, Bailey-Serres J, Weretilnyk E (2000) Responses to abiotic stresses. In: Gruissem W, Buchannan B, Jones R (eds) Biochemistry and molecular biology of plants. American Society of Plant Physiologists, Rockville, pp 1158–1249

Card SD, Johnson LJ, Teasdale S et al (2016) Deciphering microbial behaviour—the link between endophyte biology and efficacious biological control agents. FEMS Microbiol Ecol 92. https://doi.org/10.1093/femsec/fiw114

Cardoso IM, Kuyper TW (2006) Mycorrhizas and tropical soil fertility. Agric Ecosyst Environ 116:72–84

Chethan Kumar KV, Chandrashekar KR, Lakshmipathy R (2008) Variation in arbuscular mycorrhizal fungi and phosphatase activity associated with *Sida cordifolia* in Karnataka. World J Agric Sci 4:770–774

Cruz-Hernández A, Tomasini-Campocosio A, Pérez-Flores LJ, Fernández-Perrino FJ, Gutiérrez-Rojas M (2013) Inoculation of seed-borne fungus in the rhizosphere of *Festuca arundinacea* promotes hydrocarbon removal and pyrene accumulation in roots. Plant Soil 362:261–270

Egamberdieva D, Lugtenberg B (2014) Use of plant growth-promoting rhizobacteria to alleviate salinity stress in plants. In: Miransari M (ed) Use of microbes for the alleviation of soil stresses, vol 1. Springer, New York, pp 73–96

Egamberdieva D, Kamilova F, Validov S et al (2008) High incidence of plant growth-stimulating bacteria associated with the rhizosphere of wheat grown on salinated soil in Uzbekistan. Environ Microbiol 10:1–9

Fan QJ, Liu JH (2011) Colonization with arbuscular mycorrhizal fungus affects growth, drought tolerance and expression of stress-responsive genes in *Poncirus trifoliata*. Acta Physiol Plant 33:1533–1542

Francis R, Read DJ (1995) Mutualism and antagonism in the mycorrhizal symbiosis, with special reference to impacts on plant community structure. Can J Bot 73:1301–1309

Franken P (2012) The plant strengthening root endophyte *Piriformospora indica*: potential application and the biology behind. Appl Microbiol Biotechnol 96:1455–1464

Friesen ML (2013) Microbially mediated plant functional traits. In: Molecular microbial ecology of the rhizosphere, vol 1. Wiley, Hoboken, pp 87–102

Ganley RJ, Brunsfeld SJ, Newcombe G (2004) A community of unknown, endophytic fungi in western white pine. Proc Natl Acad Sci U S A 101:10107–10112

Gianinazzi S, Gollotte A, Binet MN, van Tuinen D, Redecker D, Wipf D (2010) Agroecology: the key role of arbuscular mycorrhizas in ecosystem services. Mycorrhiza 20:519–530

Graham JH, Eissenstat DM (1998) Field demonstration of carbon cost of citrus mycorrhizas. New Phytol 140:103–110

Griffiths H, Parry MAJ (2002) Plants response to water stress. Ann Bot 89:801–802

Hallmann J, Quadt-Hallmann A, Mahaffee WF et al (1997) Bacterial endophytes in agricultural crops. Can J Microbiol 43:895–914

Hardoim PR, van Overbeek LS, Berg G et al (2015) The hidden world within plants: ecological and evolutionary considerations for defining functioning of microbial endophytes. Microbiol Mol Biol Rev 79:293–320

Herrera-Carillo Z, Torres MS, Singh AP, Vorsa N, Gianfagna T, Meyer W, White JF Jr (2009) Phenolic, flavonoid and antioxidant profiling for cool-season grasses with and without endophyte. In: Proceedings of the 18th annual Rutgers Turfgrass symposium, 12 January 2009, Rutgers University, New Brunswick, p 43

Hodge AT, Helgason T, Fitter AH (2010) Nutritional ecology of arbuscular mycorrhizal fungi. Fungal Ecol 3:267–273

Hsiao TC, Acevedo TC (1974) Plant responses to water deficits, water-use efficiency, and drought resistance. Agric Meteorol 14:59–84

Huang WY, Cai YZ, Xing J, Corke H, Sun M (2007) A potential antioxidant resource: endophytic fungi from medicinal plants. Econ Bot 61(1):14–30

Iba K (2002) Acclimative response to temperature stress in higher plants: approaches of gene engineering for temperature tolerance. Annu Rev Plant Biol 53:225–245

Jennings DD, Ehrenshaft M, Pharr DM, Williamson JD (1998) Roles for mannitol and mannitol dehydrogenase in active oxygen-mediated plant defense. Proc Natl Acad Sci U S A 95:15129–15133

Johnson NC, Graham JH, Smith FA (1997) Functioning of mycorrhizal associations along the mutualism-parasitism continuum. New Phytol 135:575–585

Johnson JM, Alex T, Oelmuller R (2014) *Piriformospora indica*: the versatile and multifunctional root endophytic fungus for enhanced yield and tolerance to biotic and abiotic stress in crop plants. J Trop Agric 52:103–122

Khalmuratova I, Kim H, Nam Y, Oh Y, Jeong M, Choi H, You Y, Lee I, Shin J, Yoon H, You YH (2015) Diversity and plant growth promoting capacity of endophytic fungi associated with halophytic plants from the coast of Korea. Microbiology:43–44

Koide RT, Mosse B (2004) A history of research on Arbuscular mycorrhiza. Mycorrhiza 14:145–163

M'arquez LM, Redman RS, Rodriguez RJ et al (2007) A virus in a fungus in a plant: three-way symbiosis required for thermal tolerance. Science 315:513–515

Malinowski MP, Belesky DP (2006) Ecological importance of *Neotyphodium* spp. grass endophytes in agroecosystems. Grassl Sci 52:1–14

Malinowski M, Beleski DP, Lewis GC (2005) Abiotic stresses in endophytic grasses. In: Robert C, West CP, Spiers DE (eds) Neotyphodiumin cool season grasses. Blackwell, Ames, pp 187–199

Manoharachary C (2001) Arbuscular mycorrhizal fungi and their SEM aspects. In: Reddy SM, Ram Reddy S, Singarachary MA, Girisham S (eds) Bioinoculants for sustainable agriculture and forestry. Proceeding of national symposium, India, Scientific Publishers, pp 37–42

Mei C, Flinn BS (2010) The use of beneficial microbial endophytes for plant biomass and stress tolerance improvement. Recent Pat Biotechnol 4(1):81–95

Mercado-Blanco J (2015) Life of microbes inside the plant. In: Lugtenberg B (ed) Principles of plant-microbe interactions-microbes for sustainable agriculture. Springer, Dordrecht, pp 25–32

Miller RM, Jastrow JD (1990) Hierarchy of root and mycorrhizal fungal interactions with soil aggregation. Soil Biol Biochem 22:579–584

Murphy BR, Doohan FM, Hodkinson TR (2015) Fungal root endophytes of a wild barley species increase yield in a nutrient-stressed barley cultivar. Symbiosis 65:1–7

Nanda AK, Andrio E, Marino D, Pauly N, Dunand C (2010) Reactive oxygen species during plant-microorganism early interactions. J Integr Plant Biol 52(2):195–204

Pirozynski KA, Malloch DW (1975) The origin of land plants: a matter of mycotrophism. Biosystems 6:153–164

Porcel R, Ruiz-Lozano JM (2004) Arbuscular mycorrhizal influence on leaf water potential, solute accumulation and oxidative stress in soybean plants subjected to drought stress. J Exp Bot 55:1743–1750

Porcel R, Barea JM, Ruiz-Lozano JM (2003) Antioxidant activities in mycorrhizal soybean plants under drought stress and their possible relationship to the process of nodule senescence. New Phytol 157:135–143

Redecker D (2000) Specific PCR primers to identify arbuscular mycorrhizal fungi within colonized roots. Mycorrhiza 10:73–80

Redman RS, Dunigan DD, Rodriguez RJ (2001) Fungal symbiosis: from mutualism to parasitism, who controls the outcome, host or invader? New Phytol 151:705–716
Redman RS, Rossinck MR, Maher S, Andrews QC, Schneider WL et al (2002a) Field performance of cucurbit and tomato plants infected with a nonpathogenic mutant of *Colletotrichum magna* (teleomorph: *Glomerella magna*; Jenkins and Winstead). Symbiosis 32:55–70
Redman RS, Sheehan KB, Stout RG et al (2002b) Thermotolerance generated by plant/fungal symbiosis. Science 298:1581
Redman RS, Kim YO, Woodward CJDA et al (2011) Increased fitness of rice plants to abiotic stress via habitat adapted symbiosis: a strategy for mitigating impacts of climate change. PLoS One e14823:6
Richardson MD, Chapman GW, Hovel CS, Bacon CW (1992) Sugar alcohols in endophyte-infected tall fescue. Crop Sci 32:1060–1061
Rodriguez RJ, Henson J, Van Volkenburgh E et al (2008) Stress tolerance in plants *via* habitat-adapted symbiosis. ISME J 2:404–416
Rodriguez RJ, White JF, Arnold AE et al (2009) Fungal endophytes: diversity and functional roles. New Phytol 182:314–330
Ruiz-Lozano JM (2003) Arbuscular mycorrhizal symbiosis and alleviation of osmotic stress, New perspectives for molecular studies. Mycorrhiza 13:309–317
Ruiz-Sánchez M, Aroca R, Muñoz Y, Armada E, Polón R, Ruiz-Lozano JM (2010) The arbuscular mycorrhizal symbiosis enhances the photosynthetic efficiency and the antioxidative response of rice plants subjected to drought stress. J Plant Physiol 167:862–869
Schardl CL, Leuchtmann A, Spiering M (2004) Symbiosis of grasses with seed borne fungal endophytes. Annu Rev Plant Biol 55:315–340
Schulz BJE (2006) Mutualistic interactions with fungal root endophytes. In: Schulz BJE, Boyle CJC, Sieber TN (eds) Microbial root endophytes, vol 9. Springer, Berlin/Heidelberg
Schulz B, Rommert AK, Dammann U, Aust HJ, Strack D (1999) The endophyte–host interaction: a balanced antagonism? Mycol Res 10:1275–1283
Schussler A, Schwarzott D, Walker C (2001) A new fungal phylum, the Glomeromycota: phylogeny and evolution. Mycol Res 105:1413–1421
Sharma P, Dubey RS (2005) Drought induces oxidative stress and enhances the activities of antioxidant enzyme in growing rice seedling. Plant Growth Regul 46:209–221
Shrivastava S, Verma A (2014) From *Piriformospora indica* to rootonic: a review. Afr J Microbial Res 8(32):2984–2992
Simon L, Bousquet J, Levesque RC, Lalonde M (1993) Origin and diversification of endomycorrhizal fungi and coincidence with vascular land plants. Nature 363:67–69
Simpson D, Daft MJ (1990) Interactions between water-stress and different mycorrhizal inocula on plant growth and mycorrhizal development in maize and sorghum. Plant Soil 121:179–186
Sinha S, Raghuwanshi R (2016) Synergistic effects of Arbuscular Mycorrhizal Fungi and mycorrhizal helper bacteria on physiological mechanism to tolerate drought in *Eclipta prostrata* (L.) L. J Pure Appl Microbiol 10(2):1117–1129
Smith SE, Read DJ (2008) Mycorrhizal symbiosis, 3rd edn. Elsevier, New York
Strobel G, Daisy B (2003) Bioprospecting for microbial endophytes and their natural products bioprospecting for microbial endophytes and their natural products. Microbiol Mol Biol Rev 67:491–502
Studer C, Hu Y, Schmidhalter U (2007) Evaluation of the differential osmotic adjustments between roots and leaves of maize seedlings with single or combined NPK-nutrient supply. Funct Plant Biol 34:228–236
Stürmer SL (2012) A history of the taxonomy and systematics of arbuscular mycorrhizal fungi belonging to the phylum Glomeromycota. Mycorrhiza 22:247–258
Sylvia DM, Williams SE (1992) Vesicular-arbuscular mycorrhizae and environmental stress. In: Beihlenfalvay GJ and Linderman RG (eds) Mycorrhizae in sustainable agriculture, Am Soc Agron special publication 54, American Society of Agronomy, Madison, 101–124

Tanaka A, Christensen MJ, Takemoto D et al (2006) Reactive oxygen species play a role in regulating a fungus-perennial rye grass mutualistic interaction. Plant Cell 18:1052–1066
Tian CJ, Kasiborski B, Koul R, Lammers PJ, Bücking H, Shachar-Hill Y (2010) Regulation of the nitrogen transfer pathway in the arbuscular mycorrhizal symbiosis: gene characterization and the coordination of expression with nitrogen flux. Plant Physiol 153:1175–1187
Torres MS, Singh AP, Shah S, Herrera-Carrillo Z, Gianfagna T, White JF Jr, Vorsa N (2009) LC–MS–MS identification and quantification of phenolics in symbiotic tall fescue. In: Proceedings of the 18th annual Rutgers Turfgrass symposium, 12 January 2009, Rutgers University, New Brunswick, p 54
Treseder KK, Turner K (2007) Glomalin in ecosystems. Soil Sci Soc Am J 71:1257–1266
Vega FE, Simpkins A, Aime MC et al (2010) Fungal endophyte diversity in coffee plants from Colombia, Hawai, Mexico and Puerto Rico. Fungal Ecol 3:122–138
Wang W, Vinocur B, Altman A (2003) Plant responses to drought, salinity and extreme temperatures: towards genetic engineering for stress tolerance. Planta 218:1–14
Wu QS, Zou YN, He XH (2011) Difference of hyphal and soil phosphatase activities in drought-stressed mycorrhizal trifoliate orange *(Poncirus trifoliata)* seedlings. Sci Hort 129:294–298
Yuan ZL, Zhang C, Lin F (2009) Role of diverse non-systemic fungal endophytes in plant performance and response to stress: progress and approaches. J Plant Growth Regul 29:9112–9119
Zhu JK (2001a) Plant salt tolerance. Trends Plant Sci 6:66–71
Zhu JK (2001b) Cell signaling under salt, water and cold stresses. Curr Opin Plant Biol 4:401–406
Zhu X, Song F, Liu S (2011) Arbuscular mycorrhiza impacts on drought stress of maize plants by lipid peroxidation, proline content and activity of antioxidant system. J Food Agric Environ 9:583–587

Chapter 16
Role of Fungi As Biocontrol Agents for the Control of Plant Diseases in Sustainable Agriculture

D. U. Gawai

Abstract Biological control is the process which decreases the inoculum density of the pathogenic microbes, present in dormant state by the other microbes. Generally, it involves either the naïve or genetically modified microbes which reduce the effect of pests, pathogen, and diseases. The plant disease is controlled by the pesticides, which are now extensively used. Due to excessive use of pesticides, socioeconomic and environmental pollution issues have been resulted, which demand the alternative method to reduce content of chemical pesticides. Biological control is an eco-friendly method employed to control the plant diseases, with the aim of developing a sustainable system in agriculture. Biological control mechanism involves the interaction among the antagonists and pathogens, which aid in selection and manipulation to develop an effective control system. Currently, this approach is employed when no other alternative is available. Emergent of fungal antagonistic has made it a promising biological control strategy to control the plant diseases. The major factors which hinder the efficiency of the biocontrol agents to control the plant diseases need to be considered during the formulation of biocontrol procedure, biocontrol agent, and its application time.

Keywords Biocontrol · Fungal antagonists · Sustainable agriculture · Disease control

16.1 Introduction

Biological control is the process which reduces the number of microbes or pathogens by other microorganisms, without the external intervention of humans (Cook and Baker 1983). In 1967, Beirner stated that biological control is the controlling of one microbe by the other. This can be stated either as a large community of pest (DeBach 1964) or as an inhibitor of severe pest damage irrespective of the pest inhabitants (Cook and Baker 1983). It largely depends on the understanding of

D. U. Gawai (✉)
Botany Research Laboratory and Plant disease Clinic, PG Department of Botany, Science College Nanded, Nanded, Maharashtra, India

P. Gehlot, J. Singh (eds.), *Fungi and their Role in Sustainable Development: Current Perspectives*, https://doi.org/10.1007/978-981-13-0393-7_16

biological interactions with ecosystem and microbes up to cellular and molecular level, which are more complex and difficult to control in comparison to physical as well as chemical methods. Additionally, this method is a highly stable and long-lasting process (Baker and Cook 1974). This method is claimed to be a valuable alternative approach to regulate the plant diseases, in which one microbe inhibits the proliferation and infection caused by the other microbes (Cook 1993; Baker 1987). Being an eco-friendly approach, thus in few cases, this approach is predominantly used to save the plant from pathogenic microbes (Cook 1993). The approach employs natural predators which have the ability to eradicate and control the growth of pest as well as pathogens. This method explores the antagonism potential of microbes, which makes it an eco-friendly approach to control the plant disease. Thus, by calculating the cost of hazardous pesticides and other chemical agents, biological control is an efficient and eco-friendly method for controlling plant diseases, which can be used worldwide.

16.2 Significance of Biological Control

Biological control provides protection to the plant throughout its cultivation period. The biological agents proliferate rapidly in soil and leave no residue. Being non-toxic, it is safer for humans and plants. This approach is not limited for controlling the disease; additionally, it also enhances the growth (especially root) and yield of the crop. Due to easy handling and manufacturing, it can be used in combination with bio-fertilizers. Moreover, it is a cheap, safe, and eco-friendly method.

16.3 Importance of Biological Control

Chemical pesticides were used to enhance crop yield, but extensive use affects the nontargeted organism and surrounding environment. Thus, the current scenario demands the eco-friendly approach for controlling the pest, as chemical pesticides being not suitable for cultivation of crop. Bacteria, fungi, nematodes, protozoans, and virus have been extensively studied because of advantageous characteristics. Overexploitation of fungicides has resulted in gathering of the toxic molecules which are harmful to the environment and humans, but pathogenic microbes have adapted themselves by getting resistant to it. In order to overcome this global problem related to chemical control, alternative approaches are being exploited. Additionally, this biological control approach is highly effective for sustainable agriculture and is a vital component of integrated pest management (IPM) program.

16.4 Need for Biological Control in India

In order to meet the demand of the growing population, the yield of crop should increase to 250 million tonnes by 2020. The extensive use of fertilizer and pesticides has resulted in environmental pollution (especially, soil pollution). Over-usage of these agrochemicals and rumors created by the pesticide rivals have significantly reformed the attitude of consumers to use pesticides in their agricultural land. Controlling the large proportion of pest and disease has elevated the usage of these hazardous chemicals for proper management. Generation of resistant against fungicide and pesticide is emerging as new problem. Thus, there is a need to employ eco-friendly pesticides as they are less toxic and have low residual problem and low level of resistance. Thus, biological control approach should be used in collaboration, as efficiency of one approach varies with time, location, and environmental conditions.

16.5 Microbial Biocontrol Agents

Currently, different biocontrol agents are available, where *Agrobacterium* sp., *Bacillus* spp., and *Pseudomonas* spp. are bacterial agents and *Aspergillus* spp., *Ampelomyces* sp., *Candida* sp., *Coniothyrium* sp., *Gliocladium* sp., and *Trichoderma* spp. are fungal agents (Papavizas 1985; Koumoutsi et al. 2004; Mavrodi et al. 2002; Atehnkeng et al. 2008; Gilardi et al. 2008). Among them, the most versatile fungal agent belongs to *Trichoderma* sp. for controlling the growth of pathogenic fungi. In a previous study, conidial suspension of *Trichoderma* sp. was found to be effective against *Sclerotium rolfsii* causing the damping-off, root rot, and seed rot disease in mung bean and sunflower, and moreover, it also increased the plant growth (Yaqub and Shahzad 2008), whereas *Bacillus cereus* is used as a biocontrol agent for preventing the damping-off and root rot, which is also a food toxicant and is related to *B. anthracis*, causative agent of anthrax, a major biowar threat. Numerous studies have claimed *Trichoderma* spp. to be a valuable biocontrol agent for controlling the plant disease. Presently, commercial *Trichoderma* products are used as biopesticides which amend the soil and increase the plant growth (Papavizas 1985; Chet 1987; Harman et al. 2004; Vinale et al. 2008). In 1934, Weindling showed the biocontrol potential of *Trichoderma lignorum* (viride) against *Rhizoctonia solani*, a fungal pathogen. Further, *Trichoderma lignorum* (viride) also showed mycoparasitic activity against *Phytophthora*, *Pythium*, *Rhizopus*, and *Sclerotium rolfsii* (Wells 1988). Current fungal biocontrol agents are founded on the observation of *Trichoderma* spp., which has drawn the attention on fungus, as a biocontrol model (Chet et al. 1981, Chet 1993).

16.6 Efficacy of Microbial Biocontrol Agents

In addition to properties discussed above, there are few amendments which enhance the efficiency of this method. First, inappropriate usage of this technique should be prevented, which is mostly because of improper knowledge. Second, one should be able differentiate failure which is cause by low-quality inoculum. Moreover, inefficacy occurs because the compost/fertilizers containing biocontrol agents are not of superior quality as available in registered plant products. To improve the efficiency of the biocontrol agents, the strain should be assessed and verified against the targeted disease plus optimum condition should also be noted. Specific substrates and carriers also aid in enhancing the efficacy of the agents. Exploration of effective strains will also improve the quality of biocontrol agents and lessen the required amount.

16.7 Mass Production of Biocontrol Agents

Mass production of the biocontrol agents is required to meet the commercial demand. There is no effective method for the mass production of these biocontrol agents at industrial level, as the production of these biocontrol agents requires continuous resource which should be readily available. *Trichoderma* spp. have been reported to grow on various solid substrates such as coffee husk, saw dust, sorghum grain, waste of tea leaf, wheat grain and bran, etc. In 2004, Zaidi and Singh inoculated *Trichoderma harzianum* over the presoaked and sterilized Jhangora seed powder for 12 days. After this, the powder was diluted with talcum powder comprising 1% CMC (carboxymethylcellulose) for obtaining the biocontrol agent of desired concentration, which can be commercialized.

16.8 Commercial Products of Biocontrol Agents

Commercially available biocontrol products which control the plant disease are a new prospect. But it started in 1979, when *Agrobacterium radiobacter* strain K 84 was enlisted in EPA (United States Environmental Protection Agency) list for controlling crown gall disease in plant. Later on, *Trichoderma harzianum* ATCC 20476, the first fungal strain, was enlisted in EPA list for controlling the plant diseases. Presently, 14 bacteria and 12 fungi strains have been recorded by EPA which aid in controlling the plant disease (Fravel 2005). The majority of these biocontrol agents are commercially marketed (Table 16.1).

Table 16.1 List of crop diseases controlled by various biocontrol agents

Crop disease	Pathogen	Biocontrol agents
Blight of *Sesamum*	*Phytophthora* sp.	*T. harzianum*
		T. viride
Root rot of *Sesamum*	*M. phaseolina*	*Trichoderma* sp.
		Gliocladium sp.
Root rot chilli	*S. rolfsii*	*T. harzianum*
Dieback of chilli	*Colletotrichum capsici*	*T. viride*
		T. harzianum
Wilt of eggplant	*F. solani*	*T. viride*
		T. koningii
Damping-off of eggplant	*P. aphanidermatum*	*T. viride*
		T. koningii
Wilt of tomato	*F. oxysporum*	*T. harzianum*
	f.sp. lycopersici	
Root knot of tomato	*Meloidogyne incognita*	*T. harzianum*
	M. javanica	
Wilt of okra	*Pythium* spp.	*A. niger*
Leaf blight of sunflower	*Alternaria helianthi*	*T. virens*
Wilt of pigeon pea	*Fusarium udum*	*T. viride*
		T. hamatum
		T. harzianum
		T. koningii
Wilt of chickpea	*F. oxysporum*	*T. viride*
	f.sp. ciceri	*T. harzianum*
		T. virens
Dry root rot of soybean	*M. phaseolina*	*T. viride*
		T. harzianum
Stem rot of groundnut	*Sclerotium rolfsii*	*T. harzianum*
Damping-off of mustard	*Pythium aphanidermatum*	*T. harzianum*
		T. viride
Root rot of mung bean	*M. phaseolina*	*T. harzianum*
		T. viride

16.9 Advantages of Biological Control

Biological control is an eco-friendly approach, as it is nontoxic to plants and a non-targeted microbe, decreases the pesticide accumulation in food, regulates the activity of natural predators, and increases the microbial diversity in managed system. This process is less prominent but more stable and long-lasting, in comparison to physical and chemical controls (Baker and Cook 1974). Some of the advantages of biological controls are listed below:

16.9.1 Biocontrol Agents Are Host Specific

These agents are highly specific to host and pathogen, which save the beneficial microbes.

16.9.2 Nontoxic to Plants

As the agents do not liberate the toxic material, they are safe and nontoxic against other forms of life.

16.9.3 Application by Conventional Methods

The normal method of dusting and spraying can be used with conventional equipment's because of the versatile nature of the pathogens.

16.9.4 Ability to Multiply in Their Target Host

These agents possess the ability to proliferate within the host system.

16.9.5 Production Technology Available

Up to cottage scale, the mass multiplication of biocontrol agents can be performed by the conventional approaches.

16.10 Disadvantages of Biocontrol Agents

16.10.1 High Cost of Production

The major limiting factor of this approach is its cost as its production and registration are highly expensive in comparison to the chemical agents.

16.10.2 Additional Control Measures

Different control measures are needed to be taken against the nontargeted microbes.

16.10.3 Time of Application

The time of application is critical to be determined as it is difficult to judge the specificity of age and stage of targeted pathogen.

16.10.4 Mortality

Environmental factors aid in the high mortality rate as biocontrol agents take time to generate response against the targeted microbes, which is largely dependent on its incubation time.

16.10.5 Viability

Viability of the biocontrol agents needs to be maintained as they have short shelf life. Small change in temperature amends the microbial concentration. The exposure of UV rays from sunlight decreases virulence of the agents.

16.10.6 Difficulty in Mass Production

The mass production of the whole microbe in in vivo condition is an expensive and laborious job.

16.10.7 Legal Protection

The registration of data is an expensive procedure, and further it doesn't provide any legal protection.

16.11 Future Prospects

Biological control method is believed to be an effective method for controlling pest and increasing the crop yield. Although biological control is naturally taking place for centuries, this industrially regulated biological control is in its early stages. This approach emerged as an effective method as it is recorded to be safe, and this time is considered to be a period of modern biocontrol (Waage and Greathead 1988). On assessing the current scenario of plant yield, biocontrol approach has showed the high potency, but still it needs to be exploited more. The experiments are limited to laboratory, very less investigation has been carried out for formulating the commercial biological agents. Moreover, this approach is not commercially used because of lack of awareness among the farmers. So, there is a need to provide awareness about this concept along with its benefit. This approach requires more extensive research, as in vitro result is not reproduced when conducted in natural conditions. Physiological and ecological factors are considered to be responsible for hindering the bioagent's efficacy. Advancement in molecular technique is the available option for enhancing the potency of these biocontrol agents. Different methods have been employed for augmenting the efficiency of these biocontrol agents, which involves mutation and protoplasm fusion via chemical agents. However, there is a need for mass production of these biological agents, whose action mechanisms require deep understanding, plus environmental factors which aid in the fast growth of these biocontrol agents. The main challenge is the cost associated with the formulation and safety assessment during the production of commercial biocontrol agents. If these agents are applied to a large piece of land, the selection pressure of these agents decreases, and it shortens the shelf life of the product. Advancement in agriculture process has introduced new challenges which need to be overcome promptly, to meet the future demand. Modern techniques used worldwide have increased the crop yield including India, but widespread use of chemical pesticide and fertilizer has emerged as a major environmental issue. Therefore, biocontrol approach has emerged as the promising alternative approach, which can aid in achieving the agricultural goals. Development in this field will not only provide and ensure an alternative during failure but also facilitate a sustainable management system.

References

Atehnkeng J, Ojiambo PS, Ikotum T, Sikora RA, Cotty PJ, Bandyopadhyay R (2008) Evaluation of atoxigenic isolates of *Aspergillus flavus* as potential biocontrol agents for aflatoxin in maize. Food Addit Contam Part A 25:1266–1273

Baker KF (1987) Evolving concepts of biological control of plant pathogens. Annu Rev Phytopathol 25:67–85

Baker KF, Cook RJ (1974) Biological control of plant pathogens. W. H. Freeman and Co, San Francisco 433 pp. (Book, reprinted in 1982, Amer. Phytopathol. Soc., St. Paul, Minnesota)

Beirner BP (1967) Biological control and its potential. World Rev Pest Control 6(1):7–20

Chet I (1987) Trichoderma – application, mode of action and potential as a biocontrol agent of soil borne plant pathogenic fungi. In: Chet I (ed) Innovative approaches to plant disease control. Wiley, New York, pp 137–160

Chet I (1993) Biological control of soil-borne plant pathogens with fungal antagonists in combination with soil treatments. In: Hornby D (ed) Biological control of soil borne plant pathogens. CABI Publishers, London, p 15

Chet I, Harman GE, Baker R (1981) *Trichoderma hamatum*: its hyphal interaction with *Rhizoctonia solani* and *Pythium* spp. Microbial Biol 7:29–38

Cook RJ (1993) Making greater use of introduced microorganisms for biological control of plant pathogens. Annu Rev Phytopathol 31:53–80

Cook RJ, Baker KF (1983) The nature and practice of biological control of plant pathogens. American Phytopathological Society, St. Paul 539 pp

DeBach P (1964) Biological control of insect pests and weeds. Reinhold, New York, p 844

Fravel DR (2005) Commercialization and implementation of bio control. Annu Rev Phytopathol 43:337–359

Gilardi G, Manker DC, Garibaddi A, Gullino ML (2008) Efficacy of the biocontrol agents *Bacillus subtilis* and *Ampelomyces quisqualis* applied in combination with fungicides against powdery mildew of Zucchini. J Plant Dis Prot 115:208–213

Harman GE, Howell CR, Viterbo A, Chet I, Lorito M (2004) *Trichoderma* species-opportunistic, avirulent plant symbionts. Nat Rev Microbiol 2:43–56

Koumoutsi A, Chen XH, Henne A, Liesegang H, Hitzeroth G, Franhe P, Vater J, Borris R (2004) Structural and functional characterization of gene clusters directing nonribosomal synthesis of bioactive cyclic lipopeptides in Bacillus amyloliquefaciens strain FZB42. J Bactriol 186:1084–1096

Mavrodi DV, Mavrodi OV, McSpaddenss-Gardener BB, Landa BB, Weller DM, Thomashow LS (2002) Identification of differences in genome content among phID-positive Pseudomonas fluorescens strains by using PCR based substractive hybridization. Appl Environ Microbiol 68:5170–5776

Papavizas GC (1985) Trichoderma and gliocladium: biology, ecology, and potential for biocontrol. Ann Rev Phytopathol 23:923

Vinale FK, Sivasithamparam LE, Ghisalberti R, Marra LS, Lorito M (2008) *Trichoderma*-plant-pathogen interactions. Soil Biol Biochem 40:1–10

Waage JK, Greathead DJ (1988) Biological control: challenges and opportunities. Philos Transac Royal Soc B Biol Sci 318:1189

Weindling R (1934) Studies on lethal principle effective in the parasitic action of *Trichoderma lignorum* on *Rhizoctonia solani* and other soil fungi. Phytopathology 24:1153–1179

Wells DH (1988) Trichoderma as a biocontrol agent. In: Mukerji KG, Garg KL (eds) Biocontrol and plant diseases. CRC Press, Florida, p 73

Yaqub F, Shahzad S (2008) Effect of seed pelleting with *Trichoderma* spp., and *Gliocladium virens* on growth and colonization of roots of sunflower and mung bean by *Sclerotium rolfsii*. Pak J Bot 40:947–963

Zaid WN, Singh US (2004) Development of improved technology for the mass multiplication and delivery of fungal (*Trichoderma*) and bacterial (*Pseudomonas*) bio agents. J Mycol Plant Pathol 34:732–745

Chapter 17
Fungal Disease Management in Chickpea: Current Status and Future Prospects

Sarvjeet Kukreja, Neha Salaria, Kajal Thakur, and Umesh Goutam

Abstract Chickpea (*Cicer arietinum* L.) is one of the most important leguminous crops grown predominantly in tropical and temperate areas. The beneficial effects of chickpea on soil health and human health are well recognized. The area under chickpea production in India is 9.6 million ha with an average production of 8.8 million tons. Yield of chickpea is largely affected by exposure to both abiotic and biotic stresses. It has been observed that insects and diseases cause around 50–100% yield loss in chickpea in temperate regions. Among various diseases caused by a variety of pathogens, fungal diseases have shown to have devastating effects on the chickpea yield. Among various fungal diseases, the diseases caused by *Fusarium oxysporum* (fusarium wilt) and *Ascochyta rabiei* (Ascochyta blight) are the most common root and foliar diseases, respectively, causing severe loss to crop yield. Identification of pathogen, variation in genome and pathology, development of disease-resistant varieties, and organizing the resistant genes in different regions play an important role in fungal disease management. Implication of conventional methods is both time-consuming and slow process. Development of new genomic tools and resources aid the employment of genomics selection in improvement of chickpea. This will provide greater insight to the breeding program to work with high precision and accuracy to develop resistant chickpea cultivars.

Keywords Chickpea · Fungal disease · Disease managements

S. Kukreja
MRM PG College, Ganganagar, Rajasthan, India

Lovely Professional University, Phagwara, Punjab, India

N. Salaria · K. Thakur · U. Goutam (✉)
Lovely Professional University, Phagwara, Punjab, India

P. Gehlot, J. Singh (eds.), *Fungi and their Role in Sustainable Development: Current Perspectives*, https://doi.org/10.1007/978-981-13-0393-7_17

17.1 Introduction

Chickpea (*Cicer arietinum* L.) ranks third among food legume production in the world. According to FAOSTAT (2013), the global chickpea area cultivated was 13.5 million ha with production of 13.1 million tons. More than 54 countries are chickpea producing, but major part is contributed by the developing countries. About 70% of world chickpea production is contributed by India only, making it the largest chickpea producer. Net chickpea production in India is 8.8 million tons covering an area of 9.6 million ha.

Chickpea is known to be attacked by a variety of pathogens, viz., fungus, bacteria, viruses, mycoplasma, and nematodes. More than 172 pathogens infecting chickpea have been reported worldwide (Nene et al. 1996). Among these pathogens, fungal species are predominantly associated with the major chickpea diseases. Fungal diseases can be hazardous due to production of persistent mycotoxins. Therefore, inhibiting the fungal growth is necessary to combat the fungal infection (Yan et al. 2015). Major fungal diseases leading to high yield loss worldwide include Ascochyta blight, fusarium wilt, dry root rot, Botrytis gray mold, and wet root rot. Other minor diseases with local economical importance having limited distribution may become significant with the changing cultural practices.

Conventional methods for disease management are time-consuming and involve slow process. Therefore, there is a need to use new approaches such as genomics-assisted breeding for improvement of chickpea. Genomics of chickpea is currently in initial and formative stages; however, the basis for rapid development of genomic tools is present and progressing rapidly. Genetic reference maps are available and are based on both interspecific and intraspecific crosses. The general lack of polymorphism within the cultivated species initially hindered map development, but the development and use of SSRs and other codominant types of markers have generally overcome that obstacle. Recently, Cevik et al. (2015) explored the genetic relationships among 23 cultivated chickpea and 2 genotypes of *Cicer reticulatum* using simple sequence repeat (SSR) markers. Genotypes of *C. reticulatum* were found to be different from the cultivated chickpea. ICC 4958, "microsperma," or desi chickpea was found to be the closest cultivar to *C. reticulatum* genetically. Aggarwal et al. (2015) studied the genetic diversity in 125 in Indian chickpea cultivars (42 were resistant and 13 were susceptible) to fusarium wilt and Ascochyta blight using 40 ISSR primers. Genetic diversity revealed more variability among miscellaneous cultivars in comparison to resistant and susceptible cultivars. Genetic variability of cultivars can be increased by exploiting the available diverse germplasm. Genetic analysis of chickpea can give important indications for understanding species relationships and may help in developing resistant cultivars (Table 17.1).

Table 17.1 Pathogen distribution and QTLs identified for resistance to chickpea fungal diseases

Trait	Pathogen	Distribution	Name of population	Markers/QTL(s)	References
Ascochyta blight	*Ascochyta rabiei*	Australia, Canada, Latin America, Southern Europe, the United States, and West Asia	FLIP84-92C/PI 599072	UBC733b, UBC181a, Dia4	Santra et al. (2000)
			Lasseter/ICC1 2004	TS45, TA146, TA130	Flandez-Galvez et al. (2003b)
			Lasseter/ PI527930	CS5b650, GA2, OPB17c560	Collard et al. (2003)
			ILC 1272/ILC 3279	Ta20, TA72, ar1	Udupa and Baum (2003)
			PI 359075/ FLIP84-92C	GA16, GA24, GAA47, Ta46	Cho et al. (2004)
			Cr5-10/ILC72	OPAI09746, UBC881621	Cobos et al. (2006)
			LC3279/WR315	TA194	Iruela et al. (2007)
			ICCV96029 9/ CDC Frontier	TA64, TS54, TA176	Taran et al. (2007)
			ICCV96029/ CDC Luna	TR19, TS54	Anbessa et al. (2009)
			ICCV 96029/ CDC Corinne	TA132, TS45	
			ICCV 96029/ Amit	TA64	
			ICC 12004/ Bivanij	TA125, TA72, GA26	Kanouni et al. (2009)
			C 214/ILC 3279	AB-Q-SR-4-1, AB-Q-SR-4-2, AB-Q-APR-6-1, AB-Q-APR-6-2, AB-Q-APR-4-1, AB-Q-APR-5B	Sabbavarapu et al. (2013)
			Lasseter/ ICC3996	SNP-40000185	Stephens et al. (2014)
			S95362 9/ Howzat	TA146, TA72	

(continued)

Table 17.1 (continued)

Trait	Pathogen	Distribution	Name of population	Markers/QTL(s)	References
Fusarium wilt	*Fusarium oxysporum*	India, Iran, Peru, Syria, Ethiopia, Mexico, Spain, Tunisia, Turkey, and the United States	WR-315/C-104	CS-27, UBC170	Tullu (1996)
			ICC-4958 9/ PI498777	CS-27, UBC-855	Ratnaparkhe et al. (1998)
			JG-62 9/ Surutato-77	CS-27	Tullu et al. (1999)
			ICC-4958 9/PI 498777	CS-27, UBC-170	Tekeoglu et al. (2000)
			ICC-4958 9/PI 498777	CS27, TA96, TA27	Winter et al. (2000)
			CA2156 9/JG62	OPJ 20600	Rubio et al. (2003)
			WR-315/C-104	TA96, CS27A	Sharma et al. (2004)
			CA2139/JG62	OPJ20(600), TR59	Cobos et al. (2005)
			JG62 9/Vijay	TA110, TA96, H1B06y	Gowda et al. (2009)
			C 214/WR 315	FW-Q-APR-6-1, FW-Q-APR-6-2	Sabbavarapu et al. (2013)
			K850/WR315	GSSR 18-TC14801	Jingade and Ravikumar (2015)
Botrytis gray mold	*Botrytis cinerea*	India, Bangladesh, Nepal, Pakistan, and Australia	JG62 9/ICCV2	SA14, TA25, TA159	Anuradha et al. (2011)
Dry root rot	*Rhizoctonia bataticola*	Australia, Ethiopia, Iran, Pakistan, Bangladesh, and Nepal			Singh and Sharma (2002)
Wet root rot	*R. solani*	India, the Philippines, Sierra Leone, and Malaysia			Nene (1979)

17.2 Ascochyta Blight

Ascochyta blight (AB) is caused by *Ascochyta rabiei* (Pass.) Labrousse and is the major constraint in chickpea production worldwide. The disease has been reported in major chickpea-growing countries (Kaiser et al. 2000; Pande et al. 2005). In 1911, the first blight epidemic occurred in former Punjab province of British India (Kaiser et al. 2000). Since then, researchers have been focusing on the pathogen and its biology, spread, survival, and control. Chickpea is infected specifically by *A.*

rabiei, but other related pathogens also attack lentil and pea. Alternative hosts, such as pea, alfalfa, etc., are important for survival of pathogen as they are rarely attacked and the infections remain latent (Wiese et al. 1991). The pathogen can lead to heavy losses up to 100% in favorable environmental conditions by reducing the seed yields and quality (Duzdemir et al. 2014). Many reports suggest the significant economic losses in different geographical regions including Australia, Canada, America, and West Asia due to Ascochyta blight (Pande et al. 2005).

17.2.1 Symptoms and Disease Cycle

Plants can be attacked at any growth stage by *A. rabiei*, but its flowering to early podding stages is most susceptible to the infection. Symptoms of blight develop on aerial parts of the plant. When the infection is seed borne, the seedlings show presence of brown necrotic lesions at the stem. Increases in the size of lesion causes stem girdling. Due to this girdling stem breaks and causes plant death. Necrotic lesions act as a site for production of pycnidium. The shape of lesion varies from circular (leaves and pods) to elongated (petiole, stem, and branches).

Younger leaves are targeted by airborne conidia and ascospores which produce water-soaked necrotic spots. Splash and waterborne conidia disperse and spread the infection to foliage tissue on the same or neighboring plants (Fig. 17.1). Rapid spread of infection to aerial parts leads to death of the plant. Pod infection via testa and cotyledons affects the seed quality. Infected seeds with cankers and pycnidia appear discolored and shriveled (Pande et al. 2005).

17.2.2 Management of Ascochyta Blight (AB)

Cultural practices have proved to be effective in plant disease control as it reduces the inoculums sources. Planting disease-free seed, crop rotation, removing stubble, and sowing seeds deep into the soil are the cultural practices adopted for reducing AB epidemic. Potassium fertilizers also have the potency to enhance yield and retard AB infection in soils with high nitrogen content (Kader et al. 1990). In Australia, foliar fungicides such as chlorothalonil and mancozeb provided AB resistance in susceptible cultivars (Bretag et al. 2003). Propineb, Bordeaux mixture, chlorothalonil, zineb, dithianon, propiconazole, penconazole, sulfur, and thiabendazole are the foliar fungicides being used in management of AB. Secondary spread of AB infection can be controlled by applying fungicides on the infected crop (Pande et al. 2005). According to Reddy and Singh (1990), the application of fungicide before flowering provided the best protection against AB. Fungicides have proved as effective AB control agents, but the requirement of repeated application makes them uneconomical.

Marker-assisted selection or marker-aided selection (MAS) aims at selection of a genetic trait of interest such as productivity, disease resistance, etc. indirectly with

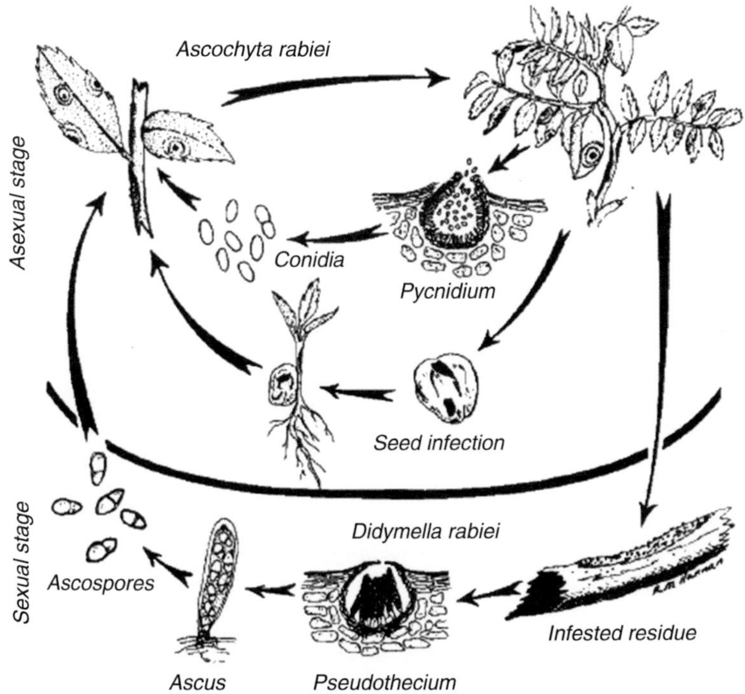

Fig. 17.1 Disease cycle of Ascochyta blight (Wiese et al. 1995)

the help of a marker (morphological, biochemical, or one based on DNA/RNA variation). MAS is being used in plant breeding for improving disease resistance and enhancing the quality (Goutam et al. 2015). MAS has proved to be efficient in selection for traits that are otherwise difficult to measure or show delayed expression during development. Quantitative trait loci (QTLs)-linked molecular markers providing resistance to AB have been discovered and are being used. Deoxyribonucleic acid (DNA) markers also encouraged the use of exotic sources of disease resistance. These markers increase the diversity in sources for pyramiding of resistant genes and reduce the time for development of resistant cultivars. Santra et al. (2000) identified two major QTLs (QTL 1 and QTL 2) for AB resistance and mapped these QTLs. QTL 1 and QTL 2 are found to be linked with microsatellite (GAA 47, ubc 733, ubc 181) and ISSR (Ta 72A, Ta2, Ts 54, and Ta 146) markers, respectively (Tekeoglu et al. 2002). Collard et al. (2003) developed a linkage map using the F2 generation of interspecific cross between Lasseter (susceptible) and PI 527930 (resistant). Seedling and stem resistance were studied using 83 molecular markers (RAPD, ISSR, STMS, and RGA). The study reported five markers for stem resistance, four of which were bearing resistance to both seedling and stem. QTLs mapped to LG 4 suggested the role of this linkage group for resistance to AB in the *Cicer* genome.

In Germany a DAF marker OPS06-1 was mapped along with the other DAF markers (Rakshit et al. 2003). Another study by Cho et al. (2004) suggested the potential of *Ar*19 (or *Ar*21d) on LG2 + 6 in complete resistance to pathotype I along with partial resistance to pathotype II and QTL on LG4A in resistance to pathotype II. Two QTLs for AB resistance were screened using an intraspecific population derived from ILC3279 × WR315 (Iruela et al. 2006). The frequency of single nucleotide polymorphism (SNP) in expressed and genomic sequences of the chickpea genome was determined using an interspecific cross between *C. arietinum* and *C. reticulatum*. These SNPs were utilized to increase marker density at ABR-*QTL*1 by designing cleavage-amplified polymorphic site (CAPS) and derived CAPS (dCAPS) markers (Rajesh et al. 2005). Castro et al. (2013) compared the effectiveness of phenotypic and MAS for selecting blight-resistant genotypes. Their findings suggested that higher frequency of AB resistant alleles can be achieved by utilizing MAS in earlier generations of breeding programs. Varshney et al. (2014) performed marker-assisted backcrossing to introgress AB resistance targeting QTL I and QTL II in C 214, an elite cultivar of chickpea. Seven AB-resistant lines were found using 43 SSR markers.

17.3 Fusarium Wilt

Fusarium wilt caused by fungus, *Fusarium oxysporum f.* sp. *ciceris* (FOC), is another important disease of chickpea more prevalent in lower latitudes (0–30°N) having relatively dryer and warmer environment. The disease was first reported from India (Singh 2003). Nowadays, wilt has been reported worldwide including India, Iran, Peru, Syria, Ethiopia, Mexico, Spain, Tunisia, Turkey, and the United States. The disease can cause yield losses of 10–90% (Landa et al. 2004). It can cause complete yield loss in grain when the pathogen attacks in the vegetative and reproductive stages of the crop (Navas-Cortes et al. 2000). Hanif et al. (1999) reported that in Pakistan the area under chickpea cultivation on irrigated land has reduced from 50 to 10 percent. Fusarium wilt incidence varied from 14 to 32% in different states of India as reported by Dubey et al. (2010). Chickpea seedling, flower, and pods get infected by pathogen, and the rate of infection increases with increase in temperature and water stress (Grewal 1969; Chaudhry et al. 2007). Foc is a facultative saprophyte, which can be either soil or seed borne. The fungus lives in seeds and dead plant materials in the form of chlamydospore. Survival period of pathogen in absence of host can extend up to 6 years (Singh 2003).

Eight races of the pathogen of fusarium wilt disease are diagnosed (0, 1A, 1B/C, 2, 3, 4, 5, and 6) by pathogenesis tests on ten chickpea cultivars (Haware and Nene 1982). 1A, 2, 3, 4, 5, and 6 races result in the wilting; 0 and 1B/C races induce chlorosis in plants (Jimenez-Diaz et al. 1991). Moreover, two separate clusters are diagnosed for fusarium wilt disease. The first one is a linkage group (LG2 chromosome F or G) which is a gene cluster including five genes (foc 1, foc 2, foc 3, foc 4, and foc 5) and genes that are effective on resistance to pathotypes of fusarium wilt. Another one is LG3 chromosome C or D which includes genes that are effective on resistance to chlorosis pathotype (race zero). Genes that are effective on resistance to 6 and 1B/C races are not yet located (Sharma and Muehlbauer 2007).

17.3.1 Symptoms and Disease Cycle

After about 30 days of sowing, wilting of young seedlings is apparent after infection, which leads to the death of adult plant. Infected seedlings do not lose their original color and fall down on the ground. Seedling and flowering stages are the most susceptible for infection among other vegetative stages. Petiole drooping, brown vascular bundles, and faded and dried leaves are characteristic symptoms, which lead to the plant death (Prasad and Padwick 1939; Westerlund et al. 1974). Leslie and Summerell (2006) studied the mechanism of fungal infection and reported invasion of vascular system via the roots. After entering the vascular system, the fungus produces cell wall-degrading enzymes, which disturb the plant transport system. Discoloration of the tissues starts from the roots and ends on aerial parts of the plant, followed by yellowing and wilting of the foliage, leading to necrosis. Infected pods from the wilted are normal, but size, shape, and color of seed are altered.

17.3.2 Management of Wilt

Crop rotation and use of pesticide are inefficient in wilt management as the pathogen is soilborne in nature. The ability of pathogen to survive and remain latent for long period of time without suitable host makes its management very difficult (Haware et al. 1996). Nowadays, application of chemical pesticides is limited because of their environmental pollutions and health risks. So, using genetic resistance and cultivating resistant genotypes are the most suitable and possible methods for control of fusarium wilt disease. Selecting the resistant genotypes by phenotypical methods is a complicated and time-consuming method. So, using DNA-based molecular markers is a major tool for the selection of the resistant cultivars, facilitating the process (Lindhout 2002; Sharma et al. 2005).

Genes for fusarium wilt resistance *foc 1*, *foc 4*, and *foc 3* (Tullu 1996; Mayer et al. 1997; Tullu et al. 1998, 1999; Winter et al. 2000; Sharma et al. 2004) have been mapped in the *Cicer* genome. In addition, 19 resistance gene analogs (RGAs) have been mapped in different populations (Huttel et al. 2002; Rajesh et al. 2002; Flandez-Galvez et al. 2003a, b). Varshney et al. (2014) performed backcrossing targeting *foc 1* using 40 SSR markers and identified 3 wilt-resistant lines. In another study, 24 indigenous and 46 exotic chickpea accessions from Pakistan were evaluated for wilt resistance. In molecular characterization for wilt response, 5 RAPD and 15 SSR markers were identified. Only in TA194 SSR marker, 85% linkage to wilt resistance was reported (Ahmad et al. 2014). In another study by Farahani et al. (2015), SCAR molecular markers were used to detect wilt resistance in chickpea genotypes. The DNA of 42 chickpea genotypes was isolated, and polymerase chain reaction was conducted by CS-27 and OPM-20 molecular markers. Flip 06-152c was the only resistant genotype found among the five.

17.4 Botrytis Gray Mold (BGM)

BGM is a major fungal disease caused by *Botrytis cinerea*, which can result in complete yield loss in favorable environmental conditions (Pande et al. 2002, 2006). During unfavorable environmental conditions, the pathogen inhabits on crop debris and seeds for its survival. The disease is distributed worldwide, but its occurrence in India, Bangladesh, Nepal, Pakistan, and Australia is of major concern (Pande et al. 2002). BGM was first reported from India by Shaw and Ajrekar (1915) and later by Butler and Bisby (1931). Argentina faced the first BGM epidemic with crop loss of more than 90% (Carranza 1965). The first epidemic in India occurred in the year 1978–1979, in which 20,000 ha crop was destroyed (Haware 1998). Nepal bears crop loss of 15% every year (Joshi 1992). After only 7 years of first documentation of BGM in Bangladesh in 1981, it becomes devastating with 100% crop loss (Bakr and Ahmed 1992).

17.4.1 Symptoms and Disease Cycle

B. cinerea is an important fungal pathogen of chickpea, as it causes seedling disorders and soft rot (Cother 1977; Bretag and Mebalds 1987; Burgess et al. 1997). The infection occurs on all aerial parts, but tips and growing flowers are most susceptible (Haware 1998; Bakr et al. 2002). Grey or brown to light brown lesions with hairy sporophores and hyaline spores can be seen on the aerial parts of plant (Haware and McDonald 1992; Haware 1998). Firstly water-soaked lesions 10–30 mm long appear on the stem near ground and then spread to other parts causing rotting of the affected area (Knights and Siddique 2002). Breakage of branches occurs at rotting point, and other aerial parts become a rotting mass (Bakr et al. 2002; Pande et al. 2002). Infection severity is largely affected by the environmental conditions and amount of inoculums, which consequently hampers pod yield. Size and color of seeds depend on the site of infection. Seeds get covered with fungal mat on pod invasion by fungus (Haware 1998; Bakr et al. 2002; Knights and Siddique 2002).

17.4.2 Management of Botrytis Gray Mold (BGM)

B. cinerea can remain latent up to 5 years on plant seed (Laha and Grewal 1983; Haware et al. 1986; Meeta et al. 1986). Disease transmission via seed can be limited by using healthy seeds. Reduced level of BGM in chickpea has achieved using manipulating sowing dates, erect cultivar, and lower plant densities (Haware 1998). The agricultural research led to the development of new biological approaches. *Trichoderma* sp. having rhizospheric origin at a concentration of 10^7conidia/mL was found to be effective against BGM (Mukherjee et al. 1995). High antagonistic potential was observed in extracts (*Allium* and *Capsicum spp*.,) and oils (*Cymbopogon martini*, *Thymus zygis*, *Cinnamomum zeylanicum*, and *Syzygium aromaticum*) of some plants against *B. cinerea* (Wilson et al. 1997). Oils of oregano, thyme, dictamnus, and marjoram possess fungicidal activity against *B. cinerea* even at low concentrations (Daferena et al. 2003). The nonchemical methods are weak and do not provide complete resistance.

Fungicides such as captan, carbendazim, chlorothalonil, mancozeb, thiabendazole, thiophanate-methyl, thiram, triadimefon, triadimenol, and vinclozolin were found effective when sprayed on foliar tissues after sowing or on appearance of initial symptoms (Knights and Siddique 2002; Pande et al. 2002; Davidson et al. 2004). Rashid et al. (2014) studied the effect of individual and combination of fungicides against BGM. Combined formulations of fungicides were found to be more effective in avoiding the fungicide resistance along with reduction in the yield losses. Host-plant resistance (HPR) becomes preferable choice for management of BGM; fungicides require repeated application to control infection.

HPR identification is being done by developing techniques like controlled environment and field screening. Only the cultivars possessing moderate resistance are being identified; hence its adoption of resistant germplasm is not possible (Pande et al. 2006). Isenegger et al. (2008) studied the allelic diversity of *B. cinerea* at nine microsatellite loci. Their findings suggest that genetic diversity, genotype flow, and the evolutionary potential of *B. cinerea* can result in breakdown of host resistance. Kaur et al. (2013) evaluated wild *Cicer* relatives (*C. judaicum* and *C. pinnatifidum*) for resistance against AB and BGM. Resistant cultivars including GL 29029, GL29206, GL29212, and GL29081 were crossed with BG 256 (susceptible cultivar). Six SSR markers (Ta 2, Ta 110, Ta 139, CaSTMS 7, CaSTMS 24, and Tr 29) were found to have polymorphic bands between resistant and the susceptible cultivars.

17.5 Dry Root Rot

Macrophomina phaseolina (*Rhizoctonia bataticola*) causes dry root rot in chickpea. It is a serious problem and has been reported from Australia, Ethiopia, Iran, Pakistan, Bangladesh, Nepal, and several other countries (Nene et al. 1991; Singh and Sharma 2002). Although it is present in all growing regions of India, it is most prevalent in Central and South India, as rainfed conditions are used for growing crop. At flowering and podding stage, environmental conditions like temperature and moisture content of soil affect the severity of the infection (Gurha et al. 2007).

17.5.1 Symptoms and Disease Cycle

The disease is more severe in sandy poor soils and generally appears at flowering and podding stage. The petioles and leaflets droop only at the top of plant. Tap roots turn black and show sign of rotting and are devoid of lateral roots. The plant is uprooted, while some fragments especially the lower one are left in soil. Sometimes a grayish mycelium is easily visible on the tap roots. Dead roots are brittle and show shredding of bark. The tip of the root is easily broken when touched. Minute sclerotia can be seen with the arid of handles on the uncovered woody root parts and internal region of the bark or when the collar part is sliced vertically. The disease is both seed- and soilborne. Many plants from the family *Leguminosae* are the favorable host of pathogen. Deficient soil moisture is favorable for disease development (Singh and Sharma 2002).

17.5.2 Management of Dry Root Rot

Deep plowing and removal of infected host debris from the soil reduce disease severity. Moisture stress conditions should be avoided. Timely sowing of early maturing varieties is a good option to escape the hot weather conditions during maturity of the disease (Gurha et al. 2007).

More than 10,000 chickpea germplasm accessions and cultivated genotypes were screened for resistance to dry root rot along with wilt, and several resistant sources have been identified. These resistant sources have been utilized in disease resistance breeding programs, and consequently varieties resistant or tolerant to multiple diseases have been developed. Treating seeds with fungicides such as captan or thiram is helpful in reducing the disease. Seed treatment with biocontrol agent *T. viride* has reported to be beneficial in managing the disease (Gurha et al. 2007).

17.6 Wet Root Rot

Wet root rot is caused by *R. solani* Kuhn which is of minor importance and is reported worldwide. Conditions like no till or reduced till favor pathogen growth and disease development. Cool and wet conditions of soil are optimum, but infection can occur at wide temperature range of soil.

17.6.1 Symptoms and Disease Cycle

The field symptoms are drying plants scattered throughout the field. In highly moist soils, the disease is reported even at seedling stage, whereas in irrigated conditions, pathogen may infect the crop at the later stages. Wilting of aerial parts like petioles occurs after yellowing of the infected seedlings. Crumpling of seedling does not occur immediately after infection. Appearance of brown lesions starts from the stem at collar region and moves downward to lower parts of plant. Rotting of stem and root can be seen below the brown lesions. Pink mycelial growth is visible, but sclerotia are not usually seen.

17.6.2 Management of Wet Root Rot

As wet root rot is of minor importance, hence its control measures are not studied extensively. Cultural practices like avoiding use of rich soil, low soil moisture, etc. can be used to reduce the frequency of infection. Using fungicides like captan, thiram, bentate, and PCNB for treating seeds before sowing is recommended.

17.7 Conclusion

Chickpea is the third most important legume crop produced worldwide. The yield gets reduced due to both abiotic and biotic stresses. Biotic stresses due to pathogens can cause complete yield loss. Fungal pathogens have shown devastating effects on the crop yield. The fungal diseases of major importance include AB, BGM, fusarium wilt, and root rot. Fungal disease management depending on the identification of pathogen, genetic, and pathological variability has led to the shift from cultural practices to chemical and biological control methods. Conventional methods are tedious and slow. Therefore, the need for novel, rapid, and effective control methods has resulted in the development of new genetic approaches. Recent advances in the development of genomic resources have made it possible to design resistant genotypes. The *Cicer* genomics is still being explored, and the new molecular approaches will provide an insight to chickpea genomics.

References

Agarwal G, Sabbavarapu MM, Singh VK, Sheelamary MTS, Gaur PM, Varshney RK (2015) Identification of a non-redundant set of 202 in silico SSR markers and applicability of a select set in chickpea (*Cicer arietinum* L.). Euphytica 205:381–394

Ahmad F, Gaur PM, Croser JS (2014) Chickpea (*Cicer arietinum* L.) genetic resources, chromo some engineering and crop improvement. Taylor and Francis, Boca Raton, pp 229–267
Anbessa Y, Taran B, Warkentin TD, Tullu A, Vandenberg A (2009) Genetic analyses and conservation of QTL for ascochyta blight resistance in chickpea (*Cicer arietinum L.*). Theor Appl Genet 119:757–765
Anuradha C, Gaur P, Pande S, Kishore G, Ganesh M, Kumar J, Varshney R (2011) Mapping QTL for resistance to botrytis grey mould in chickpea. Euphytica 182:1–9
Bakr MA, Ahmed F (1992) Botrytis gray mold of chickpea in Bangladesh. In: Haware MP, Faris DG, Gowda CLL (eds) Botrytis gray mold of chickpea summary proceedings of the BARI/ ICRISAT Working Group meeting. ICRISAT, Patancheru, pp 10–12
Bakr MA, Rahman ML, Ahmed AU (2002) Manifestation of Botrytis grey mould in chickpea in Bangladesh. In: Bakr MA, Siddique KHM, Johansen C (eds) Integrated management of Botrytis grey mould of chickpea in Bangladesh and Australia. Bangladesh Agricultural Research Institute, Gazipur, pp 63–69
Bretag TW, Mebalds MI (1987) Pathogenicity of fungi isolated from *Cicer arietinum* (chickpea) grown in north-western Victoria. Aust J Exp Agric 27:141–148
Bretag TW, Meredith KA, Knights E, Materne MA (2003) Control of ascochyta blight in chickpeas using disease resistance and foliar fungicides. In: Proceedings of the 8th international congress of plant pathology. (Christchurch, New Zealand). (International Society for Plant Pathology, UK), p 291
Burgess DR, Bretag T, Keane PJ (1997) Seed-to-seedling transmission of *Botrytis cinerea* in chickpea and disinfestations of seed with moist heat. Aust J Exp Agric 37:223–229
Butler EJ, Bisby GR (1931) The fungi of India, Science monograph no. 1, vol XVIII. ICAR, New Delhi, p 237
Carranza JM (1965) Wilt of chickpea (*Cicer arietinum* L.) caused by *B. cinerea* (In Spanish). Revista de la Facultad de Agronomia, Universidad Nacional de la Plata 41:135–138
Castro P, Rubio J, Madrid E, Fernández-Romero MD, Millán T, Gil J (2013) Efficiency of marker-assisted selection for ascochyta blight in chickpea. J Agric Sci:1–12
Cevik S, Unyayar S, Ergul A (2015) Genetic relationships between cultivars of *Cicer arietinum* and its progenitor grown in turkey determined by using the SSR markers. Turk J Field Crop 20:109–114
Chaudhry MA, Ilyas MB, Muhammad F, Ghazanfar MU (2007) Sources of resistance in chickpea germplasm against fusarium wilt. Mycopathology 5:17–21
Cho S, Chen W, Fred JM (2004) Pathotype-specific genetic factors in chickpea (*Cicer arietinum* L.) for quantitative resistance to ascochyta blight. Theor Appl Genet 109:733–739
Cobos M, Rubio J, Strange R, Moreno M, Gil J, Millan T (2006) A new QTL for Ascochyta blight resistance in a RIL population derived from an interspecific cross in chickpea. Euphytica 149:105–111
Cobos MJ, Fernandez MJ, Rubio J, Kharrat M, Moreno MT, Gil J, Millan T (2005) A linkage map of chickpea (*Cicer arietinum L.*) based on populations from Kabuli 9 Desi crosses, location of genes for resistance to fusarium wilt race 0. Theor Appl Genet 110:1347–1353
Collard BCY, Pang ECK, Ades PK, Taylor PWJ (2003) Preliminary investigation of QTLs associated with seedling resistance to ascochyta blight from *Cicer echinospermum*, a wild relative of chickpea. Theor Appl Genet 107:719–729
Cother EJ (1977) Identification and control of root-rot fungi in *Cicer arietinum* (chickpea). Plant Dis Rep 61:736–740
Daferena DJ, Ziogas BN, Polissiou MC (2003) The effectiveness of plant essential oils on the growth of *Botrytis cinerea, Fusarium* sp. and *Clavibacter michiganensis* sub sp. *michiganensis*. Crop Prot 22:39–44
Davidson JA, Pande S, Bretag TW, Lindbeck KD, Kishore GK (2004) Biology and management of *Botrytis* spp. in legume crops. In: Elad Y, Williamson B, Tudzynski P, Delen N (eds) Botrytis, biology, pathology and control. Kluwer Academic Publishers, Dordrecht, pp 295–318

Dubey SC, Singh SR, Singh B (2010) Morphological and pathogenic variability of Indian isolates of *Fusarium oxysporum* f. sp. ciceris causing chickpea wilt. Phytopathology 43:174–190
Duzdemir O, Selvi B, Yanar Y, Yildirimi A (2014) Sources of resistance in chickpea (*Cicer arietinum* L.) land races against A*scochyta rabiei* causal agent of ascochyta blight disease. Pak J Bot 46:1479–1483
Farahani S, Mori B, Maleki M, Saberi S (2015) Using SCAR molecular marker to detect resistance genes to *Fusarium oxysporum* f. sp. ciceris in chickpea cultivars and lines. Biol Forum 7:1369–1376
Flandez-Galvez H, Ades PK, Ford R, Pang EC, Taylor PW (2003a) An intraspecific map of chickpea (*Cicer arietinum* L.) genome based on sequence tagged microsatellite site and resistance gene analog markers. Theor Appl Genet 106:1447–1452
Flandez-Galvez H, Ades R, Ford R, Pang E, Taylor P (2003b) QTL analysis for ascochyta blight resistance in an intraspecific population of chickpea (*Cicer arietinum L.*). Theor Appl Genet 107:1257–1265
Goutam U, Kukreja S, Yadav R, Salaria N, Thakur K, Goyal A (2015) Recent trends and perspectives of molecular markers against fungal diseases in wheat. Front Microbiol 6:1–14
Gowda SJM, Radhika P, Kadoo NY, Mhase LB, Gupta VS (2009) Molecular mapping of wilt resistance genes in chickpea. Mol Breed 24:177–183
Grewal JS (1969) Important fungal disease of *Cicer arietinum* in India. Pulse improvement project seminar report held at Karaj Agricultural College, University of Tehran & USDA, pp 35–40
Gurha SN, Mukesh S, Shubha T, Udit N (2007) Prospects of ecofriendly management of wilt and dry root rot in chickpea. In: Ahmad S, Narani U (eds) Ecofriendly management of plant disease. Daya Publishing House, New Delhi, pp 215–221
Hanif M, Jamil FF, Haq I (1999) Effect of various sowing depths on wilt incidence of chickpea in wilt sick field in Pakistan. ICPN 6:11–12
Haware MP (1998) Diseases of chickpea. In: Allen DJ, Lenne JM (eds) The pathology of food and pasture legume. ICARDA, CAB International, Wallingford, pp 473–516
Haware MP, McDonald D (1992) Integrated management of botrytis gray mold of chickpea. In: Haware MP, Faris DG, Gowda CLL (eds) Botrytis gray mold of chickpea. ICRISAT, Patancheru, pp 3–6
Haware MP, Nene YL (1982) Races of *Fusarium oxysporum* f. sp. *ciceri*. Plant Dis 66:809–810
Haware MP, Nene YL, Mathur SB (1986) Seedborne diseases of chickpea, Technical bulletin no. 1. Danish Government Institute of Seed Pathology for Developing Countries/ICRISAT, Copenhagen/Patancheru
Haware MP, Nene YL, Natarajan M (1996) Survival of *Fusarium oxysporum* f. sp. *ciceri* in the soil in the absence of chickpea. Phytopathol Mediterr 35:9–12
Huttel B, Santra D, Muehlbauer FJ, Kahl G (2002) Resistance gene analogues of chickpea (*Cicer arietinum* L.), isolation, genetic mapping and association with a Fusarium resistance gene cluster. Theor Appl Genet 105:479–490
Iruela M, Castro P, Rubio J, Cubero JI, Jacinto C, Millan T, Gil J (2007) Validation of a QTL for resistance to ascochyta blight linked to resistance to fusarium wilt race 5 in chickpea (*Cicer arietinum L.*). Eur J Plant Pathol 119:29–37
Iruela M, Rubio J, Barro F, Cubero JI, Millan T, Gil J (2006) Detection of two quantitative trait loci for resistance to ascochyta blight in an intra-specific cross of chickpea (*Cicer arietinum* L.), development of SCAR markers associated with resistance. Theor Appl Genet 112:278–287
Isenegger DA, Macleod WJ, Ford R, Taylor PWJ (2008) Genotypic diversity and migration of clonal lineages of *Botrytis cinerea* from chickpea fields of Bangladesh inferred by microsatellite markers. Plant Pathol 57:967–973
Jimenez-Diaz RM, Singh KB, Trapero-Casas A, Trapero-Casas JL (1991) Resistance in kabuli chickpeas to fusarium wilt. Plant Dis 75:914–918
Jingade P, Ravikumar RL (2015) Development of molecular map and identification of QTLs linked to fusarium wilt resistance in chickpea. J Genet 94:723–729

Joshi S (1992) Botrytis gray mold of chickpea in Nepal. In: Haware MP, Faris DG, Gowda CLL (eds) Botrytis gray mold of chickpea. ICRISAT, Patancheru, pp 12–13

Kader DAA, El-Wakil A, Tohami MR, Ghoniem MI (1990) Effect of some agricultural practices and chemical control on the incidence of ascochyta blight of chickpea. Egypt J Phytopathol 21:31–43

Kaiser WJ, Coca FW, Vega S (2000) First report of ascochyta blight of chickpea in Latin America. Plant Dis 84:102–106

Kanouni H, Taleei A, Peyghambari SA, Okhovat SM, Baum M, Abang M (2009) QTL analysis for ascochyta blight resistance in chickpea (*Cicer arietinum* L.) using microsatellite markers. J Agric Res 25:109–127

Kaur L, Sirari A, Kumar D, Sandhu JS, Singh S, Kapoor K, Singh I, Gowda CLL, Pande S, Gaur P, Sharma M, Imtiaz M, Siddique KHM (2013) Combining ascochyta blight and Botrytis grey mould resistance in chickpea through interspecific hybridization. Phytopathol Mediterr 52:157–165

Knights EJ, Siddique KHM (2002) Manifestation of *Botrytis cinerea* on chickpeas in Australia. In: Workshop proceedings integrated management of Botrytis grey mould of chickpea in Bangladesh and Australia. Bangladesh Agricultural Research Institute, Gazipur, pp 70–77

Laha SK, Grewal JS (1983) Botrytis blight of chickpea and its perpetuation through seed. Indian Phytopathol 36:630–634

Landa BB, Navas-Cortes JA, Jimenez-Diaz RM (2004) Integrated management of fusarium wilt of chickpea with sowing date, host resistance and biological control. Phytopathology 94:946–960

Leslie JF, Summerell BA (2006) The Fusarium laboratory manual. Blackwell Publishing, Iowa

Lindhout P (2002) The perspectives of polygenic resistance in breeding for durable disease resistance. Euphytica 124:217–226

Mayer MS, Tullu A, Simon CJ, Kumar J, Kaiser WJ, Kraft JM, Muehlbauer FJ (1997) Development of a DNA marker for fusarium wilt resistance in chickpea. Crop Sci 37:625–1629

Meeta M, Bedi PS, Kumar K (1986) Chemical control of gray mold of gram caused by *Botrytis cinerea* in Punjab. J Res 23:435–438

Mukherjee PK, Haware MP, Jayanthi S (1995) Preliminary investigations in integrated biocontrol of botrytis gray mold of chickpea. Indian Phytopathol 48:141–149

Navas-Cortés JA, Hau B, Jiménez-Díaz RM (2000) Yield loss in chickpeas in relation to development of fusarium wilt epidemics. Phytopathology 90:1269–1278

Nene YL (1979) Group discussion on the resistance to soilborne diseases of legumes. In: Proceedings of the consultants. ICRISAT, Hyderabad

Nene YL, Reddy MV, Haware MP, Ghanekar AM, Amin KS (1991) Field diagnosis of chickpea diseases and their control, Information bulletin no.28. International Crops Research Institute for the Semi Arid Tropics, Patancheru

Nene YL, Sheila VK, Sharma BS (1996) A world list of chickpea and pigeonpea pathogens, 5th edn. International Crops Research Institute for the Semi-Arid Tropics, Patencheru

Pande S, Galloway JJ, Gaur PM, Siddique KHM, Tripathi HS, Taylor P, MacLeod MWJ, Basandrai AK, Bakr A, Joshi S, Krishna KG, Isenegger DA, Rao NJ, Sharma M (2006) Botrytis grey mould of chickpea, a review of biology, epidemiology, and disease management. Aust J Agric Res 57:1137–1150

Pande S, Siddique KHM, Kishore GK, Bayaa B, Gaur PM, Gowda CLL, Bretag TW, Crouch JH (2005) Ascochyta blight of chickpea (*Cicer arietinum* L.), a review of biology, pathogenicity, and disease management. Aust J Agric Res 56:317–332

Pande S, Singh G, Rao JN, Bakr MA, Chaurasia PCP, Joshi S, Johansen C, Singh SD, Kumar J, Rahman MM, Gowda CLL (2002) Integrated management of botrytis gray mold of chickpea, Information bulletin no. 61. ICRISAT, Patancheru

Prasad N, Padwick GW (1939) The genus *Fusarium* 11. A species of *Fusarium* as a cause of wilt of gram (*C. arietinum* L.). Indian Agric Sci 9:731–735

Rajesh PN, McPhee K, Muehlbauer FJ (2005) Detection of polymorphism using CAPS and dCAPS markers in two chickpea genotypes. Int Chickpea Pigeonpea Newsl 12:4–6

Rajesh PN, Tullu A, Gil J, Gupta VS, Ranjekar PK, Muehlbauer FJ (2002) Identification of an STMS marker for the double-podding gene in chickpea. Theor Appl Genet 105:604–607
Rakshit S, Winter P, Tekeoglu M, Juarez- Munoz M, Pfaff T, Benko-Iseppon AM, Muehlbauer FJ, Kahl G (2003) DAF marker tightly linked to a major locus for ascochyta blight resistance in chickpea (*Cicer arietinum* L.). Euphytica 132:23–30
Rashid MH, Hossain AM, Kashem MA, Kumar S, Rafii MY, Latif MA (2014) Efficacy of combined formulations of fungicides with different modes of action in controlling botrytis gray mold disease in chickpea. Sci World J 2014:639–646
Ratnaparkhe MB, Santra DK, Tullu A, Muehlbauer FJ (1998) Inheritance of inter-simple sequence-repeat polymorphisms and linkage with a fusarium wilt resistance gene in chickpea. Theor Appl Genet 96:348–353
Reddy MV, Singh KB (1990) Management of Ascochyta blight of chickpea through integration of host plant tolerance and foliar spraying of chlorothalonil. Indian J Plant Protect 18:65–69
Rubio J, Moussa E, Kharrat M, Moreno MT, Millan NT, Gil J (2003) Two genes and linked RAPD markers involved in resistance to *Fusarium oxysporum* f. sp. ciceris race 0 in chickpea. Plant Breed 122:188–191
Sabbavarapu MM, Sharma M, Chamarthi SK, Rathore A, Thudi M, Gaur PM, Pande S, Singh S, Kaur L, Varshneyet RK (2013) Molecular mapping of QTLs for resistance to fusarium wilt (race 1) and ascochyta blight in chickpea (*Cicer arietinum* L.). Euphytica 193:121–133
Santra DK, Singh G, Kaiser WJ, Gupta VS, Ranjekar PK, Muehlbauer FJ (2000) Molecular analysis of *Ascochyta rabiei* (Pass.) Labr., the pathogen of ascochyta blight in chickpea. Theor Appl Genet 102:676–682
Sharma KD, Muehlbauer FJ (2007) Fusarium wilt of chickpea, physiological specialization, genetics of resistance and resistance gene tagging. Euphytica 157:1–14
Sharma KD, Chen W, Muehlbauer FJ (2005) Genetics of chickpea resistance to Wve races of *Fusarium* wilt and a concise set of race differentials for *Fusarium oxysporum* f. sp. *ciceris*. Plant Dis 89:385–390
Sharma KD, Winter P, Kahl G, Muehlbauer FJ (2004) Molecular mapping of *Fusarium oxysporum* f. sp. *ciceris* race 3 resistance gene in chickpea. Theor Appl Genet 108:1243–1248
Shaw FJ, Ajekar M (1915) The genus Rhizoctonia in India. Dep Agric Bot Surv 7:177–194
Singh G, Sharma YR (2002) Fungal diseases of pulses. In: Gupta VK, Paul YS (eds) Diseases of field crops. Indus Publishing, New Delhi, pp 155–192
Singh SD (2003) Soil borne diseases of chickpea. ICRISAT, Patancheru
Stephens A, Lombardi M, Cogan NOI, Forster JW, Hobson K, Materne M, Kaur S (2014) Genetic marker discovery, interspecific linkage map construction and quantitative trait locus analysis of ascochyta blight resistance in chickpea (*Cicer arietinum* L.). Mol Breed 33:297–313
Taran B, Warkentin TD, Tullu A, Vandenberg A (2007) Genetic mapping of aschochyta blight resistance in chickpea using a simple sequence repeat linkage map. Genome 50:26–34
Tekeoglu M, Rajesh N, Muehlbauer J (2002) Integration of sequence tagged microsatellite sites to the chickpea genetic map. Theor Appl Genet 105:847–854
Tekeoglu M, Tullu A, Kaiser WJ, Muehlbauer FJ (2000) Inheritance and linkage of two genes that confer resistance to fusarium wilt in chickpea. Crop Sci 40:1247–1251
Tullu A (1996) Genetics of fusarium wilt resistance in chickpea, Dissertation. Washington State University, Pullman, WA
Tullu A, Kaiser WJ, Kraft JM, Muehlbauer FJ (1999) A second gene for resistance to race 4 of fusarium wilt in chickpea and linkage with a RAPD marker. Euphytica 109:43–50
Tullu A, Muehlbauer FJ, Simon CJ, Mayer MS, Kumar J, Kaiser WJ, Kraft JM (1998) Inheritance and linkage of a gene for resistance to race 4 of fusarium wilt and RAPD markers in chickpea. Euphytica 102:227–232
Udupa SM, Baum M (2003) Genetic dissection of pathotype specific resistance to ascochyta blight disease in chickpea (*Cicer arietinum L.*) using microsatellite markers. Theor Appl Genet 106:1196–1202

Varshney RK, Mohan SM, Gaur PM, Chamarthi SK, Singh VK, Srinivasan S, Swapna N, Sharma M, Singh S, Kaur L, Pande S (2014) Marker-assisted backcrossing to introgress resistance to fusarium wilt race 1 and ascochyta blight in C 214, an elite cultivar of chickpea. Plant Genome 7:1–11
Westerlund FV, Campbell RN, Kimble KA (1974) Fungal root rot and wilt of chickpea in California. Phytopathology 64:432–436
Wiese MV, Kaiser WJ, Smith LJ, Muehlbauer FJ (1995) Ascochyta blight of chickpea. University of Idaho College of Agriculture, CIS 886 (revised)
Wilson CL, Solar JM, El-Ghaouth A, Wisniewski ME (1997) Rapid evaluation of plant extracts and essential oils for antifungal activity against *Botrytis cinerea*. Plant Dis 81:204–210
Winter P, Benko-Iseppon AM, Huttel B, Ratnaparkhe M, Tullu A, Sonnante G, Pfa VT, Tekeoglu M, Santra D, Sant VJ, Rajesh PN, Kahl G, Muehlbaue FJ (2000) A linkage map of the chickpea (*Cicer arietinum* L.) genome based on recombinant inbred lines from a *C. arietinum* X *C. reticulatm* cross, localization of resistance genes for fusarium wilt races 4 and 5. Theor Appl Genet 101:1155–1163
Yan J, Yuan S, Jiang L, Ye X, Bun T, Wu Z (2015) Plant antifungal proteins and their applications in agriculture. Appl Microbiol Biotechnol 99:4961–4981

Chapter 18
Recent Approaches for Late Blight Disease Management of Potato Caused by *Phytophthora infestans*

Umesh Goutam, Kajal Thakur, Neha Salaria, and Sarvjeet Kukreja

Abstract Potato (*Solanum tuberosum*) is the fourth most produced noncereal crop worldwide. Among various biotic stresses, late blight caused by *Phytophthora infestans* is the most devastating disease. It affects both potato foliage in the field and tuber in the storage which can absolutely destroy a crop, producing a 100% crop loss. The occurrence and rigorousness of late blight caused by *Phytophthora* can be reduced by adopting effective and durable control methods. The use of conventional control methods (cultural practices and fungicides) was limited due to their inefficiency and non-biodegradable nature. Control of the disease has been achieved up to a great extent through the use of fungicides, but their extensive application is harmful for the environment. Therefore, there is an urgent need to find alternative eco-friendly crop protection methods. The use of microorganisms as biological control agents owing to their different modes of actions (i.e. antagonistic effects or induction of plant defence mechanisms) has proved to be a potential approach. Another economical and eco-friendly remedial measure for plant diseases being adopted involves the use of nanoparticles against plant pathogens. Providing genetic resistance against pests and diseases is another crop protection approach. Multiple resistance (R) genes have been introduced in potato varieties to provide durable resistance to late blight. Genetic modification using cisgenes is preferable as it is a feasible and highly efficient approach with low risks and high societal acceptability. Accumulation of new virulent *P. infestans* strains decreases the effectiveness of R-genes. Therefore, the loss of function in susceptible gene via gene silencing is the emerging approach which helps in exploring plant-pathogen interactions and provides potential strategies for disease control. RNA silencing without altering the plant genome overcomes the risk associated with transgenic plants.

Keywords Late blight disease · *Phytophthora infestans* · Potato · Disease management

---

U. Goutam (✉) · K. Thakur · N. Salaria
Lovely Professional University, Phagwara, Punjab, India

S. Kukreja
MRM PG College, Ganganagar, Rajasthan, India

P. Gehlot, J. Singh (eds.), *Fungi and their Role in Sustainable Development: Current Perspectives*, https://doi.org/10.1007/978-981-13-0393-7_18

18.1 Introduction

Late blight of potato, which is caused by *P. infestans* (Mont.) de Bary, is the major bottleneck in potato production in Ethiopia (Bekele and Yaynu 1996) and other parts of the world (Fry and Goodwin 1997b). It is the best-known, highly studied and most destructive disease of all potato-producing countries (Jones 1998). The estimated annual loss due to late blight is more than $5 billion; hence it is considered as threat for global food security (Latijnhouwers et al. 2004, Haverkort et al. 2008). Late blight was responsible for the Irish potato famine in the 1840s (Mercure 1998). The disease caused yield losses ranging from 31 to 100% in Ethiopia depending on the variety used (HARC 2007). Good cultural practices such as removing the cull piles which can act as potential inoculum and proper harvest and postharvest approaches aid in reduction of pathogen growth, reproduction, survival and dispersal (Garrett and Dendy 2001). The use of fungicides like metalaxyl in controlling the disease was found to boost potato yield in various East African countries (Nsemwa et al. 1992, Rees et al. 1992). Management approaches to combat late blight in potato-growing regions have changed significantly due to migration of fungicide-resistant isolates of pathogen from one region to another. This necessitated the modification in old chemical practices and employment of cultural measures to solve the above-mentioned problem. Potato late blight is probably the most studied plant disease in the world; yet relatively fewer alternative management options other than synthetic chemical fungicides are available.

Application of fungicides has been the most adopted blight control method worldwide, but production of new highly aggressive strains renders the use of fungicides (CIP 1989, Powelson and Ingils 1998). Excellent control of the late blight disease was achieved through the use of the phenylamide fungicides, like Ridomil, across the sub-Saharan region (Dekker 1984, Fekede et al. 2013, Binyam et al. 2014a). According to Mesfin and Gebremedhin (2007), the failure of Ridomil in giving perfect control of the disease in some countries of the sub-Saharan region and in some cases the intensive frequency of usage (Davidse et al. 1981; Schiessendoppler et al. 2003) lead to the development of an integrated disease management strategy involving resistant and susceptible varieties and fungicide sprays. The best management of late blight and high marginal rate of return were obtained on plots treated with combinations of all tested potato varieties and 0.75 kg ha^{-1} Ridomil application followed by 1.5 kg ha^{-1} Ridomil application (Binyam et al. 2014b). Fontem and Aighew (1993) reported that application of fungicides for late blight management increases potato tuber yield by more than 50%.

Many dominant genes resistant (R-genes) to potato late blight have been reported and used in potato breeding to produce resistant varieties. The generation of new virulent strains limits the use of R-genes. A new class of resistance has been introduced which involves loss of function of a susceptibility gene (S-gene). The S-gene plays an important role in infection as it encodes the product required by pathogen for colonization. Impaired S-genes lead to recessive resistance traits, while dominant R-genes provide recognition-based resistance. Sun and his coworkers selected 11 S-genes in *Arabidopsis thaliana* and silenced orthologous genes in the blight-susceptible potato cultivar Desiree. Complete resistance to the *P. infestans* isolate Pic99189 was achieved by silencing of five genes, and the silencing of the sixth S-gene reduced late blight susceptibility (Sun et al. 2016a, b). As we know late blight is the most studied and well-known disease of potato in this world. In addition there are so many important research works which are done about this disease all over the world. However, these research results are not compiled in a well-structured review paper. With this perspective, this paper is initiated to review on late blight of potato's biology, economic importance and its management approaches.

18.2 The Pathogen

The causal organism of potato late blight is *P. infestans* (Mont.) de Bary (Jones 1998). The genus *Phytophthora* belongs to the Oomycetes, which are unrelated to true fungi (Shaw and Khaki 1971). It contains some heterothallic species (A1 and A2 mating types) (Fry et al. 1993). The mycelium produces branched sporangiophores that produce lemon-shaped sporangia at their tips. At the places where sporangia are produced, the sporangiophores form swellings that are characteristic for these fungi (Agrios 2005). *P. infestans* is the best known, most studied and still among the destructive of all potato diseases of the species of *Phytophthora* (Jones 1998). *P. infestans* requires two mating types, A1 and A2, to produce a sexual spore known as an oospore (Kirk 2009). When the two mating types grow adjacently, the female hypha grows through the young antheridium (male reproductive cell) and develops into a globose oogonium (female reproductive cell) above the antheridium. The antheridium then fertilizes the oogonium, which develops into a thick-walled and hardy oospore. Oospores germinate through germ tube that produces a sporangium, although at times the germ tube directly forms mycelium. Sporangia germinate almost entirely by releasing three to eight zoospores at temperatures up to 12 or 15 °C, whereas at above 15 °C, sporangia may germinate directly by producing a germ tube (Agrios 2005).

18.3 Development, Epidemiology and Life Cycle of Late Blight

P. infestans can survive in living host tissue, such as in seed tubers, cull piles and volunteer potatoes (Shinners et al. 2003), on other solanaceous plants and in the soil (Kirk et al. 2013). In the spring, the disease transmission from infected tubers is via airborne spores. Infected seed potatoes are also important sources of the disease. Night temperatures of 10–16 °C, moist conditions due to fog, dew and rain and high relative humidity provide the optimum environment for blight infection and development (Kirk 2009; Kirk et al. 2013). Temperatures above 30 °C slow or stop the growth of the fungus in the field but do not kill it, and the fungus can start to sporulate again when the temperature becomes favourable, provided, of course, that the relative humidity (near 100%) is sufficiently high (Agrios 2005). Water-soaked lesions varying in shapes and colours from light to dark green appear as the first visible symptoms in field on the lower leaves where humidity is high (Kirk et al. 2013). In favourable environment, infection spreads to the upper leaves and then carried into the neighbouring fields through wind (Martin et al. 1994; Kirk et al. 2013). On the underside of the leaf, a white, mildew-like growth at the edge of the lesion is the characteristic of fungus. This white growth is only found in late blight and hence can be differentiated from other foliar diseases. In dry conditions, existing lesions stop enlarging, turn black, curl and wither, and no oomycetes appear on the underside of the leaves. When the conditions get favourable, oomycetes resume their activities, and the disease once again develops rapidly (Agrios 2005).

Tubers may get infected early or late till harvest by *P. infestans* when they come in contact with sporangia. The affected tubers at first showed more or less irregular purplish-black or brownish blotches, and when cut open, the affected tissue appears water soaked, dark and somewhat reddish brown and extends 5–15 mm into the flesh of the tuber. Infected tubers may be subsequently covered with sporangiophores and spores of the pathogen, or infected tubers may be subsequently invaded by secondary fungi and bacteria, causing soft rots and giving the rotting potatoes a putrid, offensive odour (Agrios 2005). The susceptibility of the cultivar, temperature and time duration after the initial infection affect the extent of rotting in tuber (Martin et al. 1994; Kirk et al. 2013).

18.4 Sustainable Late Blight Management Approaches

A number of management techniques of late blight have been developed and used. An integrated disease management approach can prove to be most effective control. The most important measures are cultural controls, the use of resistant cultivars and chemical controls.

18.4.1 Cultural Control

Various control methods have been used for late blight management. Cultural practices provide foremost protection against late blight (Kirk 2009; Kirk et al. 2013). These methods reduce the inoculum by decreasing the survival rate, reproduction and dispersal of the pathogen. Survival of *P. infestans* to initiate an epidemic can be reduced through avoidance of introducing late blight into a field by planting only disease-free seed tubers, destroying all cull and volunteer potatoes, avoiding frequent or night-time overhead irrigation and good soil coverage (Draper et al. 1994). The use of disease-free seeds, removal of infected plants and soil management are the cultural practices being used (Garrett and Dendy 2001). Late blight can be controlled by removing infected potatoes and using proper harvesting and storage practices (Davis et al. 2009). The proper selection of seed sources helps in blight control by limiting the entry of pathogen strains (Kirk 2009). When partially blighted leaves and stems are surviving at harvest time, it is necessary to remove the above-ground parts of potato plants or destroy them by chemical sprays (herbicides) or mechanical means to prevent the tubers from becoming infected (Agrios 2005).

Favourable environmental conditions for the pathogen should be avoided. Selection of field and implementation of proper irrigation practices can help in reducing the probability of infection and disease development. Since wet conditions are favourable for infection, the use of sprinkler irrigation should be minimized. Morning-irrigated potatoes are more susceptible to infection than midday or evening irrigated (Carlson 1994). Removal of weeds and alternative late blight hosts helps in disease control as it doesn't allow restriction of air movement within the canopy (Kirk 2009; Kirk et al. 2013).

Mulching stimulates plant root growth, increases nutrient uptake, decreases evaporation from the soil, increases soil water-holding capacity, reduces surface water run-off, facilitates drainage, regulates soil temperature and provides a high level of nutrients for soil microbes (Aryantha et al. 2000). Alfalfa meal, cotton waste, soybean meal, wheat straw, chicken manure and urea have been reported for activity against *Phytophthora*. Organic matter on decomposition produces ammonia and volatile organic acids which can kill *Phytophthora* and stimulate antagonistic microorganisms in the soil (Lazarovits et al. 2001). These amendments may be phytotoxic to the plant roots which decreases the pathogen colonization (Erwin and Ribeiro 1996). Fertilizers are also used for blight management. Enhanced plant vigour and disease resistance have been achieved with the use of fertilizers (Erwin and Ribeiro 1996). Tubers should be stored properly in a pathogen-unfavourable dry environment (Kirk 2009). Keeping a check on stored potatoes and removal of infected tubers also help in blight management (Stone 2009).

18.4.2 Chemical Control

Application of fungicides is the most frequently adopted control method globally (CIP 1989). According to Beckerman (2008), fungicides can only slow or stop the development of new symptoms but cannot cure existing symptoms. Therefore, its application should be before disease development. Several broad-spectrum and systemic fungicides are used for blight management. Bordeaux mixture has proved to be effective against various species of *Phytophthora*. It adheres well to foliage but can be toxic to some plants owing to copper in its active site (Brown et al. 1998). Bordeaux mixture is a combination of copper sulphate and calcium hydroxide but its use is limited as the application in fields is labour intensive and because of the run-off due to rainfall in tropical areas (Erwin and Ribeiro 1996). Another class of fungicides is systemic phenylamides (acylanilides) which include furalaxyl

(Fongarid), metalaxyl (Ridomil) and benalaxyl (Galben). Metalaxyl is a xylem-translocated compound, and it can't be used as a foliar spray for root diseases. Its application as a soil drench has been reported as very effective (Guest et al. 1995). In Ethiopia, the increase in yield due to the use of fungicide was reported (Mesfin and Gebremedhin 2007). Ridomil MZ spray followed by Dithane M-45 (mancozeb) has been reported to be effective in late blight control.

Binyam et al. (2014a) reported that decreased use of Ridomil provided better blight management. Fungicide mixtures containing different broad-spectrum fungicides having diverse strategies to combat with this pathogen can be employed to reduce resistance (Thind 2015). Bhaik and Trivedi (2015) tested Zorvec™, a new fungicide in South Asia, against late blight in different climatic conditions. It was reported to be effective in providing durable protection from this disease as well as increased yield when compared with commercial counterparts. The excessive use of fungicide led to the development of resistance as in case of phenylamides. Phosphonates are the other type of fungicide active against the Peronosporales. It is xylem and phloem translocated, with both downward and upward movement in the host (Ouimette and Coffey 1990). It gets oxidized to phosphate by soil microbes and is a little toxic to mammals. Phosphonates at concentrations less than threshold for inhibition of mycelial growth decrease the virulence of the pathogen and lead to production of stress metabolites. These metabolites act as elicitors for host defence (Guest et al. 1995). Phosphonates can be applied either as a drench, foliar spray, stem-canker paint or trunk injection for direct systemic control. No resistant *Phytophthora* isolates have been reported yet (Guest et al. 1995). Despite being non-eco-friendly, the efficacy of fungicides is appealing to resource-poor farmers in almost all developing countries (Forbes et al. 1997). The need for repeated applications due to UV-light degradation and erosion by wind and rainwater limits the use of fungicides. The new strains develop resistance to fungicides like metalaxyl and are unaffected. Therefore, the search for an economic and eco-friendly disease management technique is required (Drenth and Guest 2004).

18.4.3 Biological Control

Using biological agents to control or suppress the growth of *Phytophthora* provides an economic and environmentally friendly approach. Actinomycetes and fungi such as *Trichoderma* spp., *Penicillium funiculosum* and *Chaetomium globosum* have been reported as efficient biocontrol agents (You et al. 1996; Chambers and Scott 1995, Fang and Tsao 1995; Heller and Theilerhedtrich 1994). All these agents have been effective against *P. cinnamomi* (El-Tarabily et al. 1996). Myrothecium, a soil-dwelling genus, was reported as an efficient biological agent against *P. palmivora* and *P. katsurae* which cause leaf rot in coconut. A range of endophytic fungi have been reported which protect cocoa against fungal pathogens, including *Phytophthora*. Their direct antagonism helps in providing the resistance (Arnold et al. 2003).

Therefore, new biocontrol agents need to be identified to provide eco-friendly control method.

In in vitro conditions, *Bacillus subtilis* var. *amyloliquefaciens* was reported to provide 35.6% inhibition against the *P. infestans*. Similar results were observed in glasshouse and field conditions, and also tuber yield was high in comparison with control plants (Keerthana et al. 2015). Wang et al. (2016) reported fungus *Purpureocillium lilacinum* as an efficient and effective biocontrol agent against plant pathogens including *Phytophthora*. Genome sequencing of fungal isolates was carried out, and the sequences were compared with other species. The genome of fungus *P. lilacinum* is rich in genes encoding proteases, carbohydrate-active enzymes and secondary metabolites. Leucinostatins, one of the important metabolites secreted by this fungus, inhibited the growth of late blight pathogen in bioassays.

18.4.4 Host-Plant Resistance

Resistance against late blight in host plant plays a vital role in integrated management of late blight as it is durable and economic and decreases the possibility of fungicide resistance (Hakiza 1999; Mukalazi et al. 2001). In case of *P. infestans*, locating host resistance becomes more difficult as the host range is narrow. Resistance to *Phytophthora* species in different hosts is of non-specific nature. Invasion of host by the pathogen leads to the production of many antifungal agents such as phytoalexins. Phytoalexins provide resistance to the host against pathogen, but their inhibitory mechanisms are non-specific. They can induce physical or chemical pathways irrespective of the pathogen nature. The elevated levels of phytoalexins are reported in the presence of compounds (carbohydrates, lipids, amino acids and proteins) of pathogen origin. Production of phytoalexin by *Phytophthora* infection is well documented, and Bion (acibenzolar-*S*-methyl), an analogue of salicylic acid, is found to induce SAR in host and enhance the resistance against *Phytophthora* (Erwin and Ribeiro 1996, Ali et al. 2000).

The use of resistant varieties is among the most effective and environmentally safe means of managing the disease. Variations in resistance to late blight among different potato varieties have been demonstrated by several researchers (Njualem et al. 2001). Cultivars having high levels of resistance can allow them to be grown without chemical protection even in the wettest growing seasons (Fry 1978). Early-maturing varieties are usually susceptible to the disease with exception of some cultivars. Some varieties have useful foliage resistance but poor tuber-blight resistance. Yet, others have good tuber-blight resistance but poor foliage-blight resistance. Ideally, a variety should have good resistance to both foliage and tuber blight (Anonymous 2007). However, no potato varieties are fully resistant to late blight (ATTRA 2004). Most resistant varieties are not immune to late blight but possess varying degrees of resistance to various races of the pathogen (Popokova 1972). However existing potato varieties possess race specific, which can be invaded by other compatible *P. infestans* races making varieties susceptible to the pathogen in a very short period (Shtienberg et al. 1994). Generally resistant potato varieties and improved cultural practices can reduce late blight (FAO 2008).

So, R-gene introduction in potato cultivars from wild *Solanum* spp. is considered to be the most efficient and eco-friendly approach to combat this pathogen. More than 20 functional R-genes were cloned from wild *Solanum* spp. such as *S. demissum*, *S. bulbocastanum*, *S. papita*, *S. venturii*, etc. (Li et al. 2011; Kim et al. 2012). The R-genes (R1, R2, R3a and R3b) from *S. demissum* are race specific and mapped on chromosome 5, 4 and 11, respectively (Leonards-Schippers et al. 1992; Li et al. 1998; Huang et al. 2005; Kim et al. 2012). Twenty-four quantitative trait loci (QTL) for late blight resistance were identified by linkage study. Candidate gene approach was used to identify the first diagnostic marker for quantitative resistance to late blight. This approach targets the known genes or cluster of genes, and their functional studies are already conducted on the model organisms. This limits the use of novel genes and pathways in providing resistance to pathogen. GWAS (genome-wide association studies) biased on single-nucleotide polymorphisms (SNPs) present over the whole genome aid the discovery of more markers associated with resistance (Goutam et al. 2015; Mosquera et al. 2016).

18.4.5 *Biotechnological Approaches*

18.4.5.1 Systemic Acquired Resistance (SAR)

The search for novel crop protection strategies has led the attention towards increasing plant immunity. Many crops get affected by the pathogens due to their weak immunity. Susceptible plants can be protected by stimulating defence mechanism which results in alteration of innate immunity. Systemic acquired resistance (SAR) is trigged by plants due to primary infection caused by pathogens, which aid the potential immune system in succeeding infections (Oostendorp et al. 2001; Prime-A-Plant Group 2006). SAR inducers such as elicitors can act as boosters of crop innate immunity and may provide a durable solution against pathogenic microbes (Conrath 2011; Bektas and Eulgem 2015). The metabolic adaptations in SAR resistance do not affect the genome of the host but modify the expression profile by altering the enzymes and proteins which regulate the gene expression (Zhang 2008; Fu and Dong 2013). Thus, SAR inducers can be seen as efficient eco-friendly alternative to conventional pesticides.

In 2012, three research groups confirmed the role of SAR in inducing resistance against biotic stress in plants and provided insight to the long-term metabolic memory after primary infection (Luna and Ton 2012; Pieterse 2012; Rasmann et al. 2012; Slaughter et al. 2012). A non-proteinogenic amino acid, β-aminobutyric acid (BABA), is a potent inducer of SAR in plants. BABA is reported as an effective SAR inducer against many plant pathogens including *P. infestans* (Cohen 2002; Baider and Cohen 2003; Ton and Mauch-Mani 2004; Ton et al. 2005; Andreu et al. 2006; Dubreuil-Maurizi et al. 2010; Worrall et al. 2012; Janus et al. 2013).

18.4.5.2 RNA Silencing

Immune system of plants is a complex association of different mechanisms that aid prevention of pathogen colonization in host (Jones and Dangl 2006). Defence is prompted when receptors present on plasma membrane of host recognize apoplastic effectors also known as pathogen-associated molecular patterns (PAMPs). PAMP-triggered immunity (PTI) is initiated due to these effectors which comprise cascade of signal transduction reactions. Elicitins are well-known example of apoplastic effectors released by several *Phytophthora* spp., which are recognized by ELR (receptor for elicitin response) present in potato (Du et al. 2015). Another class of effectors, i.e. Avr (avirulence), is detected by R-genes (resistance genes) that form the second line of defence in host and result in ETI (effector-triggered immunity). ETI can be invaded easily as resistance provided by R-genes is race specific and pathogen is evolving day by day (Vleeshouwers et al. 2011). So instead of using a single gene, pyramiding of different R-genes can be employed to improve the durability of resistance (Zhu et al. 2013). Both PTI and ETI rely on recognition of effector molecules from pathogen to instigate signal transduction. There another class of resistance is introduced recently which depends upon loss of function of S-gene (susceptibility gene). S-genes are class of gene which benefits pathogen instead of host that encodes these genes. Hence, dysfunctional S-genes develop a phenotype which resembles to healthy and resistant plant phenotype. The well-reported S-gene providing resistance against powdery mildew in silenced mutant is PMR6 (Powdery Mildew Resistance 6) in *Arabidopsis* (Eckardt 2002). S-genes are categorized in three classes: (1) required in early infection, (2) encode negative regulators for host defence, and (3) required for sustainability of infection. CesA3 gene involved in cellulose synthesis, DMR1 (Downy Mildew Resistant 1) gene providing resistance against *Hyaloperonospora parasitica* (Ellis et al. 2002, Huibers et al. 2013) and *Fusarium graminearum* and *F. culmorum*, which cause Fusarium ear blight (FEB) disease (Brewer et al. 2014) and SWEET proteins required by *Xanthomonas oryzae* as for carbon source (Streubel et al. 2013) are other examples of S-genes providing resistance in plants against pathogens. Sun et al. (2016a, b) selected 11 S-genes from *A. thaliana* and silenced their orthologous counterparts in late blight-susceptible potato cultivar Desiree. The complete resistance from isolate pic99189 of *P. infestans* was conferred by silencing of only five genes.

RNA silencing is a well-known mechanism which regulates the gene expression in eukaryotes. It initiates mRNA degradation through sequence-specific process, but it can inhibit expression of gene at transcriptional level by RNA-directed DNA methylation or at translation also known as PTGS (post-transcriptional gene silencing). The dsRNA formed in host due to pathogen replication or by RNA-dependent RNA polymerase activity are the targets of silencing machinery. Many transgenic plants were produced which target pathogen using RNA silencing. In transcriptional gene silencing (TGS), the host plant is supplemented with the extra copies of gene, and it silences the indigenous locus of gene. Some reports suggest the more stable and efficient method of silencing in *Phytophthora* is TGS (Judelson and Tani 2007). To study the mechanism of silencing in *Phytophthora*, Judelson and Tani (2007) silenced gene and gene cluster having sequence similarity to NIFC1. Elevated resistance was observed in case of tomato when dsRNA construct was introduced in host. This construct was homologous to mannose 6-phosphate reductase encoding gene which plays a significant role in the growth of *O. aegyptiaca*. Similar findings were reported in the case of *Blumeria graminis* (Nowara et al. 2010) and *F. verticillioides* (Tinoco et al. 2010) when dsRNA was expressed in barley and tobacco, respectively. In the case of *P. infestans*, Ah-Fong et al. (2008) targeted in f1 for producing transformants and found the presence of both PGTS and TGS methods.

18.5 Conclusion

Potato is the one of the most important commercial food crops worldwide. The crop losses due to biotic and abiotic stresses affect the quality and hamper the economy. Late blight caused by *P. infestans* causes the complete yield loss. The conventional late blight control methods available are not fully effective as mentioned above. Biotechnology is emerging with new options which can be promising in the future. Therefore, there is a need to find new control methods which are efficient, eco-friendly and economic.

References

Agrios GN (2005) Plant pathology, 5th edn. Academic, New York/London, p 922

Ah Fong AMV, Bromann-Chung CA, Judelson HS (2008) Optimization of transgenic mediated silencing in *Phytophthora infestans* and its association with small-interfering RNAs. Fungal Genet Biol 45:1197–1205

Ali Z, Smith I, Guest DI (2000) Combinations of potassium phosphonate and Bion (acibenzolar-S-methyl) reduce root infection and dieback of *Pinus radiata*, *Banksia integrifolia* and *Isopogon cuneatus* caused by *Phytophthora cinnamomi*. Australas Plant Pathol 29:59–63

Andreu AB, Guevara MG, Wolski EA, Daleo GR, Caldiz DO (2006) Enhancement of natural disease resistance in potatoes by chemicals. Pest Manag Sci 62:162–170

Anonymous (2007) Knowing your enemy-blight. Available on http://sarvari-trust.org/lateblight.html/05/07/2014

Arnold AE, Majia LC, Kyllo D, Rojas EI, Maynard Z, Robbins N, Herre EA (2003) Fungal endophytes limit pathogen damage in a tropical tree. Proc Natl Acad Sci U S A 100:15649–15654

Aryantha IP, Cross R, Guest DI (2000) Suppression of *Phytophthora* cinnamomi in potting mixes amended with uncomposted and composted animal manures. Phytopathology 90:775–782

ATTRA (Appropriate Technology Transfer to Rural Area) (2004) Organic alternatives for late blight control in potatoes. National Sustainable Agriculture Information Service

Baider A, Cohen Y (2003) Synergistic interaction between BABA and Mancozeb in controlling *Phytophthora infestans* in potato and tomato and *Pseudoperonospora cubensis* in cucumber. Phytoparasitica 31:399–409

Beckerman J (2008) Understanding fungicide mobility. Purdue Extension BP-70-W

Bekele K, Yaynu H (1996) Tuber yield loss assessment of potato cultivars with different levels of resistance to late blight. In: Bekele E, Abdulahi A, Yemane A (eds) "Proceedings of the 3rd

annual conference, of Crop Protection Society of Ethiopia" 18–19 May, Ethiopia, CPSE, Addis Abeba, pp 149–152
Bektas Y, Eulgem T (2015) Synthetic plant defense elicitors. Front Plant Sci 5:804
Bhaik A, Trivedi R (2015) Zorvec™ a new age fungicide for the management of late blight in potato in South asia. 3rd international symposium on Phytophthora, Bengaluru, p 69
Binyam T, Temam H, Tekalign T (2014a) Efficacy of reduced dose of fungicide sprays in the Management of Late Blight (*Phytophthora infestans*) disease on selected potato (*Solanum tuberosum L.)* varieties Haramaya, Eastern Ethiopia. J Biol Agric Health Care 4(20):46–52
Binyam T, Temam H, Tekalign T (2014b) Tuber yield loss assessment of potato (*Solanum tuberosum* L.) varieties due to late blight (*Phytophthora infestans*) and its management Haramaya, eastern Ethiopia. J Biol Agric Health Care 4(23):45–54
Brewer HC, Hawkins ND, Hammond-Kosack KE (2014) Mutations in the Arabidopsis homoserine kinase gene DMR1 confer enhanced resistance to *Fusarium culmorum* and *F. graminearum.* BMC Plant Biol 14:3–17
Brown A, Thomson-Dans C, Marchant N (eds) (1998) Western Australia's threatened Flora. CALM, Como, pp 1–220
Carlson H (1994) Potato pest management guidelines. University of California Statewide IPM Project. Available on www.ipm.ucdavis.edu/PMG/r607101211.html
Chambers SM, Scott ES (1995) In vitro antagonism of *Phytophthora cinnamomi* and *Phytophthora citricola* by isolates of *Trichoderma spp.* and *Gliocladium virens*. J Phytopathol 143:471–477
CIP (International Potato Centre) (1989) Fungal diseases of potato. Report of planning conference on fungal diseases of the potato. CIP, Lima, p 216
Cohen YR (2002) β-amino butyric acid-induced resistance against plant pathogens. Plant Dis 86:448–457
Conrath U (2011) Molecular aspects of defense priming. Trends Plant Sci 16:524–531
Davidse LC, Looijen D, Turkensteen LJ, Van der Wal D (1981) Occurrence of metalaxyl resistant strains of *Phytophthora infestans* in Dutch potato fields. Neth J Plant Pathol 87:65–68
Davis RM, Nunez J, Aegerter BJ (2009) Potato late blight. Statewide IPM Program, Agriculture and natural resources, University of California
Dekker J (1984) The development of resistance to fungicides. Prog Pestic Biochem Toxicol 4:165–218
Draper MA, Secor GA, Gudmestad NC, Lamey HA, Preston D (1994) Leaf blight diseases of potato. Late blight. North Dakota State University Agriculture and University Extension.1084
Drenth A, Guest DI (2004) Managing Phytophthora diseases. In: Drenth A, Guest DI (eds) Diversity and Management of Phytophthora in Southeast Asia, Austalian Centre for International Agricultural Research (ACIAR) Monograph 114. Australian Centre for International Agricultural, Canberra
Du J, Verzaux E, Chaparro-Garcia A, Bijsterbosch G, Keizer LCP, Zhou J, Liebrand TWH, Xie C, Govers F, Robatzek S, van der Vossen EAG, Jacobsen E, Visser RGF, Kamoun S, Vleeshouwers VGAA (2015) Elicitin recognition confers enhanced resistance to *Phytophthora infestans* in potato. Nate Plants 1:15034
Dubreuil-Maurizi C, Trouvelot S, Frettinger P, Pugin A, Wendehenne D, Poinssot B (2010) β-amino butyric acid primes an NADPH oxidase–dependent reactive oxygen species production during grape vine-triggered immunity. Mol Plant Microbe Interact 23:1012–1021
Eckardt NA (2002) Plant disease susceptibility genes? Plant Cell Online 14(9):1983–1986
Ellis C, Karafyllidis I, Wasternack C, Turner JG (2002) The Arabidopsis mutant cev1 links cell wall signalling to jasmonate and ethylene responses. Plant Cell 14:1557–1566
El-Tarabily KA, Sykes ML, Kurtboke ID, Hardy GES, Barbosa AM, Dekker RFH (1996) Synergistic effects of a cellulase-producing Micromonospora and an antibiotic-producing *Streptomyces violascens* on the suppression of *Phytophthora cinnamomi* root rot of *Banksia grandis*. Can J Bot 74:618–624
Erwin DC, Ribeiro OK (1996) *Phytophthora* diseases worldwide. APS Press, St Paul

Fang JG, Tsao PH (1995) Efficacy of *Penicillium funiculosum* as a biological control agent against *Phytophthora* root rots of azalea and citrus. Phytopathology 85:871–878

FAO (Food and Agriculture Organization) (2008) Potato world: Africa-International Year of the Potato 2008. http://www.potato2008.org/en/world/africa.html

Fekede G, Amare A, Nigussie D (2013) Management of Late Blight (*Phytophthora* infestans) of potato (Solanum tuberosum) through potato cultivars and fungicides in Hararghe highlands, Ethiopia. Int J Life Sci 2(3):130–138

Fontem DA, Aighew B (1993) Effect of fungicides on late blight control and yield loss of potato in the western highlands of Cameroon. Int J Pest Manag 39:152–155

Forbes GA, Escobar XC, Ayala CC, Revelo J, Ordoñez ME, Fry BA, Doucett K, Fry WE (1997) Population genetic structure of *Phytophthora infestans* in Ecuador. Phytopathology 87:375–380

Fry WE (1978) Quantification of general resistance of potato cultivars and fungicide effects for integrated control of late blight. Phytopathology 68:1650–1655

Fry WE, Goodwin SB (1997) Resurgence of the Irish potato famine fungus. Bioscience 47:363–371

Fry WE, Goodwin SB, Dyer AT, Matuszak JM, Drenth A, Tooley PW, Sujkowski LS, Koh YJ, Cohen BA, Spielman LJ, Deahl KL, Inglis DA, Sandlan KP (1993) Historical and recent migrations of *Phytophthora infestans*: chronology, pathways, and implications. Plant Dis 77:653–661

Fu ZQ, Dong X (2013) Systemic acquired resistance: turning local infection into global defense. Annu Rev Plant Biol 64:839–863

Garrett KA, Dendy SP (2001) Cultural practices in potato late blight management. In: Fernandez-Northcoted N (ed) "Complementing resistance to late blight (Phytophthora infestans) in the Andes" Proceedings of GILB Latin American workshop I, 13–16 February, Cochabamba, Bolivia, pp 107–113

Goutam U, Kukreja S, Yadav R, Salaria N, Thakur K, Goyal A (2015) Recent trends and perspectives of molecular markers against fungal diseases in wheat. Front Microbiol 6:1–14

Guest DI, Pegg KG, Whiley AW (1995) Control of *Phytophthora* diseases of tree crops using trunk-injected phosphonates. Hortic Rev 17:299–330

Hakiza JJ (1999) The importance of resistance to late blight in potato breeding in Africa (Abstract). In: Proceedings of the global initiative on late blight conference, Quito, Ecuador, p 4

HARC (Holetta Agricultural Report Center) (2007) Impact of farmers' Selected IDM options on potato late blight control and yield. In: African Crop Science conference proceedings, Egypt: El-Minia 8, pp 2091–2094

Haverkort AJ, Boonekamp PM, Hutten R, Jacobsen E, Lotz LA, Kessel GJ, Visser RG, van der Vossen EA (2008) Societal costs of late blight in potato and prospects of durable resistance through cisgenic modification. Potato Res 51:47–57

Heller WE, Theilerhedtrich R (1994) Antagonism of *Chaetomium globosum*, *Gliocladium virens* and *Trichoderma viride* to four soil-borne *Phytophthora* species. J Phytopathol 141:390–394

Huang S, van der Vossen EAG, Kuang H, Vleeshouwers VGAA, Zhang N, Borm TJA, van Eck HJ, Baker B, Jacobsen E, Visser RGF (2005) Comparative genomics enabled the cloning of the *R3a* late blight resistance gene in potato. Plant J 42(2):251–261

Huibers RP, Loonen AEHM, Gao D, Ackerveken GVD, Visser RGF, Bai Y, Damme EJMV (2013) Powdery mildew resistance in tomato by impairment of SlPMR4 and SlDMR1. PLoS One 8(6):e67467

Janus Ł, Milczarek G, Arasimowicz-Jelonek M, Abramowski D, Billert H, Floryszak-Wieczorek J (2013) Normoergic NO-dependent changes, triggered by a SAR inducer in potato, create more potent defense responses to *Phytophthora infestans*. Plant Sci 211:23–34

Jones GD (1998) The epidemiology of plant diseases, 3rd edn. Kluwer Academic Publishes, London, pp 371–388

Jones JDG, Dangl JL (2006) The plant immune system. Nature 444(7117):323–329

Judelson HS, Tani S (2007) Transgene-induced silencing of the zoosporogenesis-specific NIFC gene cluster of *Phytophthora infestans* involves chromatin alterations. Eukaryot Cell 6(7):1200–1209

Keerthana U, Nagendran K, Raguchander T, Rajendran L, Karthikeyan G (2015) Biological management of potato late blight disease using Bacillus subtilis var. amyloliquefaciens. In: 3rd International symposium on Phytophthora, Bengaluru

Kim H, Lee H, Jo K, Mahdi Mortazavian SM, Huigen DJ, Evenhuis B, Kessel G, Visser RGF, Jacobsen E, Vossen JH (2012) Broad spectrum late blight resistance in potato differential set plants *MaR8* and *MaR9* is conferred by multiple stacked R genes. Theor Appl Genet 124:923–935

Kirk W (2009) Potato late blight alert for the Midwest. Field crop advisory team alert current news articles

Kirk W, Wharton P, Hammerschmidt R, Abu-el Samen F, Douches D (2013) Late blight. Michigan State University extension bulletin E-2945. East Lansing. Available on http://www.potatodiseases.org/lateblight.html

Latijnhouwers M, Ligterink W, Vleeshouwers VG, VanWest P, Govers F (2004) A Gα subunit controls zoospore mobility and virulence in the potato late blight pathogen *Phytophthora infestans*. Mol Microbiol 51:925–936

Lazarovits G, Tenuta M, Conn KL (2001) Organic amendments as a disease control strategy for soil borne diseases of high-value agricultural crops. Australs Plant Pathol 30:111–117

Leonards-Schippers C, GieVers W, Salamini F, Gebhardt C (1992) The R1 gene conferring race-specific resistance to *Phytophthora infestans* in potato is located on potato chromosome V. Mol Gen Genet 233:278–283

Li X, van Eck HJ, Rouppe van der Voort JNAM, Huigen DJ, Stam P, Jacobsen E (1998) Autotetraploids and genetic mapping using common AFLP markers: the R2 allele conferring resistance to *Phytophthora infestans* mapped on potato chromosome 4. Theor Appl Genet 96:1121–1128

Li G, Huang S, Guo X, Li Y, Yang Y, Guo Z, Kuang H, Rietman H, Bergervoet M, Vleeshouwers VGAA, van der Vossen E, Qu D, Visser RGF, Jacobsen E, Vossen JH (2011) Cloning and characterization of *R3b*; members of the R3 superfamily of late blight resistance genes show sequence and functional divergence. Mol Plant-Microbe Interact 24(10):1132–1142

Luna E, Ton J (2012) The epigenetic machinery controlling transgenic rational systemic acquired resistance. Plant Signal Behav 7:615–618

Martin AD, Gary AS, Neil CG, Arthur HL, Duane P (1994) Leaf blight diseases of potato. North Dakota State University Agriculture and University Extension

Mercure P (1998) Early blight and late blight of potato. University of Connecticut, Integrated Pest Management. Available on www.hort.uconn.edu/IPM/VEG/HTMS/BLTPOT.HTML

Mesfin T, Gebremedhin G (2007) Impact of farmers' selected IDM options on potato late blight control and yield. In: African Crop Science Conference Proceedings, vol 8, pp 2091–2094

Mosquera T, Alvarez MF, Jiménez-Gómez JM, Muktar MS, Paulo MJ, Steinemann S, Li J, Draffehn A, Hofmann A, Lübeck J, Strahwald J, Tacke E, Hofferbert H-R, Walkemeier B, Gebhardt C (2016) Targeted and untargeted approaches unravel novel candidate genes and diagnostic SNPs for quantitative resistance of the potato (*Solanum tuberosum L.*) to *Phytophthora infestans* causing the late blight disease. PLoS One 11(6):e0156254

Mukalazi J, Adipala E, Sengooba T, Hakiza JJ, Olanya M, Kidanemariam H (2001) Metalaxyl resistance, mating type and pathogenicity of *Phytophthora infestans* in Uganda. Crop Prot 20:379–388

Njualem DK, Demo P, Mendoza HA, Koi JT, Nana SF (2001) Reaction of some potato genotypes to late blight in Cameroon. Afr Crop Sci J 1(1):209–213

Nowara D, Gaya A, Lacommeb C, Shawb J, Ridoutc C, Douchkova D, Hensela G, Kumlehna J, Schweizera P (2010) HIGS: host-induced gene silencing in the obligate biotrophic fungal pathogen *Blumeria graminis*. Plant Cell 22:3130–3141

Nsemwa LTH, Mwalyego FM, Kabungo DA (1992) A review of past crop protection research in the southern highlands of Tanzania and prospects for the future. In: Proceedings of an international conference on agricultural research, training and technology transfer in the southern highlands of Tanzania: past achievements and future prospects. 5–9 October, Uyole Agricultural Centre, Mbeya Tanzania

Oostendorp M, Kuntz W, Dietrich B, Staub T (2001) Induced disease resistance in plants by chemicals. *Eur J Plant Pathol* 107:19–28

Ouimette DG, Coffey MD (1990) Symplastic entrance and phloem translocation of phosphonate. Pestic Biochem Physiol 38:18–25

Pieterse CMJ (2012) Prime time for transgenic rational defense. Plant Physiol 158:545–545

Popokova KV (1972) Late Blight of Potato. Moscow. In: Abate T (ed) (1985) Review of crop production research in Ethiopia. Proceedings of the first Ethiopian crop protection symposium. Institute of Agricultural Research, Addis Ababa, Ethiopia

Powelson M, Inglis DA (1998) Potato late blight: live on the internet. American Phytopathological Society, St. Paul. Available on www.apsnet.org/online/feature/lateblit/

Prime-A-Plant Group, Conrath U, GJM B, Flors V, García-Agustín P, Jakab G, Mauch F, Newman M, CMJ P, Poinssot B, Pozo MJ, Pugin A, Schaffrath U, Ton J, Wendehenne D, Zimmerli L, Mauch-Mani B (2006) Priming: getting ready for battle. Mol Plant-Microbe Interact 19(10):1062–1071

Rasmann S, De Vos M, Casteel CL, Tian D, Halitschke R, Sun JY, Agrawal AA, Felton GW, Jander G (2012) Herbivory in the previous generation primes plants for enhanced insect resistance. Plant Physiol 158:854–863

Rees DJ, Ekpere JA, Kamasho JA, Gillings SI, Mussei ANK, Temu AEM (1992) Adapting research to farmers' circumstances in the southern highlands of Tanzania. In: Proceedings of an international conference on agricultural research, training and technology transfer in the southern highlands of Tanzania: past achievements and future prospects, 5–9 October Uyole Agricultural Centre, Mbeya Tanzania

Schiessendoppler E, Molnar U, Glauninger J, Olaniya M, Bekele K (2003) Characterization of *Phytophthora infestans* populations in Sub-Saharan Africa (SSA) as a basis for simulation modelling and integrated disease management. Ages, Vienna

Shaw DS, Khaki IA (1971) Genetical evidence for diploidy in *Phytophthora*. Genet Res 17:165–167

Shinners CT, Bains P, McLaren D, Thomson J (2003) Commercial potato production – disease management. Western Potato Council. Available on http://www.gov.mb.ca/agriculture//crops/potatoes/bda04s07

Shtienberg D, Raposo R, Bergerson SN, Legard DE, Dyer AT, Fry WE (1994) Inoculation of cultivar resistance reduced spray strategy to suppress early and late blight on potato. Plant Dis 78:23–26

Slaughter A, Daniel X, Flors V, Luna E, Hohn B, Mauch-Mani B (2012) Descendants of primed Arabidopsis plants exhibit resistance to biotic stress. Plant Physiol 158:835–843

Stone A (2009) Organic Management of Late Blight of potato and tomato (*Phytophthora infestans*). Sustainable Agriculture Research and Education. Oregon State University. Available on http://www.extension.org/pages/18361/

Streubel J, Pesce C, Hutin M, Koebnik R, Boch J, Szurek B (2013) Five phylogenetically close rice SWEET genes confer TAL effector-mediated susceptibility to *Xanthomonas oryzae pv. oryzae*. New Phytol 200:808–819

Sun K, Wolters A, Vossen JA, Rouwet ME, Loonen AEHM, Jacobsen E, Visser RGF, Bai Y (2016a) Silencing of six susceptibility genes results in potato late blight resistance. Transgenic Res 25(5):731–742

Sun K, Wolters AMA, Loonen AEHM, Huibers RP, van der Vlugt R, Goverse A, Jacobsen E, Visser RGF, Bai Y (2016b) Down-regulation of Arabidopsis DND1 orthologs in potato and tomato leads to broad-spectrum resistance to late blight and powdery mildew. Transgenic Res 25:123–138

Thind TS (2015) Relevance of fungicides in the present day crop protection and the way ahead. J Mycol Plant Pathol 45(1):4–12

Tinoco MLP, Dias BBA, Dall'Astta RC, Pamphile JA, Aragão FJL (2010) *In vivo* trans-specific gene silencing in fungal cells by *in planta* expression of a double-stranded RNA. BMC Biol 8:27

Ton J, Mauch-Mani B (2004) β-amino-butyric acid-induced resistance against necrotrophic pathogens is based on ABA-dependent priming for callose. Plant J 38:119–130

Ton J, Jakab G, Toquin V, Flors V, Iavicoli A, Maeder MN, Métraux JP, Mauch-Mani B (2005) Dissecting the β-aminobutyric acid-induced priming phenomenon in Arabidopsis. Plant Cell 17:987–999

Vleeshouwers VG, Raffaele S, Vossen JH, Champouret N, Oliva R, Segretin ME, Rietman H, Cano LM, Lokossou A, Kesse G, Pel MA, Kamoun S (2011) Understanding and exploiting late blight resistance in the age of effectors. Annu Rev Phytopathol 49:507–531

Wang G, Liu Z, Lin R, Li E, Mao Z, Ling J, Yang Y, Yin WB, Xie B (2016) Biosynthesis of antibiotic Leucinostatins in bio-control fungus *Purpureocillium lilacinum* and their inhibition on *Phytophthora* revealed by genome mining. PLoS Pathog 12(7):e1005685. https://doi.org/10.1371/journal.ppat.1005685

Worrall D, Holroyd GH, Moore JP, Glowacz M, Croft P, Taylor JE, Paul ND, Roberts MR (2012) Treating seeds with activators of plant defense generates long-lasting priming of resistance to pests and pathogens. New Phytol 193:770–778

You MP, Sivasithamparam K, Kurboke DJ (1996) Actinomycetes in organic mulch used in avocado plantations and their ability to suppress *Phytophthora cinnamomi*. Biol Fertil Soils 22:237–242

Zhang X (2008) The epigenetical and scape of plants. Science 320:489–492

Zhu S, Duwal A, Su Q, Vossen JH, Visser RGF, Jacobsen E (2013) Vector integration in triple R gene transformants and the clustered inheritance of resistance against potato late blight. Transgenic Res 22:315–325

Part III
Fungi for sustainable Industrial and Environmental Aspects

Chapter 19
Recent Advances and Industrial Applications of Microbial Xylanases: A Review

Sunny Dhiman and Gunjan Mukherjee

Abstract Xylanase being a hydrolytic enzyme catalyses the hydrolytic breakdown of 1,4-β-D-xylosidic linkages in xylan which is an important constituent of hemicellulose. Xylanases are hemicellulases required for depolymerization of xylans which are the second most bountiful polysaccharide occurring in nature after cellulose having plant origin. A broad range of organisms have been reported to produce xylanases that include several fungi, bacteria, protozoans, crustaceans, marine algae, insects, snails, gastropods, arthropods, several seeds and plants. Filamentous fungi have been documented to be the useful producers of xylanase because of ease of cultivation, extracellular secretion of enzymes, higher yield and industrial aspect. Fungal xylanases from *Aspergillus* species and *Trichoderma* species have been widely studied and characterized and are commercially utilized in bakery and food processing industries. Microbial xylanases have been reported to be single-chain glycoproteins having molecular masses usually 8–145 kDa and exhibit maximum activity in temperature range 40–60 °C. Thermostable xylanases are ideally suited for use in industrial applications because of numerous advantages over thermolabile xylanase such as ability to work in broad temperature range, better substrate utilization and ability to tolerate high temperature in processes as well as better shelf life. Xylanases have widespread utilization in diverse industries such as food industry, textile industry and in pulp and paper industry. Xylanases have emerged to be extremely beneficial in terms of enhancing the production of numerous fruitful products. Over the years the advancements in molecular tools and techniques have enabled the better understanding of regulatory mechanisms heading xylanase production, underlying mechanism of action of xylanases as well as more precise knowledge of xylanase gene. Such advancements have paved the way for better utilization of enzymes in a much broader sense in commercial sector. Xylanases have tremendous industrial applications in commercial sector either on their own or by associating with different enzymes in numerous processes like processing of pulp and fibres; saccharification of agricultural, industrial and municipal wastes;

S. Dhiman · G. Mukherjee (✉)
Department of Biotechnology, University Institute of Biotechnology, Chandigarh University, Mohali, Punjab, India

P. Gehlot, J. Singh (eds.), *Fungi and their Role in Sustainable Development: Current Perspectives*, https://doi.org/10.1007/978-981-13-0393-7_19

flour improvement for bakery products; pretreatment of forage crops and lignocellulosic biomass; as well as an alternate to treating the textile-cellulosic waste with sulphuric acid.

Keywords Xylanase · Xylan · Hemicellulases · Saccharification · Glycoproteins · Immobilization

19.1 Introduction

Xylanase (E.C 3.2.1.8) is an enzyme belonging to glucanase family and has a quite an expanding group of enzymes which can hydrolyse the 1,4-β-D-xylosidic linkages in xylan. Being a complex molecule, the depolymerization of xylan necessitates a cooperative action of different enzymes for its thorough disintegration. β-1,4-Endoxylanase, β-xylosidase, α-L-arabinofuranosidase, α-glucoronidase, acetyl xylan esterase and phenolic acid esterase occupy a significant place amongst the xylanases that have been reported extensively. According to Sharma and Kumar (2013), xylanase is a significant industrial enzyme that causes the random disintegration of xylan by its endo-1,4-xylanase activity and produces xylose, xylooligosaccharides and xylobiose. Amongst these xylanases, endo-1,4- xylanases (1,4-β-D-xylan xylanohydrolase, E.C.3.2.1.8) catalyse xylan depolymerization by randomly hydrolysing xylan backbone. Whistler and Masek (1955) reported xylanase for the first time. According to Bastawde (1992), the importance of xylanases is discovered over 100 years ago by Hoppe-Seyler. Xylanases were primarily termed pentosanases and were recognized by International Union of Biochemistry and Molecular Biology (IUBMB) for the first time in 1961. The systematic name of xylanases is endo-β-1,4-xylanase, but more widely accepted and universally used synonyms of the enzyme include xylanase, endo-1,4-β-D-xylanase, endoxylanase, β-1,4-D-xylan xylanohydrolase, β-xylanase and β-1,4-xylanase, respectively. A number of key factors govern the yield of xylanases in fermentation that include accessibility of substrate, rate and extent of disentangling of the xylooligosaccharides, etc. Omar et al. (2008) reported that xylanase holds significant importance due to its proficiency of degrading the plant cell wall constituents. Xylanase has attained paramount industrial importance credited to their multidimensional and multifunctional role in fermentation processes and numerous other industries.

19.2 Substrates for Xylanase

Hemicelluloses are the major substrates for xylanases. The word hemicelluloses points to polysaccharides occurring in plant cell wall that are confederated with cellulose and glucans. Hemicelluloses are the second most bountiful constituents in cell wall of plants behind cellulose (Nakamura 2003). Hemicelluloses consist of complex of polymeric carbohydrates that include xylan, glucomannan, xyloglucan, galactoglucomannan and arabinogalactan (Shallom and Shoham 2003). Classification of such polymeric carbohydrates relies upon the type of sugar entities present in their structure. Xylan is a major hemicellulose composed of xylose units interlinked through β-1,4-glycosidic linkage. Xylan has been reported to be the second most abundant polysaccharide behind cellulose that is loaded with a huge potential of being converted into a majority of products having paramount importance. Saha (2003) reported that xylan holds for nearly one third of entire replenishable organic carbon present on our planet. Xylan has a complex structure and heterogenic nature. Thus on account of this, xylanase plays a prodigious role in the combination of hydrolytic enzymes obligatory for the thorough disintegration of xylan (Takahashi et al. 2013). Xylan is a heterogenic polysaccharide comprising of xylose units interlinked through β-1,4-glycosidic bonds. Whistler and Richards (1970) reported that xylan major chain is built up of β-xylopyranose units. Xylan predominantly occurs in the secondary cell wall, and it constitutes the most part of the polymeric fraction of plant cell wall in association with lignin and cellulose. Xylan provides integrity to the cell wall by virtue of its association between lignin and cellulose through covalent and noncovalent bonds (Motta et al. 2013).

19.2.1 Structural Framework and Distribution of Xylan

Xylan being a convoluted heteropolysaccharide has an eminently branched structure that differs remarkably amidst distinctive plant species. On the grounds of common substituents occurring on the xylan backbone, xylans have been categorized into linear homoxylan, arabinoxylan, glucuronoxylan and glucuronoarabinoxylan. However as a matter of fact, there occurs microheterogeneity in each category of the xylan in relevance to the extent and characteristic of branching. The occurrence of side chains in the substituted forms of xylan determines several aspects that include physical configuration, degree of solubility, mode of the enzyme action, etc. (Motta et al. 2013). On account of its intricate structure and heterogenic nature, the thorough depolymerization of xylan necessitates a cooperative action of several enzymes (Subramaniyan and Prema 2002). Xylan has been reported to be distributed in a vast variety of tissues and cells and occurs in the major fraction of plant species. Singh et al. (2003) reported that xylan is known to exist up to significant levels in hardwoods of angiosperms (15–30%) and softwoods of gymnosperms (7–10%) and also in annual plants (<30%). Wood xylan predominantly occurs as O-acetyl-4-O-methylglucuronoxylan in hardwoods and in softwoods as arabino-4-O-methylglucuronoxylan. However xylans occurring in grasses and annual plants are generally arabinoxylans (Kulkarni et al. 1999). Xylan has a similarity to other polysaccharides having plant origin with respect to vast polydiversity and polymolecularity (Sunna and Antranikian 1997). The extent of polymerization in xylans also has significant variability, for instance, hardwood xylans generally consist of 150–200 β-xylopyranose units, whereas softwood xylans generally consist of 70–130 β-xylopyranose units (Kulkarni et al. 1999). D-Xylans hold for 20–35% of entire dry weight in hardwood and annual plants and predominantly exist as the most prevalent non-cellulosic polysaccharides (Velkova et al. 2007).

19.2.2 Enzymatic Disintegration of Xylan

On account of its complex structure and heterogenic nature, thorough disintegration of plant xylan necessitates the cooperative involvement of a multienzyme hydrolytic complex having broad spectrum of activities and diverse approaches of action. The main enzymes participating in multienzyme hydrolysis of xylan include β-xylosidase, endoxylanase, acetyl xylan esterase, arabinofuranosidase, glucuronidase, galactosidase and feruloyl esterase. These enzymes act in a collegial fashion to depolymerize xylan into its monomeric units (Belancic et al. 1995). Endoxylanases hold foremost place amongst all xylanases on account of their direct participation in the cleavage of glycosidic bonds as well as in liberating short xylooligosaccharides (Verma and Satyanarayana 2012).

Xylanase randomly hydrolyses xylans into xylooligosaccharides, whereas β-xylosidase dislodges xylose units from the non-reducing ends of xylooligosaccharides. Despite this, thorough disintegration needs the activity of acetyl esterase to dislodge the acetyl substituents from the D-xylose backbone (β-1,4-linked) of xylan (Coughlan and Hazlewood 1993).

19.3 Xylanase Sources

Xylanase sources range from microorganisms like bacteria, fungi, protozoans, actinomycetes, crustaceans, marine algae, insects, snails, gastropods, arthropods, several seeds and plants that tears the glycosidic linkages in xylans thus resulting in hemi acetyls and glycans (Motta et al. 2013). Certain invertebrate organisms such as earthworms also possess a broad spectrum of fibrolytic microbes in their gut. Thus it is quite possible that a few amongst these organisms might be the producers of novel endoxylanases (Park et al. 2007). Rumen is one of the most fascinating sources of xylanases wherein the effective hydrolysis of plant polysaccharides commences. Filamentous fungi have been documented to be the useful xylanase producers because of ease of cultivation, extracellular secretion of enzymes, higher yield and industrial aspect. Fungal xylanases from *Aspergillus* species and *Trichoderma* species have been widely and thoroughly studied and characterized and are commercially utilized in bakery and food processing industries. Certain microorganisms such as *Aspergillus* and *Penicillium* have been documented to efficiently synthesize xylanases in the presence of cheaper hemicellulosic substrates such as rice bran, bagasse corn stalk, corn cob, wheat bran and rice straw. White rot fungus is extensively utilized in manufacturing pharmaceutical products, in food industries and in the manufacturing of cosmetics (Qinnhe et al. 2004). White rot basidiomycetes conducts its depolymerization action on lignocellulosic materials through the enzymatic action of several xylanase enzymes. Several bacterial strains including *Bacillus* sp. have also been reported to be the efficient producers of thermostable xylanases having good stability at varying range of temperature, pH, presence of metal ions, etc. (Battan et al. 2007).

19.3.1 Fungal Xylanases

Fungal xylanases from genera *Aspergillus*, *Trichoderma*, *Pichia* and *Fusarium* species have been considered as the efficient and producers of novel xylanases (Absul et al. 2005). Other efficient fungal species producing xylanases include *Streptomyces* (Kansoh et al. 2001) *Aspergillus kawachii* (Ito et al. 2000) and *Cunninghamella subvermispora* (Ferraz et al. 2004). Haq et al. (2002) documented *Aspergillus niger* as the most potent xylanase producer.

19.3.2 Bacterial Xylanases

Bacterial species producing high activity xylanases at alkaline pH and higher temperature are Bacillus sp. (Subramaniyan et al. 2001). Main bacterial genera that are efficient producers of novel xylanases are *Bacillus halodurans* (Ebrahimi 2010), *Saccharopolyspora pathunthaniensis* (Verma and Satyanarayana 2012), *Thermotoga* sp. (Yoon et al. 2004), *Bacillus circulans* D1(Bocchini et al. 2005), *Pseudomonas* sp. XPB-6 (Sharma and Chand 2012), *Stenotrophomonas maltophilia* (Raj et al. 2013) and *Bacillus* genus (Kaur et al. 2015). Certain filamentous bacterial strains documented to be the efficient producers of endoxylanases, xylanase and polygalacturonase are *Streptomyces* sp., *S. roseiscleroticus* and *S. cuspidosporus* (Maheswari and Chandra 2000).

19.3.3 Thermophilic Xylanases

A good proportion of xylanase-producing thermophilic and hyperthermophilic microorganisms have been reported and successfully explored from a vast variety of sources including thermal springs, self-heating decaying organic debris, hot pools and terrestrial and marine solfataric fields (Sunna and Bergquist 2003). Family 10 xylanases have been successfully explored from a diverse range of thermophilic and hyperthermophilic organisms, including *Thermoascus aurantiacus* (Khasin et al. 1993) and *Nonomuraea flexuosa* (Hakulinen et al. 2003). Xylanases from *Dictyoglomus thermophilum* and *Nonomuraea flexuosa* have been reported amongst the most stable xylanases and possess temperature optima of 80 °C and 85 °C, respectively. Certain hyperthermophilic archaea have also been reported to be the efficient producers of novel xylanases, e.g. *Pyrodictium abyssi* (Andrade et al. 1999), *Thermofilum* strains (Andrade et al. 1999), *Pyrococcus furiosus* (Cady et al. 2001), *Thermococcus zilligii* (Cady et al. 2001), *Sulfolobus solfataricus* (Cannio et al. 2004) and *T. leycettanus* (Wang et al. 2016) (Tables 19.1, 19.2, and 19.3).

Table 19.1 List of xylanase-producing microorganisms

Microorganisms	References
Fungi	
Trichoderma harzianum	Sanghvi et al. (2010)
Aspergillus niger	Subbulakshmi and Priya (2014)
Trichoderma reesei SAF3	Kar et al. (2006)
Marasmius sp.	Ratanachomsri et al. (2006)
Aspergillus terreus UL 4209	Chidi et al. (2008)
Fusarium solani F7	Gupta et al. (2009)
Aspergillus awamori	Teixeira et al. (2010)
Penicillium citrinum	Ghoshal et al. (2011)
Aspergillus usamii	Zhou et al. (2011)
Trichoderma sp.	Norazlina et al. (2013)
Cladosporium sp.	Patel and Prajapati (2014)
Penicillium crustosum	Mushimiyimana and Padmavathi (2015)
Aspergillus sp.	Thomas et al. (2016)
Aspergillus nidulans	Gabriela et al. (2016)
Rhizopus oryzae SN5	Pandey et al. (2016)
Aureobasidium pullulans NRRL Y-2311-1	Yegin (2016)
Bacteria	
Sclerotinia sclerotiorum S2	Ellouze et al. (2008)
Bacillus cereus	Roy and Habib (2009)
Bacillus pumilus	Monisha et al. (2009)
Bacillus sp. *YJ6*	Yin et al. (2010)
Bacillus sp.	Azeri et al. (2010) and Bahri et al. (2011)
Streptomyces sp. *P12–137*	Coman and Bahrim (2011)
Pseudomonas sp. *XPB-6*	Sharma and Chand (2012)
Colletotrichum graminicola	Zimbardi et al. (2013)
Bacillus genus	Kaur et al. (2015)

Table 19.2 Thermophilic organisms for family 10 xylanase

S. No.	Name of the organism	Temperature	Reference
1.	*Clostridium thermocellum*	70 °C	Herbers et al. (1995)
2.	*Rhodothermus marinus*	65 °C	Karlsson et al. (2004)
3.	*Thermotoga* sp.	105 °C	Shi et al. (2013)
4.	*Thermoascus aurantiacus* RCKK	45 °C	Jain et al. (2014)
5.	*Geobacillus stearothermophilus* KIBGE-IB29	60 °C	Bibi et al. (2014)
6.	*Caldicellulosiruptor* sp.	75 °C	Meng et al. (2015)

Table 19.3 Thermophilic organisms for family 11 xylanase

S. No.	Name of the organism	Temperature	Reference
1.	*Dictyoglomus thermophilum*	70 °C	McCarthy et al. (2000)
2.	*Paecilomyces variotii*	60 °C	Kumar et al. (2000)
3.	*Thermomyces lanuginosus*	60 °C	Singh et al. (2003)
4.	*Chaetomium thermophilum*	70 °C	Ahmed et al. (2012)
5.	*Anoxybacillus flavithermus* TWXYL3	65 °C	Ellis and Magnuson (2012)
6.	*Humicola insolens* Y1	50 °C	Shi et al. (2015)

19.4 Classification of Xylanases

Xylanases can be categorized into two separate groups: (1) one possessing low molecular mass (<30 kDa) and basic pI and (2) another one possessing high molecular mass (>30 kDa) and acidic pI (Wong et al. 1998). Also on the basis of primary sequence homology, xylanases have been categorized into two distinct families (family F or 10 and family G or 11). Xylanase under family F has lower pI values as compared to family G. Earlier in a study, family 10 xylanases have been reported to be more complex and diverse having high molecular mass (>30 kDa) (Ducros et al. 2000). In contrast family 11 xylanases are much simpler, have consistency in their structure, have greater specificity for xylan and possess low molecular mass (>20 kDa) in comparison to family 10 xylanases. Woodward (1984) reported that three distinct types of xylanases are known for their involvement in xylan degradation.

19.4.1 Endo-1,4-β-xylanase (1,4-β-D-Xylan Xylanohydrolase)

Endo-1,4-β-xylanase (1,4-β-D-xylan xylanohydrolase; EC 3.2.1.8) disunites the glycosidic linkages in the xylan backbone thus leading to reduced substrate polymerization. Xylan is not degraded randomly. As a matter of fact, the xylan hydrolysis relies upon the attributes of the substrate molecule, i.e. upon the chain length, presence of substituents and extent of branching (Li et al. 2000). Xylanase categorizes themselves into two distinct types on the grounds of type of end products of the reaction: (a) non-debranching or non-arabinose liberating and (b) branching or arabinose liberating.

(a) Non-debranching endoxylanases: Non-debranching or non-arabinose liberating endoxylanases are those endoxylanases that can be compartmentalized into two distinct variants, one giving off xylose and xylobiose as the end products and the other one giving off xylooligosaccharides as the end product.
(b) Branching endoxylanases: Branching or arabinose liberating endoxylanases can be compartmentalized into two discrete groups: group 1 that possess the capability to hydrolyse branching points hereby giving off xylooligosaccharides and arabinose as the end products and group 2 that cleaves off the xylan and branching points thereby giving off principally xylobiose, xylose and arabinose, respectively.

19.4.2 *Exo-1, 4-β-xylanase (β-1,4-D-Xylan Xylohydrolase)*

These enzymes discharge the single xylose units from the non-reducing end of the xylan chain.

19.4.3 *β-D-Xylosidase (1,4-β-D-Xylan Xylohydrolase)*

β-D-Xylosidases (1,4-β-D-xylan xylohydrolase; EC 3.2.1.37) are exoglycosidase that actively depolymerizes short xylooligosaccharides to give off xylose. β-D-Xylosidases can be classified on the grounds of their relative fondness towards xylobiose and larger xylooligosaccharides, respectively. Octavio et al. (2006) documented that a huge fraction of bacteria and fungi efficiently produce such xylanases. They may show their existence in the culture broth surrounding the cell, in alliance with the cell, or they may also exist in both. β-D-Xylosidases play a significant role in lowering the end product inhibition of endoxylanases which is a rate-limiting factor in xylan depolymerization (Andrade et al. 2004).

19.5 Mechanism of Xylanase Action

Numerous stereotypes have been put forward to elucidate the action mechanism of xylanases. Subramaniyan and Prema (2002) reported that xylanase action eventually results in the hydrolysis of xylan that may cause retention or inversion of the anomeric centre of the reducing sugar monomer of the xylan thus giving an intimation of one or two chemical transition states being involved. Transfer of glycosyl eventually causes nucleophilic substitution at the saturated carbon of the anomeric centre and commences with either retention or inversion of the anomeric configuration. A great majority of hydrolytic enzymes like xylanases and cellulases that are well recognized for hydrolysing polysaccharides eventually result in the hydrolysis of their corresponding substrates with the retention of the C1 anomeric configuration. Double displacement mechanism has been reported to be directly indulged in the anomeric retention of product (Clarke et al. 1993).

19.6 Xylanase Production

The main driving force behind search for novel xylanases is the broader range of its tremendous industrial applications. Both solid-state fermentation (SSF) system and submerged fermentation (SMF) system can be successfully utilized. Xylanase production can be efficiently carried out using solid-state cultivation systems and submerged cultures methods. Submerged fermentation (SMF) technique has been the method of choice by most of researchers because of easier regulation of various process parameters such as pH, temperature of medium, degree of aeration as well as several environmental factors indispensable for the optimal growth of microorganisms. However as a matter of fact, solid-state fermentation has procured significant attention and acceptance from the researchers worldwide over the years and has been successfully and widely utilized for xylanase synthesis (Haltrich et al. 1996). This is credited to numerous economic and engineering advantages. Submerged fermentation (SMF) system has been preferred over solid-state fermentation (SSF) system for products involving large-scale production because the yield of enzyme is higher (about 90%) and also more cost-effective as compared to solid-state fermentation (Gouda 2000). SMF has the advantage over solid-state fermentation system in extracting a large fraction of purified enzymes. Wheat bran came out to be the best carbon source in the studies conducted on xylanase production by *Stenotrophomonas maltophilia* after using commercial xylans and different agro-industrial residues (Raj et al. 2013). *Bacillus arseniciselenatis* DSM-15340 resulted in a thermoalkalophilic cellulose-free xylanase in significant level, while it was grown in solid-state conditions by utilizing economically accessible agro-residual substrate wheat bran. Thus it could be efficiently utilized for production of xylanase on large scale by utilizing such agro-residual substrates (Kamble and Anandrao 2012). SSF conditions are exclusively favourable for the fungal growth since these organisms possess the ability to grow at rather low water activities in contrast to most of the bacteria and yeast that do not grow and proliferate efficiently in such culture environment. Mushimiyimana et al. (2015) reported that xylanolytic enzymes are efficiently produced by fungi under submerged conditions. Microbes such as *Trichoderma*, *Aspergillus*, *Phanerochaete*, *Streptomyces*, *Clostridia*, *Ruminococcus*, *Chytridiomycetes*, *Bacillus* and *Fibrobacteres* are loaded with huge potential to efficiently produce xylanase enzymes (Qinnhe et al. 2004).

19.7 Characterization and Purification

Xylanases from distinct sources vary significantly in their temperature and pH optima for the maximal activity. Concentrated and pure enzymes exhibit enhanced activity and reduced risk of inhibitory substances thus putting forward themselves as ideal materials having tremendous potential for use in industrial applications. Standard column chromatography, size exclusion chromatography and ion exchange chromatography are principally utilized techniques for purification of xylanases. Dean et al. (1991) reported that low molecular weight of xylanases has rendered their successful segregation from other proteins utilizing ultra filtration technique. Widjaja et al. (2009) purified cellulase-free xylanase from *Aspergillus niger* and *Trichoderma reesei* using ion exchange chromatography. The effective utilization of xylanase for the treatment of pulp fibres demands cellulase free xylanase. Goulart et al. (2005) successfully produced cellulase-free xylanase by utilizing *Rhizopus stolonifer* cultured on wheat bran. This cellulase-free xylanase exhibited optimum activity at pH 6.0 and temperature 45 °C, respectively. For maximal xylanase activity, pH 5.5 and temperature 60 °C were documented as most appropriate conditions by Huang and Penner (1991). Coelho and Carmona (2003) documented that xylanases are significantly thermostable within the pH range 4.5–10.5. Camacho and Aguilar (2003) documented a molecular weight of 22 kDa for xylanase from *Aspergillus* sp. Sardar et al. (2000) reported a molecular weight of 24 kDa for purified xylanase upon SDS-PAGE. Yasinok et al. (2010) reported the 186-fold purification of xylanase from *Bacillus pumilus* SB-M13A by hydrophobic interaction.

19.8 Xylanase Immobilization

Pioneer immobilization reaction was executed for introduction of reactive groups onto inert glass surface so as to increase the accessible surface area for immobilization. Therefore activation of glass beads was undertaken. Free xylanase exhibited optimum activity at pH 5.0 and 35 °C temperature. Crude enzyme was immobilized onto glass beads by physical adsorption binding. Immobilized enzyme can be reused two to three times under assay conditions. The free and immobilized xylanase activity was assayed at different pH of buffer (0.1 M) ranging from 4.0 to 8.0 and at various temperatures (35–65 °C) to determine the optimum activity under reaction conditions. The immobilized xylanase was tested for its reusability using 1 g of immobilized support repeatedly up to four times and percent relative activity determined (Kumar et al. 2014).

19.9 Industrial Aspects and Applications of Xylanases

Xylans and xylanases have witnessed significant rise in their biotechnological values and have shown remarkable rise in their use. The end products of xylan degradation, furfural and xylitol have gained remarkable utility in industrial applications (Parajo et al. 1998). The industrial utilization of xylanase commenced in the 1980s. Initially xylanases were used in animal feed preparation. In the successive years, they were significantly utilized in the food, textile and paper industries, respectively. Xylanases are predominantly employed in food industry to quicken the baking process of cookies, crackers, cakes as well as various other foods since they depolymerize the polysaccharides in the dough into corresponding monomeric units.

19.9.1 Bioprocessing of Fibres

Modern research centres on substituting the hazardous chemicals with the commercial enzymes that can precisely act upon the non-cellulosic and hemicellulosic impurities are still sustaining the quality as well as upholding the production yields of textile industries (Dhiman et al. 2008). Treatment with enzyme can apparently strengthen the water soaking characteristics of fibres by eradicating the complex impurities present in the primary cell wall. (Saha 2000) reported that utilization of pure and thermostable xylanase for pretreatment of low quality jute fibers for selectively removing xylan is entrancing. Plant fibres such as linen can be effectively processed by utlizing the xylanolytic enzyme complex. This method has an advantage that the step involving usage of strong bleaching step is bypassed as lignin won't face oxidation which would have darkened the fibres (Csiszar et al. 2001). Xylanase precisely acts upon the hemicellulosic impurities and effectively eradicates them. Such enzymatic treatment do not create any harm to the fibre in terms of loss in fibre strength (Dhiman et al. 2008).

19.9.2 Biobleaching of Pulp and Paper

Treatment of the paper pulp with xylanase effectively hydrolyses the hemicellulosic chain amidst cellulose and lignin thereby eliminating the loosely held lignin from the desired cellulose. Xylanases have turned out to be a valuable as well as cost-effective asset for mills to have an edge over a vast number of bleaching benefits (Bajpai 2012). In this context, it lessens the discharge of organochlorine pollutants, for instance, dioxin, ultimately resulting in chlorineless bleaching without posing any detrimental harm to the paper's strength (Li and Hardin 1998). The major utilization of xylanases in commercial sector is in cellulose pulp bleaching. The usage of enzymes commenced in this context ever since peroxidases were employed for degrading the lignin (Sandrim et al. 2005). Dhillon and Khanna (2000) reported chemical process is preferred over enzymatic hydrolysis method for paper production in several countries, including Brazil. The routinely used method is popularly called as the kraft process. The paper production process involves chemical pulping as the pioneer step which is characterized by the breakdown of fibres and removal of majority of lignin fraction (Hong et al. 1989). Pulp bleaching can be represented as a purification process which involves the attributes such as destruction as well as the alteration or solubilization of the coloured organic matters, lignin and other inadmissible leftovers on the fibres (Madlala et al. 2001). The efficacy as well as efficiency of microbial xylanase in context of bleaching process has been well studied for *Streptomyces galbus* (Kansoh and Nagieb (2004), *Bacillus pumilus* (Duarte et al. 2003), etc. The optimum pH of bacterial xylanases in common is somewhat uplifted compared to optimum pH of fungal xylanases (Khasin et al. 1993), which is a desirable asset in majority of paper and pulp industries.

19.9.3 Significant Role in Improving the Animal Feed

Xylanase is well recognized to play a significant role in uplifting the quality of animal feed. Xylanase treatment lessens the viscosity of the fodder thereby rendering the fodder readily digestible by the animal gut. It significantly elevates propagation of the pancreatic enzymes into the food thereby boosting the overall absorption of the nutrients. Gilbert and Hazlewood (1993) documented utility of xylanase in enhancing the digestibility of ruminant feeds and also in speeding up the composting process. Xylanases have been well documented for their utilization in animal feed in association with cellulases, amylases, galactosidases, glucanases, lipases, pectinases, proteases and phytases. Twomey et al. (2003) reported that these enzymes depolymerize arabinoxylans present in the constituents of the feed thereby rendering the raw material less viscous. Young fowl and swine synthesize endogenous enzymes in comparatively lesser amount as compared to adults, so that food supplements loaded with exogenous enzymes should uplift their performance as livestock. Furthermore, such diet has been found to cut down the undesirable leftovers in the excreta (nitrogen, zinc, copper and phosphorus), an effect that might play a productive role in scaling down of the environmental contamination (Polizeli et al. 2005). Addition of xylanase to feed comprising of low viscosity foods like maize and sorghum may enhance the digestion and absorption of nutrients in the foremost part of the digestive tract eventually resulting in an improved use of energy (Van Paridon et al. 1992).

19.9.4 Pharmaceutical and Chemical Applications of Xylanase

The end product of xylan hydrolysis, xylitol a polyalcohol, is as an artificial sweetener that is significantly used in candies, chewing gums and several other food items (Parajo et al. 1998). Xylanases are sometimes added as a cocktail (mixture) of enzymes comprising of proteases, hemicellulases and various other enzymes as a dietary supplement or as a measure to cure weak digestion. Xylitol being a noncarcinogenic sweetener is highly suited for individuals suffering with diabetes and obesity. Xylitol is also recommended in cases such as lipid metabolism disorder and respiratory infections, for the prevention of osteoporosis as well as for persons suffering with kidney and parental lesions. A vast variety of commercially available products such as candies, chewing gums and several other food products are known to have xylitol as the artificial sweetener. The advancements in xylitol production technology have facilitated a way for its utilization in a broad sense in the food, odontological and pharmaceutical (Nigam and Singh 1995). The production of eco-friendly biological fuels like bioethanol is witnessing a significant hike as the other available energy sources are depleting continuously. Moreover most of the fuels currently in use generate high levels of toxic aerosols and other polutants that pose several health hazards. The products of xylan hydrolysis can be efficiently transformed into important biological fuels such as ethanol (Sun and Cheng 2002).

19.9.5 Applications in Recycling of Waste Paper

Xylanases have also proved their utility in recycling of waste paper. The recycling is principally accomplished via two-staged processes, i.e. pulping and beating. Stage one essentially consists of the sundering of fibre or fibre dissemination. The entire exercise is called as hydrating process. The foremost step in enzymatic treatment primarily consists of prefatory soaking of paper followed by enzyme incubation. However stage two essentially consists of mechanical shearing of pulp, subsequent heating of pulp with a purpose of disaggregation of fibres and deactivating the enzyme. Xylanase treatment accounts for dislodging numerous reducing sugars from waste paper pulp. The release of reducing sugars is directly correlated to temperature probably on grounds that elevated temperature hydrolyses the majority of xylans held amidst the pulp fibres. Enzymatic treatment eventually promotes the swelling of pulp fibres, which aids in further processing of the pulp material as well as significantly upgrades its physical properties (Kenealy and Jeffries 2003).

19.9.6 Applications of Xylanases in Food, Bread and Drinks Production

The industrial applications of xylanases have witnessed a significant rise in last few decades credited to their potent efficacy in bread making process (Butt et al. 2008). The usage of starch- as well as non-starch-hydrolysing enzymes is a common practice in bread making industry for uplifting the quality and texture of bread (Javier et al. 2007). Like other hemicellulases, xylanases act in a similar manner thereby depolymerizing the hemicellulose existent in wheat flour thus assisting in uniform circulation of water thereby rendering the dough more softer as well as easy to knead. Xylanases assist in delaying the crumbing process during bread baking process thereby letting the dough to grow (Polizeli et al. 2005). Xylanase utilization in baking industry has significantly aided in a significant rise in bread volumes, better absorption of water as well as enhanced resistance to fermentation (Camacho and Aguilar 2003). Butt et al. (2008) reported that xylanases effectively transform the hemicelluloses that are water insoluble into a soluble form that actively binds to the water in dough thereby reducing the firmness in dough, enhances the volume and generates finer crumbs with increased uniformity. Various enzymes like xylanases, cellulases and proteases enhance the firmness of the gluten network thereby uplifting the worthiness of bakery products (Gray and Bemiller 2003). Xylanases are highly commended for use in biscuit industry with a purpose to make cream crackers lighter as well as to uplift the texture, uniformity and palatability of the wafers. The major desirable aspects of xylanases in food industry are endurance and ability to show optimum activity in acidic pH range. The advancements in molecular tools and techniques have paved the way for more and more uses of xylanases. Xylanases play a significant role in beer production process. They effectively depolymerize arabinoxylans to lower xylooligosaccharides thereby rendering the beer less viscous thus eliminating its muddy appearance to significant levels (Dervilly et al. 2002).

19.9.7 Other Important Applications of Xylanases

Xylanases along with other hydrolases can be efficiently employed for the synthesis of important biofuels like ethanol by utilizing lignocellulosic biomass (Ahring et al. 1999). Xylanase in association with pectinase, amylase and carboxymethylcellulase can be efficiently utilized for clarification of juices. Xylanases may also be employed to enhance the extraction of coffee, plant oils and starches. Xylanases may also be successfully capitalized for boosting the nutritional aspects of agricultural silage and grain feed (Malathi and Devegowda 2001). Xylanases also have significant use in rye baking wherein the addition of xylanase prompts the dough to become more soft and sloppy (Harbak and Thygesen 2002).

19.10 Future Prospects

The industrial importance of xylanase is well established. Amongst different hydrolytic enzymes, xylanase has attained widespread commercial importance credited to its enormous potential applications in food, in feed and in pharmaceutical industries. The surplus availability of hemicellulosic biomass especially xylan becomes a major factor in xylanase production by various microorganisms. Fungal xylanases from *Aspergillus* species and *Trichoderma* species have been widely studied and characterized and are commercially utilized in bakery and food processing industries. Xylanase production economics is governed by several key factors that include accessibility of substrate and rate and extent of disentangling of the xylooligosaccharides besides several other decisive factors including inoculum size, pH, temperature, inducers, medium additives, aeration, activators and inhibitors. Submerged fermentation (SMF) system is the most promising technique used worldwide for xylanase production. This is credited to ease of control over various key process parameters, to the higher yield of enzyme (about 90%) and because of being cost-effective as compared to solid-state fermentation. The present review focuses on the various microbial sources for novel xylanase production, the range of available substrates that can be successfully utilized to meet the industrial demands of xylanases and the widespread industrial applications of microbial xylanases.

The prospects of xylan hydrolysis by xylanase from fungal species such as *Aspergillus* and *Trichoderma* and bacterial species like *Bacillus* sp. look quite promising. Thus future studies to increase the xylan hydrolysis rate as well as to assure enhanced process control for increased yield of xylanase would be envisaged.

In conclusion, xylanase is an industrially important enzyme that is loaded with huge potentials for use in commercial sector in various processes such as processing of pulp and fibres; saccharification of agricultural, industrial and municipal wastes; flour improvement for bakery products; manufacturing of several food products; and enhanced bleaching of cellulose pulps that is mainly used in food and pharmaceutical industry. Since the range of applications of this enzyme is very broad, so there is always a scope for novel xylanase with better and improved characteristics, which may be utilized for various industrial applications.

References

Absul MG, Ghule JE, Shaikh H et al (2005) Enzymatic hydrolysis of delignified bagasse polysaccharides. Carbohydr Polym 62:6–10

Ahmed S, Imdad SS, Jamil A (2012) Comparative study for the kinetics of extracellular xylanases from Trichoderma harzianum and Chaetomium thermophilum. Electron J Biotechnol 15:3

Ahring BK, Licht D, Schimd AS et al (1999) Production of ethanol from wet oxidised wheat straw by *Thermoanaerobacter mathranii*. Bioresour Technol 68:3

Andrade CMMC, Pereira N, Antranikian G (1999) Extremely thermophilic microorganisms and their polymer-hydrolytic enzymes. Rev Microbiol 30:287–298

Andrade SV, Polizeli MLTM, Terenzi HF et al (2004) Effect of carbon source on the biochemical properties of the β-xylosidase produced by *Aspergillus versicolor*. Process Biochem 39:1931–1938

Azeri C, Tamer AU, Oskay M (2010) Thermoactive cellulase-free xylanase production from alkaliphilic *Bacillus* strains using various agro-residues and their potential in biobleaching of kraft pulp. Afr J Biotechnol 9:63–72

Bahri DO, Coskun A, Ozcan N et al (2011) Some properties of a new thermostable xylanase from alkaliphilic and thermophilic *Bacillus* sp. J Anim Vet Adv 10:138–143

Bajpai P (2012) Biotechnology for pulp and paper processing. Springer 3:42–45

Bastawde KB (1992) Xylan structure microbial xylanases, and their mode of action. World J Microbiol Biotechnol 8:353–368

Battan B, Sharma J, Dhiman SS et al (2007) Enhanced production of cellulase-free thermostable xylanase by *Bacillus pumilus* ASH and its potential application in paper industry. Enzym Microbiol Technol 41:733–739

Belancic A, Scarpa J, Peirano A et al (1995) Purification and properties of two of the enzymes. J Biotechnol 41:71

Bibi Z, Ansari A, Zohra RR et al (2014) Production of xylan degrading endo-1, 4-b-xylanase from thermophilic *Geobacillus stearothermophilus* KIBGEIB29. J Radiat Res Appl Sci 7:478–485

Bocchini DA, Oliveira OMM, Gomes E et al (2005) Use of sugarcane bagasse and grass hydrolysates as carbon sources for xylanase production by *Bacillus circulans* D1 in submerged fermentation. Process Biochem 40:3653–3659

Butt MS, Nadeem MT, Ahmad Z et al (2008) Xylanases and their applications in baking industry. Food Technol Biotechnol 46:22–31

Cady SG, Bauer MW, Callen W et al (2001) Beta-endoglucanase from *Pyrococcus furiosus*. Methods Enzymol 330:346–354

Camacho NA, Aguilar OG (2003) Production, purification and characterization of a low molecular mass xylanase from *Aspergillus* sp. and its application in bakery. Appl Biochem Biotechnol 104:159–172

Cannio R, Di Prizito N, Rossi M et al (2004) A xylan-degrading strain of Sulfolobus solfataricus: isolation & characterization of the xylanase activity. Extremophiles 8:117–124

Chidi SB, Godana B, Ncube I et al (2008) Production, purification and characterization of cellulase free xylanase from *Aspergillus terreus* UL 4209. Afr J Biotechnol 7:3939–3948

Clarke AJ, Bray MR, Strating H. (1993) Xylanases mechanism of action. In: ACS symposium American Chemical Society, vol 27

Coelho GD, Carmona EC (2003) Xylanolytic complex from *Aspergillus giganteus:* production and characterization. J Basic Microbiol 43:269–277

Coman G, Bahrim G (2011) Optimization of xylanase production by *Streptomyces sp.* P12–137 using response surface methodology and central composite design. Ann Microbiol 61:773–779

Coughlan M, Hazlewood G (1993) β-1,4-D-xylan-degrading enzyme systems: biochemistry, molecular biology and applications. Biotechnol Appl Biochem 17:259–289

Csiszar E, Urbánszki K, Szakas G (2001) Biotreatment of desized cotton fabric by commercial cellulase and xylanase enzymes. J Mol Catal 11:1065–1072

Dean JFD, Gross KC, Anderson JD (1991) Ethylene biosynthesis inducing xylanase III: product characterisation. Plant Physiol 96:571

Dervilly G, Leclercq C, Zimmerman D et al (2002) Isolation and characterization of high molecular mass water-soluble arabinoxylans from barley malt. Carbohydr Polym 47:143

Dhillon A, Khanna S (2000) Production of a thermostable alkali tolerant xylanase from *Bacillus circulans* AB16 grown on wheat straw. World J Microbiol Biotechnol 27:325–327

Dhiman SS, Sharma J, Battan B (2008) Pretreatment processing of fabrics by alkalothermophilic xylanase from *Bacillus stearothermophilus*. Enzym Microbial Technol 43:262–269

Duarte MCT, Silva DA, Gomes EC et al (2003) Xylan-hydrolyzing enzyme system from *Bacillus pumilus* CBMAI 0008 and its effects on *Eucalyptus grandis kraft* pulp for pulp bleaching improvement. Bioresour Technol 88:9–15

Ducros V, Charnock SJ, Derwenda U et al (2000) Substrate specificity in glycoside hydrolase family10 structure and kinetic analysis of *S. lividans* xylanase. J Biol Chem 275:230–236
Ebrahimi M (2010) Engineering thermostable xylanase enzyme mutant from Bacillus halodurans. Afr J Biotechnol 9:8110–8117
Ellis JT, Magnuson TS (2012) Thermostable and alkali stable xylanases produced by the thermophilic bacterium *Anoxybacillus flavithermus* TWXYL3. Int Sch Res Netw ISRN Microbiol 2012:1–8
Ellouze O, Fattouch S, Mestiri F et al (2008) Optimization of extracellular xylanase production by *Sclerotinia sclerotiorum* S2 using factorial design. Indian J Biochem Biophys 45:404–405
Ferraz A, De Souza-Cruz PB, Freer J, Siika-Aho M (2004) Extraction and determination of enzymes produced by *Ceriporiopsis subvermispora* during biopulping of *Pinus taeda* wood chips. Enzym Microbial Technol 34:228–234
Gabriela PMA, Mariana BO, Ronaldo APN et al (2016) Characterization and biotechnological application of recombinant xylanases from *Aspergillus nidulans*. Int J Biol Macromol 91:60–67
Ghoshal G, Kamble A, Shivhare US et al (2011) Optimization of culture conditions for the production of xylanase in submerge fermentation by *Penicillium citrinum* using response surface methodology. Int J Res Rev Appl Sci 6:132–137
Gilbert HJ, Hazelwood GP (1993) Bacterial cellulases and xylanases. J Gen Microbiol 139:187–194
Gouda MK (2000) Purification and partial characterization of cellulose free xylanase produced in solid state and submerged fermentation by *Aspergillus tamarii*. Adv Food Sci 22:31–37
Goulart AJ, Carmona EC, Monti R (2005) Partial purification and properties of cellulose free alkaline xylanase produced by *Rhizopus stolonifer* in solid state fermentation. Braz Arch Biol Technol 48:327–333
Gray JA, Bemiller (2003) Bread staling: molecular basis and control. Rev Food Sci Food Saf 2:1–21
Gupta VK, Gaur R, Yadava SK, Darmwal NS (2009) Optimization of xylanase production from free and immobilized cells of Fusarium solani F7. Bioresources 4:932945
Hakulinen N, Turunen O, Janis J et al (2003) Three-dimensional structures of thermophilic beta-1,4-xylanases from *Chaetomium thermophilum* and *Nonomuraea flexuosa*. Eur J Biochem 270:1399–1412
Haltrich D, Nidetzky B, Kulbe KD et al (1996) Production of fungal xylanases. Bioresour Technol 58:137–161
Haq I, Khan A, Butt WA et al (2002) Effect of carbon and nitrogen sources on xylanase production by mutant strain of *Aspergillus niger* GCBMX-45. J Biol Technol 2:143–144
Harbak L, Thygesen HV (2002) Safety evaluation of a xylanase expressed in *Bacillus subtilis*. Food Chem Toxicol 40(1):1–8
Herbers K, Wilke I, Sonnewald U (1995) A thermostable xylanase from *Clostridium thermocellum* expressed at high levels in the apoplast of transgenic tobacco has no detrimental effects and is easily purified. Nat Biotechnol 13:63–66
Hong Q, Shin NH, Chang H (1989) Effect of oxygen extraction on organic chlorine contents in bleach plant effluents. TAPPI J 72:157–161
Huang XL, Penner MH (1991) Apparent substrate inhibition of the *Trichoderma reesei* cellulase system. J Agri Food Chem 39:2096–2100
Ito K, Ogasawara H, Sugimoto T et al (2000) Purification and properties of acid stable xylanases from *Aspergillus kawachii*. Biosci Biotechnol Biochem 56:547–550
Jain KK, Dey TB, Kumar S, Kuhad RC (2014) Production of thermostable hydrolases (cellulases and xylanase) from *Thermoascus aurantiacus* R C K K: a potential fungus. Bioproc Biosyst Eng 38:1–10
Javier PFI, Oscar G, Sanz-Aparicio J and Diaz P (2007). Xylanases: molecular properties and applications. In: Industrial enzymes: structure, function and applications . Polaina, J., A. P. MacCabe (Eds.), Springer, Dordrecht. pp. 65– 82
Kamble RD, Anandrao RJ (2012) Isolation, purification, and characterization of xylanase produced by a new species of *Bacillus* in solid state fermentation. Int J Microbiol 2:2–8
Kansoh AL, Nagieb ZA (2004) Xylanase and mannanase enzymes from *Streptomyces galbus* and their use in biobleaching of softwood kraft pulp. J Microbiol 85:103–114

Kansoh AL, Ali MA, El-Gammal AA (2001) Xylanolytic activities of *Streptomyces sp.* 1, taxonomy, production, partial purification and utilization agricultural wastes. Acta Microbiol Immunol 48:39–52

Kar S, Mandal A, Mohapatra PKD et al (2006) Production of cellulose free xylanase by *Trichoderma reesei* SAF3. Braz J Microbiol 37:462–464

Karlsson EN, Hachem MA, Ramchuran S et al (2004) The modular xylanase Xyn10A from *Rhodothermus marinus* is cell-attached, and its C-terminal domain has several putative homologues among cell-attached proteins within the phylum Bacteroidetes. FEMS Microbiol Lett 241:233–242

Kaur A, Chopra C, Joshi A et al (2015) Bioprocessing, biochemical characterisation and optimization of solid state fermentation of a new thermostable xylanase producing strain belonging to Bacillus genus. J Chem Pharm Res 7:266–276

Kenealy RW, Jeffries TW (2003) Enzyme processes for pulp and paper: a review of recent developmenst. In: Goodell B, Nicholas DD, Schultz TB (eds) Wood deterioration and preservation: Advances in our changing world, Chapter 12. ACS symposium series 845 American Chemical Society, Washington, DC, pp 210–239

Khasin A, Alchanati I, Shoham Y (1993) Purification and characterization of a thermostable xylanase from *Bacillus stearothermophilus* T-6. Appl Environ Microbiol 59:1725–1730

Kulkarni N, Shendye A, Rao M (1999) Molecular and biotechnological aspects of xylanases. Fed Eur Microbiol Soc 23:411–456

Kumar PR, Eswaramoorthy S, Vithayathil PJ, Viswamitra MA (2000) The tertiary structure at 1.59 Å resolution and the proposed amino acid sequence of a family-11 xylanase from the thermophilic fungus Paecilomyces varioti Bainier. J Mol Biol 295:581–593

Kumar A, Guleria A, Kumar D et al (2014) Isolation, immobilization and characterization of xylanase from a new isolate *Bacillus atrophaeus* E8. Internat jour of scientific resear 3:33–35

Li Y, Hardin JR (1998) Enzymatic scouring of cotton surfactants, agitation and selection of enzyme. Textile Colour Chem 30:23–28

Li K, Azadi P, Collins R et al (2000) Relationships between activities of xylanases and xylan structures. Enzym Microbiol Biotechnol 27:89–94

Madlala A, Bissoon S, Singh S et al (2001) Xylanase-induced reduction of chlorine dioxide consumption during elemental chlorine-free bleaching of different pulp types. Biotechnol Lett 23:345–351

Maheswari U, Chandra TS (2000) Production and potential applications of xylanase from a new strain of a Streptomyces cuspidosporus. World J Microbiol Biotechnol 16:257–263

Malathi V, Devegowda G (2001) In vitro evaluation of nonstarch polysaccharide digestibility of feed ingredients by enzymes. Poult Sci 80:302

McCarthy AA, Morris DD, Bergquist PL et al (2000) Structure of XynB, a highly thermostable β-1,4-xylanase from *Dictyoglomus thermophilum* Rt46B.1, at 1.8 Å resolution. Acta Crystallogr 56:1367–1375

Meng DD, Ying Y, Chen XH et al (2015) Distinct roles for carbohydrate-binding modules of GH10 and GH11 xylanases from *Caldicellulosiruptor sp.* F32 in thermostability and catalytic efficiency. Appl Environ Microbiol 81:2006–2014

Monisha R, Uma MV, Krishna Murthy V (2009) Partial purification and characterization of Bacillus pumilus xylanase from soil source. Kathmandu Univ J Sci Eng Technol 5:137–148

Motta FL, Andrade CCP, Santana MHA (2013) A review of xylanase production by the fermentation of xylan: classification, characterization and applications. In: Chandel AK, da Silva SS (eds) Sustainable degradation of lignocellulosic biomass – techniques, applications and commercialization. INTECH, Croatia, pp 251–275

Mushimiyimana I, Padmavathi T (2015) Xylanase enzyme production by *Penicillium crustosum* using sugar beet peel substrate by response surface methodology. Res J Pharm, Biol Chem Sci 6:1145–1149

Nakamura S (2003) Structure and function of a multiplidomain alkaline xylanase from alkaliphilic *Bacillus* sp. strain 41M-1. Catal Surv Jpn 7:157164

Nigam P, Singh D (1995) Processes for fermentative production of xylitol a sugar substitute. Process Biochem 30:117–124
Norazlina I, Meenalosani N, KuHalim KH (2013) Production of xylanase by *Trichoderma* sp. via solid state culture using sugarcane bagasse. Int J Energy Sci 3:99–103
Octavio L, Corral, Francisco VO (2006) Xylanases overview. Adv Agric Food Biotechnol:305–322
Omar AW, Khataibeh MJ, Abu-Alruz K (2008) The use of xylanases from different microbial origin in bread making and their effects on bread quality. J Appl Sci 8:672–676
Pandey AK, Edgard G, Negi S (2016) Optimization of concomitant production of cellulase and xylanase from *Rhizopus oryzae* SN5 through EVOP-factorial design technique and application in Sorghum Stover based bioethanol production. Renew Energy 98:51–56
Parajo JC, Domingues H, Domingues JM (1998) Biotechnological production of xylitol and fundamentals of its biosynthesis. Bioresour Technol 65:191–201
Park S, Rancour DM, Bednarek SY (2007) Protein domain-domain interactions and requirements for the negative regulation of *Arabidopsis* CDC48/p97 by the plant ubiquitin regulatory X (UBX) domain-containing protein, PUX1. J Biol Chem 282:5217–5224
Patel K, Prajapati K (2014) Xylanase production by *Cladosporium* sp. from agricultural waste. Int J Curr Res Acad Rev 2:84–90
Polizeli MLT, Rizzatti M, Monti R et al (2005) Xylanases from fungi: properties and industrial applications. Appl Microbiol Biotechnol 67:577–591
Qinnhe C, Xiaoyu Y, Tiangui N et al (2004) The screening of culture condition and properties of xylanase by white-rot fungus *Pleurotus ostreatus*. Process Biochem 39:1561–1566
Raj A, Sharad K, Sudheer KS (2013) A highly thermostablexylanase from stenotrophomonas maltophilia: purification and partial characterization. Enzym Res:2–8
Ratanachomsri U, Sriprang R, Sornlek W et al (2006) Thermostable xylanase from Marasmius sp.: purification and characterization. J Biochem Mol Biol 39:105
Roy N, Habib MR (2009) Isolation and characterization of xylanase producing strain of *Bacillus cereus* from soil. Iran J Microbiol 1:49–53
Saha BC (2000) A-L-arabinofuranosidases biochemistry, molecular biology and application in biotechnology. Biotechnol Adv 18:403–423
Saha BC (2003) Hemicellulose bioconversion. J Ind Microbiol Biotechnol 30:279–291
Sandrim VC, Rizzatti ACS, Terenzi HF et al (2005) Purification and biochemical characterization of two xylanases produced by *Aspergillus caespitosus* and their potential for kraft pulp bleaching. Process Biochem 40:1823–1828
Sanghvi GV, Koyani RD, Rajput KS (2010) Thermostable xylanase production and partial purification by solid-state fermentation using agricultural waste wheat straw. Mycology 2:106–112
Sardar M, Roy I, Gupta MN (2000) Simultaneous purification and immobilization of *Aspergillus niger* xylanase on the reversibly soluble polymer. Enzym Microb Technol 27:672–679
Shallom D, Shoham Y (2003) Microbial hemicellulases. Curr Opin Microbiol 6:219–228
Sharma PK, Chand D (2012) Production of cellulase free thermostable xylanase from *Pseudomonas* sp. XPB-6. Res J Biol Sci 1:31–41
Sharma M, Kumar A (2013) Xylanases overview. British Biotechnol J 3:1–28
Shi H, Zhang Y, Li X et al (2013) A novel highly thermostable xylanase stimulated by Ca2+ from *Thermotoga thermarum*: cloning, expression and characterization. Biotechnol Biofuels 6:26
Shi P, Du Y, Yang H et al (2015) Molecular characterization of a new alkaline-tolerant xylanase from *Humicola insolens* Y1. Biomed Res Int 2015:1–7
Singh S, Madlala AM, Prior BA (2003) *Thermomyces lanuginosus*: properties of strains and their hemicellulases. FEMS Microbiol Rev 27:3–16
Subbulakshmi S, Priya R (2014) Production and purification of enzyme xylanase by *Aspergillus niger*. Int J Curr Microbiol App Sci 3:664–668
Subramaniyan S, Prema P (2002) Biotechnology of microbial xylanases: enzymology, molecular biology and application. World Appl Sci J 4:277–283
Subramaniyan S, Sandhia GS, Prema P (2001) Control of xylanase production without protease activity in *Bacillus* sp. by the selected nitrogen source. Biotechnol Lett 23:369–371

Sun Y, Cheng J (2002) Hydrolysis of lignocellulosic materials for ethanol production. Bioresour Technol 83:1–11

Sunna A, Antranikian G (1997) Xylanolytic enzymes from fungi and bacteria. Crit Rev Biotechnol 17(1):39–67

Sunna A, Bergquist PL (2003) A gene encoding a novel extremely thermostable 1,4beta-xylanase isolated directly from an environmental DNA sample. Extremophiles 7(1):63–70

Takahashi Y, Kawabata H, Murakamli S (2013) Analysis of functional xylanases in xylan degradation by *Aspergillus niger* E-1 and characterization of the GH family 10 xylanase XynVII. Springer Plus 2:447

Teixeira RSS, Siqueira FGA, de Souza MV et al (2010) Purification and characterization studies of a thermostable β-xylanase from *Aspergillus awamori*. J Ind Microbiol Biotechnol 37:1041–1051

Thomas L, Parmeswaran B, Pandey A (2016) Hydrolysis of pretreated rice straw by an enzyme cocktail comprising acidic xylanase from *Aspergillus* sp. for bioethanol production. Renew Energy 98:9–15

Twomey LN, Pluske JR, Rowe JB et al (2003) The effects of increasing levels of soluble non-starch polysaccharides and inclusion of feed enzymes in dog diets on faecal quality and digestibility. Anim Feed Sci Technol 108:71–82

Van Paridon PA, Booman JCP, Selten GCM et al (1992) Application of fungal endoxylanase in poultry diets. Elsevier:371–378

Velkova ZI, Gochev VK, Kostov G, Atev A (2007) Optimization of nutritive media composition for xylanase production by *Aspergillus awamori*. Bulg J Agri Sci 13:651–656

Verma D, Satyanarayana T (2012) Molecular approaches for ameliorating microbial xylanases. Bioresour Technol 17:360–367

Wang X, Huang H, Xie X et al (2016) Improvement of the catalytic performance of a hyperthermostable GH10 xylanase from *Talaromyces leycettanus* JCM12802. Bioresour Technol 222:277–284

Whistler R, Masek E (1955) Enzymatic hydrolysis of xylan. J Am Chem Soc 77:1241–1243

Whistler RL, Richards EL (1970) Hemicelluloses. The carbohydrates. Academic, New York, pp 447–469

Widjaja A, Lestari E, Tanjung A, Alfian W et al (2009) Optimized production of xylanase from fungal strains and its purification strategies. J Appl Sci Environ Sanitation 4:219–232

Wong KKY, Tan LUL, Saddler JN (1998) Multiplicity of β-1,4-xylanase in microorganisms. Funct Appl Microbiol Rev 52:305

Woodward J (1984) Xylanases: functions, properties and applications. Top Enzym Ferment Biotechnol 8:9–30

Yasinok AE, Biran S, Kocabas A et al (2010) Xylanase from a soil isolate, *Bacillus pumilus*: gene isolation, enzyme production, purification, characterization and one-step separation by aqueous-two-phase system. World J Microbiol Biotechnol 26:1641–1652

Yegin S (2016) Single-step purification and characterization of an extreme halophilic, ethanol tolerant and acidophilic xylanase from *Aureobasidium pullulans* NRRL Y-2311-1 with application potential in the food industry. Food Chem 221:67–75

Yin LJ, Lin HH, Chiang YI et al (2010) Bioproperties and purification of xylanase from *Bacillus* sp. YJ6. J Agri Food Chem 58:557–562

Yoon H, Han NS, Kim CH (2004) Expression of *Thermotoga maritime* endo-β 1,4-xylanase gene in *E. coli* and characterisation of the recombinant enzyme. Agric Chem Biotechnol 47:157–160

Zhou C, Wang Y, Liu Z et al (2011) Acidic xylanase II from *Aspergillus usamii*: efficient expression in *Pichia pastoris* and mutational analysis. Afr J Biotechnol 10:11631–11639

Zimbardi S, Cesar S, Luana P et al (2013) Optimization of β-Glucosidase, β-Xylosidase and xylanase production by *Colletotrichum graminicola* under solid-state fermentation and application in raw sugarcane trash saccharification. Int J Mol Sci 14:2875–2902

Chapter 20
Bio-valorization of Dairy Whey for Bioethanol by Stress-Tolerant Yeast

Deepansh Sharma, Mehak Manzoor, Parul Yadav, Jagdip Singh Sohal, Gajender Kumar Aseri, and Neeraj Khare

Abstract Ethanol as a fuel has been used during the course of mankind industrial and social history. But due to the tax burden on ethanol and the cheaper cost of kerosene oil, it quickly substitutes ethanol. Ethanol is obtained by anaerobic fermentation by preferably yeast using sugars. Various different agro-industrial residues were used for the production of bioethanol at pilot scale. One important substrate, i.e., lactose, mainly present in milk and recognized as a huge unexplored waste remains from all the different kinds of cheese produced worldwide by the dairy processing sector. The global production of cheese whey is over 160 million tons production per year, showing a 1–2% annual growth rate. During fermentation of sugar to ethanol, yeast strains have to be capable to endure certain physiological stress and still growing actively at economically and in principle suitable standards. The future of ethanol production using stress-tolerant yeast to make the process more economically viable is very important. Utilization of high gravity substrate like concentrated cheese whey required the yeast strains with better and higher osmotolerant strains.

Keywords Bioethanol · Whey · Yeast · Stress tolerance · Dairy industry

20.1 Historical Background and Introduction

Today, ethanol is widely used for the formulation of the medicine and production of chemical products for household and household things. Initially the historical internal combustion automobile engines in fact ran on ethanol. Today, cars running on ethanol are more common than prior to earlier. But the olden times of the convoluted relationship amid ethanol and man goes much more back than the history of our contemporary automobile industry and car occupied culture. The exhaustion of the global petroleum resource and inclined levels of greenhouse gases have recently

D. Sharma (✉) · M. Manzoor · P. Yadav · J. S. Sohal · G. K. Aseri · N. Khare
Amity Institute of Microbial Technology, Amity University, Jaipur, Rajasthan, India

P. Gehlot, J. Singh (eds.), *Fungi and their Role in Sustainable Development: Current Perspectives*, https://doi.org/10.1007/978-981-13-0393-7_20

raised extensive attention in developing fermentation processes for commercial production of substitute biofuels like bioethanol (Aristidou and Penttila 2000).

Ethanol as a fuel has been used during the course of mankind industrial and social history. But due to the tax burden on ethanol and the cheaper cost of kerosene oil, it quickly substitutes ethanol (Morris 1993). Due to the coming out of petroleum fuel as alternative fuel, the use of ethanol as fuel was decreased considerably. This exclusion was ended by the early 1940s; the production of ethanol was accelerated again when it was used for the period of the Second World War for fuel and other chemical synthesis like manufacturing of synthetic rubber. During early and mid of the Second World War, about 600 million gallons of ethanol was produced yearly in the USA (Morris 1993). There was a decline in the demand of the ethanol after the First World War, since it became more costly to produce than fossil fuels (Demirbas et al. 2009; Balat and Balat 2009; Solomon et al. 2007).

The promotion of the bioethanol industrial production was resumed in the early of 1980, mainly to revive the farming area at a time of saturate supply of agricultural products (Johnson and Rosillo-Calle 2007). The USA rebuilds its ethanol industry more steadily than Brazil and is these days the leader in its production and practice. Presently ethanol is the key biofuel used globally, and its demand and use are progressively increased (Bastos 2007).

Ethanol is obtained by anaerobic fermentation by preferably yeast using sugars. Various different agro-industrial residues were used for the production of bioethanol at pilot scale. One important substrate, i.e., lactose, mainly present in milk and recognized as a huge unexplored waste remains from all the different kinds of cheese produced worldwide by the dairy processing sector. Lactose can be either hydrolyzed or fermented which leads to ethanol as an end product (Sansonetti et al. 2009). Bioethanol is produced using various types of agro-food residues; first-generation bioethanol production is obtained from sugar and starch containing sugarcane, beet root, corn, wheat, and other agriculture crops, whereas second-generation ethanol production aims to utilize the lignocellulose waste such as plant biomass and agricultural residues (Naik et al. 2010). Ethanol obtained from sugarcane, corn, and other agriculture crops is used as a gasoline alternative or supplement for automobile usage in various countries (Walter et al. 2008).

India is the one of the largest producers of sugarcane after Brazil. Every ton of sugarcane yields 740 kg of juice (135 kg sucrose and 605 kg water) and 130 kg of dry bagasse (Da Rosa 2009). In India, typically sugarcane bagasse and other by-products such as molasses are used for ethanol production at industrial scale, but limited availability and alternative uses of molasses and bagasse have directed to search for other feasible substrates. Additionally lack of industrial feasibility for ethanol production from lignocellulosic biomass and inadequate obtainability of food grains to meet the food requirements of growing population in country like India, ethanol production from cheese whey seems to be the only promising alternative substrate for commercial biofuel production. Currently India is the second largest milk producer around the world (FAO 2012). In countries like India, ethanol production from cheese whey will help improve the socioeconomic condition of the farmers also.

Quickly diminishing fossil fuels and rising environmental pollution have led to a global exploration for alternative fuels. Ethanol is recognized a potential renewable fuel as of its diverse and well-recognized advantages (Brethauer and Wyman 2010; Galbe and Zacchi 2002). In emergent countries such as India, which has a large inhabitant and a restricted area under cultivation, distraction of agriculture crops to ethanol production is not viable or sustainable. In India, ethanol is commonly obtained from the fermentation of molasses, but molasses has inadequate accessibility and various other alternative uses; the Government of India is sturdily advocating research for the production of second-generation ethanol.

20.2 Global Whey Production and Problems

The global production of cheese whey is over 160 million tons production per year, showing a 1–2% annual growth rate (Smithers 2008). Whey is characterized with high range of biochemical oxygen demand (BOD) and chemical oxygen demand (COD), 30–50 g L^{-1} and 60–80 g L^{-1}, respectively. Lactose moieties are mostly accountable for the elevated BOD and COD (Ghaly and Kamal 2004). Protein revitalization diminishes the COD of cheese whey simply by about 10 g L^{-1} (Domingues et al. 1999; Siso 1996). The global cheese whey generation is estimated about 160 million tons per year (anticipated as ninefold the cheese produced) (Guimarães et al. 2010). In India, various indigenous fermented dairy foods are prepared and consumed. Chhana (a type of milk preparation) is in huge demand for the Rasogolla, Sandesh, and other local preparations. Currently Indian milk sector produces 1.2 million tons of cheese and chhana per year, which consequences in a yearly production of 8.0 million tons of cheese whey. Whey constitutes majorly all the nutrients present in the milk commonly. Dumping of whey holds a severe trouble to the dairy processing units as of the elevated organic content. Consequently, the removal of whey is expensive and ecologically challenging matter for dairy dispensation sector. Thus, cheese whey is a vital environmental crisis as of the high volumes generated and its elevated organic matter substance (Marwaha and Kennedy 1988; Mawson 1994). The valorization of cheese whey has been a dispute for dairy industries the generation of of cheese globally. With rising cheese utilization and production, the amount of produced whey also raises (Rajeshwari et al. 2000, De Wit 2001).

Various strategies have been suggested for efficient removal of whey. Whey has been significantly consumed in obtaining of cheese whey powder, paste, lactose, and fermentation of lactose to lactic acid, food additives, alcoholic beverages, etc. But still the potential consumption of whey has not been completely explored. The Government of India took the idea and promulgated the Environmental Protection Act 1986 that makes it compulsory to pretreat dairy whey prior to release in local water bodies like rivers and ponds.

20.3 Biotechnological Valorization of Cheese Whey

Lactose is a significant and important substrate for the fermentation and biotechnological industrial productions. All of the abovementioned solutions utilize chemically pure lactose, while others use whey as an energy source. The biotechnological transformation of lactose to the production of lactic acid using suitable species of *Lactobacillus* has the dual benefit of alleviating a pollution problem and, at the same time, producing a marketable product (Polat 2009. The fermentation of whey lactose to ethanol, using selected yeasts, has been frequently reported and demonstrated by various reports (Guimarães et al. 2010). Generally yeasts incorporate lactose aerobically, but lactose-hydrolyzing yeasts are very uncommon and include *Kluyveromyces lactis*, *Kluyveromyces marxianus*, and *Candida pseudotropicalis* preferred strains. Being a waste discharge and environmental concern, whey represents an advantage over feedstocks such as corn for ethanol production. Different commercial products are produced by microorganism (e.g., PHB, bacteriocins, biosurfactants, and bioethanol).

20.4 Global Production of Whey and Associated Problems

Around 50% of cheese whey obtained is discarded aforementioned to any waste treatment and causes widespread ecological harm, chiefly due to its BOD (Gonzalez-Siso 1996). Many dairy industries pick up a part of the whey proteins by means of ultrafiltration for utilization as food additives or in various other milk preparations. On the other hand, cheese whey permeate obtained from this method still contains just about 85–95% of the whey lactose; the lactose is generally accountable for its high BOD value (Vienne and Stockar 1985). So, there is strong encouragement for the improvement of a process for cheese whey treatment that can obtain a biotechnological valorization from the lactose (González-Siso 1996). In the past decade, major research was on ethanol production which has been determined by the rising demand for cleaner and renewable energy (Rana et al. 2013). Additionally, the ethanol obtained from lactose permeate is utilized for food, beverage formulations, and pharmaceutical preparations and in cosmetic formulations (Guimarães et al. 2010).

Historically, the disposal of cheese mainly included discharge into rivers, lakes or local water bodies, animal feed, and direct feeding to ruminants. A different alternative would be to release the whey into lagoons for oxidation or into domestic sewage system, but the elevated BOD and COD of cheese whey typically lead to a surplus organic matter to the system (Kosikowski 1979; Smithers 2008). Release of cheese whey by these ways provides no valuable product and is expensive and demands high manpower for the production companies, who usually bear all the straight costs of handling and transport. In array to build up integrated solutions for the whey disposal problem, it must be recognized as a source and not only as a waste matter, in view of its huge prospective as a substrate for value-added products.

The first stride in most measures for cheese whey valorization comprises of the recovery of the protein part. Huge amount of surplus whey jointly with the requirement for inexpensive and commonly available substrates and, beyond that, the rapid advances in microbial biotechnology are expected to prompt further utilization of whey lactose as fermentation substrate to attain value supplementary product developments 30–50 g L^{-1}.

20.5 Whey to Biofuel

Ethanol production globally has immensely improved since the oil crises in the late 1970s. The bioethanol market grew 39 billion liters of bioethanol in 2006 and is predictable to attain 100 billion liters till 2015 (Licht 2003). In point of fact, the USA and Brazil are the largest producers of global share of ethanol production of ethanol. Brazil started a national program for ethanol production, i.e., National Alcohol Program, and became the one of the largest nations for bioethanol production (RFA 2010). The Brazil national program has worked out searches for new sugarcane breeds containing high amount of sugar per hectare of crop. It could be possible as because the sugarcane genome was sequenced in 90% in 2003, the various research groups were capable to create varieties of transgenic cane with higher sugar content along with high productivity, better resistance to dryness and adjustment to diverse climatic conditions, and better resistance to low-nutrient and nutrient-rich soils. A significant transformation also occurred concerning the fermentation with better substrate to ethanol conversion, i.e., 88–90% (Souza 2006; Orellana and Bonalume Neto 2006).

Presently, over 95% of ethanol obtained in the USA produces from corn, and the remaining 5% is coming from other grains such as wheat, barley, cheese whey, and certain beverage remains (Solomon et al. 2007). In 2007, grain-based substrates were utilized for ethanol plants, and their production capacity was about 1.4 million metric tons (MMT). On the other hand, the fuel vs. food issue comes out, and the utilization of the grains for ethanol production has restricted the projects working on grain-based ethanol production; research and production on ethanol now constrained to "nonfood ethanol" (ethanol produced from nonfood crops) were backed by the organizations like Chinese government (Fang et al. 2010). Various strategies for the production of bioethanol were developed based on nonfood agriculture crops and agro-industrial waste, like cassava, sweet sorghum, and sweet potato, dairy waste like cheese whey, and others (Li and Chan-Halbrendt 2009; Wu et al. 2010).

Further one of the common substrates available for bioethanol production nowadays, other than in the neighborhood accessible crops, easily reached for bioethanol production is cheese whey, obtained during cheese making with 90% of the amount of the milk used (Siso 1996). Commonly used yeast strains for bioethanol production are naturally *Kluyveromyces marxianus* strains able of fermenting lactose as compared to the conventional industrial strain of *S. cerevisiae* as it is not having lactose breakdown enzymes (Guimarães et al. 2010). Though fermentation of raw whey comprising about 5% lactose produces only 2.5% ethanol, which is quiet distant from cost-effectively realistic due to the elevated cost of distillation, still when a lactose-fermenting yeast is employed, the addition of glucose to increase the preliminary sugar concentration elicits catabolite repression, hindering the strain from conversion of the lactose to the ethanol during the fermenting process (Wang et al. 1987).

A number of microorganisms are capable of breaking down lactose to ethanol during fermentation such as *Kluyveromyces marxianus*. Yeast *K. marxianus* is known for its high metabolic range and its considerable extent of intraspecific polymorphism (Lane et al. 2011). Furthermore not only lactose fermentation, yeast strain of *K. marxianus* has other attractive attributes for various industrial fermentations, like thermotolerance nature, effective growth rate, and substrate conversion factor, and regularly ferments an extensive range of carbohydrates like pentoses, hexoses, and many disaccharides (Lane and Morrissey 2010). Various lactose-fermenting strains were reported earlier for bioethanol production. *K. marxianus* UFV-3, isolated from dairy site in Southeastern Brazil, is capable to ferment lactose to bioethanol with a significant yield in environment of concentrated whey permeate under oxygen-limiting conditions (Silveira et al. 2005). *K. marxianus* fermentative performance is mostly due to its improved expression of key enzymes occupied in lactose breakdown (Diniz et al. 2012). On the other hand, additional factors that may influence the fermentative ability of *K. marxianus*, like temperature, pH, and biomass-generated concentration of substrate, have not been studied or established significantly.

Ethanolic fermentation is an attractive substitute for the bioremediation of the polluting dairy whey those leftovers after removal of the whey proteins using yeast (Rogosa et al. 1947; Webb and Whittier 1948; Whittier 1944). The transformation of the lactose in dairy whey into bioethanol is barely reasonably competitive with the at present recognized processes, utilizing sugarcane and cornstarch as raw substrates, or with rising second-generation technologies utilizing lignocellulosic biomass as substrate. But, being a waste effluent represents a benefit of cheese whey over food feedstocks, like corn, for ethanol production. Furthermore, the ease of use of varied solutions for whey bioremediation is important, so that each dairy industry can assess, according to its own problems, and customize accordingly. In conclusion, ethanol production from whey is potable and so can find appropriate markets such as food and beverages, pharmaceutical formulations, and personnel hygiene to cosmetic preparations. It is remarkable that the amounts of lactose available are rather significant (Becerra et al. 2001; Siso 1996). For those grounds, the quantity of lactose accessible for ethanol production may be as elevated as 4 million tons per year, which, bearing in mind a transformation competence of 85%, could capitulate about 2.3 million m^3 of bioethanol. This is approximately 3.5% of the total global production of bioethanol in 2008, which was approximately 65 million m^3 (RFA 2009).

Straight fermentation of whey to bioethanol is usually not cost-effectively reasonable because of the little lactose content yields and low ethanol yield (2–3% v/v), resulting in high investment in the distillation of ethanol. So it's appropriate to start the fermentation with initial high concentration of lactose which can be achieved by ultrafiltration and/or reverse osmosis methods, in order to achieve high ethanol yield after fermentation (Oda and Nakamura 2009). During the last decade, various authors have worked on the production of bioethanol from whey lactose, typically referring the yeasts *K. fragilis*, *K. marxianus*, and *C. pseudotropicalis* as major lactose-hydrolyzing strains (Kurtzman and Fell 1998; Lachance 1998). In addition to scientific information, there are some examples of industrial units that produce ethanol using cheese whey, typically using *Kluyveromyces* strains.

Examples of established industrial plants are there which are producing the bioethanol from the cheese whey fermentation in countries like New Zealand, the USA, Denmark, and Ireland (Lyons and Cunningham 1980; Pesta et al. 2007; Siso 1996). Carbery Milk Products (at present designated Carbery Group) from Cork, Ireland, was in progress of large-scale ethanol production from cheese whey from 1978. The strategy was anticipated for the production of clean ethanol, but from 2005, the plant has also been producing fuel-grade ethanol for E85 and E5 blends in Ireland (Doyle 2005; Ling 2008; Ling 2008). At present, the Carbery Group produced 11 cylindro-conical fermentation systems, utilizing compressed air for agitation and aeration. The whey is fermented in batch cycle for 12–20 h or accordingly the initial concentration of the yeast cells depending upon the activity. The yeast biomass is recovered at the end of the process and repitched a number of periods before it is discarded. Ethanol yields at the end of process are on average in the sort 2.5–4.2% (v/v) after continuous distillation (Pesta et al. 2007; Ling 2008; Hamilton 1998; Thiele 2005). Since 2007, Anchor has also been providing fuel-grade ethanol to a fossil fuel-based industry such as New Zealand for E10 blend (Ling 2008). The substrate is de-proteinated cheese whey (Hamilton 1998), which is concentrated from 4% to 8% lactose content by reverse osmosis before the fermentation starts (Gibson 2006). The process of fermentation is running for about 24 h using yeast strains of *Kluyveromyces* spp., which yields an ethanol titer of approximately 4%, after distillation (Gibson 2006). In the USA, the Milbrew process was industrialized in 1972 to obtain single-cell protein and bioethanol using cheese whey by *K. fragilis* (Detroy and Julian 1982; Lyons and Cunningham 1980; Pesta et al. 2007). The ethanol concentration during the process could be improved or optimized by altering the fermentation parameters like as the rate of aeration (Lyons and Cunningham 1980). In Denmark, the Dansk Gaerings method was industrialized in the late 1970s for the whey to biofuel production in a continuous fermentation mode (Lyons and Cunningham 1980; Pesta et al. 2007). In recent times, one dairy industry Theo Müller announced the building of a plant in its Leppersdorf plant (Germany) targeting to produce 10 million liters of ethanol from cheese whey valorization.

Whey arises in huge amounts in the dairy industry during cheese production. Most of the whey is processed to whey powder or lactose, but it is also an interesting source for microbial production of β-D-galactosidase, ethanol, and other compounds (Pesta et al. 2007; Guimarães et al. 2010). *K. marxianus* is known to produce ethanol from whey-borne lactose into valuable products since this yeast species displays fast lactose utilization via respiratory and fermentative metabolism and is thus appropriate for diverse applications (Fonseca et al. 2008; Löser et al. 2011). The metabolism of *K. marxianus* can be controlled by the available oxygen since it is a Crabtree-negative yeast (van Urk et al. 1990; van Dijken et al. 1993; Wardrop et al. 2004). Moreover, the thermal tolerance of *K. marxianus* (Brady et al. 1997; Kourkoutas et al. 2002) is another advantage in large-scale processes since cultivation at high temperature possibly allows the use of non-sterile culture media. Besides microbial production of ethanol from whey, conversion of whey-borne lactose into ethyl acetate by *K. marxianus* could be an interesting alternative. Ethyl acetate is an environmentally friendly solvent due to its microbial degradability (Chan and Su 2008), and ethyl acetate is more valuable than ethanol. The prices of both compounds are subjected to considerable market fluctuations to be explained by their oil-based syntheses, but a market analysis demonstrates that the price of ethyl acetate is always ca. 30 % higher than the market price of ethanol. Another advantage over ethanol is its high volatility which allows its process-integrated recovery in biotechnological processes (Löser et al. 2011; Urit et al. 2011).

20.6 High-Throughput Strategies for Bioethanol Production

20.6.1 Ethanol Production by Stress-Tolerant Strategies Using Whey

During fermentation of sugar to ethanol, yeast strains have to be capable to endure certain physiological stress and still growing actively at economically and in principle suitable standards (Marullo et al. 2009; Zhao and Bai 2009). For that reason, improving the multi-stress tolerance of yeast strains is vital to their effectiveness in commercially viable ethanol production. In *S. cerevisiae*, stress tolerance is usually controlled by multiple loci extensively scattered all through the cellular genome (Attfield 1997; Causton et al. 2001; Zhao and Bai 2009; Zhang et al. 2002). In heredity and reproduction of yeast strains, high-throughput screening methods and efficient screening technique may yield mutant strains with the preferred traits (Zuzuarregui 2004). Growth-containing trait-relevant inhibitors were generally functional in the early selection of preferred tolerance phenotypes (Limtong et al. 2007; Shi et al. 2009; Sridhar et al. 2002).

The tolerant capability of yeast strains under the circumstances of high sugar concentration, elevated temperature, and high ethanol accumulation (Ansanay-Galeote et al. 2001; Osho 2005) during fermentation is an increasingly vital trait that magnetizes a lot of researchers as there has been a diverse range of profit which could be used during the utilization of osmotolerant, ethnotolerant, and thermophilic yeasts for economically viable ethanol production. Thermotolerant yeasts are competent of expansion and fermentation through the summer months in nontropical nations and in tropical temperatures as well. The cooling costs in fermentation in the course of ethanol production are costly; therefore, by exploring thermotolerant yeasts, the cost of cooling and continuous distillation can be controlled. In addition, high saccharification and fermentation rates, continuous ethanol distillation, and the chances of contamination reduction have encouraged exploring for routes to thermotolerant yeasts.

The capability of yeast to counter efficiently to these circumstances is vital not merely for bioethanol production but also to maintain the fermentation robustness of yeast for exploring the inconsequent fermentations. Throughout bioethanol production, cells reside in a multifarious environment, and our understanding of stress responses under such conditions is limited. The initiation of methods competent of decisive genomic and proteomic alterations within the cell is expected immensely to advance our information of yeast stress responses through commercial ethanol fermentation.

20.6.2 Osmotic Stress

Osmotic stress may be understood as any circumstances when there is a disproportion of intracellular and extracellular osmolarities, adequate to cause a harmful alteration in cell physiology (Csonka and Hanson 1991; Beney et al. 2000; Tamás and Hohmann 2003). Certainly, yeast cells may undergo high osmotic stress when there is a condition of low external osmotic potential, such as cell present in deionized water which results in an influx of water molecules to cell cytoplasm consequential in hypoosmotic stress (Dihazi et al. 2001). On the other hand, osmotic stress may also upshot when exposed to a growth condition when high solute concentrations like elevated sugar concentration or high gravity fermentations which lead to hyperosmotic stress exemplify by the loss of cytoplasmic water (Klipp et al. 2005).

Yeast commonly presents two different types of the "response" cell which may pertain on exposure to the osmotic confronts, i.e., osmotolerance and osmoadaptation. Innate osmotic resistance of a yeast cell in an environment is called osmotolerance and characterizes the means by which cells keep their viability in the existence of toxic solute concentrations.

When cells adjust their active physiological state in view to maintain their viability, they represent the state of osmoadaptation in unfavorable water potential (Poolman and Glaasker 1998; Yancey et al. 1982; Wegmann 1986; Blomberg and Adler 1992; Hernandez-saavedra et al. 1995). There are various yeast strains available for the production of the bioethanol from various substrates under sugar or osmotolerant conditions in literature (Table 20.1).

20.6.3 Ethanol Toxicity

The most important reason of ethanolic fermentation is the production of alcohol from various fermentable sugars by yeast strain such as *Saccharomyces cerevisiae* and *Kluyveromyces marxianus* yeast. Though a vital product of fermentation is a properly viable and healthy yeast biomass which can be used inconsequent fermentations (Stewart 2001), as fermentation advances, the ethanol yield enhances, and yeast cells are vulnerable to gradually increasing toxic intensity of ethanol (Briggs

Table 20.1 Various reported osmotolerant yeast strains in literature

S. No.	Yeast strain	Sugar stress (concentration)	References
1.	*Candida guilliermondii* NRRL Y-2075	30 gL^{-1}	Schirmer-Michel et al. (2008)
2.	*S. cerevisiae C2*	20 gL^{-1}	Ali et al. (2014)
3.	*Saccharomyces kluyveri K-6*	40 gL^{-1}	Brooks (2008)
4.	*C. zemplinina*	60 gL^{-1}	Tofalo et al. (2009)
5.	*S. cerevisiae NFRI3062*	40 gL^{-1}	Watanabe et al. (2010)
6.	*Saccharomyces cerevisiae ITV-01*	100–250 gL^{-1}	Ortiz-Muñiz et al. (2010)
7.	*S. cerevisiae OVB 11*	31 gL^{-1}	Yadav et al. (2011)
8.	*C. bombi*	70 gL^{-1}	Ruyters et al. (2015)
9.	*S. paradoxus*	48 gL^{-1}	Mukherjee et al. (2014)
10.	*Zymomonas mobilis*	200 gL^{-1}	Cazetta et al. (2007)
11.	*S. cerevisiae* S_1	50 gL^{-1}	Balakumar and Arasaratnam (2014)

Table 20.2 Various reported ethanol-tolerant yeast strains in literature

S. No.	Yeast strain	Ethanol tolerance (v/v)	Reference
1.	*Saccharomyces cerevisiae UVNR56*	15%	Thammasittirong et al. (2013)
2.	*Saccharomyces cerevisiae W303*	7.5%	Stanley et al. (2010)
3.	*S. cerevisiae 0271*	12%	Osho (2005)
4.	*S. cerevisiae strain X4 (CECT13014)*	15%	García-Martínez et al. (2011)
5.	*S. paradoxus*	14%	Mukherjee et al. (2014)
6.	*Kluyveromyces delphensis*	6%	Tsegaye (2016)
7.	*H. uvarum*	12%	Nwachukwu et al. (2006)
8.	*S. cerevisiae K-701*	20%	Yamaoka et al. (2014)
9.	*S. cerevisiae W303-1A*	13%	Abe et al. (2009)
10.	*S. cerevisiae TA*	14%	Ali et al. (2014)
11.	*Saccharomyces cerevisiae ITV-01*	10%	Ortiz-Muñiz et al. (2010)

et al. 2004). Under favorable conditions, end product i.e. ethanol yield lies in the array of 3–6%, however under high gravity fermentation process yields the ethanol up to a concentration of 4–10% (Table 20.2) (Briggs et al. 2004).

20.6.4 Thermotolerance

Tolerance to elevated temperature and fermentation sugar concentrations are significant properties of producer industrial strain. The fermentation potential of *S. cerevisiae* at elevated temperatures is quiet less because of increased fluidity in cytoplasmic

Table 20.3 Various reported thermotolerant yeast strains in literature

S. No.	Yeast strains	Temperature (°C)	Reference
1.	*Kluyveromyces marxianus DMKU3-1042*	40	Nonklang et al. (2008)
2.	*K. marxianus CECT10875*	42	Ballesteros et al. (2004)
3.	*H. polymorpha*	48	Voronovsky et al. (2009)
4.	*Debaryomyces hansenii*	40	Menon et al. (2010)
5.	*I. orientalis TTK316*	40	Kitagawa et al. (2010)
6.	*C. glabrata Cgrd1*	42	Watanabe et al. (2010)
7.	*K. marxianus NBRC1777*	48	Yanase et al. (2010)
8.	*P. kudriavzevii HOP-1*	40	Oberoi et al. (2012)
9.	*K. marxianus IMB3*	45	Pessani et al. (2011)
10.	*S. cerevisiae TJ14*	42	Prasetyo et al. (2011)
11.	*K. marxianus CHY1612*	45	Kang et al. (2012)
12.	*K. marxianus UFV-3*	37	Silveira et al. (2005)
13.	*S. cerevisiae LBM-1*	42	de Souza et al. (2012)
14.	*S. cerevisiae F34*	45	Shi et al. (2009)

membranes. In industrial ethanol production, elevated temperature the saving of energy achieved during the drop in cooling costs and overheating troubles usually faced in geographical with high ambient temperature.

The huge literature on yeasts mostly consists of research on that yeast which is growing in the array of 30 ± 35 °C. Physiological divergence in yeast strains has been considered only infrequently, and the exploit of the terms "thermophilic" or "thermotolerant," in recitation yeasts, has not been very clear (Arthur et al. 1978). In recent times, thermophilic yeast strains were mainly indefinite, and those usually competent of growth at higher than 40 °C have been typically recognized as thermotolerant (Krouwel and Braber 1979; Hacking et al. 1984; Hughes et al. 1984; Anderson et al. 1986a; Szczodrak and Targonski 1988; D'Amore et al. 1988, 1989; Lee et al. 1993; Kida et al. 1992; Morimura et al. 1997). Further thermotolerant yeasts which fit in to the genus *Kluyveromyces* were observed for ethanol production at or above 40 °C, and to have a maximal optimal growth temperature of 49 °C (Hughes et al. 1984) or still up to 52 °C (Banat et al. 1992) would classify them as thermophilic strain. As a result, the fermentation vessel's temperature rises to higher than 40 °C, leading to reduced ethanol production yield (Ohta et al. 1988). Rising temperature leads to increase in saturated esterified fatty acids like palmitic and palmitoleic acids in yeast cytoplasmic membrane (Van Uden 1984; Suutari et al. 1990). Related membrane changes have also been revealed to arise in the ethanol-producing bacterium *Zymomonas mobilis* (Benschoter and Ingram 1986). Furthermore, some membrane changes typically lead to the production of heat shock proteins (Hsps) which may play a significant role in providing thermal and ethanol tolerance (Michel and Starka 1986). In literature various reports related to the production of bioethanol by thermotolerant strain have been mentioned (Table 20.3).

20.7 Future Challenges and Conclusion

The future of ethanol production using stress-tolerant yeast to make the process more economically viable is very important. Utilization of high gravity substrate like concentrated cheese whey required the yeast strains with better and higher osmotolerant strains. In tropical countries or in summer season, the strain with better confrontation abilities will be the organisms of choice for industrial production. Developments in molecular biology and system biology will help the investigators to develop the more efficient osmotolerant, ethnotolerant, and thermotolerant strains in the near future.

References

Abe H, Fujita Y, Takaoka Y, Kurita E, Yano S, Tanaka N, Nakayama KI (2009) Ethanol-tolerant saccharomyces cerevisiae strains isolated under selective conditions by over-expression of a proofreadingdeficient DNA polymerase δ. J Biosci Bioeng 108(3):199–204

Ali M et al (2014) Stress-dependent proteolytic processing of the actin assembly protein Lsb1 modulates a yeast prion. J Biol Chem 289(40):27625–27639

Anderson PJ, McNeil K, Watson K (1986) High-efficiency carbohydrate fermentation to ethanol at temperatures above 40 C by kluyveromyces marxianus var. marxianus isolated from sugar mills. Appl Environ Microbiol 51(6):1314–1320

Ansanay-Galeote V, Blondin B, Dequin S, Sablayrolles JM (2001) Stress effect of ethanol on fermentation kinetics by stationary-phase cells of Saccharomyces cerevisiae. Biotechnol Lett 23(9):677–681

Aristidou A, Penttilä M (2000) Metabolic engineering applications to renewable resource utilization. Curr Opin Biotechnol 11(2):187–198

Arthur H, Watson K, McArthur CR, Clark-Walker GD (1978) Naturally occurring respiratory deficient Candida slooffii strains resemble petite mutants. Nature 271(5647):750

Attfield PV (1997) Stress tolerance: the key to effective strains of industrial baker's yeast. Nat Biotechnol 15(13):1351–1357

Balakumar S, Arasaratnam V (2014) Enhanced production of ethanol by high gravity glucose fermentation at temperatures above 40 oC by saccharomyces cerevisiae S1 using a soya flour supplemented medium. J Natl Sci Found Sri Lanka 42(2):111–117

Balat M, Balat H (2009) Recent trends in global production and utilization of bio-ethanol fuel. Appl Energy 86(11):2273–2282

Ballesteros M, Oliva JM, Negro MJ, Manzanares P, Ballesteros I (2004) Ethanol from lignocellulosic materials by a simultaneous saccharification and fermentation process (SFS) with Kluyveromyces marxianus CECT 10875. Process Biochem 39(12):1843–1848

Banat IM, Nigam P, Marchant R (1992) Isolation of thermotolerant, fermentative yeasts growing at 52 C and producing ethanol at 45 C and 50 C. World J Microbiol Biotechnol 8(3):259–263

Bastos VD (2007) Etanol, alcoolquímica e biorrefinarias. BNDES Setorial 25:5–38

Becerra M, Baroli B, Fadda AM, Mendez JB, Siso MG (2001) Lactose bioconversion by calcium-alginate immobilization of Kluyveromyces lactis cells. Enzym Microb Technol 29(8):506–512

Beney L, de Maranon IM, Marechal PA, Gervais P (2000) Influence of thermal and osmotic stresses on the viability of the yeast Saccharomyces cerevisiae. Int J Food Microbiol 55(1):275–279

Benschoter AS, Ingram LO (1986) Thermal tolerance of Zymomonas mobilis: temperature-induced changes in membrane composition. Appl Environ Microbiol 51(6):1278–1284

Blomberg A, Adler L (1992) Physiology of Osmotolerance in Fungi1. Adv Microb Physiol 33:145–212 Academic Press
Brady D, Nigam P, Marchant R, Singh D, McHale AP (1997) The effect of Mn 2+ on ethanol production from lactose using Kluyveromyces marxianus IMB3 immobilized in magnetically responsive matrices. Bioprocess Biosyst Eng 17(1):31–34
Brethauer S, Wyman CE (2010) Continuous hydrolysis and fermentation for cellulosic ethanol production. Bioresour Technol 101(13):4862–4874
Briggs DE, Boulton CA, Brookes PA, Stevens R (2004) Native african beers. In: Stevens R (ed) Brewing: science and practice. CRC Press, Cambridge, pp 589–605
Brooks AA (2008) Ethanol production potential of local yeast strains isolated from ripe banana peels. Afr J Biotechnol 7(20):3752–3755
Causton HC, Ren B, Koh SS, Harbison CT, Kanin E, Jennings EG, Lee TI, True HL, Lander ES, Young RA (2001) Remodeling of yeast genome expression in response to environmental changes. Mol Biol Cell 12(2):323–337
Cazetta ML, Celligoi MAPC, Buzato JB, Scarmino IS (2007) Fermentation of molasses by zymomonas mobilis: effects of temperature and sugar concentration on ethanol production. Bioresour Technol 98(15):2824–2828
Chan WC, Su MQ (2008) Biofiltration of ethyl acetate and amyl acetate using a composite bead biofilter. Bioresour Technol 99(17):8016–8021
Csonka LN, Hanson AD (1991) Prokaryotic osmoregulation: genetics and physiology. Annu Rev Microbiol 45(1):569–606
D'Amore TONY, Panchal CJ, Stewart GG (1988) Intracellular ethanol accumulation in Saccharomyces cerevisiae during fermentation. Appl Environ Microbiol 54(1):110–114
D'Amore T, Celotto G, Russell I, Stewart GG (1989) Selection and optimization of yeast suitable for ethanol production at 40 C. Enzym Microb Technol 11(7):411–416
Da Rosa AV (2009) Fundamentals of renewable energy processes. Academic, Burlington
Demirbas MF, Balat M, Balat H (2009) Potential contribution of biomass to the sustainable energy development. Energ Conver Manag 50(7):1746–1176
de Souza CJ, Costa DA, Rodrigues MQ, dos Santos AF, Lopes MR, Abrantes AB et al (2012) The influence of presaccharification, fermentation temperature and yeast strain on ethanol production from sugarcane bagasse. Bioresour Technol 109:63–69
De Wit JN (2001) Lecturer's handbook on whey and whey products. European Whey Products Association, Brussels
Detroy RW, St Julian G (1982) Biomass conversion: fermentation chemicals and fuels. CRC Crit Rev Microbiol 10(3):203–228
Dihazi H, Kessler R, Eschrich K (2001) Phosphorylation and inactivation of yeast 6-phosphofructo-2-kinase contribute to the regulation of glycolysis under hypotonic stress. Biochem 40(48):14669–14678
Diniz RH, Silveira WB, Fietto LG, Passos FM (2012) The high fermentative metabolism of Kluyveromyces marxianus UFV-3 relies on the increased expression of key lactose metabolic enzymes. Antonie Van Leeuwenhoek 101(3):541–550
Domingues L, Dantas MM, Lima N, Teixeira JA (1999) Continuous ethanol fermentation of lactose by a recombinant flocculating Saccharomyces cerevisiae strain. Biotechnol Bioeng 64(7):692
Doyle A (2005) Another step in biofuel supply. Ir Farmers J Interact 2005:12–16
Fang X, Shen Y, Zhao J, Bao X, Qu Y (2010) Status and prospect of lignocellulosic bioethanol production in China. Bioresour Technol 101(13):4814–4819
FAO W (2012) IFAD. The state of food insecurity in the world, 65
Fonseca GG, Heinzle E, Wittmann C, Gombert AK (2008) The yeast Kluyveromyces marxianus and its biotechnological potential. Appl Microbiol Biotechnol 79(3):339–354
Galbe M, Zacchi G (2002) A review of the production of ethanol from softwood. Appl Microbiol Biotechnol 59(6):618–628
García-Martínez T, Bellincontro A, Peinado RA, Mauricio JC, Mencarelli F, Moreno JJ (2011) Discrimination of sweet wines partially fermented by two osmo-ethanol-tolerant yeasts by gas chromatographic analysis and electronic nose. Food Chem 127(3):1391–1396

Ghaly AE, Kamal MA (2004) Submerged yeast fermentation of acid cheese whey for protein production and pollution potential reduction. Water Res 38(3):631–644
Gibson A (2006) Ethanol from whey. Sustainable energy conference, 26–27 July, Palmerston North, New Zealand
Guimarães PM, Teixeira JA, Domingues L (2010) Fermentation of lactose to bio-ethanol by yeasts as part of integrated solutions for the valorisation of cheese whey. Biotechnol Adv 28(3):375–384
Hacking AJ, Taylor IWF, Hanas CM (1984) Selection of yeast able to produce ethanol from glucose at 40 C. Appl Microbiol Biotechnol 19(5):361–363
Hamilton R (1998) The manufacture of ethanol from whey. Chemical processes in New Zealand: New Zealand Institute of Chemistry
Hernandez-Saavedra NY, Ochoa JL, Vazquez-Dulhalt R (1995) Osmotic adjustment in marine yeast. J Plankton Res 17(1):59–69
Hughes DB, Tudroszen NJ, Moye CJ (1984) The effect of temperature on the kinetics of ethanol production by a thermotolerant strain of kluveromyces marxianus. Biotechnol Lett 6(1):1–6
Johnson FX, Rosillo-Calle F (2007) Biomass, livelihoods and international trade. Stockholm Environment Institute, Stockholm
Kang HW, Kim Y, Kim SW, Choi GW (2012) Cellulosic ethanol production on temperature-shift simultaneous saccharification and fermentation using the thermostable yeast Kluyveromyces marxianus CHY1612. Bioprocess Biosyst Eng 35(1-2):115–122
Kida K, Kume K, Morimura S, Sonoda Y (1992) Repeated-batch fermentation process using a thermotolerant flocculating yeast constructed by protoplast fusion. J Ferment Bioeng 74(3):169–173
Kitagawa T, Tokuhiro K, Sugiyama H, Kohda K, Isono N, Hisamatsu M, … Imaeda T (2010) Construction of a β-glucosidase expression system using the multistress-tolerant yeast Issatchenkia orientalis. Appl Microbiol Biotechnol 87(5): 1841–1853
Klipp E, Nordlander B, Krüger R, Gennemark P, Hohmann S (2005) Integrative model of the response of yeast to osmotic shock. Nat Biotechnol 23(8):975
Kosikowski FV (1979) Whey utilization and whey products1. J Dairy Sci 62(7):1149–1160
Kourkoutas Y, Dimitropoulou S, Kanellaki M, Marchant R, Nigam P, Banat IM, Koutinas AA (2002) High-temperature alcoholic fermentation of whey using Kluyveromyces marxianus IMB3 yeast immobilized on delignified cellulosic material. Bioresour Technol 82(2):177–181
Krouwel PG, Braber L (1979) Ethanol production by yeast at supraoptimal temperatures. Biotechnol Lett 1(10):403–408
Kurtzman CP, Fell JW (eds) (1998) The yeasts, a taxonomic study. Elsevier, Amsterdam
Lachance MA (1998) Kluyveromyces van der Walt emends. In: Van der Walt (ed) The Yeasts, 4th edn, pp 227–247
Lane MM, Morrissey JP (2010) Kluyveromyces marxianus: a yeast emerging from its sister's shadow. Fungal Biol Rev 24(1):17–26
Lane MM, Burke N, Karreman R, Wolfe KH, O'Byrne CP, Morrissey JP (2011) Physiological and metabolic diversity in the yeast Kluyveromyces marxianus. Antonie Van Leeuwenhoek 100(4):507–519
Lee C, Yamakawa T, Kodama T (1993) Rapid growth of a thermotolerant yeast on palm oil. World J Microbiol Biotechnol 9(2):187–190
Li SZ, Chan-Halbrendt C (2009) Ethanol production in (the) People's Republic of China: potential and technologies. Appl Energy 86:S162–S169
Licht FO (2003) World ethanol markets: the outlook to 2012: an FO Licht special study. FO Licht
Limtong S, Sringiew C, Yongmanitchai W (2007) Production of fuel ethanol at high temperature from sugar cane juice by a newly isolated Kluyveromyces marxianus. Bioresour Technol 98(17):3367–3374
Ling KC (2008) Whey to ethanol. Is there a biofuel role for dairy cooperatives? Rural Cooperatives, Washington, DC

Löser C, Urit T, Nehl F, Bley T (2011) Screening of Kluyveromyces strains for the production of ethyl acetate: design and evaluation of a cultivation system. Eng Life Sci 11(4):369–381
Lyons TP, Cunningham JD (1980) Fuel alcohol from whey. Am Dairy Rev 42(11):42A–42E
Marullo P, Mansour C, Dufour M, Albertin W, Sicard D, Bely M, Dubourdieu D (2009) Genetic improvement of thermo-tolerance in wine Saccharomyces cerevisiae strains by a backcross approach. FEMS Yeast Res 9(8):1148–1160
Marwaha SS, Kennedy JF (1988) Whey – pollution problem and potential utilization. Int J Food Sci Technol 23(4):323–336
Mawson AJ (1994) Bioconversions for whey utilization and waste abatement. Bioresour Technol 47(3):195–203
Menon V, Prakash G, Rao M (2010) Enzymatic hydrolysis and ethanol production using xyloglucanase and Debaromyces hansenii from tamarind kernel powder: galactoxyloglucan predominant hemicellulose. J Biotechnol 148(4):233–239
Michel GP, Starka JIRI (1986) Effect of ethanol and heat stresses on the protein pattern of Zymomonas mobilis. J Bacteriol 165(3):1040–1042
Morimura S, Ling ZY, Kida K (1997) Ethanol production by repeated-batch fermentation at high temperature in a molasses medium containing a high concentration of total sugar by a thermotolerant flocculating yeast with improved salt-tolerance. J Ferment Bioeng 83(3):271–274
Morris D (1993) Ethanol: a 150 year struggle toward a renewable future. Institute for Local Self-Reliance, Washington, DC
Mukherjee V, Steensels J, Lievens B, Van de Voorde I, Verplaetse A, Aerts G et al (2014) Phenotypic evaluation of natural and industrial saccharomyces yeasts for different traits desirable in industrial bioethanol production. Appl Microbiol Biotechnol 98(22):9483–9498
Naik SN, Goud VV, Rout PK, Dalai AK (2010) Production of first and second generation biofuels: a comprehensive review. Renew Sust Energ Rev 14(2):578–597
Nonklang S, Abdel-Banat BM, Cha-aim K, Moonjai N, Hoshida H, Limtong S et al (2008) High-temperature ethanol fermentation and transformation with linear DNA in the thermotolerant yeast Kluyveromyces marxianus DMKU3-1042. Appl Environ Microbiol 74(24):7514–7521
Nwachukwu IN, Ibekwe VI, Nwabueze RN, Anyanwu BN (2006) Characterisation of palm wine yeast isolates for industrial utilisation. Afr J Biotechnol 5(19):1725–1728
Oberoi HS, Babbar N, Sandhu SK, Dhaliwal SS, Kaur U, Chadha BS, Bhargav VK (2012) Ethanol production from alkali-treated rice straw via simultaneous saccharification and fermentation using newly isolated thermotolerant Pichia kudriavzevii HOP-1. J Ind Microbiol Biotechnol 39(4):557–566
Oda Y, Nakamura K (2009) Production of ethanol from the mixture of beet molasses and cheese whey by a 2-deoxyglucose-resistant mutant of Kluyveromyces marxianus. FEMS Yeast Res 9(5):742–748
Ohta K, Wijeyaratne SC, Hayashida S (1988) Temperature-sensitive mutants of a thermotolerant yeast, Hansenula polymorpha. J Ferment Technol 66(4):455–459
Orellana C, Neto RB (2006) Brazil and Japan give fuel to ethanol market. Nat Biotechnol 24:232
Ortiz-Muñiz B, Carvajal-Zarrabal O, Torrestiana-Sanchez B, Aguilar-Uscanga MG (2010) Kinetic study on ethanol production using saccharomyces cerevisiae ITV-01 yeast isolated from sugar cane molasses. J Chem Technol Biotechnol 85(10):1361–1367
Osho A (2005) Ethanol and sugar tolerance of wine yeasts isolated from fermenting cashew apple juice. Afr J Biotechnol 4(7):660–662
Pessani NK, Atiyeh HK, Wilkins MR, Bellmer DD, Banat IM (2011) Simultaneous saccharification and fermentation of Kanlow switchgrass by thermotolerant Kluyveromyces marxianus IMB3: the effect of enzyme loading, temperature and higher solid loadings. Bioresour Technol 102(22):10618–10624
Pesta G, Meyer-Pittroff R, Russ W (2007) Utilization of whey. Utilization of by-products and treatment of waste in the food industry. Springer, New York, pp 193–207
Poolman B, Glaasker E (1998) Regulation of compatible solute accumulation in bacteria. Mol Microbiol 29(2):397–407

Polat Z (2009) Integrated approach to whey utilization through natural zeolite adsorption/desorption and fermentation. Graduate School of Engineering, Izmir

Prasetyo J, Naruse K, Kato T, Boonchird C, Harashima S, Park EY (2011) Bioconversion of paper sludge to biofuel by simultaneous saccharification and fermentation using a cellulase of paper sludge origin and, thermotolerant Saccharomyces cerevisiae TJ14. Biotechnol Biofuels 4(1):35

Rajeshwari KV, Balakrishnan M, Kansal A, Lata K, Kishore VVN (2000) State-of-the-art of anaerobic digestion technology for industrial wastewater treatment. Renew Sust Energ Rev 4(2):135–156

Rana S, Tiwari R, Arora A, Singh S, Kaushik R, Saxena AK, Nain L (2013) Prospecting Parthenium sp. pretreated with Trametes hirsuta, as a potential bioethanol feedstock. Biocatal Agric Biotechnol 2(2):152–158

Renewable Fuels Association (2009) Ethanol industry statistics. Online: http://www.ethanolrfa.org/industry/statistics

RFA-Renewable Fuels Association (2010) The industry-statistics

Rogosa M, Browne HH, Whittier EO (1947) Ethyl alcohol from whey. J Dairy Sci 30(4):263–269

Ruyters S, Mukherjee V, Verstrepen KJ, Thevelein JM, Willems KA, Lievens B (2015) Assessing the potential of wild yeasts for bioethanol production. J Ind Microbiol Biotechnol 42(1):39–48

Sansonetti S, Curcio S, Calabrò V, Iorio G (2009) Bio-ethanol production by fermentation of ricotta cheese whey as an effective alternative non-vegetable source. Biomass Bioenergy 33(12):1687–1169

Schirmer-Michel ÂC, Flôres SH, Hertz PF, Matos GS, Ayub MAZ (2008) Production of ethanol from soybean hull hydrolysate by osmotolerant candida guilliermondii NRRL Y-2075. Bioresour Technol 99(8):2898–2904

Shi DJ, Wang CL, Wang KM (2009) Genome shuffling to improve thermotolerance, ethanol tolerance and ethanol productivity of Saccharomyces cerevisiae. J Ind Microbiol Biotechnol 36(1):139–147

Silveira WB, Passos FJV, Mantovani HC, Passos FML (2005) Ethanol production from cheese whey permeates by Kluyveromyces marxianus UFV-3: a flux analysis of oxido-reductive metabolism as a function of lactose concentration and oxygen levels. Enzym Microb Technol 36(7):930–936

Siso MG (1996) The biotechnological utilization of cheese whey: a review. Bioresour Technol 57(1):1–11

Smithers GW (2008) Whey and whey proteins—from 'gutter-to-gold'. Int Dairy J 18(7):695–704

Solomon BD, Barnes JR, Halvorsen KE (2007) Grain and cellulosic ethanol: history, economics, and energy policy. Biomass Bioenergy 31(6):416–425

Souza RR (2006) Oportunidades e desafios para o mercado mundial de álcool automotivo. Universidad Federal de Río de Janeiro, Río de Janeiro

Sridhar M, Sree NK, Rao LV (2002) Effect of UV radiation on thermotolerance, ethanol tolerance and osmotolerance of Saccharomyces cerevisiae VS 1 and VS 3 strains. Bioresour Technol 83(3):199–202

Stanley D, Fraser S, Chambers PJ, Rogers P, Stanley GA (2010) Generation and characterisation of stable ethanol-tolerant mutants of Saccharomyces cerevisiae. J Ind Microbiol Biotechnol 37(2):139–149

Stewart GG (2001) Yeast management–the balance between fermentation efficiency and beer quality. Tech Q:39–46

Suutari M, Liukkonen K, Laakso S (1990) Temperature adaptation in yeasts: the role of fatty acids. Microbiology 136(8):1469–1474

Szczodrak J, Targoński Z (1988) Selection of thermotolerant yeast strains for simultaneous saccharification and fermentation of cellulose. Biotechnol Bioeng 31(4):300–303

Tamás MJ, Hohmann S (2003) The osmotic stress response of saccharomyces cerevisiae. In: Hohmann S, Mager WH (eds) Yeast stress responses. Chapman & Hall, New York, pp 121–200

Tofalo R, Chaves-López C, Di Fabio F, Schirone M, Felis GE, Torriani S et al (2009) Molecular identification and osmotolerant profile of wine yeasts that ferment a high sugar grape must. Int J Food Microbiol 130(3):179–187
Thammasittirong SNR, Thirasaktana T, Thammasittirong A, Srisodsuk M (2013) Improvement of ethanol production by ethanol-tolerant saccharomyces cerevisiae UVNR56. SpringerPlus 2(1):583
Thiele JH (2005) Estimate of the energy potential for fuel ethanol from putrescible waste in New Zealand. Waste Solutions Ltd., Dunedin
Tsegaye Z (2016) Isolation, identification and characterization of ethanol tolerant yeast species from fruits for production of bio-ethanol. Int J Mod Chem Appl Sci 3:437–443
Urit T, Löser C, Wunderlich M, Bley T (2011) Formation of ethyl acetate by Kluyveromyces marxianus on whey: studies of the ester stripping. Bioprocess Biosyst Eng 34(5):547–559
Van Dijken JP, Weusthuis RA, Pronk JT (1993) Kinetics of growth and sugar consumption in yeasts. Antonie Van Leeuwenhoek 63(3–4):343–352
Van Uden N (1984) Temperature pro®les of yeasts. In: Rose AH, Tempest DW (eds) Advances in microbial physiology, vol 25. Academic Press, London, pp 195–248
Van Urk H, Voll WL, Scheffers WA, Van Dijken JP (1990) Transient-state analysis of metabolic fluxes in Crabtree-positive and Crabtree-negative yeasts. Appl Environ Microbiol 56(1):281–287
Vienne P, Von Stockar U (1985) Metabolic, physiological and kinetic aspects of the alcoholic fermentation of whey permeate by Kluyveromyces fragilis NRRL 665 and Kluyveromyces lactis NCYC 571. Enzym Microb Technol 7(6):287–294
Voronovsky AY, Rohulya OV, Abbas CA, Sibirny AA (2009) Development of strains of the thermotolerant yeast Hansenula polymorpha capable of alcoholic fermentation of starch and xylan. Metab Eng 11(4–5):234–242
Wang CJ, Jayanata Y, Bajpai RK (1987) Effect of multiple substrates in ethanol fermentations from cheese whey. J Ferment Technol 65:249–253
Walter A, Rosillo-Calle F, Dolzan P, Piacente E, da Cunha KB (2008) Perspectives on fuel ethanol consumption and trade. Biomass Bioenergy 32(8):730–748
Wardrop FR, Liti G, Cardinali G, Walker GM (2004) Physiological responses of Crabtree positive and Crabtree negative yeasts to glucose upshifts in a chemostat. Ann Microbiol 54(1):103–114
Watanabe T, Srichuwong S, Arakane M, Tamiya S, Yoshinaga M, Watanabe I, Yamamoto M, Ando A, Tokuyasu K, Nakamura T (2010) Selection of stress-tolerant yeasts for simultaneous saccharification and fermentation (SSF) of very high gravity (VHG) potato mash to ethanol. Bioresour Technol 101(24):9710–9714
Webb BH, Whittier EO (1948) The utilization of whey: a review. J Dairy Sci 31(2):139–164
Wegmann K (1986) Osmoregulation in eukaryotic algae. FEMS Microbiol Lett 39(1–2):37–43
Whittier EO (1944) Lactose and its utilization: a review. J Dairy Sci 27(7):505–537
Wu CZ, Yin XL, Yuan ZH, Zhou ZQ, Zhuang XS (2010) The development of bioenergy technology in China. Energy 35(11):4445–4450
Yadav KS, Naseeruddin S, Prashanthi GS, Sateesh L, Rao LV (2011) Bioethanol fermentation of concentrated rice straw hydrolysate using co-culture of saccharomyces cerevisiae and pichia stipitis. J Chem Technol Biotechnol 102(11):6473–6478
Yamaoka C, Kurita O, Kubo T (2014) Improved ethanol tolerance of Saccharomyces cerevisiae in mixed cultures with Kluyveromyces lactis on high-sugar fermentation. Microbiol Res 169(12):907–914
Yanase S, Hasunuma T, Yamada R, Tanaka T, Ogino C, Fukuda H, Kondo A (2010) Direct ethanol production from cellulosic materials at high temperature using the thermotolerant yeast Kluyveromyces marxianus displaying cellulolytic enzymes. Appl Microbiol Biotechnol 88(1):381–388
Yancey PH, Clark ME, Hand SC, Bowlus RD, Somero GN (1982) Living with water stress: evolution of osmolyte systems. Science 217(4566):1214–1222
Zhang YX, Perry K, Vinci VA, Powell K, Stemmer WP, del Cardayré SB (2002) Genome shuffling leads to rapid phenotypic improvement in bacteria. Nature 415(6872):644–646

Zhao XQ, Bai FW (2009) Mechanisms of yeast stress tolerance and its manipulation for efficient fuel ethanol production. J Biotechnol 144(1):23–30
Zuzuarregui A (2004) Analyses of stress resistance under laboratory conditions constitute a suitable criterion for wine yeast selection. Antonie Van Leeuwenhoek 85(4):271–280

Chapter 21
Nutritional Quality Attributes of Edible Gasteroid Wild Mushroom *Astraeus hygrometricus*

Mundamoole Pavithra, Kandikere R. Sridhar, and Ammatanda A. Greeshma

Abstract The gasteroid mushroom *Astraeus hygrometricus* is one of the traditionally consumed wild mushrooms in Southwest India. Owing to insufficient information on nutritional attributes of this mushroom, the present study addresses proximal qualities, minerals, amino acids, protein bioavailability and fatty acid composition in uncooked and pressure-cooked samples. The proximal features include moderate quantity of protein, low total lipid, high crude fibre and high carbohydrate contents. In uncooked and cooked mushroom, potassium, iron and zinc were higher than NRC-NAS recommended pattern, while the Na/K ratio was favourable (<1). Six essential amino acids (EAAs) (His, Ile, Leu, Lys, Thr and Val) in uncooked and cooked samples surpassed the FAO-WHO stipulated pattern, while the ratio of total EAAs/total amino acids favourably increased on cooking. The EAA score for many amino acids (Ile, Leu, Lys, Phe + Tyr and Val) was higher in cooked than in uncooked samples. The in vitro protein digestibility was significantly higher in uncooked than cooked samples. The protein efficiency ratios were favourable (>2) with highest in cooked samples (2.3–2.9). Among the fatty acids, palmitic and oleic acids were high. Overall, the wild tender *A. hygrometricus* is nutritionally valuable either uncooked or cooked state with sufficient quantity of protein, high carbohydrates, high fibre, low fat with adequate quantity of some essential minerals and several EAAs. In addition to good nutritional potential, it is also known for several bioactive components and antioxidant properties that serve as indigenous natural nutraceutical source.

Keywords Amino acids · Ectomycorrhizae · Fatty acids · Minerals · Protein bioavailability · Proximal qualities

M. Pavithra · K. R. Sridhar (✉) · A. A. Greeshma
Department of Biosciences, Mangalore University, Mangalagangotri, Mangalore, Karnataka, India

P. Gehlot, J. Singh (eds.), *Fungi and their Role in Sustainable Development: Current Perspectives*, https://doi.org/10.1007/978-981-13-0393-7_21

21.1 Introduction

Wild mushrooms of forest origin are natural indigenous sources of nutrition and medicine for humans throughout the world (Boa 2004). Nearly 80% of the forest dwellers are dependent on non-timber forest products to meet up to 25–50% of their annual nutritional requirements (Singh 2011). Either wild or cultivated mushrooms serve as natural functional ingredients of food almost similar to cereals and beverages (Barros et al. 2008). The amino acid composition of wild edible mushrooms is comparable to animal food products (Gruen and Wong 1982). Their nutritional (proteins, fibre, fatty acids and vitamins) and bioactive components (ascorbic acid, carotenoids, phenolics and tocopherols) are capable to combat several human ailments (e.g. cardiovascular, inflammatory and cancers) (Fürst and Kuhn 2000; Mattila et al. 2001; Fang et al. 2002; Kruger and Mann 2003; Lindequist et al. 2005; Soobrattee et al. 2005; Barros et al. 2007a, b, c, 2008). Beyond nutrition and health, the value of wild and cultivated mushrooms expanded towards biomedical and industrial applications (e.g. chitosan, cosmetics, glucans and nanoparticles) (Manzi and Pizzoferrato 2000; Wu et al. 2004; Hyde et al. 2010; Arun et al. 2014). Similar to plant- and animal-derived food products, edible macrofungi have potential to meet the protein-energy demand (Boa 2004; Sudheep and Sridhar 2014). Generally, the wild mushrooms are rich in protein and low in fats and possess other value-added nutrients (e.g. vitamins) compared to commercially cultivated mushrooms (Barros et al. 2008). Many wild edible mushrooms are used to treat diabetes, obesity, anaemia, mumps, fever, protein deficiency and infant malnutrition in Nigeria (Sylvester et al. 2014). High content of fibre in mushrooms will also be an added advantage to treat those suffering from hyperacidity and constipation (Bora and Kawatra 2014).

The Western Ghats of India with chain of mountains along the southwest has been recognized as one of the hotspots of biodiversity and encompasses approximately 750 species of mushrooms (Thiribhuvanamala et al. 2011). Some of the recent inventories in Western Ghats and west coast of India accounted wild edible mushrooms ranging from 9 to 25 (Ghate et al. 2014; Karun and Sridhar 2014, 2016; Pavithra et al. 2015; Ghate and Sridhar 2016a, b; Greeshma et al. 2016). In spite of several recent studies addressed the diversity of wild mushrooms in the Western Ghats, evaluation of nutritional and bioactive attributes are scanty (e.g. Sudheep and Sridhar 2014; Karun et al. 2016). Similarly, a few studies have focused on the ectomycorrhizal fungi of the Western Ghats (Natarajan et al. 2005; Pavithra et al. 2015). Many edible ectomycorrhizal wild mushrooms are in association with a variety of tree families (e.g. Anacardiaceae, Betulaceae, Dipterocarpaceae, Ericaceae, Fagaceae, Moraceae, Myrtaceae, Phyllanthaceae, Pinaceae and Ulmaceae) (Sanmee et al. 2003; Natarajan et al. 2005; Pavithra et al. 2015). *Astraeus hygrometricus* is one of the common traditionally consumed highly prized popular edible mushrooms in the foothill of the Western Ghats as well as coastal regions of the Southwest India as in Northern Thailand (Sanmee et al. 2003; Pavithra et al. 2015). It is also most widely consumed mushroom in the eastern lateritic parts of India (Manna et al.

2014). Its usual geographic locations include mainly lateritic soils in association with several tree species (Pavithra et al. 2015). Owing to its edibility, tribals and local dwellers collect and market them during the wet season. As there is no specific or in-depth outlook on nutritional attributes on *A. hygrometricus*, the current study focuses to document the proximal qualities, mineral composition, amino acid composition, protein bioavailability and fatty acid profile in comparison with other wild edible mushrooms.

21.2 Mushroom

Partially hypogeous fruit bodies of gasteroid mushroom *Astraeus hygrometricus* (Pers.) Morgan were collected from five locations (~50 m apart) during monsoon season (June–July 2015) of Karkala forests of the Western Ghats, India (13°12′N, 74°58′E; 65–90 m asl). They were transported to the laboratory in cool packs and processed within 3–4 h of collection. Debris (soil, roots and mycelial mass) were removed, rinsed in water and drained. Slight abrasion peels out the outermost layer of fruit bodies and removal of outer layer is traditional practice prior to cook. Each replicate was divided into two groups, and tenderness was ascertained by white interior in cut fruit bodies (those with black interior were discarded) (Fig. 21.1). The first group of each replicate was oven dried (50–55 °C) until the moisture attains below 5%. The second group was pressure-cooked with distilled water (1:1 v/v) in a household pressure cooker (Deluxe stainless steel, TTK Prestige™, Prestige Ltd., Hyderabad, India; capacity, 6.5 L) and oven dried. The dried samples were milled in Wiley Mill (mesh # 30) and refrigerated in air-tight containers for analysis.

21.3 Nutritional Assessment

21.3.1 Proximal and Mineral Analysis

Moisture content of uncooked and cooked mushroom powder was assessed gravimetrically (AOAC 1995). Proximal qualities such as crude protein, total lipids, crude fibre, ash, total carbohydrates and calorific value of uncooked and cooked mushroom samples were evaluated based on standard protocols. The crude protein content was assessed by micro-Kjeldahl method (N × 6.25) (Humphries 1956). Total lipids content was extracted in petroleum ether (60–80 °C) using Soxhlet apparatus according to AOAC (1995) method to evaluate gravimetrically. The crude fibre and ash contents were determined gravimetrically (AOAC 1995). Total carbohydrates were estimated based on the method by Sadasivam and Manickam (2008). The calorific value was calculated according to Ekanayake et al. (1999):

Fig. 21.1 (**a**) Partially hypogeous tender sporocarps of *Astraeus hygrometricus* just below the lateritic soil partially exposed on surface scratching, (**b**) excavated tender sporocarps with debris (soil, roots and mycelia), (**c**) mature and spent sporocarps, (**d**) cleaned sporocarps, (**e**) cut open sporocarp with details of white interior (edible and used for nutritional assessment) and (**f**) cut open mature sporocarp with spore mass (inedible)

$$\text{Calorific value}\left(\text{kJ}/100\,\text{g}\right) = \left(\text{Crude protein} \times 16.7\right) + \left(\text{Total lipids} \times 37.7\right) + \left(\text{Carbohydrate} \times 16.7\right) \quad (21.1)$$

Mineral composition was assessed based on the protocol by Ramamurthy and Kannan (2009). This method is commonly employed to determine the major and minor elements in environmental samples. The moisture free samples (particle size, >40 μm) were subjected to field-emission scanning electron microscope-energy dispersive spectrometer (SEM-EDS) analysis (FESEM Carl Zeiss, Oxford Instruments, USA) with voltage of 15 kV. The SEM images and corresponding EDS spectrum generated for samples were dependent on the properties (shape, shell and size) to express minerals in percentage. The ratio of Na/K and Ca/P were calculated.

21.3.2 *Amino Acids and Protein Assessment*

The protocols by Hofmann et al. (1997, 2003) were employed to analyse the amino acids in mushroom samples. Alkaline-extracted samples were used for tryptophan, and oxidized samples were used for sulphur amino acid analysis. Procedure proposed by Brand et al. (1994) was followed for derivatization process of esterification with trifluoroacetylation. The amino acids in reaction vials dried in CH_2Cl_2 served as standards. The mushroom samples were hydrolysed in HCl and evaporated in rotary evaporator (Büchi Laboratoriumstechnik AG RE121, Switzerland) with a diaphragm vacuum pump (MC2C, Vacuubrand GmbH, Germany) followed by measurements of GC-C-IRMS/MS. The gas chromatograph (Hewlett-Packard 58590 II, Germany) connected through a combustion interface to the IRMS system (GC-C-II to MAT 252, Finnigan MAT, Germany) was employed for the measurements of GC-C-IRMS/MS for the isotopic determination of nitrogen through a transfer line with a mass spectrometer (GCQ, Finnigan MAT, Germany) for qualitative and quantitative determination of the amino acids.

The ratio between total essential amino acids (TEAAs) and total amino acids (TAAs) was calculated (TEAAs/TAAs)

The in vitro protein digestibility (IVPD) was assessed using enzymes (pepsin, trypsin and α-chymotrypsin) as per the procedure outlined by Akeson and Stahmann (1964) and protein content was assessed by micro-Kjeldahl method (Humphries 1956):

$$\text{IVPD}(\%) = (\text{Protein in digest/Protein in defatted flour}) \times 100 \quad (21.2)$$

The essential amino acid score (EAAS) was determined according to FAO-WHO EAA requirement pattern for adults (FAO-WHO 1991):

$$\text{EAAS} = (\text{Amino acid in the test protein:mg/g}) / (\text{FAO WHO EAA reference pattern:mg/g}) \quad (21.3)$$

The protein digestibility corrected amino acid score (PDCAAS) for adult was calculated.

(FAO-WHO 1991):

$$\text{PDCAAS} = (\text{EAA in test protein:mg/g}) / (\text{FAO-WHO EAA reference pattern:mg/g}) \times \text{IVPD}(\%) \quad (21.4)$$

Protein efficiency ratio (PER) was determined based on the amino acid composition of uncooked and cooked mushroom samples based on Alsmeyer et al. (1974):

$$\text{PER}_1 = -0.684 + 0.456 \times \text{Leu} - 0.047 \times \text{Pro} \quad (21.5)$$

$$\text{PER}_2 = -0.468 + 0.454 \times \text{Leu} - 0.105 \times \text{Tyr} \quad (21.6)$$

$$\text{PER}_3 = -1.816 + 0.435 \times \text{Met} + 0.78 \times \text{Leu} + 0.211 \times \text{His} - 0.944 \times \text{Tyr} \quad (21.7)$$

21.3.3 Fatty Acid Analysis

The total lipids of uncooked and cooked mushroom samples by Soxhlet method was used for analysis of fatty acid methyl esters (FAMEs) according to Padua-Resurreccion and Benzon (1979). The gas chromatograph (GC-2010, Shimadzu, Japan) combined with an autoinjector (AOI) with capillary column (BPX-70) was used. The elutants were detected on flame ionization detector (FID), and the amplified signals were transferred and monitored using GC Solutions software (http://www.shimadzu.eu/products/software/labsolutions/gcgcms/default.aspx). Analytical conditions of autosampler, injection port settings, column oven settings and column information of the gas chromatograph were maintained according to Nareshkumar (2007). The quantity of FAMEs was evaluated on comparison of the peaks of spectra and retention time (RT) with peaks, RT and hits of known compounds stocked in National Institute of Standards and Technology (NIST) library. The ratio of total unsaturated fatty acid and total saturated fatty acids (TUFA/TSFA) was calculated.

21.3.4 Data Analysis

The significance of proximal properties, mineral composition, amino acid composition, in vitro protein digestibility and FAMEs between uncooked and cooked mushroom samples were assessed by *t*-test using Statistica version #8.0 (Statsoft Inc. 2008).

21.4 Nutritional Qualities

The tender fruit bodies of wild mushroom *A. hygrometricus* possess required traits as they are large with sufficient wet biomass similar to the baby potatoes (Fig. 21.1). One of the important features is the tender fruit bodies tend to sink in water, while mature or old fruit bodies float. Such physical features of this mushroom facilitate

Table 21.1 Proximal characteristics of uncooked and cooked tender *A. hygrometricus* on dry mass basis ($n = 5$; mean±SD)

	Uncooked	Cooked
Moisture (%)	2.71 ± 0.31	2.93 ± 0.22
Crude protein (g/100 g)	16.80 ± 0.006	17.30 ± 0.46
Total lipids (g/100 g)	3.55 ± 0.24*	2.49 ± 0.06
Crude fibre (g/100 g)	14.58 ± 0.38	15.91 ± 0.22**
Ash (g/100 g)	18.43 ± 0.17**	15.53 ± 0.46
Total carbohydrates (g/100 g)	46.17 ± 0.76	48.42 ± 0.52*
Calorific value (kJ/100 g)	1185.4 ± 4.18	1191.3 ± 6.75

Asterisk across the cooked and uncooked mushroom are significantly different (*t*-test: $*p < 0.05$; $**p < 0.01$)

handling during large-scale processing for value-added food preparation. The local dwellers after collection of mushroom clean them to remove debris, process and consume on the same day or subsequent day. Cleaned mushrooms will be stored by wrapping in wet cloth to preserve at room temperature or in refrigerator for a couple of days.

21.4.1 Proximal Features

Moisture content of mushroom samples did not significantly differed, which ranged from 2.7% to 2.9%. The crude protein content of edible mushrooms is usually higher than the cereals and comparable to those of legume seeds. Pressure-cooking of fruit bodies of *A. hygrometricus* has not resulted in significant loss of crude protein ($p > 0.05$), which is an added nutritional advantage (Table 21.1). The quantity of crude protein was higher in our study (16.8–17.3%) compared to young (14.0%) as well as mature (14.7%) fruit bodies of *A. hygrometricus* landrace from the Northern Thailand (Sanmee et al. 2003), so also from the report from Central India (Singh 2011) and wood-inhabiting wild mushroom *Auricularia auricula-judae* (Karun et al. 2017). However, its content was lower than other wild edible mushrooms of the Western Ghats (soil-inhabiting *Agaricus abruptibulbus*, termite mound-inhabiting *Termitomyces globulus* and *Termitomyces umkowaan*) (Sudheep and Sridhar 2014; Karun et al. 2017).

Generally lipid content is lower in edible mushrooms, which is advantageous in human nutrition. Total lipid content was significantly higher in uncooked than in cooked samples of *A. hygrometricus* in our study ($p < 0.05$) (Table 21.1). Its content is comparable or higher than the young fruit bodies of Northern Thailand (2.5–3.6% vs. 2.7%) (Sanmee et al. 2003) as well as *A. abruptibulbus* of the Western Ghats (Sudheep and Sridhar 2014) and higher compared to *A. hygrometricus* reported from Central India (Singh 2011). Although lipid content of *A. hygrometricus* is higher than wild *A. auricula-judae*, it was lower than two termitomycetes (*T. globulus* and T. *umkowaan*) of the Western Ghats (Sudheep and Sridhar 2014;

Karun et al. 2017). Overall, the lipid content in *A. hygrometricus* is comparable with many wild edible mushrooms reported from the Indian Subcontinent (Kavishree et al. 2008).

Mushrooms are also known for high crude fibre content, and it was significantly higher in cooked than uncooked samples of *A. hygrometricus* ($p < 0.01$) (Table 21.1). Its range is higher than young and mature *A. hygrometricus* of Northern Thailand (10.8–12.3%) (Sanmee et al. 2003), so also the wild *A. abruptibulbus* and *T. globulus*, while lower than *A. auricula-judae* and *T. umkowaan* of the Western Ghats (Sudheep and Sridhar 2014; Karun et al. 2017). High fibre content in these mushrooms results in several advantages such as improvement of digestibility by trapping less proteins as well as carbohydrates (Balogun and Fetuga 1986). Besides, high fibre content is known for lowering the blood cholesterol as well as lowering the risks of bowel cancers (Anderson et al. 1995; Slavin et al. 1997).

The ash content in mushrooms tends to decrease in cooked samples as minerals drain in water used for cooking. The ash content of *A. hygrometricus* was significantly decreased in pressure-cooked samples ($p < 0.01$) (Table 21.1). Its content in uncooked as well as cooked samples is lower than tender *A. hygrometricus* of Northern Thailand (15.5–18.4% vs. 27.6%) (Sanmee et al. 2003), higher than *A. auricula-judae* and *T. umkowaan*, while it is comparable to *A. abruptibulbus* and *T. globulus* (Sudheep and Sridhar 2014; Karun et al. 2017).

As seen in crude protein and fibre, the carbohydrate content of mushrooms was also in high quantity. The total carbohydrates were significantly increased on cooking of *A. hygrometricus* ($p < 0.05$) (Table 21.1). Its quantity (46.2–48.4%) could be comparable to young fruit bodies (44.9%) of Northern Thailand landrace (Sanmee et al. 2003), but higher than the report from Central India (Singh 2011). Carbohydrate content in uncooked as well as cooked samples is also comparable to *T. globulus* and *T. umkowaan*, while lower than *A. abruptibulbus* and *A. auricula-judae* (Sudheep and Sridhar 2014; Karun et al. 2017). High carbohydrate diet is known to combat the intestinal cancer as well as lowers glycaemic index helping in prevention of the type II diabetes (Venn and Mann 2004).

Similar to crude protein, cooking did not significantly alter the calorific value of *A. hygrometricus* ($p > 0.05$) (Table 21.1). However, the calorific value of uncooked and cooked *A. hygrometricus* was lower than other wild mushrooms of the Western Ghats (1185–1191 vs. 1241–1593 kJ/100 g) (Sudheep and Sridhar 2014; Karun et al. 2017).

21.4.2 Minerals Profile

While evaluating minerals by SEM-EDS, other components like carbon, nitrogen and oxygen were also obtained. Among the ten minerals of *A. hygrometricus* assessed, the potassium was highest followed by phosphorus and sulphur

Table 21.2 Mineral composition of uncooked and cooked tender *A. hygrometricus* on dry mass basis (mg/100 g) ($n = 5$; mean±SD)

	Uncooked	Cooked	Infants/children[a]	Adults[a]
Na	69.9 ± 0.65***	40.08 ± 1.18	120–400	500
K	3216.51 ± 29.82*	3045.88 ± 89.38	500–1600	1600–2000
Ca	249.43 ± 1.92	240.46 ± 7.06	600–800	800
Mg	169.82 ± 1.57**	150.29 ± 4.41	60–170	280–350
P	539.41 ± 5**	480.93 ± 14.11	500–800	800
Fe	439.52 ± 4.08***	170.33 ± 5.00	10	10–15
Zn	219.76 ± 2.04***	140.27 ± 4.12	5–10	12–15
S	509.45 ± 4.72**	440.9 ± 12.94		
Cu	329.64 ± 3.06	BDL	0.6–2	1.5–3
Si	349.62 ± 3.24	BDL		
Na/K ratio	0.02	0.01		
Ca/P ratio	0.46	0.50		

Asterisk across the cooked and uncooked mushroom are significantly different (*t*-test: *$p < 0.05$; **$p < 0.01$; ***$p < 0.001$)
BDL below detectable level
[a]NRC-NAS (1989) recommended pattern

(Table 21.2). Except for calcium, rest of the minerals were significantly decreased on cooking. The contents of potassium, iron and zinc surpassed the NRC-NAS (1989) recommended pattern for infants/children as well as adults, while magnesium and phosphorus were comparable to the NRC-NAS pattern only for infants/children. The phosphorous content (480.9–539.4 mg/100 g) and sulphur content (440.9–509.5 mg/100 g) were almost similar to young fruit bodies of *A. hygrometricus* of Thailand, while magnesium (150.3–169.8 mg/100 g) and calcium (240.5–249.4 mg/100 g) contents were comparable with mature fruit bodies (Sanmee et al. 2003). Potassium and zinc contents are higher than young and mature fruit bodies of *A. hygrometricus* (Sanmee et al. 2003). Iron, zinc and phosphorus contents in *A. hygrometricus* were more than *A. abruptibulbus* as well as *T. globulus* of the Western Ghats (Sudheep and Sridhar 2014).

Low sodium and high potassium contents in uncooked and cooked *A. hygrometricus* reflected in low ratio of Na/K (0.01–0.02), such low ratio (<1) is advantageous to those suffering from high blood pressure and hypertension (Yusuf et al. 2007). As seen in *A. hygrometricus*, the Na/K ratio was favourable (<1) in many wild mushrooms of the Western Ghats (*A. abruptibulbus*, *A. auricula-judae*, *T. globulus* and *T. umkowaan*) (Sudheep and Sridhar 2014; Karun et al. 2017). However, the Ca/P ratio in *A. hygrometricus* (<1) is not as favourable as in other wild mushrooms of the Western Ghats (>1) (e.g. *A. abruptibulbus*, *A. auricula-judae* and *T. globulus*) (Shills and Young 1988).

Table 21.3 Amino acid composition of uncooked and cooked tender *A. hygrometricus* in comparison with soybean, wheat and FAO-WHO (1991) recommended pattern for adults (g/100 g protein) ($n = 5$; mean±SD)

	Uncooked	Cooked	Soybean[a]	Wheat[b]	FAO-WHO[c]
Essential amino acid (EAA)					
His	2.7±0.17	2.69±0.01	2.50	1.9–2.6	1.90
Ile	4.61±0.01	5.02±0.01***	4.62	3.4–4.1	2.80
Leu	6.65±0.006	7.14±0.02***	7.72	6.5–7.2	6.60
Lys	7.7±0.02	9.35±0.02***	6.08	1.8–2.4	5.80
Met	1.67±0.01	1.9±0.01***	1.22	0.9–1.5	2.50[d]
Cys	0.72±0.02***	0.17±0.01	1.70	1.6–2.6	
Phe	2.91±0.006	3.97±0.06***	4.84	4.5–4.9	6.30[e]
Tyr	2.36±0.03	2.43±0.02*	1.24	1.8–3.2	
Thr	4.88±0.01*	4.85±0.02	3.76	2.2–3.0	3.40
Trp	BDL	BDL	3.39	0.7–1.0	1.10
Val	5.75±0.006	5.88±0.02**	4.59	3.7–4.5	3.50
Non-essential amino acid					
Ala	9.33±0.02***	8.47±0.01	4.23	2.8–3.0	
Arg	4.96±0.04	5.36±0.03**	7.13	3.1–3.8	
Asp	6.16±0.006***	4.65±0.006	11.30	3.7–4.2	
Glu	11.5±0.006***	10.92±0.01	16.90	35.5–36.9	
Gly	9.25±0.01	9.68±0.006***	4.01	3.2–3.5	
Pro	6.37±0.02**	6.21±0.01	4.86	11.4–11.7	
Ser	8.32±0.02***	7.43±0.02	5.67	3.7–4.8	
TEAAs/TAAs ratio[f]	0.40	0.43	–	–	

Asterisk across the cooked and uncooked mushroom are significantly different (*t*-test: *$p < 0.05$; **$p < 0.01$; ***$p < 0.001$)
BDL Below detectable level
[a]Bau et al. (1994)
[b]USDA (1999)
[c]FAO-WHO (1991) pattern
[d]Phe + Tyr
[e]Met + Cys
[f]Total essential amino acids/total amino acids

21.4.3 Amino Acid Composition

Similar to other wild mushrooms of the Western Ghats (e.g. *A. abruptibulbus* and *T. globulus*), glutamic acid was the major amino acid in *A. hygrometricus* (Sudheep and Sridhar 2014) (Table 21.3). The lysine was the major EAA, which is higher than soybean, wheat and FAO-WHO (1991) recommended pattern. Pressure-cooking has significantly enhanced the EAA contents of isoleucine, leucine, lysine, methionine, phenylalanine ($p < 0.001$), valine ($p < 0.01$) and tyrosine ($p < 0.05$). Among them, isoleucine, leucine, lysine, phenylalanine + tyrosine and valine were higher compared to FAO-WHO (1991) recommended pattern. The ratio of TEAAs/TAAs increased in cooked *A. hygrometricus* similar to *A. auricula-judae*, *T. globulus* and

Table 21.4 In vitro protein digestibility (IVPD) ($n = 5$, mean±SD), essential amino acid score (EAAS), protein digestibility corrected amino acid score (PDCAAS) and protein efficiency ratio (PER) of uncooked and cooked tender *A. hygrometricus*

	Uncooked	Cooked
IVPD	37.60 ± 1.98*	27.28 ± 1.10
EAAS		
His	1.42	1.42
Ile	1.65	1.79
Leu	1.01	1.08
Lys	1.33	1.61
Met + Cys	0.96	0.83
Phe + Tyr	0.84	1.02
Thr	1.44	1.43
Val	1.64	1.68
PDCAAS		
His	53.39	38.74
Ile	62.04	48.83
Leu	37.98	29.46
Lys	50.01	43.92
Met + Cys	36.10	22.64
Phe + Tyr	31.58	27.83
Thr	54.14	39.01
Val	61.66	45.83
PER		
PER1	2.05	2.28
PER2	2.30	2.52
PER3	2.44	2.85

Asterisk across the cooked and uncooked mushroom for IVPD is significantly different (t-test: $*p < 0.01$)

T. umkowaan of the Western Ghats (Sudheep and Sridhar 2014; Karun et al. 2017). Such increase of TEAAs/TAAs ratio in *A. hygrometricus* after cooking reveals its nutritional value.

21.4.4 Protein Bioavailability

The IVPD is an important index of protein bioavailability, which was higher in uncooked than in cooked *A. hygrometricus* (Table 21.4) as seen in other wild mushrooms of the Western Ghats (e.g. *A. auricula-judae* and *T. umkowaan*) (Karun et al. 2017). However, cooking has substantially increased the IVPD in other wild mushrooms (*A. abruptibulbus* and *T. globulus*) (Sudheep and Sridhar 2014). Increased EAA score between uncooked and cooked mushrooms denotes improvement in nutritional quality. Cooking *A. hygrometricus* resulted in increased the EAA score

Table 21.5 Fatty acid methyl esters (FAMEs) of uncooked and cooked tender *A. hygrometricus* by Soxhlet extraction (g/100 g lipid) ($n = 5$, mean±SD)

	Uncooked	Cooked
Saturated fatty acid		
Palmitic acid	12.45 ± 0.006*	11.85 ± 0.006
Lauric acid	2.55 ± 0.01	ND
Eicosanoic acid	ND	3.68 ± 0.006
Heptadecanoic acid	ND	0.93 ± 0.006
Docosanoic acid	ND	3.56 ± 0.01
Unsaturated fatty acid		
Oleic acid	7.49 ± 0.02*	5.69 ± 0.01
Total saturated fatty acids (TSFA)	15.00 ± 0.01	20.02 ± 0.006*
Total unsaturated fatty acids (TUFA)	7.49 ± 0.02*	5.69 ± 0.01
TUFA/TSFA ratio	0.50	0.28

Asterisk across the cooked and uncooked mushroom for FAMEs, TSFA and TUFA is significantly different (*t*-test: *$p < 0.001$)
ND Not detectable

of isoleucine, leucine, lysine, phenylalanine + tyrosine and valine. The PDCAAS is another authentic parameter used worldwide to ascertain the protein quality of foodstuffs. It was substantially high in all EAAs present in uncooked *A. hygrometricus* and indicates the possibility of more benefits on consumption of uncooked or partially thermally processed mushroom. The PER is another useful parameter which helps in evaluation of the quality of foodstuffs, and those that possess PER >2 are of high quality suitable for human consumption (Friedman 1996). The PER was higher in cooked (2.28–2.85) than uncooked (2.05–2.44) *A. hygrometricus* which reveals superior protein quality. The PER of *A. hygrometricus* is higher than *T. umkowaan*, while comparable or lower than *A. auricula-judae* (Karun et al. 2017).

21.4.5 Fatty Acid Composition

The fatty acid analysis yielded five saturated and one unsaturated fatty acids in *A. hygrometricus* (Table 21.5). The uncooked samples were dominated by palmitic acid, which was significantly higher than cooked samples ($p < 0.001$). The second dominant fatty acid was oleic acid, which was also significantly higher in uncooked samples ($p < 0.001$). The palmitic acid was highest in wild mushroom like *A. abruptibulbus*, while the stearic acid was highest in *A. auricula-judae* (Sudheep and Sridhar 2014; Karun et al. 2017). Moreover, uncooked samples showed significantly higher TUFA than cooked samples of *A. hygrometricus* ($p < 0.001$). According to the Department of Health (1994), cardiac diseases will be prevented by the foodstuff possessing TUFA/TSFA ratio ≥0.45. This ratio is favourable in uncooked *A. hygrometricus* (0.5) which strongly suggests its use in preparation of foodstuffs or fortified food using uncooked mushroom. As seen in *A. hygrometricus*, uncooked wild mushrooms such as *A. auricula-judae* and *T. globulus* of the Western Ghats also showed favourable TUFA/TSFA ratio (Sudheep and Sridhar 2014; Karun et al. 2017).

21.5 Outlook

This study has demonstrated that tender *A. hygrometricus* (uncooked and cooked) possess valuable nutritional qualities and possible to utilize depending on the nutritional requirements either the mushroom or its blend for fortified foods. Moreover, tender *A. hygrometricus* exhibited several bioactive components (total phenolics, tannins, flavonoids, vitamin C, phytic acid, lycopene and β-carotene) as well as antioxidant properties (total antioxidant activity, Fe^{+} chelating capacity, reducing power and radical-scavenging activity) (Pavithra et al. 2016). Except for the total phenolics, the rest of the bioactive components including the antioxidant potential were higher in uncooked than in cooked samples which further supports the importance of uncooked tender fruit bodies to derive maximum nutraceutical benefits. As *A. hygrometricus* occur in lateritic soils and ectomycorrhizal with tree species in pure stands as well as mixed stands, the nutritional and bioactive properties might vary depending on sampling in pure or mixed stands. Tree species support mycorrhizal association of *A. hygrometricus* which needs special concern towards their conservation. The sampling and marketing of *A. hygrometricus* are entirely based on the traditional knowledge of rural folk (or tribals) of the Western Ghats, and benefits of future technological advances (e.g. IPR and patents) need to be shared for their financial security and community development.

Acknowledgements Authors are grateful to Mangalore University for the permission to carry out this research. KRS is grateful to the University Grants Commission, New Delhi, for the award of UGC-BSR Faculty Fellowship.

References

Akeson WR, Stahmann MA (1964) A pepsin pancreatin digest index of protein quality. J Nutr 83:257–261

Alsmeyer RH, Cunningham AE, Happich ML (1974) Equations predict PER from amino acid analysis. Food Technol 28:34–38

Anderson JW, Johnstone BM, Cook-Newell ME (1995) Meta-analysis of the effects of soy protein intake on serum lipids. N Engl J Med 333:276–282

AOAC (1995) Official methods of analysis, 16th edn. Association of Official Analytical Chemists, Washington, DC

Arun G, Eyini M, Gunasekaran P (2014) Green synthesis of silver nanoparticles using the mushroom fungus *Schizophyllum commune* and its biomedical applications. Biotechnol Bioprocess Eng 19:1083–1090

Balogun AM, Fetuga BL (1986) Chemical composition of some underexploited leguminous crop seeds in Nigeria. J Agric Food Chem 34:189–192

Barros L, Baptista P, Correia DM, Casal S, Oliveira B, Ferreira ICFR (2007a) Fatty acid and sugar compositions, and nutritional value of five wild edible mushrooms from Northeast Portugal. Food Chem 105:140–145

Barros L, Calhelha RC, Vaz JA, Ferreira ICFR, Baptista P, Estevinho LM (2007b) Antimicrobial activity and bioactive compounds of Portuguese wild edible mushrooms. Eur Food Res Technol 225:151–156

Barros L, Ferreira MJ, Queiros B, Ferreira ICFR, Baptista P (2007c) Total phenols, ascorbic acid, (β carotene and lycopene) in Portuguese wild edible mushrooms and their antioxidant activities. Food Chem 103:413–419

Barros L, Ferreira MJ, Queiros B, Ferreira ICFR, Baptista P (2008) Wild and commercial mushrooms as source of nutrients and nutraceuticals. Food Chem Toxicol 46:2742–2747

Bau HM, Vallaume CF, Evard F, Quemener B, Nicolas JP ad Mejean L. (1994) Effect of solid state fermentation using *Rhizophus oligosporus* sp. T-3 on elimination of antinutritional substances and modification of biochemical constituents of defatted rape seed meal. J Sci Food Agric 65:315–322

Boa E (2004) Wild edible fungi a global overview of their use and importance to people. FAO, Rome

Bora P, Kawatra A (2014) Study on nutritional evaluation and composition of oyster mushrooms (*Pleurotus florida*). Food Sci Res J 5:56–58

Brand WA, Tegtmeyer AR, Hilkert A (1994) Compound-specific isotope analysis, extending towards 15N/14N and 13C/12C. Org Geochem 21:585–594

Department of Health (1994) Nutritional aspects of cardiovascular disease, Reports on Health and Social Subjects, vol 46. HMSO, London

Ekanayake S, Jansz ER, Nair BM (1999) Proximate composition, mineral and amino acid content of mature *Canavalia gladiata* seeds. Food Chem 66:115–119

Fang YZ, Yang S, Wu G (2002) Free radicals, antioxidants, and nutrition. Nutrition 18:872–879

FAO-WHO (1991) Protein quality evaluation, Reports of a Joint FAO-WHO Expert Consultation. Food and Nutrition Paper # 51, Food and Agriculture Organization of the United Nations. FAO, Rome

Friedman M (1996) Nutritional value of proteins from different food sources – a review. J Agric Food Chem 44:6–29

Fürst P, Kuhn KS (2000) Fish oil emulsions: what benefits can they bring? Clin Nutr 19:7–14

Ghate SD, Sridhar KR (2016a) Contribution to the knowledge on macrofungi in mangroves of the Southwest India. Pl Biosys 150:977–986

Ghate SD, Sridhar KR (2016b) Spatiotemporal diversity of macrofungi in the coastal sand dunes of Southwestern India. Mycosphere 7:458–472

Ghate SD, Sridhar KR, Karun NC (2014) Macrofungi on the coastal sand dunes of South-western India. Mycosphere 5:144–151

Greeshma AA, Sridhar KR, Pavithra M, Ghate SD (2016) Impact of fire on the macrofungal diversity of scrub jungles of Southwest India. Mycology 7:15–28

Gruen EH, Wong MW (1982) Distribution of cellular amino acids, protein and total inorganic nitrogen during fruit body development in *Flammulina veluptipes*. Can J Bot 60:1330–1341

Hofmann D, Jung K, Bender J, Gehre M, Schüürmann G (1997) Using natural isotope variations of nitrogen in plants an early indicator of air pollution stress. J Mass Spectrom 32:855–863

Hofmann D, Gehre M, Jung K (2003) Sample preparation techniques for the determination of natural 15N/14N variations in amino acids by gas chromatography-combustion-isotope ratio mass spectrometry (GC-C-IRMS). Isot Environ Health Stud 39:233–244

Humphries EC (1956) Mineral composition and ash analysis. In: Peach K, Tracey MV (eds) Modern methods of plant analysis, vol 1. Springer, Berlin, pp 468–502

Hyde KD, Bahkali AH, Moslem MA (2010) Fungi – an unusual source for cosmetics. Fungal Divers 43:1–9

Karun NC, Sridhar KR (2014) A preliminary study on macrofungal diversity in an arboratum and three plantations of the southwest coast of India. Curr Res Environ Appl Mycol 4:173–187

Karun NC, Sridhar KR (2016) Spatial and temporal diversity of macrofungi in the western Ghat forests of India. Appl Ecol Environ Res 14:1–21

Karun NC, Sridhar KR, Niveditha VR, Ghate SD (2016) Bioactive potential of two wild edible mushrooms of the western Ghats of India. In: Watson RR, Preedy VR (eds) Fruits, vegetables, and herbs: bioactive foods in health promotion. Elsevier, Oxford, pp 344–362

Karun NC, Sridhar KR, Ambarish CN (2017) Nutritional potential of *Auricularia auricula-judae* and *Termitomyces umkowaan* – the wild edible mushrooms of southwestern India. In: Gupta VK, Treichel H, Shapaval V, De Oliveira LA, Tuohy MG (eds) Microbial functional foods and nutraceuticals. Wiley, New Jersey, pp 281–301

Kavishree S, Hemavathy J, Lokesh BR, Shashirekha MN, Rajarathnam S (2008) Fat and fatty acids of Indian edible mushrooms. Food Chem 106:597–602

Kruger CL, Mann SW (2003) Safety evaluation of functional ingredients. Food Chem Toxicol 41:793–805

Lindequist U, Niedermeyer THJ, Julich WD (2005) The pharmacological potential of mushrooms. eCAM 2:285–299

Manna S, Roy D, Roy A (2014) Trial relation to spatio-temporal variation of wild mushrooms in eastern lateritic part of India. Ethnobot Res Appl 12:15–24

Manzi P, Pizzoferrato L (2000) Beta-glucans in edible mushrooms. Food Chem 68:315–318

Mattila P, Könkö K, Eurola M, Pihlava JM, Astola J, Vahteristo L, Hietaniemi V, Kumpulainen J, Valtonen M, Piironen V (2001) Contents of vitamins, mineral elements, and some phenolic compounds in cultivated mushrooms. J Agric Food Chem 49:2343–2348

Nareshkumar S (2007) Capillary gas chromatography method for fatty acid analysis of coconut oil. J Plant Crop 35:23–27

Natarajan K, Senthilarasu G, Kumaresan V, Riviere T (2005) Diversity in ectomycorrhizal fungi of a dipterocarp forest in western Ghats. Curr Sci 88:1893–1895

NRC-NAS (1989) Recommended dietary allowances. National Academy Press, Washington, DC

Padua-Resurreccion AB, Banzon JA (1979) Fatty acid composition of the oil from progressively maturing bunches of coconut. Philip J Coconut Stud 4:1–15

Pavithra M, Greeshma AA, Karun NC, Sridhar KR (2015) Observations on the *Astraeus* spp. of Southwestern India. Mycosphere 6:421–432

Pavithra M, Sridhar KR, Greeshma AA, Tomita-Yokotani K (2016) Bioactive potential of the wild mushroom *Astraeus hygrometricus* in the southwest India. Mycology 7:191–202

Ramamurthy N, Kannan S (2009) SEM-EDS analysis of soil and plant (*Calotropis gigantea* Linn.) collected from an industrial village, Cuddalore Dt, Tamil Nadu, India. Rom J Biophys 19:219–226

Sadasivam S, Manickam A (2008) Biochemical methods. New Age International Pvt Ltd, New Delhi

Sanmee R, Dell B, Lumyong P, Izumori K, Lumyong S (2003) Nutritive value of popular wild edible mushrooms from northern Thailand. Food Chem 82:527–532

Shills MEG, Young VR (1988) Modern nutrition in health and disease. In: Neiman DC, Buthepodorth DE, Nieman CN (eds) Nutrition. WmC Brown, Dubugue, pp 276–282

Singh N (2011) Wild edible plants: a potential source of nutraceuticals. Int Phar Sci Res 2:216–225

Slavin J, Jacobs DR, Marquart L (1997) Whole-grain consumption and chronic disease: protective mechanisms. Nutr Cancer 27:14–21

Soobrattee MA, Neergheen VS, Luximon-Ramma A, Aruoma OI, Bahorun T (2005) Phenolics as potential antioxidant therapeutic agents: mechanism and actions. Mutat Res 579:200–213

StatSoft (2008) Statistica version # 8. StatSoft, Tulsa

Sudheep NM, Sridhar KR (2014) Nutritional composition of two wild mushrooms consumed by the tribals of the western Ghats of India. Mycology 5:64–72

Sylvester EC, Ezejiofor AJ, Nnedinma EA (2014) Survey and proximate analysis of edible mushrooms in Enugu State, Nigeria. Ann Exp Biol 2:52–57

Thiribhuvanamala G, Prakasam V, Chandrasekar G, Sakthivel K, Veeralakshmi S, Velazhahan R Kalaiselvi G (2011) Biodiversity, conservation and utilization of mushroom flora from the Western Ghats region of India. Proceedings of the 7th international conference on mushroom biology and mushroom products. pp 155–164

USDA (1999) Nutrient data base for standard reference release # 13, food group 20: cereal grains and pasta. Agriculture Handbook # 8–20, Agricultural Research Service, US Department of Agriculture, USA

Venn BJ, Mann JI (2004) Cereal grains, legumes and diabetes. Eur J Clin Nutr 58:1443–1461

Wu T, Zivanovic S, Draughon FA, Sams CE (2004) Chitin and chitosan-value-added products from mushroom waste. J Agric Food Chem 29:7905–7910

Yusuf AA, Mofio BM, Ahmed AB (2007) Proximate and mineral composition of *Tamarindus indica* Linn 1753 seeds. Sci World J 2:1–4

Chapter 22
Biotechnology of Fungal Lipases

Gautam Kumar Meghwanshi and Abhishek Vashishtha

Abstract Fungal lipases are incomparable biocatalysts which find applications in a number of catalytic processes which are of industrial potential. Many fermentation strategies have been designed for easy cultivation and scale-up of fungal lipases. However, two major methods are used for their production: submerged fermentation (SmF) and solid-state fermentation (SSF). Both the methods are employed at industrial scale, and their use is mainly dependent on the fungal product to be produced. A variety of strategies have been developed for their purification and characterization in order to know the optimal conditions for their applications. Fungal lipases are exceptional with considerable variations in their properties such as substrate specificity (chemo-, regio-, and enantiospecificity), temperature tolerance, pH tolerance, and organic solvent stability. Based on the variations in the properties, a wide array of lipases is available which can be used as per the specific need of any industrial process. Presently fungal lipases find many industrial applications such as oleochemical, food and feed, dairy, agrochemical, biomedical, biodiesel, detergent, pulp and paper, biosensors, waste remediation, leather processing, polymer synthesis, etc.

Keywords Fungal lipase · Production · Purification · Properties · Applications

22.1 Introduction

Lipases (triacylglycerol ester hydrolase E.C. 3.1.1.3) are serine hydrolases, which carry out the hydrolysis and synthesis of esters of glycerol and long-chain fatty acids under aqueous and microaqueous conditions, respectively (Jaeger and Eggert 2002) (Fig. 22.1).

Lipases were first reported as early as 1856 by Claude Bernard (Petersen and Drablos 1994). Earlier, lipases from animal and plant sources were used for both research purposes and industrial applications (Vulfsn 1994; Ghosh et al. 1996;

G. K. Meghwanshi (✉) · A. Vashishtha
Department of Microbiology, M.G.S. University, Bikaner, India

P. Gehlot, J. Singh (eds.), *Fungi and their Role in Sustainable Development: Current Perspectives*, https://doi.org/10.1007/978-981-13-0393-7_22

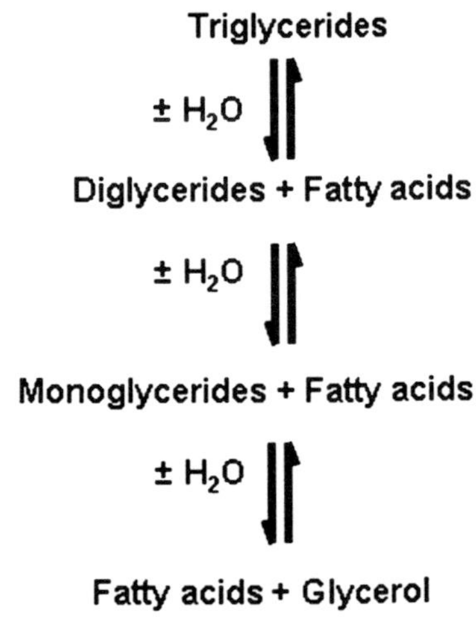

Fig. 22.1 Atypical lipase reaction: lipase catalyzed hydrolysis and synthesis of triglycerides

Dunhaupt et al. 1991; Rubin and Dennis 1997a, b). However, gradually microbial lipases have replaced them to a great extent and now are preferred for both research and industrial applications due to their multifold applications, easy extraction procedures, and unlimited supply (Saxena et al. 1999).

Industrial lipases are usually esterase enzymes that hydrolyze fats and oils, sequentially into the diglycerides, monoglycerides, and finally to glycerol and free fatty acids. Lipases from different sources exhibit substantial differences in their reaction specificities, which are known as enzyme specificities. Therefore, with reference to fatty acid, some lipases such as the *Mucor miehei* lipase have the preference for short-chain fatty acids (C_2, C_4, C_6, C_8, and C_{10}), and some like *Geotrichum candidum* lipase have the affinity for fatty acids like linoleic, oleic, and linolenic which are unsaturated fatty acids, whereas a lot are nonspecific and randomly remove the fatty acids from the triacylglycerols, e.g., many bacterial lipases (Godtfredsen 1990; Meghwanshi et al. 2006). With reference to three carbons of the glycerol backbone of the triacylglycerols, the lipases generally show specificity and attack the fatty acids at either 1 or 3 position or at both these positions; however, the fatty acid at the 2 position is not attacked. An exception to these, the *Geotrichum* sp. lipase has specificity for the 2 position (Asahara et al. 1993).

As reviewed by Ghosh et al. (1996), lipases from several organisms have been isolated, purified, and studied. Generally lipases are acidic glycoproteins with molecular weights lying between 20,000 and 60,000 Da. Some lipases are known to form aggregates in solution and this may account for the high molecular weights reported in some cases. The majority of lipases have been shown to contain 2–15% carbohydrates, the major glycoside residue in all cases being mannose.

Lipases do not require any cofactor for their activity. Salts may affect lipase activity significantly by influencing the ionization of the fatty acids produced. Generally, Ca^{++} ions activate enzyme activity by removing the calcium soaps of fatty acids, which are inhibitory (Liu et al. 1973). Diethyl *p*-nitrophenyl phosphate in most cases causes complete inhibition of enzyme activity, thus confirming the nature of lipases as serine hydrolases (Patkar and Bjorkling 1994).

Lipases are preferred catalysts of biochemists because of their potential to carry out different types of syntheses in nonaqueous or micro-aqueous media. For example, they carry out alcoholysis, acidolysis, aminolysis, interesterification, and transesterification reactions (Vulfson 1994). Lipases exhibit unique regio-, enantio-, and chemoselectivities, which allow them to catalyze the production of novel biochemicals which find applications as fine chemicals, biomedical, and agrochemicals.

Lipases bind to the water-oil interface and catalyze hydrolysis at this interface. This binding not only places the lipases close to the substrate but also increases the catalytic power of the lipases, a phenomenon called interfacial activation. Most lipases are poor catalysts in the absence of an interface such as a micelle. A conformational change in the lipase probably causes the interfacial activation (Verger 1997).

The simplest lipase-catalyzed reaction is the hydrolysis of esters in water or biphasic mixtures of water and an organic solvent. However, stability and activity of lipases in media such as organic solvents, reverse micelles, and supercritical fluids have pushed lipase catalysis to the forefront of enzymology (Balca et al. 1996).

The major producers of commercial lipases among fungi are *Aspergillus niger*, *Candida cylindracea*, *Humicola lanuginosa*, *Mucor miehei*, *Rhizopus arrhizus*, *R. delemar*, *R. japonicus*, *R. niveus*, and *R. oryza* (Godtfredsen 1990).

Currently, lipases find applications in different industrial sectors including oleochemistry, detergent formulations, textile, leather, nutrition, pulp and paper, pharmaceuticals, agrochemical, polymer, biosensor, waste treatment, biofuels, and others (Saxena et al. 1999, 2005).

22.2 Taxonomic Distribution of Fungal Lipase Producers

A lot of work has been carried out on fungal lipases, and many reports have been published on lipase-producing strains of fungi (Lazar and Schroeder 1992; Cordova et al. 1998; Abbas et al. 2002; Tan et al. 2003; Burkert et al. 2004). The major fungal groups involved in lipase production are as follows.

22.2.1 Mucorales

Among *Mucorales*, *Mucor hiemalis*, *M. miehei*, *M. lipolytica*, *Rhizopus japonicus*, *R. delemar*, *R. arrhizus*, *R. nigricans*, *R. nodosus*, *R. microsporus*, and *R. chinensis* have been studied in great detail for lipase production and applications (Sztajer and Zboinska 1988). Lipases from *Mucor* spp. have been reported to be 1,3-regiospecific in nature. The thermophilic mold *Mucor pusillus* produces a thermostable extracellular lipase. A lipase-producing strain of *M. miehei* has been reported to produce two isoenzymes with slightly different isoelectric points but a high degree of antigenic identity. Lipase from *M. miehei* immobilized on a resin (Lipozyme™) has been commercialized by Novo Industries, Denmark. Lipases from *Rhizopus mehei* is 1,3-regiospecific in nature and therefore is a good candidate for the synthesis of structured glycerides like having food and pharmaceutical applications (Saxena et al. 2005). Lipase from *R. japonicus* is used for the production of hard butter suitable for chocolate manufacture by 1,3-regiospecific interesterification of palm oil with methyl stearate. *Rhizopus* species lipases have been shown to have maximum activity toward medium-chain fatty acids (C_8–C_{10}), and their molecular weight ranges from 40 to 45 kDa. In the case of *R. delemar*, extracellular and intracellular lipase isoenzymes have been discovered. Lipase from *R. oligosporus* is used to produce characteristic flavor in tempeh (Nahas 1988).

22.2.2 Entomophthorales

Within the order *Entomophthorales*, lipase production has been reported in *Entomophthora opiculata*, *E. coronata*, *E. thaxteriana*, *E. virulenta*, *Basidiobolus* spp., and *Conidiobolus* spp. (Lazar and Schroder 1992).

22.2.3 Endomycetales

In *Endomycetales*, the main lipase producers belong to the genus *Geotrichum*. As early as 1972, the imperfect fungus *Geotrichum candidum* lipase was considered to be responsible for acid formation in dairy products by lipolyzing fat. The *G. candidum* lipase features a clear-cut specificity toward fatty acids with a *cis* double bond at C_9, and therefore this enzyme has been used in the structural analysis of triglycerides (Litchfield 1972). Another importance of this lipase is that it is the only known lipase showing 2-regiospecificity for triglycerides (Asahara et al. 1993).

22.2.4 Eurotiales

In the order *Eurotiales*, the main lipase producers belong to the genus *Aspergillus* and *Penicillium*. The most widely studied species of *Aspergillus*, viz., *A. niger*, is reported to produce intracellular and extracellular lipases which are 1,3-regiospecific (Okumura et al. 1976). *A. oryzae* was found to be an efficient host for the heterologous expression of the lipase from *Rhizomucor miehei* and *Humicola lanuginosa* (Shimada et al. 1994). Besides, these two aspergilla, viz., *A. terreus* and *A. carneus*, have been well studied for lipase production. The extracellular lipase of *A. terreus* is a thermostable lipase (molecular weight 41 kDa) which is capable of esterification and transesterification in organic solvents (Gosh et al. 1996; Yadav et al. 1997; Parmar et al. 1998; Yadav et al. 1998). The lipase of *A. carneus* which is also an extracellular enzyme has a molecular weight of 45 kDa. This lipase has been employed in various esterification and transesterification reactions. It has been employed in the synthesis of sorbitol monooleate with high efficiency, giving 94% yields in 8 h at 45 °C. Other *Aspergillus* spp. reported for lipase production include *A. fumigates* and *A. nidulans* (http://www.uniprot.org/taxonomy).

Penicillum spp. are well-known lipase producers. The lipase from *Penicillium roqueforti* has been shown to play an important role in characteristic flavor development in blue cheese (Eitenmiller et al. 1970). Lipolytic activity has also been detected in *P. camemberti*, the white surface mold of Brie, and Camembert cheese. Lipase with a clear-cut specificity for butyric acid has been isolated from a variety of strains of *Penicillium* such as *P. crustosum*, *P. cyclopium*, and *P. roqueforti* (Lazer and Schroder 1992). The lipase from *P. cyclopium* has been reported to have much higher activity toward di- and monoglycerides than triglycerides.

Besides the abovementioned major groups, there are some other fungal genera, which have also been investigated for lipase production. These include *Eurotium*, *Talaromyces*, *Neurospora*, *Puccnia*, and *Ustilago* (Lazar and Schroder 1992). Lipase from a thermophilic fungus *Humicola lanuginosa* DSM 3819 has also been studied, which is a suitable detergent additive active at alkaline pH and is stable toward anionic surfactants (Huge-Jensen and Gormsen 1988).

22.2.5 *Saccharomycetales*

Lipase production in yeasts has been studied in different genera which include *Pichia*, *Hansenula*, *Saccharomyces*, and *Candida*. The lipases from different species of *Candida* have been well studied and characterized. The well-known lipolytic species of *Candida* include *C. rugosa*, *C. antarctica*, *C. curvata*, *C. tropicalis*, *C. valida*, *C. utilis*, and *C. sake* (Stead 1986; Lazar and Schroder 1992). Lipase *from C. antarctica* was patented as a useful tool for interesterification of palm oil mid fraction and soya bean oil in continuous processes (Michiyo 1988), while lipase from *C. cylindracea* is one of the most thoroughly investigated lipases for its remarkable specificities in nonaqueous media. *Candida rugosa* is a producer of a highly active lipase frequently used for enzymatic splitting of fats (Valero et al. 1991; Chamorro et al. 1998).

In the genus *Saccharomyces*, S. *cerevisiae* and *S. lipolytica* have been reported to produce lipase. Other yeasts like *Debaryomyces hansenii*, *Kluyveromyces lactis* (*Candida sphaerica*), *Lodderomyces elongisporus* (*Saccharomyces elongisporus*), *Candida albicans*, *C. glabrata*, and *Yarrowia lipolytica* have been studied for lipase production (Lazar and Schroder 1992). A list of commercially available fungal lipases is given in Table 22.1.

Table 22.1 List of commercially available fungal lipases

Fungi	Abbreviation	Other name	Commercial source
Aspergillus niger	–	Amano A	Amano
Geotrichum candidum	GCL	Amano lipase GC	Sigma, Amano
Humicola lanuginosa	HLL	*Thermomyces lanuginosus*	Boehringer Mannheim (Chiraazyme L-8), Novo Nordisk (SP524, Lipase^{R})
Penicillium camemberti	PcamL	*Penicillium cyclopium*	Amano (Lipase G)
Penicillium roqueforti	–		Fluka
Rhizomucor javanicus	RJL, Amano lipase FAP	*Mucor javanicus*	Amano (LipaseM), Nagase Sangyo, Osaka Saiken, Fluka
Rhizomucor miehei	RML	*Mucor miehei*	Boehringer Mannheim (Chiraazyme^{R}L-9), Amano (MAP), Novo Nordisk (SP523,LIPozyme^{R}), Fluka, Sigma
Candida antarctica A	CAL-A	–	Boehringer Mannheim (Chiraazyme^{R}L-5), Novo Nordisk (SP526)
Candida lipolytica	CLL	–	Amano (Lipase R), Fluka
Candida antarctica B	CAL-B, SP-435[b]	–	Boehringer Mannheim (Chiraazyme^{R}L-20, Novo Nordisk (SP255 or Novozyme 435), Sigma
Candida rugosa	CRL	*Candida cylindracea*	Amano (Lipase AY) Boehringer Mannheim (Chiraazyme^{R}L-3)
Candida cylindracea	CCL	Lipase OF, Amano lipase AY	Boehringer Mannheim, Amano, Meito Sangyo

22.3 Lipase Production and Regulation

The production of lipases in fungi and yeasts is strongly influenced by culture conditions (Sztajer and Maliszewska 1988; Hass and Joerger 1995). Lipase producers can be divided into three groups on the basis of mechanism of lipase induction:

(a) Group I: In this group, lipase is produced in the presence of an inducer which may be a triglyceride, fatty acid, or any other lipid (Shimada et al. 1992).
(b) Group II: These organisms do not require lipids for lipase production, but their incorporation in the culture medium increases the level of lipase production (Ohnishi et al. 1994).
(c) Group III: In this group lipases are produced only constitutively, i.e., there is no need of any inducer (Chander and Klostermeyer 1983).

Production of lipases in different fungi is affected by nutritional and other physicochemical parameters (Rapp 1995; Gordillo et al. 1995).

22.3.1 Nutritional Factors

22.3.1.1 Lipid Substrates

Olive oil, which is a triglyceride of oleic acid, has been used as a standard substrate for lipase production from different microorganisms including fungi, e.g., *Aspergillus niger* (Macris et al. 1996), *Ophiostoma piliferum* (Rapp 1995), and *Mucor hiemalis* (Akhter et al. 1980). Gordillo et al. (1995) reported that oleic acid induced maximum lipase production in *C. rugosa.* Similarly, oleic acid has been used for lipase production from *Candida rugosa* (Gordillo et al. 1998). However olive oil is a costly substrate so researchers have used cheaper oils for lipase production, viz., sunflower oil for *Rhizopus delemar* (Espinosa et al. 1990) and *Aspergillus carneus* (Saxena et al. 2003), corn oil for *A. terreus* (Kaushik et al. 2010), and tung oil for *Trichosporon fermentas* WU-C12 (Chen et al. 1992) lipase production. Shimada et al. (1992) reported that extracellular lipases I and II of *Geotrichum candidum* were induced by long-chain fatty acids.

Besides oils, others lipids have also been reported as substrates for lipase production from different fungi, e.g., lecithin and cholesterol in *Rhizopus japonicus*.

22.3.1.2 Carbon Source

Different sugars, polysaccharides, and sugar alcohols have been reported to affect lipase production (Macrae and Hammond 1985). Akhter et al. (1983) reported that replacement of olive oil by glucose in the growth medium caused 1.5- to 4-fold increase in the mycelial lipase activity and as much as 3-fold decrease in soluble activity in *Rhizopus delemar* (Haas and Baily 1993). Ibrahim et al. (1987) reported that sorbitol acted as a better alternative to olive oil for producing lipase from *Humicola lanuginosa.* Use of dextran as a carbon source gave maximum lipase production by *Rhizopus delemar* (Espinosa et al. 1990).

22.3.1.3 Nitrogen Source

The production of microbial lipases is influenced by the types and concentrations of nitrogen sources used in the production medium (Haas and Joerger 1995). Corn steep liquor has been reported as the best nitrogen source for lipase production from *Humicola lanuginosa* (Ibrahim et al. 1987), in *Penicillium cyclopium* (Iwai et al. 1975).

Ammonium sulfate supported maximum lipase production in *Hendersonula toruloidea* (Odibo et al. 1995) whereas casamino acids and yeast extract in *Rhizopus delemar* CDBB H313 (Espinosa et al. 1990). However, one of the studies on solid-state fermentation of *Aspergillus niger* using gingelly oil showed that use of different inorganic and organic nitrogen sources did not affect lipase yield (Kamini et al.

1998). In case of *Candida rugosa*, urea and ammonium nitrate supported maximum lipase production (Benjamin and Pandey 1996).

22.3.1.4 Detergents and Ions

The positive effects of detergents (e.g., Tween-80, Tween-20, etc.) on lipase production are thought to be due to induction of enzyme synthesis by a substrate analog possessing the ester bond (Novotny et al. 1994). Studies have shown that the fatty acid chain length has an effect on the lipase production, e.g., polyoxyethylene sorbate compounds containing fatty acids with a shorter chain (C-12) were found to enhance lipase-mediated synthesis in comparison to those with a long-chain fatty acid in case of *Geotrichum candidum* (Iwai et al. 1975). In case of *Humicola lanuginosa*, high levels of lipase were distinctly observed on addition of Tween-80 to sorbitol corn steep liquor medium (Ibrahim et al. 1987).

Presence of Tween-80 in the medium enhanced the release of lipase from *Rhizopus delemar* CDBB H313 by 12-fold, where it acted both as an inducer and a surfactant (Espinosa et al. 1990). Odibo et al. (1995) reported stimulation of lipase production in *Hendersonula toruloidea* by sodium dodecyl sulfate.

Work carried out by different researches has shown that inclusion of ions into the medium results in enhanced lipase production from different fungi. For instance, Ca^{2+}, Sn^{2+}, Mn^{2+}, and Fe^{3+} enhanced lipase yield from *Humicola lanuginosa* (Ibrahim et al. 1987). Similarly, Ca^{2+} improved lipase production from *Fusarium solani* f. sp. *lini* SUF402 (Hoshino et al. 1992) and *Ophiostoma piliferum* (George et al. 1999).

22.3.2 Physical Factors

22.3.2.1 Temperature

The influence of temperature on lipase production is mainly related to the growth of organism. Among fungi, most of the lipase production studies have been conducted with mesophilic fungi which have a growth optimum between 25 and 37 °C (Lazar and Schroder 1992). Maximum lipase production by *Mucor heimalis* (Akhter et al. 1980), *Candida cylindracea* (Brahimi-Horn et al. 1990), *Fusarium oxysporum* (Hoshino et al. 1992), *Geotrichum candidum* (Shimada et al. 1992), and *Candida rugosa* (Benjamin and Pandey 1996) occurred between 28 and 30 °C, whereas in case of *Humicola lanuginosa*,, high lipase yields were obtained at 45 °C (Ibrahim et al. 1987).

22.3.2.2 pH

Lipase production by different fungi is greatly influenced by the pH of the production medium (Macrae and Hammond 1985). Fungi are generally known to produce lipase at lower pH values of 5.5–6.2. In this reference, *Rhizopus delemar* (Espinosa et al. 1990), *Rhizopus chinensis* (Venkata Rao and Lakshmanan 1991), *Fusarium oxysporum* (Hoshino et al. 1992), *Hendersonula toruloidea* (Odibo et al. 1995), and *Candida rugosa* (Gordillo et al. 1995) have been reported to produce lipase optimally in the pH range of 5.5–6.2. pH has been reported to affect the type of lipase being produced by the fungal culture, viz., *Geotrichum candidum* produced two lipases with opposite positional specificity, on changing the pH of the medium, one at initial pH of 6.0 and other at initial pH of 7.5 (Asahara et al. 1993). In case of *Humicola lanuginosa*, maximum lipase was produced when pH of the medium was kept between 7 and 8 (Ibrahim et al. 1987). A highly alkaline lipase from *Aspergillus carneus* was reported to be produced at pH 8.0 (Saxena et al. 2003).

22.3.2.3 Incubation Period

A large variation in optimal incubation period has been reported for different fungi (Lazar and Schroder 1992). Lipase production reached a maximum at 24 h in case of *Candida paraliplytica* (Ota et al. 1968); 72 h for *Candida rugosa*; 96 h in *Rhizopus japonicus*, *Humicola lanuginosa*, and *Rhizopus delemar* (Espinosa et al. 1990); 116 h for *Mucor hiemalis* (Akhter et al. 1983); 120 h for *Hendersonula toruloidea* (Odibo et al. 1995); 220 h for *Fusarium oxysporum* (Odibo et al. 1995); and 5 days for *Aspergillus niger* (Kamini et al. 1998).

22.4 Purification of Lipases

Although lipases can be used in their crude form in many industries like detergent, oleochemical, leather, bioremediation of waste, biodiesel, etc., however purified lipases are required in certain studies like the primary amino acid sequence and three-dimensional structure determination of lipases and molecular weight determination and for their kinetic studies.

A combination of several procedures has been used for lipase purification, which are presented in Table 22.2.

22.5 Assay Procedures

A number of assays have been developed to determine lipase activity, which are summarized in Table 22.3.

Table 22.2 Purification strategies used for fungal lipases

Fungi	Purification strategy	Purif. yield/ fold	References
A. niger MYA135	(i) Electroelution	47% yield 8.4-fold	Romero et al. (2012)
	(ii) DEAE Sepharose anion exchange chromatography	53.4% 16.6-fold	
A. niger NCIM 1207	Ammonium sulfate precipitation, phenyl Sepharose, sephacryl S-100 gel chromatography	54% yield	Mhetras et al. (2009)
A. oryzae	Ammonium sulfate precipitation, hydrophobic interaction, gel filtration, DEAE Sepharose, hydroxylapatite adsorption	5.5% yield 1242-fold	Toida et al. (1995)
Penicillium Camemberti	Ammonium sulfate precipitation and ethanol, DEAE Toyopearl, phenyl Toyopearl 650 M,	26% yield 210-fold	Yamasuchi and Mase (1991)
Mucor miehei	Lipase A: DE-cellulose 52, con-A Sepharose, affinity chromatography	57% yield 15-fold	Huge-Jensen et al. (1987)
	Lipase B: DEAE Sephadex, Phenyl Sepharose	32% yield Eightfold	Huge-Jensen et al. (1987)
Fusarium heterosporum	Ammonium sulfate precipitation, Sephadex gel filtration, isoelectric focusing, column chromatography	38% yield 21.15-fold	Shimada et al. (1993a, b)
Fusarium sp. YM −30	Ultrafiltration, ammonium sulfate, DEAE Toyopearl 650M, Phenyl Toyopearl 650 M	32% yield 17.07-fold	Mase et al. (1995)
Neurospora Crassa	Saline extract of conidia, charcoal treatment, dialysis Sephadex G-100 chromatography	2% yield 26-fold	Kundu et al. (1987)
A. terreus	Ammonium sulfate, acetone precipitation, gel filtration	100% yield 11.01-fold	Yadav et al. (1998)
A. carneus	Ammonium sulfate precipitation, Octyl Sepharose column		Saxena et al. (2003)
Penicillium chrysogenum	Ultrafiltration, Phenyl Sepharose, mono Q, HR5/5, and PD-10 column Sephadex (G-25) chromatography	44% 30-fold	Ferrer et al. (2000)
Penicillium cyclopium	$NH_4.SO_4$ precipitation, DEAE Sepharose, Sephadex (G-75) gel filtration chromatography	30% 590-fold	Chahinian et al. (2000)
Trichoderma viride	Ammonium sulfate precipitation, DEAE cellulose, and gel permeation chromatography	46% 134-fold	Kashmiri et al. (2006)
Candida paralipolytica	Ammonium sulfate precipitation, CM-Sephadex C-50,DEAE Sephadex G-25	32% yield 132-fold	Ota et al. (1970)
C. rugosa	Ethanol precipitation, anion exchange (mono Q) chromatography	18% yield 6-fold	Veeraragavan and Gibbs (1989)
Kurtzmanomyces sp. I-11	Ammonium sulfate fractionation, anion exchange chromatography on DEAE, gel filtration chromatography (Sephadex G-50)	16.6% yield 3140-fold	Kakugawa et al. (2002a, b)

Table 22.3 Lipase assay methods

Substrate	Reaction product	Method	Final product	O.D. (nm)
Plate assay				
Glycerides (triolein, tributyrin)	Free fatty acids	Colored indicators		
		Phenol red	–	–
		Methyl red	–	–
		Victoria blue	–	–
		Rhodamine	–	–
Spectroscopic				
1,2 diglycerides	Glycerol	Enzymatic conversion	Quinone	550
Glycerides (triolein)	Free fatty acids	Enzymatic conversion	NAD	240
Fluorescence				
Glycerides	Free fatty acids	Complex formation	11-(dansylamino) undecanoic acid	Ex. 350 nm
				Em. 500 nm
Titrimetry				
Glycerides	Free fatty acids	pH determination	Endpoint/pH-stat method	–

22.6 Properties of Fungal Lipases

Fungal lipases show great diversity with respect to their biochemical properties, viz., pH optima and tolerance, temperature optima and tolerance, molecular weights, and substrate specificities.

22.6.1 Substrate Specificity

Substrate specificity of lipases is guided by their structure, molecular properties, and factors responsible for enzyme-substrate interactions (Jensen 1983).

The specificities of lipases can be divided into five major types:

- Fatty acid specificity
- Positional specificity
- Lipid class specificity
- Alcohol specificity
- Stereochemical specificity

The diversity in fungi with regard to substrate specificity can be described as follows.

22.6.2 *Fatty Acid Specificity*

Lipases show varying preferences for fatty acids with reference to their degrees of saturation and carbon chain length (Jensen 1983; Macrae and Hammond 1985). Their preference has been exploited in production of cheese flavors in food products where release of short-chain fatty acids is contributing to the characteristic flavors (Haas and Joeger 1995). A study carried out with *Mucor miehei* lipase showed that the enzyme displayed similar preferences for the release of C4 and C6 fatty acids from their corresponding glycerides at both acidic (5.3) and alkaline (8.0) pHs; however, longer-chain fatty acids were released more slowly at acidic pH (Moskowitz et al. 1977). *Rhizopus* species lipases exhibited maximum activity toward medium-chain fatty acids (C_8–C_{10}). Alcohol chain length specificity of lipase from *Rhizopus arrhizus* in ester synthetic reactions follows bell-shaped distribution in the number of carbon atoms centered at C_4 and C_2 (Langrand et al. 1990), whereas binomial distribution consisting of two superimposed bell-shaped distribution centered at C_8–C_{10} was observed for *Rhizopus delemar* (Iwai et al. 1980). Two types of fatty acid specificities have been reported from *Geotrichum candidum* lipases. One lipase is specific for long-chain polyunsaturated fatty acids, whereas another lipase, i.e., lipase B, is highly specific for *cis*-Δ9 unsaturated fatty acids (Charton and Macrae 1992; Holmquist et al. 1997). *Candida rugosa* lipase showed specificity for long-chain polyunsaturated fatty acids (Hoshino et al. 1990).

Lipase from *Aspergillus carneus* showed good activity on all the triglycerides ranging from C_4 to C_{18} with a preference for C_{12} (Saxena et al. 2003). Lipase from *A. niger* MYA 135 showed highest affinity for long-chain fatty acid esters, viz., *p*-nitrophenyl stearate (C18) (Romero et al. 2012), whereas lipase from *A. niger* CICC 4009 showed affinity for medium-chain fatty acid esters such as p-nitrophenyl caprate (C10) and *p*-nitrophenyl caprylate (C8). On the other side, lipases from other *A. niger* strains showed affinity for long-chain fatty acid esters; for instance, lipase from *A. niger* F044 and *A. niger* MYA 135 showed highest affinity for esters of palmitic acid (C16) and stearic acid (C18), respectively (Romero et al. 2012).

22.6.3 *Positional Specificity*

Lipases can be divided into two groups based on their positional specificity. The first group contains lipases that are positionally nonspecific, i.e., can release fatty acids from *sn*-1, *sn*-2, or *sn*-3 position of triglycerides. Second group displays positional or regional specificity, which is of two kinds, *sn*-1,3 specific or *sn*-2 specific (Jensen 1983), and the reaction products are mainly di- and monoglycerides.

sn-1,3-specific lipases preferentially release fatty acid residues from the terminal positions of the glycerol backbone but not from the central carbon atom, whereas *sn*-2-specific lipases preferentially release fatty acids from the central carbon atom. Examples of *sn*-1,3-specific lipases have been reported from *Rhizopus arrhizus*, *R. delemar*, *R. oryzae*, *R. niveus*, *Aspergillus niger*, *A.carneus*, and *Mucor miehei* (Okumura et al. 1976; Hass and Joerger 1995; Saxena et al. 2003). In nature *sn*-2-specific lipases are very rare; however it has been reported from *Geotrichum candidum*, which is capable of hydrolyzing oleic and linoleic acids from the *sn*-2 position of the glyceride molecules (Jensen et al. 1986).

Rogalska et al. (1993) surveyed the regioselectivity of 25 lipases or acylglycerol molecules. Some lipases, e.g., from *Rhizomucor miehei* and *Candida rugosa*, showed high selectivity toward *sn*-1 position, whereas regioselectivity of lipases from other fungi varied with the triglyceride type, e.g., *Candida antarctica* lipase B showed *sn*-3 selectivity for trioctanoin but *sn*-1 selectivity with triolein. In another study by Okumura et al. (1976), it was found that lipases from *Aspergillus niger* and *Rhizopus delemar* released the fatty acids from *sn*-3 position first and then from the *sn*-1 position. The fatty acid at *sn*-2 position is attacked only after its spontaneous intramolecular transfer to one of the terminal positions via isomerization reactions, whereas lipases from *Penicillium cyclopium* and *Geotrichum candidum* attacked the *sn*-3 and *sn*-2 positions with similar preferences. The order of attack, i.e., change in regiospecificity, is reversed for esterification reactions. Stadler et al. (1995) also reported considerable changes as well as reversals in selectivity using analogs of triacylglycerides with ether or alkyl groups at *sn*-2 position.

22.6.4 Lipid Class Specificity

Lipid class specificity is well observed in *Penicillium* group. Okumura et al. (1980) reported that lipase from a strain of *P. cyclopium* displayed unique and strictly defined lipid class selectivity showing highest activity on monoglycerides and much lower toward di- and triglycerides. A novel lipase hydrolyzing mono- and diacylglycerol was isolated from culture filtrate of *Penicillium camemberti* by Yamaguchi and Mase (1991). The enzyme was separated into two forms, A and B enzymes. The B-enzyme was specific for hydrolysis of mono- and diglycerides, and triglycerides were found to be completely inert as substrates. Lipase of *Fusarium* sp. YM-30 (Mase et al. 1995) and *Aspergillus oryzae* (Toida et al. 1995) also showed similar type of lipid class selectivity. In contrast to this, *M. miehei* lipase hydrolyzed triglycerides more rapidly than mono- and diglycerides.

22.6.5 Stereospecificity

Stereospecificity is a novel characteristic of lipases. A large number of researchers have reported stereoselective nature of lipase catalysis on substrates such as straight-chain secondary alcohols, carboxylic acids, acetonoids, oxazolidones, inositols, glycidyl esters, β-blockers, and several esters of ibuprofens. This property of lipase has found immense use in the pharmaceutical sector in the resolution of chiral drugs. Lipases from *Candida cylindracea* and *C. antarctica* are most widely used for asymmetric transformations of synthetic chemicals. In addition to general synthetic utility of lipases in organic chemistry, they also exhibit stereoselectivity in reactions with triglycerides (Macrae and Hammond 1985; Saxena et al. 1999). Primary ellenic alcohols with axial chirality have been resolved using *Candida rugosa* lipase using the hydrolysis mode (Cross et al. 2004). Raminelli et al. (2007) catalyzed the enzymatic resolution of racemic phenylpropanols in high enantiomeric excesses (up to 99%) and good yield (40–45%) using lipase from *Candida antarctica* (Novozyme 435). Immobilized lipases from *Rhizomucor miehei* and *Mucor javanicus* were used for catalyzing the resolution of racemic mixture of ketoprofen esters with good enantiomeric excess (ee_p were 86.2% and 99.2%, respectively) (Zhang et al. 2015). In another investigation both *Candida rugosa* and *Candida antarctica* lipases B (CAL-B) were shown to catalyze the preferential reaction of (*R*)-enantiomer of different alcohols in transesterification reactions with diketoestes (Nanda and Scott 2004).

22.6.6 Alcohol Specificity

Alcohol selectivity is another form of substrate specificity of lipases and is extensively utilized in kinetic resolution of primary and secondary alcohols (Lalonde et al. 1995). Two factors are important for determining the effect of alcohol moiety of a substrate on lipase activity. These are steric hindrance and electrophilic induction (Brokkeroff and Jensen 1974). Triglycerides are bulky in the glycerol end, and thus steric hindrance with triglycerides may be greater than those with monohydric alcohols. Bulkiness in the alcohol moiety can thus inhibit ester hydrolysis; similarly presence of an electrophilic moiety in the alcohol part of an ester facilitates nucleophilic attach on the carbonyl carbon of the ester bond.

The substrate specificities of lipases are thus of a broad and varied nature. They can essentially be used for the synthesis of novel organic compounds of desired configurations and modify the already existing ones.

22.6.7 pH Tolerance

The activity of lipases is pH-dependent. Some lipases have shown good tolerance over a wide range of pH values. The ideal lipase would be the one, which would be active over the full pH range of 0–14; such a lipase may never be available in nature. Nevertheless, the lipase studied have usually shown profound activity at neutral pH or near the neutral pH range of 6.0–7.5, with considerable activity at acidic pH down to 4.0 to alkaline pH up to 8.0.

An investigation by Yadav et al. (1998) on different fungal lipases has shown that extracellular lipases of *Aspergillus niger* and *Rhizopus* spp. were particularly optimally active at acidic pHs. *Aspergillus niger* showed optimal activity in pH range of 4.5–5.6 and *Rhizopus delemar* at pH 5.6 (Commenil et al. 1995). *Botrytis cinerea, Candida curvata, Geotrichum candidum, Penicillium cyclopium*, and *P. roqueforti* were active in near neutral pH range with optimum at pH 6.0. Lipases from *Candida rugosa, A. oryzae, M. miehei* (Toida et al. 1995), and *Neurospora crassa* (Kundu et al. 1987) were also active in the neutral pH range, with optimum at pH 7.0. On the other hand, lipases from *Fusarium heterosporum* (Shimada et al. 1993) and *Rhizopus japonicus* (Aisaka and Terada 1979) were active in the acidic pH range, with optimum at 5.5–6.0 and 5.0, respectively, whereas lipases from *Fusarium* sp.YM-30 (Mase et al. 1995) and *Yarrowia lipolytica* (Vakhlu and Kaur 2006) showed good activity in the neutral to alkaline pH range with optima between 7.0–8.0 and 9.0, respectively. Yadav et al. (1998) further reported that the lipase from *Aspergillus terreus* was active in a broad pH range from 3.0 to 11.0 with optimum at pH 6.0, whereas *F. globulosum* (Gulati et al. 2005) and *Aspergillus carneus* (Saxena et al. 2003) lipases were active in the neutral to alkaline pH range and showed pH optima, respectively, at pH 10.0 and 9.0.

22.6.8 Temperature Tolerance

On the basis of optimum temperature of catalysis, lipases can be divided into three groups:

(a) Psychrophilic lipases
(b) Mesophilic lipases
(c) Thermophilic lipases

- *Psychrophilic lipases*: So far very few reports on psychrophilic lipases are available from fungi. Lipases from *Mucor racemosus* and *M. caseolyticus* exhibited maximum activity at 22 °C and 20 °C, respectively (Ghosh et al. 1996). *Fusarium globulosum* lipase although is not a psychrophilic lipase, however, showed 81% activity at 20 °C and 71% activity at 10 °C of the optimal activity (Gulati et al. 2005).

- *Mesophilic lipases:* A number of fungal lipases with optimal activity in mesophilic temperature range have been reported. Lipases from *Aspergillus niger*, *A. oryzae* (Toida et al. 1995), and *Neurospora crassa* (Kundu et al. 1987) showed temperature optimum of 30 °C. *Penicillium roqueforti* (Eitenmiller et al. 1970) lipase exhibited optimal activity at 35 °C, whereas lipases from *Fusarium* sp. YM-30 (Mase et al. 1995) and *Yarrowia lipolytica* (Destain et al. 1997) showed temperature optimum of 37 °C. *Botrytis cinerea* lipase showed temperature optimum of 38 °C (Comménil et al. 1995), whereas *Rhizopus japonicus* lipase showed good activity in the temperature range of 35–40 °C (Nahas 1988). The lipase from *Mucor* sp. showed temperature optimum of 40 °C (Savitha et al. 2007). However, the maximal stability of the lipase was observed at 30 °C up to 1 h (Savitha et al. 2007). Lipase from *Fusarium heterosporum* (Shimada et al. 1993) showed high activity in temperature range of 40–50 °C with no clear-cut mention of a single temperature optimum.
- *Thermophilic lipases*: Lipases exhibiting optimal activity at 50 °C have been reported from *Geotrichum candidum* (Mladenoska 2014) and *Trichosporon asteroides* (Dharmsthiti and Ammaranond 1997). One the other hand, *Candida curvata* lipase showed temperature optima in the range of 50–60 °C (Montet et al. 1985). Salleh et al. (1993) reported two lipases from a thermophilic strain of *Rhizopus oryzae*, one extracellular and the other intracellular. The temperature optimum of extracellular lipase was 45 °C, and that of intracellular was 37 °C. Thus both mesophilic and thermophilic lipases were produced by the same fungal strain. The lipase from *F. solani* FS1 showed optimal activity at 45 °C (Maia et al. 1999). The lipase produced by *Fusarium* sp. and *F. oxysporum* f. sp. *vasinfectum* showed optimum activity at 42 and 45 °C, respectively (Hoshino et al. 1992). A highly thermophilic lipase has been reported from *Kurtzmanomyces* sp. I-11, showing optimal activity at 75 °C (Kakugawa et al. 2002).

22.7 Applications of Fungal Lipases

Fungal lipases find applications in a number of industries as already mentioned. However, here only few of the more specialized and important applications are described.

22.7.1 Applications in Food Industry

22.7.1.1 Desaturation of Saturated Fatty Acids

Fatty acid selective lipases from fungi have been used in triacylglycerol hydrolysis for the production of low SAFA (saturated fatty acid) oil. Reduction in blood cholesterol levels by altering the fat composition has been a burning issue for many years. Presently, the total fat intake is recommended to be within the prescribed limits, especially saturated fats; hence, there is a great interest in lowering the saturated fat contents in the edible products. It has been shown for the *Fusarium oxysporum* lipase that it can lower the level of saturated fatty acids in oils via selective hydrolysis. Both intracellular and extracellular lipases from this fungus have been ascribed to exhibit high affinity of hydrolysis for saturated fatty acids (Hoshino et al. 1992).

The intracellular lipase from *Fusarium oxysporum* has been shown to preferentially remove the saturated fatty acids from cottonseed oil and groundnut oil (Joshi and Dhar 1987). A second process for producing low SAFA oil involves a two-step process based on the high selectivity of *Geotrichum candidum* lipase B (Charton and Macrae 1992; Holmquist et al. 1997). The lipase B has a characteristic preference for unsaturated fatty acids with a *cis*-Δ-9 double bond, e.g., oleic or linoleic acid. These fatty acids can be selectively removed from oils and can be reesterified with glycerol molecules to produce unsaturated fatty acid-rich oils.

Free fatty acids can be reesterified with glycerol to give a corresponding triglycerides using *Rhizomucor miehei* lipase. This reaction is carried out by immobilized lipase from *R. miehei* at 60 °C and continuous water removal. Under these conditions the product comprised of >95% (w/w) triglycerides (Mc Neil et al. 1991).

22.7.1.2 Enrichment of Docosahexaenoic Acid (DHA) and Eicosapentaenoic Acid (EPA) from Fish Oil

It is now well established that polyunsaturated fatty acids (PUFA) are important for human nutrition and health. Among PUFA ω-3 and ω-6 family fatty acids are essential for human metabolism, but they cannot be interconverted in the body.

These acids can be obtained from a number of sources, e.g., fish oils, marine algae, and some microorganisms. Among these, fish oils are more significant commercially, although the levels of PUFA in fish oils are only low to moderate. Lipases can be employed in concentration process for PUFA in fish oils by adopting two approaches, viz., hydrolysis of fish oil glycerides and (trans)esterification of fish oil fatty acids with glycerol to produce. Hydrolysis of fish oil triglycerides leads to partial glycerides enriched in EPA and DHA. Lipase from *Geotrichum candidum* has been used for this purpose (Shimada et al. 1994). Lipases having 1,3-regiospecificity are less commonly used because they are unable to hydrolyze fatty acids present at the sn-2 position of triglycerides; however, *Rhizomucor miehei* lipase has been reported to enrich long-chain PUFA (LCPUFA) from anchovy oil (Uston et al. 1997). An increase of 40% has been reported (Uston et al. 1997).

Another important lipase in this respect is that from *Candida rugosa* which exhibited low affinity for DHA than EPA (Tanaka et al. 1992; McNeill et al. 1996). This property of the *C. rugosa* lipase was used for the production of LCPUFA fractions enriched in DHA than EPA. In case of Chilean fish oil, the ratio of DHA to EPA was raised finally to approximately 5:1 from the initial ration of 1:1 (McNeill et al. 1996). The products of the enrichment process can be reesterified using immobilized *Rhizomucor miehei* lipase to generate triacylglycerols enriched in these fatty acids (Moore and McNeill 1996).

22.7.1.3 Separation of Conjugated Linoleic Acid (CLA) Isomers

Conjugated linoleic acids are compounds derived from linoleic acid ($C_{18:2}$, n-6) having conjugated double bonds. These double bonds are present in either *cis* or *trans* forms. Lipase from *Geotrichum candidum* especially the B-isoform has characteristic selectivity, i.e., it shows very high preference for fatty acids with a *cis* Δ 9 bond. Therefore it is used in the isolation of the two isomeric forms of CLA, viz., C-9, t-11 CLA and C-12, t-10 CLA (Haas et al. 1999)

Lipases from *Penicillium* spp. like *P. camemberti* and *P. cyclopium* M1 are specific for mono- and diacylglycerols (Yamaguchi and Mase 1991; Ibrik et al. 1998) which can be used for the production of mono- and diglycerides by fatty acid esterification, whereas lipases from *Penicillium* spp. UZLM-4 (Gulomova et al. 1996), *P. cyclopium* (Ibrik et al. 1998), *P. expansum*, and *P. roqueforti* (Millqvist Fureby et al. 1997) are specific for triacylglycerols and can be used for the production of 1,2-diglycerides by triglycerides hydrolysis or alcoholysis.

For the production of 1,3-diglycerides, 1,3-specific lipases from *Rhizmucor miehei* (Rosu et al. 1999) or *Rhizopus arrhizus* have been used. These lipases synthesize 1,3-diglycerides by esterification of free fatty acids and glycerol. Experimental data show that diglyceride yields between 60% and 85% have been obtained using both fatty acids and ethyl esters as acyl donor (Fureby 1995).

22.7.1.4 2-Monoglycerides

Lipase from *Rhizopus delemar* has been reported to catalyze the hydrolysis of palm oil to produce 2 monopalmitin (Holmberg and Osterberg 1988). The reaction was carried out at 35 °C with 80% degree of conversion obtained (on a molar basis). Similarly lipase from *Rhizopus arrhizus* was used for tripalmitin hydrolysis at 35 °C with over 95% molar yield of 2-monopalmitin (Holmberg and Osterberg 1988).

22.7.1.5 1(3)-Monoglycerides

Lipase G from *Penicillium camemberti* gave 90% pure monoglycerides with 76% degree of conversion (on molar basis) by esterification of fatty acids with glycerol (Yamaguchi and Mase 1991). It is worth to describe here that the *Penicillium* sp. lipase described above is a different lipase from a triglyceride-specific lipase of other *Penicillium* spp. used 1,2-diglyceride production (Gulomova et al. 1996).

22.7.1.6 Separation of Lipids and Fatty Acids

The property of some lipases to prefer or select particular fatty acids or acyl moieties has been exploited to enrich selective fatty acids or their esters on oils/fats via selective hydrolysis, esterification, and transesterification (Mukherjee 1995, 1998).

Lipases from *C. rugosa*, *R. miehei*, and *R. arrhizus* (Mukherjee et al. 1993; Jachmanian et al. 1996), *Penicillium cyclopium* and *Penicillium* sp. (lipase G) (Mukherjee et al. 1993), and those from *Rhizopus* sp. (Mukherjee and Kiewitt 1998) have closely resembling substrate specificities in esterification with *n*-butanol. These lipases exhibit strong preference for unsaturated fatty acids with first double bond from the carboxyl end at an even number carbon, i.e., *cis*-4, e.g., all-*cis*-4,7,10,13,16,19-docosahexaenoic acid (*n*-3 22:6); *cis*-6, e.g., γ-linolenic (all-*cis*-6,9,12-octadecatrienoic. *n*-6 18:3), petroselinic (*cis*-6-octadecanoic *cis*-6 18:1), and stearidonic (all-*cis*-6,9,12,15-octaecaretraenoic, *n*-3 18:4); or *cis*-8, e.g., dihomo-γ-linolenic (all-*cis*-8,11,14-ecosatrienoic, *n*-6 20:3) acid.

In esterification studies of common and unusual fatty acids with *n*-butanol, it was found that lipases from *C. rugosa*, *R. miehei*, and *R. arrhizus* (Jachmanian et al. 1996) and *Rhizopus* sp. (Mukherjee and Kiewitt 1998) exhibited strong affinity for fatty acids with hydroxyl groups, e.g., 12-hydroxystearic acid and ricinoleic (12-hydroxy-*cis*-9-octadecenoic), and epoxy groups, e.g., cyclopentenyl fatty acids and *trans*-9,10-epoxystearic acid having saturated alkyl chains, e.g., hydnocarpic [11-(cyclopent-2-en-1-yl) undecanoic] and chaulmoorgric [13-(cyclopent-2-en-1-yl) tridecanoic] acid, whereas a cyclopentenyl fatty acid having a *cis*-6 olefinic bond, i.e., gorlic [13-(cyclopent-2-en-1-yl) tridec-6-enoic] acid, is selectively preferred by many lipases (Jachmanian et al. 1996).

22.7.1.7 Enhancing the Polyunsaturated Fatty Acid Contents in Oils

Lipases from *R. miehei*, *C. rugosa*, and *G. candida* have the ability to discriminate between γ-linolenic acid and other fatty acids present in evening primrose seed oil and borage seed oil. These lipases therefore can be used to enrich γ-linolenic acid in the source oil, via selective esterification of n-butanol with the other fatty acids present. The γ-linolenic acid is thus obtained as a concentrate in the unesterified fatty acid fraction (Hills et al. 1990a, b; Syed Rahmatullah et al. 1994).

R. miehei and *C. rugosa* lipases have been reported to carry out chemoselective hydrolysis of the acyl moieties other than the γ-linolenoyl moieties in evening primrose oil or borage oil leading to their selective enrichment in these processed oil derivatives (Syed Rahmatullah et al. 1994).

22.7.2 Detergent Aid, Leather Processing, and Paper Industry

The detergent industry is the one which consumes the biggest amount of lipase. Lipases assist in fatty stain removal from fabrics and overall fat removal from the utensils used for storage and preparation of oily foods. So they make an important part of laundry detergents and household dishwashers to remove fat-containing stains (Jaeger and Reetz 1998). The first and most widely used detergent lipase is from *Thermomyces* sp. overexpressed in a recombinant strain of *Aspergillus oryzae* (Lipolase, Novozymes). Detergent lipases have specific prerequisite for their use in detergent formulations, viz., broad substrate specificity, i.e., should be active against all sorts of fats containing short-, medium-, or long-chain fatty acid (both saturated and unsaturated) containing glycerides, activity and stability at alkaline pH and temperatures above 40 °C, and compatibility with different components in a detergent, including metal ions, surfactants, oxidants, and proteases (Hasan et al. 2006; Liu et al. 2009; Grbavčić et al. 2011).

Another application of these enzymes is in the removal of "pitch" (hydrophobic components of wood, namely, triglycerides and waxes) from pulp produced in the paper industry. This enzymatic pitch control method has been used in large scale since early 1990s (Jaeger and Reetz 1998; Hasan et al. 2006). Lipases can also be used to remove fats and grease from animal skins and hides. This process generates a higher-quality product (more uniform color and a cleaner appearance) when compared to leather that is manufactured by traditional methods (Hasan et al. 2006).

22.7.3 Synthesis of Fine Chemicals (Therapeutics, Agrochemicals, Fragrances)

Stereoselectivity of lipases is an important parameter for their use in the synthesis of a single enantiomer of a stereoactive compound, as usually only one of the enantiomers is of biological use and the other is not (or less active). Stereoselective lipases are therefore used for the production of pharmaceuticals, pesticide, and agrochemicals with high enantiomeric purity. Lipases offer several advantages over traditional chemical routes, such as mild reaction environment, high efficiency, selectivity, and easy separation from the reaction mixture (Hasan et al. 2006; Manoel et al. 2012; Jaeger and Eggert 2002). Machado et al. 2011 performed the kinetic resolution of (R,S)-1,2-isopropylidene glycerol ester derivatives (solketal) using different lipases.

Solketal is an important auxiliary for the synthesis of optically active isoforms of many biologically active compounds, such as glycerophospholipids, *β*-blockers, prostaglandins, and leukotrienes. Cunha et al. (2006) used Novozyme 435 (*Candida antarctica* lipase expressed in *Aspergillus oryzae*) for the kinetic resolution of (±)-1,2-*O*-isopropylidene-3,6-di-*O*-benzyl-myoinositol (precursor of chiral myo-inositol derivatives) (Cunha et al. 2010). Similar results were obtained by Manoel et al. (2012) with Novozyme 435 for the kinetic resolution of DL-1,3,6-tri-*O*-benzyl-myoinositol where the *O*-acylated L enantiomorph was obtained in up to >99% ee with conversions up to >49% (Manoel et al. 2012). Lipases are presently used by many pharmaceutical industries over the world for the preparation of chiral intermediates at commercial scale (Hasan et al. 2006).

Organic esters from short-chain fatty acids are among the most important natural fragrances and flavors. Examples of these esters include butyl butyrate, isoamyl acetate (banana flavor), ethyl valerate (green apple flavor), and butyl acetate (pineapple flavor) (Mendes et al. 2012). The use of lipases results in better-quality products suitable to fragrance and flavor industry (Rodrigues and Fernandez-Lafuente 2012).

22.7.4 Production of Biodiesel and Biolubricants

Biodiesel is gaining an edge over the petroleum-based diesel. Nonspecific lipases are becoming the choicest biocatalysts for biodiesel synthesis. These enzymes would be substituting conventional alkali/acid catalyst-based processes that generate undesirable by-products, which are difficult to separate from the leftover glycerol, produce highly alkaline/acidic waste, and use high-quality raw materials (Gog et al. 2012; de Sousa et al. 2010). Lipases catalyze both the transesterification of triacylglycerols and alcohols, as well as the esterification of free fatty acids and alcohols to yield monoalkyl esters. They can synthesize biodiesel with oils from different origins, including waste and non-edible oils, and also enable easy separation from the by-product, glycerol (Robles-Medina et al. 2009; Fjerback et al. 2009). The enzymatic biodiesel production has been carried out by several research groups using free, immobilized, and whole cells (de Sousa et al. 2010; Shimada et al. 1999; Ban et al. 2001; K¨ose et al. 2002; Corrêa et al. 2011). These reports show good results with conversion ranges above 90%. However, enzyme's high cost remains a barrier for enzymatic production of biodiesel (Gog et al. 2012). Efforts have been focused on the development of low-cost enzyme preparations and more stable and active biocatalysts, in order to improve conversion yields in a shorter period of time and to recycle the enzyme for as much batches as possible. Biolubricants are another group of great industrial interest in which lipases can be used as biocatalysts. The patent EP 2657324 A1 (daSilva et al. 2013) describes an enzymatic process for production of biolubricant from methyl ricinoleate (biodiesel from castor oil) and/or from a mixture of methyl oleate and linoleate (biodiesel from Jatropha oil) by transesterification with trimethylolpropane. Conversions of 80–99% of the ester (castor oil and Jatropha oil biodiesel) were achieved and the product showed good properties of viscosity, viscosity index (VI), pour point, and oxidation stability. Aguieiras et al. have investigated biolubricant synthesis from oleic acid and methyl ricinoleate using immobilized commercial lipases (Aguieiras et al. 2011). Novozyme 435 showed the best performance, and the synthesized product exhibited good values of pour point viscosity and viscosity index.

22.7.5 *Pretreatment of Lipid-Rich Wastewaters*

Wastewaters from dairies, slaughterhouses, and fish processing contain high levels of fats and proteins with low biodegradability that may cause serious environmental damage if not properly treated (Rosa et al. 2009; Cammarota and Freire 2006). An enzymatic hydrolytic step before further biological treatment can reduce the particles diameter, increasing their surface area and favoring organic matter assimilation by the microbial consortium (Valladao et al. 2011). Thus, the application of lipases in wastewater treatment may improve biological degradation of fatty wastewaters, thereby accelerating this process (Cammarota and Freire 2006). Nevertheless, this added pretreatment procedure becomes economically unfeasible if there are high costs associated with the commercial enzymatic preparations. Therefore, the development of low-cost enzyme preparations becomes essential.

22.7.6 *Production of Biodegradable Polymers*

Another field for lipases application is the synthesis of useful and biodegradable biopolymers like polyesters, produced from renewable natural resources (Kadokawa and Kobayashi 2010). Lipase-catalyzed polymerization reactions are classified into two major polymerization modes: ring-opening polymerization (Uyama and Kobayashi 1993) and polycondensation. The latter may include polycondensation of dicarboxylic acids or their derivatives with diols and polycondensation of oxyacids or their esters (Kadokawa and Kobayashi 2010). This polymerization approach has been successfully employed by several works through a range of strategies. *Rhizomucor miehei* lipase was used in the polymerization of bis(2,2,2-trifluoroethyl) sebacate and aliphatic diols (Kadokawa and Kobayashi 2010; Uyama and Kobayashi 1993). Mahapatro et al. (2004) have studied the polymerization of 12-hydroxydodecanoic acid catalyzed by Novozyme 435 and obtained conversion of 91% in oligomers of high molecular weight. Polyesters of ricinoleic acid with 72% of conversion were synthesized by Bódalo et al. using immobilized lipase from *Candida rugosa* (Bódalo et al. 2008). These results indicate the applicability of lipases for polymerization reactions through different pathways. Taking into account the wide range of applications and the importance of lipases in biotechnology, many efforts have been directed toward understanding how these enzymes work at the molecular and atomic level. Since the BRIDGE-T lipase project (1990–1994) until today, numerous three-dimensional structures of lipases from several different organisms have been reported, shedding light onto the mechanism used by these enzymes during catalysis.

References

Abbas H, Hiol A, Deyris V, Comeau L (2002) Isolation and characterization of an extracellular lipase from *Mucor* sp strain isolated from palm fruit. Enzym Microb Technol 31:968–975

Aguieiras ECG, Veloso CO, Bevilaqua JV, Rosas DO, da Silva MAP, Langone MAP (2011) Estolides synthesis catalyzed by immobilized lipases. Enzym Res. https://doi.org/10.4061/2011/432746

Aisaka K, Terada O (1979) Production of lipoprotein lipase and lipase by *Rhizopus japonicus*. Agric Biol Chem 43(10):2125–2129

Akhtar MW, Mirza AQ, Nawazish MN, Chughtai MID (1983) Effect of triglycerides on the production of lipids and lipase by *Mucor heimalis*. Can J Microbiol 29:664–669

Akhter MW, Miraz AQ, Chughati MDI (1980) Lipase induction in *Mucor hiemalis*. Appl Environ Microbiol 18:257–263

Asahara T, Matori M, Ikemoto M, Ota Y (1993) Production of 2 types of lipases with opposite positional specificity by *Geotrichum* sp. FO401B. Biosci Biotechnol Biochem 57:390–394

Balcao VM, Paiva AL, Malcata FX (1996) Bioreactors with immobilized lipases: state of the art. Enzym Microb Technol 18:392–416

Ban K, Kaieda M, Matsumoto T (2001) Whole cell biocatalyst for biodiesel fuel production utilizing *Rhizopus oryzae* cells immobilized within biomass support particles. Biochem Eng J 8(1):39–43

Benjamin S, Pandey A (1996) Optimization of liquid media for lipase production by *Candida rugosa*. Bioresour Technol 55:167–170

Brahimi-Horn MC, Guglielmino ML, Elling L, Sparrow LG (1990) The esterase profile of a lipase from *Candida cylindracea*. Biochim Biophys Acta 1042(1):51–54

Brockerhoff H, Jensen R (1974) Lipolytic enzymes. Academic, New York 1974

Burkert JF, Maugeri F, Rodrigues MI (2004) Optimization of extracellular lipase production by *Geotrichum* sp. using factorial design. Bioresour Technol 91(1):77–84

Bódalo A, Bastida J, Máximo MF, Montiel MC, Gómez M, Murcia MD (2008) Production of ricinoleic acid estolide with free and immobilized lipase from *Candida rugosa*. Biochem Eng J 39(3):450–456

Cammarota MC, Freire DMG (2006) A review on hydrolytic enzymes in the treatment of waste water with high oil and grease content. Bioresour Technol 97(17):2195–2210

Chahinian H, Vanot G, Ibrik A, Rugani N, Sarda L, Comeau LC (2000) Production of extracellular lipases by *Penicillium cyclopium* purification and characterization of partial acylglycerol lipase. Biosci Biotechnol Biochem 64(2):215–222

Chamorro S, Sanchez-Montero JM, Alcantara AR, Sinisterra JV (1998) Treatment of *Candida rugosa* lipase with short-chain polar organic solvents enhances its hydrolytic and synthetic activities. Biotechnol Lett 20:499–505

Chander H, Klostermeyer H (1983) Production of lipase by *Fusarium solani* under various growth conditions. Sci Aliment 3:274–285

Charton E, Macrae AR (1992) Substrate specificities of lipases A and B from *Geotrichum candidum* CMICC 335426. Biochim Biophys Acta 1123(1):59–64

Chen J, Ishu T, Shimura S, Kirimura K, Usami S (1992) Lipase production by *Trichosporn fermentas* WU-G2 , a newly isolated yeast. J Ferment Bioeng 73:412–414

Comménil P, Belingheri L, Sancholle M, Dehorter B (1995) Purification and properties of an extracellular lipase from the fungus *Botrytis cinerea*. Lipids 30(4):351–356

Cordova J, Nemmaoui M, Ismaili-Alaoui M, Morin A, Roussos S, Raimbault M, Benjilali B (1998) Lipase production by solid state fermentation of olive cake and sugar cane bagasse. J Mol Catal B Enzym 5:75–78

Corrêa INS, Souza SL, Catran M, Bernardes OL, Portilho MF, Langone MAP (2011) Enzymatic biodiesel synthesis using a byproduct obtained from palm oil refining. Enzym Res. https://doi.org/10.4061/2011/814507

Cross H, Marriot R, Gorgan G (2004) Enzymatic esterification of lavandulol – a partial kinetic resolution of (S)-lavandulol and preparation of optically enriched (R)-lavandulyl acetate. Biotechnol Lett 26(5):457–460

Cunha AG, da Silva AAT, da Silva AJR, Tinoco LW, Almeida RV, De Alencastro RB, Simas ABC, Freire DMG (2010) Efficient kinetic resolution of (±)-1,2-O-isopropylidene-3,6-diO-benzyl-myo-inositol with the lipase B of *Candida antarctica*. Tetrahedron Asymmetry 21(24):2899–2903

da Silva JA, Freire DMG, Habert A, Soares V (2013) Process for the production of bio-lubricant from methyl biodiesel and bio-lubricant obtained by said process. EP2657324A1, October 30, 2013

de Maia M MD, de Morais MMC, de Morais MA Jr, Melo EHM, de Lima Filho JL (1999) Production of extracellular lipase by the phytopathogenic fungus *Fusarium solani* FS1. Rev Microbiol 30(4):304–309

de Sousa JS, Cavalcanti-Oliveira ED, Aranda DAG, Freire DMG (2010) Application of lipase from the physic nut (*Jatropha curcas* L.) to a new hybrid (enzyme/chemical) hydro esterification process for biodiesel production. J Mol Catal B Enzym 65(1–4):133–137

Destain J, Roblain D, Thonart P (1997) Improvement of lipase production from *Yarrowia lipolytica*. Biotechnol Lett 19(2):105–107

Dharmsthiti S, Ammaranond P (1997) Purification and characterization of lipase from a raw-milk yeast (Trichosporon asteroides). Biotechnol Appl Biochem 26(2):111–116

Dunhaupt A, Lang S, Wagner F (1991) Properties and partial purification of a *Pseudomonas cepacia* lipase. In: Alberghina L, Schmid RD, Verger R (eds) Lipases: structure, mechanism and genetic engineering, GBF monographs, vol 16. VCH, Weinheim 389pp

Eitenmiller RR, Vakil JR, Shahani KM (1970) Production and properties of *Penicillium roqueforti* lipase. J Food Sci 35:130–133

Espinosa E, Sanchez S, Farres A (1990) Nutritional factors affecting lipase production by *Rhizopus delemar* CDBB H313. Biotechnol Lett 12:209–214

Ferrer M, Plou FJ, Nuero OM, Reyes F, Ballesteros A (2000) Purification and properties of a lipase from *Penicillium chrysogenum* isolated from industrial wastes. J Chem Technol Biotechnol 75(7):569–576

Fjerbaek L, Christensen KV, Norddahl B (2009) A review of the current state of biodiesel production using enzymatic transesterification. Biotechnol Bioeng 102(5):1298–1315

Fureby AM (1995) Aspectson lipase-catalyzed preparation of partialacyl glycerols. Ph.D. thesis, Lund University, Lund. 46 pp

George E, Tamerler C, Martinez A, Martinez MJ, Keshavarz T (1999) Influence of growth medium composition on the lipolytic enzyme activity of *Ophiostoma piliferum* (Cartapip™). J Chem Technol Biotechnol 74:137–140

Ghosh PK, Saxena RK, Gupta R, Yadav RP, Davidson S (1996) Microbial lipases production and applications. Sci Prog 79:119–157

Godtfredsen SE (1990) Microbial lipases. In: Fogarty WM, Kelly ET (eds) Microbial enzymes and biotechnology. Elsevier, Amsterdam 255pp

Gog A, Roman M, Tos M, Paizs C, Irimie FD (2012) Biodiesel production using enzymatic transesterification: current state and perspectives. Renew Energy 39(1):10–16

Gordillo MA, Obradors N, Montesinos L, Valero F, Lafuente J, Solà C (1995) Stability studies and effect of the initial oleic acid concentration on lipase production by *Candida rugosa*. Appl Micobiol Biotechnol 43:38–41

Gordillo MA, Sanz A, Sánchez A, Valero F, Montesinos JL, Lafuente J, Solà C (1998) Enhancement of *Candida rugosa* lipase production by using different control fed-batch operational strategies. Biotechnol Bioeng 60:156–168

Grbavĉić S, Bezbradica D, Izrael-Živković L, Avramović N, Milosavić N, Karadžić I, Knežević-Jugović Z (2011) Production of lipase and protease from an indigenous *Pseudomonas aeruginosa* strain and their evaluation as detergent additives: compatibility study with detergent ingredients and washing performance. Bioresour Technol 102(24):11226–11233

Gulati R, Isar J, Kumar V, Prasad AK, Parmar VS, Saxena RK (2005) Production of a novel alkaline lipase by *Fusarium globulosum* using neem oil and its applications. Pure Appl Chem 77:251–262

Gulomova K, Ziomek E, Schrag JD, Davranov K, Cygler M (1996) Purification and characterisation of a *Penicillium* sp. lipase which discriminates against diglycerides. Lipids 31:379–384

Haas MJ, Bailey DG (1993) Glycerol as a carbon source for lipase production by the fungus *Rhizopus delemar*. Food Biotechnol 7(1):49–73

Haas MJ, Joerger RD (1995) Lipases of the genera *Rhizopus* and *Rhizomucor* versatile catalyst in nature and the laboratory. In: Hui YH, Khachatourians GG (eds) Food biotechnology. VCH Publications, Wyoming 549pp

Haas MJ, Kramer JKG, McNeill GP, Scott K, Foglia TA, Sehat N, Fritsche J, Mossoba MM, Yurawecz MP (1999) Lipase-catalysed fractionation of conjugated linoleic acid isomers. Lipids 34:979–987

Hasan F, Shah AA, Hameed A (2006) Industrial applications of microbial lipases. Enzym Microb Technol 39(2):235–251

Hills MJ, Kiewitt I, Mukherjee KD (1990a) Lipase from *Brassica napus* L discriminates against cis4 and cis-6 unsaturated fatty acids and secondary and tertiary alcohols. Biochim Biophys Acta 1042:237–240

Hills MJ, Kiewitt I, Mukherjee KD (1990b) Enzymatic fractionation offatty acids: enrichment of clinolenic acid and docosahexaenoic acid by selective esterification catalysed by lipases. J Am Oil Chem Soc 67:561–564

Holmberg K, Osterberg E (1988) Enzymatic preparation of monoglycerides in microemulsion. J Am Oil Chem Soc 65:1544–1548

Holmquist M, Tessier DC, Cygler M (1997) Identification of residues essential for differential fatty acyl specificity of *Geotrichum candidum* lipases I and II. Biochemistry 36(48):15019–15025

Hoshino T, Yamane T, Shimizu S (1990) Selective hydrolysis of fish oil by lipase to concentrate n3 polyunsaturated fatty acid. Agric Biol Chem 54:1459–1467

Hoshino T, Sasaki T, Watanabe Y, Nagasawa T, Yamane T (1992) Purification and some characteristic of extracellular lipase from *Fusarium oxysporum*. Biosci Biotechnol Biochem 56:660–664

Huge-Jensen B, Gormsen E (1988) Enzymatic detergent additives. EP Patent 258 068, 2 March 1988

Huge-Jensen B, Galluzzo DR, Jensen RG (1987) Partial purification and characterization of free and immobilized lipases from *Mucor miehei*. Lipids 22(8):559–565

Ibrahim CO, Nishio N, Nagai S (1987) Fat hydrolysis and estérification by a lipase from *Humicola lanuginose*. Agric Biol Chem 51(8):2153–2159

Ibrik A, Chahinian H, Rugani N, Sarda L, Comeau LC (1998) Biochemical and structural characterisation of triacylglycerol lipase from *Penicillium cyclopium*. Lipids 33:377–384

Iwai M, Okumura S, Tsujisaka Y (1975) The comparison of the properties of two lipases from *Penicillium cyclopium*. Agric Biol Chem 39:1063–1070

Iwai M, Shimada Y, Tsujiska Y (1980) Purification of *Rhizopus delemar* lipase by its binding with phospholipids. J Biochem 88:533–538

Jachmanian I, Schulte E, Mukherjee KD (1996) Substrate selectivity in esterification of less common fatty acids catalysed by lipases from different sources. Appl Microbiol Biotechnol 44:563–567

Jaeger KE, Eggert T (2002) Lipases for biotechnology. Curr Opin Biotechnol 13:390–397

Jaeger K, Reetz MT (1998) Microbial lipases form versatile tools for biotechnology. Trends Biotechnol 16(9):396–403

Jensen RG (1983) Detection and determination of lipase (acyglycerol hydrolase) activity from various sources. Lipids 18(9):650–657

Jensen RG, DeJong FA, Clark RM (1986) Determination of lipase specificity. Lipids 18(3):239–252

Joshi S, Dhar DN (1987) Specificity of fungal lipase in hydrolytic cleavage of oil. Acta Microbiol Hung 34(2):111–114

Kadokawa J, Kobayashi S (2010) Polymer synthesis by enzymatic catalysis. Curr Opin Chem Biol 14(2):145–153

Kakugawa K, Shobhayashi M, Suzuki O, Miyakawa T (2002a) Purification and characterization of a lipase from the glycolipid producing yeast *Kurtzmanomyces* sp. I-11. Biosci Biotechnol Biochem 66(5):978–985

Kakugawa K, Tamai M, Imamura K, Miyamoto K, Miyoshi S, Morinaga Y, Suzuki O, Miyakawa T (2002b) Isolation of yeast *Kurtzmanomyces* sp. I-11, novel producer of mannosyl erythritol lipid. Biosci Biotechnol Biochem 66:188–191

Kamini NR, Mala JGS, Puvanakrishnan P (1998) Lipase production from *Aspergillus niger* by solid state fermentation using ginglly oil cake. Process Biochem 35:501–511

Kashmiri MA, Adnan A, Butt BW (2006) Production, purification and partial characterization of lipase from *Trichoderma viride*. Afr J Biotechnol 5(10):878–882

Kaushik R, Marwah RG, Gupta P, Saran S, Saso L, Parmar VS, Saxena RK (2010) Optimization of lipase production from *Aspergillus terreus* by response surface methodology and its potential for synthesis of partial glycerides under solvent free conditions. Indian J Microbiol 50(4):456–462

Köse O, Tüter M, Aksoy HA (2002) Immobilized *Candida antarctica* lipase-catalyzed alcoholysis of cotton seed oil in a solvent-free medium. Bioresour Technol 83(2):125–129

Kundu M, Basu J, Guchhait M, Chakrabarti P (1987) Isolation and characterization of an extracellular lipase the from conidia of *Neurospora crassa*. J Gen Microbiol 133:149–153

Lalonde JJ, Govardhan C, Khaalaf N, Martinez AG, Visuri K, Margolin AL (1995) Cross-linked crystals of *Candida rugosa* lipase: highly efficient catalysts for the resolution of chiral esters. J Am Chem Soc 117:6845–6852

Langrand G, Rondon N, Triantaphylides C (1990) Short chain flavor esters synthesis by microbial lipases. Biotechnol Lett 12:581–596

Lazar G, Schroder FR (1992) Degradation of lipid by fungi. In: Winekelmann G (ed) Microbial degradation of natural products. VCH, Weiheim 267pp

Litchfield C (1972) Analysis of triglycerides. Academic, New York

Liu WH, Beppu T, Arima K (1973) Effect of various inhibitors on lipase action of thermophilic fungus *Humicola lanuginosa*. S-38. Agric Biol Chem 37:2487–2492

Liu R, Jiang, Mou H, Guan H, Hwang HM, Li X (2009) A novel low-temperature resistant alkaline lipase from a soda lake fungus strain *Fusarium solani* N4-2 for detergent formulation. Biochem Eng J 46(3):265–270

Macrae AR, Hammond RC (1985) Present and future applications of lipases. Biotechnol Genet Eng Rev 3:193–217

Macris JB, Kourentzi E, Hatzinikolaou DG (1996) Studies on localization and regulations of lipase production by *Aspergillus niger*. Process Biochem 31:807–812

Mahapatro A, Kumar A, Gross RA (2004) Mild, solventfree ω-hydroxy acid polycondensations catalyzed by *Candida antarctica* lipase B. Biomacromolecules 5(1):62–68

Manoel EA, Pais KC, Cunha AG, Coelho MAZ, Freire DMG, Simas ABC (2012) On the kinetic resolution of sterically hindered myo-inositol derivatives in organic media by lipases. Tetrahedron Asymmetry 23(1):47–52

Mase T, Matsumiya Y, Akiba T (1995) Purification and characterization of a new lipase from *Fusarium* sp. YM-30. Biosci Biotechnol Biochem 59:1771–1772

McNeill GP, Shimizu S, Yamane Y (1991) Further improvements in the yield of monoglycerides during enzymatic glycerolysis of fats and oils. J Am Oil Chem Soc 68:6–10

McNeill GP, Ackman RG, Moore SR (1996) Lipase catalyzed enrichment of long-chain polyunsaturated fatty acids. J Am Oil Chem Soc 73:1403–1407

Meghwanshi GK, Agrawal L, Dutt K, Saxena RK (2006) Characterization of 1,3-regiospecific lipases from new *Pseudomonas* and *Bacillus* isolates. J Mol Catal B Enzym 40:127–131

Mendes AA, Oliveira PC, de Castro HF (2012) Properties and biotechnological applications of porcine pancreatic lipase. J Mol Catal B Enzym 78:119–134

Mhetras NC, Bastawde KB, Gokhale DV (2009) Purification and characterization of acidic lipase from *Aspergillus niger* NCIM 1207. Bioresour Technol 100:1468–1490

Michiyo I (1988) Positionally non-specific lipase. Patent CA 1339844 C, 28 April 1998

Millqvist FA, Adlercreutz P, Mattiasson B (1997) Preparation of diglycerides by lipase-catalyzed alcoholysis of triglycerides. Enzym Microb Technol 20:198–206

Mladenoska I (2014) Isolation and purification of lipases from *Geotrichum Candidum* grown on a sunflower oil waste as a carbon source. Chem Eng Trans 42:49–54

Montet D, Ratomahenina R, Pina M, Graille J, Galzy P (1985) Purification and characterization of a lipase from *Candida curvata* Lodder and Kreger – van Rij CBS 570. Eur J Lipid Sci Technol 87(5):181–185

Moore SR, McNeill GP (1996) Production of triglycerides enriched in long-chain n-3 polyunsaturated fatty acids from fish oil. J Am Oil Chem Soc 73:1409–1414

Moskowitz GJ, Cassaign R, West IR, Shen T, Feldman LI (1977) Hydrolysis of animal fat and vegetable oil with *Mucor miehei* esterase. J Agric Food Chem 25:1146–1150

Mukherjee KD (1995) Fractionation of fatty acids and other lipids via lipase-catalyzed reactions. J Fr Oleagineaux Corps Gras Lipides 2:365–368

Mukherjee KD (1998) Lipid biotechnology. In: Akoh CC, Min DB (eds) Food lipids chemistry, nutrition, and biotechnology. Marcel Dekker, New York 589pp

Mukherjee KD, Kiewitt I (1998) Structured triacylglycerols resembling human milk fat by transesterification catalyzed by papaya (*Carica papaya*) latex. Biotechnol Lett 20(6):613–616

Mukherjee KD, Kiewitt I, Hills MJ (1993) Substrate specificities of lipases in view of kinetic resolution of unsaturated fatty acids. Appl Microbiol Biotechnol 40(4):489–493

Nahas E (1988) Control of lipase production *Rhizopus oligosporus* under various growth conditions. J Gen Microbiol 134:227–233

Nanda S, Scott AI (2004) A highly efficient chemoselective synthesis of 3,5-diketoesters by lipase-catalyzed transesterification: application to the resolution of secondary alcohols. J Mol Catal B Enzym 30:1–12

Novotny C, Dolezalova L, Lieblova J (1994) Dimorphic growth and lipase production in lipolytic yeasts. Folia Microbiol 39(1):71–73

Odibo FJC, Okereke UO, Oyeka CA (1995) Influence of culture condition on the production of lipase of *Hendersonula toruloidea*. Bioresour Technol 54:81–83

Ohnishi K, Yoshida Y, Sekiguchi J (1994) Lipase production of *Aspergillus oryzae*. J Ferment Bioeng 77(5):490–495

Okumura S, Iwai M, Tsujiskaka Y (1976) Positional specificities of four kinds of microbial lipases. Agric Biol Chem 40:655–660

Ota Y, Miyairi S, Yamada K (1968) Sterol requirement for the lipase production by *Candida cylindracea*. Agric Biol Chem 32(12):1476–1478

Ota Y, Nakamiya T, Yamada K (1970) Lipase from *Candida paralipolytica*. Agric Biol Chem 34(9):1368–1374

Parmar VS, Pati HN, Yadav RP, Kumar A, Bisht KS, Gupta R, Davidson S, Poonam, Saxena RK (1998) Utility of a novel lipase from *Aspergillus terreus* in deacetylation reactions. Biocatal Biotransformation 16:17–25

Patkar SA, Bjorkling F (1994) Lipase inhibitors. In: Woolley P, Petersen SB (eds) Lipases-their structure, biochemistry and application. Cambridge University Press, Cambridge 207pp

Petersen SB, Drablos F (1994) A sequence analysis of lipases, esterases and related proteins. In: Woolley P, Petersen SB (eds) Lipases – their structure, biochemistry and application. Cambridge University Press, Cambridge 23pp

Raminelli C, Kagohara E, Pellizari VH, Comasseto JV, Andrade LH, Porto ALM (2007) Biotransformations of mannich bases and propiophenones by Brazilian microorganisms and enzymatic resolution of phenylpropanols by lipase from *Candida antarctica* (Novozyme 435). Enzym Microb Technol 40:362–369

Rapp P (1995) Production, regulation and some properties of lipase activity from *Fusarium oxysporum* f. sp. *vasinfectum*. Enzym Microb Technol 17:832–838

Robles-Medina A, González-Moreno PA, Esteban-Cerdán L, Molina-Grima E (2009) Biocatalysis: towards ever greener biodiesel production. Biotechnol Adv 27(4):398–408

Rodrigues RC, Fernandez-Lafuente R (2012) Lipase from *Rhizomucor miehei* as an industrial biocatalyst in chemical process. J Mol Catal B Enzym 64(1–2):1–22

Romero CM, Pera LM, Loto F, Vallejos C, Castro G, Baigori MD (2012) Purification of an organic solvent tolerant lipase from *A. niger* MYA135 and its applications in ester synthesis. Biocatal Agric Biotechnol 1:25–31

Rosa DR, Duarte ICS, Saavedra NK, Varesche MB, Zaiat M, Cammarota MC, Freire DMG (2009) Performance and molecular evaluation of an anaerobic system with suspended biomass for treating wastewater with high fat content after enzymatic hydrolysis. Bioresour Technol 100(24):6170–6176

Rosu R, Yasui M, Iwasaki Y, Yamane T (1999) Enzymatic synthesis of symmetrical 1,3-diacylglycerols by direct esterification in solvent-free system. J Am Oil Chem Soc 76:839–843

Rubin B, Dennis EA (1997a) Lipases, part a: biotechnology. Methods Enzymol 284:1–440

Rubin B, Dennis EA (1997b) Lipases, part B: enzyme characterization and utilization. Methods Enzymol 286:1–600

Salleh AB, Musani R, Basri M, Ampon K, Yunus WMZ, Razak CNA (1993) Extra-and intracellular lipases from a thermophilic *Rhizopus oryzae* and factors affecting their production. Can J Microbiol 39:976–981

Savitha J, Srividya S, Jagat R, Payal P, Priyanki S, Rashmi GW, Roshini KT, Shantala YM (2007) Identification of potential fungal strain(s) for the production of inducible, extracellular and alkalophilic lipase. Afr J Biotechnol 6(5):564–568

Saxena RK, Ghosh PK, Gupta R, Davidson WS, Bradoo S, Gulati R (1999) Microbial lipases: potential biocatalyst for the future industry. Curr Sci 77:101–115

Saxena RK, Davidson WS, Sheoran A, Giri B (2003) Purification and characterization on an alkaline thermostable lipase from *Aspergillus carneus*. Process Biochem 39:239–247

Saxena RK, Agarwal L, Meghwanshi GK (2005) Diversity of fungal and yeast lipases: present in future scenario for the 21st century. In: Satyanaryana T, Johri BN (eds) Microbial diversity: current perspectives and potential applications. I.K. International Pvt. Ltd, New Delhi 796pp

Shimada Y, Sugihara A, Nagao T, Tominaga Y (1992) Induction of *Geotrichum candidum* lipase by long-chain fatty acids. J Ferment Bioeng 74:77–80

Shimada Y, Koga C, Sugihara A, Takada N, Tsunasawa S, Tominaga Y (1993a) Purification and characterization of a novel solvent-tolerant lipase from *Fusarium heterosporum*. J Ferment Bioeng 75:349–352

Shimada Y, Sugihara A, Tominaga Y (1993b) Microbial lipase: structure and production. In: Murooka Y (ed) Recombinant microbes for industrial and agricultural applications. CRC Press, Boca Raton 359pp

Shimada Y, Maruyama K, Okazaki S, Nakamura M, Sugihara A, Tominaga Y (1994) Enrichment of polyunsaturated fatty acids with *Geotrichum candidum* lipase. J Am Oil Chem Soc 71(9):951–954

Shimada Y, Watanabe Y, Samukawa T, Sugihara A, Noda H, Fukuda H, Tominaga Y (1999) Conversion of vegetable oil to biodiesel using immobilized *Candida antarctica* lipase. J Am Oil Chem Soc 76(7):789–793

Stead R (1986) Microbial lipases: their characteristic, role in food spoliage and industrial uses. J Dairy Sci 53:481–505

Syed Rahmatullah MSK, Shukla VKS, Mukherjee KD (1994) c-Linolenic acid concentrates from borage and evening primrose oil fatty acids via lipase-catalyzed esterification. J Am Oil Chem Soc 71:563–567
Sztajer H, Zboinska C (1988) Microbial lipases in biotechnology. Acta Biotechnol 8:169–175
Sztazer H, Maliszewska I (1988) The effect of culture condition on lipolytic productivity of microorganisms. Biotechnol Lett 10:199–204
Tan T, Zhang M, Wang B, Ying C, Deng L (2003) Screening of high lipase producing *Candida* sp. and production of lipase by fermentation. Process Biochem 39(4):459–465
Tanaka Y, Hirano J, Funada T (1992) Concentration of docosahexaenoic acid in glyceride by hydrolysis of fish oil with *Candida cylindracea* lipase. J Am Oil Chem Soc 69:1210–1214
Toida J, Kondoh K, Fukuzawa M, Ohnishi K, Sekiguchi J (1995) Purification and characterization of a lipase from *Aspergillus oryzae*. Biosci Biotechnol Biochem 59(7):1199–1203
Uston G, Goner S, Arer G, Türkay S, Erciyes AT (1997) Enzymatic hydrolysis of anchovy oil: production of glycerides enriched in polyunsaturated fatty acids. Appl Biochem Biotechnol 68:171–185
Uyama H, Kobayashi S (1993) Enzymatic ring-opening polymerization of lactones catalyzed by lipase. Chem Lett 22:11497–11150
Vakhlu J, Kour A (2006) Yeast lipases: enzyme purification, biochemical properties and gene cloning. Electron J Biotechnol 9(1):69–85
Valero F, del-Rio JL, Poch M, Solà C (1991) Fermentation behaviour of lipase production by *Candida rugosa* growing on different mixtures of glucose and olive oil. J Ferment Bioeng 72:399–401
Valladao ABG, Cammarota MC, Freire DMG (2011) Performance of ananaerobic reactor treating poultry abattoir wastewater with high fat content after enzymatic hydrolysis. Environ Eng Sci 28(4):299–307
Veeraragavan K, Gibbs BF (1989) Detection and partial purification of two lipases from *Candida rugosa*. Biotechnol Lett 11:345–348
Venkata Rao P, Lakshmanan CM (1991) Lipase enzyme technology and its potential application in the oils and fats industry. Ind Chem Eng 33:7–29
Verger R (1997) Interfacial activation of lipases: facts and artifacts. Trends Biotechnol 15:32–38
Vulfson EN (1994) Industrial application of lipases. In: Woolley P, Petersen SB (eds) Lipases-their structure, biochemistry and application. Cambridge University Press, Cambridge 271pp
Yadav RP, Saxena RK, Gupta R, Davidson S (1997) Production of biosurfactant from sugar alcohols and natural triglycerides by *Aspergillus terreus* lipase. J Sci Ind Res 56:479–482
Yadav RP, Saxena RK, Gupta R, Davidson WS (1998) Purification and characterization of a regiospecific lipase from *Aspergillus terreus*. Biotechnol Appl Biochem 28:243–249
Yamaguchi S, Mase T (1991) High yield synthesis of monoglyceride by monoacylglycerol and diacylglycerol lipase from *Penecillium camembertii* U-150. J Ferment Bioeng 72:162–167
Zhang WW, Jia JQ, Wang N (2015) Improved activity of lipase immobilized in microemulsion based organogels for (R,S)-ketoprofen ester resolution: long term stability and reusability. Biotechnol Rep 7:1–8

Chapter 23
Biopigments: Fungal Pigments

Sharmila Tirumale and Nazir Ahmad Wani

Abstract Synthetic dyes are frequently used in different fields such as for food industry, paper and agricultural industry and science and technology. But due to the adverse toxicological side effects of synthetic pigments used in the industries, now research is focused on the products from natural resources. Microbial compounds are natural-coloured substances produced by microorganisms, especially fungi and bacteria. Biopigments from natural resources can replace the synthetic dyes used in pharma industries. Most of the microbes reported to produce carotenoids belong to *Myxococcus* spp. Other organisms include spp. of *Serratia*, *Streptomyces* and *Agrobacterium*. The red-coloured basidiomycetous yeast *Xanthophyllomyces dendrorhous*, green alga *Haematococcus Pluvialis* and *Agrobacterium aurantiacum* are known to produce astaxanthin, an orange-red pigment. Other organisms such as *Serratia marcescens*, *Vibrio psychoerythrus*, *Rugamonas rubra*, *Streptoverticillium rubrireticuli* and other eubacteria produce prodigiosin, a red pigment used in various applications. Astaxanthin from *Xanthophyllomyces* sp., arpink red from *Penicillium* sp. and riboflavin from *Ashbya* sp. and pigments from *Monascus* spp. are used in many food industries. Other pigment-producing fungi are *chaetomium cupreum*, *Penicillium aculeatum*, *Fusarium chlamydosporum*, etc. Fungi produce an interesting class of pigmented secondary metabolites, called azaphilones. Recently many pharmaceutical industries are using microbial pigments in their products. Microbial pigments produced by pharmaceutical industry may act as antibiotics, anticancer, antiproliferative and immunosuppressive compounds.

Keywords Fungi · Pigments

S. Tirumale (✉) · N. A. Wani
Department of Microbiology and Biotechnology, Bangalore University, Benguluru, India

P. Gehlot, J. Singh (eds.), *Fungi and their Role in Sustainable Development: Current Perspectives*, https://doi.org/10.1007/978-981-13-0393-7_23

23.1 Introduction

Pigments change the colour of reflected or transmitted light and is capable of absorbing light in the visible range (400–700 nm). The word pigment has originated from Latin and was initially used to denote a colour, but later its usage extended to indicate the coloured compounds. Pigments are used in food and pharmaceutical industries. Natural organic pigments have been known to mankind for a long time.

23.2 Global Colour Market

The demand for biopigments with special reference to the food colourants is increasing day by day through the world. It is estimated that the market for food biopigments is 35 million USD in the late 1980s, 250 million in 2000 and 600 million in 2011. The colour market, both synthetic and natural, was estimated at $1.8 billion in 2010, of which natural colour market was estimated at $0.66 billion of total. The growth of total colour usage is about 4.5% per year, but the growth of the natural colours is higher, i.e. 6.7% per year. Europe stands first in holding the largest share of the natural colour market, which is followed by North America and Asia Pacific (Natural Colorants Market – Global Industry Trends, Share, Analysis and Forecast, 2012–2012). According to another report on food colour market in combination with the market expertise (Mintel and Leatherhead Food Research 2011), there is an inclination for the use of natural colours from artificial or synthetic colour all over the world.

According to the report of food colours, Market, Technical and Regulatory Insights Report 2013, the global food colour market was appearing to be about $1.55 billion in 2011 which had shown a growth of 13% from 2007. The industrialization and thereby consumption of processed foods are increasing in the Asian countries. Therefore it is certain that the biopigment market will grow to about 1.5 billion USD by 2020 and also that the international market of the natural colours is in a phase of extraordinary expansion. There are no reliable published statistics on the size of the colour market; however, on a global scale, a reasonable estimate of $940 million for the year 2002 is segmented as shown in Fig. 23.1.

23.3 Natural Pigments

Pigments are classified into three main classes such as synthetic pigments, plant-derived pigments and microbial pigments. Most of the natural compounds are used as colourants, flavours and fragrances. Biopigments are produced from natural sources which can replace synthetic dyes (Palanichamy et al. 2011). Biopigments that are produced from living organisms can be successfully used in place of

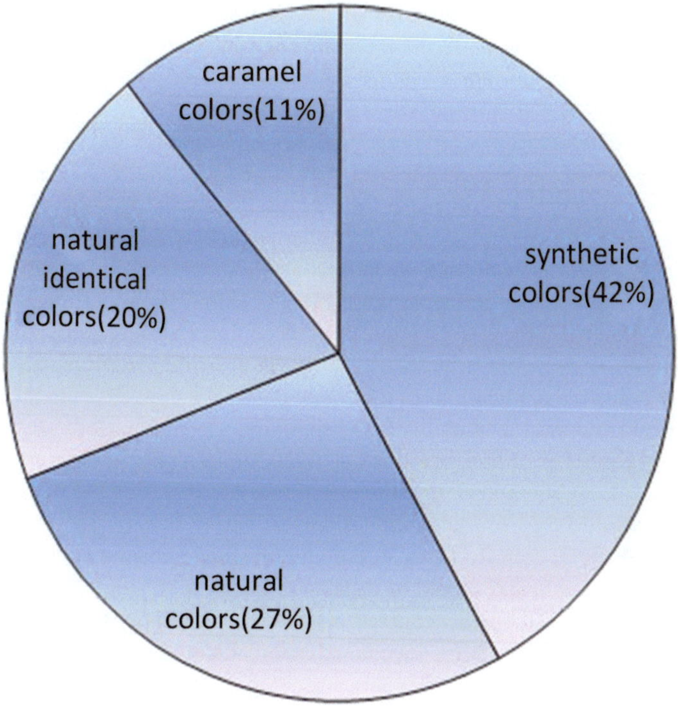

Fig. 23.1 Biopigments: Fungal pigments

synthetic dyes in the industries and laboratories. As it has been suggested that synthetic pigments have various negative effects, now research is being oriented on the production of metabolites from natural resources. This is also evident from the banning of synthetic dyes and their imports from other countries by Europe (Natural Colorants Market – Global Industry Trends, Share, Analysis and Forecast, 2012–2018). Microorganisms and herbs are the two major sources of natural dyes. Microbial pigments are more stable against light and other environmental factors; they are water soluble and are available throughout the year. Thus, there is worldwide interest for the development and production of pigments and colours from microbes.

23.3.1 *Microbial Pigments*

"Biopigments" or microbial pigments are natural-coloured substances produced by microorganisms, especially fungi and bacteria. Microorganisms produce a variety of pigments by fermentation process with higher pigment yields and production of lower residues compared to that of the plant and animal sources (Duran et al. 2012). A number of microorganisms are also reported for the production of carotenoids with different colours. Most of the microbes reported to produce carotenoids belong to *Myxococcus* spp. Other organisms include spp. of *Serratia*, *Streptomyces* and *Agrobacterium*. The red yeast *Xanthophyllomyces dendrorhous*, alga *Haematococcus Pluvialis* and *Agrobacterium aurantiacum* are known to produce astaxanthin, an orange-red pigment. Other organisms such as *Serratia marcescens*, *Vibrio psychoerythrus*, *Rugamonas rubra*, *Streptoverticillium rubrireticuli* and other eubacteria produce prodigiosin, a red pigment used in various applications.

Microbial pigments can increase the marketability of the products of various industries especially food processing and pharmaceutical industries. They possess various biological activities such as antioxidative, anticancerous, antiproliferative, antimicrobial, immunosuppressive and biodegradability properties (Venil and Lakshmanaperumalsamy 2009). Because of these properties, recently, the production of microbial pigments is increasing due to their medicinal properties and food value. The industrial applications of microbial pigments, their environmental stability, high yield and their availability make them industrially feasible for cultivation technology (Kim et al. 1999a, b; Parekh et al. 2000). In addition, their production is independent of environmental conditions and seasons as that of the plants and animals and is produced by fermentation processes.

A microorganism used for pigment production should contain sufficient amount of carbon and nitrogen and should be capable of giving high yield. Also it should be non-toxic and non-pathogenic coupled with easy separation from cell biomass. Extensive research has not been done on microorganisms for pigment production.

23.3.2 Fungal Pigments

The term "fungi" refers to a group of eukaryotic organisms whose members include yeasts and moulds and also the more familiar (macroscopic) mushrooms. The group is considered sufficiently distinct from other groups of organisms such as plants, animals or bacteria. While between 80,000 and 1,00,000 species of fungi are known to date, some estimates of total numbers suggest that as many as 1.5 million species may exist. Fungi are important source of numerous secondary metabolites, including organic acids; polyenes and polyynes; phenolic compounds such as quinones, antraquinones and xanthones; mono- to triterpenes; polysaccharides; proteins; etc. (Hashimo et al. 1994; Hellwig et al. 2005; Hobbs 2003).

Fungi are also known to produce various pigments with different structures and biological activities. In contrast to higher plants, fungi are more useful for industrial production of pigments because of their fast growth rate, easier culture techniques, ability to grow on cheap substrates, independent of weather conditions, production of different shades of colours, high yield of the products and the feasibility of bioprocess development (Zhao et al. 1998). The important criteria for an ideal pigment-producing microbial cell are that it should be non-toxic and non-pathogenic to humans (Duffose et al. 2014). Recently, β-carotene, a carotenoid by the fungus *Blakeslea trispora*, has been produced by M/s DSMTM in the Netherlands and has been approved as safe for food use (Britton et al. 2004). The microbial pigments, like astaxanthin, arpink red and riboflavin, are already being used in many food items (Duffose 2006). Other pigment-producing fungi are *chaetomium cupreum* (Soumya et al. 2013), *Penicillium aculeatum*, *Fusarium chlamydosporum*, etc.

23.3.3 Fungal Azaphilones

Fungi produce an interesting and significant class of secondary metabolites, called azaphilones. Azaphilones are: "Structurally diverse class of fungal secondary metabolites (polyketide derivatives), namely pigments" (Strudikova et al. 2000; Dong et al. 2006). They have affinity for ammonia, and they react with amino group present in biomolecules to form red or purple *vinylogous* ɣ*-pyridones* due to the exchange of pyrane oxygen for nitrogen and thus form water-soluble pigments (Stadler et al. 1995). This reaction can occur with ammonia alone as found in monascorubramine and rubropunctamine (Akihisa et al. 2005) or with the side chain of a macrocyclic polypeptide, as in the case of chlorofusin (Duncan et al. 2001). Azaphilones have various significant properties like their natural origin, yellow-red spectra, thermostability and easy derivatization. The coloured azaphilones are produced by numerous species of ascomycetous and basidiomycetous fungi (Gill and Steglich 1987) including genera *Aspergillus*, *Penicillium*, *Chaetomium*, *Talaromyces*, *Pestalotiopsis*, *Phomopsis*, *Emericella* and *Epicoccum* as well as *Monascus* and *Hypoxylon*, where they are responsible for the bright yellow, red or green colours produced by either fruiting bodies or mycelia (Mapari et al. 2006a, b). Similar compounds were also identified in *Aspergillus ustus*, *Cochliobolus lunatus*, *Talaromyces* sp. and *Emericella falconensis*. Elaborative research has been carried out which reveals that they show a wide spectrum of biological activities such as antimicrobial, antifungal, antiviral, antioxidant, cytotoxic, nematicidal and anti-inflammatory activities. These potent biological activities of azaphilones may be due to the presence of vinylogous ɣ-pyridones (Park et al. 2005; Osmanova et al. 2010). Azaphilones also exhibit different industrial applications such as food and cosmetic.

23.4 Cultivation of Pigment-Producing Fungi

23.4.1 Culture Media Used for Pigment Production

The production of different pigments is often affected by different cultivation media (Mapari et al. 2006a, b). The different concentration of carbon and nitrogen sources in the growth media results in enhanced pigment production (Mapari et al. 2009). For efficient pigment production, fungi are grown on natural media – potato dextrose agar (PDA), corn extract agar (CEA), Jowar extract agar (JEA), rice extract agar (REA) and also two synthetic media – Sabouraud dextrose agar (SDA) medium and Czapek-Dox agar (CZA) medium.

23.4.2 Extraction of Pigments

Media selection depends on whether the pigments produced are extracellular or intracellular. Some fungi produce pigments in extracellular liquid medium, and some of them produce intracellular pigments in the mycelia. There are a number of methods available for extraction of pigments. Pigment extraction generally depends upon the nature of pigments and is usually done by polar and non-polar solvents either individually or in combinations. Ultrasonic and hydrochloric acid and also supercritical CO_2 extraction techniques are used. But there is a need to develop a suitable extraction method with maximum pigment yield from microbial sources, which can be applied on a commercial scale.

23.4.3 Intracellular Pigment Extraction

The intracellular pigments are produced from mycelium of the fungi. The organisms are cultured in potato dextrose broth under submerged fermentation conditions for 7–10 days at 28 °C. For intracellular pigment, mycelium is separated from broth by homogenizing the mycelia with sea sand several times until the debris of the mycelia will become colourless. The extracts are collected, filtered and evaporated to dryness in a rotary vacuum evaporator. Organisms such as *Fusarium aculeatum* produce reddish-orange, and *Penicillium chlamydosporum* produces deep red-coloured pigment.

23.4.4 Extracellular Pigment Extraction

The extracellular pigments are produced and released into the broth by the fungi. The fungi are cultured in potato dextrose broth under submerged fermentation conditions on rotary shaker at 150 rpm for 10–15 days at 28 °C. For extracellular pigment, mycelium is separated from broth by filtration. Then the extracellular pigments are extracted from the filtrate by using different solvents in the ratio of 1:1. The extracts are collected, filtered and evaporated to dryness in a rotary vacuum evaporator. The extracellular pigment-producing organisms are *Chaetomium cupreum* which produces reddish-purple pigment, *Mycelium sterilia* producing deep orange pigments and species of *Trichoderma*, *Penicillium*, *Aspergillus* and *Alternaria* producing different coloured pigments.

23.4.5 Biological Applications of Pigments

Microbial pigments are used in pharmaceutical, food and cosmetic applications and as new drugs for the treatment of various human diseases (Table 23.1).

Table 23.1 Some of the significant pigment-producing microorganisms and their bioactivities

S. No.	Microorganisms	Pigments	Biological activity	References
1.	*Agrobacterium aurantiacum*	Astaxanthin	Antioxidant, anticancer, anti-inflammatory, photoprotectant	Reyes et al. (1996)
2.	*Haematococcus pluvialis*	Astaxanthin	Antioxidant, anticancer, anti-inflammatory, photoprotectant	Reyes et al. (1996)
3.	*Bradyrhizobium* spp.	Canthaxanthin	Anticancer, antioxidant	Chew et al. (1998) and Duffose (2006)
4.	*Pseudoalteromonas denitrificans*	Cycloprodigiosin	Anticancer, anti-plasmodial	Kim et al. (1999a, b)
5.	*Corynebacterium insidiosum*	Indigoidine	Antimicrobial	Starr (1958) and Cude et al. (2012)
6.	*Serratia marcescens*	Prodigiosin	Anticancer, DNA cleavage, immunosuppressant	Feher et al. (2008) and Deorukhar et al. (2007)
7.	*Streptomyces echinoruber*	Rubrolone	Anti-inflammatory	Iacobucci and Sweeney (1981)
8.	*Cyanobacteria*	Scytonemin	Anti-inflammatory, anti-proliferative	Stevenson et al. (2002)
9.	*Sphingobacterium Multivorum*	Zeaxanthin	Protection against photodamage	Hammond and White (1970)
10.	*Monascus* spp.	Ankaflavin	Antitumour, anti-inflammatory	Hsu et al. (2011)
11.	*Fusarium sporotrichioides*	Lycopene	Antioxidant, anticancer	Di Mascio et al. (1989) and Giovannucci et al. (2002)
12.	*Monascus* spp.	Monascorubramin, rubropunctatin	Anticancer, antimicrobial, anticancer	Blanc et al. (1994)
13.	*Blakeslea trispora*	β-carotene	Antioxidant, anticancer, suppression of cholesterol synthesis	Casta et al. (2005), Dufosse (2009) and Cerdaolmedo (2001)
14.	*Fusarium sporotrichioides*	β-carotene	Antioxidant, anticancer, suppression of cholesterol synthesis	Casta et al. (2005), Dufosse (2009) and Cerdaolmedo (2001)
15.	*Neurospora crassa*	β-carotene	Antioxidant, anticancer, suppression of cholesterol synthesis	Casta et al. (2005), Dufosse (2009), Lopes et al. (2009) and Cerdaolmedo (2001)

(continued)

Table 23.1 (continued)

S. No.	Microorganisms	Pigments	Biological activity	References
16.	*Phycomyces*	β-carotene	Antioxidant, anticancer, suppression of cholesterol synthesis	Casta et al. (2005), Dufosse (2009) and Cerdaolmedo (2001)
17.	*Phaffia rhodozyma*	Astaxanthin	Antioxidant, anticancer,, anti-inflammatory, photoprotectant	Florencio et al. (1998) and Flores-Cotera and Sanchez (2001)
18.	*Xanthophyllomyces*	Astaxanthin	Antioxidant, anticancer, anti-inflammatory, photoprotectant	Florencio et al. (1998) and Flores-Cotera and Sanchez (2001)
19.	*Rhodotorula* spp.	Torularhodin	Antioxidant, antimicrobial	Sakaki et al. (2000) and Ungureanu and Ferdes (2012)
20.	*Haloferax alexandrines*	Canthaxanthin	Antioxidant, anticancer	Mathews-Roth (1982), Chew et al. (1998) and Duffose (2006)

23.5 Pharmaceutical Applications

Pharmaceutical industry uses pigments as colouring agents in their products. Pigments are used as anticancer, antidiabetic and even as efficient antimicrobials. Because of these properties, these can be exploited for the treatment of cancer, diabetes and other infectious diseases. They are also proved to be efficient antioxidants and could be applied to bring down the level of free radicals. Fungal pigments are also used as antibacterial, antifungal and antiprotozoa (*Leishmania braziliensis*).

Epicoccum nigrum is a non-mycotoxigenic fungus and is known to be a potential producer of polyketide pigment orevactaene having good antioxidant property (Mapari et al. 2008). Gerber and Ammar (1979) have found fusarubin from *Fusarium solani* which are known to possess antibiotic properties. The studies of Blanc et al. (1994) have shown that monascorubramine from *Monascus* spp. is known to exhibit good antimicrobial activity. Andersen et al. (1991) and Prathumpai et al. (2006) have described the antimicrobial activities of anthraquinone pigment of *Penicillium oxalicum* (antifungal and antiviral) and naphthoquinone pigment of *Cordyceps unilateralis* (antibacterial and trypanocidal), respectively. Canthaxanthin produced by *Monascus roseus* is reported to be a good antioxidant (Mathews and Roth 1982). Pigments produced by *Monascus* spp. (ankaflavin, rubropunctatin and monascorubramine), *Monascus roseus* (canthaxanthin), *Fusarium sporotrichioides* (lycopene and β-carotene), *Blakeslea trispora* (lycopene and β-carotene) and *Cordyceps unilateralis* (naphthoquinone) are known to possess good anticancer activity (Tuli et al. 2014). Pigments such as ankaflavin from *Monascus* spp. also exhibit anti-inflammatory activities (Hsu et al. 2011). Fatima et al. (2014) have clearly indicated the greater potentiality of fungal pigments in cancer therapy. They have successfully designed the fungal pigments based on computation drug designing which have shown promising results as aromatase inhibitors which can stop the creation of oestrogen, as a result of which the development of breast cancer cell scan be reduced. Some examples of such type of pigments are given below.

23.5.1 Anthocyanin

Anthocyanins are flavonoids and water-soluble pigments. They exhibit anticancer and antioxidant activity (Katsube et al. 2003; Martin et al. 2003) and decrease and modulate immune response. Anthocyanin exhibits anticancer activity by interfering with cell cycle, apoptosis, inflammation, angiogenesis, invasion and metastasis process. The antioxidant activity of anthocyanin is due to phenolic hydroxyl groups that donate a hydrogen atom or an electron to a free radical.

23.5.2 Prodigiosin

Prodigiosin is produced by different organisms such as *Vibrio psychroerythrus*, *S. marcescens*, *Pseudomonas magnesiorubra* and other eubacteria. This pigment was first reported from *S. marcescens*. *Streptomyces* is also known to produce this pigment. It is a potential pigment having various industrial applications (Furstner 2003). It shows immunosuppressive activity (Kim et al. 1989) and anticancer activity on about 60 cell lines of human tumour cells derived from different malignant tissues (Pandey et al. 2007). Prodigiosin is also used for the treatment of diabetes mellitus (Hwanmook et al. 2006).

23.5.3 Violacein

Violacein, is a violet pigment and indole derivative, isolated mainly from bacterium *Chromobacterium violaceum*, which exhibits various pharmaceutical activities such as antitumoural, antiprotozoan (Matz et al. 2004), anticancer, (Ferreira et al. 2004; Kodach et al. 2006), antiviral (Sánchez et al. 2006), antibacterial (Lichstein and van De Sand 1946; Nakamura et al. 2003) and antioxidant activities (Konzen et al. 2006).

23.5.4 Red Yeast Rice (RYR)

Red yeast rice (RYR) is a fermented rice product produced traditionally by fermenting cooked rice kernels with yeast *Monascus*. It is produced by different species of *Monascus* such as *M. pilosus*, *M. ruber* and *M. purpureus*. These *Monascus* spp. produce compounds with polyketide structure with yellow, orange and red pigments. Angkak is produced from *Monascus ruber* with anticholesterol activity. RYR is proved to contain many active constituents resembling statins in its structure, unsaturated fatty acid, vitamins, sterols and B complex (Wang et al. 1997). Various studies also reported that RYR and statins decrease blood glucose levels in diabetes patients (Chang et al. 2006).

23.6 Food Industry

Microbial pigments produced especially by some fungi such as β-carotene from the fungus *Blakeslea trispora* in Europe and *Monascus* pigments in Asia are now in use in the food industry (Wissgott and Bortlik 1996; Lampila et al. 1985). *Monascus* red pigments, generally produced as *Monascus* fermented rice (MFR) powder, improve the organoleptic characteristics of the food products. These pigments are useful in maintaining cholesterol level because they contain monocolins (Vidyalakshmi et al. 1999). Some strains of *Aspergillus* spp. such as *A. glaucus*, *A. cristatus* and *A. repens* are found to produce yellow and red hydroxyanthraquinoid pigment (HAQN), such as emodin (yellow), physcion (yellow), questin (yellow to orange-brown), erythroglaucine (red), catenarin (red) and rubrocristin (red). These are known to be used as food colourants. This was the first report on hydroxyanthraquinoid pigment as food colourant (Caro et al. 2012). Similarly there are various other microbial pigments that are known to possess good potentiality and are still under research stage.

23.6.1 β-Carotene

β-carotene, also known as provitamin A, is a yellowish carotenoid pigment and acts as antioxidant. Following microbes are mainly used for β-carotene production.

23.6.2 **Blakeslea trispora**

Some strains of this mould produce high level of β-carotene. *B. trispora* strains are of two types: (+) mating type and (−) mating type (−). Strains obtained by the specific ratio of above two mating types produce β-carotene (Dufosse 2006). Today *B. trispora* fungal β-carotene is produced by two industries located in Russia, Ukraine and Spain.

23.6.3 Mucor circinelloides

Some strains of *M. circinelloides*, when exposed to the blue light, get activated due to significant change in structural genes of β-carotene and hence produce it to high level.

23.6.4 Phycomyces blakesleeanus

Phycomyces spp. are known to enhance carotenogenic potential when grown on solid substrates or in liquid media (Garton et al. 1951).

23.7 Textile Industry

Textile industries use large amount of synthetic dyes. Synthetic dyes are widely available at an economical price and produce a wide variety of colours, but they may cause skin allergy and other harms to human body. In addition, they may release hazardous chemicals during their manufacture. Due to this reason, there is more demand on natural or microbial pigments. The use of natural pigments in the textile industry is eco-friendly and harmless. Studies of Sharma et al. (2012) suggest that pigments produced by *Curvularia lunata*, *Trichoderma virens* and *Alternaria alternata* can be used to dye the textiles such as silk and wool. Further these dyes have no adverse effect on the tensile strength of the fabrics and were found to be non-toxic to the human skin. Gupta et al. (2013) have reported the use of pigments from T*richoderma* spp. for dyeing silk and wool. For making the natural dyes for textiles commercially successful for any particular fibre, the appropriate and standardized techniques of dyeing need to be adopted. Therefore to obtain newer shade with acceptable colour fastness, behaviour and reproducible colour yield, appropriate scientific dyeing techniques and procedures are yet to be developed.

References

Akihisa et al (2005) Anti-tumor-initiating effects of monascin, an azaphilonoid pigment from the extract of Monascus pilosus fermented rice (red-mold rice). Chem Biodivers 2(10):1305–1309

Andersesen DO, Weber ND, WoodS G, Hughes BG, Murray BK, North JA (1991) In vitro virucidal activity of selected anthraquinones and anthraquinones derivatives. Antivir Res 16(2):185–196

Blanc PJ, Loret MO, Santerre AL, Pareilleux A, Prome JC, Laussac JP, Goma G (1994) Pigments of Monascus. J Food Sci 59:862–865

Britton G et al (2004) Carotenoids- handbook. Birkhauser. ISBN:3-7643-6180-8

Caro Y, Anamale L, Fouillaud M, Laurent P, Petit T, Dufosse L (2012) Natural hydroxyanthraquinoid pigments as potent food grade colorants: an overview. Nat Prod Bioprospect 2:174–193

Casta FTM, Justo GZ, Duran N, Noguerira PA, Lopes SCP (2005) The use of violacein in its free and encapsulated form in polymeric systems against malaria. Brazilian Patent PIBr.056399-0
Cera-Olmedo E (2001) Phycomyces and the biology of light and color. FEMS Microbiol Rew 25:503–512
Chang JC, Wu MC, Liu IM, Cheng JT (2006) Plasma glucose-lowering action of Hon-Chi in streptozotocin-induced diabetic rats. Horm Metab Res 38:76–81
Chew BP, Park JS, Wong MW, Wong TS (1998) A comparison of the anticancer activities of beta-carotene, canthaxanthin and astaxanthin in mice in vivo. Anticancer Res 19(3a):1849–1853
Cude WN, Mooney J, Tavanaei AA, Hadden MK, Frank AM, Gulvik CA (2012) Production of the antimicrobial secondary metabolite indigoidine contributes to competitive surface colonization by the marine roseobacter Phaeobacter sp. strain Y4I. Appl Environ Microbiol 78:4771–4780
Deorukhar AA, Chander R, Ghosh SB, Sainis KB (2007) Identification of a red pigment bacterium producing a potent anti-tumor N-alkylated prodigiosin as Serratia Marcescens. Res Microbiol 158(5):399–404
Di-Mascio P, Kaiser S, Sies H (1989) Lycopene as the most efficient biological carotenoid singlet oxygen quencher. Arch Biochem Biophys 274(2):532–538
Dong J et al (2006) New nematicidal azaphilones from the aquatic fungus Pseudohalonectria adversaria. Microbiol Lett 264(1):65–79
Dufosse L (2006) Microbial production of food grade pigments. Food Technol Biotechnol 44(3):313–321
Dufosse L (2009) Pigments. Microb Encylopedia Microbiol 4:457–471
Dufosse L, Fouillaud M, Caro Y, Mapari SAS, Sutthiwong N (2014) Filamentous fungi are large scale production of pigments and colorants for the food industry. Curr Opin Biotechnol 26:56–61
Duncan SJ et al (2001) Isolation and structure elucidation of Chlorofusin, a novel p53-MDM2 antagonist from a Fusarium sp. J Am Chem Soc 123(4):554–560
Duran M, Ponezi AN, Faljoni-Alario A, Teixerira MF, Justo GJ, Duran N (2012) Potential application of violacein: a microbial pigment. Med Chem Res 21(7):1524–1532
Fatima N, Kalsoom A, Mumtaz A, Muhammad SA (2014) Computational drug designing of fungal pigments as potential aromatase inhibitors. Bangladesh J Pharmacol 9:575–579
Feher D, Barlow RS, Lorenzo PS, Hemscheidt T (2008) A 2-substituted prodiginine, 2-(p-hydroxybenzyl) prodigiosin from Pseudoalteromonas rubra. J Nat Prod 71(11):1970–1972
Ferreira CV, Bos CL, Versteeg HH, Justo GZ, Durán N, Peppelenbosch MP (2004) Molecular mechanism of violacein mediated human leukemia cell death. Blood 104:1459–1467
Florencio JA, Soccol CR, Furlanetto LF, Biofilm TMB, Krieger N, Baron M, Fontana JD (1998) A factorial approach for a sugarcane juice based low cost culture medium: increasing the astaxanthin production by the red yeast Phaffia rhodozyma. Bioprocess Eng 19:161–164
Flores-Cotera LB, Sanchez S (2001) Copper but not iron limitation increases astaxanthin production by Phaffia rhodozyma in a chemically defined medium. Biotechnol Lett 23:793–797
Furstner A (2003) Chemistry and biology of roseophilin and the prodigiosin alkaloids: a survey of the last 2500 years. Chem Int Ed Engl 42:3582–3603
Garton GA, Goodwin TW, Lijinsky W (1951) General conditions governing //-carotene synthesis by the fungus Phycomyces blakesleeanus Burgeff. Biochem J 48:154–163
Gerber NN, Ammar MS (1979) New antibiotic pigments related to fusarubin from Fusarium solani (Mart.) Sacc. II. Structure elucidations. J Antibiot 32:685–688
Gill M, Steglich W (1987) Pigments of fungi (macromycetes). Prog Chem Org Nat Prod 51:1
Giovannucci E, Rimm EB, Liu Y, Stampfer MJ, Willett WC (2002) A prospective study of tomato products, lycopene and prostate cancer risk. J Nat Cancer Inst 94(5):391–398
Gupta C, Sharma D, Aggarwal S, Nagpal N (2013) Pigment production from *Trichoderma* spp. for dyeing of silk and wool. IJSN 4(2):351–355
Hammond RK, White DC (1970) Inhibition. Inhibition of carotenoid hydroxylation in *Staphylococcus aureus* by mixed-function inhibitors. J Bacteriol 103:607–610
Hashimoto T, Tahara S, Takaoka S, Tori M, Asakawa Y (1994) Structures of daldinins A.apprx.C, three novel azaphilone derivatives from ascomycetous fungus Daldinia concentrica. Chem Pharm Bull 42:2397–2379

Hellwig V, Ju YM, Rogers JD, Fournier J, Stadler M (2005) Hypomiltin, a novel azaphilone from Hypoxylon hypomiltum, and chemotypes in Hypoxylon sect. Hypoxylon as inferred from analytical HPLC profiling. Mycol Prog 4:39–54

Hobbs C (2003) Medicinal mushrooms: an exploration of tradition, health and culture. Botanica Press, Summertown

Hsu LC, Hsu YW, Liang YH, Kuo YH, Pan TM (2011) Anti-tumor and anti-inflammatory properties of ankaflavin and monaphilone A from Monascus purpureus NTU 568. J Agri Food Chem 59(4):1124–1130

http://www. Transparencymarketresearch.com/natural-colours-market.html (n.d.). Food additives market- global industry analysis, size, growth, trends and forecast 2012–2018. Transparency Market Research – Publications, New York

Hwanmook K, Sangbae H, Changwoo L, Kihoon L, Sehyung P, Youngkook K (2006) Use of prodigiosin for treating diabetes mellitus. US Patent 6638968

Iacobucci GA, Sweeney LG (1981) Process for enhancing the sunlight stability of rubrolone US Patent 4:285–985

Katsube N, Iwashita K, Tsushida T, Yamaki K, Kobori M (2003) Induction of apoptosis in cancer cells by Bilberry (Vaccinium myrtillus) and the anthocyanins. J Agric Food Chem 51:68–75

Kim H, Han SB, Lee OW, Lee K, Park S, Kim Y (1989) Use of prodigiosin for treating diabetes mellitus. US Patent 6 (1989) 638, 968 B1. 33

Kim CH, Kim SW, Hong SI (1999a) An integrated fermentation separation process for the production of red pigment by Serratia sp. KH-95. Process Biochem 35:485–490

Kim HS, Hayashi HJ, Shibata Y (1999b) Cycloprodigiosin hydrochloride obtained from Pseudoalteromonas denitrificans in a potent antimalarial agent. Biol Pharm Bull 22(5):532–534

Kodach LL, Bos CL, Durán N, Peppelenbosch MP, Ferreira CV, Hardwick JCH (2006) Violacein synergistically increases 5- fluorouracil cytotoxicity, induces apoptosis and inhibits akt mediated signal transduction in human colorectal cancer cells. Carcinogenesis 27:508–516

Konzen M, De Marco D, Cordova CA, Vieira TO, Antonio AV, Creezynski-Pasa TB (2006) Antioxidant properties of violacein: possible relation on its biological function. Bioorg Med Chem 14(24):8307–8313

Lampila LE, Wallen SE, Bullerman LB (1985) A review of factors affecting biosynthesis of carotenoids by the order Mucorales. Mycopathologia 90:65–80

Lichstein HC, van De Sand VF (1946) The antibiotic activity of violacein, prodigiosin, and phthiocol. J Bacteriol 52:145–146

Mapari SAS, Meyer AS, Thrane UJ (2006a) Agric Food Chem 54:7027

Mapari SAS, Meyer AS, Thrane U (2006b) Colorimetric characterization for comparative analysis of fungal pigments and natural food colorants. J Agric Food Chem 54(19):7028–7035

Mapari SAS, Hansen ME, Meyer AS, Thrane U (2008) Computerized screening for novel producers of Monascus-like food pigments in Penicillium species. J Agric Food Chem 56:9981–9989

Mapari SAS, Meyer AS, Thrane U, Frisvad JC (2009) Identification of potentially safe promising fungal cell factories for the production of polyketide natural food colorants using chemotaxonomic rationale. Microb Cell Factories 8:24

Martin S, Giannone G, Andriantsitohaina R, Martinez MC (2003) Delphinidin, an active compound of red wine, inhibits endothelial cell apoptosis via nitric oxide pathway and regulation of calcium homeostasis. Br J Pharm 39:1095–1102

Mathews-Roth MM (1982) Antitumor activity of β-carotene, canthaxanthin and phytoene. Oncology 39(1):33–37

Matz C, Deines P, Boenigk J, Arndt H, Eberl L, Kjelleberg S et al (2004) Impact of violacein producing bacteria on survival and feeding of Bacterivorous nanoflagellates. Appl Environ Microbiol 70:1593–1599

Mintel/Leatherhead Food Research (2011) http://www. Food coloors, market, technical and regulatory insights

Nakamura Y, Sawada T, Morita Y, Tamiya E (2003) Isolation of a psychrotrophic bacterium from the organic residue of a water tank keeping rainbow trout and antibacterial effect of violet pigment produced from the strain. Biochem Eng J 12:79–86

Osmanova N, Schultze W, Ayoub N (2010) Azaphilones: a class of fungal metabolites with diverse biological activities. Phytochem Rev 9:315
Palanichamy V, Hundet A, Mitra B, Reddy N (2011) Optimization of cultivation parameters for growth and pigment Production by streptomyces spp. isolated from marine sediment and rhizosphere soil. Int J Plant Anim Envir Sci 1(3):158–170
Pandey R, Chander R, Sainis KB (2007) Prodigiosins A novel family of immunosuppressants with anticancer activity. Ind J Biochem Biophys 44:295–302
Parekh S, Vinci VA, Strobel RJ (2000) Improvement of microbial strains and fermentation processes. Appl Microbiol Biotechnol 54:287–301
Park et al (2005) Antifungal activity against plant pathogenic fungi of chaetoviridins isolated from *Chaetomium globosum*. FEMS Microbiol Lett 252(2):309–313
Prathumpai W, Phimmakong K, Srikitikulchai P, Wongsa P (2006) Kinetic study of naphthoquinone and key metabolite production of C unilateralis BCC 1869. Thai J Biotechnol 7(2):39–43
Reyes FG, Valim MF, Vercesi AE (1996) Effect of organic synthetic food colours on mitochondrial respiration. Food Addit Contam 13(1):5–11
Sakaki H, Nakanishi T, Satonaka KY, Miki W, Fujita T, Komemushi S (2000) Properties of high to rularhodin producing mutant of Rhodotorula glutinis cultivated under oxidative stress. J Biosci Bioeng 89:203–205
Sánchez C, Brana AF, Méndez C, Salas JA (2006) Re- evaluation of the violacein biosynthetic pathway and its relationship to indolocarbazole biosynthesis. Chem Bio Chem 7:1231–1240
Sharma D, Gupta C, Aggarwal S, Nagpal N (2012) Pigment extraction from fungus for textile dyeing. Indian J Fibre Text Res 37:68–73
Soumya K, Narasimha Murthy K, Sreelatha GL, Srinivas C, Sharmila T (2013) Influence of growth factors on pigmentation of *Chaetomium cupreum SS – 02* and the antibacterial efficacy of the pigment against *Ralstonia solanacearum*. Int J Adv Res 1(10):212–219
Stadler M et al (1995) Novel bioactive azaphilones from fruit bodies and mycelial cultures of the ascomycete Bulgaria inquinans (Fr.). Nat Prod Lett 7(1):7–14
Starr MP (1958) The blue pigment of *Corynebacterium insidiosum*. Arch Mikrobiol 30:25–334
Stevenson CS, Capper EA, Roshak AK (2002) Scytonemin- a marine natural product inhibitor of kinases key in hyperproliferation inflammatory diseases. Inflamm Res 51(12):112–114
Strudikova M et al (2000) Mikrobialna produkcia farbnych azaphilonovych metabolitov. Chem List 94:105–110
Tuli HS, Chaudhary P, Beniwal V, Sharma AK (2014) Microbial pigment as natural color sources: current trends and future perspectives. J Food Sci Technol. https://doi.org/10.1007/s13197-014-1601-6
Ungureanu C, Ferdes M (2012) Evaluation of antioxidant and antimicrobial activities of torularhodin. Adv Sci Lett 18(1):50–53
Venil CK, Lakshmanaperumalsamy P (2009) An insightful overview on microbial pigment, prodigiosin. Electron J Biol 5(3):49–61
Vidyalakshmi R, Paranthaman R, Murugesh S, Singaravadivel K, Carroll PO (1999) Microbial bioconversion of rice broken to food grade pigments naturally exciting colours. World Ingred 12:39–42
Wang J, Lu Z, Chi J (1997) Multicenter clinical trial of the serum lipid-lowering effects of a Monascus purpureus (red yeast) rice preparation from traditional Chinese medicine. Curr Ther Res 58(12):64–78
Wissgott U, Bortlik K (1996) Prospects for new natural food colorants. Trends Food Sci Technol 7.298–302
Zhao DH, Lu L, Qin HL (1998) Studies on the properties of a microbial blue pigment. Food Ferment Ind 5:21–24

Chapter 24
Bioethanol Production from Renewable Biomass by Yeast

Ajay Kumar, Ranjan Deb, and Joginder Singh

Abstract Bioethanol produced from renewable biomass such as lignocellulosic materials like potato peel waste, wood, and agricultural and forest residues has the potential to replace the fuel extracted from nonrenewable resources. The main step involved is hydrolysis which enables the hemicelluloses and cellulose present in the renewable biomass to be converted into monomeric sugars. The process included a basic pretreatment of the renewable biomass with different concentrations acid, base, ionic liquid, etc. to get the maximum concentration of reducing sugars and nonreducing sugars. Media optimization is a basic step which can be achieved by adding different concentrations of yeast extract, peptone, KH_2PO_4, $MgSO_4.7H_2O$, $(NH_2)_2$ SO_4, and glucose and subjected to fermentation. Media is supplemented by the addition of organic and inorganic nitrogen source for better fermentation and high yield. FTIR is used to analyze the structure of pretreated biomass, while HPLC is used to determine the concentration of reducing sugars. Formation of ethanol and by-products such as acetic acid is analyzed by GC-MS method.

Keywords Potato peel waste · Bioethanol · Hydrolysis · Pretreatment · Optimization · Fermentation · Lignocellulosic biomass · FTIR · GC-MS · HPLC · Ethanol · Yeast

24.1 Introduction

Bioethanol production is the need of the society. In most developed countries, more than 90% of total energy comes from nonrenewable fuel resources, and countries all over the world import that fuel resources for the increasing need of energy (Duruyurek et al. 2015). This correlation causes energy crisis throughout the world because of their nonrenewability as predicted by the International Energy Agency: the consumption of energy from fossil fuels will double in the next 20 years. The drastic uses of fossil fuels have stimulated and led to economic development and

A. Kumar (✉) · R. Deb · J. Singh
Department of Bioengineering and Biosciences, Lovely Professional University, Phagwara, Punjab, India

P. Gehlot, J. Singh (eds.), *Fungi and their Role in Sustainable Development: Current Perspectives*, https://doi.org/10.1007/978-981-13-0393-7_24

have boosted the progress of human civilization since the twentieth century. Increasing consumption of fossil fuels has caused grave concern. This has led to increasing the demand for a power supply which has led to a brisk consumption of fossil fuels and resulted in global power cataclysm (Wang et al. 2016). The International Energy Agency has provided that the 80% of the primary sources of fuel are coal, oil, and natural gas. The global oil demand elevates by a range of 1.6% per year. Fossil fuel is limited and nonrenewable and poses a major challenge for the society of the twenty-first century as they are being consumed consequently. The production of oil is predicted to fall in the next 10–100 years (Sheikh et al. 2016). There is a high demand for contemporary and recyclable energy sources because the global energy crisis has led people to find new and sustainable sources. A growing concern about the price hikes and environmental problems has arisen due to the usage of petrol and diesel. So due to this, we have a necessity that we look for each and every probable source of energy. A lot of studies in the field of renewable energy are focused on various alternatives, but among these, biofuels produced by renewable plant products using fermentation technology is the most efficiently preferred. The popularity of nonconventional energy is increasing yearly, including bioethanol, hydroelectric power, firepower, wind energy, solar energy, biological hydrogen production, fuel cell, etc., which was the possible impact as discussed during the International Energy Meeting in 2005. Bioethanol is used as an alternative for gasoline in most of the European countries. Bioethanol amalgamates easily with natural gasoline which adds up to 10% resulting in demand of high quantities of bioethanol. Many countries all over the world have either developed or are developing process technologies in order to mix ethanol and gasoline. The worldwide demand to lessen oil importation has led to increased bioethanol productivity and consumption. It has led to improvised air quality of the world, and it has led to economic development of rural areas. Worldwide production of ethanol has progressed considerably to 51,000 million liters. The USA and Brazil are the key manufacturers of ethanol. Studies show that 73% of the ethanol on an average is produced throughout the world which is analogous to bioethanol, 17% ethanol to beverages, and 10% to industrially used ethanol (Arapoglou et al. 2010). For consumption of fuel from nonconventional resources, the EU directive (2003/30/EC) for bioethanol requires member states to acquire legislation. Two percent of the total fuel expenses covered the consumption in 2005. It has increased to 5.75% in 2010 and more (Berna 1998). The yearly bioethanol production was 2155 million liters in 2008 in EU (Arapoglou et al. 2010). Some of the nonconventional fuels, such as bioethanol, biodiesel, biohydrogen, etc., are extracted from a wide variety of biomass as sugarcane, corn, switch grass, algae, etc. which can substitute all fuels derived from petroleum. Sugarcane is the primary feedstock consumed in tropical countries such as Brazil and India. Corn has good starch content, and in countries like North America and Europe, it is used as biomass for ethanol production. Fermentation employed with technology can be used for rigorous ethanol fuel production in countries with rich agricultural economy. Feedstocks are 20–55% of the total production costs over estimated costs of 0.40/l ethanol produced. In recent years research has been conducted for better consumption of lignocellulosic biomass as primary feedstock.

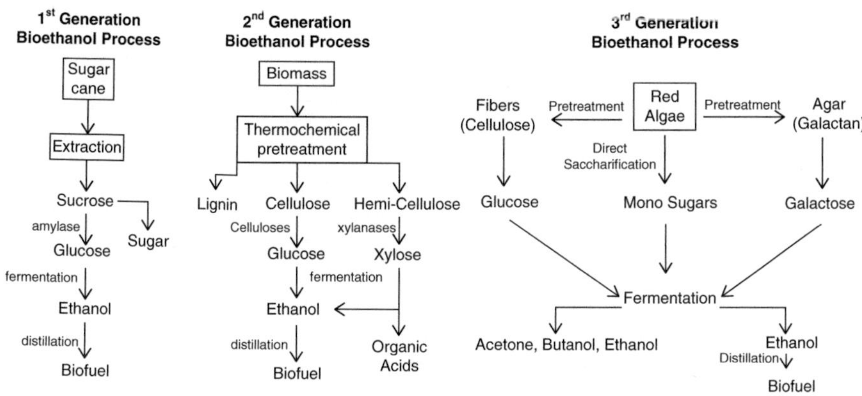

Fig. 24.1 Representation of method and steps involved in different generation of biofuels

Table 24.1 Illustration of the different generation of biofuel

First-generation biofuels	Second-generation biofuels	Third-generation biofuels	Fourth-generation biofuel
Substrate: Seeds, grains, or sugars	**Substrate:** Lignocellulosic biomass	**Substrate:** Algae, seaweeds	Genetically modified carbon-negative crops (biofuel from high solar efficiency cultivations)
Bioethanol or butanol by fermentation of starch (wheat, barley, corn, potato) or sugars (sugarcane, sugar beet, etc.)	Bioethanol or butanol by enzymatic hydrolysis	Biodiesel from algae	
Biodiesel by transesterification of plant oils (rapeseed, soybeans, sunflower, palm, coconut, Jatropha, used cooking oil, animal fats, etc.)	Methanol, Fischer-Tropsch gasoline and diesel, mixed alcohol, dimethyl ether, and green diesel by thermochemical processes	Bioethanol and biobutanol from algae and seaweeds	
	Biomethane by anaerobic digestion	Hydrogen from green algae and microbes	

Lignocellulosic biomass is vitally important because it contributes for bioenergy accountable for sustainable development (Sanchez and Cardona 2008).

On the basis of the feedstock used for the production of biofuel, there are four different generations of biofuels as shown in Fig. 24.1 and Table 24.1.

Conversion into secondary carriers of energy is feasible due to abundance of resources which can be utilized without any high-cost investment. Some of the rich biomass include agricultural residues, wood, solid waste, and medicated crops. Renewable resources such as biofuels have a wide scope of decreasing the fossil fuel burning and CO_2 production. Furthermore with the rapid increase of consumption of nonrenewable resources and rising environmental concern, the worldwide demand of bioethanol has risen. Several new technologies with pretreatment processes are improvised constantly, and better processes yielding high amount of ethanol are discovered. Issues like global warming and climatic changes have increased attention toward bioethanol production. Bioethanol is regarded to be the purest form of liquid fuel which can be a substitute to gasoline due to its economic, environmental, and strategic qualities. Agricultural and forest residues, solid municipal waste, and woody and herbaceous crops act as good cellulosic biomass for ethanol production. Bioethanol, produced from agricultural biomass, is pure burning, sustainable, carbon neutral, and nonconventional fuel (Song et al. 2013). Agriculture and energy have a good correlation over a long period of time. The primary source of bioenergy is agriculture. The agricultural waste comprises mainly of lignocellulosic biomass, which can range up to 50% of agricultural production, and they are considered as no cost product and accessible and are in abundance for production of bioethanol. Most of the plants store starch/cellulose, and they can be utilized for producing ethanol. Plant cell walls comprise primarily of glycoproteins, carbohydrate polymers, minerals, and lignin. The sugar sources for biofuels production are primarily obtained from carbohydrate components like hemicelluloses, cellulose, and pectin (Taylor 2008). Sugarcane, sugar beets, and maize (corn) are the common sources used for ethanol production. The sources from where ethanol can be produced are corn, sugar beet, sweet sorghum, and sweet potato or from the abundant cheap cellulosic feedstocks like wheat straw, switchgrass, wood, etc. known as cellulosic ethanol. In order to prevent the use of staple crops, we are looking for waste by-products from sugar industries like sugarcane, molasses, etc. which are considered as cheaper sources of ethanol. The worldwide production of bioethanol ranged to 92 billion liters in 2014. Large amounts of industrial wastes act as a key substrate for production of value-added products and help in cost reduction with both economic and environmental management. Properties like availability, abundance, and rich carbon and nitrogen content have driven focus toward bioethanol production from industrial wastes. Ethanol, biodiesel, and methanol are considered as substitute to fuels obtained from petroleum. The production of fuel ethanol throughout the world was estimated to be 91 billion liters in 2013. The primary generation crops as wheat, cassava, and sugarcane are responsible for present ethanol production which is commercialized worldwide with installation of 650 plants and a maximum storage of 100 billion liters. Corn-based ethanol is followed up by ethanol from sugarcane with production yield of 60 billion liters and 20 billion liters in 2012 correspondingly. Various concerns about the limited land use and environment are raised by the primary generation crop users for production of bioethanol. Second and third-generation feedstocks like lignocellulosic biomass and algae have been a good substitute but are still not cost-effective as first-generation crops. Hence, the agricultural wastes for ethanol production have been considered as primary source of carbon. Factors like abundance, availability, biodegradability, agro-industrial wastes, and high carbon and nitrogen content have driven focus on production of ethanol and also contributed to waste management both economically and environmentally.

Several industries reported production of bioethanol which included the waste materials of several fruit peels, potato peels, and wastes from olive oil industry. Production of ethanol from potatoes and potato peels has been studied consequently. Due to the abundance in availability of ethanol throughout the world, it has been used for fuel ethanol production (FEP). Worldwide potato production was reported to be about 325 million tons in 2007. This signifies that it can be produced at 1200–7200 million liters of fuel ethanol annually. According to Minal and Deshpande (2010), the second most consumed food is potato. Complex pretreatments are not required as potatoes have high starch content. Being a high-value crop, 5% to 20% end up as waste potato by-products from cultivation of potato which can be suitably utilized for bioethanol production. Eighteen percent of the potatoes are generated as biowaste in industries such as in potato chip industry during the processing of potato chips. Potato peel waste being a rich carbon source can be used in enriched growth media for fermentation and production of ethanol. Peels are basically removed during processing. Lignocellulosic biomass such as potato peel is produced as waste in potato processing industries in a high amount which can be used for ethanol production. The worldwide potato production was estimated to be 350 million tons in 2012. During the industrial processing of potatoes, approximately 40% of the potatoes are wasted, principally as a peel. Two million tons has been predicted to be produced by industries as potato peel per year. Studies have reported that the industrial potato waste including potato peels, mash, and pulp and potato processing wastewater has the potential for the production of bioethanol, lactic acid, glucoamylase, and pullulan. Moreover, PPW shows antioxidant and antimicrobial properties as it has been reported.

Sugarcane bagasse comprises 60% of the total bioethanol production. Industries are looking for a generation of ethanol from cheap and abundant agricultural feedstock for providing maximum supply. Potato peel is a zero-value by-product (Richelle et al. 2015). It contains a substantial quantity of cellulose, starch, hemicellulose, lignin, and fermentable sugars (Arapoglou et al. 2010). Factors like low-cost, high-abundance, and high-carbohydrate content correspond for the PPW as the ideal source for bioethanol production. Optimization has to be done during processing. The cellulose, hemicelluloses, and lignin comprise PPW which has to metabolically convert to fermentable sugar form by enzymatic hydrolysis (Zabed et al. 2016).

24.1.1 *Pretreatment of Biomass*

Renewable biomass such as potato peel waste (PPW) subjected to pretreatment converts the carbohydrate polymers into fermentable monomers. The cellulose in PPW is coated by hemicelluloses forming a cellulose-hemicellulose complex. It acts a chemical barrier by preventing access of enzymes in the complex naturally (Brodeur et al. 2011). Lignin encompasses the cellulose-hemicellulose complex and thus acts as a tangible barrier for the conversion of lignocellulosic biomass (LCB) into ethanol (Agustini et al. 2012).

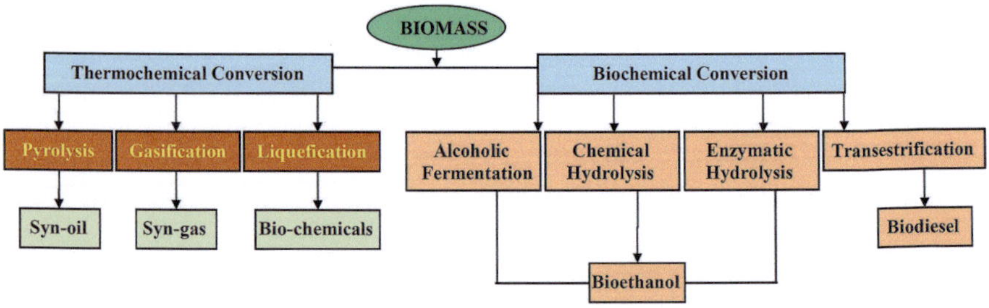

Fig. 24.2 Classification of biomass processing technologies

The pretreatment method is required to alter the lignocellulosic matrix so that enzymes can access the cellulose easily. Pretreatment is done to depolymerize lignin, increase the porosity of matrix, dissolve the hemicellulose, and reduce the crystallinity of cellulose to be degraded easily (Yuk et al. 2014) (Fig. 24.2).

24.1.2 Methods of Pretreatment

Pretreatment methods are generally identified as chemical, physical, biological, and physiochemical. The physical pretreatment works on the biomass by increasing the pore size and decreasing the degree of polymerization of cellulose matrix and hydrolysis of hemicelluloses and lignin. For chemical pretreatment, acids and primary alkalis can be used. It results in degradation of lignin, decreasing of the polymerization, and decrystallization of cellulose (Taha et al. 2016; Rao et al. 2016; Szczodrak and Fiedurek 1996). H_2SO_4, HCl, H_3PO_4, and HNO_3 can be used during acid pretreatment of biomass. NaOH is the most used alkali, while H_2SO_4 is mostly used for acids. Acid pretreatment is applied to dissolve the hemicellulosic fraction in the PPW and makes cellulose more accessible to enzymes. Organic acids like fumaric and maleic acid are used as alternate acids which enhance hydrolysis of celluloses and reduction in inhibitor production (Wyman et al. 2005).

Chemical solvents like alkaline H_2O_2, dioxane, ozone, glycerol, organosol vents, phenol, or ethylene glycol are capable of disrupting cellulose structure and promoting hydrolysis. Mineral acids (like H_2SO_4, HCl, hydrazine), ammonia-based solvents and aprotic solvents (DMSO), metal complexes (ferric sodium tartrate, cadoxen, cuoxan), and wet oxidation have also been reported for decrystallization of cellulose and its dissociation from lignin and hemicelluloses (Mosier et al. 2005).

Physicochemical pretreatment methods include a wide number of techniques like a steam explosion, ammonia fiber explosion (AFEX), soaking aqueous ammonia (SAA), wet oxidation, CO_2 explosion, etc. The advantages of these methods are the increase of the surface area accessibility of PPW for the enzyme, decrystallization of cellulose, and removal of the lignin and hemicelluloses. It is responsible for effecting of both physical and chemical conditions (Alvira et al. 2010; Mosier et al. 2005).

Biological pretreatment of PPW can be done using microorganisms like fungi which include brown rot, white rot, and soft rot fungi (Sarkar et al. 2012). The most efficient are the white rot fungi. Such kind of treatment disrupts the structure of lignin and thus separating it from the lignocellulosic matrix. White rot and soft rot fungi work both on cellulose and lignin, while brown rot degrades cellulose (Prasad et al. 2007). Forest agricultural residues and crops as lignocellulosic biomass offer vast benefits but are hindered by technological and economic hindrances (Amado et al. 2014). The fungi secrete enzymes to degrade lignin in biomass during the biological pretreatment process. The common disadvantages include lower rates of hydrolysis and longer pretreatment time compared to other methodologies. Currently, new biological pretreatments can be designed with other pretreatments and develop more sophisticated methods for instant hydrolysis (Brodeur et al. 2011).

24.1.3 *Physical Pretreatment Methods*

Mechanical communication is a pretreatment method of grinding, chipping, and milling of fine particle size of the material. Chipping converts into 10–30 mm particles, while milling and grinding convert into 0.2–0.4 mm. It is easy to handle and results in decrystallization of cellulose, depolymerization, and an increased surface area (Alvira et al. 2010; Elgharbawy and Moniruzzaman 2016). Irradiation is the technique of treatment of biomass with high-energy radiation including gamma beam rays, electron beam, ultrasound, pulsed electric field, UV, and microwave heating (Zheng et al. 2009).

24.1.4 *Chemical Treatment Methods*

It involves dilute acid treatments in which concentrations of acids are kept below 4%. It works at a high temperature of 180^0 C or 120^0 C with longer incubation time (20–30 min). Both organic (fumaric and maleic acids) and inorganic acids (HCl, H_2SO_4, H_3PO_4) can be used for pretreatment (Alvira et al. 2010; Mosier et al. 2005). Concentrated acid treatment exposed to higher concentrations of 70–77% can be done coupled with 40–100 ° C. Generally inorganic acids like H_2SO_4 and H_3PO_4 are used (Alvira et al. 2010).

Basic treatment involves bases such as NaOH, $Ca(OH)_2$, KOH, and NH_4OH. It is performed at room temperature, and hydrolysis time can range from several hours to days. It results in amorphous cellulose (Alvira et al. 2010; Mosier et al. 2005). Ozonolysis includes treatment by ozone gas usually at room temperatures. The reaction time can range to several hours. It helps in selective degradation of lignin with no effects on cellulose. Inhibitor formation is low and formation of furfural is negligible.

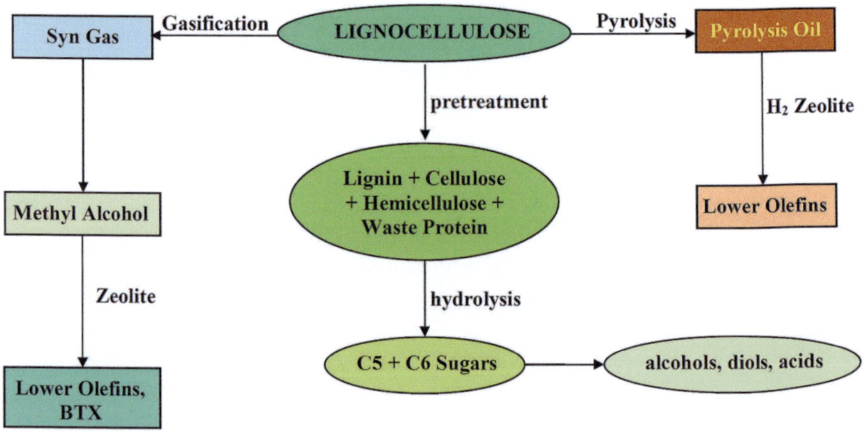

Fig. 24.3 Strategies for primary conversion of lignocellulose. (Modified from Sheldon 2014)

In ionic liquid pretreatment, ILs comprise large organic cations and inorganic anions. Ionic liquids remain in a liquid state at room temperature (<100 °C). Toxic gases are not formed. Carbohydrates and lignin are dissolved. Hydrolysis in a single step can be employed in ionic liquids that maintain cellulose stability and activity. It is done at lower temperatures (Elgharbawy and Moniruzzaman 2016; Paulov 2017).

Organo-solvents are a mixture of aqueous and organic solvents like methanol, ethanol, acetone, and ethylene glycol. The solvents are combined with acids like HCl, H_2SO_4, and oxaloacids for the breakdown of hemicellulose bonds (Alvira et al. 2010; Nahyun 2015). Figure 24.3 shows the strategies for conversion of lignocellulose into fermentable sugars.

24.1.5 Physicochemical Pretreatment Methods

Uncatalyzed steam explosion includes exposure to steam at T = 160–260 ° C and a pressure of 0.69–4.83 MPa. The reaction time can range from seconds to hours. No extra chemical is added (Paulov 2017). The acid-catalyzed steam explosion is done by addition of H_2SO_4 or HCl to steam explosion at a temperature range of 160–220 °C. It removes hemicelluloses. It has no effects on the environment and is a hazardless process (Mosier et al. 2005; Paulov 2017).

LHW (liquid hot water) involves treatment of biomass with rapid decompression. At elevated temperatures the pressure is changed to maintain water in a liquid state at higher temperatures (>160 °C). The pH is set in between 4 and 7 and the reaction time is 15 min. The main disadvantage involves high water and energy demand and hemicelluloses degradation (Mosier et al. 2005; Paulov 2017).

Ammonia fiber explosion (AFEX) involves treatment with anhydrous ammonia liquid at 60–120 °C and above 3 MPa for 30–60 min, followed by rapid decompression (Alvira et al. 2010; Mosier et al. 2005; Paulov 2017). Ammonia recycling percolation (ARP) includes the addition of aqueous ammonia at a concentration ranging from 5 to 15% which passes through a packaged reactor comprising of biomass at 140–210 °C for up to 90 min with the rate of percolation about 5 mL/min. Advantages include no inhibitor formation (Mosier et al. 2005; Paulov 2017).

Soaking aqueous ammonia (SAA) is a substitute to ARP and is done at a lower temperature (30–75 °C). An oxidative pretreatment of biomass is done using oxygen or air as a catalyst. The oxidation is performed for 10–15 min at 170–200 °C and 10–12 bars. Formation of low inhibitors and energy requirements are low. Solubilization of lignin and hemicelluloses is easily done (Alvira et al. 2010; Mosier et al. 2005; Paulov 2017).

In CO_2 explosion, CO_2 is used as a supercritical fluid for pretreatment of biomass, i.e., LCB. Main advantages involve effective lignin degradation, lower sugar degradation, increased digestibility, increased surface area, and exposure at a lower temperature (Alvira et al. 2010; Paulov 2017). Different lignocellulosic biomass has different physical and chemical properties; hence different pretreatment technologies coupled with multiple conditions are used according to the nature of the biomass. Certain factors like digestibility of cellulose, toxic compound generation, consumption of energy in the downstream processing, and treatment of biomass with water pay a huge role in the treatment process (Galbe and Zacchi 2007). Different pretreatment techniques like AFEX, wet oxidation, and water treatment such as of hot water are used for yielding good sugar content by pretreating agricultural residues, e.g., corn stover (Mosier et al. 2005). The woody biomass compared to LCB is tougher to hydrolyze, and H_2SO_4 or SO_2 are generally used for pretreating biomass, hence increasing the hydrolysis of hemicelluloses portion. Woody biomass seems to be efficiently affected by catalyzed steam pretreatment undertaken in rigorous work in pilot and demonstration plants in industries (Ropars et al. 1992; Zacchi 2006).

Optimization is the sole factor for high waste-high yield ratio. Methods like response surface methodology and artificial neural network are statistical methods that can be used for optimization and experiment design methodology. RSM is a statistical method which calculates the effect of multiple factors individually or in combination in a minimal number of experiments (Izmirlioglu and Demirci 2016).

Artificial neural networks (ANN) are an artificial intelligence technique commonly used for modeling and optimizing complex phenomena involving a large number of process variables (Sebayang et al. 2017). Survey of recent literature has shown that the commercial ethanol production from PPW is low in yield. However hydrothermal pretreatment has shown considerable yield.

24.1.6 *Pyrolysis*

It is one of the most promising thermochemical processes for the conversion of biomass into gases, liquids, vapors, char, and ash. It is an endothermic reaction that involves the thermal cracking or decomposition of biomass at an elevated temperature (220–315 °C for hemicellulose, 315–400 °C for cellulose, 160–900 °C for lignin) in the absence of oxygen (Yang et al. 2007).

Pyrolytic reaction using cellulose

$$3C_6H_{10}O_5 \rightarrow C_3H_8O + 8H_2O + CH_4 + 2CO + 2CO_2 + 7C$$

where C_3H_8O + $8H_2O$ is the liquid bio-oil; $8H_2O$ is the water pyrolysis; $CH_4 + 2CO + 2CO_2$ is the combustible gas; and 7C is the biochar.

There are two different modes of pyrolysis, but the major focus nowadays is on fast pyrolysis because it yields liquid products that can be stored and transported easily according to the requirement. In order to maximize the liquid yield during fast pyrolysis, there are three main variables that need to be focused more (Bridgwater et al. 1999):

(i) Reaction temperature.
(ii) Biomass heating rate.
(iii) Vapor residence time.

Generally moderate temperature of approximately 500 °C and short residence time are required for high yield. Under these conditions the yield of liquid bio-oil is 75%, whereas 13 and 12% of combustible gas and biochar, respectively, are obtained (Bridgwater et al. 1999). Whereas in case of slow pyrolysis in which conditions of low temperature and long residence time are maintained, the yield is 305, 35% of liquid bio-oil, combustible gas, and biochar, respectively, are obtained.

24.1.7 *Gasification*

It is the thermal conversion of biomass by partial oxidation at elevated temperatures to produce permanent gases (carbon monoxide, carbon dioxide, hydrogen, and methane), char, and water, condensable as minor products. The low-energy density of the gases and poorer qualities in terms of heating value (4–7 MJ m^{-3}), produced during gasification process, is suitable for turbine and engine operations but not for pipeline transportation, whereas if oxygen is used, then a better quality gas can be obtained which have higher heating value (10–18 MJm^{-3}) (Bridgwater et al. 1995).

Gasification occurs in three sequential steps that include drying, pyrolysis, and gasification. Firstly, the moisture content is removed by drying, and then pyrolysis occurs during which feedstock is heated at 300–500 °C in the absence of oxygen which gives rise to liquids, tar, gases, and solid char residues followed by gasification which includes the partial oxidation of pyrolysis gases, tar, and solid char in the presence of air to produce permanent gases (carbon monoxide, carbon dioxide, and hydrogen). There are various factors that affect the gas composition produced during gasification such as feed composition (industrial waste, agricultural residues, energy crops, food processing industry), water content, temperature, and the type of

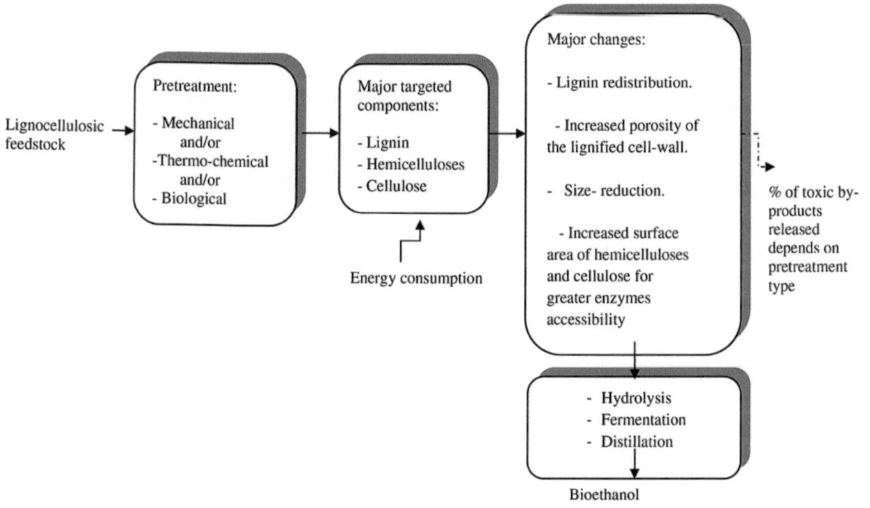

Fig. 24.4 Pretreatment upstream process (Limayem and Ricke 2012)

pyrolysis product used for gasification (Bridgwateret al. 1995). Pretreatment upstream process has been shown in Fig. 24.4.

Pretreatment Kinetics

The kinetic model for cellulose hydrolysis is suggested by Saeman (1945) and later on verified by several researchers (Jacobsen and Wyman 2000; Bhandari et al. 1984). It follows the first-order reaction kinetics:

$$\text{Cellulose} \xrightarrow{k_1} \text{Glucose} \xrightarrow{k_2} \text{Degradation products}$$

$$d[C]/dt = -k_1 \times [C]$$

$$d[G]/dt = k_2 \times [G] - k_1 \times [C]$$

where $d[C]/dt$ represents the reaction rate of cellulose, $d[G]/dt$ is the reaction rate of glucose, C is the mass of cellulose, G is the mass of glucose, and k_1 and k_2 are reaction rate constants for reaction of cellulose and glucose, respectively.

The Arrhenius form of the rate constants in these equations is as follows:

$$\mathrm{k} = \mathrm{A}\exp^{(-\mathrm{E}/\mathrm{RT})} \tag{24.1}$$

in which A_0 is a constant, C is the acid concentration in weight percent, m has a value close to 1.0, E is the activation energy, R is the universal gas constant, and T is the absolute temperature.

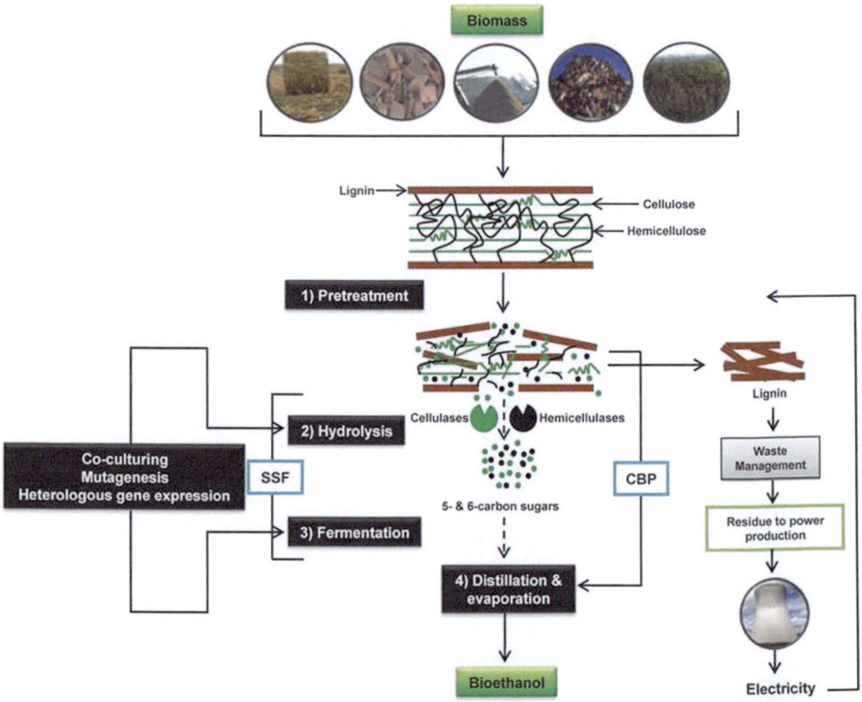

Fig. 24.5 Schematic picture for the conversion of lignocellulosic biomass to ethanol, including the major steps. Simultaneous saccharification and fermentation (SSF) in consolidated bioprocessing (CBP), all bioconversion steps are minimized to one step in a single reactor using one or more microorganisms. (Source: Dashtban et al. 2009)

The most common description of cellulase adsorption is the Langmuir isotherm (Zhang and Lynd 2004). The Langmuir isotherm may be represented as:

$$E_{\alpha} = \frac{W_{\max} K_P E_f}{1 + K_P E_f} \tag{24.2}$$

where Eα is adsorbed cellulase (mg or μmol cellulase/L), W_{max} is the maximum cellulase adsorption = A_{max}*S (mg or μmol cellulase/L), Amax is the maximum cellulase adsorption per g cellulose (mg or μmol cellulase/g cellulose), S is cellulose concentration (g cellulose/L), E_f is free cellulase (mg or μmol cellulase/L), and K_P is the dissociation constant in terms of L/g cellulose.

Schematic pictorial representation for the conversion of lignocellulosic biomass to ethanol is shown in the Fig. 24.5.

24.2 By–Product Formation

Pretreatment is an important step for converting renewable biomass such as LCB into ethanol, but due to this process, lots of lignocellulosic by-products are formed. It acts as inhibitors during fermentation process if the yield of by-products is high (Cavka and Jönsson 2013). The by-products that may be produced are acetic acids, formic acids, sugar acids, furfural and levulinic acids, and hydroxymethylfurfural (HMF) (Jönsson et al. 2013). However, pre-treatment inhibitors can be categorized majorly into three groups, such as furfural, phenols and aliphatic acids. Certain processes like biological, physical, and physiochemical methods are used for detoxification. The type and quantity of inhibitors depend on the pretreatment method used and according to the detoxification method used. When strongly inhibiting hydrolysates are fermented, inhibitors accumulate in the fermentation broth due to recirculation of streams or when the microorganism is susceptible to low inhibitor concentration. The process of detoxification should be cheap and easy to operate in the process (Palmqvist 2000). Roadmap for conversion of lignocellulosic biomass to fuels and chemicals is shown in the Fig. 24.6, and production of bioethanol from different biomass is represented by Table 24.2.

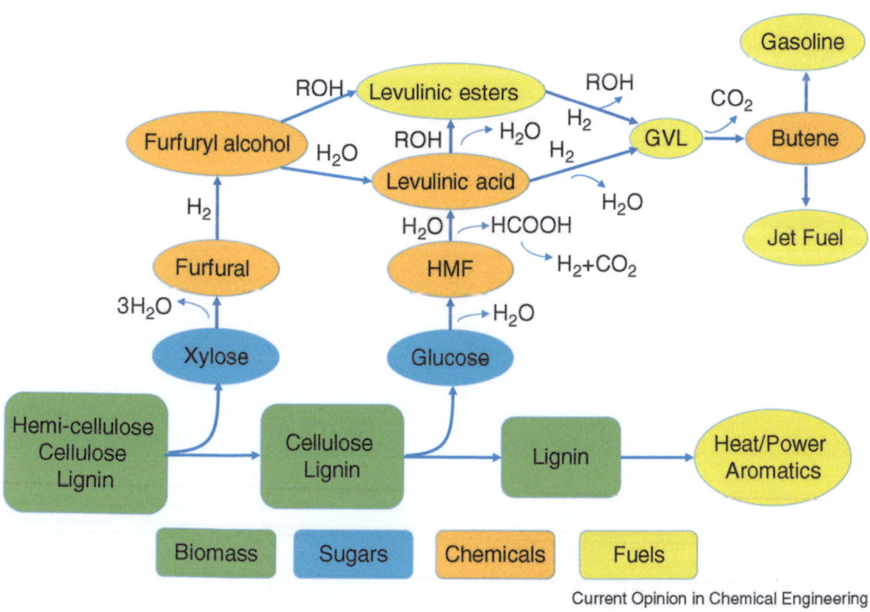

Fig. 24.6 Roadmap for conversion of lignocellulosic biomass (green) to fuels (yellow) and chemicals (orange), passing through the intermediate formation of furfural and levulinic acid from C5 and C6 sugars (blue) (Wettstein et al. 2012)

Table 24.2 Production of bioethanol from different biomass

Strain species	Temperature (°C)	pH value	Carbon source and concentration (g/L)	Nitrogen source and concentration (g/L)	Incubation time (h)	Concentration of ethanol produced (g/L)	References
Saccharomyces cerevisiae 424A(LNH-ST)	30	5.5	Corn Stover (this CS contains 34.1% cellulose, 20.4% xylan, 3.3% arabinan, and 2.3% protein on a dry weight basis)	Complex media. YEP (5 g/L yeast extract and 10 g/L Pepton)	72	wa0.51 g EtOH/g sugar	Lau and Dale. (2009)
Saccharomyces cerevisiae (ATCC 24859)	30 °C	5.5	Waste potato mash	Poultry meal	48	30.99	Izmirlioglu and Demirci (2012)
Saccharomyces cerevisiae D5A.	38 C	4.8	Rice straw	Aqueous Ammonia	120	12.7 g of ethanol/L	Ko et al. 2009
Saccharomyces cerevisiae	35	6	Rice straw	Aqueous Ammonia	72	7.61 g/L ethanol	El-Zawawy et al. 2011
Saccharomyces cerevisiae var. bayanus	32	5	Potato peel waste (PPW)	Aqueous Ammonia	48	7.6 g/L	Arapoglou et al. (2010)
Yeast *Kluyveromyces marxianus* K21	40 °C	4.2–4.5	Taro waste	Bone meal (BM), chicken meal (CM), fish meal (FM), soy meal (SM), corn gluten meal (CGM), and dried distillers grains with soluble (DDGS)	22	(48.98 g/L)	Wu et al. (2016)
Saccharomyces cerevisiae MP 3013	30	5.5	Corn cob	$(NH_4)_2SO_4$	72	24 g/L	Cai et al. (2016)

Saccharomyces cerevisiae 3013	30	6	Corn stalk	$(NH_4)_2SO_4$	72	22.6 g/L	Li et al. (2016)
Scheffersomyces stipitis NRRL Y-7124	30	6.5	Sugarcane bagasse	$(NH_4)_2SO_4$	72	7.34 g/L	Dussán et al. (2016)
Scheffersomyces shehatae UFMG HM 52.2.	30	6.5	Sugarcane bagasse	$(NH_4)_2SO_4$	72	18 g/L	Dussán et al. (2016)
Scheffersomyces stipitis CBS5774	37	4.5	Sweet sorghum	$(NH_4)_2SO_4$	21	84 g/L	Barcelos et al. (2016)

24.3 Role of Microorganisms

Various microorganisms such as bacteria, yeast, and fungi are used for the major conversion steps (detoxification, hydrolysis, and fermentation) during ethanol production. For the production of ethanol, microorganism such as bacteria, yeast, and fungi can be used from various sources of biomass. In this experiment yeast (*Saccharomyces cerevisiae*) was used during the course of the project for the fermentation of glucose to ethanol.

In pretreatment, microorganisms play a major role during pretreatment of biomass. To delignify the lignocellulosic network, microorganisms are used which partially release cellulose and hemicelluloses from the matrix. The microorganism plays a major role in hydrolysis. Carbohydrate polymers such as cellulose can be both enzymatically and chemically hydrolyzed, but microorganisms are widely used for this purpose for their cheap availability, requiring less energy, and mild environment compared to acid hydrolysis. Some of the bacteria used for hydrolysis are *Bacillus*, *Cellulomonas*, and *Thermomonospora* and fungal genera include *Aspergillus* and *Schizophyllum*. In the fermentation process, microorganisms have played a vital role; many yeast and bacteria can ferment soluble sugars in an anaerobic condition and produce ethanol. Filamentous fungi also produce ethanol but at low rates.

Many microorganisms can ferment sugar and produce ethanol, but their efficiency and feasibility are not equal. Only a few microorganisms are used for large-scale production of ethanol, and the widely used microorganisms for ethanol fermentation is yeast, that is, *Saccharomyces cerevisiae*, and it is suited for fermentation of lignocellulosic biomass and effectively breaks six carbon sugars, but due to the lack of enzymes which can convert xylose to xylulose fermentation of pentose, it is hardly done by this yeast. Gram-negative bacteria are commonly used for ethanol production. There are many factors which are responsible for the performance of any microorganisms as ethanol fermenter. Factors responsible for microorganism performance are osmotic tolerance, pH range, temperature range, growth rate, ethanol tolerance, and inhibitor. Some ethanologenic microorganism characteristics are high ethanol yielding, high ethanol productivity, and good growth in simple and inexpensive media.

Bioethanol Production Kinetics

Ethanol production from the lignocellulosic biomass follows the Luedeking-Piret model (Krishnan et al. 1999). Most of the studies for ethanol production were carried out using free cells of *Saccharomyces cerevisiae* in batch and fed batch processes (Laopaiboon et al. 2007). The ethanol yield (Y_{ps}) was calculated as the actual ethanol produced and expressed as g ethanol per g sugar utilized (g g^{-1}).

$$\frac{dx}{dt} = \mu x \tag{24.3}$$

$$\mu = (\mu_{\max} s) / (K_s + s) \tag{24.4}$$

$$\frac{dp}{dt} = \alpha \frac{dx}{dt} + \beta x \tag{24.5}$$

$$\frac{ds}{dt} = -(\mu_{\max} / Y_{XS} + q_p / Y_{PS} + m_s) x \tag{24.6}$$

where μ = *specific growth rate* h^{-1}; K_s, Monod constant for growth on glucose or xylose (g/L); S, substrate concentration (g/L); x, cell dry weight (g/L); μ_m, maximum specific growth rate (/h); P, ethanol concentration (g/L); q_p, specific rate of product formation; Y_{xs}, true biomass yield from substrate; Y_{ps}, true product yield from substrate; and α and β are constant.

$$\frac{dV}{dt} = F \tag{24.7}$$

$$\frac{d(xV)}{dt} = Fx_i + \mu xV - k_d xV \tag{24.8}$$

$$\frac{F}{V} = D \tag{24.9}$$

$$\frac{dx}{dt} = x(\mu - D) \tag{24.10}$$

$$\frac{ds}{dt} = D(s_i - s) - \left(\frac{\mu}{Y_{XS}} + \frac{q_P}{Y_{PS}} + m_s \right) x \tag{24.11}$$

where m_s is the maintenance coefficient (h^{-1}); *F*, volumetric flow rate of entering feed; *V*, liquid volume; *D*, dilution rate; and k_d, specific death rate constant.

Bioethanol Recovery

Once ethanol has been produced by batch, fed batch, or CSTR mode, it must be recovered because higher concentration of ethanol is toxic for the cell. For the recovery of the ethanol from the fermentation broth, different kinds of unit operations such as distillation, gas stripping, liquid-liquid extraction, and pervaporation can be used to get rid of product inhibition. Table 24.3 shows different unit operations used for bioethanol recovery.

Table 24.3 Bioethanol recovery

Methods	Principle	Advantage	Disadvantage
Distillation	Boiling occurs when the vapor pressure of a liquid exceeds the ambient pressure	Traditional method	Expensive to perform
Gas stripping	Heating of effluent, purging with gas, condensation of solvent/water vapors	Simple to perform, low chance of clogging or fouling	Low selectivity, no complete removal of solvents, more energy required compared to membrane-based processes
Liquid-liquid extraction	Contact of water – Immiscible solvent with fermentation broth, recovery of acetone/butanol / isopropanol by distillation	High capacity, high selectivity, low chance of clogging or fouling	Expensive to perform, possible formation of emulsions
Pervaporation	Selective diffusion of solvents across a nonporous membrane, recovery of evaporated vapors by applying vacuum or sweep gas	High selectivity compared to membrane evaporation, simple to perform	Lower membrane flux compared to membrane evaporation, possible clogging and fouling

24.4 Metabolic Engineering of Yeast

Ethanol fermentation is normally carried out at 30 °C and at pH 5.0. To facilitate the uptake of hemicellulosic sugar for the ethanol production by yeast, metabolic engineering is one of the vital tools (Zaldivar et al. 2001; Van Vleet and Jeffries 2009) as shown in Fig. 24.7. Yeast takes up xylose poorly. So to enable this organism for the utilization of pentose sugar, three main strategies has been approached:

(i) Insertion of bacterial isomerase gene from *E. coli*, *Bacillus subtilis*, or *Thermus thermophilus*.
(ii) Insertion of pentose utilization gene XR (xylose reductase) and XDH (xylitol dehydrogenase) from *Pichia stipitis*.
(iii) Improvement of xylulose consumption can be demonstrated by insertion of XKS1 gene encoding xylulokinase from *S. cerevisiae* and heterologous gene XR and XDH from *P. stipites* which resulted in hybrid yeast capable to grow on xylose through the PPP (pentose phosphate pathway).

Mutant yeast requires tolerating high concentration of both glucose and ethanol for a successful fermentation which can be achieved through metabolic engineering (Alper et al. 2006). Thus metabolic engineering enables *S. cerevisiae* to ferment xylose and arabinose to ethanol (Jeffries and Jin 2004).

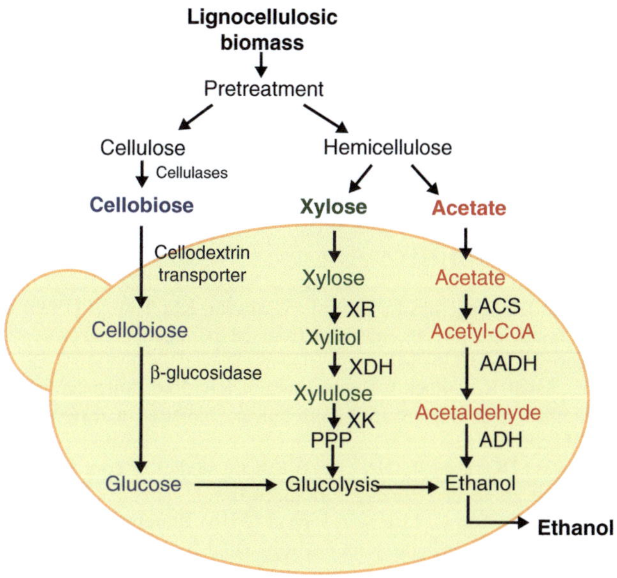

Fig. 24.7 Schematic overview of biofuel production through co-utilization of cellobiose, xylose, and acetic acid from lignocellulosic biomass by engineered yeast. *XR* xylose reductase, *XDH* xylitol dehydrogenase, *XK* xylulose kinase, *ACS* acetyl-CoA synthetase, *AADH* acetylating acetaldehyde dehydrogenase, *ADH* alcohol dehydrogenase (Wei et al. 2015)

24.5 Conclusions

Unicellular fungus such as yeast has the potential to produce bioethanol and fine chemicals by utilization of lignocellulosic biomass such as residue of agriculture and forestry. Pretreated lignocellulosic biomass produces cellulose and hemicellulose which are hydrolyzed by enzyme to produce monomeric sugars. Thus ethanol is produced by the fermentation of sugars by yeast mainly *S. cerevisiae* strains. The metabolic engineering of yeast strains is promising strategies for improved ethanol production.

References

Agustini L, Efiyanti L, Faulina SA, Santoso E (2012) Isolation and characterization of cellulase- and xylanase- producing microbes isolated from Tropical Forests in Java and Sumatra, 3(3):154–167

Alper H, Moxley J, Nevoigt E, Fink GR, Stephanopoulos G (2006) Engineering yeast transcription machinery for improved ethanol tolerance and production. Science 314(5805):1565–1568

Alvira P, Ballesteros M, Negro MJ (2010) Bioresource technology pretreatment technologies for an efficient bioethanol production process based on enzymatic hydrolysis: a review. Bioresour Technol 101(13):4851–4861. https://doi.org/10.1016/j.biortech.2009.11.093

Amado IR, Franco D, Sánchez M, Zapata C, Vázquez JA (2014) Optimisation of antioxidant extraction from Solanum tuberosum potato peel waste by surface response methodology. Food Chem 165:290–299. https://doi.org/10.1016/j.foodchem.2014.05.103

Arapoglou D, Varzakas T, Vlyssides A, Israilides C (2010) Ethanol production from potato peel waste (PPW). Waste Manag 30(10):1898–1902. https://doi.org/10.1016/j.wasman.2010.04.017

Barcelos CA, Maeda RN, Santa Anna LMM, Pereira N (2016) Sweet sorghum as a whole-crop feedstock for ethanol production. Biomass Bioenergy 94:46–56

Berna G (1998) Integrated biomass system. Office for Official Publications of the EC, Luxembourg, p 27

Bridgwater AV, Meier D, Radlein D (1999) An overview of fast pyrolysis of biomass. Org Geochem 30(12):1479–1493

Bridgwater AV, Elliott DC, Fagernäs L, Gifford JS, Mackie KL, Toft AJ (1995) The nature and control of solid, liquid and gaseous emissions from the thermochemical processing of biomass. Biomass Bioenergy 9(1–5):325–341

Brodeur G, Yau E, Badal K, Collier J, Ramachandran KB, Ramakrishnan S (2011) Chemical and physicochemical pretreatment of lignocellulosic biomass: a review. 2011. https://doi.org/10.4061/2011/787532

Bhandari N, Macdonald DG, Bakhshi NN (1984) Kinetic studies of corn stover saccharification using sulphuric acid. Biotechnol Bioeng 26(4):320–327

Cai D, Dong Z, Wang Y, Chen C, Li P, Qin P et al (2016) Biorefinery of corn cob for microbial lipid and bio-ethanol production: an environmental friendly process. Bioresour Technol 211:677–684

Cavka A, Jönsson LJ (2013) Bioresource technology detoxification of lignocellulosic hydrolysates using sodium borohydride. Bioresour Technol 136:368–376. https://doi.org/10.1016/j.biortech.2013.03.014

Dashtban M, Schraft H, Qin W (2009) Fungal bioconversion of lignocellulosic residues; opportunities and perspectives. Int J Biol Sci 5(6):578

Duruyurek M, Dusgun C, Gulhan MF, Selamoglu Z (2015) Production of bioethanol from waste potato. Turk J Agric Food Sci Technol 3(5):331–334

Dussán KJ, Silva DD, Perez VH, da Silva SS (2016) Evaluation of oxygen availability on ethanol production from sugarcane bagasse hydrolysate in a batch bioreactor using two strains of xylose-fermenting yeast. Renew Energy 87:703–710

Elgharbawy AA, Moniruzzaman M (2016) Ionic liquid pretreatment as emerging approaches for enhanced enzymatic hydrolysis of lignocellulosic biomass. Biochem Eng J 109(May):252–267. https://doi.org/10.1016/j.bej.2016.01.021

El-Zawawy WK, Ibrahim MM, Abdel-Fattah YR, Soliman NA, Mahmoud MM (2011) Acid and enzyme hydrolysis to convert pretreated lignocellulosic materials into glucose for ethanol production. Carbohydr Polym 84(3):865–871

Galbe M, Zacchi G (2007) Pretreatment of lignocellulosic materials for efficient bioethanol production, (July), pp 41–65

Izmirlioglu G, Demirci A (2016) Ethanol production in biofilm reactors from potato waste hydrolysate and optimization of growth parameters for Saccharomyces cerevisiae. Fuel 181:643–651. https://doi.org/10.1016/j.fuel.2016.05.047

Jacobsen SE, Wyman CE (2000) Cellulose and hemicellulose hydrolysis models for application to current and novel pretreatment processes. In: Twenty-first symposium on biotechnology for fuels and chemicals (pp. 81–96). Humana Press, Totowa, NJ

Jeffries TW, Jin YS (2004) Metabolic engineering for improved fermentation of pentoses by yeasts. Appl Microbiol Biotechnol 63(5):495–509

Jönsson LJ, Alriksson B, Nilvebrant N (2013) Bioconversion of lignocellulose: inhibitors and detoxification. Biotechnol Biofuels 6:1–10

Ko JK, Bak JS, Jung MW, Lee HJ, Choi IG, Kim TH, Kim KH (2009) Ethanol production from rice straw using optimized aqueous-ammonia soaking pretreatment and simultaneous saccharification and fermentation processes. Bioresour Technol 100(19):4374–4380

Krishnan MS, Ho NW, Tsao GT (1999) Fermentation kinetics of ethanol production from glucose and xylose by recombinant saccharomyces 1400 (pLNH33). Appl Biochem Biotechnol 78(1–3):373–388

Laopaiboon L, Thanonkeo P, Jaisil P, Laopaiboon P (2007) Ethanol production from sweet sorghum juice in batch and fed-batch fermentations by Saccharomyces cerevisiae. World J Microbiol Biotechnol 23(10):1497–1501

Lau MW, Dale BE (2009) Cellulosic ethanol production from AFEX-treated corn Stover using Saccharomyces cerevisiae 424A (LNH-ST). Proc Natl Acad Sci 106(5):1368–1373

Li P, Cai D, Luo Z, Qin P, Chen C, Wang Y et al (2016) Effect of acid pretreatment on different parts of corn stalk for second generation ethanol production. Bioresour Technol 206:86–92

Limayem A, Ricke SC (2012) Lignocellulosic biomass for bioethanol production: current perspectives, potential issues and future prospects. Prog Energy Combust Sci 38(4):449–467

Minal D, Deshpande RS (2010) Biofuel from cellulosic agricultural waste. Int J Chem Eng Res 2(2):197–217

Mosier N, Wyman C, Dale B, Elander R, Lee YY, Holtzapple M, Ladisch M (2005) Features of promising technologies for pretreatment of lignocellulosic biomass. Bioresour Technol 96:673–686. https://doi.org/10.1016/j.biortech.2004.06.025

Nahyun P (2015) Organosolv pretreatment with various catalysts for enhancing enzymatic hydrolysis of pitch pine (Pinus rigida). Bioresour Technol 101(18):7046–7053. https://doi.org/10.1016/j.biortech.2010.04.020

Palmqvist E (2000) Fermentation of lignocellulosic hydrolysates . II: inhibitors and mechanisms of inhibition. Bioresour Technol 74:25–33

Paulov L (2017) Advanced fermentation processes, (April). https://doi.org/10.1201/b15426-6

Prasad S, Singh A, Joshi HC (2007) Ethanol as an alternative fuel from agricultural, industrial and urban residues. Resour Conser Recycl 50:1–39. https://doi.org/10.1016/j.resconrec.2006.05.007

Rao LV, Goli JK, Gentela J, Koti S (2016) Bioconversion of lignocellulosic biomass to xylitol: an overview. Bioresour Technol 213:299. https://doi.org/10.1016/j.biortech.2016.04.092

Richelle A, Ben Tahar I, Hassouna M, Bogaerts P (2015) Macroscopic modeling of bioethanol production from potato peel wastes in batch cultures supplemented with inorganic nitrogen. Bioprocess Biosyst Eng 38(9):1819–1833. https://doi.org/10.1007/s00449-015-1423-6

Ropars M, Marchal R, Pourquid J, Vandecasteele R (1992) Large-scale enzymatic hydrolysis of agricultural lignocellulosic biomass . Part 1 : pretreatment procedures. Bioresour Technol 42:197–204

Saeman JF (1945) Kinetics of wood saccharification. Hydrolysis ofcellulose and decomposition of sugars in dilute acid at high temperature. Ind Eng Chem 37:43–52

Sanchez OJ, Cardona CA (2008) Trends in biotechnological production of fuel ethanol from different feedstocks. Bioresour Technol 99(13):5270–5295

Sarkar N, Ghosh SK, Bannerjee S, Aikat K (2012) Bioethanol production from agricultural wastes: an overview. Renew Energy 37(1):19–27. https://doi.org/10.1016/j.renene.2011.06.045

Sebayang AH, Masjuki HH, Ong HC, Dharma S, Silitonga AS, Kusumo F, Milano J (2017) Optimization of bioethanol production from sorghum grains using artificial neural networks integrated with an ant colony. Ind Crop Prod 97:146–155. https://doi.org/10.1016/j.indcrop.2016.11.064

Sheikh RA, Al-Bar OA, Soliman YMA (2016) Biochemical studies on the production of biofuel (bioethanol) from potato peel wastes by Saccharomyces cerevisiae: effects of fermentation periods and nitrogen source concentration. Biotechnol Biotechnol Equip 30(3):497–505. https://doi.org/10.1080/13102818.2016.1159527

Sheldon RA (2014) Green and sustainable manufacture of chemicals from biomass: state of the art. Green Chem 16(3):950–963

Song L, Yu H, Ma F, Zhang X (2013) Biological pretreatment under non-sterile conditions for enzymatic hydrolysis of corn stover. BioResources 8(3):3802–3816

Szczodrak J, Fiedurek JAN (1996) Technology for conversion of lignocellulosic biomass to ethanol. Biomass Bioenergy 10(95):367–375

Taha M, Foda M, Shahsavari E, Aburto-medina A, Adetutu E, Ball A (2016) ScienceDirect commercial feasibility of lignocellulose biodegradation : possibilities and challenges. Curr Opin Biotechnol 38:190–197. https://doi.org/10.1016/j.copbio.2016.02.012

Taylor G (2008) Biofuels and the biorefinery concept. Energy Policy 36(12):4406–4409

Van Vleet JH, Jeffries TW (2009) Yeast metabolic engineering for hemicellulosic ethanol production. Curr Opin Biotechnol 20(3):300–306

Wang F, Jiang Y, Guo W, Niu K, Zhang R, Hou S et al (2016) An environmentally friendly and productive process for bioethanol production from potato waste. Biotechnol Biofuels 9(1):50. https://doi.org/10.1186/s13068-016-0464-7

Wei N, Oh EJ, Million G, Cate JH, Jin YS (2015) Simultaneous utilization of cellobiose, xylose, and acetic acid from lignocellulosic biomass for biofuel production by an engineered yeast platform. ACS Synth Biol 4(6):707–713

Wettstein SG, Alonso DM, Gürbüz EI, Dumesic JA (2012) A roadmap for conversion of lignocellulosic biomass to chemicals and fuels. Curr Opin Chem Eng 1(3):218–224

Wu WH, Hung WC, Lo KY, Chen YH, Wan HP, Cheng KC (2016) Bioethanol production from taro waste using thermo-tolerant yeast Kluyveromyces marxianus K21. Bioresour Technol 201:27–32

Wyman CE, Dale BE, Elander RT, Holtzapple M, Ladisch MR, Lee YY (2005) Coordinated development of leading biomass pretreatment technologies. Bioresour Technol 96:1959–1966. https://doi.org/10.1016/j.biortech.2005.01.010

Yang H, Yan R, Chen H, Lee DH, Zheng C (2007) Characteristics of hemicellulose, cellulose and lignin pyrolysis. Fuel 86(12):1781–1788

Yuk IH, Zhang JD, Ebeling M, Berrera M, Gomez N, Werz S, Szperalski B (2014) Effects of copper on cho cells: insights from gene expression analyses. Biotechnology:1–34. https://doi.org/10.1002/btpr

Zabed H, Sahu JN, Boyce AN, Faruq G (2016) Fuel ethanol production from lignocellulosic biomass: an overview on feedstocks and technological approaches. Renew Sust Energ Rev 66:751–774. https://doi.org/10.1016/j.rser.2016.08.038

Zacchi G (2006) Bio-ethanol – the fuel of tomorrow from the residues of today. Trends Biotechnol 24(12). https://doi.org/10.1016/j.tibtech.2006.10.004

Zaldivar J, Nielsen J, Olsson L (2001) Fuel ethanol production from lignocellulose: a challenge for metabolic engineering and process integration. Appl Microbiol Biotechnol 56(1):17–34

Zhang YHP, Lynd LR (2004) Toward an aggregated understanding of enzymatic hydrolysis of cellulose: noncomplexed cellulase systems. Biotechnol Bioeng 88(7):797–824

Zheng Y, Pan Z, Zhang R (2009) Overview of biomass pretreatment for cellulosic ethanol production. Int J Agric Biol Eng 2(3):51–68. https://doi.org/10.3965/j.issn.1934-6344.2009.03.051-068

Chapter 25
Fungi Inhabiting in Hypersaline Conditions: An Insight

Abhishek Vashishtha and Gautam Kumar Meghwanshi

Abstract Until the last decade of the last century, it was believed that the contamination of food stuff preserved with salt is the only source of the halotolerant and halophilic fungi. After this era, it got established that many hypersaline lakes, solar salterns, etc. are natural habitats of such fungi. Many phylogenetically unrelated fungi were reported to grow in salty water with more than 30% NaCl concentration. A huge diversity of fungi from all the three major groups, viz., halophilic and halotolerant, xerotolerant, and sporadic, have been isolated all over the globe. Different species of *Cladosporium*, *Aspergillus*, *Penicillium*, *Emericella*, and *Eurotium* are some representative types of black yeast-like melanized fungi found in hypersaline conditions.

Many indigenous species like *Aureobasidium pullulans*, *Debaryomyces hansenii*, *Hortaea werneckii*, and *Wallemia ichthyophaga* have also been isolated universally from the natural hypersaline environments. *D. hansenii*, *H. werneckii*, *and W. ichthyophaga* have been considered as model organisms for learning eukaryotic halotolerance with *Hortaea werneckii* being the best studied. *D. hansenii* is ubiquitous fungi found in the oceans globally. *W. ichthyophaga* was reported in 2005 only and is most halophilic fungi described to date. For proper survival and to combat the ill effects of salinity, fungi inhabiting in hypersaline conditions employ various strategies like ion exchange, accumulation of compatible osmolytes, alterations in membrane fluidity, and many more. Their special characteristics make them a potent organism with ample biotechnological potential.

Keywords *Cladosporium* · *Debaryomyces hansenii* · *Hortaea* · *Hypersalinity* · *Osmolytes* · *Penicillium* · *Wallemia*

A. Vashishtha (✉) · G. K. Meghwanshi
Department of Microbiology, M.G.S. University, Bikaner, India

P. Gehlot, J. Singh (eds.), *Fungi and their Role in Sustainable Development: Current Perspectives*, https://doi.org/10.1007/978-981-13-0393-7_25

25.1 Introduction

Hypersaline environments, due to their high salinity, exposure to high and low temperature, low oxygen conditions, and in some cases high pH values, constitute the typical example of an extremophilic environment. From the historical times, saturated solution (crystalline) of sodium chloride (NaCl) is believed to be hostile to most forms of life and that is why it had been used for centuries as a food preservative. But now hypersaline water bodies, food preserved with the high concentration of salt, and solar salterns are known to harbor a diverse group of halophilic organisms. Halophiles are a group of microorganisms that live in saline environments and in many cases require a high concentration of various salts to survive (Gunde-Cimerman et al. 2009). Halophilic organisms are found in all the three domains of life, *Archaea*, *Bacteria*, and *Eukarya*, and over a wide range of extreme conditions (salinity, temperature, light intensity, pressure, oxygen, and nutrient conditions). Over the years, such extreme environment was typically known to be populated exclusively by prokaryotes (Oren 2002), and it was believed that halophilic fungi could survive only as a contaminant in the food stuff that use a high concentration of salt or in the environment with low water activity (a_w). It is in the year 2000, when Gunde-Cimerman et al. (2000) first time reported the presence of fungi in solar salterns. After this report, many new species were discovered from the natural hypersaline environments, and species that were considered as food contaminants have been studied worldwide. Prior to the year 2000, scarce reports describing the isolation of fungi from moderately saline natural environments, such as saline soil (Guiraud et al. 1995), salt marshes (Newell 1996), and sea water (Kohlmeyer and Kohlmeyer 1991), were available. Research has shown that pre-crystallization and crystallization ponds have an abundant diversity of halophilic and halotolerant fungi (Zalar et al. 2007, 2008).This knowledge proved a handful to develop sufficient understanding of the contaminants of the food preserved with salt, especially about the mycotoxigenic fungi. Based upon the population and diversity assessment studies, fungi nowadays are recognized as an essential part of the native microbial communities in natural hypersaline habitats.

25.2 Fungi Inhabiting in Hypersaline Environments

The story of halophilic fungi starts with the belief that the fungi which inhabit on substrates with $a_w \leq 0.85$ (corresponding to 17% NaCl) generally have a xerophilic phenotype (Northolt et al. 1995). Such phenotype is dependent upon the water potential of the medium rather than the chemical composition. In recent years, regardless of geographical locality, stable and consistent composition of mycobiota has been reported in hypersaline environments worldwide. They include phylogenetically unrelated groups of fungi with minor impact on local environment. The abundance and diversity of these fungi depends upon a_w, dissolved oxygen, nutrients like phosphorus and nitrogen, pH, etc. (Gunde-cimerman et al. 2009). Most of the natural inhabitant fungi in hypersaline environments show a typical halophilic behavior, different from that of majority of prokaryotes. However, it is noteworthy here that many halophilic fungi can grow equally well in freshwater environments and do not require high salt concentration for their viability (Plemenitas et al. 2008) and can be isolated regularly at 10% or above salinity. Such groups of fungi are termed as "Extremely Halotolerant Fungi" (Plemenitas et al. 2008). Black yeast-like fungi and associated species from the melanized filamentous genera *Alternaria* and *Cladosporium* (Plemenitas et al. 2008) are good examples of extremely halotolerant fungi. Those which have an obligate requirement of salt for their growth are termed as halophilic (Gunde-Cimerman et al. 2009). Halophilic organisms can grow even in 3 M salinity or more. Two species of the genus *Wallemia*, viz., *W. ichthyophaga* and *W. muriae*, represent halophilic fungi.

25.3 Main Groups of Halophilic Fungi

Fungi isolation from hypersaline water samples is generally done by filtration of the water, microbial baiting, and spreading of biofilms covering the brine. Selective media with high fractions of NaCl (from 17% to 32%) or sugars (from 50% to 70%), i.e., having low a_w, are normally used for the isolation process. The incubation period may vary from few days to several weeks (Butinar et al. 2011; Gunde-Cimerman and Zalar 2014). In addition to hypersaline water, halophilic fungi have been obtained on wood immersed in brine (Zalar et al. 2005a), tropical microbial mats (Cantrell et al. 2006, 2011, 2013), and surface of halophytic plants. Some were reported from Arctic glacial ice and other water-based Arctic environments (Gostincar et al. 2010).

In spite of very high diversity, certain genera and species of fungi dominate in almost all of the hypersaline environments. Such fungi can be divided in three major groups: (*i*) halophilic and halotolerant fungi, those which are primarily isolated on selective growth media containing various concentrations of NaCl; (*ii*) xerotolerant fungi, which can grow on saline media and sugar; and (*iii*) sporadic fungal isolates that can contaminate low a_w media.

Presently about 140 orders of fungi are known; however, only a few of them are capable of growing in medium with very little a_w. This number is restricted to few species belonging to small number of genus. About 25 species of halophilic fungi, since 1998, have been reported from hypersaline environments. They include three species of the genus *Wallemia* (*W. muriae*, *W. sebi*, and *W. ichthyophaga*).These species were put into new class *Wallemiomycetes* and order *Wallemiales* (Zalar et al. 2005b). In addition, two new species of the genus *Emericella* (*E. appendiculata* and *E. stella-maris*) and four from frequently spotted varieties of *Aureobasidium pullulans* were isolated (Zalar et al. 2008). Twelve new species of the genus *Cladosporium*, namely, *C. fusiforme*, *C. dominicanum*, *C. psychrotolerans*, *C. halotolerans*, *C. herbarioides*, *C. ramotenellum*, *C. salinae*, *C. subinflatum*, *C. subtilissimum*, *C.*

spinulosum, *C. tenellum*, and *C. velox*, were also identified (Schubert et al. 2007). Zalar et al. (1999) isolated black yeast *Trimmatostroma salinum* from hypersaline water. Moreover, two novel species of *Candida*, i.e., *C. galli* and *C. pseudorugosa* (Butinar et al. 2005a), and *Eurotium halotolerans* were obtained universally from hypersaline waters of different salterns (Butinar et al. 2005b).

Some fungal species belonging to genera like *Penicillium*, *Aspergillus*, *Eurotium*, and *Cladosporium* with adaptive xerotolerance/halotolerance have been reported frequently in hypersaline waters around the world. Meristematic melanized yeast-like fungi, non-melanized yeasts (Butinar et al. 2005b), and various related species of the genus *Cladosporium* (Zalar et al. 2007) and filamentous genera *Scopulariopsis*, *Wallemia*, and *Alternaria* (Gunde-Cimerman et al. 2005; Zalar et al. 2005b) are normal mycobiotic inhabitants of global hypersaline waters. Members of classes *Ascomycota* like *Dothideales*, *Capnodiales*, and *Eurotiales* with halophilic and halotolerant properties have also been reported in hypersaline conditions. Members of order *Dothideales* and *Capnodiales* show xerotolerance and have tendency to grow as cryptoendolithic or epilithic at elevated and low temperatures (Selbmann et al. 2005) and in hypersaline coastal areas (Butinar et al. 2005c).Such fungi are virtually absent in nonnatural environments. *Phaetotheca triangularis*, *Hortaea werneckii*, *Trimmatostroma salinum*, and the halotolerant *Aureobasidium pullulans* are examples of fungi found in such environment.

Members of genus *Cladosporium* (order *Capnodiales*) are important extremophilic fungi with cosmopolitan distribution. Some species of this genus have been isolated consistently from the salty lakes across the world (Gunde-Cimerman et al. 2000).In the earlier times, many members of *Cladosporium* were considered as airborne contaminants, but later they were reported as most common and frequent fungal taxa, especially *Cladosporium sphaerospermum*, in hypersaline circumstances (Park et al. 2004). Now it is known that a complex of eight new species of the genus *Cladosporium* have either a narrow or wide ecological distribution (Zalar et al. 2007).

Many species of order *Eurotiales* (division *Ascomycota*) show halotolerance and xerotolerance as persistent plesiomorphic phenomena. *Aspergillus sydowii*, *Aspergillus niger*, *Penicillium chrysogenum*, and *Eurotium amstelodami* are main members of this group and have been recovered from biofilms, baits, and brines. *Aspergillus candidus* was observed in abundance. Some members of this order like *Aspergillus flavus*, *A. tubingensis*, *A. versicolor*, *Penicillium steckii*, *P. citrinum*, and *Eurotium herbariorum* are isolated from natural brine environments in fairly high frequencies however in lower mean counts. In fact about 60 different species of *Penicillium* and *Aspergillus* have been recovered from the oligotrophic and eutrophic hypersaline conditions around the world. This clearly indicates the diversity of these two genera in such condition. Two members of *Emericella*, viz., *A. versicolor* and *A. sydowii* (division *Ascomycota*), were also isolated from the hypersaline environments. Many of the above stated species are producers of mycotoxins and some other secondary metabolites. Thus, evaporated brine salt can have high concentration of foodborne fungi and their by-products.

Many species of order *Saccharomycetales* (division *Ascomycota*) have been reported in hypersaline environments. Such yeasts have hemi-ascomycetous affinities, and genera *Debaryomyces*, *Candida*, *Pichia*, and *Metschnikowia* are representative genus of such yeasts.

Candida tropicalis, *C. parapsilosis*, *C. glabrata*, and *C. krusei* were reported from hypersaline water of the Dead Sea. Among all of these, *C. parapsilosis* was the only species known formerly as foodborne halotolerant yeast; others however were never known for their halotolerance (Butinar et al. 2005b).

Three orders, viz., Trichonosporales, Sporidiales, and Wallemiales, of the division Basidiomycota have been reported with halophilic and halotolerant properties. The representative of the order Trichonosporales includes *Trichosporon mucoides* (Scorzetti et al. 2002), while *Rhodosporidium babjevae*, *Rhodosporidium sphaerocarpum*, and *Rhodotorula laryngis* are the halophilic representatives of the order *Sporidiales* (Butinar et al. 2005b). Order *Wallemiales* is represented by *Wallemia sebi*, *Wallemia muriae*, and *Wallemia ichthyophaga*. This is noteworthy here that the entire order *Wallemiales* (genus *Wallemia*) is either halophilic/xerophilic or xerotolerant (Zalar et al. 2005b) and its members have been isolated from the hypersaline lakes throughout the world (Wasser et al. 2003; Zalar et al. 2005b).

Natural saline environments, despite the little mycorrhizal affinity of the halophytes (Brundrett 1991), also harbor many arbuscular mycorrhizal fungi (Harisnaut et al. 2003; Yamato et al. 2008). It is however noteworthy that the average density of their spores in saline areas is low (Carvalho et al. 2001). Aliasgharzadeh et al. (2001) found *Glomus intraradices*, *G. versiforme*, and *G. etunicatum* to be the most predominant species of arbuscular mycorrhizal fungi in the severely saline soils. With the increase in soil salinity, no significant decrease in the concentration of the spores of these arbuscular mycorrhizal fungi was observed. Moreover, Aliasgharzadeh et al. (2001) observed a moderately high spore number (mean of 100/10 g soil). The higher fungal spore density in saline soils can be attributed to the reason that sporulation is stimulated by salt stress (Tressner and Hayes 1971) which means that in extremely saline environments, arbuscular mycorrhizal fungi may produce spores at low root-colonization levels (Aliasgharzadeh et al. 2001).

25.4 Representative Classes

Saccharomyces cerevisiae is the most studied eukaryotic microorganism as far as salt tolerance is concerned. However, *S. cerevisiae* principally cannot be placed among list of salt-sensitive organisms. It can grow up to 1.2 M NaCl salinity without any specific adaptation to hypersaline conditions. Recent studies have shown that *Hortaea werneckii*, *Debaryomyces hansenii*, and *Wallemia ichthyophaga* are better representative organisms for the study of halotolerance in eukaryotes than *S. cerevisiae*. These organisms have been reported in hypersaline environments worldwide and can grow up to 5.0 M (Plemenitas et al. 2008), 3.0 M (Prista et al. 2005), and

5.2 M NaCl (Zalar et al. 2005b) concentration, respectively. They adopt different strategies to combat the salinity stress.

25.4.1 Hortaea werneckii

Hortaea werneckii is the best studied extremely halotolerant eukaryotic model organism (Lenassi et al. 2013; Ali et al. 2016a). It was initially recognized as the causative agent of *Tinea nigra*, a typical sporadic asymptomatic infection of salty human hands and soles (Bonifaz et al. 2008). Now this is an established fact that *H. werneckii* is a dominant species in hypersaline and seawater-related environments (Zalar et al. 1999). Recurrent peaks of *H. werneckii* are primarily correlated with the high values of nitrogen. *H. werneckii* is the only species of genus *Hortaea*. It has no known sexual stage.

H. werneckii has a thick cell wall and is heavily melanized. At high salinities, it exhibits a meristematic, isodiametric kind of thallus enlargement, which leads to the formation of highly resistant cell clumps (Sterflinger 1998). This offers great resistance to high salinity, vagaries in temperature and desiccation. On the contrary at lower salinities, *H. werneckii* shows a characteristically polymorphic morphology (Sterflinger et al. 1999). *H. werneckii* can grow in a comprehensive growth optima ranging from 1.0 to 5.0 M NaCl (Gunde-Cimerman and Zalar 2014; Plemenitaš et al. 2014).

The genome of *H. werneckii* is quite bulky (51.6 Mb) with 23,333 reported genes. A heterothallic mating locus is present in the genome (Lenassi et al. 2013). *H. werneckii* has distinctive mechanisms of adaptation toward hypersaline environments that have not been seen with either salt-sensitive or moderately salt-tolerant fungi (Plemenitas et al. 2008).The most significant physiological characteristics are the properties and composition of plasma membrane and transport across it (Turk et al. 2007; Plemenitaš et al. 2016), osmolyte composition and accumulation of ions (Kogej et al. 2006), and structure and melanization of the cell wall (Kogej et al. 2007).

Salt stress in halophilic *H. werneckii* is known to cause an increase in the length of fatty acid chain and unsaturation with an increase in cis18:2$\Delta^{9,12}$ fatty acids and decrease in C16:0 fatty acids resulting in high plasma membrane fluidity over a wide range of salinities. Synthesis of these fatty acids remains under the control of two copies of desaturases genes *HwODE12* and two copies of gene *HwELO1*. The increase in NaCl concentration up-regulates the expression of these genes except *HwELO1B* (Gostincar et al. 2010).

The membrane fluidity in *H. werneckii* is controlled by sterol biosynthesis which in turn is regulated by 3-hydroxy-3-methylglutaryl I coenzyme A reductase (HmgR). Two isoforms of HmgR, mitochondrial *HwHmg1P* and *HwHmg2P* (confined to endoplasmic reticulum), have been identified (Vaupotic and Plemenitas 2007). These workers observed that the activities of these enzymes are high during the hypo- and hypersaline stress conditions while low in optimal salinity conditions. This clearly indicates that HmgR activities are regulated at transcriptional level.

To get adopted in its natural environment, *H. werneckii* is known to maintain low concentration of Na^+. This is due to continued exclusion of Na^+. *H. werneckii* is also known to accumulate glycerol (both by endogenous synthesis and uptake from the medium). Glycerol acts as a key compatible solute and helps to combat the ill effects of hypersaline conditions. Glycerol accumulation is accomplished by enhanced expression of two salt-induced GPD1 genes that encode glycerol-P dehydrogenase (Lenassi et al. 2011).

25.4.2 Debaryomyces hansenii

Debaryomyces hansenii and *Candida famata* (anamorphic state of *D. hansenii*) are halo-, osmo-, and xerotolerant yeasts which were previously known to be contaminants of salt-preserved food with low a_w (Prista et al. 1997). It is common yeast involved in the spoilage of brine-preserved food, frozen food, and other low a_w products. *D. hansenii* has been reported from many habitats with low water activities like sea water, meat, cheese, wine, beer, fruit, and soil (Barnett et al. 2000) as well as in high-sugar products (Marth 1978). In fact it is ubiquitous in the oceans and hypersaline habitats worldwide (Gunde-Cimerman et al. 2005). *D. hansenii* has also been observed in Antarctic soils (Atlas et al. 1978), glacial waters (Bridge et al. 2007), overcooled saline cryopegs in permafrost (Gilichinsky et al. 2005), and frost from Antarctic glaciers (Di Menna 1966).

D. hansenii is oleaginous yeast that can accumulate lipids up to 70% of their dry biomass (Breuer and Harms 2006). It plays a major a role in meat fermentation and is known to produce lytic enzymes and alditols (Breuer and Harms 2006). On the basis of different properties, present-day taxonomy distinguishes two varieties of *D. hansenii*: *D. hansenii* var. *fabryi* and *D. hansenii* var. *hansenii*.

All species of *Debaryomyces* are perfect haploid yeasts which reproduce vegetatively through multilateral budding. The sexual reproduction in *Debaryomyces* proceeds via heterogamous conjugation, i.e., the conjugation of two cells of dissimilar form or size (Breuer and Harms 2006). This conjugation commonly leads to diplophase followed by meiosis and ascospore formation (Forrest et al. 1987). The asci comprise one to four globular, spherical, lenticular smoothy or ovoidal or warty ascospores. Isogamous conjugation may also occur in some cases (Nakase et al. 1998).

D. hansenii is resistant toward penconazole and intermediately resistant toward benomyl and cycloheximide. Another interesting feature of *D. hansenii* is that it shows high tolerance toward biocide chlorine dioxide (ClO_2), a powerful biocide (Ramirez-Orozco et al. 2001). *D. hansenii* is haploid yeast with a substantial level of chromosome length polymorphism (Corredor et al. 2003). Genome of *D. hansenii* is available at http://cbi.labri.fr/Genolevures/ (Dujon et al. 2004). Thus, this provides new possibilities for an integrated approach toward the understanding of halophily/halotolerance. *D. hansenii* has a very small genome size, 12.2 Mb, as compared to 12.5 Mb of *Saccharomyces cerevisiae* and 14.1 Mb of *Schizosaccharomyces pombe*. However, the calculated gene number for *D. hansenii* is about 1000–2000 higher than in *S. cerevisiae* and *S. pombe*, and thus it represents a type of genetic uniqueness. As compared to *S. cerevisiae*, *D. hansenii* has significantly higher ratio of genes per kb and coding DNA (genes per kb ratio and coding DNA fraction in *S. cerevisiae* is 0.46 and 0.71, respectively, while values for *D. hansenii* are 0.57 and 0.79, respectively). These values represent the maximal limits for the genome of yeast (Axelson-Fisk and Sunnerhagen 2006).

Two separate groups working on the molecular biology of *D. hansenii* found linear plasmids titled as pDHL2 (9.2 kB), pDHL1 (8.4 kB), and pDHL3 (15.0 kB) (Gunge et al. 1993) and pDH1A and B (Cong et al. 1994). Both the groups reported that these plasmids were stable under osmotic pressure and in NaCl-free culture medium plasmids got cured. Salt dependency of the plasmid was later linked to the growth temperature (Fukuda et al. 2004).

25.4.3 Wallemia ichthyophaga

Wallemia ichthyophaga, a basidiomycete, is the utmost halophilic fungus identified to date. This fungus thrives in saturated salt solution and requires at least 10% NaCl. It shows its growth at an a_w of 0.77, corresponding to 5.2 M NaCl. *Wallemia* is a cosmopolitan genus of halophilic fungi that can be found in a variety of environments having low a_w (Zalar et al. 2005b). Species falling under the genus *Wallemia* are one of the most significant food pollutants of salty, sweet, and dry food. About 20 strains of *W. ichthyophaga* have been isolated from the hypersaline solar salterns and bittern's and salted meat. Two of the *Wallemia* species (*W. ichthyophaga* and *W. muriae*) require supplementary solutes for their growth and thus are considered as xerophilic/halophilic.

Physiological position of *Wallemia* remained unclear for many years, and it was placed in different positions. It was only after the genome sequencing that its correct phylogenetic position was depicted in basidiomycetes (Padamsee et al. 2012). A phylogenomic based study has shown that *Wallemiomycetes* is sister group of 495 million-year-old *Agaricomycotina* (Zajc et al. 2013). Since, species under the genus *Wallemia*, present in hypersaline water, were properly recognized after 2005, a little is known about their ecology and the mechanism of adaptation in low a_w (Zalar et al. 2005b). The sexual behavior of this species is also unclear. Until now no fruiting bodies or teleomorphs have been reported in any of the *Wallemia* spp. The genome of this fungus contains no mating type locus, and it seems that its entire species are asexual (Zajc et al. 2013).

W. ichthyophaga is also known to produce yet unidentified toxic metabolite with hemolytic activity. It is observed that high concentration of salt and sugar stimulates the synthesis of this metabolite. *W. ichthyophaga* like *Hortaea werneckii* forms covered multicellular meristematic clumps at all salinities. Clumps form a *Sarcina*-like structure which at high salinities shows fourfold increase in size. The cell wall thickens by threefold. As a result functional cell volume decreases significantly (Zalar et al. 2005a, b). The cover of the clumps is an extracellular polymeric substance matrix synthesized in a salinity-dependent phenomenon – more at suboptimal, lower salinity. Unlike other two species of the genus *Wallemia*, *W. ichthyophaga* does not exhibit hyphal growth in broth cultures.

Similar to *Hortaea werneckii*, *W. ichthyophaga* also uses accumulation of compatible organic solutes as a strategy to combat the turgor pressure. Moreover it maintains low intracellular Na^+ concentration. A mixture of polyolsis used to serve this purpose, glycerol being the key solute (Hohmann et al. 2007). Glycerol-3-phosphate dehydrogenase is the key enzyme involved in glycerol biosynthesis in *W. ichthyophaga*. This enzyme is coded by the gene *WiGPD1*. *W. ichthyophaga* is enriched with hydrophobins which contains unusually high ratio of acidic amino acids. Accumulation of high concentration of acidic amino acids is an adaptation toward hypersalinity (Zajc et al. 2013).

25.5 Mechanism for Living in Hypersalinity Conditions

Fungi exposed to hypersaline conditions experience two types of stresses: (a) ionic stress and (b) osmotic stress. Ionic stress results in the entry of ions (e.g., Na^+) into the cytoplasm, thus building up the concentration of ions inside the cytoplasm. This leads to breakdown of the cellular proteins and membrane system. The hyperosmotic stress results in efflux of water from the cell. This leads to the reduction in turgor pressure and hence dehydration of cytoplasm (Gunde-Cimerman and Zalar 2014). Fungi adapted to life in hypersaline (low a_w) conditions exhibit various strategies to combat such stresses.

25.5.1 Compatible Salt Strategy

Many organisms (both eukaryotes and prokaryotes) accumulate some compatible solutes so as to low down the concentration of Na^+ below toxic levels. Halophilic bacteria are known to accumulate compatible solutes such as glycine betaine, ectoine, proline, and glutamate, while the buildup of glycerol and other similar organic compounds is well-known strategy in alga and fungi. Fungi are known to accumulate various types of polyols and amino acids and their derivatives (Hohmann 2002).

Glycerol is an exclusive osmolyte used by the salt-sensitive *Saccharomyces cerevisiae*; however, other yeasts and fungi can accumulate and/or produce different polyols like ribitol, erythritol, and polyols from the environment, such as xylitol, galactitol, arabitol, sorbitol, and mannitol (Hohmann 2002).

Debaryomyces hansenii like *Saccharomyces cerevisiae* also accumulates glycerol as main compatible solute. Arabitol, trehalose, glutamic acid, and alanine are other compatible solutes present in low amounts in *D. hansenii* (Gunde-Cimerman et al. 2009). In hypersaline conditions, *D. hansenii* starts to accumulate more glycerol, but in moderate conditions of salinity, trehalose is accumulated more than glycerol (González et al. 2005). Glycerol-3-phosphate dehydrogenase is the key enzyme involved in glycerol biosynthesis. In high salinity conditions, the concentration of this enzyme along with glycerol-3-phosphatase uplifts. This escalation can be attributed to the increased expression of two *GPD1* genes which encodes for glycerol-3-phosphate dehydrogenase and *GPP2* encoding glycerol-3-phosphatase (Thomé 2005; Gori et al. 2005; Lenassi et al. 2011). In *D. hansenii*, glutamate acts a protective shield to tolerate high salt content. This has come to notice that when grown in high salinity, activity of NADP-glutamate dehydrogenase in *D. hansenii* increases (Alba-Lois et al. 2004).

Similar to *D. hansenii* and *Saccharomyces cerevisiae*, glycerol acts as the most important compatible solute in *Hortaea werneckii* (Petrovic et al. 2002). *H. werneckii* in saline conditions accumulates glycerol along with a mixture of erythritol and polyols mannitol and arabitol. The amount of these organic compounds depends upon the salinity and the fungal growth phase. During high salinity conditions, the level of erythritol gradually increases during the exponential phase and reaches to its maximum in the stationary phase, while polyol concentration uplifts during the exponential phase and diminishes steeply in the stationary phase. At high salinities, level of glycerol remains high during all the phases. Mycosporines, another class of compatible solutes, also gets accumulated in *H. werneckii* (Oren and Gunde-Cimerman 2007). They steeply accumulate up to 1.0 M NaCl but decrease at higher NaCl concentrations (Kogej et al. 2006).

Similarly, glycerol acts as the main osmotically compatible solute in *Wallemia ichthyophaga*. *W. ichthyophaga* detects salinity changes in the environment by employing a high-osmolarity glycerol (HOG) signaling pathway which has been conserved evolutionarily (Konte et al. 2016) and works equally in *Saccharomyces cerevisiae*. The accumulated concentration of glycerol in *W. ichthyophaga* is regulated by the salinity (i.e., glycerol concentration builds up with the increase of salinity and decreases with the decline of salinity). Moreover minor quantities of arabitol and mannitol were also detected (Gunde-Cimerman et al. 2009; Zajca et al. 2014).

Thus it can be concluded that all the three model halophilic fungi use accumulate ion of compatible solute strategy to combat the hypersaline conditions with glycerol as the main solute. Besides accumulation of glycerol, its transport and retention play a key role in adaptation to salt stress.

25.5.2 Ion Homeostasis Strategy

In order to survive, cells inhabiting in natural saline systems try to maintain a low water potential than their surroundings by maintaining a transport equation between Na^+ and K^+ ions.

As far as ion homeostasis is concerned, *Debaryomyces hansenii* is the most studied halotolerant fungi. Many studies have proven that high concentration of Na+ is not toxic for *D. hansenii* and, when grown in high salinity (NaCl), it develops better and accumulates more Na^+ than *Saccharomyces cerevisiae* (Prista et al. 1997; Gonzalez-Hernandez et al. 2004).

Some worker reported that the presence of high concentration of Na^+ can shield the cells against other stress factors (Almagro et al. 2000; Papouskova and Sychrova 2007). This suggests that *D. hansenii* is an efficient intruder of Na^+. In *D. hansenii* two genes *DhENA1* and *DhENA2* have been reported to code for ENA ATPases. These genes work for salt extrusion. Expression of *DhENA1* boosts up in the presence of high NaCl concentrations, whereas *DhENA2* expression requires a high pH (Almagro et al. 2001; Rodríguez-Navarro and Benito 2010). This fungus employs a different but somewhat similar low- and high-affinity K^+ and Na^+ efflux practices, cation/cation exchange system, and sodium-glycerol symporter. Mortensen et al. (2006) reported that *D. hansenii* efficiently manages its NaCl stress by maintaining the pK(a,i) close to its pHi homeostasis level. Plasma and internal membrane of *D. hansenii* harbors K^+ and Na^+ transporters (Montiel and Ramos 2007; Prista et al. 2007; Plemenitas et al. 2016).The cation transporters check the intracellular gathering of Na^+ to avoid the lethal effects of Na^+ accumulation while keeping high K^+/Na^+ ratio which is required for necessary functioning of the cells.

A proton pump similar to *S. cerevisiae* operated with the help of H+-ATPase (coded by *DhPMA1*) exists in *D. hansenii* (Prista et al. 2005). When grown in 4.5 M salinity, contrary to *D. hansenii*, *H. werneckii* is reported to maintain low levels of Na^+ and K^+. This suggests that *H. werneckii* efficiently extrudes Na^+ ions and prevents their influx. An ENA-like ATPase coded by *HwENA1* and *HwENA2* genes regulates the exclusion process during the high salt and pH conditions in *H. werneckii*. These two genes work in close harmony to maintain ion homeostasis. *HwENA1* expression advances at high pH, while in salt stress conditions, *HwENA2* is expressed at higher level.

W. ichthyophaga is reported to maintain moderately small intracellular concentration of Na^+ and K^+ at constant salinities; however, hyperosmotic shock leads to the buildup of these cations inside the cell (Lenassi et al. 2011).This can be concluded that as compared to other two fungi, *W. ichthyophaga* maintains an intermediate intracellular concentration of Na^+ and K^+.

These findings reveal that mechanism of ion homeostasis has evolved independently in all the abovementioned halophilic fungi.

25.6 Applications of Halophilic Fungi

The studies related to understanding the molecular mechanism of fungal adaptive response in hypersaline conditions have expanded the horizon of understanding the mechanism of salinity tolerance in other eukaryotic cells (Gunde-Cimerman and Plemenita 2006; Gostincar et al. 2011). This has also helped to combat the problems arising from the escalation of salinization of agriculture lands. Extremophiles are considered as a source of chemically diverse and often novel metabolites. Fungi have proven as a rice reservoir of halotolerant genes that can be transferred to increase the halotolerance in plants (Gostinar et al. 2012). Stress tolerance in mutant *S. cerevisiae* improved considerably after the heterologous insertion of the genes from *Debaryomyces hansenii* and *H. werneckii* (Vaupoti et al. 2007). Halotolerance and drought tolerance in *Arabidopsis thaliana* increased after insertion of coding sequence of ApHal2 (3′-phosphoadenosine-5′-phosphatase protein) from halotolerant yeast *A. pullulans*. Under moderate and severe saline conditions, the root length in transformed *Arabidopsis thaliana* plants increased to more than twice and have 5–10% larger leaf length. Genome analysis clearly demonstrates that halophilic fungi especially *H. werneckii* and *W. ichthyophaga* are valuable gene pool that could be used to improve stress tolerance in economically and agriculturally important plants (Gasparic et al. 2013). Many species of halophilic *Aspergillus*, *Sterigmatomyces*, etc. are rich source of antibacterial (Rani et al. 2013) and antioxidant compounds active in hypersaline conditions. Many enzymes that are active in saline conditions were also isolated from these fungi (Ali et al. 2016b). Gunny et al. (2015) isolated halophilic cellulase from *Aspergillus terreus*. Halophilic fungi are also used for the production of pigments, antibiotics, many enzymes, and retinal proteins. *Debaryomyces hansenii* is used industrially to synthesize D-D-arabinitol, xylitol, riboflavin, and pyruvic acid (Breuer and Harms 2006). Recently, Jiang et al. (2016) reported efficient biodegradation of phenol by halophilic fungi. They are also used for the bioremediation of toxic waste and to preserve food materials (Pavitra et al. 2017). Some are involved in melanin biosynthesis. α-Amylase enzyme isolated from halophilic fungi *Aspergillus gracilis* performs better at high NaCl concentrations (Imran et al. 2014) and can be used for many purposes including waste water remediation, laundry, and textile, pharmaceutical, and food industries.

25.7 Conclusion

For a long time, *Saccharomyces cerevisiae* acted as a model organism to study salt tolerance in eukaryotes. Nowadays, *S. cerevisiae* is not considered as a typical example of halophilic or halotolerant eukaryote as neither it can adopt its self in hypersaline conditions nor it is very salt sensitive. Halophilic fungi were previously thought to be contaminants of food preserved in high concentration of salt (low a_w). From last three decades, a large number of halophilic fungi have been isolated from hypersaline lakes, solar salterns, and other natural saline environments. Surprisingly, a huge diversity of many halophilic and halotolerant fungi (new and previously known) were discovered from these locations.

Most of the fungal species discovered in hypersaline conditions show extreme halotolerance, i.e., they do not require salt for viability but can adjust themselves in full salinity range – fresh water to saturated NaCl solutions. Some representatives of halophilic fungi have been reported in polythermal Arctic glaciers too.

Three fungi, namely, *Hortaea werneckii*, *Debaryomyces hansenii*, and *Wallemia ichthyophaga*, have been isolated from almost all the saline habitats. In comparison with *S. cerevisiae* that can grow up to 1.2 M NaCl, *H. werneckii* can grow up to

5.0 M NaCl and *D. hansenii* in up to 3.0 M NaCl, while *W. ichthyophaga* in up to 5.2 M NaCl (saturation level) and requires at least 1.5 M NaCl for growth. They are considered as eukaryotic models to study the mechanism of adaptation in hypersaline conditions. They belong to different taxonomic groups and practices different strategies to manage the problems of osmotic and ionic stress. This indicated that the halophilic and halotolerant fungi evolved independently in unrelated groups.

Almost all the members of halophilic fungi adopt various strategies like ion exchange, accumulation of compatible solutes, alterations in membrane fluidity, and many more to cope the ill effects of extreme saline conditions. Nowadays these fungi have been explored for their biotechnological applications also.

References

Alba-Lois L, Segal C, Rodarte B, Valdes-Lopez V, DeLuna A, Cardenas R (2004) NADP-glutamate dehydrogenase activity is increased under hyperosmotic conditions in the halotolerant. Curr Microbiol 48(1):68–72

Ali I, Prasongsuk S, Akbar A, Aslam M, Lotrakul P, Punnapayak H, Rakshit SK (2016a) Hypersaline habitats and halophilic microorganisms. Maejo Int J Sci Technol 10(03):330–345

Ali I, Siwarungson N, Punnapayak H, Lotrakul P, Prasongsuk S, Bankeeree W, Rakshit SK (2016b) Screening of potential biotechnological applications from obligate halophilic fungi, isolated from a man-made solar saltern located in Phetchaburi Province, Thailand. Pak J Bot 46(3):983–988

Aliasgharzadeh N, Saleh RN, Towfighi H, Alizadeh A (2001) Occurrence of arbuscular mycorrhizal fungi in saline soils of the Tabriz Plain of Iran in relation to some physical and chemical properties of soil. Mycorrhiza 11:119–122

Almagro A, Prista C, Castro S, Quintas C, Madeira-Lopes A, Ramos J, Loureiro-Dias MC (2000) Effects of salts on *Debaryomyces hansenii* and *Saccharomyces cerevisiae* under stress conditions. Int J Food Microbiol 56:191–197

Almagro A, Prista C, Benito B, Loureiro-Dias MC, Ramos J (2001) Cloning and expression of two genes coding for sodium pumps in the salt-tolerant yeast *Debaryomyces hansenii*. J Bacteriol 183:3251–3255

Atlas RM, Di Menna ME, Cameron RE (1978) Ecological investigations of yeasts in Antarctic soils. Antarct Res Ser 30:27–34

Axelson-Fisk M, Sunnerhagen P (2006) Comparative genomics and gene finding in fungi. In: Sunnerhagen P, Piskur J (eds) Topics in current genetics: comparative genomics. Using fungi as models. Springer, Heidelberg, pp 1–28

Barnett JA, Payne RW, Yarrow D (2000) In: Barnett JA, Payne RW, Yarrow D (eds) Yeasts: characteristics and identification, 3rd edn. Cambridge University Press, Cambridge

Bonifaz A, Badali H, de Hoog GS, Cruz M, Araiza J, Cruz MA, Ponce RM (2008) Tinea nigra by *Hortaea werneckii*, a report of 22 cases from Mexico. Stud Mycol 61:77–82

Breuer U, Harms H (2006) *Debaryomyces hansenii* – an extremophilic yeast with biotechnological potential. Yeast 23(6):415–437

Bridge P, Spooner B, Roberts P (2007) List of non-lichenized fungi from the Antarctic region. http://www.antarctica.ac.uk/Resources/BSD/Fungi/Speciespublic2.html#Lich

Brundrett M (1991) Mycorrhizas in natural ecosystem. Adv Ecol Res 21:300–313

Butinar L, Santos S, Spencer-Martins I, Oren A, Gunde-Cimerman N (2005a) Yeast diversity in hypersaline habitats. FEMSMicrobiol Lett 244:229–234

Butinar L, Sonjak S, Zalar P, Plemenita A, Gunde-Cimerman N (2005b) Melanized halophilic fungi are eukaryotic members of microbial communities in hypersaline waters of solar salterns. Bot Mar 1:73–79
Butinar L, Zalar P, Frisvad JC, Gunde-Cimerman N (2005c) The genus *Eurotium* – members of indigenous fungal community in hypersaline waters of salterns. FEMS Microbiol Ecol 51:155–166
Butinar L, Frisvad JC, Gunde-Cimerman N (2011) Hypersaline waters: a potential source of food-borne toxigenic aspergilla and penicillia. FEMS Microbiol Ecol 77:186–199
Cantrell SA, Casillas L, Molina M (2006) Characterization of fungi from hypersaline environments of solar salterns using morphological and molecular technique. Mycol Res 110:962–970
Cantrell SA, Dianese JC, Fell J, Gunde-Cimerman N, Zalar P (2011) Unusual fungal niches. Mycologia 103:1161–1174
Cantrell SA, Tkavc R, Gunde-Cimerman N, Zalar P, Acevedo M, Báez-Félix C (2013) Fungal communities of young and mature hypersaline microbial mats. Mycologia 105:827–836
Carvalho LM, Correia PH, Martins-Loucao A (2001) Arbuscular mycorrhizal fungal propagules in a salt marsh. Mycorrhiza 14:165–170
Cong YS, Yarrow D, Li YY, Fukuhara H (1994) Linear DNA plasmids from *Pichia etchellsii, Debaryomyces hansenii* and *Wingea robertsiae*. Microbiologica 140:1327–1335
Corredor M, Davila AM, Casarégola S, Gaillardin C (2003) Chromosomal polymorphism in the yeast species *Debaryomyces hansenii*. Antonie Van Leeuwenhoek 84:81–88
Di Menna ME (1966) Yeasts in Antarctic soils. Antonie Van Leeuwenhoek 32:29–38
Dujon B, Sherman D, Fischer G, Durrens P, Casaregola S, Lafontaine I et al (2004) Genome evolution in yeasts. Nature 430:35–44
Forrest SI, Robinow CF, Lachance MA (1987) Nuclear behaviour accompanying ascus formation in *Debaryomyces polymorphus*. Can J Microbiol 33:967–970
Fukuda K, Jin-Shan C, Kawano M, Sudo K, Gunge N (2004) Stress responses of linear plasmids from *Debaryomyces hansenii*. FEMS Microbiol Lett 237:243–248
Gasparic MB, Lenassi M, Gostinar C, Rotter A, Plemenita A, Gunde-Cimerman N, Gruden K, Že J (2013) Insertion of a specific fungal 3′-phosphoadenosine-5′-phosphatase motif into a plant homologue improves halotolerance and drought tolerance of plants. PLoS One https://doi.org/10.1371/journal.pone.0081872
Gilichinsky D, Rivkina E, Bakermans C, Shcherbakova V, Petrovskaya L, Ozerskaya S, Ivanushkina N, Kochkina G, Laurinavichuis K, Pecheritsina S, Fattakhova R, Tiedje JM (2005) Biodiversity of cryopegs in permafrost. FEMS Microbiol Ecol 53:117–128
Gonzalez-Hernandez JC, Cardenas-Monroy CA, Pena A (2004) Sodium and potassium transport in the halophilic yeast *Debaryomyces hansenii*. Yeast 21:403–412
González-Hernández JC, Jiménez-Estrada M, Pena A (2005) Comparative analysis of trehalose production by *Debaryomyces hansenii* and *Saccharomyces cerevisiae* under saline stress. Extremophiles 9:7–16
Gori K, Mortensen HD, Arneborg N, Jespersen L (2005) Expression of the GPD1 and GPP2 orthologous and glycerol retention during growth of *Debaryomyces hansenii* at high NaCl concentrations. Yeast 22:1213–1222
Gostinar C, Gunde-Cimerman N, Turk M (2012) Genetic resources of extremotolerant fungi: a method for identification of genes conferring stress tolerance. Bioresour Technol 111:360–367
Gostincar C, Grube M, de Hoog S, Zalar P, Gunde-Cimerman N (2010) Extremotolerance in fungi: evolution on the edge. FEMS Microbiol Ecol 7:12–11
Gostincar C, Lenassi M, Gunde-Cimerman N, Plemenita A (2011) Fungal adaptation to extremely high salt concentrations. Adv Appl Microbiol 77:71–96
Guiraud P, Steiman R, Seigle-Murandi F, Sage L (1995) Mycoflora of soil around the Dead Sea. Syst Appl Microbiol 18:318–322
Gunde-Cimerman N, Plemenita A (2006) Ecology and molecular adaptations of the halophilic black yeast *Hortaea werneckii*. Rev Environ Sci Biotechnol 5:323–331

Gunde-Cimerman N, Zalar P (2014) Extremely halotolerant and halophilic fungi inhabit brine in solar salterns around the globe. Food Technol Biotechnol 52(2):170–179

Gunde-Cimerman N, Zalar P, de Hoog GS, Plemenita A (2000) Hypersaline waters in salterns – natural ecological niches for halophilic black yeasts. FEMS Microbiol Ecol 32:235–240

Gunde-Cimerman N, Frisvad JC, Zalar P, Plemenitas A (2005) Halotolerant and halophilic fungi. In: Deshmukh SK, Rai MK (eds) Biodiversity of fungi – their role in human life. Oxford and IBH Publishing Co. Pvt. Ltd, New Delhi, pp 69–128

Gunde-Cimerman N, Amos J, Plemenitas A (2009) Halotolerant and halophilic fungi. Mycol Res. https://doi.org/10.1016/j.mycres.2009.09.002

Gunge N, Fukuda K, Morikawa S, Murakami K, Takeda M, Miwa A (1993) Osmophilic linear plasmids from the salt-tolerant yeast *Debaryomyces hansenii*. Curr Genet 23:443–449

Gunny AAN, Arbain D, Jamal P, Gumba RE (2015) Improvement of halophilic cellulase production from locally isolated fungal strain. Saudi J Biol Sci 22(4):476–483

Harisnaut P, Poonsopa D, Roengmongkol K, Charoensataporn R (2003) Salinity effects on antioxidant enzymes in mulberry cultivar. Sci Asia 29:109–113

Hohmann S (2002) Osmotic stress signaling and osmoadaptation in yeasts. Microbiol Mol Biol Rev 66:300–372

Hohmann S, Krantz M, Nordlander B (2007) Yeast osmoregulation methods. Enzymologia 428:29–45

Imran A, Akbar A, Yanwisetpakdee B, Prasongsuk S, Lotrakul P, Hunsa Punnapayak H (2014) Purification, characterization, and potential of saline waste water remediation of a polyextremophilic α-Amylase from an obligate halophilic *Aspergillus gracilis*. BioMed Res Inter. 7 pages. Article ID 106937

Jiang Y, Shang Y, Yang K, Wan H (2016) Phenol degradation by halophilic fungal isolate JS4 and evaluation of its tolerance of heavy metals. Appl Microbiol Biotechnol 100(4):1883–1890

Kogej T, Gostinar C, Volkmann M, Gorbushina AA, Gunde-Cimerman N (2006) Mycosporines in extremophilic fungi – novel complementary osmolytes? Environ Chem 3(2):105–110

Kogej T, Stein M, Volkmann M, Gorbushina AA, Galinski EA, Gunde-Cimerman N (2007) Osmotic adaptation of the halophilic fungus *Hortaea werneckii*: role of osmolytes and melanization. Microbiology 153:4261–4273

Kohlmeyer J, Kohlmeyer BV (1991) Illustrated key to the filamentous higher marine fungi. Bot Mar 34:1–61

Konte T, Terpitz U, Plemenitas A (2016) Reconstruction of the high-osmolarity glycerol (HOG) signaling pathway from the halophilic fungus *Wallemia ichthyophaga* in *Saccharomyces cerevisiae*. Front Microbiol 7:901

Lenassi M, Zajc J, Gostincar C, Gorjan A, Gunde-Cimerman N, Plemenitas A (2011) Adaptation of the glycerol-3-phosphate dehydrogenase Gpd1 to high salinities in the extremely halotolerant *Hortaea werneckii* and halophilic *Wallemia ichthyophaga*. Fungal Biol 115(10):959–970

Lenassi M, Gostinčar C, Jackman S, Turk M, Sadowski I, Nislow C (2013) Whole genome duplication and enrichment of metal cation transporters revealed by de novo genome sequencing of extremely halotolerant black yeast *Hortaea werneckii*. PLoS One 8:e71328

Marth EH (1978) Dairy products. In: Beuchat LR (ed) Food and beverages mycology. AVI, Westport, pp 145–172

Montiel V, Ramos J (2007) Intracellular Na^+ and K^+ distribution in *Debaryomyces hansenii*. Cloning and expression in *Saccharomyces cerevisiae* of DhNHX1. FEMS Yeast Res 7:102–109

Mortensen HD, Gori K, Siegumfeldt H, Nissen P, Jespersen L, Arneborg N (2006) Intracellular pH homeostasis plays a role in the NaCl tolerance of *Debaryomyces hansenii* strains. Appl Microbiol Biotechnol 71:713–719

Nakase T, Suzuki M, Phaff HJ, Kurtzman CP (1998) *Debaryomyces* Lodder and Kreger-van Rij Nom. Cons. In: Cletus Kurtzman P, Fell JW (eds) The yeasts – a taxonomic study, 4th edn. Elsevier Science Publishers, Amsterdam, pp 157–173

Newell SY (1996) Established and potential impacts of eukaryotic mycelial decomposers in marine/terrestrial ecotones. J Exp Mar Biol Ecol 200:187–206

Northolt MD, Frisvad JC, Samson RA (1995) Occurrence of foodborne fungi and factors for growth. In: Samson RA, Hoekstra ES, Frisvad JC, Filtenborg O (eds) Introduction to food-borne fungi. CBS, Delft, pp 243–250

Oren A (2002) Halophilic microorganisms and their environments. In: COLE Series 5. Kluwer Academic Publishers, Dordrecht, 575 pp

Oren A, Gunde-Cimerman N (2007) Mycosporines and mycosporine- like amino acids: UV protectants or multipurpose secondary metabolites? FEMS Microbiol Let 269:1–10

Padamsee M, Kumar TKA, Riley R, Binder M, Boyd A, Calvo AM, Furukawa K, Hesse C, Hohmann S, James TY (2012) The genome of the xerotolerant mold Wallemia sebi reveals adaptations to osmotic stress and suggests cryptic sexual reproduction. Fungal Genet Biol 49:217–226

Papouskova K, Sychrova H (2007) The co-action of osmotic and high temperature stresses results in a growth improvement of *Debaryomyces hansenii* cells. Int J Food Microbiol 118:1–7

Park HG, Managbanag JR, Stamenova EK, Jong SC (2004) Comparative analysis of common indoor *Cladosporium* species based on molecular data and conidial characters. Mycotaxon 89:441–451

Pavitra S, Anuradha S, Nupur M (2017) Introduction to halophiles. Inter J Multidis App Stud 04(1):43–60

Petrovic U, Gunde-Cimerman N, Plemenitas A (2002) Cellular responses to environmental salinity in the halophilic black yeast *Hortaea werneckii*. Mol Microbiol 45:665–672

Plemenitas A, Vaupotic T, Lenassi M, Kogej T, Gunde-Cimerman N (2008) Adaptation of extremely halotolerant black yeast *Hortaea werneckii* to increased osmolarity: a molecular perspective at a glance. Stud Mycol 61:67–75

Plemenitaš A, Lenassi M, Konte T, Kejžar A, Zajc J, Gostinčar C, Gunde-Cimerman N (2014) Adaptation to high salt concentrations in halotolerant/halophilic fungi: a molecular perspective. Front Microbiol 5:199

Plemenitas A, Konte T, Cene Gostinčar C, Gunde Cimerman N (2016) Transport systems in Halophilic Fungi. Adv Exp Med Biol 892:307–325

Prista C, Almagro A, Loureiro-Dias MC, Ramos J (1997) Physiological basis for the high salt tolerance of *Debaryomyces hansenii*. Appl Environ Microbiol 63:4005–4009

Prista C, Loureiro-Dias MC, Montiel V, Garcıa R, Ramos J (2005) Mechanisms underlying the halotolerant way of *Debaryomyces hansenii*. FEMS Yeast Res 5:693–701

Prista C, González-Hernández JC, Ramos J, Loureiro-Dias MC (2007) Cloning and characterization of two Kþ transporters of *Debaryomyces hansenii*. Microbiology 153:3034–3043

Ramirez-Orozco M, Hernandez-Saavedra NY, Ochoa JL (2001) *Debaryomyces hansenii* growth in nonsterile seawater ClO_2 –peptone-containing medium. Can J Microbiol 47:676–679

Rani MHS, Ramesh T, Subramanian J, Kalaiselvam M (2013) Production and characterization of melanin pigment from halophilic black yeast *Hortaea werneckii*. Int J Pharma Res Rev 2(8):9–17

Rodríguez-Navarro A, Benito B (2010) Sodium or potassium efflux ATPase a fungal, bryophyte, and protozoal ATPase. Biochim Biophys Acta 1798(10):1841–1853

Schubert K, Groenewald JZ, Braun U, Dijksterhuis J, Starink M, Hill CF (2007) Biodiversity in the *Cladosporium herbarum* complex (Davidiellaceae, Capnodiales), with standardization of methods for *Cladosporium* taxonomy and diagnostics. Stud Mycol 58:105–156

Scorzetti G, Fell JW, Fonseca A, Statzell-Tallman A (2002) Systematics of basidiomycetous yeasts: a comparison of large subunit D1/D2 and internal transcribed spacer rDNA regions FEMS Yeast Res 2:495–517

Selbmann L, de Hoog GS, Mazzaglia A, Friedmann EI, Onofri S (2005) Fungi at the edge of life: cryptoendolithic black fungi from Antarctic deserts. Stud Mycol 51:1–32

Sterflinger K (1998) Temperature and NaCl-tolerance of rock-inhabiting meristematic fungi. Antonie Van Leeuwenhoek 74:271–281

Sterflinger K, de Hoog GS, Haase G (1999) Phylogeny and ecology of meristematic ascomycetes. Stud Mycol 43:5–22

Thomé PE (2005) Heterologous expression of glycerol 3-phosphate dehydrogenase gene (DhGPD1) from the osmotolerant yeast *Debaryomyces hansenii* in *Saccharomyces cerevisiae*. Curr Microbiol 71:713–719

Tressner HD, Hayes JA (1971) Sodium chloride tolerance of terrestrial fungi. Appl Microbiol 22:210–213

Turk M, Abramovi Z, Plemenita A, Gunde-Cimerman N (2007) Salt stress and plasma-membrane fluidity in selected extremophilic yeasts and yeast-like fungi. FEMS Yeast Res 7(4):550–557

Vaupoti T, Gunde-Cimerman N, Plemenita A (2007) Novel 3′-phosphoadenosine-5′-phosphatases from extremely halotolerant *Hortaea werneckii* reveal insight into molecular determinants of salt tolerance of black yeasts. Fungal Genet Biol 44:1109–1122

Vaupotic T, Plemenitas A (2007) Osmoadaptation-dependent activity of microsomal HMG-CoA reductase in the extremely halotolerant black yeast *Hortaea werneckii* is regulated by ubiquitination. FEBS Lett 581(18):3391–3395

Wasser SP, Grishkan I, Kis-Papo T, Buchalo AS, Volz PA, Gunde-Cimerman N, Zalar P, Nevo E (2003) Species diversity of the dead sea. In: Nevo E, Oren A, Wasser SP (eds) Fungal life in the dead sea. ARG Ganter Verlag K.-G, Ruggell, pp 203–270

Yamato M, Ikeda S, Iwase K (2008) Community of arbuscular mycorrhizal fungi in coastal vegetation on Okinawa Island and effect of the isolated fungi on growth of sorghum under salt-treated conditions. Mycorrhiza 18:241–249

Zajc J, Liu Y, Dai W, Yang Z, Hu J, Gostin Ar C, Gunde-Cimerman N (2013) Genome and transcriptome sequencing of the halophilic fungus *Wallemia ichthyophaga*: Haloadaptations present and absent. BMC Genomics 14:617

Zajca J, Kogeja T, Galinskib EA, Ramosc J, Gunde-Cimerman N (2014) Osmoadaptation strategy of the most halophilic fungus, *Wallemia ichthyophaga*, growing optimally at salinities above 15% NaCl. Appl Environ Microbiol 80(1):247–256

Zalar P, de Hoog GS, Gunde-Cimerman N (1999) *Trimmatostroma salinum*, a new species from hypersaline water. Stud Mycol 43:57–62

Zalar P, de Hoog GS, Schroers HJ, Frank JM, Gunde-Cimerman N (2005a) Taxonomy and phylogeny of the xerophilic genus *Wallemia* (Wallemiomycetes and Wallemiales, cl Et ord nov). Antonie Van Leeuwenhoek 87:311–328

Zalar P, Kocuvan MA, Plemenita A, Gunde-Cimerman N (2005b) Halophilic black yeasts colonize wood immersed in hypersaline water. Bot Mar 48:323–326

Zalar P, de Hoog GS, Schroers HJ, Crous PW, Groenewald JZ, Gunde-Cimerman N (2007) Phylogeny and ecology of the ubiquitous saprobe *Cladosporium sphaerospermum* with descriptions of seven new species from hypersaline environments. Stud Mycol 58:157–183

Zalar P, Frisvad JC, Gunde-Cimerman N, Varga J, Samson RA (2008) Four new species of *Emericella* from the Mediterranean region of Europe. Mycologia 100:779–795

Chapter 26
Carbon Sequestration and the Significance of Soil Fungi in the Process

Santosh Kumar Mehar and S. Sundaramoorthy

Abstract With the advent of industrial revolution, the release of carbon dioxide in the air has increased its concentration in the atmosphere to alarming levels. The value that was approximately 277 ppm in the beginning has crossed the 400 ppm mark in the recent times. The increase in the level of CO_2 can be controlled by two pronged approach: first the causes for the rise in the CO_2 be identified and managed, and second, the factors which can act as sink to reduce the increase in the CO_2 concentration in atmosphere be identified and used in proper manner. Terrestrial component has proven to be acting as a sink to carbon and has in the past kept the concentration of CO_2 in check by sequestering it in various forms in the soil. Soil fungi are found to be the main regulatory component in the process, by accumulating carbon in their biomass and by producing recalcitrant decomposition products that have very long residence time in the soil, ranging from years to centuries.

Keywords Carbon sequestration · Soil fungi · Carbon sink

26.1 Introduction

From the beginning of the industrial era (1750), there has been vast increase in the concentration of carbon dioxide in the atmosphere. The value has increased from less than 280 (277 parts per million (ppm)) at the beginning of the industrial era (Joos and Spahni 2008) to ~400 (399.4 ppm) 2 years back from now as reported by Dlugokencky and Tans (2016). Scripps (2013) reported that the observations from Mauna Loa station which keeps track of the direct measurement of atmospheric CO_2 concentration went above 400 ppm in May 2013. During the year 2015, global monthly average from March to May was greater than 400 ppm, and the same trend was recorded again in November 2015 (Dlugokencky and Tans 2016). Initially the rise in atmospheric CO_2 concentration was caused by the release of carbon in the atmosphere due to reduction in forest cover and other land use change activities.

S. K. Mehar (✉) · S. Sundaramoorthy
Department of Botany, J.N.V. University, Jodhpur, Rajasthan, India

P. Gehlot, J. Singh (eds.), *Fungi and their Role in Sustainable Development: Current Perspectives*, https://doi.org/10.1007/978-981-13-0393-7_26

Since the 1920s, fossil fuel burning by mankind became the major cause of carbon emission in the atmosphere, and its comparative share has continuously increased ever since.

For the year 2016, the figures of the carbon budget of the entire world reported by Le Quéré et al. (2016) provide the overall statistics related to the factors responsible for the change in the concentration of CO_2 and have reference to the initiation of the industrial era. The report numerically analyses the contribution of CO_2 in atmosphere from different anthropogenic activities, resulting increase in the concentration of atmospheric CO_2 and change in the stored quantities of carbon in the land and ocean reservoirs with increase in the CO_2 concentration. The different components of CO_2 budget are presented below:

1. Combustion and oxidation of fossil fuel and production of cement, which is referred as E_{FF}
2. The emission from anthropogenic actions on land resulting in change of how land is used, referred as E_{LUC}
3. Increase in the atmospheric CO_2 concentration, referred as G_{ATM}
4. The uptake of CO_2 by the CO_2 sinks, viz. the oceans referred as S_{OCEAN} and land referred as S_{LAND}

Le Quéré et al. (2016) suggest that the emissions at global level and their differential contribution among the components, viz. land, ocean and atmosphere, remain in balance, as the sum of E_{FF} of E_{LUC} remains equal to G_{ATM}, S_{OCEAN} and S_{LAND}.

The figures of the global carbon budget when averaged for the 10-year duration (2006–2016) point to the facts that during the period, more than 90% of the emissions were effected by burning of fossil fuel and industrial activities and less than 10% were due to change in land use. When emissions were divided among the different components, atmosphere, ocean and land contributed 44%, 26% and 30%, respectively (Le Quéré et al. 2016). Among all the components except for land use change, rest of the components have increased in quantity since 1959. Besides, there has been difference in the increase in concentration of CO_2 and the terrestrial CO_2 sink. Le Quéré et al. (2016) have however not included the net contribution of CO_2 released when reactive carbon-containing gases get chemically oxidized, which is a source different from the burning of fossil fuels (e.g. CH_4 emissions from anthropogenic activities, industrial processes, biogenic emission resulting due to change in vegetation, fire and wetlands). Besides, carbon monoxide and certain volatile compounds like isoprene and terpene also contribute to rise in carbon concentration (Gonzalez-Gaya et al. 2016). E_{FF} component of the equation includes CO emissions but not fugitive anthropogenic CH_4 emissions, which when included, will add to E_{FF}. Regnier et al. (2013) have proposed that anthropogenic disturbance in the cycling of carbon in terrestrial freshwater bodies, such as estuaries, modifies lateral transition from terrestrial system to the open area and CO_2 from rivers and lakes to the atmosphere, and the net air to sea anthropogenic CO_2 flux in the coastal areas also contributes to the carbon budget.

26.2 Terrestrial Component in Global Carbon Cycle

Terrestrial component is an integral part of the global carbon cycle. During the process of photosynthesis, carbon is brought into the terrestrial biosphere; thereafter it is released back through various processes such as plant respiration, microbial respiration or decomposition, fire and a variety of human land use practices. When photosynthesis rate is more than the sum of plant respiration, microbial respiration and land use release, there is a net sink of carbon from the atmosphere into the ecosystem. The gain or loss of carbon in the terrestrial ecosystem is reflected by small changes in the atmospheric concentration of CO_2. But gradually, on a longer time scale (from decades to centuries), the cycling of carbon in the terrestrial ecosystem significantly influences the CO_2 level in the atmosphere, concomitantly becoming significant contributory factor to the amount of CO_2 in the atmosphere.

For long, the scientific belief has been that the terrestrial component is a net sink of atmospheric carbon. During the past few decades, the terrestrial ecosystem has absorbed significant quantities of atmospheric CO_2, emitted mainly from sources such as burning of fossil fuels and to some point from production of cement. This has kept atmospheric CO_2 concentration at a lower level than it would otherwise have been. Terrestrial biosphere has continued to keep under check the fossil fuel emissions from entering into the atmosphere. This fact is further highlighted by the observations of Prentice et al. (2001) that during the 1990s, approximately 6 petagrams of carbon per year (PgC $year^{-1}$) were released from burning of the fossil fuels and also production of cement. Houghton (2000) has reported that during the same time period, another 1–2 PgC $year^{-1}$ was being released through land use practices, mainly deforestation in the tropics. Together both of these factors should have caused 7–8 Pg rise in the carbon content of the atmosphere on annual basis. However, it has been reported by Bousquet et al. (2000) and Schimel et al. (2001) that nearly half of this carbon only accumulated in the atmospheric component, which is known as airborne fraction. The other half is supposed to be absorbed in the oceans and terrestrial biosphere. Thus research work has stressed the importance of both the oceanic and terrestrial systems in reducing the release of CO_2 in the atmosphere. This fact has been pointed out in the report of the Intergovernmental Panel on Climate Change (IPCC), wherein it was mentioned that the terrestrial component of the biosphere absorbed nearly 1.4 PgC $year^{-1}$ (approximately 22% of the anthropogenic emission) during the 1990s and nearly 0.2 PgC $year^{-1}$ (about 4% of the anthropogenic emission) during the 1980s.

26.3 Organic and Inorganic Carbon Pools

There are five global carbon pools, viz. atmospheric, geologic, biotic, oceanic and soil. Of the five global carbon pools, soil C pool is the third largest. Its magnitude of 2500 Pg to 1 m depth is more than three times of the atmospheric pool which stands at 760 Pg and four times of the biotic pool (620 Pg). The other pools which are greater than soil C pool are geologic (4130 Pg) and oceanic pools (38,400 Pg; Lal 2008). The soil carbon pool consists of two major and distinct components: a SOC (soil organic carbon) pool of about 1550 Pg and a SIC (soil inorganic carbon) pool of 950 Pg, both to 1 m depth (Batjes 1996; Eswaran et al. 1993). The SOC pool comprises of organic C assembly, dead or alive, and includes microbial biomass (Jenny 1980). Lal (2008) suggests the SOC pool consists of different constituents which are given below:

1. Small amount of plant and animal tissues from the original biomass input
2. Products of the biological and chemical decomposition of the biomass addition to the soil
3. Living and dead microbial cells
4. Degradation of soil organisms
5. The interaction of products of any or all of the substances mentioned above

The SOC is also referred as SOM (soil organic matter) or humus. Humus is a dark brown or black amorphous material, and highly decomposed component of the SOM, having large surface area, high reactivity and high affinity for the clay fraction and high charge density (Lal 2008). Humic substances are dark-coloured organic macromolecules which are rich in phenolic compounds, derived from plant remains and microbial metabolism (Perveen et al. 2014). Humic substances are highly dynamic, being formed from plant and animal residues and simultaneously decomposed by the microbial processes (Allison 2012; Waring et al. 2013).

The SIC pool is more important in soils of dry climates and includes elemental C, primary (calcite and dolomite) and secondary carbonates. The source of primary carbonates is weathering of the parent material, while the secondary carbonates derive from the dissolution of CO_2 in the soil air and due to reactions of weak carbonic acids with Ca^{+2}, Mg^{+2} and other cations which are brought into the system (Lal 2008).

For the accumulation of SOC, a positive imbalance between the input into the soil organic matter stock and outputs from it is required. Carbon build-up can increase due to inputs from the photosynthesis carried out by the autotrophic components of the ecosystem, and also when there is reduction in the loss of C from the ecosystem, or due to both of the processes working together. Although certain processes such as decomposition, runoff, leaching and erosion cause loss of C from a particular area, they can result in addition of C inputs at some other places, mainly the later three processes.

26.4 Regulatory Factors for Carbon Sequestration

The meaning of carbon sequestration is transferring of CO_2 in the atmosphere to other long-lived pools and storing it securely so that it is not remitted immediately. Hence, carbon sequestration in soil means increase in the SOC and SIC reservoirs.

According to Jastrow et al. (2007), carbon sequestration can occur only when a positive disequilibrium continues for long periods of time; as a result the system ultimately achieves a new higher steady state. Since the carbon derived from the atmosphere is ultimately cycled back to it, carbon storage in soil is a dynamic process. Luo et al. (2003) suggest that the residence time of carbon in the soil (τ) is the major determining factor of the ability of the system to sequester carbon. When there is increase in τ, SOC could get sequestered even without any increase in the inputs. On the other hand, if τ remains undisturbed, then only the increase in the inputs gets sequestered as long as the processes controlling the τ do not get saturated (Six et al. 2002). According to Freeman et al. (2001), under the extremes of environmental factors such as moisture, temperature, pH and nutrient availability, the decomposer activity is limited which has effect on τ. Under such conditions, inputs are sequestered without any difficulty but in relatively undecomposed forms, such as in boreal peat deposit. However, the carbon accumulated in this manner is susceptible to loss from the system when the conditions turn favourable for the decomposer community, allowing the process of decomposition to progress. Otherwise under normal conditions of the environment, two other mechanisms, viz. biochemical change and physicochemical protection, have major influence on the stabilization of SOC and thus greater influence on τ. When carbonaceous material is transformed biochemically, SOC is altered by both living organisms and abiotic forces into forms which are very hard to decompose and get retained in the system due to sorption with colloidal particles of soil. This transformation process is known as humification which results in the formation of humic substances. Besides, SOC could also get protected physicochemically from biochemical forces when it gets stabilized onto the soil surface, undergoes complexation with different mineral in soil, enters into aggregates and gets deposited in pores which are not accessible to the decomposer community and even to the secreted enzymes (Jastrow et al. 2007).

26.5 Soil Fungi: An Important Component of Soil Microflora and Their Contribution in Sequestration of Carbon vis-a-vis Other Microbes

Soil microorganisms are the biotic controlling factors for SOC turnover and thereby significant in the regulation of τ. According to Six et al. (2006), microbially derived organic matter pool referred to as MOM builds up from the assimilation of carbon by the microbial biomass. The build-up of MOM depends on the total microbial biomass of the system, and its dynamics is influenced by the growth efficiency of the soil microbes. Among the soil microorganisms, bacteria and fungi are the most dynamic and effective in terms of the regulation of the MOM. The two groups are very different in their biochemical approach and the effect they have on stabilization of soil organic carbon.

During decomposition of organic matter in the soil, microorganism partitions the substrate carbon taken up by them into biomass build-up and respiration. The amount of biomass carbon which is produced from each unit of substrate carbon utilized in the metabolism is known as microbial growth efficiency (MGE). This also determines how much of the carbon can be lost as CO_2 from the soil. Estimations of MGE are useful for knowing the growth patterns of soil microorganisms and for estimating the SOM budgets at the ecosystem level. In the Century SOM model, Parton et al. (1987) observed that more than half of the carbon (55%) assimilated by the microbes could be lost during the decomposition of nonlignin SOM pool as CO_2. This estimation had the basis that fungi mainly decompose surface litter, have comparatively higher MGE when compared with bacteria and lead to carbon stabilization during decomposition. Further, the model also indicated based on sensitivity analysis that comparatively small change in MGE could have significant effect on the dynamics of carbon and nitrogen, because soil communities dominated by fungi are reported to consume lesser quantities of substrate carbon per unit of the biomass measured.

Similar to MGE, carbon utilization efficiency (CUE) is also used for comparing relative efficiency of different microbes in the organic matter decomposition. In the opinion of Payne (1970), there is no difference in the CUE of bacteria and fungi, but in the opinion of other researchers (Adu and Oades 1978), fungal CUE is greater than bacterial. The significance of comparisons of CUE in estimates of sequestration of carbon stems from the fact that those organisms which have comparatively lower CUE respire a greater part of carbon metabolized by them, meaning thereby for metabolizing similar quantities of carbon, since bacteria have lower CUE, they have to respire a greater portion of metabolized carbon as CO_2 and thereby will contribute very less to the newly stabilized soil organic carbon pools in comparison to fungi. Their relative approach to decomposition also makes fungi better contributors to carbon sequestration, which is exemplified by the difference in the extracellular enzymes produced by fungi and bacteria.

The source of carbon for soil decomposers is primarily the litter that falls in the soil. Plant litter comprises of different amounts of various kinds of organic compounds, which vary with plant species, and the plant parts such as leaves, stem, root and bark. Some of the compounds are readily lost by dissolution and leaching from the source litter, the process being partly assisted by the activity of rapidly growing opportunistic microorganisms. In contrary to this, the large macromolecules, which include cellulose, hemicelluloses and lignin, degrade more slowly. During the process of decomposition of litter, also there is marked difference among the microorganisms in their capability, and the approach, e.g. fungi, can grow inside the litter even before it falls.

Decomposition of cellulose can be carried out by many species of bacteria and fungi. The degradation of cellulose needs to be carried out outside the microbial cell, where the insoluble polymer is broken down to monomers or oligomer units consisting of a few glucose molecules, e.g. cellobiose. These are then taken up by the microbial cells and get metabolized. Eriksson et al. (1990) reported approximately 74 species of fungi that are capable of degrading cellulose. According to Ander and Eriksson (1977), *Phanerochaete chrysosporium* (previously *Sporotrichum pulverulentum* Novabranova) is one of the important wood-decay fungi. Highley

(1988) has observed many species of brown rot fungi that are capable of solubilizing microcrystalline cellulose.

While white rot fungi are capable of fully mineralizing lignin to carbon dioxide and water, many brown rot fungi mainly decompose cellulosic and hemicellulosic parts of the wood and modify the lignin molecules only partially. Crawford (1981) proposed that soft rot fungi only soften the wood, by breaking down middle lamella of the cells. Majority of these fungal species belong to deuteromycetes and ascomycetes and are particularly involved in the decay of moist wood. However, Nilsson and Daniel (1989) observed that the soft rot fungi are able to degrade lignin also.

The process of decomposition of litter and ultimately the transformation of various forms of carbon in the soil is typically related to and dependent on the availability of other nutrients. For instance, lignin degradation is repressed if the substrate has high N levels. Besides, it has been observed that the degradation of lignin in many species of fungi, namely, *P. chrysosporium*, *Coriolus hirsutus*, *Lentinus edodes* and *Polyporus* sp., is influenced when another carbon source apart from lignin is available, in which case the degradation of lignin was stimulated. According to Berg and McClaugherty (2008), in comparison to glucose, cellulose has a stronger stimulating effect on lignin degradation.

Large amounts of phenoloxidases, peroxidises and laccases are produced by fungi, which carry out the decomposition of lignin by promotion of condensation reactions. In comparison to this, bacteria produce cellulases and lipases, required for attacking nonlignite material. When both of the microbial groups, viz. fungi and bacteria, are compared with regard to their contribution in the formation of humus, fungal-mediated decomposition contributes more in the formation of humus due to the fact that the monomeric units arising out of degradation of lignitic material are ingredients of the humic substances.

There are inherent differences in the decomposition capabilities of bacteria and fungi also. While both of them can degrade cellulose, lignin is more recalcitrant, and its complete degradation is carried out only by some selected types of fungi which are capable of producing lignin peroxidase (ten Have and Teunissen 2001). Till date, there is no report of bacteria which can initiate degradation of lignin (Jastrow et al. 2007), and it is reported that they are capable of mineralizing only the partially altered intermediate products of fungal degradation. Thus, when some source of carbon is given as input in the soil, fungi and bacteria carry out selective degradation resulting in the decomposition of the labile portion first, which leads to gradual increase in the recalcitrance in the carbon remaining as the time passes. The resulting more recalcitrant carbon can only be decomposed at a very slow rate, and the process is carried out by the activity of specific extracellular enzymes. Shen et al. (2002) reported that there is one more limitation in the decomposition of recalcitrant C, even by the extracellular enzymes, which arises due to the fact that the extracellular enzymes have a very limited lifetime in the soil (only a few days). Besides the short lifetime of the enzymes, their activity is also hampered by being sorbed to minerals which restricts their access to the target substrate. As a result, decomposition gets restricted to regions near the source of the release of the enzymes only (George et al. 2005).

Martin and Haider (1979) observed that when soils have greater biomass of fungi in comparison to bacteria, the decomposition of organic matter derived microbially (microbial-derived organic matter (MOM)) will be much slower. Simpson et al. (2004) proposed this due to the fact that fungal metabolic products are more resistant to chemical degradation and get preferentially secluded from degradation by their interaction with the clay minerals and the soil aggregates.

Going into the detail of the process of extracellular enzymatic decomposition by bacteria reveals the causes why bacteria have limited degradation power. There are only a small number of sites on the surface of the substrate available for sorption for the extracellular enzymes; therefore, when the sites get saturated by the enzymes, additional enzyme molecules must diffuse to longer distances, more distant from the source, i.e. the bacterial cell which secreted them, and thus will return significantly less products to the cell, which released them (Schimel and Weintraub 2003). The negative feedback loop thus created is simulated by reverse Michelis-Menton relations and is the regulatory force for the production of extracellular enzymes by the microorganisms. Vetter et al. (1998) have proposed a model about how the extracellular enzymes are produced by the bacterial cells, which proposes that foraging distance of the extracellular enzymes produced by the bacterial cells is on the order of 10–50 μm. The review of the model clarifies that as the distance increases, benefits to the bacterial cells for production of enzymes are much lower than the cost of producing them. In comparison to this, fungi have longer foraging distances particularly in undisturbed soils. The hyphal nature of fungi allows them to form bridge between the litter molecules and the soil. These hyphal networks become a large component of the soil fabric (Haynes and Beare 1997).

Fungi contribute more to C stabilization by virtue of difference in their metabolism and formation of biomass also. Cell wall of fungal cells consists of chitin- and melanin-like molecules, while the membranes of bacterial cells are predominantly phospholipidic in nature. Frey et al. (2001) observed that after the death of bacteria, bacterial phospholipids are metabolized rapidly by other bacterial cells. In comparison to this, the fate of the fungal cell walls is different. Chitin and melanin residues are quite recalcitrant and persist in the soil longer (Holland and Coleman 1987; Guggenberger et al. 1999). This is also confirmed by the study carried out by Suberkropp and Weyers (1996), where they studied the growth of microbes on leaf litter. Their observations suggest that carbon stored by fungal cells was 26 times greater than that stored by bacterial cells. Besides this, the sequestration of carbon in soil depends not only on how much of the bacterial and fungal by-products get accumulated in the soil, and the chemical composition of the by-products, but also the extent to which these substances remain protected in the soils. ^{14}C-labelled cell wall's decomposition was studied in grassland soil (Nakas and Klein 1979), and it was observed that fungal material decomposition was much slower than the bacterial, as hyphae of fungi having high melanin content decay more slowly than those having low melanin content (Martin and Haider 1986).

The residence time of carbon in soil is determined by the chemical modification of organic matter in the soil to forms which can resist microbial decomposition better, get adsorbed to the soil particles or both. The biochemical alteration takes place in two different stages: first, the plant resides are broken down into particulate organic matter, which is later degraded to small molecules. Similarly, the carbon assimilated in the bodies of soil microbes gets decomposed after the death of the organism. Condensation and polymerization reactions follow the first stage, during which the small molecules are polymerized to create larger molecules. Both of these stages are biologically mediated.

The process of condensation and polymerization makes use of some of the molecules which were released during the stage of decomposition to create new compounds. Since for the formation of new compounds, the different monomers released during decomposition are used, the resulting compounds do not have a well-defined specific structure and composition. The new condensates thus comprise of variety of chemically modified and unmodified compounds held together by hydrophobic interaction and hydrogen bonding. According to Chefetz et al. (2002), they have greater amounts of fatty and carboxylic acid carbon and lesser quantities of polysaccharide carbon, in comparison to the plant litter. Probably due to these biochemical properties or enhanced capacity for sorption, these are quite hard to decompose.

Saprotrophic fungi carry out the decomposition of litter in the soil, and mycorrhizal fungi are the other kind of soil fungi, not directly involved in the process of decomposition. The arbuscular mycorrhizal (AM) fungi form symbiotic association between higher plants and the species of fungal phylum *Glomeromycota*. This beneficial association, which is mutual for both the partners, is believed to have assisted the invasion of dry lands ca 450 Ma ago by the plants and, according to Smith and Read (2008), is adopted by the majority of the land plants. The AM symbiosis with plants is reported to contribute to fluxes of carbon between the plants and the atmosphere. Carbon is sequestered in the soil by AM fungi when the photosynthesis products from host are transferred initially to the intraradical fungal hyphae, and later to the extraradical hyphae, after which it is released into the soil (Parniske 2008). Since the AM fungi obtain their carbohydrates from the plants, they create a sink (demand for carbohydrates) through this association, draining 4–20% of carbon from the host plant, and thereby influence carbon sequestration in soil. Of the two kinds of AM fungal hyphae (intraradical and extraradical), the extraradical hyphae incorporate greater proportions of carbon. Although the life span of extraradical hyphae is relatively short, the AM fungi play significant role in carbon sequestration by virtue of the volume of hyphal biomass formed, time needed for the turnover of the fungal biomass which is influenced by various abiotic and biotic factors and also the stabilization of soil aggregates mediated by these fungal hyphae (Zhu and Miller 2003). There are reports (Leake et al. 2004) that the Glomeromycota fungi of the grassland soils represent major portion of fungal biomass in the soil, and in these systems up to 30% of the carbon in microbial biomass derives from AM fungi. Similar to this in prairie soils, the extraradical hyphae are reported to be as high as 28 m/cm^3 soil and have an annual turnover of 26%.

As mentioned earlier, the fungal hyphal biomass consists of recalcitrant compounds that reduce the rate of turnover of soil organic carbon. Chitin is the main constituent of the extraradical hyphae which is quite resistant to decomposition. Due to this the slower turnover of extraradical hyphae results in the accumulation of hyphal residues in soil. The duration of time for which the chitinous residues of the cell wall can persist in the soil could be up to 49 years (for protein- and chitin-derived by-products). AM fungi also produce glomalin, a substance similar to glycoproteins, which is quite stable in soil (Wright and Upadhyaya 1998). Radiocarbon dating study has shown that glomalin extracts have quite long residence time in soil (6–42 years), much longer than the residence time of AM fungal hyphae (Rillig et al. 2001). Carbon in glomalin form is reported to constitute up to 5% of the soil carbon in tropical soil, much greater than the microbial biomass carbon. On the basis of close correlation among the glomalin content in soil, hyphal length and soil aggregate stability, Wright and Upadhyaya (1996) propose it is a proof that the glomalin is capable of influencing the carbon storage in soil indirectly by affecting the soil aggregate stability. An estimate by Zhu and Miller (2003) suggests that with up to 30 cm depth in soil with 1.2 g/cm^3 bulk density and 50% carbon content in the dry hyphae, the AM fungal hyphae mediated soil organic carbon could be up to 900 kg/ha.

26.6 Protected and Unprotected Organic Matter Pool

On the basis of rate of death, soil microbial biomass is divided into two parts, viz. protected and unprotected. According to van Veen et al. (1985), the fraction of microbial biomass having a death rate around 0.5% on a daily basis is considered as protected pool and the fraction with death rate around 70% on similar daily basis is counted as unprotected. Protected and unprotected microbial biomass fractions in soil can be isolated by separation of microbial biomass into light fraction (LF) which is associated with plant residues and the fraction associated with silt and clay particles. According to Chotte et al. (1998), due to association with clay, the fraction is more sheltered from predation in soil. Further, Swanston et al. (2004) have pointed to the fact that LF-associated microbial biomass is metabolically less active. Several mechanisms have been proposed regarding the protection of microbial biomass in the soil, some of which are given below:

1. Microbial growth is promoted in clay when pH is maintained in the optimum limits (Stotzky and Rem 1966).
2. Clay adsorbs certain metabolites which have inhibitory effect on the microbial growth (Martin et al. 1976).
3. When microbes interact with clay, they get protection from desiccation in soil (Bushby and Marshall 1977).
4. The specific location of microbial cells in the soil; microbes residing in small pores get protection from predators like protozoa (Rutherford and Juma 1992).

Therefore, since several researchers concluded that most of the mechanisms related to the protection of microbial biomass in soil are related directly or indirectly with the clay content of the soil, clay content of the soil is principally important for the protection of microbial biomass in the soil.

26.7 Effect of Soil Management Practices

The kind of management practice followed has significant effect on the microbial biomass dynamics in the soil. According to Holland and Coleman (1987) and Allison et al. (2005), systems with relatively lower level of disturbance (abandoned agricultural soils and no-tillage systems) hold comparatively more MOM, with the following reasons for such conditions: reduced disturbance, increased fungal dominance in the system and holding greater part of the microbial biomass. A critical evaluation of the hypothesis that fungal-dominated systems are more efficient in comparison to bacteria-dominated systems suggests certain assumptions must be satisfied for this to be true, such as MGE should be greatest in communities with only fungi, lower for communities having assemblage of fungi and bacteria and still lower (or lowest) for communities dominated by bacteria only. However, these arguments are not observed as such, and there is significant overlapping in the range of MGE reported (Six et al. 2006). Although the unequivocal evidence for fact that fungi have greater MGE in comparison to bacteria is still lacking, in natural ecosystems fungi are biggest contributors of the soil microbial biomass and so also in no-tillage systems (Holland and Coleman 1987; Frey et al. 1999). Besides this, there are inherent differences in the C/N ratio of fungi and bacteria. Fungi have C/N ratio of 10, while the figure is 4 for bacterial cells (Sylvia et al. 2005). These points to the fact that for a particular amount of soil microbial biomass, carbon contained in the fungal biomass is significantly greater per unit of nitrogen than the bacterial biomass. Therefore, much more carbon is reserved in the fungal biomass than in the bacterial cells, provided both of these have the same level of protection.

Agricultural management practices include a variety of operations on land, which vary significantly in the kind and intensity of soil disturbance and thus exert different effects on the microbial biomass. It has been observed that microbial biomass is higher in undisturbed systems and decreases with increase in the disturbance in soil. Certain kinds of management practices have been found to favour increase in the level of microbial biomass, which are briefly summarized below:

1. Practices with greater carbon inputs (Schnurer et al. 1985)
2. Reduction in tillage (Beare et al. 1992; Frey et al. 1999)
3. Crop residue retention in the soil (Gupta et al. 1994)
4. Organic amendments (Hassink et al. 1991)

Systems which include the practice of crop rotations on the piece of land influence the amount of soil microbial biomass, activity of soil microbes and ratios of fungal vs bacterial biomass. According to Acosta-Martinez et al. (2003), the activity of soil enzymes increases in soils under crop rotation in comparison to monocropping practised continuously. Besides this, the kind of crop included in the rotation also has influence on the microbial biomass as legume cover crops have been found to promote the soil microbial biomass (Lupwayi et al. 1999). Microbial-derived carbohydrates in the soil are reported to be influenced more by crop rotation in comparison to plant-derived carbohydrates (Six et al. 2006). Six et al. further concluded that crop rotation practice increases carbon sequestration when compared with continuous crop management or crop rotations which have fallow periods. When cropping is intensive, it increases soil carbon along with increase in the microbial activity and biomass also; concomitantly its influence on microbial community composition increases the level of soil fungi.

The kind of tillage practiced will affect the amount of stored carbon in the soil. Ogle et al. (2003) and Six et al. (2002) propose that no-till practice (NT) and also minimum tillage show greater carbon storage when compared with conventional tillage (CT). But Peterson et al. (1998) had argued that the differences get eliminated during the fallow periods. Overall, these observations suggest that multiple management practices need to be practiced for increasing the storage of carbon in the soil. While adopting the management practice, the fact that the influence of NT and CT occurs in the upper layer of soil only (upper 5 cm) should not be left out of consideration. This is important because within this zone only, in NT soils significantly greater influence of fungal component has been indicated, summarized on the basis of fungal to bacterial ratio (Frey et al. 1999), mycorrhizal colonization (Drijber et al. 2000), greater activity of soil enzymes (Acosta-Martınez et al. 2003) and total contents of carbon and nitrogen (Feng et al. 2003).

Frey et al. (1999) analysed fungal to bacterial biomass ratios in tillage comparison experiments were carried over for longer duration of times, along with different climate gradients. They found the increase in fungal biomass and also in the ratio of biomass of fungi to bacteria, when tillage was reduced. This change was found to be positively correlated with the moisture content of the soil. This indicates that microbial communities become dominated by fungi when the tillage disturbance is reduced in the soil, which is important for the decomposition of residues and also the cycling of soil nutrients in the upper layers of the soil. Bailey et al. (2002) therefore argue that agroecosystems with reduced disturbance, which have greater resemblance to the natural systems, are mostly found to be dominated by fungi. The reduced disturbance favours the growth and activity of fungi allowing the establishment and preservation of the hyphal networks (Wardle 1995). Such systems therefore need less support in the form of inputs to sustain the decomposition of organic material and the cycling of nutrients.

26.8 Conclusion

Increase in the concentration of atmospheric carbon is a lurking danger which can have enormous impact on life of the planet earth. Terrestrial component has been proven to be an important sink of the carbon where the variety of soil microorganism carry out the sequestration of carbon initially assimilated in the bodies of the photosynthetic organisms. During the decomposition of litter, fungi transform it to forms that are very hard to degrade further. Both as the part of the fungal biomass, and also the end products of fungal decomposition of litter, fungi transform the organic matter into more recalcitrant forms. The production of recalcitrant material by fungi during the decomposition of litter greatly increases the residence time of carbon in the soil which makes the terrestrial component a valuable sink of carbon, which results in lowering of the concentration of carbon dioxide in the atmosphere. Soil microbial communities act to accumulate the carbon in their biomass, the decomposition intermediates and the end products also have variable residence lifetimes in the soil, but all these are influenced greatly by anthropogenic effect such as disturbance of soil and other kinds of management practices.

Sequestration of carbon in the terrestrial ecosystem can greatly help in reducing CO_2 concentration of atmosphere, but what is needed is adoption of management practices which can promote the sequestration of carbon in the soil.

Acknowledgement The authors gratefully acknowledge the financial assistance received from UGC-CAS.

References

Adu JK, Oades JM (1978) Utilization of organic materials in soil aggregates by bacteria and fungi. Soil Biol Biochem 10:117–122

Acosta-Martınez V, Zobeck TM, Gill TE, Kennedy AC (2003) Enzyme activities and microbial community structure in semiarid agricultural soils. Biol Fertil Soils 38:216–227

Allison SD (2012) A trait-based approach for modelling microbial litter decomposition. Ecol Lett 15:1058–1070

Allison VJ, Miller RM, Jastrow JD, Matamala R, Zak DR (2005) Changes in soil microbial community structure in a tallgrass prairie chronosequence. Soil Sci Soc Am J 69:1412–1421

Ander P, Eriksson K-E (1977) Selective degradation of wood components by white-rot fungi. Physiol Plant 41:239–248

Bailey VL, Smith JL, Bolton H Jr (2002) Fungal-to-bacterial ratios in soils investigated for enhanced C sequestration. Soil Biol Biochem 34:997–1007

Batjes NH (1996) Total carbon and nitrogen in the soils of the world. Eur J Soil Sci 47:151–163

Beare MH, Parmelee RW, Hendrix PF, Cheng W, Coleman DC, Grossley DA Jr (1992) Microbial and faunal interactions and effects on litter nitrogen and decomposition in agroecosystems. Ecol Monogr 62:569–591

Berg B, McClaugherty C (2008) Plant litter. Decomposition, humus formation, carbon sequestration. Springer, Berlin

Bousquet P, Peylin P, Ciais P, Quere CL, Friedlingstein P, Tans PP (2000) Regional changes in carbon di oxide fluxes of land and oceans since 1980. Science 290:1342–1346

Bushby HVA, Marshall KC (1977) Water status of rhizobia in relation to their susceptibility to desiccation and to their protection by montmorillonite. J Gen Microbiol 99:19–27

Chefetz B, Tarchitzky J, Deshmukh AP, Hatcher PG, Chen Y (2002) Structural characterization of soil organic matter and humic acids in particle-size fractions of an agricultural soil. Soil Sci Soc Am J 66:129–141

Chotte JL, Ladd JN, Amato M (1998) Sites of microbial assimilation, and turnover of soluble and particulate ^{14}C-labelled substrates decomposing in a clay soil. Soil Biol Biochem 30:205–218

Crawford RL (1981) Lignin biodegradation and transformation. Wiley, New York

Dlugokencky E, Tans P (2016) Trends in atmospheric carbon dioxide, National Oceanic & Atmospheric Administration, Earth System Research Laboratory (NOAA/ESRL), available at: http://www.esrl.noaa.gov/gmd/ccgg/trends/global.html. Last accessed 28 Oct 2016

Drijber RA, Doran JW, Parkhurst AM, Lyon DJ (2000) Changes in soil microbial community structure with tillage under long-term wheat-fallow management. Soil Biol Biochem 32:1419–1430
Eriksson K-E, Blanchette RA, Ander P (1990) Microbial and enzymatic degradation of wood and wood components. Springer, Berlin
Eswaran H, Van Den Berg E, Reich P (1993) Organic carbon in soils of the world. Soil Sci Soc Am J 57:192–194
Feng Y, Motta AC, Reeves DW, Burmester CH, van Santen E, Osborne JA (2003) Soil microbial communities under conventional-till and no-till continuous cotton systems. Soil Biol Biochem 35:1693–1703
Freeman C, Ostle N, Kang H (2001) An enzymic "latch" on a global carbon store – a shortage of oxygen locks up carbon in peatlands by restraining a single enzyme. Nature 409:149
Frey SD, Elliott ET, Paustian K (1999) Bacterial and fungal abundance and biomass in conventional and no-tillage agroecosystems along two climatic gradients. Soil Biol Biochem 31:573–585
Frey SD, Gupta V, Elliott ET, Paustian K (2001) Protozoan grazing affects estimates of carbon utilization efficiency of the soil microbial community. Soil Biol Biochem 33:1759–1768
George TS, Richardson AE, Simpson RJ (2005) Behaviour of plant-derived extracellular phytase upon addition to soil. Soil Biol Biochem 37:977–988
Gonzalez-Gaya B, Fernandez-Pinos MC, Morales L, Mejanelle L, Abad E, Pina B, Duarte CM, Jimenez B, Dachs J (2016) High atmosphere-ocean exchange of semivolatile aromatic hydrocarbons. Nat Geosci 9:438–442
Guggenberger G, Frey SD, Six J, Paustian K, Elliott ET (1999) Bacterial and fungal cell-wall residues in conventional and no-tillage agroecosystems. Soil Sci Soc Am J 63:1188–1198
Gupta VVSR, Roper MM, Kirkegaard JA, Angus JF (1994) Changes in microbial biomass and organic matter levels during the first year of modified tillage and stubble management practices on a red Earth. Aust J Soil Res 32:1339–1354
Hassink J, Lebbink G, Van Veen JA (1991) Microbial biomass and activity of a reclaimed-polder soil under a conventional and a reduced-input farming system. Soil Biol Biochem 23:507–514
Haynes RJ, Beare MH (1997) Influence of six crop species on aggregate stability and some labile organic matter fractions. Soil Biol Biochem 29:1647–1653
Highley TL (1988) Cellulolytic activity of brown-rot and white-rot fungi on solid media. Holzforschung-Int J Biol Chem Phys Technol Wood 42:211–216
Holland EA, Coleman DC (1987) Litter placement effects on microbial and organic matter dynamics in an agroecosystem. Ecology 68:425–433
Houghton RA (2000) A new estimate of global sources and sinks of carbon from land use change. Eos 81:S281
Jastrow JD, Amonette JE, Bailey VL (2007) Mechanisms controlling soil carbon turnover and their potential application for enhancing carbon sequestration. Clim Chang 80:5–23
Jenny H (1980) The soil resource: origin and behavior. Springer, New York
Joos F, Spahni R (2008) Rates of change in natural and anthropogenic radiative forcing over the past 20,000 years. Proc Natl Acad Sci 105:1425–1430
Lal R (2008) Sequestration of atmospheric CO_2 in global carbon pools. Energy Environ Sci 1:86–100
Le Quéré C, Andrew RM, Canadell JG, Sitch S, Korsbakken JI, Peters GP, Manning AC, Boden TA, Tans PP, Houghton RA, Keeling RF, Alin S, Andrews OD, Anthoni P, Barbero L, Bopp L, Chevallier F, Chini LP, Ciais P, Currie K, Delire C, Doney SC, Friedlingstein P, Gkritzalis T, Harris I, Hauck J, Haverd V, Hoppema M, Goldewijk KK, Jain AK, Kato E, Körtzinger A, Landschützer P, Lefèvre N, Lenton A, Lienert S, Lombardozzi D, Melton JR, Metzl N, Millero F, Monteiro PMS, Munro DR, Nabel JEMS, Nakaoka S, Brien KO, Olsen A, Omar AM, Ono T, Perrot D, Poulter B, Rödenbeck C, Salisbury J, Schuster U, Schwinger J, Séférian R, Skjelvan I, Stocker BD, Sutton AJ, Takahashi T, Tien H, Tilbrook B, Laan-Luijkx V, Werf V, Viovy N, Walker AP, Wiltshire AJ, Zaehle S (2016) Global carbon budget 2016. Earth Syst Sci Data 8:605–649

Leake J, Johnson D, Donnelly D, Muckle G, Boddy L, Read D (2004) Networks of power and influence: the role of mycorrhizal mycelium in controlling plant communities and agroecosystem functioning. Can J Bot 82:1016–1045

Luo Y, White LW, Canadell JG, DeLucia EH, Ellsworth DS, Finzi A, Lichter J, Schlesinger WH (2003) Sustainability of terrestrial carbon sequestration: a case study in Duke Forest with inversion approach. Glob Biogeochem Cycles 17:1021

Lupwayi NZ, Rice WA, Clayton GW (1999) Soil microbial biomass and carbon dioxide flux under wheat as influenced by tillage and crop rotation. Can J Soil Sci 79:273–280

Martin JP, Haider K (1979) Biodegradation of ^{14}C-labeled model and cornstalk lignins, phenols, model phenolase humic polymers, and fungal melanins as influenced by a readily available carbon source and soil. Appl Environ Microbiol 38:283–289

Martin JP, Haider K (1986) Influence of mineral colloids on turnover rates of soil organic carbon. In: Huang PM, Shnitzer M (eds) Interactions of soil minerals with natural organics and microbes, SSSA Special Pub No. 17. SSSA, Madison, pp 283–304

Martin JP, Filip Z, Haider K (1976) Effect of montmorillonite and humate on growth and metabolic activity of some actinomycetes. Soil Biol Biochem 8:409–413

ten Have R, Teunissen PJ (2001) Oxidative mechanisms involved in lignin degradation by white-rot fungi. Chem Rev 101:3397–3414

Nakas JP, Klein DA (1979) Decomposition of microbial cell components in a semi-arid grassland soil. Appl Environ Microbiol 38:454–460

Nilsson T, Daniel G (1989) Chemistry and microscopy of wood decay by some higher ascomycetes. Holzforschung-Int J Biol Chem Phys Technol Wood 43:11–18

Ogle SM, Breidt FJ, Eve M, Paustian K (2003) Uncertainty in estimating land use and management impacts on soil organic carbon storage for US agricultural lands between 1982–1997. Glob Chang Biol 9:1521–1542

Parniske M (2008) Arbuscular mycorrhiza: the mother of plant root endosymbioses. Nat Rev Microbiol 6:763–775

Parton WJ, Schimel DS, Cole CV, Ojima DS (1987) Analysis of factors controlling soil organic matter levels in Great Plains grasslands. Soil Sci Soc Am J 51:1173–1179

Payne WJ (1970) Energy yields and growth of heterotrophs. Ann Rev Microbiol 24:17–52

Peterson GA, Halvorson AD, Havlin JL, Jones OR, Lyon DJ, Tanaka DL (1998) Reduced tillage and increasing cropping intensity in the Great Plains conserves soil C. Soil Tillage Res 47:207–218

Perveen N, Barot S, Alvarez G, Klumpp K, Martin R, Rapaport A, Herfurth D, Louault F, Fontaine S (2014) Priming effect and microbial diversity in ecosystem functioning and response to global change: a modeling approach using the SYMPHONY model. Glob Chang Biol 20:1174–1190

Prentice IC, Farqygar GD, Fasham JR, Goulden ML, Heimann M, Jaramillo VJ, Kheshgi HS, Quere CL, Scholes RJ, Wallace DWR (2001) The carbon cycle and atmospheric carbon di oxide. In: Houghton JT, Ding Y, Griggs DJ, Noguer M, van der Linden PJ, Dai X, Maskell K, Johnson CA (eds) Climate change 2001: the scientific basis (contribution of working group I to the third assessment report of the intergovernmental panel on climate change). Cambridge University Press, Cambridge

Regnier P, Friedlingstein P, Ciais P, Mackenzie FT, Gruber N, Janssens IA, Laruelle GG, Lauerwald R, Luyssaert S, Andersson AJ, Arndt S, Arnosti C, Borges AV, Dale AW, Gallego-Sala A, Goddéris Y, Goossens N, Hartmann J, Heinze C, Ilyina T, Joos F, La Rowe DE, Leifeld J, Meysman FJR, Munhoven G, Raymond PA, Spahni R, Suntharalingam P, Thullner M (2013) Anthropogenic perturbation of the carbon fluxes from land to ocean. Nat Geosci 6:597–607

Rillig MC, Wright SF, Nichols KA, Schmidt WF, Torn MS (2001) Large contribution of arbuscular mycorrhizal fungi to soil carbon pools in tropical forest soils. Plant Soil 233:167–177

Rutherford PM, Juma NG (1992) Influence of soil texture on protozoa-induced mineralization of bacterial carbon and nitrogen. Can J Soil Sci 72:183–200

Schimel JP, Weintraub MN (2003) The implications of exoenzyme activity on microbial carbon and nitrogen limitation in soil: a theoretical model. Soil Biol Biochem 35:549–563

Schimel DS, House JI, Hibbard KA, Bousquet P, Ciais P, Peylin P, Braswell BH, Apps MJ, Baker D, Bondeau A, Canadell J, Chukina G, Cramer W, Denning AS, Field CB, Friedlingstein P, Goodale C, Heimann M, Houghton RA, Melillo JM, Moore B III, Murdiyarso D, Noble I, Pacala SW, Prentice IC, Raupach MR, Rayner PJ, Scholes RJ, Steffen WL, Wirth C (2001) Recent patterns and mechanisms of carbon exchange by terrestrial ecosystems. Nature 414:169–172

Schnurer J, Clarholm M, Rosswall T (1985) Microbial biomass and activity in an agricultural soil with different organic matter contents. Soil Biol Biochem 17:611–618

Scripps: The Keeling Curve (2013) Available at: http://keelingcurve.ucsd.edu/. Last accessed 7 Nov 2013

Shen S, Tu S-I, Taylor RW (2002) Interactions of enzymes with clays and applications in bioremediation. In: Soil mineralogy with environmental applications. Soil Science Society of America, Madison, pp 795–817

Simpson RT, Frey SD, Six J, Thiet RK (2004) Preferential accumulation of microbial carbon in aggregate structures of no tillage soils. Soil Sci Soc Am J 68:1249–1255

Six J, Conant RT, Paul EA, Paustian K (2002) Stabilization mechanisms of soil organic matter: implications for C-saturation of soils. Plant Soil 241:155–176

Six J, Frey SD, Thiet RK, Batten KM (2006) Bacterial and fungal contributions to carbon sequestration in agroecosystems. Soil Sci Soc Am J 70:555–569

Smith SE, Read DJ (2008) Mycorrhizal symbiosis. Academic, London

Suberkropp K, Weyers H (1996) Application of fungal and bacterial production methodologies to decomposing leaves in streams. Appl Environ Microbiol 62:1610–1615

Swanston C, Homann PS, Caldwell BA, Myrold DD, Ganio L, Sollins P (2004) Long-term effects of elevated nitrogen on forest soil organic matter stability. Biogeochemistry 70:229–252

Sylvia DM, Fuhrmann JJ, Hartel PG, Zuberer DA (2005) Principles and applications of soil microbiology, 2nd edn. Pearson Education, Upper Saddle River

Stotzky G, Rem LT (1966) Influence of clay minerals on microorganisms. I. Montmorillonite and kaolinite on bacteria. Can J Microbiol 12:547–563

van Veen AJ, Ladd JN, Amato M (1985) Turnover of carbon and nitrogen through the microbial biomass in a sandy loam and a clay soil incubated with 14C(U) glucose and 15N(NH4)2SO4 under different moisture regimes. Soil Biol Biochem 17:747–756

Vetter YA, Deming JW, Jumars PA, Krieger-Brockett BB (1998) A predictive model of bacterial foraging by means of freely released extracellular enzymes. Microb Ecol 36:75–92

Wardle DA (1995) Impacts of disturbance on detritus food webs in agro-ecosystems of contrasting tillage and weed management practices. Adv Ecol Res 26:107–185

Waring BG, Averill C, Hawkes CV (2013) Differences in fungal and bacterial physiology alter soil carbon and nitrogen cycling: insights from meta-analysis and theoretical models. Ecol Lett 16:887–894

Wright SF, Upadhyaya A (1996) Extraction of an abundant and unusual protein from soil and comparison with hyphal protein of arbuscular mycorrhizal fungi. Soil Sci 161:575–586

Wright SF, Upadhyaya A (1998) A survey of soils for aggregate stability and glomalin, a glycoprotein produced by hyphae of arbuscular mycorrhizal fungi. Plant Soil 198:97–107

Zhu Y-G, Miller RM (2003) Carbon cycling by arbuscular mycorrhizal fungi in soil–plant systems. Trends Plant Sci 8:407–409

Chapter 27
Fungal Infections and Intervention in Diabetic Complication

Amarish Kumar Sharma and Anjana Rana Sharma

Abstract Prevalence of diabetes has globally increased including our homeland, India. India ranks among top nations in the world map, battling diabetes. Diabetes mellitus is commonly associated with fungal infections; however extensive and ubiquitous data availability of fungal infections in diabetic patients is sparse. Many studies has been undertaken to profile different fungal organism, its types and their associated infections and its specific treatments in diabetes mellitus patients. Fungal infections associated with foot are common in general diabetic population, but its consequences for people whose peripheral neurological or vascular status has been compromised, for them, are life-threatening. Skin infections and disorders are prominent in underdiagnosed diabetic mellitus patients. Factors like hyperglycaemia and advanced glycation end products could lead to complications like ulcerations, skin lesions, diabetic foot, etc. Therefore pathophysiology of early skin disorders and dermo-cosmetic management becomes important screening factor for diabetic mellitus patients. Therefore effective precautionary measurements and education become a prime importance to diagnose, screen and control the fungal disease associated with diabetes mellitus patients.

Keywords Diabetes mellitus · Immunization · Infections · Vaccines

A. K. Sharma (✉)
School of Bioengineering & Biosciences, Lovely Professional University, Phagwara, Punjab, India
e-mail: amarish.19824@lpu.co.in

A. R. Sharma
Gyan Vihar University, Jaipur, Rajasthan, India

P. Gehlot, J. Singh (eds.), *Fungi and their Role in Sustainable Development: Current Perspectives*, https://doi.org/10.1007/978-981-13-0393-7_27

27.1 Diabetes Mellitus (DM)

Diabetes mellitus is a metabolic syndrome which becomes chronic under hyperglycaemic conditions and mostly affects the body's ability to metabolize the energy input which comes through food intake, in a balanced manner. It is characterized by availability of significantly high amount of blood sugar, which may be due to insufficient response of insulin hormone to the availability of blood sugar in the systemic circulation or insufficient amount of pancreatic cells to secrete required insulin hormone, respectively.

The symptoms of onset of diabetes mellitus are characterized by high blood sugar level along with frequent urination and enhanced level of thirst and appetite, respectively. If DM is left untreated for long, it may cause several acute complications which included non-ketotic hyperosmolar coma, diabetic ketoacidosis or death. Long-term complications associated with diabetes are cardiovascular disease, kidney failure, foot ulcers, fungal skin infections and eye damage.

The onset of Diabetes mellitus depends on two factors. (a) Either the pancreas is not producing sufficient amount of insulin hormone to meet up the glucose uptake in blood circulation or (b) the body cells are not responding to the presence of Insulin (Mohan et al. 2007) (Fig. 27.1).

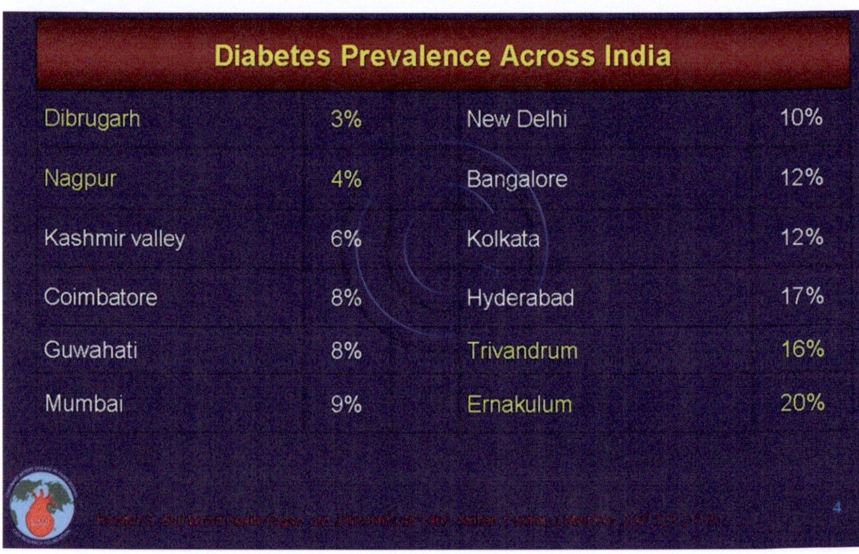

Fig. 27.1 Diabetes prevalence across India (Reddy et al. 2006)

27.2 Diabetes and Its Types

World Health Organization recognizes diabetes mellitus in three major types:

27.3 Diagnosis

Clinically a patient is termed to be diabetic, when he is having a fasting plasma glucose level in excess of 120 mg/dL and having post-prandial plasma glucose level in excess of 200 mg/dL. A normal person has glucose level in the range of 100 mg/dL.

27.3.1 Type 1 Diabetes

Type 1 diabetes is due to autoimmune destruction of pancreatic beta cells, which leads to under-secretion of required insulin for glucose uptake. It was earlier termed as insulin-dependent diabetes mellitus (IDDM) or juvenile-onset diabetes, which accounts for 10% of total diagnosed cases of diabetes till date. Risk factors associated with this disorder is generally due to autoimmune, genetic and environmental factors. Type 1 diabetes is characterized by recurrent loss of pancreatic cells, i.e. islets of Langerhans, an insulin-producing cellular centre, due to T cell-mediated autoimmunogenic attack, leading to progressive decrease in insulin levels in body fluids. The principle treatment of type 1 diabetes is careful profiling of blood glucose level at regular interval of time and replenishing.

27.3.2 Type 2 Diabetes

Type 2 diabetes is also termed as non-insulin-dependent diabetes (NIDDM). It accounts for more than 90% of all screened diabetic cases yet diagnosed. It occurs primarily due to lethargic and sedentary lifestyle, a gift of increasing urbanization, which ultimately leads to obesity and stress, and due to genetic manifestation.

Type 2 diabetes is a metabolic disorder which mainly happens due to the pathological condition where insulin gets desensitized to the presence of glucose in the body fluids, i.e. insulin becomes resistant to glucose present in body fluids, which leads to hyper-accumulation of blood sugar in the body fluids which results in many disease outcomes. Some of the common risk factors associated with type 2 diabetes are cardiovascular disease, retinopathy, impaired glucose tolerance, obesity, dementia, kidney failure and physical inactivity.

27.4 Mechanism of Action (Fig. 27.2)

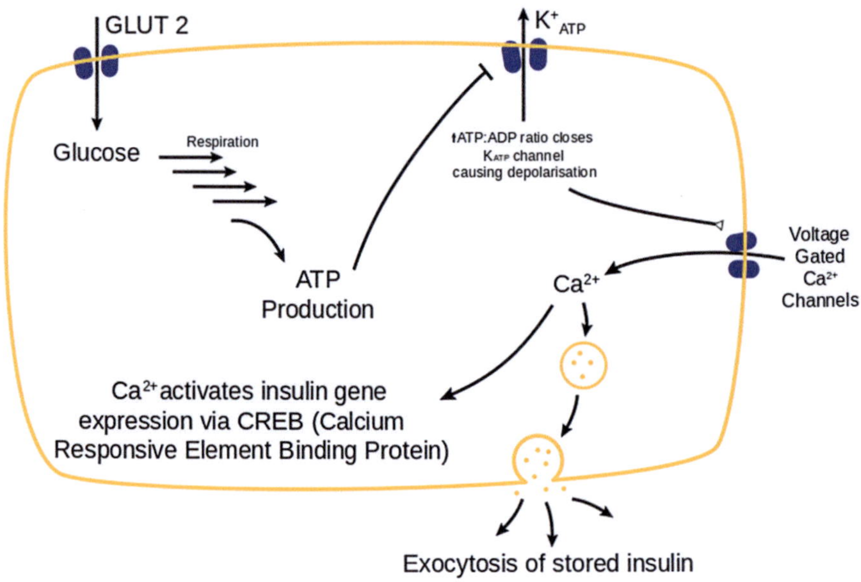

Fig. 27.2 Mechanism of action of pancreatic beta cells for the instantaneous release of insulin. Production of insulin is steadily constant, which is triggered by food intake chiefly glucose source

Table 27.1 WHO diabetes diagnostic criteria (Vijan 2010)

Diabetes diagnostic criteria as per WHO recommendations				
Metabolic condition	Post-prandial glucose – 2 h	Glucose fasting	HbA_{1C}	
Unit	mmol/l(mg/dl)	mmol/l(mg/dl)	mmol/mol	DCCT %
Normal	<7.8 (<140)	<6.1 (<110)	<42	<6.0
Glucose tolerance-impaired	≥7.8 (≥140)	<7.0 (<126)	42–46	6.0–6.4
Fasting glycaemia-impaired	<7.8 (<140)	≥6.1(≥110) and <7.0(<126)	42–46	6.0–6.4
Diabetes mellitus	≥11.1 (≥200)	≥7.0 (≥126)	≥48	≥6.5

27.5 Gestational Diabetes (Occurring During Pregnancy)

Gestational diabetes occurs during pregnancy of women. Its symptoms are moreover similar to that of type 2 diabetes that involved insulin resistance, which may happen due to pregnancy hormones. It occurs temporarily in pregnant women and ranges in between 2% and 10% of all pregnancies. However, after pregnancy, almost 10% of women affected with gestational diabetes are prone to have type 2 diabetes mellitus. Gestational diabetes is temporary and treatable but requires careful medical attention and full dietary control.

If kept untreated, gestational diabetes can be dangerous to both pregnant mother and the developing foetus. The risk factors associated to baby include macrosomia, congenital heart disease and respiratory distress syndrome. Enhanced level of bilirubin in blood could lead to destruction of red blood cells (World Health Organization 2006) (Table 27.1).

27.6 Pathophysiology of Diabetes

Blood glucose is entirely regulated by the coordinated and synergistic activity of pancreatic hormones, namely, insulin, glucagon and somatostatin. Insulin is the principle hormone which regulated the glucose homeostasis in blood circulation in human system, especially in adipose and muscle cell, respectively. Insulin acts on the blood glucose molecules via insulin growth factor-1 (IGF-1). There could be two possible reasons for the onset of diabetic condition. First, either the insulin is not produced in prerequisite, or it is insensitive to the presence of glucose in the body system. Therefore we can conclude that the deficiency of insulin or the insulin receptor insensitivity for sugar molecule in blood vasculature plays a vital role in the onset of the metabolic disorder (Barrett 2010).

Glucose is available to the body tissues for energy metabolism through intestinal absorption of digested food, breakdown of stored glucose in the form of glucagon in liver and through production of glucose through gluconeogenesis. Insulin plays a pivotal role in regulating the glucose level either through inhibiting the breakdown of glycogen or through the process of gluconeogenesis, or it can enhance the conversion of glucose into its storage form i.e. into triacylglycerides or glycogen (Gardner and Greenspan 2007).

Insulin secretion is activated by beta cells of islets of Langerhans in pancreas, in coordinated and feedback response to the rising levels of blood glucose in vasculature, typically post-prandial. In hypoglycaemic condition, insulin secretion is reduced to a great extent, simultaneously breaking down the glycogen into glucose to saturate the blood with normal glucose concentration for cellular energy metabolism (American Diabetes Association 2014).

27.7 Epidemiology

Epidemiology refers to the statistical, time-based, prevalence and incident-based study for most of the non-communicable disease. Non-communicable disease such as type 2 diabetes has shown a dramatic increase in past decade. The World Health Organization has estimated that almost 9% of total world population was affected with diabetes mellitus, and more than 90% of them were specifically affected with type 2 diabetes.

Five million deaths occur every year mostly due to type 2 diabetes. Deaths happen mostly due to cardiovascular disease for the above five million deaths, which signifies the closeness of both the disease. It is also estimated that more six million deaths is expected globally by 2030. It has been correlated with many studies and reports that type 2 diabetes is very closely associated with obesity and cardiovascular disease, and most of the populations from middle income and developing countries are affected with this metabolic disorder, where urbanization and rapid changing sedentary lifestyle are more prevalent (Table 27.2).

Table 27.2 Global estimates of diabetes prevalence for 2013 and projections for 2035 (Guariguata et al. 2014)

	2013			2035			
Region	Population (20–79 years) millions	Number of people with diabetes (20–79 years) millions	Comparative diabetes prevalence (20–79 years) millions	Population (20–79 years) millions	Number of people with diabetes (20–79 years) millions	Comparative diabetes prevalence (20–79 years) millions	Increase in the number of people with diabetes
AFR	407.8	19.8	5.7	775.5	41.4	6	109.1
EUR	658.7	56.3	6.8	668.7	68.9	7.1	22.4
MENA	347.5	34.6	10.9	583.7	67.9	11.3	96.2
NAC	334.9	36.7	9.6	404.5	50.4	12.3	37.3
SACA	300.5	24.1	8.2	394.2	38.5	8.2	59.8
SEA	883.2	72.1	8.7	1216.9	123	9.4	70.6
WP	1613.2	138.2	8.1	1818.2	201.8	8.4	46
World	4572.9	381.8	8.3	5861.7	591.9	8.8	55

AFR Africa, *EUR* Europe, *MENA* Middle East and North Africa, *NAC* North America and Caribbean, *SACA* South and Central America, *SEA* South East Asia, *WP* Western Pacific

27.8 Fungal Infections and Diabetes

Diabetes mellitus raises a growing concern to world fraternity due to its serious complications that ends with devastating effects of disease. A projected rise from 172 million in 2000 to 367 million in 2030 is hypothesized looking into the prevalence of the disease among world populations. Glucose insensitivity is the immediate stage between normal glucose sensitivity and diabetes. Either the insulin gets desensitized in presence of blood sugar, or there is insufficient secretion of insulin due to pancreatic cell loss due to autoimmunity. It results in ketoacidosis and hyperglycaemia causing dysfunction of the immune system. In early stages, if left untreated properly, such infections may get complicated and can prove even more lethal. Hence it becomes urgently important to recognize signs and symptoms of such infections and treat the patients effectively through a diabetologist or dermatologist (Muller et al. 2005).

In general, there is an acute increase in morbimortality due to onset of infectious disease, which is more serious and/or frequent in patients suffering with diabetes mellitus. The hyperglycaemic environment due to diabetic condition favours immune dysfunction (suppression of antioxidant system and humoral immunity, damage of functional activity of neutrophil), dysmotility of urinary system and gastrointestinal system and angiopathy. Many of these complications are prominent in diabetic patients such as foot infection, malignant external otitis, gangrenous cholecystitis and rhinocerebral mucormycosis. Fungal foot infections in populations affected with peripheral neurological or vascular disorder is compromised by the diabetes sequel and could lead to more serious complications. Skin disorders are often neglected and occasionally underdiagnosed in metabolic disorder especially type 1 and 2 diabetes, respectively, which results in cutaneous infection and/or dry skin and pruritus infections. Ulceration and ulcerations are the most common skin disorders associated under hyperglycaemia and advanced glycation end products condition, which can end up in major complications. Limited data is available regarding early-stage skin disorders in diabetic patients, even as diabetic's skin disorders are consistent in the literature. Morbidity due to diabetes can be extensively controlled through early-stage treatment (skin hydration, orthotic devices), disease control and public awareness (Venmans 2007).

Research studies showed that myco-infections caused due to *Candida* and intertrigo in men are highly indicative of diabetic conditions. Moreover, glossitis, paronychia and onychomycosis are more prominent and frequent (Saifullah 2009). Recent studies have proved that dermatophyte infections like *Epidermophyton floccosum* and *Trichophyton mentagrophytes* and *T. rubrum* are commonly associated with diabetes (Lauder and Binienda 2005) (Tables 27.3, 27.4, 27.5 and 27.6).

Table 27.3 Clinical classification of fungal system (Kapur 2016)

Site of infection	Classification	Example
Dermal and subcutaneous tissue	Subcutaneous	Sporotrichosis
Multiple internal organ	Systemic; opportunistic	Candidiasis, cryptococcosis, mucormycosis
Deep inside the epidermis and nails	Cutaneous	Dermatophytosis
Topical skin surface and hair	Superficial	Malasseziasis (tinea versicolor)

Table 27.4 As per data of collected surface for diabetic and nondiabetic patients, frequency of identified species of *Candida* species (Volpato et al. 2013)

Species of *Candida* spp.	Type of collected surface											
	Mucosa		Sensitive film		Nonsensitive film		Sensor		Device		Total	
	D	ND	D	ND	D	ND	D	ND	D	ND	D	ND
Albicans	93	110	97	84	109	115	65	70	56	52	420	431
Krusei	4	5	2	1	–	4	1	1	2	2	9	13
Parapsilosis	6	9	4	7	4	5	2	2	5	4	21	27
Guilliermondii	1	15	4	16	1	14	–	13	2	14	8	72
Tropicalis	18	25	10	28	11	32	13	31	–	20	52	136
Total	122	164	117	136	125	170	81	117	65	92	510	679

D diabetic, *ND* nondiabetic

Table 27.5 Fungal infections in diabetes (Bansal et al. 2008; Nowakowska et al. 2004)

Possible increased incidence cases	Pulmonary mucormycosis	(a) Infection of biliary tract
	Invasive aspergillosis	(b) Postoperative peritonitis
	Dermatophytoses	
	Candidiasis	
Increased incidence cases	Candidiasis	(a) Cutaneous, (b) prostatic abscess and (c) peritonitis: patients undergoing peritoneal dialysis
Most prominent and increased incidence cases	Mucormycosis candidiasis aspergillosis	(a) Rhino cerebral, (b) cutaneous, (c) vulvovaginal, (d) candiduria, (e) ascending pyelonephritis, (f) otitis

Table 27.6 Synthetic drugs for fungal treatment

Antifungal agents	Chemical structure	Mechanism of action (treatment)
Clotrimazole	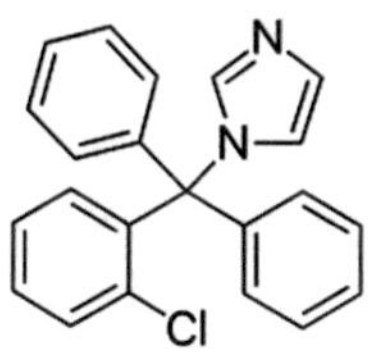	Fungal cell permeability is altered as clotrimazole binds to phospholipids in the cell membrane and ergosterol biosynthesis is subsequently inhibited which is responsible for cell membrane production. This leads to loss of intracellular elements resulting into cell death (Sravanthi et al. 2014)
Miconazole	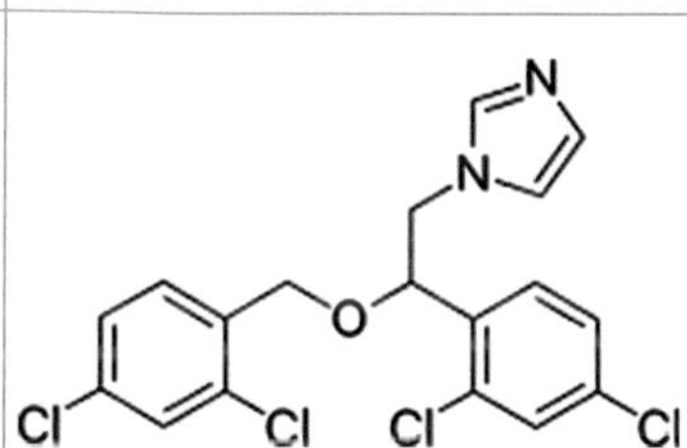	Miconazole inhibits the biosynthesis of ergosterol, an important component of fungal wall biosynthesis. Cytochrome-P450 is an important element in the conversion of lanosterol to ergosterol through its interaction with 14-alpha-demethylase. Miconazole acts by hampering the endogenous respiration and inhibiting the yeast transformation to its mycelial form. It also inhibits the purine uptake thereby impairing he phospholipid biosynthesis (Duret et al. 2006)
Ketoconazole	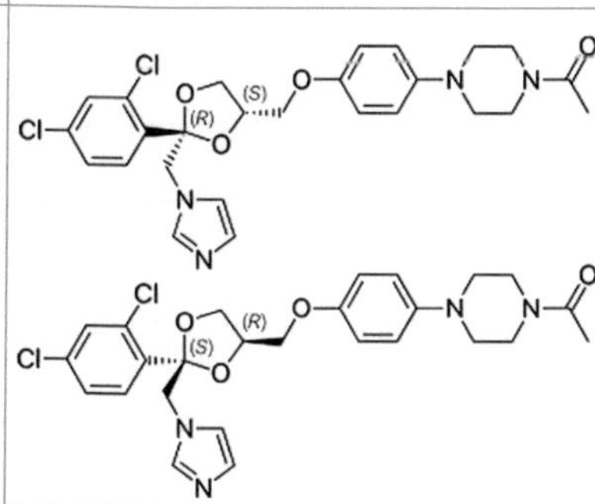	A synthetic imidazole antifungal drug, sold as oral formulation as well as formulations for topical administration. This drug interferes with the fungal biosynthesis of ergosterol, an important constituent of fungal cell membrane. It inhibits the enzyme cytochrome P450 14 alpha-demethylase (Loose et al. 1983)

(continued)

Table 27.6 (continued)

Antifungal agents	Chemical structure	Mechanism of action (treatment)
Econazole	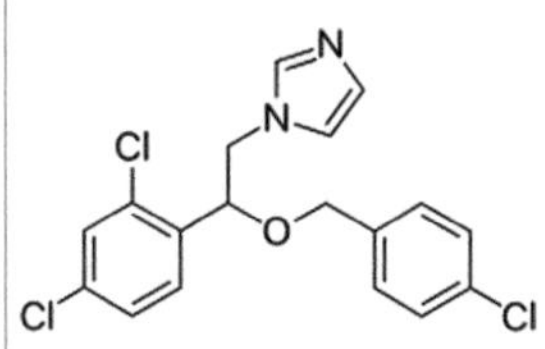	An antifungal drug sold under the name of Spectrazole, highly effective to treat skin infections. It interacts with 14-alpha-demethylase, an essential enzyme to convert lanosterol to ergosterol, which plays an important role in fungal cell membrane synthesis.
		The drug inhibits endogenous respiration, interacts with membrane phospholipids and impairs its biosynthesis (Thienpont et al. 1975)
Itraconazole	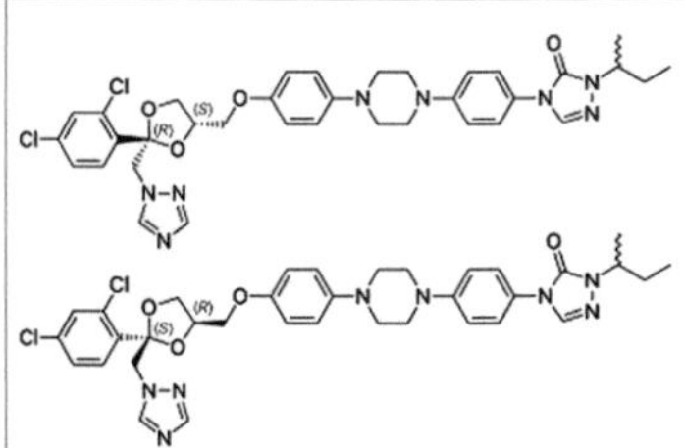	It is a triazole antifungal agent effective against antifungal infections, and mostly it is orally administered. It inhibits the fungal mediated biosynthesis of ergosterol via inhibition of lanosterol 14 alpha-demethylase. It has been observed to inhibit sonic-hedgehog pathway and angiogenesis. It has also been investigated to be a potent anticancer agent (Trevor et al. 2010)
Micafungin	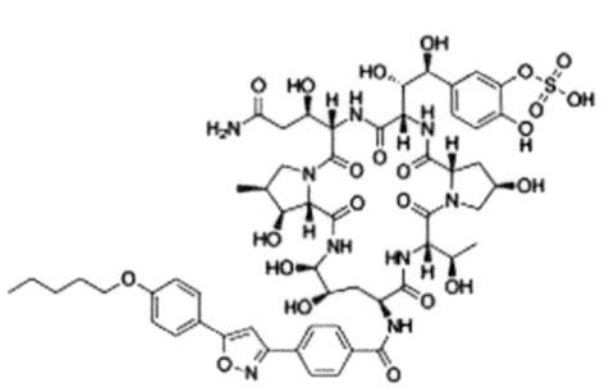	It is a potent antifungal secondary metabolite derived from fungi as an essential defence mechanism for competition of food nutrients and survivability. It inhibits the production of beta-1,3 glucan, an essential constituent of fungal cell wall. It is administered through intravenous route (Kohno et al. 2004)
Voriconazole	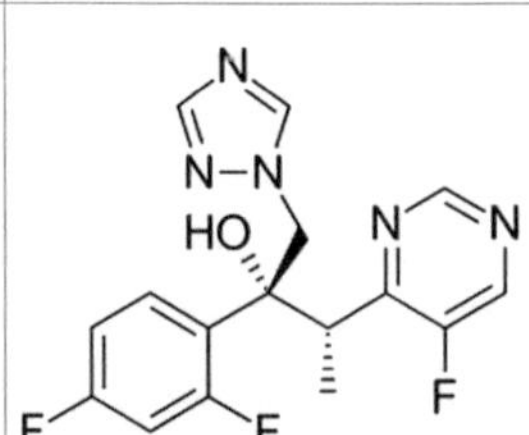	A triazole antifungal medication generally used to treat patients who are immune-compromised and suffering with serious fungal infections. It inhibits the action of cytochrome P450(CYP 450)-dependent 14 alpha-lanosterol demethylase. 14 alpha-lanosterol demethylase demethylation results in the depletion of ergosterol, an important component for fungal cell wall synthesis (Girmenia et al. 2005)

(continued)

Table 27.6 (continued)

Antifungal agents	Chemical structure	Mechanism of action (treatment)
Griseofulvin	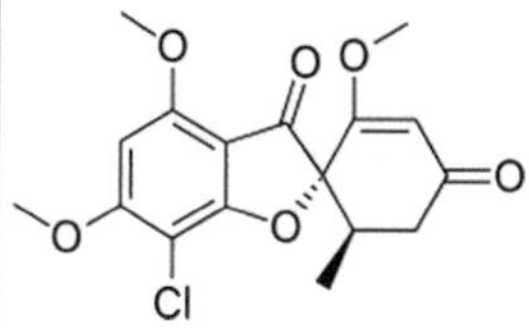	This drug is fungistatic which inhibits or restricts fungal cell mitosis and nuclear acid synthesis. Binding to alpha- and beta-tubulin, it subsequently interferes the cellular function of spindle and cytoplasmic microtubules (Grove et al. 1952)
Nystatin	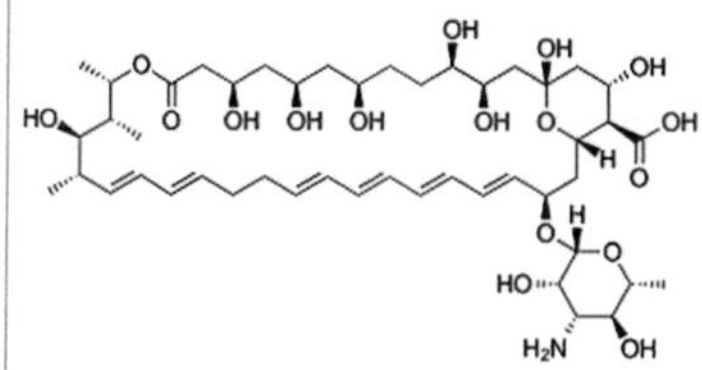	Commercially known as Mycostatin, it is a potent antifungal medication to treat chronic candidiasis. It actively binds with ergosterol, an essential fungal cell wall component and forms pores in the cell membrane that results into potassium ions leakage which finally ends up in acidification and fungus death (Akaike and Harata 1994)
Amphotericin B	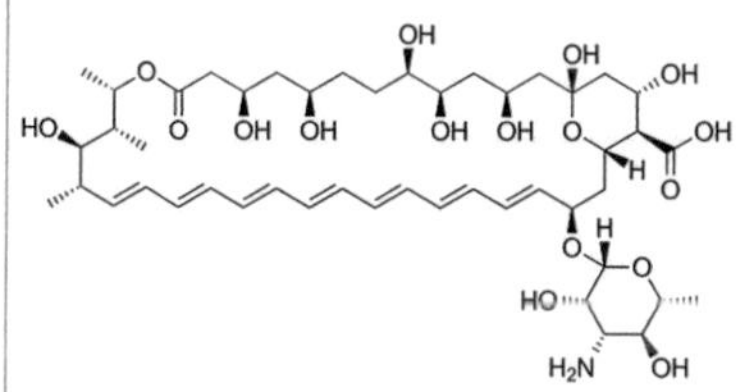	It is a polyene, extensively used as antifungal agent generally to treat aspergillosis, candidiasis and cryptococcosis. It is also used to treat leishmaniasis. It irreversibly binds with ergosterol-forming pores in the membrane which causes in rapid leakage of essential monovalent ions such as K^+, Na^+ and H^+ and results in subsequent disintegration of fungal cell membrane and ends up with cell death (Mesa-Arango et al. 2012)
Fluconazole	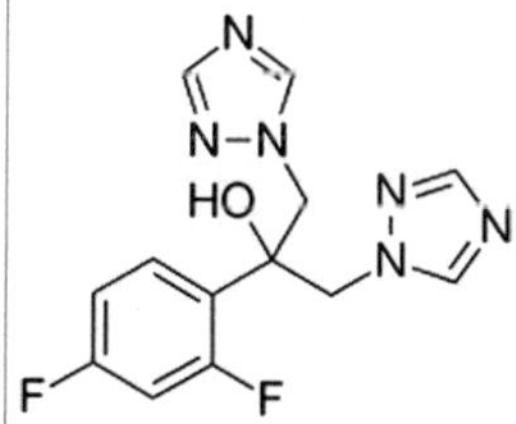	It is an antifungal drug specifically used for the treatment of candidiasis, cryptococcosis and dermatophytosis. It is a triazole class of antifungal agent which inhibits the expression of cytochrome P450 enzyme 14α-demethylase. The biosynthetic conversion from lanosterol to ergosterol is inhibited, an essential component of fungal cytoplasmic membrane (Udy et al. 2014)

27.9 Physiopathology of Fungal Infections in Diabetes

Fungi are eukaryotic aerobic cells, more complex than bacteria, which grow best at 25–30 °C. Fungi, which generally cause cutaneous and subcutaneous diseases, grow poorly at temperatures more than 37 °C. A myco-infection usually associated with DM usually appears on the skin, as the microorganism live on skin protein, keratin. This protein makes up the skin, hair and nail. Exhaustive research has already been conducted to understand the pathogenesis of immune dysfunction in diabetes mellitus. Leukocyte chemotaxis and its adherence is one of the phagocytic mechanisms, which is progressively impaired under high blood glucose concentration and diabetic acidosis (Peleg et al. 2007).

Leucocyte function and several other immune functional responses are supressed and extensively compromised. Diabetic patients are very much susceptible to urinary tract infection, skin infection and many more. Urinary tract infection is considered to be most troublesome and problem creating because urine rich in glucose concentration supports an enriched culture medium. If the diabetic patient has developed autonomic neuropathy, which could end up in urinary retention from poor bladder emptying. This condition may cause, Pyelonephritis, a chronic kidney infection, due to urinary obstructions. Urinary bladder infection due to hyperglycaemic complication could lead to devastating renal papillary necrosis. Mucormycosis is a dreaded fungal infection, due too poorly controlled diabetes. This fatal infection originates in the nasopharynx and communicates rapidly to the orbit and brain (Madhu et al. 2015).

Opsonization and phagocytosis of non-self-microorganism are promoted by antibody-mediated beta cellularis complement system. The products of complement system also induce signals for the B-lymphocyte activation and antibody production. Some studies claimed that increased glycation in diabetes mellitus is associated with interleukin-1 and interleukin-6 secretion by monocytes and mononuclear cells (Geerlings and Hoepelman 1999).

27.10 Pulmonary Fungal Infections and Diabetes

Functions of respiratory epithelium and ciliary motility along with alterations of host immune defence system in the entire body are the characterizations of pulmonary infections in diabetic or hyperglycaemic condition. They are further characterized by serious clinical manifestations, longer durations and chronic complications and increased mortality. It has been observed that the mortality rate of diabetic patients with chronic pulmonary infections with end-stage renal disease is tenfold than in general complications.

Mucor infections are more prone in patients having metabolic disorder especially type 2 diabetes mellitus. The decrease ability of bronchoalveolar macrophages to inhibit generation of fungal spore's results in enhanced pulmonary infections characterized with haematogenous disseminations and ends up in patient's death. Hemoptysis could occur in case of vascular invasion. Diabetic patients undergone with renal transplants are moreover observed to be affected with mucormycosis. Proper diagnosis and hyperglycaemic control and acidosis and broad-spectrum antibiotic treatment of amphotericin B with surgical debridement could be the most acceptable and beneficial treatment done (Szalai et al. 2006).

Diabetic patients could suffer from fungal pneumonia caused by *Aspergillus*. Intracavitary mass or mycetoma or chronic pneumonia could be the manifestations of fungal pneumonia. Intracavitary pulmonary colonization is caused by *Aspergillus niger* and *Aspergillus flavus*. Fever and cough are the common symptoms in chronic *Aspergillus pneumonia*. These symptoms are more often sedentary and show their presence in late course of 1–6 months. The antibiotic therapy of choice is amphotericin B or rimantadin. Diabetic patients are more prone to cryptococcal pneumonia and coccidioidomycosis compared to their nondiabetic counterpart. Diabetic patients have a high carrier rate of fungal infections caused by *Candida albicans*, including oral candidiasis and vulvovaginal candidiasis, but fungal urinary tract infection and fungal pneumonia are not usual complications and need serious medical attention (Koziel and Koziel 1995).

27.11 Cryptococcosis and Diabetes

Diabetic patients with underlying immunodeficiency are prone to opportunistic fungal infections such as *Cryptococcus neoformans*, which rarely affects hosts with normal defences. It existence is documented as a yeast which is encapsulated and surrounded by a polysaccharide capsule and has a widespread reach. The infection is routed to lungs through oral inhalation of spores. Patients with T cell immunized hyperglycaemic infections contained within the lungs, whereas it rapidly spreads to the other sites, notably the central nervous system.

Cryptococcosis is diagnosed by isolating the organism from a sterile topical body site, by cryptococcal capsular antigen testing or by histopathology. HIV patients, who are also infected with *C. neoformans*, have 70% positive blood cultures (Mohan et al. 2007). The diagnosis can also be made by examining the tissue culture, body fluids, sputum and cerebrospinal fluids microscopically (Haron et al. 1993) (Table 27.7).

27.12 Treatment

Treatment is usually initiated with an amphotericin B formulation, with or without flucytosine. HIV patients with cryptococcal meningoencephalitis infection are usually administered with fluconazole. Cryptococcosis microbes are generally spread by bird droppings, especially pigeon droppings. Patient with weakened immune

Table 27.7 Predisposing factors of cryptococcosis in diabetic complication (Park et al. 2009)

Predisposing factors of cryptococcosis
HIV infection
Corticosteroids
Solid organ transplantation
Malignancies
CD4+ T cell lymphopenia
Connective tissue diseases or disease of immunological disorder
Monoclonal immunoglobulin (antibody)
Diabetes mellitus
Chronic pulmonary diseases or lung cancer
Renal failure
Cirrhosis
Pregnancy cases

system are more prone to get infected, so they are advised to avoid areas contaminated with bird droppings (World Health Organization 2011).

27.13 Staphylococcal Infection and Diabetes

Why many of the people suffering with metabolic disorder may not realize that the culprit behind the wound or any septic infection could be because of *Staphylococcus* infection. The *Staphylococcus* bacterium most often lives on the skin without causing any real harm to the body. But when there is a skin abrasion or puncture to the dermal layer and left untreated, it could lead to sepsis and can progressively cause drastic drop in blood pressure.

There is high risk of staphylococcal infection in diabetic patients having symptoms such as high nasal carriage rate. Diabetic patients are more prone to both community-acquired and nosocomial pneumonia. Diabetic patients infected with *Staphylococcus aureus* are estimated to up to 30% compared to healthy individual, whose probability ranges up to 10–15%. It has been observed that glycosylated haemoglobin is directly related to rate of nasal carriage due to *Staphylococcus aureus* infection (Lipsky et al. 1987).

27.14 Treatment

Generally *Staphylococcus* treatment is either surgically treated or through antibiotic application. Patients who are surgically treated may have to undergo antibiotic treatment as well. Surgical treatment involves incision and drainage of pus and removal of sources of infection, respectively. Many antibiotics which are available for fungal treatment are nafcillin, cefazolin, dicloxacillin, doxycycline, etc. (Bansal et al. 2008).

27.15 Candidiasis and Diabetes

Mucosal and deep-tissue infections, which are termed as chronic human fungal pathogens, are majorly caused by *Candida* species. They are opportunistic pathogens that live as normal commensals in the living system. They are the basic cause of oral infections and were identified in 1840. In the recent past, infections concerned to *Candida* have increased to a great extent. Prevalent incidence of systemic mycoses caused by *Candida* species is an important cause of morbidity and mortality in hospitalized patients. Mycoses caused by these fungi are generally superficial, as with cutaneous and mucosal infections and of higher severity in case of invasive candidiasis (Bansal et al. 2007).

Candida organisms are thin walled, microscopic sized (4–7 μm) ovoid or spherical yeast cells that reproduce and colonize mainly through budding. Other morphologic forms, such as pseudo-hyphae and hyphae are also observed in clinical specimens for most *Candida* species except *Candida* glabrata. Almost 150 *Candida* species have been identified and documented previously; however, very few are considered important human pathogens. The pathogenic species include *Candida albicans*, *Candida tropicalis*, *Candida parapsilosis*, *C. glabrata* (formerly classified as *Torulopsis glabrata*), *C. krusei*, *C. guilliermondii*, *Candida lusitaniae*, *Candida kefyr*, *Candida rugosa*, *Candida dubliniensis* and *Candida stellatoidea*.

Candida species constitute the microbiota of various anatomical sites through one of the main transmission mechanisms known as endogenous candidaemia under conditions of susceptibility and weakness of host, which finally behave as opportunistic pathogens. Exogenous mechanism is another mode of transmission, which occurs mainly through hands of health professionals who care for patients. Healthcare materials, such as contaminated catheters and intravenous solutions, are also indicated as other causes of spread of infection.

Nosocomial candidaemia, one of the dreaded infections, results in 61% mortality compared with 12% among cases and control studies well matched for age group, sex type, service type, underlying diagnosis and duration of hospital stay (Colombo et al. 2007) (Fig. 27.3).

27.16 Treatment

Treatment of candidiasis is mostly based on infections anatomical location, the current immune status of the patient and/or other underlying disease reported. The patient's susceptibility to antifungal drugs, specifically the susceptibility towards *Candida* species and the specific *Candida* species responsible for infections, is the

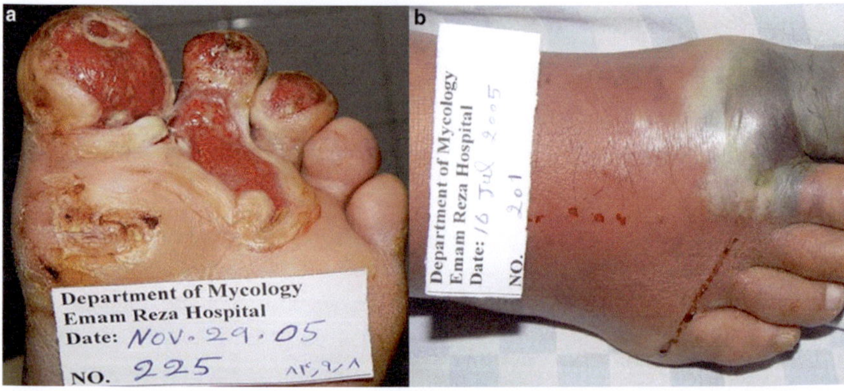

Fig. 27.3 (**a**) A typical deep ulceration caused by *Candida albicans* in a diabetic patient; (**b**) Simultaneous gangrene of the toe and interdigital space with superimposed infection due to *Trichophyton mentagrophytes* in a diabetic patient (Saba et al. 2012)

Table 27.8 Candidiasis and antifungal treatment strategies (Coelho et al. 2008)

Chemical class	Antifungal drug	Cellular target
Azole	Miconazole	Biosynthesis of ergosterol
	Ketoconazole	
	Voriconazole	
	Fluconazole	
Echinocandins	Caspofungin	Synthesis of glucan
Pyrimidine and polyenes	Flucytosine; amphotericin B	DNA and RNA synthesis; ergosterol

major criterion to be taken into consideration while deciding the treatment and manage *Candida* infections. The latest and most accepted recommendations for the treatment of *Candida* infections include miconazole, ketoconazole, caspofungin along with fluconazole as well as amphotericin B formulated with lipid molecules (Coelho et al. 2008) (Table 27.8).

27.17 Aspergillosis and Diabetes

Aspergillus are filamentous fungi most often found in dark, moist and cool areas in homes. Upon ingestion through respiratory tract, *Aspergillus* forms balls in bronchiole cavities in human lungs or in sinuses but in rare cases. The patients, who are victim of aspergillosis, are severely immune-suppressed and are neutropenic. Also patients with bone marrow transplant and/or solid organ transplant and patients with HIV-AIDS are affected with granulomatous disease. In addition to the above, patients suffering with chronic liver disease are at a higher risk for aspergillosis infection. Immunocompetent patients rarely develop such kind of infection. They show positive symptoms generally in systemic and pulmonary abnormalities such as fibrotic lung disease, especially when he patient is on corticosteroids.

27.18 Treatment

The most effective treatment for allergic bronchopulmonary aspergillosis infections is the application of itraconazole, a potent antifungal medication. Corticosteroids may also be beneficial. Voriconazole is an antifungal medication especially used for the treatment of invasive aspergillosis. Other effective anti-mycotic medications for the treatment of aspergillosis are micafungin, caspofungin, amphotericin B, posaconazole, etc. (Addrizzo-Harris et al. 1997) (Table 27.9).

27.19 Fungal Foot Infections and Diabetes

People affected with diabetes mellitus most often suffers with fungal skin diseases like athlete's foot (tinea pedis) and nail infections (onychomycosis). Onychomycosis, the nails becomes discoloured, brittle and thickened in appearance. Fungal foot infections are irritant and unsightly, risking bacterial entry via damaged skin. Eighty-four percent of patients suffering with type 1 diabetes are prone to fungal foot infections. *Trichophyton rubrum* was found in 69.2% and 0.53.1% of type 2 diabetes mellitus, and 37.9% of nondiabetic family members had fungal foot infections. Higher level of glycosylated haemoglobin and impaired sweating are some common symptoms of fungal infections in type 2 diabetic patients.

Identifications of fungal foot infections are more often based on clinical symptoms. Whenever the nails look abnormal, having change in colour and dystrophy, fungal infections should be suspected. Some common symptoms like appearance of yellow or white streaks alongside of the nails may be due to onychomycosis. It could also happen that there is occurrence of scaling under the distal nail and the nails may become discoloured, thickened and opaque. These symptoms may be due to distal onycholysis and hyperkeratosis. Superficial white onychomycosis could lead to development of small white flaky patches, and small pits may appear of the superficial surface of nails, and finally the mails become rough in textures and crumbles easily. Total dystrophic onychomycosis could lead to complete destruction of nails. Fungal infections mainly in the inflamed regions of foot commonly between the interdigital areas in toes and on the sole are mainly due to the presence of tinea pedis (Chadwick 2013).

27.20 Treatment

Oral terbinafine is found to be safe and effective for oral itraconazole therapy for the treatment of onychomycosis in people with diabetes mellitus. However, for optimal treatment of tinea pedis in diabetes, efficacy results were found to be poor as there was no clinical evidence (Crawford and Hollis 2007).

Table 27.9 Invasive fungal infections in diabetic complication (Meltem et al. 2015)

Serial no	Sex/age	Type of DM/years of diagnosis	Osteomyelitis	Response to antibacterial therapy	Surgery, type of surgery	Ulcer microbiology	Ulcer mycobiology	Treatment response
1.	F/79	Type 2/10	Yes	Yes	Yes, below knee amputation	Hafnia alvei	*Candida lipolytica*	Fluconazole therapy progressed to amputation
2.	F/53	Type 2/11	Yes	Yes	Yes, transmetatarsal amputation	*P. aeruginosa*	*C. glabrata*	Amphotericin B therapy rapidly progressed to amputation
3.	M/63	Type 2/10	Yes	No	Yes, above knee amputation	*P. aeruginosa*	*C. glabrata*	Amputation performed without interpreting culture
4.	M/49	Type 2/new	Yes	No	Yes, finger amputation	*P. aeruginosa*	*C. glabrata*	Amputation performed without resulting culture
5.	M/64	Type 2/16	Yes	Yes	Yes, Ray's amputation	*P. aeruginosa*	*C. glabrata*	Amputation performed without resulting culture

27.21 Conclusion

Diabetic patient skin is more susceptible to fungal infections than to nondiabetics. In order to avoid severe complications in the latent phase, these infections require immediate observation and quick diagnosis. Specific consideration is advised with regard to elevated frequency of fungal infections which can lead to the development of obvious disease in diabetic individuals.

References

Addrizzo-Harris DJ, Harkin TJ, McGuinness G, Naidich DP, Rom WN (1997) Pulmonary aspergilloma and AIDS: a comparison of HIV-infected and HIV-negative individuals. Chest 111(3):612–618

Akaike N, Harata N (1994) Nystatin perforated patch recording and its applications to analyses of intracellular mechanisms. Jpn J Physiol 44(5):433–473

American Diabetes Association (n.d.) Insulin basics. Retrieved 24 Apr 2014

Bansal E, Garg A, Bhatia S, Attri AK, Chander J (2008) Spectrum of microbial flora in diabetic foot ulcers. Indian J Pathol Microbiol 51(2):204

Barrett KE (2010) Ganong's review of medical physiology, 24th edn. McGraw-Hill Medical, New York

Chadwick P (2013) Fungal infection of the diabetic foot: the often ignored complication. Diabet Foot Can 1(2):20–24

Coelho LM, Aquino-Ferreira R, Maffei CML, Martinez-Rossi NM (2008) In vitro antifungal drug susceptibilities of dermatophytes microconidia and arthroconidia. J Antimicrob Chemother 62(4):758–761

Colombo AL, Guimaraes T, Silva LR, de Almeida Monfardini LP, Cunha AKB, Rady P, Alves T, Rosas RC (2007) Prospective observational study of candidemia in Sao Paulo, Brazil: incidence rate, epidemiology, and predictors of mortality. Infect Control Hosp Epidemiol 28(5):570–576

Crawford F, Hollis S (2007) Topical treatments for fungal infections of the skin and nails of the foot. The Cochrane Library

Duret C, Daujat-Chavanieu M, Pascussi JM, Pichard-Garcia L, Balaguer P, Fabre JM, Vilarem MJ, Maurel P, Gerbal-Chaloin S (2006) Ketoconazole and miconazole are antagonists of the human glucocorticoid receptor: consequences on the expression and function of the constitutive androstane receptor and the pregnane X receptor. Mol Pharmacol 70(1):329–339

Gardner DGS, Greenspan D (2007) Greenspan's basic & clinical endocrinology. McGraw-Hill Medical, New York

Geerlings SE, Hoepelman AI (1999) Immune dysfunction in patients with diabetes mellitus (DM). Pathog Dis 26(3–4):259–265

Girmenia C, Moleti ML, Micozzi A, Iori AP, Barberi W, Foà R, Martino P (2005) Breakthrough *Candida krusei* fungemia during fluconazole prophylaxis followed by breakthrough zygomycosis during caspofungin therapy in a patient with severe aplastic anemia who underwent stem cell transplantation. J Clin Microbiol 43(10):5395–5396

Grove JF, MacMillan J, Mulholland TPC, Rogers MT (1952) 762. Griseofulvin. Part IV. Structure. J Chem Soc (resumed), 3977–3987

Guariguata L, Whiting DR, Hambleton I, Beagley J, Linnenkamp U, Shaw JE (2014) Global estimates of diabetes prevalence for 2013 and projections for 2035. Diabetes Res Clin Pract 103(2):137–149

Haron E, Vartivarian S, Anaissie E, Dekmezian R, Bodey GP (1993) Primary Candida pneumonia: experience at a large cancer center and review of the literature. Medicine 72(3):137–142

Kapur A (2016) Diabetes and chronic infections. J Ceylon Coll Phys 46:1–2

Kohno S, Masaoka T, Yamaguchi H, Mori T, Urabe A, Ito A, Niki Y, Ikemoto H (2004) A multicenter, open-label clinical study of micafungin (FK463) in the treatment of deep-seated mycosis in Japan. Scand J Infect Dis 36(5):372–379

Koziel H, Koziel MJ (1995) Pulmonary complications of diabetes mellitus. Pneumonia Infect Dis Clin N Am 9(1):65–96

Lauder N, Binienda J (2005) Case study: postsexual penile ulcer as a symptom of diabetes. Clin Diabetes 23(4):191–192

Lipsky BA, Pecoraro RE, Chen MS, Koepsell TD (1987) Factors affecting staphylococcal colonization among NIDDM outpatients. Diabetes Care 10(4):483–486

Loose DS, Kan PB, Hirst MA, Marcus RA, Feldman D (1983) Ketoconazole blocks adrenal steroidogenesis by inhibiting cytochrome P450-dependent enzymes. J Clin Investig 71(5):1495

Madhu SV, Saboo B, Makkar BM, Reddy GC, Jana J, Panda JK, Singh J, Setty N, Rao PV, Chawla R, Sahay RK (2015) RSSDI clinical practice recommendations for management of type 2 diabetes mellitus, 2015. Int J Diabetes Dev Countries 35(1):1–71

Meltem T, Uysal S, Oztürk A, Arda B, Metin D, Pullukcu H (2015) Rare cause of the diabetic foot: invasive fungal infections. Mycoses 58:89

Mesa-Arango AC, Scorzoni L, Zaragoza O (2012) It only takes one to do many jobs: amphotericin B as antifungal and immunomodulatory drug. Front Microbiol 3:286

Mohan V, Sandeep S, Deepa R, Shah B, Varghese C (2007) Epidemiology of type 2 diabetes: Indian scenario. Indian J Med Res 125(3):217

Muller LMAJ, Gorter KJ, Hak E, Goudzwaard WL, Schellevis FG, Hoepelman AIM, Rutten GEHM (2005) Increased risk of common infections in patients with type 1 and type 2 diabetes mellitus. Clin Infect Dis 41(3):281–288

Nowakowska D, Kurnatowska A, Stray-Pedersen B, Wilczyński J (2004) Species distribution and influence of glycemic control on fungal infections in pregnant women with diabetes. J Infect 48(4):339–346

Park BJ, Wannemuehler KA, Marston BJ, Govender N, Pappas PG, Chiller TM (2009) Estimation of the current global burden of cryptococcal meningitis among persons living with HIV/AIDS. AIDS 23(4):525–530

Peleg AY, Weerarathna T, McCarthy JS, Davis TM (2007) Common infections in diabetes: pathogenesis, management and relationship to glycaemic control. Diabetes Metab Res Rev 23(1):3–13

Reddy KS, Prabhakaran D, Chaturvedi V, Jeemon P, Thankappan KR, Ramakrishnan L, Mohan BM, Pandav CS, Ahmed FU, Joshi PP, Meera R (2006) Methods for establishing a surveillance system for cardiovascular diseases in Indian industrial populations. Bull World Health Organ 84(6):461–469

Saba F, Hadi SMM, Rabe'eh F, Javad NM, Monavvar A, Mohsen G, Mohammad M, Ali N, Mojtaba M Abdolmajid F (2012) Mycotic infections in diabetic foot ulcers in Emam Reza hospital, Mashhad, 2006–2008. Jundishapur J Microbiol, 2011(1, Winter), 0–0

Saifullah MG (2009) Diabetic's skin: a storehouse of infections. J Pak Assoc Derma 19:34–37

Sravanthi V, Mayure VK, Donthineni K, Meher CP, Finny GJ (2014) A review on candidiasis: types and medications. PharmaTutor 2(9):36–45

Szalai G, Fellegi V, Szabo Z, Vitez LC (2006) Mucormycosis mimicks sinusitis in a diabetic adult. Ann N Y Acad Sci 1084(1):520–530

Thienpont D, Van Cutsem J, Van Nueten JM, Niemegeers CJ, Marsboom R (1975) Biological and toxicological properties of econazole, a broad-spectrum antimycotic. Arzneimittelforschung 25(2):224–230

Trevor AJ, Katzung BG, Masters SB, Kruidering-Hall M (2010) Pharmacology examination & board review. McGraw-Hill Medical, New York

Udy AA, Jarrett P, Stuart J, Lassig-Smith M, Starr T, Dunlop R, Wallis SC, Roberts JA, Lipman J (2014) Determining the mechanisms underlying augmented renal drug clearance in the critically ill: use of exogenous marker compounds. Crit Care 18(6):657

Venmans LMAJ (2007) Diabetes and infections: towards an optimal treatment strategy in primary care. Doctoral dissertation, Utrecht University

Vijan S (2010) Type 2 diabetes. Ann Intern Med 152(5):ITC3–ITC1

Volpato FC, Pires JR, Martinez IDRDC, Orrico SRP, Costa MPD, Spolidório DMP, Gonçalves A (2013) Prevalence of Candida spp. during radiographic examination in diabetes mellitus patients. Rev Odontol UNESP 42(1):13–19

World Health Organization (2006) Definition and diagnosis of diabetes mellitus and intermediate hyperglycemia: report of a WHO/IDF consultation. World Hearth Org

World Health Organization (2011) Rapid advice: diagnosis, prevention and management of cryptococcal disease in HIV-infected adults, adolescents and children: December 2011

Part IV
Bioconversion Technologies of Fungi

Chapter 28
Arbuscular Mycorrhizal Fungi: Effects on Secondary Metabolite Production in Medicinal Plants

Devendra K. Pandey, Prabhjot Kaur, and Abhijit Dey

Abstract Medicinal plants are used by 80% of the world population for their primary health care. The medicinal value of plants is primarily attributed to the secondary metabolite content such as terpenoids, alkaloids, and phenolics. These compounds play a crucial role in plant defense, are merchandised valued for their therapeutic applications and ecological role, and are also used as flavoring agents. Arbuscular mycorrhizal fungi (AMF) or *Glomeromycota* is known to form a symbiotic relationship with many terrestrial plants. AM fungi–plant consortium enhanced the production of plant terpenoids, alkaloids, and phenolics, which are valuable to human health. The potential role of arbuscular mycorrhiza (AM) symbiosis in amplification of the secondary metabolite content has attained enormous recognition for sustainable cultivation of medicinally important crops. AMF–plant symbiosis not only improves the growth and nutrients but also exerts a synergistic effect on accumulation of bioactive compounds with medicinal importance. Current studies have also recognized AM-mediated modulation of morphology, biochemistry, and gene expression in medicinal as well as in the industrial important plants. This chapter provides an appraisal on contemporary finding in the area of AMF investigation with a marked emphasis on the yield of pharmaceutically important plant-derived secondary metabolites.

Keywords Secondary metabolites · Medicinal plants · Nutrient uptake · AM fungi

---

D. K. Pandey (✉) · P. Kaur
Department of Biotechnology, Lovely Faculty of Technology and Sciences, Lovely Professional University, Phagwara, Punjab, India

A. Dey
Department of Life Sciences, Presidency University, Kolkata, India

P. Gehlot, J. Singh (eds.), *Fungi and their Role in Sustainable Development: Current Perspectives*, https://doi.org/10.1007/978-981-13-0393-7_28

28.1 Introduction

The use of medicinal plants as therapeutic agents against various diseases has been notable for thousands of years (Gopal 2001). The World Health Organization (WHO) appraises that more than 80% of global population utilize plants and plant products for their primary health care (Cordell 1995). There is a huge demand for traditional herbal medicine which has increased several folds globally during the recent past. The primary reason behind this boon is thought to be the introduction of traditional knowledge-based drugs like Taxol (anticancer), Artemisinin (antimalarial), Forskolin (antihypertensive) (Ghisalberti 1993), and many others in the Western market.

Fransworth et al. (1985) estimated that at least 119 compounds derived from 90 plant species are considerably significant drugs currently practiced in one or more countries and 77% of these are obtained mostly from botanicals used in traditional medicine. The traditional Chinese medicine (TCM), the Indian system of medicine (ISM), the Japanese Kampo, and African folklore are still practiced by major population of Asia and Africa. The botanical-based treatments have remained a major part of TCM, and even today it comprises approximately 50% of total drugs used in China. Of the 5500 medicinal plants used in TCM, about 300–500 are commonly used in day-to-day prescriptions (Corizier 1974). One such drug that has been used in China for the past hundreds of years is *Ephedra sinica* (Ma huang), the active principle of which is a potent sympathomimetic amine, ephedrine. This medicine in the form of various salts is now used in Western medicine to treat bronchial asthma.

The Indian system of medicine (ISM), mainly constituting of Ayurveda, Siddha, and Unani, is considered as one of the traditional system of medicine with thoroughly documented plant-based therapeutics. In India and abroad also, Ayurveda is being practiced by a vast population. In ISM, the remedies are primarily based on plants and plant products that usually work on the human system to boost immunity, resistance, and strength in order to cure the diseases (Swami Tirtha 1998). Some of the classical examples are as follows: (a) reserpine, an indole alkaloid as a potent antihypertensive and tranquilizing agent developed from the Indian medicinal plant *Rauwolfia serpentina* (Warrier et al. 1996), and (b) forskolin, a highly oxygenated labdane diterpenoid from the roots of Indian medicinal plant *Coleus forskohlii* exhibiting potent antihypertensive, antithrombotic, and positive ionotropic properties (Rastogi and Meharotra 1990). The Kampo system of medicine is practiced by a considerable population in Japan, where about 70% of physicians prescribe one or another Kampo formulations (Ishibashi 2002). Some of the bioactive isolates from the plants used in Japanese Kampo are baicalein (from *Scutellaria baicalensis*) (Huang 1999), glycyrrhizin (from *Glycyrrhiza uralensis*) (Tang and Eisenbrand 1992), and saikosaponin I (from *Bupleurum falcatum*) (Shibata et al. 1923).

In most of the African countries, up to 80% of population depends entirely on botanicals for therapeutics (Suzuki 2002; David 2000), many of which have been documented in the form of an African pharmacopeia issued by a scientific technical research commission of the organization of African Unity in 1984. A few promising drugs developed from African medicinal plants include (a) caffeine from the coffee tree *Coffea arabica* (from the highlands of southwest Ethiopia), a purine alkaloid exhibiting potent CNS activity and positive inotropic activity and activates lipolysis, and (b) ouabain (γ-strophanthin) isolated from *Strophanthus gratus* (from tropical West Africa) – a cardiac glycoside used to treat heart problems in acquits cardiac insufficiency (Hostettmann et al. 2000).

Due to the devoted work from the researchers, good proportion of promising drugs, numerous therapeutic leads, and many novel pharmacologically potent drugs have been developed from botanicals (Fig. 28.1) (Phillipson 1999). E. Merck, in the year 1826, manufactured an analgesic morphine, which was derived from opium poppy on commercial level, which was marked as the dawn of commercialization of plant-based drugs (Galbley and Thiericke 1999). Reserpine, the chief constituent from the roots of the Indian medicinal plant *R. serpentina*, introduced by CIBA (USA) in 1953 is another classical example of plant-derived drug. The following (Fig. 28.1) are the examples of plant-derived drugs which are under commercialization.

The widespread skepticism on synthetic medicines and worry of their usage due to adverse side effects such as toxicity, teratogenicity, and carcinogenicity which steered to the noticeable change in the status of herbal drugs as alternative medicine. The recent boon in the demand for herbal drugs is attributed to the introduction of promising plant-based drugs, viz., (1) Taxol (anticancer), (2) artemisinin (antimalarial), (3) forskolin (antihypertensive), etc. in the Western market. The present annual global market for herbal medicine is 62.0 billion US dollars. However, it is very sad to note the Indian share amounts to a meager 1%.

28.2 Plant Secondary Metabolites

According to Verpoorte (1999): "Secondary metabolites are compounds which act as a defensive role in the interaction of the organism with its environment for survival in the ecosystem and are restricted to particular taxonomic group". Three major secondary metabolites present in plants such as triterpenoids, alkaloids and phenolics.

28.2.1 Terpenoids

Terpenoids or isoprenoids are the largest class of secondary metabolites represented by nearly 27,000 compounds. Triterpenoids are classified on the basis of number of isoprenoid units present in any terpenoidal compound. For example, the monoterpenes are made up of two isoprenoid units (citronellal in lemon and menthol from peppermint), sesquiterpenes are made up of three isoprenoid units (zingiberene and artemisinin), diterpenes are made up of four isoprenoid units (gibberellic acid), triterpenes are made up of six isoprenoid units (ursolic, oleanolic, and betulinic acid), and tetraterpenes are made up of eight isoprenoid units (lanosterol, stigmasterols, diosgenin, lupeol) (Gershenzon and Kreis 1999).

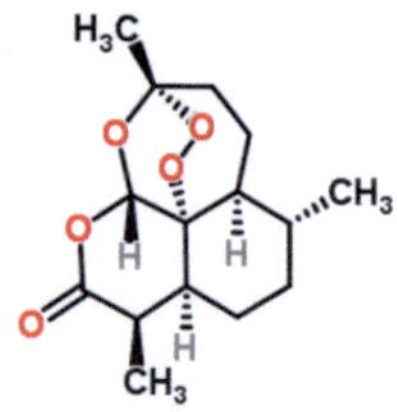

Artemisinin

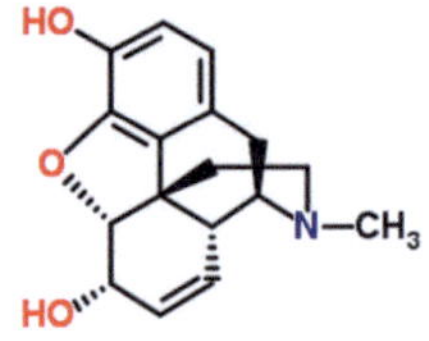

Morphine

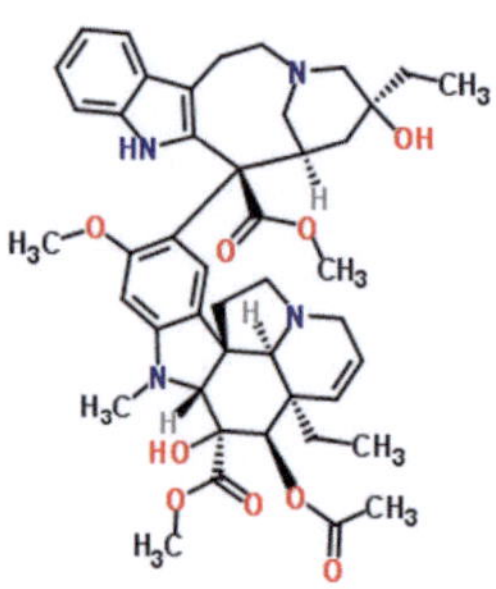

Vinblastine

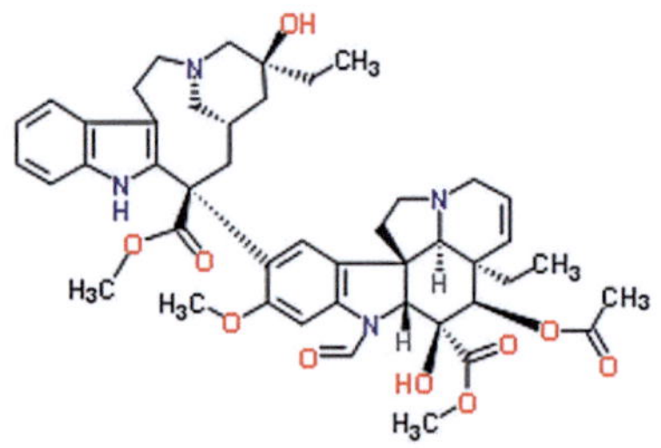

Vincristine

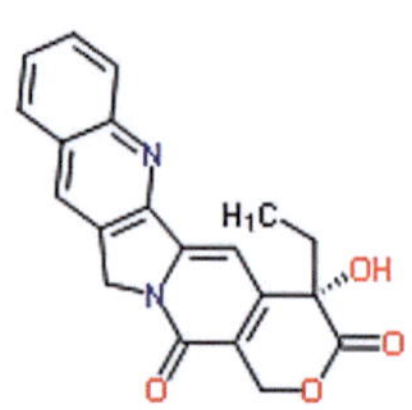

Camptothecin

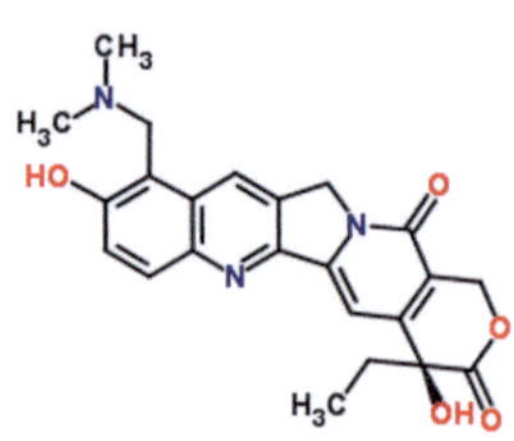

Topotecan (camptothecin derivatives)

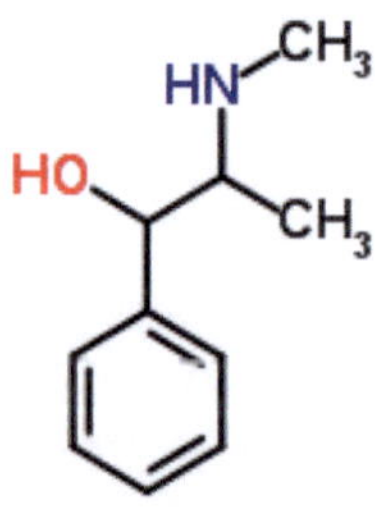

Ephedrine

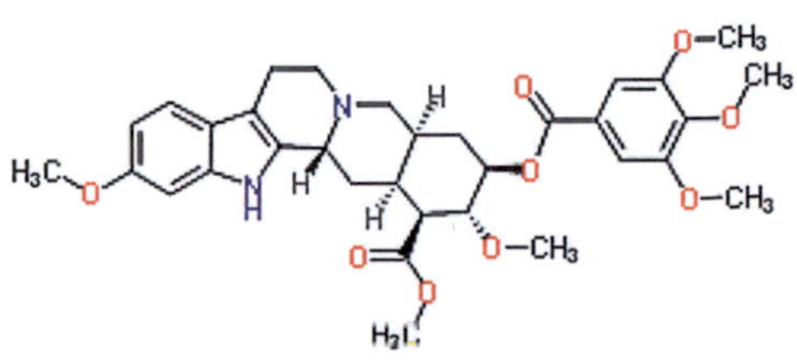

Reserpine

Fig. 28.1 Potent plant-based drugs derived from various traditional medicinal systems of the world

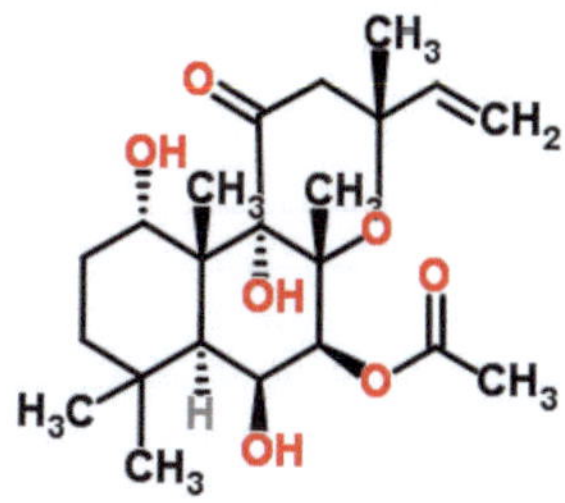

Forskolin

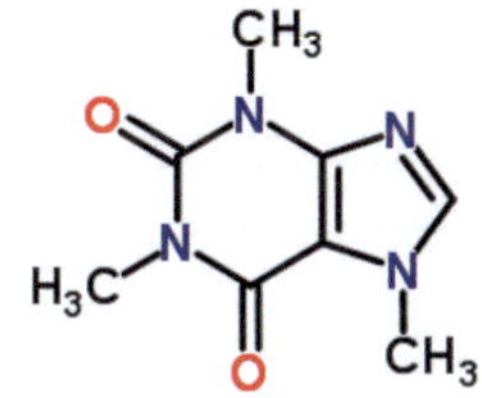

Caffine

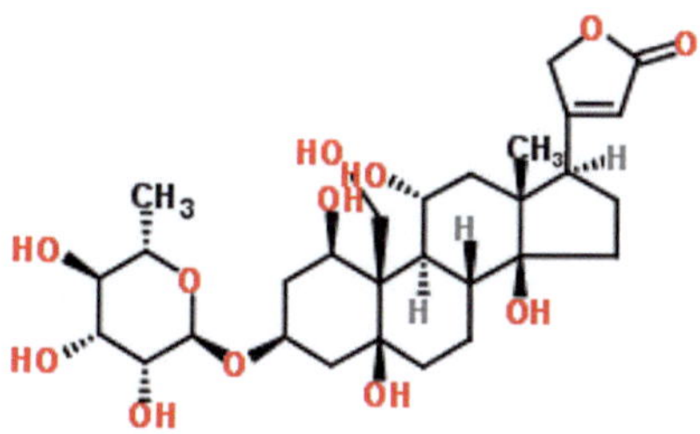

Ouabaine

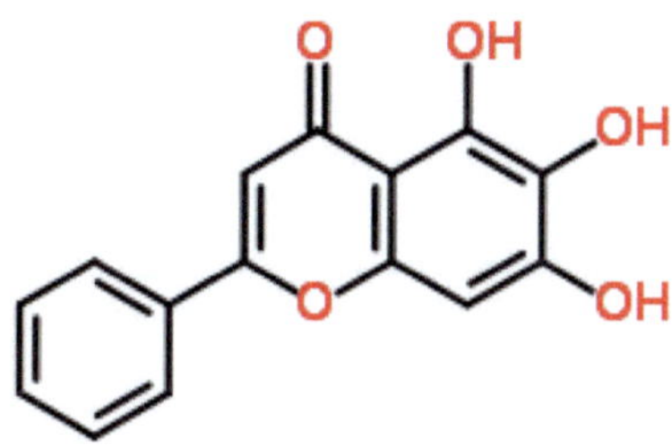

baicalein

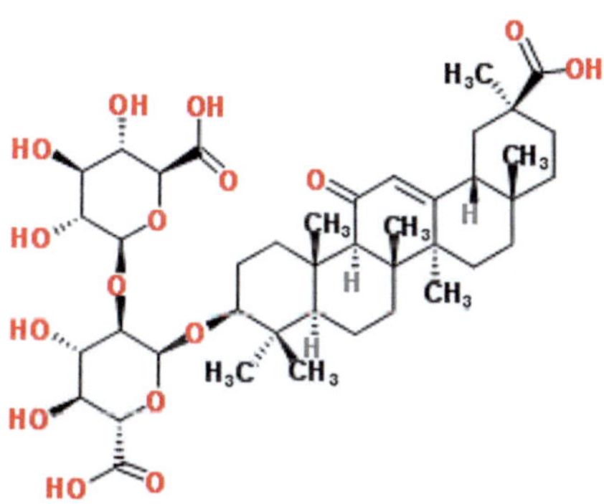

Glycyrrhizin

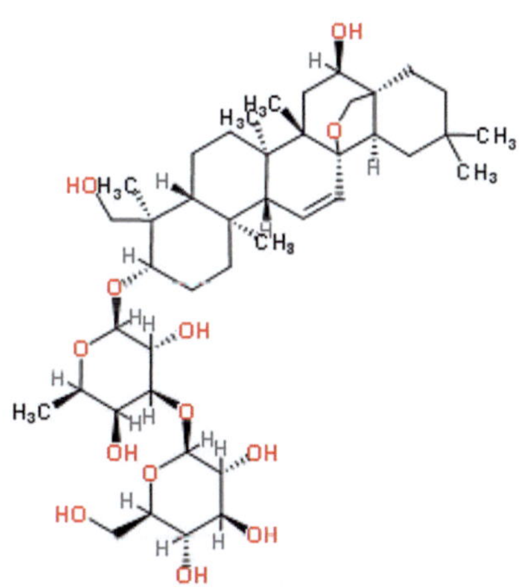

Saikosaponin

Fig. 28.1 (continued)

28.2.2 Alkaloids

The alkaloids contain nitrogen besides carbon, oxygen, and hydrogen, commonly as part of a cyclic system, and are numerous among plants with diverse pharmacological properties depending on the specific alkaloid structure. The well-known examples of potent alkaloids are reserpine, ajmaline, vincristine, colchicines, mescaline, nicotine, cocaine, and morphine. Alkaloid biosynthesis in plants is from amino acids, terpenes, and aromatic compounds (Herbert 2001).

28.2.3 Plant Phenolics

Phenolic compounds are comprised of carbon, hydrogen, and oxygen atoms arranged as an aromatic ring having a hydroxyl substituent. Among plant polyphenols, the flavonoids form the largest group followed by phenolic quinones, lignans, xanthones, coumarins, and monocyclic phenols (Croteau et al. 2000).

28.2.4 Transport, Storage, and Turnover of Plant Secondary Metabolites

The biosynthesis of secondary metabolites is restricted in plants to various compartments, and their accumulation, storage, and release occur in highly specialized organs or tissues such as in glands, bark, roots, stem, and flowers (Croteau et al. 2000). In the vacuole the high polar water-soluble compounds are generally accumulated (Wink 1999; Boller and Wiemken 1986), whereas low polar compounds are restricted in other parts of plants such as resin ducts, lactifers, glandular hairs, trichomes, thylakoid membranes, or the cuticle and defend plants from abiotic and biotic stress (Wiermann 1981). In the peppermint monoterpenes, essential oils are accumulated in glandular trichomes during the development of leaves (McGarvey and Croteau 1995).

28.3 Biotic and Abiotic Factors Influencing Accumulation of Secondary Metabolites

In order to become a key player at the global level and to provide efficacious herbal remedy to the society, it is absolutely necessary to consider certain regulatory aspects. The primary aspect is the identification of high bioactive secondary metabolite-producing plants. The collection of secondary metabolites in a plant species depends principally on (1) soil nature, (2) climatic condition, and (3) altitude.

Microbes also play a pivotal role in accumulation of secondary metabolites in herbal plants. Some important microbes such as plant growth-promoting rhizobacteria (PGPR) and *mycorrhiza* influence the bioactive phytochemical accumulation in herbal plants. Detailed investigation in this aspect will help in identifying the high-yielding plant source as well as will be useful in optimizing the parameters in which the medicinal plants may be cultivated under stringent and unfavorable conditions. Arbuscular mycorrhizal fungi (AM fungi) exist in mutually beneficial symbiotic coexistence with the higher land plants (Smith and Read 1997). Clark and Zeto (2000) and Augé (2001) have indicated that AMF improves plant nutrient assimilation specially phosphorus along with nitrogen, zinc, iron, and copper as well as water relations. Furthermore, AMF ameliorate cultivation condition and abiotic stresses (Jeffries et al. 2003) through high underground root biomass and improved absorption of soil nutrients and secretion of different enzymes by AMF colonized roots and/or hyphae (Marschner 1995; Smith and Read 1997).

28.3.1 Arbuscular Mycorrhizal Fungi (AMF)

AMF is symbiotic fungi that form a beneficial relation with the roots of a diverse group of terrestrial plants (Fig. 28.1). The AM fungi are classified under the order *Glomales* under the *Zygomycota* (Redecker et al. 2000), but currently it is moved to a new phylum, the *Glomeromycota* (Schüßler et al. 2001). These fungi are widespread soil-borne fungi, whose origin and occurrence have been established more than 450 million years (Redecker et al. 2000). AMF association with higher plants can be seen in temperate, tropical, and arctic regions but was found to be absent in waterlogged conditions (Smith and Read 1997). These fungi have a synergistic effect on plant growth performance in comparison with any microbes by an increase in biomass of the root system of the plant and also by enhancing the root surface area for absorption of minerals and water (Leake et al. 2004). Thus, it can be inferred that arbuscular mycorrhizal (AM) symbiosis with terrestrial plants is of significant utility in forest ecology, land reformation, improved growth, and better yield of plants in low-input systems (Sieverding et al. 1991).

28.3.2 Agronomic and Ecological Roles of AMF

The economically and medicinally important crops are known to be colonized by AMF (Sieverding et al. 1991). AMF improves the phosphorus nutrition by increasing the root biomass and surface area and by improving the P uptake (Koide 1991). AMF act as phosphate solubilizers by producing organic acids that increased the availability of insoluble mineral phosphates (Lapeyrie 1988) and enhanced the phosphorus uptake by host plants. Moreover, AMF also improves the uptake of macronutrients and micronutrients (Clark and Zeto 2000; Smith et al. 2004).

Marschner (1998) and Hodge and Campbell (2001) on their studies on mycorrhiza established that the better plant nutrition is dependent on (i) enhanced root surface via extra-radical hyphae that extend beyond root depletion zone, (ii) decomposition of organic substances, and (iii) enhanced microbial consortium in the rhizosphere zone.

Contemporary studies reveal that AMF is an ideal tool for sustainable agriculture, landscape resurrection, and horticulture via decomposition of organic material, seedling establishment, enhanced pathogenic resistance, herbivore tolerance, soil stability, heavy metal detoxification, water stresses/cold temperature resistance, and reducing desertification (Jeffries et al. 2003; Hart and Trevors 2005). The function of AMF to their hosts in nutrient suppressive soil (less P) environment is high whereas it is less under P-sufficient conditions (Koide and Schreiner 1992), and plant growth rates can be reduced by AM colonization when available P is present (Peng et al. 1993).

28.3.3 Mycorrhizal Association with Medicinal Plants

Inoculation of crop plants with AMF is known to enhance growth, nutritional, and secondary metabolite contents of many medicinally and industrially significant crops which is attributed to enhanced uptake of nutrients, production of growth-promoting factors, biotic and abiotic stress tolerance, and synergistic interaction with PGPR such as nitrogen-fixing rhizobacteria and PSB (Bagyaraj and Varma 1995; Rajan et al. 2000). Enhanced mineral nutrition, i.e., N, P, Fe, Cu, Zn, and B, helps in the synthesis of chlorophyll thus enhancing photosynthetic rate (Bian et al. 2001; Feng et al. 2002; Toussaint 2007). In addition, AMF inoculation on medicinal plants is responsible for reduction in the intensity of disease caused by bacterial and fungal pathogens by morphological modulations such as thickening of the cell wall, stronger vascular bundles, etc. Moreover AMF inoculation leads to physiological alterations in host such as enhanced levels of P, phenolics, sulfur-containing amino acids, etc. (Boby and Bagyaraj 2003).

AM fungi that have a symbiotic relationship with medicinal plants belong to several families of angiosperms and gymnosperms. AM fungi are associated with *Adhatoda vasica* and *Datura somnifera* (Wei and Wang 1989). Several authors reported symbiotic relation of AMF with plants that belong to Araliaceae such as *Panax ginseng* (Zhang et al. 1990; Xing et al. 2000, 2003; Li 2003a; Cho et al. 2009; Ren et al. 2007; Zhang et al. 2011). There were many plants that belong to Labiatae family having association with AMF such as *Salvia miltiorrhiza* (He et al. 2009a, b; Wang and He 2009; Ma et al. 2009; He et al. 2010; Meng and He 2011), *Schizonepeta tenuifolia* (Wei and Wang 1991), *Bupleurum scorzonerifolium* (Teng and He 2005), *Pogostemon cablin* (Arpana et al. 2008), *Coleus forskohlii* (Sailo and Bagyaraj 2005), *Salvia officinalis* (Nell et al. 2009), *Mentha arvensis* (Gupta et al. 2002; Karagiannidis et al. 2011), *Ocimum basilicum* (Copetta et al. 2006; Toussaint et al. 2007; Lee and Scagel 2009; Prasad et al. 2011; Rasouli-Sadaghianil et al. 2010), and *Origanum* sp. (Khaosaad et al. 2006; Morone Fortunato and Avato 2008; Karagiannidis et al. 2011). Mycorrhization was also observed in plants that belong to Umbelliferae such as *Anethum graveolens*, *Trachyspermum ammi* (Kapoor et al. 2002a), *Coriandrum sativum* (Kapoor et al. 2002b; Farahani et al. 2008), *Foeniculum vulgare* (Kapoor et al. 2004), and *Angelica dahurica* (Cao and Zhao 2007; Zhao et al. 2009; Zhao and He 2011). AMF was reported in several plants that belong to Liliaceae such as *Gloriosa superba* L. (Yadav et al. 2013; Pandey et al. 2014), *Aloe barbadensis* (Gong et al. 2002; Mamta et al. 2012; Pandey and Banik 2009; Burni et al. 2013), *Ophiopogon japonicus* (Pan et al. 2008), *Paris polyphylla var.* (Zhou et al. 2009, 2010), and *Allium sativum* (Borde et al. 2009). Several authors reported mycorrhizal plants belonging to Leguminosae such as *Pueraria lobata*, *Astragalus membranaceus*, *Glycyrrhiza inflata*, *Castanospermum australe*, and *Prosopis laevigata* (Wang et al. 2006; Liu and He 2008, 2009; He et al. 2009b; Liu et al. 2007; Abu-Zeyad et al. 1999; Rojas-Andrade et al. 2003). AMF reported in plants belong to the Asteraceae family (Binet et al. 2011; Asrar and Elhindi 2010; Jurkiewicz et al. 2010; Araim et al. 2009; Kapoor et al. 2007; Rapparini et al. 2008; Chaudhary et al. 2008; Awasthi et al. 2011; Huang et al. 2011; Guo et al. 2006; Zhang et al. 2010, 2011; Asrar and Elhindi 2010). AMF is also reported in *Forsythia suspense*, *Taxus chinensis var.*, *Pinellia ternata*, *Catharanthus roseus*, *Valeriana officinalis*, *Atractylodes macrocephala*, *Camptotheca acuminate*, *Coix lacryma-jobi var.*, *Ginkgo biloba*, and *Gentiana manshurica* (Wang et al. 1998, 2010; Qi et al. 2002, 2003; Zhang et al. 2004; Wu and Wei 2008; Li 2003b; Huang et al. 2003; Zhao et al. 2006, 2007; Yu et al. 2010; Lu and He 2005, 2008; Lu et al. 2008a, b, 2011; Ren et al. 2008; Chen et al. 2009a, b, 2010; Guo et al. 2010; Shen et al. 2011; Zubek et al. 2012; Rosa-Mera et al. 2011; Nell et al. 2010; Geneva et al. 2010) (Fig. 28.2 and Table 28.1).

28.3.4 Synergistic Effects of AMF on Secondary Metabolites Accumulation in Medicinal Plants

There is alteration in the secondary metabolite accumulation due to chemical and biological events taking place during the AMF–host interaction (Akiyama and Hayashi 2002; Allen et al. 1982; Barrios 2007; Cai et al. 2008). AMF is known to enhance contents of secondary metabolites such as terpenoids, alkaloids, and phenolics in many economically and industrially significant crops such as production of flavonoids, cyclohexanone derivatives and apocarotenoids, phytoalexins, phenolic compounds, triterpenoids, and glucosinolates in herbal, and medicinal important plants colonized by AMF has been reported (Gianinazzi et al. 2010; Hadwiger et al. 1986; Harris et al. 2001; Harrison, 1999; Heet al. 2009c; Huang et al. 2004; Janardhan and Abdul-Khaliq 1995, Jie et al. 2007; Loomis and Corteau 1972; Maier et al. 1999; Paterson and Simmonds 2003; Shah et al. 1980; Silva et al. 2008; Singh et al. 2013; Smith and Read 2008; Szakiel and Paczkoski 2011a, b; Toussaint et al.

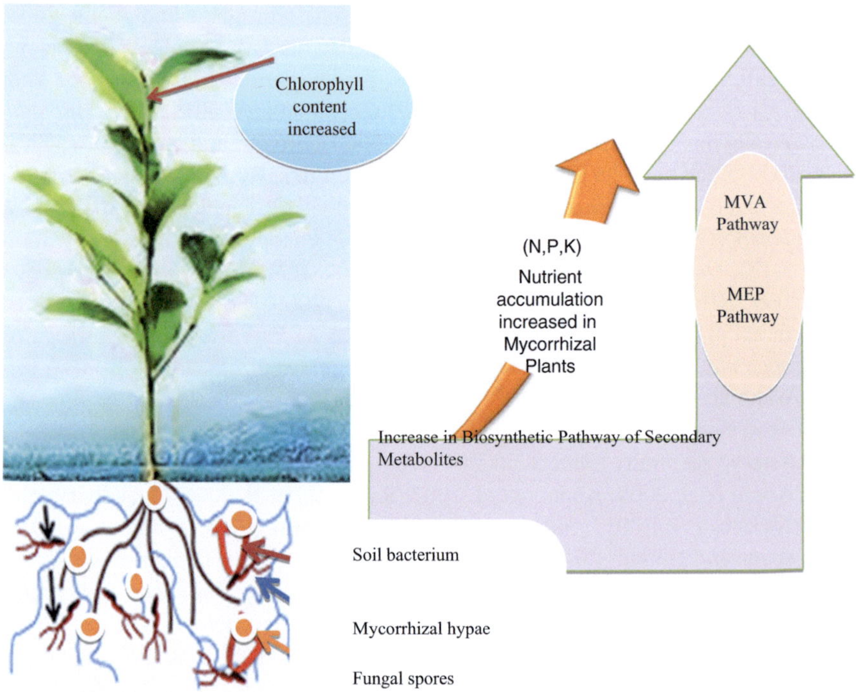

Fig. 28.2 Symbiotic association of AMF with plants

2004; Volpin et al. 1994; Wink 1997; Wu et al. 2010; Xiao et al. 2011; Yang et al. 2008; Zeng et al. 2007; Zubek et al. 2010, 2013).

28.3.4.1 Plant Terpenoid Accumulation Associated with AMF

The influence of mycorrhization on the terpenoid production in different plants has been presented in Table 28.2. The enhancements in triterpenoids are both qualitative and quantitative due to modifications in morphology of various aromatic plants due to (atractylol, anethole, thymol, artemisinin, stevioside, rebaudioside A, β-caryophyllene, p-cymene, geraniol, patchoulol, valerenic acid, and glycyrrhizic acid) in AMF-colonized plants (Kapoor et al. 2007; Mandal et al. 2013, 2015; Morone Fortunato and Avato 2008; Khaosaad et al. 2006; Lu et al. 2011; Farahani et al. 2008; Wei and Wang 1991; Arpana et al. 2008; Geneva et al. 2010; Liu et al. 2007; Jurkiewicz et al. 2010). Furthermore many researchers reported that increase in terpenoids (thymol, geraniol, anethol, forskolin) in aromatic and medicinal plants is due to P availability (Torelli et al. 2000; Kapoor et al. 2002a; Strack et al. 2003; Krishna et al. 2005; Sailo and Bagyaraj 2005; Bagheri et al. 2014; Guo et al. 2006; Zhang et al. 2010, 2011) and transcription of gene responsible for terpenoid

Table 28.1 Medicinal plants that have been investigated for AM effects

Species	Families	Major marker compounds	Medicinal value
Datura stramonium	*Solanaceae*	Hyoscyamine	Cures rheumatism, cough, relieving pain
Panax ginseng	Araliaceae	Ginsenosides	Immunodulatory, increase vigorness
Panax notoginseng	Araliaceae	Ginsenosides	Improves vital energy
Salvia miltiorrhiza	Labiatae	*Salvinorin*	Regularize menstruation, hepatoprotective
Schizonepeta tenuifolia	Labiatae	*Pulegone*, limonene, and menthofuran	Headache and improves digestion and cures vomiting
Bupleurum scorzonerifolium	Labiatae	Saikosaponin	Effective against cough and cold; regularize menstruation 1
Pogostemon cablin	Labiatae	Pogostol	Cures cold and diarrhea
Coleus forskohlii	Labiatae	Forskolin	Antihypertensive and cures placebo, asthma, and glaucoma
Salvia officinalis	Labiatae	*Salvinorin*	Anti-inflammatory agent
Gloriosa superba L.	Liliaceae	Colchicine	Abortifacient, anti-gout, anticancer properties
Gentiana manshurica	Gentianaceae	Amarogentin, swertiamarin	Cures acute icterohepatitis, fever, and convulsions
Aloe barbadensis	Liliaceae	Barbaloin	Bactericidal effects and treat inflammation, improves immunity
Ophiopogon japonicas	Liliaceae	Ophiopojaponin	Cures cough and insomnia and digestive problem
Paris polyphylla var.	Liliaceae	Polyphyllin	Cures mouth ulcer
Allium sativum	Liliaceae	*Allicin*, ajoene	Cures hypertension, potent drugs for cardiovascular and immune systems
Ginkgo biloba	Ginkgoales	Ginkgolic acid, bilobalide	Cures hypertension
Coix lacryma-jobi var.	*Poaceae*	Palmitic, stearic, oleic, and linoleic acid	Cures diarrhea and lung problem
Camptotheca acuminata	Nyssaceae	Camptothecin and *piperine*	Potent drug against tumors
Atractylodes macrocephala	*Poaceae*	Astragaloside	Treatment of digestive disorders
Atractylodes lancea	Asteraceae	Astragaloside	Treatment of digestive disorders, rheumatism, cold, and nyctalopia
Artemisia annua	Asteraceae	Artemisinin	Cures malaria
Echinacea purpurea	Asteraceae	Cichoric acid	Potent antimicrobial and antiviral properties

(continued)

Table 28.1 (continued)

Species	Families	Major marker compounds	Medicinal value
Arnica montana	Asteraceae	*Helenalin* and chamissonolid	Effective against bruises, swellings, and mouth ulcers
Inula ensifolia	Asteraceae	Borneol, β-caryophyllene, p-cymene, and bornyl acetate	Potent anti-inflammatory and antiviral compound
Artemisia umbelliformis	Asteraceae	Artemisinin	Antimicrobial effect
Tagetes erecta		Carotenoids, scopoletin, ferulic acid	Cures inflammation, cold, and cough
Phellodendron amurense	Rutaceae	Nexrutine, quercetin, berberine	Treating digestive disorders
Phellodendron chinense	Rutaceae	Nexrutine, quercetin, berberine	Potent drug to cure dysentery
Citrus aurantium	Rutaceae	Limonin glucoside and phlorin	Potent digestive agent and improves appetite
Pueraria lobata	Leguminosae	Quercetin, tryptamine, apigenin 5-hydroxytryptamine	Cures fever and antioxidant activity
Astragalus membranaceus	Leguminosae	Astragaloside	Immunomodulatory and diabetes II
Glycyrrhiza inflata	Leguminosae	Glycyrrhizin	Cures throat problem
Castanospermum austral	Leguminosae	1-epilexine	Anti-HIV, diabetes II
Prosopis laevigata	Leguminosae	Juliflorine, patulitrin	Immunomodulatory
Mentha arvensis	Lamiaceae	Eucalyptol, terpinolene, linalool, pulegol, menthol, menthofuran, menthyl acetate	Cures fever, cough and cold
Ocimum basilicum	Lamiaceae	Eugenol, linalool	Fever, cold, and antivenom
Origanum sp.	Lamiaceae	Carvacrol, thymol, cosmocide, vicenin-2, rosmarinic acid	Improves appetite and stomach ailment
Origanum vulgare	Lamiaceae	Carvacrol, thymol, cosmocide, vicenin-2, rosmarinic acid	Cures digestive problems
Origanum onites	Lamiaceae	Carvacrol, thymol, cosmocide, vicenin-2, rosmarinic acid	Cures digestive problems
Coleus forskohlii	Lamiaceae	Forskolin	Anti-inflammatory activity
Forsythia suspense	*Oleaceae*	Phillyrin, pinoresinol, phillygenin, lariciresinol, and forsythiaside	Detoxification, cold and cough

(continued)

Table 28.1 (continued)

Species	Families	Major marker compounds	Medicinal value
Taxus chinensis var.	*Taxaceae*	10-deacetylbaccatin, baccatin III, cephalomannine and paclitaxel paclitaxel	Chemotherapy drug used to cure tumor and cancer
Pinellia ternate	*Araceae*	β-cubebene, atractylon, methyl eugenol, and δ-cadinene	Treating cough and vomiting
Anethum graveolens	Umbelliferae	Alpha-phellandrene, apiole, dill ether, limonene, geraniol	Cures flatulence, anti-spasm effect
Trachyspermum ammi	Umbelliferae	Thymol	Cures gaseous distention, cholera and diarrhea
Coriandrum sativum	Umbelliferae	Caffeic acid, chlorogenic acid, quercetin, kaempferol, rhamnetin, and apigenin	Improve appetite and cures analgesia, detoxifying
Foeniculum vulgare	Umbelliferae	Anethole, fenchone, and methyl chavicol	Promoting appetite and potent drug to cure digestive problem
Angelica dahurica	Umbelliferae	Ferulic acid and ligustilides	Treating headache, toothache, and nasosinusitis
Hypericum perforatum	Garcinia	Hypericin	Antimicrobial effect, diminishing inflammation
Catharanthus roseus	*Apocynaceae*	Vinblastine, vincristine, vindoline, and catharanthine	Antiproliferation and antitumor and anticancer effect
Valeriana officinalis	Valerianaceae	Valerenic acid	Sedatives and tranquilizers, treating insomnia
Hypericum perforatum,	Valerianaceae	Hypericin, naphthodianthron, and hyperforin	Antidepressant activity and potent anti-inflammatory properties, antispasmodic, analgesic, antiseptic, carminative, cholagogic, diaphoretic, and vasodilator properties

biosynthetic pathways (Floß et al. 2008; Mandal et al. 2014, 2015). Adams et al. (2004) investigated that composition of essential oil (EO) levels in Vetiver roots is altered in the presence of unidentified bacteria PGPR, and AMF were suggested to be involved in the modulation of essential oil (EO) accumulation. Copetta et al. (2006) and Freitas et al. (2004) reported that AM fungal root colonization enhances the essential oil content (linalool and geraniol, menthol, menthone, carvone, and pulegone) in *Ocimum basilicum* and *Mentha arvensis*, respectively. Moreover, many authors reported increase in terpenoids such as menthol, menthone, carvone, pulegone, carvacrol, and thymol in aromatic plants (Gupta et al. 2002; Karagiannidis et al. 2011; Khaosaad et al. 2006; Morone Fortunato and Avato 2008; Rasouli

Table 28.2 Effects of AM symbiosis on secondary metabolism of medicinal plants under the conditions of artificial inoculation

Secondary metabolite		Medicinal plant	AMF	Change in secondary metabolite
Terpene	–	*Schizo tenuifolia*	*G. epigae*, *G. mossae*	Significant enhancement
Terpene	Atractylol	*Atractylodes macrocephala*	*Glomus mosseae*	Significant increase
Terpene	Menthol, menthone, carvone, and pulegone	*Mentha arvensis*	*G. fasciculatum*, *G. etunicatum*, *G. lamellosum*	Significant enhancement
Terpene	Carvacrol, thymol	*Origanum* sp.	*G. mosseae*	Significant improvement
Terpene	Thymol	*Origanum vulgare*	*G. viscosum*	No change
Terpene	Linalool and geraniol	*Ocimum basilicum*	*Gigaspora margarita*, *Gigaspora rosea*	Significant increase
Terpene	Eugenol	*Ocimum sanctum*	*Glomus fasciculatum*	Relatively increased
Terpene	Anethole	*Anethum graveolens*	*G. macrocarpum*, *G. fasciculatum*	90% of the control
Terpene	Thymol	*Trachyspermum ammi*	*G. macrocarpum*, *G. fasciculatum*	72% of the control
Terpene	β-caryophyllene, p-cymene, geraniol	*Coriandrum sativum*	*Glomus hoi*	Significant improvement
Terpene	Anethol	*Foeniculum vulgare*	*G. macrocarpum*, *G. fasciculatum*	78% of the control
Terpene	Sesquiterpenes Artemisinin	*Artemisia annua*	*G. macrocarpum*, *G. fasciculatum*	Significant improvement
Terpene	Patchoulol	*Pogostemon cablin*	*A. laevis*, *G. mosseae*, *S. calaspora*	Significant improvement
Terpene	Atractylol	*Atractylodes lancea*	*G. mosseae*:*G. mosseae*, *G. aggregatum*, *G. versiforme*, *G. intraradices*	No change
Terpene	Menthol, menthone, carvone, and pulegone	*Mentha spicata*	*G. etunicatum*, *G. mosseae*	No change
Terpene	Thymol	*Trachyspermum ammi*	*G. fasciculatum*	51.21%
Terpene	Geraniol, linalool	*Coriandrum sativum*	*G. macrospermum*, *G. fasciculatum*	28% and 43% over control
Terpene	Anethol	*Foeniculum vulgare*	*G. macrospermum*, *G. fasciculatum*	Significant increase

(continued)

Table 28.2 (continued)

Secondary metabolite		Medicinal plant	AMF	Change in secondary metabolite
Terpene	Artemisinin	*Artemisia annua*	*G. macrospermum*, *G. fasciculatum*, *G. mosseae*, *G. versiforme*	15.2–32.8% over the control
Terpene	Sesquiterpene lactones	*Arnica montana*	*G. geosporum*, *G. constrictum*	Either increase or decrease
Terpene (triterpenoid)	Glycyrrhizic acid bornyl acetate, 1,8-cineole, α-thujones, β-thujones	*Glycyrrhiza inflata*	*G. mosseae*, *G. versiforme*	Significant increase
Triterpene saponins		*Dioscorea* spp.	*G. clarum*, *G. etunicatum*	42% over the control
Terpene	Valerenic acid	*Valeriana officinalis*	*G. intraradices*	Relative quantity increase
Terpene (diterpenoid)	Stevioside, rebaudioside-A	*Stevia rebaudiana*	*Rhizophagus fasciculatus*	Significant increase
		Stevia rebaudiana	*G. intraradices*	Upregulates genes for biosynthesis
Phenolics	Rosmarinic, caffeic acids	*Ocimum basilicum*	*G. caledonium*, *G. mosseae*	Significant increase
Phenolics		*Ocimum basilicum*	*Glomus intraradices*	No change
Phenolics	Diobulbinone	*Dioscorea* spp.	*G. clarum*, *G. etunicatum*	Significant increase
Phenolics	Phenolics	*Echinacea purpurea*	*Glomus intraradices*	Significant increase in roots
Phenolic acids	Caffeic, chlorogenic acid	*Arnica montana*	*G. geosporum*, *G. constrictum*,	Significant increase in roots
Phenolics	Caffeic acid	*Viola tricolor*	*Rhizophagus irregularis*	Significant influence
Phenolics			*Funneliformis mosseae*	No response
Phenolics	Salvianolic acid B	*Salvia miltiorrhiza*	*G. mosseae*, *G. aggregatum*	Significant increase
Phenolics	Gallic acid	*Libidibia ferrea*	*Claroideoglomus etunicatum*	21% over the control
Phenolics			*A. longula*	No response
Phenolics			*G. albida*	No response

(continued)

Table 28.2 (continued)

Secondary metabolite		Medicinal plant	AMF	Change in secondary metabolite
Phenolics	Rutin, quercetin, and kaempferol	*Catharanthus roseus*	*Glomus species*	Significant increase
Phenolics	Caffeic acid, rosmarinic acid, and luteolin	*Mentha spicata*	*G. etunicatum*, *G. mosseae*	Significant increase
Phenolics	Coumaric and ferulic acids	*Myracrodruon urundeuva*	*A. longula*	81.03%
Phenolics	Thymol derivative	*Inula ensifolia*	*G. intraradices*, *G. clarum*	Either increase or decrease
Phenolics	Gallic acid, chlorogenic acid, catechin, hydroxyl benzoic acid	*Valeriani jatamansi*	*G. intraradices*	Significant increase
Flavonoid		*Salvia miltiorrhiza,*	*G. mosseae*	Significant increase
Flavonoid		*Astragalus membranaceus*	*G. mosseae*	Significant increase
Flavonoid		*Myracrodruon urundeuva*	*A. longula*	57.5% over control
		Viola tricolor	*Rhizophagus irregularis*	Significant influence
Flavonoid	Imperatorin, total coumarins	*Angelica dahurica*	*Glomus species*	Significant increase
Flavonoid	Total phenolic, rosmarinic acid	*Salvia officinalis*	*G. mosseae*, *G. intraradices*	No change in leaves and roots
Flavonoid	Valerenic acid	*Valeriana officinalis*	*G. mosseae*, *G. intraradices*	Significant increase
Tannins		*Valeriana jatamansi*	*G. intraradices*	Significant increase
Alkaloid	Castanospermine	*Castanospermum australe*	*G. intraradices*, *G. margarita*	Significant improvement with *G. intraradices*
Alkaloid	Trigonelline	*Prosopis laevigata*	*G. rosea*	1.8-fold increase in roots
Alkaloid	Vinblastine alkaloid	*Catharanthus roseus*	*Glomus species*	Significant increase
Alkaloid	Hyoscine, hyoscyamine	*Datura stramonium*	*G. mosseae*, *G. epigaeum*	Significant increase

(continued)

Table 28.2 (continued)

Secondary metabolite		Medicinal plant	AMF	Change in secondary metabolite
Alkaloid	Berberine, jatrorrhizine, palmatine	*Phellodendron amurense*	*G. mosseae*, *G. etunicatum*,*G. versiforme*, *G. diaphanum*	Significant increase
Alkaloid	Berberine	*Phellodendron Chinese*	*A. laevis*, *A. mellea*	Significant increase in the tree bark
Alkaloid	Vindoline, vinblastine, vincristine, catharanthine, ajmalicine, and serpentine	*Catharanthus roseus*	*G. etunicatum* and *G. intraradices*	Significant increase
Alkaloid	Nicotine, anabasine, and nornicotine	*Nicotiana tabacum*	*G. etunicatum*, *G. intraradices*	Significant increase
Alkaloid	L-ephedrine, guanosine	*Pinellia ternate*	*G. intraradices*, *G. mosseae*	Significant increase in the tuber
Alkaloid	Camptothecin	*Camptotheca acuminate*	*A. mellea*, *G. intraradices*	Significant increase
Alkaloid	Forskolin	*Coleus forskohlii*	*A. laevis*, *G. monosporum*, *S. calospora*	Significant increase
	Alliin	*Allium sativum*	*G. fasciculatum*	Significant increase
Others	Hypericin, pseudohypericin	*Hypericum perforatum*	*G. intraradices*, *G. mosseae*	Significant increase
	Oleoresin	*Zingiber officinale*	*S. heterogama*, *G. decipiens*, *A. koskei*, *Entrophospora colombiana*	Significant increase

Sadaghianil et al. 2010; Chaudhary et al. 2008). Moreover, studies on anatomical and pharmacognosy suggested that enhanced production of essential oils was due to improvement in the number of glandular hairs in leaves in AM-inoculated plants (Guerrieri et al. 2004). Furthermore, plant growth regulator levels are also known to be amended in both arbuscule-containing cells and whole tissues of mycorrhizal plants (Allen et al. 1980).

28.3.4.2 Plant Alkaloid Accumulation Associated with AMF

The effect of arbuscular mycorrhizae (AM) on accumulation and biosynthesis of alkaloids has been attributed to mycorrhization in various medicinal plants such as *Castanospermum australe* (Abu-Zeyad et al. 1999), *Catharanthus roseus* (Rosa-Mera et al. 2011; Andrade et al. 2013), *Datura stramonium* (Wei and Wang (1989), *Phellodendron amurense* (Fan et al. 2006), and *Camptotheca acuminate* (Zhao et al. 2006; Yu et al. 2010). Liu et al. (2007) investigated that enhanced production of alkaloid in medicinal plants are linked with the higher biomass production in AM-inoculated plants which are usually connected to the nutritional benefits of mycorrhization. AM fungi directly controlled the expression of different genes associated with the plant defense system, with consequences for whole-plant fitness and its response to biotic stresses (Pozo and Azco'n-Aguilar 2007; Fan et al. 2006; Zhou and Fan 2007). It was reported in many medicinal plants that mycorrhization enhances alkaloid contents such as ephedrine, camptothecin, and alliin (Guo et al. 2010; Zhao et al. 2006; Yu et al. 2010; Borde et al. 2009). Studies performed by El-Sayed and Verpoorte (2007) and Roepke et al. (2010) reported the improved production of several pharmacologically important monoterpene indole alkaloids (MIAs), such as vinblastine, vincristine, ajmalicine, vindoline, catharanthine, and serpentine in *Catharanthus roseus* and pyridine alkaloids (PAs) in *Nicotiana species.*

28.3.4.3 Plant Phenolic Accumulation Associated with AMF

The effect of AM on the production of phenolics in various plants has been summarized in Table 28.2. The enhancements of phenolics such as anthraquinone glycosides and flavanoids are both qualitative and quantitative (Araim et al. 2009; Silva et al. 2014; Rosa-Mera et al. 2011; Bagheri et al. 2014; De Sousa et al. 2013; He et al. 2009a, b, c; Meng and He 2011; Zhao et al. 2009; Zhao and He 2011; Nell et al. 2009). It was observed by many researchers that the enhancement of phenolics in medicinal plants was due to increase in the biomass of plants due to better nutrition and expression of genes related to the defense system of plants. Jurkiewicz et al. (2010) reported increase in the caffeic and chlorogenic acid in *Arnica Montana*. Furthermore there was improved production of phenolics in *Viola tricolor* (Zubek et al. 2015), *Salvia* spp. (Yang et al. 2017), and *Aloe vera* (Pandey and Banik 2009; Mamta et al. 2012). There were significant increases in the content of rosmarinic and caffeic acids (Jugran et al. 2015; Zubek et al. 2015; Jurkiewicz et al. 2010; Toussaint et al. 2007; Lee and Scagel 2009) and diobulbinone (Lu et al. 2015).

Structure of Secondary Metabolites Affected by AM

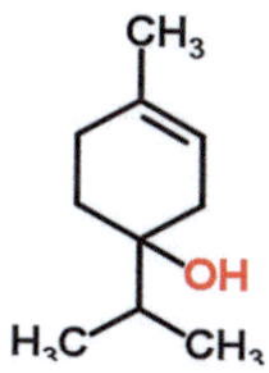
4-terpinenol

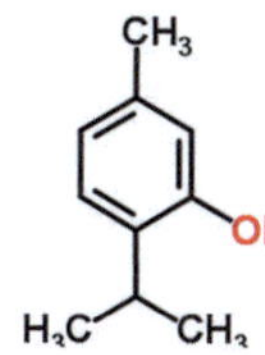
Thymol

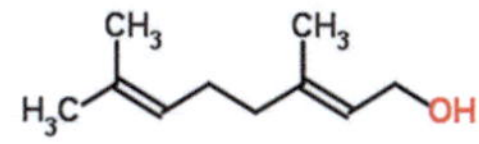
Geraniol

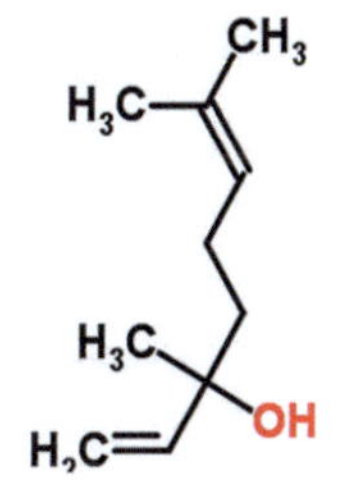
Linalool

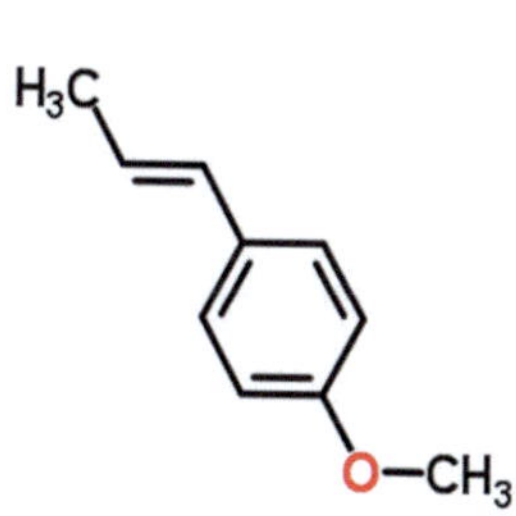
trans-Anethole

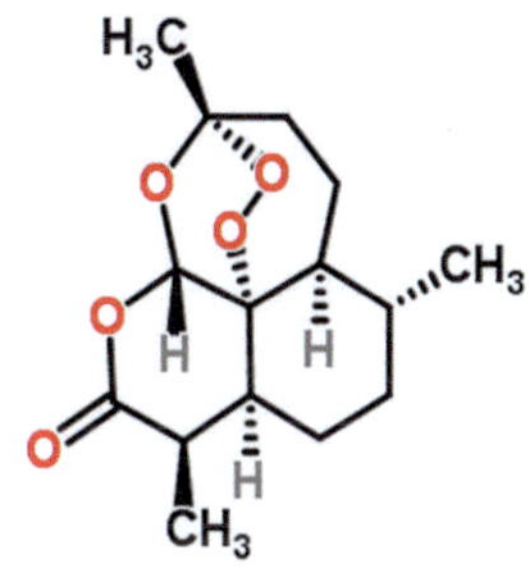
Artemisinin

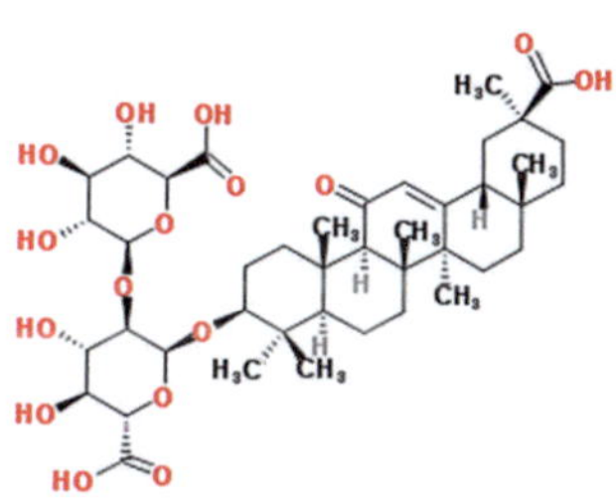
Glycyrrhizin

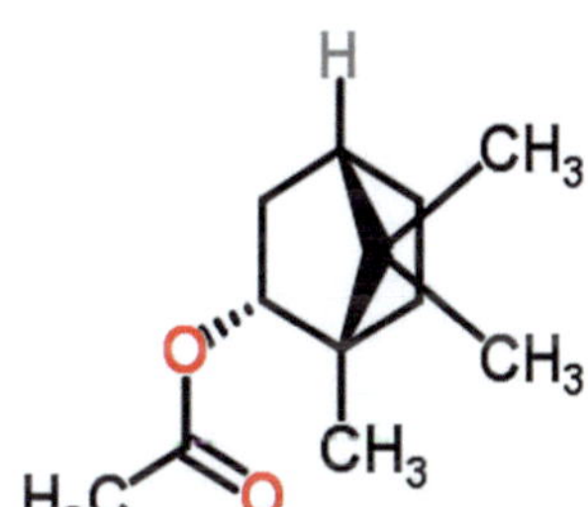
l-Bornyl acetate

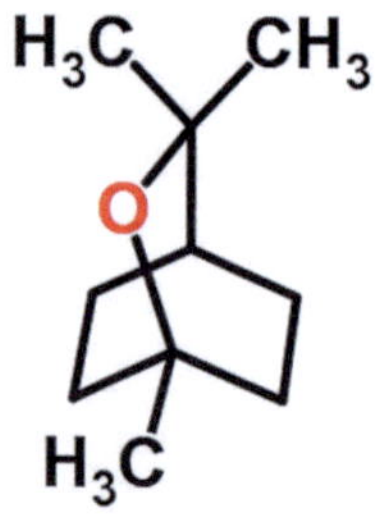

(±)-Eucalyptol

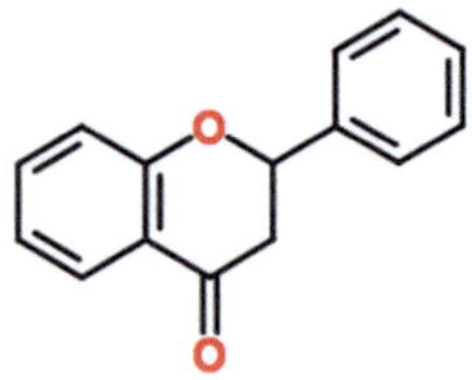

FLAVANONE

A. Triterpene

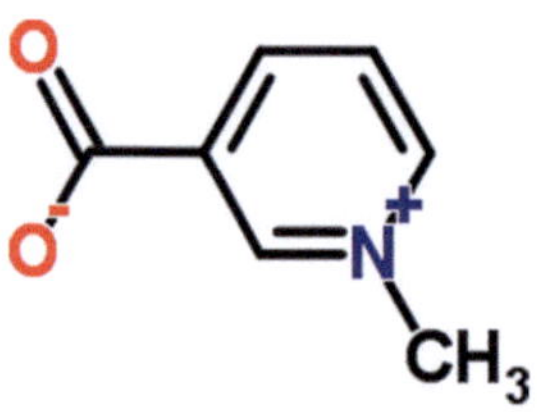

Trigonelline

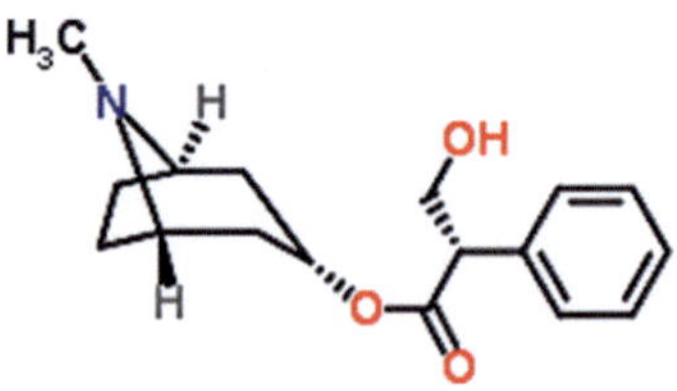

(S)-(-)-atropine

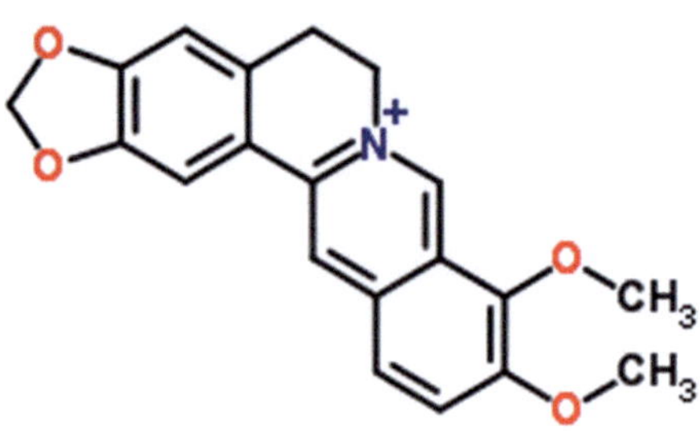

Jatrorrhizine

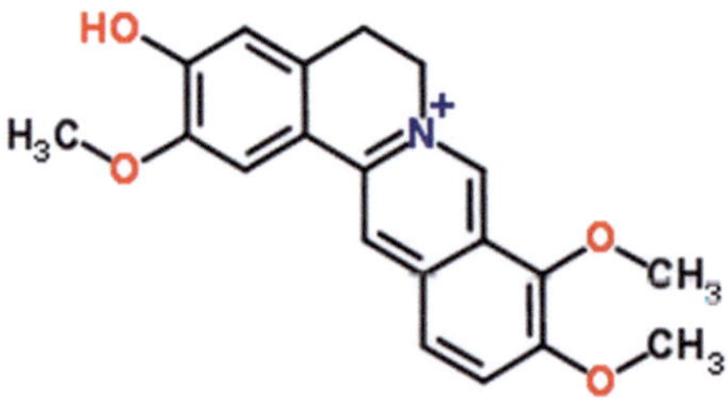

Berberine

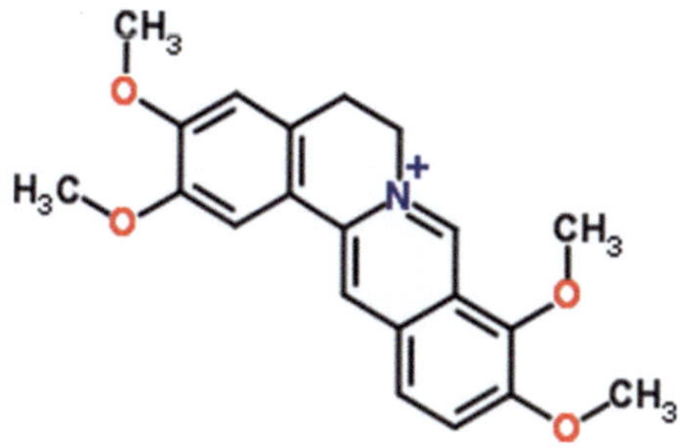

palmatine

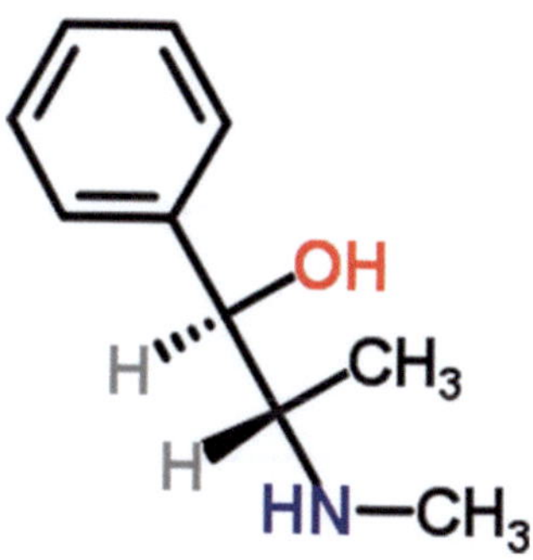

L-(-)-Ephedrine

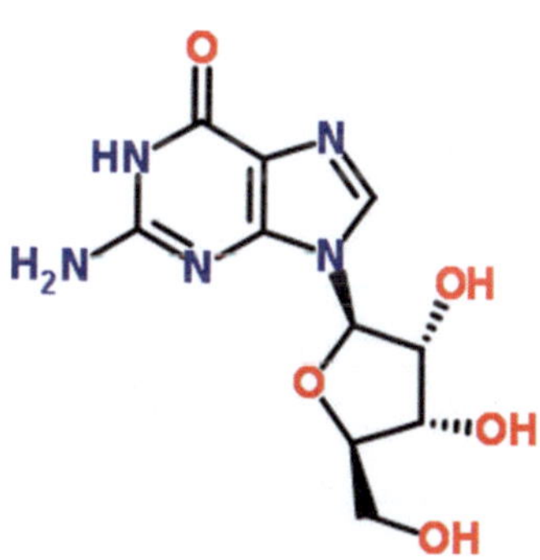

Guanosine

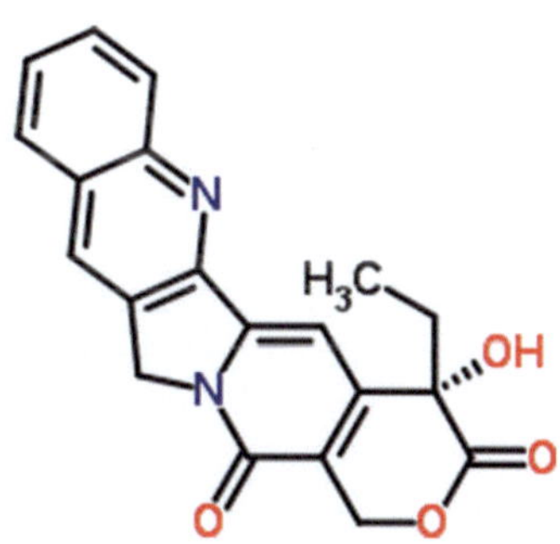

Camptothecin

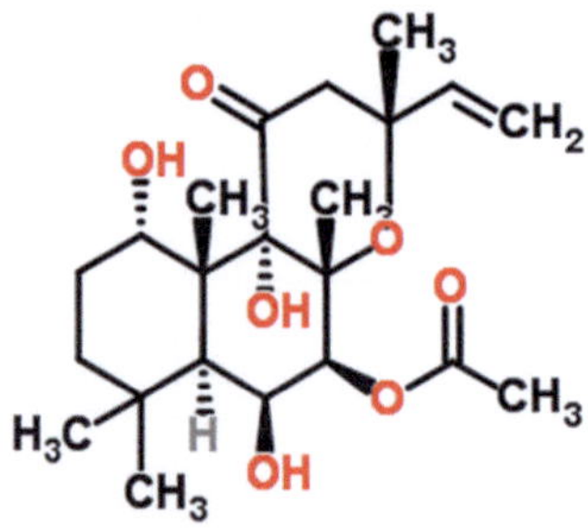

Forskolin

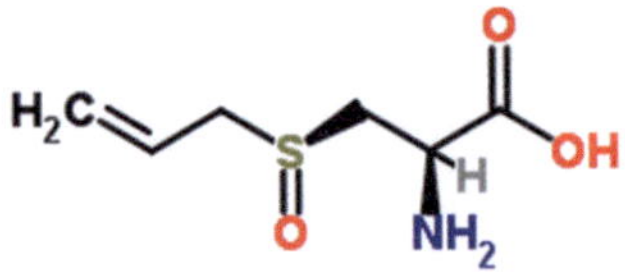

(+)-L-Alliin

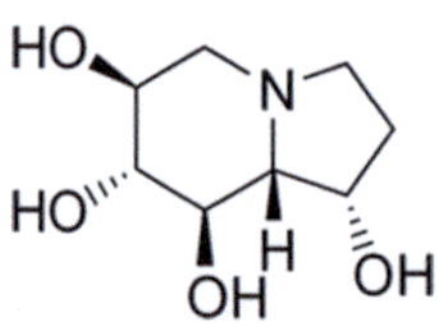

Castanospermine

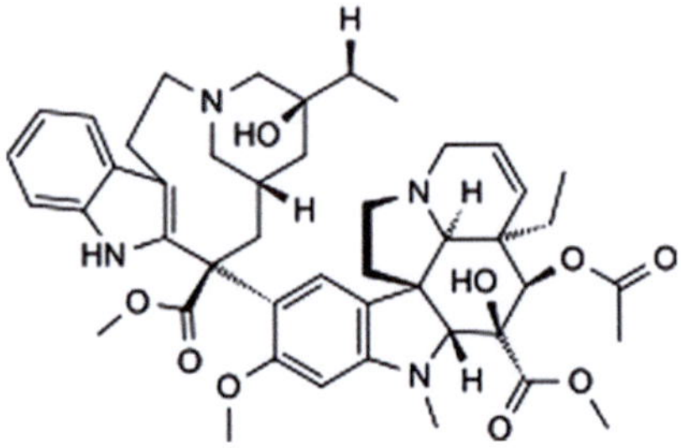

Vinblastine

B. Alkaloids

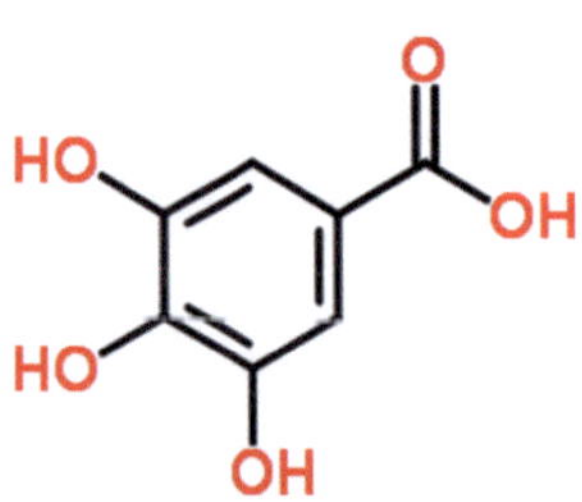

Gallic acid

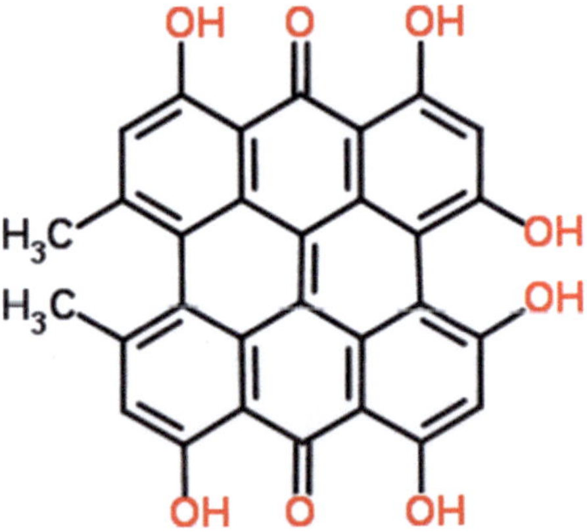

Hypericin

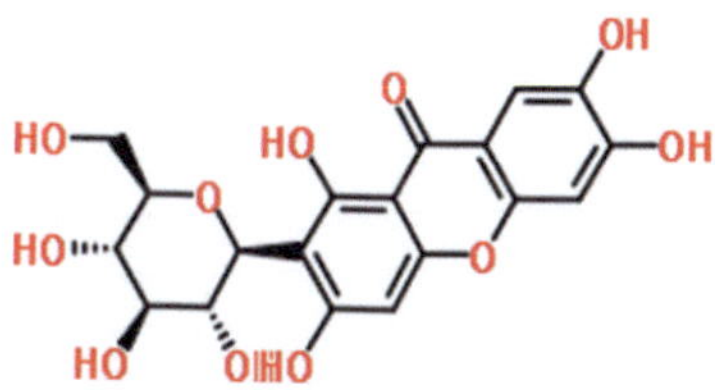

Mangiferin

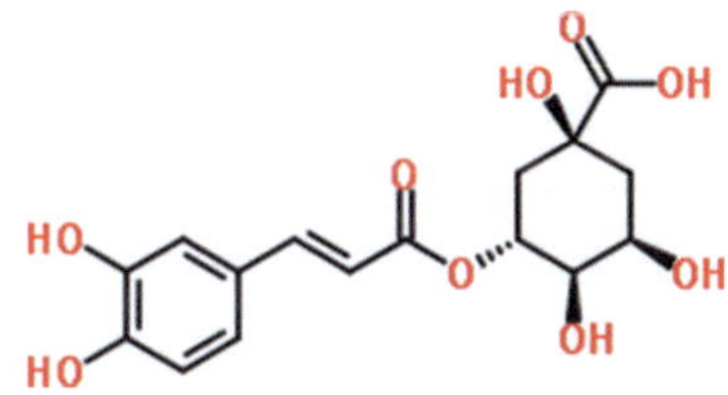

Chlorogenic acid

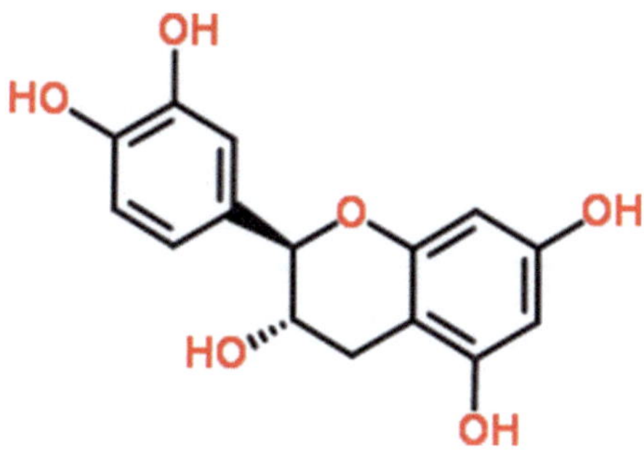

D-(+)-Catechin

C. Flavanoids and Phenolic Compounds

28.4 Conclusion

Now it is well-established that symbiotic association between medicinal plants and AM fungi can be exploited for the enhanced production of plant secondary metabolites. These secondary metabolites not only play an important role in uplifting the defense system of plants but also can be used for curing human ailments. In production of herb-based materials, there is a need to implement mycorrhizal technology to explore the role of AM fungi in the cultivation of medicinal plants. Thus future research priorities should emphasize on (1) isolation, selection, and screening of promising and effective AM fungal strains which can be adapted to natural habitat where medicinal and aromatic plants have to be grown, (2) strategies for the production of seedlings of medicinal and aromatic plants with mycorrhiza, (3) intensive cultivation of medicinal plants and the production of medicinally important phytochemicals as well as plant parts rich in medicinally important compounds, and (4) strategies by which AM fungi modulate the contents of bioactive principles in medicinal plants.

References

Abu-Zeyad R, Khan AG, Khoo C (1999) Occurrence of arbuscular mycorrhiza in *Castanospermum australe* A. Cunn. & C. Fraser and effects on growth and production of castanospermine. Mycorrhiza 9:111–117

Adams RP, Habte M, Park S, Dafforn MR (2004) Preliminary comparison of vetiver root essential oils from cleansed (bacteria-and fungus-free) versus non-cleansed (normal) vetiver plants. Biochem Syst Ecol 32(12):1137–1144

Akiyama K, Hayashi H (2002) Arbuscular mycorrhizal fungus-promoted accumulation of two new triterpenoids in cucumber roots. Biosci Biotechnol Biochem 66(4):762–769

Allen MF, Moore TS, Christensen M (1980) Phytohormone changes in *Bouteloua gracilis* infected by vesicular-arbuscular mycorrhizae. I. Cytokinin increases in the host plant. Can J Bot 58:371–374

Allen MF, Moore TS, Christensen M (1982) Phytohormone changes in *Bouteloua gracilis* infected by vesicular-arbuscular mycorrhizae. II. Altered levels of gibberellin-like substances and abscisic acid in the host plant. Can J Bot 60:468–471

Andrade SAL, Malik S, Sawaya ACHF, Bottcher A, Mazzafera P (2013) Association with arbuscular mycorrhizal fungi influences alkaloid synthesis and accumulation in Catharanthus roseus and Nicotiana tabacum plants. Acta Physiol Plant 35(3):867–880

Araim G, Saleem A, Arnason JT, Charest AC (2009) Root colonization by an arbuscular mycorrhizal (AM) fungus increases growth and secondary metabolism of purple coneflower, *Echinacea purpurea* (L.) Moench. J Agric Food Chem 57:2255–2258

Arpana J, Bagyaraj DJ, Prakasa Rao EVS, Parameswaran TN, Abdul Rahiman BA (2008) Symbiotic response of patchouli [*Pogostemon cablin* (Blanco) Benth.] to different arbuscular mycorrhizal fungi. Adv Environ Biol 2(1):20–24

Asrar AWA, Elhindi KM (2010) Elhindi. Alleviation of drought stress of marigold (*Tagetes erecta*) plants by using arbuscular mycorrhizal fungi. Saudi J Biol Sci 18(1):93–98

Augé RM (2001) Water relations, drought and vesicular-arbuscular mycorrhizal symbiosis. Mycorrhiza 11(1):3–42

Awasthi A, Bharti N, Nair N, Singh R, Shukla AK, Gupta MM, Darokar MP, Kalra A (2011) Synergistic effect of *Glomus mosseae* and nitrogen fixing *Bacillus subtilis* strain Daz26 on artemisinin content in *Artemisia annua* L. Appl Soil Ecol 49:125–130

Bagheri S, Ebrahimi MA, Davazdahemami S, Minooyi J (2014) Terpenoids and phenolic compounds production of mint genotypes in response to mycorrhizal bio-elicitors. Tech J Eng Appl Sci 4:339–348

Bagyaraj D J, Varma A (1995) Interaction between arbuscular mycorrhizal fungi and plants. In Advances in Microbial Ecology Springer, Boston pp 119–142

Barrios E (2007) Soil biota, ecosystem services and land productivity. Ecol Econ 64:269–285

Bian XJ, Hu L, Li XL, Zhang FS (2001) Effect of VA mycorrhiza on the turfgrass quality and mineral nutrient uptakes. Acta Pratacul Sin 10(3):42–46

Binet MN, Van Tuinen D, Deprêtre N, Koszela N, Chambon C, Gianinazzi S (2011) Arbuscular mycorrhizal fungi associated with *Artemisia umbelliformis* Lam, an endangered aromatic species in Southern French Alps, influence plant P and essential oil contents. Mycorrhiza 21:523–535

Boby VU, Bagyaraj DJ (2003) Biological control of root-rot of Coleus forskohlii Briq. using microbial inoculants. World J Microbiol Biotechnol 19(2):175–180

Boller T, Wiemken A (1986) Dynamics of vacuolar compartmentation. Annu Rev Plant Physiol 37(1):137–164

Borde M, Dudhane M, Jite PK (2009) Role bioinoculant (AM fungi) increasing in growth, flavor content and yield in *Allium sativum* L. under field condition. Not Bot Hortic Agrobot Cluj 37(2):124–128

Cai BY, Jie WG, Ge JP, Yan XF (2008) Molecular detection of the arbuscular mycorrhizal fungi in the rhizosphere of *Phellodendron amurense*. Mycosystema 27(6):884–893

Cao DX, Zhao JL (2007) The investigation of arbuscular mycorrhizal fungi and soil factors from the rhizospere of medicinal plant *Angelica dahurica*. Acta Agric Boreali-Sin 22:47–50

Copetta A, Lingua G, Berta G (2006) Effects of three AM fungi on growth, distribution of glandular hairs, and essential oil production in *Ocimum basilicum* L. var. Genovese. Mycorrhiza 16:485–494

Chaudhary V, Kapoor R, Bhatnagar AK (2008) Effectiveness of two arbuscular mycorrhizal fungi on concentrations of essential oil and artemisinin in three accessions of *Artemisia annua* L. Appl Soil Ecol 40:174–181

Chen LT, Guo QS, Liu ZY (2009a) Colonization pattern and dynamic change of arbuscular mycorrhizal fungi in *Pinellia ternate*. Guizhou Agric Sci 37(2):37–39

Chen LT, Liu ZY, Guo QS, Zhu GS (2009b) Advances in studies on arbuscular mycorrhizas in medicinal plants. Chin Tradit Herb Drugs 40(1):156–160

Chen LT, Guo QS, Liu ZY (2010) Arbuscular mycorrhiza of cultivated and wild *Pinellia ternate*. Chin J Chin Mater Med 35(4):405–410

Cho EJ, Lee DJ, Wee CD, Kim HL, Cheong YH, Cho JS, Sohn BK (2009) Effects of AM FUNGI inoculation on growth of *Panax ginseng* C.A. Meyer seedlings and on soil structures in mycorrhizosphere. Sci Hortic 122(4):633–637

Clark RB, Zeto SK (2000) Mineral acquisition by arbuscular mycorrhizal plants. J Plant Nutr 23(7):867–902

Cordell GA (1995) Changing strategies in natural products chemistry. Phytochem 40(6):1585–1612

Corizier R (1974) In: Kleinman VS et al. (eds) Medicine in Chinese Culture, Department of Health, Education and Welfare Publications, pp 26

Croteau R, Kutchan TM, Lewis NG (2000) Natural products (secondary metabolites). Biochem Mol Biol 24:1250–1319

David S (2000) The history of WWII medicine. http://home.att.net/~steinert/wwii.htm

De la Rosa-Mera CJ, Ferrera-Cerrato R, Alarcón A, de Jesús Sánchez-Colín M, Muñoz-Muñiz OD (2011) Arbuscular mycorrhizal fungi and potassium bicarbonate enhance the foliar content of the vinblastine alkaloid in Catharanthus roseus. Plant Soil 349(1–2):367–376

El-Sayed M, Verpoorte R (2007) Catharanthus terpenoid indole alkaloids: biosynthesis and regulation. Phytochem Rev 6(2–3):277–305

Fan JH, Yang GT, Mu LQ, Zhou JH (2006) Effect of AM fungi on the content of berberine, jatrorrhizine and palmatine of *Phellodendron amurense* seedings. Prot For Sci Technol 5:24–26

Farahani HA, Lebaschi MH, Hamidi A (2008) Effects of arbuscular mycorrhizal fungi, phosphorus and water stress on quantity and quality characteristics of coriander. J Adv Nat Appl Sci 2(2):55–59

Feng G, Zhang F, Li X, Tian C, Tang C, Rengel Z (2002) Improved tolerance of maize plants to salt stress by arbuscular mycorrhiza is related to higher accumulation of soluble sugars in roots. Mycorrhiza 12(4):185–190

Floß DS, Hause B, Lange PR, Küster H, Strack D, Walter MH (2008) Knock-down of the MEP pathway isogene 1-deoxy-d-xylulose 5-phosphate synthase 2 inhibits formation of arbuscular mycorrhiza-induced apocarotenoids, and abolishes normal expression of mycorrhiza-specific plant marker genes. Plant J 56(1):86–100

Fransworth NR, Akerele O, Bingel AS, Soejarto DD, Guo Z (1985) Drugs from medicinal plants. Bull WHO 63:965–981

Freitas MSM, Martins MA, Vieira IJC (2004) Yield and quality of essential oils of Mentha arvensis in response to inoculation with arbuscular mycorrhizal fungi. Pesq Agropec Bras 39(9):887–894

Galbley S, Thiericke R (1999) Drug Discovery FROM Nature. Springer, Berlin

Geneva MP, Stancheva IV, Boychinova MM, Mincheva NH, Yonova PA (2010) Effects of foliar fertilization and arbuscular mycorrhizal colonization on *Salvia officinalis* L. growth, antioxidant capacity, and essential oil composition. J Sci Food Agric 90:696–702

Gershenzon J, Kreis W (1999) Biochemistry of terpenoids: monoterpenes, sesquiterpenes, diterpenes, sterols, cardiac glycosides and steroid saponins. Biochem Plant Sec Met 2:222–299

Ghisalberti EL (1993) Detection and isolation of bioactive natural products. In: Colegate SM, Molyneux RJ (eds) Bioactive natural products: detection, isolation, and structural determination. CRC Press, Boca Raton, pp 9–57

Gianinazzi S, Gollotte A, Binet MN, Tuinen DV, Redecker D, Wipf D (2010) Agroecology: the key role of arbuscular mycorrhizas in ecosystem services. Mycorrhiza 20:519–530

Gong MQ, Wang FZ, Chen Y (2002) Study on application of arbuscular-mycorrhizas in growing seedling of Aloe vera. J Chin Med Mater 25(1):1–3

Gopal RM (2001) J Med Aromat Plant Sci, 22/4A & 23/1A, 572

Guerrieri E, Lingua G, Digilio MC, Massa N, Berta G (2004) Do interactions between plant roots and the rhizosphere affect parasitoid behaviour? Ecol Entomol 29(6):753–756

Guo LP, Wang HG, Hang LQ (2006) Effects of arbuscular mycorrhizae on growth and essential oil of *Atractylodes lancea*. Chin J Chin Mater Med 31(8):1491–1495

Guo QS, Chen LT, Liu ZY (2010) Study on influence of arbuscular mycorrhizal fungi on *Pinellia ternata* yield and chemical composition. Chin J Chin Med 35(3):333–338

Gupta ML, Prasad A, Ram M, Kumar S (2002) Effect of the vesicular-arbuscular mycorrhizal (VAM) fungus *Glomus fasciculatum* on the essential oil yield related characters and nutrient acquisition in the crops of different cultivars of menthol mint (*Mentha arvensis*) under field conditions. Bioresour Technol 81(1):77–79

Hadwiger A, Neimann H, Kaebisch A, Bauer H, Tamura T (1986) Appropriate glucosylation of the FMS gene product is a prerequisite for its transforming potency. EMBO J 5:689–694

Harris JC, Cottrell S, Plummer S, Lloyd D (2001) Antimicrobial properties of *Allium sativum* (garlic). Appl Microbiol Biotechnol 57:282–286

Harrison M (1999) Molecular and cellular aspects of the arbuscular mycorrhizal symbiosis. Annu Rev Plant Biol 50:361–389

Hart MM, Trevors JT (2005) Microbe management: application of mycorrhyzal fungi in sustainable agriculture. Front Ecol Environ 310:533–539

Herbert RB (2001) The biosynthesis of plant alkaloids and nitrogenous microbial metabolites. Nat Prod Rep 18(1):50–65

He XL, Li J, Gao AX, Zhao LL, Zao JL (2009a) Effects of different host plants on the development of AM fungi in the rhizospere of *Salvia miltiorrhiza*. J Hebei Univ 29(5):533–537

He XL, Li J, He C (2009b) Effects of AM fungi on the chemical components of *Salvia miltiorrhiza* Bge. Chin Agric Sci Bull 25(14):182–185

He XL, Liu T, Zhao LL (2009c) Effects of inoculating AM fungi on physiological characters an nutritional components of *Astragalus membranaceus* under different N application levels. Chin J Appl Ecol 20(9):2118–2122

He XL, Wang LY, Ma J, Zhao LL (2010) AM fungal diversity in the rhizosphere of *Salvia miltiorrhiza* in Anguo city of Hebei province. Biodivers Sci 18(2):187–194

Hodge A, Campbell CDFAH (2001) An arbuscular mycorrhizal fungus accelerates decomposition and acquires nitrogen directly from organic material. Nature 413:297–299

Hostettmann K, Marston A, Ndjoko K, Wolfender JL (2000) The potential of African plants as a source of drugs. Curr Org Chem 4(10):973–1010

Huang KC (1999) The pharmacology of Chinese herbs: a brief history of Chinese medicine, 2nd edn. CRC Press, Boca Raton, pp 1–16

Huang YF, Li HH, Chen HY, Li Y (2003) Preliminary study on the mycorrhiza inoculation on the seeding of camptotheca acuminate. Guangdong For Sci Technol 19(1):40–42

Huang LQ, Chen ML, Xiao PG (2004) The modern biological basis and model hypothesis on the research of genuineness of Chinese herbal medicine. Chin J Chin Mater Med 29(6):494–496

Huang JH, Tan JF, Jie HK, Zeng RS (2011) Effects of invoculating arbuscular mycorrhizal fungi on *Artemisia annua* growth and its officinal components. Chin J Appl Ecol 22(6):1443–1449

Ishibashi A (2002) In: Complementary and alternative medicine in Japan, SY19-4

Janardhanan KK, Abdul-Khaliq K (1995) Influence of vesicular arbuscular mycorrhizal fungi on growth and productivity of German chamomile in alkaline usar soil. In: Adholeya A, Singh S (eds) Mycorrhizae: biofertilizers for the future. Tata Energy Research Institute, New Delhi, pp 410–412

Jeffries P, Gianinazzi S, Peretto S, Turnau K, Barea JM (2003) The contribution of arbuscular mycorrhizal fungi in sustainable maintenance of plant health and soil fertility. Biol Fertil Soils 37:1–16

Jie WG, Cai BY, Ge JP, Yan XF (2007) Identification of arbuscular mycorrhizal fungi of *Phellodendren amurense* Rupr. Biotechnology 17(6):32–35

Jugran AK, Bahukhandi A, Dhyani P, Bhatt ID, Rawal RS, Nandi SK, Palni LMS (2015) The effect of inoculation with mycorrhiza: AM on growth, phenolics, tannins, phenolic composition and antioxidant activity in *Valeriana jatamansi* Jones. J Soil Sci Plant Nutr 15(4):1036–1049

Jurkiewicz A, Ryszka P, Anielska T, Waligórski P, Białońska D, Góralska K, Michael MT, Turnau K (2010) Optimization of culture conditions of *Arnica montana* L.: effects of mycorrhizal fungi and competing plants. Mycorrhiza 20:293–306

Kapoor R, Giri B, Mukerji KG (2002a) Glomus macrocarpum: a potential bioinoculant to improve essential oil quality and concentration in Dill (Anethum graveolens L.) and Carum (*Trachyspermum ammi* (Linn.) Sprague). World J Microbiol Biotechnol 18(5):459–463

Kapoor R, Giri B, Mukerji KG (2002b) Mycorrhization of coriander (*Coriandrum sativum* L.) to enhance the concentration and quality of essential oil. J Sci Food Agric 82:339–342

Kapoor R, Giri B, Mukerji KG (2004) Improved growth and essential oil yield and quality in *Foeniculum vulgare* mill on mycorrhizal inoculation supplemented with P-fertilizer. Bioresour Technol 93:307–311

Kapoor R, Chaudhary V, Bhatnagar AK (2007) Effects of arbuscular mycorrhiza and phosphorus application on artemisinin concentration in *Artemisia annua* L. Mycorrhiza 17:581–587

Karagiannidis N, Thomidisa T, Lazari D, Filotheou EP, Karagiannidou C (2011) Effect of three Greek arbuscular mycorrhizal fungi in improving the growth, nutrient concentration, and production of essential oils of oregano and mint plants. Sci Hortic 129:329–334

Khaosaad T, Vierheilig H, Nell M, Zitterl-Eglseer K, Novak J (2006) Arbuscular mycorrhiza alter the concentration of essential oils in oregano (*Origanum sp.*, Lamiaceae). Mycorrhiza 16:443–446

Koide RT (1991) Nutrient supply, nutrient demand and plant response to mycorrhizal infection. New Phytol 117:365–386

Koide RT, Schreiner RP (1992) Regulation of the vesicular-arbuscular mycorrhizal symbiosis. Annu Rev Plant Physiol Plant Mol Biol 43:557–581

Krishna H, Singh S, Sharma RR, Khawale RN, Grover M, Patel VB (2005) Biochemical changes in micropropagated grape (Vitis vinifera L.) plantlets due to arbuscular-mycorrhizal fungi (AMF) inoculation during ex vitro acclimatization. Sci Hort 106(4):554–567

Lapeyrie F (1988) Oxalate synthesis from soil bicarbonate by fungus *Paxillus involutus*. Plant Soil 110:3–8

Leake J, Johnson D, Donnelly D, Muckle G, Boddy L, Read (2004) Networks of power and influence: the role of mycorrhizal mycelium in controlling plant communities and agroecosystem functioning. Can J Bot 82:1016–1045

Lee J, Scagel CF (2009) Chicoric acid found in basil (*Ocimum basilicum* L.) leaves. Food Chem 115(2):650–656

Li CX (2003a) Effect of vesicular-arbuscular mycorrhizal fungi on production of *Ginseng*. J Chin Med Mater 26(7):475–476

Li CX (2003b) Effects of infecting vesicular-arbuscular mycorrhiza on growth and development of *Coix Lachryma-jobi* L. J Shanxi Agric Univ 23(4):351–353

Liu T, He XL (2008) Research on the formation course of arbuscular mycorrhizae from *Astragalus membranaceus* (Fisch.) Bunge seedlings. J Hebei For Orc Res 23(3):311–314

Liu SL, He XL (2009) Effects of AM fungi on growth of *Glycyrrhiza inflata* Bat under water stress. J Nucl Agric Sci 23(4):692–696

Liu JN, Wu LJ, Wei SL, Xiao X, Su CX, Jiang P, Song ZB, Wang T, Yu ZL (2007) Effects of arbuscular mycorrhizal fungi on the growth, nutrient uptake and glycyrrhizin production of licorice (*Glycyrrhiza uralensis* Fisch). Plant Growth Regul 52(1):29–39

Loomis WD, Corteau R (1972) Essential oil biosynthesis. Rec Adv Phytochem 6:147–185

Lu YQ, He XL (2005) Effects of AM fungi on the chemical composition and growth amount of *Atractylodes macrocephala koidz* seedling on different N levels. J Hebei Univ 25(6):650–653

Lu YQ, He XL (2008) Effects of AM fungi on photosynthetic pigment of *Atractylodes macrocephala* under different nitrogen levels. Acta Agric Bor Occi Sin 17(4):314–321

Lu YQ, Cui Y, He XL (2008a) Effects of AM fungi on biomass and nitrogen content of *Atractylodes macrocephala* under different nitrogen levels. J Henan Agric Sci 4:94–96

Lu YQ, He XL, Li LZ (2008b) Effects of AM fungi on leaf protective enzymes of *Atractrlodes macrocephala* under different nitrogen levels. Hubei Agric Sci 47(6):659–660

Lu YQ, Wang DX, Lu XL, Li LM, Li Y, He XL (2011) Effects of AM fungi on physiological character and nutritional component of *Atractylodes macrocephala* under different N levels. Acta Bot Bor Occi Sin 31(2):351–356

Lu FC, Lee CY, Wang CL (2015) The influence of arbuscular mycorrhizal fungi inoculation on yam (*Dioscorea* spp.) tuber weights and secondary metabolite content. Peer J 3:e1266

Ma J, He XL, Jiang ZM, Wang LY (2009) Influence of soil factors on arbuscular mycorrhizal fungal colonization of *Salvia miltiorrhiza*. Acta Agric Bor Occi Sin 18(5):194–198

Maier W, Schmidt J, Wray V, Walter MH, Strack D (1999) The arbuscular mycorrhizal fungus, *Glomus intraradices*, induces the accumulation of cyclohexenone derivatives in tobacco roots. Planta 207:620–623

Mamta G, Rahi P, Pathania V, Gulati A, Singh B, Bhanwra RK, Tewari R (2012) Comparative efficiency of phosphate-solubilizing bacteria under greenhouse conditions for promoting growth and aloin-A content of Aloe barbadensis. Arch Agron Soil Sci 58(4):437–449

Mandal S, Evelin H, Giri B, Singh VP, Kapoor R (2013) Arbuscular mycorrhiza enhances the production of stevioside and rebaudioside-A in *Stevia rebaudiana* via nutritional and non-nutritional mechanisms. Appl Soil Ecol 72:187–194

Mandal A, Mandal S, Park MH (2014) Genome-wide analyses and functional classification of proline repeat-rich proteins: potential role of eIF5A in eukaryotic evolution. PLoS One 9(11):e111800

Mandal S, Upadhyay S, Singh VP, Kapoor R (2015) Enhanced production of steviol glycosides in mycorrhizal plants: a concerted effect of arbuscular mycorrhizal symbiosis on transcription of biosynthetic genes. Plant Physiol Biochem 89:100–106

Marschner H (1995) Mineral nutrition of higher plants, 2nd edn. Elsevier Science, San Diego

Marschner H (1998) Mineral nutrition of higher plants. Academic, London

McGarvey DJ, Croteau R (1995) Terpenoid metabolism. Plant Cell 7(7):1015

Meng JJ, He XL (2011) Effects of AM fungi on growth and nutritional contents of *Salvia miltiorrhiza* Bge. under drought stress. J Agric Univ Hebei 34(1):51–61

Morone Fortunato I, Avato P (2008) Plant development and synthesis of essential oils in micropropagated and mycorrhiza inoculated plants of *Origanum vulgare* L. ssp. *hirtum* (Link) Ietswaart. Plant Cell Tissue Organ 93:139–149

Nell M, Vötsch M, Vierheilig H, Steinkellner S, Zitterl-Eglseer K, Franz C, Novak J (2009) Effect of phosphorus uptake on growth and secondary metabolites of garden (*Salvia officinalis* L.). J Sci Food Agric 89:1090–1096

Nell M, Wawrosh C, Steinkellner S, Vierheilig H, Kopp B, Lössl A, Franz C, Novak J, Zitterl-Eglseer K (2010) Root solonization by symbiotic arbuscular mycorrhizal fungi increases sesquiterpenic acid concentrations in *Valeriana of ficinalis* L. Planta Med 76:393–398

Pan PL, Chen DQ, Chen YT, Zhou FM (2008) The research on the sift and germinate of AM FUNGI spore of *Ophiopogon japonicas*. Mod Chin Med 10(10):13–14

Pandey DK, Banik RM (2009) The influence of dual inoculation with Glomus mossae and Azotobacter on growth and barbaloin content of Aloe vera. Am Eu J Sust Agric 3(4):703–714

Pandey DK, Banik RM, Dey A, Panwar J (2014) Improved growth and colchicines concentration in Gloriosa superb in mycorrhizal inoculation supplemented with phosphorus – fertilizer. Afr J Tradit Complement Altern Med 11(2):439–446
Peng S, Eissenstat DM, Graham JH, Williams K, Hodge NC (1993) Growth depression in mycorrhizal citrus at high-phosphorus supply: analysis of carbon costs. Plant Physiol 101:1063–1071
Petersen M, Simmonds MSJ (2003) Rosmarinic acid. Phytochemistry 62:121–125
Phillipson JD (1999) New drugs from nature—It could be yew. Phytother Res 13(1):2–8
Pozo MJ, Azcón-Aguilar C (2007) Unraveling mycorrhiza-induced resistance. Curr Opin Plant Biol 10(4):393–398
Prasad A, Kumar S, Khaliq A, Pandey A (2011) Heavy metals and arbuscular mycorrhizal (AM) fungi can alter the yield and chemical composition of volatile oil of sweet basil (*Ocimum basilicum* L.). Biol Fertil Soils 47(8):853–861
Qi GH, Zhang LP, Yang WL, Lu XR, Li CL (2002) Effects of arbuscular mycorrhizal fungi on growth and disease resistance of replanted ginkgo (*Ginkgo biloba* L.) seedlings. J Hebei For Orch Res 17(1):58–61
Qi GH, Zhang LP, Yang WL, Lv GY (2003) The effects of abruscular mycorrhiza fungi on ginkgo (*Ginkgo biloba* L.) in the field. Hebei Fruits 19(1):40–42
Rajan SK, Reddy BJD, Bagyaraj DJ (2000) Screening of arbuscular mycorrhizal fungi for their symbiotic efficiency with Tectona grandis. Forest Ecol Manag 126(2):91–95
Rapparini F, Llusia J, Penuelas J (2008) Effect of arbuscular mycorrhizal (AM) colonization on terpene emission and content of *Artemisia annua* L. Plant Biol 10:108–122
Rasouli-Sadaghianil MH, Hassani A, Barin M, Danesh YR, Sefidkon F (2010) Effects of arbuscular mycorrhizal (AM) fungi on growth, essential oil production and nutrients uptake in basil. J Med Plant Res 4(21):2222–2228
Rastogi PR, Meharotra BN (1990) (eds) In: Compendium of Indian medicinal plants, publications and information directorate, CSIR, New Delhi, vol.l, p 339
Redecker D, Morton JB, Bruns TD (2000) Ancestral lineages of arbuscular mycorrhizal fungi (Glomales). Mol Phylogenet Evol 14:276–284
Ren JH, Liu RX, Li YL (2007) Study on arbuscular mycorrhizae of *Panax notoginseng*. Microbiology 34(2):224–227
Ren JH, Zhang JF, Liu RX, Li YQ (2008) Study on arbuscular mycorrhizae in *Taxus chinensis* var. mairei. Acta Bot Bor Occi Sin 28(7):1468–1473
Roepke J, Salim V, Wu M, Thamm AM, Murata J, Ploss K, Wilhelm B, De Luca V (2010) Vinca drug components accumulate exclusively in leaf exudates of Madagascar periwinkle. Proc Natl Acad Sci 107(34):15287–15292
Rojas-Andrade R, Cerda-García-Rojas CM, Frías-Hernández JT, Dendooven L, Olalde-Portugal V, Ramos Valdivia AC (2003) Changes in the concentration of trigonelline in a semi-arid leguminous plant (*Prosopis laevigata*) induced by an arbuscular mycorrhizal fungus during the presymbiotic phase. Mycorrhiza 13:49–52
Sailo GS, Bagyaraj DJ (2005) Influence of different AM-fungi on the growth, nutrition and forskolin content of *Coleus forskohlii*. Mycol Res 109(7):795–798
Schüßler A, Schwarzott D, Walker C (2001) A new fungal phylum, the Glomeromycota, phylogeny and evolution. Mycol Res 105:1413–1421
Singh R, Soni SK, Kalra A (2013) Synergy between *Glomus fasciculatum* and a beneficial *Pseudomonas* in reducing root diseases and improving yield and forskolin content in *Coleus forskohlii* Briq. under organic field conditions. Mycorrhiza 23:35–44
Shah V, Bhat SV, Bajwa BS, Domacur H, De SNJ (1980) The occurrence of forskolin in Labiatae. Planta Med 39:183–185
Shen XL, Guo QS, Liu ZY, Zhu GS, Liu YX (2011) Colonization progress of arbuscular mycorrhizae on tissue-cultured plantlets of *Pinellia ternata*. Chin J Chin Mater Med 36:93–96
Shibata K, Iwata S, Nakamura M (1923) Baicalin, a new flavone-glucuronic acid compound from the roots of Scutellaria baicalensis. Acta Phytochim 1:105–139

Sieverding E, Friedrichsen J, Suden W (1991) Vesicular-arbuscular mycorrhiza management in tropical agrosystems. Sonderpublikation der GTZ (Germany)

Silva MFD, Pescador R, Rebelo RA, Stürmer SL (2008) The effect of arbuscular mycorrhizal fungal isolates on the development and oleoresin production of micropropagated *Zingiber officinale*. Braz J Plant Physiol 20(2):119–130

Silva FA, Ferreira MR, Soares LA, Sampaio EV, Maia LC (2014) Arbuscular mycorrhizal fungi increase gallic acid production in leaves of field grown *Libidibia ferrea* (Mart. ex Tul.) LP Queiroz. J Med Plant Res 8(36):1110–1115

Smith SE, Read DJ (1997) Mycorrhizal symbiosis. Academic, London

Smith SE, Read DJ (2008) Mycorrhizal symbiosis, 3rd edn. Academic, London

Smith SE, Smith FA, Jakobsen I (2004) Functional diversity in arbuscular mycorrhizal (AM) symbioses: the contribution of the mycorrhizal P uptake pathway is not correlated with mycorrhizal responses in growth or total P uptake. New Phytol 162:511–524

de Sousa OM, da Silva Campos MA, de Albuquerque UP, da Silva FSB (2013) Arbuscular mycorrhizal fungi (AMF) affects biomolecules content in *Myracrodruon urundeuva* seedlings. Ind Crop Prod 50:244–247

Strack D, Fester T, Hause B, Schliemann W, Walter MH (2003) Arbuscular mycorrhiza: biological, chemical, and molecular aspects. J Chem Ecol 29(9):1955–1979

Suzuki Y (2002, May 29) In Current status and future directions of alternative medicine, WHO, Geneva, Switzerland

Swami Tirtha SS (1998) In: Uniyal RC et al (eds) Ayurvedic Encyclopedia, Natural Secrets to Healing, Prevention, and Longevity: history of Ayurvedic Tree, 1st edn. New Delhi, Sai Satguru Publications

Szakiel A, Pączkowski H (2011a) Influence of environmental biotic factors on the content of saponins in plants. Phytochem Rev 10:493–502

Szakiel A, Pączkowski H (2011b) Influence of environmental abiotic factors on the content of saponins in plants. Phytochem Rev 10:471–491

Burni T, Saadia N, Tabassum Y, Sakina B (2013) Arblscular mycorrhizal stuidies in "Aloe vera (l). burm. f." biologically active and potential medicinal plant. Wudpecker J Agric Res 2(1):039–042

Tang W, Eisenbrand G (1992) In: Chinese drugs of plant origin, SpingerVerlag, Berlin, p 127

Teng HR, He XL (2005) Effects of different AM fungi and N levels on the flavonoid content of *Bupleuruin scorzonerifolium* Willd. J Shanxi Agric Sci 4:53–54

Torelli A, Trotta A, Acerbi L, Arcidiacono G, Berta G, Branca C (2000) IAA and ZR content in leek (Allium porrum L.), as influenced by P nutrition and arbuscular mycorrhizae, in relation to plant development. Plant Soil 226(1):29–35

Toussaint JP (2007) Investigating physiological changes in the aeria parts of AM plants: what do we know and where should we be heading? Mycorrhiza 17:349–353

Toussaint JP, St-Arnaud M, Charest C (2004) Nitrogen transfer and assimilation between the arbuscular mycorrhizal fungus *Glomus intraradices* Schenck & Smith and Ri T-DNA roots of *Daucus carota* L. in an in vitro compartmented system. Can J Microbiol 50:251–260

Toussaint JP, Smith FA, Smith SE (2007) Arbuscular mycorrhizal fungi can induce the production of phytochemicals in *sweet basil* irrespective of phosphorus nutrition. Mycorrhiza 17:291–297

Verpoorte R (1999) Secondary metabolism. In: Verpoorte R, Alfermann AW (eds) Metabolic engineering of plant secondary metabolism. Klumer Academic Publishers, Dordrecht, pp 1–29

Volpin H, Elkind Y, Okon Y, Kapulnik Y (1994) A vesicular arbuscular mycorrhizal fungus (*Glomus intraradices*) induces a defense response in alfalfa roots. Plant Physiol 104:683–689

Wang LY, He XL (2009) The resource and spatio-temporal distribution of AM fungi from *Salvia miltiorrhiza* in Anguo. J Agric Univ Hebei 32(6):73–79

Wang Q, Li HQ, Du YR, Li Y, Li HW (1998) Isolation and identification of VA mycorrhizal fungi on *Radix gentianae*. Biotechnology 8(2):19–22

Wang Q, He XL, Chen TS, Dou WF (2006) Ecological research of arbuscular mycorrhizal fungi in rhizosphere of *Puerraria lobata*. J Hebei Univ 26(4):420–425

Wang DX, Lu YQ, He XL (2010) Effects of AM fungi on growth and physiological characters of *Atractylodes macrocephala* under different P-applied levels. Acta Bot Bor Occi Sin 30(1):136–142

Warrier PK, Nambiar VPK, Ramankutti C (1996) (eds), In Indian medicinal plants: a compendium of 500 species, Orient Longmann, Hyderabad, vol 4. p 409

Wei GT, Wang HG (1989) Effects of VA mycorrhizal fungi on growth, nutrient uptake and effective compounds in Chinese medicinal herb *Datura stramonium* L. Sci Agric Sin 22(5):56–61

Wei GT, Wang HG (1991) Effect of vesicular-arbuscular mycorrhizal fungi on growth, nutrient uptake and synthesis of volatile oil in *Schizonepeta tenuifolia* Briq. Chin J Chin Mater Med 16(3):139–142

Wiermann R (1981) Secondary plant products and cell and tissue differentiation. In: The biochemistry of plants, vol. 7. Academic, New York, pp 85–116

Wink M (1997) Compartmentation of secondary metabolites and xenobiotics in plant vacuoles. Adv Bot Res 25:141–169

Wink M (1999) Introduction: biochemistry, role and biotechnology of secondary metabolites. In: Wink M (ed) Biochemistry of plant secondary metabolism. Sheffield Academic Press Ltd, Sheffield, pp 1–16

Wu QC, Wei QA (2008) Arbuscular mycorrhizae of *Ginkgo biloba* and its correlation with soil available phosphorus. J Yangtze Univ 5(3):49–52

Wu QS, Liu W, Zhai HF, Ye XF, Zhao LJ (2010) Influences of AM fungi on growth and root antioxidative enzymes of *Trifoliate orange* seedlings under salt stress. Acta Agric Univ Jiangxiensis 32(4):759–762

Xing XK, Li Y, Yolande D (2000) Ten species of vAM fungi in five ginseng fields of Jilin province. J Jilin Agric Univ 22(2):41–46

Xing XK, Li Y, Wang Y, Zhang MP (2003) Foundation of dual cultural system of ginseng VA mycorrhiza fungi. J Jilin Agric Univ 25(2):154–157

Xiao WJ, Yang G, Chen ML, Guo LP, Wang M (2011) AM and its application in plant disease prevention of Chinese medicinal herbs cultivation. Chin J Chin Med 36(3):252–257

Yang G, Guo LP, Huang LQ, Chen M (2008) Inoculation methods of AM fungi in medicinal plant. Resour Sci 30(5):778–785

Yang Y, Ou X, Yang G, Xia Y, Chen M, Guo L, Liu D (2017) Arbuscular mycorrhizal fungi regulate the growth and phyto-active compound of *Salvia miltiorrhiza* seedlings. Appl Sci 7(1):68

Yadav K, Aggarwal A, Singh N (2013) Arbuscular mycorrhizal fungi (AMF) induced acclimatization, growth enhancement and colchicine content of micropropagated Gloriosa superba L. plantlets. Ind Crop Prod 45:88–93

Yu Y, Yu T, Wang Y, Yan XF (2010) Effect of inoculation time on camptothecin content in arbuscular mycorrhizal *Camptotheca acuminate* seedlings. Chin J Plant Ecol 34(6):687–694

Zeng Y, Guo LP, Hang LQ, Zhou J, Sun YZ (2007) AM and its application in TCM cultivation. World Sci Technol/Moder TCM Mater Med 9(6):83–87

Zhang MC, Jing YJ, Ma J (1990) The changing of microbial ecological types after the improvement of ginseng soil. J Jilin Agric Univ 12(4):42–46

Zhang Y, Xie LY, Xiong BQ, Zeng M, Yu D (2004) Correlation between the growth of arbuscular mycorrhizal fungi in the rhizosphere and the flavonoid content in the root of *Ginkgo biloba*. Mycosystema 23(1):133–138

Zhang J, Liu DH, Guo LP, Jin H, Zhou J, Yang G (2010) Effects of four AM fungi on growth and essential oil composition in rhizome of *Atractylodes lancea*. World Sci Technol/Moder TCM Mater Med 12(5):779–782

Zhang J, Liu DH, Guo LP, Jin H, Yang G, Zhou J (2011) Effects of arbuscular mycorrhizae fungi on biomass and essential oil in rhizome of *Atractylodes lancea* in different temperatures. Chin Tradit Herb Drugs 42(2):372–375

Zhao JL, He XL (2011) Effects of AM fungi on drought resistance and content of chemical components in *Angelica dahurica*. Acta Agric Bor Occi Sin 20(3):184–189

Zhao X, Wang BW, Yan XF (2006) Effect of arbuscular mycorrhiza on camptothecin content in *Camptotheca acuminate* seedlings. Acta Ecol Sin 26(4):1057–1062

Zhao PJ, An F, Tang M (2007) Effects of arbuscular mycorrhiza fungi on drought resistance of *Forsythia suspense*. Acta Bot Bor Occi Sin 27(2):396–399

Zhao JL, Deng HY, He XL (2009) Effects of AM fungi on the quality of trueborn *Angelica dahurica* from Hebei province. Acta Agric Boreali-Sin 24:299–302

Zhou JH, Fan JH (2007) Effects of AM fungi on the berberine content in *Phellodendron chinense* seedings. North Hortic 12:25–27

Zhou N, Xia CL, Jiang B, Bai ZC, Liu GN, Ma XK (2009) Arbuscular mycorrhiza in *Paris polyphylla* var. yunnanensis. Chin J Chin Med 34(14):1768–1772

Zhou N, Zou L, Wang GZ, Jiang B (2010) Primary explore to relation of arbuscular mycorrhizae and its secondary metabolite steroidal saponin in *Paris polyphylla*. Chin J Exp Tradit Med Formulae 16(16):85–88

Zubek S, Stojakowska A, Anielska T, Turnau K (2010) Arbuscular mycorrhizal fungi alter thymol derivative contents of *Inula ensifolia* L. Mycorrhiza 20:497–504

Zubek S, Mielcarek S, Turnau K (2012) Hypericin and pseudohypericin concentrations of a valuable medicinal plant *Hypericum perforatum* L. are enhanced by arbuscular mycorrhizal fungi. Mycorrhiza 22:149–156

Zubek S, Błaszkowski J, Seidler-Łożykowska K, Bąba W, Mleczko P (2013) Arbuscular mycorrhizal fungi abundance, species richness and composition under the monocultures of five medicinal plants. Acta Sci Pol-Hortoru 12:127–141

Zubek S, Rola K, Szewczyk A, Majewska ML, Turnau K (2015) Enhanced concentrations of elements and secondary metabolites in *Viola tricolor* L. induced by arbuscular mycorrhizal fungi. Plant Soil 390(1–2):129–142

Chapter 29
Endophytic Fungi: Carrier of Potential Antioxidants

Jyoti Goutam, Ranjana Singh, Rajyoganandh S. Vijayaraman, and Mukesh Meena

Abstract The asymptomatic association of fungi with plants is termed as endophytes. These plant-associated mycoflora are a promising source of bioactive natural products. Metabolites released by endophytes not only possess many important functions but also supply antioxidant compounds, which are expected to fight disease due to its anti-aging properties. Currently, hectic lifestyle and stressful environment have become the prime cause for the generation of excessive free radicals in the human body. These free radicals create a destructive process in the body cells which leads to various chronic diseases and deleterious effects. Antioxidants are the chemical moieties that engulf free radicals which are followed by delaying cell damages and health disorders. Antioxidant moieties are generally synthesized by both plants and other microorganisms to survive adverse situations such as harmful radiations and abiotic and biotic stress. Hence, they are beneficial to both plants and animals which fed on the plant, thereby decreasing the reactive oxygen species level which are elevated in their normal metabolism process. Collectively, they help us to properly detoxify the body from these harmful molecules. This overview will discuss about antioxidants and highlight the different antioxidant compounds that have been derived from endophytic fungi. Although synthetic antioxidant compounds are being used, but due to their side effects and less bioavailability, they are not widely accepted. Therefore endophytes could prove to be a natural resource for sustainable antioxidant.

J. Goutam (✉)
Mycopathology and Microbial Technology Laboratory, Centre of Advance Study in Botany, Institute of Science, Banaras Hindu University, Varanasi, India

R. Singh
Centre of Advance Study in Physics, Department of Physics, Institute of Science, Banaras Hindu University, Varanasi, India

R. S. Vijayaraman
Department of Microbiology, Sri Ramachandra Medical College, Porur, Chennai, India

M. Meena
Centre for Transgenic Plant Development, Department of Biotechnology, Faculty of Science, Jamia Hamdard, New Delhi, India

Mycopathology and Microbial Technology Laboratory, Centre of Advance Study in Botany, Institute of Science, Banaras Hindu University, Varanasi, India

P. Gehlot, J. Singh (eds.), *Fungi and their Role in Sustainable Development: Current Perspectives*, https://doi.org/10.1007/978-981-13-0393-7_29

Keywords Endophytic fungi · Antioxidant compounds · Free radicals

29.1 Introduction

Endophytic fungi are the prominent components of plant microecosystem. Besides possessing a variety of potentials, these endophytic fungi have become an alternative source for antioxidant moieties. Antioxidants can be derived from natural sources as well as obtained synthetically. Among them, compound pestacin isolated from endophytes is considered as "keystone" in the research field, due to its extraordinary higher antioxidant potential than trolax, a synthetic antioxidant. This has motivated the researcher to extract and study some more compounds from endophytes. Although synthetic antioxidants such as butylated hydroxyanisole [BHA], butylated hydroxytoluene [BHT], and propyl gallate are being used in food and cosmetics industries, they do have their own drawbacks, such as probable carcinogenicity when administered *in vivo*, the concern of public safety, and economic values. Therefore compounds isolated from plant and other natural products are gaining momentum in cosmetics and food additives industry. Since endophytic fungi are an equally alternative source for such kind of products, exploring endophytic fungi for antioxidant compounds is becoming important day by day. They are considered to have strong protective mechanism against generation of free radicals which leads to several disorders such as aging, cancer, atherosclerosis, coronary heart ailment, and diabetes (Halliwell et al. 1994), neurodegenerative disorders such as Alzheimer's and Parkinson's disease, and rheumatoid arthritis (Valko et al. 2007). Generation of free radicals and its deleterious effect in normal body function are the key element of oxidative stress. The brain is highly vulnerable to oxidative stress as it undergoes excessive physiological activities. The unsaturated lipids are targeted and oxidized. This makes antioxidants an important component in treating brain disease. This article is being focused on the antioxidant potential of endophytic fungi. Few important compounds that have been explored from endophytes have been discussed here.

29.2 Why Do Plants Synthesize Antioxidants?

Plants have been the major sources of natural products since the beginning of research, and it could also be stated that they are the natural craftsmen of molecules created in inexhaustible orders. Since plants are sedentary organisms, they entertain a number of changes in order to adapt to unfavorable conditions. These changes occur due to the formation of different important chemicals. These chemicals are accepted as the valuable resource of plants since they maintain their age and health. Endophytes and plants have a symbiotic relationship, where the endophytes acquire benefit in the form of nutrition and in return synthesize certain compounds that help the plant in metabolism and protect it from unfavorable conditions. These compounds produced by the endophytes present hidden repertoires of known and unknown medicinal significance.

Heinig et al. (2013) gave the concept of horizontal gene transfer which suggests that endophytes and the host plant contribute to the co-production of bioactive molecules. Some endophytes have been known to possess superior biosynthetic capabilities, owing to their presumable gene recombination with the host while residing and reproducing inside the healthy plant tissues (Li et al. 2005). Taxol, jesterone, ambuic acid, torreyanic acid, pestaloside, pestalotiopsins, and 2-a-hydroxydimeniol are few examples of such compounds (Strobel and Daisy 2003). These bioactive molecules synthesized by plants can be used for the treatment of several human diseases. Apart from plants, endophytes which are being in the symbiotic relationship with plants are also considered to be a significant source of antioxidants (Huang et al. 2007).

29.3 Synthetic Antioxidants vs. Natural Antioxidants

Edible antioxidants are the best way of acquiring antioxidants in the stressful environment. Studies suggest that antioxidants added in food materials have diverse positive effect (Willis et al. 2009). Synthetic phenolic antioxidants such as butylated hydroxyanisole [BHA], butylated hydroxytoluene [BHT], and propyl gallate are used as food additives to inhibit the oxidation of food materials. In addition to this, people are consuming synthetic antioxidants in the form of vitamins, colorants, flavoring agents (spice and herb), and preservatives. Long-term antioxidant stress may result in weak immunological response and development of asthma, allergies, and obesity which have become serious concerns for public health. Moreover, the quest for safer supplements that are devoid of negative effects has motivated the researchers to search for consumer-friendly natural antioxidants derived from plants and other natural sources (Table 29.1).

After discussing natural antioxidants and their role in body defense system, it can be suggested that non-enzymatic antioxidants are also added in the form of natural dietary components.

29.4 Structure of Antioxidant Compounds and Their Role in Decreasing Oxidation

Chemical format of molecules suggest that, compounds that possess any aromatic ring bonded to hydroxyl groups (–OH) as substituents act as effective antioxidants; these include phenols, flavonoids, etc. A phenolic compound develops resonance and delocalizes and converts itself into phenoxide ion which further loses an electron and forms a corresponding free radical. Phenolic rings, which donate H to free radical during self-association, become a free radical. Later on, they are stabilized by internal delocalization of electron within the aromatic ring (Brewer et al. 2011). Due to the presence of –OH group, they also act as an anti-inflammatory and

Table 29.1 Natural antioxidants

1.	Enzymatic antioxidant (breaking down and removing free radicals): Superoxide dismutase (SOD) converts superoxide ion into dioxygen and hydrogen peroxide; this enzyme exists in three forms: cytosolic Cu, Zn-SOD, mitochondrial Mn-SOD, and extracellular SOD (Landis and Tower 2005). Catalase converts hydrogen peroxide to water and has the highest turnover rate (Mates et al. 1999).Glutathione peroxidase and glutathione reductase both involve themselves in secondary enzyme defense system glucose-6-phosphate dehydrogenase regenerate NADPH thereby creating a reducing environment (Gamble and Burke 1984)
2.	Nonenzymatic antioxidants (interrupting free radical chain reactions): Minerals such as selenium, copper, iron, zinc, and manganese (acting as cofactor for antioxidants); vitamins such as vitamin A, vitamin C, and vitamin E; carotenoids such as β-carotene, lycopene, lutein, and zeaxanthin; and volatile oils found in nature such as eugenol, carvacrol, safrole, thymol, menthol, 1, 8-cineole, α-terpineol, p-cymene, cinnamaldehyde, myristicin, and piperine (Shan et al. 2005). Polyphenols: Phenolic acids (hydrocinnamide and hydrobenzoic acid). Flavonoids: They are scavengers of superoxide ions (Robak and Gryglewski 1998). Examples include flavones (luteolin, apigenin, tangeretin), flavanols (quercetin, kaemferol, myricetin, isorhamnetin, pachypodo), flavanones (hesteretin, naringenin, eriodictyol), isoflavanones (genistein, daidzein, glycitein), anthocyanin C (cyanidin, delphinidin, malvidin, pelargonidin, peonidin, petunidin), gingerol, and curcumin
3.	Others: Transition metal-binding proteins such as albumin, ceruloplasmin, haptoglobin, and transferrin
	Nonprotein antioxidants include bilirubin, uric acid (Ames et al. 1981), and ubiquinol (coenzyme Q) (Stocker et al. 1987; Papas et al. 1998)

contain antimicrobial properties. Hence, phenolic compounds have radical-scavenging and antioxidant activity (Waterman and Mole 1994). Among the diverse groups, flavonoids, tannins, and phenolic acids are the main phenolic compounds (Koes et al. 1994; Burns et al. 2001; Rababah et al. 2005). Antioxidants are not only eaters of free radicals but also suppress generation of superoxide and possess chemopreventive effects (Lippman et al. 1994).

Antioxidant compound uses the different mechanism to reduce the populations of free radicals as it chelates metal ions, so they become unable to generate reactive species. They also aid in quenching O_2, preventing formation of peroxides which break the auto-oxidative chain reactions, thus reducing localized O_2 concentrations (Nawar et al.1996).

29.5 Structure and Functional Role of Antioxidant Compound Associated with Different Endophytic Fungi

29.5.1 Flavipin

Flavipin is a well-known natural product that was isolated from endophytes belonging to *Chaetomium* sp. associated with leaves of *Ginkgo biloba.* (Ye et al. 2013). Apart from it, in previous studies, *Epicoccum nigrum*, *Aspergillus flavipes*, and *Aspergillus terreus* also showed the production of flavipin (Raistrick and

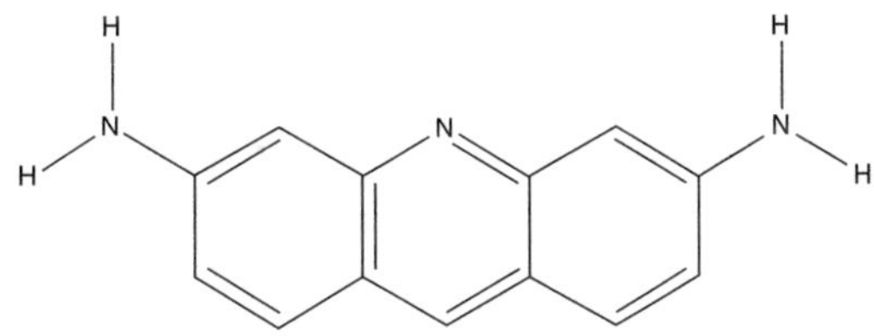

Fig. 29.1 Flavipin (A) (Raistrick and Rudman 1956) https://www.aspergillus.org.uk/content/flavipin

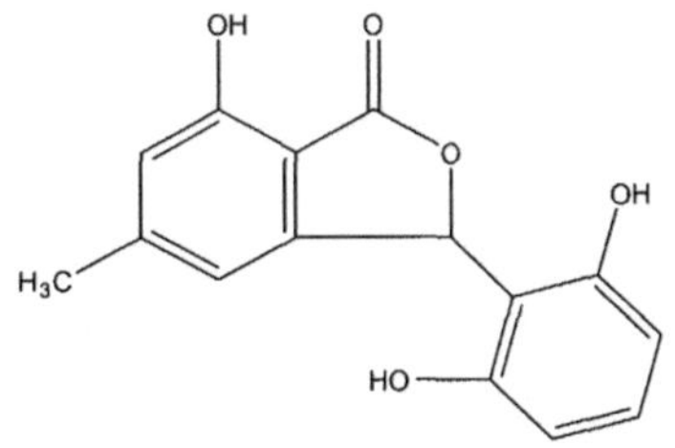

Fig. 29.2 Isopestacin (B) (Harper et al. 2003)

Rudman 1956; Bamford et al. 1961). Earlier this compound has been reported to possess antifungal activity against potent pathogens like *Botrytis allii* and *Aspergillus flavipes*. It has also exhibited activity against nematodes (soybean cyst) (Nitao et al. 1999). Flavipin produced by *Chaetomium* sp. exhibited antifungal activity against plant pathogenic fungi *Fusarium graminearum.* In 2013 Ye et al. studied the antioxidant behavior of this compound and found that the compound had an IC_{50} value of 0.73 μg mL − 1 against *Fusarium graminearum* (Fig. 29.1).

Flavipin (acroflavin) (1) is chemically known as 1, 2-benzenedicarboxaldehyde-3, 4, 5-trihydroxy-6-methyl. Combinatorial chemistry has modified this compound and is now sold as acriflavine by several companies all over the world. Apart from its medical usage, this compound is also being used in the laboratory as an intercalating dye.

29.5.2 *Isopestacin*

Isopestacin (B) is an important compound obtained from endophytic fungus *Pestalotiopsis microspora* associated with combretaceaous plant *Terminalia morobensis*. This plant was collected from Sepik River drainage of Papua New Guinea (Womersley 1995) (Fig. 29.2).

From the bioactivity reported by Strobel (2002) and Harper (2003), isopestacin (crude methylene extract) was found to be moderately antimycotic, as the growth of *Pythium ultimum*, a plant pathogenic oomycete, was completely inhibited at 40 mg/ml at 48 h. Apart from antimycotic and antioxidant activity, an antimalarial potential of a compound was also reported. Isopestacin was recognized as an antioxidant on the basis of structural similarity with flavonoids. Its structure was confirmed by electron spin resonance spectroscopy technique.

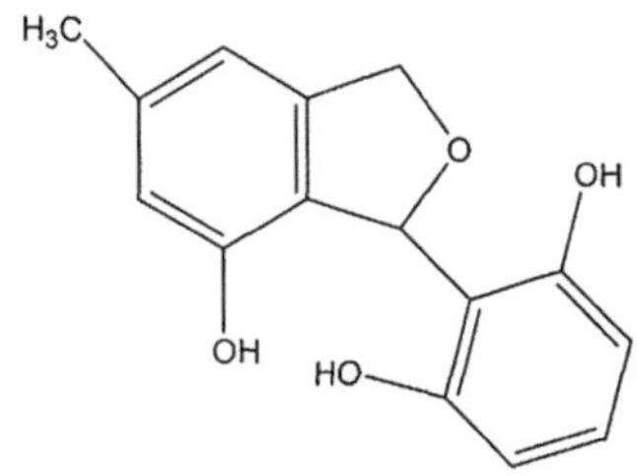

Fig. 29.3 Isopestacin (C) (Strobel et al. 2002)

29.5.3 Pestacin

The chemical name of pestacin (C) is (1, 5, 7-trisubstituted) 1, 3-dihydro isobenzofuran. The compound is isolated from endophytic fungi *Pestalotiopsis microspora* possessing antioxidant and antimycotic activities. Antioxidant properties of pestacin (C) were analyzed by known antioxidant assay TOSC. This assay is based on the ability of the compound to inhibit a free radical initiator's oxidation of alfa-keto-g methiolbutyric acid to release ethylene gas (Dugas et al. 2000). Further, decrease in production of ethylene gas was measured by gas chromatography. By comparing vitamin E derivative, trolox, it was exhibited that 18:8 0:9 mM was the concentration of solution that exerted the same effect. Antioxidant effect of pestacin was observed 11 times greater than the vitamin E derivative trolox (a potential active antioxidant) (Fig. 29.3).

29.5.4 Graphislactone A

In 1968 graphislactone (D) was extracted from botralin which is a fungal metabolite (Overeem and Van Dijkman 1968). Graphislactone is a well-known free radical antioxidant isolated from a mycobiont of lichens of the genus *Graphis*, hence termed as such. This detail was first reported by a Japanese researcher at Kobe Pharmaceutical University in 1997. Different derivatives of *Graphis* sp. are also isolated from sponge-derived mycobiont of lichen (Tanahanshi et al. 2003). *Microsphaeropsis olivacea*, an endophytic fungus, was found to be a producer of graphislactone and botralin (Hormazabala et al. 2005). Chemically graphislactone A belongs to phenolic benzopyranones. Graphislactones A–H and the structurally related ulocladol are highly oxygenated resorcylic lactones produced by lichens and fungi (Altemollar et al. 2009) (Fig. 29.4).

In 2005 Song demonstrated that graphislactone A (D) has stronger free radical-scavenging and antioxidant activity as compared to standard butylated hydroxytoluene (BHT) and ascorbic acid when co-assayed in the study. An endophytic fungus *Cephalosporium* sp. IFB-E001 was isolated from *Trachelospermum jasminoides (wine)*. Among different derivatives of graphislactone, graphislactone A acts both as an antioxidant and a scavenger of free radicals (Song et al. 2005),

Fig. 29.4 Graphislactone A (D) (Song et al. 2005)

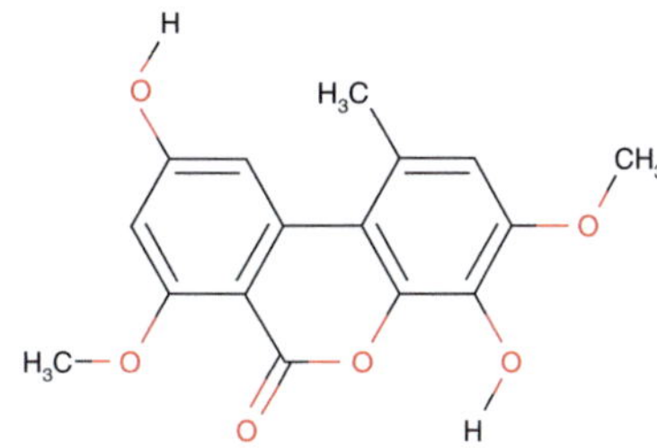

whereas graphislactones (again A PLZ CHECK) A, G, and H were found to be active against the SW1116 cell line with IC_{50} 8.5, 21, and 12 mg/mL, respectively (Zhang et al. 2005).

29.6 Natural Polyphenolic Components Derived from Endophytic Fungi

Polyphenols are the structural classes that are characterized by multiple structural units of phenols. Polyphenols are commonly found in our daily dietary products, generally vegetables and fruits with associated phytochemicals. They are often produced among plants as a response to light, stress, injury, etc. (Valentine et al. 2003). Polyphenols exist in versatile categories such as vitamins, flavonoids, etc. Under *in vivo* treatments, polyphenols have shown their significant role in heart diseases. Although, there are lots of controversies regarding their bioavailability, *in vivo* studies found that they impart their effect on plasma membrane, transcription factor, and enzymes.

Endophytes associated with five Sudanese medicinal plants *Calotropis procera*, *Catharanthus roseus*, *Euphorbia prostrata*, *Vernonia amygdalina*, and *Trigonella foenum-graecum* were screened for their antioxidant behavior. On the basis of total phenolic content and total antioxidant capacity, 21 endophytic fungi showed positive results. Natural phenolic antioxidants act as reducing agents as they absorb light in the ultraviolet regions (100–400 nm) and chelate transition metals (Fig. 29.5).

29.6.1 Depsidones

Despidones (E) are the polyphenolic compounds that are prevalent in different species of family named as Ericaceae, Lamiaceae, and Papaveraceae. Despidones are actually esters that contain depsides and cyclic ether. Depsides are known to have multipotential activities such as antibiotic, anti-HIV, antioxidant, and antiproliferative activity (Neamati et al. 1997; Nielsen et al. 1998; Kumar and Müller 1999; Reynertson et al. 2006). In addition to this, depsidones and its derivatives have been

Fig. 29.5 Despidones (E) (Bauer et al. 2012)

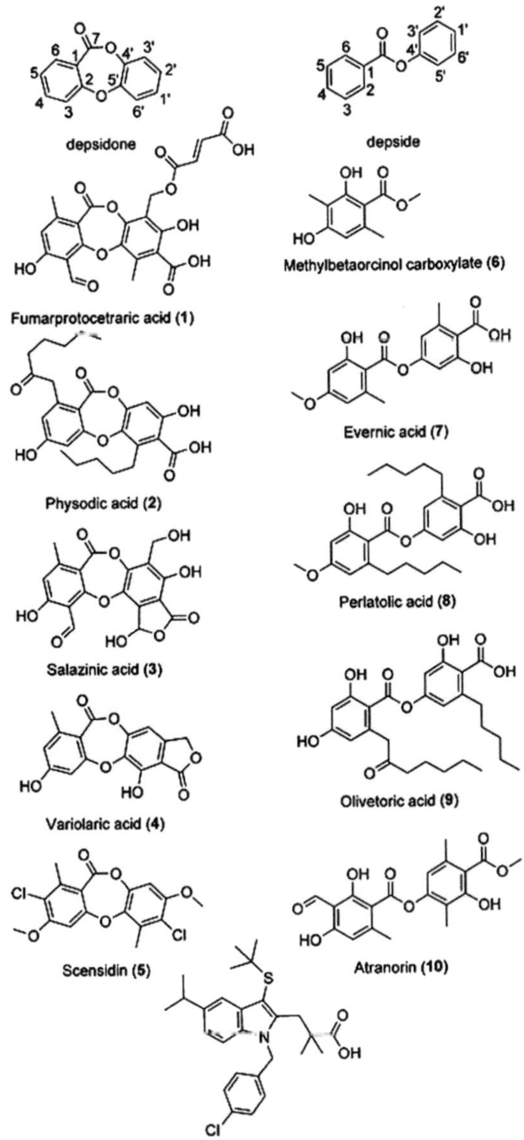

isolated from the endophytic fungus *Corynespora cassiicola* L36. Among them corynesidone B (DPPH scavenging 2, 2-diphenyl-1-picrylhydrazyl) had showed the best activity with an IC_{50} value of 22.4 lM comparable to ascorbic acid (IC_{50} 21.2 lM). Lang et al. (2007) have discovered new depsidone (excelsione) from unknown endophytes of endemic tree *Knightia excelsa*.

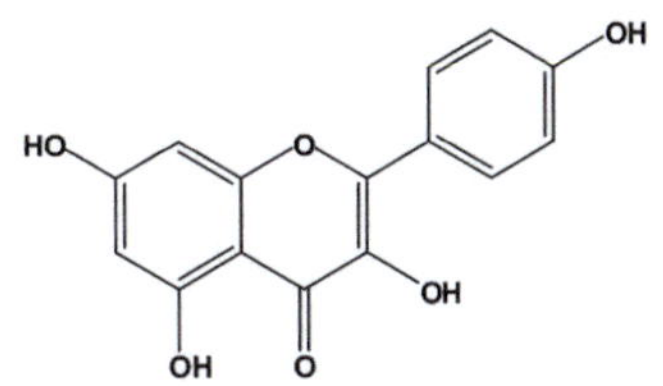

Fig. 29.6 Kaempferol (F) (Smith et al. 2006)

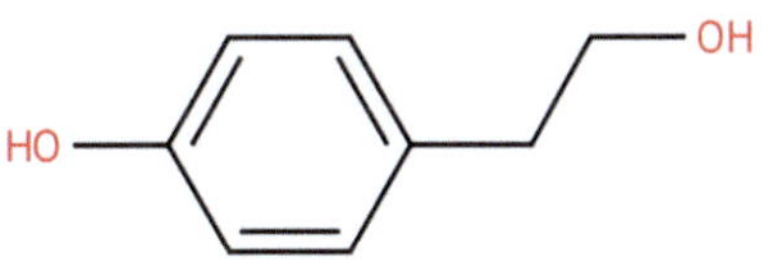

Fig. 29.7 Tyrosol (G) (CHEB: 1879) (Specian et al. 2012)

29.6.2 Kaempferol

Kaempferol (F) is a polyphenol antioxidant highly prevalent in vegetables, tea, broccoli, apples, and beans (Somerset and Johannot 2008). Dietary intake of this antioxidant may be very helpful in maintaining the antioxidant defense system of the body thereby decreasing the susceptibility to harmful diseases such as cancer, diabetes, etc. (Fig. 29.6).

Although, extraction of kaempferol from different sources is rapid, it has many disadvantages; therefore, isolating it from natural sources could be more effective.

Chen and Chen (2013) discussed that kaempferol has wide spectrum ability to kill cancer in apoptosis, angiogenesis, and metastasis and inflammation stages. Kaempferol has been be isolated from endophytes such as *Mucor fragilis* (*Sinopodophyllum hexandrum*), etc.

Kaempferol is not only responsible for apoptosis promotion but also modifies a host of cellular signaling pathways (Ramos 2007) and is also found to be less toxic to normal cells (Zhang et al. 2008). Also, it has chemopreventive activity (Huang et al. 2014). It has been found to prevent arteriosclerosis by inhibiting the oxidation of low-density lipoprotein and the formation of platelets in the blood. Collectively it could be said that kaempferol is a potent initiator of apoptosis in cancer cells (Fig. 29.7).

29.6.3 Tyrosol

Tyrosol (G) is a phenylated alcohol which is present in a variety of oils and wine. Generally, it is applied topically to remove blemishes of skin such as scars, pimples, and acne. It is also used to treat dry skin, rashes, and eczema. Tyrosol also possesses antioxidant activity, a property used in the treatment of atherosclerosis. The antioxidant content of the LDL particle is critical for its protection. Thus, phenolics, which

Fig. 29.8 Pestalachlorides (H) (Li et al. 2008)

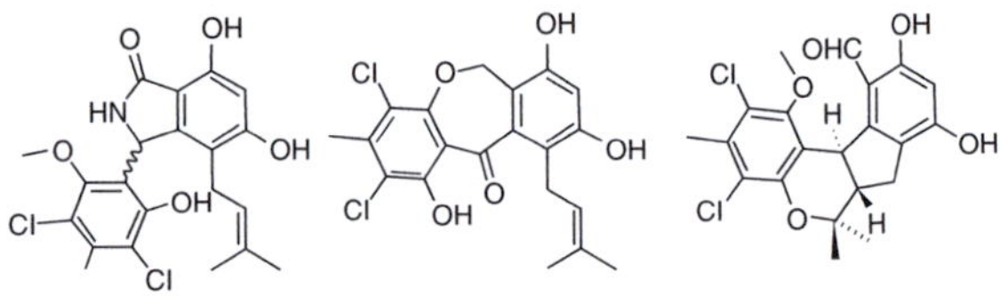

Fig. 29.9 Chaetopyranin (I) (Wang et al. 2006)

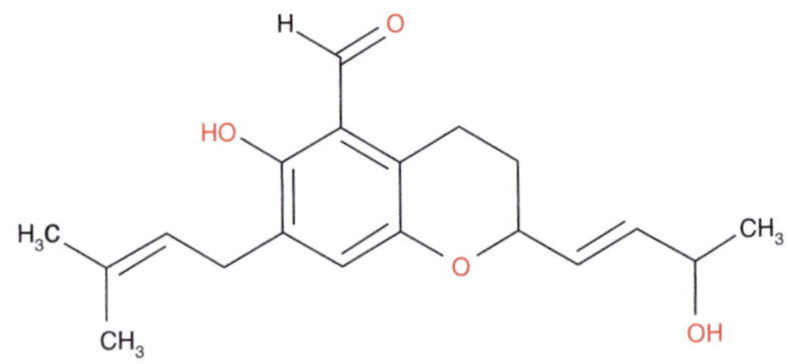

are able to bind LDL, could be effective in preventing lipid peroxidation and atherosclerotic processes. Tyrosol is one among them. Specian et al. (2012) isolated tyrosol from endophytic fungus *Diaporthe helianthi* derived from *Luehea divaricata.* Tyrosol was also found to be produced by *Ceratocystis adiposa*, a phytopathogen (Lopez et al. 2007). Tyrosol in association with nectriapyrone was extracted from endophyte *Glomerella cingulata* (Guimares et al. 2008). This compound has an aromatic ring that is linked with two hydroxyl groups at ortho position that makes it stronger than α-tocopherol (Fig. 29.8).

29.6.4 Pestalachloride

Pestalachlorides (H) A–C (1–3) are three new chlorinated benzophenone derivatives that have been isolated from the endophytic fungus, *Pestalotiopsis adusta*. The structures of these compounds were determined by NMR spectroscopy, and the structures of 1 and 3 were further confirmed by X-ray crystallography. Pestalachlorides 1 and 2 are considered to be a good antifungal agent (Li et al. 2008).

29.6.5 Chactopyranin

The basic structure of chaetopyranin (I) is chromenol (I) (chromene carrying one or more hydroxyl substituents). It is chemically known as 3, 4-dihydro-2H-chromene substituted by a hydroxyl group at position 6, a 3-hydroxybut-1-en-1-yl at position 2, a formyl group at position 5, and a prenyl group at position 7 (Fig. 29.9) (Wang et al. 2006).

These two compounds have been isolated from an endophytic fungus *Chaetomium globosum*, associated with *Polysiphonia urceolata*, and are found to possess antioxidant activity. The former compound also exhibits anticancer activity (Wang et al. 2006). Chaetopyranin also showed antioxidant activity.

29.7 Conclusions

The contribution of endophytes due to outstanding bioactivities led to continued research of endophytically derived chemical moieties. Studies revealed that antioxidant bioactivities are prevalent in these microbial metabolites. Antioxidants are vital additives required in the food industry to maintain the healthy system of the human body, and they are used in the manufacture of processed food and natural food ingredients. Additionally, they are widely used in cosmetics and dermatology also. Apart from natural plant extract, endophytes are also the producer of many good antioxidants. Hence, the call for sustainable and eco-friendly antioxidant production is motivating cosmetics and pharmaceutical industries to continue their exploration.

References

Altemoller M, Gehring T, Cudaj J, Podlech J, Goesmann H, Feldmann C, Rothenberger A (2009) Total synthesis of Graphislactones A C D and H of ulocladol and of the originally proposed and revisited structures of graphislactones E and F. Eur J Org Chem: 2130–2140

Ames BN, Cathcart R, Schwiers E, Hochstein P (1981) Uric acid provides an antioxidant defense in humans against oxidant- and radical-caused aging and cancer: a hypothesis. Proc Nat AcadSci USA 73:6858–6862

Bamford PC, Norris GLF, Ward G (1961) Flavipin production by Epicoccum species. Trans Brit MycolSoc 44:354–356

Bauer J, Waltenberger B, Noha SM, Schuster D, Rollinger JM, Boustie J, Chollet-Krugler M, Stuppner H, Werz O (2012) Discovery of depsides and depsidones from lichen as potent inhibitors of microsomal prostaglandin E2 synthase-1 using pharmacophore models. Chem Med Chem 7(12):2077–2081

Brewer MS (2011) Natural antioxidants: sources, compounds, mechanisms of action, and potential applications. Compr Rev Food Sci Food Saf 10:221–247

Burns J, Gardner PT, Matthews D, Duthie GG, Lean MEJ, Crozier A (2001) Extraction of phenolics and changes in antioxidant activity of red wines during vinification. J Agric Food Chem 49:5797–5808

Chen AY, Chen YC (2013) A review of the dietary flavonoid, kaempferol on human health and cancer chemoprevention. Food Chem 138:2099–2107

Dugas AJ, Castaneda-Acosta J, Bonin GC, Price KL, Fischer NH, Winston GWJ (2000) Evaluation of the total peroxyl radical-scavenging capacity of flavonoids: structure-activity relationships. Nat Prod 63:327–331

Gamble PE, Burke J (1984) Effect of water stress on the chloroplast antioxidant system. Plant Physiol 76:615–621

Guimares DO, Borges WO, Kawano CY, Riebeiro PH, Goldman GH, Nomizo A, Thiemann OH, Oliva G, Lopes NP, Pupo MT (2008) Biological activities from extracts of endophytic fungi

isolated from *Viguiera arenaria* and *Tithonia diversifolia*. FEMS Immun Med Microbiol 52:134–144
Halliwell B (1994) Free radicals, antioxidants, and human disease: curiosity, cause, or consequence? Lancet 344:721–724
Harper JK, Arif AM, Ford EJ, Strobel GA, Porco JA, Tomer DP, Oneill KL, Heider EM, Grant DM (2003) Pestacin: a 1, 3-dihydro isobenzofuran from Pestalotiopsis microspora possessing antioxidant and antimycotic activities. Tetrahedron 59:2471–2476
Heinig S, Scholz S, Wein J (2013) Getting to the bottom of Taxol biosynthesis by fungi. Fungal Divers 60:161–170
Hormazabala E, Schmeda-Hirschmanna G, Asudillo L, Rodriguez J, Theodulozb C (2005) Metabolites from *Microsphaeropsis olivacea*, an endophytic fungus of *Pilgerodendron uviferum*. Naturforschung 60:11–21
http://www.ebi.ac.uk/chebi/searchId.do?chebiId=CHEBI:1879
Huang WY, Cai YZ, Xing J, Corke H, Sun MA (2007) Potential antioxidant resource: endophytic fungi from medicinal plants. Econ Bot 61:14–30
Huang JX, Zhang J, Zhang XR, Zhang K, Zhang X, He XR (2014) Mucor fragilis as a novel source of the key pharmaceutical agents podophyllotoxin and kaempferol. Pharm Biol 10:1237–1243
Koes RE, Quattrocchio F, Mol JNM (1994) The flavonoid biosynthetic pathway in plants: func. evol. Bioessays 16:123
Kumar KC, Müller K (1999) Lichen metabolites. Antiproliferative and cytotoxic activity of gyrophoric, usnic, and diffractaic acid on human keratinocyte growth. J Nat Prod 62:821–823
Landis GN, Tower J (2005) Superoxide dismutase evolution and life span regulation. Mech Ageing Dev 126:365–379
Lang G, Cole ALJ, Blunt JW, Robinson WT, Munro MHG (2007) Excelsione, a depsidone from an endophytic fungus isolated from the New Zealand endemic tree Knightia excels. J Nat Prod 70:310–311
Li Y, Song YC, Liu JY, Ma YM, Tan RX (2005) Anti-helicobacter pylori substances from endophytic fungal cultures. W J Microbiol Biotechnol 21:553–558
Li EW, Jiang LH, Guo LD, Zhang H, Che YS (2008) Pestalachlorides A C, antifungal metabolites from the plant endophytic fungus *Pestalotiopsis adusta*. Bioorg Med Chem 16:7894–7899
Lippman SM, Benner SE, Hong WK (1994) Cancer chemoprevention. J Clin Oncol 12:851–873
Lopez OG, Trigos A, Fernandez FJ, Morales MJ, Castaneda GS (2007) Tyrosol and tryptophol produced by Ceratocystis adipose. World J Microbiol Biotechnol 23:1473–1477
Mates JM, Perez-Gomez C, De Castro IN (1999) Antioxidant enzymes and human diseases. Clin Biochem 32:595–603
Nawar WF (1996) Lipids. In: Fennema O (ed) Food Chem, 3rd edn. Marcel Dekker, New York, pp 225–320
Neamati N, Hong H, Mazumder A, Wang S, Sunder S, Nicklaus MC, Milne GW, Proksa B, Pommier Y (1997) Depsides and depsidones as inhibitors of HIV-1 integrase; discovery of novel inhibitors through 3D database searching. J Med Chem 40:942–951
Nielsen J, Nielsen PH, Frisvad JC (1998) Fungal depside, guisinol, from a marine derived strain of *Emericella unguis*. Phytochemistry 50:263–265
Nitao JK, Meyer SLF, Chitwood DJ (1999) In vitro assays of Meloidogyne incognita and Heteroderaglycines for detection of nematode-antagonistic fungal compounds. J Nematol 31:172–183
Overeem JC, Van Dijkman A (1968) Total synthesis of graphislactone. Rec Trav Chim Pays-Bas 87:940–944
Papas AM (ed) (1998) Antioxidant status, diet, nutrition and health. CRC Press, Boca Raton
Rababah TM, Ereifej KI, Howard L (2005) Effect of ascorbic acid and dehydration on concentrations of total phenolics, antioxidant capacity, anthocyanins, and color in fruits. J Agric Food Chem 53:4444–4447
Raistrick H, Rudman P (1956) Flavipin, A crystalline metabolite of *Aspergillus flavipes* Thom & Church and *Aspergillus terreus* thom. Studies in the biochemistry of micro-organisms. Biochem J 63(3):395–406

Ramos S (2007) Effects of dietary flavonoids on apoptotic pathways related to cancer chemoprevention. J Nutr Biochem 18:427–442

Reynertson KA, Wallace AM, Adachi S, Gil RR, Yang H, Basile MJ, Darmiento J, Weinstein IB, Kennelly EJ (2006) Bioactive depsides and anthocyanins from jaboticaba (Myrciaria cauliflora). J Nat Prod 69:1228–1230

Robak J, Gryglewski RJ (1988) Flavonoids are scavengers of superoxide anions. Biochem Pharmacol 37:837–841

Shan B, Cai YZ, Sun M, Corke H (2005) Antioxidant capacity of 26 spice extracts and characterization of their phenolic constituents. J Agric Food Chem 53:7749–7759

Smith JA, Maloney DJ, Clark DE, Xu Y, Hecht SM, Lannigan DA (2006) Influence of rhamnose substituents on the potency of SL0101, an inhibitor of the Ser/Thr kinase, RSK. Bioorg Med Chem 14:6034–6042

Somerset SM, Johannot L (2008) Dietary flavonoid sources in Australian adults. Nutr Cancer 60:442–449

Song YC, Huang WY, Sun C, Wang FW, Tan RX (2005) Characterization of Graphislactone A as the antioxidant and free radical-scavenging substance from the culture of *Cephalosporium* sp. IFB-E001, an endophytic fungus in *Trachelospermum jasminoides*. Biol Pharm Bull 28:506–509

Specian V, Sarrgiotto HM, Pamphile AJ, Clemente E (2012) Chemical characterization of bioactive compounds from endophytic fungus Diaporthe Helianthi isolated from Luehea Divaicata. Braz J Microbiol: 1174–1182

Strobel GA, Daisy B (2003) Bioprospecting for microbial endophytes and their natural products. Microbiol Mol Biol Res 67:491–502

Stocker R, Yamamoto Y, McDonagh AF, Glazer A, Ames BN (1987) Bilirubin is an antioxidant of possible physiological importance. Science 235:1043–1046

Strobel G, Ford E, Worapoang J, Karper JK, Arif AM, Grant DM, Fung PC, Ming WCR (2002) Isopestacin, an isobenzofuranone from *Pestalotiopsis microspora*, possessing antifungal and antioxidant activities. Phytochemistry 60:179–183

Tanahashi T, Takenaka Y, Nagakura N, Hamada N (2003) 6H-dibenzo[b,d]pyran-6-one derivatives from the cultured lichen mycobionts of Graphis spp. and their biosynthetic origin. Phytochemistry 62:71–75

Valentine IK, Maria VK, Bruno B (2003) Phenolic cycle in plants and environment. J Mol Cell Biol 2:13–18

Valko M, Leibfritz D, Moncol J, Cronin MTD, Mazur M, Telser J (2007) Free radicals and antioxidants in normal physiological functions and human diseases. Int J Biochem Cell Biol 39:44–84

Wang S, Li XM, Teuscher F, Li DL, Diesel A, Ebel R, Proksch P, Wang BG (2006) Chaetopyranin, a benzaldehyde derivative, and other related metabolites from Chaetomiumglobosum, an endophytic fungus derived from the marine red alga *Polysiphonia urceolata*. J Nat Prod 69(11):1622–1625

Waterman PG, Mole S (1994) Analysis of phenolic plant metabolites. Alden Press Limited, Oxford

Willis LM, Shukitt-Hale B, Joseph JA (2009) Recent advances in berry supplementation and age-related cognitive decline. Curr Clin Nutr Metabol Care 12:91–94

Womersley JS (1995) Handbooks of the Flora of Papua New Guinea. Melbourn e University Press, Melbourne

Ye Y, Xiao Y, Ma L, Li H, Xie Z, Wang M, Ma H, Tang H, Liu J (2013) Flavipin in *Chaetomium globosum* CDW7, an endophytic fungus from *Ginkgo biloba*, contributes to antioxidant activity biotechnological products and process engineering. Appl Microbiol Biotechnol 97:7131–7139

Zhang HW, Huang WY, Song YC, Chen JR, Tan RX (2005) Four 6H-dibenzo [b, d] pyran-6-one derivatives produced by the endophyte *Cephalosporium acremonium* IFB-E007. Helv Chim Acta 88:2861–2864

Zhang Y, Chen AY, Li M, Chen C, Yao Q (2008) *Ginkgo biloba* extract kaempferol inhibits cell proliferation and induces apoptosis in pancreatic cancer cells. J Surg Res 148:17–23

Chapter 30
Current Perspectives of Endophytic Fungi in Sustainable Development

Rashmi Mishra and V. Venkateswara Sarma

Abstract Endophytic fungi inhabit plant tissues asymptomatically and confer a high diversity. They are polyphyletic in nature and primarily belong to the division *Ascomycota*. Endophytes are an important component of sustainable development in community ecology as they support the rich biodiversity and bioremediation of organic pollutants, wastewater, poisonous gases, industrial sewage, and heavy metals or in agricultural sufficiency. Endophytes are known to produce different kinds of secondary metabolites, and many of them are similar to what the host plants produce. Around 80% of the world's population, mostly those in the developing countries, still rely on herbal medicines for their primary healthcare. Usage of plant sources for commercial production of bioactive compounds requires sacrifice of several trees to obtain satisfactory amounts. This is more so if they were to be isolated from endemic and endangered plant species. Such an aspect poses a grave threat to biodiversity conservation and forces us to go for a judicious usage of plant resources. Hence, production of bioactive compounds by the microorganisms that are intrinsic to the host plant tissues gives us hope. This is because microbial fermentation is comparably cheaper and economically stable and offers a sustainable source to pharmaceutical industries. In this chapter current perspectives of endophytic fungi in sustainable usage such as bioactive compounds of therapeutic use, biotransformations, plant defense and protection mechanisms, biocontrol, and crop production and in sustainable development including nutrient recycling and ecosystem functioning are reviewed and discussed.

Keywords Endophytic fungi · Nutrient recycling · Antibiotics · Biotransformation · Biocontrol · Bioactive compounds

---

R. Mishra · V. Venkateswara Sarma (✉)
Department of Biotechnology, School of Life Sciences,
Pondicherry University, Kalapet, Pondicherry, India

P. Gehlot, J. Singh (eds.), *Fungi and their Role in Sustainable Development: Current Perspectives*, https://doi.org/10.1007/978-981-13-0393-7_30

30.1 Introduction

Sustainable development entails sustaining the natural system's ability to produce resources and at the same time meeting our day-to-day needs that result in overall human development. It takes into account the future needs without exploiting the finite resources exhaustibly. Such an aspect requires a comprehensive approach that not only meets the present needs but also does not undermine the natural ecosystem, thus allowing the future generations to enjoy the privileges to meet their needs (WCED 1987). The whole idea is to ensure a better society by this process, through a planned resource usage for present and future generations. Though this concept appears to be primarily focusing on future needs, it also targets to balance the competing needs of today and tomorrow.

30.2 Microbes and Sustainable Development

Microbial world, an integral part of our ecosystem, is indispensable for our sustenance. Microbes are beneficial to plants, animals, and human beings. Microbiota of the biosphere has the capability to clean the environment. They are involved in biogeochemical cycles, nutrient recycling, waste degradation, environmental remediation, etc. Recycling of core elements of ecosystem like carbon, phosphorus, nitrogen, oxygen, sulfur, etc., occurs through biogeochemical cycles. They are efficient decomposers as well as scavengers; thus they can clean and maintain the integrity of the biosphere. Life without microbes is not possible when compared to higher life forms. The microbial world and its diversity are exploited in sustainable development. When it comes to application in industries and commercial purposes, their role is worth millions of rupees. Besides having great economic value, microbes are still an untapped resource as they provide us the potent pharmaceuticals, novel chemicals, and improved technologies for biotechnological applications. They are a treasure of potent biomolecules, and they contribute greatly in maintaining dynamic equilibrium and sustainable development.

When it comes to organic farming, agricultural waste is considered to be one of the important components of farming. Agricultural inputs as part of organic amelioration provide nutrients to the crops. Microbes secrete extracellular enzymes such as ligninases, cellulases, laccases, xylanases, and amylases which break down the complex compounds into the simpler ones that are water soluble (Hankin et al. 1975). *Trichoderma* and *Phanerochaete* are the most comprehensively studied fungi responsible for lignocellulolytic degradation (Tiquia et al. 2002). Other fungi involved in cellulolytic degradation of composting materials are *Penicillium*, *Fusarium*, *Aspergillus*, *Rhizopus*, *Chaetomium*, *Alternaria*, and *Cladosporium* (Lynch et al. 1981). In addition, bacteria are involved in cellulose degradation, and many species including those belonging to *Cytophaga*, *Bacillus*, *Cellulomonas*, *Pseudomonas*, *Klebsiella*, and *Azomonas* are commonly involved in aerobic decomposition of substrates (Nakasaki et al. 1985; Strom 1985a, b). However, the present chapter deals with the filamentous fungi that thrive endophytically inside the living leaves.

30.3 Fungal Endophytes

Depending upon the mode of life, niches that they occupy, and interactions with the hosts, fungi can be classified into saprophytes, pathogenic fungi, mycorrhizal and rhizosphere fungi, endophytes, etc. Endophytic fungi also play a lead role in the sustainable development including recycling of biological wastes, providing nutrition, food, beverages, health, pharmaceuticals, and agriculture. Endophytic fungi inhabiting plant tissues asymptomatically confer a high diversity. They are polyphyletic in nature and primarily belong to the division *Ascomycota* (Aly et al. 2011). Around a million endophytic fungal species have been estimated by researchers (Strobel and Daisy 2003). In the first place, Bary (1866) acquainted the world with the concept of endophytes as organisms dwelling inside living tissues. The most widely used definition for endophytes is that of Petrini, who defined endophytes as, "all organisms inhabiting plant organs that at some time in their life, can colonize internal plant tissues without causing apparent harm to the host" (Petrini 1991). Another highly acknowledged definition was provided by Bacon and White (2000), who defined the endophytes as "microbes that colonize living, internal tissues of plants without causing any immediate, overt negative effects." Endophytes are an important component of sustainable development in community ecology as they support the rich biodiversity and bioremediation of organic pollutants, wastewater, poisonous gases, industrial sewage, and heavy metals or in agricultural sufficiency. Endophytes are known to produce different kinds of secondary metabolites, and many of them are similar to what the host plant produce.

30.4 Endophytes As Secondary Metabolite Producers

During 1981 to 2010, natural products constituted almost half (50%) of the drugs introduced in the market. Out of these, approximately three-fourths were anti-infective in nature (Newman and Cragg 2012). Plants are a wonderful source of naturally occurring bioactive compounds and have been exploited for ages to treat many diseases in our ancient societies (Davis 1995). However reports suggest that only 10–15% of present-day higher plant species are explored for potent metabolites (Bisht et al. 2006). Out of these only 6% seem to have been tested for potent activity (Verpoorte 2000). It has to be reminded that even in the present times, around 80% of the world's population, mostly those in the developing countries, still rely on herbal medicines for their primary healthcare (Gurib-Fakim 2006).

We get many clues from ethnobotanical knowledge, and at the same time, it has been exploited in the screening of natural products with biological activity (Strobel and Daisy 2003; Ashforth et al. 2010), from traditional plants with medicinal properties. Ever since penicillin was discovered (Fleming 1929), massive efforts were taken for bioprospecting and production of such bioactive chemical entities with a renewed vigor.

One of the main disadvantages of usage of plant sources for commercial production of bioactive compounds is that numerous trees have to be sacrificed to obtain satisfactory amounts. This is more so if these natural bioactive metabolites were to be isolated from endemic and endangered plant species. Such an aspect poses a grave threat to biodiversity conservation and forces us to go for a judicious usage of plant resources. Hence, plant availability is considered as a limiting factor in the commercial utilization of some natural products. Often, a large quantity of plants is required to produce enough amounts of the bioactive compounds for clinical use. Also in many cases, compounds have been isolated from endangered or highly endemic plants. These are the major concerns to be taken into account with regard to the biodiversity conservation. Although plant tissue culture offers an alternative solution, the production cost is high and hence uneconomical (McAlpine et al. 1999). Against this background, the production of bioactive compounds by the microorganisms that are intrinsic to the host plant tissues gives us hope. This is because microbial fermentation is comparably cheaper and economically stable. Consequently, these microbes can complement and facilitate the needs of pharmaceutical industries concurrently offering a sustainable source.

Fungal endophytes possess structurally and biologically unique potent secondary metabolites and play a myriad role in the realm of health, ecology, and industry. Mycological experts have recommended cutting-edge research in endophytes for their role in pharmaceuticals, medicine, plant pathogen resistance, agriculture, genetics/genomics, ecology, and other areas. Thus a concept of sustainability could be interlinked to the interaction between fungal endophytes and their hosts from an ecological point of view. In terms of sparing the destruction of numerous trees, they prove to be an alternative source in the discovery of new drugs. The endophytes as a treasure of bioactive compounds have been well recognized (Casella et al. 2013). The propensity of endophytic fungi to produce myriad forms of bioactive metabolites is unparalleled, and the credit goes to the exclusive niches provided by the host tissues (Tejesvi et al. 2006). The coevolution of endophytes with their host plants increases the chances of genetic recombination which would credit the endophytes with host-related bioactive compounds (Stierle et al. 1993; Zhang et al. 2006). The diversity of chemical compounds produced by endophytes has made drug discovery research from endophytes to take preeminence when compared to their ecological diversity (Tejesvi et al. 2007). The endophytic fungi have higher chances of getting more diverse and potent secondary metabolites when compared to more explored fungal groups such as genera belonging to soil fungi. Specialized habitats in which endophytes dwell make them well adapted to harsh environmental conditions to develop different mechanisms to survive and flourish. That might be one of the possible explanations for their richness in diversity in metabolite production and its use in biotechnology (Bills and Polishook 1992).

An array of secondary metabolites are secreted by endophytes (Tan and Zou 2001; Schulz et al. 2002; Strobel 2003; Prado et al. 2013), falling under different discrete structural groups including tetralones, benzopyranones, enniatines, cytochalasines, xanthones, terpenoids, steroids, chinones, phenols, and isocoumarins (Schulz et al. 2002). These compounds possess antimicrobial potential. In the present age, there is a growing need of these compounds and hence a flip side to plant materials has to be hunted for addressing the needs (Heinig et al. 2013).

30.4.1 Antibiotics from Fungal Endophytes

Secondary metabolite like colletotric acid, isolated from the endophytic fungus *Colletotrichum gloeosporioides*, dwelling in *Artemisia annua* (Zou et al. 2000), a Chinese traditional herb, was shown to have activity against pathogenic plant fungi and human pathogenic bacteria (Lu et al. 2000). This Chinese traditional herb has already been reported to produce artemisinin (an antimalarial drug). *Pestalotiopsis* sp., an endophyte of *Rhizophora mucronata*, a mangrove, produced pestalotiopen A, exhibiting activity against *Enterococcus faecalis* (Hemberger et al. 2013). A novel phenolic compound, 4-(2, 4, 7-trioxa-bicyclo [4.1.0] heptan-3-yl) phenol, was isolated from *Pestalotiopsis mangiferae* associated with *Mangifera indica*. The compound exhibits activity against *Bacillus subtilis* and *K. pneumoniae* (MICs 0.039 μg/ml), *E. coli* and *Micrococcus luteus* (MICs 1.25 μg/ml), and *P. aeruginosa* (MIC 5.0 μg/ml) (Subban et al. 2013).

Phomopsis spp. occurring as endophytes on different host plants produce several chemically diverse bioactive compounds. *Phomopsis longicolla*, associated with mint plant *Dicerandra frutescens*, was found to produce dicerandrol A, B, and C with antimicrobial activity exhibiting zones of inhibition of 11, 9.5, and 8.0 mm against *B. subtilis* and 10.8, 9.5, and 7.0 mm against *S. aureus*. Similarly *Phomopsis longicolla* strain C81, associated with a seaweed *Bostrychia radicans*, produced dicerandrol C active against *S. aureus* and *Staphylococcus saprophyticus* (Wagenaar and Clardy 2001). *Phomopsis longicolla* S1B4, isolated from a plant in Hadong-gun, Gyeongnam Province, South Korea, was found to produce several antibiotics such as dicerandrol A, dicerandrol B, dicerandrol C, deacetyl phomoxanthone B, and fusaristatin A (Lim et al. 2010). Some species produce phomoxanthone A, from the host *Costus* sp., resident of Costa Rica rain forest, that is effective against *Bacillus megaterium* (Elsässer et al. 2005). Phomol, from the *Erythrina crista-galli*, a medicinal plant, is effective against *Arthrobacter citreus* and *Corynebacterium insidiosum* (Weber et al. 2004).

Species belonging to the genus *Colletotrichum* are also well-known producers of several antibacterial compounds that are aromatic steroids in nature and show potent activity against *E. coli* and *B. megaterium* (Zhang et al. 2009b). Similar antibacterial activity was also shown by the genus *Phoma* associated with the host *Salsola oppositifolia* (Qin et al. 2010). Further exploration of endophytic fungi for antimicrobial compounds is still ongoing.

30.4.2 Anticancer Compounds from Fungal Endophytes

Paclitaxel (Taxol), a well-known compound with anticancerous property, is a plant metabolite. Optimization for its large-scale production in cell culture has been accomplished (Malik et al. 2011; Cusido et al. 2014). *Taxomyces andreanae*, an endophyte on *Taxus brevifolia*, was found to produce taxol (Strobel et al. 1993), a discovery that led to more fungal endophyte isolates from a variety of *Taxus* spp. (Strobel et al. 1996; Zhang et al. 2009a). Further research on these lines led to identification of numerous other species belonging to *Alternaria*, *Fusarium*, *Penicillium*, *Aspergillus*, *Cladosporium*, *Monochaetia*, *Pestalotia*, *Pestalotiopsis*, *Pithomyces*, *Xylaria*, etc., from yew or non-*Taxus* plants, which fell under the category of microbial taxane-producing varieties (Heinig et al. 2013).

Pestalotiopsis microspora, an endophytic fungus on the host plant *Torreya taxifolia*, belonging to Taxaceae (Lee et al. 1996), was found to produce torreyanic acid, a selectively cytotoxic quinone dimer with the property of an anticarcinogenic compound. Similarly, *Trametes hirsuta*, a novel endophyte, secretes podophyllotoxin, an aryl tetralin lignan (Puri et al. 2006), and is considered to be a promising one. As an important precursor of anticancer drugs such as etoposide, teniposide, and etopophos phosphate, podophyllotoxin production from the endophytes has been thoroughly investigated (Kusari et al. 2008; Kour et al. 2008). Podophyllotoxin has been reported to be produced from *Aspergillus fumigatus* (Kusari et al. 2009), *Fusarium oxysporum* (Kour et al. 2008), and *Phialocephala fortinii* (Eyberger et al. 2006) isolated from *Juniperus communis*, *Juniperus recurva*, and *Podophyllum peltatum*, respectively.

Camptothecin, another precursor of anticancer drug topotecan and irinotecan (Shaanker et al. 2008), is a pentacyclic quinolone alkaloid. Camptothecin was first reported to be synthesized by *Entrophospora infrequens*, an endophytic fungus of *Nothapodytes foetida*, by Puri et al. (2005). Later, this compound has been found to be secreted by different endophytic fungi on various hosts including *Xylaria* sp. isolated from *Camptotheca acuminata* (Liu et al. 2010), *Fusarium solani* from *Apodytes dimidiata* (Shweta et al. 2010), and *Neurospora* sp. from *Nothapodytes nimmoniana* (Rehman et al. 2008).

Mimusops elengi, one of the Indian medicinal plants, has been found to accommodate an endophytic fungus that actively produced ergoflavin, which is a dimeric xanthene and is very effective in cancer (Deshmukh et al. 2009). Rubrofusarin B and aurasperone A were isolated from *Aspergillus niger*, an endophyte of the host *Cynodon dactylon.* Cytotoxic activity in colon cancer cell lines has been investigated as they are strong coinhibitors of xanthine oxidase (XO), and promising results were observed (Song et al. 2004). As an antineoplastic agent, L asparagine (LA) can be exploited to tumor cells and acute lymphoblastic leukemia. This enzyme was found to be secreted by several genera of the endophytes such as *Aspergillus*, *Cladosporium*, *Fusarium*, and *Penicillium*, isolated from green, brown, and red algae (Murali 2011). Compounds from dissimilar families of plants have also been found. For example, Berberidaceae, harboring an endophyte producing podophyllotoxin (Eyberger et al. 2006), and Meliaceae, harboring *Fusarium* sp., capable of producing an anticancer drug precursor, rohitukine, have also been reported (Kumara et al. 2012). *Catharanthus roseus* was reported to harbor the endophytic fungus *Fusarium oxysporum*, which has the ability to produce vincristine (Kumara et al. 2012).

30.4.3 Antioxidants from Fungal Endophytes

Secondary metabolites produced by fungal endophytes also have antioxidant properties. Since antioxidant compounds are commonly found in naturally occurring plant and plant products, they also harbor the endophytes that can produce antioxidant compounds due to their interaction within (Pandey et al. 2014). Pestacin (1,3-dihydroisobenzofuran) and isopestacin (isobenzofuranone) are the bioactive compounds showing antioxidant properties of scavenging superoxide and hydroxyl free radicals (Harper et al. 2003). These compounds are found to be produced by *Pestalotiopsis microspora*, an endophyte of the host *Terminalia morobensis* (Strobel et al. 2002). These compounds are structurally similar to natural flavonoids.

30.4.4 Antimicrobial Compounds from Fungal Endophytes

Endophytic fungi are one of the most attractive microbes for antimicrobial compounds. They have been proven to be a novel source of various bioactive compounds. Species belonging to the genera *Phyllosticta*, *Nodulisporium*, and *Xylaria* isolated from *Dipterocarpus* trees are found to have antimicrobial activity against microbial pathogens *Staphylococcus aureus*, *Pseudomonas aeruginosa*, *Bacillus subtilis*, and *Escherichia coli* (Sutjaritvorakul et al. 2011). Compounds such as hypericin and emodin produced by endophytes isolated from Indian medicinal plants have been shown to be having activity against bacterial pathogens, namely, *Klebsiella pneumoniae*, ssp. *ozaenae*, *Staphylococcus aureus* ssp. *aureus*, *Pseudomonas aeruginosa*, and *Escherichia coli*, and fungal pathogens such as *Candida albicans* and *Aspergillus niger* (Kusari et al. 2008). Tropical plants usually inhabited by the endophytic fungi belonging to the genus *Xylaria* have been shown to be having broad-spectrum antimicrobial activity (Liu et al. 2008).

Aspergillus fumigatus CY018, a leaf endophytic fungus of *Cynodon dactylon*, was found to produce several metabolites such as asperfumin, fumigaclavine C, asperfumoid, fumitremorgin C, helvolic acid, and physcion, which are antifungal in nature and also inhibit the growth of *Candida albicans* (Liu et al. 2004). Similarly other antifungal compounds such as sordaricin (Pongcharoen et al. 2008) and multiplolides A and B (Boonphong et al. 2001) were found to be isolated from the genus *Xylaria*. *Melilotus dentatus*, a grassland plant, inhabiting the coastal area of the Baltic Sea, Ahrenshoop, Germany, was investigated for endophytic secondary metabolite production. Several potential compounds like polyketide metabolite 5-methoxy-7-hydroxyphthalide (7-hydroxyphthalide, 4-hydroxyphthalide, (3R,4R)-*cis*-4-hydroxymellein, and (3R,4R)-*cis*-4-hydroxy-5-methylmellein) along with few steroids like ergosterol and 5α, 8α-epidioxyergosterol were identified from an unidentified endophytic fungus belonging to *Ascomycota* (Hussain et al. 2009). *S. aureus*, a pathogenic bacterium, was found to be inhibited by *Trichophaea abundans*, *Diaporthe phaseolorum*, and *Fusarium redolens* from the hosts *Pinus* sp., *Picrorhiza* sp., and *Artemisia* sp., respectively, with their fermented broth. Similarly, extracts of *Chaetomium globosum* from *Artemisia* sp. and *Phomopsis* sp. from *Nothapodytes* sp. have been found to be effective against *E. coli* and *S. aureus* with IC50 value of 50 μg/ml. In the same way, endophytic fungal broth of *Fusarium tricinctum*, *Alternaria* sp., and *Gibberella avenacea* inhabiting the host *Artemisia annua* inhibited fungal pathogen *Candida albicans* (Qadri et al. 2013). The screening of fungal crude extracts for antimicrobial activity against various pathogens has been reviewed by Martinez-Klimova et al. (2016). The above examples show a great potential of endophytic fungi in producing antimicrobial compounds and are encouraging to further explore more of them.

30.4.5 Antimycobacterial Compounds from Fungal Endophytes

Tuberculosis (TB), caused by *Mycobacterium* sp., mainly *Mycobacterium tuberculosis*, is a deadly disease. It is mainly a lung-infecting disease but also targets the central nervous system, skeletal tissues, as well as the lymphatic system. *M. tuberculosis* has been found to affect one-third of the world's population (Koul et al. 2011) and has caused approximately 1.5 million deaths (Organization 2011). Genomic plasticity is one of the well-known features of *M. tuberculosis* (Domenech et al. 2001). Multidrug resistance is one of the primary concerns with this organism as it develops frequent mutations in more than one of the target genes (Rattan et al. 1998). Endophytic fungi are also a potential source of novel antitubercular compounds.

A group in Thailand investigated the extracts of endophytic fungi from local medicinal plants against *M. tuberculosis* and found it to be effective with MIC of 0.0625–200 μg/ml. These fungal extracts have also shown activity against *Plasmodium falciparum*, antiviral activity against herpes simplex virus type I, anti-proliferative activity against carcinoma cell lines, etc. (Wiyakrutta et al. 2004). Drug discovery and biological diversity of fungal endophytes showed a magnificent correlation vis-à-vis novel bioactive compound production. Another antimycobacterial compound, 3-nitropropionic acid, was obtained from a species of *Phomopsis* belonging to six Thai medicinal plants. This compound is known to modulate the virulence and fatty acid catabolism of *Mycobacterium* and exhibited effective inhibition against *Mycobacterium tuberculosis* with MIC of 3.3 μM

(Muñoz-Elías and McKinney 2005; Chomcheon et al. 2005). A *Phomopsis* sp. from leaves of *Tectona grandis* L., in Northern Thailand, was found to produce compounds phomoxanthones A and B that showed significant "in vitro" antitubercular activities when compared to the existing drugs such as isoniazid and kanamycin sulfate (Isaka et al. 2001).

Coniothyrium cereale, an endophytic fungus, from marine green alga *Enteromorpha* sp., was found to produce (−)-trypethelone, which was found to be effective against *Mycobacterium phlei*, *S. aureus*, and *E. coli* (Elsebai et al. 2011). *Chaetomium globosum* strain IFB-E036, isolated from *Cynodon dactylon*, was found to secrete chaetoglobosins A and B, having activity against *Micrococcus luteus* and *Mycobacterium smegmatis* (Ge et al. 2011). Compounds such as 4-methoxycinnamaldehyde, biscogniazaphilones A and B, 5-hydroxy-3,7,4-trimethoxyflavone, *N-trans*-feruloyl-3-*O*-methyldopamine, 4-methoxy-*trans*-cinnamic acid, and methyl 3, 4-methylene dioxycinnamate isolated from *Biscogniauxia formosana*, residing inside *Cinnamomum* sp., have been shown to be having antimycobacterial activities against *M. tuberculosis* strain H37Rv in vitro (Cheng et al. 2012). *Fusarium* sp. BCC 14842, isolated from bamboo leaves in Thailand, was found to produce javanicin, javanicin (181), 3-*O*-methylfusarubin, and 5-hydroxy-3-methoxydihydrofusarubin A, which showed antimycobacterial activity (Kornsakulkarn et al. 2011)

Mycobacterium tuberculosis which causes the dreaded disease tuberculosis in human beings is still prevalent in virulent form in several countries and is in need of novel bioactive compounds for a successful treatment. Endophytic fungi show a promise for this disease also.

30.4.6 Other Compounds from Fungal Endophytes

Antidiabetic activity: Endophytic fungi associated with *Salvadora oleoides* were investigated to test their ability to lower lipids and act against diabetes. Wistar albino rats, suffering from diabetes, were treated for antidiabetic and hypolipidemic activities, and endophytic fungi have been observed to significantly reduce the blood glucose level (Dhankhar et al. 2013).

Volatile organic compounds (VOCs): Endophytes are excellent producers of volatile organic compounds (VOCs). *Muscodor albus*, a xylariaceaous endophyte of host *Cinnamomum zeylanicum*, is known to produce 28 different VOCs. The component of these VOCs is a mixture of five gases, i.e., alcohols, acids, esters, ketones, and lipids (Strobel et al. 2001). The effective compounds from these VOCs are 1-butanol and 3-methyl-acetate that were shown to be able to inhibit bacterial and fungal pathogens. Strobel et al. (2001) also reported the production of VOCs from *Muscodor crispans* isolated from *Ananas ananassoides, such as* 1-butanol, 3-methyl-; 1-butanol, 3-methyl-, acetate; propanoic acid, 2-methyl-; propanoic acid, 2-methyl-, 2-methylbutyl ester; and ethanol, effective against a citrus pathogen *Xanthomonas axonopodis* pv. *citri*. VOCs from *Muscodor crispans* are also known to kill human pathogens such as *Yersinia pestis*, *Mycobacterium tuberculosis*, and *Staphylococcus aureus* (Deshmukh et al. 2015). Similarly, mixture of fungal VOCs are reported to be lethal to several plant and human pathogens such as three drug-resistant strains of *Mycobacterium tuberculosis* (Mitchell et al. 2010).

Immunosuppressive compounds, viz., subglutinols A and B, isolated from the endophytic fungus, *Fusarium subglutinans*, from the host *T. wilfordii*, have been reported to be effective against autoimmune diseases (Lee et al. 1995).

Anti-inflammatory compound, ergoflavin, is basically a pigment isolated from an endophytic fungus, inhabiting the leaves of *Mimusops elengi* (bakul), an Indian medicinal plant, which has been shown to be having anti-inflammatory properties (Deshmukh et al. 2009).

More precise strategies have to be adopted to isolate endophytic strains that have potential to produce novel drugs. These diverse and endemic species of fungal endophytes hold a great promise to humankind with their novel chemical entities (Strobel and Daisy 2003). Arnold and Lutzoni (2007) suggested that tropical forests are one of the biodiversity-rich sites for endophytic fungal exploration. Careful selection of hosts for endophytic fungal isolation is an important step in this endeavor.

Genomics of plants and metabolite production by endophytic fungi need to be correlated to give us a better insight of the biosynthetic pathways in host plants. Molecular techniques in the identification of endophytic fungi and understanding the bioactivities both at the genotype and strain level are required. There is an immense natural wealth in these tiny organisms, but what is needed is an interdisciplinary effort to tap and exploit the resources. More funds and collaborations are needed to explore a plethora of chemical entities.

To achieve ideal laboratory conditions for optimum production of secondary metabolites, a thorough understanding of host-endophyte relationship has to be explored at various levels including molecular and genetic, biogenetic gene cluster regulation, and effects due to changes in the environment. Future research, incorporating advanced molecular techniques, may increase our knowledge on endophytic fungal biodiversity and the regulation of fungal secondary metabolism.

30.5 In Vitro Biotransformation by Fungal Endophytes

Biotransformation can be defined as "a chemical process of modification of a compound by a biological system into another compound, which makes it more bioavailable for further metabolism." Biotransformation is considered as a substitute for biological synthesis of contemporary pharmaceutical compounds (Azerad 1999).

This technology not only favors "a new drug metabolism" but also "facilitates its scale-up of production." Through this process chemicals can be modified by changing the functional groups or carbon skeleton. The advantages of this technique are its cost-effectiveness, less energy needs, and moderate experimental conditions (Suresh et al. 2006) in addition to facilitating a better structural elucidation that can give accuracy in regio- and stereospecificities (Gao et al. 2013).

Plant metabolites get biotransformed by the endophytes thriving in plants during secondary metabolism. Such fungal endophytes could be exploited as biocatalysts. For various biotechnological applications, by adding stereo specificity to biotransformation processes, the fungal endophytes can cut down the complicated chemical steps of purification of enantiomers during synthesis. *Phomopsis* sp., *Glomerella cingulata*, *Diaporthe phaseolorum*, and *Aspergillus fumigatus* have been found to be capable of stereoselective biotransformation of thioridazine (THD), a neuroleptic drug, into (R) and (S) configurations from its racemic mixture of the enantiomers (Borges et al. 2009a), and thioridazine sulfoxides biotransformation to stereoselective mixture from racemic was also investigated. In the same way, juglone, from the endophytic fungus *Paraconiothyrium variabile*, was biotransformed into (4S)-isosclerone through one-step enantioselection (Prado et al. 2013). They can also act on the core moiety and may lead to more bioactive compounds with improved properties.

Biomolecules such as alkaloids, lignans, monoterpenes, diterpenes, and triterpenes are also exploited by this technique through stereospecific and stereoselective reactions to form new potential pharmaceutics for the benefit of different industries. A biotransformed product, "3,4-dimethyl-2-(4-hydroxy-3,5-dimethoxyphenyl)-5-methoxy-tetrahydrofuran" from tetrahydrofuran lignan, (-)-grandisin, produced by *Phomopsis* sp., an endophytic fungus of *Viguiera arenaria*, has been found to be having trypanocidal activity. The natural precursor of this new compound has a similar activity against *Trypanosoma cruzi*, a parasite that causes Chagas disease (Figueiredo et al. 1996)

Similarly, root and shoot endophytes of *Aphelandra tetragona* are found to biotransform benzoxazinones, 2-benzoxazolinone (BOA) and 2-hydroxy-1,4-benzoxazin-3-one (HBOA), into different benzoxazinones that are produced during plant defense against pests in which toxic BOA (benzoxazolinone, 2-benzoxazolinone) or the less toxic lactams such as HBOA (2-hydroxy-1,4-benzoxazin-3-(2H)-one) are formed as a result of hydroxamic acid metabolism. *Fusarium sambucinum*, an endophyte, has been reported to detoxify these toxic compounds into *N*-(2-hydroxyphenyl) acetamide, *N*-(2-hydroxyphenyl) acetamide, and analogues (Zikmundova et al. 2002).

Basic enzymatic reactions such as oxidation and hydroxylation may modify a structure. A fungus isolated from *Camellia sinensis* has been shown to carry out biotransformation of (?)-catechin and (-)-epicatechin, which are C4-flavans, into dihydroflavan derivatives through the stereoselective biooxidation (Agusta et al. 2005; Shibuya et al. 2005). Also an endophytic fungus belonging to *Coelomycetes* has been reported to convert the alkaloid berberine into its 7-*N*-oxide (Agusta et al. 2014). Endophytic fungi residing inside the inner bark of *Taxus yunnanensis* were found to biotransform taxoids through deacetylation, hydroxylation, and epoxidation (Zhang et al. 1998). Endophytes also produce volatiles as one of their bioactive compounds (Abrahão et al. 2013). These compounds play a major role in making the plants to repel and attract insects and restrict the pathogens. *Penicillium canescens*, an endophyte of pigeon pea, which has deacetylation ability, has been shown to be involved in the biotransformation of astragalosides to astragaloside IV (Yao et al. 2014).

Volatile compounds (VOCs) from endophytes also possess various properties such as vanillin, an essential oil constituent, having antimicrobial properties, alkanolides having antifungal and antiviral properties, eugenol having antioxidant properties, nootkatone having somatic fat-reducing properties, 2-(E)-hexenal having blood pressure regulation, and 1,8-cineole having anti-inflammatory properties (Berger 2009). "Thioridazine (THD)," a phenothiazine neuroleptic drug, has also been investigated for biotransformation with the help of endophytic fungi through stereoselective kinetics, and it was concluded that biotransformation reactions are a kind of biomimicry of mammalian metabolism (Doble et al. 2004).

30.6 Endophytic Fungi in Plant Defense and Improved Performance

Reports suggest that microbes modulate the host defense system across a spectrum of phytopathogens. The strategy includes either directly targeting the pathogen or its lysis or it may be indirectly actuating the mechanism of host defense or growth promotion. By harboring fungal endophytes, the host plants are well equipped to tackle biotic and abiotic stresses that can impart fitness benefits. They can play a key role in the agriculture activities and in the restoration of ecosystems (Rodriguez et al. 2008; Aly et al. 2011; Franken 2012; Johnson et al. 2013; Card et al. 2016). One of the finest examples of this is shown in *Piriformospora indica*, a root-dwelling endophyte, which enhances plant growth promotion traits by modulating the phytohormones similar to reinforced nutrient uptake and translocation. Evidences to show host protection by endophytic fungi from diseases and in limiting the damages caused by pathogenic microorganisms have been provided by various workers (Arnold et al. 2003; Ganley et al. 2008; Mejía et al. 2008).

Pathogenic infections, nutritional deficits, and external environmental changes can act as external stimuli and generally induce secondary metabolite production in an organism (Strohl 2000). Macia-Vicente (2009) reported that endophytic fungi, *Fusarium equiseti* and *Pochonia chlamydosporia*, having a dual antagonistic system, were introduced into the roots of barley. They found that the pathogenic disease in the host was curtailed along with an enhanced growth. The host plants get multiple benefits through an association with fungal endophytes that promote plant growth (Yong et al. 2009), improve counteraction against multiple stress (Malinowski et al. 2004), and shield from diseases and insects (Wilkinson et al. 2000; Tanaka et al. 2005; Vega et al. 2008). Endophytes also arm the hosts with plant defense mechanisms that can avert herbivory (Zhang et al. 2012) or cut down the population of juveniles of pests (Jaber and Vidal 2009).

30.7 Plant Protection Mechanisms Bestowed by Fungal Endophytes

30.7.1 Antibiotic Production

Endophytes can control and prevent phytopathogens by secreting antibiotics or hydrolytic enzymes (Strobel 2003; Berg and Hallmann 2006), limiting the population of insects (Azevedo et al. 2000) and nematodes (Hallmann et al. 1998) from affecting the host plants. Fungal endophytes can produce secondary metabolites that are antifungal and antibacterial in nature to inhibit the growth of pathogens (Gunatilaka 2006). An array of antibiotics including alkaloids, terpenoids, aromatic compounds, and polypeptides are produced by endophytic fungi that can restrict the plant pathogens.

Induced resistance can be classified into two types: systemic acquired resistance (SAR), which is pathogen induced, has pathogenesis-related (PR) proteins getting accumulated. The second one is induced systemic resistance (ISR), which is induced by nonpathogenic microbes and is devoid of accumulation of pathogenesis-related (PR) proteins (Vallad and Goodman 2004; Tripathi et al. 2008). In induced systemic resistance, the host defense mechanism is activated by the bioactive metabolites secreted by endophytes that can mobilize the host defense mechanism (Kloepper and Ryu 2006). For example, induced systemic resistance is effected by *Fusarium solani*, which is a root endophyte of tomato, which acts against *Septoria lycopersici*, a tomato foliar pathogen, in triggering the expression of PR genes, PR5 and PR7 (Kavroulakis et al. 2007).

Pathogenesis-related gene expression could also be linked to the fungal endophyte-induced ISR. These PR proteins, as enzymes like chitinases and ß-1, 3-glucanases (Fukuda and Shinshi 1994), may act to lyse the breaching cells, boosting infection resistance, or induce localized cell death.

30.7.2 Stimulation of Secondary Metabolite Production

When fungal endophytes colonize a host, the plant secretes hydrolases to limit its growth. Thus when endophytes are co-cultured with the hosts, they elicit secondary metabolite production in hosts in the same manner that they stimulate plant resistance in the host plants under natural conditions.

In a study it was reported that *Euphorbia pekinensis* growth could be enhanced by endophytic fungi *Fusarium* spp. E4 and E5 through increased terpenoid content (Yong et al. 2009). Similarly, a 1.8-fold increase in yield of paclitaxel was recorded when a fungal endophyte culture supernatant was added to *Taxus cuspidata* culture suspension (Li and Tao 2009).

30.8 Endophytes As Biocontrol Agents

Employment of one living organism to control the population of an unwanted organism is generally referred to as biocontrol. Biocontrol agents are those which help in achieving such a trait. BCAs may include insects, pathogens, or grazing animals to suppress the unwanted population. This method reestablishes a balance of natural enemies with their hosts. Since this is an eco-friendly natural method and does not involve the use of any chemicals or machinery, it is sustainable, economically safer, and harmless to beneficial organisms or human beings or their health. Biological control is considered as one of the most sustainable and everlasting solutions (Bateman 2002). Fungal endophytes exploit the mechanisms such as mycoparasitism, antibiosis, and competition to mitigate the pathogenic infection by directly proliferating within the host or indirectly via induced intrinsic resistance responses to the host (Aneja et al. 2005; Bailey et al. 2006). By understanding the patterns of host-endophyte ecology and the intrinsic mechanisms involved in interactions among the pathogens, endophytes, and hosts, one can draw a well-thought-out biocontrol program (Herre et al. 2007).

Some of the well-known endophytic fungal genera implicated in biocontrol include *Acremonium* and *Heteroconium* species (Joost 1995; Narisawa et al. 2000; Jäschke et al. 2010). Occasionally the biocontrol action could also be attributed to toxins produced by endophytes, e.g., the *Acremonium zeae*/maize interaction (Poling et al. 2008). In another example, it has been shown that endophytic fungi limit the pathogenic damage caused in *T. cacao* (Arnold et al. 2003; Evans et al. 2003; Mejia et al. 2003; Holmes et al. 2004; Rubini et al. 1980; Tondje et al. 2006).

Similar claims on other endophytic fungi in combating pathogenic damage have been reported by other researchers also (Herre et al. 2007). In a greenhouse experiment, it was shown that when pods were treated with *C. gloeosporioides*, in different farms, symptoms of black pod disease were reduced considerably (Arnold and Herre 2003). Introduction of *C. rosea* in the host curtailed the sporulation of *M. roreri*, the most common epiphytic mycoparasite in cacao, leading to control of a part of its life cycle under field conditions (Mejía et al. 2008). Antibiotic yields from *C. rosea* (Berry and Deacon 1992; Hajlaoui et al. 2001) and biocontrol of *Botrytis cinerea* in roses (Morandi et al. 2000) have been well-documented. *Gaeumannomyces graminis*, a well-known plant pathogen causing take-all disease, which affects the roots of grasses and cereals, in temperate climates and *Fusarium oxysporum*, a primary causative agent of wilt of lentils, have been shown to be controlled by root endophytic fungi such as *Piriformospora indica*, *Sebacina vermifera*, and *Trichoderma* species (Ghahfarokhi and Goltapeh 2010; Kari Dolatabadi et al. 2011).

Endophytic fungi have also been shown to be acting as growth inhibitors of potential pathogens such as *Rhizoctonia solani*, by *Trichoderma* species, *including Trichoderma harzianum*, *Trichoderma viride*, and *Trichoderma aureoviride* (Shalini and Kotasthane 2007). Also they may weaken the severity of the diseases such as

Pseudocercosporella herpotrichoides in hosts colonized by the endophyte *P. indica*. This endophytic fungus has also been implicated in the increased yield of tomato and acting against *Verticillium dahliae* (Fakhro et al. 2010). Volatile organic compounds (VOCs), secreted by fungal endophytes, act as potent agents against various fungi and bacteria that are pathogenic to human beings and plants (Strobel 2006). *Pythium ultimum*, *Phytophthora infestans*, and *Phytophthora capsici* are some of the plant pathogens belonging to oomycetes that causes havoc to important crops. Kim et al. (2007) proved, in dual culture tests, that fungal endophytes isolated from different vegetable crops exhibit anti-oomycete property.

In addition, it has also been shown that during antagonistic communication between interspecies, fungal endophytes affect the plant growth. Yet another example of mitigating the pathogen, *Ustilago maydis*, through the secretion of pathogenic growth-limiting metabolites by *Fusarium verticillioides*, an endophyte in *Zea mays*, was shown by Rodriguez et al. (2012). Bioactive metabolites produced by endophytes such as ergot and indole diterpene alkaloids, swainsonine, lolines, and peramine are known to determine herbivory (Panaccione et al. 2014). While screening the endophytes for antimicrobial and herbicidal compounds, it was found that endophytes can also synthesize compounds that could be directed against weedy plants (Schulz et al. 1999). Besides, it was also found in this study that the number of endophytes producing these herbicidal substances was three times higher than that of soil microbes and twice that of plant pathogenic microorganisms. Insects also seem to influence the endophytic fungi to produce bioactive compounds in the host plants (Kusari et al. 2013). Thus, insecticidal substances also can be found among those produced by the endophytes.

It has been shown that the important cacao pathogens such as *Phytophthora palmivora*, *Moniliophthora roreri*, *and Moniliophthora perniciosa* could be controlled by fungi isolated from leaves and pods of the host *Theobroma cacao* (Bailey et al. 2006; Mejía et al. 2008). Similarly, problems posed by *Puccinia recondite* f.sp. *tritici*, responsible for pustule development in leaf rust, could be mitigated by *Chaetomium* sp., a foliar endophyte of wheat (Dingle and Mcgee 2003). Subsequently it has been shown that various species of *Chaetomium* can prevent the growth of *P. tritici-repentis* (Istifadah et al. 2006).

The above account shows that the endophytic fungi can give beneficial effects when introduced in the fields. Gimenez et al. (2007) reported that expected pattern of metabolite production may significantly vary when the host is not natural to the introduced endophytes.

The effect of introduction of endophytic fungi into the hosts and the chemical biology of host-endophyte-microbiota interactions need to be fully understood. Assessment of the desired results such as increase in crop yield, disease protection, host fitness, and resistance toward pathogens needs to be looked into. Thus future studies are wanting on the aforesaid aspects.

30.9 Fungal Endophytes for Sustainable Crop Production

Agriculture has seen a paradigm change in recent times. Nowadays the emphasis is more on the new technologies, maximizing production, mechanization, increased chemical usage, and government policies that favor the needs of an ever-increasing population, which is expected to reach 9.7 billion by 2050 (DESA 2015).

Intensive agricultural methods, although they have met several targets and have achieved positive effects in reducing risks in farming, are unsustainable and expensive. Depletion of top layer of our soils, decrease and contamination of ground water, and biotic and abiotic stresses in agriculture are some of the drawbacks to consider. These agrochemical inputs are economically not viable and are not environmentally friendly. As of now, these practices have made agricultural yields unpredictable.

Modern agricultural practices involve introduction of chemical fertilizers and agrochemicals to control pests and weeds. The world is facing a grave threat due to diminished water supply and increased soil salinity for crop production (Egamberdieva et al. 2008; Egamberdieva and Lugtenberg 2014). Microbes such as nitrogen-fixing bacteria and mycorrhizal fungi are a boon for plant growth (Berendsen et al. 2012; Santoyo et al. 2016). The discovery and application of endophytes such as AR1 and AR37 to protect ryegrass against the Argentine stem weevil, in New Zealand, have proved that fungal endophytes can be a reliable alternative in agriculture (Easton et al. 2001; Rodriguez et al. 2009; White et al. 2002).

Endophytic association in the host plants induces a better nutrient uptake. Seeds coated with endophytic fungi can help seed germination as part of a symbiotic relationship. These mutualist microbes help in degrading the cellulose of cuticle and facilitate the availability of carbon required for the plant germination and development. They also secrete metabolites that can act as plant growth modulators which can help in germination of crop seeds (Bhagobaty et al. 2010).

The best approach for introducing the fungal endophytes in agricultural fields is still being explored. Addition as soil inoculant or as seed dressing seems to be the most suitable way of their introduction in the fields. This strategy has been found to be successful in the case of sugarcane (Da Silva et al. 2012). Similarly, other significant roles such as promoting plant growth and combating biotic and abiotic stresses by some endophytic microbes, under laboratory conditions, have been proven (Baltruschat et al. 2008; Waller et al. 2005; Hubbard et al. 2014).

When plants are under environmentally stressed conditions, the endophytic fungi, dwelling in roots of the hosts, facilitate the proliferation by limiting the levels of ethylene, which causes growth inhibition in host plants. The endophytic fungi are strong contenders in the process of remediation of long-standing soils for agriculture. Approximately 70% of significant increase in seed germination was reported with fungal application and enhanced seedling (Chen et al. 2013).

Field-scale inoculation of endophytes may not be much effective when compared to inoculating the crops at each planting time in case vertical transmission fails to occur (O'Callaghan 2016). Robinson et al. (2016) found that, in wheat, when potential endophytes were incubated with sterile excised embryos, true vertical transmission was not observed. Thus introducing the potential fungal endophytes, in the fields, as seed dressings, would support successful establishment of seed-adhering microorganisms in the endosphere.

Selection of potential endophytes and understanding their behavior in field conditions are critical before they are introduced in the fields. Furthermore, it is very important to understand the full life cycles of fungal endophytes as it is necessary to predict the chances of its conversion into a pathogenic form from endophytic stage under stress or abiotic conditions (Redman et al. 2001).

Continuous use of agrochemicals made us to become dependent on them. While there are multiple benefits of endophytic fungi, we should always remember that some endophytic traits may be hazardous to other organisms such as *Epichloë* which are toxic to vertebrates (Schardl et al. 2004). Therefore, a thorough investigation is needed to check for any negative impact it may have on other life forms. A better understanding on host and endophytic interaction need to be explored before selecting a suitable candidate organism in the field. Appropriate usage of these wonderful microbes holds a promise and lays a platform for sustainable development. During plant senescence, endophytes promptly access available nutrients being entrenched into the host tissue (Rodriguez et al. 2008), (Aly et al. 2011). Some fungal endophytes also mediate the nutrient recycling process in the ecosystem by initiating the biodecay of dead/dying plant (Strobel and Daisy 2003; Zhang et al. 2006; Vega et al. 2010; Aly et al. 2011; Boberg et al. 2011).

30.10 Fungal Endophytes and Remediation

Industrialization and urbanization have made the modern agriculture vulnerable to elevated levels of heavy metals in the soils. Soil, being the uppermost layer of the earth crust, supports life and is essential for primary production, biogeochemical cycles, and biodiversity maintenance. However, phytoextraction of the heavy metals from the environment throws a challenge. In phytoremediation, plants have natural ability to extract unwanted harmful chemicals from different components of ecosystem such as soil, water, and air with an advantage of being more cost-effective and more sustainable when compared to the conventional methods. Phytoremediation is one of the sustainable, economic, and environment-friendly technologies that promises to reduce the deleterious effects of heavy metals. Heavy metal concentration in plant tissues and the magnitude of plant biomass is directly proportional to the effectiveness of phytoremediation. Plants are hyperaccumulators, and their potential for detoxification accelerates due to efficient root-to-shoot transport system. Plants play a major role in treating and eliminating the heavy metals, excess nutrients, and radionuclides in several forms of phytoextraction, rhizofiltration, phytostabilization, and phytovolatilization. The presence of these heavy metals impairs the health of the flora and affects its metabolism. Host plants exploit endophytic microorganisms to extract the nutrients, and the bioavailability of metal to plant is enhanced by the secondary metabolites produced by the microbes (Rajkumar et al. 2010). Endophytic microorganisms have been screened and investigated for their roles and efficacy in phytoremediation. By adopting different metabolic pathways, the endophytes have proved to be a valuable resource in bioremediating pollutants and biotransforming the organic substances (Stępniewska and Kuźniar 2013).

Fungi play a crucial role, in different ways, in the sustainable development including metal-fungal interactions or fungal-clay interactions, mineral formation or rock and mineral transformations, and elemental recycling or bioweathering. When compared to bacteria, mycelial growth of fungi provides them an advantage to be more adapted and acclimatized to the environment and be more affirmative in strategies for such solutions. Higher biomass and ability to sequester or chelate metals make endophytic fungi better candidates for heavy metal tolerance and bioremediation (Aly et al. 2011). Fungal endophytes make use of various tolerance mechanisms, e.g., sequestering and precipitating extracellular metals, binding of metal to fungal cell walls, and sequestering intracellularly along with complexation, compartmentation, and volatilization (Fomina et al. 2007). Phytoremediation in combination with endophytic fungi is still an unexplored area and hence needs to be more exploited. *Portulaca*, a plant known for hyperaccumulating heavy metals, accommodates numerous fungal endophytes including genera such as *Paecilomyces*, *Penicillium*, *Trichoderma*, *Fusarium*, *Aspergillus*, *Cladosporium*, and *Lasiodiplodia* (Deng et al. 2012). Fungal taxa which are recurrently isolated and are found to be effective in tolerating heavy metal stresses are those belonging to *Penicillium* spp. and *Trichoderma* spp. (Babu et al. 2014b, a; Khan et al. 2014). *Solanum nigrum*, a cadmium hyperaccumulator, harbors an endophytic fungus *Microsphaeropsis* sp. LSE10, which is reported to show biosorption capacity of Cd up to 247.5 mg/g, the highest when compared with other adsorbents (Xiao et al. 2010). *Exophiala pisciphila*, a dark septate endophytic strain, has the ability to accumulate lead over 20% of cadmium and lead up to 5% of the dry weight (Zhang et al. 2008).

Endophytic fungi are great producers of enzymes that are economically important, and these enzymes help them in growth and development within a host. In one such case, ligninolytic enzymatic system can bioremediate and modulate a broad spectrum of organic pollutants. *Monotospora* sp. strain W823, an endophytic isolate of *Cynodon dactylon*, produces potential laccases comparable with white-rot fungi belonging to different species of *Trametes*. *Pestalotiopsis palmarum* BM-04, an endophyte, is shown to be having the potential to grow in extremes like 50,000 ppm of extra-heavy crude oil and 2000 mM sodium chloride (Naranjo-Briceño et al. 2013). When wheat bran or lignin peroxidase was used as substrate with extra-heavy crude oil as sole carbon and energy source, this strain has overproduced laccase in higher amounts. Thus, *Pestalotiopsis palmarum* showed its potential to be a promising candidate for bioremediation in harsh conditions of halo regions and crude oil-contaminated regions.

Endophytic fungi are also investigated for their ability to degrade plastic polyester polyurethane (PUR). Russell et al. (2011) reported that two fungal isolates of *Pestalotiopsis microspora* used plastic polyester polyurethane as a sole carbon source under aerobic and anaerobic conditions. Polycyclic aromatic hydrocarbons (PAHs) are one of the major concerns related to biodegradation. The endophyte *Phomopsis liquidambari* proved to degrade PAHs and showed growth in hydroxybenzoic acid, using it as a sole carbon source. *Phomopsis liquidambari*, in soil, releases significant amounts of mineral N by exploiting the organic N of soil. Therefore they have the potency to degrade N-heterocyclic molecules, e.g., indole into much simpler structures (Chen et al. 2013; Dai et al. 2010).

Xiao et al. (2014) reported degradation of phenolic acids and significant increment in germination of watermelon seedlings cultured in soil, when treated with the endophytic fungus *Ceratobasidium stevensii*. *Festuca arundinacea* and *Festuca pratensis*, two grass species, were experimentally proved to remediate an age-old petroleum-contaminated soil when infected by endophytic fungi. Infected grass contained higher root and shoot biomass when compared to non-infected grass. Thus remediation by degradation of TPH from oil-contaminated soils by endophytic fungi-infested grasses showed significant results (Soleimani et al. 2010). Dark septate endophytes colonize the host and curtail the damaging effects of exaggerated heavy metal supplements and boost root shoot growth in maize under heavy metal stress conditions. They also enhance the metal resistance of host plants in multimetal-contaminated conditions. For example, *Exophiala pisciphila* showed mutual symbiosis with its host and exemplifies an efficient strategy to host survival under higher heavy metal stress (Li et al. 2011).

Trichoderma H8 and rhizosphere *Aspergillus* G16, associated with *Acacia auriculiformis*, supported the host growth when associated with mustard in soils contaminated with Cd and Ni. *Trichoderma* enhanced the fresh weight by 167% in the Cd-Ni-contaminated soils, whereas 44% enhancement of plant yields was observed in the case of *Aspergillus* G16 inoculation in Cd-Ni-contaminated soils. When inoculated together, it has resulted in 178% more plant yields in the same conditions (Jiang et al. 2008). Endophyte *Lasiodiplodia* sp. MXSF31, dwelling in stems of *Portulaca oleracea*, increased the biomass production of canola (*Brassica napus* L.) and Cd extraction in the Cd and Pb-contaminated soils by canola (Deng et al. 2014). The Cd-induced membrane injury to host was minimized by *Penicillium janthinellum LK5*, a gibberellin-producing endophyte. Also oxidative stress was diminished through reducing the electrolytes and lipid peroxidation and enhancing the reduced glutathione content and catalase activities (Khan et al. 2014). The isolates FT2G59 (*Penicillium*) and FT2G7 (*P. columnaris*) associated with *Dysphania ambrosioides*, a hyperaccumulator from two Pb-Zn contaminated sites, exhibited better metal tolerance (Li et al. 2016). Thus, fungal endophytism as a symbiosis can counteract the stress of heavy metals in plants.

Thus, we can safely conclude here that fungal endophytes play a key role in enhancing the phytoremediation processes. Elimination of environmental contaminants by endophytic fungi has significantly contributed to this momentum. Endophytic fungi that are metal resistant or capable of degrading organic contaminants should be employed on a large scale in the polluted areas, for sustainable development.

On the other hand, very few studies have been carried out on the role of endophytic fungi with organic-contaminant-degrading capability and/or metal resistance. Their role in helping flora in phytoremediation of the ecosystem is very less known. Laboratory studies related to endophyte-assisted phytoremediation have to be tested in the field on the large scale to prove its practical worthiness. Endophytic ecology and their diversity in the host plants dwelling in the contaminated sites need to be explored at a deeper level for application part. Bioengineering of endophytes to tackle slow degrading pollutants has to be considered to enhance their biodegradative capabilities against a wide range of stubborn contaminants.

30.11 Enzymatic Potential of Fungal Endophytes

In the present era, there is a need to be completely independent in energy for our needs by reducing our dependence on fossil oil. For instance, ethanol obtained mainly from fermentation of sugarcane and corn could be used in a mixture with petrol as fuel for vehicles. Cellulolytic and hemicellulolytic enzymes can degrade bagasse, a lignocellulosic raw material from sugarcane industry, and can achieve production of second-generation (2G) ethanol. Recently a group reported that endophytic fungi such as *Aspergillus niger*, *Trichoderma atroviride*, *Alternaria* sp., *Annulohypoxylon stigyum*, and *Talaromyces wortmannii*, isolated from paper of spoiled book, were found to produce hydrolytic enzymes that decompose sugarcane biomass at industrial level (Robl et al. 2013). Several endophytic fungi are also responsible for producing commercially beneficial enzymes. For example, *Neotyphodium lolii* and *Epichloë festucae* are found to produce −1, 6-glucanases which can degrade fungal cell wall (Bryant et al. 2007). Interspecies symbiosis also plays an important role in secreted proteins. For instance, *Poa ampla*, a fungal endophyte on interaction with endophyte *Neotyphodium* sp., expressed a novel fungal chitinase (Li et al. 2004). Similarly, isolation, purification, and characterization of two extracellular β-1, 6-glucanases from *Acremonium* OXF C13 were also reported (Martin et al. 2006). Endophytic isolates of *Colletotrichum musae*, an agriculturally important genus, are also found to produce several fungal phosphatases (Macchcroni Jr and Azevedo 1998). The effects of the carbon and nitrogen sources, initial pH, and incubation temperature on laccase production by the endophytic fungus *Monotospora* sp. were evaluated (Wang et al. 2006). The ability to produce different kinds of enzymes like cellulase, mannanase, proteinase, and xylanase is investigated in trees from Thailand (Lumyong et al. 2002). Recently, *Cercospora kikuchii*, an endophytic fungus of host *Tithonia diversifolia*, was reported to produce lipase that has chemoselectivity in organic synthesis (Nogueira et al. 2008). Borges et al. 2009b summarized the enzymes produced from several fungal endophytes and explained how these endophytes are prolific source of commercially important enzymes. Diverse array of fungal endophytes produces wide variety of enzymes with significant implications to their lifestyle.

30.12 Nutrient Cycling in Terrestrial Ecosystems by Endophytic Fungi

Endophytic fungi also play a significant role in nutrient cycling in different ecosystems. Endophytic phyllosphere fungi affect litter degradation and nutrient exchange as they act as saprotrophs later in their life cycle concomitantly affecting the abundance and richness of other decomposer organisms (Saikkonen et al. 2015). Evidences support that during early senescence, endophytes act as saprotrophs and prevent occupation of other saprophytes to the newly abscised leaves, thus acting as latent saprotrophs. The endophytic mode of lifestyle aids the endophytes to be the first to utilize the host resources and thus paves a path for competitive advantage for available space and nutrients over the other saprotrophic fungi, which may colonize the plant litter later in time (Wilson 1993; Saikkonen et al. 2015). Healthy leaves of grasses harbor non-systemic endophytes that play a role in senescence and abscission of leaf (Zabalgogeazcoa et al. 2013). Endophytic fungi also affect the litter produced by the host by affecting the growth, reproduction, and stress tolerance level of the host (Hoveland 1993). Evidences support that host species shows competitive dominance due to systemic fungal endophytes within thus resulting in prevention of species invasions with decreased richness of flora in successional grasslands. A number of studies suggest that the systemic endophytes can, by enhancing the competitive dominance of their host species, prevent species invasions (Saikkonen et al. 2013) and reduce plant species richness in successional grasslands (Clay and Holah 1999). Presence of endophyte also affects the saprotrophs invasion of host in positive as well as negative aspect. Prior colonization and allelochemicals production by endophytes can competitively exclude saprophytic fungi. On the other side, they can positively facilitate the invasion of latent saprotrophs into the host, for example, through the salicylic acid (SA) and the jasmonic acid (JA) pathways, i.e., phytohormone signaling pathways (Lemons et al. 2005; Faeth and Shochat 2010). Thus evidences support that endophytes do play a significant role in nutrient cycling directly by involving as early successional saprophytes or indirectly by affecting the litter formation and invasion of saprophytes.

30.13 Conclusions

The search for new drugs has researchers looking into fungal endophytic role to produce diverse metabolites with various activities such as antibacterial, antiviral, anticancer, antioxidant, antidiabetic, and immunosuppressive compounds. The mutualistic relationship between the endophytes and their hosts requires further in-depth analysis (Abdalla and Matasyoh 2014).

Endophytes have evolved mechanisms that allow them to compete with other microorganisms for the microhabitats inside plants. Therefore, endophytic microorganisms are a good resource to search for antimicrobial and antifungal agents.

The plausible reason for looking into the endophytes for antimicrobial and antifungal agents might be the coevolution of these microbes with their hosts to survive in the microhabitats while competing with other microorganisms (Martinez-Klimova et al. 2016). One of the major concerns, while culturing the endophytes, is to mimic the conditions of strong physiological bonds that it has with the hosts which perhaps is responsible for inducing the metabolite production. Some endophytes are host-specific, while some are tissue-specific.

Also the modification of the crude metabolites to be active in biological systems and making them commercially available is still one of the important areas of future research.

Researchers link the idea of importance of ethnobotanical medicinal knowledge to the more obvious chances of finding bioactive molecules (Castillo et al. 2002; Alvin et al. 2014). Considering the ethnobotanical knowledge as a prime factor in selecting a host medicinal plant can result in the isolation of fungi that can actively produce bioactive compounds. Application of pharmaceutical and agricultural industries could be a great option for exploiting bioactive metabolites. These efforts require an interdisciplinary approach with experts from different fields in biology and chemistry to meet the targets.

Future holds a bright scope for the discovery and utilization of new bioactive compounds of pharmaceutical interest from the fungal endophytes. With the advancement of latest technologies, upgraded and sophisticated instruments, and encouragement of interdisciplinary sciences, now it has become possible to achieve the targets. Last but not least, by employing the genome mining techniques with the help of recombinant molecular techniques, we can raise the chances of discovery of compounds of medical interest from endophytic fungi.

References

Abdalla MA, Matasyoh JC (2014) Endophytes as producers of peptides: an overview about the recently discovered peptides from endophytic microbes. Nat Prod Bioprospect 4:257–270

Abrahão MRE, Molina G, Pastore GM (2013) Endophytes: recent developments in biotechnology and the potential for flavor production. Food Res Int 52:367–372

Agusta A, Maehara S, Ohashi K, Simanjuntak P, Shibuya H (2005) Stereoselective oxidation at C-4 of flavans by the endophytic fungus *Diaporthe* sp. isolated from a tea plant. Chem Pharm Bull 53:1565–1569

Agusta A, Wulansari D, Nurkanto A, Fathoni A (2014) Biotransformation of protoberberine alkaloids by the endophytic fungus Coelomycetes AFKR-3 isolated from yellow moonseed plant (Archangelisia flava (L.) Merr.). Procedia Chem 13:38–43

Alvin A, Miller KI, Neilan BA (2014) Exploring the potential of endophytes from medicinal plants as sources of antimycobacterial compounds. Microbiol Res 169:483–495

Aly AH, Debbab A, Proksch P (2011) Fungal endophytes: unique plant inhabitants with great promises. Appl Microbiol Biotechnol 90:1829–1845

Aneja M, Gianfagna TJ, Hebbar PK (2005) *Trichoderma harzianum* produces nonanoic acid, an inhibitor of spore germination and mycelial growth of two cacao pathogens. Physiol Mol Plant Pathol 67:304–307

Arnold AE, Herre EA (2003) Canopy cover and leaf age affect colonization by tropical fungal endophytes: ecological pattern and process in *Theobroma cacao* (Malvaceae). Mycologia 95:388–398

Arnold AE, Lutzoni F (2007) Diversity and host range of foliar fungal endophytes: are tropical leaves biodiversity hotspots. Ecology 88:541–549

Arnold AE, Mejía LC, Kyllo D, Rojas EI, Maynard Z, Robbins N, Herre EA (2003) Fungal endophytes limit pathogen damage in a tropical tree. Proc Natl Acad Sci 100:15649–15654

Ashforth EJ, Fu C, Liu X, Dai H, Song F, Guo H, Zhang L (2010) Bioprospecting for antituberculosis leads from microbial metabolites. Nat Prod Rep 27:1709–1719

Azerad R (1999) Microbial models for drug metabolism. In: Biotransformations. Springer, Berlin, pp 169–218

Azevedo JL, Maccheroni W Jr, Pereira JO, de Araújo WL (2000) Endophytic microorganisms: a review on insect control and recent advances on tropical plants. Electron J Biotechnol 3:15–16

Babu AG, Shim J, Bang K-S, Shea PJ, Oh B-T (2014a) *Trichoderma virens* PDR-28: a heavy metal-tolerant and plant growth-promoting fungus for remediation and bioenergy crop production on mine tailing soil. J Environ Manag 132:129–134

Babu AG, Shim J, Shea PJ, Oh B-T (2014b) *Penicillium aculeatum* PDR-4 and *Trichoderma* sp. PDR-16 promote phytoremediation of mine tailing soil and bioenergy production with sorghum-sudangrass. Ecol Eng 69:186–191

Bacon CW, White JF (2000) Microbial endophytes. Marcel Deker, NewYork

Bailey BA, Bae H, Strem MD, Roberts DP, Thomas SE, Crozier J, Samuels GJ, Choi I-Y, Holmes KA (2006) Fungal and plant gene expression during the colonization of cacao seedlings by endophytic isolates of four *Trichoderma* species. Planta 224:1449–1464

Baltruschat H, Fodor J, Harrach BD, Niemczyk E, Barna B, Gullner G, Janeczko A, Kogel K, Schäfer P, Schwarczinger I (2008) Salt tolerance of barley induced by the root endophyte *Piriformospora indica* is associated with a strong increase in antioxidants. New Phytol 180:501–510

Bateman R (2002) Best-bet solutions for cocoa diseases. Gro-Cocoa Newsl 1:4–5

Berendsen RL, Pieterse CMJ, Bakker PAHM (2012) The rhizosphere microbiome and plant health. Trends Plant Sci 17:478–486

Berg G, Hallmann J (2006) Control of plant pathogenic fungi with bacterial endophytes. In: Microbial root endophytes. Springer, Berlin, pp 53–69

Berger RG (2009) Biotechnology of flavours—the next generation. Biotechnol Lett 31:1651–1659

Berry LA, Deacon JW (1992) Video-analysis of *Gliocladium roseum* in relation to mechanism of antagonism of plant pathogens. Bull OILB/SROP 15:64–66

Bhagobaty RK, Joshi SR, Kumar R (2010) *Penicillium verruculosum* RS7PF: a root fungal endophyte associated with an ethno-medicinal plant of the indigenous tribes of Eastern India. Afr J Microbiol Res 4:766–770

Bills GF, Polishook JD (1992) Recovery of endophytic fungi from *Chamaecyparis thyoides*. Sydowia 44:1–12

Bisht D, Owais M, Venkatesan K (2006) Potential of plant-derived products in the treatment of mycobacterial infections. Mod Phytomedicine Turn Med Plants into Drugs, 293–311

Boberg JB, Ihrmark K, Lindahl BD (2011) Decomposing capacity of fungi commonly detected in *Pinus sylvestris* needle litter. Fungal Ecol 4:110–114

Boonphong S, Kittakoop P, Isaka M, Pittayakhajonwut D, Tanticharoen M, Thebtaranonth Y (2001) Multiplolides A and B, new antifungal 10-membered lactones from *Xylaria multiplex*. J Nat Prod 64:965–967

Borges KB, de Souza Borges W, Durán-Patrón R, Pupo MT, Bonato PS, Collado IG (2009a) Stereoselective biotransformations using fungi as biocatalysts. Tetrahedron Asymmetry 20:385–397

Borges W d S, Borges KB, Bonato PS, Said S, Pupo MT (2009b) Endophytic fungi: natural products, enzymes and biotransformation reactions. Curr Org Chem 13:1137–1163

Bryant MK, May KJ, Bryan GT, Scott B (2007) Functional analysis of a β-1, 6-glucanase gene from the grass endophytic fungus *Epichloë festucae*. Fungal Genet Biol 44:808–817
Card S, Johnson L, Teasdale S, Caradus J (2016) Deciphering endophyte behaviour-the link between endophyte biology and efficacious biological control agents. FEMS Microbiol Ecol 92:fiw114
Casella TM, Eparvier V, Mandavid H, Bendelac A, Odonne G, Dayan L, Duplais C, Espindola LS, Stien D (2013) Antimicrobial and cytotoxic secondary metabolites from tropical leaf endophytes: isolation of antibacterial agent pyrrocidine C from Lewia infectoria SNB-GTC2402. Phytochemistry 96:370–377
Castillo UF, Strobel GA, Ford EJ, Hess WM, Porter H, Jensen JB, Albert H, Robison R, Condron MAM, Teplow DB (2002) Munumbicins, wide-spectrum antibiotics produced by Streptomyces NRRL 30562, endophytic on *Kennedia nigricans*. Microbiology 148:2675–2685
Chen Y, Wang H-W, Li L, Dai C-C (2013) The potential application of the endophyte *Phomopsis liquidambari* to the ecological remediation of long-term cropping soil. Appl Soil Ecol 67:20–26
Cheng M-J, Wu M-D, Yanai H, Su Y-S, Chen I-S, Yuan G-F, Hsieh S-Y, Chen J-J (2012) Secondary metabolites from the endophytic fungus *Biscogniauxia formosana* and their antimycobacterial activity. Phytochem Lett 5:467–472
Chomcheon P, Wiyakrutta S, Sriubolmas N, Ngamrojanavanich N, Isarangkul D, Kittakoop P (2005) 3-Nitropropionic acid (3-NPA), a potent antimycobacterial agent from endophytic fungi: is 3-NPA in some plants produced by endophytes? J Nat Prod 68:1103–1105
Clay K, Holah J (1999) Fungal endophyte symbiosis and plant diversity in successional fields. Science (80-) 285:1742–1744
Cusido RM, Onrubia M, Sabater-Jara AB, Moyano E, Bonfill M, Goossens A, Pedreño MA, Palazon J (2014) A rational approach to improving the biotechnological production of taxanes in plant cell cultures of *Taxus* spp. Biotechnol Adv 32:1157–1167
Da Silva MF, de Souza Antônio C, de Oliveira PJ, Xavier GR, Rumjanek NG, de Barros Soares LH, Reis VM (2012) Survival of endophytic bacteria in polymer-based inoculants and efficiency of their application to sugarcane. Plant Soil 356:231–243
Dai C, Tian L, Zhao Y, Chen Y, Xie H (2010) Degradation of phenanthrene by the endophytic fungus *Ceratobasidum stevensii* found in *Bischofia polycarpa*. Biodegradation 21:245–255
Davis EW (1995) Ethnobotany: an old practice, a new discipline. Dioscorides Press, Portland
Deng Z, Wang W, Tan H, Cao L (2012) Characterization of heavy metal-resistant endophytic yeast *Cryptococcus* sp. CBSB78 from rapes (*Brassica chinensis*) and its potential in promoting the growth of *Brassica* spp. in metal-contaminated soils. Water Air Soil Pollut 223:5321–5329
Deng Z, Zhang R, Shi Y, Tan H, Cao L (2014) Characterization of Cd-, Pb-, Zn-resistant endophytic *Lasiodiplodia* sp. MXSF31 from metal accumulating *Portulaca oleracea* and its potential in promoting the growth of rape in metal-contaminated soils. Environ Sci Pollut Res 21:2346–2357
DESA UN (2015) Population division
Deshmukh SK, Mishra PD, Kulkarni-Almeida A, Verekar S, Sahoo MR, Periyasamy G, Goswami H, Khanna A, Balakrishnan A, Vishwakarma R (2009) Anti-inflammatory and anticancer activity of ergoflavin isolated from an endophytic fungus. Chem Biodivers 6:784–789
Deshmukh SK, Verekar SA, Bhave SV (2015) Endophytic fungi: a reservoir of antibacterials. Front Microbiol 5:715
Dhankhar S, Dhankhar S, Parkash Yadav J (2013) Investigations towards new antidiabetic drugs from fungal endophytes associated with *Salvadora oleoides* Decne. Med Chem (Los Angeles) 9:624–632
Dingle J, Mcgee PA (2003) Some endophytic fungi reduce the density of pustules of *Puccinia recondita* f. sp. *tritici* in wheat. Mycol Res 107:310–316
Doble M, Kruthiventi AK, Gaikar VG (2004) Biotransformations and bioprocesses. CRC Press, Boca Raton
Domenech P, Barry CE, Cole ST (2001) *Mycobacterium tuberculosis* in the post-genomic age. Curr Opin Microbiol 4:28–34

Easton HS, Christensen MJ, Eerens JPJ, Fletcher LR, Hume DE, Keogh RG, Lane GA, Latch GCM, Pennell CGL, Popay AJ (2001) Ryegrass endophyte: a New Zealand Grassland success story. In: Proceedings of the conference-New Zealand Grassland Association, pp 37–46

Egamberdieva D, Lugtenberg B (2014) Use of plant growth-promoting rhizobacteria to alleviate salinity stress in plants. In: Use of microbes for the alleviation of soil stresses, vol 1. Springer, New York, pp 73–96

Egamberdieva D, Kamilova F, Validov S, Gafurova L, Kucharova Z, Lugtenberg B (2008) High incidence of plant growth-stimulating bacteria associated with the rhizosphere of wheat grown on salinated soil in Uzbekistan. Environ Microbiol 10:1–9

Elsässer B, Krohn K, Flörke U, Root N, Aust H, Draeger S, Schulz B, Antus S, Kurtán T (2005) X-ray structure determination, absolute configuration and biological activity of phomoxanthone A. Eur J Org Chem 2005:4563–4570

Elsebai MF, Natesan L, Kehraus S, Mohamed IE, Schnakenburg G, Sasse F, Shaaban S, Gütschow M, König GM (2011) HLE-inhibitory alkaloids with a polyketide skeleton from the marine-derived fungus *Coniothyrium cereale*. J Nat Prod 74:2282–2285

Evans HC, Holmes KA, Thomas SE (2003) Endophytes and mycoparasites associated with an indigenous forest tree, *Theobroma gileri*, in Ecuador and a preliminary assessment of their potential as biocontrol agents of cocoa diseases. Mycol Prog 2:149–160

Eyberger AL, Dondapati R, Porter JR (2006) Endophyte fungal isolates from *Podophyllum peltatum* produce podophyllotoxin. J Nat Prod 69:1121–1124

Faeth SH, Shochat E (2010) Inherited microbial symbionts increase herbivore abundances and alter arthropod diversity on a native grass. Ecology 91:1329–1343

Fakhro A, Andrade-Linares DR, von Bargen S, Bandte M, Büttner C, Grosch R, Schwarz D, Franken P (2010) Impact of *Piriformospora indica* on tomato growth and on interaction with fungal and viral pathogens. Mycorrhiza 20:191–200

Figueiredo AC, Almendra MJ, Barroso JG, Scheffer JJC (1996) Biotransformation of monoterpenes and sesquiterpenes by cell suspension cultures of *Achillea millefolium* L. ssp. *millefolium*. Biotechnol Lett 18:863–868

Fleming A (1929) On the antibacterial action of cultures of a penicillium, with special reference to their use in the isolation of *B. influenzae*. Br J Exp Pathol 10:226

Fomina M, Charnock JM, Hillier S, Alvarez R, Gadd GM (2007) Fungal transformations of uranium oxides. Environ Microbiol 9:1696–1710

Franken P (2012) The plant strengthening root endophyte *Piriformospora indica*: potential application and the biology behind. Appl Microbiol Biotechnol 96:1455–1464

Fukuda Y, Shinshi H (1994) Characterization of a novel cis-acting element that is responsive to a fungal elicitor in the promoter of a tobacco class I chitinase gene. Plant Mol Biol 24:485–493

Ganley RJ, Sniezko RA, Newcombe G (2008) Endophyte-mediated resistance against white pine blister rust in *Pinus monticola*. For Ecol Manag 255:2751–2760

Gao F, Zhang J-M, Wang Z-G, Peng W, Hu H-L, Fu C-M (2013) Biotransformation, a promising technology for anti-cancer drug development. Asian Pac J Cancer Prev 14:5599–5608

Ge HM, Zhang Q, Xu SH, Guo ZK, Song YC, Huang WY, Tan RX (2011) Chaetoglocins A–D, four new metabolites from the endophytic fungus *Chaetomium globosum*. Planta Med 77:277–280

Ghahfarokhi RM, Goltapeh ME (2010) Potential of the root endophytic fungus *Piriformospora indica; Sebacina vermifera* and *Trichoderma* species in biocontrol of take-all disease of wheat *Gaeumannomyces graminis var. tritici* in vitro. J Agric Technol 6:11–18

Gimenez C, Cabrera R, Reina M, Gozalez-Coloma A (2007) Fungal endophytes and their role in plant protection. Curr Org Chem 11:707–720

Gunatilaka AAL (2006) Natural products from plant-associated microorganisms: distribution, structural diversity, bioactivity, and implications of their occurrence⊥. J Nat Prod 69:509–526

Gurib-Fakim A (2006) Medicinal plants: traditions of yesterday and drugs of tomorrow. Mol Asp Med 27:1–93

Hajlaoui MR, Dip D, Cherif M (2001) Contribution to Sclerotinia blight caused by *Sclerotinia sclerotiorum* (Lib.) de Bary. Al-Awamia 104:85–101

Hallmann J, Quadt-Hallmann A, Rodrıguez-Kábana R, Kloepper JW (1998) Interactions between *Meloidogyne incognita* and endophytic bacteria in cotton and cucumber. Soil Biol Biochem 30:925–937

Hankin L, Poincelot RP, Anagnostakis SL (1975) Microorganisms from composting leaves: ability to produce extracellular degradative enzymes. Microb Ecol 2:296–308

Harper JK, Arif AM, Ford EJ, Strobel GA, Porco JA, Tomer DP, Oneill KL, Heider EM, Grant DM (2003) Pestacin: a 1, 3-dihydroisobenzofuran from *Pestalotiopsis microspora* possessing antioxidant and antimycotic activities. Tetrahedron 59:2471–2476

Heinig U, Scholz S, Jennewein S (2013) Getting to the bottom of Taxol biosynthesis by fungi. Fungal Divers 60:161–170

Hemberger Y, Xu J, Wray V, Proksch P, Wu J, Bringmann G (2013) Pestalotiopens A and B: stereochemically challenging flexible sesquiterpene-cyclopaldic acid hybrids from *Pestalotiopsis* sp. Chem Eur J 19:15556–15564

Herre EA, Mejía LC, Kyllo DA, Rojas E, Maynard Z, Butler A, Van Bael SA (2007) Ecological implications of anti-pathogen effects of tropical fungal endophytes and mycorrhizae. Ecology 88:550–558

Holmes KA, Schroers H-J, Thomas SE, Evans HC, Samuels GJ (2004) Taxonomy and biocontrol potential of a new species of *Trichoderma* from the Amazon basin of South America. Mycol Prog 3:199–210

Hoveland CS (1993) Importance and economic significance of the Acremonium endophytes to performance of animals and grass plant. Agric Ecosyst Environ 44:3–12

Hubbard M, Germida JJ, Vujanovic V (2014) Fungal endophytes enhance wheat heat and drought tolerance in terms of grain yield and second-generation seed viability. J Appl Microbiol 116:109–122

Hussain H, Krohn K, Draeger S, Meier K, Schulz B (2009) Bioactive chemical constituents of a sterile endophytic fungus from *Meliotus dentatus*. Rec Nat Prod 3:114–117

Isaka M, Jaturapat A, Rukseree K, Danwisetkanjana K, Tanticharoen M, Thebtaranonth Y (2001) Phomoxanthones A and B, novel xanthone dimers from the endophytic fungus *Phomopsis* species. J Nat Prod 64:1015–1018

Istifadah N, Saleeba JA, McGee PA (2006) Isolates of endophytic *Chaetomium* spp. inhibit the fungal pathogen *Pyrenophora tritici-repentis* in vitro. Botany 84:1148–1155

Jaber LR, Vidal S (2009) Interactions between an endophytic fungus, aphids and extrafloral nectaries: do endophytes induce extrafloral-mediated defences in *Vicia faba*? Funct Ecol 23:707–714

Jäschke D, Dugassa-Gobena D, Karlovsky P, Vidal S, Ludwig-Müller J (2010) Suppression of clubroot (*Plasmodiophora brassicae*) development in *Arabidopsis thaliana* by the endophytic fungus *Acremonium alternatum*. Plant Pathol 59:100–111

Jiang M, Cao L, Zhang R (2008) Effects of *Acacia* (*Acacia auriculaeformis* A. Cunn)-associated fungi on mustard (*Brassica juncea* (L.) Coss. *var. foliosa* Bailey) growth in Cd-and Ni-contaminated soils. Lett Appl Microbiol 47:561–565

Johnson LJ, Koulman A, Christensen M, Lane GA, Fraser K, Forester N, Johnson RD, Bryan GT, Rasmussen S (2013) An extracellular siderophore is required to maintain the mutualistic interaction of *Epichloë festucae* with *Lolium perenne*. PLoS Pathog 9:e1003332

Joost RE (1995) Acremonium in fescue and ryegrass: boon or bane? A review. J Anim Sci 73:881–888

Kari Dolatabadi H, Mohammadi Goltapeh E, Mohammadi N, Rabiey M, Rohani N, Varma A (2011) Biocontrol potential of root endophytic fungi and *Trichoderma* species against fusarium wilt of lentil under in vitro and greenhouse conditions. J Agric Sci Technol 14:407–420

Kavroulakis N, Ntougias S, Zervakis GI, Ehaliotis C, Haralampidis K, Papadopoulou KK (2007) Role of ethylene in the protection of tomato plants against soil-borne fungal pathogens conferred by an endophytic *Fusarium solani* strain. J Exp Bot 58:3853–3864

Khan AL, Waqas M, Hussain J, Al-Harrasi A, Lee I-J (2014) Fungal endophyte *Penicillium janthinellum* LK5 can reduce cadmium toxicity in *Solanum lycopersicum* (Sitiens and Rhe). Biol Fertil Soils 50:75–85

Kim HY, Choi GJ, Lee HB, Lee SW, Kim HK, Jang KS, Son SW, Lee SO, Cho KY, Sung ND, Kim J-C (2007) Some fungal endophytes from vegetable crops and their anti-oomycete activities against tomato late blight. Lett Appl Microbiol 44:332–337

Kloepper JW, Ryu CM (2006) Bacterial endophytes as elicitors of induced systemic resistance. In: Microbial root endophytes. Springer, Berlin, pp 33–52

Kornsakulkarn J, Dolsophon K, Boonyuen N, Boonruangprapa T, Rachtawee P, Prabpai S, Kongsaeree P, Thongpanchang C (2011) Dihydronaphthalenones from endophytic fungus *Fusarium* sp. BCC14842. Tetrahedron 67:7540–7547

Koul A, Arnoult E, Lounis N, Guillemont J, Andries K (2011) The challenge of new drug discovery for tuberculosis. Nature 469:483–490

Kour A, Shawl AS, Rehman S, Sultan P, Qazi PH, Suden P, Khajuria RK, Verma V (2008) Isolation and identification of an endophytic strain of *Fusarium oxysporum* producing podophyllotoxin from *Juniperus recurva*. World J Microbiol Biotechnol 24:1115–1121

Kumara PM, Zuehlke S, Priti V, Ramesha BT, Shweta S, Ravikanth G, Vasudeva R, Santhoshkumar TR, Spiteller M, Shaanker RU (2012) *Fusarium proliferatum*, an endophytic fungus from *Dysoxylum binectariferum* Hook. f, produces rohitukine, a chromane alkaloid possessing anti-cancer activity. Antonie Van Leeuwenhoek 101:323–329

Kusari S, Lamshöft M, Zühlke S, Spiteller M (2008) An endophytic fungus from *Hypericum perforatum* that produces hypericin. J Nat Prod 71:159–162

Kusari S, Lamshöft M, Spiteller M (2009) *Aspergillus fumigatus* Fresenius, an endophytic fungus from *Juniperus communis* L. Horstmann as a novel source of the anticancer pro-drug deoxy-podophyllotoxin. J Appl Microbiol 107:1019–1030

Kusari S, Pandey SP, Spiteller M (2013) Untapped mutualistic paradigms linking host plant and endophytic fungal production of similar bioactive secondary metabolites. Phytochemistry 91:81–87

Lee JC, Lobkovsky E, Pliam NB, Strobel G, Clardy J (1995) Subglutinols A and B: immuno-suppressive compounds from the endophytic fungus *Fusarium subglutinans*. J Org Chem 60:7076–7077

Lee JC, Strobel GA, Lobkovsky E, Clardy J (1996) Torreyanic acid: a selectively cytotoxic quinone dimer from the endophytic fungus *Pestalotiopsis microspora*. J Org Chem 61:3232–3233

Lemons A, Clay K, Rudgers JA (2005) Connecting plant–microbial interactions above and below-ground: a fungal endophyte affects decomposition. Oecologia 145:595–604

Li Y-C, Tao W-Y (2009) Paclitaxel-producing fungal endophyte stimulates the accumulation of taxoids in suspension cultures of *Taxus cuspidata*. Sci Hortic (Amsterdam) 121:97–102

Li HM, Sullivan R, Moy M, Kobayashi DY, Belanger FC (2004) Expression of a novel chitinase by the fungal endophyte in *Poa ampla*. Mycologia 96:526–536

Li T, Liu MJ, Zhang XT, Zhang HB, Sha T, Zhao ZW (2011) Improved tolerance of maize (Zea mays L.) to heavy metals by colonization of a dark septate endophyte (DSE) *Exophiala pisciphila*. Sci Total Environ 409:1069–1074

Li X, Li W, Chu L, White JF Jr, Xiong Z, Li H (2016) Diversity and heavy metal tolerance of endophytic fungi from *Dysphania ambrosioides*, a hyperaccumulator from Pb–Zn contaminated soils. J Plant Interact 11:186–192

Lim C, Kim J, Choi JN, Ponnusamy K, Jeon Y, Kim S-U, Kim JG, Lee C (2010) Identification, fermentation, and bioactivity against *Xanthomonas oryzae* of antimicrobial metabolites isolated from *Phomopsis longicolla* S1B4. J Microbiol Biotechnol 20:494–500

Liu JY, Song YC, Zhang Z, Wang L, Guo ZJ, Zou WX, Tan RX (2004) *Aspergillus fumigatus* CY018, an endophytic fungus in *Cynodon dactylon* as a versatile producer of new and bioactive metabolites. J Biotechnol 114:279–287

Liu X, Dong M, Chen X, Jiang M, Lv X, Zhou J (2008) Antimicrobial activity of an endophytic *Xylaria* sp. YX-28 and identification of its antimicrobial compound 7-amino-4-methylcoumarin. Appl Microbiol Biotechnol 78:241–247

Liu K, Ding X, Deng B, Chen W (2010) 10-Hydroxycamptothecin produced by a new endophytic *Xylaria* sp., M20, from *Camptotheca acuminata*. Biotechnol Lett 32:689–693

Lu H, Zou WX, Meng JC, Hu J, Tan RX (2000) New bioactive metabolites produced by *Colletotrichum* sp., an endophytic fungus in *Artemisia annua*. Plant Sci 151:67–73

Lumyong S, Lumyong P, McKenzie EHC, Hyde KD (2002) Enzymatic activity of endophytic fungi of six native seedling species from Doi Suthep-Pui National Park, Thailand. Can J Microbiol 48:1109–1112

Lynch JM, Slater JH, BENNETT JA, Harper SHT (1981) Cellulase activities of some aerobic micro-organisms isolated from soil. Microbiology 127:231–236

Maccheroni W Jr, Azevedo JL (1998) Synthesis and secretion of phosphatases by endophytic isolates of *Colletotrichum musae* grown under conditions of nutritional starvation. J Gen Appl Microbiol 44:381–387

Maciá-Vicente JG, Jansson HB, Talbot NJ, Lopez-Llorca LV (2009) Real-time PCR quantification and live-cell imaging of endophytic colonization of barley (*Hordeum vulgare*) roots by *Fusarium equiseti* and *Pochonia chlamydosporia*. New Phytol 182:213–218

Malik S, Cusidó RM, Mirjalili MH, Moyano E, Palazón J, Bonfill M (2011) Production of the anticancer drug taxol in *Taxus baccata* suspension cultures: a review. Process Biochem 46:23–34

Malinowski DP, Zuo H, Belesky DP, Alloush GA (2004) Evidence for copper binding by extracellular root exudates of tall fescue but not perennial ryegrass infected with *Neotyphodium* spp. endophytes. Plant Soil 267:1–12

Martin KL, Unkles SE, McDougall BM, Seviour RJ (2006) Purification and characterization of the extracellular β-1, 6-glucanases from the fungus *Acremonium* strain OXF C13 and isolation of the gene/s encoding these enzymes. Enzym Microb Technol 38:351–357

Martinez-Klimova E, Rodríguez-Peña K, Sánchez S (2016) Endophytes as sources of antibiotics. Biochem Pharmacol 134:1–17

McAlpine JB, Pazoles C, Stafford A (1999) Phytera's strategy for the discovery of novel anti-infective agents from plant cell cultures. In: Bioassay methods in natural product research and drug development. Springer, Houten, pp 159–166

Mejia LC, Rojas EI, Maynard Z, Arnold AE, Kyllo D, Robbins N, Herre EA (2003) Inoculation of beneficial endophytic fungi into *Theobroma cacao* tissues. In: Proceedings of the 14th International Cocoa Research conference, Accra, II, pp 669–705

Mejía LC, Rojas EI, Maynard Z, Van Bael S, Arnold AE, Hebbar P, Samuels GJ, Robbins N, Herre EA (2008) Endophytic fungi as biocontrol agents of *Theobroma cacao* pathogens. Biol Control 46:4–14

Mitchell AM, Strobel GA, Moore E, Robison R, Sears J (2010) Volatile antimicrobials from *Muscodor crispans*, a novel endophytic fungus. Microbiology 156:270–277

Morandi MAB, Sutton JC, Maffia LA (2000) Effects of host and microbial factors on development of *Clonostachys rosea* and control of *Botrytis cinerea* in rose. Eur J Plant Pathol 106:439–448

Muñoz-Elías EJ, McKinney JD (2005) *Mycobacterium tuberculosis* isocitrate lyases 1 and 2 are jointly required for in vivo growth and virulence. Nat Med 11:638–644

Murali TS (2011) L-asparaginase from marine derived fungal endophytes of seaweeds. Mycosphere 2:147–155

Nakasaki K, Sasaki M, Shoda M, Kubota H (1985) Change in microbial numbers during thermophilic composting of sewage sludge with reference to CO2 evolution rate. Appl Environ Microbiol 49:37–41

Naranjo-Briceño L, Pernía B, Guerra M, Demey JR, Sisto Á, Inojosa Y, González M, Fusella E, Freites M, Yegres F (2013) Potential role of oxidative exoenzymes of the extremophilic fungus *Pestalotiopsis palmarum* BM-04 in biotransformation of extra-heavy crude oil. Microb Biotechnol 6:720–730

Narisawa K, Tokumasu S, Hashiba T, Ohki T (2000) Suppression of clubroot and Verticillium yellows in Chinese cabbage in the field by the root endophytic fungus, *Heteroconium chaetospira*. Plant Pathol 49:141–146

Newman DJ, Cragg GM (2012) Natural products as sources of new drugs over the 30 years from 1981 to 2010. J Nat Prod 75:311–335

Nogueira SCP, Sandrim VC, Guimarães LHS, Jorge JA, Terenzi HF, Polizeli MLTM (2008) Evidence of thermostable amylolytic activity from *Rhizopus microsporus* var. rhizopodiformis using wheat bran and corncob as alternative carbon source. Bioprocess Biosyst Eng 31:329–334

O'Callaghan M (2016) Microbial inoculation of seed for improved crop performance: issues and opportunities. Appl Microbiol Biotechnol 100:1–18

Organization WHO (2011) World Health Organization global tuberculosis control. World Health Organization, Geneva

Panaccione DG, Beaulieu WT, Cook D (2014) Bioactive alkaloids in vertically transmitted fungal endophytes. Funct Ecol 28:299–314

Pandey PK, Singh S, Yadav RNS, Singh AK, Singh MCK (2014) Fungal endophytes: promising tools for pharmaceutical science

Petrini O (1991) Fungal endophytes of tree leaves. In: Microbial ecology of leaves. Springer, New York, pp 179–197

Poling SM, Wicklow DT, Rogers KD, Gloer JB (2008) *Acremonium zeae*, a protective endophyte of maize, produces dihydroresorcylide and 7-hydroxydihydroresorcylides. J Agric Food Chem 56:3006–3009

Pongcharoen W, Rukachaisirikul V, Phongpaichit S, Kühn T, Pelzing M, Sakayaroj J, Taylor WC (2008) Metabolites from the endophytic fungus *Xylaria* sp. PSU-D14. Phytochemistry 69:1900–1902

Prado S, Buisson D, Ndoye I, Vallet M, Nay B (2013) One-step enantioselective synthesis of (4S)-isosclerone through biotransformation of juglone by an endophytic fungus. Tetrahedron Lett 54:1189–1191

Puri SC, Verma V, Amna T, Qazi GN, Spiteller M (2005) An endophytic fungus from *Nothapodytes foetida* that produces camptothecin. J Nat Prod 68:1717–1719

Puri SC, Nazir A, Chawla R, Arora R, Riyaz-ul-Hasan S, Amna T, Ahmed B, Verma V, Singh S, Sagar R (2006) The endophytic fungus *Trametes hirsuta* as a novel alternative source of podophyllotoxin and related aryl tetralin lignans. J Biotechnol 122:494–510

Qadri M, Johri S, Shah BA, Khajuria A, Sidiq T, Lattoo SK, Abdin MZ, Riyaz-Ul-Hassan S (2013) Identification and bioactive potential of endophytic fungi isolated from selected plants of the Western Himalayas. Springerplus 2:1

Qin S, Hussain H, Schulz B, Draeger S, Krohn K (2010) Two new metabolites, Epoxydine A and B, from *Phoma* sp. Helv Chim Acta 93:169–174

Rajkumar M, Ae N, Prasad MNV, Freitas H (2010) Potential of siderophore-producing bacteria for improving heavy metal phytoextraction. Trends Biotechnol 28:142–149

Rattan A, Kalia A, Ahmad N (1998) Multidrug-resistant *Mycobacterium tuberculosis*: molecular perspectives. Emerg Infect Dis 4:195

Redman RS, Dunigan DD, Rodriguez RJ (2001) Fungal symbiosis from mutualism to parasitism: who controls the outcome, host or invader? New Phytol 151:705–716

Rehman S, Shawl AS, Kour A, Andrabi R, Sudan P, Sultan P, Verma V, Qazi GN (2008) An endophytic *Neurospora* sp. from *Nothapodytes foetida* producing camptothecin. Appl Biochem Microbiol 44:203–209

Robinson RJ, Fraaije BA, Clark IM, Jackson RW, Hirsch PR, Mauchline TH (2016) Endophytic bacterial community com-position in wheat (*Triticum aestivum*) is determined by plant tissue type, developmental stage and soil nutrient availability. Plant Soil 405:381–396. https://doi.org/10.1007/s11104-015-2495-4

Robl D, da Silva Delabona P, Mergel CM, Rojas JD, dos Santos Costa P, Pimentel IC, Vicente VA, da Cruz Pradella JG, Padilla G (2013) The capability of endophytic fungi for production of hemicellulases and related enzymes. BMC Biotechnol 13:94

Rodriguez RJ, Henson J, Van Volkenburgh E, Hoy M, Wright L, Beckwith F, Kim Y-O, Redman RS (2008) Stress tolerance in plants via habitat-adapted symbiosis. ISME J 2:404–416

Rodriguez RJ, White JF Jr, Arnold AE, Redman RS (2009) Fungal endophytes: diversity and functional roles. New Phytol 182:314–330

Rodriguez EAE, Jonkers W, Corby KH, May G (2012) Interactions between *Fusarium verticillioides, Ustilagomaydis*, and *Zea mays*: an endophyte, a pathogen, and their shared plant host. Fungal Genet Biol 49:578–587

Rubini MR, Silva-Ribeiro RT, Pomella AWV, Maki CS, Araújo WL, Dos Santos DR, Azevedo JL, Cruz R, Romero JV, Cortina Guerrero H (1980) Diversity of endophytic fungal community of cacao (*Theobroma cacao* L.) and biological control of *Crinipellis perniciosa*, causal agent of Witches' Broom Disease. CATIE, Turrialba

Russell JR, Huang J, Anand P, Kucera K, Sandoval AG, Dantzler KW, Hickman D, Jee J, Kimovec FM, Koppstein D, Marks DH, Mittermiller PA, Núñez SJ, Santiago M, Townes MA, Vishnevetsky M, Williams NE, Vargas MP, Boulanger LA, Bascom-Slack C, Strobel SA (2011) Biodegradation of polyester polyurethane by endophytic fungi. Appl Environ Microbiol 77:6076–6084

Saikkonen K, Ruokolainen K, Huitu O, Gundel PE, Piltti T, Hamilton CE, Helander M (2013) Fungal endophytes help prevent weed invasions. Agric Ecosyst Environ 165:1–5

Saikkonen K, Mikola J, Helander M (2015) Endophytic phyllosphere fungi and nutrient cycling in terrestrial ecosystems. Curr Sci 109:121–126

Santoyo G, Moreno-Hagelsieb G, del Carmen Orozco-Mosqueda M, Glick BR (2016) Plant growth-promoting bacterial endophytes. Microbiol Res 183:92–99

Schardl CL, Leuchtmann A, Spiering MJ (2004) Symbioses of grasses with seedborne fungal endophytes. Annu Rev Plant Biol 55:315–340

Schulz B, Römmert A-K, Dammann U, H-Jür A, Strack D (1999) The endophyte-host interaction: a balanced antagonism? Mycol Res 103:1275–1283

Schulz B, Boyle C, Draeger S, Römmert A-K, Krohn K (2002) Endophytic fungi: a source of novel biologically active secondary metabolites. Mycol Res 106:996–1004

Shaanker RU, Ramesha BT, Ravikanth G, Gunaga R, Vasudeva R, Ganeshaiah KN (2008) Chemical profiling of Nothapodytes nimmoniana for camptothecin, an important anticancer alkaloid: towards the development of a sustainable production system. In: Bioactive molecules and medicinal plants. Springer, Berlin, pp 197–213

Shalini S, Kotasthane AS (2007) Parasitism of *Rhizoctonia solani* by strains of *Trichoderma* spp. EJEAF Chem 6:2272–2281

Shibuya H, Agusta A, Ohashi K, Maehara S, Simanjuntak P (2005) Biooxidation of (+)-catechin and (−)-epicatechin into 3, 4-dihydroxyflavan derivatives by the endophytic fungus *Diaporthe* sp. isolated from a tea plant. Chem Pharm Bull 53:866–867

Shweta S, Zuehlke S, Ramesha BT, Priti V, Kumar PM, Ravikanth G, Spiteller M, Vasudeva R, Shaanker RU (2010) Endophytic fungal strains of *Fusarium solani*, from *Apodytes dimidiata* E. Mey. ex Arn (Icacinaceae) produce camptothecin, 10-hydroxycamptothecin and 9-methoxycamptothecin. Phytochemistry 71:117–122

Soleimani M, Afyuni M, Hajabbasi MA, Nourbakhsh F, Sabzalian MR, Christensen JH (2010) Phytoremediation of an aged petroleum contaminated soil using endophyte infected and non infected grasses. Chemosphere 81:1084–1090

Song YC, Li H, Ye YH, Shan CY, Yang YM, Tan RX (2004) Endophytic naphthopyrone metabolites are co-inhibitors of xanthine oxidase, SW1116 cell and some microbial growths. FEMS Microbiol Lett 241:67–72

Stępniewska Z, Kuźniar A (2013) Endophytic microorganisms – promising applications in bioremediation of greenhouse gases. Appl Microbiol Biotechnol 97:9589–9596

Stierle A, Strobel G, Stierle D (1993) Taxol and taxane production by *Taxomyces andreanae*, an endophytic fungus of Pacific yew. Sci York Then Washington 260:214

Strobel GA (2003) Endophytes as sources of bioactive products. Microbes Infect 5:535–544

Strobel G (2006) Harnessing endophytes for industrial microbiology. Curr Opin Microbiol 9:240–244

Strobel G, Daisy B (2003) Bioprospecting for microbial endophytes and their natural products. Microbiol Mol Biol Rev 67:491–502

Strobel G, Stierle A, Stierle D, Hess WM (1993) Taxomyces andreanae, a proposed new taxon for a bulbilliferous hyphomycete associated with Pacific yew (*Taxus brevifolia*). Mycotaxon 47:71–80

Strobel G, Yang X, Sears J, Kramer R, Sidhu RS, Hess WM (1996) Taxol from *Pestalotiopsis microspora*, an endophytic fungus of Taxus wallichiana. Microbiology 142:435–440

Strobel GA, Dirkse E, Sears J, Markworth C (2001) Volatile antimicrobials from *Muscodor albus*, a novel endophytic fungus. Microbiology 147:2943–2950

Strobel G, Ford E, Worapong J, Harper JK, Arif AM, Grant DM, Fung PCW, Chau RMW (2002) Isopestacin, an isobenzofuranone from *Pestalotiopsis microspora*, possessing antifungal and antioxidant activities. Phytochemistry 60:179–183

Strohl WR (2000) The role of natural products in a modern drug discovery program. Drug Discov Today 5:39–41

Strom PF (1985a) Effect of temperature on bacterial species diversity in thermophilic solid-waste composting. Appl Environ Microbiol 50:899–905

Strom PF (1985b) Identification of thermophilic bacteria in solid-waste composting. Appl Environ Microbiol 50:906–913

Subban K, Subramani R, Johnpaul M (2013) A novel antibacterial and antifungal phenolic compound from the endophytic fungus *Pestalotiopsis mangiferae*. Nat Prod Res 27:1445–1449

Suresh B, Ritu T, Ravishankar GA (2006) Biotransformations as applicable to food industries. Food Sci Technol York-Marcel Dekker 148:1655

Sutjaritvorakul T, Whalley AJS, Sihanonth P, Roengsumran S (2011) Antimicrobial activity from endophytic fungi isolated from plant leaves in Dipterocarpus forest at Wiang Sa district Nan province, Thailand. J Agric Technol 7:115–121

Tan RX, Zou WX (2001) Endophytes: a rich source of functional metabolites. Nat Prod Rep 18:448–459

Tanaka A, Tapper BA, Popay A, Parker EJ, Scott B (2005) A symbiosis expressed non-ribosomal peptide synthetase from a mutualistic fungal endophyte of perennial ryegrass confers protection to the symbiotum from insect herbivory. Mol Microbiol 57:1036–1050

Tejesvi MV, Mahesh B, Nalini MS, Prakash HS, Kini KR, Subbiah V, Shetty HS (2006) Fungal endophyte assemblages from ethnopharmaceutically important medicinal trees. Can J Microbiol 52:427–435

Tejesvi MV, Nalini MS, Mahesh B, Prakash HS, Kini KR, Shetty HS, Subbiah V (2007) New hopes from endophytic fungal secondary metabolites. Bol Soc Quím Méx 1:19–26

Tiquia SM, Richard TL, Honeyman MS (2002) Carbon, nutrient, and mass loss during composting. Nutr Cycl Agroecosyst 62:15–24

Tondje PR, Hebbar KP, Samuels G, Bowers JH, Weise S, Nyemb E, Begoude D, Foko J, Fontem D (2006) Bioassay of *Geniculosporium* species for *Phytophthora megakarya* biological control on cacao pod husk pieces. Afr J Biotechnol 5:648

Tripathi S, Kamal S, Sheramati I, Oelmuller R, Varma A (2008) Mycorrhizal fungi and other root endophytes as biocontrol agents against root pathogens. In: Mycorrhiza. Springer, Heidelberg, pp 281–306

Vallad GE, Goodman RM (2004) Systemic acquired resistance and induced systemic resistance in conventional agriculture. Crop Sci 44:1920–1934

Vega FE, Posada F, Aime MC, Pava-Ripoll M, Infante F, Rehner SA (2008) Entomopathogenic fungal endophytes. Biol Control 46:72–82

Vega FE, Simpkins A, Aime MC, Posada F, Peterson SW, Rehner SA, Infante F, Castillo A, Arnold AE (2010) Fungal endophyte diversity in coffee plants from Colombia, Hawai'i, Mexico and Puerto Rico. Fungal Ecol 3:122–138

Verpoorte R (2000) Pharmacognosy in the new millennium: leadfinding and biotechnology. J Pharm Pharmacol 52:253–262

Wagenaar MM, Clardy J (2001) Dicerandrols, new antibiotic and cytotoxic dimers produced by the fungus *Phomopsis longicolla* isolated from an endangered mint. J Nat Prod 64:1006–1009

Waller F, Achatz B, Baltruschat H, Fodor J, Becker K, Fischer M, Heier T, Hückelhoven R, Neumann C, von Wettstein D (2005) The endophytic fungus *Piriformospora indica* reprograms barley to salt-stress tolerance, disease resistance, and higher yield. Proc Natl Acad Sci U S A 102:13386–13391

Wang JW, Wu JH, Huang WY, Tan RX (2006) Laccase production by *Monotospora* sp., an endophytic fungus in Cynodon dactylon. Bioresour Technol 97:786–789

Weber D, Sterner O, Anke T, Gorzalczancy S, Martino V, Acevedo C (2004) Phomol, a new anti-inflammatory metabolite from an endophyte of the medicinal plant *Erythrina crista-galli*. J Antibiot (Tokyo) 57:559–563

WCED (1987) Our common future by World Commission on Environment and Development (WCED). Oxford University Press, New York, p 383

White JF, Belanger F, Meyer W, Sullivan RF, Bischoff JF, Lewis EA (2002) Clavicipitalean fungal epibionts and endophytes-development of symbiotic interactions with plants. Symbiosis 33:201–213

Wilkinson HH, Siegel MR, Blankenship JD, Mallory AC, Bush LP, Schardl CL (2000) Contribution of fungal loline alkaloids to protection from aphids in a grass-endophyte mutualism. Mol Plant-Microbe Interact 13:1027–1033

Wilson D (1993) Fungal endophytes: out of sight but should not be out of mind. Oikos 68:379–384

Wiyakrutta S, Sriubolmas N, Panphut W, Thongon N, Danwisetkanjana K, Ruangrungsi N, Meevootisom V (2004) Endophytic fungi with anti-microbial, anti-cancer and anti-malarial activities isolated from Thai medicinal plants. World J Microbiol Biotechnol 20:265–272

Xiao X, Luo S, Zeng G, Wei W, Wan Y, Chen L, Guo H, Cao Z, Yang L, Chen J (2010) Biosorption of cadmium by endophytic fungus (EF) *Microsphaeropsis* sp. LSE10 isolated from cadmium hyperaccumulator Solanum nigrum L. Bioresour Technol 101:1668–1674

Xiao J, Zhang Q, Gao YQ, Shi XW, Gao JM (2014) Antifungal and antibacterial metabolites from an endophytic Aspergillus sp. associated with *Melia azadirachta*. Nat Prod Res 28:1388–1392

Yao M, Liu J-Z, Jin S, Jiao J, Gai Q, Wei Z, Fu Y, Zhao J (2014) A novel biotransformation of astragalosides to astragaloside IV with the deacetylation of fungal endophyte *Penicillium canescens*. Process Biochem 49:807–812

Yong YH, Dai CC, Gao FK, Yang QY, Zhao M (2009) Effects of endophytic fungi on growth and two kinds of terpenoids for Euphorbia pekinensis. Chin Tradit Herb Drug 40:1136–1139

Zabalgogeazcoa I, Gundel PE, Helander M, Saikkonen K (2013) Non-systemic fungal endophytes in *Festuca rubra* plants infected by Epichloë festucae in subarctic habitats. Fungal Divers 60:25–32

Zhang J, Zhang L, Wang X, Qiu D, Sun D, Gu J, Fang Q (1998) Microbial transformation of 10-deacetyl-7-epitaxol and 1β-hydroxybaccatin I by fungi from the inner bark of *Taxus yunnanensis*. J Nat Prod 61:497–500

Zhang HW, Song YC, Tan RX (2006) Biology and chemistry of endophytes. Nat Prod Rep 23:753–771

Zhang Y, Zhang Y, Liu M, Shi X, Zhao Z (2008) Dark septate endophyte (DSE) fungi isolated from metal polluted soils: their taxonomic position, tolerance, and accumulation of heavy metals in vitro. J Microbiol 46:624–632

Zhang P, Zhou P-P, Yu L-J (2009a) An endophytic taxol-producing fungus from *Taxus media*, *Cladosporium cladosporioides* MD2. Curr Microbiol 59:227–232

Zhang W, Draeger S, Schulz B, Krohn K (2009b) Ring B aromatic steroids from an endophytic fungus, *Colletotrichum* sp. Nat Prod Commun 4:1449–1454

Zhang XX, Li CJ, Nan ZB, Matthew C (2012) *Neotyphodium* endophyte increases *Achnatherum inebrians* (drunken horse grass) resistance to herbivores and seed predators. Weed Res 52:70–78

Zikmundova M, Drandarov K, Bigler L, Hesse M, Werner C (2002) Biotransformation of 2-benzoxazolinone and 2-hydroxy-1, 4-benzoxazin-3-one by endophytic fungi isolated from *Aphelandra tetragona*. Appl Environ Microbiol 68:4863–4870

Zou WX, Meng JC, Lu H, Chen GX, Shi GX, Zhang TY, Tan RX (2000) Metabolites of *Colletotrichum gloeosporioides*, an endophytic fungus in *Artemisia mongolica*. J Nat Prod 63:1529–1530

Chapter 31
Marine Fungi for Sustainable Development

V. Venkateswara Sarma

Abstract Marine fungi are an ecologically unified group but are taxonomically diverse. They occur on wood, sediments, algae (seaweeds), dead corals, calcareous tubes of mollusks, decaying leaves, seedlings, prop roots and pneumatophores of mangroves, intertidal grasses, and living animals. Fungi that develop reproductive propagules on natural substrata (plant or animal) are considered as obligate marine fungi, and those that are isolated from soil or water samples of marine environments which sprout in the agar plates under artificial conditions are considered as facultative marine fungi. Many in the first group have special structures such as sheaths and appendages on their propagules (spores and conidia) which help them in attaching to different substrata. The fungi of the second group are predominantly comprised of aspergilli/penicilli. Marine fungi play an important role in the nutrient regeneration cycles by degrading and decaying the dead organic matter; thus they are involved in the production of organic detritus that supports a large animal community. Besides the wood borers and bacteria, the fungi are major decomposers of woody and herbaceous substrata that enter into the marine ecosystems. These aspects have been reviewed in this chapter. Also in recent times various biotechnological applications of marine fungi have been studied including their enzymatic potential and secondary metabolite production, which also have been discussed. Finally, recommendations have been provided for future research.

Keywords Decomposition · Enzymes · Mangroves · Nutrient recycling · Secondary metabolites

---

V. Venkateswara Sarma (✉)
Department of Biotechnology, Pondicherry University, Kalapet, Pondicherry 605014, India

P. Gehlot, J. Singh (eds.), *Fungi and their Role in Sustainable Development: Current Perspectives*, https://doi.org/10.1007/978-981-13-0393-7_31

31.1 Definition

Kohlmeyer and Kohlmeyer (1979) have offered a definition for marine fungi as follows: "Those that grow and sporulate exclusively in a marine or estuarine habitat" as "obligate marine fungi" and "those from a freshwater or terrestrial milieu, able to grow and possibly also sporulate in the marine environment" as "facultative marine fungi." Jones et al. (2015) felt that the above definition is narrow and should be expanded to include all those fungi that occur in marine environment. Further, many researchers were using the term "marine-derived fungi" for those fungi that are isolated from marine sediment and water samples in the marine environment to distinguish them from obligate marine fungi. Most of the secondary metabolites reported from marine environment belonged to "marine-derived fungi" category, and there has been a debate whether they could be categorized as marine fungi or not. Against this backdrop, recently, Pang et al. (2016) proposed a revised, broad definition of a marine fungus as "any fungus that is recovered repeatedly from marine habitats because: (1) it is able to grow and/or sporulate (on substrata) in marine environments; (2) it forms symbiotic relationships with other marine organisms; or (3) it is shown to adapt and evolve at the genetic level or be metabolically active in marine environments."

31.2 Diversity of Marine Fungi

Marine fungi have special structures such as sheaths and appendages on their propagules (spores and conidia) which help them in attaching to different substrata. In fact, they are not restricted to any particular taxonomic group but are rather an ecologically and physiologically defined group. Investigations on marine fungi were carried out using two techniques, viz., (i) direct microscopic examination method where the fungi occurring on natural samples alone were considered as typical marine fungi (e.g., driftwood, mangrove wood, etc.) and (ii) culture techniques where soil samples from mud beneath the mangroves or beaches are plated on to appropriate agar media. Most of the fungi isolated by the latter technique belonged to the terrestrial fungi, viz., species of *Aspergillus* and *Penicillium*, and none of the typical marine fungi were isolated (Newell 1976). Jones et al. 2009 reported the number of typical marine fungi (obligate group) to be 530. In a recent monograph, Jones et al. (2015) documented 1112 species as marine fungi belonging to 472 genera including *Ascomycota* 805 species (in 352 genera), *Basidiomycota* 21 species (in 17 genera), *Chytridiomycota* and related phyla 26 species (in 13 genera), *Zygomycota* 3 species (in 2 genera), *Blastocladiomycota* 1 species (in 1 genus), asexual filamentous fungi 43 (in 26 genera), and marine yeasts classified under *Ascomycota* 138 species (in 35 genera) and *Basidiomycota* 75 species (in 26 genera). These fungi could be grouped under 65 orders and 129 families. Among the different families, the *Halosphaeriaceae* is the largest family of marine fungi with

141 species in 59 genera, while the most speciose genera are *Aspergillus* (47 species), *Penicillium* (39 species), and the yeast genus *Candida* (64 species). While the above numbers show the known marine fungi, Jones (2011) estimated that there could be more than 10,000 marine fungi, and the recent molecular studies are suggestive of such an estimate (Jones et al. 2015).

31.3 Role of Marine Fungi in the Environment

Marine fungi occur on different substrata including wood, sediments, algae (seaweeds), dead corals, calcareous tubes of mollusks, decaying leaves, seedlings, prop roots and pneumatophores of mangroves, intertidal grasses, and living animals (e.g., those growing in the guts of crustaceans) (Kohlmeyer and Kohlmeyer 1979). These fungi play a key role in the production of organic detritus and support large animal communities (Kohlmeyer and Kohlmeyer 1979), including commercial fisheries where the detritus acts as breeding and nursery ground (Jones and Alias 1997). They also play an important role in nutrient regeneration cycles as decomposers of dead and decaying organic matter in marine ecosystems (Fell and Master 1980; Hyde et al. 1998). Besides the woodborers and bacteria, the fungi form major decomposers of woody and herbaceous substrata entering marine ecosystems. Marine fungi can colonize any untreated piece of wood that is submerged for a certain period in marine or estuarine waters. This fact has kindled interest in many workers to conduct research on fungi on man-made structures in the sea, mangrove wood, and intertidal wood (Kohlmeyer and Kohlmeyer 1979; Hyde et al. 2000). Wood in the open ocean (drift wood) and decaying/decomposing substrata in mangrove habitats are the two predominant substrata that accommodate marine fungi. While the aerial parts of the mangrove plants are usually occupied by the terrestrial fungi, the marine fungi colonize lower parts of mangroves where their trunks and roots are permanently or intermittently submerged in water. At the high-tide mark where an interface with reference to exposure to water could be indicated, marine and terrestrial fungi overlap. Kohlmeyer and Kohlmeyer (1979) opined that the mangrove trees are fascinating study objects for the mycological investigations. Earlier mangroves were considered to be the second most important hosts for marine fungi after driftwood (Hyde et al. 2000). But after several studies in the Southeast Asia and other countries, it has been found that the mangroves seem to be dominating as far as marine fungal diversity is concerned, even outcompeting driftwood (Sarma and Hyde 2001). Hence this environment is dealt with in more detail with regard to the fungi and their role in decomposing the dead organic matter in mangroves in this chapter.

A major contribution of mangroves is generating leaf detritus and other biomass and supplying nutrients for primary production. The converted nutrients also are a major energy input for fisheries. The biomass is recycled by woodborers, fungi, and bacteria thus maintaining ecological balance through the detritus cycle. Detritus organisms including different microbes that are dependent on the mangroves would be lost if any damage to this ecosystem occurs as that would reflect on the loss of microbes also. Fell and Master (1980) suggested that fungi rather than bacteria are considered to play a major role in nutrient regeneration cycles maintaining C/N ratios.

Maintenance of detrital-based food webs in the coastal environment depends on the availability of mangrove leaf litter. As one of the great productive ecosystems, mangroves support a high abundance and rich diversity, which is suggestive of high leaf production, leaf fall, and rapid breakdown of the detritus. In order to conserve and sustainably manage the mangrove ecosystem, it is necessary to understand the key processes involved in the production and breakdown of mangrove litter (Ashton et al. 1999).

Nutrient recycling starts with leaf senescence and leaf fall. The leaves are needed to be degraded into fragments and small bits and pieces, which would finally be converted into smaller compounds or elements. During this course of decomposition and aging, the chemistry of mangrove detritus changes significantly. Leaf breakdown could be considered as a weight loss due to physical fragmentation (caused by abiotic factors), animal feeding, microbial activity, and leaching. Losses ranging from 14% to 40% of the initial dry mass of the mangrove leaves brought about in 3–28 days (Cundell et al. 1979; Steinke et al. 1993) due to very rapid leaching of dissolved organic matter (DOM), including organic carbon, nitrogen, and tannins, have been well documented. However, the remaining leaf biomass, particulate organic matter (POM), is decomposed more slowly and is highly dependent on the action of bacterial and fungal communities that develop rapidly on the leaves and increase the absolute nitrogen content of the leaf litter and also tame the leaf litter for invertebrates as feed (Ashton et al. 1999).

Heald and Odum (1970) and Odum and Heald (1972) have determined the detritus production to be 3 metric tons (dry weight) per acre per year from mangrove leaf fall alone in a South Florida estuary. According to these authors, leaves are greater contributors to the food web than mangrove twigs, bark, and leaf scales, and hence mangrove leaves as a source of detritus are of great interest to investigate the role of microorganisms in the breakdown of leaves (Kohlmeyer and Kohlmeyer 1979). Exploitation of environmental resources is more efficient in diverse communities, which possess greater structural and functional versatility than individual species. Further, a rich biodiversity would sustain the productivity and stabilize community performance under different environmental conditions. This is more so in the case of detrital organisms of which fungi have been considered to play a greater role in nutrient recycling than bacteria (Fell and Master 1980).

Fell and Master (1973, 1975) have investigated the breakdown of leaves in mangroves. They found that senescent leaves still attached harbor a number of parasitic and saprobic terrestrial fungi (Kohlmeyer and Kohlmeyer 1979). During the 1st week of submergence, the lower fungi occurred along with a few other primary invaders, mostly *Hyphomycetes*. During the second and third week, the obligate marine fungi, such as *Lulworthia* sp. and *Zalerion varium*, were observed with most of the lower fungi disappearing during this time (Kohlmeyer and Kohlmeyer 1979). Fell and Master (1973, 1975, 1980) have shown the role of fungi in *Rhizophora mangle* leaf degradation. In contrast, Benner and Hodson (1985) concluded that fungi had minimal participation in the mineralization of the lignocellulosic component of leaves of *Rhizophora mangle*. Further, Cundell et al. (1979) reported that fungi colonized the mangrove (*Rhizophora mangle*) leaves only after 4 weeks of submergence although Newell et al. (1987) found fungal colonization within 2 weeks of submergence on the same plant. Although cellulolytic activity of phylloplane fungi of *Bruguiera gymnorrhiza* was proved under lab conditions, most of the initial mass losses in mangroves were due to leaching of dissolved organic matter (DOM) (Singh and Steinke 1992).

Raghukumar et al. (1994, 1995) have studied the litter degradation by fungi by using litterbags (both in the field and in the laboratory conditions). In these studies, lower fungi such as *Halophytophthora* spp. and thraustochytrids were found to colonize the leaves initially followed by a few species of *Hyphomycetes*, e.g., *Acremonium* sp., *Aspergillus* spp., *Cladosporium herbarum*, *Cirrenalia basiminuta*, *Fusarium moniliforme*, and *Penicillium* sp. Kuthubutheen (1981) has found a greater diversity of fungi in senescent leaves than young or mature green leaves. However, he has not studied the leaf litter-degrading fungi. The fungi recorded were *Pestalotiopsis* spp., *Zygosporium* spp., *Aspergillus* sp., *Cladosporium oxysporum*, *Penicillium* sp., *Cephalosporium* sp., *Corynespora cassiicola*, *Curvularia* sp., *Phyllosticta* sp., *Parasympodiella* sp., *Botryodiplodia theobromae*, and *Drechslera* spp. Ananda and Sridhar (2004) reported leaf litter fungi from west coast of Indian mangroves along Karnataka coast which included mostly *Hyphomycetes*. Kuthubutheen (1981) showed that mostly terrestrial fungi are recorded in the aerial parts of mangroves. The zeal to investigate the marine fungi in mangroves has dominated, and we have a wealth of information on the same (Sarma and Hyde 2001). However, the mycota on the aerial parts of the mangroves is less investigated. Hence, in the future, more time has to be spent on the mycota of aerial parts of the mangroves also. Sarma and Vittal (2001) have concluded mentioning that it would be interesting to examine whether there is any distinct mycota on aerial parts of mangroves that is not seen in other land plants.

Other than the leaves, the woody litter also gets degraded in the mangrove environment albeit slowly over several months or years of time. All this woody litter once again is degraded by fungi in addition to bacteria and tiny faunal organisms. Such degraded and decomposed material becomes available in the inorganic form for the plant growth thus maintaining the nutrient regeneration cycles. Several studies have been conducted on the diversity and ecological observations including frequency of occurrence, vertical and horizontal distribution, host and substrate specificity, etc. (Sarma and Hyde 2001; Sarma and Vittal 2000, 2001). Many marine fungi have shown host specificity on *Rhizophora* spp. and *Nypa fruticans*. In fact, more than 100 species have been reported on each of these hosts, and several of them have been found to be host specific (Hyde et al. 2000). If any damage occurs to the host species, then it would directly affect the marine fungal diversity also as many of them are host specific. Hence we need a sustainable development of the mangrove environment also to make sure that the plant diversity and the dependent organisms are not lost. Among the various factors, the rainfall and time and exposure to sea water and depth of the canals or tributaries in the estuaries/deltas seem to affect the fungal diversity in mangroves. Though more than 200 obligate marine fungi have been reported from mangroves, only a few are dominant and have repeated occurrence. These are known as "core group fungi" based on their percentage occurrence and subsequent conversion into frequency groupings based on their percentage occurrence. These core group fungi vary from one country to another country or within the same country from one coast to another coast with a provision for overlapping of them here and there. In an excellent review, Jones (2000) has reviewed the factors affecting the biodiversity of marine fungi. The frequency of occurrence of marine fungi on the woody litter in mangroves has been reviewed by Sarma and Hyde (2001), and readers are suggested to consult these reviews for more details on diversity and ecology of marine fungi in mangroves.

Most of the fungi occurring in mangroves are dependent on the plant hosts. If anything happens to these host plants, we may lose the mycota depending on them. Hence it is important to conserve the mangroves. Also it is important to make surveys on the mycota growing on the plant litter. While all the fungi colonizing mangrove litter can be said to be halotolerant, we do not know the adaptations these fungi have in addition to growing in high concentrations of salt amended in the artificial media. Such studies would kindle the interest of those interested in salt related physiological changes in the microorganisms.

31.4 Potential in Biotechnological Applications

Lignocellulosic materials in marine environments have been found to support a wide range of fungi capable of wood decay. Marine fungi naturally adapted to aquatic conditions could be expected to grow easily under submerged fermenting conditions (liquid media) with increased enzyme production. Lignocellulose-degrading enzymes such as xylanases and lignin-modifying enzymes (LMEs) have found their use in waste bioconversion, biopulping, biobleaching, and bioremediation technologies. The source of lignocellulose (a heteropolymer that comprises cellulose, hemicellulose, and lignin) in marine environments is driftwood, mangrove wood, wood from shoreline plants, and other halophytes. White rot fungi (basidiomycetes) are known to be capable of producing a battery of ligninolytic enzymes, e.g., laccases, lignin peroxidases, and manganese-dependent peroxidases. Some marine ascomycetous fungi have also been shown to be efficient in producing these enzymes (Raghukumar et al. 1996; Raghukumar et al. 1999; Sarma 2007). Ligninolytic and xylanolytic enzymes from terrestrial fungi are used in various biotechnological applications including biobleaching, biopulping, and bioremediation. There are a few reports that are available to show that marine fungi also have similar capabilities. For example, decolorization of paper mill bleach plant effluents (Raghukumar et al. 1996) and synthetic dye decolorization by a marine basidiomycete, viz., *Flavodon flavus* (Raghukumar et al. 1999), have been reported. The same fungus has also been reported to be having a potential in simultaneous detoxification and decolorization of molasses spent wash by immobilized cells (Raghukumar et al. 2004) and in removal of the colored pollutants in the paper industry (Raghukumar

2002). In another report it has been shown that the crude culture filtrate of a marine fungal isolate has thermostable, cellulose-free alkaline xylanase activity of 2457 U mg^{-1} protein that could be used in biobleaching of paper pulp (Raghukumar et al. 2004). These few reports suggest possible applications for marine fungi, which need further extensive screening and intensive studies (Sarma 2007).

31.5 Secondary Metabolites

Well-known antibiotics from terrestrial fungi are penicillin, cyclosporin, and griseofulvin. Most of the efforts from terrestrial environments did not yield fruitful results but instead showed a recurrence of known compounds. Nowadays there is a shift in drug search program as new environments are being explored. One such environment is marine environment. Little information is available on cultural aspects and/or biotechnological potential of obligate marine fungi. The past half-century has witnessed considerable effort toward the isolation and characterization of metabolites and toxins from fungi, bacteria, and actinomycetes. The oceans represent the largest habitat of the world as they represent ¾ of the earth's surface. It is only in the recent two decades that more attention has been paid to the flora and fauna of the sea. Several structurally and pharmacologically interesting compounds have been isolated from algae, sponges, mollusks, ascidians, and other relatively higher organized groups. When compared to these organisms, the marine microbes in general and marine fungi in particularly were poorly investigated until recently (Liberra and Lindequist 1995).

Due to difficulties in isolation and maintenance of typical marine fungi, researchers have screened taxa that are more closely related to or identical with terrestrial fungi that occur in maritime environments (Sarma 2007). For example, Höller et al. (2000) reported strains isolated from algae and sponges in the marine environments were terrestrial fungi but produce several bioactive compounds and these belonged to species of *Aspergillus* and *Penicillium*. A marine fungal strain identified as *Penicillium* sp. produced a neuritogenic compound (epolactaene), which can treat various neurodegenerative diseases involving dementia (Kakeya et al. 1997). *Emericella variecolor* isolated from Caribbean Sea, Venezuela, produces varitriol that showed potency to cure some renal, central nervous system, and breast cancer cell lines (Malmstrøm et al. 2002). *Wardomyces anomalus* has been reported to be producing xanthone derivatives that showed antioxidant and tyrosine kinase inhibitor activities (Abdel-Lateff et al. 2003). Imhoff (2016) considers that though there is a huge phylogenetic diversity of marine fungi and their almost ubiquitous distribution, only few genera, e.g., *Penicillium* and *Aspergillus*, have been studied for secondary metabolites, and there is a need to investigate the secondary metabolites of less studied species which deserve special attention.

Some reports on bioactive compounds from typical marine fungi are also available. These include isoculmorin produced by *Kallichroma tethys*, siccayne produced by *Halocyphina villosa*, helicascolides (lactones) produced by *Helicascus kanaloanus*, cyclic lipopeptide having antifungal activity produced by *Halorosellinia oceanica*, cathestatins produced by *Microascus longirostris*, obionene having antibiotic activity and culmorin produced by *Leptosphaeria oraemaris*, and auranticins having antibiotic activity and produced by *Preussia aurantiaca* (Bugni and Ireland 2004). Though many drugs would not have reached the market, conceptually they have been proven to be having therapeutic values. Literature shows that *Pyrenomycetes* and *Loculoascomycetes* have been shown to be attractive for screening for novel natural products. Most of the marine fungi, interestingly, belong to these two groups (Huang and Kaneko 1996). Hence typical marine fungi also could be considered as a potential source for novel compounds.

Cultivation-dependent studies show that marine macroorganisms, like sponges or algae, are a rich source for biologically active fungi (Zhang et al. 2009; Paz et al. 2010; Imhoff et al. 2011; Rateb and Ebel 2011; Wiese et al. 2011). Fungal strains isolated from sponges have been shown to be producing a considerable proportion (28%) of new compounds followed by those obtained from algae (27%) (Bugni and Ireland 2004). Up to only 15 fungal metabolites were reported until 1992 (Fenical and Jensen 1993), and around 270 compounds were reported by 2002 (Bugni and Ireland 2004). Between 2000 and 2005 approximately 100 marine fungal metabolites were classified by Saleem et al. (2007). In the following 5-year period, i.e., from 2006 to 2010, totally 690 natural products from fungi isolated from marine habitats have been reported (Rateb and Ebel 2011). These studies revealed that almost half of the compounds (nearly 50%) belonged to polyketides and their isoprene hybrids, while terpenoids, alkaloids, and peptides contributed to 14–20%. Strains belonging to the fungal genera *Penicillium* and *Aspergillus* produced most of the new compounds. Strains belonging to other genera such as *Acremonium*, *Emericella*, *Epicoccum*, *Exophiala*, *Paraphaeosphaeria*, *Phomopsis*, and *Halarosellinia*, to mention a few, were less common. Compounds of considerable interest isolated from marine fungi include ulocladol, halimide, avrainvillamide, pestalone, and the halovirs A–E.

In addition to the above, cytotoxic secondary metabolites from marine fungi have been reported by Lee et al. (2011), Sun et al. (2012), Xia et al. (2012), Gao et al. (2010), Li et al. (2011), and Zheng et al. (2012). Similarly, antimicrobial activity against various pathogenic bacteria and fungi by producing different secondary metabolites by marine fungi was reported, in recent times, by Li et al. (2012), Klaiklay et al. (2012), Cohen et al. (2011), Yang et al. (2011), Elsebai et al. (2011), Gao et al. (2011), and Almeida et al. (2011). However, in almost all of the above studies that have been researched, the fungi belonged to the marine-derived fungi category and not typical marine fungi.

31.6 Future Research

It is well established that the fungi play key roles in nutrient regeneration cycles by degrading plant and animal substrata in marine environments particularly in coastal environments such as mangroves. In the future efforts should be made to correlate the chemistry of the host substrata and the fungi thriving on these hosts to check what kind of roles the host chemistry plays. Similarly, the literature surveyed in the present chapter shows that most of the fungi investigated for secondary metabolites belonged to the marine-derived fungi and not the typical marine fungi. Efforts should be made to investigate with typical marine fungi also for secondary metabolites. Similarly, there is a need to screen more typical marine fungi for their enzymatic potential. There is a need for more isolations of marine fungi and their deposition in culture collection centers. The same would be of help to test their efficacy in various biotechnological applications and in the sustainable use of marine fungi.

References

Abdel-Lateff A, Klemke C, König GM, Wright AD (2003) Two new xanthone derivatives from the algicolous marine fungus Wardomyces anomalus. J Nat Prod 66:706–708

Almeida C, Kehraus S, Prudêncio M, König GM (2011) Marilones A–C, phthalides from the sponge-derived fungus Stachylidium sp. Beilstein J Org Chem 7:1636–1642

Ananda K, Sridhar KR (2004) Diversity of filamentous fungi on decomposing leaf and woody litter of mangrove forests in the southwest coast of India. Curr Sci 87:1431

Ashton EC, Hogarth PJ, Ormond R (1999) Breakdown of mangrove leaf litter in a managed mangrove forest in peninsular Malaysia. In: Diversity and function in mangrove ecosystems. Springer, Dordrecht, pp 77–88

Benner R, Hodson RE (1985) Microbial degradation of the leachable and lignocellulosic components of leaves and wood from Rhizophora mangle in a tropical mangrove swamp. Mar Ecol Prog Ser 23:221

Bugni TS, Ireland CM (2004) Marine-derived fungi: a chemically and biologically diverse group of microorganisms. Nat Prod Rep 21:143–163

Cohen E, Koch L, Thu KM, Rahamim Y, Aluma Y, Ilan M, Yarden O, Carmeli S (2011) Novel terpenoids of the fungus aspergillus insuetus isolated from the Mediterranean sponge Psammocinia sp. collected along the coast of Israel. Bioorg Med Chem 19:6587–6593

Cundell AM, Brown MS, Stanford R, Mitchell R (1979) Microbial degradation of Rhizophora mangle leaves immersed in the sea. Estuar Coast Mar Sci 9:281IN1–286IN4

Elsebai MF, Kehraus S, Lindequist U, Sasse F, Shaaban S, Gütschow M, Josten M, Sahl H-G, König GM (2011) Antimicrobial phenalenone derivatives from the marine-derived fungus Coniothyrium cereale. Org Biomol Chem 9:802–808

Fell JW, Master IM (1973) Fungi associated with the degradation of mangrove (Rhizophora mangle L.) leaves in South Florida. Estuar Microb Ecol 455–465

Fell JW, Master IM (1975) Phycomycetes (Phytophthora spp. nov. and Pythium sp. nov.) associated with degrading mangrove (Rhizophora mangle) leaves. Can J Bot 53:2908–2922

Fell JW, Master IM (1980) The association and potential role of fungi in mangrove detrital systems. Bot Mar 23:257–263

Fenical W, Jensen PR (1993) Marine microorganisms: a new biomedical resource. In: Pharmaceutical and bioactive natural products, Springer, p 419–457

Gao S-S, Li X-M, Du F-Y, Li C-S, Proksch P, Wang B-G (2010) Secondary metabolites from a marine-derived endophytic fungus Penicillium chrysogenum QEN-24S. Mar Drugs 9:59–70

Gao S-S, Li X-M, Li C-S, Proksch P, Wang B-G (2011) Penicisteroids A and B, antifungal and cytotoxic polyoxygenated steroids from the marine alga-derived endophytic fungus Penicillium chrysogenum QEN-24S. Bioorg Med Chem Lett 21:2894–2897

Heald EJ, Odum WE (1970) The contribution of mangrove swamps to Florida fisheries

Höller U, Wright AD, Matthee GF, Konig GM, Draeger S, Aust H-J, Schulz B (2000) Fungi from marine sponges: diversity, biological activity and secondary metabolites. Mycol Res 104:1354–1365

Huang LH, Kaneko T (1996) Pyrenomycetes and Loculoascomycetes as sources of secondary metabolites. J Ind Microbiol 17:402–416

Hyde KD, Jones EBG, Leaño E, Pointing SB, Poonyth AD, Vrijmoed LLP (1998) Role of fungi in marine ecosystems. Biodivers Conserv 7:1147–1161

Hyde KD, Sarma VV, Jones EBG (2000) Morphology and taxonomy of higher marine fungi. Mar Mycol Pract Approach 172–204

Imhoff JF (2016) Natural products from marine fungi—still an underrepresented resource. Mar Drugs 14:19. https://doi.org/10.3390/md14010019

Imhoff JF, Labes A, Wiese J (2011) Bio-mining the microbial treasures of the ocean: new natural products. Biotechnol Adv 29:468–482

Jones EBG (2000) Marine fungi: some factors influencing biodiversity. Fungal Divers 4:53–73

Jones EBG, Alias SA (1997) Biodiversity of mangrove fungi. Biodivers Trop microfungi. Hong Kong Univ Press, Hong Kong, pp 71–92

Jones EBG, Sakayaroj J, Suetrong S, Somrithipol S, Pang KL (2009) Classification of marine Ascomycota, anamorphic taxa and Basidiomycota. Fungal Divers 35:187

Jones EBG, Suetrong S, Sakayaroj J, Bahkali AH, Abdel-Wahab MA, Boekhout T, Pang K-L (2015) Classification of marine Ascomycota, Basidiomycota, Blastocladiomycota and Chytridiomycota. Fungal Divers 73:1–72

Kakeya H, Onozawa C, Sato M, Arai K, Osada H (1997) Neuritogenic effect of epolactaene derivatives on human neuroblastoma cells which lack high-affinity nerve growth factor receptors. J Med Chem 40:391–394

Klaiklay S, Rukachaisirikul V, Tadpetch K, Sukpondma Y, Phongpaichit S, Buatong J, Sakayaroj J (2012) Chlorinated chromone and diphenyl ether derivatives from the mangrove-derived fungus Pestalotiopsis sp. PSU-MA69. Tetrahedron 68:2299–2305

Kohlmeyer J, Kohlmeyer E (1979) Submarine lichens and lichenlike associations. Mar mychology high fungi. Academic, New York, pp 70–78

Kuthubutheen AJ (1981) Fungi associated with the aerial parts of Malaysian mangrove plants. Mycopathologia 76:33–43

Lee Y-M, Li J, Zhang P, Hong J-K, Lee C-O, Jung J-H (2011) A cytotoxic fellutamide analogue from the sponge-derived fungus aspergillus versicolor. Bull Kor Chem Soc 32:3817–3820

Li H, Huang H, Shao C, Huang H, Jiang J, Zhu X, Liu Y, Liu L, Lu Y, Li M (2011) Cytotoxic norsesquiterpene peroxides from the endophytic fungus Talaromyces flavus isolated from the mangrove plant Sonneratia apetala. J Nat Prod 74:1230–1235

Li D, Xu Y, Shao C-L, Yang R-Y, Zheng C-J, Chen Y-Y, Fu X-M, Qian P-Y, She Z-G, de Voogd NJ (2012) Antibacterial bisabolane-type sesquiterpenoids from the sponge-derived fungus Aspergillus sp. Mar Drugs 10:234–241

Liberra K, Lindequist U (1995) Marine fungi – a prolific resource of biologically active natural products? Pharmazie 50:583–588

Malmstrøm J, Christophersen C, Barrero AF, Oltra JE, Justicia J, Rosales A (2002) Bioactive metabolites from a marine-derived strain of the fungus Emericella variecolor. J Nat Prod 65:364–367

Newell SY (1976) Mangrove fungi: the succession in the mycoflora of red mangrove (Rhizophora mangle L.) seedlings

Newell SY, Miller JD, Fell JW (1987) Rapid and pervasive occupation of fallen mangrove leaves by a marine zoosporic fungus. Appl Environ Microbiol 53:2464–2469

Odum WE, Heald EJ (1972) Trophic analyses of an estuarine mangrove community. Bull Mar Sci 22:671–738

Pang K-L, Overy DP, Jones EBG, Calado MDL, Burgaud G, Walker AK, Johnson JA, Kerr RG, Cha H-J, Bills GF (2016) 'Marine fungi' and 'marine derived fungi' in natural product chemistry research: toward a new consensual definition. Fungal Biol Rev 30:163–175

Paz Z, Komon-Zelazowska M, Druzhinina IS, Aveskamp MM, Shnaiderman A, Aluma Y, Carmeli S, Ilan M, Yarden O (2010) Diversity and potential antifungal properties of fungi associated with a Mediterranean sponge. Fungal Divers 42:17–26

Raghukumar C (2002) Bioremediation of coloured pollutants by terrestrial versus facultative marine fungi

Raghukumar S, Sharma S, Raghukumar C, Sathe-Pathak V, Chandramohan D (1994) Thraustochytrid and fungal component of marine detritus. IV. Laboratory studies on decomposition of leaves of the mangrove *Rhizophora apiculata* Blume. J Exp Mar Bio Ecol 183:113–131

Raghukumar S, Sathe-Pathak V, Sharma S, Raghukumar C (1995) Thraustochytrid and fungal component of marine detritus. III. Field studies on decomposition of leaves of the mangrove Rhizophora apiculata. Aquat Microb Ecol 9:117–125

Raghukumar C, Chandramohan D, Michel FC, Redd CA (1996) Degradation of lignin and decolorization of paper mill bleach plant effluent (BPE) by marine fungi. Biotechnol Lett 18:105–106

Raghukumar C, D'souza TM, Thorn RG, Reddy CA (1999) Lignin-modifying enzymes of Flavodon flavus, a basidiomycete isolated from a coastal marine environment. Appl Environ Microbiol 65:2103–2111

Raghukumar C, Mohandass C, Kamat S, Shailaja MS (2004) Simultaneous detoxification and decolorization of molasses spent wash by the immobilized white-rot fungus *Flavodon flavus* isolated from a marine habitat. Enzym Microb Technol 35:197–202

Rateb ME, Ebel R (2011) Secondary metabolites of fungi from marine habitats. Nat Prod Rep 28:290–344

Saleem M, Ali MS, Hussain S, Jabbar A, Ashraf M, Lee YS (2007) Marine natural products of fungal origin. Nat Prod Rep 24:1142–1152

Sarma VV (2007) Marine fungi and their biotechnological potential – an appraisal. In: Proceedings of the international conference on "Applied bioengineering" held on 5–7 December, 2007 at Sathyabama University, Chennai, India, pp 16–22

Sarma VV, Hyde KD (2001) A review on frequently occurring fungi in mangroves. Fungal Divers 8:1–34

Sarma VV, Vittal BPR (2000) Biodiversity of mangrove fungi on different substrata of *Rhizophora apiculata* and *Avicennia* spp. from Godavari and Krishna deltas, east coast of India. In: Hyde KD, Ho WH, Pointing SB (eds) Aquatic mycology across the millennium, vol 5. Fungal Diversity Press, Hong Kong, pp 23–41

Sarma VV, Vittal BPR (2001) Biodiversity of manglicolous fungi on selected plants in the Godavari and Krishna deltas, east coast of India. Fungal Divers 6:115–130

Singh N, Steinke TD (1992) Colonization of decomposing leaves of Bruguiera gymnorrhiza (Rhizophoraceae) by fungi, and in vitro cellulolytic activity of the isolates. South Afr J Bot 58:525–529

Steinke TD, Holland AJ, Singh Y (1993) Leaching losses during decomposition of mangrove leaf litter. South Afr J Bot 59:21–25

Sun L-L, Shao C-L, Chen J-F, Guo Z-Y, Fu X-M, Chen M, Chen Y-Y, Li R, de Voogd NJ, She Z-G (2012) New bisabolane sesquiterpenoids from a marine-derived fungus Aspergillus sp. isolated from the sponge Xestospongia testudinaria. Bioorg Med Chem Lett 22:1326–1329

Wiese J, Ohlendorf B, Blümel M, Schmaljohann R, Imhoff JF (2011) Phylogenetic identification of fungi isolated from the marine sponge Tethya aurantium and identification of their secondary metabolites. Mar Drugs 9:561–585

Xia X, Zhang J, Zhang Y, Wei F, Liu X, Jia A, Liu C, Li W, She Z, Lin Y (2012) Pimarane diterpenes from the fungus Epicoccum sp. HS-1 associated with Apostichopus japonicus. Bioorg Med Chem Lett 22:3017–3019

Yang G, Sandjo L, Yun K, Leutou AS, Kim G-D, Choi HD, Kang JS, Hong J, Son BW (2011) Flavusides A and B, antibacterial cerebrosides from the marine-derived fungus Aspergillus flavus. Chem Pharm Bull 59:1174–1177

Zhang Y, Mu J, Feng Y, Kang Y, Zhang J, Gu P-J, Wang Y, Ma L-F, Zhu Y-H (2009) Broad-spectrum antimicrobial epiphytic and endophytic fungi from marine organisms: isolation, bioassay and taxonomy. Mar Drugs 7:97–112

Zheng C-J, Shao C-L, Guo Z-Y, Chen J-F, Deng D-S, Yang K-L, Chen Y-Y, Fu X-M, She Z-G, Lin Y-C (2012) Bioactive hydroanthraquinones and anthraquinone dimers from a soft coral-derived Alternaria sp. fungus. J Nat Prod 75:189–197

Chapter 32
Fungi-Mediated Biodeterioration of Household Materials, Libraries, Cultural Heritage and Its Control

Bhupendra Koul and Hina Upadhyay

Abstract Fungi are cosmopolitan in distribution. The fungal deterioration of paper materials (books, manuscripts, journals and files), wood (household furniture, library and museum furniture), textiles (household and museum specimens) and cultural heritage (storage spaces of museums, monument walls, ceilings, statues and wall paintings) is a serious problem throughout the world. It has been observed that fungi of the class *Ascomycetes*, followed by *Deuteromycetes* and *Zygomycetes* are mostly responsible for such type of damages. Moreover, fungal spores are allergenic and may produce mycotoxins. They enter the human body through inhalation or dermal contact and may cause severe diseases such as air-tract infections, mycosis, asthma and immune system problems. Several physical and chemical methods and specimen treatment regimens, depending on the type and intensity of infection, have been suggested from time to time. The objective of this chapter is to create awareness on the aforementioned fungal biodeterioration phenomena and also to deal with precautionary and protective control measures to prolong the shelf-life of the household objects, museum specimens, historic monuments, objects and archaeological sites.

32.1 Introduction

Microbial deterioration and degradation of living and nonliving materials is an inevitable phenomenon observed in human society. Thus, biodeterioration is often defined as an undesirable change in the properties of materials caused by the activities of microorganisms which are the primary deteriogens (Allsopp 2011). Because of their biodeteriorative potential, the microorganisms, such as bacteria, archaea,

---

B. Koul
School of Bioengineering and Biosciences, Lovely Professional University, Phagwara, Punjab, India

H. Upadhyay (✉)
School of Agriculture, Lovely Professional University, Phagwara, Punjab, India
e-mail: hina.18745@lpu.co.in

P. Gehlot, J. Singh (eds.), *Fungi and their Role in Sustainable Development: Current Perspectives*, https://doi.org/10.1007/978-981-13-0393-7_32

fungi, lichens and insects, influence each and every aspect of human life, which covers public health issues, that is, change in water, food and feed, and also influence our cultural heritage (Sterflinger and Piñar 2013; Upadhyay and Jain 2014; Palla and Barresi 2017). Molecular techniques along with conventional methodologies are required for the studies on microflora inhabiting the cultural heritage sites (Tarsitani et al. 2014). Most of the industries including textile, paper, leather and pharmaceuticals are influenced by biodeterioration and biodegradation in one way or the other. Moreover, it has direct connection with our economy as it is related to the sustainable conservation of their quality (Agrawal and Dhawan 1985; Upadhyay 2001; Upadhyay and Jain 2005).

Among the microorganisms that cause biodeterioration, major damage is caused by fungi to household objects, archives, museum collections, books in libraries, wooden materials, textiles, leather objects, paintings, metal objects, statues and other irreplaceable objects of historical importance. Fungi (moulds and mildews) are found both in outdoor and indoor environments. They are simple, plant-like, achlorophyllous organisms which are heterotrophic, and they break down dead organic matter of the host (plant and animal tissues) by releasing enzymes from hyphal tips (saprophytic). Some fungi only stain the wood, while others have the potential to completely destroy the wood cellulose and lignin (wood crumbles to a powder). In a nutshell, fungi have a parasitic relationship with their host. They speedily grow in a congenial environment and disperse spores (sometimes airborne) to extend their range and effect.

32.2 Factors Influencing Biodeterioration

There are some factors which are interdependent and responsible for the deterioration of household, museum specimens, library materials and cultural heritage. These can be clustered under **[A]** abiotic factors, i.e. (I) *physical*: (i) temperature, (ii) humidity, (iii) water, (iv) dust and dirt, (II) *chemical*: (i) pH, (ii) pollution of atmosphere, (III) *physiochemical*: (i) light and (IV) *disasters*, and **[B]** biotic factors, i.e. (i) microorganisms, (ii) insects, (iii) rodents and (iv) anthropogenic factors. Controlling any factor can be used for preventing the growth of microorganisms.

32.2.1 Abiotic Factors

32.2.1.1 Physical Factors

32.2.1.1.1 Temperature

The majority of fungi can grow at temperature within the range of 10–40 °C, and a faster growth is observed at room temperature that is between 15 and 30 °C. Interestingly, some of the fungi are highly heat-resistant and can tolerate a temperature up to 50 °C. They are termed as thermophilic (Cooney and Emerson 1964). High temperature and low humidity cause cellulose fibres of the paper to dehydrate, due to which the paper becomes brittle and loses its flexibility and even tends to crumble on touch. On the other hand, high temperature and high humidity favour the growth of moulds. High-power electric bulbs too generate heat and increase the room temperature. Extreme variations in temperature ($\leq$5 °C in winter and $\geq$45 °C in summer) also affect the physical conditions of the libraries and, indirectly, biodeterioration.

32.2.1.1.2 Humidity

Humidity indicates the moisture amount of the atmosphere. It is essential for the development of mould, but it is the microclimate or relative humidity (RH) at the surface of the object rather than the ambient conditions in the storage or display areas that is the critical factor for fungal growth. This determines the equilibrium moisture content of the substrate, which in turn provides the necessary moisture to mould spores to germinate. Moisture is the major cause of deterioration (physical, chemical and biological) of household, library materials and cultural heritage. When there is high humidity, the papers absorb more moisture because of the absorbency property. Although a reasonable amount of humidity is essential for the flexibility of paper, in prolonged humid conditions, paper becomes soggy and weak. Furthermore, this weakens the adhesive which makes the book binding to become loose and causes sticking of pages, spreading of ink, paper-yellowing and staining. Some species of fungi can survive below 60% RH, but majority requires more than 70%. Moreover, light, temperature, oxygen and pH also influence the degree of fungal infestation.

32.2.1.1.3 Water

Water exists in all the three forms as solid, liquid and gas. It affects the hygroscopic substances to undergo dimensional changes. Water is damaging for stored household objects, libraries and cultural heritage sites. It comes from natural calamities, human negligence, leaky ceilings, faulty plumbing and ventilators and through open doors and windows during rains. The presence of water supports the growth of fungus or mildews and leads to staining of papers, smearing of ink and rottening of fabric and leathers and causes rusting of iron furniture.

32.2.1.1.4 Dust and Dirt

Dust consists of minute and dry particles of any matter. It is airborne and settles down on any surface of the object. These airborne particles of the dust include larger particles (size larger than 100 μm), medium-sized particles (particles in the range of 1–100 μm), and small particles (size less than 1 μm). Dust on the surface of household objects and ancient art objects has a heterogenous and variable composition. It usually contains chemical particles of diverse origin and nature, insect eggs, microorganism spores and flower pollens. During humid conditions (due to its hygroscopic nature), dust gets transformed into dirt which sticks to the surface of the books and is difficult to remove. Dust and dirt are responsible for both physical and chemical deterioration of the library collections. Dust provides a nutritive layer for the microorganisms, which can survive for a long time. It also acts as a nucleus for the moisture, and this moisture facilitates (provides humidity) the growth of fungus and its chemical action (formation of acids). Since dust and dirt are solid particles of variable size and hardness, they exert abrasion on the surface of the books. Microorganisms, including fungi and bacteria, are carried in the house, libraries and museums by the residents and visitors, respectively, which are diffused by breathing (bacteria) or by dust particles attached to the shoes. Their effect depends upon the receptive quality of the microenvironment and is certainly more harmful in wet conditions whether inside the house, libraries or museums.

32.2.1.2 Chemical Factors

In paper manufacturing, sometimes low cellulose-containing fibres and chemical compounds like alum, rosin, etc. are used for paper sizing. With the passage of time, these chemical compounds cause acidic effect which facilitate chemical deterioration of the papers.

32.2.1.2.1 pH

Fungi have variable pH requirements but normally grow best in acidic conditions (pH 5.5–6.0). Some fungi like *A. niger* grow in highly acidic environment (pH 2) and even below that (Smith and Onions 1983).

32.2.1.2.2 Pollution of the Atmosphere

The presence of oxides of carbon (CO_2), sulphur (SO_2), nitrogen (NO_2) and hydrogen sulphide (H_2S) and moisture affects the papers, library materials, fabrics and walls of the monuments. The oxides of nitrogen, sulphur and ozone majorly deteriorate the library materials. Among these, SO_2 is most damaging to cellulose fibres of paper and cloth. A common observation in libraries is the presence of brown and brittle edges of books, which is caused by SO_2. The second most abundant pollutant is NO_2 which comes from automobile exhausts and combines with oxygen (O_2) and water (H_2O) to form nitric acid (HNO_3). This strong acid affects the clothes, dyes in ink, papers and leather goods. In humid conditions, ozone damages organic materials and fades the colours of fabrics and book covers and also deteriorates the book-binding materials such as leather, gelatin and glue.

32.2.1.3 Physiochemical Factors

32.2.1.3.1 Light

The ultraviolet radiation of sunlight causes photochemical degradation of paper, in the presence of air (oxygen). Due to this, some portion of cellulose is oxidized to oxycellulose, and the long cellulose chains break down making the paper weak and brittle. The formation of oxycellulose is responsible for the fading of paper colour and paper-yellowing. Paper-yellowing is also caused by artificial lights (fluorescent tube light) which radiate a high amount of ultraviolet radiation. Altogether, the extent of photochemical damage is directly proportional to the intensity of light and duration of exposure and inversely proportional to the distance from the source of light.

32.2.1.3.2 Disasters

Disasters are unexpected and can be natural or man-made. In libraries, archives and museums, there are chances of fire as the materials are generally organic in nature. Items that do not directly come in contact with flames can be spoiled by soot and smoke. The heat of the fire causes the bindings to shrink and warp and melting of plastic base materials. Water, if used to extinguish fires, will cause massive damage, as it will cause the documents to absorb water, swell, warp and become enormously susceptible to physical damage. Dyes and ink may fade and book pages may stick together. Leather-bound books may extremely distort and change their shape. Besides fire, high-intensity winds, cyclones, floods and earthquakes are also agents of deterioration for the library collections as well as cultural heritage.

32.2.2 *Biotic*

32.2.2.1 Microorganisms

32.2.2.1.1 Fungi

Fungi are a large heterogeneous group of plant-like organisms. The fungal spores are prevalent in the soil, water and air and may remain in a dormant state for long periods. These spores germinate and grow during favourable moisture and temperature. Generally, the fungi grow at 63–100% relative humidity and 15–35 °C temperature. The fungi (mould or mildew) appear as brown or black vegetative growth on paper, wood, leather, textiles, etc. They consume cellulose of the wood and thrive on the nutrients present in leather, adhesives, binding threads, etc. and cause weakening, staining and discolouration of papers (Table 32.1).

Table 32.1 List of common biodeteriorating fungi

S. No.	Fungal species	Materials affected	Symptoms of biodeterioration
1.	*Aspergillus niger*, *A. glaucus*, *A. flavus*, *A. restrictus*, *Alternaria* sp., *Curvularia* sp., *Trichoderma* sp., *Fusarium* sp., *Cladosporium* sp., *Penicillium* sp.	Paper, cotton, textiles and herbarium specimens	Stains and discolouration: caused by surface growth of fungi on paper, leather, textile, etc. Disfigurement: changes in the structure of a painting due to fungus
2.	*A. flavus*, *A. niger*, *A. ochraceous*, *Fusarium* sp., *Mucor* sp. and dry rot fungi	Wooden objects	
3.	*A. flavus*, *A. niger*, *Cladosporium* sp.	Leather objects	Impenetrations: the invasion of the lumen of certain natural fibres by fungi Changes in mechanical properties: loss of weight and strength of wood or reduction of breaking strength of cotton fibres attacked by fungi Changes in chemical properties: breakdown of cellulose microorganisms, foxing on paper, etc. Development of odour: mildewing on paper

32.2.2.1.2 Bacteria

Besides fungi, different types of bacteria also decompose cellulose in paper and binding textiles.

32.2.2.2 Insects

There are several insects which badly damage the household objects, library materials and museums, but silverfish, cockroaches, booklice, bookworms and termites are the notorious ones.

32.2.2.2.1 Silverfish

Silverfishes are wingless and silvery or grey in colour and are about 8–10 mm in length. They prefer dust and dirt and are nocturnal. They eat the gum of postage stamps, envelopes, etc. and bore holes in photos, papers, catalogue cards and cardboard boxes. They lay eggs in the dark spaces on the catalogue cabinets, library racks and drawers.

32.2.2.2.2 Cockroaches

Cockroaches are brown or blackish-brown-coloured arthropods which are found worldwide. They feed on sewage, kitchen waste, paper leaves, fabrics and other organic materials. They are cursorial and nocturnal and are frequently found in libraries, museums, washrooms and kitchens. They inhabit the damp corners, wall and floor crevices, behind and beneath the almirahs and in wooden cupboards. They stain the papers with a dark brown fluid which they excrete.

32.2.2.2.3 Bookworms or Book Beetles

The bookworms are a menace to the library archives. They lay eggs on the edges of the books and on the book binding and finally make tunnels in it.

32.2.2.2.4 Book Lice

They are grey or white in colour and grow on dark, dusty, damp and warm areas. They damage the books by eating the binding glue. They also feed on the fungus which grows in between the edges of inner cover of the library books.

32.2.2.2.5 Termites or White Ants

Termites eat and damage the wood, papers and any type of cellulosic material. They grow under wet or damp conditions. In the tropical climate, the termites cause much damage to the library materials, museum objects and buildings. They may be located by the mud encrustation on the attacked materials. Although the termites inhabit the soil, their presence in the homes, libraries, furniture showrooms, etc. can be noticed by the mud tunnels on the walls, book cases and furniture, respectively. The wood-inhabiting termites live above the ground and enter the building through cracks.

32.2.2.3 Rodents

Rodents, like mice, rats and squirrels that are found in the homes, libraries and museums, enter through dry drains, open doors, windows, ventilators and chimneys. In libraries, they eat and damage materials made up of paper, cloth, leather, glue, etc. These animals are cursorial and hide in dark corners.

32.2.2.4 Anthropogenic Factors

Apart from the aforementioned factors, a serious cause of deterioration of food materials, library books and museum objects is our casual attitude. The librarians in charge of the documentary heritage are directly responsible for the overall maintenance and conservation of the collections. But sometimes they are not acquainted with handling, storing and careful maintenance of the collections so as to prevent damage and aid preservation. Inappropriate storage, defective repairment, rough handling, marking by ball pen, mutilation, reading books while having food, vandalism, etc. cause deterioration of books by human beings.

32.3 Fungi-Mediated Health Hazards and Biodeterioration of Indoor Objects

Some common fungi including *Aspergillus versicolor*, *Aspergillus nidulans*, *Alternaria alternata*, *Cladosporium cladosporioides*, *Eurotium* sp., *Exophiala* sp., *Penicillium funiculosum*, *Penicillium chrysogenum*, *Rhodotorula* sp., *Stachybotrys chartarum* and others are a serious threat to public health in the indoor environments (Samet and Spengler 2003; Khan 2009; Khan et al. 2009; Ayanbimpe et al. 2010; Khan and Karuppayil 2010, 2011) (Fig. 32.1). These deposit on humans, food items, pets, and household items and cause severe allergies (Fig. 32.2). The spores enter the body through inhalation and skin contact and cause several diseases such as air-tract infections, mycosis, immune system issues and asthma (Nielsen 2003). *Alternaria alternata*, *Penicillium* and *Cladosporium* species are known to cause asthma, while *Aspergillus* species cause aspergillosis (De Hoog et al. 2000; Salo et al. 2006). *Aspergillus* being a ubiquitous fungus can severely infect patients with weak immune system. The common fungi, which grow on eatables, even under refrigeration are *Penicillium*, *Aspergillus* and *Claviceps*. Fungal contamination of food takes place due to the production of mycotoxins that are the potent cause of diseases. Mycotoxins may cause severe and long-lasting illnesses, induce cancers and damage the vital organs such as the brain, liver and kidneys. Several fungi like *Fusaria*, *Trichothecium*, *Cephalosporium*, etc. contaminate the grains and cause headaches, dizziness, chills, vomiting, blurred vision and diarrhoea. The moulds which spread on nuts, corns, millets and figs produce aflatoxins. These toxins produce symptoms like jaundice and loss of appetite immediately, and with repeated exposure, they become carcinogenic. These fungi are known to produce a 'tremorgen' which cause 'staggers' in cattle and sheep. Ochratoxin A (mycotoxin), a secondary metabolite produced by *Aspergillus* and *Penicillium* species, contaminates foods and beverages such as coffee, wine and beer. It hinders protein synthesis by competition with the amino acid phenylalanine (its structural analogue) and also augments the production of oxygen free radicals. Its various toxic effects include cytotoxicity, carcinogenicity, genotoxicity, mutagenicity, teratogenicity and urinary

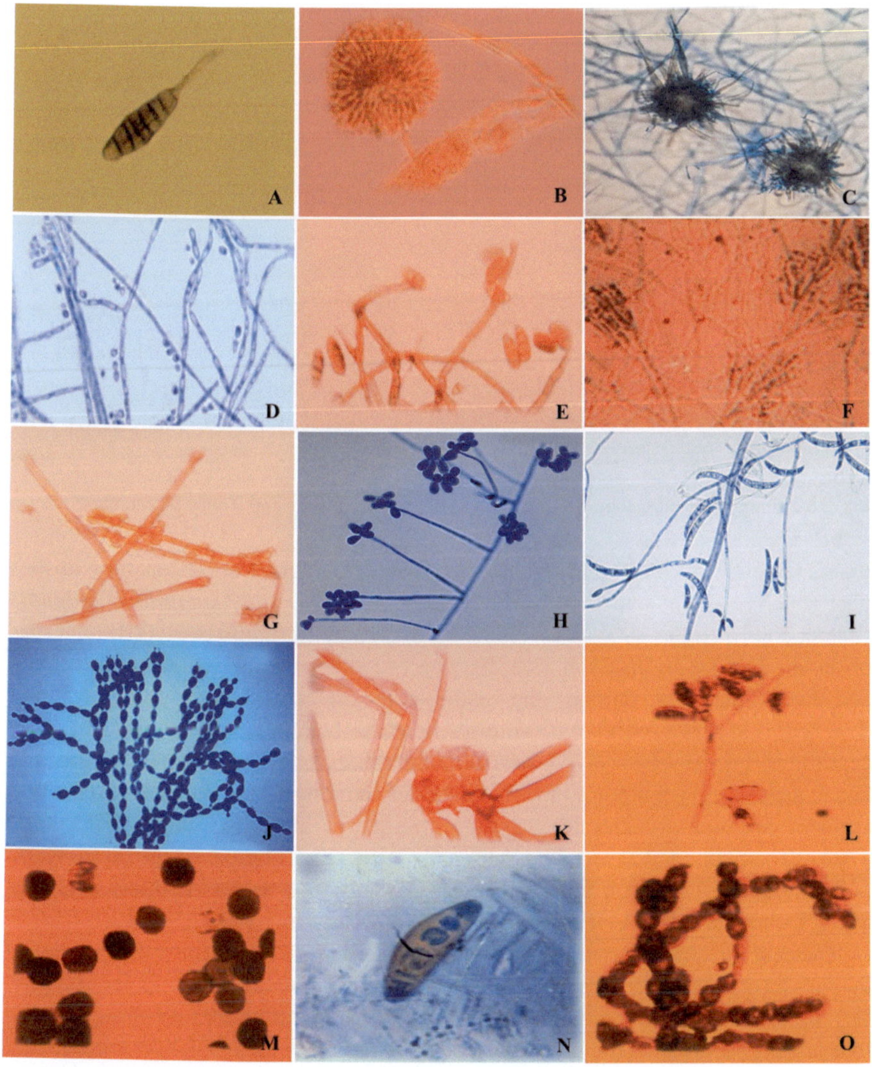

Fig. 32.1 Common fungi found in indoor environment. (**a**) *Alternaria* sp., (**b**) *Aspergillus* sp., (**c**) *Chaetomium* sp., (**d**) *Cladosporium* sp., (**e**) *Curvularia* sp., (**f**) *Penicillium* sp., (**g**) *Botrytis* sp. (**h**) *Trichothecium* sp., (**i**) *Fusarium* sp., (**j**) *Monilia* sp., (**k**) *Rhizopus* sp., (**l**) *Drechslera* sp., (**m**) *Nigrospora* sp., (**n**) *Helminthosporium* sp. and (**o**) *Torula* sp.

tract tumours. Ergot alkaloids (EA) are mycotoxins that are produced by several fungi which cause disease both in plants and animals. They act mostly on the nervous systems of animals.

Apart from food spoilage, the fungi do affect the household materials. In moist conditions, fungi develop on both natural and synthetic materials (Khan and Karuppayil 2012). They also colonize the air filters and ventilation ducts of cooling

Fig. 32.2 Fungi-mediated damage to [frame **a**] household objects and [frame **b**] eatables

plants (Noris et al. 2011). Inorganic materials often get colonized as they absorb dust which serves as nutrition for *Aspergillus versicolor* and *Aspergillus fumigatus* (Samet and Spengler 2003). In a study conducted by Ebrahimi et al. (2011), it was observed that *Aspergillus* spp. and *Zygomycota* affect the fabrics, *Aspergillus* spp. and *Penicillium* spp. damage the leather-made objects, while *Zygomycota* and *Aspergillus* spp. are found to be dominant on wooden objects. Wood is highly susceptible to fungal attack. *Cladosporium* and *Penicillium* spp. (*P. brevicompactum* and *P. expansum*) are reported to infest wooden building materials. Kiln-dried wood surfaces are more vulnerable to fungi (Sailer et al. 2010). *Aspergillus* spp., *Trichoderma* spp. and *Penicillium* spp. infest acylated wood-based furniture, wood polyethylene composites, plywood and other modified wood products (Thacker 2004; Doherty et al. 2011).

The prefabricated gypsum board containing gypsum as the inner wall material favours the growth of *Stachybotrys chartarum* (Breum et al. 1999). Paper and glue that are used to decorate the indoor surfaces are excellent growth substrates for the indoor fungi. Polyurethane coatings used in composites for insulation are infested by *Penicillium* spp., *Trichoderma harzianum* and *Paecilomyces variotii* (Yazicioglu et al. 2004). *A. versicolor*, *Cladosporium* and *Penicillium* species grow on fibre glass insulation and tiles (Erkara et al. 2008). *Aspergillus* and *Penicillium* grow superficially on painted walls, but *Aureobasidium pullulans* deteriorates the paint (O'Neill 1988; Shirakawa et al. 2002; Lugauskas et al. 2003). *Alternaria* spp., *Cladosporium* spp. and *Aspergillus* spp. also attack the acrylic painted surfaces (Shirakawa et al. 2011).

32.4 Biodeterioration in Libraries, Museums and Cultural Heritage

Fungi-mediated biodeterioration is commonly observed in public libraries, museum collections and monuments (Figs. 32.3 and 32.4). The infections are mostly airborne and are subjected to seasonal variations (Kaarakainen et al. 2009). Poor ventilation in indoor environment leads to a humid microenvironment that is congenial for some fungal species. A high relative humidity (>70%), temperature (>15 °C), a neutral to acid pH and the presence of organic nutritive sources are the favourable conditions for fast growth and reproduction of moulds, which attack museum objects (Gorbushina et al. 2004). These fungi present in storage sites of museums may be highly allergic to museum professionals and users (Milanesi et al. 2006). It has been observed that fungi affecting the library books mostly belong to the class *Ascomycetes* (*Monilia* sp., *Penicillium* sp., *Trichothecium* sp., *Aspergillus flavus*, *Aspergillus fumigatus*, *Aspergillus niger*, *Chaetomium* sp., *Chrysosporium* sp., *Cladosporium* sp. and *Paecilomyces* sp.) followed by *Deuteromycetes* (*Alternaria* sp., *Cladosporium herbarum*, *Curvularia lunata*, *Fusarium* sp.) and *Zygomycetes* (*Rhizopus nigricans*) (Upadhyay 2005; Upadhyay and Jain 2005; Pinzari and Montanari 2011; Upadhyay and Jain 2012). Table 32.2 describes the effect of library-associated fugal species on different types of papers.

Both the fungi and bacteria are involved in 'foxing' (isolated, rusty brownish-red spots) phenomenon of papers (De Paolis and Lippi 2008; Michaelsen et al. 2009, 2010). Proteolytic fungi of the class *Ascomycetes* (*Gymnoascus* and *Chaetomium*) as well as mitosporic fungi (*Aspergillus*, *Aureobasidium*, *Acremonium*, *Epicoccum*, *Verticillium* and *Trichoderma*) damage the historic objects and documents made from parchment (Strzelczyk 2004). Fungi-mediated damage in libraries and museum

Fig. 32.3 Fungi-mediated deterioration of library books

Fig. 32.4 Fungi-mediated deterioration of cultural heritage

Table 32.2 Growth of fungi on four different kinds of paper

S. No.	Fungal species	Humid conditions				
		Fungal class	Newspaper	Book paper	Filter paper	Glossy paper
1.	*Aspergillus flavus*	*Ascomycetes*	++++	++++	+++	+
2.	*Aspergillus fumigatus*		+++	+++	+++	−
3.	*Aspergillus niger*		++++	++++	++++	+
4.	*Monilia* sp.		−	+	−	−
5.	*Penicillium* sp.		++++	+++	+	++
6.	*Trichothecium* sp.		+++	++	+++	−
7.	*Chaetomium* sp.		+++	++++	+++	+
8.	*Alternaria* sp.	*Deuteromycetes*	+	+++	+	−
9.	*Cladosporium herbarum*		++	++	++	+
10.	*Curvularia lunata*		+++	++++	+++	+
11.	*Fusarium* sp.		−	+	−	−
12.	*Rhizopus nigricans*	*Zygomycetes*	+	+	+	−

− No growth
\+ Slight growth
++ Moderate growth
+++ Abundant
++++ Highly abundant growth

paintings may occur because of mechanical stress, production of staining compounds and enzymatic action (Santos et al. 2009; Sterflinger 2010; López-Miras et al. 2013). The filamentous fungi that damage papers and oil paintings on canvas can dissolve cellulose fibres (through cellulolytic enzymes), dissolve adhesives and inks or damage the oil binders. Several fungi such as *Aspergillus* spp., *Alternaria* spp., *Cladosporium* spp., *Cephalosporium* spp., *Fusarium oxysporum*, *Curvularia lunata*, *Chaetomium* spp., *Mycelia sterilia*, *Penicillium Stachybotrys*, *Phoma* spp., *Mucor* spp., *Rhizopus* spp., *Trichoderma* spp. and *Verticillium* spp. have been reported to degrade the paper paintings (Shrivastava 2015). Fungal species belonging to the *Eurotium* genera (such as *Eurotium halophilicum*) are observed as monospecific infections inside compact shelves (Montanari et al. 2012). During excessive moisture in the libraries, moulds that are adapted to the wet conditions produce toxic compounds (*Stachybotrys* spp.), strong odours (*Trichoderma* spp.) and coloured stains (*Chaetomium* spp., *Monoascus* spp. and *Epicoccum* spp.).

Fungi are extremely erosive to stone objects, buildings and monuments (Sterflinger 2000; Scheerer et al. 2009). These fungi form coloured stains, patinas and biofilms on the surfaces of the stone monuments due to the production of organic acids such as oxalic, citric, acetic, fumaric, succinic and gluconic (Farrooq et al. 2015). The rock art caves of Lascaux (France) are a live example of the historical objects suffering from serious fungal invasions (Bastian and Alabouvette 2009). Black fungi have been reported to penetrate inside the marble, limestone and antique glasses (Piñar et al. 2013). This phenomenon is known as bio-pitting. They have also been reported in rock surfaces of caves and catacombs (Saarela et al. 2004). Thus, the stone-inhabiting fungi can be grouped into two categories: (a) found in moderate or humid climates and (b) found in arid and semiarid environments. There are two major groups of stone-inhabiting and stone-dwelling fungi. In the category (a), the fungal communities on rock are dominated by hyphomycetes (*Cladosporium*, *Epicoccum*, *Alternaria*, *Phoma* and *Aureobasidium*) that form hyphal networks (mycelia) in the porous spaces of the stones (Sterflinger and Prillinger 2001; Rosling et al. 2009). In the category (b), rocks are dominated by black fungi of the genera *Sarcinomyces*, *Hortaea*, *Exophiala*, *Coniosporium*, *Trimmatostroma*, *Capnobotryella* and *Knufia* which form small black colonies on and inside the stones, in association with lichens (Sterflinger 2005).

32.5 Strategies to Prevent/Control Fungal Growth

The treatment of libraries and museum objects against biodeterioration follows two general courses of action:

(a) *Preventive measures* – includes all indirect approaches that aim at increasing the life expectancy of undamaged or damaged components of the cultural property. It includes all the good housekeeping practices, dusting, caretaking, periodical supervision and prevention of damage by physical, chemical and biological factors.

(b) *Control measures* – consists of all forms of direct actions which aim at increasing the life expectancy of damaged or undamaged materials of cultural property. It includes fumigation, deacidification, repairing, mending, lamination and other works depending on physical condition of the individual material. The general measures are as follows:

1. The libraries and museum should be built on well-examined area after studying its soil microflora.
2. The interior design for the houses, libraries and museums should have adequate number of ceiling fans and exhaust fans to facilitate proper air circulation.
3. Wood to be used for furniture should be well seasoned and chemically treated to avoid insect pests.
4. Plantation should not be close to the library or museum buildings.
5. Such buildings must be constructed away from the traffic site to get rid of the dust.
6. The window panes should have lemon yellow-, green-coloured glass panes or acrylic plastic sheets as they filter out UV rays to a greater extent.
7. An ideal room temperature (20–25 °C) and relative humidity (45–55%) are essential for preservation of books and other records.
8. Air conditioning of the libraries maintains an optimal temperature and humidity for the storage of documents.
9. If air conditioning is not possible, then use humidifiers in dry climate and dehumidifiers in wet seasons. High humidity can also be maintained (minimized) by the use of silica gel (dehydrating agents).
10. Weekly cleaning of library bookshelves and remote corners is recommended.
11. Important books and journals should be kept in specially designed cupboards.
12. The visitors and regular readers should avoid licking of fingers to turn the book pages.
13. Books should not be kept open on the reading table, with face downwards, and must be stored in closed almirahs and not on open racks.
14. Of all the biotic and abiotic factors, humidity, temperature and food are the three most important biological factors which influence the fungal pollution.

32.5.1 Chemical Treatment

Most of the materials require some kind of chemical protection (fumigation) against biological agencies of deterioration.

In tropical countries where high relative humidity continues for a long period of the year, the use of fungicides becomes a necessity for the prevention of moulds. At the same time, it should be considered that the fungicide does not volatilize too readily and must not stain the object. The growth of fungus may be arrested by cleaning the fungicide. Hueck Vander Plas (1966) and Allsopp and Allsopp (1983) have proposed a list of biocides for various types of material, but for the application to art object, only few are recommended as they need certain characteristics which will not tarnish the aesthetic beauty of the object.

In countries with high humidity conditions, the new acquisitions for large museums are fumigated before its display. One of the safest fumigants is thymol (Plenderlieth and Werner 1972; Landi 1984). The new object must be segregated immediately if suspected with mould or insect infestation. The closed spaces around the objects in drawers, in showcases or in the polythene bags should be constantly fumigated with repellents such as naphthalene, paradichlorobenzene or vapona strips (Hueck 1972). Ethylene oxide and propylene oxide are used for both bacteria and fungi (Perkin 1978). For non-proteinaceous materials, formalin is very effective. The items can be fumigated in a specialized fumigation chamber with a mixture of 40% ethylene oxide and 60% carbon dioxide (wt/vol.). Among the gases, hydrocyanic acid (HCN) is also widely used for large-scale fumigation, especially for wooden objects. One percent solution of polyhexamethylene guanidine in water has been proven safe for paper. It is a polymer compound more stable than monomer. When heated for a short time or exposed to sunlight, it does not change its properties. It has pH 8–10 and very stable in storage. For leather objects 2% solution of pentachlorophenol is recommended (Werner 1980). Sodium salt of orthophenylphenol and pentachlorophenol laureate can be used as solutions in alcohol. A 0.2% of trinitrophenol completely inhibited the fungi. Sharma and Sharma (1980) observed that trinitrophenol was antifungal at 0.01% concentration and exhibited complete inhibition of fungi at 0.2%. ß-naphthol and ß-hydroxynaphthaldehyde were also highly effective at 0.01% concentration. A 2% solution of sodium pentachlorophenol in ethyl alcohol is very effective for wood and textiles (Mori and Arai 1977; Walston 1984). The use of quaternary ammonium compounds (QAC) has an advantage because their solutions are colourless, they do not interact with fibre and they are soluble in water and organic solvents. They are highly active bactericide and fungicide, and the range of their activity is very wide (Zaitseva 1978). Kowalik (1980) suggested some ideal preservatives for the protection of valuable wooden materials: zinc chloride ($ZnCl_2$), ammonium fluoride $(NH_4)_2\ F_2$ 4% aq., sodium fluoride (NaF) 4% aq., ammonium fluoride $(NH_4)_2F_2$ 0.07 2%, zinc silicofluoride 10%, magnesium silicofluoride ($Mg\ Si\ F_4$) 10%, sodium silicofluoride ($Na_2\ SiF_4$) 10%, ammonium silicofluoride 10% and borax and boric acid 1–2%. Mori and Arai (1977) recommended 4–13 g/m^2 of paraformaldehyde as fungicide for wooden objects. The antifungal substances like polyhexamethylene guanidine have been proven safe for paper and may be used for restoration of library materials of any value (Nysukha 1980). Lollar (1956) suggested the use of *P*-nitrophenol (between 0.2% and 0.4%) for leather objects. Sequeira et al. (2017) has reported the use of calcium propionate and parabens for the complete protection of paper materials from fungal growth. Interestinly, out of five fungal species tested (*Cladosporium cladosporioides*, *Aspergillus niger*, *Penicillium Chrysogenum*, *Penicillium cory-*

lophilum and *Chaetomium globosum*), only *P. corylophilum* was able to grow little more than the untreated controls.

32.5.2 Use of Fungicidal Paper

When the materials are delicate, fungicidal paper can be used for wrapping and keeping it in between the layers of costumes or paper objects. Fungicidal paper is prepared by dipping white blotting sheets in 10% EtOH solution of thymol or ortho-phenylphenol. Object which is to be disinfected should be placed between the disinfected saturated sheets and then sealed in polythene bags. The duration of sterilization may vary from 3 to 7 days.

Nowadays, neem leaves are used in protecting paper and textile material from fungi and insects. It contains quercetin which is known to have antibacterial and antifungal action (Prasad 1981–1983; Dutta 1985–1987). Aranyanak (1988) and Zhong (1988) also suggested the use of pepper (*Piper nigrum*) as an insecticide for paper material.

32.6 Conclusions

It is impossible to eradicate fungi, since they are abundant. However, we can effectively control their presence in household objects, museum collections, libraries and monuments. If the relative humidity is kept moderate (>60%), fungi-mediated deterioration may be prevented. Sufficient light, ventilation and regular cleaning of rooms also do not allow any fungal types to grow. Maintenance of museums and libraries must be done by trained staff. The use of dehumidifiers from time to time can also be helpful in preventing the growth of fungi. Ethnographical collections contain a large number of objects made of organic materials. The method for treating inorganic objects of metal, ceramics and stone is very much similar to those used for the conservation of archaeological objects in general. But, in the field of biodeterioration of organic objects, very little work has been done and requires further research, as some of them are very delicate and must be handled very carefully. They require a very systematic and scientific study because fungi and insects are a menace to every ethnographical museum in the world.

References

Agrawal OP, Dhawan S (1985) Technical note −2 Control of biodeterioration in museums (NRLC)

Allsopp D (2011) Worldwide wastage: the economics of biodeterioration. Microbiol Tod 38:150–153

Allsopp C, Allsopp D (1983) An updated survey of commercial products used to protect materials against biodeterioration. Int Bio Bull 19(3):99–146
Aranyanak C (1988) Conservation of ancient Thai books, conservation of far eastern art, II C ed. Mills, Smith and Yamasaki: 22–29
Ayanbimpe GM, Wapwera SD, Kuchin D (2010) Indoor air mycoflora of residential dwellings in Jos metropolis. Afr Health Sci 10:172–176
Bastian F, Alabouvette C (2009) Lights and shadows on the conservation of a rock art cave: the case of Lascaux cave. Int J Speleol 38:55–60
Breum NO, Nielsen BH, Lyngbye M, Midtgard U (1999) Dustiness of chopped straw as affected by lignosulfonate as a dust suppressant. Ann Agric Environ Med 6:133–140
Cooney DG, Emerson R (1964) Thermophilic fungi: an account of their biology, activities, and classification. W. H. Freeman & Co., San Francisco/London, p Xii+188
De Hoog GS, Guarro J, Figueras MJ, et al (2000) Atlas of clinical fungi, 2nd edn. Centraalbureauvoor Schimmelcultures/Universitat Rovirai Virgili, Utrecht/Reus, 1124
De Paolis MR, Lippi D (2008) Use of metabolic and molecular methods for the identification of a bacillus strain isolated from paper affected by foxing. Microbiol Res 16:121–131
Doherty WOS, Mousavioun P, Fellows CM (2011) Valueadding to cellulosic ethanol: lignin polymers. Ind Crop Prod 33(2):259–276
Dutta PK (1987) Use of neem oil in conservation. Conserv Cult Prop India 18-20:98–100
Ebrahimi A, Karimi S, Lotfalian S, Majidi F (2011) Allergenic fungi in deteriorating historic objects of Shahrekord museum, in Iran. JJM 4(4):261–265
Erkara IP, Asan A, Yilmaz V, Pehlivan S, Okten SS (2008) Airborne Alternaria and Cladosporium species and relationship with meteorological conditions in Eskisehir City, Turkey. Environ Monit Assess 144(1–3):31–41
Farooq M, Hassan M, Gull F (2015) Mycobial deterioration of stone monuments of Dharmarajika, Taxila. J Microbiol Experimentation 2:36. https://doi.org/10.15406/jmen.02.00036
Gorbushina AA, Heyrman J, Dornieden T et al (2004) Bacterial and fungal diversity and biodeterioration problems in mural painting environments of St. Martins church (Greene-Kreiensen, Germany). Int Biodeter Biodegr 53:13–24
Hueck HJ (1972) Textile pests and their control. In: Leene JE (ed) Textile conservation (Butterworths):76–77
Hueck Vander Plas E (1966) Survey of commercial products used to protect materials against biological deterioration. Int Biod Bull 2(2):69–120
Kaarakainen P, Rintala H, Vesalainen A, Hyvarinen A, Nevalainen A, Meklin T (2009) Microbial content of house dust samples determined with qPCR. Sci Total Environ 407:4673–4680
Khan AAH (2009) Studies on indoor fungi and their control. Thesis, Swami Ramanand Teerth Marathwada University, Nanded
Khan AAH, Karuppayil SM (2010) Potential natural disinfectants for indoor environments. Int J Clin Aromather 7:1–5
Khan AAH, Karuppayil SM (2011) Practices contributing to biotic pollution in airconditioned indoor environments. Aerobiologia 27:85–89
Khan AAH, Karuppayil S (2012) Fungal pollution of indoor environments and its management. Saudi J Biol Sci 19(4):405–426
Khan AAH, Karuppayil SM, Chary M, Kunwar IK, Waghray S (2009) Isolation, identification and testing of allergenicity of fungi from air-conditioned indoor environments. Aerobiologia 25:119–123
Kowlik R (1980) Microbiodeterioration of library materials part-2. Microbiodecomposition of basic organic library materials. 4:135–219
Landi S (1984) The textile conservators manual (Butterworth): 200
Lollar RM (1956) The chemistry of liming in O'Flaherty. In: Reddy F, Lollar RM (eds) The chemistry and technology of leather, vol 1. Reinhold Publishing Co, New York, pp 297–322
López-Miras, Sanchez M, Rodrıguez, Noguera J, Galiano F, Ettenauer J, Sterflinger K, Pinar G (2013) Contribution of the microbial communities detected on an oil painting on canvas to its biodeterioration. PLoS 8(11):1–13

Lugauskas A, Levinskaite L, Peciulyte D (2003) Micromycetes as deterioration agents of polymeric materials. Int Biodeterior Biodegrad 52(4):233–242

Michaelsen A, Piñar G, Montanari M, Pinzari F (2009) Biodeterioration and restoration of a 16th-century book using a combination of conventional and molecular techniques: a case-study. Int Biodeterior Biodegrad 63:161–168

Michaelsen A, Piñar G, Pinzari F (2010) Molecular and microscopical investigation of the microflora inhabiting a deteriorated Italian manuscript dated from the 13th-century. Microb Ecol 60:69–80

Milanesi C, Baldi F, Vignani R (2006) Fungal deterioration of medieval wall fresco determined by analyzing small fragments containing copper. Int Biodeterior Biodegrad 57:7–13

Montanari M, Melloni V, Pinzari F, Innocenti G (2012) Fungal biodeterioration of historical library materials stored in compactus movable shelves. Int Biodeterior Biodegrad 75:83–88

Mori H, Arai H (1977) Proceedings of Ist, ISCRP Japan:1–16

Nielsen KF (2003) Mycotoxin production by indoor molds. Fungal Genet Biol 39(2):103–117

Noris F, Siegel JA, Kinney KA (2011) Evaluation of HVAC filters as a sampling mechanism for indoor microbial communities. Atmos Environ 45(2):338–346

Nysukha IP (1980) Some special cases of biological deterioration of books. 5(1–4): 171–182

O'Neill TB (1988) Succession and interrelationships of microorganisms on painted surfaces. Int Biodeterior 24(4–5):373–379

Palla F, Barresi G (2017) Biotechnology and conservation of cultural heritage. Springer, Switzerland, pp 1–67

Perkin JJ (1978) Principles and methods of sterilization in health sciences, 2nd edn. Charls C. Thomas Publ: 57

Piñar G, Garcia-Valles M, Gimeno-Torrente D, Fernandez-Turiel JL, Ettenauer J, Sterflinger K (2013) Microscopic, chemical, and molecular-biological investigation of the decayed medieval residential dwellings in Jos. Niger Ann Nat Sci 6(2):52–58

Pinzari F, Montanari M (2011) Mould growth on library materials stored in compactus-type shelving units. In: Abdul-Wahab Al-Sulaiman SA (ed) Sick building syndrome in public buildings and workplaces. Elsevier, Burlington

Plenderleileith HJ, Warner AE (1972) The conservation of antiquities and works of art, 2nd edn. Oxford University Press, London

Prasad LK (1981–83) Role of neem leaves in protecting textile materials and paper documents, conservation of cultural property in India, 14–15: 90–94

Rosling A, Finlay RD, Gadd GM (2009) Geomycology. Fungal Biol Rev 23:91–93

Saarela M, Alakomi HL, Suihko ML, Maunuksela L, Raaska L, Mattila-Sandholm T (2004) Heterotrophic microorganisms in air and biofilm samples from roman catacombs with special emphasis on actinobacteria and fungi. Int Biodeterior Biodegrad 54:27–37

Sailer MF, Van Nieuwenhuijzen EJ, Knol W (2010) Forming of a functional biofilm on wood surfaces. Ecol Eng 36(2):163–167

Salo PM, Arbes SJ Jr, Sever M et al (2006) Exposure to Alternaria alternata in US homes is associated with asthma symptoms. J Allergy Clin Immun 118:892–898

Samet JM, Spengler JD (2003) Indoor environments and health: moving into the 21st century. Am J Public Health 93(9):1489–1493

Santos A, Cerrada A, García S, San Andrés M, Abrusci C, Marquina D (2009) Application of molecular techniques to the elucidation of the microbial community structure of antique paintings. Microb Ecol 58:692–702

Scheerer S, Ortega-Morales O, Galarde C (2009) Microbial deterioration of stone monuments – an updated overview. Adv Microbiol 66:97–139

Sequeira SO, Phillips AJL, Cabrita EJ, Macedo MF (2017) Antifungal treatment of paper with calcium propionate and parabens: short-term and longterm effects. Int Biodeterior Biodegrad 120:203–215

Sharma OP, Sharma KD (1980) Changes in properties of finished Indian leather in store due to fungal infestation. Int Biod Bull 16(3):73–78

Shirakawa MA, Gaylarde CC, Gaylarde PM, John V, Gambale W (2002) Fungal colonization and succession on newly painted buildings and the effect of biocide. FEMS Microbiol Ecol 39(2):165–173

Shirakawa MA, Loh K, John VM, Silva MES, Gaylarde CC (2011) Biodeterioration of painted mortar surfaces in tropical urban and coastal situations: comparison of four paint formulations. Int Biodeterior Biodegrad 65(5):669–674

Shrivastava S (2015) Biodeterioration of art objects on paper and their conservation. Res J Recent Sci 4:42–46

Sterflinger K (2000) Fungi as geologic agents. Geomicrobiol J 17:97–124

Sterflinger K (2005) Black yeasts and meristematic fungi: ecology, diversity and identification. In: Rosa C, Gabor P (eds) Yeast handbook, Biodiversity and Ecophysiology of Yeasts, vol 1. Springer, New York, pp 501–514

Sterflinger K (2010) Fungi: their role in deterioration of cultural heritage. Fungal Biol Rev 24:47–55

Sterflinger K, Piñar G (2013) Microbial deterioration of cultural heritage and works of art. Appl Microbiol Biotechnol 97:9637–9646

Sterflinger K, Prillinger H (2001) Molecular taxonomy and biodiversity of rock fungal communities in an urban environment (Vienna, Austria). Antonie Van Leeuwenhoek 80:275–286

Strzelczyk AB (2004) Observations on aesthetic and structural changes induced in polish historic objects by microorganisms. Int Biodeterior Biodegrad 53:15–16

Tarsitani G, Moroni C, Cappitelli F, Pasquariello G, Maggi O (2014) Microbiological analysis of surfaces of Leonardo Da Vinci's Atlantic codex: biodeterioration risk. Int J Microbiol 2014: 1–7. Article ID 214364

Thacker PD (2004) Airborne mycotoxins discovered in moldy buildings. Environ Sci Technol 38(15):282A

Upadhyay H (2001) Biodeteriorating components with special reference to fungi inside the library environment at Gwalior (MP). Thesis, Jiwaji University Gwalior

Upadhyay H, Jain AK (2005) Fungal flora inside the library environment of Jiwaji University Gwalior. Indian J Aerobiol 18(1):6–11

Upadhyay H, Jain AK (2012) Isolation and identification of paper deteriorating fungi from book samples of Gwalior University library. Biosci Biotech Res Commun 5(2):162–167

Upadhyay H, Jain AK (2014) Fungal deterioration of books in libraries. In: Dhavan S, Virendra Nath AK (eds) Biodeterioration of cultural property-7, International council for biodeterioration of cultural property. Sukriti Nikunj, pp 169–178

Walston S (1984) Techniques for minimum environmental damage to wooden objects moved from Papua New Guinea to Australia with particular reference to seasoning,7th ICOM Trienniel meeting copenhagan. 2:56

Werner AE (1980) Insect infestation and its control the conservation of cultural materials in Humid climate, UNESCO

Yazicioglu M, Asan A, Ones U, Vatansever U, Sen B, Ture M, Bostancioglu M, Pala O (2004) Indoor airborne fungal spores and home characteristics in asthmatic children from Edirne region of Turkey. Allergol Immunopathol 32(4):197–203

Zaitseva GA (1978) Dermestidae beetles injurious to museum objects and protection against them ICOM 5th Triennel meeting 2: 1–8

Zhong ZB (1988) The preservation of ancient Chinese paper. The conservation of Far Eastern Arted. Mills, Smith and Yamasaki:19–21

Chapter 33
Co-cultivation Strategies to Induce De Novo Synthesis of Novel Chemical Scaffolds from Cryptic Secondary Metabolite Gene Clusters

Dharmesh Harwani, Jyotsna Begani, and Jyoti Lakhani

Abstract Recent developments in genomic mining have revealed that many microbes possess the huge potential to produce structurally diverse secondary metabolites of industrial and therapeutic use. Many microbial gene clusters have been identified to be cryptic or orphan or silent under standard laboratory growth conditions. In the last few years, several tools and techniques have been developed to trigger/induce these cryptic/silent/orphan biosynthetic pathways. In addition to the tools that require prior understanding of the gene sequences of the studied microbes, several other tools and techniques have been designed, independent to genome or genome sequences. One of the approaches is a "co-culturing" which involves the cultivation of two (or more) microbes (interspecies interactions) in the same closed and restricted environment. Microbial co-cultures can be seen as the natural communities where they interact with each other through cell to cell contact or signaling molecules. Such interactions lead to the induction of valuable chemical scaffolds (secondary metabolites/natural products) which may escort to the novel drug discovery. The present chapter emphasizes on the co-culture strategies involving fungus-fungus and fungus-bacterium interactions that lead to the production of diverse metabolite structures from otherwise cryptic/silent/orphan biosynthetic gene clusters. The various strategies, highlighting variable methods of performing intra- and interspecies co-culture experiments and the biological activities as well as diversity of thus produced metabolites, are presented.

Keywords Fungi · Co-culture · Bacteria · Interspecies · Secondary metabolites · Cryptic genes · Genetic clusters

D. Harwani (✉) · J. Begani
Department of Microbiology, Maharaja Ganga Singh University, Bikaner, Rajasthan, India
e-mail: dharmesh@mgsubikaner.ac.in

J. Lakhani
Department of Computer Science, Maharaja Ganga Singh University, Bikaner, Rajasthan, India

P. Gehlot, J. Singh (eds.), *Fungi and their Role in Sustainable Development: Current Perspectives*, https://doi.org/10.1007/978-981-13-0393-7_33

33.1 Natural Product Discovery Using Co-cultures

Nature provides an important reservoir of microorganisms that produce potentially beneficial and structurally diverse compounds, many of which are yet to be explored (Zhu et al. 2011). Microorganisms have been proved to produce many metabolites of therapeutic use (Cragg and Newman 2013), and approximately, more than 170,000 natural products have been identified (Seyedsayamdost and Clardy 2014; Chapman and Hall 2015). The best examples include β-lactam, penicillin, macrolides such as antibiotics (erythromycin), immunosuppressive drugs (cyclosporin), antifungals (amphotericin B), depsipeptides (fusafungin), the cholesterol-lowering (lovastatin), etc. Drugs approved by the US Food and Drug Administration (FDA) relating to antitumor compounds contain 33% of natural product which are known to be derivatives of bacterial and fungal metabolites (Giddings and Newman 2013). Thus, bacteria and fungi are considered to be valuable sources of drugs and other therapeutic compounds (Pearce et al. 2009; Berdy 2012).

Natural conditions where microbes produce secondary metabolites can be mimicked artificially in laboratory in vitro by culturing two or three different microorganisms (interspecies and intraspecies) (De Roy et al. 2013). Co-culture or mixed fermentation constitutes an artificial community to study secondary metabolite induction. The idea of co-culture to find new scaffolds of chemical structure has increased greatly nowadays. Nonetheless, comparing methods which are completely based on gene-specific expression for bioprospecting novel chemical scaffolds, co-culture technique seems to be more explicit. Depending upon the interaction, secondary metabolites may result from the activation of various genes functioning in completely unrelated pathways. This makes co-culture method a different approach from other molecular methods in which predominant polyketide synthase and non-ribosomal peptide synthetase gene cluster pathways are generally observed (Brakhage and Schroeckh 2011).

33.2 Microbial Community Dynamics

The interactions between bacteria-bacteria, bacteria-fungi, and fungi-fungi leading to inhibition of growth have been extensively reported in fungal and bacterial soil communities leading to the synthesis of antibiotics (Park et al. 2009; An et al. 2013; Sullivan et al. 2013). The coexistence of many microbial communities sharing the same niche may affect the organism's morphology, development, and synthesis of secondary metabolites (Rico-Gray 2001; Sandland et al. 2007). The invention of penicillin in a co-culture of *Staphylococcus* sp. and *Penicillium* sp. (Sir Alexander Fleming in 1928) (Fleming 1929) is the best-known example of such interaction.

33.3 Mycobiome and Microbiome

The mycobiome refers to the overall fungal microbiota localized in a given environment (Ghannoum et al. 2010). About 70,000 fungal species are already known, and 1.5×10^6 species have been speculated to be present in nature (Hawksworth 1991). The number above represents a hope to the researchers for the exploitation and optimization of novel secondary metabolites out of them. The number of metabolites from actinomycetes and fungi is numerous and accounts for different 19,000 molecules which represent 60% of total bioactive microbial secondary metabolites (Berdy 2005). Fungi are one of the most diverse group of microorganisms, and among these terrestrial fungi, entomopathogenic, endophytic, coprophilous, marine and fungi native to freshwater represent the maximum diversity in ecosystems. *Penicillium*, *Trichoderm*, *Aspergillus*, and *Fusarium* have been known to produce the highest number of secondary metabolites. Using co-cultures of fungi-fungi, the production of secondary metabolites can be induced (Keller et al. 2005; Scherlach and Hertweck 2009). Fungal partners produce secondary metabolites to support their growth when there is a competition for space and nutrients. Fungal colonies develop an interconnecting network of multicellular mycelium (Donnelly and Boddy 2001) which in close propinquity come into contact and interact in space (Rayner 1988). Fungal mycelium of fungus inhibits the growth of other fungus for the purpose of utilization of available resources (exploitation competition). While mycelial mesh contacts each other, the antagonism of this type may occur in proximity or at a distance (Falconer et al. 2008). The mechanism linked to this phenomenon is not obvious, but at distance, it necessitates the requirement of compounds which diffuses from one place to another (Peiris et al. 2008) such as trisporic acids in *Mucorales* (Schachtschabel et al. 2008). However, individual hyphae must contact each other before a reaction is stimulated (Ikediugwu and Webster 1970) that can result in the production of secondary metabolites (phenolic and quinonic compounds) (Griffith et al. 1994). The metabolome and different enzymes have also been observed to be induced. Few examples to quote are the secretion of manganese-dependent peroxidase (White and Boddy 1992), phenoloxidases, and peroxidases (Score et al. 1997) and lignin-degrading enzymes (Chi et al. 2007) and laccase (Wei et al. 2010). Confrontation or barrier zones or zone lines are also formed in competitive co-culture interactions (Campbell 1932). These zone lines and their pattern signify high metabolic activity, which may further be used to search for the novel secondary metabolites (Bohni et al. 2013). Approximately 10 billion microorganisms can be found in a gram of soil which may include 1000 and 10,000 diverse genera and species of unknown prokaryotes (Rosselló-Mora and Amann 2001). In such communities, dynamic mutualistic interactions exist, in which bacterial cells use fungal hyphae for shelter and for nutrient access. The best example of this kind of interaction is the oxalate-carbonate pathway which epitomizes a unique fungal and bacterial interaction (Martin et al. 2012). These microbiomes are presided by dynamic multidimensional interactions and chemical communication which lead to the bioactive, secondary metabolite induction.

33.4 Distinction Between Primary and Secondary Metabolites

Fats, proteins, carbohydrates, and nucleic acids are categorized in primary metabolites, whereas secondary metabolites are used for defense and communication and categorized as molecules of adaptation (Dewick 2009; O'Brien and Wright 2011). Due to the lack of evolutionary basis, there is a long debate between differential existence of primary and secondary metabolites, and importantly, many metabolites do not fit in either category (Firn and Jones 2009). In addition, vitamin C (ascorbic acid) and heparin (blood anticoagulant) make the demarcation between primary and secondary metabolite to some extent difficult. The classification of secondary metabolites becomes more difficult if containing moieties are from more than one class. Unfortunately, there is no general consensus for the division of the different types of secondary metabolites. However, the six different classes are most commonly used, namely, polyketides and fatty acids, phenylpropanoids and aromatic amino acids, terpenoids, alkaloids, peptides, proteins, and specialized class of carbohydrates (Hanson 2003). A huge and distinct class of secondary metabolites is represented by polyketides and fatty acids; affiliation of this class comprised of macrolide antibiotics and statins (Kirakosyan et al. 2004). Phenylpropanoids are synthesized using the shikimate pathway which is exclusive to plants and microorganisms. The class comprises of folic acid, salicylic acid (anti-inflammatory and analgesic), myristicin (hallucinogenic), and resveratrol (antioxidant) (Corder et al. 2003; Dewick 2009). The head-to-tail linkage of isoprene units constitutes another large group of secondary metabolites, the terpenoids, which consisted of hemiterpenes, monoterpenes, sesquiterpenes, diterpenes, sesterterpenes, triterpenes, and tetraterpenes. Compounds such as menthol, steroids, and the anticancer drug taxol are included in members of terpenoids (Dewick 2009; Hanson 2003). Conceivably the most famous class of secondary metabolites constitutes the nitrogen-containing alkaloids which include cocaine, atropine, nicotine, caffeine, morphine, strychnine, etc. Between the primary and secondary metabolites, the gray zone comprises peptides and proteins. Non-ribosomal peptide synthetases (NRPS) produce secondary metabolites which are derived from amino acids instead of the typical ribosomal peptide biosynthesis sequences. The typical examples of NRPS enzymes comprised of the immunosuppressant cyclosporine and beta-lactam antibiotics (Kalb et al. 2013).

33.5 Secondary Metabolites in Fungi

Fungi occur in complex biotic ecosystems with microbes, insects, or plants and are well-known for the production of secondary metabolites of industrial and therapeutic use. Fungi produce various chemically and biologically diverse secondary metabolites which include polyketides, terpenoids, steroids, alkaloids, and peptides. The biological activities of these special compounds include antitumor, antifungal, phytotoxic, and antibacterial activities which have already contributed in the better understanding of natural product synthesis and discovery (Hanson 2003). Recent reports on fungal biotic interactions indicated that many novel metabolites can be produced using co-cultures of fungi with fungi or bacteria (Cueto et al. 2001; Bertrand et al. 2013a). The interactions and combinatorial pathways may either change the yields of secondary metabolites or could generate the analogues of known metabolites (Degenkolb et al. 2002; Oh et al. 2007). Up till recently the main focus of various research laboratories was to discover novel metabolites with potentially useful properties from monocultures of fungi. The reports pertaining to the study, the origin, and the biological role of the fungal metabolites and how they interact with biotic systems are scanty (Zuck et al. 2011; Bertrand et al. 2013a, b; Li et al. 2014). Today, co-cultivation method has turned into one of the main players in the discovery of novel secondary metabolites relevant to pharmaceutical or industrial applications (Brakhage 2013; Moody 2014).

33.6 Mixed Fermentations

To increase the libraries of diverse natural chemical scaffolds, mixed fermentation is a simple and effectual way (Pettit 2009). The culturing of fungi on solid media has permitted to study the phenotype and metabolic changes that occur during mycelial interaction (Sonnenbichler et al. 1989; Woodward and Boddy 2008). The studies pertaining to the confrontation of fungi and bacteria in fungus-fungus and fungus-bacterium interactions have been studied extensively (Schoeman et al. 1996; Wald et al. 2004; Mela et al. 2011; Martin et al. 2012; Cheng et al. 2013; Ola et al. 2013). There are incredible opportunities and ways to develop a huge number of inter-/intraspecies combinations because of the existence of more than a million fungal microbiota (Bass and Richards 2011), which could lead to the identification of novel bioactive natural products. Even for bacteria and other microorganisms, it holds true that their exact biosynthetic potential is not known and yet to be explored. It is highly difficult to observe the dynamics of the interacting individuals in the liquid media, but the induction of the same can be monitored quite decently. A good example of this method includes fermenting cultures of therapeutic mushrooms to produce secondary metabolites. Previously, the metabolites were extracted from the fruiting bodies with high production costs. But today, liquid culture has become a well-known method for the recovery of higher amount of metabolites used therapeutically (Tang et al. 2007). Mixed fermentation methods such as fungus-fungus (Zhu and Lin 2006), bacterium-fungus (Mendes et al. 2013; Rateb et al. 2013), and bacterium-bacterium (Kumar et al. 2013) have been well documented in literature.

33.7 Genetic Regulation of Secondary Metabolite Production in Fungi

In fungi the genes coding for metabolites are positioned in large clusters. These gene clusters contain the important biosynthetic genes encoding the multi-modular enzymes responsible for the production of the secondary metabolites. The number of gene clusters in genome producing secondary metabolite varies considerably between different fungal species. In *Aspergillus* sp., the genome might contain 50 or more secondary metabolite clusters (von Döhren 2009; Burmester et al. 2011). Importantly, there are a huge number of gene clusters that can code for secondary metabolites in *Aspergillus* sp., but only few genes are expressed, and only few metabolites are known. This clearly indicates that the biosynthesis genes producing secondary metabolite are silent under standard laboratory conditions and thus are referred as cryptic or silent or orphan (Gross 2007; Chiang et al. 2011; Brakhage 2013). It is hypothesized that the development of unconventional methods may activate these cryptic biosynthetic pathways which may lead to the production of novel bioactive compounds. It is a well-established fact that co-culturing of microorganisms can direct the activation of otherwise silent/cryptic gene clusters, but intriguingly the mechanisms for the same have not been worked out very well. Microorganisms are known to synthesize transcriptional regulators and epigenetic modifiers, exemplified by a study in which *Aspergillus fumigatus* is grown in a mixed system, resulting in the activation of the silent pathway (Konig et al. 2013). Genetic alterations (mutation) and subsequent expression of otherwise silent/cryptic gene clusters (Charusanti et al. 2012) or horizontal gene transfers are the major key factors which may result in the identification of yet to be characterized chemical motifs (Kurosawa et al. 2008). The regulatory pathways directing the synthesis of complex secondary metabolites have been known to induce in response to stress factors such as pH and temperature and are also dependent on the accessibility of carbon, iron, and nitrogen sources, etc. in close vicinity. Optimization and characterization of these environmental conditions may also turn on the expression of secondary metabolites (Sorensen et al. 2012). Unfortunately, the induction is relevant to few genetic clusters, and de novo production of only few metabolites has been observed (Brakhage 2013). But the approach known as interspecies crosstalk has proven to be a success in de novo production of secondary metabolites (Brakhage and Schroeckh 2011; Bertrand et al. 2013a; Marmann et al. 2014; Schroeckh et al. 2014).

33.8 Induction of Cryptic Secondary Metabolite Gene Clusters

In various environments fungi and bacteria live in close proximity and compete for the available resources such as food and shelter (Frey-Klett et al. 2011). Under these conditions, antagonism resulting from their interactions augments the production of secretary metabolites. Few examples of such interactions include pestalone, a potent antibiotic against methicillin-resistant *Staphylococcus aureus* (MRSA) and vancomycin-resistant *Enterococcus faecium*, which has been identified when *Thalassopia* sp. (gram-negative bacterium) and *Pestalotia* (marine fungus) (Cueto et al. 2001) are co-cultured. Marine microbiota is considered to be a huge source of diverse secondary metabolites, but the bio-prospection of more than 90% of the well-known natural compounds and their derivatives has been done from soil microbiota (Subramani and Aalbersberg 2012). The bacterial genus *Streptomyces*, *Pseudomonas*, and *Bacillus* are reported to have the most important role as fungal co-symbiont (Bouizgarne 2011). *Bacillus subtilis* (gram-positive bacteria) is known to augment the production of macrocarpon, 2-carboxymethylamino benzoic acid, and citreoisocoumarinol when co-cultured with *Fusarium tricinctum* (Ola et al. 2013). Similarly, lateropyrone, cyclic depsipeptides of the enniatin type, and fusaristatin were observed to be produced 78-fold higher (Ola et al. 2013) in a co-culture as compared to the fungal monoculture. The co-cultivation of *Emericella parvathecia* (marine fungus) and *Salinispora arenicola* (actinomycete) directs a hundredfold production of emericellamides by the fungus (Marmann et al. 2014; Oh et al. 2007). The co-cultivation of *Thalassopia* sp. (marine α-proteobacterium) and *Libertella* sp. (fungal partner) led to the production of libertellenones, the diterpenoids were not produced in a *Libertella* monoculture or by the bacterial culture alone (Oh et al. 2005), and there is a direct role of cell to cell interaction for libertellone production. *Aspergillus fumigates* has been observed to produce at least 226 potentially bioactive metabolites (Frisvad et al. 2009) such as gliotoxins, pseurotins, and fumagillins. Co-culturing *A. fumigatus* with *Streptomyces peucetius* produced formylxanthocillin analogs such as fumiformamide and diformamide (Zuck et al. 2011). *A. fumigatus* when co-cultured with *Streptomyces bullii* bacteria produced ergosterol and diketopiperazine alkaloids, brevianamide, spirotryprostatin, 6-methoxy spirotryprostatin, fumitremorgin, verruculogen, and 11-*O*-methylpseurotin (Rateb et al. 2013). Antifungal iturins were observed to be produced in a co-culture of *Pseudomonas aeruginosa* and *Aspergillus fumigatus* (Moree et al. 2013). Co-cultures of *Sphingomonas* and *A. fumigatus* strains which are procured from coalmine drainage induced the production of a new diketopiperazine, glionitrin A (Park et al. 2009), having a strong antibiotic activity against MRSA and cytotoxic activity against human cancer cell lines. *Fusarium pallidoroseum* and *Saccharopolyspora erythraea* interactions were observed to result in novel tetramic acid analog, equisetin (Whitt et al. 2014).

33.9 Diversity of Novel Chemical Scaffolds from Cryptic Secondary Metabolite Gene Clusters

In co-culture experiments, polyketide compounds have been observed the most. This could be attributed to the higher structural diversity of polyketide compounds and the high abundance of organisms producing these compounds (Hertweck 2009). It is a well-established fact the co-cultivation of microbes augments the production of secondary metabolites because of specific genetic alterations resulting in the expression of silent/cryptic gene clusters (Charusanti et al. 2012). In addition to that, horizontal gene transfer may also lead to the production of previously undetected or yet to be characterized novel chemical structures of therapeutic use. The activation of silent polyketide synthetase (PKS) gene clusters via co-culturing augments the induction of a huge number of polyketides in *Aspergillus nidulans* and *A. fumigatus* (Schroeckh et al. 2009; Konig et al. 2013). The compounds which have been identified so far in *A. nidulans* are 2,4-dihydroxybenzoic acid derivatives (Cannell 1998; Seidel 2005; Wolfender et al. 2006; Challal et al. 2012). A sulfonated trimer of 2,4-dihydroxybenzoic acid was reported to be induced when *Bionectria ochroleuca* and *Trichophyton rubrum* (Sticher 2008) were co-cultured. Furthermore, the upregulation of the polyketide zearalenone (mycotoxin) has been observed in co-cultures of *Fusarium culmorum* and *Alternaria tenuissima* (Muller et al. 2012). Additionally, various polyketides including linear skeletons (Tokimoto et al. 1987; Kossuga et al. 2013), aromatic monocycles (Cueto et al. 2001; Notz et al. 2002; Nonaka et al. 2011), and multiple aromatic fused cycles (Zhang et al. 2008; Li et al. 2011; Onaka et al. 2011; Chagas et al. 2013) have also been reported to be produced in a mixed system. In literature, there are numerous reports clearly indicating that the activation of silent gene clusters is directly involved in the production of non-ribosomal peptide synthase (NRPS) (Schwarzer et al. 2003). Different types of peptides, depsipeptides (Wang et al. 2013), iturins (Moree et al. 2013), and lipoaminopeptides (Degenkolb et al. 2002), have been reported also. NRPS have been known to be implicated in the synthesis of hydroxamic acid compounds such as siderophores (Jonkers et al. 2012). Cytochalasins are the best example of the activation of hybrid NRPS-PKS gene clusters during the co-culture of two *Aspergillus* sp. (Fujii et al. 2013). The biosynthesis of the pseurotin compound class necessitates the involvement of PKS-NRPS gene (Maiya et al. 2007), as well as the production of terpenes has also well documented (Smanski et al. 2012). The compounds, namely, sesquiterpenes (Sarker and Nahar 2012) involving *Fusarium* sp. (Muller et al. 2012), diterpenes (Cho and Kim 2012), and triterpenes, are observed in a mixed fermentation of *Aspergillus* sp. with *Streptomyces* sp. (Rateb et al. 2013). Geris and Simpson 2009 also observed that the biosynthetic pathway on terpene synthases leads to meroterpenoid induction. Moreover, various alkaloids such as procyanines (Angell et al. 2006), pyrazines (Zhu et al. 2009), pyrrols (Straight et al. 2007; Onaka et al. 2011), marinamides (Zhu et al. 2013), and picolinic acid derivatives (Jonkers et al. 2012) have also been shown to be induced during co-culture. The activation of N-formyl alkaloids (co-culture of *A. fumigatus* with *S. peucetius*) (Zuck et al. 2011), free fatty acids (Combes et al. 2012), and glycosides (Kurosawa et al. 2008) is an unusual example of co-culture activation of metabolites. Co-culturing *Gloeophyllum abietinum* with *Heterobasidion annosum* induced the production of antibiotics oospoglycol, oosponol, and melledonal A and C (Sonnenbichler et al. 1994). The purification of taxol, a potent anticancer agent by *Paraconiothyrium* sp., was also induced in a co-culture of *Alternaria* sp. and *Phomopsis* sp. (endophytes of *Taxus* trees) (Soliman and Raizada 2013). Various studies have observed improved bactericidal or bacteriostatic activities of crude or purified extracts from mixed fermentations as compared to those from analogous monocultures (Losada et al. 2009). Several selective studies on known natural antimicrobial compounds have revealed that the induction of secondary metabolites can be augmented using co-cultures (Sonnenbichler et al. 1994; Muller et al. 2012).

33.10 Conclusion

It is widely accepted that microbes possess huge possibilities for the production of diverse secondary metabolites. Majority of these biosynthetic gene clusters are cryptic, and indeed novel methodologies are required to awake them. Awakening gene clusters, producing therapeutically and industrially important metabolites, is challenging, but many combinatorial approaches such as culturing the unculturables, co-culture, data mining, synthetic chemistry, etc. have yielded promising outcomes. These developments also include the induction of metabolites through cryptic pathways, combining various genetic engineering approaches to produce polymorphism of novel chemical scaffolds. Various examples have been furnished in the present chapter which clearly emphasize that microbial consortia, if co-cultured in an artificial environment, can induce or activate the production of diverse range of novel and yet unidentified chemical motifs. Due to the complex nature and the diverse chemistry of induced metabolome from co-cultures, the variability must be accurately measured to underscore the compound of interest. The mechanisms of regulation of secondary metabolites by biosynthetic gene cluster are poorly established, nevertheless, the microbial mixed fermentation approaches constitute purposeful methods to underscore new structures of secondary metabolites with pertinent biological activity. And undoubtedly, these promising approaches will add new dimensions in the field of drug research and the discovery of novel skeletons of therapeutic use.

References

An S, Couteau C, Luo F, Neveu J, DuBow MS (2013) Bacterial diversity of surface sand samples from the Gobi and Taklamakan deserts. Microb Ecol 66:850–860

Angell S, Bench BJ, Williams H, Watanabe CMH (2006) Pyocyanin isolated from a marine microbial population: synergistic production between two distinct bacterial species and mode of action. Chem Biol 13:1349–1359

Bass D, Richards TA (2011) Three reasons to re-evaluate fungal diversity on earth and in the ocean. Fungal Biol Rev 25:159–164

Berdy J (2005) Bioactive microbial metabolites: a personal view. J Antibiot 58:1–26

Berdy J (2012) Thoughts and facts about antibiotics: where we are now and where we are heading. J Antibiot 65:385–395

Bertrand S, Schumpp O, Bohni N, Bujard A, Azzollini A, Monod M, Gindro K, Wolfender JL (2013a) Detection of metabolite induction in fungal co-cultures on solid media by high throughput differential ultra-high pressure liquid chromatography-time-of-flight mass spectrometry fingerprinting. J Chromatogr A 1292:219–228

Bertrand S, Schumpp O, Bohni N, Bujard A, Azzollini A, Monod M et al (2013b) Detection of metabolite induction in fungal co-cultures on solid media by high-throughput differential ultra-high pressure liquid chromatography–time-of-flight mass spectrometry fingerprinting. J Chromatogr 292:219–228

Bohni N, Schumpp O, Schnee S, Bertrand S, Gindro K et al (2013) Targeted isolation of induced and bioactive metabolites from fungal co-cultures. Planta Med 79:SL41

Bouizgarne B (2011) Bacteria for plant growth promotion and disease management, in Bacteria. In: Maheshwari DK (ed) Agrobiology: disease management. Springer Verlag, Heidelberg, pp 15–47

Brakhage AA (2013) Regulation of fungal secondary metabolism. Nat Rev Microbiol 11:21–32

Brakhage AA, Schroeckh V (2011) Fungal secondary metabolites – strategies to activate silent gene clusters. Fungal Genet Biol 48:15–22

Burmester A, Shelest E, Glöckner G, Heddergott C, Schindler S, Staib P, Heidel A, Felder M, Petzold A, Szafranski K, Feuermann M, Pedruzzi I, Priebe S, Groth M, Winkler R, Li W, Kniemeyer O, Schroeckh V, Hertweck C, Hube B, White TC, Platzer M, Guthke R, Heitman J, Wöstemeyer J, Zipfel PF, Monod M, Brakhage AA (2011) Comparative and functional genomics provide insights into the pathogenicity of dermatophytic fungi. Genome Biol 12:R7

Campbell WG (1932) The chemistry of the white rots of wood. III. The effect on wood substance of *Ganoderma applanatum* (Pers.) Pat., *Fomes fomentarius* (Linn.) Fr., *Polyporus adustus* (Willd.) Fr., *Pleurotus ostreatus* (Jacq.) Fr., *Armillaria mellea* (Vahl.) Fr., *Trametes pini* (Brot.) Fr., and *Polystictus abietinus* (Dicks.) Fr. Biochem J 26:1829–1838

Cannell RJP (1998) Natural products isolation, 1st edn. Humana Press, Totowa

Chagas F, Dias L, Pupo M (2013) A mixed culture of endophytic fungi increases production of antifungal polyketides. J Chem Ecol 39:1335–1342

Challal S, Bohni N, Buenafe OE, Esguerra CV, de Witte PAM et al (2012) Zebrafish bioassay-guided micro fractionation for the rapid in vivo identification of pharmacologically active natural products. Chimia 66:229–232

Chapman and Hall (2015) The dictionary of natural products online [Online]. CRC Press. Available at: http://dnp.chemnetbase.com

Charusanti P, Fong NL, Nagarajan H, Pereira AR, Li HJ et al (2012) Exploiting adaptive laboratory evolution of *Streptomyces clavuligerus* for antibiotic discovery and overproduction. PLoS One 7:e33727

Cheng YF, Jin W, Mao SY, Zhu WY (2013) Production of citrate by anaerobic fungi in the presence of co-culture methanogens as revealed by 1H NMR spectrometry. Asian-Australas J Anim Sci 26:1416–1423

Chi Y, Hatakka A, Maijala P (2007) Can co-culturing of two white-rot fungi increase lignin degradation and the production of lignin-degrading enzymes? Int Biodeterior Biodegrad 59:32–39

Chiang YM, Chang SL, Oakley BR, Wang CCC (2011) Recent advances in awakening silent biosynthetic gene clusters and linking orphan clusters to natural products in microorganisms. Curr Opin Chem Biol 15:137–143

Cho JY, Kim MS (2012) Induction of antifouling diterpene production by Streptomyces cinnabarinus PK209 in co-culture with marine-derived Alteromonas sp. KNS-16. Biosci Biotechnol Biochem 76:1849–1854

Combes A, Ndoye I, Bance C, Bruzaud J, Djediat C et al (2012) Chemical communication between the endophytic fungus *Paraconiothyrium variabile* and the phytopathogen *Fusarium oxysporum*. PLoS One 7:e47313

Corder R, Crozier A, Kroon PA (2003) Drinking your health? It's too early to say. Nature 426:119

Cragg GM, Newman DJ (2013) Natural products: a continuing source of novel drug leads. Biochim Biophys Acta Gen Subj 1830:3670–3695

Cueto M, Jensen PR, Kauffman C, Fenical W, Lobkovsky E, Clardy J (2001) Pestalone, a new antibiotic produced by a marine fungus in response to bacterial challenge. J Nat Prod 64:1444–1446

De Roy K, Marzorati M, Van den Abbeele P, Van de Wiele T, Boon N (2013) Synthetic microbial ecosystems: an exciting tool to understand and apply microbial communities. Environ Microbiol 16:1472–1481

Degenkolb T, Heinze S, Schlegel B, Strobel G, Grafe U (2002) Formation of new lipoaminopeptides, acremostatins A, B, and C, by co-cultivation of *Acremonium* sp. Tbp-5 and *Mycogone rosea* DSM 12973. Biosci Biotechnol Biochem 66:883–886

Dewick PM (2009) Medicinal natural products: a biosynthetic approach, 3rd edn. Wiley, Chichester

Donnelly DP, Boddy L (2001) Mycelial dynamics during interactions between *Stropharia caerulea* and other cord-forming, saprotrophic basidiomycetes. New Phytol 151:691–704

Falconer RE, Bown JL, White NA, Crawford JW (2008) Modelling interactions in fungi. J R Soc Interface 5:603–615

Firn RD, Jones CG (2009) A Darwinian view of metabolism: molecular properties determine fitness. J Exp Bot 60:719–726

Fleming A (1929) On the antibacterial action of cultures of a *Penicillium*, with special reference to their use in the isolation of *B. influenzae*. Br J Exp Pathol 10:226–236

Frey-Klett P, Burlinson P, Deveau A, Barret M, Tarkka M, Sarniguet A (2011) Bacterial-fungal interactions: hyphens between agricultural, clinical, environmental, and food microbiologists. Microbiol Mol Biol Rev 75:583–609

Frisvad JC, Rank C, Nielsen KF, Larsen TO (2009) Metabolomics of *Aspergillus fumigatus*. Med Mycol 47:S53–S71

Fujii R, Minami A, Gomi K, Oikawa H (2013) Biosynthetic assembly of cytochalasin backbone. Tetrahedron Lett 54:2999–3002

Geris R, Simpson TJ (2009) Meroterpenoids produced by fungi. Nat Prod Rep 26:10631094

Ghannoum MA, Jurevic RJ, Mukherjee PK, Cui F, Sikaroodi M et al (2010) Characterization of the oral fungal microbiome (mycobiome) in healthy individuals. PLoS Pathog 6:e1000713

Giddings LA, Newman D (2013) Microbial natural products: molecular blueprints for antitumor drugs. J Ind Microbiol Biotechnol 40:1181–1210

Griffith GS, Rayner ADM, Wildman HG (1994) Extracellular metabolites and mycelial morphogenesis of *Hypholoma fasciculare* and *Phlebia radiata* (Hymenomycetes). Nova *Hedwigia* 59:311–329

Gross H (2007) Strategies to unravel the function of orphan biosynthesis pathways: recent examples and future prospects. Appl Microbiol Biotechnol 75:267–277

Hanson JR (2003) Natural products: the secondary metabolites. Royal Society of Chemistry, Cambridge

Hawksworth DL (1991) The fungal dimension of biodiversity: magnitude, significance, and conservation. Mycol Res 95:641–655

Hertweck C (2009) The biosynthetic logic of polyketide diversity. Angew Chem Int Ed 48:4688–4716

Ikediugwu FEO, Webster J (1970) Antagonism between *Coprinus heptemerus* and other coprophilous fungi. Trans Br Mycol Soc 54:181–204

Jonkers W, Rodriguez Estrada A, Lee K, Breakspear A, May G et al (2012) Metabolome and transcriptome of the interaction between *Ustilago maydis* and *Fusarium verticillioides* in vitro. Appl Environ Microbiol 78:3656–3667

Kalb D, Lackner G, Hoffmeister D (2013) Fungal peptide synthetases: an update on functions and specific signatures. Fungal Biol Rev 27:43–50

Keller NP, Turner G, Bennett JW (2005) Fungal secondary metabolism from biochemistry to genomics. Nat Rev Microbiol 3:937–947
Kirakosyan A, Sirvent TM, Gibson DM, Kaufman PB (2004) The production of hypericins and hyperforin by in vitro cultures of St. John's wort (*Hypericum perforatum*). Biotechnol Appl Biochem 39:71–81
Konig CC, Scherlach K, Schroeckh V, Horn F, Nietzsche S et al (2013) Bacterium induces cryptic meroterpenoid pathway in the pathogenic fungus *Aspergillus fumigatus*. Chem Biol Chem 14:938–942
Kossuga MH, Ferreira AG, Sette LD, Berlinck RGS (2013) Two polyketides from a co-culture of two marine-derived fungal strains. Nat Prod Commun 8:721–724
Kumar RS, Moorthy IMG, Baskar R (2013) Modeling and optimization of glutamic acid production using mixed culture of *Corynebacterium glutamicum* NCIM2168 and *Pseudomonas reptilivora* NCIM2598. Prep Biochem Biotechnol 43:668–681
Kurosawa K, Ghiviriga I, Sambandan TG, Lessard PA, Barbara JE et al (2008) Rhodostreptomycins, antibiotics biosynthesized following horizontal gene transfer from *Streptomyces padanus* to *Rhodococcus fascians*. J Am Chem Soc 130:1126–1127
Li C, Zhang J, Shao C, Ding W, She Z et al (2011) A new xanthone derivative from the coculture broth of two marine fungi (strain No. E33 and K38). Chem Nat Compd 47:382–384
Li C, Wang J, Luo C, Ding W, Cox DG (2014) A new cyclopeptide with antifungal activity from the co-culture broth of two marine mangrove fungi. Nat Prod Res 28:616–621
Losada L, Ajayi O, Frisvad JC, Yu J, Nierman WC (2009) Effect of competition on the production and activity of secondary metabolites in *Aspergillus* species. Med Mycol 47:S88S96
Maiya S, Grundmann A, Li X, Li S-M, Turner G (2007) Identification of a hybrid PKS/NRPS required for pseurotin A biosynthesis in the human pathogen *Aspergillus fumigatus*. Chem Biol Chem 8:1736–1743
Marmann A, Aly A, Lin W, Wang B, Proksch P (2014) Co-cultivation-a powerful emerging tool for enhancing the chemical diversity of microorganisms. Mar Drugs 12:1043–1065
Martin G, Guggiari M, Bravo D, Zopfi J, Cailleau G et al (2012) Fungi, bacteria and soil pH: the oxalate-carbonate pathway as a model for metabolic interaction. Environ Microbiol 14:2960–2970
Mela F, Fritsche K, de Boer W, van Veen JA, de Graaff LH et al (2011) Dual transcriptional profiling of a bacterial/fungal confrontation: *Collimonas fungivorans* versus *Aspergillus niger*. ISME J 5:1494–1504
Mendes R, Garbeva P, Raaijmakers JM (2013) The rhizosphere microbiome: significance of plant beneficial, plant pathogenic, and human pathogenic microorganisms. FEMS Microbiol Rev 37:634–663
Moody SC (2014) Microbial co-culture: harnessing inter-microbial signaling for the production of novel antimicrobials. Future Microbiol 9:575–578
Moree WJ, Yang JY, Zhao X, Liu WT, Aparicio M et al (2013) Imaging mass spectrometry of a coral microbe interaction with fungi. J Chem Ecol 39:1045–1054
Muller MEH, Steier I, Köppen R, Siegel D, Proske M et al (2012) Co-cultivation of phytopathogenic *Fusarium* and *Alternaria* strains affects fungal growth and mycotoxin production. J Appl Microbiol 113:874–887
Nonaka K, Abe T, Iwatsuki M, Mori M, Yamamoto T et al (2011) Enhancement of metabolites productivity of *Penicillium pinophilum* FKI-5653, by co-culture with *Trichoderma harzianum* FKI-5655. J Antibiot 64:769–774
Notz R, Maurhofer M, Dubach H, Haas D, Defago G (2002) Fusaric acid-producing strains of *Fusarium oxysporum* alter 2,4-diacetylphloroglucinol biosynthetic gene expression in *Pseudomonas fluorescens* CHA0 in vitro and in the rhizosphere of wheat. Appl Environ Microbiol 68:2229–2235
O'Brien J, Wright GD (2011) An ecological perspective of microbial secondary metabolism. Curr Opin Biotechnol 22:552–558

Oh DC, Jensen PR, Kauffman CA, Fenical W (2005) Libertellenones A-D: induction of cytotoxic diterpenoid biosynthesis by marine microbial competition. Bioorg Med Chem 13:5267–5273

Oh DC, Kauffman CA, Jensen PR, Fenical W (2007) Induced production of emericellamides A and B from the marine-derived fungus *Emericella* sp. incompeting co-culture. J Nat Prod 70:515–520

Ola ARB, Thomy D, Lai D, Brötz-Oesterhelt H, Proksch P (2013) Inducing secondary metabolite production by the endophytic fungus fusarium tricinctum through coculture with *Bacillus subtilis*. J Nat Prod 76:2094–2099

Onaka H, Mori Y, Igarashi Y, Furumai T (2011) Mycolic acid-containing bacteria induce natural-product biosynthesis in *Streptomyces* species. Appl Environ Microbiol 77:400–406

Park HB, Kwon HC, Lee CH, Yang HO (2009) Glionitrin A, an antibiotic-antitumor metabolite derived from competitive interaction between abandoned mine microbes. J Nat Prod 72:248–252

Pearce C, Eckard P, Gruen-Wollny I, Hansske FG (2009) Microorganisms: their role in the discovery and development of medicines. Natural product chemistry for drug discovery. The Royal Society of Chemistry, Cambridge, pp 215–241

Peiris D, Dunn W, Brown M, Kell D, Roy I et al (2008) Metabolite profiles of interacting mycelial fronts differ for pairings of the wood decay basidiomycete fungus, *Stereum hirsutum* with its competitors *Coprinus micaceus* and *Coprinus disseminatus*. Metabolomics 4:52–62

Pettit R (2009) Mixed fermentation for natural product drug discovery. Appl Microbiol Biotechnol 83:19–25

Rateb ME, Hallyburton I, Houssen W, Bull A, Goodfellow M et al (2013) Induction of diverse secondary metabolites in *Aspergillus fumigatus* by microbial co-culture. RSC Adv 3:14444–14450

Rayner A (1988) Life in a collective – lessons from the fungi. New Sci 120:49–53

Rico-Gray V (2001) Interspecific interaction. Wiley, Chichester

Rosselló-Mora R, Amann R (2001) The species concept for prokaryotes. FEMS Microbiol Rev 25:39–67

Sandland GJ, Rodgers JK, Minchella DJ (2007) Interspecific antagonism and virulence in hosts exposed to two parasite species. J Invertebr Pathol 96:43–47

Sarker SD, Nahar L (2012) Natural products isolation, 3rd edn. Humana Press, Totowa

Schachtschabel D, David A, Menzel KD, Schimek C, Wöstemeyer J et al (2008) Cooperative biosynthesis of trisporoids by the (+) and (−) mating types of the zygomycete *Blakeslea trispora*. Chem Biol Chem 9:3004–3012

Scherlach K, Hertweck C (2009) Triggering cryptic natural product biosynthesis in microorganisms. Org Biomol Chem 7:1753–1760

Schoeman MW, Webber JF, Dickinson DJ (1996) The effect of diffusible metabolites of *Trichoderma harzianum* on in vitro interactions between basidiomycete isolates at two different temperature regimes. Mycol Res 100:1454–1458

Schroeckh V, Scherlach K, Nützmann HW, Shelest E, Schmidt Heck W et al (2009) Intimate bacterial–fungal interaction triggers biosynthesis of archetypal polyketides in *Aspergillus nidulans*. Proc Natl Acad Sci 106:14558–14563

Schroeckh V, Nützmann HW, Brakhage AA (2014) Fungal actinomycete interactions wakening of silent fungal secondary metabolism gene clusters via inter organismic interactions. In: Osbourn RJ Goss A, Carter GT (eds) Natural products: discourse, diversity, and design. Wiley, Hoboken, pp 147–158

Schwarzer D, Finking R, Marahiel MA (2003) Nonribosomal peptides: from genes to products. Nat Prod Rep 20:275–287

Score AJ, Palfreyman JW, White NA (1997) Extracellular phenoloxidase and peroxidase enzyme production during interspecific fungal interactions. Int Biodeterior Biodegrad 39:225–233

Seidel V (2005) Initial and bulk extraction. In: Sarker SD, Latif Z, Gray AI (eds) Natural products isolation, 2nd edn. Humana Press, Totowa, pp 27–46

Seyedsayamdost MR, Clardy J (2014) Natural products and synthetic biology. ACS Synth Biol 3:745–747

Smanski MJ, Peterson RM, Huang SX, Shen B (2012) Bacterial diterpene synthases: new opportunities for mechanistic enzymology and engineered biosynthesis. Curr Opin Chem Biol 16:132–141
Soliman SSM, Raizada MN (2013) Interactions between co-habitating fungi elicit synthesis of taxol from an endophytic fungus in host Taxus plants. Front Microbiol 4:3
Sonnenbichler J, Bliestle IM, Peipp H, Holdenrieder O (1989) Secondary fungal metabolites and their biological activities, I. Isolation of antibiotic compounds from cultures of *Heterobasidion annosum* synthesized in the presence of antagonistic fungi or host plant cells. Biol Chem Hoppe Seyler 370:1295–1304
Sonnenbichler J, Dietrich J, Peipp H (1994) Secondary fungal metabolites and their biological activities. Investigations concerning the induction of the biosynthesis of toxic secondary metabolites in basidiomycetes. Biol Chem Hoppe Seyler 375:71–79
Sorensen JL, Akk E, Thrane U, Giese H, Sondergaard TS (2012) Production of fusarielins by *Fusarium*. Int J Food Microbiol 160:206–211
Sticher O (2008) Natural product isolation. Nat Prod Rep 25:517–554
Straight PD, Fischbach MA, Walsh CT, Rudner DZ, Kolter R (2007) A singular enzymatic megacomplex from *Bacillus subtilis*. Proc Natl Acad Sci 104:305–310
Subramani R, Aalbersberg W (2012) Marine actinomycetes: an ongoing source of novel bioactive metabolites. Microbiol Res 167:571–580
Sullivan TS, McBride MB, Thies JE (2013) Soil bacterial and archaeal community composition reflects high spatial heterogeneity of pH, bioavailable Zn, and Cu in a metalliferous peat soil. Soil Biol Biochem 66:102–109
Tang YJ, Zhu LW, Li HM, Li DS (2007) Submerged culture of mushrooms in bioreactors – challenges, current state-of-the-art, and future prospects. Food Technol Biotechnol 45:221–229
Tokimoto K, Fujita T, Takeda Y, Takaishi Y (1987) Increased or induced formation of antifungal substances in cultures of *Lentinus edodes* by the attack of *Trichoderma* spp. Proc Jpn Acad, Ser B 63:277–280
von Döhren H (2009) A survey of nonribosomal peptide synthetase (NRPS) gene in *Aspergillus nidulans*. Fungal Genet Biol 46(suppl 1):S45–S52
Wald P, Pitkanen S, Boddy L (2004) Interspecific interactions between the rare tooth fungi *Creolophus cirrhatus*, *Hericium erinaceus* and *H. coralloides* and other wood decay species in agar and wood. Mycol Res 108:1447–1457
Wang J, Lin W, Wray V, Lai D, Proksch P (2013) Induced production of depsipeptides by coculturing *Fusarium tricinctum* and *Fusarium begoniae*. Tetrahedron Lett 54:2492–2496
Wei F, Hong Y, Liu J, Yuan J, Fang W et al (2010) Gongronella sp. induces overproduction of laccase in Panus rudis. J Basic Microbiol 50:98–103
White NA, Boddy L (1992) Extracellular enzyme localization during interspecific fungal interactions. FEMS Microbiol Lett 98:75–79
Whitt J, Shipley SM, Newman DJ, Zuck KM (2014) Tetramic acid analogues produced by coculture of *Saccharopolyspora erythraea* with *Fusarium pallidoroseum*. J Nat Prod 77:173–177
Wolfender JL, Queiroz EF, Hostettmann K (2006) The importance of hyphenated techniques in the discovery of new lead compounds from nature. Expert Opin Drug Discovery 1:237–260
Woodward S, Boddy L (2008) Interactions between saprotrophic fungi. In: Boddy L, Frankland JC, Van West P (eds) Ecology of saprotrophic basidiomycetes. Academic, London, pp 125–141
Zhang CL, Zheng BQ, Lao JP, Mao LJ, Chen SY et al (2008) Clavatol and patulin formation as the antagonistic principle of *Aspergillus clavatonanicus*, an endophytic fungus of *Taxus mairei* Appl Microbiol Biotechnol 78:833–840
Zhu F, Hu G, Chen X, Yuan Y, Huang M et al (2009) Studies on the secondary metabolites produced by mixed cultures of two epiphytic fungi isolated from a fruit of mangrove *Avicennia marina*. Zhongshan Daxue Xuebao, Ziran Kexueban 48:79–82
Zhu F, Lin Y (2006) Marinamide, a novel alkaloid and its methyl ester produced by the application of mixed fermentation technique to two mangrove endophytic fungi from the South China Sea. Sci Bull 51(12):1426–1430

Zhu F, Qin C, Tao L, Liu X, Shi Z et al (2011) Clustered patterns of species origins of nature derived drugs and clues for future bioprospecting. Proc Natl Acad Sci 108:12943–12948

Zhu F, Chen GY, Wu JS, Pan JH (2013) Structure revision and cytotoxic activity of marinamide and its methyl ester, novel alkaloids produced by co-cultures of two marine-derived mangrove endophytic fungi. Nat Prod Res 27:1960–1964

Zuck KM, Shipley S, Newman DJ (2011) Induced production of N-formyl alkaloids from *Aspergillus fumigates* by co-culture with *Streptomyces peucetius*. J Nat Prod 74:1653–1657

Part V
Modern Biotechnological Interventions of Fungi

Chapter 34
Computational Approaches to Understand the Genome and Protein Sequences of Fungi

Atul Kumar Upadhyay and Gaurav Sharma

Abstract Fungi are large group of organism with tremendous diversity and economical importance. Application of bioinformatics approaches to understand the development and growth of organisms has a great scope. Bioinformatics is one of the rapidly emerging branches of science, which helps in understanding biological systems by using computer softwares and tools. The huge amount of data generated in life sciences on a daily basis from several projects is the major driving force for the growth and development of bioinformatics. Bioinformatics, also known as computational biology, is used to analyze genes, proteins, and genomes. Computational tools of genome, transcriptome, or exome analysis are very essential to make a meaning from this tremendous amount of data. In this chapter, I have described the bioinformatics approaches (databases and tools) that can be used in the better understanding of the fungi genes, proteins, and genomes. I have also discussed about the implication of next-generation sequencing technology (NGS) tools on fungi genetics and genomes. Application of these tools and databases to understand the fungi genome and transcriptome will have tremendous effect on development, improvement, and sustainable cultivation of fungi.

Keywords Bioinformatics · Computational biology · Fungi · Genome · Protein · Sequences

A. K. Upadhyay (✉)
School of Bioengineering and Biosciences, Lovely Professional University, Phagwara, India
e-mail: atul.20436@lpu.co.in

G. Sharma
Department of Natural sciences, Sant Baba Bhag Singh University, Jalandhar, India

P. Gehlot, J. Singh (eds.), *Fungi and their Role in Sustainable Development: Current Perspectives*, https://doi.org/10.1007/978-981-13-0393-7_34

34.1 Introduction

Fungi are one of the most diverse groups of organisms. To understand the basic biology of fungi for its sustainable growth, development, and improvement, computational biology approaches play a crucial role. To facilitate the data mining of genomes of fungi, many tools have been developed. Therefore, bioinformatics tools can be helpful in the prediction of a gene or protein, along with detailed sequence and structure analysis. Computational tools are very much useful in complex jobs such as structural modeling, molecular simulations, metabolic pathway analysis, etc. This chapter emphasizes on different tools and databases of bioinformatics and application of these tools and databases with special focus on fungi.

34.1.1 Introduction to Proteins

Proteins are a group of complex organic macromolecules that contain carbon, hydrogen, oxygen, nitrogen, and sulfur. These molecules are translated from the genes present in the genome of each organism. Proteins are essential for each and every living cell to function and survive. These biomolecules are the structural building blocks of cells, found as enzymes, hormones, immunoglobulin, receptors, etc., and are very diverse in nature. They are involved in several functions of the cell and play essential roles. One category of proteins that misfold is the precursor of several neurodegenerative diseases. The misfolded proteins, which are precursors for amyloids, are another type of cellular toxic protein forms, which lead to several neurodegenerative diseases (Abrahamson et al. 1992).

34.1.2 Protein Structure

Proteins are made up of amino acids which acquire a complex structure and are categorized into class, fold, superfamily, family, etc. (Jones and Thornton 1995). They undergo folding, which is a spontaneous process and depends on several factors like pH, temperature, concentration of proteins, etc. On the basis of structural-type proteins, molecules can be classified into two groups, viz., globular and fibrous protein. The most common class is that of globular proteins which may be compact, soluble, or spherical in shape. The other type is fibrous proteins, which are generally elongated and insoluble protein molecules. A protein fold can be visualized in a hierarchical manner into primary, secondary, tertiary, and quaternary structure based on the level of interactions within and across protein subunits.

The primary structure is the simple arrangement of amino acids, bonded together through peptide bonds within a protein. There are 20 different kinds of amino acids with a variable side chain commonly known as R group. The R group of the amino acid can be categorized into hydrophobic or hydrophilic depending on the chemical groups present. This R group determines several properties of each amino acid like hydrophobicity, polarity, allowed phi-psi angles, type of interactions with the surrounding environment, etc. and ultimately decides the structure of the protein. Due to the delicate balance in stability and a multitude of interactions, a small change (mutation) within the protein sequence could result in loss of function of a protein and/or improper folding of the protein. This can lead to severe diseases like cancer, amyloidogenesis, and neurodegenerative diseases.

Secondary structure is the second level in our understanding of the protein structure, where local folding of a polypeptide chain is observed as ordered and repetitive structural elements. These elements possess specific intramolecular hydrogen-bonding patterns. Depending on the local hydrogen-bonding pattern, protein molecules can have two major groups of secondary structures. The first and the most common group is the α-helix. The structure of the α-helix is a spiral shape secured by the presence of regular hydrogen bonds of each residue within the helix with the consecutive fourth residue. The second most common secondary structure group in protein structures is the β-pleated sheet.

Tertiary structure refers to the overall shape of the protein made up of a single folded polypeptide chain with one or more secondary structures. There are different types of bonds and forces required to give a protein molecule its stable and compact tertiary structure. The shape and interior of a protein is majorly determined by hydrophobic interactions.

Arrangement of more than two polypeptide chains in 3D space provides quaternary structure of proteins. Every polypeptide chain is referred to as a monomeric subunit. The quaternary structure of the protein may consist of the same type of subunit or different types of subunits referred as homo-oligomer and hetero-oligomer, respectively. Protein subunits interact through their interfaces, and within many protein complexes, the orientations of subunits may vary and hence protein monomeric subunits may have several interfaces.

34.2 Computational Tools and Databases for Protein Sequence Analysis

Bioinformatics approaches are powerful tools to perform analysis over large datasets and provide an early understanding of interesting targets for detailed experimental characterization. For instance, tools like BLAST (McGinnis and Madden 2004) can search homologues for given query protein sequence in sequence databases containing millions of entries, which is practically impossible to perform manually. Combined knowledge of protein structure and bioinformatics tools can help to extract better knowledge of biochemical and molecular biology of fungi and could devise improved future research prospective.

With the introduction of high-throughput experimental techniques in structural biology, thousands of structures of proteins are available complexed with diverse set of ligands and interacting partners. It is now crucial to use computational means to gain further insights in the field of protein structural biology. In this chapter, several bioinformatics tools and resources that can be used to understand the development and sustainable growth of fungi have been discussed in detail.

34.2.1 BLAST

Basic Local Alignment Search Tool (BLAST) is a very popular tool and frequently used for searching sequence homologues. This tool can be used to search and compare homologous sequences from various databases. BLAST has several variants, viz., Blastn, Blastp, Blastx, and tBlastn. It is a highly useful and efficient tool in the field of molecular biology. Similarities between sequences from a vast array of sequence databases are computed. It is possible to infer the function of newly sequenced proteins, predict new members of protein families, and give a chance to explore evolutionary relationships between homologous sequences identified at high sequence similarities between proteins by using BLAST and its variants. These set of homologues are further used to determine level of evolutionary conservation of each residue, which in turn hints about the importance of these residues structurally and functionally. In BLAST, the parameters can be further customized to provide flexibility to the user to tune their searches according to query sequence.

34.2.2 PSI-BLAST

PSI-BLAST stands for Position-Specific Iterative-BLAST. In this the first search is as normal blast followed by formation of PSSM (position-specific scoring matrix) profile from the homologous sequences. PSSM retains the information about extent of conservation of particular position as well as position in the alignment and stores it in the form of a matrix. This matrix is further used to score each hit obtained in subsequent iterations. Based on the extent of divergence required in finding related sequences, the number of iterations is decided.

34.2.3 HMMER

HMMER is a searching tool based on Hidden Markov Model (HMM), which is a statistical model that works on assumption that the system being modeled is in Markov process. HMMER (Eddy 2011) is used for homologous protein sequence searches from databases. It is also helpful in proper alignment of protein sequences. It implements methods using probabilistic models, which do not consider the penultimate state of the events, called profile Hidden Markov Models (profile HMMs).

34.2.4 Domain Identification Algorithm (DIAL)

DIAL (Pugalenthi et al. 2005) is an in-house program for the identification of structural domains basis of the three-dimensional coordinates of a protein. In the algorithm, firstly, secondary structural states are assigned with reference of hydrogen bonds of main chain. For this assigning of secondary structural states, SSTRUC program is used. Then, average C-alpha distances between secondary structures, also referred as proximity indices, are calculated for all secondary structure pairs in the protein. The calculated proximity indices are used to perform clustering with the help of KITSCH program within PHYLIP package (Felsenstein 1989). On the basis of extent of compactness between secondary structures, they are treated as tertiary structural clusters of the protein. These tertiary clusters include super-secondary structures as well as structural domains.

34.2.5 The Pfam Protein Families Database

The Pfam database (Finn et al. 2013) is a repository of protein families. Each family has a well-curated multiple sequence alignment along with a Hidden Markov Model (HMM) of all the members of that family. Proteins typically possess one or more functional elements generally known as functional domains. Numerous combinations of these domains, with different number of repeat, result in diverse proteins found in nature. Hence, the knowledge of protein domain gives prominent indication of their function. Pfam majorly is grouped into two prominent classes: Pfam-A and Pfam-B. Pfam-A contains well-curated entries with their representative multiple sequence alignments.

34.2.6 UniProt

UniProt is a database of protein sequences having very reliable information about each entry in this database. It is a large, comprehensive, and freely accessible protein sequence database and high-quality functional information (Apweiler et al. 2004). In 2002, Swiss Institute of Bioinformatics (SIB), European Bioinformatics Institute (EBI), and Protein Information Resource (PIR) came together to form the UniProt consortium. Several entries of this database are being derived from multiple genome sequencing projects. All entries are annotated with information about the biological function of proteins derived from the research literature. EBI maintains a huge repository of bioinformatics databases along with allied services. SIB hosts ExPASy (Expert Protein Analysis System) is a web server hosting-related tools and databases for proteomics study. One of the oldest databases of protein sequences is Atlas of Protein Sequence and Structure (APSS), which was created by Margaret Dayhoff in 1965. PIR is an advanced version of APSS.

34.2.7 Gene Ontology (GO)

The Gene Ontology (GO) (Ashburner et al. 2000) is a database of experimental annotation terms and description of gene products across different databases. The first time it came into existence was in 1998, and it is a result of collaborative work on three popular model organism, viz., *Drosophila*, *Saccharomyces*, and mouse.

The GO database provides ontologies to annotate gene products in three non-overlapping domains of molecular biology. This database contains information in terms of molecular function (MF), biological process (BP), and cellular component (CC). MF describes activities at the molecular level such as catalytic or binding activities. MF annotation tells about the activities of the molecule rather than the entities (molecules or complexes) that perform the actions. Examples of individual molecular function could be "kinase activity" and the more specific would be "6-phosphofructokinase activity," which represents a subtype of kinase activity. BP describes biological phenomenon accomplished by one or more ordered assemblies of molecular functions. Examples of biological functions could be processes such as cell death. For the subcellular locations and structures, one can refer to CC. Examples of cellular components include "nuclear inner membrane."

The GO Consortium (GOC) has recently included many specialized databases dedicated for plants, animals, and fungal genomes. There are three major objectives of GOC:

1. To create and maintain GO ontologies
2. Annotations of gene products by providing links among the ontologies with the genes and gene products in the different databases
3. The development of tools and software for the creation, maintenance, and use of ontologies

GO serves as a platform where curators can agree on stating how and why a specific term is used and how to consistently apply it.

34.2.8 *Database for Annotation, Visualization, and Integrated Discovery (DAVID)*

DAVID (Dennis et al. 2003) is an enrichment tool for function annotation and analysis, composed of four main modules: Annotation Tool, Gene Ontology Charts, KEGG Charts, and Domain Charts. This is an automated too for annotation of gene lists for their respective functions. This tool has module, the Gene Ontology Charts module, for calculation of the distribution of differentially expressed genes. KEGG Charts show the distribution of differentially expressed genes among KEGG biochemical pathways. Domain Charts display the distribution of differentially expressed genes among Pfam protein domains.

34.3 Protein Structure Databases and Tools

34.3.1 *Protein Data Bank (PDB)*

The PDB is a collection of the three-dimensional structural data of variety of biological molecules, such as proteins and nucleic acids (Bernstein et al. 1977).The structural data is generally obtained by X-ray crystallography or NMR spectroscopy. All the data are freely accessible on the web. The PDB is a very important resource, which is freely available to everyone, and it is very useful to researchers, especially to structural biologists. There are several derived databases developed using the contents of PDB that categorize the data differently. For example, both SCOP (Murzin et al. 1995) and CATH (Knudsen and Wiuf 2010) classify structures according to type of structure adopted by protein, while GO (Harris et al. 2004) categorizes structures on the basis of functions.

34.3.2 *Molecular Modeling of Protein Structures*

Molecular modeling of protein structure is describing the structure with the help of computer. The cases were sequence identity is higher than 40%, referred as closely related protein sequences, the alignment is almost always correct. Alignment becomes difficult when identity is less than 30% (twilight zone). While selecting the template, other than sequence identity, few more criteria are taken care such as biological proximity, conformational state of the target, resolution of the protein, etc.

34.3.3 *Steps in Comparative Modeling*

34.3.3.1 Identification of Template

The first step in comparative modeling is to select a template structure related to the target (query) sequences. Using different sequence and structural databases and programs, comparative modeling can be performed. Templates can be found by performing sequence searches in structure databases such as Protein Data Bank (PDB) (Bernstein et al. 1977) and Structural Classification of Proteins (SCOP) (Murzin et al. 1995) using query sequence. Template selection is a critical step of comparative modeling. Template is selected on the basis of source organism, resolution of structure, and B-factor values of the template structure.

34.3.3.2 Target-Template Alignment

The most crucial step of comparative modeling is an alignment of target and the template. After successful selection of template(s), the target and template sequence should be aligned. Special care should be taken if the identity is less than 30% (twilight zone) as alignment becomes difficult in such cases. The most commonly used tool for alignment is CLUSTALW (Larkin et al. 2007).

34.3.3.3 Model Building

With the perfect target-template alignment, next step is to build 3D model for the query sequence. Most widely used method for molecular modeling of protein sequence is by MODELLER software (Sali and Blundell 1993). MODELLER has been used for comparative protein structure modeling. MODELLER computes comparative protein structure modeling of a query sequence with agreement with spatial restraints as observed in the template structure that include the distances and dihedral angles in the target sequence, extracted from its alignment with the template structures, bond length and bond angle preferences, statistical preferences for dihedral angles, and non-bonded interatomic distances. At default parameters, 20 models get generated by MODELLER, which can be changed as per requirement. Overall template selection and alignment accuracy usually have a larger impact on the model accuracy. The Discrete Optimized Protein Energy (DOPE) is an atomic distance-dependent statistical potential based on a physical reference state that accounts for the finite size and spherical shape of proteins (Shen and Sali 2006). The best model is selected based on the DOPE score or molpdf values from incorrect conformations. The lower the DOPE score, the better the model.

34.3.3.4 Model Evaluation

The assessment of the quality of model of a protein is an important step of comparative modeling. Thus, finding the accuracy of the generated model is an essential step. An important evaluation measure is to have good stereochemistry, for which programs like PROCHECK (Laskowski et al. 1993) are used. Other programs like HARMONY (Pugalenthi et al. 2006) and PROSA (Wiederstein and Sippl 2007) are tools to examine the compatibility between the sequence and structure of a protein. It works by providing scores to each residue by taking care of their local environments. VERIFY3D (Eisenberg et al. 1997) analyzes the compatibility of an atomic model with its own amino acid sequence.

34.3.3.5 Molecular Dynamics Study of Proteins

Molecular dynamics (MD) is a computational approach used for simulation of physical movements of atoms and molecules. This was first time used by Alder and Wainwright (Alder and Wainwright 1959). In this method, the atoms and molecules are allowed to interact for a period of time. MD simulations are one of the routinely used computational experiments to gain information regarding the fluctuations and conformational changes of proteins and nucleic acids. MD simulations are now used on regular basis to understand the structure, dynamics, and thermodynamics of biological molecules and their complexes. They are also used in understanding the structural details from X-ray crystallography and from NMR experiments. Some of the important steps of Global MD are (a) creating of GROMACS topology file from the PDB file, (b) adding solvent water around the protein, (c) running energy minimization, (d) equilibrating the water around the protein, (e) running the production simulation, (f) analysis of the simulation, (g) comparing fluctuations, (h) analyzing the distance and hydrogen bonds, (i) viewing the trajectory, and (j) inferring conclusion(s).

To start MD simulation, we need to define a force field, which requires the definition of a potential function. Force fields are the parameters, which are used to describe the potential energy of a system of particles. It includes both bonded and non-bonded terms. Principal component analysis (PCA) provides considerable insight into the nature of conformational changes in a protein.

34.3.3.6 GROMACS

GROMACS (Pandini et al. 2013) is a high-throughput and highly parallel open-source molecular simulation toolkit. This software can handle wide classes of biomolecules, such as proteins, nucleic acids, and lipids. There are several supported force fields in this toolkit with specific conditions and molecule such as AMBER

(Case et al. 2005) and CHARMM (Brooks et al. 2009) that are used for the intramolecular interactions and GROMOS (Scott et al. 1999) and OPLS (Jorgensen and Tirado-Rives 1988) that are used in coarse-grained force fields. Computations of the force fields have three different components: bonded, non-bonded, and special interactions, which are the restraints of the position, distance, or angle.

34.4 Genomic Tools

In this section, we have reviewed different computational tools and databases used in the analysis of genomic and transcriptomics data from different organisms with especial case of fungi.

34.4.1 SOAPdenovo

There is a rapid increase in number of de novo genome sequencing by using next-generation sequencing (NGS). One of the biggest challenges is to assemble these short reads, generated from different NGS platforms, into a complete or draft genome. There are many tools for such assembly, but accuracy, precision, and selection of appropriate assembler require experience and expertise. SOAPdenovo (Luo et al. 2012) has been successfully used in many genome projects for the assembly of genomes.

34.4.2 MAKER

MAKER (Cantarel et al. 2008) is a genome annotation pipeline, and it can be easily configured to any operating system. MAKER has so many software such as SNAP, Exonerate, Augustus, etc. MAKER first identifies and masks repeat, aligns ESTs and proteins to a genome to generate accurate gene models, and produces ab initio gene predictions. It also automatically synthesizes these data into gene annotations with the help of evidence-based quality values. The outputs of MAKER can be directly loaded into a GMOD database and can also be viewed in the Apollo genome browser (Lee et al. 2009).

34.4.3 RSEM

RSEM (Li and Dewey 2011) is a user-friendly software package for quantifying gene and isoform abundances from single-end or paired-end RNA-Seq data. RSEM gives result in terms of abundance of transcripts at 95% credibility and helps in visualization of these files. It can also simulate RNA-Seq data without any reference genome. Thus, RSEM along with de novo transcriptome assembly also enables accurate transcript quantification for species without sequenced genome. Mapping of reads to reference genome or transcriptome can be done effectively by RSEM.

34.4.4 SAMTOOLS

SAMTOOLS (Heng et al. 2009) have many utilities to deal with alignments in the SAM format. Some of these utilities involve sorting, merging, indexing, and generating alignment in a per-position format from SAM formats. SAM format is able to store all the alignment information generated by various alignment programs. SAM is very compressed and compact in file size and allows most of the operations on alignment without loading the whole alignment into the memory.

34.4.5 JELLYFISH

JELLYFISH (Marçais and Kingsford 2011) is a fast and memory efficient k-mer counting algorithm. It is works on multithreaded and multiprocessors. It has a lock-free hash table which is optimized for counting k-mers till 31 bases in length. This algorithm includes following steps:

- Generation of a fast k-mer hash table.
- Updates the lock-free hash table.
- Less memory usage for a hash entry.
- Encodes space-efficient keys.
- Efficient merging of intermediate hash tables.
- Analysis of running time.

34.4.6 Augustus

Augustus (Stanke and Morgenstern 2005) is a gene prediction software for eukaryotic genome sequences. It works on the basis of a generalized Hidden Markov Model. It calculates probability distributions for the various sections of genomic sequences such as introns, exons, intergenic regions, etc. The probabilistic model of Augustus considers following points while assigning genes in genomic sequences:

- The splice sites and gene sequences around it
- To identify the branch point region of sequences
- The translation start site and base before it
- The coding and noncoding regions on the sequence

- The first coding bases of a gene
- The lengthwise distribution of single exons
- Identification of initial and internal exons
- To find terminal exons and intergenic regions
- The distribution of the number of exons per gene
- The distribution of introns according to its size

34.5 Specialized Databases for Fungi

Even though most of the commonly used databases have protein and DNA sequences and structures from all the taxon of living organism, but having specialized databases and tools help us in performing the analysis more efficiently and smoothly. There are many databases exclusively dedicated to fungi genome and protein sequences. Some of these are MycoBank database (Robert et al. 2013), Fungal Barcoding database (http://www.fungalbarcoding.org/), FungiDB (http://fungidb.org/), and Ensembl Fungi (http://fungi.ensembl.org/).

34.5.1 MycoBank Database

MycoBank database is an online repository of all fungal taxonomic novelties published. The main aim of this database is to provide an open access updated data of fungal taxonomy (including new names) to the mycological community. This database was first time launched in 2004. There are experts in MycoBank to check the newly proposed nomenclature of fungi. Experts check the validity, legitimacy, and linguistic correctness of the proposed names. MycoBank also provides links to other databases containing information regarding DNA data, reference specimens, living cultures, and pleomorphic names.

34.5.2 Fungal Barcoding Database

The aim of this database is to facilitate updated information of fungi in form of barcoding, which in turn helps in finding collaborations among researchers. DNA barcoding is a technique for identification of species by using short standardized segments of the genome of that organism. Fungal barcoding database promotes the DNA barcoding of fungi and many related organisms. With an estimate fungi number varies from 1.5 millions to 13.5 millions. Because of huge complexity, fungal species are particularly suitable for DNA barcoding technique to identify different species.

34.6 Conclusion

In the era of post genomics, there is vast number of protein and genomic sequences, which needs to be annotated, characterized, and classified to various structural and functional classes. Recent developments in high-throughput technologies have revolutionized the developments in genomics, RNA profiling, proteomics, and the study of small molecules from metabolic process of organisms. These "-omics" technologies can provide a dynamic profile of a species at different conditions and stages of life cycle. Genomics is a field of science dedicated to the study of sequencing and analysis of genomes. Sequencing genome of an organism, whether of small organism like microbes, fungi, or large and complex organisms like plants, is not an expensive task any more. As a result of this, at affordable pricing, complete genomes of several complex plants and fungi are sequenced in recent years with an objective to understand the whole biological system. The knowledge of complete genome and proteome is paving new dimensions to understand the functioning of genomes, its architecture and engineering of the desired genes, and pathways for well-being of human and environment.

It is highly cumbersome to elucidate structures of all these sequences and assign their functions by experimental methods, leading to computational approaches. These methods exploit sequence information for automatic annotation transfer to hypothetical sequences. In this book chapter, I have discussed various computational tools and databases employed in the research work of fungi, to understand the molecular mechanism in detail.

References

Abrahamson M, Jonsdottir S, Olafsson I, Jensson O, Grubb A (1992) Hereditary cystatin C amyloid angiopathy: identification of the disease-causing mutation and specific diagnosis by polymerase chain reaction based analysis. Hum Genet 89:377–380

Alder BJ, Wainwright TE (1959) Studies in molecular dynamics. I. General method. J Chem Phys 31:459

Apweiler R, Bairoch A, Wu CH, Barker WC, Boeckmann B, Ferro S, Gasteiger E, Huang H, Lopez R, Magrane M et al (2004) UniProt: the Universal Protein knowledgebase. Nucleic Acids Res 32:D115–D119

Ashburner M, Ball CA, Blake JA, Botstein D, Butler H, Cherry JM, Davis AP, Dolinski K, Dwight SS, Eppig JT et al (2000) Gene ontology: tool for the unification of biology. The Gene Ontology Consortium. Nat Genet 25:25–29

Bernstein FC, Koetzle TF, Williams GJ, Meyer EF, Brice MD, Rodgers JR, Kennard O, Shimanouchi T, Tasumi M (1977) The Protein Data Bank: a computer-based archival file for macromolecular structures. J Mol Biol 112:535–542

Brooks BR, Brooks CL, Mackerell AD, Nilsson L, Petrella RJ, Roux B, Won Y, Archontis G, Bartels C, Boresch S et al (2009) CHARMM: the biomolecular simulation program. J Comput Chem 30:1545–1614

Cantarel BL, Korf I, Robb SMC, Parra G, Ross E, Moore B, Holt C, Sánchez Alvarado A, Yandell M (2008) MAKER: an easy-to-use annotation pipeline designed for emerging model organism genomes. Genome Res 18:188–196

Case DA, Cheatham TE, Darden T, Gohlke H, Luo R, Merz KM, Onufriev A, Simmerling C, Wang B, Woods RJ (2005) The Amber biomolecular simulation programs. J Comput Chem 26:1668–1688

Dennis G, Sherman BT, Hosack DA, Yang J, Gao W, Lane HC, Lempicki RA (2003) DAVID: database for annotation, visualization, and integrated discovery. Genome Biol 4:P3

Eddy SR (2011) Accelerated profile HMM searches. PLoS Comput Biol 7:e1002195

Eisenberg D, Lüthy R, Bowie JU (1997) VERIFY3D: assessment of protein models with three-dimensional profiles. Methods Enzymol 277:396–404

Felsenstein J (1989) PHYLIP – phylogeny inference package (version 3.2). Cladistics 5:164–166

Finn RD, Mistry J, Tate J, Coggill P, Heger A, Pollington JE, Gavin OL, Gunasekaran P, Ceric G, Forslund K et al (2013) The Pfam protein families database. Nucleic Acids Res 38:D211–D222

Harris MA, Clark J, Ireland A, Lomax J, Ashburner M, Foulger R, Eilbeck K, Lewis S, Marshall B, Mungall C et al (2004) The Gene Ontology (GO) database and informatics resource. Nucleic Acids Res 32:D258–D261

Heng Li1, Handsaker B, Wysoker A, Fennell T, Jue Ruan NH, Marth G, Goncalo Abecasis RD (2009) The sequence alignment/map format and SAMtools. Bioinformatics 25:2078–2079

Jones S, Thornton JM (1995) Protein-protein interactions: a review of protein dimer structures. Prog Biophys Mol Biol 63:31–65

Jorgensen WL, Tirado-Rives J (1988) The OPLS [optimized potentials for liquid simulations] potential functions for proteins, energy minimizations for crystals of cyclic peptides and crambin. J Am Chem Soc 110:1657–1666

Knudsen M, Wiuf C (2010) The CATH database. Hum Genomics 4:207–212

Larkin MA, Blackshields G, Brown NP, Chenna R, McGettigan PA, McWilliam H, Valentin F, Wallace IM, Wilm A, Lopez R et al (2007) Clustal W and Clustal X version 2.0. Bioinformatics 23:2947–2948

Laskowski RA, MacArthur MW, Moss DS, Thornton JM (1993) PROCHECK: a program to check the stereochemical quality of protein structures. J Appl Crystallogr 26:283–291

Lee E, Harris N, Gibson M, Chetty R, Lewis S (2009) Apollo: a community resource for genome annotation editing. Bioinformatics 25:1836–1837

Li B, Dewey CN (2011) RSEM: accurate transcript quantification from RNA-Seq data with or without a reference genome. BMC Bioinformatics 12:323

Luo R, Liu B, Xie Y, Li Z, Huang W, Yuan J, He G, Chen Y, Pan Q, Liu Y et al (2012) SOAPdenovo2: an empirically improved memory-efficient short-read de novo assembler. Gigascience 1:18

Marçais G, Kingsford C (2011) A fast, lock-free approach for efficient parallel counting of occurrences of k-mers. Bioinformatics 27:764–770

McGinnis S, Madden TL (2004) BLAST: at the core of a powerful and diverse set of sequence analysis tools. Nucleic Acids Res 32:W20–W25

Murzin AG, Brenner SE, Hubbard T, Chothia C (1995) SCOP: a structural classification of proteins database for the investigation of sequences and structures. J Mol Biol 247:536–540

Pandini A, Fornili A, Fraternali F, Kleinjung J (2013) GSATools: analysis of allosteric communication and functional local motions using a structural alphabet. Bioinformatics 29:2053–2055

Pugalenthi G, Archunan G, Sowdhamini R (2005) DIAL: a web-based server for the automatic identification of structural domains in proteins. Nucleic Acids Res 33:W130–W132

Pugalenthi G, Shameer K, Srinivasan N, Sowdhamini R (2006) HARMONY: a server for the assessment of protein structures. Nucleic Acids Res 34:W231–W234

Robert V, Vu D, Amor ABH, van de Wiele N, Brouwer C, Jabas B, Szoke S, Dridi A, Triki M, Ben DS et al (2013) MycoBank gearing up for new horizons. IMA Fungus 4:371–379

Sali A, Blundell TL (1993) Comparative protein modelling by satisfaction of spatial restraints. J Mol Biol 234:779–815

Scott WRP, Hünenberger PH, Tironi IG, Mark AE, Billeter SR, Fennen J, Torda AE, Huber T, Krüger P, van Gunsteren WF (1999) The GROMOS biomolecular simulation program package. J Phys Chem A 103:3596–3607

Shen M-Y, Sali A (2006) Statistical potential for assessment and prediction of protein structures. Protein Sci 15:2507–2524

Stanke M, Morgenstern B (2005) AUGUSTUS: a web server for gene prediction in eukaryotes that allows user-defined constraints. Nucleic Acids Res 33:W465–W467

Wiederstein M, Sippl MJ (2007) ProSA-web: interactive web service for the recognition of errors in three-dimensional structures of proteins. Nucleic Acids Res 35:W407–W410

Chapter 35
Bioprospecting of Endophytic Fungi for Bioactive Compounds

C. Ganesh Kumar and Poornima Mongolla

Abstract Higher plants that host microflora that reside asymptomatically in their tissues for some time of their life and do not result in apparent symptoms of diseases are called as endophytes. These endophytic microbes may provide protection and survival strategies in their host plants with production of a repertoire of chemically diverse and structurally unprecedented secondary metabolites and drug leads exhibiting an array of biological activities such as antimicrobial, anticancer, antiviral, antioxidants, insecticidal, antimalarial, antiparasitics, antidiabetic, immunosuppressants, etc. The advancements made in fungal cultivation, separation of metabolites, spectroscopic techniques for metabolite characterization, and *in vitro* bioassays for activity evaluation have disclosed the huge potential of endophytic fungi as an unexploited bioresource of novel chemical scaffolds. The present book chapter highlights the chemical potential of endophytic fungi isolated from diverse niche environments and their contribution from a drug discovery perspective.

Keywords Extrolites · Bioactives · Scaffold · Antimicrobial · Antioxidant · Anticancer · Antiviral · Antidiabetic · Immunosuppressants

35.1 Introduction

Mankind has used natural products and/or chemicals derived from plants, fungi, bacteria, and other living organisms since antiquity and in folklore medicine to prepare a diverse range of formulations to treat various diseases and illnesses. In this regard, the fungi have played an important role in producing useful secondary metabolites for treating diseases of plants, animals, and human. Fungi are eukaryotic organisms that differ from bacteria and other prokaryotes in many ways. Endophytic fungi among all other fungi are the ones being widely explored. Plants act as a potential reservoir for indigenous fungi (Lu et al. 2012), bacteria (Patel et al.

C. Ganesh Kumar (✉) · P. Mongolla
Medicinal Chemistry and Biotechnology Division, CSIR-Indian Institute of Chemical Technology, Hyderabad, Telangana, India
e-mail: cgkumar@iict.res.in

P. Gehlot, J. Singh (eds.), *Fungi and their Role in Sustainable Development: Current Perspectives*, https://doi.org/10.1007/978-981-13-0393-7_35

2012), or actinomycetes (Cao et al. 2004) principally known as endophytes which colonize the plant living tissues internally and can function obligatively or facultatively without causing any harmful effects or symptoms to the host externally and instead contribute beneficial attributes to the host plant (Rosenbluth and Martinez-Romero 2006). These endophytes are responsible for nutrient assimilation, generation of defense system, and synthesis of secondary metabolites (Petrini 1991; Suto et al. 2002). Endophytes interact with higher plants and transfer information, and also many biochemical pathways are evolved which results in the production of various novel bioactive compounds with potential applications in medicine and biotechnology (Strobel et al. 1999a; Gangadevi and Muthumary 2008a). The research on fungal endophytes carried out over for several decades has resulted in adequate information with new additions on day-to-day basis. Endophytes synthesize various bioactive agents that help in the defense response of plants against pathogens, while others were explored to identify novel drugs from a drug discovery perspective.

35.2 Relationship of Fungal Endophytes with Host Plant, Occurrence, and Biodiversity

In the recent years, endophytic fungi are recognized to play significant role in affecting the secondary metabolite types and the quantities synthesized through fungus-host interaction, which signifies the knowledge generated on the endophytic fungi and medicinal plants relationships (Faeth and Fagan 2002). Although the understanding on these exact relationships is very limited, endophytic fungi have played a key role in the plant micro-ecosystem with regard to growth and development of host plants.

Different host-endophyte relationships such as mutualism, symbiosis can be ascribed due to the host specificity, recurrence, selectivity, or preference (Zhou and Hyde 2001; Schulz and Boyle 2005; Cohen 2006; Arnold 2007; Purahong and Hyde 2011). These interactions are controlled at the genetic level, involving the genes from the endophytic fungi and its host which modulates the environment (Moricca and Ragazzi 2008); these interactions remain asymptomatic till a balance exists between plant defense and fungal virulence.

The term host preference represents the distinctive occurrence of an endophyte with respect to a particular host and indicates the divergence in the composition of endophytic community and relationships among the different host plants (Suryanarayanan and Kumaresan 2000). The fungal endophytic diversity varies based on different plant parts and the geographical distribution of the host plant in different environments including temperate (Ganley et al. 2004), tropic (Mohali et al. 2005), aquatic (Krzic et al. 2006; Debbab et al. 2011), and xerophytic (Suryanarayanan et al. 2005). Fungal endophytes were isolated almost from every plant groups such as lichens (Li et al. 2007), palm (Rungjindamai et al. 2008), grasses (Shankar and Shashikala 2010), sea grasses (Thirunavukkarasu et al. 2011), large trees (Sutjaritvorakul et al. 2011), and medicinal plants (Gautam 2013).

Among the isolated fungal classes, deuteromycetes were highly prevalent (Khan et al. 2010), while other isolated forms belonged to ascomycetes (Rodrigues and Samuels 1994) and basidiomycetes (Rungjindamai et al. 2008). Endophytic fungal diversity in medicinal plants has been investigated on the basis of their number, types, colonization rates, and isolation strategy.

35.3 Bioactive Compounds Derived from Fungal Endophytes

Endophytic fungi protect its host from pathogenic microflora by disallowing the pathogens to develop a systemic relationship with the host plant (Jalgaonwalal et al. 2011) and are practiced as a new approach for controlling plant diseases. A diverse range of metabolites has been isolated from endophytic fungi emphasizing their potential ecological role. Estimates made, until 2003, indicated 4000 bioactive compounds from some fungal species, viz., *Penicillium*, *Fusarium*, and *Acremonium*, playing a functional role on different aspects; however, few reports exist from endophytes (Dreyfuss and Chapela 1994; Padhi et al. 2013).

35.4 Antibiotics from Fungal Endophytes

Fungal secondary metabolites have been first isolated and differentiated for industrial application (Schulz 2002). Some natural products are harmful such as mycotoxins, while others are beneficial such as antibiotics to mankind (Strobel et al. 2004). For a mycologist to study fungal ecology, it is clear that secondary metabolites play an important role *in vivo* and are important for many metabolic interactions between their plant host and fungal host, playing a functional role in signaling, regulation, and defense of the symbiosis. A diverse array of natural products were synthesized by endophytic fungi which exhibited antimicrobial activity against various pathogenic microflora, and these can find commercial use in pharmaceuticals, medicine, and agriculture (Castillo et al. 2002, 2003; Ezra et al. 2004; Park et al. 2005; Wang et al. 2007). *Colletotrichum gloeosporioides*, an endophytic fungus, produced three new antimicrobial metabolites, whose structures were elucidated based on different spectroscopic techniques as 6-isoprenylindole-3-carboxylic acid, 3β,5α-dihydroxy-6β-acetoxy-ergosta-7,22-diene, and 3β,5α-dihydroxy-6β-phenylacetyloxy-ergosta-7,22-diene. These metabolites exhibited antimicrobial activity against different human pathogenic bacteria as well as fungistatic activity against crop pathogenic fungi (Lu et al. 2000). *Streptomyces* sp. (MSU-2110) isolated from endophytic *Monstera* sp. produced a peptide antibiotic, Coronamycin, which exhibited antimicrobial activity (MIC, 2 μg/mL) against *Cryptococcus neoformans* as well as promising antimalarial activity (IC_{50} value of 9 ng/mL) against *Plasmodium falciparum*, while chloroquine showed IC_{50} value of 7 ng/mL (Ezra et al. 2004).

35.5 Anticancer Agents from Fungal Endophytes

Cancer causes 8.8 million deaths worldwide in 2015 (World Health Organization 2017a), a disease manifested by unregulated cell proliferation, which results in abnormal cell multiplication and uncontrolled tissue growth (American Cancer Society 2009). Several plant-derived compounds such as taxol (*Taxus baccata*, *T. brevifolia*, *T. canadensis*), vinblastine and vincristine (*Catharanthus roseus*), camptothecin (*Camptotheca acuminata*), and its derivatives like topotecan and irinotecan, etoposide obtained from epipodophyllotoxin, an isomer of podophyllotoxin (roots of *Podophyllum peltatum*) have played an important role in the development of clinically useful anticancer drugs (Firakova et al. 2007; Debbab et al. 2011; Sheela 2012). Some examples of endophytic fungi producing anticancer activity are shown in Table 35.1.

Paclitaxel (Taxol) is the major bioactive compound having a tetracyclic diterpenoid structure obtained from the bark of the Pacific yew tree, *Taxus brevifolia* Nutt. (Taxaceae family) (Wani et al. 1971). This compound exhibited diverse bioactivities with unique structures. It was the world's first billion dollar anticancer drug. Bristol-Myers Squibb received approval in December 1992 from US Food and Drugs Administration (US-FDA) to commercialize pactitaxel under the trade name Taxol for treatment of ovarian and breast cancer. It generated a renewed attention in medicine as a blockbuster drug for clinical use in the treatment of various cancers due to its distinctive mode of action. In addition, paclitaxel finds extensive use in biomedical research as a microtubule stabilizer (Strobel et al. 1993, 1996a, b, 1997; Miao et al. 2009; Deng et al. 2009; Gond et al. 2014). Its mode of action against various cancer types is due to the stabilization of the microtubules and disruption of their dynamic equilibrium (Wang et al. 1999). It is also reported to be effective against non-cancerous conditions like polycystic kidney diseases (Visalakchi and Muthumary 2010). *Taxus* trees are rare and few in number and produce small quantities (0.01–005%) of Taxol which cannot meet the full regimen (2 g of purified Taxol) for antitumor treatment as well as fulfill the current global demand (Stierle et al. 1993). A major breakthrough on the identification of Taxol (paclitaxel)-producing fungal endophytes has opened new avenues for investigating the anticancer properties of its metabolites. This served as an alternative strategy to meet the growing demand of Taxol (Gangadevi and Muthumary 2008a). The first Taxol-producing fungal endophyte reported was *Taxomyces andreanae* from the Pacific yew, *Taxus brevifolia* (Stierle et al. 1993). Since then a number of endophytic fungal genera have demonstrated Taxol production which are listed in Table 35.2.

Based on chemical inhibitors, gene discovery, and gene expression studies, it was confirmed that the biosynthesis of fungal paclitaxel is through shikimic acid pathway (Soliman et al. 2011). An endophytic fungus, *Pestalotiopsis microspora* CP-4 isolated from the tree *Torreya taxifolia*, produced a selective cytotoxic quinone dimer, torreyanic acid, which was more cytotoxic (five to ten times) to cell lines sensitive to protein kinase C agonists, which resulted in apoptotic cell death (Lee et al. 1996).

Table 35.1 Some examples of bioactive compounds produced by endophytic fungi exhibiting anticancer activity

Name of the Endophyte	Plant source	Compound Name	Structure	References
Phyllosticta spinarum, Bartalinia robillardoides	*Taxus baccata*	Taxol	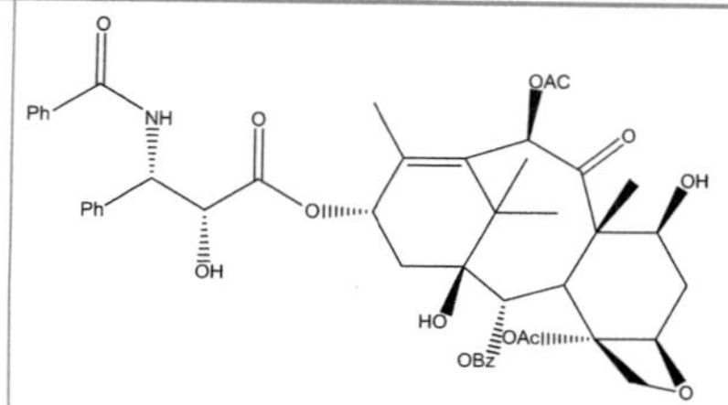	Gangadevi and Muthumary (2008a)
Pestalotiopsis microspora	*Torreya taxifolia*	Torreyanic acid	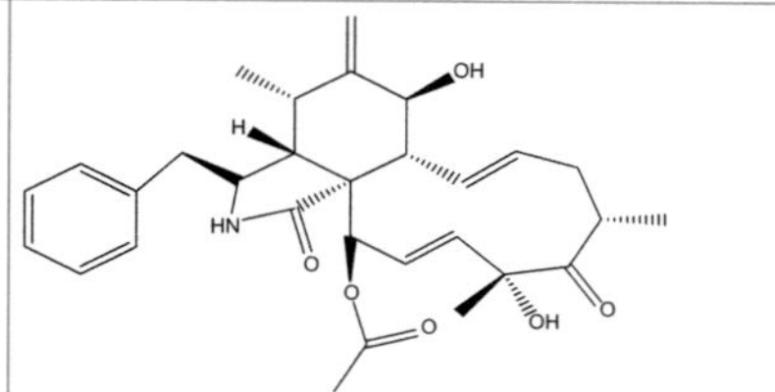	Lee et al. (1996)
Fusarium solani	*Camptotheca acuminate*	Camptothecin	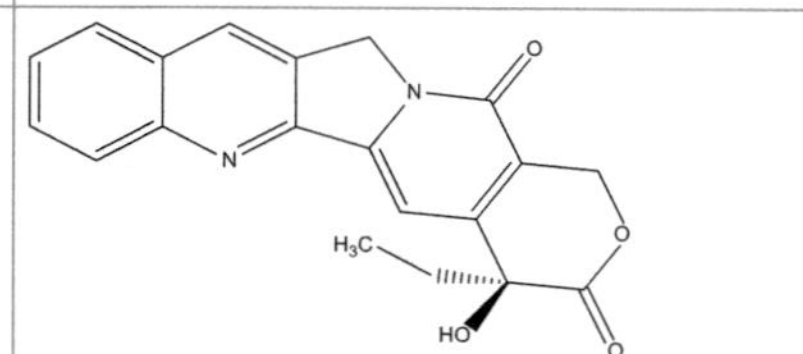	Kusari et al. (2009a)

(continued)

Table 35.1 (continued)

Name of the Endophyte	Plant source	Compound Name	Structure	References
Aspergillus fumigatus	*Podophyllum peltatum*	Podophyllotoxin	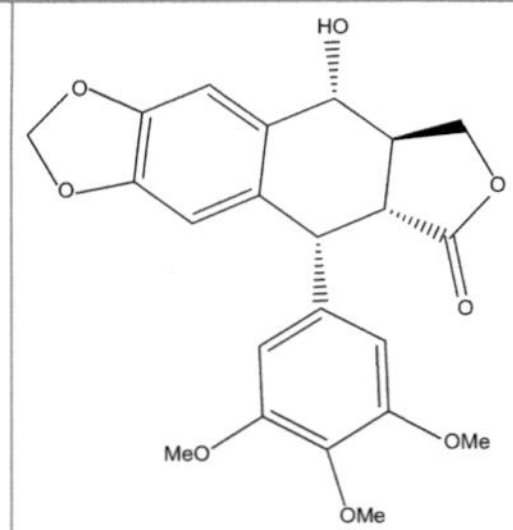	Eyberger et al. (2006)
Unidentified fungus XG8D	*Xylocarpus granatum*	Merulin A and C	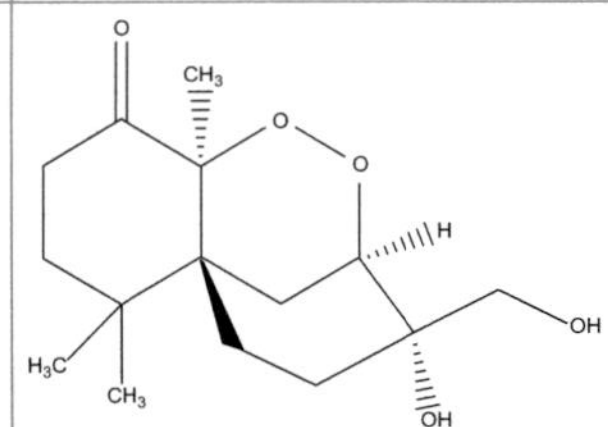	Chokpaiboon et al. (2010)

Table 35.2 Some examples of various endophytic fungal genera producing taxol

Endophytic fungus	Host plant	Taxol yield (μg L^{-1})	Reference
Alternaria sp. strain C-9	*Corylus avellana* (hazelnut)	n.s.	Michalczyk et al. (2014)
Aspergillus candidus strain MD3	*Taxus × media* (yew)	112 μg g^{-1} (dry wt. of mycelium)	Zhang et al. (2009b)
Aspergillus niger var. *taxi* strain HD86–9	*Taxus cuspidata* (yew)	273.46	Zhao et al. (2009)
Bartalinia robillardoides Tassi strain AMB-9	*Aegle marmelos* Correa ex Roxb.	187.6	Gangadevi and Muthumary (2008a)
Botryodiplodia theobromae	n.s.	118.7 μg L^{-1}	Muthumary and Sashirekha (2007)
Cladosporium cladosporioides strain MD2	*Taxus media* (yew)	800	Zhang et al. (2009c)
Chaetomella raphigera strain TAC-15	*Terminalia arjuna*	79.6 μg L^{-1}	Gangadevi and Muthumary (2009a)
Colletotrichum sp.	*Maguireothamnus speciosus*	6 ng/L	Strobel et al. (1999a)
Colletotrichum capsici	*Capsicum annuum* L. (Chilli)	687 μg L^{-1}	Kumaran et al. (2011)
Colletotrichum gloeosporioides strain JGC-9	*Justicia gendarussa*	163.4 μg L^{-1}	Gangadevi and Muthumary (2008b)
Colletotrichum gloeosporioides (Penz.) Sacc.	*Plumeria acutifolia*	57.54 μg L^{-1}	Nithya and Muthumary (2009)
Colletotrichum gloeosporioides (Penz.)	*Tectona grandis*	n.s.	Senthilkumar et al. (2013)
Fusarium arthrosporioides strain F-40	*Taxus cuspidata* (yew)	131 μg L^{-1}	Li et al. (2008c)
Fusarium solani	*Taxus celebica*	1.6 μg L^{-1}	Chakravarthi et al. (2008)
Fusarium redolens strain TBPJ-B	*Taxus baccata* subsp. *wallichiana* (yew)	66 μg L^{-1}	Garyali et al. (2013)
Fusarium sp. strain 53	*Taxus wallichiana* var. *mairei* (yew)	270 μg g^{-1} (dry wt. of mycelium)	Jian et al. (2015)
Gliocladium sp.	*Taxus baccata* (yew)	10 μg L^{-1}	Sreekanth et al. (2009)
Grammothele lineata	*Corchorus olitorius* (jute)	382.2 μg L^{-1}	Das et al. (2017)
Lasiodiplodia theobromae	*Morinda citrifolia*	245 μg L^{-1}	Pandi et al. (2011)
Metarhizium anisopliae strain H-27	*Taxus chinensis* (yew)	846.1	Liu et al. (2009)
Ozonium sp. strain BT2	*Taxus chinensis* var. *mairei* (yew)	n.s.	Guo et al. (2006)

(continued)

Table 35.2 (continued)

Endophytic fungus	Host plant	Taxol yield (μg L^{-1})	Reference
Ozonium sp. strain EFY-21	*Taxus chinensis* var. *mairei* (yew)	n.s.	Zhou et al. (2007)
Paraconiothyrium variabile	*Taxus baccata* (English yew)	7.11 μg L^{-1}	Somjaipeng et al. (2015)
Periconia sp.	*Torreya grandifolia* (yew)	30–831 ng/L	Li et al. (1998)
Pestalotiopsis sp. strain W-1f-1	*Wollemia nobilis* (wollemi pine)	172 ng/L	Strobel et al. (1997)
Pestalotiopsis guepinii strain W-1f-2	*Wollemia nobilis* (wollemi pine)	485 ng/L	Strobel et al. (1997)
Pestalotiopsis breviseta	*Ervatamia divaricata*	0.064 mg/L	Kathivaran and Raman (2010)
Pestalotiopsis menezesiana	n.s.	274.9 mg/L	Muthumary and Sashirekha (2007)
Pestalotiopsis microspora strain CP-4	*Taxodium distichum* (bald cypress)	50–1487 ng/L	Li et al. (1996)
Pestalotiopsis microspora strain NE-32	*Taxus wallichiana* (Himalayan yew)	60–70 μg L^{-1}	Strobel et al. (1996a)
Pestalotiopsis microspora	*Maguireothamnus speciosus*	114 ng/L	Strobel et al. (1999a)
Pestalotiopsis microspora strain KS-15	*Taxus wallichiana* (Himalayan yew)	203.4 ng/L	Shrestha et al. (2001)
Pestalotiopsis pauciseta strain CHP-11	*Cardiospermum halicacabum*	113.3 μg/L	Gangadevi et al. (2008)
Pestalotiopsis pauciseta strain VM1	*Tabebuia pentaphylla* Hemsl.	208.6 μg/L	Vennila and Muthumary (2011)
Pestalotiopsis terminaliae	*Terminalia arjuna*	211.1 μg/L	Gangadevi and Muthumary (2009b)
Pestalotiopsis uvicola	n.s.	118.7 μg/L	Muthumary and Sashirekha (2007)
Pestalotiopsis versicolor	*Taxus cuspidata* (yew)	478 μg/L	Kumaran et al. (2010)
Phoma betae	*Ginkgo biloba*	795 μg L^{-1}	Kumaran et al. (2012)
Phoma sp.	*Calotropis gigantea*	76.13 μg L^{-1}	Hemamalini et al. (2015)
Phoma medicaginis strain 218	*Taxus chinensis* var. *mairei* (yew)	1215 μg L^{-1}	Jian et al. (2017)
Phomopsis sp. strain BKH 27	*Taxus cuspidata* (Japanese yew)	418 μg L^{-1}	Kumaran and Hur (2009)
Phomopsis sp. strain BKH 30	*Ginkgo biloba*	372 μg L^{-1}	Kumaran and Hur (2009)
Phomopsis sp. strain BKH 35	*Larix leptolepis*	334 μg L^{-1}	Kumaran and Hur (2009)

(continued)

Table 35.2 (continued)

Endophytic fungus	Host plant	Taxol yield (μg L^{-1})	Reference
Phyllosticta spinarum	*Cupressus* sp. (conifer)	235 μg L^{-1}	Kumaran et al. (2008a)
Phyllosticta citricarpa	*Citrus medica*	265 μg/L	Kumaran et al. (2008b)
Phyllosticta melochiae	*Melochia corchorifolia*	274 μg/L	Kumaran et al. (2008c)
Phyllosticta tabernaemontanae	*Wrightia tinctoria*	461 μg/L	Kumaran et al. (2009)
Seimatoantlerium tepuiense strain B(V) L3 (epiphyte)	*Maguireothamnus speciosus*	263 ng/L	Strobel et al. (1999a)
Sporormia minima strain KS-12	*Taxus wallichiana* (Himalayan yew)	15.7 ng/L	Shrestha et al. (2001)
Stemphylium sedicola SBU-16	*Taxus baccata* (English yew)	6.9 μg/L	Mirjalili et al. (2012)
Taxomyces andreanae	*Taxus brevifolia* (Pacific yew)	24–50 ng/L	Stierle et al. (1993)
Trichothecium sp. strain KS-13	*Taxus wallichiana* (Himalayan yew)	165.7 ng/L	Shrestha et al. (2001)
Tubercularia sp. strain TF5	*Taxus mairei* (yew)	n.s.	Wang et al. (2000)
Xylaria sp.	*Maguireothamnus speciosus*	242 ng/L	Strobel et al. (1999a)

n.s. not specified

Alkaloids are potent anticancer agents and their presence was established in endophytic fungi. Three novel cytochalasin alkaloids exhibiting antitumor activity were produced by the endophytic fungus *Rhinocladiella* (Wagenaar et al. 2000). Camptothecin and 10-hydroxycamptothecin are pentacyclic quinoline alkaloids which act as precursors for the synthesis of chemotherapeutic agents such as topotecan and irinotecan, which function as topoisomerase I inhibitors (Meng et al. 2003; Uma et al. 2008). Camptothecin has been reported to find wide usage for skin disease treatment in China (Guo et al. 2008). An endophytic fungus belonging to the family of Phycomycetes produced camptothecin, when cultured in Sabouraud broth (Puri et al. 2005). Several endophytes such as *Entrophospora infrequens* (Amna et al. 2006), *Neurospora* sp. (Rehman et al. 2008), *Fusarium solani* (Kusari et al. 2009a; Shweta et al. 2010), and *Xylaria* sp. M20 (Liu et al. 2010b) were reported to produce camptothecin and other analogs of camptothecin. An endophytic fungus strain PM0651480 isolated from leaves of *Mimusops elengi* (Sapotaceae family) produced ergoflavin that exhibited promising anticancer activities (Deshmukh et al. 2009). Naturally, vinblastine and vincristine are terpenoid indole alkaloids produced by *Catharanthus roseus*. Vincristine was derived from endophytic *Fusarium oxysporum* to *Catharanthus roseus* (Zhang et al. 2000; Tung et al. 2002), while vinblastine was obtained from *Alternaria* sp. isolated from *Catharanthus roseus* (Guo and Kunming 1998). Both vinblastine and vincristine were produced by endophytic

Fusarium oxysporum AA-CRL-6 isolated from *Catharanthus roseus* with yields of 76 and 67 μg, respectively (Kumar et al. 2013). They function by interfering with the microtubule and mitotic spindle (Zhao et al. 2010).

Lignans are shikimic acid pathway-originated secondary metabolites displaying cytotoxic and other biological activities having significant medical importance (Gordaliza et al. 2004). Especially in cancer chemotherapy, lignans showed vast structural and biological diversity (Korkina 2007). Podophyllotoxins are aryl tetralin lignans (2,7′ cyclolignan) of commercial interest and have been primarily derived from the roots and rhizomes of *Podophyllum* (*Sinopodophyllum*) *hexandrum* and *P. peltatum*. Cultivation of *Podophyllum* sp. has been unsuccessful due to lack of proper climatic conditions (Moraes et al. 2001; Lee and Xiao 2003). In view of the short supply and growing demand, there has been a search for alternative routes to derive podophyllotoxins (Bhattacharyya and Chattopadhyay 2015; Ardalani et al. 2017). Commercially, the total chemical synthesis of podophyllotoxins has not been feasible due to the presence of four chiral centers along with γ-lactone and a high degree of oxygenation (Damayanti and Lown 1998; Berkovitz et al. 2000). The microbial-derived lignans are bioactive in nature exhibiting promising antioxidant, anticancer, and radioprotective properties (Arora et al. 2005a, b, c). The podophyllotoxin derivatives are widely used in the synthesis of topoisomerase inhibitors for cancer chemotherapy. Phenylpropanoids produced by endophytes have gained renewed interest in medicine as anticancer and antioxidant agents (Korkina 2007). *Trametes hirsuta* isolated from *Podophyllum hexandrum* (Puri et al. 2006), *Aspergillus fumigatus* from *Juniperus communis* L. (Kusari et al. 2009b), *Fusarium oxysporum* from *Juniperus recurva* (Kour et al. 2008), and *Phialocephala fortinii* from *Podophyllum peltatum* (Eyberger et al. 2006) were identified as important fungal endophytic sources of podophyllotoxin. The endophytic *Penicillium brasilianum* isolated from *Melia azedarach* (root bark) was reported to be involved in the phenylpropanoid amide biosynthesis (Fill et al. 2010).

Curvularia lunata, isolated from endophytic *Niphates olemda*, produced cytoskyrins exhibiting antibacterial and potential anticancer activities (Brady and Clardy 2000; Jadulco et al. 2002). Brefeldin A produced by endophytic *Phoma medicaginis*, *Medicago sativa* and *Medicago lupulina*, caused apoptosis (Weber et al. 2004b). Secalonic acid D, a secondary metabolite from the mangrove endophytic fungus (ZSU44) exhibited promising anticancer activities against HL60 and K562 cell lines with IC_{50} values of 0.38 and 0.43 μM, respectively, as well as caused cell arrest of G(1) phase and induced leukemia cell apoptosis through GSK-3β/β-catenin/c-Myc pathway (Zhang et al. 2009a).

An unidentified endophytic fungus XG8D produced merulin A, B, and C. However, merulins A and C showed promising cytotoxicity against BT474 (breast cancer) cell line (IC_{50}: 19.60 and 5.57 μM) and against SW620 (colon) cell line (IC_{50}: 19.05 and 14.57 μM) (Chokpaiboon et al. 2010). Endophytic *Fusarium* sp. strain ZZF41, isolated from mangrove tree *Kandelia candel*, produced 5-*O*-methyl-2′-methoxy-3′-methylalpinumisoflavone, with significant cytotoxicity against Hep-2 (IC_{50}: 4 μM) and HepG2 (IC_{50}: 11 μM) cancer cell lines (Huang et al. 2010). *Diaporthe* sp. HLY-1, an endophytic fungus to mangrove *Kandelia candel*, produced mycoepoxydiene, which induced cell cycle arrest at G2/M phase and apoptosis in HeLa cells (Wang et al. 2010a).

Expansols A and B were two new polyphenols possessing both phenolic bisabolane sesquiterpenoid and diphenyl ether units reported to be produced by endophytic fungus *Penicillium expansum* 091006. Expansol A exhibited moderate cytotoxic activity (IC_{50}: 15.7 μM) against HL-60 cell line, while expansol B showed pronounced cytotoxicity against A549 (IC_{50}: 1.9 μM) and HL-60 (IC_{50}: 5.4 μM) cell lines (Lu et al. 2010). Allantopyrone A and islandic acid-II methyl ester were two bioactive compounds reported from endophytic fungus *Allantophomopsis lycopodina* KS-97 which exhibited cytotoxicity on HL60 cells (IC_{50}: 0.32 and 6.55 μM), and these compounds also induced apoptosis (Shiono et al. 2010). *Paecilomyces* sp. endophytic to the bark of mangrove produced new depsidones, named as paeciloxocins A and B. Paeciloxocin A exhibited significant cytotoxicity against HepG2 (IC_{50}: 2.69 μM), as well as arrested *Curvularia lunata* and *Candida albicans* (Wen et al. 2010). Phomoxanthones A and B obtained from endophytic *Phomopsis* sp. BCC 1323 (Isaka et al. 2001) and photinides A–F were benzofuranone-derived γ-lactones derived from endophytic fungus *Pestalotiopsis photiniae* (Ding et al. 2009) and also exhibited cytotoxicity to various cancer cell lines.

Among the active endophytes, the genus *Chaetomium* produces different cytotoxic metabolites such as globosumones A–C (Bashyal et al. 2005); globosuxanthone, a dihydroxanthenone (Wijeratne et al. 2006); chaetoglobosins (Ding et al. 2006; Qin et al. 2009; Li et al. 2014); cochliodinols (Debbab et al. 2009); chaetomugilin A (Qin et al. 2009); etc. The cytotoxic effect on Potoroo kidney cell line (PtK2) showed cell enlargement, loss of actin fibers, and failure of cell division after nuclear changes (Suryanarayanan et al. 2009), while cytochalasin analogs inhibited actin polymerization (Yahara et al. 1982).

Several solanapyrones A, D, E, F, and G, produced by endophytic *Alternaria* sp., functioned as DNA polymerase inhibitors (Mizushina et al. 2002). Endophytic *Nigrospora* sp. produced epoxyexserohilone (Cutler et al. 1991), nigrosporolide (Harwood et al. 1995), phomalactone (Kim et al. 2001), aphidicolin (Lopes and Pupo 2011), griseofulvin (Zhao et al. 2012), bostrycin (Zhang et al. 2013a), and nigrosphaerin A (Metwaly et al. 2014). Aphidicolin (tetracyclic diterpene-tetraol) inhibited nuclear DNA synthesis in eukaryotes and interfered with α-enzyme of DNA polymerase. It also inhibited S phase of cell cycle (Spadari et al. 1982). Endophytic *Fusarium* sp. secreted apicidin, a novel cyclic tetrappetide which functions as histone deacetylase inhibitor (Han et al. 2000), while enniatins are inhibitors of Pdr5p, a major multidrug efflux pumps present in *Saccharomyces cerevisiae* that confers multidrug resistance (Hiraga et al. 2005).

Endophytic *P. microspora* produced torreyanic acid, a selectively cytotoxic quinine dimer which induced apoptosis followed by cell death (Lee et al. 1996). Endophytic *Aspergillus niger* IFB-E003 to *Cyndon dactylon* produced various naphtho-γ-pyrones such as rubrofusarin B, fonsecinone A, asperpyrone B, and aurasperone A. Among them, rubrofusarin B showed cytotoxicity to colon cancer cell line SW1116 (IC_{50} value: 4.5 μg/ml), while aurasperone A showed inhibition of xanthine oxidase (IC_{50} value: 10.9 μM) (Song et al. 2004). These spirobisnaphthalenes exhibited several medicinal properties including antimicrobial (Cai et al.

2009), cytotoxic (Chu et al. 1995; Seephonkai et al. 2002), and as DNA gyrase inhibitors (Sakemi et al. 1995). Tetracyclic cyclopiane diterpenes such as conidiogenones B–D and F were isolated from endophytic *Penicillium chrysogenum* QEN-24S present in algae, *Laurencia* sp. Conidiogenone B showed promising antimicrobial activity against MRSA, *P. fluorescens*, *P. aeruginosa*, and *S. epidermidis* with MIC value of 8 μg/mL (Gao et al. 2011). Endophytic *Halorosellinia* sp. produced anthraquinone SZ-685C which stopped proliferation of different cancer cell lines such as human breast cancer, prostate cancer, glioma, and hepatoma (IC_{50}: 3.0–9.6 μM) and induced apoptosis through Akt/FOXO pathway (Xie et al. 2010). *Aspergillus parasiticus*, an endophyte to *Sequoia sempervirens*, produced four sequoiatones C–F, which were cytotoxic to brine shrimp (Stierle et al. 2001). Gliocladicillins A and B were produced by *Cordyceps*-colonizing fungi which induced tumor cell apoptosis through both extrinsic and intrinsic pathways in melanoma B16 cells (Chen et al. 2009). Endophytic fungus *Cryptosporiopsis* sp. produced (R)-5-hydroxy-2-methylchroman-4-one which exhibited significant cytotoxicity against human leukemia cell line, HL-60 (IC_{50}: 4 μg/mL) and also significantly induced G2 cell cycle arrest of HL-60 (Zilla et al. 2013). Two naphthoquinones, namely, anhydrofusarubin and methyl ether of fusarubin, produced by endophytic fungus *Cladosporium* sp. exhibited potential cytotoxicity against human leukemia cells (K-562) (IC_{50}: 3.97 and 3.58 μg/mL) (Khan et al. 2015). Further, an endophytic *Chaetomium arcuatum* strain SAF-2 isolated from fruit of *Semecarpus anacardium* produced two bioactive compounds such as eugenetin and 6-hydroxyeugenin, which showed cytotoxicity against A549, HeLa, MCF-7, MBA-MB-231, and COLO205 cell lines (Kumar et al. 2017).

35.6 Antimicrobial Compounds from Fungal Endophytes

Antimicrobial metabolites produced by endophytes are low molecular weight extrolites not essential for growth and produced at low concentrations against pathogenic invasion (Strobel and Daisy 2003; Guo et al. 2006). Thus endophytes are a promising resource of antimicrobial compounds to counteract the serious threat from human drug-resistant and plant pathogens (Tan and Zou 2001; Yu et al. 2010). On the basis of literature survey, it was found that endophytes produced several structurally diverse classes of antimicrobial compounds such as alkaloids, terpenoids, phenols, quinines, flavonoids, and steroids (Guo et al. 2000). Monoterpene preaustinoids A1, A2, and B1 produced by *Penicillium* sp., endophytic to *Melia azedarach*, showed moderate bacteriostatic activity against *E. coli*, *S. aureus*, *P. aeruginosa*, and *Bacillus* sp. (Geris dos Santos and Rodrigues-Fo 2002, 2003). Similarly, four other monoterpenes such as preaustinoid B2, preaustinoid A3, austinolide, and isoaustinone were identified which were produced due to α-ketol rearraangements and Bayer-Villiger oxidations involved in the biosynthetic process (Fill et al. 2007).

Endophytic *Rhizoctonia* sp. Cy064 to *Cynodon dactylon* produced rhizoctonic acid exhibiting anti-*Helicobacter pylori* activity, the causative agent for peptic ulcer (Ma et al. 2004). Periconicins A and B produced by endophytic fungus *Periconia* sp.

exhibit antibacterial activity (Kim et al. 2004). Naphthodianthrone derivative, hypericin produced by *Hypericum perforatum*, exhibited antimicrobial activity against various human pathogens as well as antifungal activity against *Aspergillus niger* and *C. albicans* (Kusari et al. 2008). *Xylaria* sp. YX-28 endophytic to *Ginkgo biloba* produced 7-amino-4-methylcoumarin which showed antimicrobial activity against various pathogens (Liu et al. 2008). Cylarphthalide A, (-)-5-carboxylmellein, and (-)-5-methylmellein, produced by endophytic *Xylaria* sp. GDG-102 from *Sophora tonkinensis*, exhibited significant antibacterial activity (Zheng et al. 2017).

Curvularide B produced by endophyte *Curvularia geniculata* exhibited antifungal activity against *Candida albicans* (Chomcheon et al. 2010). Endophytic *Talaromyces* sp. produced antimicrobial extrolites such as 7-epiaustdiol, stemphyperylenol, and secalonic acid A. It was observed that 7-epiaustdiol showed significant inhibitory activity against *P. aeruginosa* (MIC, 6.25 μg/mL). Stemphyperylenol inhibited *Sarcina ventriculi* with MIC value of 3.21 μg/mL, while secalonic acid A inhibited all the tested microbes (Liu et al. 2010a). Pyrrocidine C produced by *Lewia infectoria* SNB-GTC2402 exhibited antibacterial activity against *S. aureus* ATCC 29213 (MIC, 2 μg/mL) (Casella et al. 2013). Equisetin derived from endophytic *Fusarium* sp. showed antimicrobial activities against *Bacillus subtilis* UBC 344 (MIC, 8 μg/mL), *Staphylococcus aureus* ATCC 43300 (MIC, 16 μg/mL), and methicillin-resistant *Staphylococcus aureus* ATCC 33591 (MRSA, MIC, 16 μg/mL) (Ratnaweera et al. 2015). Chaetoglobosin A, a cytochalasan produced by endophytic *Chaetomium globosum*, exhibited antimicrobial activities against *Bacillus subtilis* (MIC, 16 μg/mL), *Staphylococcus aureus* (MIC, 32 μg/mL), and MRSA (MIC, 32 μg/mL) (Dissanayake et al. 2016).

Yu et al. (2010) compiled an excellent review on the antimicrobial metabolites produced by endophytes which cater to different classes such as alkaloids, peptides, steroids, terpenoids, phenols, quinines, and flavonoids. A significant number of antimicrobial compounds have been reported from endophytes, however, they occupy a small portion of the total endophytic community identified. Endophytic *Aspergillus fumigatus* CY018 produced asperfumin which inhibited *Candida albicans* (Liu et al. 2004). Endophytic *Fusarium redolens* Dzf2 isolated from *Dioscorea zingiberensis* rhizomes produced beauvericin which exhibited significant antimicrobial activity (Xu et al. 2010). Beauvericin produced by endophytic *Epicoccum nigrum* showed remarkable antibacterial activity against *Bacillus cereus* (MIC, 3.12 μg/mL) and *Salmonella typhimurium* (MIC, 6.25 μg/mL) (Dzoyem et al. 2017). *Microdiplodia hawaiiensis* CZ315 endophytic to *Garcinia mangostana* produced bioactive compounds with antimicrobial activity against *S. aureus*, *B. subtilis*, *M. luteus*, *E. coli*, *S. typhi*, and *P. aeruginosa* (MIC, 25–200 μg/mL) (Radji et al. 2011). Different endophytic fungi isolated from *Ocimum sanctum* (Tulsi) showed high antimicrobial activity against *Candida albicans* and *Pseudomonas aeruginosa* (Pavithra et al. 2012). Endophytic *Diaporthe helianthi* (anamorphic *Phomopsis helianthi*) isolated from *Luehea divaricata* produced 2(-4-hydroxyphenyl)-ethanol (tyrosol) which exhibited antimicrobial activity against *E. coli*, *S. aureus*, *S. typhi*, *Enterococcus hirae*, and *Xanthomonas axonopodis* pv. *phaseoli* (Specian et al. 2012). Endophytic *Eurotium amstelodami* isolated from *Ipomea pes-carprae* L. produced tetrahydroauroglaucin and flavoglaucin which showed antimicrobial activity against *B. subtilis*, *S. aureus*, *P. aeruginosa*, *and C. albicans* (Chaipackdee et al. 2013). Endophytic *Phoma* sp. produced bioactive compounds such as sclerodin, 8,9-dihydro-3,5,7-trihydroxy-1,8,8,9-tetramethyl-5-(2-oxopropyl)-4*H*-phenaleno[1,2-*b*]furan-4,6(5*H*)-dione, atrovenetinone, and sclerodione. Among them atrovenetinone showed good antibacterial activity toward *Eurotium repens*. Furthermore, atrovenetinone and sclerodione displayed very strong antifungal activity toward *Ustilago violacea* (Hussain et al. 2015). Some examples of endophytic fungi producing antimicrobial activity are shown in Table 35.3.

35.7 Antifungal Compounds from Endophytic Fungi

Cryptocin and cryptocandin were antifungal metabolites produced by endophytic *Cryptosporiopsis* cf. *quercina* (Strobel et al. 1999b; Li et al. 2000). Cryptocandin, a lipopeptide antimycotic produced by *Cryptosporiopsis* cf. *quercina*, exhibited promising antifungal activity (MIC values of 0.03–0.07 μg/ml) against *Candida albicans*, *Trichophyton mentagrophytes*, and *Trichophyton rubrum* along with plant pathogenic fungi such as *Sclerotinia sclerotiorum* and *Botrytis cinerea* (Strobel et al. 1999b), while cryptocin, a tetramic acid, showed antimycotic activity against *Pyricularia oryzae* (Li et al. 2000). Pestaloside produced by *Pestalotiopsis microspora* showed antimycotic activity (Lee et al. 1995b). Ambuic acid, a functionalized cyclohexenone, was produced by endophytic *Pestalotiopsis* spp. and *Monochaetia* sp. which exhibited antifungal activity against *Pythium ultimum* (MIC, 7.5 μg/ml) (Li et al. 2001). Jesterone and hydroxy-jesterone (cyclohexenone epoxides) were produced by *Pestalotiopsis jesteri* which exhibited antimycotic activity against oomycetous fungi (Li and Strobel 2001). Endophytic *Pestalotiopsis adusta* produced three new chlorinated benzophenone derivatives, Pestalachlorides A–C. Pestalachloride A showed antifungal activity against *Fusarium culmorum* (IC_{50}: 0.89 μM), while Pestalachloride B was active against *Gibberella zeae* (IC_{50}: 0.89 μM) (Li et al. 2008a).

Altomare et al. (2000) isolated two alpha pyrones antifungal compounds named as fusapyrone and deoxyfusapyrone from *Fusarium semitectum* which showed very potential activity against a number of pathogenic or mycotoxogenic filamentous fungi such as *Alternaria alternata*, *Aspergillus flavus*, *Botrytis cinerea*, *Cladosporium cucumerinum*, *Phoma tracheiphila*, *and Penicillium verrucosum*. Culture extracts of *P. guepinii*, *Phomopsis* sp., and *Guignardia* sp. showed very active antifungal activity against *S. cerevisiae*, *Geotrichum* sp., *Cladosporium elatum*, *Mycotypha* sp., and *Penicillium canadensi* (Rodrigues et al. 2000). *Candida albicans*, *Cryptococcus neoformans*, and *Aspergillus fumigatus* are the major pathogenic fungi which cause disease in human beings. *Streptomyces* sp. produced bioactive polyenes having a broad spectrum activity against various *Aspergillus* and *Candida* species (Hay 2003). Amphotericin B, nystatin, and natamycin are main polyenes derived from *Streptomyces* sp. which find wide use for treatment of coccidioidal meningitis, cutaneous dermatophytes, and histoplasmosis and other mycotic diseases caused by *Candida albicans*, *Cryptococcus neoformans*, and *Aspergillus fumigatus* (Gupte et al. 2002; Iznaga et al. 2004; Augustine et al. 2005; Gohel et al. 2006). Oocydin A,

Table 35.3 Some examples of bioactive compounds produced by endophytic fungi exhibiting antimicrobial activity

Name of the Endophyte	Plant source	Compound name	Structure	References
Cladosporium sp.	*Quercus variabilis*	Brefeldin	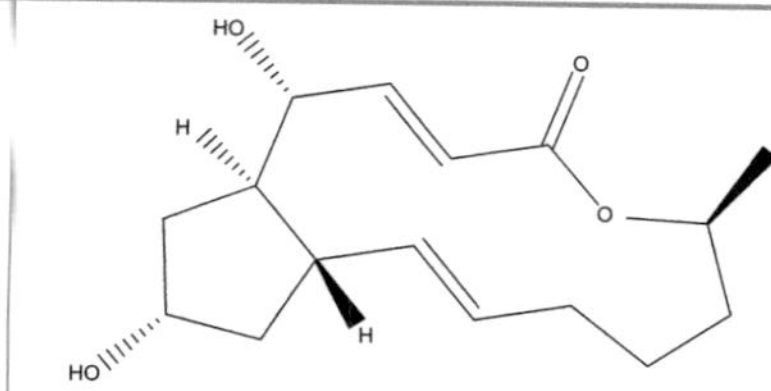	Wang et al. (2007)
Xylaria sp. YX-28	*Ginkgo biloba*	7-amino-4-methylcoumarin	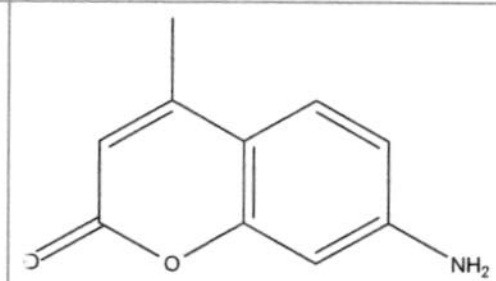	Liu et al. (2008)
Xylaria sp. F0010	*Abies holophylla*	Griseofulvin	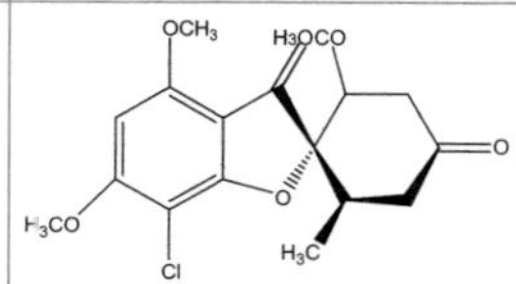	Park et al. (2005)

(continued)

Table 35.3 (continued)

Name of the Endophyte	Plant source	Compound name	Structure	References
Pestalotiopsis adusta	Unidentified plant	Pestalachloride A	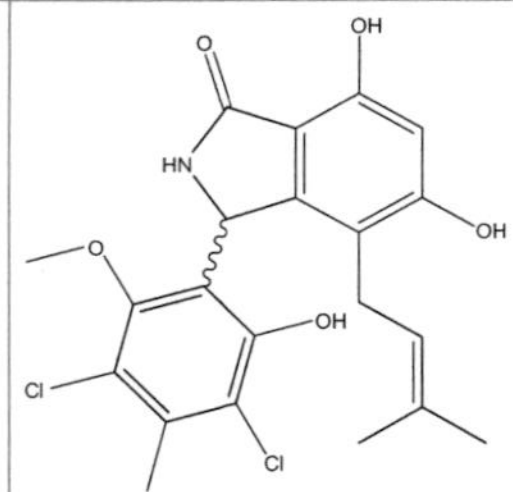	Li et al. (2008a)
		Pestalachloride B	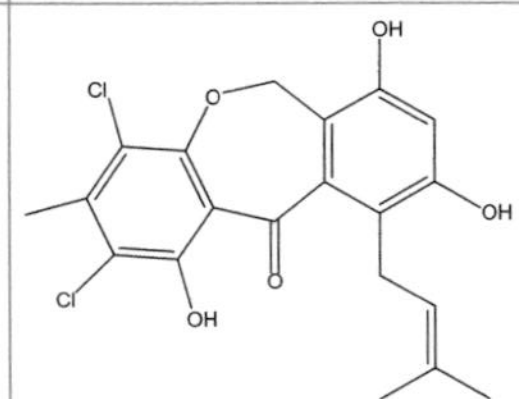	

Phomopsis cassia	*Cassia spectabilis*	Ethyl 2,4-dihydroxy-5,6-dimethylbenzoate	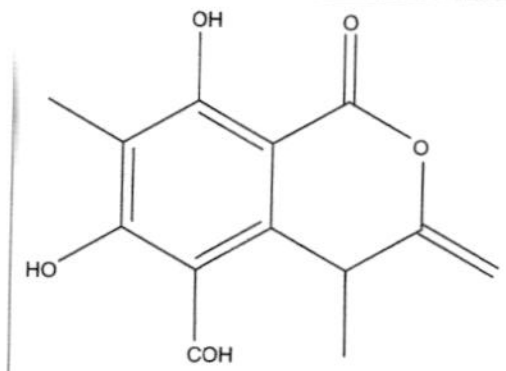	Silva et al. (2005)
		Phomopsilactone	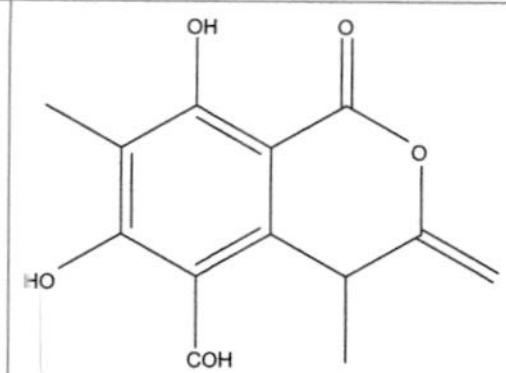	
Phomopsis sp.	*Erythrina crista*	Phomol	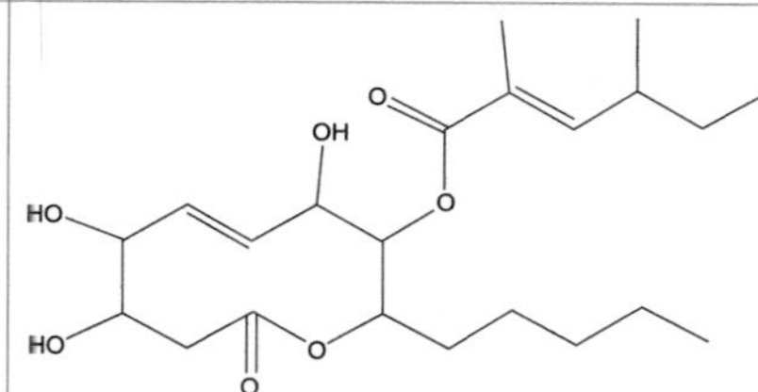	Weber et al. (2004a)
Curvularia lunata	*Niphates olemda*	Cytoskyrin A	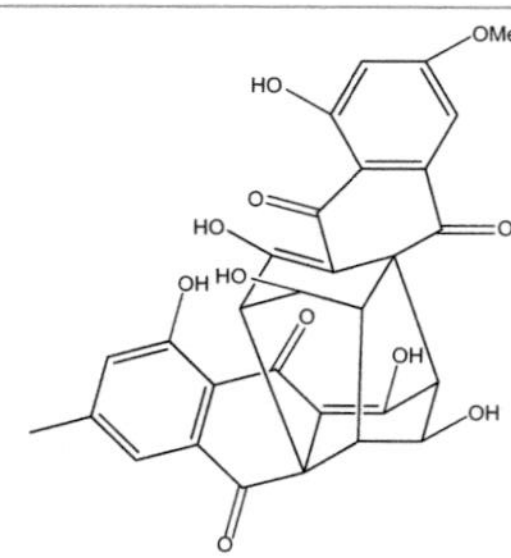	Jadulco et al. (2002)

a chlorinated macrocyclic lactone derived from an endophytic *Serratia marcescens* isolated from *Rhyncholacis pedicillata*, finds use in agriculture for control of oomycetous phytopathogenic fungi such as *Pythium ultimatum*, *Phytophthora parasitica*, *Phytophthora cinnamomi*, and *Phytophthora citrophora* (Strobel et al. 1999c).

The endophytic *Xylaria* sp. produced various antifungal metabolites such as griseofulvin (Park et al. 2005), cytochalasin D (Cafeu et al. 2005), multiplolides (Boonphong et al. 2001), 7-amino-4-methylcoumarin (Liu et al. 2008), sordaricin (Pongcharoen et al. 2008), phomenone (Silva et al. 2010), and xyolide (Baraban et al. 2013) which exhibited promising antifungal activity against various plant pathogenic fungi. Multiplolides A and B produced by *Xylaria multiplex* BCC 1111 exhibited antifungal activity against *Candida albicans* (IC_{50}: 7 and 2 μg/mL) (Boonphong et al. 2001). Endophytic *Xylaria* sp. produced 2-hexyl-3-methylbutanodioic acid and cytochalasin D with antifungal activity against *Cladosporium cladosporioides* and *C. sphaerospermum* (Cafeu et al. 2005). Griseofulvin produced by *Xylaria* sp. F0010 exhibited antimycotic activity against *Magnaporthe grisea*, *Corticium sasaki*, *Puccinia recondita*, and *Blumeria graminis* f. sp. *hordei*, with MIC values ranging between 50 and 150 μg/mL (Park et al. 2005). *Xylaria* sp. YX-28 endophytic to *Ginkgo biloba* produced 7-amino-4-methylcoumarin which showed antimycotic activity against *Candida albicans* (MIC, 15 μg/mL), *Penicillium expansum* (MIC, 40 μg/mL), and *Aspergillus niger* (MIC, 25 μg/mL) (Liu et al. 2008). *Xylaria* sp. PSU-D14 produced sordaricin which exhibited antifungal activity against *Candida albicans* ATCC90028 (MIC, 32 μg/mL) (Pongcharoen et al. 2008). Xyolide, a new nonenolide, (4*S*,7*S*,8*S*,9*R*)-4-*O*-succinyl-7,8-dihydroxy-9-heptyl-nonen-9-olide, was purified from endophytic *Xylaria feejeensis* strain E6912B which showed antimycotic activity against *Pythium ultimum* with MIC value of 425 μM (Baraban et al. 2013). Endophytic fungi such as *Plectophomella* sp., *Physalospora* sp., and *Crataegus monogyna* produced different bioactive compounds exhibiting antifungal activities such as (-)-mycorrhizin A, cytochalasins E and K, and radicinin, respectively. It was observed that mycorrhizin A showed significant antifungal activity toward *Ustilago violacea* and *Eurotium repens*. Cytochalasins E and K and radicinin showed strong antifungal activity against *Eurotium repens* and *Mycotypha microspora* (Hussain et al. 2014).

Colletotrichum gloeosporioides Mc-7 endophytic to *Michelia champaca* produced 2-phenylethyl 1*H*-indol-3-yl-acetate with antifungal activity against *Cladosporium cladosporioides* and *C. sphaerospermum* (Chapla et al. 2014). Fusidilactones produced by endophytic *Fusidium* sp. exhibited promising antimycotic activity against *Eurotium repens* and *Fusarium oxysporum* (Krohn et al. 2002). *Acremonium zeae* endophyte to maize produced pyrrocidines A and B, which showed significant antimycotic activity against *A. flavus* and *Fusarium verticillioides* (Wicklow et al. 2005). *Phomopsis cassia* endophytic to *Cassia spectabilis* produced two metabolites such as ethyl 2,4-dihydroxy-5,6-dimethylbenzoate and phomopsilactone exhibiting strong antifungal activity against *Cladosporium cladosporioides* and *C. sphaerospermum* (Silva et al. 2005). *Phomopsis* sp. ZSU-H76 a mangrove endophytic fungus to *Excoecaria agallocha* produced phomopsins A, B, and C along with cytosporones B and C. However, only cytosporone B and C exhibited antimycotic activity against *Candida albicans* and *Fusarium oxysporum* (MIC, 32–64 mg/

mL) (Huang et al. 2008b). *Cladosporium* sp. I(R)9-2 endophytic to *Quercus variabilis* produced brefeldin A which displayed promising antifungal activity (Wang et al. 2007). *Chaetomium globosum* endophytic to leaves of *Ginkgo biloba* produced azaphilone derivatives such as chaetomugilin A and D along with chaetoglobosins A and C which exhibited antifungal activity (Qin et al. 2009). Griseofulvin produced by endophytic *Nigrospora* sp. LLGLM003 exhibited good antifungal activity against *B. cinerea* and *Colletotrichum orbiculare* (EC_{50}: 20 and 0.49 μg/ml) (Zhao et al. 2012). Cladosporin, chaetoatrosin A, and chaetoviridin A produced by endophytic *Chaetomium globosum* F211_UMNG were found to be active against *Fusarium oxysporum* (Fierro-Cruz et al. 2017). *Bacillus amyloliquefaciens* strain BUZ-14 showed antifungal activity for the biocontrol of different *Penicillium* spp. such as *P. digitatum* and *P. italicum* in oranges and *P. expansum* in apples (Calvo et al. 2017). Ericin was identifed as a lantibiotic produced by *B. velezensis* RC 218 exhibiting antifungal activity against *Fusarium graminearum* head blight (Palazzini et al. 2016). Endophytic *Bacillus velezensis* CC09 produced antifungal compounds for the biocontrol of wheat powdery mildew disease (Cai et al. 2017).

Endophytic *Muscodor albus* (Strobel et al. 2001; Worapong et al. 2001), *M. vitigenus* (Daisy et al. 2002a, b), *M. roseus* (Worapong et al. 2002), *M. yucatanensis* (Gonzalez et al. 2009), *M. crispans* (Mitchell et al. 2010), *M. cinnamoni* (Suwannarach et al. 2010), *M. fengyangensis* (Zhang et al. 2010), and *M. camphora* (Meshram et al. 2017) produce volatile antimicrobials which are mixtures of volatile organic compounds (VOCs) including various alcohols, acids, esters, ketones, and lipids that function synergistically by arresting the growth of a wide variety of plant pathogenic fungi and bacteria. These VOCs find use in mycofumigation for treating buildings, soils, agricultural produce, and human wastes (Strobel 2006a, b; Mitchell et al. 2010; Strobel 2011; Alpha et al. 2015). The volatile antimicrobials produced by *Muscodor albus* have also been used for the biocontrol of green mold and sour rot of stored lemon (Mercier and Smilanick 2005), seedling diseases of sugar beet and *Verticillium* wilt of eggplant (Stinson et al. 2003), root-knot nematode on tomato (Grimme et al. 2007), and postharvest fungal apple decay (Ramin et al. 2007). A mixture of VOCs such as acetic acid, ethyl ester, propanoic acid, 2-methyl-, methyl ester and acetic acid, and 2-methylpropyl ester was produced by *Muscodor albus* MOW12 (Banerjee et al. 2014). A volatile DNA-methylating agent N-methyl-N-nitrosoisobutyramide was produced by *Muscodor albus* which finds use as a potential mycofumigant (Hutchings et al. 2017). A total of 132 sponge-associated Antarctic strains were characterized, which displayed antimicrobial activity against *Burkholderia cepacia* complex (Bcc) bacteria, which are opportunistic pathogens causing infections in the lungs of cystic fibrosis (CF) patients (Papaleo et al. 2012). The antimicrobial activity by Antarctic bacteria was demonstrated to be due to the production of microbial volatile organic compounds (mVOCs) (Romoli et al. 2011). Headspace solid-phase microextraction coupled with gas chromatography-mass spectrometry (HS-SPME-GC-MS) analysis revealed that some bacterial genera, *Arthrobacter* sp. strain TB23 (Fondi et al. 2012; Orlandini et al. 2014), *Burkholderia cepacia* (Papaleo et al. 2013), *Psychrobacter* sp. (Broekaert et al. 2013), *Pseudoalteromonas* sp. TB41 (Romoli et al. 2014), and *Gillisia* sp. strain CAL575 (Maida et al. 2014) synthesized a mixture of (at least) 30 different mVOCs that may be responsible for the inhibition of the growth of Bcc bacteria.

Moreover, it was also demonstrated that the synthesis of mVOCs was strongly dependent on the growth medium composition (Papaleo et al. 2013). O-anisaldehyde has been identified as the most abundant volatile produced by *Bacillus atrophaeus* CAB-1 which exhibited highest inhibition on the mycelial growth of *Botrytis cinerea* (Zhang et al. 2013b). Endophytic *Bacillus velezensis* ZSY-1 produced pyrazine (2,5-dimethyl), benzothiazole, phenol (4-chloro-3-methyl), and phenol-2,4-bis (1,1-dimethylethyl) which showed significant antifungal activity (Gao et al. 2017). Volatile organic compounds produced by *Bacillus amyloliquefaciens* CPA-8 were demonstrated for antifungal activity against postharvest fruit pathogen decays of cherry caused by *Monilinia laxa*, *M. fructicola*, and *Botrytis cinerea* (Gotor-Vila et al. 2017).

35.8 Antimycobacterial Compounds from Fungal Endophytes

Tuberculosis (TB) is potential, dreadful, and infectious disease caused by *Mycobacterium tuberculosis*. It infects the lungs and also attacks the central nervous system, lymphatic system, and skeletal tissue. Some common symptoms observed in patients affected with TB include chronic cough with blood-tinged sputum, fever, night sweats, and weight loss. It is transmitted through air when infected individuals cough, sneeze, or spit, causing the bacterium to spread from their throat or lungs. Latent TB is prevalent in approximately one-third of the current world population (Koul et al. 2011). In 2015, there were approximately 10.4 million cases of active TB, and this disease caused approximately 1.8 million deaths (World Health Organization 2017b). The standard first-line therapy for tuberculosis involves a 2-month regimen using a combination of rifampicin, isoniazid, ethambutol, and pyrazinamide, followed by an additional 4-month regimen using a combination of rifampicin and isoniazid (Grosset 1996). Secondary metabolites produced by endophytic fungi exhibiting antimycobacterial activity were tested for antitubercular assay against *M. tuberculosis* strains H37Rv or H37Ra using microplate alamar blue assay method (Collins and Franzblau 1997). 3-Nitropropionic acid (3-NPA) produced by endophytic *Phomopsis* sp. strain usia5 exhibited antimycobacterial activity against *M. tuberculosis* H37Ra with MIC value of 3.3 μM. The mode of action of 3-NPA for antimycobacterial activity is due to the inhibition of mycobacterial succinate dehydrogenase (Chomcheon et al. 2005). Endophytic *Periconia* sp. produced piperine, (5-(3, 4-methylenedioxyphenyl)-1-piperidinopent-2, 4-dien-1-one), which exhibited strong antimycobacterial activity against *M. tuberculosis* (MIC, 1.74 μg/mL) and *M. smegmatis* (MIC, 2.62 μg/mL) (Verma et al. 2011). Piperine is an efflux pump inhibitor of the NorA (Khan et al. 2006; Kumar et al. 2008) and MdeA (Mirza et al. 2011) present in *S. aureus*. It is also known to function as inhibitor of Rv1258c, a putative multidrug efflux pump of *M. tuberculosis* (Sharma et al. 2010). Some examples of endophytic fungi producing antitubercular compounds are shown in Table 35.4.

Table 35.4 Examples of endophytic fungi producing antitubercular compounds

Endophyte producer/(host plant)	Compound	Active against	MIC values	References
Phomopsis sp. BCC 1323 (*Tectona grandis* L.)	Phomoxanthones A and B	*M. tuberculosis* H37Ra	0.5 and 6.25 μg/mL, respectively	Isaka et al. (2001)
Microsphaeropsis sp. BCC 3050 (*Dirinaria applanata*; lichen)	Preussomerins	*M. tuberculosis* H37Ra	3.12–50 μg/mL	Seephonkai et al. (2002)
Phomopsis sp. strain usia5 (*Urobotrya siamensis*)	3-Nitropropionic acid	*M. tuberculosis* H37Ra	3.3 μM	Chomcheon et al. (2005)
Dothideomycete sp. LRUB20 (*Leea rubra* Blume ex. Spreng)	2-hydroxymethyl-3-methylcyclopent-2-enone, asterric acid, and 2,4-dinitrophenylhydrazone derivative	*M. tuberculosis* H37Ra	200 μg/mL	Chomcheon et al. (2006)
Phomopsis sp. PSU-D15 (*Garcinia dulcis* (Roxb.) Kurz)	Phomoenamide	*M. tuberculosis* H37Ra	6.25 μg/mL	Rukachaisirikul et al. (2008)
Periconia sp. (*Piper longum* L.)	Piperine, (5-(3, 4-methylenedioxyphenyl)-1-piperidinopent-2, 4-dien-1-one)	*M. tuberculosis*	1.74 μg/mL	Verma et al. (2011)
		M. smegmatis	2.62 μg/mL	
Fusarium sp. DZ-27 (*Kandelia candel* (L.) Druce; mangrove)	Cd^{2+} and Cu^{2+} complexes of fusaric acid	*M. tuberculosis* H37Rv	10 μg/mL	Pan et al. (2011)
		M. bovis strain BCG	4 μg/mL	
Diaporthe sp. P133 (*Pandanus amaryllifolius*)	Diaportheone A and B	*M. tuberculosis* H37Rv	100.9 and 3.5 μM, respectively	Bungihan et al. (2011)
Biscogniauxia formosana BCRC 33718 (*Cinnamomum* sp.)	Biscogniazaphilones A and B	*M. tuberculosis* H37Rv	5.12 and 2.52 μg/mL, respectively	Cheng et al. (2012)
Nigrospora sp. (*Fragaria virginiana*)	Linoleic acid derivatives and (+)-abscisic acid	*M. tuberculosis* H37Ra	n.s.	Clark et al. (2013)
Phoma sp. NRRL 46751 (*Saurauia scaberrinae*)	Phomapyrrolidones B and C	*M. tuberculosis* H37Rv	5.9 and 5.2 μg/mL, respectively	Wijeratne et al. (2013)

(continued)

Table 35.4 (continued)

Endophyte producer/(host plant)	Compound	Active against	MIC values	References
Fusarium heterosporum NRRL 50135 (*Asplenium* sp.; fern)	Pyrrolocins A, B, and C	*M. tuberculosis* H37Ra	26.3, 112.9, and 56.4 μM, respectively	Jadulco et al. (2014)
Seimatosporium sp. (*Hypericum perforatum*)	(-)-Avenaciolide	*M. tuberculosis* H37Ra	n.s.	Clark et al. (2014)
Unidentified endophyte (*Heracleum maximum*; cow parsnip)	Phomopsolide A and 6(E)-phomopsolide A	*M. tuberculosis* H37Ra	n.s.	Clark et al. (2015)
Penicillium sp. (isolate HL4–159-41B) and *Coniothyrium* sp. (isolate HL6–097-027B) (*Aralia nudicaulis*)	Palitantin and botrallin, craterellin C, and mycosporulone	*M. tuberculosis* H37Ra	84.2, 104, 433, and 90.3 μM, respectively	Li et al. (2015)
Penicillium sp. CAM64 (*Garcinia nobilis*)	Penialidin A, penialidin B, penialidin C, citromycetin, p-hydroxy-phenyl-glyoxalaldoxime, and brefeldin A	*M. smegmatis*	62.5, 62.5, 15.6, 31.2, 62,5, and 250 μg/mL, respectively	Jouda et al. (2016)
Fusarium solani A2 (*Glycyrrhiza glabra*)	3,6,9-trihydroxy-7-methoxy-4,4-dimethyl-3,4-dihydro-1H-benzo[g]isochromene-5,10-dione (**1**), fusarubin (**2**), 3-O-methylfusarubin (**3**), and javanicin (**4**)	*M. tuberculosis* H37Rv	256, 8, 64, and 32 μg/mL, respectively	Shah et al. (2017)
Unidentified endophyte TC2-085 (*Geum macrophyllum*)	Phomopsolide A	*M. tuberculosis* H37Ra	35 μM	Mullin et al. (2017)
Gliomastix sp. ZSDS1-F7 (sponge *Phakellia fusca*)	Macrophorin A, 4′-oxomacrophorin and 7-deacetoxyyanuthone A	*M. tuberculosis* H37Ra	22.1, 2.44, and 17.5 μM, respectively	He et al. (2017)

35.9 Antiviral Activities of Endophytes

Human immunodeficiency virus (HIV), influenza, and hepatitis C virus (HCV) are the most perilous viral diseases affecting human with high mortality rates (Roy 2017). Endophytic fungi were reported to produce various antiviral agents, for example, cytonic acids A and B produced by endophytic fungus *Cytonaema* sp., which function as novel human cytomegalovirus (hCMV) protease inhibitors (Guo et al. 2000). Hinnuliquinone is an isoprenyl-indolyl-2,5-dihydroxyquinone pigmented compound isolated from a fungus endophytic to *Quercus coccifera* leaves, which acts as a potent human immunodeficiency virus type 1 (HIV-1) protease inhibitor (Singh et al. 2004). Dihydroiso-coumarin (*R*)-(-)-mellein produced by endophytic fungi such as *Pezicula livida*, *Plectophomella* sp., and *Cryptosporiopsis malicoticis* inhibited hepatitis C virus (HCV) protease activity (IC_{50}: 35 mM) (Krohn et al. 1997; Florke et al. 2006). Pestalotheol C isolated from endophytic fungus *Pestalotiopsis theae* displayed inhibition on HIV-1_{LAI} replication in C8166 cells (EC_{50}: 16.1 μM) (Li et al. 2008b). Pestalofficiols F, G, H, J, and K, new isoprenylated chromone derivatives, purified from endophytic fungus *Pestalotiopsis fici* showed inhibitory effects on HIV-1 replication in C8166 cells (Liu et al. 2009). An endophytic fungus *Penicillium sclerotiorum* PSU-A13 produced (+)-sclerotiorin which exhibited anti-HIV-1 integrase activity (IC_{50}: 14.5 μg/mL) and protease activity (IC_{50}: 62.7 μg/mL) (Arunpanichlert et al. 2010). Oblongolide Z produced by endophyte *Phomopsis* sp. BCC 9789 isolated from *Musa acuminata* exhibited anti-HSV-1 activity (IC_{50}: 14 μM) (Bunyapaiboonsri et al. 2010). Emerimidine A and B produced by endophytic fungus *Emericella* sp. HK-Z exhibited moderate inhibitory activity against influenza virus H_1N_1 (emerimidine A, IC_{50}, 42.07 mg/mL; emerimidine B, IC_{50}, 62.05 mg/mL) (Zhang et al. 2011). Altertoxins V, I, II, and III produced by endophytic *Alternaria tenuissima* QUE1Se exhibited potent anti-HIV activity (Bashyal et al. 2014). Some examples of endophytic fungi producing antiviral activity are shown in Table 35.5.

35.10 Antioxidants from Fungal Endophytes

Natural antioxidants are present in medicinal plants, fruits, and vegetables. Nevertheless, metabolites produced by endophytes are reported to exhibit novel naturally produced antioxidants. Microbial polysaccharides are also studied as natural antioxidants. These compounds are effective against damage due to ROS and oxygen-derived free radicals derived from oxygen (Huang et al. 2007; Seifried et al. 2007). Antioxidants showed therapeutic promise for treatment of various ROS-linked diseases such as atherosclerosis, hypertension, cancer, etc. (Valko et al. 2007). Pestacin and isopestacin produced by endophytic fungus *Pestalotiopsis microspora* displayed potent antioxidant activity. Pestacin's proposed antioxidant activity is due to breakdown of reactive C-H bond and through OH abstraction. The

Table 35.5 Some examples of bioactive compounds produced by endophytic fungi exhibiting antiviral activity

Name of the Endophyte	Plant source	Compound name	Structure	References
Cytonaema sp.	Unidentified plant	Cytonic acid A	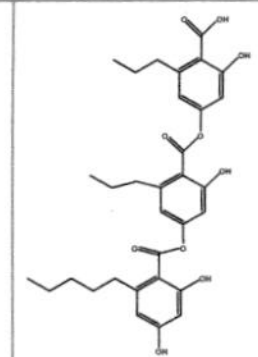	Guo et al. (2000)
		Cytonic acid B	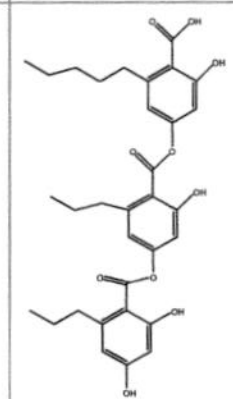	
Pullularia sp.	*Quercus coccifera*	Pullularins A	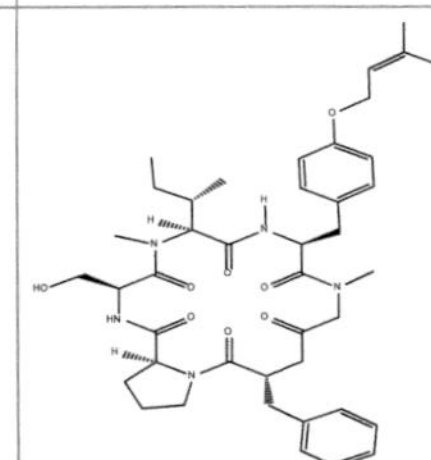	Isaka et al. (2007)
Pestalotiopsis theae	Unidentified tree	Pestalotheol C	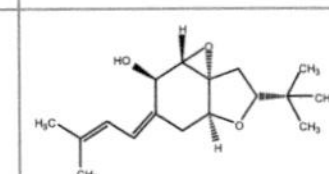	Li et al. (2008b)

Xylaria mellisii (BCC 1005)	Thai medicinal plant	Mellisol	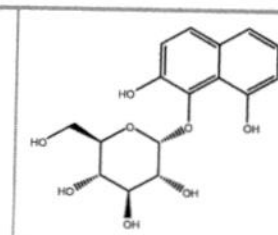	Pittayakhajonwut et al. (2005)
Penicillium sclerotiorum	*Eugenia jambolana Lam*	Sclerotiorin	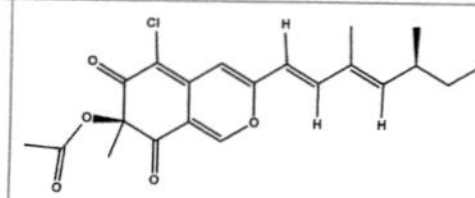	Arunpanichlert et al. (2010)
Emericella sp.	*Aegiceras corniculatum*	Emerimidine A	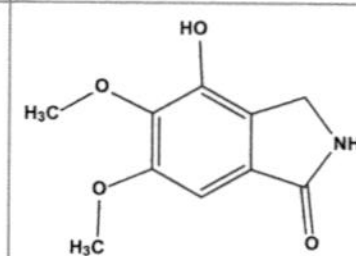	Zhang et al. (2011)
		Emerimidine B	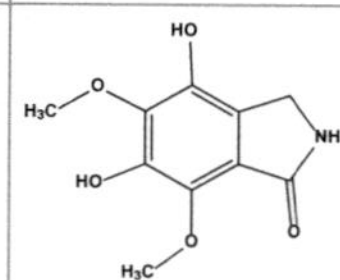	

antioxidant activity is believed to 11 times greater than trolox (Strobel et al. 2002; Harper et al. 2003).

Phyllosticta sp. an endophyte fungus from *Guazuma tomentosa* exhibited strong antioxidant activity (Srinivasan et al. 2010). Endophytic fungi isolated from *Scapania verrucosa* exhibited promising antioxidant activity, and the major antioxidant constituents were due to total phenolics (Huang et al. 2007, 2008a; Zeng et al. 2011). Endophytic fungus *Pseudocercospora* sp. ESL 02 produced bioactive compounds exhibiting DPPH radical scavenging activity such as terreic acid (IC_{50}: 0.22 mM/L) and 6-methylsalicylic acid (IC_{50}: 3.87 mM/L) (Prihantini and Tachibana, 2017). Some examples of endophytic fungi exhibiting antioxidant activity are shown in Table 35.6.

Table 35.6 Some examples of bioactive compounds produced by endophytic fungi exhibiting antioxidant activity

Name of the Endophyte	Plant source	Compound name	Structure	References
Pestalotiopsis microspora	*Terminalia morobensis*	Pestacin	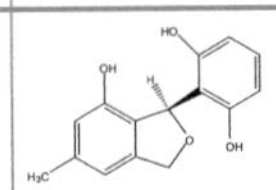	Harper et al. (2003):
		Isopestacin	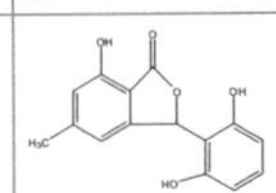	Strobel et al. (2002)
Cephalosporium sp. IFB-E001	*Trachelospermum jasminoides*	Graphislactone A	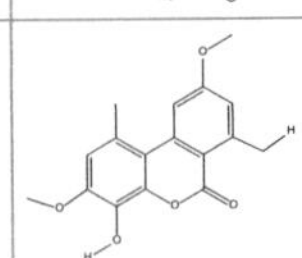	Hormazabal et al. (2005); Song et al. (2005)
Microsphaeropsis olivacea	*Pilgerodendron uviferum*			
Corynespora cassiicola L36	*Lindenbergia philippensis* (Cham.) Benth.	Corynesidone B	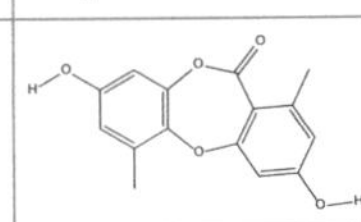	Chomcheon et al. (2009)
Corynespora cassiicola L36	*Lindenbergia philippensis* (Cham.) Benth.	Corynesidones A	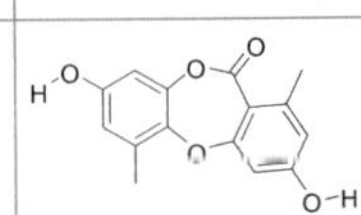	Chomcheon et al. (2009)
Corynespora cassiicola L36	*Lindenbergia philippensis* (Cham.) Benth.	Corynether A	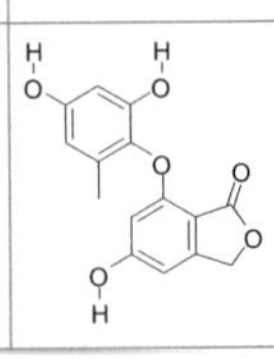	Chomcheon et al. (2009)

35.11 Insecticidal Activities from Fungal Endophytes

Systemic and foliar *Clavicipetalean* grass fungal endophytes produced toxic alkaloids that protected the hosts against insect and vertebrate herbivores. Mechanisms involving the production of toxic repellent compounds by endophytic fungus have also been reported (Webber and Gibbs 1984). A number of endophytic entomopathogenic and nematophagous fungi function as biocontrol agents, which serve as leads for new insecticides or nematicides. Phomopsolides A and B produced by endophytic *Phomopsis* sp. prevent the boring and feeding activities in elm bark beetles (Grove 1985). Endophytic *Claviceps purpurea* and *C. chaetomium* produced insecticidal metabolites against cotton aphids (Zhang et al. 2010). Penicinoline, produced by endophytic *Penicillium* sp., exhibited strong insecticidal activity at 1000 ppm with 100% mortality of sucking pest *Aphis gossypii* (Shao et al. 2010). Some examples of endophytic fungi exhibiting insecticidal activity are shown in Table 35.7.

Table 35.7 Some examples of bioactive compounds produced by endophytic fungi exhibiting insecticidal activity

Name of the Endophyte	Plant source	Compound name	Structure	References
Nodulisporium sp.	*Bontia daphnoides*	Nodulisporic acid	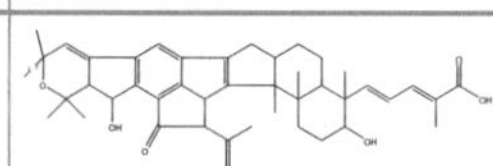	Demain (2000)
Muscodor vitigenus	*Paullinia paullinioides*	Naphthalene	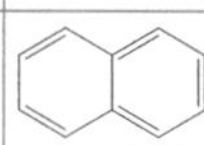	Daisy et al. (2002a)
Penicillium sp.	*Acanthus ilicifolius*	Penicinoline	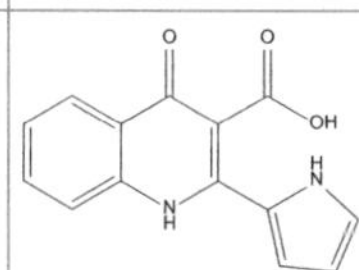	Shao et al. (2010)
Eupenicillium sp.	*Murraya paniculata*	Alantryphenone (R=Benzyl) Alantryleunone (R=Isobutyl)	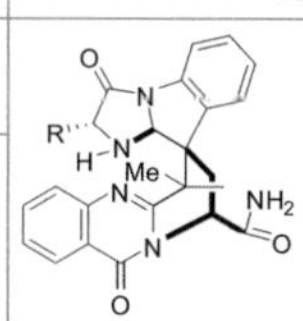	Fábio et al. (2005)
Eupenicillium sp.	*Murraya paniculata*	Alantrypinene	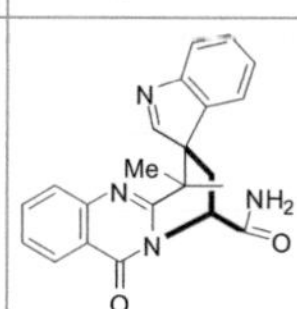	Fábio et al. (2005)

35.12 Antidiabetic Agents from Fungal Endophytes

Endophytes are reported to produce many antidiabetic metabolites along with other bioactives. Demethyl asterriquinone B-1 (DMAQ B-1, L-783,281) is a non-peptidyl fungal metabolite, produced by a tropical endophytic fungus *Pseudomassaria* sp. (ATCC 74411) that acted as an orally active insulin mimetic agent which significantly lowered the blood glucose levels in diabetic mouse model. This insulin mimetic agent functions as an efficient insulin substitute for diabetic therapy (Zhang et al. 1999; Salituro et al. 2001). 8-hydroxy-6,7-dimethoxy-3-methylisocoumarine produced by an endophytic fungus, *Xylariaceae* sp. QGS 01, displayed α-glucosidase inhibitory activity (IC_{50}: 41.75 μg/mL (Indrianingsih and Tachibana 2017). Some examples of endophytic fungi producing antidiabetic compounds are depicted in Table 35.8.

35.13 Fungal Endophytes Producing Immunosuppressive Compounds

Immunosuppressive agents prevent allograft rejection in organ transplantation and find use for the treatment of various autoimmune diseases. Phomal, a polyketide lactone, was identified as an anti-inflammatory metabolite produced by endophytic *Phomopsis* sp. (Weber et al. 2004a). Endophytic *Pestalotiopsis leucothes* isolated from *Tripterygium wilfordii* produced immunomodulatory compounds (Kumar et al. 2005). Collutelin A was isolated from endophytic *Colletotrichum dematium* which exhibited strong immunosuppressive activity (Ren et al. 2008). Ergoflavin, a

Table 35.8 Some examples of bioactive compounds produced by endophytic fungi exhibiting antidiabetic activity

Name of the Endophyte	Plant source	Compound name	Structure	References
Aspergillus sp.	*Salvadora oleoides* Decne.	2, 6-di-tert-butyl-p-cresol	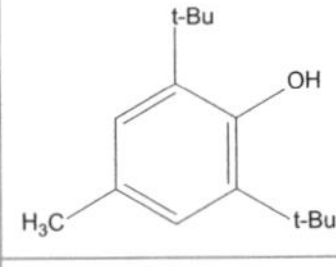	Dhankhar and Yadav (2013)
Phoma sp.	*Salvadora oleoides decne*	Phenol	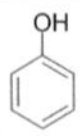	Dhankhar and Yadav (2013)
Pseudomassaria sp.	Unindentified plant	Demethyl asterriquinone B	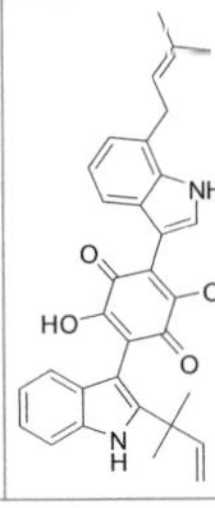	Salituro et al. (2001); Strobel (2002)

Table 35.9 Some examples of bioactive compounds produced by endophytic fungi exhibiting immunosuppressive activity

Name of the Endophyte	Plant source	Compound name	Structure	References
Penicillium sp.	*Limonium tubiflorum*	7-methyl anthraquinone	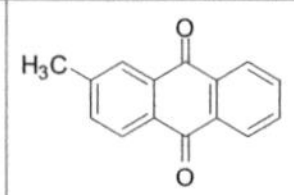	Aly et al. (2011)
Penicillium sp.	*Limonium tubiflorum*	1-chloro-2,4-dihydroxy-5-methoxy-7-methyl anthraquinone	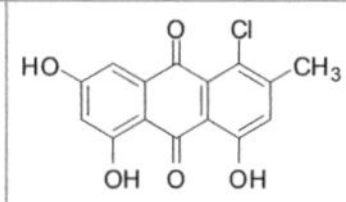	Aly et al. (2011)
Penicillium sp.	*Limonium tubiflorum*	11-β-methoxy curvularin	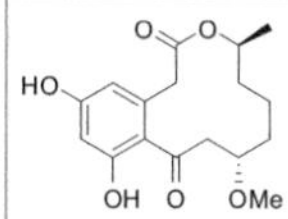	Aly et al. (2011)
Penicillium sp.	*Limonium tubiflorum*	11-α-methoxy curvularin	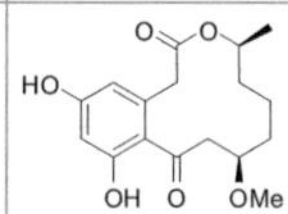	Aly et al. (2011)
Fusarium subglutinans	*Tripterygium wilfordii*	Subglutinol A	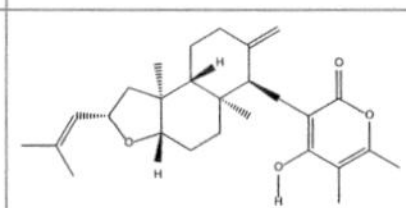	Lee et al. (1995a)
		Subglutinol B	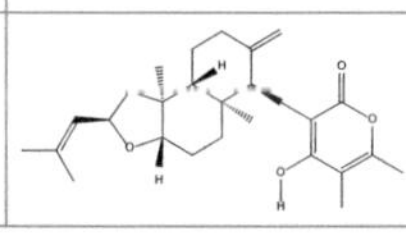	

dimeric xanthene produced by an endophytic fungus strain PM0651480 isolated from *Mimusops elengi*, showed potent anti-inflammatory activity by inhibiting the production of cytokines such as TNF-α and IL-6 by lipopolysaccharide (LPS)-stimulated human monocytic cell line, THP-1 (Deshmukh et al. 2009). Epicoccins M and R and ent-epicoccin G were reported from endophytic *Epicoccum nigrum* that showed promising anti-inflammatory activities (Wang et al. 2010b). Nuclear factor kappa B (NF-κB) regulates various inflammatory and immune responses. Endophytic *Penicillium* sp. from *Limonium tubiflorum* produced various NF-κB inhibitors (IC_{50}: 1.6–10.1 μM) (Aly et al. 2011). Endophytic fungus *Phomopsis* produced phomol and mevinic acid exhibiting strong anti-inflammatory activity (Weber et al. 2004a). Some examples of bioactive compounds produced by endophytic fungi exhibiting immunosuppressive activity are depicted in Table 35.9.

35.14 Other Biological Activities of Endophytes

Monoamine oxidase (MAO) is an enzyme which catalyzes the oxidation of monoamines acting as neurotransmitters and is pharmacologically important since it functions in the brain and peripheral tissues (Ramsay 2012). MAO exists in two isoforms:

MAO-A and MAO-B, which are encoded by distinct genes (Shih et al. 1999; Tipton et al. 2004). MAO-A exhibits specific activity in catecholamines and other biogenic amines, such as norepinephrine and epinephrine; conversely, MAO-B shows preference for hydrophobic substrates containing benzylamine and 2-phenylethylamine (Orhan 2016). Although the two isoforms are related to major depressive and neuropsychiatric disorders, MAO-A is mainly associated with depression and anxiety, whereas MAO-B is a major molecular target to treat Alzheimer's and Parkinson's diseases (Yamada and Yasuhara 2004; Pacher and Kecskeméti 2004; Youdim et al. 2006). Endophytic *Talaromyces wortmannii* LGT-4 produced secovironolide which showed weak MAO inhibitory activity (Ding et al. 2015).

Xanthine oxidase (XO) catalyzes the oxidation of hypoxanthine to xanthine which is further catalyzed into uric acid, having functional role in gout. From a clinical perspective, allopurinol inhibited XO which finds use for gout treatment. Many endophytes produced compounds exhibiting strong XO inhibitory activity; alternariol isolated from endophytic *Alternaria brassicicola* showed XO inhibitory activity with IC_{50} value of 15.5 μM. Rubrofusarin B and aurasperone A produced by endophytic *Aspergillus niger* also showed strong co-inhibition of xanthine oxidase (Song et al. 2004; Gu 2009). Xyloketal A produced by mangrove fungus *Xylaria* sp. no. 2508 inhibited acetylcholine esterase at 1.5×10^{-6} mol/L (Lin et al. 2001). Endophytic *Pullularia* sp. produced pullularins A, B, and C showing promising antimalarial activity against *Plasmodium falciparum* K1 (K1, refer to multidrug-resistant strain) (Isaka et al. 2007). Endophytic *Xylaria* sp. produced lactones with promising anti-*Plasmodium falciparum* activity (Romero et al. 2008). Endophytic *Aspergillus terreus* produced butyrolactone V exhibiting antimalarial activity (IC_{50}: 17.95 μM) (Haritakun et al. 2010). Eremophilane-type sesquiterpenoids compounds produced by endophytic *Xylaria* exhibited antimalarial activity (IC_{50}: 8.1–13.0 μM) (Isaka et al. 2010). Mycoepoxydiene, deacetylmycoepoxydiene, phomoxydienes A and C, and cytosporones E and P produced by endophytic fungus *Phomopsis* sp. BCC 45011 displayed antimalarial activities with IC_{50} values of 2.57, 2.41, 3.52, 3.26, 2.02, and 3.65 μg/mL, respectively (Kornsakulkarn et al. 2015). Endophytic *Penicillium janthinellum* to *Melia azedarach* produced a polyketide citrinin exhibiting 100% anti-*Leishmania* activity (Marinho et al. 2005). Endophytic *Phomopsis phaseoli* produced 3-hydroxypropionic acid exhibiting nematicidal activity (Schwarz et al. 2004).

35.15 Conclusions

In conclusion, it is clear that endophytic fungi is a largely untapped bioresource with novel and interesting biological and chemical diversity, which is poorly investigated. Literature survey showed that a number of endophytic fungal strains have been investigated, nevertheless a significant number of novel strains remain unexplored inhabiting various niches which need to be discovered and systematic studies need to be performed to structurally characterize the bioactive metabolites and also perform biological activities. Further, the extreme halophilic (salt stress) conditions prevailing in the mangrove environment favor the associated microflora to secrete an array of secondary metabolites; however, by stimulating some of the silent genes in these microbial strains may enable the biosynthesis of unprecedented secondary metabolites.

In view of the growing drug resistance problem in cancer and pathogenic bacteria, there is an urgent need to find new drug leads or therapies to combat these diseases. One strategy that has been practised in the recent past to augment the diversity of secondary metabolite production is the OSMAC (one strain many compounds) approach, in which different culture media are used along with minor changes in the physical (temperature, pressure, light, etc.) or chemical (addition of chemicals, pH, etc.) parameters during the fungal cultivation. Another approach would be the co-cultivation of two or more fungal strains, or mixed fungal and bacterial strains, which mimics a very competitive natural environment and thus favors enhanced production of new bioactive secondary metabolites. Another recent strategy would be the regulation of the fungal biosynthetic gene expression or silencing based on epigenetic modification using DNA methyltransferase (DNMT) and histone deacetylase (HDAC) inhibitors.

References

Alpha CJ, Campos M, Jacobs-Wagner C, Strobel SA (2015) Mycofumigation by the volatile organic compound-producing fungus *Muscodor albus* induces bacterial cell death through DNA damage. Appl Environ Microbiol 81:1147–1156

Altomare C, Perrone G, Zonno MC, Evidente A, Pengue R, Fanti F, Polonelli L (2000) Biological characterization of fusapyrone and deoxyfusapyrone, two bioactive secondary metabolites of *Fusarium semitectum*. J Nat Prod 63:1131–1135

Aly AH, Debbab A, Clements C, Ebel RAE, Orlikova B, Diederich M, Wray V, Lin WH, Proksch P (2011) NF kappa B inhibitors and antitrypanosomal metabolites from endophytic fungus *Penicillium* sp. isolated from *Limonium tubiflorum*. Bioorg Med Chem 19:414–421

American Cancer Society (2009) Cancer facts and figures. American Cancer Society, Atlanta

Amna T, Puri SC, Verma V, Sharma JP, Khajuria RK, Musarrat J, Spiteller M, Qazi GN (2006) Bioreactor studies on the endophytic fungus *Entrophospora infrequens* for the production of an anticancer alkaloid camptothecin. Can J Microbiol 52:189–196

Ardalani H, Avan A, Ghayour-Mobarhan M (2017) Podophyllotoxin: a novel potential natural anticancer agent. Avicenna J Phytomed 7:285–294

Arnold AE (2007) Understanding the diversity of foliar endophytic fungi: progress, challenges, and frontiers. Fungal Biol Rev 21:51–66

Arora R, Gupta D, Chawla R, Sagar R, Sharma AK, Prasad J, Singh S, Kumar R, Sharma RK (2005a) Radioprotection by plant products: present status and future prospects. Phytother Res 19:1–22

Arora R, Chawla R, Sagar R, Prasad J, Singh S, Kumar R, Sharma A, Singh S, Sharma RK (2005b) Evaluation of radioprotective activities of *Rhodiola imbrricata* Edgew – a high altitude plant. Mol Cell Biochem 273:209–223

Arora R, Chawla R, Puri SC, Sagar RK, Singh S, Kumar R, Sharma AK, Prasad J, Singh S, Kaur G, Chaudhary P, Qazi GN, Sharma RK (2005c) Radioprotective and antioxidant properties of low altitude *Podophyllum hexandrum* (LAPH). J Environ Pathol Toxicol Oncol 24:253–269

Arunpanichlert J, Rukachaisirikul V, Sukpondma Y, Phongpaichit S, Tewtra-kul S, Rungjindamai S, Sakayarojd J (2010) Azaphilone and isocoumarin derivatives from the endophytic fungus *Penicillium sclerotiorum* PSU–A13. Chem Pharm Bull 58:1033–1036

Augustine SK, Bhavsar SP, Kapadnis BP (2005) A non-polyene antifungal antibiotic from *Streptomyces albidoflavus* PU23. J Biosci 30:201–211

Banerjee D, Pandey A, Jana M, Strobel G (2014) *Muscodor albus* MOW12 an endophyte of *Piper nigrum* L. (Piperaceae) collected from North East India produces volatile antimicrobials. Indian J Microbiol 54:27–32

Baraban EG, Morin JB, Phillips GM, Phillips AJ, Strobel SA, Handelsman J (2013) Xyolide, a bioactive nonenolide from an Amazonian endophytic fungus, *Xylaria feejeensis*. Tetrahedron Lett 54:4058–4060

Bashyal BP, Wijeratne EM, Faeth SH, Gunatilaka AA (2005) Globosumones A-C, cytotoxic orsellinic acid esters from the Sonoran desert endophytic fungus *Chaetomium globosum*. J Nat Prod 68:724–728

Bashyal BP, Wellensiek BP, Ramakrishnan R, Faeth SH, Ahmad N, Gunatilaka AAL (2014) Altertoxins with potent anti-HIV activity from *Alternaria tenuissima* QUE1Se, a fungal endophyte of *Quercus emoryi*. Bioorg Med Chem 22:6112–6116

Berkovitz DB, Choi S, Maeng JH (2000) Enzyme assisted asymmetric total synthesis of (−)-podophyllotoxin and (−)-picropodophyllotoxin. J Org Chem 65:847–860

Bhattacharyya D, Chattopadhyay S (2015) Characterization of podophyllotoxin biosynthetic pathway and future prospect of podophyllotoxin production from *Podophyllum hexandrum* Royle. Chem Biol Lett 2:12–21

Boonphong S, Kittakoop P, Isaka M, Pittayakhajonwut D, Tanticharoen M, Thebtaranonth Y (2001) Multiplolides A and B, new antifungal 10-membered lactones from *Xylaria multiplex*. J Nat Prod 64:965–967

Brady SF, Clardy J (2000) CR377 a new pentaketide antifungal agent isolated from an endophytic fungus. J Nat Prod 63:1447–1448

Broekaert K, Noseda B, Heyndrickx M, Vlaemynck G, Devliegere F (2013) Volatile compounds associated with *Psychrobacter* spp. and *Pseudoalteromonas* spp., the dominant microbiota of brown shrimp (*Crangon crangon*) during aerobic storage. Int J Food Microbiol 63:1447–1448

Bungihan ME, Tan MA, Kitajima M, Kogure N, Franzblau SG, Dela Cruz TEE, Takayam H, Nonato MG (2011) Bioactive metabolites of *Diaporthe* sp. P133, an endophytic fungus isolated from *Pandanus amaryllifolius*. J Nat Med 65:606–609

Bunyapaiboonsri T, Yoiprommarat S, Srikitikulchai P, Srichomthong K, Lumyong S (2010) Oblongolides from the endophytic fungus *Phomopsis* sp. BCC 9789. J Nat Prod 73:55–59

Cafeu MC, Silva GH, Teles HL, Bolzani VS, Araújo AR, Young MCM, Pfenning LH (2005) Antifungal compounds of *Xylaria* sp., an endophytic fungus isolated from *Palicourea marcgravii* (Rubiaceae). Quim Nova 28:991–995

Cai X, Shan T, Li P, Huang Y, Xu L, Zhou L, Wang M, Jiang W (2009) Spirobisnaphthalenes from the endophytic fungus Dzf12 of *Dioscorea zingiberensis* and their antimicrobial activities. Nat Prod Commun 4:1469–1472

Cai X-C, Liu C-H, Wang B-T, Xue Y-R (2017) Genomic and metabolic traits endow *Bacillus velezensis* CC09 with a potential biocontrol agent in control of wheat powdery mildew disease. Microbiol Res 196:89–94

Cao L, Qiu Z, Dai X, Tan H, Lin Y, Zhou S (2004) Isolation of endophytic actinomycetes from roots and leaves of banana (*Musa acuminata*) plants and their activities against *Fusarium oxysporum* f. sp. *cubense*. World J Microbiol Biotechnol 20:501–504

Casella TM, Eparvier V, Mandavid H, Bendelac A, Odonne G, Dayan L, Duplais C, Espindola LS, Stein D (2013) Antimicrobial and cytotoxic secondary metabolites from tropical leaf endophytes: isolation of antibacterial agent pyrrocidine C from *Lewia infectoria* SNB-GTC2402. Phytochemistry 96:370–377

Castillo U, Strobel GA, Ford EJ, Hess WM, Porter H, Jensen JB, Albert H, Robison R, Condron MA, Teplow DB, Stevens D, Yaver D (2002) Munumbicins, wide spectrum antibiotics produced by *Streptomyces munumbi*, endophytic on *Kennedia nigriscans*. Microbiology 148:2675–2685

Castillo U, Harper JK, Strobel GA, Sears J, Alesi K, Ford E, Lin J, Hunter M, Maranta M, Ge H, Yaver D, Jensen JB, Porter H, Robinson R, Millar D, Hess WM, Condron M, Teplow D (2003) Kakadumycins, novel antibiotics from *Streptomyces* sp. NRRL 30566, an endophyte of *Grevillea pteridifolia*. FEMS Microbiol Lett 224:183–190

Chaipackdee V, Pornpakakul S, Chokratin P (2013) Bioactive compounds produced by endophytic fungi isolated from beach morning glory (*Ipomoea pes-caprae* Linn.). Asian J Microbiol Biotechnol Environ Sci 15:479–482

Chakravarthi BVSK, Das P, Surendranath K, Karande AA, Jayabaskaran C (2008) Production of paclitaxel by *Fusarium solani* isolated from *Taxus celebica*. J Biosci 33:259–267

Chapla VM, Zeraik ML, Leptokarydis IH, Silva GH, Bolzani VS, Young MCM, Pfenning LH, Araújo AR (2014) Antifungal compounds produced by *Colletotrichum gloeosporioides*, an endophytic fungus from *Michelia champaca*. Molecules 19:19243–19252

Chen Y, Guo H, Du Z, Liu XZ, Che Y, Ye X (2009) Ecology-based screen identifies new metabolites from a *Cordyceps*-colonizing fungus as cancer cell proliferation inhibitors and apoptosis inducers. Cell Prolif 42:838–847

Cheng M-J, Wu M-D, Yanai H, Su Y-S, Chen I-S, Yuan G-F, Hsieh S-Y, Chen J-J (2012) Secondary metabolites from the endophytic fungus *Biscogniauxia formosana* and their antimycobacterial activity. Phytochem Lett 5:467–472

Chokpaiboon S, Sommit D, Teerawatananond T, Muangsin N, Bunyapaiboonsri T, Pudhom K (2010) Cytotoxic nor-chamigrane and chamigrane endoperoxides from a basidiomycetous fungus. J Nat Prod 73:1005–1007

Chomcheon P, Wiyakrutta S, Sriubolmas N, Ngamrojanavanich N, Isarangkul D, Kittakoop P (2005) 3-Nitropropionic acid (3-NPA), a potent antimycobacterial agent from endophytic fungi: is 3-NPA in some plants produced by endophytes? J Nat Prod 68:1103–1105

Chomcheon P, Sriubolmas N, Wiyakrutta S, Ngamrojanavanich N, Chaichit N, Mahidol C, Ruchirawat S, Kittakoop P (2006) Cyclopentenones, scaffolds for organic syntheses produced by the endophytic fungus mitosporic *Dothideomycete* sp. LRUB20. J Nat Prod 69:1351–1353

Chomcheon P, Wiyakrutta S, Sriubolmas N, Ngamrojanavanich N, Kengtong S, Mahidol C, Ruchirawat S, Kittakoop P (2009) Aromatase inhibitory, radical scavenging, and antioxidant activities of depsidones and diaryl ethers from the endophytic fungus *Corynespora cassiicola* L36. Phytochemistry 70:407–413

Chomcheon P, Wiyakrutta S, Aree T, Sriubolmas N, Ngamrojanavanich N, Mahidol C, Ruchirawat S, Kittakoop P (2010) Curvularides A– E: antifungal hybrid peptide–polyketides from the endophytic fungus *Curvularia geniculata*. Chem Eur J 16:11178–11185

Chu M, Truumees I, Patel M, Blood C, Das PR, Puar MS (1995) Sch 50673 and Sch 50676, two novel antitumor fungal metabolites. J Antibiot 48:329–331

Cohen SD (2006) Host selectivity and genetic variation of *Discula umbrinella* isolates from two oak species: analyses of intergenic spacer region sequences of ribosomal DNA. Microbial Ecol 52:463–469

Collins LA, Franzblau SG (1997) Microplate Alamar blue assay versus BACTEC 460 system for high-throughput screening of compounds against *Mycobacterium tuberculosis* and *Mycobacterium avium*. Antimicrob Agents Chemother 41:1004–1009

Clark TN, Ellsworth KT, Li H, Johnson JA, Gray CA (2013) Isolation of the plant hormone (+)-abscisic acid as an antimycobacterial constituent of the medicinal plant endophyte *Nigrospora* sp. Nat Prod Commun 8:1673–1674

Clark TN, Bishop AI, McLaughlin M, Calhoun LA, Johnson JA, Gray CA (2014) Isolation of (-)-avenaciolide as the antifungal and antimycobacterial constituent of a *Seimatosporium* sp. endophyte from the medicinal plant *Hypericum perforatum*. Nat Prod Commun 9:1495–1496

Clark TN, Ellsworth KT, Jean S, Webster D, Robichaud GA, Johnson JA, Gray CA (2015) Isolation of phomopsolide A and 6(*E*)-phomopsolide A as antimycobacterial products from an unidentified endophyte of the Canadian medicinal plant *Heracleum maximum*. Nat Prod Commun 10:1647–1648

Cutler HG, Hoogsteen K, Littrell RH, Arison BH (1991) Epoxyexserohilone, a novel metabolite from *Nigrospora sphaerica*. Agric Biol Chem 55:2037–2042

Calvo H, Marco P, Blanco D, Oria R, Venturini ME (2017) Potential of a new strain of *Bacillus amyloliquefaciens* BUZ-14 as a biocontrol agent of postharvest fruit diseases. Food Microbiol 63:101–110

Daisy B, Strobel GA, Castillo U, Ezra D, Sears J, Weaver D, Runyon JB (2002a) Naphthalene, an insect repellent, is produced by *Muscodor vitigenus*, a novel endophytic fungus. Microbiology 148:3737–3741

Daisy B, Strobel GA, Ezra D, Castillo U, Baird G, Hess WM (2002b) *Muscodor vitigenus anam.* sp. nov., an endophyte from *Paullinia paullinioides*. Mycotaxon 84:39–50

Damayanti Y, Lown JW (1998) Podophyllotoxins: current status and recent developments. Curr Med Chem 5:205–252

Das A, Rahman MI, Ferdous AS, Amin A, Rahman MM, Nahar N, Uddin MA, Islam MR, Khan H (2017) An endophytic basidiomycete, *Grammothele lineata*, isolated from *Corchorus olitorius*, produces paclitaxel that shows cytotoxicity. PLoS One 12:e0178612. https://doi.org/10.1371/journal.pone.0178612

Debbab A, Aly AH, Ebel RAE, Müller WEG, Mosaddak M, Hakiki A, Ebel R, Proksch P (2009) Bioactive secondary metabolites from the endophytic fungus *Chaetomium* sp. isolated from *Salvia officinalis* growing in Morocco. Biotechnol Agron Soc Environ 13:229–234

Debbab A, Aly AH, Proksch P (2011) Bioactive secondary metabolites from endophytes and associated marine derived fungi. Fungal Divers 49:1–12

Demain AL (2000) Microbial natural products: a past with a future. In: Wrigley SK, Hayes MA, Thomas R, Chrystal EJT, Nicholson N (eds) Biodiversity: new leads for pharmaceutical and agrochemical industries. The Royal Society of Chemistry, Cambridge, UK, pp 3–16

Deng BW, Liu KH, Chen WQ, Ding XW, Xie XC (2009) *Fusarium solani*, Tax-3, a new endophytic taxol-producing fungus from *Taxus chinensis*. World J Microbiol Biotechnol 25:139–143

Deshmukh SK, Mishra PD, Kulkarni-Almeida A, Verekar S, Sahoo MR, Periyasamy G, Goswami H, Khanna A, Balakrishnan A, Vishwakarma R (2009) Anti-inflammatory and anticancer activity of ergoflavin isolated from an endophytic fungus. Chem Biodivers 6:784–789

Ding G, Song YC, Chen JR, Xu C, Ge HM, Wang XT, Tan RX (2006) Chaetoglobosin U, a cytochalasan alkaloid from endophytic *Chaetomium globosum* IFB-E019. J Nat Prod 69:302–304

Ding G, Zheng Z, Liu S, Zhang H, Guo L, Che Y (2009) Photinides A-F, cytotoxic benzofuranone-derived γ-lactones from the plant endophytic fungus *Pestalotiopsis photiniae*. J Nat Prod 72:942–945

Ding H-E, Yang Z-D, Sheng L, Zhou S-Y, Li S, Yao X-J, Zhi K-K, Wang Y-G, Zhang F (2015) Secovironolide, a novel furanosteroid scaffold with a five-membered B ring from the endophytic fungus *Talaromyces wortmannii* LGT-4. Tetrahedron Lett 56:6754–6757

Dhankhar S, Yadav JP (2013) Investigations towards new antidiabetic drugs from fungal endophytes associated with *Salvadora oleoides* Decne. Med Chem 9:624–632

Dissanayake RK, Ratnaweera PB, Williams DE, Wijayarathne CD, Wijesundera RLC, Andersen RJ, de Silva ED (2016) Antimicrobial activities of endophytic fungi of the Sri Lankan aquatic plant *Nymphaea nouchali* and chaetoglobosin a and C, produced by the endophytic fungus *Chaetomium globosum*. Mycol: Int J Fungal Biol 7:1–8. https://doi.org/10.1080/21501203.2015.1136708

Dreyfuss MM, Chapela IH (1994) Potential of fungi in the discovery of novel, low-molecular weight pharmaceuticals. In: Gullao VP (ed) The discovery of natural products with therapeutic potential. Butterworth-Heinemann, London, pp 49–80

Dzoyem JP, Melong R, Tsamo AT, Maffo T, Kapche DGWF, Ngadjui BT, McGaw LJ, Eloff JN (2017) Cytotoxicity, antioxidant and antibacterial activity of four compounds produced by an endophytic fungus *Epicoccum nigrum* associated with *Entada abyssinica*. Rev Bras 27:251–253

Eyberger L, Dondapati R, Porter JR (2006) Endophyte fungal isolates from *Podophyllum peltatum* produce podo-phyllotoxin. J Nat Prod 69:1121–1124

Ezra D, Castillo UF, Strobel GA, Hess WM, Porter H, Jensen J, Condron M, Teplow DB, Sears J, Maranta M, Hunter M, Weber B, Yaver D (2004) Coronamycins, peptide antibiotics produced by a verticillate *Streptomyces* sp. (MSU-2110) endophytic on *Monstera* sp. Microbiology 150:785–793

Fábio A, Barros P, Filho ER (2005) Four spiroquinazoline alkaloids from *Eupenicillium* sp. isolated as an endophytic fungus from leaves of *Murraya paniculata* (Rutaceae). Biochem Syst Ecol 33:257–268

Faeth SH, Fagan WF (2002) Fungal endophytes: common host plant symbionts but uncommon mutualists. Integr Comp Biol 42:360–368

Fierro-Cruz JE, Jimenez P, Coy-Barrera E (2017) Fungal endophytes isolated from *Protium heptaphyllum* and *Trattinnickia rhoifolia* as antagonists of *Fusarium oxysporum*. Rev Argent Microbiol 49:255–263

Fill TP, Pereira GK, Geris Dos Santos RM, Rodrigues–Fo E (2007) Four additional meroterpenes produced by *Penicillium* sp. found in association with *Melia azedarach*. Possible biosynthetic intermediates to austin. Z Naturforsch 62b:1035–1044

Fill TP, da Silva BF, Rodrigues-Fo E (2010) Biosynthesis of phenylpropanoid amides by an endophytic *Penicillium brasilianum* found in root bark of *Melia azedarach*. J Microbiol Biotechnol 20:622–629

Firakova S, Sturdíkova M, Muckova M (2007) Bioactive secondary metabolites produced by microorganisms associated with plants. Biologia 62:251–257

Florke U, Krohn K, Zia-Ullah HH (2006) An orthorhombic modification of (*R*) – (-) -8-hydroxy-3-methyl-3, 4-dihydro-1*H*-2benzopyran-1-one [(*R*)-(-)-mellein]. Acta Crystallogr E 62:03539–03541

Fondi M, Orlandini V, Maida I, Perrin E, Papaleo MC, Emiliani G, de Pascale D, Parrilli E, Tutino ML, Michaud L, Lo Giudice A, Fani R (2012) Draft genome sequence of the volatile organic compound-producing Antarctic bacterium *Arthrobacter* sp. strain TB23, able to inhibit cystic fibrosis pathogens belonging to the *Burkholderia cepacia* complex. J Bacteriol 194:6334–6335

Gangadevi V, Muthumary J (2008a) Taxol, an anticancer drug produced by an endophytic fungus *Bartalinia robillardoides* Tassi, isolated from a medicinal plant, *Aegle marmelos* Correa ex Roxb. World J Microbiol Biotechnol 24:717–724

Gangadevi V, Muthumary J (2008b) Isolation of *Colletotrichum gloeosporioides*, a novel endophytic taxol-producing fungus from the leaves of a medicinal plant, *Justicia gendarussa*. Mycol Balc 5:1–4

Gangadevi V, Muthumary J (2009a) A novel endophytic taxol-producing fungus *Chaetomella raphigera* isolated from a medicinal plant, *Terminalia arjuna*. Appl Biochem Biotechnol 158:675–684

Gangadevi V, Muthumary J (2009b) Taxol production by *Pestalotiopsis terminaliae*, an endophytic fungus of *Terminalia arjuna* (Arjun tree). Biotechnol Appl Biochem 52:9–15

Gangadevi V, Murugan M, Muthumary J (2008) Taxol determination from *Pestalotiopsis pauciseta*, a fungal endophyte of a medicinal plant. Sheng Wu Gong Cheng Xue Bao 24:1433–1438

Ganley RJ, Brunsfeld SJ, Newcombe GA (2004) A community of unknown, endophytic fungi in western white pine. Proc Nat Acad Sci 101:10107–10112

Gao SS, Li XM, Zhang Y, Li CS, Wang BG (2011) Conidiogenones H and I, two new diterpenes of cyclopiane class from a marine-derived endophytic fungus *Penicillium chrysogenum* QEN-24S. Chem Biodivers 8:1748–1753

Gao Z, Zhang B, Liu H, Han J, Zhang Y (2017) Identification of endophytic *Bacillus velezensis* ZSY-1 strain and antifungal activity of its volatile compounds against *Alternaria solani* and *Botrytis cinerea*. Biol Control 105:27–39

Garyali S, Kumar A, Reddy MS (2013) Taxol production by an endophytic fungus, *Fusarium redolens*, isolated from Himalayan yew. J Microbiol Biotechnol 23:1372–1380

Gautam AK (2013) Diversity of fungal endophytes in some medicinal plants of Himachal Pradesh, India. Arch Phytopathol Plant Protect 47:537–544

Geris dos Santos RM, Rodrigues-Fo E (2002) Meroterpenes from *Penicillium* sp. found in association with *Melia azedarach*. Phytochemistry 61:907–912

Geris dos Santos RM, Rodrigues-Fo E (2003) Further meroterpenes produced by *Penicillium* sp., an endophyte obtained from *Melia azedarach*. Z Naturforsch 58c:663–669

Gohel V, Singh A, Vimal M, Ashwini P, Chhatpar HS (2006) Bioprospecting and antifungal potential of chitinolytic microorganisms. Afr J Biotechnol 5:54–72

Gond SK, Kharwar RN, White JF Jr (2014) Will fungi be the new source of the blockbuster drug taxol? Fungal Biol Rev 28:77–84

Gonzalez MC, Anaya AL, Glenn AE, Macias-Rubalcava ML, Hernandez-Bautista BE, Hanlin RT (2009) *Muscodor yucatanensis*, a new endophytic ascomycete from Mexican chakah, *Bursera simaruba*. Mycotaxon 110:363–372

Gordaliza M, Garcia PA, Del Corral JMM, Castro MA, Gómez-Zurita MA (2004) Podophyllotoxin: distribution, sources, applications and new cytotoxic derivatives. Toxicon 44:441–459

Gotor-Vila A, Teixido N, Di Francesco A, Usall J, Ugolini L, Torres R, Mari M (2017) Antifungal effect of volatile organic compounds produced by *Bacillus amyloliquefaciens* CPA-8 against fruit pathogen decays of cherry. Food Microbiol 64:219–225

Grimme E, Zidack NK, Sikora RA, Strobel GA, Jacobsen BJ (2007) Comparison of *Muscodor albus* volatiles with a biorational mixture for control of seedling diseases of sugar beet and root-knot nematode on tomato. Plant Dis 91:220–225

Grosset J (1996) Current problems with tuberculosis treatment. Res Microbiol 147:10–16

Grove JF (1985) Metabolic products of *Phomopsis oblonga*, part 2. Phomopsolide A and B, tiglic esters of two 6-substituted 5,6-dihydro-5-hydroxypyran-2-ones. J Chem Soc Perkin Trans 1:865–869

Gu W (2009) Bioactive metabolites from *Alternaria brassicicola* ML-P08, an endophytic fungus residing in *Malus halliana*. World J Microbiol Biotechnol 25:1677–1683

Guo B, Kunming LH (1998) A middle vinblastine fungi isolated. J Yunnan Univ 20:214–215

Guo B, Dai JR, Ng S, Huang Y, Leong C, Ong W, Carte BK (2000) Cytonic acids A & B: novel tridepside inhibitors of hCMV protease from the endophytic fungus *Cytonaema* species. J Nat Prod 63:602–604

Guo BH, Wang YC, Zhou XW, Hu K, Tan F, Miao ZQ, Tang KX (2006) An endophytic taxol-producing fungus BT2 isolated from *Taxus chinensis* var. *mairei*. Afr J Biotechnol 5:875–877

Guo B, Wang Y, Sun X, Tang K (2008) Bioactive natural products from endophytes: a review. Biochem Microbiol 44:136–142

Gupte M, Kulkarni P, Ganguli BN (2002) Antifungal antibiotics. Appl Microbiol Biotechnol 58:46–57

Han JW, Ahn SH, Park SH, Wang SY, Bae GU, Seo DW, Kwon HK, Hong S, Lee HY, Lee YW, Lee HW (2000) Apicidin, a histone deacetylase inhibitor, inhibits proliferation of tumor cells *via* induction of p21WAF1/Cip1 and gelsolin. Cancer Res 60:6068–6074

Haritakun R, Rachtawee P, Chanthaket R, Boonyuen N, Isaka M (2010) Butyrolactones from the fungus *Aspergillus terreus* BCC 4651. Chem Pharm Bull 58:1545–1548

Harper JK, Arif AM, Ford EJ, Strobel GA, Tomer DP, Grant DM, Porco J, Oneill K (2003) Pestacin: a 1,3-dihydro isobenzofuran from *Pestalotiopsis microspora* possessing antioxidant and antimycotic activities. Tetrahedron 59:2471–2476

Harwood JS, Cutler HG, Jacyno JM (1995) Nigrosporolide, a plant growth-inhibiting macrolide from the mould *Nigrospora sphaerica*. Nat Prod Lett 6:181–185

Hay RJ (2003) Antifungal drugs used for systemic mycoses. Dermatol Clin 21:577–587

He W-J, Zhou X-J, Qin X-C, Mai Y-X, Lin X-P, Liao S-R, Yang B, Zhang T, Tu Z-C, Wang J-F, Liu Y (2017) Quinone/hydroquinone meroterpenoids with antitubercular and cytotoxic activities produced by the sponge-derived fungus *Gliomastix* sp. ZSDS1-F7. Nat Prod Res 31:604–609

Hemamalini V, Mukesh Kumar DJ, Rebecca AIN, Srimathi S, Muthumary J, Kalaichelvan PT (2015) Isolation and characterization of taxol producing endophytic *Phoma* sp. from *Calotropis gigantea* and its anti-proliferative studies. J Acad Ind Res 3:645–649

Hiraga K, Yamamoto S, Fukuda H, Hamanaka N, Oda K (2005) Enniatin has a new function as an inhibitor of Pdr5p, one of the ABC transporters in *Saccharomyces cerevisiae*. Biochem Biophys Res Commun 328:1119–1125

Hormazabal E, Schmeda–Hirschmann G, Astudillo L, Rodriguez J, Theoduloz C (2005) Metabolites from *Microsphae-ropsis olivacea*, an endophytic fungus of *Pilgerodendron uviferum*. Z Naturforsch C 60:11–21

Huang WY, Cai YZ, Xing J, Corke H, Sun M (2007) Potential antioxidant resource: endophytic fungi isolated from traditional Chinese medicinal plants. Econ Bot 61:14–30

Huang WY, Cai YZ, Hyde KD, Corke H, Sun M (2008a) Biodiversity of endophytic fungi associated with 29 traditional Chinese medicinal plants. Fungal Divers 33:61–75

Huang Z, Cai X, Shao C, She Z, Xia X, Chen Y, Yang J, Zhou S, Lin Y (2008b) Chemistry and weak antimicrobial activities of phomopsins produced by mangrove endophytic fungus *Phomopsis* sp. ZSU-H76. Phytochemistry 69:1604–1608

Huang Z, Yang J, She Z, Lin Y (2010) Isoflavones from the mangrove endophytic fungus *Fusarium* sp. (ZZF41). Nat Prod Commun 5:1771–1773

Hussain H, Kliche-Spory C, Al-Harrasi A, Al-Rawahi A, Abbas G, Green IR, Schulz B, Krohn K, Shah A (2014) Antimicrobial constituents from three endophytic fungi. Asian Pac J Trop Med 7:S224–S227

Hussain H, John M, Al-Harrasi A, Shah A, Hassan Z, Abbas G, Rana UA, Green IR, Schulz B, Krohn K (2015) Phytochemical investigation and antimicrobial activity of an endophytic fungus *Phoma* sp. J King Saud Univ – Sci 27:92–95

Hutchings ML, Alpha-Cobb CJ, Hiller DA, Berro J, Strobel SA (2017) Mycofumigation through production of the volatile DNA-methylating agent N-methyl-N-nitrosoisobutyramide by fungi in the genus *Muscodor*. J Biol Chem 292:7358–7371

Indrianingsih AW, Tachibana S (2017) α-glucosidase inhibitor produced by an endophytic fungus, *Xylariaceae* sp. QGS 01 from *Quercus gilva* Blume. Food Sci Human Wellness 6:88–95

Isaka M, Jaturapat A, Rukseree K, Danwisetkanjana K, Tanticharoen M, Thebtaranonth Y (2001) Phomoxanthones A and B, novel xanthone dimers from the endophytic fungus *Phomopsis* species. J Nat Prod 64:1015–1018

Isaka M, Berkaew P, Intereya K, Komwijit S, Sathitkunanon T (2007) Antiplasmodial and antiviral cyclohexadepsipeptides from the endophytic fungus *Pullularia* sp. BCC 8613. Tetrahedron 63:6855–6860

Isaka M, Chinthanom P, Boonruangprapa T, Rungjindamai N, Pinruan U (2010) Eremophilane-type sesquiterpenes from the fungus *Xylaria* sp. BCC 21097. J Nat Prod 73:683–687

Iznaga Y, Lemus M, González L, Garmendía L, Nadal L, Vallín C (2004) Antifungal activity of actinomycetes from Cuban soils. Phytother Res 18:494–496

Jadulco R, Brauers G, Edrada RA, Ebel R, Wray V, Sudarsono V, Proksch P (2002) New metabolites from sponge-derived fungi *Curvularia lunata* and *Cladosporium herbarum*. J Nat Prod 65:730–733

Jadulco RC, Koch M, Kakule TB, Schmidt EW, Orendt A, He H, Janso JE, Carter GT, Larson EC, Pond C, Matainaho TK, Barrows LR (2014) Isolation of pyrrolocins A–C: cis- and trans-decalin tetramic acid antibiotics from an endophytic fungal-derived pathway. J Nat Prod 77:2537–2544

Jalgaonwalal RE, Mohite BV, Mahajan RT (2011) Evaluation of endophytes for their antimicrobial activity. Afr J Biotechnol 10:103–107

Jian Z, Wang H, Wang H, Meng L, Xu G (2015) Isolation and identification of an endophytic fungus producing paclitaxel from *Taxus wallichiana* var. *mairei*. Nutr Hosp 32:2932–2937

Jian Z, Meng L, Hu X (2017) An endophytic fungus efficiently producing paclitaxel isolated from *Taxus wallichiana* var. *mairei*. Medicine (Baltimore) 96:e7406. https://doi.org/10.1097/MD.0000000000007406

Jouda JB, Mawabo IK, Notedji A, Mbazoa CD, Nkenfou J, Wandji J, Nkenfou CN (2016) Anti-mycobacterial activity of polyketides from *Penicillium* sp. endophyte isolated from *Garcinia nobilis* against *Mycobacterium smegmatis*. Int J Mycobacteriol 5:192–196

Kathivaran G, Raman VS (2010) *In vitro* taxol production, by *Pestalotiopsis breviseta* – a first report. Fitoterapia 81:557–564

Khan IA, Mirza ZM, Kumar A, Verma V, Qazi GN (2006) Piperine, a phytochemical potentiator of ciprofloxacin against *Staphylococcus aureus*. Antimicrob Agents Chemother 50:810–812

Khan R, Shahzad S, Choudhary MI, Khan SA, Ahmad A (2010) Communities of endophytic fungi in medicinal plant *Withania somnifera*. Pak J Bot 42:1281–1287

Khan MIH, Sohrab MH, Rony SR, Tareq FS, Hasan CM, Mazid MA (2015) Cytotoxic and antibacterial naphthoquinones from an endophytic fungus, *Cladosporium* sp. Toxicol Rep 3:861–865

Kim JC, Choi GJ, Park JH, Kim HT, Cho KY (2001) Activity against plant pathogenic fungi of phomalactone isolated from *Nigrospora sphaerica*. Pest Manag Sci 57:554–559

Kim S, Shin DS, Lee T, Oh KB (2004) Periconicins two new fusicoccane diterpenes produced by an endophytic fungus *Periconia* sp. with antibacterial activity. J Nat Prod 67:448–450

Korkina LG (2007) Phenylpropanoids as naturally occurring antioxidants: from plant defense to human health. Cell Mol Biol 53:15–25

Kornsakulkarn J, Somyong W, Supothina S, Boonyuen N, Thongpanchang C (2015) Bioactive oxygen-bridged cyclooctadienes from endophytic fungus *Phomopsis* sp. BCC 45011. Tetrahedron 71:9112–9116

Koul A, Arnoult E, Lounis N, Guillemont J, Andries K (2011) The challenge of new drug discovery for tuberculosis. Nature 469:483–490

Kour A, Shawl AS, Rehman S, Sultan P, Qazi PH, Suden P, Khajuria RK, Verma V (2008) Isolation and identification of an endophytic strain of *Fusarium oxysporum* producing podophyllotoxin from *Juniperus recurva*. World J Microbiol Biotechnol 24:1115–1121

Krohn K, Bahramsari R, Florke U, Ludewig K, Kliche-Spory C, Michel A, Aust HJ, Draeger S, Schulz B, Antus S (1997) Dihydroisocoumarins from fungi: isolation, structure elucidation, circular dichroism and biological activity. Phytochemistry 45:313–320

Krohn K, Biele C, Drogies KH, Steingröver K, Aust H-J, Draeger S, Schulz B (2002) Fusidilactones, a new group of polycyclic lactones from an endophyte, *Fusidium* sp. Eur J Org Chem 14:2331–2336

Krzic NS, Pongrac P, Klemenc M, Kladnik A, Regvar M, Gagerscik A (2006) Mycorrhizal colonisation in plants from intermittent aquatic habitats. Aquat Bot 85:331–336

Kumar DSS, Lau CS, Wan JMF, Yang D, Hyde KD (2005) Immunomodulatory compounds from *Pestalotiopsis leucoth*? (HKUCC 10197), an endophytic fungus of *Tripterygium wilfordii*. Life Sci 78:147–156

Kumar A, Khan IA, Koul S, Koul JL, Taneja SC, Ali I, Ali F, Sharma S, Mirza ZM, Kumar M, Sangwan PL, Gupta P, Thota N, Qazi GN (2008) Novel structural analogues of piperine as inhibitors of the NorA efflux pump of *Staphylococcus aureus*. J Antimicrob Chemother 61:1270–1276

Kumar A, Patil D, Rajamohanan PR, Ahmad A (2013) Isolation, purification and characterization of vinblastine and vincristine from endophytic fungus *Fusarium oxysporum* isolated from *Catharanthus roseus*. PLoS One 8:e71805. https://doi.org/10.1371/journal.pone.0071805

Kumar CG, Godumagadda NR, Mongolla P, Pombala S, Badrinarayan P, Reddy CR, Das B, Ramakrishna KVS, Sridhar B, Kamal A (2017) Novel bioactive compounds from an endophytic *Chaetomium arcuatum* strain SAF-2 isolated from *Semecarpus anacardium*. Chapter 3. In: Bhukya B, Tangutur AD (eds) Microbial biotechnology: technological challenges and developmental trends. Apple Academy Press, Waretown, pp 63–87

Kumaran RS, Hur BK (2009) Screening of species of the endophytic fungus *Phomopsis* for the production of the anticancer drug taxol. Biotechnol Appl Biochem 54:21–30

Kumaran RS, Muthumary J, Hur BK (2008a) Production of taxol from *Phyllosticta spinarum*, an endophytic fungus of *Cupressus* sp. Eng Life Sci 8:438–446

Kumaran RS, Muthumary J, Hur BK (2008b) Taxol from *Phyllosticta citricarpa*, a leaf spot fungus of the angiosperm *Citrus medica*. J Biosci Bioeng 106:103–106

Kumaran RS, Muthumary J, Hur BK (2008c) Isolation and identification of taxol, an anticancer drug from *Phyllosticta melochiae* Yates, an endophytic fungus of *Melochia corchorifolia* L. Food Sci Biotechnol 17:1246–1253

Kumaran RS, Muthumary J, Hur BK (2009) Isolation and identification of an anticancer drug, taxol from *Phyllosticta tabernaemontanae*, a leaf spot fungus of an angiosperm, *Wrightia tinctoria*. J Microbiol 47:40–49

Kumaran RS, Kim HJ, Hur BK (2010) Taxol producing fungal endophyte, *Pestalotiopsis* species isolated from *Taxus cuspidata*. J Biosci Bioeng 110:541–546

Kumaran RS, Jung H, Kim HJ (2011) *In vitro* screening of taxol, an anticancer drug produced by the fungus, *Colletotrichum capsici*. Eng Life Sci 11:264–271

Kumaran RS, Choi YK, Lee S, Jeon HJ, Jung H, Kim HJ (2012) Isolation of taxol, an anticancer drug produced by the endophytic fungus, *Phoma betae*. Afr J Biotechnol 11:950–960

Kusari S, Lamshoft M, Zuhlke S, Spiteller M (2008) An endophytic fungus from *Hypericum perforatum* that produces hypericin. J Nat Prod 71:159–162

Kusari S, Zühlke S, Spiteller M (2009a) An endophytic fungus from *Camptotheca acuminata* that produces camptothecin and analogues. J Nat Prod 72:2–7

Kusari S, Lamshoft M, Spiteller M (2009b) *Aspergillus fumigates* Fresenius, an endophytic fungus from *Juniperus communis* L. Horstmann as a novel source of the anticancer pro-drug deoxypodophyllotoxin. J Appl Microbiol 107:1019–1030

Lin Y, Wu X, Feng S, Jiang G, Luo J, Zhou SL, Vrijmoed LP, Jones EBG, Krohn K, Steingrover K, Zsila F (2001) Five unique compounds: Xyloketals from mangrove fungus *Xylaria* sp. from the South China sea coast. J Org Chem 66:6252–6256

Lee KH, Xiao Z (2003) Lignans in treatment of cancer and other diseases. Phytochem Rev 2:341–362

Lee J, Lobkovsky E, Pliam NB, Strobel GA, Clardy J (1995a) Subglutinols A & B: immunosuppressive compounds from the endophytic fungus *Fusarium subglutinans*. J Org Chem 60:7076–7077

Lee JC, Yang X, Schwartz M, Strobel GA, Clardy J (1995b) The relationship between an endangered North American tree and an endophytic fungus. Chem Biol 2:721–727

Lee JC, Strobel GA, Lobkovsky E, Clardy J (1996) Torreyanic acid: a selectively cytotoxic quinone dimer from the endophytic fungus *Pestalotiopsis microspora*. J Org Chem 61:3232–3233

Lopes AA, Pupo MT (2011) Biosynthesis of aphidicolin proceeds via the mevalonate pathway in the endophytic fungus *Nigrospora sphaerica*. J Braz Chem Soc 22:460–462

Li JY, Strobel GA (2001) Jesterone and hydroxy-jesterone antioomycete cyclohexenenone epoxides from the endophytic fungus *Pestalotiopsis jesteri*. Phytochemistry 57:261–265

Li JY, Strobel GA, Sidhu RS, Hess WM, Ford E (1996) Endophytic taxol-producing fungi from bald cypress, *Taxodium distichum*. Microbiology 142:2223–2226

Li JY, Sidhu RS, Ford EJ, Hess WM, Strobel GA (1998) The induction of taxol production in the endophytic fungus *Periconia* sp. from *Torreya grandifolia*. J Ind Microbiol Biotechnol 20:259–264

Li JY, Strobel GA, Harper JK, Lobkovsky E, Clardy J (2000) Cryptocin, a potent tetramic acid antimycotic from the endophytic fungus *Cryptosporiopsis* cf. *quercina*. Org Lett 2:767–770

Li JY, Harper JK, Grant DM, Tombe BO, Bashyal B, Hess WM, Strobel GA (2001) Ambuic acid, a highly functionalized cyclohexenone with antifungal activity from *Pestalotiopsis* spp. and *Monochaetia* sp. Phytochemistry 56:463–468

Li W, Zhou J, Lin Z, Hu Z (2007) Study on fermentation condition for production of huperzine A from endophytic fungus 2F09P03B of *Huperzia serrata*. Chin Med Biotechnol 2:254–259

Li E, Jiang L, Guo L, Zhang H, Che Y (2008a) Pestalachlorides A-C, antifungal metabolites from the plant endophytic fungus *Pestalotiopsis adusta*. Bioorg Med Chem 16:7894–7899

Li E, Tian R, Liu S, Chen X, Guo L, Che Y (2008b) Pestalotheols A–D, bioactive metabolites from the plant endophytic fungus *Pestalotiopsis theae*. J Nat Prod 71:664–668

Li CT, Li Y, Wang QJ, Sung CK (2008c) Taxol production by *Fusarium arthrosporioides* isolated from yew, *Taxus cuspidata*. J Med Biochem 27:454–458

Li H, Xiao J, Gao YQ, Tang JJ, Zhang AL, Gao JM (2014) Chaetoglobosins from *Chaetomium globosum*, an endophytic fungus in *Ginkgo biloba*, and their phytotoxic and cytotoxic activities. J Agric Food Chem 62:3734–3741

Li H, Doucet B, Flewelling AJ, Jean S, Webster D, Robichaud GA, Johnson JA, Gray CA (2015) Antimycobacterial natural products from endophytes of the medicinal plant *Aralia nudicaulis*. Nat Prod Commun 10:1641–1642

Liu JY, Song YC, Zhang Z, Wang L, Guo ZJ, Zou WX, Tan RX (2004) *Aspergillus fumigatus* CY018, an endophytic fungus in *Cynodon dactylon* as a versatile producer of new and bioactive metabolites. Biotechnology 114:279–287

Liu X, Dong M, Chen X, Jiang M, Lv X, Zhou J (2008) Antimicrobial activity of an endophytic *Xylaria* sp. YX-28 and identification of its antimicrobial compound 7-amino-4-methylcoumarin. Appl Microbiol Biotechnol 78:241–247

Liu K, Ding X, Deng B, Chen W (2009) Isolation and characterization of endophytic taxol-producing fungi from *Taxus chinensis*. J Ind Microbiol Biotechnol 36:1171–1177

Liu F, Cai XL, Yang H, Xia XK, Guo ZY, Yuan J, Li MF, She ZG, Lin YC (2010a) The bioactive metabolites of the mangrove endophytic fungus *Talaromyces* sp. ZH-154 isolated from *Kandelia candel* (L.) Druce. Planta Med 76:185–189

Liu K, Ding X, Deng B, Chen W (2010b) 10-Hydroxycamptothecin produced by a new endophytic *Xylaria* sp. M20, from *Camptotheca acuminate*. Biotechnol Lett 32:689–693

Lu H, Zou WX, Meng JC, Hu J, Tan RX (2000) New bioactive metabolites produced by *Colletotrichum* sp., an endophytic fungus in *Artemisia annua*. Plant Sci 151:67–73

Lu Z, Zhu H, Fu P, Wang Y, Zhang Z, Lin H, Liu P, Zhuang Y, Hong K, Zhu W (2010) Cytotoxic polyphenols from the marine-derived fungus *Penicillium expansum*. J Nat Prod 73:911–914

Lu Y, Chen C, Chen H, Zhang J, Chen W (2012) Isolation and identification of endophytic fungi from *Actinidia macrosperma* and investigation of their bioactivities. Evid-Based Complement Alternat Med 2012:382742. https://doi.org/10.1155/2012/382742

Ma YM, Li Y, Liu JY, Song YC, Tan RX (2004) Anti-*Helicobacter pylori* metabolites from *Rhizoctonia* sp. Cy064, an endophytic fungus in *Cynodon dactylon*. Fitoterapia 75:451–456

Maida I, Fondi M, Papaleo MC, Perrin E, Orlandini V, Emiliani G, De Pascale D, Parrilli E, Tutino ML, Michaud L, Giudice AL (2014) Phenotypic and genomic characterization of the Antarctic bacterium *Gillisia* sp. CAL575, a producer of antimicrobial compounds. Extremophiles 18:35–49

Marinho AMR, Rodrigues-Filho E, Moitinho MDLR, Santos LS (2005) Biologically active polyketides produced by *Penicillium janthinellum* isolated as an endophytic fungus from fruits of *Melia azedarach*. J Braz Chem Soc 16:280–283

Meng LH, Liao ZY, Pommier Y (2003) Non-camptothecin DNA topoisomerase I inhibitors in cancer therapy. Curr Top Med Chem 3:305–320

Mercier J, Smilanick JL (2005) Control of green mold and sour rot of stored lemon by biofumigation with *Muscodor albus*. Biol Control 32:401–407

Meshram V, Kapoor N, Chopra G, Saxena S (2017) *Muscodor camphora*, a new endophytic species from *Cinnamomum camphora*. Mycosphere 8:568–582

Metwaly AM, Kadry HA, El-Hela AA, Mohammad A-EI, Ma G, Cutler SJ, Ross SA (2014) Nigrosphaerin A a new isochromene derivative from the endophytic fungus *Nigrospora sphaerica*. Phytochem Lett 7:1–5

Miao Z, Wang Y, Yu X, Guo B, Tang K (2009) A new endophytic taxane production fungus from *Taxus chinensis*. Appl Biochem Microbiol 45:81–86

Michalczyk A, Cieniecka-Roslonkiewicz A, Cholewinska M (2014) Plant endophytic fungi as a source of paclitaxel. Herba Pol 60:22–33

Mirjalili MH, Farzaneh M, Bonfill M, Rezadoost H, Ghassempour A (2012) Isolation and characterization of *Stemphylium sedicola* SBU-16 as a new endophytic taxol-producing fungus from *Taxus baccata* grown in Iran. FEMS Microbiol Lett 328:122–129

Mirza ZM, Kumar A, Kalia NP, Zargar A, Khan IA (2011) Piperine as an inhibitor of the MdeA efflux pump of *Staphylococcus aureus*. J Med Microbiol 60:1472–1478

Mitchell AM, Strobel GA, Moore E, Robison R, Sears J (2010) Volatile antimicrobials from *Muscodor crispans*, a novel endophytic fungus. Microbiology 156:270–277

Mizushina Y, Kamisuki S, Kasai N, Shimazaki N, Takemura M, Asahara H, Linn S, Yoshida S, Matsukage A, Koiwai O, Sugawara F, Yoshida H, Sakaguchi K (2002) A plant phytotoxin, Solanapyrone A, is an inhibitor of DNA polymerase b and l. J Biol Chem 277:630–638

Mohali S, Burgess TI, Wingfield MN (2005) Diversity and host association of the tropical tree endophyte *Lasiodiplodia theobromae* revealed using simple sequence repeat markers. Forest Pathol 35:385–396

Moraes RM, Burandt C, Ganzera M, Li X, Khan I, Canel C (2001) The American mayapple revisited – *Podophyllum peltatum* – still a potential cash crop? Econ Bot 54:471–476

Moricca S, Ragazzi A (2008) Fungal endophytes in Mediterranean oak forests: a lesson from *Discula quercina*. Phytopathology 98:380–386

Mullin T, Clark T, Johnson J, Gray C (2017) Isolation of phomopsolide A as the antimycobacterial constituent of an unidentified endophyte of the Canadian medicinal plant *Geum macrophyllum*. Can J Undergrad Res 2:69–71

Muthumary J, Sashirekha S (2007) Detection of taxol, an anticancer drug, from selected coelomycetous fungi. Indian J Sci Technol 1:1–10

Nithya K, Muthumary J (2009) Growth studies of *Colletotrichum gloeosporioides* (Penz.) Sacc. – a taxol producing endophytic fungus from *Plumeria acutifolia*. Indian J Sci Technol 2:14–19

Orhan IE (2016) Potential of natural products of herbal origin as monoamine oxidase inhibitors. Curr Pharm Des 22:268–276

Orlandini V, Maida I, Fondi M, Perrin E, Papaleo MC, Bosi E, de Pascale D, Tutino ML, Michaud L, Lo Giudice A, Fani R (2014) Genomic analysis of three sponge-associated *Arthrobacter* Antarctic strains, inhibiting the growth of *Burkholderia cepacia* complex bacteria by synthesizing volatile organic compounds. Microbiol Res 169:593–601

Pacher P, Kecskemeti V (2004) Trends in the development of new antidepressants. Is there a light at the end of the tunnel? Curr Med Chem 11:925–943

Padhi L, Mohanta YK, Panda SK (2013) Endophytic fungi with great promises: a review. J Adv Pharm Educ Res 3:152–170

Patel HA, Patel RK, Khristi SM, Parikh K, Rajendran G (2012) Isolation and characterization of bacterial endophytes from *Lycopersicon esculentum* plant and their plant growth promoting characteristics. Nepal J Biotechnol 2:37–52

Pan JH, Chen Y, Huang YH, Tao YW, Wang J, Li Y, Peng Y, Dong T, Lai XM, Lin YC (2011) Antimycobacterial activity of fusaric acid from a mangrove endophyte and its metal complexes. Arch Pharm Res 34:1177–1181

Pandi M, Kumaran RS, Choi YK, Kim HJ, Muthumary J (2011) Isolation and detection of taxol, an anticancer drug produced from *Lasiodiplodia theobromae*, an endophytic fungus of the medicinal plant *Morinda citrifolia*. Afr J Biotechnol 10:1428–1435

Papaleo MC, Fondi M, Maida I, Perrin E, Lo Giudice A, Michaud L, Mangano S, Bartolucci G, Romoli R, Fani R (2012) Sponge-associated microbial Antarctic communities exhibiting antimicrobial activity against *Burkholderia cepacia* complex bacteria. Biotechnol Adv 30:272–293

Papaleo MC, Romoli R, Bartolucci G, Maida I, Perrin R, Fondi M, Orlandini V, Mengoni A, Emiliani G, Tutino ML, Parrilli E, de Pascale D, Michaud L, Lo Giudice A, Fani R (2013) Bioactive volatile organic compounds from Antarctic (sponges) bacteria. New Biotechnol 30:824–838

Park JH, Choi GJ, Lee HB, Kim KM, Jung HS, Lee SW, Jang KS, Cho KY, Kim JC (2005) Griseofulvin from *Xylaria* sp. strain F0010, an endophytic fungus of *Abies holophylla* and its antifungal activity against plant pathogenic fungi. J Microbiol Biotechnol 15:112–117

Pavithra N, Sathish L, Ananda K (2012) Antimicrobial and enzyme activity of endophytic fungi isolated from Tulsi. J Pharm Biomed Sci 16:1–6

Palazzini JM, Dunlap CA, Bowman MJ, Chulze SN (2016) *Bacillus velezensis* RC 218 as a biocontrol agent to reduce *Fusarium* head blight and deoxynivalenol accumulation: genome sequencing and secondary metabolite cluster profiles. Microbiol Res 192:30–36

Petrini O (1991) Fungal endophytes in the tree leaves. In: Andrews JH, Hirano SS (eds) Microbial ecology of leaves. Springer, New York, pp 179–197

Pittayakhajonwut P, Suvannakad R, Thienhirun S, Prabpai S, Kongsaeree P, Tanticharoen M (2005) An anti-*Herpes simplex* virus-type 1 agent from *Xylaria mellisii* (BCC 1005). Tetrahedron Lett 46:1341–1344

Pongcharoen W, Rukachaisirikul V, Phongpaichit S, Kühn T, Pelzing M, Sakayaroj J, Taylor WC (2008) Metabolites from the endophytic fungus *Xylaria* sp. PSU-D14. Phytochemistry 69:1900–1902

Prihantini AI, Tachibana S (2017) Antioxidant compounds produced by *Pseudocercospora* sp. ESL 02, an endophytic fungus isolated from *Elaeocarpus sylvestris*. Asian Pac J Trop Biomed 7:110–115

Purahong W, Hyde KD (2011) Effects of fungal endophytes on grass and non-grass litter decomposition rates. Fungal Divers 47:1–7

Puri SC, Verma V, Amna T, Qazi GN, Spiteller M (2005) An endophytic fungus from *Nothapodytes foetida* that produces camptothecin. J Nat Prod 68:1717–1719

Puri SC, Nazir A, Chawla R, Arora R, Riyaz-ul-Hasan S, Amna T, Ahmed B, Verma V, Singh S, Sagar R, Sharma A, Kumar R, Sharma RK, Qazi GN (2006) The endophytic fungus *Trametes hirsuta* as a novel alternative source of podophyllotoxin and related aryl tetralin ligans. J Biotechnol 122:494–510

Qin JC, Zhang YM, Gao JM, Bai MS, Yang SX, Laatsch H, Zhang AL (2009) Bioactive metabolites produced by *Chaetomium globosum*, an endophytic fungus isolated from *Ginkgo biloba*. Bioorg Med Chem Lett 19:1572–1574

Radji M, Sumiati A, Rachmayani R, Elya B (2011) Isolation of fungal endophytes from *Garcinia mangostana* and their antibacterial activity. Af J Biotechnol 10:103–107

Ramin AA, Prange RK, Gordon-Braun P, Delong JM (2007) Biofumigation of postharvest fungal apple decay with *Muscodor albus* volatiles. J Appl Sci Res 3:307–310

Ramsay RR (2012) Monoamine oxidases: the biochemistry of the proteins as targets in medicinal chemistry and drug discovery. Curr Top Med Chem 12:2189–2209

Ratnaweera PB, de Silva ED, Williams DE, Andersen RJ (2015) Antimicrobial activities of endophytic fungi obtained from the arid zone invasive plant *Opuntia dillenii* and the isolation of equisetin, from endophytic *Fusarium* sp. BMC Comp Alt Med 15:220. https://doi.org/10.1186/s12906-015-0722-4

Rehman S, Shawl AS, Kour A, Andrabi R, Sudan P, Sultan P, Verma V, Qazi GN (2008) An endophytic *Neurospora* sp. from *Nothapodytes foetida* producing camptothecin. Appl Biochem Microbiol 44:203–209

Ren Y, Strobel GA, Graff JC, Jutila M, Park SG, Gosh S, Teplow D, Condron M, Pang E, Hess WM, Moore E (2008) Colutellin A, an immunosuppressive peptide from *Colletotrichum dematium*. Microbiology 154:1973–1879

Rodrigues KF, Samuels GJ (1994) *Letendraeposis palmarum*, a new genus and species of local ascomycetes. Mycologia 86:254–258

Rodrigues KF, Hesse M, Werner C (2000) Antimicrobial activities of secondary metabolites produced by endophytic fungi from *Spondias mombin*. J Basic Microbiol 40:261–267

Romero JC, Barria OE, Arnold AE, Rios CL (2008) Activity against *Plasmodium falciparum* of lactones isolated from the endophytic fungus *Xylaria* sp. Pharm Biol 46:1–4

Romoli R, Papaleo MC, de Pascale D, Tutino ML, Michaud L, Lo Giudice A, Fani R, Bartolucci G (2011) Characterization of the volatile profile of Antarctic bacteria by using solid-phase microextraction-gas chromatography–mass spectrometry. J Mass Spectrom 46:1051–1059

Romoli R, Papaleo MC, de Pascale D, Tutino ML, Michaud L, Lo Giudice A, Fani R, Bartolucci G (2014) GC–MS volatolomic approach to study the antimicrobial activity of the Antarctic bacterium *Pseudoalteromonas* sp. TB41. Metabolomics 10:42–51

Rosenbluth M, Martinez-Romero E (2006) Bacterial endophytes and their interactions with hosts. Mol Plant-Microbe Interact 19:827–837

Roy BG (2017) Potential of small-molecule fungal metabolites in antiviral chemotherapy. Antivir Chem Chermother 25:20–52

Rukachaisirikul V, Sommart U, Phongpaichit S, Sakayaroj J, Kirtikara K (2008) Metabolites from the endophytic fungus *Phomopsis* sp. PSU-D15. Phytochemistry 69:783–787

Rungjindamai N, Pinruan U, Choeyklin R, Hattori T, Jones EBG (2008) Molecular characterization of basidiomycetous endophytes isolated from leaves, rachis and petioles of the oil palm, *Elaeis guineensis*, in Thailand. Fungal Divers 33:139–161

Sakemi S, Inagaki T, Kaneda K, Hirai H, Iwata E, Sakakibara T, Yamauchi Y, Norcia M, Wondrack LM, Sutcliffe JA, Kojima N (1995) CJ-12,371 and CJ-12,372, two novel DNA gyrase inhibitors. Fermentation, isolation, structural elucidation and biological activities. J Antibiot 48:134–142

Schulz B (2002) Endophytic fungi: a sources of novel metabolites. In: British Mycological Society, international symposium proceedings on bioactive fungal metabolites – impact and exploitation at University of Wales

Schulz B, Boyle C (2005) The endophytic continuum. Mycol Res 109:661–686

Schwarz M, Kopcke B, Weber RWS, Sterner O, Anke H (2004) 3-Hydroxypropionic acid as a nematicidal principle in endophytic fungi. Phytochemistry 65:2239–2245

Seephonkai P, Isaka M, Kittakoop P, Palittapongarnpim P, Kamchonwongpaisan, Tanticharoen M, Thebtaranonth Y (2002) Evaluation of antimycobacterial, antiplasmodial and cytotoxic activities of preussomerins isolated from the lichenicolous fungus *Microsphaeropsis* sp. BCC 3050. Planta Med 68:45–48

Seifried HE, Anderson DE, Fisher EI, Milner JA (2007) A review of the interaction among dietary antioxidants and reactive oxygen species. J Nutr Biochem 18:567–579

Senthilkumar N, Murugesan S, Mohan V, Muthumary J (2013) Taxol producing fungal endophyte, *Colletotrichum gleospoiroides* (Penz.) from *Tectona grandis* L. Curr Biotica 7:8–15

Singh SB, Ondeyka JG, Tsipouras N, Ruby C, Sardana V, Schulman M, Sanchez M, Pelaez F, Stahlhut MW, Munshi S, Olsen DB, Lingham RB (2004) Hinnuliquinone, a C2-symmetric dimeric non-peptide fungal metabolite inhibitor of HIV-1 protease. Biochem Biophys Res Commun 324:108–113

Shah A, Rather MA, Hassan QP, Aga MA, Mushtaq S, Shah AM, Hussain A, Baba SA, Ahmad Z (2017) Discovery of anti-microbial and anti-tubercular molecules from *Fusarium solani*: an endophyte of *Glycyrrhiza glabra*. J Appl Microbiol 122:1168–1176

Shankar NB, Shashikala J (2010) Diversity and structure of fungal endophytes in some climbers and grass species of Malnad region, Western Ghats, Southern India. Mycosphere 1:265–274

Shao CL, Wang CY, Gu YC, Wei MY, Pan JH, Deng DS, She ZG, Lin YC (2010) Penicinoline, a new pyrrolyl 4-quinolinone alkaloid with an unprecedented ring system from an endophytic fungus *Penicillium* sp. Bioorg Med Chem Lett 20:3284–3286

Sharma S, Kumar M, Sharma S, Nargotra A, Koul S, Khan IA (2010) Piperine as an inhibitor of Rv1258c, a putative multidrug efflux pump of *Mycobacterium tuberculosis*. J Antimicrob Chemother 65:1694–1701

Sheela C (2012) Endophytic fungi: novel sources of anticancer lead molecules. Appl Microbiol Biotechnol 95:47–59

Shih JC, Chen K, Ridd MJ (1999) Monoamine oxidase: from genes to behavior. Annu Rev Neurosci 22:197–217

Shiono Y, Yokoi M, Koseki T, Murayama T, Aburai N, Kimura K (2010) Allantopyrone A, a new alpha-pyrone metabolite with potent cytotoxicity from an endophytic fungus, *Allantophomopsis lycopodina* KS-97. J Antibiot 63:251–253

Shrestha K, Strobel GA, Shrivastava SP, Gewali MB (2001) Evidence for paclitaxel from three new endophytic fungi of Himalayan yew of Nepal. Planta Med 67:374–376

Shweta S, Zuehlke S, Ramesha BT, Priti V, Kunar PM, Ravikanth G, Spiteller M, Vasudeva R, Shaanker RU (2010) Endophytic fungal strains of *Fusarium solani*, from *Apodytes dimidiata* E. Mey. ex Arn (Icacinaceae) produce camptothecin, 10-hydroxycamptothecin and 9-methoxycamptothecin. Phytochemistry 71:117–122

Silva GH, Teles LH, Trevisan CH, Bolzani SV, Young MC, Haddad R, Araujo RA (2005) New bioactive metabolites produced by *Phomopsis cassiae*, an endophytic fungus in *Cassia spectabilis*. J Braz Chem Soc 16:1463–1466

Silva GH, de Oliveira CM, Teles HL, Pauletti PM, Castro-Gamboa I, Silva DHS, Bolzani VS, Young MCM, Costa-Neto CM, Pfenning LH, Berlinck RGS, Araujo AR (2010) Sesquiterpenes from

Xylaria sp., an endophytic fungus associated with *Piper aduncum* (Piperaceae). Phytochem Lett 3:164–167
Salituro GM, Pelaez F, Zhang BB (2001) Discovery of a small molecule insulin receptor activator. Recent Prog Horm Res 56:107–126
Soliman SSM, Tsao R, Raizada MN (2011) Chemical inhibitors suggest endophytic fungal paclitaxel is derived from both mevalonate and non-mevalonate-like pathways. J Nat Prod 74:2497–2504
Somjaipeng S, Medina A, Kwasna H, Ortiz JO, Magan N (2015) Isolation, identification, and ecology of growth and taxol production by an endophytic strain of *Paraconiothyrium variabile* from English yew trees (*Taxus baccata*). Fungal Biol 119:1022–1031
Song YC, Li H, Ye YH, Shan CY, Yang YM, Tan RX (2004) Endophytic naphthopyrone metabolites are co-inhibitors of xanthine oxidase, SW1116 cell and some microbial growths. FEMS Microbiol Lett 241:67–72
Song YC, Huang WY, Sun C, Wang FW, Tan RX (2005) Characterization of graphislactone A as the antioxidant and free radical scavenging substance from the culture of *Cephalosporium* sp. IFB-E001, an endophytic fungus in *Trachelospermum jasminoides*. Biol Pharm Bull 28:506–509
Spadari S, Sala F, Pedrali-Noy G (1982) Aphidicolin: a specific inhibitor of nuclear DNA replication in eukaryotes. Trends Biochem Sci 7:29–32
Specian V, Sarragiotto MH, Pamphile JA, Clemente E (2012) Chemical characterization of bioactive compounds from the endophytic fungus *Diaporthe hellianthi* isolated from *Luehea divaricata*. Braz J Microbiol 43:1174–1182
Sreekanth D, Syed A, Sarkar S, Sarkar D, Santhakumari B, Ahmad A, Khan MI (2009) Production, purification and characterization of taxol and 10DAB III from a new endophytic fungus *Gliocladium* sp. isolated from the Indian yew tree, *Taxus baccata*. J Microbiol Biotechnol 19:1342–1347
Srinivasan L, Jagadish K, Shenbhagaraman R, Muthumary J (2010) Antioxidant activity of endophytic fungus *Phyllosticta* sp. isolated from *Guazuma tomentosa*. J Phytology 2:37–41
Stierle A, Strobel G, Stierle D (1993) Taxol and taxane production by *Taxomyces andreanae*, an endophytic fungus of Pacific yew. Science 260:214–216
Stierle AA, Stierle DB, Bugni T (2001) Sequoiatones, C-F constituents of the redwood endophytes. J Nat Prod 64:1350–1353
Strobel G (2002) Microbial gifts from rain forests. Can J Plant Pathol 24:14–20
Strobel G (2006a) Harnessing endophytes for industrial microbiology. Curr Opin Microbiol 9:240–244
Strobel G (2006b) *Muscodor albus* and its biological promise. J Ind Microbiol Biotechnol 33:514–522
Strobel G (2011) *Muscodor* species-endophytes with biological promise. Phytochem Rev 10:165–172
Strobel G, Daisy B (2003) Bioprospecting for microbial endophytes and their natural products. Microbiol Mol Biol Rev 67:491–502
Strobel GA, Stierle A, Stierle D, Hess WM (1993) *Taxomyces andreanae* a proposed new taxon for a bulbilliferous hyphomycete associated with Pacific yew. Mycotaxon 47:71–78
Strobel G, Yang X, Sears J, Kramer R, Sidhu RS, Hess WM (1996a) Taxol from *Pestalotiopsis microspora*, an endophytic fungus of *Taxus wallichiana*. Microbiology 142:435–440
Strobel GA, Hess WM, Ford E, Sidhu RS, Yang X (1996b) Taxol from fungal endophytes and issue of biodiversity. J Ind Microbiol 17:417–423
Strobel GA, Hess WM, Li JY, Ford E, Sears J, Sidhu RS, Summerell B (1997) *Pestalotiopsis guepinii*, a taxol producing endophyte of the Wollemi pine, *Wollemia nobilis*. Aust J Bot 45:1073–1082
Strobel GA, Ford E, Li JY, Sears J, Sidhu RS, Hess WM (1999a) *Seimatoantlerium tepuiense* gen. nov. a unique epiphytic fungus producing taxol from the Venezuelan-Guayana system. Syst Appl Microbiol 22:426–433

Strobel GA, Miller RV, Miller C, Condron M, Teplow DB, Hess WM (1999b) Cryptocandin, a potent antimycotic from the endophytic fungus *Cryptosporiopsis* cf. *quercina*. Microbiology 145:1919–1926

Strobel G, Li JY, Sugawara F, Koshino H, Harper J, Hess WM (1999c) Oocydin A, a chlorinated macrocyclic lactone with potent anti-oomycete activity from *Serratia marcescens*. Microbiology 145:3557–3564

Strobel GA, Dirkse E, Sears J, Markworth C (2001) Volatile antimicrobials from *Muscodor albus*, a novel endophytic fungus. Microbiology 147:2943–2950

Strobel G, Ford E, Worapong J, Harper JK, Arif AM, Grant DM, Fung PC, Chau RMW (2002) Isopestacin, an isobenzofuranone from *Pestalotiopsis microspora*, possessing antifungal and antioxidant activities. Phytochemistry 60:179–183

Strobel G, Daisy B, Castillo U, Harper J (2004) Natural product from endophytic microorganisms. J Nat Prod 67:257–268

Stinson AM, Zidack NK, Strobel GA, Jacobsen BJ (2003) Effect of mycofumigation with *Muscodor albus* and *Muscodor roseus* on seedling diseases of sugarbeet and *Verticillium* wilt of eggplant. Plant Dis 87:1349–1354

Suryanarayanan TS, Kumaresan V (2000) Endophytic fungi of some halophytes from an estuarine mangrove forest. Mycol Res 104:1465–1467

Suryanarayanan TS, Thirunavukkarasu N, Hriharan GN, Balaji P (2005) Occurrence of non-obligate microfungi inside lichen thalli. Sydowia 57:20–130

Suryanarayanan TS, Thirunavukkarasu N, Govindarajulu MB, Sasse F, Jansen R, Murali TS (2009) Fungal endophytes and bioprospecting. Fungal Biol Rev 23:9–19

Sutjaritvorakul T, Whalley AJS, Sihanonth P, Roengsumran S (2011) Antimicrobial activity from endophytic fungi isolated from plant leaves in dipterocarpous forest at Viengsa district Nan province, Thailand. J Agric Technol 7:115–121

Suto M, Takebayashi M, Saito K, Tanaka M, Yokota A, Tomita F (2002) Endophytes as producers of xylanase. J Biosci Bioeng 93:88–90

Suwannarach N, Bussaban B, Hyde KD, Lumyong S (2010) *Muscodor cinnamoni*, a new endophytic species from *Cinnamomum bejolghota*. Mycotaxon 114:15–23

Tan RX, Zou WX (2001) Endophytes: a rich source of functional metabolites. Nat Prod Rep 18:448–459

Thirunavukkarasu N, Suryanarayanan TS, Murali TS, Ravishankar JP, Gummadi SN (2011) L-asparaginase from marine derived fungal endophytes of seaweeds. Mycosphere 2:147–155

Tipton KF, Boyce S, O'Sullivan J, Davey GP, Healy J (2004) Monoamine oxidases: certainties and uncertainties. Curr Med Chem 11:1965–1982

Tung CY, Yang DB, Gou M (2002) A preliminary study on the condition of the culture and isolate of endophytic fungus producing vincristine. J Chuxiong Norm Univ 6:39–41

Uma SR, Ramesha BT, Ravikanth G, Rajesh PG, Vasudeva R, Ganeshaiah KN (2008) Chemical profiling of *N. nimmoniana* for camptothecin, an important anticancer alkaloid: towards the development of a sustainable production system. In: Ramawat KG, Merillion J (eds) Bioactive molecules and medicinal plants. Springer, Berlin, pp 198–210

Vennila R, Muthumary J (2011) Taxol from *Pestalotiopsis pauciseta* VM1, an endophytic fungus of *Tabebuia pentaphylla*. Biomed Prev Nutr 1:103–108

Verma VC, Lobkovsky E, Gange AC, Singh SK, Prakash S (2011) Piperine production by endophytic fungus *Periconia* sp. isolated from *Piper longum* L. J Antibiot 64:427–431

Visalakchi S, Muthumary J (2010) Taxol (anticancer drug) producing endophytic fungi: an overview. Int J Pharm Biosci 1:1–9

Valko M, Leibfritz D, Moncol J, Cronin MTD, Mazur M, Telser J (2007) Free radicals and antioxidants in normal physiological functions and human disease. Int J Biochem Cell Biol 39:44–84

Wagenaar M, Corwin J, Strobel GA, Clardy J (2000) Three new chytochalasins produced by an endophytic fungus in the genus *Rhinocladiella*. J Nat Prod 63:1692–1695

Wang LG, Liu XM, Kreis W, Budman DR (1999) The effect of antimicrotubule agents on signal transduction pathways of apoptosis: a review. Cancer Chemother Pharmacol 44:355–361

Wang J, LI G, Lu H, Zheng Z, Huang Y, Su W (2000) Taxol from *Tubercularia* sp. strain TF5, an endophytic fungus of *Taxus mairei*. FEMS Microbiol Lett 193:249–253
Wang FW, Jiao RH, Cheng AB, Tan SH, Song YC (2007) Antimicrobial potentials of endophytic fungi residing in *Quercus variabilis* and brefeldin A obtained from *Cladosporium* sp. World J Microbiol Biotechnol 23:79–83
Wang J, Zhao B, Zhang W, Wu X, Wang R, Huang Y, Chen D, Park K, Weimer BC, Shen Y (2010a) Mycoepoxydiene, a fungal polyketide, induces cell cycle arrest at the G2/M phase and apoptosis in HeLa cells. Bioorg Med Chem Lett 20:7054–7058
Wang JM, Ding GZ, Fang L, Dai JG, Yu SS, Wang YH, Chen XG, Ma SG, Qu J, Xu S, Du D (2010b) Thiodiketopiperazines produced by the endophytic fungus *Epicoccum nigrum*. J Nat Prod 73:1240–1249
Wani MC, Taylor HL, Wall ME, Coggon P, McPhail AT (1971) Plant antitumor agents. VI. The isolation and structure of taxol, a novel antileukemic and antitumor agent from *Taxus brevifolia*. J Am Chem Soc 93:2325–2327
Webber JF, Gibbs JN (1984) Colonization of elm barks by *Phomopsis oblonga*. Trans Br Mycol Soc 82:348–352
Weber D, Sterner O, Anke T, Gorzalczancy S, Martin V, Acevedo C (2004a) Phomol, a new anti-inflammatory metabolite from an endophyte of the medicinal plant *Erythrina cristagalli*. J Antibiot 57:559–563
Weber RW, Stenger E, Meffert A, Hahn M (2004b) Brefeldin A production by *Phoma medicaginis* in dead pre-colonized plant tissue: a strategy for habitat conquest? Mycol Res 108:662–671
Wen L, Chen G, She Z, Yan C, Cai J, Mu L (2010) Two new paeciloxocins from a mangrove endophytic fungus *Paecilomyces* sp. Russ Chem Bull 59:1656–1659
Wicklow DT, Roth S, Deyrup ST, Gloer JB (2005) A protective endophyte of maize: *Acremonium zeae* antibiotics inhibitory to *Aspergillus flavus* and *Fusarium verticillioides*. Mycol Res 109:610–618
Wijeratne EMK, Turbyville TJ, Fritz A, Whitesell L, Gunatilaka AAL (2006) A new dihydroxanthenone from a plant-associated strain of the fungus *Chaetomium globosum* demonstrates anticancer activity. Bioorg Med Chem 14:7917–7923
Wijeratne EM, He H, Franzblau SG, Hoffman AM, Gunatilaka AA (2013) Phomapyrrolidones A-C, antitubercular alkaloids from the endophytic fungus *Phoma* sp. NRRL 46751. J Nat Prod 76:1860–1865
Worapong J, Strobel GA, Ford EJ, Li JY, Baird G, Hess WM (2001) *Muscodor albus* anam. nov., an endophyte from *Cinnamomum zeylanicum*. Mycotaxon 79:67–79
Worapong J, Strobel GA, Daisy B, Castillo U, Baird G, Hess WM (2002) *Muscodor roseus* anam. nov., an endophyte from *Grevillea pteridifolia*. Mycotaxon 81:463–475
World Health Organization (2017a) Cancer. Fact sheet Reviewed February 2017. http://www.who.int/mediacentre/factsheets/fs297/en/ (Retrieved 12 Aug 2017)
World Health Organization (2017b) Tuberculosis. Fact sheet Reviewed March 2017. http://www.who.int/mediacentre/factsheets/fs104/en/ (Retrieved 26 Aug 2017)
Xie G, Zhu X, Li Q, Gu M, He Z, Wu J, Li J, Lin Y, Li M, She Z, Yuan J (2010) SZ-685C, a marine anthraquinone, is a potent inducer of apoptosis with anticancer activity by suppression of the Akt/FOXO pathway. Br J Pharmacol 159:689–697
Xu L, Wang J, Zhao J, Li P, Shan T, Wang J, Li X, Zhou L (2010) Beauvericin from the endophytic fungus, *Fusarium redolens*, isolated from *Dioscorea zingiberensis* and its antibacterial activity. Nat Prod Commun 5:811–814
Yahara I, Harada F, Sekita S, Yoshihira K, Natori S (1982) Correlation between effects of 24 different cytochalasins on cellular structures and cellular events and those on actin *in vitro*. J Cell Biol 92:69–78
Yamada M, Yasuhara H (2004) Clinical pharmacology of MAO inhibitors: safety and future. Neurotoxicology 25:215–221
Youdim MBH, Edmondson D, Tipton KF (2006) The therapeutic potential of monoamine oxidase inhibitors. Nat Rev Neurosci 7:295–309

Yu H, Zhang L, Li L, Zheng C, Guo L, Li W, Sun P, Qin L (2010) Recent developments and future prospects of antimicrobial metabolites produced by endophytes. Microbiol Res 165:437–449

Zeng PY, Wu JG, Liao LM, Chen TQ, Wu JZ, Wong KH (2011) In vitro antioxidant activities of endophytic fungi isolated from the liverwort *Scapania verrucosa*. Genet Mol Res 10:3169–3179

Zhang B, Salituro G, Szalkowski D, Li Z, Zhang Y, Royo I, Vilella D, Dez MT, Pelaez F, Ruby C, Kendall RL, Mao X, Griffin P, Calaycay J, Zierath JR, Heck JV, Smith RG, Moller DE (1999) Discovery of small molecule insulin mimetic with antidiabetic activity in mice. Science 284:974–981

Zhang LB, Gou LH, Zeng SV (2000) Preliminary study on the isolation of endophytic fungus of *Catharanthus roseus* and its fermentation to produce product of therapeutic value. Chin Trad Herb Drugs 11:805–807

Zhang JY, Tao LY, Liang YJ, Yan YY, Dai CL, Xia XK, She ZG, Lin YC, Fu LW (2009a) Secalonic acid D induced leukemia cell apoptosis and cell cycle arrest of G1 with involvement of GSK-*3β/β*-catenin/c-Myc pathway. Cell Cycle 8:2444–2450

Zhang P, Zhou PP, Yu LJ (2009b) An endophytic taxol-producing fungus from *Taxus* × *media, Aspergillus candidus* MD3. FEMS Microbiol Lett 293:155–159

Zhang P, Zhou PP, Yu LJ (2009c) An endophytic taxol-producing fungus from *Taxus media, Cladosporium cladosporioides* MD2. Curr Microbiol 59:227–232

Zhang CL, Wang GP, Mao LJ, Komon-Zelazowska M, Yuan ZL, Lin FC, Druzhinina IS, Kubicek CP (2010) *Muscodor fengyangensis* sp. nov. from southeast China: morphology, physiology and production of volatile compounds. Fungal Biol 114:797–808

Zhang G, Sun S, Zhu T, Lin Z, Gu J, Li D, Gu Q (2011) Antiviral isoindolone derivatives from an endophytic fungus *Emericella* sp. associated with *Aegiceras corniculatum*. Phytochemistry 72:1436–1442

Zhang H, Deng Z, Guo Z, Chen J, Wang J, Zou K (2013a) Anthraquinone metabolites from endophytic *Nigrospora* sp BM-2 and optimization of fermentation medium for high bostrycin production. J Pure Appl Microbiol 7:2489–2494

Zhang X, Li B, Wang Y, Guo Q, Lu X, Li S, Ma P (2013b) Lipopeptides, a novel protein, and volatile compounds contribute to the antifungal activity of the biocontrol agent *Bacillus atrophaeus* CAB-1. Appl Microbiol Biotechnol 97:9525–9534

Zhao K, Ping W, Li Q, Hao S, Zhao L, Gao T, Zhou D (2009) *Aspergillus niger* var. *taxi*, a new species variant of taxol-producing fungus isolated from *Taxus cuspidata* in China. J Appl Microbiol 107:1202–1207

Zhao J, Zhou L, Wang J, Shan T, Zhong L, Liu X, Gao X (2010) Endophytic fungi for producing bioactive compounds originally from their host plants. In: Mendez-Vilas A (ed) Current research, technology and education topics in applied microbiology and microbial biotechnology. Formatex Res Center, Badajoz, pp 567–576

Zhao JH, Zhang YL, Wang LW, Wang JY, Zhang CL (2012) Bioactive secondary metabolites from *Nigrospora* sp. LLGLM003, an endophytic fungus of the medicinal plant *Moringa oleifera* Lam. World J Microbiol Biotechnol 28:2107–2112

Zheng N, Yao F, Liang X, Liu Q, Xu W, Liang Y, Liu X, Li J, Yang R (2017) A new phthalide from the endophytic fungus *Xylaria* sp. GDG-102. Nat Prod Res. https://doi.org/10.1080/14786419.2017.1311892

Zhou D, Hyde KD (2001) Host-specificity, host-exclusivity, and host-recurrence in saprobic fungi. Mycol Res 105:1449–1457

Zhou X, Wang Z, Jiang K, Wei Y, Lin J, Sun X, Tang K (2007) Screening of taxol-producing endophytic fungi from *Taxus chinensis* var. *mairei*. Appl Biochem Microbiol 43:439–443

Zilla MK, Qadri M, Pathania AS, Strobel GA, Nalli Y, Kumar S, Guru SK, Bhushan S, Singh SK, Vishwakarma RA, Riyaz-Ul-Hassan S, Ali A (2013) Bioactive metabolites from an endophytic *Cryptosporiopsis* sp. inhabiting *Clidemia hirta*. Phytochemistry 95:291–297

Chapter 36
Bioactive Compounds from Endophytic Fungi

Vijayalakshmi Selvakumar and A. Panneerselvam

Abstract Endophytic fungi are highly potential for the production of pharmaceutically valuable compounds such as anticancer, antioxidant, antimicrobial, antidiabetic and industrial enzymes, etc. Today human faces a lot of challenges for existence due to the appearance of new diseases, infections, drug resistances and imbalances in the ecosystems. Here endophytes were the potential sources for the new remedies. Endophytes are contemplated as a treasury for bioprospecting, and they assist in many forms to conquer many complications. Among the various endophytic microbes, fungi have been found most potential microorganisms which are a reservoir of largely untapped bioactive metabolites. In the future we need to seek for endophytes for their bioactive compounds.

Keywords Endophytic fungi · Bioactive compounds · Host plants · Podophyllotoxin · Camptothecine · Vinblastine · Hypericin

36.1 Introduction

Emergence of modern diseases, advancement of drug-resistant pathogenic microorganisms, advent of life aggressive viruses and authority of post-operative complications in patient with organ transplantations are some of the objection in front of examiners and researchers. This position has forced researchers to seek various natural sources for the secure, economic and dominant agents to meet the confrontation of the twenty-first century.

Endophytes are microorganisms (mostly fungi and bacteria) that possess plant hosts for all or sector of their life cycle. They conquer the internal plant tissues below the epidermal cell layers without causing any probable harm or symptomatic

V. Selvakumar (✉)
Department of Microbiology, Shrimati Indira Gandhi College, Tiruchirappalli, India

A. Panneerselvam
Department of Botany and Microbiology, A.V.V.M Sri Pushpam College, (Autonomous), Thanjavur, Tamil Nadu, India

P. Gehlot, J. Singh (eds.), *Fungi and their Role in Sustainable Development: Current Perspectives*, https://doi.org/10.1007/978-981-13-0393-7_36

infection to their host, living within the intercellular spaces of the tissues, and it seems that they may go through the living cells (Strobel 2003; Firakova et al. 2007).

Studies on endophytic fungi are desired to give basic knowledge for the appraisal of global fungal dissimilarity and dissemination. Endophytic fungi develop to be metabolically more contemporary compared to soil fungi (Schulz et al. 2002) or fungi associated with algae (Schulz et al. 2008) with regard to bioactive compounds, that they produce correlate bioactive metabolites (Mitchell et al. 2008).

Less endophytes have been beleaguered for novel antibiotics (Bérdy 2005). To date, only about 80,000–100,000 fungal species have been expressed, out of a stable estimate of 1.5 million. The latest studies of endophytic fungi from tropical and temperate forests back the high appraisal of species assortment (Marquez et al. 2007).

During the last 20 years, it has been detected that much of the wealth of microbial biodiversity with novel biochemistry and secondary stability production consists in plant tissues (Porras-Alfaro and Bayman 2011). Interest in such microorganisms, described as endophytes, increased excessively with the discovery of an endophytic fungus, from *Taxus brevifolia*, bringing forth the billion-dollar anticancer drug, Taxol (Stierle et al. 1993).

The arrival of combinatorial chemistry has shifted the investigate focus away from natural products; fungal endophytes carry on to be a basis for new drugs (Bérdy 2005; Wang et al. 2008); they produce a cluster of bioactive compounds of varied structural groups such as terpenoids, steroids, xanthones, chinones, phenols, isocoumarins, benzo pyranones, tetralones, cytochalasins and enniatins. The compounds of endophytic fungi consist of antibacterial, antiviral, antifungal and anticancer activities (Gunatilaka 2006).

In addition, more than 60% of the anticancer and 70% of the antimicrobial drugs at present in clinical use are natural products or natural product derivatives. This is not amazing in the light of their progression over millions of years in varied ecological niches and natural abode. In difference to other natural sources like plants, microorganisms are highly diverse but narrowly analysed. Studies based on estimation of microbial populations have affirmed that only about 1% of bacteria and 5% of fungi have been characterized and the rest remain undetermined for their addition to the human benefits. Plentiful bioactive molecules have been isolated from endophytic fungi since this ground-breaking discernment (Zhang et al. 2012).

The investigation of Moricca and Ragazzi (2008) reveals that the type of interaction between an endophyte and a plant is controlled by the genes of both organisms and regulated by the environment. A compassionate endophyte consists in the host tissue in a symptomless state or one that may be good to its host may turn into a pathogen in reaction to several environmental conditions. It is logical to assume that such a shift in the nature of the endophyte would also result in a change in its metabolite profile. Pathogenic fungi can reside as symptomless endophytes in plant tissues (Suryanarayanan and Murali 2006); furthermore, the species diversity of foliar endophyte congregation is known to change with the leaf age (Suryanarayanan and Thennarasan 2004). These facts point out those sampling endophytes from a plant community for bioprospecting on a single occasion may not capture the entire spectrum of endophytes and their metabolites. The effort has to be uninterrupted to obtain the whole extent of secondary metabolites.

Endophytes have been portrayed as mutualists that protect both grasses and conifers against insect herbivory, and many of those fungi yield biologically active secondary metabolites (Pelaez et al. 1998). Dreyfuss (1986) reported antibiotic activity from isolates of the endophytic *Pleurophomopsis* species and *Cryptosporiopsis* species, as well as from a sterile endophyte from *Abies alba*. Strains of the endophytic *Pezicula* species (and its anamorph *Cryptosporiopsis*) from several tedious and coniferous tree hosts produce altogether bioactive secondary metabolites in culture (Schultz et al. 1995).

Endophytic species of *Xylariaceae* frequently yield compounds with more biological activity, including cytochalasins (Dreyfuss 1986) and indole diterpenes. Many endophytes Producing toxins in culture, such compounds have been difficult to detect in plant host tissue (Henson et al. 1999). An analysis has revealed that 6-pentyl-α-pyrone and other α-pyrone analogues show antibiotic activity against the development of the fungus *Gaeumannomyces graminis* var. *tritici* (Ghisalberti and Rowland 1993). 6-Pentyl-α-pyrone has been reported to inhibit growth in vitro of a number of fungi and that it reduced the rate of damping-off in lettuce by restraining the growth of *Rhizoctonia solani* (Simon et al. 1988).

The pharmaceutical and medical concerns of new drugs are the coexistence of these prospective drugs to human tissues. Since the plant tissue where the endophytes exist is a eukaryotic system, it would appear that the secondary metabolites produced by the endophytes may have decreased cell toxicity; otherwise, mortality of the host tissue may develop. Thus, the host itself has naturally served as an alternative system for microbes having bioactive molecules with decreased toxicity leading to higher organisms (Strobel 2003).

36.2 Taxonomy of Endophytic Fungi

Mostly sac fungi, anamorphic fungi and *Basidiomycetes* class of fungi are described as endophytic fungi (Dayle et al. 2001). A lot of generic groups of fungi associating to first two classes are managing to live individually in plants. The class and species of the fungi depend upon the host plants. A number of coprophilous fungi belonging to the genera *Ascobolus*, *Coprinus*, *Delitschia*, *Gelasinospora*, etc., were isolated from plants as endophytes (Petrini and Petrini 1985).

36.3 Endophytic Fungi: A Source for Novel Bioactive Secondary Metabolites

For many years, accustomed results have been used directly as drugs or have provided the basic chemical architecture for deriving such drugs. There are at least 200,000 natural metabolites with coactive substances (Bérdy 2005). For detail, about 52% of the new chemicals offered into the merchandise worldwide between 1981 and 2002 were natural products or their by-products. Besides plants, microorganisms constitute a leading source of natural products with desirable bioactive properties. More than 20,000 bioactive metabolites of microbial origin were known by the end of 2002 (Bérdy 2005). Fungi are among the most important groups of eukaryotic organisms that are being explored for metabolites for clinical applications. Existing drugs of fungal origin include β-lactam antibiotics, griseofulvin, cyclosporine A, Taxol, ergot alkaloids and lovastatin.

In the past 5 years from 2000 to 2005, 23 NME (new molecular entity) gathered from plants and microbes for treating different human disorders such as cancer, neural disorders, contagious and cardiovascular disease, metabolic and immunological syndrome and genetic disorders were brought to the market. In the past 9 years from 1993, about 1500 metabolites have been reported from fungi to have antitumour or antibiotic activity (Pela'ez 2005).

Endophytes are the chemical compound producers inside plants. Many of them are able to synthesize bioactive compounds that can be used as budding sources of pharmaceutical leads. Endophytic fungi have been tested useful for innovative drug discovery as recommended by the chemical different of their secondary metabolites. Many endophytic fungi have been reported to produce novel antibacterial, antifungal, antiviral, anti-inflammatory, antitumour and other substance belonging to the alkaloid, steroid, flavonoid and terpenoid extracts and other structure types (De Souza et al. 2011).

36.4 Need for Research on Endophytic Fungi

Endophytic fungi are a collection of organisms which live within the living plant without causing any symptom. It has symbiotic or mutualistic position with its host. Each and every one plant species was studied for their endophytic fungal assemblages (Tan and Zou 2001). Only a small fraction of plant species have been examined. Study on endophytes is developing now; there are many aspects of endophytic organisms that could be examined. Study of biodiversity of endophytic fungi is significant to have the primary idea about the endophytic population of specific plant species.

Antibiotic inhabitance profiles lead to scare about the emergence and re-emergence of multidrug-resistant (MDR) pathogens and parasites. Once a Person infected with MDR bacteria, it is not likely to cure easily, and he/she has to spend more time in the hospital and needs to have multiple treatments of broad-spectrum antibiotics, which are less sufficient, more toxic and more expensive. Moreover, development of or alternation in antimicrobial compounds to improve bactericidal ability is a significant area of research in this modern era (Humberto et al. 2010).

Infection with ESBL-generating organisms can be life-threatening. In vitro studies would recommend that carbapenems or non-β-lactam antibiotics should be the most favourable therapy for ESBL-producing strains since they are not hydrolysed by ESBLs. The furthermost clinical practice is with carbapenems, and these remain the normal therapy for severe infections due to ESBL producers. Co-production of ESBLs and carbapenem-hydrolysing enzymes has been explained and may intimidate the usefulness of carbapenems in the future. For this basis, the progressive development of new antibiotics active against gram-negative bacilli is greatly important.

The history of endophytes dating back 4400 years to grass seeds identified in the tomb of a fifth dynasty Egyptian pharaoh and revealed by Vogl (1898) in seeds of *Lolium temulentum*, no consequent work has been undertaken in this field in Egypt. An investigation on endophytes has essentially been alarmed with parasitic *Clavicipitaceae* which take place within grasses (White et al. 1996). Mysterious or symptomless fungal infection of non-grass hosts, also termed "endophytes", has been returned from impermanent trees and shrubs (Lodge et al. 1996), angiosperm parasites (Suryanarayanan et al. 2000), bamboo (Umali et al. 1999), lichens (Petrini et al. 1990), palms (Frohlich et al. 2000) and mangroves (Suryanarayanan et al. 1998).

Around, there are near to 300,000 plant species on earth, and each creature plant is the host to many endophytes, and many of them may inhabit assured hosts. It has been predicted that there may be as many as one million dissimilar endophytic fungal taxa; thus endophytes may be hyperdiverse (Huang et al. 2007). Endophytes possibly will generate a plethora of bioactive compounds that may be concerned in the host-endophyte relationship and may provide as prospective point of supply novel natural products for development in medicine, agriculture and industry (Strobel and Daisy 2003).

36.5 Discovery of Bioactive Metabolites from Endophytic Fungi

Endophytic fungi that survive inside the tissues of living plants are an under-investigated group of microorganisms. Endophytic fungi had antibacterial, antifungal, anticancer and antiviral agents. Success of Taxol-Producing fungi improved the value of endophytes and exchanged natural products explore to endophytic fungi (Dreyfuss and Chapela 1994).

36.6 Applications of Endophytic Fungi

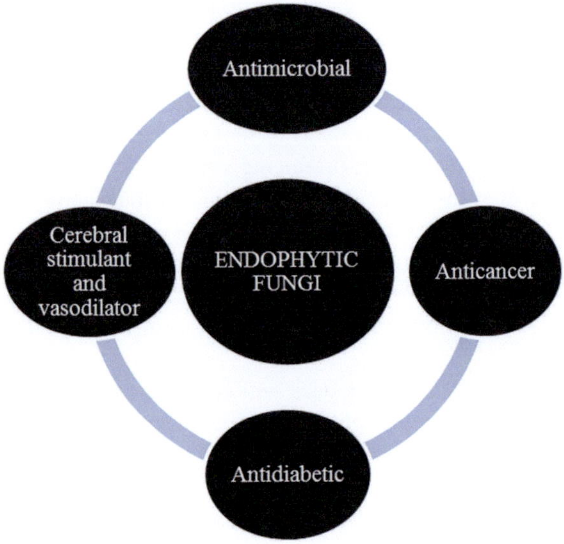

36.6.1 Antimicrobial Activity

A considerable number of information focused on antimicrobial metabolites segregated from mangrove saprophytic fungi, for instance, auranticins, which are antimicrobial depsidones, and the epimeric δ-lactones, helicascolides A and B, were acquired from fungi isolated from mangroves. Aigialomycins A–E, new 14-associated resorcylic macrolides, were separated along with an identified hypothemycin from the mangrove fungus, *Aigialus parvus* BCC 5311 (Isaka et al. 2002). Hypothemycin and aigialomycin D showed in vitro antimalarial activity with IC_{50} values of 2.2 and 6.6 μg/mL, correspondingly, while other analogues were inactive (Poch and Gloer 1991).

The antimicrobial possibility of the ethyl acetate solution of 70 endophytic fungi strains separated from leaves of Brazilian mangrove plant *Laguncularia racemosa* (L.) Gaertn. not in favour of bacteria *Bacillus subtilis*, *Enterococcus faecalis*, *Escherichia coli*, *Micrococcus luteus*, *Pseudomonas aeruginosa* and *Staphylococcus aureus* was noticed by disc diffusion method. Thirty-four (48.6%) endophytic fungi strains were possible to bring forth secondary metabolites with antimicrobial activity, and the basic extracts of *Aspergillus niger*, *Curvularia pallescens*, *Guignardia bidwellii*, *Paecilomyces variotii* and mycelia sterilia showed the best outcome. The bacterium *M. luteus* was the most susceptible, while *P. aeruginosa* was the tolerant, being only suppressed by the extracts of *Aspergillus niger* strains (Silva et al. 2011).

Marine genus *Trichoderma* generates a kind of bioactive metabolites which let in the antimycobacterial, aminolipopeptids, trichoderins (Pruksakorn 2010), antifungal, ketone, cytotoxic dipeptide, trichodermamide B (Garo et al. 2003) and antibacterial tetra hydroanthraquinone, xanthone derivatives.

Two strange pyridines, trichodin A and trichodin B, along with known compound, pyridoxatin, were infused from mycelia and culture broth of the marine fungus, *Trichoderma* actions against the gram-positive *Bacillus subtilis*, *Staphylococcus aureus* and the yeast, *Candida albicans*. The substance pyridoxatin was used against *B. subtilis*, *S. epidermis*, *Staphylococcus aureus*, *C. albicans* and *Trichophyton rubrum* (Bin Wu et al. 2014).

Several of the chemical compounds formed specific antifungal and antibacterial movement. 2-Phenylethanol and tyrosol obtained from *T. harzianum* are reported for the first time from *Trichoderma* species. The ultimate effective metabolite detached from these constrictions was 6-*n*-pentyl-α-pyrone, which showed the highest antimicrobial activity and completely reserved the growth of *Armillaria mellea* fungus at a concentration of 200 ppm. Compound sorbicillin presented cautious activity against the fungal analysis organisms.

36.6.2 Cerebral Stimulant and Vasodilator

Endophytic fungal strain QJ18 was created to produce the biometric compound gentiopicrin like its host plant *G. macrophylla*. Besides, the medicinal plant *Vinca minor* contains the alkaline vincamine, which is used in the medical industry and acts as a cerebral catalyst and vasodilator. An extract from endophytic fungus (Vm-J2) was shown to produce the same bioactive constituent, vincamine, as the host plant (Yin and Sun 2011).

36.6.3 Antibacterial Activity

A whole of 300 endophytes were isolated from many parts of plants from the National Park, Pahang. 3.3% of extracts showed dynamic (IC_{50} <0.01 μg/mL) cell toxic activity against the murine leukemic P388 cell line and 1.7% against a human chronic myeloid leukaemia cell line K562. *Sporothrix* sp. KK29FL1 isolated from *Crepe ginger* showed tough cytotoxicity in opposition to colorectal carcinoma (HCT116) and human breast adenocarcinoma (MCF7) cell lines with IC_{50}values of 0.05 μg/mL and 0.02 μg/mL, correspondingly. Antibacterial activity was demonstrated for 8% of the abstracts. T22 azaphilone, 1-hydroxy-3-methyl-anthraquinone, 1,8-dihydroxy-3-methyl-antraquinone, T39 butenolide, harzianolide and harzianopyridone were purified from *Trichoderma harzianum* (Vinale et al. 2009).

Screening of antibacterial activity in crude extracts from a few *Trichoderma* spp. was examined and determined using CYS80 medium. Most *Trichoderma* spp. had control effects against 12 of the tested pathogens. *T. reesei* and *T. viride* were greatly effectual near morbidic bacteria tested such as *E.coli* (Ec1), *E. coli* (Ec2), *Staphylococcus agalactiae* (SA), *S. aureus* (SAU), *S. pneumoniae* (SP), *Streptococcus pyogenes* (Spy), *Enterobacter faecalis* (EF), methicillin-resistant *Staphylococcus aureus*, *Klebsiella pneumoniae* (KP), *S. sonnei* (Sh.s), *P. mirabilis* (Prot. M) and *P. aeruginosa* (PA). The restriction regions invoked by extracts of these fungi aligned from 10 to 28 mm restriction zones. Diverse restriction zones aligned from 1 to 42.3 mm were recorded for obolifera, *T. koningii*, *Chaetomium cupreum* and *Penicillium sclerotiorum* against EC1 and EC2 (Imtiaj and Lee 2007; Takahashi et al. 2008). *S. pneumoniae*, *S. pyogenes* and *P. aeruginosa* were restricted by *Verticillium albo-atrum* (3 mm inhibition zone), *V. lecanii* (2 mm inhibition zone) and *V. albo-atrum* (2.3 mm inhibition zone) (Mekawey 2010). Conversely, *S. pneumoniae*, *S. pyogenes*, *E. faecalis* and *P. aeruginosa* are not controlled by *Agaricus* cf. *nigrecentulus*, *Agrocybe perfecta*, *Basidiomycetes*, *Climacodon pulcherrimus*, *Phellinus* sp. and *Tyromyces duracinus*. Extremely tiny data has been published on the inhibition of *S. agalactiae*, MRSA, *K. pneumoniae*, *S. sonnei* and *P. mirabilis* by fungi.

Screening of the antimicrobial activity of endophytic extracts revealed a considerable action opposite to the disease-causing cultures proven. The majority of the cultures inhibited the growth of *Candida albicans*, *Escherichia coli* and *Staphylococcus aureus*. Totally four different groups of fungi have been segregated out of the mess *Roccella montagnei*. Amid the quarter genera, *Aspergillus niger*'s *(A. niger)* ability to generate chitosan (1.3 g/L) on the duodecimal day of incubation occurs. Glucose acts a significant part in the efficiency of chitosan; furthermore the yield was top at 10% (1.93 g/L). Antibacterial activity unfolds that *Vibrio cholerae* was sensitive to chitosan followed by *Escherichia coli* (Logesh et al. 2012).

Aspergillus flavus displays the strongest antimicrobial activity against *Staphylococcus aureus* whereas also in *Trichoderma koningii* Oudem, with 20.2 ± 0.6 and 18.05 ± 0.8 suppressed, correspondingly. From leaves and root tissues of *Melia azedarach* L., these above two endophytic fungi were separated, and their antimicrobial activities ascertained in the crude fungal extract cultures were being briefed for the first time (Kaushal Kanwer et al. 2013).

The fungal endophytes presented considerable antibacterial activity against gram-positive bacteria as well as gram-negative bacteria. The distinct endophytes like *Aspergillus conicus* BPEF2, *Penicillium janthinellum* BPEF1 and *Phomopsis amygdali* BPEF3 solutions lack resemblance sententiously in their activity against bacterial pathogens. This distinctiveness may characterized the fact that the incidence of distinct antimicrobial compounds with distinctive solvents (Bharathidasan and Panneerselvam 2013).

Fusarium sp., *Penicillium* sp., *Alternaria* sp. and *Chaetomium* sp. are endophytic fungi separated from medicinal plants of western Himalayas. *Fusarium* sp. suppressed *S. aureus* firmly; *Chaetomium* sp. and *Phomopsis* sp. suppressed the development of *E. coli* and *S. aureus*. *Fusarium* and *Alternaria* sp. are alive against the fungal parasite *Candida albicans* (Qadri et al. 2013).

36.6.4 Anticancer Activity

Endophytic fungus *Aspergillus fumigatus* Fresenius, from *Juniperus communis* plant as an atypical producer of deoxypodophyllotoxin, is not readily accessible from economical source and massive extraction producers. It might be employed as a substrate for bioconversion studies and as a forerunner to promote anticancer drugs. Deoxypodophyllotoxin has antimicrobial action against *S. aureus*, *K. pneumoniae* and *P. aeruginosa* (Kusari et al. 2009a, b).

From the stem tissues of this plant, the endophytic fungus *Chaetomium* sp. was separated. The fungal strain was determined by PCR (a method for amplifying DNA). When tested for cytotoxicity against L5178Y mouse lymphoma cells, the crude organic extract of the fungal strain was attested to be alive (Adbessamad et al. 2009).

36.6.5 Anticancer Activity

Sclerotiorin was separated from an endophytic fungus *Cephalotheca faveolata*. It was effective and anti-proliferative against distinct cancer cells. Sclerotiorin causes apoptosis in colon cancer (HCT-116) cells through the activation of BAX and down-regulation of BCL-2; those in addition stimulated dissected caspase-3 causing apoptosis of cancer cells (Giridharan et al. 2012).

36.6.6 Antioomycetous Activity

The fungal genera *Acremonium*, *Aspergillus*, *Fusarium* and *Penicillium* are "creative species" positioned on the formulation of many bioactive metabolites (Dreyfuss and Chapela 1994). The unusual antioomycetous element, oocydin A from the endophytic strain *Serratia marcescens* from *Rhyncholacis penicillata*.

36.6.7 *Anti*-Helicobacter pylori *Activity*

From lagoons and mangroves in Venezuelan waters, *Aspergillus* and *Penicillium* were separated and predicted to provide optimal cases for the detection of new metabolites. Amidst the entire 227 isolated fungi, 61 strains of *Penicillium citrinum* antibacterial activity connected well with matter of secondary metabolites as deliberated by HPLC, and 30 isolates of *Penicillium steckii* bear very similar profiles of secondary metabolites, and 6 of these has activity against either *Vibrio parahaemolyticus* or *Staphylococcus aureus* or both (Christophersen et al. 1999).

Investigation of endophytic fungus *Rhizoctonia* sp. yielded rhizoctonic acid with anti-*Helicobacter pylori* activity, the causative bacterium of peptic ulcer (Ma et al.

2004). Rubrofusarin B, fonsecinone A, asperpyrone B and aurasperone A isolated from Aspergillus niger IFB– E003, an endophyte in Cynodon dactylon. The four compounds exhibited growth inhibitions against the disease-causing microbes with minimal inhibitory concentrations (MICs) ranging in between 1.9 and 31.2 μg/mL.

36.6.8 Antifungal Activity

Some of the freshly accepted drugs of fungal base are micafungin, an antifungal metabolite from *Coleophoma empetri*; mycophenolate, a result of *Penicillium brevicompactum* put upon for precluding renal graft; rosuvastatin from *Penicillium citrinum* and *P. brevicompactum* put upon for processing dyslipidaemias; and cefditoren pivoxil, a broad-spectrum antibiotic gotten from *Cephalosporium* sp. Derivatives of fumagillin, an antibiotic made by *Aspergillus fumigatus*, and illudin-S, a sesquiterpenoid from *Omphalotus illudens*, possess anticancer action. Fungal metabolites find crucial usage in agriculture as well. It is apt to observe that, among the microfungi, only few genera such as *Aspergillus* and *Penicillium* have been strictly screened for bioactive compounds; of the 6450 bioactive metabolites from microfungi, more than 30% are incurred from these 2 genera (Bérdy 2005).

The alive element ergokonin A was separated from *Trichoderma longibrachiatum*. The antifungal array of ergokonin A was very broad and suppressed the development of many *Candida* sp., *S. cerevisiae* and most of the filamentous fungi. During their screening process, many hundred examples of *Trichoderma* spp. were examined to attempt for new antifungal agents, but only one of these isolate tests resulted in ergokonin A (Vicente et al. 2011).

36.6.9 Antitumour Activity

Eight new indole triterpenes named shearinines D–K, along with shearinine A, paspalitrem A and paspaline, have separated from the mangrove endophytic fungus *Penicillium* sp.; shearinines D, E and (with reduced potency) G showed important in vitro trait on large affinity to calcium-excited potassium transmission. One of them is cytosporone B which shows broad traits against fungi; was refined from the fermentation broth of an endophytic fungus, *Dothiorella* sp.; and was separated from mangrove plant *Avicennia marina* at the estuary of Jiulong River, Fujian Province, China (Cheng et al. 2009).

Trichoderma longibrachiatum isolated from the ventral of *Juglans mandshurica*. The conditioned medium of *T. longibrachiatum* displayed a high antitumour capability against liver cancer cell – HepG2. In addition, there was prominent selective controlling effect against the normal liver cell strain HL-7702 and its cancer anticipated strain HepG2. The restricting burden against strain HL-7702 was only one quarter of that opposed Hepg2 at the density of IC_{50} (Li et al. 2009).

36.6.10 Tuberculosis Inhibitors

In recent times, tuberculosis was suppressed by endophytes; few of the endophytes were found to restrict *Mycobacterium aurum* and *Mycobacterium tuberculosis*, the causative organisms of tuberculosis. Phomoenamide exposed in vitro antimycobacterial process against *M. tuberculosis* H37Ra. Fungal endophytes connected with higher plants come to be a good source of new antioxidants as well (Huang et al. 2007. Cheng et al. (2009) reported one of them is cytosporone B which exhibits broad function against fungi; was refined from the fermentation broth of an endophytic fungus, *Dothiorella* sp.; and was restricted from mangrove plant *Avicennia marina* at the estuary of Jiulong River, Fujian Province, China.

36.6.11 Antidiabetic Property

An endophytic *Syncephalastrum* sp. was isolated from *Adhatoda beddomei*. The ethyl acetate crude extract of the fungi was screened for antidiabetic activity. The extract exhibited inhibition of α-amylase (Prabavathy and Valli 2013). Similarly an endophytic fungus *Syncephalastrum racemosum* isolated from the seaweed *Gracilaria corticata* showed remarkable antidiabetic property by the in vitro assay of alpha-amylase inhibition.

36.7 Bioactive Compounds

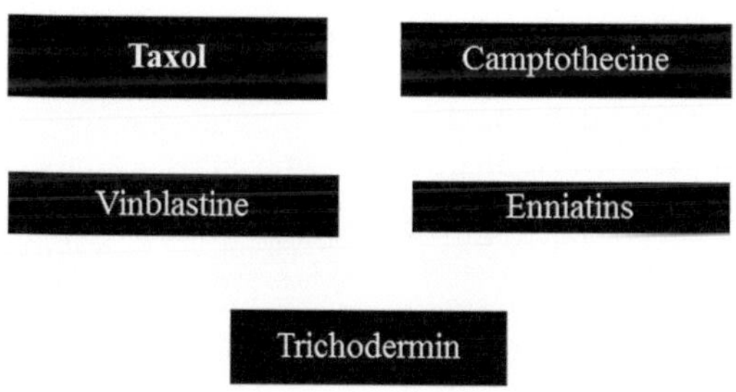

36.7.1 Paclitaxel (Taxol)

Taxol, a highly functionalized diterpenoid, is found in each yew (*Taxus*) species but was originally isolated from *Taxus brevifolia*. This compound is the world's first billion-dollar anticancer drug, and it is used for the treatment of ovarian and breast cancers, but now it is used to treat a number of other human tissue-proliferating diseases as well. Its cost makes it unavailable to most of the world's people (Stierle et al. 1993) in that yew trees might support certain endophytic microorganisms that may also synthesize Taxol. Furthermore, several other *P. microspora* isolates were obtained from bald cypress in South Carolina and were also shown to produce Taxol. This was the first indication that other endophytes than *T. andreanae* residing in plants other than *Taxus* spp. were producing Taxol. Numerous reports have shown that many of the other endophytic fungi such as *Pestalotiopsis guepini* and *Periconia* sp. also produce Taxol (Strobel et al. 1997).

Endophytic fungi Fusarium solani segregated out of *Taxus chinensis* moreover common endophytic genera including *Alternaria* and *Aspergillus* out of *Ginkgo biloba* moreover *Podocarpus* sp. individually, held published as per creators concerning of Taxol (Liu et al. 2009).

Paclitaxel (Taxol), as a prominent and greatly operated tetracyclic diterpenoid bioactive composite, established from the yelp of *Taxus brevifolia.* Yet the important supply of paclitaxel has been from the wild *Taxus* plants. However, it is found in extremely mean quantity in different segments such as the spikes, yelp and roots of *Taxus* species. Luckily, a paclitaxel-producing endophytic fungus *Taxomyces andreanae* was revealed out of the Pacific yew (*Taxus brevifolia*) in 1993 (Stierle et al. 1993). Broad study includes probing for paclitaxel-generating endophytic fungi from *Taxus* and other associated plant species. Paclitaxel was developed by microbial fermentation and genetic engineering method. As of today, relatively 19 genera of endophytic fungi such as *Alternaria*, *Aspergillus*, *Botryodiplodia*, *Botrytis*, *Cladosporium*, *Ectostroma*, *Fusarium*, *Metarhizium*, *Monochaetia*, *Mucor*, *Ozonium*, *Papulaspora*, *Periconia*, *Pestalotia*, *Pestalotiopsis*, *Phyllosticta*, *Pithomyces*, *Taxomyces* and *Tubercularia* were screened the ability to generate paclitaxel furthermore its counter path. The host of paclitaxel-producing fungi importantly such as *Taxus* (i.e. *T. baccata*, *T. cuspidata*, *T. media* and *T. yunnanensis*) and non-*Taxus* species likewise *Cardiospermum halicacabum*, *Citrus medica*, *Cupressus* sp., *Ginkgo biloba*, *Hibiscus rosa-sinensis*, *Podocarpus* sp., *Taxodium distichum*, *Terminalia arjuna*, *Torreya grandifolia* and *Wollemia nobilis*. The great number and broad range imply that both paclitaxel-producing fungi and their hosts have biological diversity. Endophytic fungi showed us a promising way that is an alternative paclitaxel-producing resource.

Phyllosticta spinarum, *Bartalinia robillardoides*, *Pestalotiopsis terminaliae*, *Botryodiplodia theobromae*, *Alternaria*, *Aspergillus*, *Botrytis*, *Cladosporium*, *Fusarium* and *Mucor* spp. have been reported as producers of Taxol (Zhao et al. 2010).

36.7.2 Camptothecin

Camptothecin is a pentacyclic quinoline alkaloid and was initially isolated from the wood of *Camptotheca acuminate*. Camptothecin acts as an antineoplastic agent. The primary action mechanism of CPT is by virtue of inhibiting the intranuclear enzyme topoisomerase-1, which is required in DNA replication and transcription during molecular events. Hycamtin (topotecan) and camtostar (irinotecan), two of the famous CPT semi-synthetic drugs, have already been in clinical use against ovarian, small lung and refractory ovarian cancers. CPT semi-synthetic drugs successfully discovered from a CPT-producing endophytic fungus Neurospora sp. from the seeds of Nothapodytes foetida. *Fusarium solani* isolated as an endophytic fungus from *Camptotheca acuminate* is able to produce CPT, 10-hydroxycamptothecin and 9-methoxycamptothecin (Kusari et al. 2009a, b).

Min and Wang (2009) reported an unidentified endophytic fungal strain XK001 which could yield 0-hydroxycamptothecin with the yield of 677 μg/L (Min and Wang 2009). Two endophytic fungi *Fusarium solani* strains MTCC9667 and MTCC9668 had the ability to generate CPT (cytidine triphosphate) and 9-methoxycamptothecin (0.45 μg/g), and the endophyte MTCC9668 might also produce 10-hydroxycamptothecin (0.08 μg/g). *Entrophospora infrequens* was able to produce camptothecin (Shweta et al. 2010).

36.7.3 Vinblastine

Vinblastine and vincristine, the terpenoid indole alkaloid derivative as of the combination of vindoline and catharanthine monomers, are anticancer agents. The primary action mechanism of vincristine is via interference with microtubule formation and mitotic spindle dynamics, disruption of intracellular transport and decreased tumour blood flow, with the latter probably as a consequence of anti-angiogenesis. Endophytic fungus *Alternaria* sp. was isolated commencing the phloem of *Catharanthus roseus* that had the capacity to produce vinblastine. Endophytic *Fusarium oxysporum* from the phloem of *C. roseus* could produce vincristine (Zhang et al. 2000). An unidentified vincristine-producing endophytic fungus from the leaves of *C. roseus* could be a potential source to produce either vinblastine or vincristine (Guo et al. 1998).

36.7.4 Enniatins

The antibiotic, insecticidal and phytotoxic activities of the enniatins have been studied extensively and appear to be linked to their ionophoric properties. Commencing G, a new compound with the structure of cyclohexapeptide, was also isolated from the culture broth of mangrove fungus *Fusarium* sp., which was collected from Thailand, and displayed antitumour activity of Heps 7402 with ED50 value of 12 μg/mL. Enniatins are cyclohexadepsipeptides and consist of three D-2-hydroxyisovaleric acid (HyIv) residues linked alternatively with L-amino acids or *N*-methyl-L-amino acids to give an 18-membered cyclic skeleton (Lin and Zhou 2003).

36.7.5 *Trichodermin*

Trichodermin was defined out of *Trichoderma harzianum*, an endophytic fungus surviving in *Ilex cornuta*, a perennial flowering shrub from East Asia. Trichodermin has been announced towards guard facing the solanaceous plant germs *Alternaria solani* and *R. solani* ex vivo *status* (Chen et al. 2007). Some of the main bioactive compounds and their activity were tabulated (Table 36.1).

Table 36.1 Name of the bioactive compound and their activity

S. no.	Name of the fungi	Bioactive compound	Activity
1.	*Acremonium zeae*	Pyrrocidines A and B	Antifungal activity against *A. flavus* and *F. verticillioides*
2.	*Aspergillus fumigatus* CY018	Asperfumoid, fumigaclavine C, fumitremorgin C, physcion and helvolic acid	Restrict *Candida albicans*
3.	*Cephalosporium* sp. IFB-E001	Graphislactone A	Free radical scavenging
4.	*Cephalotheca faveolata*	Sclerotiorin	Antibacterial
5.	*Chaetomium globosum*	Chaetoglobosins A and C	Inhibit *Mucor miehei*
6.	*Chaetomium globosum* IFB-E019	Chaetoglobosin U, C, F, E	Cytotoxic
7.	*Cladosporium* sp.	Brefeldin A	Antifungal activity
8.	*Cryptosporiopsis* cf. *quercina*	Cryptocin and Cryptocandian	Antifungal activity against
			Human pathogens
			Candida albicans
			Trichophyton sp.
			Plant pathogens
			Sclerotinia sclerotiorum
			Botrytis cinerea
			Pyricularia oryzae
9.	*Discosia* sp.	*Amylase*	Enzymatic
10.	*Entrophospora infrequens*	Camptothecin	Anticancer
11.	*Fusarium oxysporum*	Vinca alkaloids	Leukaemia
12.	*Fusarium solani*	Berberine	Anticancer
13.	*Fusarium solani* LCPANCF01	Taxol	Anticancer
14.	*Fusidium* sp.	Fusidikactones	Antifungal activity

(continued)

Table 36.1 (continued)

S. no.	Name of the fungi	Bioactive compound	Activity
15.	*Hypericum perforatum*	Hypericin and emodin	Antibacterial activity
			Staphylococcus aureus,
			Pseudomonas aeruginosa,
			Klebsiella pneumoniae
			Salmonella enterica
			Antifungal activity
			A. niger
			C. albicans
16.	*Penicillium janthinellum*	Citrinin	Antimicrobial
17.	*Penicillium* sp.	Alkaloid A, B	Bacteriostatic effect
			E.coli
			Staphylococcus aureus
			Pseudomonas aeruginosa
			Bacillus
18.	*Periconia* sp.	Periconicin A	Antibacterial
		Piperine	Inhibit *M. tuberculosis* and *M. smegmatis*
19.	*Pestalotiopsis adusta*	Pestalachlorides	Antifungal activity against
			Plant pathogens
			Fusarium culmorum
			Gibberella zeae
			Verticillium albo-atrum
20.	*Pestalotiopsis jester*	Jesterone	Antifungal activity against plant pathogens
21.	*Pestalotiopsis microspora*	Ambuic acid and pestalotiopsins A and B	Antifungal activity
		Pestaloside	Phytotoxic properties
		Taxol	Anticancer
22.	*Phomopsis cassia*	Ethyl 2,4 dihydroxy-5,6-dimethyl benzoate and phomopsis lactone	Antimycotic action facing *Cladosporium cladosporioides*
			C. sphaerospermum
23.	*Phomopsis* sp.	Cytosporone B and C	Inhibit *Candida albicans*, *F. oxysporum*
24.	*Rhizoctonia* sp.	Rhizoctonic acid	Anti-helicobacter pylori activity
25.	*Talaromyces* sp.	7-epiaustdiol	Inhibit multidrug-resistant opportunistic pathogen *P. aeruginosa*
26.	*Xylaria* sp.	Griseofulvin	To treat human and animal mycotic diseases

36.8 Conclusions

Endophytic fungi are to be a potential source of novel bioactive compounds. Endophytic fungi could be another better source for development of new drugs against cancer and multidrug-resistant bacteria. Endophytic fungi are easily cultured laboratory and fermentor. It will help to avoid harvesting plants and affecting the environmental biodiversity.

References

Adbessamad D, Aly Amal H, Edrada-Ebel RA, Mueller WEG, Mosaddak M, Hakikj A, Ebel R, Proksch P (2009) Bioactive secondary metabolites from the endophytic fungus Chaetomium sp. isolated from Salvia officinalis growing in Morocco. Biotechnol Agron Soc Environ (BASE) 13(2):229–234

Bérdy J (2005) Bioactive microbial metabolites: a personal view. J Antibiot 58:1–26

Bharathidasan R, Panneerselvam A (2013) Diversity and biotechnological application of endophytic fungi from mangrove plants of Karankadu, Ramanathapuram district, Tamil Nadu, India. Ph.D thesis, Refer to Bharathidasan University, Trichy

Chen L, Chen J, Zheng X, Zhang J, Yu X (2007) Identification and antifungal activity of the metabolite of endophytic fungi isolated from *Ilex cornuta*. Nongyaoxue Xuebao 9:143–150

Cheng Z-s, Pan J-H, Wen-cheng T, Qi-jin C, Lin Y-c (2009) Biodiversity and biotechnological potential of mangrove – associated fungi. J For Res 20(1):63–72

Christophersen C, Crescente O, Frisvad JC, Gram L, Nielsen J, Nielsen PH, Rahbaek L (1999) Antibacterial activity of marine – derived fungi. Mycopathologia 143:135–138

Dayle ES, Polans NO, Paul DS, Melvin RD (2001) Angiosperm DNA contamination by endophytic fungi: detection and methods of avoidance. Plant Mole Biol Rep 19:249–260

De Souza JJ, Vieira IJ, Rodrigues-Filho E, Braz-Filho R (2011) Terpenoids from endophytic fungi. Mol Ther 16:10604–10618

Dreyfuss MM (1986) Neue Erkienntnisse aus einem pharmakologischen Pilzscreening. Sydowia 39:22–36

Dreyfuss MM, Chapela IH (1994) Potential of fungi in the discovery of novel, low molecular weight pharmaceuticals. In: Gullo VP (ed) The discovery of natural products with therapeutic potential. Butterworth-Heinemann, Boston, pp 49–80

Firakova S, Sturdikova M, Muckova M (2007) Bioactive secondary metabolites produced by microorganisms associated with plants. Biologia 62/3:251–257

Frohlich J, Hyde KD, Petrini O (2000) Endophytic fungi associated with palm. Mycol Res 104:1202–1212

Garo E, Starks CM, Jensen PR, Fenical W, Lobkovsky E, Clardy J (2003) Trichodermamides A and B, cytotoxic modified dipeptides from the marine – derived fungus *Trichoderma virens*. J Nat Prod 66:423–426

Ghisalberti E, Rowland CJ (1993) Antifungal metabolites from *Trichoderma harzianum*. Nat Prod 56:1799

Giridharan P, Verekar SA, Khanna A, Mishra PD, Deshmukh SK (2012) Anticancer activity of sclerotiorin, isolated from an endophytic fungus *Cephalotheca faveolata* Yaguchi, Nishim. & Udagawa. Indian J Exp Biol 50(7):464–468

Gunatilaka AAL (2006) Natural products from plant-associated microorganisms: distribution, structural diversity, bioactivity, and implications of their occurrence. J Nat Prod 69:509–526

Guo B, Li H, Zhang L (1998) Isolation of the fungus producing vinblastine. J Yunnan Univ (Nat Sci Ed) 20:214–215

Henson RNA, Rugg MD, Shallice T, Josephs O, Dolan RJ (1999) Recollection and familiarity in recognition memory: an event related functional magnetic resonance imaging study. J Neurosci 19(10):3962–3972
Huang WY, Cai YZ, Xing J, Corke H, Sun M (2007) Potential antioxidant resource: endophytic fungi isolated from traditional Chinese medicinal plants. Econ Bot 61:14–30
Humberto H, Lara V, Ayala-Nunez NV, Carmen LD, Ixtepan T, Cristina RP (2010) Bactericidal effect of silver nanoparticles against multidrug-resistant bacteria. World J Microbiol Biotechnol 26:615–621
Imtiaj A, Lee T-S (2007) Screening of antibacterial and antifungal activities from Korean wild mushrooms. World J Agric Sci 3:316–321
Isaka M, Suryansestakorn C, Tanticharoen M (2002) Aigialomycins A–E, new resorcylic macrolides from the marine mangrove fungus *Aigialus parvus*. J Organomet Chem 67:1561–1566
Kaushal Kanwer S, Rao DV, Batra A (2013) In vitro antimicrobial activities of endophytic fungi isolates medicinal tree – *Melia azedarach* L. J Microbiol Res 3(1):19–24
Kusari S, Zuhlke S, Spiteller M (2009a) An endophytic fungus from *Camptotheca acuminata* that produces camptothecin and analogues. J Nat Prod 72:2–7
Kusari S, Lamshoft M, Spiteller M (2009b) *Aspergillus fumigatus* Fresenius, an endophytic fungus from *Juniperus communis* L. Horstmann as a novel source of the anticancer prodrug deoxypodophyllotoxin. J Appl Microbiol 107:1019–1030
Li MY, Xiao Q, Pan JY, Wu J (2009) Natural products from semi-mangrove flora: source, chemistry and bioactivities. Nat Prod Rep 26(2):281–298
Lin YC, Zhou SN (2003) Marine microorganisms and its metabolites. Chemical Industry Press, Beijing, pp 426–427
Liu K, Ding X, Deng B, Chen W (2009) Isolation and characterization of endophytic taxol-producing fungi from Taxus chinensis. J Ind Microbiol Biotechnol 36:1171–1177
Lodge D, Fisher PG, Sutton BC (1996) Endophytic fungi of *Manilkara bidentata* in fruticose lichens. Mycologia 82:444–451
Logesh AR, Thillaimaharani KA, Sharmila K, Kalaiselvam M, Raffi SM (2012) Production of chitosan from endolichenic fungi isolated from mangrove environment and its antagonistic activity. Asian Pac J Trop Biomed 2(2):140–143
Ma YM, Li Y, Liu JY, Song YC, Tan RX (2004) Anti-*Helicobacter pylori* metabolites from *Rhizoctonia* sp. Cy064, an endophytic fungus in *Cynodon dactylon*. Fitoterapia 75:451–456
Marquez S, Bills GF, Zabalgogeazcoa I (2007) The endophytic mycobiota of the grass *Dactylis glomerata*. Fungal Divers 27:171–195
Mekawey AAI (2010) The antibiotic properties of several strains of fungi. Aust J Basic Appl Sci 4:3441–3454
Min C, Wang X (2009) Isolation and identification of the 10-hydroxycamptothecin-producing endophytic fungi from *Camptotheca acuminata* Decne. Acta Botan Boreali-Occiden Sin 29:0614–0617
Mitchell AM, Strobel GA, Hess WM, Vargas PN, Ezra D (2008) *Muscodor crispans*, a novel endophyte from *Ananas ananasoides* in the Bolivian Amazon. Fungal Divers 31:37–43
Moricca S, Ragazzi A (2008) Fungal endophytes in Mediterranean oak forests: a lesson from *Discula quercina*. Phytopathology 98:380–386
Pela'ez F (2005) Biological activities of fungal metabolites. In: An Z (ed) Handbook of industrial mycology. Marcel Dekker, New York, pp 49–92
Pelaez F, Collado J, Arenal F, Basilio A, Cabello A, Diez Matas MT, Garcia B, Gonzalez Del Val A, Gonzalez V, Gorrochategui J, Hernandez P, Martin I, Platas G, Vicente F (1998) Endophytic fungi from plants living on gypsum soils as a source of secondary metabolites with antimicrobial activity. Mycol Res 102(6):755–761
Petrini LE, Petrini O (1985) Xylariaceous fungi as endophytes. Sydowia 38:216–234
Petrini A, Hake U, Dreyfuss M (1990) An analysis of fungal communities isolated from fruticose lichens. Mycologia 82:444–451

Poch GK, Gloer JB (1991) Auranticins A and B: two depsidones from a mangrove isolate of the fungus *Preussia aurantiaca*. J Nat Prod 54:213–217
Porras-Alfaro A, Bayman P (2011) Hidden fungi, emergent properties: endophytes and microbiomes. Annu Rev Phytopathol 49:291–315
Prabavathy D, Valli NC (2013) Antimicrobial and antidiabetic activity of an endophytic fungi isolated from Adathoda beddomei. Int J Pharm Pharm Sci 5:780–783
Pruksakorn P, Arai M, Kotoku N, Vilcheze C, Baughn AD, Moodley P, Jacobs WR, Kobayashi M (2010) Tricoderins, novel aminolipopeptiides from a marine sponge-derived *Trichoderma* sp., are active against dormant mycobacteria. Bioorg Med Chem Lett 20(12):3658–3663
Qadri M, Johri S, Shah BA, Khajuria A, Sidiq T, Lattoo SK, Abdin MZ, Hassan UI, S.R. (2013) Identification and bioactive potential of endophytic fungi isolated from selected plants of the western Himalayas. Springer Plus 2(8):2–14
Schultz B, Sucker J, Aust HJ, Krohn K, Ludewig K, Jones PG, Doring D (1995) Biologically active secondary metabolites of *Pezicula* species. Itlycol Res 99:1007–1015
Schulz B, Boyle C, Draeger S, Rommert AK, Krohn K (2002) Endophytic fungi: a source of novel biologically active secondary metabolites. Mycol Res 106:996–1004
Schulz B, Draeger S, dela Cruz TE, Rheinheimer J, Siems K, Loesgen S, Bitzer J, Schloerke O, Zeeck A, Kock I, Hussain H, Dai J, Krohn K (2008) Screening strategies for obtaining novel, biologically active, fungal secondary metabolites from marine habitats. Bot Mar 51:219–234
Shweta S, Zuehlke S, Ramesha BT, Priti V, Mohanakumar P, Ravikanth G, Spiteller M (2010) Endophytic fungal strains of Fusarium solani, from Apodytes dimidiata E. Mey.ex Arn (Icacinaceae) produce camptothecin, 10-hydroxycamptothecin and 9-methoxycamptothecin. Phytochemistry 71(1):117–122
Silva MRO, Almeida AC, Arruda FVF, Gusmao N (2011) Endophytic fungi from Brazilian mangrove plant *Laguncularia racemosa* (L.) Gaertn. (Combretaceae): their antimicrobial potential. Sci Against Microb Pathog: Commun Curr Res Technol Adv 2:1260–1266
Simon A, Dunlop RW, Ghulsalberti EL, Silvsithamparam K (1988) *Trichoderma koningii* produces a pyrone compound with antibiotic properties. Soil Biol Biochem 20:263
Stierle A, Strobel GA, Stierle DB (1993) Taxol and taxane production by Taxomyces *andreanae*, an endophytic fungus of Pacific yew. Science 260:214–216
Strobel G (2003) Endophytes as sources of bioactive products. Microbes Infect 5:535–544
Strobel G, Daisy B (2003) Bioprospecting for microbial endophytes and their natural products. Microbiol Mole Biol 67(4):491–502
Strobel G, Hess WM, Li JY, Ford E, Sears J, Sidhu RS, Summerell B (1997) *Pestalotiopsis guepinii*, a taxol producing endophyte of the Wollemi Pine, Wollemia nobilis. Aust J Bot 45:1073–1082
Suryanarayanan TS, Murali TS (2006) Incidence of *Leptosphaerulina crassiasca* in symptomLess leaves of peanut in southern India. J Basic Microbiol 46:305–309
Suryanarayanan TS, Thennarasan S (2004) Temporal variation in endophyte assemblages of *Plumeria rubra* leaves. Fungal Divers 15:197–204
Suryanarayanan TS, Kumaresan V, Johnson JA (1998) Foliar fungal endophytes from two species of the mangrove *Rhizophora*. Can J Microbiol 44:1003–1006
Suryanarayanan TS, Senthilarasu G, Muruganandam V (2000) Endophytic fungi from *Cuscuta reflexa* and its host plants. Fungal Divers 4:117–123
Takahashi JA, Monteiro de Castro MC, Souza GG, Lucas EMF, Bracarense AAP, Abreu LM, Marriel IE, Oliveira MS, Floreano MB, Oliveira TS (2008) Isolation and screening of fungal species isolated from Brazilian cerrado soil for antibacterial activity against Escherichia coli, Staphylococcus aureus, Salmonella typhimurium, Streptococcus pyogenes and Listeria monocytogenes. J Mycol Méd 18:198–204
Tan RX, Zou WX (2001) Endophytes: a rich source of functional metabolites. Nat Prod Rep 18:448–459
Umali TE, Quimio TH, Hyde KD (1999) Endophytic fungi in leaves of *Bambusa tuldoides*. Fungal Sci 14:11–18

Vicente MF, Cabello A, Platas G, Basilio A, Díez MT, Dreikorn S, Giacobbe RA, Onishi JC, Meinz M, Kurtz MB, Rosenbach M, Thompson J, Abruzzo G, Flattery A, Kong L, Tsipouras A, Wilson KE, Peláez F (2011) Antimicrobial activity of ergokonin A from *Trichoderma longibrachiatum*. J Appl Microbiol 91(5):806–813

Vinale F, Ghisalberti EL, Sivasithamparam K, Marra R, Ritieni A, Ferracane R, Woo S, Lorito M (2009) Factors affecting the production of *Trichoderma harzianum* secondary metabolites during the interaction with different plant pathogens. Lett Appl Microbiol 48:706–711

Vogl A (1898) Mehl und die anderen mehl produkte der cerealien und leguminosen. Zeitschrift Nahrungsmittle untersuchung, Hgg. Warlenkunde 21:25–29

Wang FW, Hon ZM, Wang CR, Li P, Shi DH (2008) Bioactive metabolites from *Penicillium* sp., and endophytic fungi residing in *Hopea hainanensis*. World J Biotechnol 24:2143–2147

White F Jr, Drake TE, Martin N (1996) Endophyte host associations in grasses. XXIII. A study of two species of *Balansia* that form stromata on nodes of grasses. Mycologia 88:89–97

Wu B, Oesker V, Wiese J, Schmalijohann R, Imhoff JF (2014) Two new antibiotic pyridines produced by a marine fungus, *Trichoderma* sp. Strain MF106. Mar Drugs 12:1208–1219

Yin H, Sun YH (2011) Vincamine-producing endophytic fungus isolated from *Vinca minor*. Phytomedicine 18:802–805

Zhang L, Guo B, Li H, Zeng S, Shao H, Gu S, Wei R (2000) Preliminary study on the isolation of endophytic fungus of *Catharanthus roseus* and its fermentation to produce products of therapeutic value. Chin Tradit Herb Drug 31:805–807

Zhang Y, Han T, Ming Q, Wu L, Rahman K, Qin L (2012) Alkaloids produced by endophytic fungi: a review. Nat Prod Commun 7(7):963–968

Zhao J, Zhou L, Wang J, Shan T, Zhong L, Liu X, Gao X (2010) Endophytic fungi for producing bioactive compounds originally from their host plants. Curr Res Technol Educat Top Appl Microbiol Microbial Biotechnol 1:567–576

Chapter 37
Bioactive Potential of Nonconventional Edible Wild Mushroom *Amanita*

Ammatanda A. Greeshma, Kandikere R. Sridhar, Mundamoole Pavithra, and Kaori Tomita-Yokotani

Abstract This study evaluated the bioactive components and antioxidant potential of uncooked and cooked tender edible mushroom *Amanita* sp. This mushroom is common in lateritic scrub jungles during the early monsoon season of the southwest coast of India and a delicacy for tribals and native people. Nine bioactive components of tender *Amanita* sp. (total phenolics, tannins, flavonoids, vitamin C, phytic acid, lycopene, β-carotene, trypsin inhibition and haemagglutinin) showed higher quantities in uncooked than in cooked samples, so also the five antioxidant activities (total antioxidant activity, ferrous ion-chelating capacity, reducing power and DPPH and ABTS radical-scavenging activities). It was devoid of L-DOPA, and there was no significant difference in tannins, flavonoids and phytic acid contents between uncooked and cooked samples. Bioactive principles as well as antioxidant activities were comparable or higher than many edible *Amanita* spp. In principal component analysis irrespective of thermal processing, tannin content and trypsin inhibition activity were clustered with total antioxidant activity, reducing power and DPPH radical-scavenging activity. Owing to the potential nutritional and antioxidant properties, *Amanita* sp. and its ectomycorrhizal native host tree species deserve conservation priorities. Future research needs to focus on this nonconventional wild edible source for nutraceutical potential.

Keywords Agaricales · Antioxidant potential · Bioactive compounds · Ectomycorrhizae · Nutraceuticals · Nonconventional food

---

A. A. Greeshma · K. R. Sridhar (✉) · M. Pavithra
Department of Biosciences, Mangalore University, Mangalagangotri, Mangalore, Karnataka, India

K. Tomita-Yokotani
Applied Biochemistry and Life Sciences, University of Tsukuba, Tsukuba, Ibaraki, Japan

P. Gehlot, J. Singh (eds.), *Fungi and their Role in Sustainable Development: Current Perspectives*, https://doi.org/10.1007/978-981-13-0393-7_37

37.1 Introduction

Macrofungi have become integral part of human life (nutrition and medicine) owing to their versatile nutritional and bioactive principles (Boa 2004). Several mushrooms are cultivated worldwide to fulfil mainly nutritional requirements (e.g. *Agaricus bisporus*, *Flammulina velutipes*, *Lentinus edodes*, *Pleurotus sajor-caju* and *Volvariella volvacea*). Similar to ethnic knowledge on plants, many wild macrofungi possess nonconventional nutritional and medicinal value especially in Asian countries (Aly et al. 2011; Xu et al. 2011; De Silva et al. 2013). Several macrofungi are attractive sources in development of new drugs and nutraceuticals (Mau et al. 2004; Cheung and Cheung 2005; Barros et al. 2007). Many studies have revealed the importance of macrofungi in combating lifestyle diseases, owing to their therapeutic potential such as antioxidant activity, antimicrobial activity, anticancer properties, cholesterol lowering and immunostimulatory effects (Barros et al. 2007; De Silva et al. 2013; Thatoi and Singdevsachan 2014). The potential of cultivated and wild mushrooms has expanded towards industrial applications beyond mere nutritional or health perspectives by exposing new metabolites to science (e.g. cosmetics, chitosan, glucans and nanoparticles) (Manzi and Pizzoferrato 2000; Wu et al. 2004; Hyde et al. 2010; Vikineswary and Chang 2013; Arun et al. 2014; Taofiq et al. 2016).

The Western Ghats and the west coast of India are known for wild macrofungi of nutritional, medicinal and industrial significance (Mohanan 2011; Farook et al. 2013; Senthilarasu 2014; Pavithra et al. 2015; Senthilarasu and Kumaresan 2016). Inventories of wild mushrooms in Southwest India have exposed several edible mushrooms based on the traditional knowledge of native people and tribals (Ghate et al. 2014; Senthilarasu 2014; Karun and Sridhar 2014, 2016; Pavithra et al. 2015, 2016). Some examples of the prominent traditionally consumed wild mushrooms of the Southwest India include *Amanita* sp., *Astraeus* spp., *Auricularia* spp., *Lentinus* spp., *Russula* spp. and *Termitomyces* spp. (Senthilarasu 2014; Karun and Sridhar 2014, 2016; Pavithra et al. 2015).

Tulloss (2005) estimated a range of 900–1000 species of *Amanita* that are distributed worldwide. Zhang et al. (2015) reviewed the diversity, phylogeography and population genetics of *Amanita* mushrooms. Among these, 50% of species have been described with 100 species as poisonous, 50 species as edible and the rest as unknown edibility. Recently, Sánchez-Ramírez et al. (2015) have dealt with origin, distribution and diversification of edible ectomycorrhizal clade of *Amanita* (section: Caesareae). Although many *Amanita* spp. are poisonous, several edible *Amanita* spp. are reported from Thailand (Sanmee et al. 2008), Tanzania (Härkönen et al. 1994), Tropical Africa (Buyck 1994), Turkey (Doğan 2013) and Western Ghats of India (Senthilarasu 2014). According to Sanmee et al. (2008), at least six *Amanita* spp. are edible and common in Northern Thailand during wet season (*Amanita chepangiana*, *A. hemibapha*, *A. princeps*, *A. manginiana*, *A. pseudoporphyria* and *A. sinensis*). During spatio-temporal inventory of macrofungi of the Southwest India, an ectomycorrhizal *Amanita* sp. occurred predominantly in lateritic soils, which were consumed by the native people in young stage (Karun and Sridhar 2014). It was also recovered as ectomycorrhizal in *Acacia auriculiformis* in coastal sand dunes of the Southwest India (Ghate et al. 2014; Ghate and Sridhar 2016). This edible mushroom is designated as 'motte-anabe' in Kannada language (meaning 'egg mushroom', as young sporocarps are similar to small eggs) crop up during monsoon season (early June–early August) (Karun and Sridhar 2014). Being ecto-mycorrhizal, its fruit bodies will be collected by the native people underneath the tree species like *Acacia auriculiformis*, *Anacardium occidentale*, *Hopea ponga* and *Terminalia paniculata*. The young sporocarps of this mushroom will be consumed at spherical, oval, dumble stages and partially ruptured volva stage. *Amanita* sp. occurring in the lateritic belt of Southwest India being traditionally edible, the current study aims at evaluating bioactive principles and antioxidant potential to find out its nutraceutical values.

37.2 Mushroom

Edible stages of *Amanita* sp. (young sporocarp stages) were collected from the lateritic soils of the Southwest India (Konaje village, Dakshina Kannada, Mangalore, India: 12°48′N, 74°55′E, 115 m asl) with support of local dwellers who regularly consume during monsoon season (June–August) (Fig. 37.1). Its fruit bodies are very common underneath the tree species of *Acacia auriculiformis*, *Anacardium occidentale*, *Hopea ponga* and *Terminalia paniculata*. Based on macro- and micromorphological features, although the *Amanita* sp. roughly matches with *Amanita marmorata* as reported from Hawaii (Miller Jr et al. 1996), several glaring differences support to consider it as a new species. The tender sporocarps collected and consumed by the villagers include spherical, oval, dumble shapes and just partially ruptured volva stage (Fig. 37.1a–k). Sampling was carried out in five locations with about 50 m apart in lateritic scrub jungles. The young stages of mushroom in each sample were separately rinsed in distilled water to eliminate soil, roots and other debris. They were wiped with clean cloth to eliminate moisture on the surface. Each replicate was divided into two groups: the first group was oven-dried at 50–55 °C, while the second group was separately cooked in a household pressure cooker with distilled water (1:1 v/v) followed by oven drying. The dried samples were milled in Wiley mill (mesh # 30), and the powder was refrigerated in air-tight containers for analysis.

37.3 Bioactive Components

37.3.1 Total Phenolics

The total phenolic content of mushroom flour was determined based on the method by Rosset et al. (1982). To mushroom powder (100 mg), methanol (50%, 10 ml) was added, mixed, kept in water bath (95 °C, 10 min), cooled and centrifuged (2000 rpm, 20 min) to recover supernatant. Extraction was repeated, and the pooled final volume

Fig. 37.1 Various stages of immature sporocarps (**a–k**), maturing sporocarps (**l–n**) and mature fruit bodies (**o–r**) of *Amanita* sp.: spherical (**a** and **b**) (note roots on sporocarp in **a**, arrows), beak-like protuberance (**c**), extended protrusion (**d** and **e**), dumble shape (**f** and **g**), partially ruptured volva (**h** and **i**), cut-open spherical sporocarp (**j**), cut-open beaked sporocarp (**k**), extended stipe prior to open pileus (**l** and **m**), gills and partial veil of immature fruit body (**n**), mature fruit bodies (**o** and **q**), ruptured volva of mature fruit body (**p**) and gills of mature fruit body (**r**)

of extract was made to 20 ml. Extract (0.5 ml) was diluted by equal volume of distilled water; sodium carbonate (in 0.1 N NaOH, 5 ml) was added and incubated (10 min, room temperature). Folin-Ciocalteau's reagent (diluted 1:2, 0.5 ml) was added, and absorbance was read (725 nm; UV-VIS Spectrophotometer-118, Systronics, Ahmedabad, Gujarat, India). The total phenolic content was expressed as mg tannic acid equivalents (standard) per gramme mushroom powder (mg TAEs/g).

37.3.2 Tannins

The tannin content of mushroom flour was determined based on the protocol by Burns (1971). To mushroom powder (1 g), methanol (50 ml) was added to extract tannins on a rotary shaker (28 °C, 24 h) and centrifuged (1500 rpm) to collect the supernatant. To the extract (1 ml), vanillin hydrochloride was added (5 ml: 4% in methanol +8% concentrated HCl in methanol; 1:1) and incubated (20 min, room temperature), and the absorbance was read at 500 nm. Catechin in methanol served as standard to express tannin in mg catechin equivalents (mg CEs/g).

37.3.3 Flavonoids

Total flavonoid content was determined using aluminium chloride colorimetric method (Chang et al. 2002). Mushroom flour (1 mg) was extracted in methanol (1.5 ml), and aliquot of extract (0.5 ml) was mixed with aluminium chloride (10%, 0.1 ml) and potassium acetate (1 M, 0.1 ml). The final volume was made to 3 ml in distilled water and incubated (30 min, room temperature). Quercetin dihydrate served as standard, and absorbance was measured (415 nm) to express flavonoid content in mg equivalents per gramme mushroom flour (mg QEs/g).

37.3.4 Vitamin C

The vitamin C content in mushroom flour was quantified based on the method of Roe (1954). The flour (1 g) was extracted by trichloroacetic acid (TCA, 5%, 10 ml), and aliquot of extract (0.2 ml) was made up to 1 ml in TCA (5%), mixed and added with chromogen (1 ml) (dinitrophenyl hydrazine thiourea copper sulphate solution: 5 parts of 5% thiourea +5 parts of 0.6% copper sulphate +90 parts of 2% 2,4-dinitrophenylhydrazine in H_2SO_4). The reaction mixture was incubated (boiling water bath, 10 min), cooled, added with H_2SO_4 (65%, 4 ml) and incubated (room temperature, 10 min), and absorbance was read (540 nm). Ascorbic acid served as standard to quantify vitamin C to represent in mg ascorbic acid equivalents per gramme mushroom flour (mg AAEs/g).

37.3.5 Phytic Acid

The phytic acid content of the mushroom powder was estimated based on the methodology by Deshpande et al. (1982) and Sathe et al. (1983). The flour (2 g) was extracted with sodium sulphate (10 ml, 10% in 1.2% HCl) followed by stirring (room temperature, 2 h). After centrifugation (3000 rpm, 10 min), the supernatant was made up to 10 ml (in extracting solvent). The extract (5 ml) that was mixed with ferric chloride (3 ml) (2 g in 16.3 ml of 12 N HCl, diluted to 1 L), vortexed and centrifuged (3000 rpm, 10 min). The supernatant was filtered (Whatman # 1), and the filtrate was made to 10 ml in distilled water. The free soluble phosphorous was estimated by vanadomolybdophosphoric acid method using potassium dihydrogen phosphate as reference and expressed in percentage of phytic acid:

$$\text{Total phosphorous}\left(\text{g}/100\ \text{g}\right) = \frac{M \times V \times F}{10{,}000 \times W \times V} \times \Delta A_{\text{p}}$$

where M is the average of phosphorous standard ($\mu g/\Delta A_p$); V, the original sample in ml; F, dilution factor; ΔA_p, absorbance; W, weight of sample (g); and V, sample volume (ml).

$$\text{Phytic acid}\left(\%\right) = \frac{\text{Phosphorous}\left(\text{g}/100\,\text{g}\right)}{0.282}$$

where 0.282 is the factor used to convert phosphorous content into phytic acid as it contains 28.2% phosphorous.

37.3.6 Carotenoids

The carotenoids lycopene and β-carotene in the mushroom flour were estimated based on Nagata and Yamashita (1992). Mushroom flour (~5 g) was extracted in methanol (100 ml) on centrifugation (150 rpm, 24 h) and filtered (Whatman # 1). The extraction was repeated, pooled and allowed to evaporate to dryness. The dried extract (100 mg) was dissolved in acetone-hexane (4:6, 10 ml) and filtered (Whatman # 1). The absorbance was measured (453, 505 and 663 nm). Contents of lycopene and β-Carotene were calculated:

$$\text{Lycopene}\left(\text{mg}\ \ 100/\text{ml}\right) = \left(-0.0458\,\text{A}_{663} + 0.372\,\text{A}_{505} - 0.0806\,\text{A}_{453}\right)$$
$$\beta\text{ - carotene}\left(\text{mg}\ \ 100/\text{ml}\right) = \left(0.216\,\text{A}_{663} - 0.304\,\text{A}_{505} + 0.452\,\text{A}_{453}\right)$$

37.3.7 L-DOPA

The L-DOPA (L-3,4-dihydroxyphenylalaninne) of the mushroom flour was evaluated based on Fujii et al. (1991). The flour samples were mixed in distilled water (1 ml), incubated (room temperature, 2 h) and centrifuged (1500 rpm, 10 min), and the supernatant was concentrated to dryness in a rotary evaporator. To discard the compounds with high molecular weight, the extract was dissolved in distilled water and filtered through ultrafilter overnight. The fraction was purified in ODS extraction minicolumn (C18 Sep-Pak Cartridge, Waters) with water and was evaporated to dryness. The L-DOPA was determined in HPLC (Tosoh system DP-8020; UV-8020, 280 nm; Column, Aqua 180 Mightsil; Kanto chemical Co. Inc., Japan) and LC-ESI/MS (Positive mode; Waters 181 Associates Inc., Milford, MA).

37.3.8 Trypsin Inhibition Activity

The trypsin activity was measured according to Kakade et al. (1974). The mushroom flour (1 g) was constantly stirred with NaOH (0.01 N, 50 ml, 10 min). The extract (1 ml) was diluted with distilled water (1:1), followed by addition of enzyme standard (2 mg trypsin/100 ml 0.001 M NaOH; 2 ml) and incubated in water bath (37 °C, 10 min); BAPNA (40 mg N_α-benzoyl-L-arginine 4-nitroanilide hydrochloride dissolved in dimethyl sulphoxide and made to 100 ml with Tris buffer at 37 °C; 5 ml) was added and incubated (room temperature, 10 min). To abort the reaction, acetic acid (30%; 1 ml) was added, and the absorbance was measured (410 nm). The control was prepared as per protocol without the addition of the mushroom extract. Trypsin inhibition (TIu) per mg of mushroom flour was calculated:

$$\text{TIu / mg} = \frac{\left[\left(A_{c410} - A_{s410}\right) \times 100\right] \text{per ml extract}}{\text{Mg sample per ml of extract}}$$

where A_c is absorbance of control and A_s absorbance of sample.

37.3.9 Haemagglutinin Activity

To determine the haemagglutinin activity in mushroom flour, method outlined by Occenã et al. (2007) was followed. The mushroom extract was prepared by mixing defatted flour (1 g) in NaCl (0.9%; 10 ml) and incubated (room temperature, 1 h). Centrifuged (2000 rpm, 10 min) after incubation, supernatant was collected, filtered and used as crude agglutinin. Heparinized blood samples of human blood were withdrawn and centrifuged (2000 rpm, 10 min) to separate erythrocytes. Separated erythrocytes (A^+, B^+, AB^+, O^+) were washed (1:4; chilled saline, 0.9%) repeatedly until the clear supernatant was obtained. Washed erythrocytes (4 ml) were dispensed into phosphate buffer (100 ml; 0.0006 M, pH 7.4). Trypsin (2%, 1 ml) was added, mixed and incubated (37 °C, 1 h). After incubation, trypsinized solution was repeatedly washed with saline (0.9%) to eliminate trypsin content. The erythrocytes obtained were suspended in saline (0.9%) and made to 100 ml. The round-bottomed 96-well microtiter plate was used for the assay. Initially phosphate buffer (50 μl) was allocated in well # 1–11, followed by the addition of crude agglutinin extract (50 μl) to well # 1, and mixed, and twofold serial dilution was followed up to well # 11. The suspension of erythrocytes (50 μl) was added to well # 1–11. The control was made in well # 12 as followed for sample. The contents in the wells are gently mixed and incubated (room temperature, 4 h) and observed for haemagglutination in each well. Haemagglutination unit pergramme (Hu/g) was calculated:

$$\mathrm{Hu} / g = \frac{D_{\mathrm{a}} \times D_{\mathrm{b}} \times S}{V}$$

where D_{a} is the dilution factor of extract in well #1; D_{b}, dilution factor of well containing 1 Hu that is the well in which the haemagglutination was observed; S, initial extract per gramme mushroom flour; and V, volume of extract in well #1.

37.4 Antioxidant Activities

Samples of mushroom flour (0.5 g) were extracted using methanol (30 ml) on a rotary shaker (150 rpm, 48 h). On centrifuging, the supernatant was collected in a preweighed Petri plate and allowed to evaporate (at room temperature). The weight of extract was assessed gravimetrically followed by dissolving in known quantity of methanol (1 mg/ml) to evaluate different antioxidant activities.

37.4.1 Total Antioxidant Activity

The total antioxidant activity (TAA) was evaluated by method of Prieto et al. (1999). To the methanolic extract of mushroom (1 mg/ml; 0.1 ml), reagent mixture was added (28 mM sodium phosphate +4 mM ammonium molybdate in 0.6 M sulphuric acid) and incubated (95 °C, 90 min). The absorbance was measured (695 nm), and the total antioxidant capacity was expressed in μM equivalents of ascorbic acid per gramme (μM AAEs/g).

37.4.2 Ferrous Ion-Chelating Capacity

The ferrous ion-chelating capacity of the methanolic extract was determined based on the protocol of Hsu et al. (2003). Methanol extract (1 ml) was mixed with 2 mM ferrous chloride (0.1 ml) and ferrozine (5 mM, 0.2 ml). The volume was made to 5 ml in methanol and incubated (room temperature, 10 min), and the absorbance was read (562 nm). The control was prepared similar to the protocol for sample without the addition of the extract to calculate the ferrous ion-chelating capacity:

$$\text{Ferrous ion chelating activity}(\%) = \left(1 \quad \frac{A_{s562}}{A_{c562}}\right)100$$

where A_s is absorbance of sample and A_c absorbance of control.

37.4.3 Reducing Power

The reducing power of the extract was determined by the method proposed by Oyaizu (1986) with a slight modification. The extract in different concentrations (0.2–1.0 mg/ml) was prepared phosphate buffer (0.2 M, pH 6.6) followed by the addition of potassium ferricyanide (1%, 2.5 ml) and incubated (50 °C, 20 min). After incubation, TCA (10%, 2.5 ml) was added to the mixture and centrifuged (3000 rpm, 10 min). The supernatant (2.5 ml) was mixed with equal volume of double-distilled water, and $FeCl_3$ (0.1%, 0.5 ml) was added. Absorbance was measured (700 nm) and higher absorbance indicates an increased reducing capability.

37.4.4 Free Radical-Scavenging Activity

The radical-scavenging activity of the mushroom extract was evaluated based on Singh et al. (2002). The extract of different concentrations (0.2–1.0 mg/ml) was made up to 1 ml using methanol followed by the addition of reagent (0.001 M DPPH in methanol, 4 ml). The contents were mixed well and incubated in the dark (room temperature, 20 min). The reagent without addition of extract served as control, and the absorbance was read (517 nm) to calculate the radical-scavenging activity:

$$\text{Free radical - scavenging activity}(\%) = \left(\frac{A_{c517} - A_{s517}}{A_{c517}}\right)100$$

where A_c is absorbance of control and A_s absorbance of sample.

The ABTS [(2,2′-azinobis(3-ethylbenzothiazoline-6-sulfonic acid)] cationic radical decolourization assay was carried out based on the method outlined by Adedapo et al. (2008). The stock solution ($ABTS^+$, 7.4 mM, and potassium persulphate, 2.6 mM) and the working solution (mixing stock solutions 1:1, allowed to react at room temperature, for 12 h in the dark) were prepared. The working solution was diluted with methanol to obtain suitable absorbance (1 ± 0.01 units at 734 nm). Different concentrations of mushroom extract (made up to 62 μl using absolute alcohol) were treated with $ABTS^+$ (188 μl, in dark, 30 min). The absorbance was measured (734 nm) to determine the percent inhibition:

$$\text{Inhibition percentage}(\%) = \left(\frac{\text{Assay control} - (\text{Test} - \text{Control})}{\text{Assay control}} \right) 100$$

where assay control is ethanol + ABTS reagent and control sample + ethanol + methanol.

37.5 Data Analysis

Student's *t*-test was used to ascertain the difference in bioactive components in uncooked and cooked mushroom samples (Statistica version # 8.0) (StatSoft 2008). The principal component analysis (PCA) was performed between bioactive components and antioxidant activities for uncooked as well as cooked mushroom samples separately (SPSS version 16.0: www.spss.com).

37.6 Bioactive Potential

In addition to nutritional and medicinal attributes, mushroom-derived antioxidants are gaining importance worldwide, owing to carcinogenic effects of synthetic antioxidants (e.g. BHT, BHA and TBHQ) in food industry. Antioxidant potential of many Asian wild mushrooms has been connected with total phenolic content (Cheung and Cheung 2005; Lo and Cheung 2005). Phenolic compounds of mushrooms are the principal components valuable in decreasing the incidence of cancers, cardiovascular diseases and atherosclerosis (Randhir et al. 2008; Alothman et al. 2009). The content of total phenolics in *Amanita* sp. was significantly decreased on pressure-cooking ($p < 0.05$), while the tannin content was not significantly changed on cooking ($p > 0.05$) (Fig. 37.2a, b). Total phenolic content is lower than *Amanita caesarea* and *A. citrina* while higher than *Amanita fulva*, *A. loosii*, *A. ovoidea* and *A. vaginata* (Doğan 2013; Paloi and Acharya 2013; Tripathy et al. 2014; Sharma and Gautam 2015), so also compared to some of the wild mushrooms of Southwest India (*Astraeus hygrometricus* and *Auricularia auricula-judae*) (Karun et al. 2016;

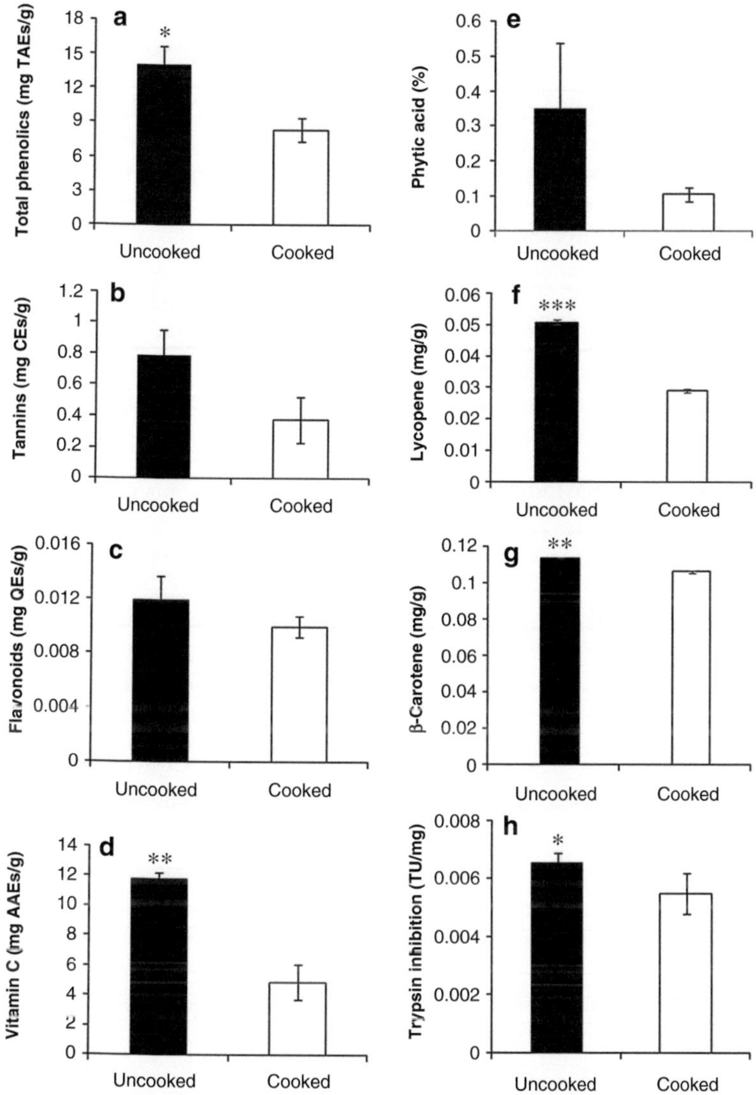

Fig. 37.2 Bioactive principles of tender fruit bodies of *Amanita* sp.: total phenolics (**a**), tannins (**b**), flavonoids (**c**), vitamin C (**d**), phytic acid (**e**), lycopene (**f**), β-carotene (**g**) and trypsin inhibition (**h**) (*t*-test: *$p < 0.05$; **$p < 0.01$; ***$p < 0.001$)

Pavithra et al. 2016). However, the total phenolic content in uncooked samples was lower than popular edible wild mushroom *Termitomyces umkowaan* of Southwest India (Karun et al. 2016). Similar to total phenolic content, tannin content surpassed the concentration in *A. hygrometricus* and *A. auricula-judae* (Karun et al. 2016; Pavithra et al. 2016). The total phenolics as well as tannins could be retained in the diet of *Amanita* sp. by following alternative thermal treatment than pressure-cooking (e.g. partial or microwave cooking).

Flavonoids consist of a variety of bioactive compounds that elicit protective effect against several human ailments (e.g. diabetes, cancers and cardiovascular diseases) (Champ 2002; Tapas et al. 2008). Similar to the tannin content, flavonoids in *Amanita* sp. also do not significantly changed on cooking ($p > 0.05$) (Fig. 37.2c). Its content was lower than many *Amanita* spp. (*A. caesarea*, *A. citrina*, *A. fulva*, *A. loosii* and *A. vaginata*) (Paloi and Acharya 2013; Tripathy et al. 2014; Sharma and Gautam 2015). Flavonoid content was also lower than the wild *A. auricula-judae* and *T. umkowaan* (Arun et al. 2016). Partial cooking of *Amanita* sp. could be beneficial in retention of maximum flavonoids.

Vitamin C serves as potent reducing agent and most essential cofactor in numerous enzymatic reactions. Its consumption in diet or supplement is known to cover the risks of cardiovascular diseases (Knekt et al. 2004; Ye and Song 2008). Vitamin C content in *Amanita* sp. was significantly decreased on cooking ($p < 0.01$), owing to its thermal sensitivity (Fig. 37.2d). Vitamin C of uncooked *Amanita* sp. is higher than *A. caesarea*, *A. citrina*, *A. fulva*, *A. loosii* and *A. vaginata* (Paloi and Acharya 2013; Tripathy et al. 2014; Sharma and Gautam 2015). Its content is substantially higher than the wild mushrooms of Southwest India (*A. hygrometricus*, *A. auricula-judae* and *T. umkowaan*) (Karun et al. 2016; Pavithra et al. 2016).

Phytic acid as an antioxidant binds to minerals such as calcium to prevent formation of kidney stones and improve heart health by preventing calcium deposits in arteries (Grases et al. 2008; Sekita et al. 2016). Its content in diets is also responsible for several benefits like absorption of glucose, reduction of renal lithiasis and prevention of dental caries (Kumar et al. 2010). The decrease in phytic acid content on cooking in *Amanita* sp. was not a significant change ($p > 0.05$) (Fig. 37.2e). Its quantity in *Amanita* sp. is substantially higher than the wild mushroom *A. hygrometricus* (Pavithra et al. 2016).

Carotenoids are known for quenching singlet oxygen and also inhibit oxidation of fats (lipid peroxidation); however its functions in humans appear to be more complex (Young and Lowe 2001). Similar to vegetables, mushrooms are also known for carotenoids, and their presence has been correlated with antioxidant activity in wild mushrooms in Portugal (Barros et al. 2007). In *Amanita* sp., lycopene ($p < 0.001$) and β-carotene ($p < 0.01$) were significantly decreased on pressure-cooking (Fig. 37.2f, g). Carotenoid content is considerably low in *Amanita* sp. compared to *A. caesarea* as well as *A. loosii* (Tripathy et al. 2014). Lycopene as well as β-carotene contents were lower than *A. vaginata* (Paloi and Acharya 2013). However, their contents were substantially higher than *A. caesarea*, *C. citrina* and *A. fulva* as reported from North India (Sharma and Gautam 2015). Carotenoid content is also higher than the wild mushroom *A. hygrometricus* of Southwest India (Pavithra et al. 2016). The present study revealed that uncooked samples of *Amanita* sp. will be more beneficial in accessing carotenoids.

The L-DOPA being nonprotein amino acid is known for its usefulness in treating Parkinson's disease (Hornykiewicz 2002). In *Amanita* sp. its content was below detectable level. However, other wild mushrooms (*A. auricula-judae* and *T. umkowaan*) occurring along with *Amanita* sp. possess good amount of L-DOPA (Karun et al. 2016).

Usually deficiency of sulphur amino acids is related to presence of trypsin inhibitors in food stuffs, owing to utilization of sulphur amino acids for synthesis of trypsin and chymotrypsin (Liener and Kakade 1980). Trypsin inhibition in *Amanita* sp. which was significantly decreased on cooking is nutritionally advantageous ($p < 0.05$) (Fig. 37.2h). However, its quantity is substantially lower compared to wild mushroom *A. hygrometricus* (Pavithra et al. 2016).

Haemagglutination activity in mushroom is due to the presence of lectins. Singh et al. (2014) identified 144 lectins from edible mushrooms; among them 38 and 30 lectins are known from poisonous and medicinal mushrooms, respectively. However, lectins in food stuffs are known as immunomodulators (Hartmann and Meisel 2007). On testing haemagglutinin activity of *Amanita* sp. on human erythrocytes (A^+, B^+, AB^+, O^+), only uncooked extract showed activity at 100 Hu/g against A^+ and B^+, while at 50 Hu/g against AB^+ and O^+ indicates cooked *Amanita* sp. is free from haemagglutinins.

37.7 Antioxidant Potential

The antioxidant potential of mushrooms is under the influence of cumulative effect of a variety of bioactive components. For a fair assessment of antioxidant potential of a specific biological material, at least two methods have to be adapted (Wong et al. 2006). The present study has followed five methods of assessment, and in all methods, uncooked samples of *Amanita* sp. showed significantly better activities than cooked samples ($p < 0.05$) (Fig. 37.3a–f).

The total antioxidant activity was significantly decreased on cooking ($p < 0.05$) (Fig. 37.3a). It is comparable to *Amanita vaginata* and *Astraeus hygrometricus* while lower than *Auricularia auricula-judae* and *Termitomyces umkowaan* (Paloi and Acharya 2013; Karun et al. 2016; Pavithra et al. 2016). According to Zhang et al. (2011), such results could be related to the amount of polysaccharides present in *Amanita* sp. Metal ions are known to deteriorate the food stuffs, which leads to cause arthritis as well as cancer, and its chelation reduces many damaging effects (Gordon 1990; Halliwell et al. 1995; Leopoldini et al. 2006; Soares et al. 2009). The ferrous ion-chelating capacity was also significantly decreased on cooking in *Amanita* sp. ($p < 0.001$) (Fig. 37.3b). Its activity is comparable to *A. vaginata* (Paloi and Acharya 2013), *A. auricula-judae* and *A. hygrometricus* while lower than *T. umkowaan* (Karun et al. 2016; Pavithra et al. 2016). The reducing power of *Amanita* sp. was significantly higher in uncooked than cooked samples ($p < 0.05$) (at highest concentration studied, $p < 0.001$) (Fig. 37.3c), which is comparable with *A. vaginata* and *T. umkowaan* (Paloi and Acharya 2013; Karun et al. 2016), while lower than *A. caesarea*, *A. citrina*, *A. fulva*, *A. ovoidea* and *Astraeus hygrometricus* (Doğan 2013; Sharma and Gautam 2015; Pavithra et al. 2016). The DPPH as well as ABTS radical-scavenging assays depict the ability of mushroom sample to serve as hydrogen donors. Results of these assays were similar to the reducing power ($p < 0.05$) (DPPH at 1 mg/ml: $p < 0.001$; ABTS at 88 μg/ml: $p < 0.001$) (Fig. 37.3d, e). The DPPH

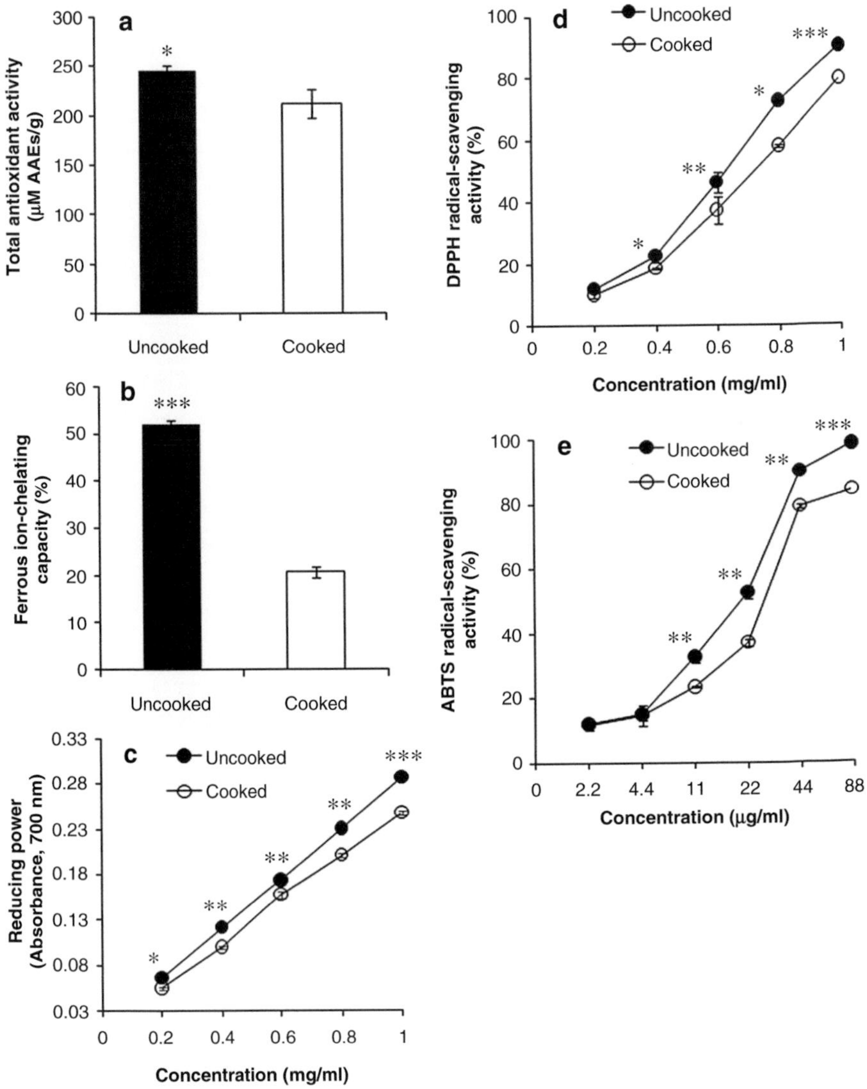

Fig. 37.3 Antioxidant activities of tender fruit bodies of *Amanita* sp.: total antioxidant activity (**a**), ferrous ion-chelating capacity (**b**), reducing power (**c**), DPPH radical-scavenging activity (**d**) and ABTS radical-scavenging activity (**e**) (*t*-test: *$p < 0.05$; **$p < 0.01$; ***$p < 0.001$)

radical-scavenging activity of *Amanita* sp. is comparable with *A. vaginata* (Paloi and Acharya 2013), while it was higher than *A. hygrometricus* and *T. umkowaan* (Karun et al. 2016; Pavithra et al. 2016). Overall, the above facts reveal that pressure-cooking reduces the antioxidant potential in *Amanita* sp. possibly due to damage of cell structure causing elimination of potent radical scavengers (Arora and Singh 2014).

37.8 Bioactive Components vs. Antioxidant Potential

The PCA of bioactive components of uncooked *Amanita* sp. versus antioxidant capacity resulted in two components with 100% variance. The rotated score plot for component 1 showed variance of 71.8% and 28.2% for component 2 (Fig. 37.4a). The bioactive principles like tannins (TaU), β-carotene (BcU) and trypsin inhibition (TiU) were clustered with antioxidant properties like total antioxidant activity (TaaU), reducing power (RpU) and DPPH radical-scavenging activity (DPPHU) at the right-hand region of the plot. The PCA for bioactive principles of cooked *Amanita* sp. against antioxidant capacity yielded two components with 100% variance. The rotated score plot for component 1 with variance is 61.2% and 38.8% for component 2 (Fig. 37.4b). The bioactive principles like tannins (TaC), vitamin C (VCC) and lycopene (LpC) clustered with reducing power (RpC) at the right-hand part of the plot, while only trypsin inhibition activity (TiC) clustered with total antioxidant activity (TaaC), DPPH (DPPHC) as well as ABTS (ABTSC) radical-scavenging activities at the top of the plot. On scrutiny of PCA results, uncooked and cooked *Amanita* sp. showed different pictures in clustering of bioactive principles with antioxidant properties. Uncooked as well as cooked samples have common features of association with tannins (Ta) and trypsin inhibition (Ti) with total antioxidant activity (Taa), reducing power (Rp) and DPPH radical-scavenging activity. Interestingly, tannin content in *Amanita* sp. was not significantly decreased on pressure-cooking ($p > 0.05$) and has likely immense impact on antioxidant potential of *Amanita* sp. In uncooked *A. hygrometricus*, also tannin contents were associated with reducing power (RpU) and DPPH radical-scavenging activity (DPPHU) (Pavithra et al. 2016). Overall, the PCA of *Amanita* sp. depicts several advantages of uncooked and cooked samples depending on the nutritional, pharmaceutical and nutraceutical requirements.

37.9 Outlook

Compared to the conventional plant-based foods and drugs, macrofungal products seem to be fairly recent and need serious concern in the future to fulfil nutritional requirements and to treat many life-threatening diseases. The west coast of India particularly lateritic scrub jungles are endowed with a variety of versatile wild mushrooms of nutritional and medicinal value (e.g. *Amanita* spp., *Astraeus* spp., *Phallus* spp. and *Termitomyces* spp.). *Amanita* sp. crop up in lateritic belts of scrub jungles for a short duration (up to 6–8 weeks) needs vigilance for sampling. As ectomycorrhizal, *Amanita* sp. is dependent on native tree species (e.g. *Hopea ponga* and *Terminalia paniculata*), and such tree species needs conservation measure. Detailed studies were performed on the phenolic compounds of edible *Amanita ovoidea* by Doğan (2013) showed antioxidant potential, and such in-depth studies are necessary for *Amanita* sp. to harness its nutritional and medicinal potential in future.

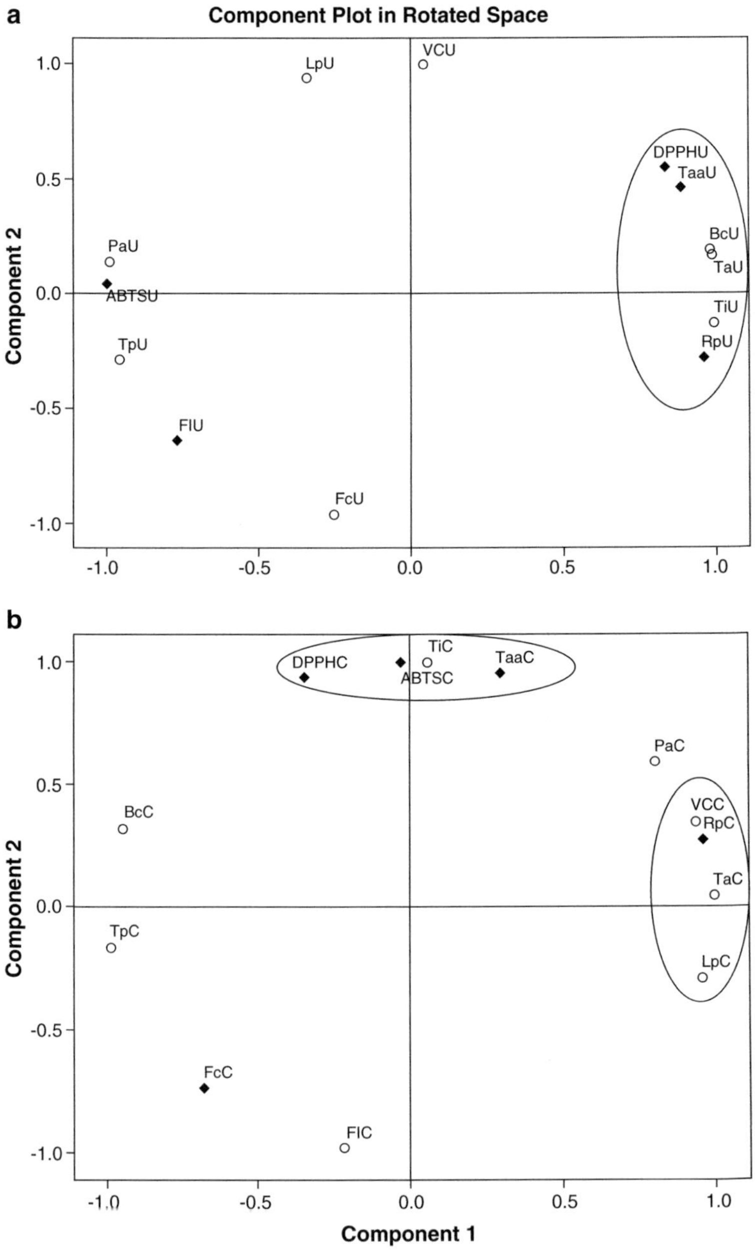

Fig. 37.4 Principal component analysis of uncooked (with suffix U) (**a**) and cooked (with suffix C) (**b**) tender fruit bodies of *Amanita* sp. [bioactive principles: total phenolics (*Tp*), tannins (*Ta*), flavonoids (*Fl*), vitamin C (*VC*), phytic acid (*Pa*), lycopene (*Lp*), β-carotene (*Bc*) and trypsin inhibition (*Ti*); antioxidant activities – total antioxidant activity (*Taa*), ferrous ion-chelating capacity (*Fc*), reducing power (*Rp*) assay, DPPH radical-scavenging activity (DPPH) and ABTS radical-scavenging activity (ABTS)]

Acknowledgements Authors are grateful to Mangalore University in accomplishing this study in the Department of Biosciences. GAA greatly acknowledges the award of INSPIRE Fellowship, Department of Science and Technology, New Delhi, Government of India. KRS is grateful to the University Grants Commission, New Delhi, India, for the award of UGC-BSR Faculty Fellowship during the tenure of this study.

References

Adedapo AA, Jimoh FO, Afolayan AJ, Masika PJ (2008) Antioxidant activities and phenolic contents of the methanol extracts of the stems of *Acokanthera oppositifolia* and *Adenia gummifera*. BMC Complement Altern Med 8:1–7

Alothman M, Bhat R, Karim AA (2009) Effects of radiation processing on phytochemicals and antioxidants in plant produce. Trends Food Sci Technol 20:201–212

Aly AH, Debbab A, Proksch P (2011) Fifty years of drug discovery from fungi – review. Fungal Divers 50:3–19

Arora B, Singh M (2014) Effect of cooking on antioxidant activity and phenolic content of various species of edible mushrooms of India. In: Proceedings of 8th international conference on mushroom biology and mushroom products, New Delhi, pp 576–581

Arun G, Eyini M, Gunasekaran P (2014) Green synthesis of silver nanoparticles using the mushroom fungus *Schizophyllum commune* and its biomedical applications. Biotechnol Bioprocess Eng 19:1083–1090

Barros L, Ferreira MJ, Queiros B, Ferreira ICFR, Baptista P (2007) Total phenols, ascorbic acid, β-carotene and lycopene in Portuguese wild edible mushrooms and their antioxidant activities. Food Chem 103:413–419

Boa E (2004) Wild edible fungi a global overview of their use and importance to people. Food and Agricultural Organization, Rome

Burns R (1971) Methods for estimation of tannins in grain sorghum. Agron J 63:511–512

Buyck B (1994) Ubwoba: les champignons comestibles de I'Ouest du Burundi. Administration Generale de la Cooperation au Developpement Bruxelles Publ # 34

Champ MM (2002) Non nutrient bioactive substances of pulses. Br J Nutr 88:307–319

Chang C, Yang M, Wen H, Chern J (2002) Estimation of total flavonoid content in propolis by two complementary colorimetric methods. J Food Drug Anal 10:178–182

Cheung LM, Cheung PCK (2005) Mushroom extracts with antioxidant activity against lipid peroxidation. Food Chem 89:403–409

De Silva DD, Rapior S, Sudarman E, Stadler M, Xu J, Alias SA, Hyde KD (2013) Bioactive metabolites from macrofungi: ethnopharmacology, biological activities and chemistry. Fungal Divers 62:1–40

Deshpande SS, Sathe SK, Salunkhe DK, Cornforth DP (1982) Effects of dehulling on phytic acid, polyphenols and enzyme –inhibitors of dry beans (*Phaseolus vulgaris* L). J Food Sci 47:1846–1850

Doğan HH (2013) Evaluation of phenolic compounds, antioxidant activities and fatty acid composition of *Amanita ovoidea* (Bull.) Link. in Turkey. J Food Compos Anal 31:87–93

Farook AV, Khan SS, Manimohan P (2013) A checklist of agarics (gilled mushrooms) of Kerala state, India. Mycosphere 4:97–131

Fujii Y, Shibuya T, Yasuda T (1991) L 3,4-dihydroxyphenylalanine as an allelochemical from *Mucuna pruriens* (L.) DC var *utilis*. Agric Biol Chem 55:617–618

Ghate SD, Sridhar KR (2016) Spatiotemporal diversity of macrofungi in the coastal sand dunes of Southwestern India. Mycosphere 7:458–472

Ghate SD, Sridhar KR, Karun NC (2014) Macrofungi on the coastal sand dunes of South-Western India. Mycosphere 5:144–151

Gordon MH (1990) The mechanism of antioxidant action in vitro. In: Hudson BJF (ed) Food antioxidants. Elsevier Applied Science, London, pp 1–18

Grases F, Sanchis P, Perello J, Isern B, Prieto RM, Fernandez-Palomeque C, Saus C (2008) Phytate reduces age-related cardiovascular calcification. Front Biosci 13:7115–7122

Halliwell B, Murcia HA, Chico S, Aruoma OI (1995) Free radicals and antioxidants in food an in vivo: what they do and how they work. CRC Crit Rev Food Sci Nutr 35:7–20

Härkönen M, Saarimäki T, Mwasumbi L (1994) Tanzania mushrooms and their uses 4. Some reddish edible and poisonous Amanita species *Karstenia,* 34:47–60

Hartmann R, Meisel H (2007) Food-derived peptides with biological activity: from research to food application. Curr Opin Biotechnol 18:163–169

Hornykiewicz O (2002) L-DOPA: from a biologically inactive amino acid to a successful therapeutic agent. Amino Acids 23:65–70

Hsu CL, Chen W, Weng YM, Tseng CY (2003) Chemical composition, physical properties and antioxidant activities of yam flours as affected by different drying methods. Food Chem 83:85–92

Hyde KD, Bahkali AH, Moslem MA (2010) Fungi – an unusual source for cosmetics. Fungal Divers 43:1–9

Kakade ML, Rackis JJ, McGhee JE, Puski G (1974) Determination of trypsin inhibitor activity of soy products, a collaborative analysis of an improved procedure. Cereal Chem 51:376–382

Karun NC, Sridhar KR (2014) A preliminary study on macrofungal diversity in an arboratum and three plantations of the southwest coast of India. Curr Res Environ Appl Mycol 4:173–187

Karun NC, Sridhar KR (2016) Spatial and temporal diversity of macrofungi in the Western Ghat forests of India. Appl Ecol Environ Res 14:1–21

Karun NC, Sridhar KR, Niveditha VR, Ghate SD (2016) Bioactive potential of two wild edible mushrooms of the Western Ghats of India. In: Watson RR, Preedy VR (eds) Fruits, vegetables, and herbs: bioactive foods in health promotion. Elsevier, Oxford, pp 344–362

Knekt P, Ritz J, Pereira MA, O'Reilly EJ, Augustsson K, Fraser GE, Goldbourt U, Heitmann BL, Hallmans G, Liu S, Pietinen P (2004) Antioxidant vitamins and coronary heart disease risk: a pooled analysis of 9 cohorts. Am J Clin Nutr 80:1508–1520

Kumar V, Sinha AK, Makkar HPS, Becker K (2010) Dietary roles of phytate and phytase in human nutrition: a review. Food Chem 120:945–959

Leopoldini M, Russo N, Chiodo S, Toscano M (2006) Iron chelation by the powerful antioxidant flavonoid quercetin. J Aric Food Chem 54:6343–6351

Liener IE, Kakade ML (1980) Protease inhibitors. In: Liener IE (ed) Toxic-constituents of plant foodstuffs, 2nd edn. Academic, New York

Lo KM, Cheung PCK (2005) Antioxidant activity of extracts from the fruiting bodies of *Agrocybe aegerita* var *alba*. Food Chem 89:533–539

Manzi P, Pizzoferrato L (2000) Beta-glucans in edible mushrooms. Food Chem 68:315–318

Mau JL, Chang CN, Huang SJ, Chen CC (2004) Antioxidant properties of methanolic extracts from *Grifola frondosa, Morchella esculenta* and *Termitomyces albuminosus* mycelia. Food Chem 87:111–118

Miller OK Jr, Hemmes DE, Wong G (1996) *Amanita marmorata* subsp *Myrtacearum* – a new subspecies in *Amanita* section Phalloideae from Hawaii. Mycologia 88:140–145

Mohanan C (2011) Macrofungi of Kerala. Handbook # 27. Kerala Forest Research Institute, Peechi, India

Nagata M, Yamashita I (1992) Simple method for simultaneous determination of chlorophyll and carotenoids in tomato fruit. Nippon Shokuhin Kogyo Gakkaish 39:925–928

Occenã IV, Majica E-RE, Merca FE (2007) Isolation of partial characterization of a lectin from the seeds of *Artocarpus camansi* Blanco. Asian J Plant Sci 6:757–764

Oyaizu M (1986) Studies on products of browning reactions: antioxidative activities of products of browning reaction prepared from glucosamine. Jap J Nutr 44:307–315

Paloi S, Acharya K (2013) Antioxidant activities and bioactive compounds of polyphenol rich extract from *Amanita vaginata* (Bull.) Lam. Int J Pharm Tech Res 5:1645–1654

Pavithra M, Greeshma AA, Karun NC, Sridhar KR (2015) Observations on the *Astraeus* spp. of Southwestern India. Mycosphere 6:421–432

Pavithra M, Sridhar KR, Greeshma AA, Tomita-Yokotani K (2016) Bioactive potential of the wild mushroom *Astraeus hygrometricus* in South-West India. Mycology 7:191–202. https://doi.org/10.1080/21501203.2016.1260663

Prieto P, Pineda M, Aguilar M (1999) Spectrophotometric quantitation of antioxidant capacity through the formation of a phosphomolybdenum complex: specific application to the determination of vitamin E. Anal Biochem 269:337–341

Randhir R, Kwon YI, Shetty K (2008) Effect of thermal processing on phenolics, antioxidant activity and health-relevant functionality of select grain sprouts and seedlings. Innov Food Sci Emerg Technol 9:355–364

Roe JH (1954) Chemical determination of ascorbic, dehydroascorbic and diketogluconic acids. In: Glick D (ed) Methods of biochemical analysis, vol 1. Inter Science Publishers, New York, pp 115–139

Rosset J, Bärlocher F, Oertli JJ (1982) Decomposition of conifer needles and deciduous leaves in two black Forest and two Swiss Jura streams. Int Rev Gesamten Hydrobiol 67:695–711

Sánchez-Ramírez S, Tulloss RE, Amalfi M, Moncalvo JM (2015) Palaeotropical origins, boreotropical distribution and increased rates of diversification in a clade of edible ectomycorrhizal mushrooms (*Amanita* section Caesareae). J Biogeogr 42:351–363

Sanmee R, Tulloss RE, Lumyong P, Dell B, Lumyong S (2008) Studies on *Amanita* (Basidiomycetes: Amanitaceae) in Northern Thailand. Fungal Divers 32:97–123

Sathe SK, Deshpande SS, Reddy NR, Goll DE, Salunkhe DK (1983) Effects of germination on proteins, raffinose oligosaccharides and antinutritional factors in the Great Northern Beans (*Phaseolus vulgaris* L.). J Food Sci 48:1796–1800

Sekita A, Okazaki Y, Katayama T (2016) Dietary phytic acid prevents fatty liver by reducing expression of hepatic lipogenic enzymes and modulates gut microflora in rats fed a high-sucrose diet. Nutrition 32:720–722

Senthilarasu G (2014) Diversity of agarics (gilled mushrooms) of Maharashtra, India. Curr Res Environ Appl Mycol 4:58–78

Senthilarasu G, Kumaresan V (2016) Diversity of agaric mycota of Western Ghats of Karnataka, India. Curr Res Environ Appl Mycol 6:75–101

Sharma SK, Gautam N (2015) Chemical, bioactive, and antioxidant potential of twenty wild culinary mushroom species. BioMed Res Int. https://doi.org/10.1155/2015/346508

Singh RP, Murthy CKN, Jayaprakasha GK (2002) Studies on antioxidant activity of pomegranate (*Punica granatum*) peel and seed extracts using in vitro methods. J Agric Food Chem 50:81–86

Singh SS, Wang H, Chan YS, Pan W, Dan X, Yin CM, Akkouh O, Tzi BN (2014) Lectins from edible mushrooms. Molecules 20:446–469

Soares SR, Carvalho-Oliveira R, Ramos-Sanchez E, Catanozi S, Silva LF, Mauad T, Gidlund M, Goto H, Garcia ML (2009) Air pollution and antibodies against modified lipoproteins are associated with atherosclerosis and vascular remodeling in hyperlipemic mice. Atherosclerosis 207:368–373

StatSoft (2008) Statistica Version # 8. StatSoft, Tulsa

Taofiq O, González-Paramás AM, Martins A, Barreiro MF, Ferreira ICFR (2016) Mushrooms extracts and compounds of cosmetics, cosmeceuticals and nutricosmetics – a review. Ind Crop Prod 90:38–48

Tapas AR, Sakarkar DM, Kakde RB (2008) Flavonoids as nutraceuticals: a review. Trop J Pharm Res 7:1089–1099

Thatoi H, Singdevsachan SK (2014) Diversity, nutritional composition and medicinal potential of Indian mushrooms: a review. Afr J Biotechnol 13:523–545

Tripathy SS, Rajoriya A, Gupta N (2014) Nutritive properties and antioxidative activity of *Amanita caesarea* and *A. loosii* wild edible mushrooms from Odisha. Int J Inno Drug Disc 4:124–129

Tulloss RE (2005) *Amanita* – distribution in the Americas, with comparison to eastern and southern Asia and notes on spore character variation with latitude and ecology. Mycotaxon 93:189–231

Vikineswary S, Chang ST (2013) Edible and medicinal mushrooms for sub-health intervention and prevention of lifestyle diseases. Tech Monitor (July–September):33–43

Wong SP, Leong LP, Koh JHW (2006) Antioxidant activities of aqueous extracts of selected plants. Food Chem 99:775–783

Wu T, Zivanovic S, Draughon FA, Sams CE (2004) Chitin and chitosan-value-added products from mushroom waste. J Agric Food Chem 29:7905–7910

Xu X, Wu Y, Chen H (2011) Comparative antioxidative characteristics of polysaccharide-enriched extracts from natural sclerotia and cultured mycelia in submerged fermentation of *Inonotus obliquus*. Food Chem 127:74–79

Ye Z, Song H (2008) Antioxidant vitamins intake and the risk of coronary heart disease: meta-analysis of cohort studies. Eur J Card Prev Rehab 15:26–34

Young AJ, Lowe GM (2001) Antioxidant and prooxidant properties of carotenoids. Arch Biochem Biophys 385:20–27

Zhang H, Wang ZY, Zhang Z, Wang X (2011) Purified *Auricularia auricula-judae* polysaccharide (AAP I-a) prevents oxidative stress in an ageing mouse model. Carbohydr Polym 84:638–648

Zhang P, Tang LP, Cai Q, Xu JX (2015) A review on the diversity, phylogeography and population genetics of *Amanita* mushrooms. Mycology 6:86–93

Chapter 38
Fungus *Monascus*-Fermented Red Yeast Rice (RYR): Natural Therapeutic Statin Source or Mycotoxin?

Seema Patel and Nadeem Akhtar

Abstract The kingdom fungi, including the genus *Monascus*, is a trove of potential pharmaceuticals. This mold *Monascus* holds paradoxical stature in the realm of healthcare. On one hand, it has been an ubiquitous ingredient of Oriental medication such as Traditional Chinese Medicine (TCM), since ages. *Monascus*-fermented product red yeast rice (RYR) is being used to ameliorate atherosclerosis, cancer, diabetes, bone ailments, and a gamut of other inflammations. On the other hand, it elaborates mycotoxins such as citrinin, which poses nephrotoxicity, hepatotoxicity, and teratogenicity risks. Also, the threat of the adverse interaction of RYR with chemotherapeutic antihypertensive drugs and the immunosuppressive effects loom over. With the unprecedented escalation in the instances of cardiovascular diseases, and the cognizance of side effects of the conventional statins, which includes myalgia, myositis, hepatic, and renal toxicities, the interest in the *Monascus*-fermented product RYR is growing. The advent in analytical techniques has characterized RYR to contain polyketides, unsaturated fatty acids, sterols, pigments, and statins. Monacolin K, which shares configurational similarity with commercial cholesterol-lowering drug lovastatin, has drawn worldwide attention. In vitro, in vivo, and clinical cohort studies have traced the biological mechanism of monacolin K to the crosstalk with human endogenous antioxidant coenzyme Q10 and the inhibition of HMG-CoA reductase, a crucial enzyme of the mevalonate pathway. As the purification of the statins from the hazardous mycotoxins of RYR, and dosage standardization, might provide with a natural source of antilipemic nutraceutical, this aspect holds relevance for exploration.

Keywords Red yeast rice · RYR · *Monascus* · Monacolin K · Lovastatin · HMG-CoA reductase · Coenzyme Q10

S. Patel (✉)
Bioinformatics and Medical Informatics Research Center, San Diego State University, San Diego, CA, USA

N. Akhtar
Department of Animal Biosciences, University of Guelph, Guelph, Ontario, Canada

P. Gehlot, J. Singh (eds.), *Fungi and their Role in Sustainable Development: Current Perspectives*, https://doi.org/10.1007/978-981-13-0393-7_38

38.1 Introduction

Red yeast rice or RYR is a mold-fermented product, with culinary as well as medicinal importance. *Monascus purpureus* from the family *Aspergillaceae* colonizes white rice, secreting a repertoire of bioactive metabolites, such as unsaturated fatty acids, polyketides (Fung et al. 2012), pigments (monascin, ankaflavin, monascorubrin, rubropunctatin, rubropunctamine, monascorubramine, etc.) (Hong et al. 2008; Patakova 2013), sterols (Venero et al. 2010), and condensed tannins (Li et al. 2005). Figure 38.1 shows the RYR production pathway. In Oriental countries, RYR is traditionally consumed as an ingredient in tofu, vinegar, meat, pastries, and wine (Liu et al. 2006). RYR is named variably in different countries, some of the popular names being “angkak” and “Xuezhikang” in China (Hong et al. 2008; Shang et al. 2012) and “red koji” in Japan (Fung et al. 2012). RYR has been a part of Traditional Chinese Medicine (TCM), administered for body rejuvenation, facilitation of blood circulation, and pH homeostasis of stomach, among other functions (Sham et al. 2014). But now, with the awareness of the deleterious effects of allelopathic drugs, and antibiotics, the Western world has also shown interest in RYR. The scientific validations of its benefits in blood pressure homeostasis have spurred its overwhelming popularity. This attention has prompted intense investigations on various aspects of RYR. Monacolins secreted by the mold *Monascus* have been identified as the therapeutic component. Monacolin K shares configurational resemblance to lovastatin (Klimek et al. 2009). Lovastatins are the mainstay therapy of hypertension, but these drugs exert adverse reactions in a majority of patients, and the intolerance increases with chronic usage. Consequently, the Food and Drug Administration (FDA) has imposed restrictions on the marketing of RYR as a dietary supplement. This chapter weighs the risk-benefit ratio of RYR as complementary and alternative medicines (CAM).

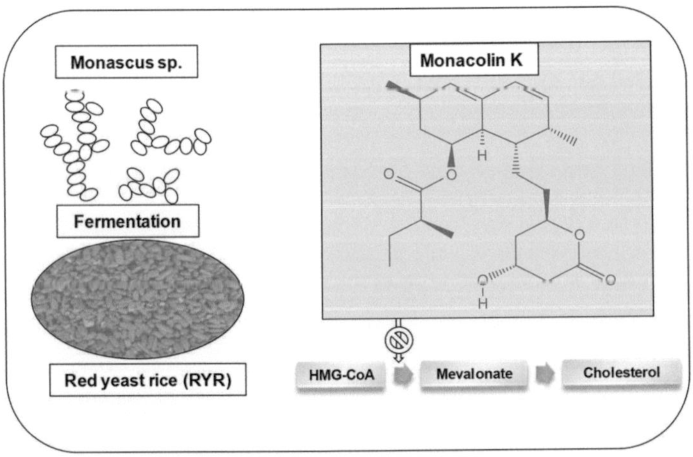

Fig. 38.1 The inhibition of cholesterol biosynthesis by RYR

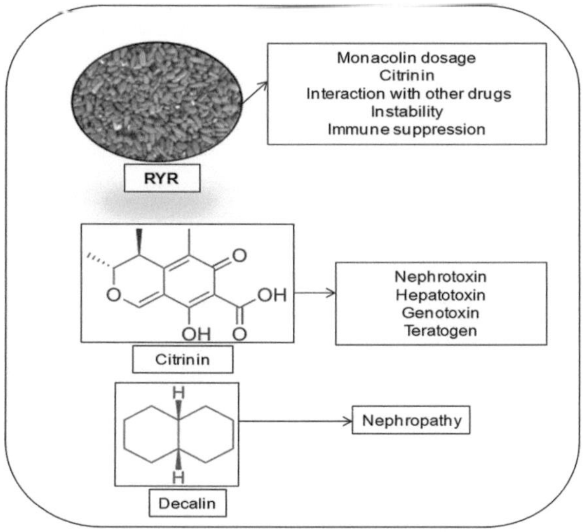

Fig. 38.2 The adverse effects of RYR intake

38.2 Hyperlipidemia and High Blood Pressure

High blood cholesterol and low-density lipoprotein (LDL) are a threat to circulatory system. These lipids adhere to the capillaries and arteries, narrowing them, raising blood pressure. So, the inhibition of cholesterol biosynthesis is an ideal target for lowering LDL level. Cholesterol synthesis is a multi-step process, with a critical enzyme being hydroxymethylglutaryl-CoA (HMG-CoA) reductase, which produces mevalonate through the mevalonate pathway (DeBose-Boyd 2008). Statins act as the inhibitors of HMG-CoA reductase (Istvan 2003). This mechanism has been illustrated in Fig. 38.2. A number of synthetic statins, including lovastatin, atorvastatin, simvastatin, etc., exist; however, they lead to detrimental effects like pain and inflammation of skeletal muscle (manifested in conditions like myopathy, myalgia, and myositis) and liver and kidney malfunctions (Golomb and Evans 2008; Chauvin et al. 2013; Khayznikov et al. 2015; Pirillo and Catapano 2015). The mevalonate pathway is critical for the synthesis of components such as coenzyme Q10, heme-A, steroids, which hamper physiological processes (Golomb and Evans 2008; Abdelbaset et al. 2014). In this regard, RYR is expected to be a natural substitute for the existing panel of statins.

38.3 RYR as Health Promoter and the Underlying Mechanisms

The most convincingly-validated properties of RYR include the anti-inflammatory, hypotensive, hypolipemic, cardioprotective, antidiabetic, anticancer, osteogenic, etc. The ameliorative mechanisms have been reviewed in details (Patel 2016).

Azaphilones (monascusazaphilones A–C), derived from the ethanol extract of RYR, inhibited nitric oxide (NO) elaboration from the macrophages (Wu et al. 2013). In hyperlipidemic rat models, Xuezhikang stabilized the lipid levels and eased the filtration stress on the kidney glomeruli (Ding et al. 2014). High blood pressure compromises the vasculature integrity, leading to a litany of pathologies like myocardial infarction, stroke, circulatory collapse, renal failure, and even sudden death (James et al. 2014). RYR extract in combination with a herb *Orthosiphon stamineus* (from the family Lamiaceae), berberine (a plant alkaloid), policosanol (a plant wax extract), folic acid, and coenzyme Q10, reduced blood pressure and corrected lipid profile, up to a period of 4 h (Trimarco et al. 2012). A number of other studies have agreed upon the blood pressure-controlling effect of RYR, when paired with antihypertensive drugs and statins (Xiong et al. 2015). A study shows that the intake of RYR by coronary artery and dyslipidemia patients led to a 30–60% reduction in the instances of cardiovascular complications (Chen et al. 2015). A randomized, double-blind, placebo-controlled study of hyperlipidemia patients revealed the antilipemic effect of nattokinase enzyme derived from the fermented soya bean (natto) plus RYR extract (Yang et al. 2009). The intake of two capsules made of the above two ingredients, twice a day, for a period of 6 months, led to the decline of blood LDL by 41%. Also, the high-density lipoprotein (HDL) level increased by 7.5% (Yang et al. 2009). Another randomized, double-blind, placebo-controlled study reported that patients receiving RYR had a significant reduction in total cholesterol by 15%, compared to placebo, after 16 weeks of the oral intake (Bogsrud et al. 2010). The intake of RYR, along with folic acid, coenzyme Q10, berberine, policosanol, and astaxanthin, for a 12-month period, by statin-intolerant elderly hypercholesteremic patients resulted in the fall of LDL (31%), total cholesterol (20%), as well as insulin resistance (10%) (Marazzi et al. 2011). In a controlled, randomized, multicenter study, the blend of RYR, policosanol, and berberine, at a level of one tablet daily for 16 weeks, improved the blood lipid profile of dyslipidemia patients (Trimarco et al. 2011). RYR was capable of LDL cholesterol reduction, in combination with bitter gourd, chlorella, soy protein, and licorice, when administered for 12 weeks (Lee et al. 2012). A blend of RYR, berberine, policosanol, folic acid, coenzyme Q10, and astaxanthin, administered to breast cancer patients on hormone replacement therapy (HRT), reduced total cholesterol as well as LDL cholesterol (Zanardi et al. 2012). A double-blind study spanning 16 weeks showed that the intake of a supplement formulated with RYR, policosanols, and artichoke leaf extract can attenuate cholesterol (Ogier et al. 2013). RYR paired with berberine and policosanol has alleviative effect on metabolic syndromes (Affuso 2012). Fruit drinks supplemented with RYR, niacin, L-carnitine, coenzyme Q10, phytosterol esters, and vitamin C lowered the total cholesterol by 14% and LDL cholesterol by 17%, during weeks 4–8 (Karl et al. 2012). RYR when blended with phytosterols reduced total cholesterol by 19% and LDL cholesterol by 33%, after 6-week intake (Feuerstein and Bjerke 2012). *Silybum marianum* (milk thistle) is a medicinal plant (Patel 2015). A double-blind, placebo-controlled trial revealed that RYR, when paired with this plant extract and octacosanol (a fatty alcohol), can improve the lipid profile while attenuating the inflammatory parameters in low-risk dyslipidemia patients (Derosa et al. 2014).

RYR and coenzyme Q10 combination lowered cholesterol in a manner comparable to atorvastatin, but without elevating creatine kinase level (Abdelbaset et al. 2014). As creatine kinase is the cause of myopathy, this side effect-free effect of the fortified RYR is encouraging. A cohort study reported that Xuezhikang, at a dose 1200–2400 mg administered for a duration of 4–12 weeks, lowers LDL in dyslipidemia patients, without causing any tolerance issue (Moriarty et al. 2014). A study found that Xuezhikang extract, when ingested at a dose of 300 mg/kg per day, significantly lowered blood coagulation activation and tissue factor expression rate in atherosclerotic rats, with superior efficacy than that of lovastatin (Li et al. 2011). RYR extract supplementation resisted the angiotensin II-caused vasoconstriction in apolipoprotein E-deficient mice (Xie et al. 2012). Further, the RYR ameliorated atherosclerotic lesions, by reducing the level of serum total cholesterol, and matrix metalloproteinase (MMP2), intercellular adhesion molecule 1 (ICAM-1), and vascular cell adhesion molecule 1 (VCAM-1) (Xie et al. 2012). ICAMs, VCAM, and MMP are central players in the process of inflammation. Xuezhikang upregulated eNOS (endothelial constitutive nitric oxide synthase) expression from immune cells, dilating the vasculature and facilitating blood flow in atherosclerotic rats (Zhu et al. 2013b). In hyperlipidemic rats, RYR delayed platelet coagulation, suggesting its role in thrombus management (Lee et al. 2013). A study found that Mediterranean diet supplemented with RYR can regulate the cholesterol level of type 2 diabetes patients (Sartore et al. 2013). Xuezhikang decreased the blood glucose level of diabetic mice by promoting insulin secretion (Wang et al. 2014). The insulin secretion was explained to be due to the protection of pancreatic cells from oxidative stress (Wang et al. 2014). In an in vitro assay, RYR exerted antiproliferation and apoptosis toward the human colon HCT-116 and HT-29 cell lines, by manipulating the enzyme HMG-CoA reductase. The enzyme-encoding genes were controlled by the transcription factor sterol response element-binding protein-2 (SREBP-2) (Hong et al. 2008). SREBP-2 crosstalks with INSIG1 (insulin-induced gene 1) and CREB (cAMP-response element-binding protein). Three cytotoxic azaphilones isolated from RYR extract exhibited selective cytotoxicity toward the human lung adenocarcinoma A549 cell line (Li et al. 2010). In another study, RYR reduced xenograft prostate gland tumor volume in severe combined immunodeficiency (SCID) mice models (Hong et al. 2011). The shrinkage of tumor was linked to the suppression of genes encoding enzymes such as AKR1C3, HSD3B2, and SRD5A1, crucial for steroidogenesis (Hong et al. 2011). The constitutive expression of these genes produces excess androgen that leads to the relapse of prostate cancer after surgery (Jernberg et al. 2013). The finding above suggests that the RYR-driven unavailability of the hormone can prevent the expression of the receptors. RYR grown on a garlic medium had anticancer activities on murine melanoma B16F10, human liver HepG2, and human colon HT-29 cancer cell lines (Park and Kim 2011). The combined treatment of RYR-derived monascuspiloin pigments and irradiation enhanced the apoptosis of human prostate cancer PC-3 cells, via the inhibition of Akt/mTOR signaling pathways (Chiu et al. 2012). In another in vitro study, this pigment inhibited androgen-dependent LNCaP cell lines by the intervention of PI3K/Akt/mTOR pathway and androgen-independent PC-3 human prostate cancer cells lines, by arresting G2/M phase cell cycle and inducing 5′ AMP-activated protein kinase (AMPK)-dependent autophagy (Chen et al. 2012). Dehydromonacolin K was highly cytotoxic for human breast adenocarcinoma MCF-7, hepatocarcinoma HepG2, and human intestinal Caco-2 cell lines (Zhu et al. 2012b). Red azaphilone pigments extracted from RYR induce cellular senescence, which lowers the viability of HepG2 cells (Wei and Popovich 2013). Bone growth-stimulating property of RYR was observed through both in vitro (rat osteosarcoma UMR 106 cell line) as well as in vivo test paradigms (Wong and Rabie 2008). This finding might lend insight to osteoporosis management approaches. Through an MTT assay, it was shown that osteoblast-like MC3T3-E1 cells undergo proliferation and differentiation, following a 24 h incubation with RYR methanol extract (Cho et al. 2010). A rat model study showed that RYR, in particular monacolin K, promotes bone mineral density (BMD) while lowering bone mineralization factors such as osteocalcin (a non-collagenous protein hormone also known as bone gamma-carboxyglutamic acid-containing protein) and tartrate-resistant acid phosphatase (Wang et al. 2015). In an in vitro experiment, RYR improved the viability of osteoblasts and the induction of morphogenetic protein-2 (BMP-2) (Wang et al. 2015). Statins as inducers of bone BMP-2 have already been proven (Wong and Rabie 2008). The reported benefits are only a fraction of the vast amount of literature published on the therapeutic spectrum of RYR. Figure 38.1 illustrates the inhibition of cholesterol biosynthesis by RYR.

38.4 Health Risks of RYR Ingestion

Despite the promising aspects of RYR, it has hazards as well, which must be clarified before its usage as a supplement or CAM. The compositional inconsistencies, instability, mycotoxins, interactions with drugs, and immune suppression are some of the conflicts associated with RYR intake. The key bioactive component, monacolin, occurs in variable amount in RYR from different sources. The analysis of 12 different RYR samples resulted in different quantities of monacolin, though the product labels claimed them to be 600 mg/capsule (Gordon et al. 2010). As the efficacy as well as the safety of any supplement hinges on the right dosage, this uncertainty of the monacolin content undermines the therapeutic relevance of RYR. Daily consumption of the standardized RYR formulation with 10 mg monacolins can lower LDL level by 20% (McCarty et al. 2015). Amount lesser or higher than this dose can prove inadequate or can trigger undesirable health consequences. RYR components are also prone to photo- and thermal instability. Monacolin K undergoes degradation during storage. A 30-day period can cause the loss of 43% of the monacolin (Hsieh et al. 2013). RYR is often contaminated with mycotoxins such as citrinin ($C_{13}H_{14}O_5$), a polyketide nephrotoxin, secreted by the pathogenic fungi like *Penicillium* and *Aspergillus*, apart from *Monascus* (Childress et al. 2013; Flajs and Peraica 2009). Also, it has been proven to be teratogenic (Flajs and Peraica

2009), hepatotoxic (Gayathri et al. 2015), and genotoxic (Klimek et al. 2009). So, safety assessment of the RYR products is important. A meta-analysis of 20 studies on RYR intake reflected the poor safety considerations, with the instances of hepatic and renal impairment being 0–5% (Gerards et al. 2015). Herb-drug interaction (HID) is a health risk, which often gets trivialized. Adverse interactions between popular herbal products and chemotherapeutics like aspirin, acetaminophen, corticosteroids, sedatives, and antidepressants have been documented. The deleterious effects can range from inflammations, high blood pressure, ulcers, hemorrhage, hepatotoxicity, and nephrotoxicity (Abebe 2002, 2003; Girard and Vohra 2011). The herb-drug combinations often cause the excess expression of P-glycoproteins, which result in drug resistance, and manipulate the crucial cytochrome P450 enzymes, leading to aberrant immune activation (Hafner-Blumenstiel 2011). The polar fractions of RYR, when not eliminated properly, have been found to react with antihypertensive drugs such as the calcium channel blockers (e.g., verapamil) (Fung et al. 2012). Also, the immunosuppressive effect of RYR has come forth. The ethyl acetate extract of RYR contains decalin (decahydronaphthalene) derivatives (Zhu et al. 2012a). These naphthalene homologs can be toxic to renal tissues (Dill et al. 2003). As a weak immunity exposes the body to a diverse array of infectious agents, the adverse immune modulation by RYR is a matter of serious concern (Fishman 2013). Figure 38.2 shows the causes and hazardous effects of RYR intake.

38.5 Hurdles, Solutions, and Scopes

As RYR is not free of health risks, precaution should be followed in its intake. Regulatory authorities should ensure the standardization of monacolin content. Mislabeling and fraudulent practices, which is prevalent in CAM drugs and dietary supplements development, should be discouraged with stringent surveillance policies. RYR production and marketing should adhere to the current good manufacturing practices (CGMP). Chromatography and mass spectrometry (MS) analyses can be recruited to identify the discrepancies. In a study, high-performance liquid chromatography (HPLC) was able to find the monacolin K and citrinin content in RYR samples (Mornar et al. 2013). A variation of chromatography, micellar electrokinetic chromatography (MEKC), was also sensitive toward these components and can measure them even in trace amounts (Nigović et al. 2013). A coupled chromatography/MS UHPLC-DAD-QToF-MS, followed by chemometric tool principle component analysis (PCA), could detect monacolins, pigments, and citrinin (Avula et al. 2014). A quartz crystal microbalance (QCM) sensor could detect lovastatin in RYR at nanomolar amount (0.030 nM) (Eren et al. 2015). A surface plasmon resonance (SPR) biosensor could detect citrinin even in nano-level (0.0017 ng/mL) (Atar et al. 2015). Variations of UHPLC such as UHPLC-Q-TOF-MS and UHPLC-QQQ-MS methods could identify the decalins at the range of 0.44–1.96 mg/kg (Zhu et al. 2013a). A NMR assay could also measure the statin content of RYR (Lachenmeier et al. 2012).

The efficacy or the toxicity of RYR depends on the mold strain, as well as the fermentation parameters (Ma et al. 2000). Apart from *M. purpureus*, other mold species such as *M. ruber* and *M. pilosus* are capable of elaborating monacolins. From the butanol fraction of the ethanolic extract of *M. pilosus*-fermented RYR, a sesquiterpene monaspilosuslin, along with other metabolites, has been isolated (Cheng et al. 2010). Hence, these and other oligoketide-synthesizing RYR-forming strains ought to be screened and assessed for monacolin. To attenuate the production of citrinin, metabolic engineering is suggested to be a promising approach. *Monascus purpureus* SM001 genome was genetically engineered to generate citrinin-free mutant strain while keeping its monacolin production ability unaffected (Jia et al. 2010). Further work in this direction can facilitate superior quality and economically-viable RYR. Higher amount of monacolin K can be recovered by solvent extraction coupled with countercurrent chromatography (CCC) (Liu et al. 2010). In the above experiment, 51.2 mg monacolin K was obtained from 300 mg RYR extract, where the product was 98.7% pure (Liu et al. 2010). Vacuum packaging is a packaging technique that relies on the removal of air while placing the item in the package and sealing. Using this method to package RYR powder at 4 °C temperature enhanced the stability of monacolin K (Jirasatid et al. 2013). The statistical optimization of ultrasound-assisted extraction improved the stability of monacolins (Zhou et al. 2015). So, from the mold strain selection to optimizing different steps of downstream processing is vital for RYR production.

One way to compensate for the adverse effects or to complement the ameliorative properties of RYR is to blend other nutrients with it. In this regard, soy protein, almonds, dietary fibers, and plant-derived stanols and sterols have been incorporated into RYR (Jia et al. 2010). In particular, the pairing with benzylisoquinoline alkaloid berberine is known to inhibit triglyceride production (Srivastava et al. 2012; McCarty et al. 2015). Based on a clinical trial, the ingestion of RYR, enriched with chlorella, soy protein, bitter gourd (*Momordica charantia*), and licorice (*Glycyrrhiza glabra*) for a period of 12 weeks, amended cholesterol imbalance and alleviated the pathological symptoms of metabolic syndrome (Lee et al. 2012). Among different diets, Mediterranean diet is regarded for healthy lipid profile restoration. It showed compatibility with RYR, in terms of cholesterol level attenuation (Dontas et al. 2007; Altomare et al. 2013; Sartore et al. 2013; Del Chierico et al. 2014; Castro-Quezada et al. 2014). A cohort study found that this combination is beneficial for dyslipidemia and type 2 diabetes patients with intolerance for statins (Sartore et al. 2013). Based on the result of cohort study, the concomitant administration of monacolin (10.82 mg) and a phenylethanoid hydroxytyrosol (9.32 mg) dosage given for 8 weeks reduced blood pressure, LDL cholesterol, and blood glucose level (Verhoeven et al. 2015).

Despite the variations in organs or systems affected, all the health issues have one origin, i.e., inflammation. It is the result of encountering stressors, extracellular acidosis, and the consequent agitated neural-endocrine-immune system (Tracy 2006). However nutrient-dense a supplement might be, it cannot negate the inflammasomes, if the stressors (such as pathogen, allergen, pollution, chemical exposure) are not eliminated. Further, any health benefits of drug-like compounds occur via the manipulation of cytochrome enzymes such as aromatase, which in turn regulate hormones. Meddling with the endocrine system is always risky, as the hormones are the messengers, the conveyors of signals. Further, the hypothalamic-pituitary-adrenal (HPA) and renin-angiotensin-aldosterone system (RAAS) overdrive result in high cortisol (a stress hormone), high angiotensin (a vasoactive peptide) and high blood pressure (Kubzansky and Adler 2010; Pacurari et al. 2014; Güder et al. 2015). So, the effect of RYR on these pivotal pathways needs to be studied. Also, the beneficial effects of a nutraceutical are dependent on age, gender, comorbidities, and a myriad of other host factors. This rule applies to the efficacy and safety of RYR too. It might be beneficial to one patient, but may not be responded to or may be perilous in another patient. Hence, these aspects must be evaluated prior to the recommendation of the RYR-based supplements or drugs. Large-scale epidemiological and consumer response studies should be conducted in this regard.

38.6 Conclusion

After going through the literature on RYR, it is clear that this mold-fermented product has health benefits but risks as well. The ameliorative effects span beyond antilipemic, encompassing anticancer, antiglycemic, and osteoprotective as well. The threats arise from the inconsistency of monacolin content, its instability on prolonged storage, co-occurring mycotoxins, interaction with chemotherapeutics, and suppressive effects on the immune system, among others. This chapter can be an interesting reference point to address the caveats in order to develop RYR as a benign dietary supplement.

References

Abebe W (2002) Herbal medication: potential for adverse interactions with analgesic drugs. J Clin Pharm Ther 27:391–401

Abebe W (2003) An overview of herbal supplement utilization with particular emphasis on possible interactions with dental drugs and oral manifestations. J Dent Hyg 77:37–46

Abdelbaset M, Safar MM, Mahmoud SS et al (2014) Red yeast rice and coenzyme Q10 as safe alternatives to surmount atorvastatin-induced myopathy in hyperlipidemic rats. Can J Physiol Pharmacol 92:481–489. https://doi.org/10.1139/cjpp-2013-0430

Affuso F (2012) A nutraceutical combination improves insulin sensitivity in patients with metabolic syndrome. World J Cardiol 4:77. https://doi.org/10.4330/wjc.v4.i3.77

Altomare R, Cacciabaudo F, Damiano G et al (2013) The Mediterranean diet: a history of health. Iran J Public Health 42:449–457

Atar N, Eren T, Yola ML (2015) A molecular imprinted SPR biosensor for sensitive determination of citrinin in red yeast rice. Food Chem 184:7–11. https://doi.org/10.1016/j.foodchem.2015.03.065

Avula B, Cohen PA, Wang Y-H et al (2014) Chemical profiling and quantification of monacolins and citrinin in red yeast rice commercial raw materials and dietary supplements using

liquid chromatography-accurate QToF mass spectrometry: chemometrics application. J Pharm Biomed Anal 100:243–253. https://doi.org/10.1016/j.jpba.2014.07.039
Bogsrud MP, Ose L, Langslet G et al (2010) HypoCol (red yeast rice) lowers plasma cholesterol – a randomized placebo controlled study. Scand Cardiovasc J 44:197–200. https://doi.org/10.3109/14017431003624123
Castro-Quezada I, Román-Viñas B, Serra-Majem L (2014) The Mediterranean diet and nutritional adequacy: a review. Nutrients 6:231–248. https://doi.org/10.3390/nu6010231
Chauvin B, Drouot S, Barrail-Tran A, Taburet A-M (2013) Drug-drug interactions between HMG-CoA reductase inhibitors (statins) and antiviral protease inhibitors. Clin Pharmacokinet 52:815–831. https://doi.org/10.1007/s40262-013-0075-4
Chen R-J, Hung C-M, Chen Y-L et al (2012) Monascuspiloin induces apoptosis and autophagic cell death in human prostate cancer cells via the Akt and AMPK signaling pathways. J Agric Food Chem 60:7185–7193. https://doi.org/10.1021/jf3016927
Chen H-HS, Neher J, Safranek S (2015) Clinical inquiry: is red-yeast rice a safe and effective alternative to statins? J Fam Pract 64:128–135
Cheng M-J, Wu M-D, Chen I-S, Yuan G-F (2010) A new sesquiterpene isolated from the extracts of the fungus Monascus pilosus-fermented rice. Nat Prod Res 24:750–758. https://doi.org/10.1080/14786410903178533
Childress L, Gay A, Zargar A, Ito MK (2013) Review of red yeast rice content and current Food and Drug Administration oversight. J Clin Lipidol, 7:117–122. https://doi.org/10.1016/j.jacl.2012.09.003
Chiu H-W, Fang W-H, Chen Y-L et al (2012) Monascuspiloin enhances the radiation sensitivity of human prostate cancer cells by stimulating endoplasmic reticulum stress and inducing autophagy. PLoS One 7:e40462. https://doi.org/10.1371/journal.pone.0040462
Cho Y-E, Alcantara E, Kumaran S et al (2010) Red yeast rice stimulates osteoblast proliferation and increases alkaline phosphatase activity in MC3T3-E1 cells. Nutr Res 30:501–510. https://doi.org/10.1016/j.nutres.2010.06.011
Del Chierico F, Vernocchi P, Dallapiccola B, Putignani L (2014) Mediterranean diet and health: food effects on gut microbiota and disease control. Int J Mol Sci 15:11678–11699. https://doi.org/10.3390/ijms150711678
DeBose-Boyd RA (2008) Feedback regulation of cholesterol synthesis: sterol-accelerated ubiquitination and degradation of HMG CoA reductase. Cell Res 18:609–621. https://doi.org/10.1038/cr.2008.61
Derosa G, Bonaventura A, Bianchi L et al (2014) A randomized, placebo-controlled study on the effects of a nutraceutical combination of red yeast rice, *silybum marianum* and octasonol on lipid profile, endothelial and inflammatory parameters. J Biol Regul Homeost Agents 28:317–324
Ding M, Si D, Zhang W et al (2014) Red yeast rice repairs kidney damage and reduces inflammatory transcription factors in rat models of hyperlipidemia. Exp Ther Med 8:1737–1744. https://doi.org/10.3892/etm.2014.2035
Dill JA, Fuciarelli AF, Lee KM et al (2003) Single administration toxicokinetic studies of decalin (decahydronaphthalene) in rats and mice. Toxicol Sci 72:210–222. https://doi.org/10.1093/toxsci/kfg027
Dontas AS, Zerefos NS, Panagiotakos DB et al (2007) Mediterranean diet and prevention of coronary heart disease in the elderly. Clin Interv Aging 2:109–115
Eren T, Atar N, Yola ML, Karimi-Maleh H (2015) A sensitive molecularly imprinted polymer based quartz crystal microbalance nanosensor for selective determination of lovastatin in red yeast rice. Food Chem 185:430–436. https://doi.org/10.1016/j.foodchem.2015.03.153
Feuerstein JS, Bjerke WS (2012) Powdered red yeast rice and plant stanols and sterols to lower cholesterol. J Diet Suppl 9:110–115. https://doi.org/10.3109/19390211.2012.682645
Fishman JA (2013) Opportunistic infections--coming to the limits of immunosuppression? Cold Spring Harb Perspect Med 3:a015669. https://doi.org/10.1101/cshperspect.a015669

Flajs D, Peraica M (2009) Toxicological properties of citrinin. Arch Ind Hyg Toxicol 60:457–464. https://doi.org/10.2478/10004-1254-60-2009-1992
Fung WT, Subramaniam G, Lee J et al (2012) Assessment of extracts from red yeast rice for herb-drug interaction by in-vitro and in-vivo assays. Sci Rep 2:298. https://doi.org/10.1038/srep00298
Gayathri L, Dhivya R, Dhanasekaran D et al (2015) Hepatotoxic effect of ochratoxin A and citrinin, alone and in combination, and protective effect of vitamin E: in vitro study in HepG2 cell. Food Chem Toxicol 83:151–163. https://doi.org/10.1016/j.fct.2015.06.009
Gerards MC, Terlou RJ, Yu H et al (2015) Traditional Chinese lipid-lowering agent red yeast rice results in significant LDL reduction but safety is uncertain – a systematic review and meta-analysis. Atherosclerosis 240:415–423. https://doi.org/10.1016/j.atherosclerosis.2015.04.004
Girard L, Vohra S (2011) Ethics of using herbal medicine as primary or adjunct treatment and issues of drug-herb interaction. In: Benzie IFF, Wachtel-Galor S (eds) Herbal Medicine: biomolecular and Clinical Aspects. 2nd edition. Boca Raton (FL): CRC Press/Taylor & Francis
Golomb BA, Evans MA (2008) Statin adverse effects: a review of the literature and evidence for a mitochondrial mechanism. Am J Cardiovasc Drugs 8:373–418
Gordon RY, Cooperman T, Obermeyer W, Becker DJ (2010) Marked variability of monacolin levels in commercial red yeast rice products: buyer beware! Arch Intern Med 170:1722–1727. https://doi.org/10.1001/archinternmed.2010.382
Güder G, Hammer F, Deutschbein T et al (2015) Prognostic value of aldosterone and cortisol in patients hospitalized for acutely decompensated chronic heart failure with and without mineralocorticoid receptor antagonism. J Card Fail 21:208–216. https://doi.org/10.1016/j.cardfail.2014.12.011
Hafner-Blumenstiel V (2011) Herbal drug-drug interaction and adverse drug reactions. Ther Umschau Rev thérapeutique 68:54–57. https://doi.org/10.1024/0040-5930/a000120
Hong MY, Seeram NP, Zhang Y, Heber D (2008) Anticancer effects of Chinese red yeast rice versus monacolin K alone on colon cancer cells. J Nutr Biochem 19:448–458. https://doi.org/10.1016/j.jnutbio.2007.05.012
Hong MY, Henning S, Moro A et al (2011) Chinese red yeast rice inhibition of prostate tumor growth in SCID mice. Cancer Prev Res (Phila) 4:608–615. https://doi.org/10.1158/1940-6207.CAPR-10-0219
Hsieh C-W, Lu Y-R, Lin S-M et al (2013) Stability of monacolin K and citrinin and biochemical characterization of red-koji vinegar during fermentation. J Agric Food Chem 61:7276–7283. https://doi.org/10.1021/jf401542q
Istvan E (2003) Statin inhibition of HMG-CoA reductase: a 3-dimensional view. Atheroscler Suppl 4:3–8
James PA, Oparil S, Carter BL et al (2014) 2014 evidence-based guideline for the management of high blood pressure in adults: report from the panel members appointed to the Eighth Joint National Committee (JNC 8). JAMA 311:507–520. https://doi.org/10.1001/jama.2013.284427
Jernberg E, Thysell E, Bovinder Ylitalo E et al (2013) Characterization of prostate cancer bone metastases according to expression levels of steroidogenic enzymes and androgen receptor splice variants. PLoS One 8:e77407. https://doi.org/10.1371/journal.pone.0077407
Jia XQ, Xu ZN, Zhou LP, Sung CK (2010) Elimination of the mycotoxin citrinin production in the industrial important strain Monascus purpureus SM001. Metab Eng 12:1–7. https://doi.org/10.1016/j.ymben.2009.08.003
Jirasatid S, Nopharatana M, Kitsubun P, Tongta A (2013) Degradation kinetics of monacolin K in red yeast rice powder using multiresponse modeling approach. J Food Eng 116:436–443. https://doi.org/10.1016/j.jfoodeng.2012.12.018
Karl M, Rubenstein M, Rudnick C, Brejda JA (2012) multicenter study of nutraceutical drinks for cholesterol (evaluating effectiveness and tolerability). J Clin Lipidol 6:150–158. https://doi.org/10.1016/j.jacl.2011.09.004

Khayznikov M, Hemachrandra K, Pandit R et al (2015) Statin intolerance because of myalgia, myositis, myopathy, or myonecrosis can in most cases be safely resolved by vitamin D supplementation. N Am J Med Sci 7:86–93. https://doi.org/10.4103/1947-2714.153919

Klimek M, Wang S, Ogunkanmi A (2009) Safety and efficacy of red yeast rice (Monascus purpureus) as an alternative therapy for hyperlipidemia. P T 34:313–327

Kubzansky LD, Adler GK (2010) Aldosterone: a forgotten mediator of the relationship between psychological stress and heart disease. Neurosci Biobehav Rev 34:80–86. https://doi.org/10.1016/j.neubiorev.2009.07.005

Lachenmeier DW, Monakhova YB, Kuballa T et al (2012) NMR evaluation of total statin content and HMG-CoA reductase inhibition in red yeast rice (Monascus spp.) food supplements. Chin Med 7:8. https://doi.org/10.1186/1749-8546-7-8

Lee I-T, Lee W-J, Tsai C-M et al (2012) Combined extractives of red yeast rice, bitter gourd, chlorella, soy protein, and licorice improve total cholesterol, low-density lipoprotein cholesterol, and triglyceride in subjects with metabolic syndrome. Nutr Res 32:85–92. https://doi.org/10.1016/j.nutres.2011.12.011

Lee C-I, Shih C-D, Lee C-L et al (2013) Effect of red mold rice on blood coagulation and anticoagulation factors in a rat model of hyperlipidemia. J Funct Foods 5:1956–1965. https://doi.org/10.1016/j.jff.2013.09.017

Li Z, Seeram NP, Lee R et al (2005) Plasma clearance of lovastatin versus Chinese red yeast rice in healthy volunteers. J Altern Complement Med 11:1031–1038. https://doi.org/10.1089/acm.2005.11.1031

Li J-J, Shang X-Y, Li L-L et al (2010) New cytotoxic azaphilones from Monascus purpureus-fermented rice (red yeast rice). Molecules 15:1958–1966. https://doi.org/10.3390/molecules15031958

Li P, Yang Y, Liu M (2011) Xuezhikang, extract of red yeast rice, inhibited tissue factor and hypercoagulable state through suppressing nicotinamide adenine dinucleotide phosphate oxidase and extracellular signal-regulated kinase activation. J Cardiovasc Pharmacol 58:307–318. https://doi.org/10.1097/FJC.0b013e3182244a2d

Liu J, Zhang J, Shi Y et al (2006) Chinese red yeast rice (Monascus purpureus) for primary hyperlipidemia: a meta-analysis of randomized controlled trials. Chin Med 1:4. https://doi.org/10.1186/1749-8546-1-4

Liu Y, Guo X, Duan W et al (2010) Accelerated solvent extraction of monacolin K from red yeast rice and purification by high-speed counter-current chromatography. J Chromatogr B Anal Technol Biomed Life Sci 878:2881–2885. https://doi.org/10.1016/j.jchromb.2010.08.028

Ma J, Li Y, Ye Q et al (2000) Constituents of red yeast rice, a traditional Chinese food and medicine. J Agric Food Chem 48:5220–5225

Marazzi G, Cacciotti L, Pelliccia F et al (2011) Long-term effects of nutraceuticals (berberine, red yeast rice, policosanol) in elderly hypercholesterolemic patients. Adv Ther 28:1105–1113. https://doi.org/10.1007/s12325-011-0082-5

McCarty MF, O'Keefe JH, DiNicolantonio JJ (2015) Red yeast rice plus berberine: practical strategy for promoting vascular and metabolic health. Altern Ther Health Med 21(Suppl 2):40–45

Moriarty PM, Roth EM, Karns A, et al (2014) Effects of Xuezhikang in patients with dyslipidemia: a multicenter, randomized, placebo-controlled study. J Clin Lipidol 8:568–575. https://doi.org/10.1016/j.jacl.2014.09.002

Mornar A, Sertić M, Nigović B (2013) Development of a rapid LC/DAD/FLD/MS n method for the simultaneous determination of monacolins and citrinin in red fermented rice products. J Agric Food Chem 61:1072–1080. https://doi.org/10.1021/jf304881g

Nigović B, Sertić M, Mornar A (2013) Simultaneous determination of lovastatin and citrinin in red yeast rice supplements by micellar electrokinetic capillary chromatography. Food Chem 138:531–538. https://doi.org/10.1016/j.foodchem.2012.10.104

Ogier N, Amiot M-J, Georgé S et al (2013) LDL-cholesterol-lowering effect of a dietary supplement with plant extracts in subjects with moderate hypercholesterolemia. Eur J Nutr 52:547–557. https://doi.org/10.1007/s00394-012-0357-x

Pacurari M, Kafoury R, Tchounwou PB, Ndebele K (2014) The renin-angiotensin-aldosterone system in vascular inflammation and remodeling. Int J Inf Secur 2014:689360. https://doi.org/10.1155/2014/689360

Patel S (2015) A promising CAM therapeutic for multiple cancers: milk thistle (Silybum). In: Emerging bioresources with nutraceutical and pharmaceutical prospects. Springer, Cham, pp 101–114

Patel S (2016) Functional food red yeast rice (RYR) for metabolic syndrome amelioration: a review on pros and cons. World J Microbiol Biotechnol 32:87. https://doi.org/10.1007/s11274-016-2035-2

Park H-J, Kim I-S (2011) Antioxidant activities and anticancer effects of red yeast rice grown in the medium containing garlic. Food Sci Biotechnol 20:297–302. https://doi.org/10.1007/s10068-011-0042-5

Patakova P (2013) Monascus secondary metabolites: production and biological activity. J Ind Microbiol Biotechnol 40:169–181. https://doi.org/10.1007/s10295-012-1216-8

Pirillo A, Catapano AL (2015) Statin intolerance: diagnosis and remedies. Curr Cardiol Rep 17:27. https://doi.org/10.1007/s11886-015-0582-z

Sartore G, Giovanni S, Burlina S et al (2013) Mediterranean diet and red yeast rice supplementation for the management of hyperlipidemia in statin-intolerant patients with or without type 2 diabetes. Evid Based Complement Alternat Med 2013:743473. https://doi.org/10.1155/2013/743473

Sham T-T, Chan C-O, Wang Y-H et al (2014) A review on the traditional Chinese medicinal herbs and formulae with hypolipidemic effect. Biomed Res Int 2014:925302. https://doi.org/10.1155/2014/925302

Shang Q, Liu Z, Chen K et al (2012) A systematic review of xuezhikang, an extract from red yeast rice, for coronary heart disease complicated by dyslipidemia. Evid Based Complement Alternat Med 2012:636547. https://doi.org/10.1155/2012/636547

Srivastava RAK, Pinkosky SL, Filippov S et al (2012) AMP-activated protein kinase: an emerging drug target to regulate imbalances in lipid and carbohydrate metabolism to treat cardiometabolic diseases: thematic review series: new lipid and lipoprotein targets for the treatment of cardiometabolic diseases. J Lipid Res 53:2490–2514. https://doi.org/10.1194/jlr.R025882

Tracy RP (2006) The five cardinal signs of inflammation: calor, dolor, rubor, tumor ... and penuria (apologies to aulus cornelius celsus, de medicina, c. A.D. 25). J Gerontol A Biol Sci Med Sci 61:1051–1052

Trimarco B, Benvenuti C, Rozza F et al (2011) Clinical evidence of efficacy of red yeast rice and berberine in a large controlled study versus diet. Med J Nutrition Metab 4:133–139. https://doi.org/10.1007/s12349-010-0043-6

Trimarco V, Cimmino CS, Santoro M et al (2012) Nutraceuticals for blood pressure control in patients with high-normal or grade 1 hypertension. High Blood Press Cardiovasc Prev 19:117–122. https://doi.org/10.2165/11632160-000000000-00000

Venero CV, Venero JV, Wortham DC, Thompson PD (2010) Lipid-lowering efficacy of red yeast rice in a population intolerant to statins. Am J Cardiol 105:664–666. https://doi.org/10.1016/j.amjcard.2009.10.045

Verhoeven V, Van der Auwera A, Van Gaal L et al (2015) Can red yeast rice and olive extract improve lipid profile and cardiovascular risk in metabolic syndrome?: a double blind, placebo controlled randomized trial. BMC Complement Altern Med 15:52. https://doi.org/10.1186/s12906-015-0576-9

Wang J, Jiang W, Zhong Y et al (2014) Xuezhikang attenuated the functional and morphological impairment of pancreatic islets in diabetic mice via the inhibition of oxidative stress. J Cardiovasc Pharmacol 63:282–289. https://doi.org/10.1097/FJC.0000000000000047

Wang YF, Liu WT, Chen CY et al (2015) Anti-osteoporosis activity of red yeast rice extract on ovariectomy-induced bone loss in rats. Genet Mol Res 14:8137–8146. https://doi.org/10.4238/2015.July.27.2

Wei Y, Popovich DG (2013) Red azaphilone pigments extracted from red yeast rice induces cellular senescence and reduces viability in HepG2 cells. Biomed Prev Nutr 3:331–337. https://doi.org/10.1016/j.bionut.2013.08.003

Wong RWK, Rabie B (2008) Chinese red yeast rice (Monascus purpureus-fermented rice) promotes bone formation. Chin Med 3:4. https://doi.org/10.1186/1749-8546-3-4

Wu M-D, Cheng M-J, Yech Y-J et al (2013) Monascusazaphilones A-C, three new azaphilone analogues isolated from the fungus Monascus purpureus BCRC 38108. Nat Prod Res 27:1145–1152. https://doi.org/10.1080/14786419.2012.715289

Xie X, Wang Y, Zhang S et al (2012) Chinese red yeast rice attenuates the development of angiotensin II-induced abdominal aortic aneurysm and atherosclerosis. J Nutr Biochem 23:549–556. https://doi.org/10.1016/j.jnutbio.2011.02.011

Xiong X, Wang P, Li X et al (2015) The effects of red yeast Rice dietary supplement on blood pressure, lipid profile and C-reactive protein in hypertension: a systematic review. Crit Rev Food Sci Nutr. https://doi.org/10.1080/10408398.2015.1018987

Yang N-C, Chou C-W, Chen C-Y et al (2009) Combined nattokinase with red yeast rice but not nattokinase alone has potent effects on blood lipids in human subjects with hyperlipidemia. Asia Pac J Clin Nutr 18:310–317

Zanardi M, Quirico E, Benvenuti C, Pezzana A (2012) Use of a lipid-lowering food supplement in patients on hormone therapy following breast cancer. Minerva Ginecol 64:431–435

Zhou G, Fu L, Li X (2015) Optimisation of ultrasound-assisted extraction conditions for maximal recovery of active monacolins and removal of toxic citrinin from red yeast rice by a full factorial design coupled with response surface methodology. Food Chem 170:186–192. https://doi.org/10.1016/j.foodchem.2014.08.080

Zhu L, Lu J-G, Li T et al (2012a) Immunosuppressive decalin derivatives from red yeast rice. J Nat Prod 75:567–571. https://doi.org/10.1021/np2007236

Zhu L, Yau L-F, Lu J-G et al (2012b) Cytotoxic dehydromonacolins from red yeast rice. J Agric Food Chem 60:934–939. https://doi.org/10.1021/jf203579f

Zhu L, Han Q-B, Ho A et al (2013a) Characterization and simultaneous determination of immunosuppressive decalins in red yeast rice by ultra-high-performance liquid chromatography hyphenated with mass spectrometry. J Chromatogr A 1303:54–61. https://doi.org/10.1016/j.chroma.2013.06.045

Zhu X-Y, Li P, Yang Y-B, Liu M-L (2013b) Xuezhikang, extract of red yeast rice, improved abnormal hemorheology, suppressed caveolin-1 and increased eNOS expression in atherosclerotic rats. PLoS One 8:e62731. https://doi.org/10.1371/journal.pone.0062731

Chapter 39
Nanoparticles from Fungi (Myconanoparticles)

G. Subashini and S. Bhuvaneswari

Abstract Myconanoparticles are solid colloidal metal particles produced from fungi in the span range approximately from 1 nm to 100 nm in dimensions and form more structure fabricating pieces of nanotechnology. Metal nanoparticles like gold, silver and platinum need picked up respectable consideration in current era due to their basic and technical enthusiasm. The nanoparticles have exclusive catalytic, electronic and visual characters discrete from the metal nanoparticles. In current era, several techniques have been intended to make nanoparticles by physical, chemical and biological methods. The chemical process utilizes chemical agents such as sodium borohydride, sodium citrate and alcohols. The physical process uses physical agents as UV rays, gamma rays, etc. Yet the biological methods utilize biological agents to portray a low cost and an environmental protective way for creation of metal nanoparticles. Nowadays, the nanoparticles synthesized by biological organisms include plants, bacteria, yeast and fungi. Among these microorganism, fungus is a well-organized system for the production of nanoparticles. Myconanotechnology is a budding field, where fungi can be used for synthesis of nanoparticles with attractive shape. The mycosynthesis of nanoparticles has monodispersity, magnitude and constancy. Furthermore, this method is ecological and inexpensively viable for the synthesis of nanoparticles. Mycosynthesized nanoparticles discovered its limitless provision for agriculture, environment, medicine and food preservation and material fabrics. This chapter highlighted on production of nanoparticles from fungi and its application to various fields in biology.

Keywords Myconanoparticles · Mycosynthesis · Fungi · Green nanotechnology

---

G. Subashini (✉) · S. Bhuvaneswari
Shrimati Indira Gandhi College, Trichy-2, India

P. Gehlot, J. Singh (eds.), *Fungi and their Role in Sustainable Development: Current Perspectives*, https://doi.org/10.1007/978-981-13-0393-7_39

39.1 Introduction

Nanotechnology is a field that is burgeoning day by day making an importance in all of human life. Nanotechnology is rising as fast-developing sciences have wide application in science and technology to creating novel material at nanolevel. Nanomaterials have lost list of applicability in improving human life and its environment. Nanotechnology is a vital area in current research dealing with make-up and synthesis particles which is about 1–100 nm in single dimension. The term nano is utilized to demonstrate one billionth of a metre or 10^{-9}. The word nano is derived from Greek word 'nanos' synonymous to dwarf meaning extremely minute. The term nanotechnology had been instituted by Taniguchi in 1974 for the first time to delineate exactness assembling of molecules at the nanoscale dimension.

Nanotechnology has a fast-growing significance in all areas such as healthcare, food industry, beauty care, environment, energy sciences, medical sciences, reography and photoelectrochemical applications.

39.2 Green Nanotechnology

'Green nanotechnology' is defined as the environmental friendly developed process that diminishes waste product which ultimately leads to atomically exact molecular manufacturing with nil waste. Green nanotechnology is of prime importance because chemical synthesis is toxic and leads to by-products that are not environmentally benign. In green nanotechnology, the nanoparticles are synthesized by various plant extracts and microbes. Green nanotechnology is the combination of green chemistry and green engineering with biological ideology used to fabricate nanosized products with particular purpose. Green nanotechnology also represents a cost-effective alternate for physical and chemical methods for nanoparticle synthesis. Presently the development of a green methodology for union of nanoparticles and its application to reduce the human and environmental risks are imperative aspects of nanotechnology. This is an especially squeezing requirement for new and effective strategy for the blend of nanoparticles.

The chemical process has lots of disadvantage such as:

- Utilization of toxic ingredients
- Production of dangerous by-products
- Use of high energy
- Risk to environment and human health

Likewise, the physical techniques, such as sputter affidavit, laser removal or microwave-helped combination, are used for union of metal nanoparticles. But it contributes high temperature, high radiation and is pressure-associated (Alzaharani

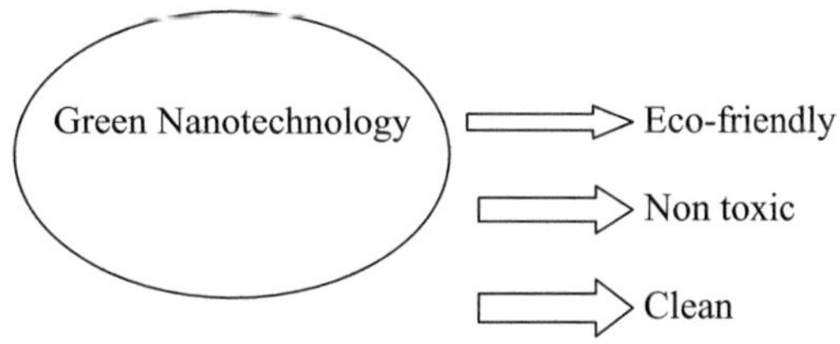

Fig. 39.1 Advantage of green nanotechnology

et al. 2015).Biological methods for nanoparticles using microorganisms have been suggested as possible eco-friendly alternates for physical and chemical methods. Biosynthesis of nanoparticles by microbes is simple, rapid and eco-friendly protocol. In green fabrication process, the nanoparticles are synthesized by microorganisms. Numerous microorganisms precipitate inorganic molecules within or outside the cell to form nanoparticles. Green technology is a combination of nanotechnology with microbial process for the manufacturing of nanoproducts (Fig. 39.1).

39.3 Myconanotechnology

Myconanotechnology refers to interconnecting nanotechnology and mycology. 'Myco' means fungi; nanotechnology is creation and exploitation of materials in size range of 1–100 nm. The mycogenic synthesis of nanoparticles is the main feature of myconanotechnology leading to an exciting novel and sensible emerging science with extensive potential owing to a wide variety of diversity in fungi. Metal nanoparticles synthesized from fungi are defined as mycogenic nanoparticles. This fungal organism developed as 'bionanofactories' incorporated nanoparticles such as silver, platinum, gold and cadmium particles.

The term 'myconanotechnology' is described about on amalgamation of nanoparticles utilizing fungi. Therefore, mycogenic nanotechnology is an incorporation of nanotechnology and mycology. Among these microbes, fungi are presented ubiquitous in nature; hence it is used in union of metal nanoparticles. The improvement of clean, non-poisonous, eco-friendly and a biocompatible technique for the combination of nanoparticles is required. The utilization of fungi for the union of nanoparticles is picking up stimulus because of the easy amalgamation of nanoparticles; in addition, the rate of accomplishment was considerably elevated (Adil et al.2015).

39.4 Myconanoparticles

Nanoparticles synthesized from fungi are called as 'myconanoparticles'. In the past, fungi have been consumed as a source of food and harnessed to ferment and conserve food and beverages. In the twentieth century, fungi are used to guard the

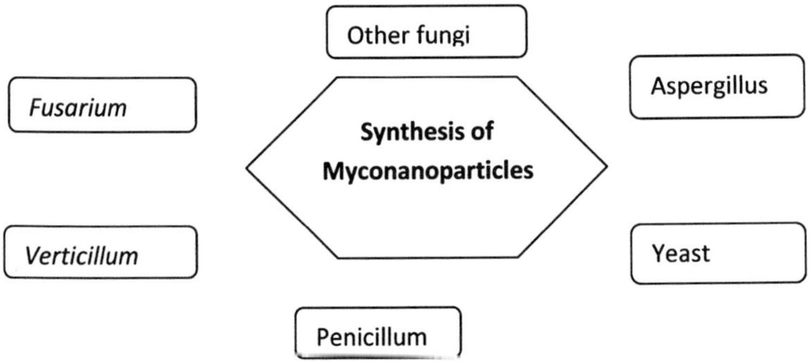

Fig. 39.2 Fungi used for synthesis of nanoparticles

human health by producing antibiotics and anticholesterol agents and used for the large-scale production of enzymes, proteins and acids in industry. In present day nanotechnology growth has lived on primarily by giving a greener other option to artificially integrated nanoparticle.

Fungi are heterotrophic, non-phototrophic and eukaryotic microbes consisting of an inflexible cell wall (Bowman and Free 2011; Duran and Nombela 2004).They have easy nutritional needs being chemoorganotrophs (Holan and Volesky 1995). Most of these fungi grow on dead organic matter. A few fungi grow as parasites on other organisms. Fungi normally nourish by secreting enzymes that assimilate food digested internally. Fungi act as 'nanofactories' for producing nanoparticles. Fungi can be used as outstanding resource of various extracellular synthesizing enzymes which control nanoparticles production (Fig. 39.2).

39.4.1 Fungi as Proficient Microorganism for Synthesizing Myconanoparticles

- Fungi produce large amount of extracellular enzymes and proteins, which catalyze the heavy metal ions producing nanoparticles. Organisms emit a lot of extracellular proteins with various capacities. The purported secretome incorporates the majority of emitted proteins into extracellular and it is utilized for modern creation of homologous and heterologous proteins.
- Fungi grow on readily existing and cheap substrates and have the capacity to produce a broad range of metabolites.
- They produce nanoparticles more rapidly than chemical synthesis.
- Fungi are easy to cultivate and subculture.
- Fungi have simple nutrition required for growth.

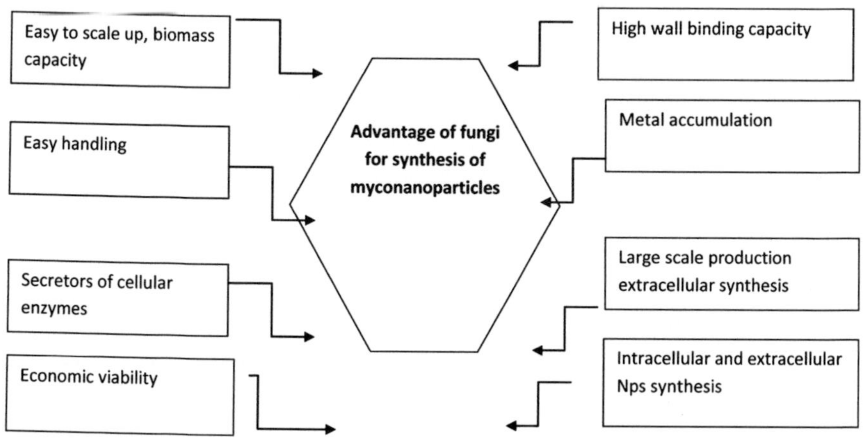

Fig. 39.3 Advantages of fungi for the synthesis of myconanoparticles among other microbes

- Fungi are simple to handle in laboratory and they require simple methods are needed for isolation.
- Nanoparticles are produced outside the cell (extracellular) which is suitable for easier downstream processing and handling biomass.
- Enzymes produced from fungi can be used to manufacture particles with defined size and shape.
- Fungal mycelia net can resist current pressure and agitation and various conditions within fermenters or other chambers.
- The majority fungi have high tolerance to metals and elevated wall binding capacity as well as ability to uptake metal intracellularly.
- Cost-effective feasibility for large-scale synthesis using a little quantity of biomass.
- Nanoparticles precipitated exterior of the cell and they can be directly used for diverse application. Therefore fungi could be considered as natural 'biofactories' for biosynthesis of nanoparticles.
- Fungi are used chiefly in the production of nanoparticles and have higher metal tolerance and ability to gather the metals.
- Fungi have the capacity to synthesize nanoparticles and nanoproducts by reducing enzymes both intracellularly and extracellularly.
- Fungi are potent candidate to examine producing nanoparticles and their use in disease detection and control.
- Fungi offer better manipulation and control over crystal growth due to its slower kinetics (Vaidyanathan et al. 2009) (Fig. 39.3).

39.4.2 Fungi in the Creation of Nanoparticles (Table 39.1)

Table 39.1 Number of fungi used in synthesis of myconanoparticles

Nanoparticles	Fungi	Mode of synthesis	Size (nm)	Shape	Application
Ag	*Alternaria alternata*	Extracellular	20–60	Spherical	Antifungal
	Aspergillus flavus	Intracellular	8.92-1.62 nm	ND	Antibacterial
	Aspergillus fumigatus	Extracellular	5–25	Spherical, triangular	ND (not determined)
	Aspergillus niger	Extracellular	15–20	ND	Antifungal
	Aspergillus terreus	Extracellular	1–20	Spherical	Antifungal
	Cladosporium cladosporioides	Extracellular	10–100 nm	Spherical rods and triangular	ND
	Fusarium acuminatum	Extracellular	5–40 nm	Spherical, triangular	Antibacterial
	Novosphingobium oryzae	Intra-/extracellular	30–90	Spherical	Antibacterial
	Penicillium sp.	Extracellular	16–40	Multishaped	Antibacterial
	Phoma glomerata	Extracellular	60–80	Spherical	Antibacterial
	Rhizopus stolonifer	Extracellular	5–50	Spherical	Antibacterial
	Trichoderma harzianum	Extracellular	30–50	Spherical	Vegetable and fruit preservation
	Trichoderma viride	Spherical	5–40	Spherical and rodlike	Vegetable and fruit preservation
	Verticillium sp. *AAT-TS-4*	Intracellular	25 ± 12	Spherical	ND
Au	*Fusarium oxysporum*	Extracellular	20–40	Multishaped	ND
	Helminthosporium solani	Extracellular	2–70	Rods, triangles	ND
	Fusarium semitectum	Extracellular	18–80	Multishaped	Antibacterial
	Rhizopus oryzae	Extra/intra	10	Multishaped	Antibacterial
	Penicillium sp.	Extracellular	30–50	Spherical	Antibacterial
	Sclerotium rolfsii	Extracellular	25	Triangle, decahedral and spherical	Antibacterial
	Lentinula edodes	Extracellular	5 50	Spherical	Antifungal
	Aspergillus fumigatus A. flavus	Extracellular	17.76–26	Triangles, hexagons, spherical	Antibacterial

(continued)

Table 39.1 (continued)

Nanoparticles	Fungi	Mode of synthesis	Size (nm)	Shape	Application
CdS	*F. oxysporum*	Extracellular	5–20	Spherical	ND
Magnetite	*Lactobacillus* sp.	Extracellular	2.5–5.5		ND
	Saccharomyces cerevisiae				
	Coriolus versicolor	Extracellular	100	Spherical	ND
	S. pombe	Extracellular	1-1.5	Hexagonal lattice	ND
CdSe	*F. oxysporum*	Extracellular	9-15	Spherical	ND
Se	A. *alternata*	Extracellular	30 ± 5	Spherical	Antibacterial
$SrCO_3$	*F. oxysporum*	Extracellular		Needle shaped	Antibacterial
Si	*F. oxysporum*	Extracellular	5–15	Quasi-spherical	Antibacterial
Ti	*F. oxysporum*	Extracellular	2-6	Quasi-spherical	Antibacterial
$BaTiO_3$	*F. oxysporum*	Extracellular	4–5	Quasi-spherical	Antibacterial
Bi_2O_3	*F. oxysporum*	Extracellular	5–8	Quasi-spherical	Antibacterial
Pt	*F. oxysporum*	Extracellular	20–60	Triangle	Antibacterial

39.5 Synthesis of Myconanoparticles

39.5.1 *General Methodology for Production of Myconanoparticles*

The production of myconanoparticles mainly depends upon fungi isolate involved. The culture techniques and media vary according to fungal species. Fungal hyphae are grown in liquid media and then cleansed with distilled water to remove growth media and put in distilled water and are incubated on shaker for 24–48 h. After incubation, the fungal mycelium is removed from the supernatant and then 1 mM ion solution is added to supernatant. The ion solution is then observed for 2–3 days for the development of nanoparticles.

An additional to general procedure is to add fungal hypha straight into 1 nm ion solution; it replaces the use of fungal filtrate. Silver nitrate is most common source of silver ions. Chloroauric acid is used as source of gold nanoparticle synthesis. Cadmium sulphide is used as resource of cadmium ion synthesis.

39.5.2 Mechanism of Intra- and Extracellular Synthesis of Myconanoparticles

The myconanoparticles can be synthesized both intra- and extracellularly in nanoscale dimension. In the intracellular nanoparticles, synthesis occurs within the fungal cell. In this process, metal salt solution is treated with fungal biomass and incubates for 24 h in the dark chamber. In extracellular nanoparticles, synthesis occurs on the surface of the fungal cell wall. In this process, the metal salt solution is treated with fungal filtrate for the formation of nanoparticles.

39.5.2.1 Mechanism of Intracellular Synthesis

- Intracellular synthesis of myconanoparticles by bioreduction process.
- The intracellular process was initiated with trapping of metal ions over fungal surface due to electrostatic interaction.
- The electrostatic interaction is initiated by enzymes present in the cell wall.
- Then the metal ions are condensed by enzymes within the cell wall.
- It directs the precipitation of metal ions and leads to the creation of nanoparticles.
- Intracellular synthesis of nanoparticles is small in size.
- Its downstream processing becomes more difficult.
- The cost of nanoparticles synthesized was increased in intracellular method (Fig. 39.4).

39.5.2.2 Mechanism of Extracellular Synthesis

In the extracellular synthesizing nanoparticle, the filtrate from the fungus is treated with metal solution. Three mechanisms are used for the extracellular synthesis of nanoparticles such as:

- Nitrate reductase action
- Electron shuttle quinines
- Both

In this process:

- The culture filtrates with specific metal solution.
- The metal ions are transported into cytoplasmic membrane and cytoplasm.
- The ions are reduced by the enzyme reductase.
- The enzyme is responsible of the reduction of Ag^+ into silver nanoparticles.

The enzyme nitrate reductase plays an important role in the synthesis of silver nanoparticles. This enzyme NADPH-dependent nitrate reductase was separated from *Fusarium oxysporum* used for creation of silver nanoparticles having 10–20 nm

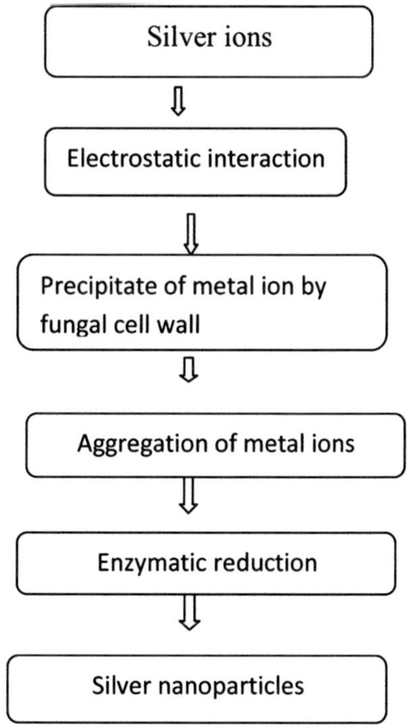

Fig. 39.4 Mechanism of intracellular synthesis of myconanoparticles

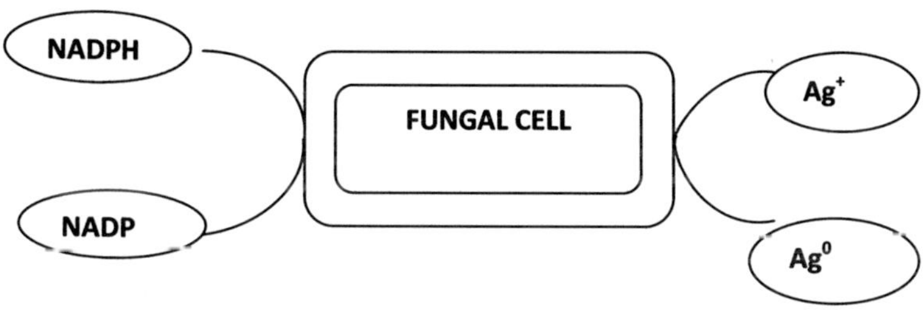

Fig. 39.5 Mechanism of extracellular synthesis of myconanoparticles

in diameter and characterized by XRD, TEM and UV-Vis absorption. The construction of silver nanoparticles needs the reduction of NADPH to $NADP^+$, and hydroxyquinoline acts as electron donors and it transports the electrons produced during reduction of nitrate to Ag^+ ions and then the electrons transfer to Ag^0. Advantages of extracellular synthesis of nanoparticles are as follows:

- Does not require broad downstream processing.
- Offers less demand.
- Financially efficient process. Thus extracellular methods of nanoparticles are mostly preferred (Fig. 39.5).

39.6 Strategies for Synthesis of Myconanoparticles

There are two processes that are used for synthesis of nanoparticles:

1. Bottom-up strategy
2. Top-down strategy

39.6.1 Top-Down Process

In top-down method, an appropriate starting material is reduced in size by physical or chemical way. This includes formation of nanosize material from massive substrate. In this process, some of fabrication techniques are used to make nanoparticles such as cutting, grinding and etching. It depends upon the nature of basal mater used for formation of nanoparticles. In top-down approach, a lot of impurities and structural defects occur in the synthesized nanoparticles (Fig. 39.7).

39.6.2 Bottom-Up Process

Bottom-up approach means from bottom to top, that is, from smaller molecules to produce larger molecules. It also refers to the building of structure of atom with atom, molecule by molecule or via self-organization. In Bottom-Up process, self or position assembly of atoms or molecules onto each other gives rise to crystal or tubes further stack onto each other, resulting in the synthesis of particles with nanodimensions. Mycosynthesis of nanoparticles is a sort of bottom-up approach. The primary response of this technique involves reduction/oxidation taking place on substrates to produce colloidal structures (Moghaddam 2010). Enzymes produced from fungi and metabolites with antioxidant and reducing properties are accountable for reduction of metal ions into individual nanoparticles, and then metal atom undergoes nucleation to produce nanoproducts (Zhang et al.2011).The creation of functional nanometre-sized objects and semiconductor QDs is a great case of bottom-up approach. The benefit of the bottom-up approach is the better potential outcomes to get nanostructures (nanorods, nanocubes, nanotubes, nanowires, nanosheets and so forth) with minor imperfections and more homogeneous chemical compositions (Thakkar et al. 2010) (Fig. 39.6).

39.7 Factors Affecting Synthesis of Myconanoparticles

The ecological conditions would control the growth and development of microorganisms. The enzyme synthesis by fungi is influenced by state in which the organisms are cultivated. Therefore, standardization studies will not only support good

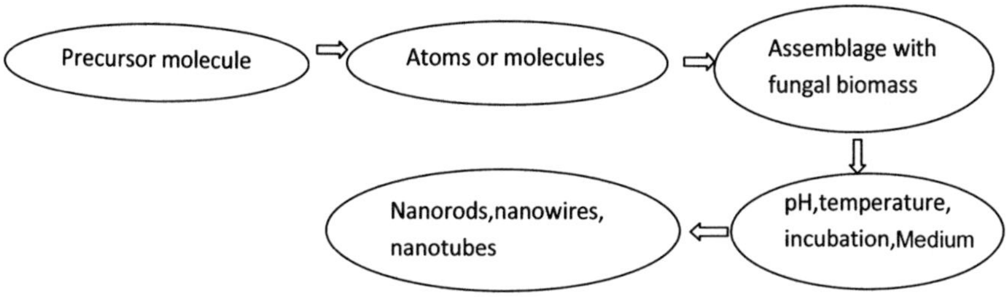

Fig. 39.6 Bottom-up strategy for synthesis of myconanoparticles

Fig. 39.7 Top-down strategy

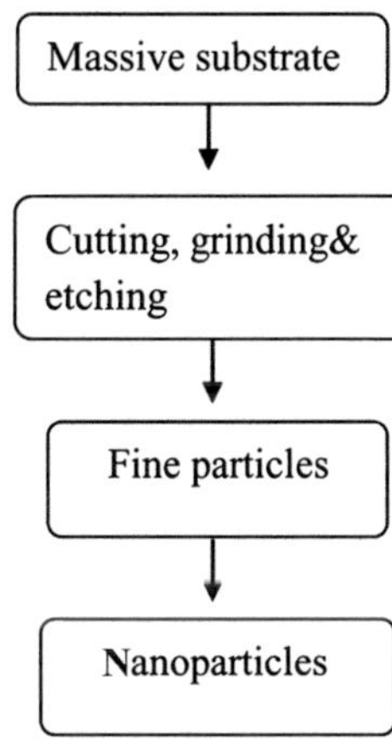

growth but also increase the product yield. Mycosynthesis of nanoparticles is influenced by several factors such as:

- Temperature
- pH
- Incubation time
- Nature of parent compound
- Metal ions
- Fungal biomass
- Colloidal association conditions that influence the size, shape and localization of nanoparticles

Silver nanoparticles produced by use of surface methodology are influenced by effect of pH, temperature, agitation rate, incubation time, silver salt concentration and weight of fungal biomass. They obtained spherical-shaped silver nanoparticles (50 nm) by plunging *F. oxysporum* in silver nitrate (3 mM; pH 6.0) solution, incubated at 25 °C with 180 rpm agitation for 96 h (Karbasian et al. 2008) (Fig. 39.8).

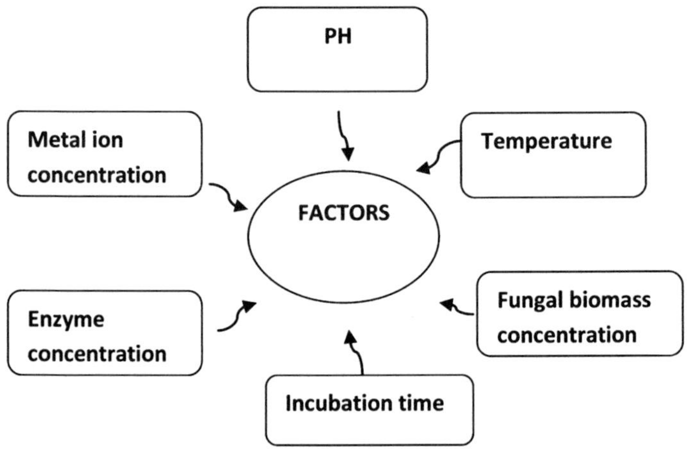

Fig. 39.8 Important factors affecting the synthesis of myconanoparticles

39.7.1 Metal Ion Concentration

The impact of metal ion concentration on the amalgamation of nanoparticles utilizing *Penicillium fellutanum* proposed that higher concentration would hamper the development of nanoparticles. The element size and monodispersity of the particles differ from desired nanosize range at elevated silver ion concentration (Kathiresan et al. 2009). The concentration of reactants influences particle size and monodispersity. For example, gold nanoparticles synthesis by *verticillium luteoalbum*, outcomes showed that when $AuCl_4$ concentration was beneath 500 mgL^{-1}, the particle was narrow and uniform. But, when the concentration was over this, the size also increased, but size fluctuated from 50 nm to several hundred nanometres.

39.7.2 pH

pH is an essential factor having prominent effect on the synthesis of nanoparticles. In order to reveal the function of pH in fabrication of nanoparticles, Gericke and Pinches (2006) showed the adjusted shape of the crystals with altering pH. At pH 3 nanoparticles created by *V. luteoalbum* were predominantly spherical in shape (\10.0 nm), and also they acquired larger nanoparticles with well-defined shapes such as hexagons, triangles, rods and spheres at pH 5.0. However, at pH 7 and 9, the particles formed included little spherical particles as well as bigger particles with unpredictable and unclear shapes. Additionally, the quantity of particles produced per cell was brought down at pH 9 than at pH 7.

39.7.3 Temperature and Incubation Time

Temperature assumes a potential part in controlling the activity of fungus and ion movement. Subsequently, it can be gathered that temperature exhibits marked effects on the growth of fungus as well as on metal uptake from the surrounding environment. Thus, the rate of development of the nanoparticles was specifically affected by the incubation temperature. Gericke and Pinches (2006) demonstrated that at lower temperatures, the larger part of circular (\10 nm)-shaped NPs framed after 1 h presentation to gold solution. Additionally expanding the incubation time frame up to 24 h leads to a reduction in the quantity of little particles, though the quantity of bigger particles was increased characterized with well-defined shape. At 50 °C, no dissimilarity was distinguished in the size and morphology of particles delivered after 1–24 h presentation to gold and not very many little circular particles were available. It is clear from this investigation that the span of the nanoparticles to a vast degree can be controlled by working at low temperature and it would be molecularly arranged at a slower rate.

39.7.4 Enzyme Concentration

The enzymes which are secreted by fungi are extremely promising for the production of predefined nanoparticles of diverse chemical composition, shape and size. Ahmad et al. (2003) noticed that the silver nanoparticle was not synthesized in the presence of *Fusarium moniliforme*, but it was formed when *F. oxysporum* was there in the silver ion solution. Protein analyses of the two fungi demonstrate that a specific reductase enzyme was formed only by *F. oxysporum* and not by *Fusarium moniliforme*. It obviously indicates that NADH-dependent reductase is used for the construction of nanoparticles.

39.8 Mycosynthesis of Different Nanoparticles

The nanoparticles are generally produced extracellularly (sometimes intracellular production does occur); they are devoid of various impurities from the cell and can be used directly. The use of eukaryotic organisms for nanoparticle synthesis was first demonstrated by the use of *Verticillium* sp. for the synthesis of gold nanoparticle. The fungi *F. oxysporum* have been the extensively used fungal species for NP synthesis.

39.8.1 Silver Nanoparticles (Ag)

Silver nanoparticles are synthesized by using many fungal species including *Trichoderma*, *Rhizoctonia*, *Pleurotus* and *Aspergillus*. Extracellular synthesis of Ag are by *T. viride*, *T. reesei*, *F. oxysporum*, *F. semitectum*, *F. solani*, *Aspergillus niger*, *A. flavus*, *A. fumigatus*, *Pleurotus ostreatus*, *Cladosporium cladosporioides*, *Penicillium brevicompactum*, *P. fellutanum*, *Rhizoctonia* sp., *Epicoccum nigrum* and *Phoma glomerata*. While intracellular union appeared by *Verticillium* species and in *Neurospora crassa*.

The mechanism of making silver nanoparticles by fungi is said to pursue the following steps:

- Entrapment of Ag + ions at the surface of the fungal cell.
- Reduction of silver ions by enzymes present in the fungus.
- The extracellular enzymes like Naphthoquinones and anthraquinones are encourage the reduction of silver ion and produce silver nanoparticles.

Jain et al. (2011) described a two-step hypothetical system for the synthesis of silver nanoparticles. In the first step, mass silver ions are reduced to silver nanoparticles by a 32 KDa protein reductase secreted by *A. flavus*. In the second step, silver nanoparticles capped by 35 kDa protein attached with the nanoparticles giving stability. Comparable results were reported that *F. oxysporum* have two extracellular proteins such as 24 and 28 kDa accountable for the synthesis of zirconium oxide nanoparticles.

The proteins, polysaccharides and organic acids from fungi were supposed to assist the formation of diverse crystal shapes and direct the growth into spherical particles (Balaji et al. 2009). Afreen and Ranganath (2011) have reported extracellular synthesis of monodispersed AgNPs by *Rhizopus stolonifer* which is cost-effective as well as eco-friendly, and these nanoparticles characterized by UV-Vis spectrophotometer, electron microscope studies and FTIR have antibacterial activity against multidrug-resistant *Pseudomonas aeruginosa* isolated from burnt patients.

The role of *phoma glomerata* in the synthesis of AgNPs has been examined and characterized by UV-Vis, SEM, FTIR confirmed capping over AgNPs. In this study they have also suggested that capping by biomolecules could serve as better candidate for drug delivery system. Furthermore, they have also investigated the enhanced antibacterial efficacy of synthesized silver nanoparticles in opposition to resistant *Escherichia coli*, *Pseudomonas aeruginosa* and *Staphylococcus aureus*.

Endophytic fungi living in symbiotic association with plants are also involved in AgNPs synthesis. The endophytic fungi *penicillum* sp isolated from curcuma leaves is a tremendous producer of silver nanoparticles where it acts against *Escherichia coli* and *Staphylococcus aureus*. Silver nanoparticles produced from endophytic fungi *Epicoccum nigrum* isolated from *Phellodendron amurense* are highly stable at varied pH and temperature.

39.8.2 Cadmium (CDTE/CDS)

The production of CdTe quantum dots by using *F. oxysporum* has been reported by Syed et al. (2013) and Ahmad et al. (2002). The nanoparticles synthesized via biological route confirmed increase of the antimicrobial activity against gram-negative and gram-positive bacteria. Prasada and Jha (2010) has reported synthesis of CdS nanoparticle using *S. cerevisiae* as rapid and low-cost green method. Participation of white-rot fungus *C. versicolor* has also been well reported for synthesis of cadmium nanoparticles.

39.8.3 Gold Nanoparticles (Au)

Gold nanoparticles were produced by using *Fusarium*, *Neurospora*, *Verticillium*, *yeast* and *Aspergillus*. Extracellular gold nanoparticles are synthesized by *F. oxysporum*, *A. niger* and cytosolic extract of *Candida albicans*. Intracellular gold nanoparticles are formed by *Verticillium* sp. and *V. luteoalbum*. The use of eukaryotic organisms for nanoparticle synthesis was first demonstrated by the use of *Verticillium* sp. for the synthesis of gold nanoparticles. The synthesized gold nanoparticles were present on the surface of the cytoplasmic membrane of the fungal organisms. Due to the formation of gold nanoparticles, the mycelial mass attains a typical purple colour demonstrating intracellular generation. TEM analyses of nanoparticles give the shape of nanoparticles like triangular, hexagonal or spherical shape formed on the cell wall, and hexagonal morphology was formed of the cytoplasmic membrane.

Rapid extracellular synthesis of AuNP in cell filtrate and intracellular synthesis in fungal biomass were achieved by *Penicillium* sp. Gold nanoparticles are synthesized both intracellularly and extracellularly in a minute via NADPH reductase enzyme (Narayanan and Sakthivel 2010). Gold nanoparticles synthesized by *Sclerotium rolfsii* have variable shapes, triangle, hexagonal and decahedral. Size, shape and state aggregation of nanoparticle are determined by various factors such as precursor salts and cellular fractions. Gold and silver nanoparticles synthesized from *F. oxysporum* have different shape at diverse cellular fractions (Deepa and Panda 2014). Synthesis of AuNPs by edible mushroom *Pleurotus florida* has been documented. The synthesized AuNPs showed anticancer activity against human lung carcinoma, human chronic myelogenous leukaemia (K562), human cervix (HeLa) and human adenocarcinoma mammary gland (MDA-MB) under in vitro conditions. Synthesis of catalytically active AuNPs by *Pleurotus florida* has been reported. The glucan content of mushroom was responsible for stability of synthesized AuNPs.

39.8.4 Other Nanoparticles

In association with gold and silver, *F. oxysporum* is able to synthesize zirconia and titanium. Elemental selenium nanoparticles were synthesized by the white-rot fungus *Phanerochaete chrysosporium*; nanosized magnetite by *Mucor javanicus*, *F. oxysporum* and *Verticillium* sp.; CdSe quantum dots by *F. oxysporum;* selenium nanoparticle by *Alternaria alternata*: strontium carbonate crystals by *F. oxysporum*; silica nanoparticle by *F. oxysporum*; titanium nanoparticle by *F. oxysporum*, *S. cerevisiae* and *A. flavus*; barium titanate nanoparticle by *F. oxysporum*; and Bi_2O_3 and platinum nanoparticles by *F. oxysporum*.

39.9 Myconanoparticles from Yeast

Yeasts are most useful in the synthesis of semiconductor nanoparticles like cadmium sulphide, lead, sulphide, antimony oxide, etc. *Candida glabrata* is the yeast which can intracellularly synthesize uniform spherically shaped peptide-bound CdS nanocrystals of size about 20 Å. They tend to form metal thiolate complex with phytochelatins which neutralize metal ions. *Schizosaccharomyces pombe* can also synthesize hexagonal crystal CdS nanopanoparticles with size 1–1.5 nm. *Torulopsis* sp. was the foremost yeast in which synthesis cubic structured PbS nanocrystals have semiconductor properties, which was intracellularly produced in the dimension of 2–5 nm in spherical structure. *S. cerevisiae* (baker's yeast) can reduce Au^{+3} to give gold nanoparticles. The reduction reaction carried out by aldehyde group present in reducing sugars occurs in the peptidoglycan layer of the cell wall. *Pichia jadinii* synthesize the gold nanoparticles intracellularly which have diverse morphologies resembling spherical, triangular, hexagonal, etc. sizes less than 100 nm in diameter. *S. cerevisiae* can also produce antimony oxide (Sb_2O_3) nanoparticles with spherical shape with size 2–10 nm at room temperature conditions.

39.10 Characterization of Nanoparticles

39.10.1 Visual Colour Change

In extracellular synthesis of silver nanoparticles, the silver ion solution is generally brown in colour. In intracellular synthesis of silver nanoparticles produced from fungi, the hyphae darken to brownish but remains clear. In both intracellular and extracellular synthesis, the browning reaction is created to the surface plasmon resonance of the metal nanoparticles. In external gold nanoparticle production, the solution colour can vary depending on the size of the gold nanoparticles; smaller particles appear pink while larger particles appear purple. Externally synthesized cadmium sulphide nanoparticles were appeared as bright yellow colour.

39.10.2 *Ultraviolet-Visible Spectrophotometry*

The Ag and Au nanoparticles were characterized in a UV-Vis spectrophotometer to identify the kinetic behaviour of nanoparticles. The scanning range for silver NP is 200–800, while for gold it was 300–600 nm in a speed of 420 nm/min.

39.10.3 *Differential Light Scattering (DLS)*

Dissolve the Ag and Au nanoparticles in distilled water (dH2O) by using a bath sonicator prior to detect the size. Dynamic light scattering size detection was performed with use of Malvern Zetasizer Nano ZS (operating with version 5.03 DTS).

39.10.4 *Scanning Electron Microscopy (SEM)*

The freeze-dried samples of AgNP and AuNP are sonicated with distilled water; this sample was taken on glass slide and allowed to dry. A thin layer of platinum was covered to make the nanoparticles conductive to SEM control at a vacuum of 10–15 torr. The speed-up voltage was kept in the range 10–20 kV (Fig. 39.9).

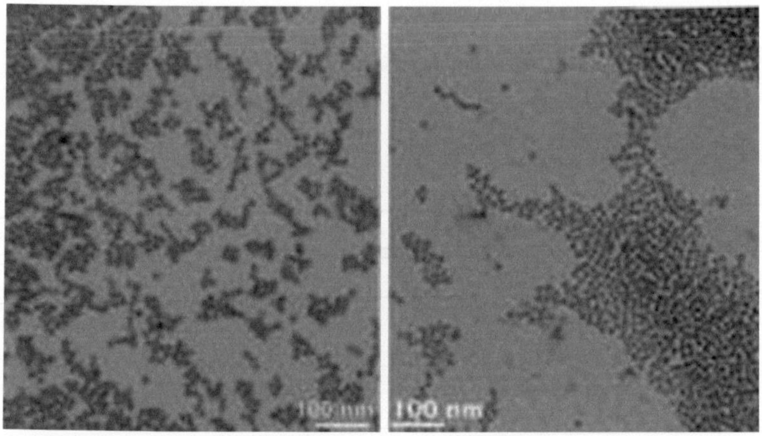

Fig. 39.9 SEM image of silver nanoparticles

39.10.5 Fourier Transmission Infrared Spectroscopy

FTIR is a chemical analytical technique that measures infrared intensity of wavelength or wave number of light. It is used to analyse the biomolecule and also the bonding interaction between them. IR spectroscopy identifies the vibration characteristics of chemical and functional groups of nanoparticles. The infrared light interacts with matter; chemical bond gives the stretch, contract and bond form. This chemical function group can absorb infrared radiation in a specific wave length and give the structure of the rest of the molecule. The silver nanoparticles formation, FTIR data calculate the interaction between Ag salts and protein molecules, which accurate for the reduction of silver ions and stable form of Ag nanoparticles formed.

39.10.6 X-Ray Diffraction Method

XRD technique is used to study phase composition, crystal structure, texture or orientation of nanoparticles. The principle of XRD is the X-rays are passed through a material, and the formed pattern gives details of size and shape of the cell. The atoms are crystal in structure arranged in a periodically array, so it can diffract light at different angle. While X-rays are passed through a crystal, it produces a diffraction pattern, and it gives the details about the atomic arrangement of crystals.

39.10.7 Atomic Force Microscopy

The AFM is used to measure the topography of nanoparticles. A nanosized tip found on a cantilever is placed over the sample, and 3D image of the nanoparticles is generated on a computer. The advantage of the AFM over SEM is the capacity to make topographical measurements for recognition and study the size and shape of nanoparticles in 3D dimensions. The AFM generally measures the nanoparticles height.

39.11 Applications of Myconanoparticles

Nanotechnology is an innovative, quick-developing science posturing important impacts on agriculture and other sectors that likely will deliver nanostructured materials (Gade et al. 2008). Even though, nanoparticles synthesized by fungus have wide applications. Nanoparticles are being produced that offer the chance to regulate pesticides, herbicides and fertilizers more effectively and safely by controlling the release of pesticides to the environment. The myconanoparticles such as silver, gold and platinum have enormous biomedical applications. Among these, silver and gold nanoparticles are widely used. Silver nanoparticles exhibit strong antimicrobial potential, whereas gold nanoparticles have their applications in drug delivery for many diseases including cancer (Rai et al. 2015).

39.11.1 Antimicrobial Activity

Myconanotechnology represents a novel concept in the development of antimicrobial nanomaterials. Silver nanoparticles produced extracellularly from *F. oxysporum* can be integrated into materials such as cloth. These clothes with silver nanoparticles are useful to prevent infection against pathogenic bacteria such as *Staphylococcus aureus*. Silver nanoparticles with 45–100 nm size synthesized by *Lecanicillium lecanii* that were encrusted on the bleached cotton fabrics have antibacterial activity against *S. aureus* and *E. coli*. Extracellular synthesis of gold nanoparticles (10 nm) by *Rhizopus oryzae* is used for the making of nanogold-bioconjugate (NGBC) structure. Double-capsulized nanosilver (1.5 nm) have antifungal effect against rose powdery mildew caused by *Sphaerotheca pannosa var. rosae* (Kim et al. 2008). The combination of AgNps with amphiphilic macromolecules exposed an effective antimicrobial agent. Silver nanoparticles have efficiency against *E. coli*, *Pseudomonas*, *Salmonella typhi* and *Vibrio cholera* (Morones et al. 2005) (Fig. 39.10).

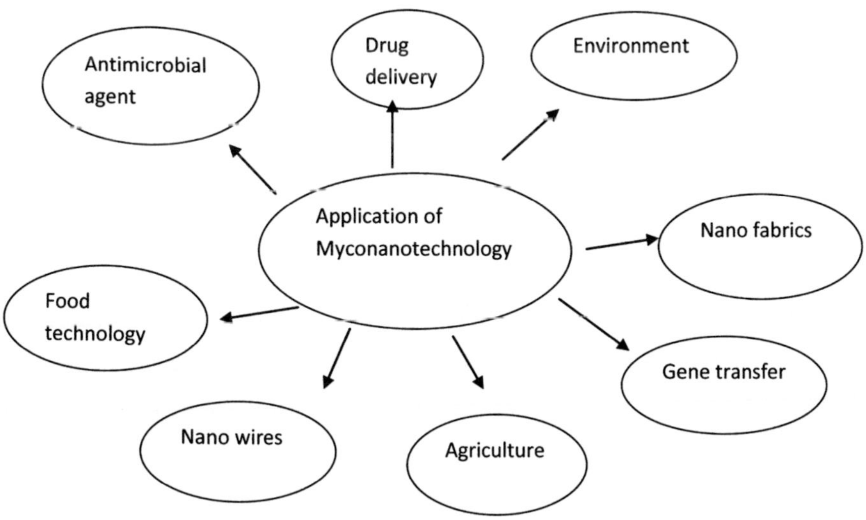

Fig. 39.10 Applications of myconanoparticles in various fields

39.11.2 Antibacterial Activity

The multidrug-resistant bacterial strains increase rapidly, and it signifies a major threat to modern medicine. The increase in antibiotic résistance is caused by the broad and inappropriate use of antibiotic in human, animals and agriculture. Using nanoparticles in nanomedicine with nanoconjugate has great potential against pathogenic bacteria, but they cause severe toxicity among people. Pathogens show resistance to antibiotics like ampicillin, vancomycin and streptomycin. When the silver nanoparticles combined with these antibiotics, it shows the antibacterial effect of mycosynthesized AgNPs against pathogenic bacteria like *Staphylococcus aureus*, *Salmonella typhi*, *E. coli*, *Pseudomonas fluorescens*, *Pseudomonas aeruginosa*, *Agrobacterium tumefaciens* and *Magnaporthe oryzae*.

39.11.2.1 Mechanism of Antibacterial Activity

- Nanoparticles penetrate the bacterial cell wall and subsequently damage the cell membrane that leads to death of the cell.
- The generation of reactive oxygen species (ROS) when AgNPs interact with respiratory enzymes after entering to the bacterial cells. The generated ROS is lethal to the cell.
- Interaction of Ag + phosphate of DNA inhibits DNA replication and causes DNA damage that leads to bacterial cell death.
- AgNPs have also shown to modulate the signalling pathway required for cell growth. AgNPs dephosphorylate the substrates on tyrosine residues, which lead to inhibition of signal cascade and thus stop the cell growth.
- Interference with nutrient uptake. The multiple targets of action could help NPs to fight effectively against different plant pathogens.

39.11.3 Antifungal Activity

- Use of nanosilver as potent antifungal agent.
- It is a more inexpensive process.
- Used to control the plant diseases.
- It is safer than synthetic fungicides.
- Strong antifungal against *Botrytis cinerea* by applying with silver nanoparticles and fluconazole.
- Antifungal against *Phoma glomerata*, *Fusarium semitectum*, *Trichoderma* and *Candida albicans* (Gajbhiye et al. 2009).
- ZnO and $ZnTiO_3$ are potent biocides against *A. niger*.
- Ag nanoparticles are antifungals against *A. altria, Macrophomina phaseolina* and *Rhizoctonia solani*.
- Cu-chitosan NP inhibits *A. altria*.

39.11.4 Antiviral Activity

- Myconanoparticles are used to control and prevent viral diseases.
- Antiviral activity against H1N1 influenza virus.
- Nanosilver with 1–10 nm binds and inhibits HIV (Lara et al. 2010).
- Silver with polyvinylpyrrolidone development of gel is used in contamination by virus.
- Antiviral against herpes simplex virus and human parainfluenza virus.
- Gold nanoparticles are used for treatment of AIDS, tuberculosis and cancer (Madhusudhan et al. 2014).

39.11.5 Food Preservation

Fayaz et al. (2009) established that AgNPs produced from *T. viride* integrated into sodium alginate thin film have antibacterial activity and develop the shelf life of carrot and pear in terms of weight loss and dissolve protein content.

39.11.6 Nanofertilizers

- Increased use of chemical fertilizer decreases soil fertility and yield crop.
- Use of nanofertilizer protects soil.
- It increases nutrient uptake and reduces soil toxicity and negative impact of chemical fertilizer.
- Nanostructure slowly delivers fertilizer according to environmental trigger.
- Nanoclay and zeolites are used as nanofertilizers.
- Urea zeolite chips are used as nitrogen fertilizer (Naderi and Dancsh-Shahraki 2013).

39.11.7 Nanoparticles as Pesticides

- Nanopesticides are organic and inorganic ingredients in different forms.
- It is active against plant pathogens, insects and pest.
- Used for formation of pesticides, insecticides and insect repellents.
- Nanosilica as nanopesticide (Barik et al. 2008).
- Nanosilica gets adsorbed into cuticular lipids by physical absorption and causing death.
- Silver, aluminium oxide, zinc and titanium dioxide control rice diseases and grasserie disease in silkworm caused by *Sitophilus oryzae* and *Bombyx mori* (Goswami etal. 2010).

39.11.8 Nanofungicides

- Fungi as plant pathogens.
- Nanofungicides control *Fusarium*, *Phoma*, *Aspergillus* and *Phytophthora* (Ingle and Rai 2011).

39.11.9 Nanosized

- Nanosilica-silver at 0.3 ppm is synthesized from fungi.
- It inhibits powdery mildews of pumpkin.
- Used to treat *Erysiphe cichoracearum*-infected leaves.
- Nanosilver as antifungal agent against rose powdery mildew.

39.11.10 Nanowires

- Nanowires are synthesized by fungi.
- Useful for creating miniature of electronic devices.
- Nanowires are used in the field of microelectonics and nanoscale electronics devices are used in agriculture.
- Gold microwires produced by *Aspergillus* and *Neurospora*.
- Gold nanoparticles are exposed to fungi.
- Then its surface treated with glutamate, aspirate and polyethylene glycol.
- Fungi with gold nanoparticle synthesizes gold microwires (Sugunan et al. 2007).

39.11.11 Nanofibrous Mats

- Nanofibrous material used from fungi.
- It has high surface area and porous structure.
- It is used as biosensors, nanocomposites and antimicrobials and in tissue engineering.
- Protects clothes and drug delivery.
- It is nanofibrous with chitosan and *T. viride* against phytopathogens.

39.11.12 Quantum Dots

Latest advanced research has been developed in the field of luminescent nanocrystals in luminous tagging by quantum dots with bio-recognition molecules (Shao et al.2011). Quantum dots are spherical, fluorescent nanocrystals consisting of semiconductor material that links the individual atoms and semiconductor solids.

39.11.13 Smart Delivery Systems

Myconanoparticles are used as smart delivery systems in agriculture. The nanomaterial has specific delivery system to target site in living organism. It was first investigated for medical (Kukowska-Latallo et al. 2005) uses only. In plants this system can be utilized, in particular to manage insect pest and diseases. Nanoparticles attached with agrochemicals could reduce damage to nontarget plant tissue and reduce level of chemicals released into environment. It is necessary for enhanced plant disease resistance, crop engineering and ecological checking. DNA coated with gold nanoparticles has used to generate transgenic plants by bombardment gene transfer method.

39.11.14 Vector Control

Banu and Balasubramanian (2014) synthesized AgNPs from entomopathogenic fungus *Beauveria bassiana* and found it effective against dengue vector *Aedes aegypti*. The adulticidal effect of AgNps synthesised from *Chrysosporium keratinophilium*, *verticillum lecanii* and *F. oxysporium* active aganist filariasis vector *culex quinquefaciatus*.

39.11.15 Wound Treatment

Sundaramoorthi et al. (2009) have explored the wound healing property of AgNPs synthesized from *Aspergillus niger* in experimental rat model (excision wound model and thermal wound model). The properties of AgNPs are better wound healing, percentage of wound contraction and period of epithelialization in dose- and time-dependent manner. In another study, *Phytophthora infestans* synthesizing AgNPs was assessed for its wound healing activity. They found 0.125% (w/w) AgNP ointments have better wound healing property as compared to standard silver sulphadiazine ointment.

39.11.16 Molecular Detection

The new PCR assay has been developed by Bansod et al. (2013) for rapid detection of pathogenic fungi *Candida* sp. from low concentrated DNA. The gold and silver nanoparticles are synthesized by *F. oxysporum* and conjugate with master DNA sample of *Candida* sp.. This bio-nanoconjugate has high specificity and sensitivity.

39.11.17 Nanoparticles as Vaccines

The development of nanovaccines is in progress. Antigens are coated with nanospheres, utilized as vaccine carriers focussing on human dendritic cells and used as nasal vaccines. Nanospheres had a present enact on human dendritic cell, initiate transcription of genes important for phagocytosis and immunity.

39.11.18 Nanoscaffolds

Nanofiber scaffolds produced from nanoparticles are used to rejuvenate CNS cells and other organs. Research is executed on a hamster axonal tissue with optic tract regenerated by nanofiber scaffold.

39.11.19 Local Treatments for Anaesthetic Toxicity

Local anaesthetics edge is occasionally extremely lethal, ranging from local neurotoxicity to cardiovascular damage and coma. Beside traditional treatment, drug-scavenging nanoparticles increase the continued existence rate from all animals in the control group (untreated) to all animals in the treated group.

39.11.20 Bioseparation

The myconanoparticles are used to prepare the nanotube membranes which work as channel for extremely semipermeable transport molecules and ions among solutions on both sides of the membranes. For example, membrane containg nanotubes are smaller inside diameter (less 1 nm) to segregate smaller molecules on the basis of molecular weight and size of molecules, while those larger in diameter (20–60 nm) are used to segregate proteins.

39.11.21 Anticancer Activity

Biosynthesized myconanoparticles have anticancer activity in opposition to diverse cancer cells that are investigated in cytotoxicology research. AuNPs give an idea about the considerable activity to human breast cancer cells, and gold nanoparticles have anticancer activity against cancer cell and yield HBL cell death. Au nanoparticles synthesized from *Penicillium brevicompactum* that can crash cancer cell lines were studied (Mishra et al. 2014).

39.11.22 Catalysis

Nanoparticles synthesized from fungi have remarkable biocatalytic properties. Mishra et al. 2014 described the catalytic power of biosynthesized gold nanoparticles from *Trichoderma viride*. This gold nanoparticle in presence of $NaBH^4$ reduces nitrophenol to 4-aminophenol and also shows antimicrobial property. Further, metal nanoparticles obtained from fungi can be used in enzyme immobilization for improved enzymatic activity.

39.12 Conclusion

Myconanoparticles have shown the way to the improvement of nonhazardous and relatively more bioactive nanoparticles. Thus, nanoparticles from fungi are currently an essential science of micro-/nanotechnology and are called as myconanotechnology. Efforts are necessary for mass-scale manufacture of nanoparticles, but synthesis from fungi reduces these efforts in large-scale production. Mycosynthesis of different nanoparticles is used as narrative antimicrobial with potential to undertake the problem of multidrug-resistant microbes. There is a spacious power of nanoparticles in the areas such as agriculture and environment.

References

Adil SF, Assal ME, Khan M (2015) Biogenic synthesis of metallic nanoparticles and prospects toward green chemistry. Dalton Trans 44:9709–9717

Afreen RV, Ranganath E (2011) Synthesis of monodispersed silver nanoparticles by *Rhizopus stolonifer* and its antibacterial activity against MDR strains of Pseudomonas aeruginosa from burnt patients. Int J Environ Sci 1(7):1582–1592

Ahmad A, Mukherjee P, Mandal D (2002) Enzyme mediated extracellular biosynthesis of CdS nanoparticles by the fungus *Fusarium oxysporum*. J Amer Chem Soc 124:12108–12109

Ahmad A, Senapati S, Khan MI (2003) Intracellular synthesis of gold nanoparticles by a novel alkalotolerant actinomycete, *Rhodococcus* species. Nanotechnology 14:824–828

Alzahrani E, Sharfalddin A, Alamodi M (2015) Microwave hydrothermal synthesis of ferric oxide doped with cobalt. Adv Nanopart 4:53–60

Balaji DS, Basavaraja S, Deshpande R, Mahesh DB, Venkataraman A (2009) Extracellular biosynthesis of functionalized silver nanoparticles by strains of *Cladosporium cladosporioides* fungus. Colloids Surf B: Biointerfaces 68(1):88–92

Bansod S, Bonde S, Tiwari V, Bawaskar M, Deshmukh S, Gaikwad S, Gade A, Rai M (2013) Bioconjugation of gold and silver nanoparticles synthesized by *F. oxysporum* and their use in rapid identification of *Candida* species by using bioconjugate nanopolymerase chain reaction. J Biomed Nanotechnol 9(12):1962–1971

Banu NA, Balasubramanian C (2014) Mycosynthesis of silver nanoparticles using *Beauveria bassiana* against dengue vector, Aedesaegypti(Diptera: Culicidae). Parasitol Res 113(8):2869–2877

Barik TK, Sahu B, Swain V (2008) Nano-silica from medicine to pest control. Parasitol Res 103:253–258

Bowman SM, Free JF (2011) The structure and synthesis of the fungal cell-wall. BioEssays 28:799–808

Deepa K, Panda T (2014) Synthesis of gold nanoparticles from different cellular fractions of *Fusarium oxysporum*. J Nanosci Nanotechnol 14:3455–3463

Duran A, Nombela C (2004) Fungal cell-wall biogenesis: building a dynamic interface with the environment. Microbiology 150:3099–3103

Fayaz AM, Balaji K, Girilal M (2009) Biogenic synthesis of silver nanoparticles and their synergistic effect with antibiotics: a study against gram-positive and gram-negative bacteria. Nanomedicine 6:103–109

Gade AK, Bonde P, Ingle AP (2008) Exploitation of *Aspergillus niger* for synthesis of silver nanoparticles. J Biobased Mater Bioener 2:243–247

Gajbhiye MB, Kesharwani JG, Ingle AP (2009) Fungus mediated synthesis of silver nanoparticles and its activity against pathogenic fungi in combination of fluconazole. Nanomed NBM 5:282–286

Gericke M, Pinches A (2006) Biological synthesis of metal nanoparticles. Hydrometallurgy 83:132–140

Goswami A, Roy I, Sengupta S, Debnath N (2010) Novel applications of solid and liquid formulations of nanoparticles against insect pests and pathogens. Thin Solid Films 519:1252–1257

Holan ZR, Volesky B (1995) Accumulation of cadmium, lead and nickel by fungal and wood biosorbents. ApplBiochemBiotechnol 53(2):133–146

Ingle AP, Rai MK (2011) Genetic diversity among Indian phytopathogenic isolates of *Fusarium semitectum* Berkeley and Ravenel. Adv Biosci Biotechnol 2:142–148

Jain N, Bhargava A, Majumdar S (2011) Extracellular biosynthesis and characterization of silver nanoparticles using *Aspergillusflavus*NJP08: a mechanism perspective. Nanoscale 3:635–641

Karbasian M, Atyabi SM, Siadat SD, Momem SB, Norouzian D (2008) Optimizing nano-silver formation by *Fusarium oxysporum* (PTCC 5115) employing response surface methodology. Am J AgricBiolSci 3:433–437

Kathiresan K, Manivannan S, Nabeel MA, Dhivya B (2009) Studies on silver nanoparticles synthesized by a marine fungus, Penicillium fellutanum isolated from coastal mangrove sediment. Colloid Surf B 7:133–137

Kim YS, Kim JS, Cho HS (2008) Twenty-eight-day oral toxicity, genotoxicity, and gender-related tissue distribution of silver nanoparticles in Sprague-Dawley rats. InhalToxico 20:575–583

Kukowska-Latallo JF, Candido KA, Cao Z, Nigavekar SS, Majoros IJ, Thomas TP, Balogh LP, Khan MK, Baker JR (2005) Nanoparticle targeting of anticancer drug improves therapeutic response in animal model of human epithelial cancer. Cancer Res 65:5317–5324

Lara HH, Ayala-Nunez NV, Ixtepan-Turrent LDC, PadilhaRodrigues C((2010) Mode of antiviral action of silver nanoparticles against HIV. J Nanobiotechnol 8:1–10

Madhusudhan A, Reddy GB, Venkatesham M (2014) Efficient pH dependent drug delivery to target cancer cells by gold nanoparticles capped with carboxymethyl chitosan. Int J MolSci 15:8216–8234

Mishra A, Kumari M, Pandey S, Chaudhry V, Gupta KC, Nautiyal CS (2014) Biocatalytic and antimicrobial activities of gold nanoparticles synthesized by T*richoderma* sp. Bioresour Technol 166:235–242

Moghaddam KM (2010) An introduction to microbial metal nanoparticle preparation method. J Young Investig 19:1–7

Morones JR, Elechiguerra JL, Camacho A (2005) The bactericidal effect of silver nanoparticles. Nanotechnology 16:2346–2353

Naderi MR, Danesh-Shahraki A (2013) Nanofertilizers and their roles in sustainable agriculture. Int J Agric Crop Sci 5:2229–2232

Narayanan KB, Sakthivel N (2010) Biological synthesis of metal nanoparticles by microbes. Adv Colloid Interf Sci 156:1–13

Prasada K, Jha AK (2010) Biosynthesis of Cds nanoparticles: an improved and rapid procedure. J Colloid Interf Sci 342(1):68–72

Rai M, Ingle AP, Gade A, Duran N (2015) Synthesis of silver nanoparticles *by Phoma gardenia* and in vitro evaluation of their efficacy against human disease causing bacteria and fungi. IETNanobiotechnol 9:71–75
Shao L, Gao Y, Feng Y (2011) Semiconductor quantum dots for biomedical applications. Sensors 11:11736–11751
Sugunan A, Melin P, Schnurer J, Hilborn JG, Dutta J (2007) Nutrition-driven assembly of colloidal nanoparticles: growing fungi assemble gold nanoparticles as microwires. Adv Mater 19:77–81
Sundaramoorthi C, Kalaivani M, Mathews DM, Palanisamy S, Kalaiselvan V, Rajasekaran A (2009) Biosynthesis of silver nanoparticles from *Aspergillus niger* and evaluation of its wound healing activity in experimental rat model. Int J PharmTech Res 1(4):1523–1529
Syed A, Saraswati S, Kundu GC, Ahmad A (2013) Biological synthesis of silver nanoparticles using the fungus Humicola sp. and evaluation of their cytotoxicity using normal and cancer cell lines. Spectrochim Acta A Mol Biomol Spectosc 114:144–147
Taniguchi N (1974) On the basic concept of nanotechnology, In: Proceedings of the international conference on production engineering, Part II, Japan society of Precision Engineering, Tokyo
Thakkar KN, Mhatre SS, Parikh RY (2010) Biological synthesis of metallic nanoparticles. Nanomedicine 6:257–262
Vaidyanathan R, Kalishwaralal K, Gopalram S, Gurunathan S (2009) Nanosilver the burgeoning therapeutic molecule and its green synthesis. Biotechnol Adv 27:924–937
Zhang X, Yan S, Tyagi RD, Surampalli RY (2011) Synthesis of nanoparticles by microorganisms and their application in enhancing microbiological reaction rates. Chemosphere 82:489–494